LINEAR ALGEBRA

Pure & Applied

LINEAR ALGEBRA
Pure & Applied

Edgar G. Goodaire
Memorial University, Canada

World Scientific
NEW JERSEY • LONDON • SINGAPORE • BEIJING • SHANGHAI • HONG KONG • TAIPEI • CHENNAI

Published in India by

Cambridge University Press India Pvt. Ltd.
under the imprint of Foundation Books
Cambridge House
4381/4 Ansari Road
Daryaganj,
New Delhi - 110 002

This reprint edition 2014

ISBN- 13: 978-93-84463-34-2

This edition of *Edgar G. Goodaire / Linear Algebra: Pure and Applied,* is published by arrangement with World Scientific Publishing Co. Pte. Ltd.

This edition is for sale in India, Pakistan, Sri Lanka, Bhutan, Nepal, Maldives, Afghanistan and Bangladesh only, not for export elsewhere.

Printed in India at Shree Maitrey Printech Pvt. Ltd., Noida

Contents

Preface

This book had its genesis in the last century and has been used at Memorial University in St. John's, Newfoundland in one form or another for at least fifteen years. I am grateful to World Scientific Publishing Co. for making the latest edition widely available.

At many universities, many students take a course in linear algebra which should more properly be called "matrix algebra" because the content is primarily matrix manipulation and solving linear systems. Like calculus, the first course in linear algebra is often a course without proofs. To be successful, a student simply matches examples to homework with almost no reading and little real understanding.

My goal here is to use an intrinsically interesting subject to show students that there is more to mathematics than numbers, and that reading to understand while challenging, can also be very satisfying. A secondary purpose is to raise the level of the typical Linear Algebra I course so that the transition to Linear Algebra II is not as difficult as it has become.

The presentation in this text is frequently interrupted by "Reading Checks," little questions a student can use to see if she really does understand the point just made and which are intended to promote active reading. Every section in this book concludes with answers to these Reading Checks followed by some "True/False" questions that test definitions and concepts, not dexterity with numbers or buttons. Every exercise set contains some "Critical Reading" exercises which include problems some of which are harder, yes, but also some that can be solved quite easily **if understood**. Each chapter concludes with review exercises and a list of key words and concepts that have been introduced in that chapter.

There are several reasons for including "pure and applied" in the title. While a pure mathematician myself and a person who, as a student, loved the abstract vector space/linear transformation approach to linear algebra, I think such treatment is less successful today because it requires more maturity than the typical sophomore possesses. This book is mainly about Euclidean n-space and matrices. Essentially every vector space is a subspace of R^n. Other standard examples, like polynomials and matrices, are mentioned, but not emphasized. In my opinion, when students are thoroughly comfortable with R^n, they have little trouble transferring concepts to general vector spaces. Moreover, most theorems in elementary linear algebra have proofs that are independent of the vector space, so why not let the student remain in his comfort zone. The emphasis on matrices and matrix factorizations, the recurring themes of projections and codes, and shorter sections on subjects like facial recognition, Markov chains, graphs and electric circuits give this book a genuinely applied flavour.

This is a book which emphasizes "pure" mathematics, understanding and rigour. Virtually nothing is stated without justification, and a large number of exercises are of the "show or prove" variety. Since students tend to find such exercises difficult, I have included an appendix called "Show and Prove" designed to help students write a mathematical proof, and another, "Things I Must Remember," which has essentially been written by my students over the years. So despite its emphasis on the concrete, this book would also serve well to support the "introduction to proof" transition course that is now a part of many degree programs.

Organization, Philosophy, Style

Many linear algebra texts begin with linear equations and matrices, which students enjoy, but which suggest incorrectly that linear algebra is about computation. This book begins with vectors, primarily in R^2 and R^3, so that I can introduce in a concrete setting some key concepts—linear combination, span, linear dependence and independence—long before the "invariance of dimension" theorem, which is often the point where students decide that linear algebra is hard. The idea of a plane being "spanned" by two vectors is not hard for beginning students; neither is the idea of "linear combination" or the fact that linear dependence of three or more vectors means that the vectors all lie in a plane. That $A\mathsf{x}$ is a linear combination of the columns of A, surely one of the most useful ideas of linear algebra, is introduced early in Chapter 2 and used time and time again. Long before we introduce abstract vector spaces, all the terminology and techniques of proof have been at play for some time. A solid early treatment of linear independence and spanning sets in the concrete setting of R^n alerts students to the fact that this course is not about computation; indeed, serious thought and critical reading will be required to succeed. Many exercise sets contain questions that can be answered very briefly, if understood. There are some writing exercises too, marked by the symbol ✍, that ask for brief explanations or, sometimes, biographical information.

In some respects, this book is at a higher level than its competitors. The theory behind Markov chains many students will find difficult. Topics like the pseudoinverse and singular value decomposition and even complex matrices and the spectral theorem are tough for most second year students. On the other hand, versions of this book have been well received and proven to work for students with average to good mathematics backgrounds.

Technology goes almost unmentioned in this book. While a great fan of technology myself, and after delivering courses via MAPLE worksheets, for instance, I have formed the opinion that teaching students how to use sophisticated calculators or computer algebra packages is more time-consuming and distracting than useful.

I try to introduce new concepts only when they are needed, not just for their own sake. Linear systems are solved by transforming to row echelon form, not **reduced** row echelon form, the latter idea introduced where it is of more benefit, in the calculation of the inverse of a matrix. The notion of the "transpose" of a matrix is introduced in the section on matrix multiplication because it is important for students to know that the **dot** product $\mathsf{u} \cdot \mathsf{v}$ of two column vectors is the **matrix** product $\mathsf{u}^T\mathsf{v}$. Symmetric matrices, whose LDU factorizations are so easy, appear for the first time in the section on LDU factorization, and reappear as part of the characterization of a projection matrix.

There can be few aspects of linear algebra more useful, practical, and interesting than eigenvectors, eigenvalues and diagonalizability. Moreover, these topics provide an excellent opportunity to discuss linear independence in a nonthreatening manner, so these ideas appear early and in a more serious vein later.

Many readers and reviewers have commented favourably on my writing which, while seemingly less formal than that of other authors is most assuredly not lacking in rigour. There is far less emphasis on computation and far more on mathematical reasoning in this book. I repeatedly ask students to explain "why." Already in Chapter 1, students are asked to show that two vectors are linearly dependent if and only if one is a scalar multiple of another. The concept of matrix inverse appears early, as well as its utility in solving matrix equations, several sections before we discuss how actually to find the inverse of a matrix. The fact that students find the first chapter quite difficult is evidence to me that I have succeeded in emphasizing the importance of asking "Why", discovering "Why," and then clearly communicating the reason "Why."

A Course Outline

To achieve economy within the first three chapters, I omit Section 2.8 on LDU factorization (but never the section on LU) and discuss only a few of the properties of determinants (Section 3.2), most of which are used primarily to assist in finding determinants, a task few people accomplish by hand any more.

Linear Algebra II is essentially Chapters 4 through 7, but there is more material here than I can ever manage in a 36-lecture semester. Thus I sometimes discuss the matrix of a linear transformation only with respect to standard bases, omitting Sections 5.4 and 5.5. The material of Section 6.1, which is centred around the best least squares solution to over-determined linear systems, may be nontraditional, but try to resist the temptation to omit it. Many exercises on this topic have numerical answers (which students like!), there are lots of calculations, but lots of theory to be reinforced too. For example, the fact that the formula $P = A(A^TA)^{-1}A^T$ works only for matrices with linearly independent columns provides another opportunity to talk about linear independence.

To do proper justice to Chapter 7—especially the unitary diagonalization of Hermitian matrices and some (remedial?) work on complex numbers and matrices—I have to cut Sections 6.4 and 6.5 on orthogonal subspaces and the pseudoinverse. When I include Section 6.5 on the pseudoinverse (a non-standard topic that students like), I bypass the first two sections of Chapter 7 and head directly to the orthogonal diagonalization of real symmetric matrices in Section 7.3 after a brief review of the concepts of eigenvalue and eigenvector.

Acknowledgements

Over the years, I have been helped and encouraged by a number of colleagues, including Ivan Booth, Peter Booth, Hermann Brunner, John Burry, Clayton Halfyard, Mikhail Kotchetov, George Miminis, Michael Parmenter and Donald Rideout. In particular, Misha Kotchetov and my friend of many years, Michael Parmenter, one of the best proof readers I have ever known, made numerous suggestions that improved this work immeasurably.

Many students have also helped me to improve this book and to make the subject easier for those that follow. In particular, I want to acknowledge the enthusiasm and assistance of Gerrard Barrington, Shauna Gammon, Ian Gillespie, Philip Johnson and Melanie Ryan.

I hope you discover that this book provides a refreshing approach to an old familiar topic with lots of "neat ideas" that you perhaps have not noticed or fully appreciated previously. If you have adopted this book at your institution for one of your courses, I am very pleased, but also most genuine in my request for your opinions, comments and suggestions.

Edgar G. Goodaire
edgar@mun.ca
St. John's, Newfoundland, Canada

To My Students

From aviation to the design of cellular phone networks, from data compression (CDs and jpegs) to data mining and oil and gas exploration, from computer graphics to Google, linear algebra is everywhere and indispensable. With relatively little emphasis on sets and functions, linear algebra is "different." Most students find it enjoyable and a welcome change from calculus. Be careful though. The answers to most calculus problems are numerical and easily confirmed with a text message. The answers to many problems in this book are **not** numerical, however, and require explanations as to **why** things are as they are. So let me begin this note to you with a word of caution: this is a book you **must read**.

For many of you, linear algebra will be the first course where many exercises ask you to explain, to answer why or how. It can be a shock to discover that there are mathematics courses (in fact, most of those above first year) where words are more important than numbers. Gone are the days when the solution to a homework problem lies in finding an identical worked example in the text. Homework is now going to require some critical thinking!

Many years ago, when this book existed just as course notes, a student came into my office one day to ask for help with a homework question. When this happens with a book of which I am an author, I am always eager to discover whether or not I have laid the proper groundwork so that the average student could be expected to make a reasonable attempt at an exercise. From your point of view, a homework exercise should be very similar to a worked example. Right? In the instance I am recalling, I went through the section with my student page by page until we found such an example. In fact, we found precisely the question I had assigned, with the complete solution laid out as an example that I had forgotten to delete when I moved it to the exercises! The student felt a little sheepish while I was completely shocked to be reminded, once again, that some students don't read their textbooks.

It is always tempting to start a homework problem right away, without preparing yourself first, but this approach isn't going to work very well here. You will find it imperative to read a section from start to finish before attempting the exercises at the end of a section. And please do more than just glance at the list of "key words" that appears at the end of each chapter. Look carefully at each word. Are you sure you know what it means? Can you produce an example or a sentence that would convince your teacher that you understand what the word means? If you can't, it's for sure you won't be able to answer a question where that word is used. Go to the back of the book, to a glossary where every technical term is defined and accompanied by examples. If you are not sure what is required when asked to prove something or to show that something is true, read the appendix "Show and Prove" that is also at the back. (You will find there the solutions to several exercises from the text itself!)

In another appendix, entitled "Things I Must Remember," I have included many important ideas that my students have helped me to collect over the years. You will also find there some ideas that are often just what you need to solve a homework problem.

I hope that you like my writing style, that you discover you like linear algebra, and that you soon surprise yourself with your ability to write a good clear mathematical proof. I hope that you do well in your linear algebra courses and all those other courses where linear algebra plays an important role. Let me know what you think of this book. I like receiving comments—good, bad and ugly—from anyone.

Edgar G. Goodaire
edgar@mun.ca
St. John's, Newfoundland, Canada

Suggested Lecture Schedule

1 The Geometry of the Plane and 3-space **11 Lectures**

1.1 Vectors and Arrows 2
1.2 Length and Direction 2
1.3 Lines, Planes and the Cross Product 4
1.4 Projections 2
1.5 Linear Dependence and Independence 1

2 Matrices and Linear Equations **18 Lectures**

2.1 The Algebra of Matrices . . . . . . $2\frac{1}{2}$
2.2 Application: Generating Codes with Matrices 1
2.3 Inverse and Transpose . . . . . . $2\frac{1}{2}$
2.4 Systems of Linear Equations 2
2.5 Application: Electric Circuits 1
2.6 Homogeneous Systems; Linear Independence 2
2.7 Elementary Matrices and LU Factorization 3
2.8 LDU Factorization 2
2.9 More on the Inverse of a Matrix 2

3 Determinants, Eigenvalues, Eigenvectors **11 Lectures**

3.1 The Determinant of a Matrix 2
3.2 Properties of Determinants 3
3.3 Application: Graphs 1
3.4 Eigenvalues and Eigenvectors . . . . . . $1\frac{1}{2}$
3.5 Similarity and Diagonalization . . . . . . $1\frac{1}{2}$
3.6 Application: Linear Recurrence Relations 1
3.7 Application: Markov Chains 1

4 Vector Spaces **14 Lectures**

4.1 The Theory of Linear Equations 3
4.2 Subspaces . . . . $3\frac{1}{2}$
4.3 Basis and Dimension . . . . $2\frac{1}{2}$
4.4 Finite-Dimensional Vector Spaces . . . . $3\frac{1}{2}$
4.5 One-sided Inverses . . . . $1\frac{1}{2}$

5 Linear Transformations **9 Lectures**

5.1 Fundamentals 3
5.2 Matrix Multiplication Revisited 1
5.3 Application: Computer Graphics 1
5.4 The Matrices of a Linear Transformation 2
5.5 Changing Coordinates 2

6 Orthogonality **12 Lectures**

6.1 Projection Matrices 2
6.2 Application: Data Fitting . . . . $1\frac{1}{2}$
6.3 The Gram-Schmidt Algorithm and QR Factorization . . . . $2\frac{1}{2}$
6.4 Orthogonal Subspaces and Complements 3
6.5 The Pseudoinverse of a Matrix 3

7 The Spectral Theorem **4 or 10 Lectures**

7.1 Complex Numbers and Matrices 2
7.2 Unitary Diagonalization 3
7.3 Real Symmetric Matrices 2
7.4 Application: Quadratic Forms, Conic Sections 1
7.5 The Singular Value Decomposition 2

Chapter 1

Euclidean n-Space

1.1 Vectors and Arrows

A *two-dimensional vector* is a pair of numbers written one above the other in a column and enclosed in brackets. For example,

$$\begin{bmatrix}1\\3\end{bmatrix}, \quad \begin{bmatrix}2\\4\end{bmatrix}, \quad \begin{bmatrix}-2\\3\end{bmatrix}, \quad \begin{bmatrix}0\\0\end{bmatrix}$$

are two-dimensional vectors. Different people use different notation for vectors. Some people underline, others use boldface type and still others arrows. Thus, in various contexts, you may well see

$$\underline{v}, \quad \mathsf{v} \quad \text{and} \quad \vec{v}$$

as notation for a vector. In this book, we will use boldface, the second form, but in handwriting the author prefers to underline.

The *components* of the vector $\mathsf{v} = \begin{bmatrix}a\\b\end{bmatrix}$ are the numbers a and b. By general agreement, vectors are *equal* if and only if they have the same first component and the same second component. Thus, if

$$\begin{bmatrix}a-3\\2b\end{bmatrix} = \begin{bmatrix}-1\\6\end{bmatrix},$$

then $a - 3 = -1$ and $2b = 6$, so $a = 2$ and $b = 3$. The vector $\begin{bmatrix}a\\b\end{bmatrix}$ is often pictured by an arrow in the plane.

1.1.1 Take any point $A(x_0, y_0)$ as the tail and $B(x_0 + a, y_0 + b)$ as the head of an arrow. This arrow, from A to B, is a picture of the vector $\begin{bmatrix}a\\b\end{bmatrix}$.

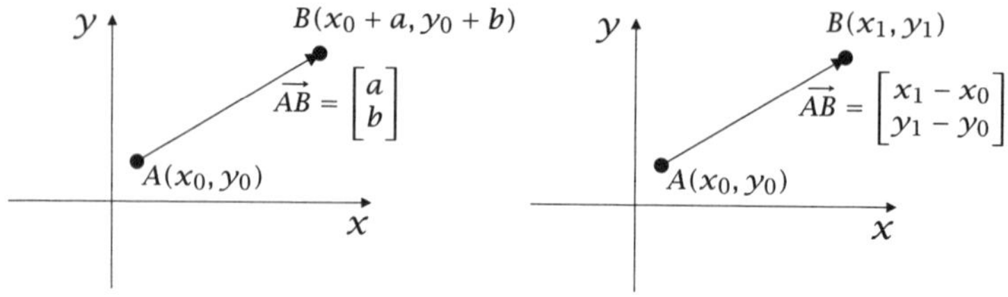

Figure 1.1: The arrow from $A(x_0, y_0)$ to $B(x_1, y_1)$ is a picture of the vector $\begin{bmatrix} x_1 - x_0 \\ y_1 - y_0 \end{bmatrix}$.

The notation $\overrightarrow{AB}$ means the vector pictured by the arrow from A to B. Thus $\overrightarrow{AB} = \begin{bmatrix} a \\ b \end{bmatrix}$ is pictured by the arrow shown on the left in Figure 1.1. It is important to distinguish between vectors, which are columns of numbers, and arrows, which are pictures.

Letting $x_1 = x_0 + a$ and $y_1 = y_0 + b$, so that the coordinates of B become (x_1, y_1), we have $a = x_1 - x_0$ and $b = y_1 - y_0$, so $\overrightarrow{AB} = \begin{bmatrix} a \\ b \end{bmatrix} = \begin{bmatrix} x_1 - x_0 \\ y_1 - y_0 \end{bmatrix}$.

1.1.2 The arrow from $A(x_0, y_0)$ to $B(x_1, y_1)$ is a picture of the vector $\overrightarrow{AB} = \begin{bmatrix} x_1 - x_0 \\ y_1 - y_0 \end{bmatrix}$.

For example, if $A = (2,3)$ and $B = (7,4)$, the vector $\overrightarrow{AB}$ is $\begin{bmatrix} 7-2 \\ 4-3 \end{bmatrix} = \begin{bmatrix} 5 \\ 1 \end{bmatrix}$.

READING CHECK 1. If $A = (-2,3)$ and $B = (1,5)$, what is the vector $\overrightarrow{AB}$?

READING CHECK 2. Suppose the arrow from A to B is a picture of the vector $\overrightarrow{AB} = \begin{bmatrix} 3 \\ -1 \end{bmatrix}$. If $A = (-4,2)$, what is B?

Notice that a vector can be pictured by many arrows, since we can place the tail at any point (x_0, y_0). Each of the arrows in Figure 1.2 is a picture of the vector $\begin{bmatrix} 1 \\ 2 \end{bmatrix}$. How do we know if two arrows are pictures of the same vector?

1.1.3 Two arrows picture the same vector if and only if they have the same length and the same direction.

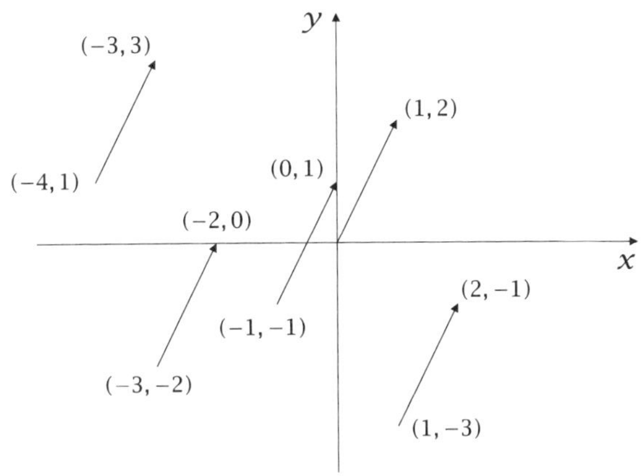

Figure 1.2: Five arrows, each one a picture of the vector $\begin{bmatrix}1\\2\end{bmatrix}$.

Scalar Multiplication

We can multiply vectors by numbers, an operation called "scalar multiplication." Almost always, "scalar"[1] means "real number," so "scalar multiplication" means "multiplication by a real number." If v is a vector and c is a scalar, we produce the *scalar multiple* cv multiplying "componentwise" in the obvious way. For example,

$$4\begin{bmatrix}-1\\3\end{bmatrix}=\begin{bmatrix}-4\\12\end{bmatrix}; \quad -2\begin{bmatrix}2\\-3\end{bmatrix}=\begin{bmatrix}-4\\6\end{bmatrix}; \quad 0\begin{bmatrix}4\\\sqrt{3}\end{bmatrix}=\begin{bmatrix}0\\0\end{bmatrix};$$
$$-\begin{bmatrix}-3\\8\end{bmatrix}=(-1)\begin{bmatrix}-3\\8\end{bmatrix}=\begin{bmatrix}3\\-8\end{bmatrix}.$$

As illustrated by the last example, −v means (−1)v.

READING CHECK 3. Is −v a scalar multiple of v?

Can you see the connection between an arrow for v and an arrow for 2v or an arrow for −v? Look at Figure 1.3. The vector 2v has the same direction as v, but it is twice as long; −v has the same length as but direction opposite that of v; $0\mathsf{v}=\begin{bmatrix}0\\0\end{bmatrix}$ is called the *zero vector* and denoted with a boldface **0**. It has length 0, no direction and is pictured by a single point.

1.1.4 Definitions. Vectors u and v are *parallel* if one is a scalar multiple of the other, that is, if u = cv or v = cu for some scalar c. They have the *same direction* if $c > 0$ and *opposite direction* if $c < 0$.

[1]In Chapter 7, scalars will be complex numbers. In general, scalars can come from any "field" (a special kind of algebraic number system). They might be just 0s and 1s, for instance.

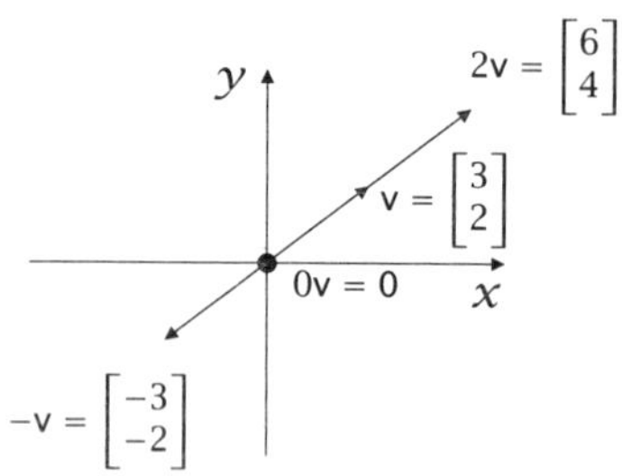

Figure 1.3

Since $0 = 0u$ for any vector u, the zero vector is parallel to any vector.

READING CHECK 4. Some students misinterpret the definition of "parallel." They claim that vectors are not parallel because "they are not scalar multiples of each other." Is this correct?

Vector Addition

We add vectors just as we multiply by scalars, componentwise:

$$\begin{bmatrix} 1 \\ -3 \end{bmatrix} + \begin{bmatrix} 2 \\ 5 \end{bmatrix} = \begin{bmatrix} 3 \\ 2 \end{bmatrix}, \quad \begin{bmatrix} -1 \\ 4 \end{bmatrix} + \begin{bmatrix} 1 \\ -4 \end{bmatrix} = \begin{bmatrix} 0 \\ 0 \end{bmatrix}, \quad \begin{bmatrix} 2 \\ 3 \end{bmatrix} + \begin{bmatrix} 0 \\ 0 \end{bmatrix} = \begin{bmatrix} 2 \\ 3 \end{bmatrix}.$$

There is a nice connection between arrows for vectors u and v and an arrow for the sum u + v.

1.1.5 **Parallelogram Rule:** If arrows for vectors u and v are drawn with the same tail O, then an arrow for u + v is the diagonal of the parallelogram whose sides are u and v with tail at O.

We illustrate in Figure 1.4 with $u = \overrightarrow{OB} = \begin{bmatrix} 3 \\ 1 \end{bmatrix}$ and $v = \overrightarrow{OA} = \begin{bmatrix} 1 \\ 2 \end{bmatrix}$. The sum of these vectors is

$$u + v = \begin{bmatrix} 3 \\ 1 \end{bmatrix} + \begin{bmatrix} 1 \\ 2 \end{bmatrix} = \begin{bmatrix} 4 \\ 3 \end{bmatrix}.$$

The parallelogram rule says that this vector can be pictured by the diagonal OC of the parallelogram $OACB$. So we have to convince ourselves that C is the point $(4, 3)$. To see why this is the case, use the fact that OA and BC are parallel and of the same length to conclude that triangles BQC and OPA are congruent, so BQ has length 1, QC has length 2 and, indeed, the coordinates of C are $(4, 3)$: $u + v = \overrightarrow{OC}$ is pictured by the arrow from O to C.

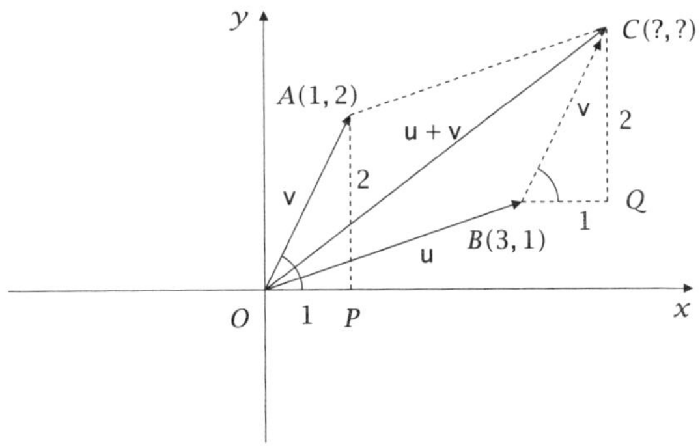

Figure 1.4: Vector u + v can be pictured by the arrow that is the diagonal of the parallelogram with sides u and v.

There is another way to picture u + v. Still with reference to Figure 1.4, we note that v = $\overrightarrow{OA} = \overrightarrow{BC}$. So $\overrightarrow{OC}$ = u + v = $\overrightarrow{OB} + \overrightarrow{BC}$: The vector $\overrightarrow{OC}$ is the third side of triangle OBC. This gives a second way to picture the sum of two vectors.

1.1.6 **Triangle Rule:** If an arrow for u goes from A to B and an arrow for v from B to C, then an arrow for u + v goes from A to C: $\overrightarrow{AB} + \overrightarrow{BC} = \overrightarrow{AC}$.

Figure 1.5: On the left, the parallelogram rule for vector addition; on the right, the triangle rule.

Subtracting Vectors

Just as 5 − 2 means 5 + (−2), we subtract vectors using the convention that

1.1.7 $$u - v = u + (-v).$$

For instance,

$$\begin{bmatrix}7\\2\end{bmatrix} - \begin{bmatrix}5\\3\end{bmatrix} = \begin{bmatrix}7\\2\end{bmatrix} + \left(-\begin{bmatrix}5\\3\end{bmatrix}\right) = \begin{bmatrix}7\\2\end{bmatrix} + \begin{bmatrix}-5\\-3\end{bmatrix} = \begin{bmatrix}2\\-1\end{bmatrix}.$$

Similarly,

$$\begin{bmatrix}-1\\3\end{bmatrix} - \begin{bmatrix}-2\\2\end{bmatrix} = \begin{bmatrix}1\\1\end{bmatrix} \quad \text{and} \quad \begin{bmatrix}2\\0\end{bmatrix} - \begin{bmatrix}0\\-3\end{bmatrix} = \begin{bmatrix}2\\3\end{bmatrix}.$$

Again, we look for a geometrical interpretation of this operation. How can we find an arrow for $u - v$ from arrows for u and v?

Since $v + (u - v) = u$—see READING CHECK 5—it follows from the triangle rule that if we represent v by the arrow from A to C and $u - v$ by an arrow from C to B, then the arrow from A to B is a picture of u: $\overrightarrow{AC} + \overrightarrow{CB} = \overrightarrow{AB}$, so $\overrightarrow{AB} - \overrightarrow{AC} = \overrightarrow{CB}$.

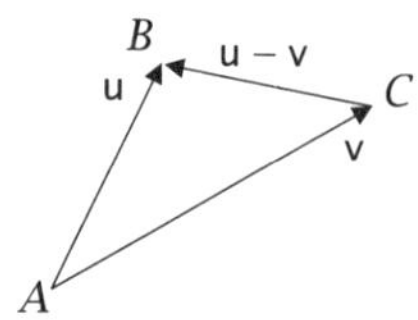

1.1.8 If vectors u and v are pictured by arrows with the same tail, then $u - v$ can be pictured by the arrow that goes from the head of v to the head of u.

Notice that $u + v$ and $u - v$ can be pictured by arrows that are the diagonals of the parallelogram with sides u and v.

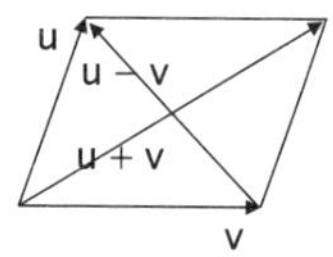

The facts that vectors can be pictured by arrows and that any two arrows with the same length and direction describe the same vector can be used to prove some familiar propositions from Euclidean geometry.

1.1.9 **Problem.** Prove that the diagonals of a parallelogram bisect each other.

Solution. Label the vertices of the parallelogram A, B, C, D and draw the diagonal AC. Let X be the midpoint of AC. We wish to show that X is also the midpoint of BD, equivalently, that the arrows from B to X and X to D have the same length and direction, that is, $\overrightarrow{BX} = \overrightarrow{XD}$. Since the arrows from A to X and from X to C have the same length and the same direction, $\overrightarrow{AX} = \overrightarrow{XC}$. Now $\overrightarrow{BX} = \overrightarrow{BA} + \overrightarrow{AX}$ and $\overrightarrow{XD} = \overrightarrow{XC} + \overrightarrow{CD}$. Since $\overrightarrow{BA} = \overrightarrow{CD}$ (the arrows have the same length and direction) and $\overrightarrow{AX} = \overrightarrow{XC}$, we have $\overrightarrow{BX} = \overrightarrow{XD}$, the desired result. ☝

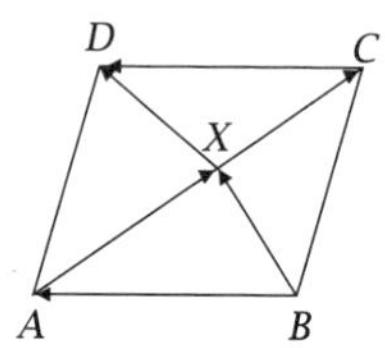

Before continuing, we record some properties of the addition and scalar multiplication of vectors. Most of these properties have names that make them easy to reference.[2]

1.1.10 Theorem (Properties of Vector Addition and Scalar Multiplication). *Let* u, v *and* w *be vectors and let* c *and* d *be scalars.*

1. *(Closure under addition)* $\mathsf{u}+\mathsf{v}$ *is a vector.*
2. *(Commutativity of addition)* $\mathsf{u}+\mathsf{v}=\mathsf{v}+\mathsf{u}$.
3. *(Associativity of addition)* $(\mathsf{u}+\mathsf{v})+\mathsf{w}=\mathsf{u}+(\mathsf{v}+\mathsf{w})$.
4. *(Zero)* $\mathsf{u}+\mathsf{0}=\mathsf{0}+\mathsf{u}=\mathsf{u}$.
5. *(Additive Inverse) There is a vector called the additive inverse of* u *or "minus* u*" and denoted* $-\mathsf{u}$ *which has the property that* $\mathsf{u}+(-\mathsf{u})=(-\mathsf{u})+\mathsf{u}=\mathsf{0}$.
6. *(Closure under scalar multiplication)* $c\mathsf{u}$ *is a vector.*
7. *(Scalar associativity)* $c(d\mathsf{u})=(cd)\mathsf{u}$.
8. *(One)* $1\mathsf{u}=\mathsf{u}$.
9. *(Distributivity)* $c(\mathsf{u}+\mathsf{v})=c\mathsf{u}+c\mathsf{v}$ *and* $(c+d)\mathsf{u}=c\mathsf{u}+d\mathsf{u}$.

We leave proofs of these simple facts to the exercises. Together, they say that vector algebra is very much like ordinary algebra.

1.1.11 Examples.

- $4(\mathsf{u}-3\mathsf{v})+6(-2\mathsf{u}+3\mathsf{v})=-8\mathsf{u}+6\mathsf{v}$.
- If $2\mathsf{u}+3\mathsf{v}=6\mathsf{x}+4\mathsf{u}$, then $6\mathsf{x}=-2\mathsf{u}+3\mathsf{v}$, so $\mathsf{x}=-\frac{1}{3}\mathsf{u}+\frac{1}{2}\mathsf{v}$.
- If $\mathsf{x}=3\mathsf{u}-2\mathsf{v}$ and $\mathsf{y}=\mathsf{u}+\mathsf{v}$, then u and v can be expressed in terms of x and y like this:
$$\begin{array}{rcl} \mathsf{x} & = & 3\mathsf{u}-2\mathsf{v} \\ 2\mathsf{y} & = & 2\mathsf{u}+2\mathsf{v} \\ \hline \mathsf{x}+2\mathsf{y} & = & 5\mathsf{u}\,, \end{array}$$
so $\mathsf{u}=\frac{1}{5}(\mathsf{x}+2\mathsf{y})$ and $\mathsf{v}=\mathsf{y}-\mathsf{u}=-\frac{1}{5}\mathsf{x}+\frac{3}{5}\mathsf{y}$. ☺

READING CHECK 5. How does $\mathsf{v}+(\mathsf{u}-\mathsf{v})=\mathsf{u}$ follow from Theorem 1.1.10? Show all steps and justify each one.

[2] We have already used some of these properties. They are so natural that it is easy to use them without realizing it.

Higher Dimensions

Most of what we have said about two-dimensional vectors applies more generally, although our ability to draw pictures becomes difficult or impossible.

1.1.12 Definition. For any positive integer $n \geq 1$, an *n-dimensional vector* is a column $\begin{bmatrix} x_1 \\ x_2 \\ \vdots \\ x_n \end{bmatrix}$ of n numbers enclosed in brackets. The numbers $x_1, x_2, \ldots, x_n$ are called the *components* of the vector.

1.1.13 Examples.

- $\begin{bmatrix} 1 \\ 2 \\ 3 \end{bmatrix}$ is a three-dimensional vector with components 1, 2 and 3;
- $\begin{bmatrix} -1 \\ 0 \\ 0 \\ 1 \\ 2 \end{bmatrix}$ is a five-dimensional vector with components -1, 0, 0, 1, 2;
- $\begin{bmatrix} 1 \\ 2 \\ 3 \\ \vdots \\ n \end{bmatrix}$ is an n-dimensional vector whose components are $1, 2, 3, \ldots, n$;
- The n-dimensional *zero vector* is the vector $\begin{bmatrix} 0 \\ 0 \\ \vdots \\ 0 \end{bmatrix}$ whose n components are all 0.

☺

1.1.14 Definition. *Euclidean n-space* is the set of all n-dimensional vectors. It is denoted R^n:

$$\mathsf{R}^n = \left\{ \begin{bmatrix} x_1 \\ x_2 \\ \vdots \\ x_n \end{bmatrix} \mid x_1, x_2, \ldots, x_n \in \mathsf{R} \right\}.$$

Euclidean 2-space,

$$\mathsf{R}^2 = \left\{ \begin{bmatrix} x \\ y \end{bmatrix} \mid x, y \in \mathsf{R} \right\}$$

is more commonly called the *Euclidean plane* or the *plane* or sometimes the *xy-plane*. Euclidean 3-space

$$\mathsf{R}^3 = \left\{ \begin{bmatrix} x \\ y \\ z \end{bmatrix} \mid x, y, z \in \mathsf{R} \right\}$$

is often called simply 3-*space*.

Vector Algebra

Two vectors $\mathsf{x} = \begin{bmatrix} x_1 \\ x_2 \\ \vdots \\ x_n \end{bmatrix}$ and $\mathsf{y} = \begin{bmatrix} y_1 \\ y_2 \\ \vdots \\ y_m \end{bmatrix}$ are *equal* if and only if $n = m$ and the corresponding components are equal: $x_1 = y_1$, $x_2 = y_2$, ..., $x_n = y_n$.

For example, $\begin{bmatrix} 1 \\ 2 \end{bmatrix}$ and $\begin{bmatrix} 1 \\ 2 \\ 0 \end{bmatrix}$ are not equal while $\mathsf{u} = \begin{bmatrix} a \\ b \\ c \end{bmatrix}$ and $\mathsf{v} = \begin{bmatrix} 3 \\ -2 \\ 7 \end{bmatrix}$ are equal if and only if $a = 3$, $b = -2$ and $c = 7$.

Addition and scalar multiplication of n-dimensional vectors are defined componentwise, as you would expect. If $\mathsf{x} = \begin{bmatrix} x_1 \\ x_2 \\ \vdots \\ x_n \end{bmatrix}$ and $\mathsf{y} = \begin{bmatrix} y_1 \\ y_2 \\ \vdots \\ y_n \end{bmatrix}$ and c is a scalar, then

$$\mathsf{x} + \mathsf{y} = \begin{bmatrix} x_1 + y_1 \\ x_2 + y_2 \\ \vdots \\ x_n + y_n \end{bmatrix} \quad \text{and} \quad c\mathsf{x} = \begin{bmatrix} cx_1 \\ cx_2 \\ \vdots \\ cx_n \end{bmatrix}.$$

1.1.15 Examples. $\begin{bmatrix} -2 \\ 4 \\ 7 \end{bmatrix} + \begin{bmatrix} 1 \\ -1 \\ 6 \end{bmatrix} = \begin{bmatrix} -1 \\ 3 \\ 13 \end{bmatrix}$; $-2\begin{bmatrix} 4 \\ -1 \\ 5 \end{bmatrix} = \begin{bmatrix} -8 \\ 2 \\ -10 \end{bmatrix}$. :)

Linear Combinations

A sum of scalar multiples of vectors, such as $-8\mathsf{u} + 6\mathsf{v}$, has a special name.

1.1.16 Definition. A *linear combination* of vectors u and v is a vector of the form $a\mathsf{u} + b\mathsf{v}$, where a and b are scalars. More generally, a *linear combination* of k vectors $\mathsf{u}_1, \mathsf{u}_2, \dots, \mathsf{u}_k$ is a vector of the form $c_1\mathsf{u}_1 + c_2\mathsf{u}_2 + \cdots + c_k\mathsf{u}_k$, where $c_1, \dots, c_k$ are scalars.

1.1.17 Examples. • $\begin{bmatrix} -5 \\ 9 \end{bmatrix}$ is a linear combination of $\mathsf{u} = \begin{bmatrix} -2 \\ 3 \end{bmatrix}$ and $\mathsf{v} = \begin{bmatrix} -1 \\ 1 \end{bmatrix}$ since $\begin{bmatrix} -5 \\ 9 \end{bmatrix} = 4\begin{bmatrix} -2 \\ 3 \end{bmatrix} - 3\begin{bmatrix} -1 \\ 1 \end{bmatrix} = 4\mathsf{u} - 3\mathsf{v}$.

- $\begin{bmatrix} 2 \\ -6 \end{bmatrix}$ is a linear combination of $\mathsf{u}_1 = \begin{bmatrix} -2 \\ 3 \end{bmatrix}$, $\mathsf{u}_2 = \begin{bmatrix} 6 \\ -5 \end{bmatrix}$ and $\mathsf{u}_3 = \begin{bmatrix} 4 \\ 5 \end{bmatrix}$ since $\begin{bmatrix} 2 \\ -6 \end{bmatrix} = 3\begin{bmatrix} -2 \\ 3 \end{bmatrix} + 2\begin{bmatrix} 6 \\ -5 \end{bmatrix} - \begin{bmatrix} 4 \\ 5 \end{bmatrix} = 3\mathsf{u}_1 + 2\mathsf{u}_2 + (-1)\mathsf{u}_3$.

- The equation $\begin{bmatrix} -11 \\ 7 \\ -4 \\ 2 \end{bmatrix} = 2\begin{bmatrix} -2 \\ 0 \\ 5 \\ 1 \end{bmatrix} - 7\begin{bmatrix} 1 \\ -1 \\ 2 \\ 0 \end{bmatrix}$ says that $\begin{bmatrix} -11 \\ 7 \\ -4 \\ 2 \end{bmatrix}$ is a linear combination of $\begin{bmatrix} -2 \\ 0 \\ 5 \\ 1 \end{bmatrix}$ and $\begin{bmatrix} 1 \\ -1 \\ 2 \\ 0 \end{bmatrix}$. ☺

1.1.18 Example. The vector $\begin{bmatrix} 1 \\ 5 \\ 2 \\ 0 \\ -1 \end{bmatrix}$ is not a linear combination of $\begin{bmatrix} 1 \\ 2 \\ -1 \\ 3 \\ 4 \end{bmatrix}$ and $\begin{bmatrix} 3 \\ 5 \\ 0 \\ 1 \\ 2 \end{bmatrix}$ because $\begin{bmatrix} 1 \\ 5 \\ 2 \\ 0 \\ -1 \end{bmatrix} = a\begin{bmatrix} 1 \\ 2 \\ -1 \\ 3 \\ 4 \end{bmatrix} + b\begin{bmatrix} 3 \\ 5 \\ 0 \\ 1 \\ 2 \end{bmatrix}$ implies

$$\begin{aligned} a + 3b &= 1 \\ 2a + 5b &= 5 \\ -a &= 2 \\ 3a + b &= 0 \\ 4a + 2b &= -1. \end{aligned}$$

The third equation says $a = -2$. Comparing with the first we get $b = 1$, but the pair $a = -2, b = 1$ does not satisfy the second equation. The required scalars a and b do not exist. ☺

1.1.19 Problem. Determine whether or not the vector $\mathsf{x} = \begin{bmatrix} -1 \\ -2 \\ 2 \end{bmatrix}$ is a linear combination of $\mathsf{u} = \begin{bmatrix} 0 \\ 1 \\ 4 \end{bmatrix}$, $\mathsf{v} = \begin{bmatrix} -1 \\ 1 \\ 2 \end{bmatrix}$ and $\mathsf{w} = \begin{bmatrix} 3 \\ 1 \\ -2 \end{bmatrix}$.

Solution. The question asks whether or not there exist scalars a, b, c so that $\mathsf{x} = a\mathsf{u} + b\mathsf{v} + c\mathsf{w}$; that is, such that

$$\begin{bmatrix} -1 \\ -2 \\ 2 \end{bmatrix} = a\begin{bmatrix} 0 \\ 1 \\ 4 \end{bmatrix} + b\begin{bmatrix} -1 \\ 1 \\ 2 \end{bmatrix} + c\begin{bmatrix} 3 \\ 1 \\ -2 \end{bmatrix}.$$

The vector on the right is $\begin{bmatrix} -b+3c \\ a+b+c \\ 4a+2b-2c \end{bmatrix}$, so the question is, are there numbers a, b, c such that

$$\begin{aligned} -\quad b + 3c &= -1 \\ a + \quad b + \quad c &= -2 \\ 4a + 2b - 2c &= \quad 2 \end{aligned} \quad ?$$

We find that $a = 1$, $b = -2$, $c = -1$ is a solution.[3] Thus $\mathsf{x} = \mathsf{u} - 2\mathsf{v} - \mathsf{w}$ is a linear combination of u, v and w.

1.1.20 Problem. Is $\begin{bmatrix} 3 \\ 0 \end{bmatrix}$ a linear combination of $\begin{bmatrix} -1 \\ 2 \end{bmatrix}$ and $\begin{bmatrix} -3 \\ 3 \end{bmatrix}$?

Solution. The question asks if there are scalars a and b such that $\begin{bmatrix} 3 \\ 0 \end{bmatrix} = a\begin{bmatrix} -1 \\ 2 \end{bmatrix} + b\begin{bmatrix} -3 \\ 3 \end{bmatrix}$. Equating corresponding components, we must have $3 = -a - 3b$ and $0 = 2a + 3b$. Adding these equations gives $a = 3$, so $-3b = 3 + a = 6$ and $b = -2$. The answer is "yes:" $\begin{bmatrix} 3 \\ 0 \end{bmatrix} = 3\begin{bmatrix} -1 \\ 2 \end{bmatrix} - 2\begin{bmatrix} -3 \\ 3 \end{bmatrix}$ is a linear combination of $\begin{bmatrix} -1 \\ 2 \end{bmatrix}$ and $\begin{bmatrix} -3 \\ 3 \end{bmatrix}$.

READING CHECK 6. Is $\begin{bmatrix} 0 \\ 0 \end{bmatrix}$ a linear combination of $\begin{bmatrix} -1 \\ 2 \end{bmatrix}$ and $\begin{bmatrix} -3 \\ 3 \end{bmatrix}$?

Actually, the answer to Problem 1.1.20 does not depend on the given vector $\begin{bmatrix} 3 \\ 0 \end{bmatrix}$: Any two-dimensional vector is a linear combination of u and v. Moreover, this is the case for any u and v that are not parallel.

---

[3] Whether or not you can find a, b and c yourself at this point isn't important. Solving systems of equations such as the one here is the subject of Section 2.4.

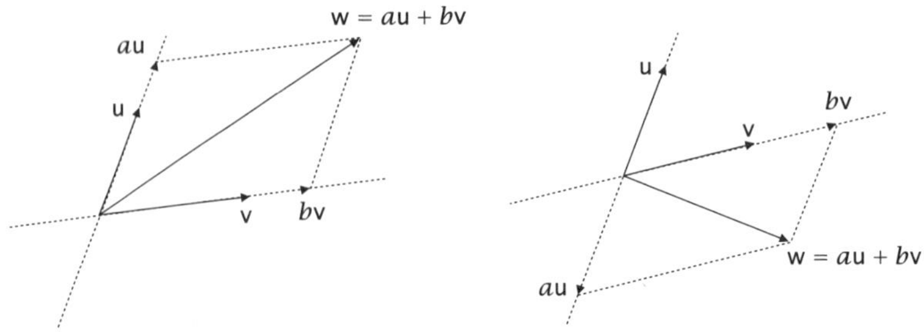

Figure 1.6: Any vector w is a linear combination of u and v.

In Figure 1.6, we show two situations of nonparallel vectors u and v and another vector w and, in each case, we show a parallelogram whose diagonal is w and whose sides are multiples of u and v. Thus the sides have the form au and bv and w is their sum au + bv, a linear combination of u and v.

So any vector in the plane is a linear combination of u and v. Certainly the converse is true: any linear combination of u and v is in the xy-plane, so we have met a very important idea.

1.1.21 If two-dimensional vectors u and v are not parallel, then the set of linear combinations of u and v is the entire xy-plane.

In fact, we can deduce a more general fact from Figure 1.6. By lifting your book from your desk and moving it around so that the vectors u and v lie in many planes, we see that

1.1.22 If three-dimensional vectors u and v are not parallel, then the set of all linear combinations of u and v is a plane in R^3.

1.1.23 Definition. The *plane spanned by vectors* u *and* v is the set of all linear combinations of u and v.

We reinforce this idea in Figure 1.7. The arrow from 0 to A is a picture of u + 2v and the arrow from 0 to B is a picture of 2u + v. Every vector in the plane shown in the figure is a linear combination of u and v.

READING CHECK 7. What vector is pictured by the arrow from 0 to C?

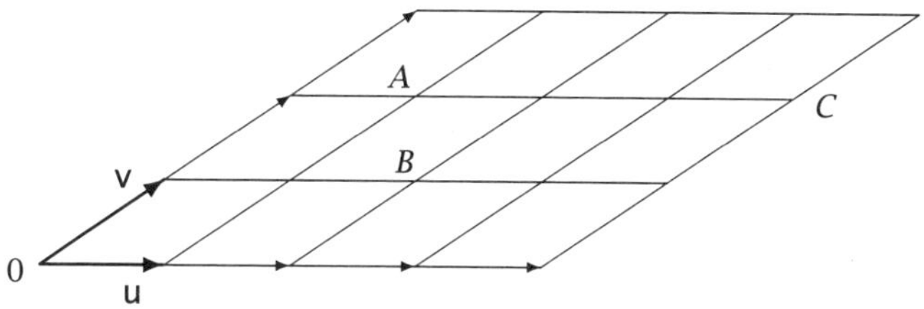

Figure 1.7: The set of all linear combinations of u and v is a plane.

The Span of Vectors

1.1.24 Definition. The *span* of vectors $\mathsf{u}_1, \mathsf{u}_2, \dots, \mathsf{u}_k$ is the set U of all linear combinations of these vectors. We refer to U as the set *spanned* by $\mathsf{u}_1, \mathsf{u}_2, \dots, \mathsf{u}_k$.

"Span" is used both as a noun and as a verb.

1.1.25 Examples. • The span of a single vector u is $\{a\mathsf{u} \mid a \in \mathsf{R}\}$ is just the set of all scalar multiples of u, which is a line, and a line passing through the origin (because $\mathsf{0} = 0\mathsf{u}$ is a scalar multiple of u).

- The *xy-plane* in R^3 is the set of all vectors of the form $\begin{bmatrix} x \\ y \\ 0 \end{bmatrix}$. This plane is spanned by $\mathsf{i} = \begin{bmatrix} 1 \\ 0 \\ 0 \end{bmatrix}$ and $\mathsf{j} = \begin{bmatrix} 0 \\ 1 \\ 0 \end{bmatrix}$ because any $\begin{bmatrix} x \\ y \\ 0 \end{bmatrix}$ can be written $x\begin{bmatrix} 1 \\ 0 \\ 0 \end{bmatrix} + y\begin{bmatrix} 0 \\ 1 \\ 0 \end{bmatrix} = x\mathsf{i} + y\mathsf{j}$.

- The span of two nonparallel vectors u and v in R^3 is the set of all linear combinations of u and v, which is a plane, and a plane containing the origin (because $\mathsf{0} = 0\mathsf{u} + 0\mathsf{v}$ is a linear combination of u and v).

- The span of the vectors $\begin{bmatrix} 1 \\ 2 \\ 3 \\ 4 \end{bmatrix}, \begin{bmatrix} -2 \\ 0 \\ 1 \\ 5 \end{bmatrix}, \begin{bmatrix} 3 \\ 1 \\ 1 \\ -4 \end{bmatrix}$ is the set of all vectors of the form

$$a\begin{bmatrix} 1 \\ 2 \\ 3 \\ 4 \end{bmatrix} + b\begin{bmatrix} -2 \\ 0 \\ 1 \\ 5 \end{bmatrix} + c\begin{bmatrix} 3 \\ 1 \\ 1 \\ -4 \end{bmatrix},$$

that is, the set of all vectors in R^4 that can be written in the form $\begin{bmatrix} a-2b+3c \\ 2a+c \\ 3a+b+c \\ 4a+5b-4c \end{bmatrix}$ for scalars a, b, c. ☺

1.1.26 Remark. The span of vectors $\mathsf{u}_1, \mathsf{u}_2, \ldots, \mathsf{u}_k$ always contains the zero vector because $\mathsf{0} = 0\mathsf{u}_1 + 0\mathsf{u}_2 + \cdots + 0\mathsf{u}_k$ is a linear combination of these vectors.

We extend the concept of "plane" to higher dimensions.

1.1.27 Definition. A *plane* in R^n, for any $n \geq 2$, is the span of two nonparallel vectors u and v.

1.1.28 Problem. Let $\mathsf{u} = \begin{bmatrix} -1 \\ 0 \\ 2 \\ 1 \end{bmatrix}$ and $\mathsf{v} = \begin{bmatrix} 2 \\ 3 \\ -4 \\ 1 \end{bmatrix}$. Describe the plane spanned by u and v. Does $\begin{bmatrix} 1 \\ 6 \\ -1 \\ 0 \end{bmatrix}$ belong to this plane?

Solution. The plane spanned by u and v is the set of linear combinations of u and v. So the plane consists of vectors of the form

$$a\mathsf{u} + b\mathsf{v} = a\begin{bmatrix} -1 \\ 0 \\ 2 \\ 1 \end{bmatrix} + b\begin{bmatrix} 2 \\ 3 \\ -4 \\ 1 \end{bmatrix} = \begin{bmatrix} -a+2b \\ 3b \\ 2a-4b \\ a+b \end{bmatrix}.$$

The vector $\begin{bmatrix} 1 \\ 6 \\ -1 \\ 0 \end{bmatrix}$ is in the plane if and only if there are numbers a and b so that

$$\begin{aligned} -a + 2b &= 1 \\ 3b &= 6 \\ 2a - 4b &= -1 \\ a + b &= 0. \end{aligned}$$

So we would need $b = 2$ and $a = -b = -2$ (equations two and four), but then the first equation is not satisfied. No such a and b exist, so the vector is **not** in the plane. ☞

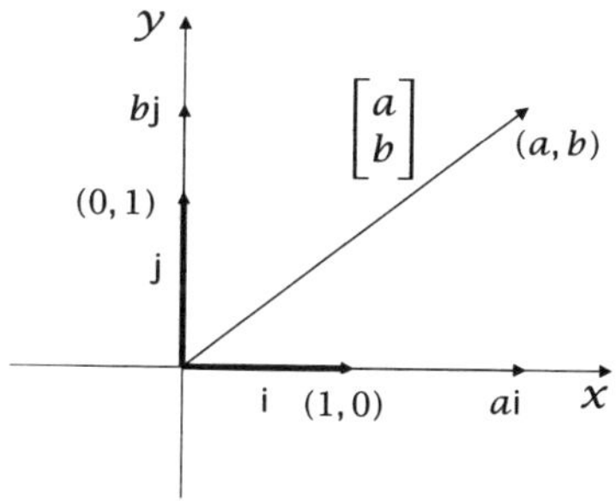

Figure 1.8: $\begin{bmatrix} a \\ b \end{bmatrix} = a\mathsf{i} + b\mathsf{j}$.

The Standard Basis Vectors

1.1.29 Definition. The *standard basis vectors* in the plane are $\mathsf{i} = \begin{bmatrix} 1 \\ 0 \end{bmatrix}$ and $\mathsf{j} = \begin{bmatrix} 0 \\ 1 \end{bmatrix}$.

The *standard basis vectors* in R^3 are $\mathsf{i} = \begin{bmatrix} 1 \\ 0 \\ 0 \end{bmatrix}$, $\mathsf{j} = \begin{bmatrix} 0 \\ 1 \\ 0 \end{bmatrix}$ and $\mathsf{k} = \begin{bmatrix} 0 \\ 0 \\ 1 \end{bmatrix}$. The *standard basis vectors* in R^n are the vectors $\mathsf{e}_1, \mathsf{e}_2, \ldots, \mathsf{e}_n$ where e_i has ith component 1 and all other components 0:

$$\mathsf{e}_1 = \begin{bmatrix} 1 \\ 0 \\ 0 \\ 0 \\ \vdots \\ 0 \end{bmatrix}, \quad \mathsf{e}_2 = \begin{bmatrix} 0 \\ 1 \\ 0 \\ 0 \\ \vdots \\ 0 \end{bmatrix}, \quad \mathsf{e}_3 = \begin{bmatrix} 0 \\ 0 \\ 1 \\ 0 \\ \vdots \\ 0 \end{bmatrix}, \quad \ldots, \quad \mathsf{e}_n = \begin{bmatrix} 0 \\ 0 \\ 0 \\ \vdots \\ 0 \\ 1 \end{bmatrix}.$$

Note that in R^2, $\mathsf{e}_1 = \begin{bmatrix} 1 \\ 0 \end{bmatrix} = \mathsf{i}$ and $\mathsf{e}_2 = \begin{bmatrix} 0 \\ 1 \end{bmatrix} = \mathsf{j}$. These are pictured in Figure 1.8. and it is ridiculously easy to see that

1.1.30

> Every two-dimensional vector is a linear combination of i and j: $\begin{bmatrix} a \\ b \end{bmatrix} = a\mathsf{i} + b\mathsf{j}$.

For example, $\begin{bmatrix} 3 \\ 4 \end{bmatrix} = 3\begin{bmatrix} 1 \\ 0 \end{bmatrix} + 4\begin{bmatrix} 0 \\ 1 \end{bmatrix} = 3\mathsf{i} + 4\mathsf{j}$.

In R^3, $\mathsf{e}_1 = \begin{bmatrix} 1 \\ 0 \\ 0 \end{bmatrix} = \mathsf{i}$, $\mathsf{e}_2 = \begin{bmatrix} 0 \\ 1 \\ 0 \end{bmatrix} = \mathsf{j}$ and $\mathsf{e}_3 = \begin{bmatrix} 0 \\ 0 \\ 1 \end{bmatrix} = \mathsf{k}$. These vectors are pictured in Figure 1.9, which also illustrates that

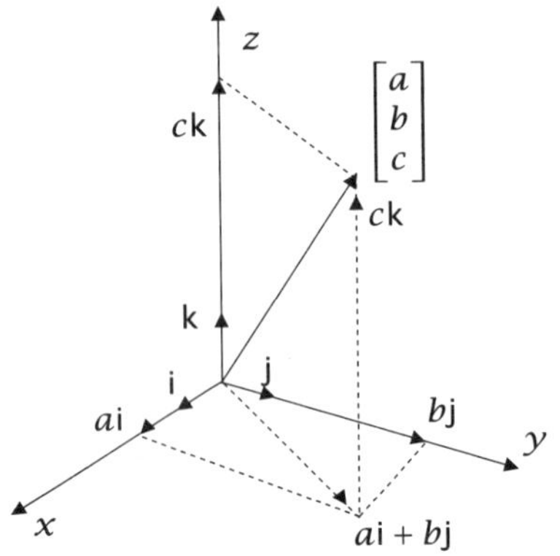

Figure 1.9: $\begin{bmatrix} a \\ b \\ c \end{bmatrix} = a\mathsf{i} + b\mathsf{j} + c\mathsf{k}$.

1.1.31

> Every vector in R^3 is a linear combination of i, j and k:
> $$\begin{bmatrix} a \\ b \\ c \end{bmatrix} = a\mathsf{i} + b\mathsf{j} + c\mathsf{k}.$$

In general, every vector in R^n is a linear combination of $\mathsf{e}_1, \mathsf{e}_2, \ldots, \mathsf{e}_n$. Another way to say this is "The standard basis vectors span R^n."

1.1.32

> Every vector in R^n is a linear combination of $\mathsf{e}_1, \mathsf{e}_2, \ldots, \mathsf{e}_n$:
> $$\begin{bmatrix} x_1 \\ x_2 \\ \vdots \\ x_n \end{bmatrix} = x_1\mathsf{e}_1 + x_2\mathsf{e}_2 + \cdots + x_n\mathsf{e}_n.$$

These facts remind the author of questions such as "Who wrote Brahms' *Lullaby*?" and "Who is buried in Grant's tomb?" To ask how to write $\begin{bmatrix} 3 \\ 5 \\ 9 \end{bmatrix}$ as a linear combination of i, j, k is to give away the answer—$\begin{bmatrix} 3 \\ 5 \\ 9 \end{bmatrix} = 3\mathsf{i} + 5\mathsf{j} + 9\mathsf{k}$.

READING CHECK 8. So who wrote Brahms' *Lullaby*?

READING CHECK 9. How long did the "Seven Years War" last?

READING CHECK 10. After whom is J. R. Smallwood Collegiate named?

READING CHECK 11. Who wrote the *William Tell* overture?

READING CHECK 12. Write $\begin{bmatrix} 3 \\ -4 \\ 5 \\ 6 \end{bmatrix}$ as a linear combination of the standard basis vectors in R^4.

Application: Facial Recognition

1.00	1.00	0.50	1.00	1.00
0.75	0.75	0.75	0.75	0.75
1.00	0.25	1.00	0.25	1.00
1.00	0.25	1.00	0.25	1.00

$$\begin{bmatrix} 1.00 \\ 0.25 \\ 1.00 \\ 0.25 \\ 1.00 \\ 1.00 \\ 0.25 \\ 1.00 \\ 0.25 \\ 1.00 \\ 0.75 \\ 0.75 \\ 0.75 \\ 0.75 \\ 0.75 \\ 1.00 \\ 1.00 \\ 0.50 \\ 1.00 \\ 1.00 \end{bmatrix}$$

Any fan of a crime show must be impressed at the speed with which good likenesses of suspects can be produced from the recollections of witnesses. Vectors provide a useful way to store and manipulate picture data. Stored in a computer, a picture is just a rectangular grid of tiny squares called "pixels." By measuring the average intensity of each pixel, for example on a scale of 0–1, 0 meaning pitch black and 1 pure white, a picture can be converted to a vector that records the intensity of the pixels. Storing pictures as vectors rather than as digital images is very efficient—much less memory is required. Here we attempt to show a robot with dark legs and lightish arms on a white background. The picture consists of 20 pixels and can be described by the shown vector. In reality, a picture would be comprised of tens of thousands of pixels so the vector would live in a Euclidean space R^n with a very large n. With the help of linear algebra, it is possible to create a relatively small number of standard vectors called "eigenfaces," linear combinations of which can be used to make a pretty good approximation to any face. For example, a face might be 20% eigenface 1, 15% eigenface 2, 60% eigenface 3 and 5% eigenface 4: Face $= 0.20\text{EF}_1 + 0.15\text{EF}_2 + 0.60\text{EF}_3 + 0.05\text{EF}_4$. Can you imagine your face as a linear combination of the faces of the people sitting around you?

Answers to Reading Checks

1. $\overrightarrow{AB} = \begin{bmatrix} 1-(-2) \\ 5-3 \end{bmatrix} = \begin{bmatrix} 3 \\ 2 \end{bmatrix}$.

2. Let B have coordinates (x, y). Then $\overrightarrow{AB} = \begin{bmatrix} 3 \\ -1 \end{bmatrix} = \begin{bmatrix} x-(-4) \\ y-2 \end{bmatrix}$. Thus $x + 4 = 3$ and $y - 2 = -1$, so $x = -1$, $y = 1$. The point B has coordinates $(-1, 1)$.

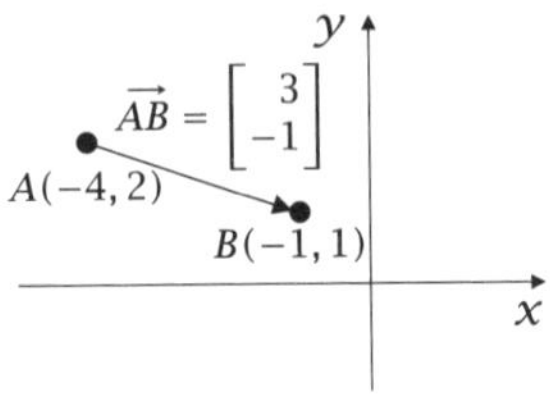

3. Sure it is! We just said that $-\mathsf{v} = (-1)\mathsf{v}$.

4. This is **not** what the definition says. "Not scalar multiples of each other" is not the same as "neither is a scalar multiple of the other." Let $\mathsf{u} = \begin{bmatrix} 0 \\ 0 \\ 0 \end{bmatrix}$ and $\mathsf{v} = \begin{bmatrix} 1 \\ 1 \\ 1 \end{bmatrix}$. These vectors are not scalar multiples of each other because $\mathsf{v} \neq c\mathsf{u}$ for any c, but they are parallel because u is a scalar multiple of v.

5.
$$\begin{aligned} \mathsf{v} + (\mathsf{u} - \mathsf{v}) &= \mathsf{v} + [\mathsf{u} + (-\mathsf{v})] && \text{definition of subtraction} \\ &= \mathsf{v} + [(-\mathsf{v}) + \mathsf{u}] && \text{commutativity} \\ &= [\mathsf{v} + (-\mathsf{v})] + \mathsf{u} && \text{associativity} \\ &= 0 + \mathsf{u} && \text{additive inverse} \\ &= \mathsf{u} && \text{zero} \end{aligned}$$

6. This question can be solved by the approach used in Problem 1.1.20, but it is easier simply to notice that $\begin{bmatrix} 0 \\ 0 \end{bmatrix} = 0\begin{bmatrix} -1 \\ 2 \end{bmatrix} + 0\begin{bmatrix} -3 \\ 3 \end{bmatrix}$. Indeed the zero vector is a linear combination of any set of vectors, with all scalars 0.

7. The arrow from 0 to C represents $4\mathsf{u} + 2\mathsf{v}$.

8. Johannes Brahms!

9. Seven years

10. Joseph R. "Joey" Smallwood, the Premier who brought Newfoundland into Confederation with Canada in 1949.

11. Gioacchino Rossini. (Don't press your luck.)

12. $\begin{bmatrix} 3 \\ -4 \\ 5 \\ 6 \end{bmatrix} = 3\begin{bmatrix} 1 \\ 0 \\ 0 \\ 0 \end{bmatrix} - 4\begin{bmatrix} 0 \\ 1 \\ 0 \\ 0 \end{bmatrix} + 5\begin{bmatrix} 0 \\ 0 \\ 1 \\ 0 \end{bmatrix} + 6\begin{bmatrix} 0 \\ 0 \\ 0 \\ 1 \end{bmatrix} = 3\mathsf{e}_1 - 4\mathsf{e}_2 + 5\mathsf{e}_3 + 6\mathsf{e}_4.$

True/False Questions

Decide, with as little calculation as possible, whether each of the following statements is true or false and, if you say "false," explain your answer. (Answers are at the back of the book.)

1. If $A = (1, 2)$ and $B = (-3, 5)$, the vector $\overrightarrow{AB} = \begin{bmatrix} 4 \\ -3 \end{bmatrix}$.

2. The vectors $\begin{bmatrix} 1 \\ 2 \end{bmatrix}$ and $\begin{bmatrix} 0 \\ 0 \end{bmatrix}$ are parallel.

3. If u and v are nonzero vectors in 3-space that are not parallel, then u and v span a plane.

4. $\begin{bmatrix} 1 \\ 2 \\ 3 \end{bmatrix}$ is a linear combination of $\begin{bmatrix} 1 \\ 0 \\ 1 \end{bmatrix}$ and $\begin{bmatrix} 2 \\ 0 \\ -1 \end{bmatrix}$.

5. $4\mathsf{u} - 9\mathsf{v}$ is a linear combination of $2\mathsf{u}$ and $7\mathsf{v}$.

6. $\begin{bmatrix} 2 \\ -3 \\ 4 \end{bmatrix} = 2\mathsf{i} - 3\mathsf{j} + 4\mathsf{k}$.

7. The zero vector is a linear combination of any three vectors $\mathsf{u}, \mathsf{v}, \mathsf{w}$.

8. The set of all linear combinations of the vectors $\begin{bmatrix} 1 \\ 2 \end{bmatrix}$ and $\begin{bmatrix} 3 \\ 1 \end{bmatrix}$ is a plane.

9. Every vector in R^3 is a linear combination of i, j and k.

10. Every vector in R^3 is a linear combination of i, j, k and $\mathsf{u} = \begin{bmatrix} 1 \\ 2 \\ 3 \end{bmatrix}$.

EXERCISES

Answers to exercises marked [BB] can be found at the Back of the Book.

1. Find the vector $\overrightarrow{AB}$ and illustrate with a picture if

   (a) [BB] $A = (-2, 1)$ and $B = (1, 4)$ **(b)** $A = (1, -3)$ and $B = (-3, 2)$
   (c) $A = (-4, -1)$ and $B = (4, 3)$ **(d)** $A = (2, 4)$ and $B = (-2, -2)$.

2. **(a)** [BB] Find B given $A = (1, 4)$ and $\overrightarrow{AB} = \begin{bmatrix} -1 \\ 2 \end{bmatrix}$.

   (b) Find A given $B = (1, 4)$ and $\overrightarrow{AB} = \begin{bmatrix} -1 \\ 2 \end{bmatrix}$.

   (c) Find A given $B = (-2, 3)$ and $\overrightarrow{AB} = \begin{bmatrix} 4 \\ 7 \end{bmatrix}$.

   (d) Find B given $A = (-9, -5)$ and $\overrightarrow{AB} = \begin{bmatrix} -2 \\ -3 \end{bmatrix}$.

3. [BB] If possible, express $\mathsf{x} = \begin{bmatrix} -2 \\ 7 \\ 4 \end{bmatrix}$ as a scalar multiple of $\mathsf{u} = \begin{bmatrix} 8 \\ -28 \\ -16 \end{bmatrix}$, as a scalar multiple of $\mathsf{v} = \begin{bmatrix} 0 \\ 0 \\ 0 \end{bmatrix}$, and as a scalar multiple of $\mathsf{w} = \begin{bmatrix} \frac{2}{7} \\ -1 \\ -\frac{4}{7} \end{bmatrix}$. Justify your answers.

4. If possible, express $\mathsf{x} = \begin{bmatrix} 6 \\ -2 \\ 8 \end{bmatrix}$ as a scalar multiple of $\mathsf{u} = \begin{bmatrix} 4 \\ -\frac{4}{3} \\ \frac{16}{3} \end{bmatrix}$, of $\mathsf{v} = \begin{bmatrix} \frac{1}{2} \\ -\frac{1}{6} \\ \frac{4}{3} \end{bmatrix}$, of $\mathsf{w} = \begin{bmatrix} -3 \\ 1 \\ -4 \end{bmatrix}$, and of $\mathsf{y} = \begin{bmatrix} 15 \\ -5 \\ 20 \end{bmatrix}$. Justify your answers.

5. Let $\mathsf{u} = \begin{bmatrix} 1 \\ 0 \\ -1 \end{bmatrix}$ and $\mathsf{v} = \begin{bmatrix} 0 \\ 0 \\ 0 \end{bmatrix}$. Is u a scalar multiple of v? Is v a scalar multiple of u? Are u and v parallel? Explain.

6. Which of the following vectors are scalar multiples of other vectors in the list? Justify your answers.

$$\mathsf{u}_1 = \begin{bmatrix} -4 \\ 2 \\ 2 \end{bmatrix}, \ \mathsf{u}_2 = \begin{bmatrix} 3 \\ -6 \\ 12 \end{bmatrix}, \ \mathsf{u}_3 = \begin{bmatrix} 6 \\ -3 \\ -3 \end{bmatrix}, \ \mathsf{u}_4 = \begin{bmatrix} 0 \\ 0 \\ 0 \end{bmatrix}, \ \mathsf{u}_5 = \begin{bmatrix} 5 \\ -10 \\ 20 \end{bmatrix}.$$

7. [BB] Given $\mathsf{u}_1, \mathsf{u}_2, \mathsf{u}_3, \mathsf{u}_4, \mathsf{u}_5$ as in Exercise 6, list all pairs of parallel vectors. In the case of parallel vectors, state, if possible, whether they have the same or opposite direction.

8. Express each of the following linear combinations of vectors as a single vector.

(a) [BB] $4\begin{bmatrix} 2 \\ -3 \end{bmatrix} + 2\begin{bmatrix} 3 \\ 1 \end{bmatrix}$ (b) $a\begin{bmatrix} -1 \\ 5 \end{bmatrix} - 3\begin{bmatrix} -a \\ 2 \end{bmatrix}$

(c) [BB] $3\begin{bmatrix} 2 \\ 1 \\ 3 \end{bmatrix} - 2\begin{bmatrix} 1 \\ 0 \\ -5 \end{bmatrix} - 4\begin{bmatrix} 0 \\ -1 \\ 2 \end{bmatrix}$ (d) $3\begin{bmatrix} 1 \\ 0 \\ -2 \end{bmatrix} - 4\begin{bmatrix} 6 \\ 1 \\ 5 \end{bmatrix} + 2\begin{bmatrix} -1 \\ 1 \\ 2 \end{bmatrix}$

(e) $7\begin{bmatrix} 2 \\ 1 \\ 3 \end{bmatrix} + 4\begin{bmatrix} 0 \\ -4 \\ 2 \end{bmatrix} - 10\begin{bmatrix} 1 \\ -\frac{1}{2} \\ 3 \end{bmatrix}$ (f) $\frac{1}{3}\begin{bmatrix} 9 \\ -4 \\ 7 \end{bmatrix} - \frac{1}{5}\begin{bmatrix} 25 \\ 6 \\ -2 \end{bmatrix} + 3\begin{bmatrix} 2 \\ 1 \\ -3 \end{bmatrix}$

(g) $\frac{2}{5}\begin{bmatrix} 5 \\ 6 \\ -3 \end{bmatrix} - \frac{17}{5}\begin{bmatrix} 0 \\ 1 \\ 2 \end{bmatrix}$ (h) $a\begin{bmatrix} -1 \\ 1 \\ 7 \end{bmatrix} + 2\begin{bmatrix} a \\ a \\ 0 \end{bmatrix} - 4\begin{bmatrix} 0 \\ -1 \\ a \end{bmatrix}$.

9. Suppose $\mathsf{u} = \begin{bmatrix} a \\ -5 \end{bmatrix}$, $\mathsf{v} = \begin{bmatrix} 1 \\ 6-b \end{bmatrix}$ and $\mathsf{w} = \begin{bmatrix} 3 \\ 1 \end{bmatrix}$. Find a and b if

(a) [BB] $2\mathsf{u} - \mathsf{v} = \mathsf{w}$; (b) $4\mathsf{u} - \mathsf{v} + 3\mathsf{w} = 0$;

(c) $2\mathsf{u} - 3\mathsf{v} + 5\mathsf{w} = 0$; (d) $6(\mathsf{u} - \mathsf{w}) = 12\mathsf{v}$.

10. Suppose x, y, u and v are vectors. Express x and y in terms of u and v, given

(a) [BB] $\mathsf{x} - \mathsf{y} = \mathsf{u}$ and $2\mathsf{x} + 3\mathsf{y} = \mathsf{v}$ **(b)** $2\mathsf{x} - 3\mathsf{y} = -\mathsf{u}$ and $-8\mathsf{x} + 13\mathsf{y} = 3\mathsf{v}$
(c) $\mathsf{x} + \mathsf{y} = 2\mathsf{u}$ and $\mathsf{x} - \mathsf{y} = \mathsf{v}$ **(d)** $3\mathsf{x} + 5\mathsf{y} = -\mathsf{u}$ and $4\mathsf{x} - \mathsf{y} = \mathsf{v}$.

11. Find a and b, if possible.

(a) [BB] $\begin{bmatrix} a-1 \\ 2b \\ 3 \\ a-b \end{bmatrix} = \begin{bmatrix} 1-b \\ -6 \\ -b \\ 8 \end{bmatrix}$ (b) $\begin{bmatrix} 2-a \\ a+2b \\ 1 \\ 2b \\ 3a \end{bmatrix} = \begin{bmatrix} a \\ b \\ 0 \\ 4 \\ -2 \end{bmatrix}$

(c) $\begin{bmatrix} a \\ -1 \\ 1 \\ b \\ 2 \end{bmatrix} + \begin{bmatrix} b \\ 0 \\ 6 \\ 1 \\ 1 \end{bmatrix} = \begin{bmatrix} 0 \\ -1 \\ 7 \\ 0 \\ 3 \end{bmatrix}$ (d) $2a\begin{bmatrix} 0 \\ -1 \\ 2 \\ 2 \end{bmatrix} - 3b\begin{bmatrix} 1 \\ 3 \\ -1 \\ 2 \end{bmatrix} = \begin{bmatrix} 6 \\ 2 \\ 26 \\ 54 \end{bmatrix}$

(e) $a\begin{bmatrix} 1 \\ -1 \\ 1 \\ 2 \end{bmatrix} + b\begin{bmatrix} -1 \\ 1 \\ 2 \\ 3 \end{bmatrix} = \begin{bmatrix} -3 \\ a+1 \\ 12 \\ 4b-1 \end{bmatrix}$

(f) $(a-b)\begin{bmatrix} -1 \\ 2 \\ 3 \\ 4 \end{bmatrix} - 3a\begin{bmatrix} 0 \\ 4 \\ 2 \\ 5 \end{bmatrix} = (a+2b)\begin{bmatrix} 1 \\ 2 \\ -1 \\ 1 \end{bmatrix}$.

12. In each case, determine whether v is a linear combination of

$$\mathsf{v}_1 = \begin{bmatrix} 3 \\ 4 \\ 5 \\ 1 \\ 2 \end{bmatrix}, \mathsf{v}_2 = \begin{bmatrix} 2 \\ 3 \\ 4 \\ 5 \\ 1 \end{bmatrix}, \mathsf{v}_3 = \begin{bmatrix} -1 \\ -2 \\ -3 \\ -4 \\ -5 \end{bmatrix}, \mathsf{v}_4 = \begin{bmatrix} 4 \\ 5 \\ 1 \\ 2 \\ 3 \end{bmatrix} \text{ and } \mathsf{v}_5 = \begin{bmatrix} 5 \\ 1 \\ 2 \\ 3 \\ 4 \end{bmatrix}.$$

Justify your answers.

(a) [BB] $\mathsf{v} = \begin{bmatrix} 1 \\ 2 \\ 3 \\ 4 \\ 5 \end{bmatrix}$ (b) $\mathsf{v} = \begin{bmatrix} 0 \\ 0 \\ 0 \\ 0 \\ 0 \end{bmatrix}$ (c) $\mathsf{v} = \begin{bmatrix} 4 \\ 6 \\ 8 \\ 10 \\ 2 \end{bmatrix}$.

13. In each case, either express p as a linear combination of u, v, w or explain why there is no such linear combination.

(a) [BB] $\mathsf{p} = \begin{bmatrix} -4 \\ 7 \\ 5 \end{bmatrix}$, $\mathsf{u} = \mathsf{i}, \mathsf{v} = \mathsf{j}, \mathsf{w} = \mathsf{k}$

(b) $\mathsf{p} = \begin{bmatrix} -1 \\ 2 \\ 4 \\ 0 \end{bmatrix}, \mathsf{u} = \begin{bmatrix} 3 \\ 7 \\ 0 \\ -4 \end{bmatrix}, \mathsf{v} = \begin{bmatrix} 0 \\ 2 \\ 0 \\ 9 \end{bmatrix}, \mathsf{w} = \begin{bmatrix} 3 \\ 1 \\ 0 \\ 5 \end{bmatrix}$

(c) $\mathsf{p} = \begin{bmatrix} 0 \\ 2 \\ 4 \\ 3 \end{bmatrix}, \mathsf{u} = \begin{bmatrix} 1 \\ -1 \\ 0 \\ 1 \end{bmatrix}, \mathsf{v} = \begin{bmatrix} -1 \\ 2 \\ 3 \\ 1 \end{bmatrix}, \mathsf{w} = \begin{bmatrix} 0 \\ 1 \\ 1 \\ 1 \end{bmatrix}$

(d) $\mathsf{p} = \begin{bmatrix} 0 \\ 3 \\ 5 \end{bmatrix}, \mathsf{u} = \begin{bmatrix} 1 \\ 2 \\ 3 \end{bmatrix}, \mathsf{v} = \begin{bmatrix} -1 \\ 0 \\ 1 \end{bmatrix}, \mathsf{w} = \begin{bmatrix} -4 \\ 1 \\ 5 \end{bmatrix}$

(e) $\mathsf{p} = \begin{bmatrix} 1 \\ 1 \\ 1 \\ 1 \\ 1 \end{bmatrix}$, $\mathsf{u} = \begin{bmatrix} 1 \\ -1 \\ 1 \\ -1 \\ 1 \end{bmatrix}$, $\mathsf{v} = \begin{bmatrix} -3 \\ 3 \\ -3 \\ 3 \\ -3 \end{bmatrix}$, $\mathsf{w} = \begin{bmatrix} -2 \\ 2 \\ -2 \\ 2 \\ -2 \end{bmatrix}$

(f) $\mathsf{p} = \begin{bmatrix} 1 \\ 2 \\ 3 \\ 4 \end{bmatrix}$, $\mathsf{u} = \begin{bmatrix} 1 \\ -1 \\ 1 \\ 0 \end{bmatrix}$, $\mathsf{v} = \begin{bmatrix} 2 \\ 0 \\ 4 \\ 3 \end{bmatrix}$, $\mathsf{w} = \begin{bmatrix} -2 \\ 2 \\ 5 \\ 1 \end{bmatrix}$.

14. In each of the following cases, find $\mathsf{u} - \mathsf{v}$ and illustrate with a picture.

(a) [BB] $\mathsf{u} = \begin{bmatrix} -2 \\ -2 \end{bmatrix}$, $\mathsf{v} = \begin{bmatrix} 4 \\ 1 \end{bmatrix}$ (b) $\mathsf{u} = \begin{bmatrix} -4 \\ 2 \end{bmatrix}$, $\mathsf{v} = \begin{bmatrix} 1 \\ -4 \end{bmatrix}$

(c) $\mathsf{u} = \begin{bmatrix} 1 \\ 1 \end{bmatrix}$, $\mathsf{v} = \begin{bmatrix} -3 \\ 4 \end{bmatrix}$ (d) $\mathsf{u} = \begin{bmatrix} 1 \\ 1 \end{bmatrix}$, $\mathsf{v} = \begin{bmatrix} 4 \\ -2 \end{bmatrix}$.

15. (a) [BB] Shown at the right are two nonparallel vectors u and v and four other vectors $\mathsf{w}_1, \mathsf{w}_2, \mathsf{w}_3, \mathsf{w}_4$. Reproduce u, v and w_1 in a picture by themselves and exhibit w_1 as the diagonal of a parallelogram with sides that are scalar multiples of u and v. Guess (approximate) values of a and b so that $\mathsf{w}_1 = a\mathsf{u} + b\mathsf{v}$.

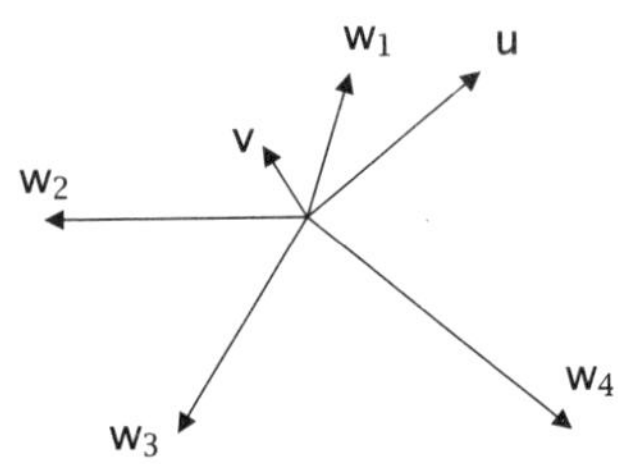

Repeat part (a) using each of (b) w_2, (c) w_3 and (d) w_4, instead of w_1.

16. Express $\begin{bmatrix} 7 \\ 7 \end{bmatrix}$ as a linear combination of each of the following pairs of vectors:

(a) [BB] $\begin{bmatrix} 2 \\ 3 \end{bmatrix}$ and $\begin{bmatrix} -1 \\ 2 \end{bmatrix}$ (b) $\begin{bmatrix} -1 \\ 1 \end{bmatrix}$ and $\begin{bmatrix} 5 \\ 2 \end{bmatrix}$

(c) $\begin{bmatrix} 11 \\ 5 \end{bmatrix}$ and $\begin{bmatrix} 3 \\ 2 \end{bmatrix}$ (d) $\begin{bmatrix} 5 \\ 5 \end{bmatrix}$ and $\begin{bmatrix} 20 \\ 2 \end{bmatrix}$.

17. The vectors $\mathsf{u} = \begin{bmatrix} 1 \\ 2 \end{bmatrix}$ and $\mathsf{v} = \begin{bmatrix} -1 \\ 1 \end{bmatrix}$ are not parallel. According to 1.1.21, any vector x in the plane is a linear combination of u and v.

Verify this statement with $\mathsf{x} = \begin{bmatrix} 1 \\ 0 \end{bmatrix}$, $\mathsf{x} = \begin{bmatrix} 0 \\ 1 \end{bmatrix}$ and $\mathsf{x} = \begin{bmatrix} 1 \\ 1 \end{bmatrix}$.

18. **(a)** [BB] Is it possible to express $\begin{bmatrix} 1 \\ 2 \end{bmatrix}$ as a linear combination of $\begin{bmatrix} -2 \\ -2 \end{bmatrix}$ and $\begin{bmatrix} 3 \\ 3 \end{bmatrix}$? Explain.

(b) Does your answer to part (a) contradict 1.1.21? Explain.

19. Let $v_1 = \begin{bmatrix} 1 \\ 2 \\ 3 \end{bmatrix}$, $v_2 = \begin{bmatrix} 4 \\ 5 \\ 6 \end{bmatrix}$ and $v_3 = \begin{bmatrix} 0 \\ 1 \\ 0 \end{bmatrix}$. Is $\begin{bmatrix} 1 \\ 0 \\ 0 \end{bmatrix}$ a linear combination of v_1, v_2, v_3 [BB]? What about $\begin{bmatrix} 0 \\ 1 \\ 0 \end{bmatrix}$? What about $\begin{bmatrix} 0 \\ 0 \\ 1 \end{bmatrix}$?

20. **(a)** [BB] Is it possible to express the zero vector as a linear combination of $\begin{bmatrix} 1 \\ 4 \end{bmatrix}$ and $\begin{bmatrix} -2 \\ -8 \end{bmatrix}$ in more than one way? Justify your answer.

    (b) Is it possible to express the zero vector as a linear combination of $\begin{bmatrix} 1 \\ 4 \end{bmatrix}$ and $\begin{bmatrix} 2 \\ -3 \end{bmatrix}$ in more than one way? Justify your answer.

21. Let $u = \begin{bmatrix} -1 \\ 0 \\ 3 \end{bmatrix}$ and $v = \begin{bmatrix} 3 \\ -1 \\ 0 \end{bmatrix}$. In each case, determine whether w is in the plane spanned by u and v.

    (a) [BB] $w = \begin{bmatrix} 0 \\ 0 \\ 0 \end{bmatrix}$ **(b)** $w = \begin{bmatrix} -7 \\ 3 \\ -6 \end{bmatrix}$ **(c)** $w = \begin{bmatrix} 1 \\ 1 \\ 1 \end{bmatrix}$

    (d) $w = \begin{bmatrix} 8 \\ -4 \\ 5 \end{bmatrix}$ **(e)** $w = \begin{bmatrix} -18 \\ 5 \\ 9 \end{bmatrix}$.

22. In each case, determine whether or not u is in the span of the other vectors.

    (a) [BB] $u = \begin{bmatrix} 1 \\ 2 \\ 3 \end{bmatrix}$, $v = \begin{bmatrix} 4 \\ 5 \\ 6 \end{bmatrix}$, $w = \begin{bmatrix} 7 \\ 8 \\ 0 \end{bmatrix}$ **(b)** $u = \begin{bmatrix} 17 \\ 19 \\ -6 \end{bmatrix}$, $v = \begin{bmatrix} 4 \\ 5 \\ 6 \end{bmatrix}$, $w = \begin{bmatrix} 7 \\ 8 \\ 0 \end{bmatrix}$

    (c) $u = \begin{bmatrix} 13 \\ 7 \end{bmatrix}$, $v = \begin{bmatrix} -5 \\ 6 \end{bmatrix}$, $w = \begin{bmatrix} 4 \\ 9 \end{bmatrix}$

    (d) $u = \begin{bmatrix} -1 \\ 0 \\ 1 \\ 1 \end{bmatrix}$, $v_1 = \begin{bmatrix} 1 \\ 0 \\ -2 \\ 1 \end{bmatrix}$, $v_2 = \begin{bmatrix} 0 \\ 1 \\ -1 \\ 1 \end{bmatrix}$, $v_3 = \begin{bmatrix} 1 \\ 3 \\ 0 \\ 1 \end{bmatrix}$

    (e) $u = \begin{bmatrix} 4 \\ 2 \\ -5 \\ 3 \end{bmatrix}$, $v_1 = \begin{bmatrix} 1 \\ 0 \\ -2 \\ 1 \end{bmatrix}$, $v_2 = \begin{bmatrix} 0 \\ 1 \\ -1 \\ 1 \end{bmatrix}$, $v_3 = \begin{bmatrix} 1 \\ 3 \\ 0 \\ 1 \end{bmatrix}$.

23. **(a)** [BB] Do the vectors $u_1 = \begin{bmatrix} -2 \\ 4 \end{bmatrix}$ and $u_2 = \begin{bmatrix} 3 \\ -6 \end{bmatrix}$ span R^2? If the answer is yes, provide a proof; otherwise, explain why the answer is no.

    (b) Answer part (a) with $u_1 = \begin{bmatrix} 1 \\ 2 \end{bmatrix}$ and $u_2 = \begin{bmatrix} 2 \\ 1 \end{bmatrix}$.

24. Do the vectors $u_1 = \begin{bmatrix} 2 \\ 1 \\ -3 \end{bmatrix}$ and $u_2 = \begin{bmatrix} 6 \\ 0 \\ 1 \end{bmatrix}$ span R^3? If the answer is yes, provide a proof; otherwise, explain why the answer is no.

25. **(a)** Suppose x is a vector in the plane spanned by nonparallel vectors u and v. Show that any scalar multiple of x lies in the same plane.

    (b) Let $x = \begin{bmatrix} x_1 \\ x_2 \\ x_3 \end{bmatrix}$ be a vector whose components satisfy the equation $ax + by + cz = 0$. Show that the components of any scalar multiple of x satisfy the same equation.

26. Let $O(0,0), A(2,2), B(4,4), C(2,-1), D(4,1), E(6,3), F(2,-4), G(6,0), H(8,2)$ be points in the xy-plane. Let $u = \overrightarrow{OA}$ be the vector pictured by the arrow from O to A and $v = \overrightarrow{OC}$ the vector pictured by the arrow from O to C.

    (a) [BB] Find u and v.

    (b) Express each of the following vectors as a linear combination of u and v.

    i. [BB] $\overrightarrow{OD}$ **v.** $\overrightarrow{AF}$
    ii. $\overrightarrow{OG}$ **vi.** $\overrightarrow{GE}$
    iii. $\overrightarrow{OH}$ **vii.** $\overrightarrow{HC}$.
    iv. $\overrightarrow{CA}$

    For example, $\overrightarrow{OE} = \begin{bmatrix} 6 \\ 3 \end{bmatrix} = 2u + v$.

27. [BB] Suppose ABC is a triangle. Let D be the midpoint of AB and E be the midpoint of AC. Use vectors to show that DE is parallel to BC and half its length.

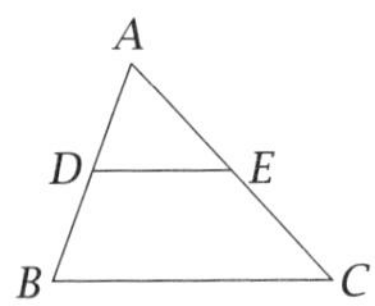

28. Suppose $ABCD$ is a quadrilateral, with sides AB and CD parallel and of the same length. Use vectors to show that $ABCD$ is a parallelogram.

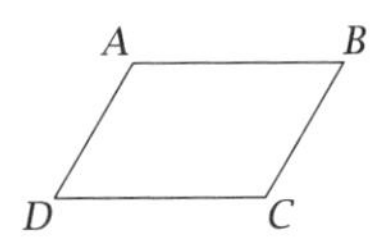

29. Let A, B and C denote the three vertices of a triangle. Let E be the midpoint of AC. Show that $\overrightarrow{BE} = \frac{1}{2}(\overrightarrow{BA} + \overrightarrow{BC})$.

30. Let $ABCD$ be an arbitrary quadrilateral in the plane. Let P, Q, R and S be, respectively, the midpoints of AB, BC, DC and AD. Use vectors to show that $PQRS$ is a parallelogram.

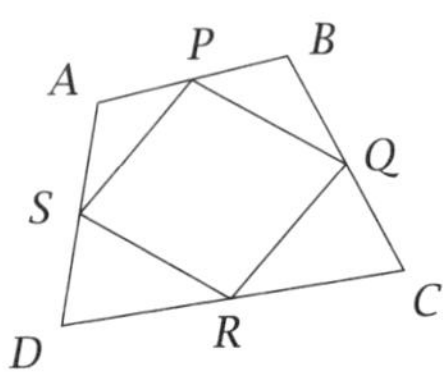

31. Problem 1.1.9 says that the diagonals of a parallelogram bisect each other. Establish the "converse;"[4] namely, that if the diagonals of a 4-sided polygon bisect each other, then that polygon is a parallelogram.

[4]The *converse* of "if $\mathcal{A}$, then $\mathcal{B}$" is the statement "if $\mathcal{B}$, then $\mathcal{A}$."

32. [BB] Let $A(1,2)$, $B(4,-3)$ and $C(-1,-2)$ be three points in the plane. Find a fourth point D such that A, B, C and D are the vertices of a parallelogram and justify your answer. Is the answer unique?

33. Given three points $A(-1,0)$, $B(2,3)$, $C(4,-1)$ in the plane, find all points D such that the four points A, B, C, D are the vertices of a parallelogram.

34. Given $P(2,-3,6)$ and $Q(2,2,-4)$, find a point R on the line segment from P to Q that is two fifths of the way from P to Q.

35. Use vectors to show that the midpoint of the line joining $A(x_1,y_1,z_1)$ to $B(x_2,y_2,z_2)$ is the point $C(\frac{x_1+x_2}{2}, \frac{y_1+y_2}{2}, \frac{z_1+z_2}{2})$.

36. [BB; 1, 2, 5] Prove all parts of Theorem 1.1.10 for two-dimensional vectors.

Critical Reading

37. Suppose c is a nonzero scalar and x is a nonzero vector in R^n. Explain why $c\mathsf{x} \neq \mathsf{0}$.

38. Suppose u and w are nonzero vectors in a (finite dimensional) vector space V neither of which is a scalar multiple of the other. Find a vector that is not a scalar multiple of u and not a scalar multiple of w.

39. **(a)** What does it mean to say that a plane is “spanned” by vectors u and v?

    (b) Let π be the plane spanned by nonparallel vectors u and v. Show that u and v are each in π.

40. Give an example of three nonzero vectors in R^3 that span a plane and explain how you found these vectors.

41. [BB] Show that any linear combination of $\begin{bmatrix}1\\ \frac{3}{2}\\ 0\end{bmatrix}$ and $\begin{bmatrix}0\\3\\6\end{bmatrix}$ is also a linear combination of $\begin{bmatrix}2\\3\\0\end{bmatrix}$ and $\begin{bmatrix}0\\1\\2\end{bmatrix}$.

42. **(a)** [BB] Given vectors u and v, show that any linear combination of $2\mathsf{u}$ and $-3\mathsf{v}$ is a linear combination of u and v.

    (b) Show that any linear combination of u and v is a linear combination of $2\mathsf{u}$ and $-3\mathsf{v}$.

43. Suppose u and v are vectors. Show that the span of u, v and $\mathsf{0}$ is the same as the span of u and v.

44. In 1.1.32, we observed that every vector in R^n is a linear combination of the standard basis vectors $\mathsf{e}_1, \mathsf{e}_2, \dots, \mathsf{e}_n$. Explain why the coefficients used in such a linear combination are unique; that is, if $x_1\mathsf{e}_1 + x_2\mathsf{e}_2 + \cdots + x_n\mathsf{e}_n = y_1\mathsf{e}_1 + y_2\mathsf{e}_2 + \cdots + y_n\mathsf{e}_n$, then $x_1 = y_1$, $x_2 = y_2$, ..., $x_n = y_n$.

45. Suppose u and v are vectors and u is a scalar multiple of v. Need v be a scalar multiple of u? Explain.

46. Let u, v and w be vectors. Show that any linear combination of u and v is also a linear combination of u, v and w.

47. Suppose u, v and w are vectors and w = 2u − 3v. Show that any linear combination of u, v and w is actually a linear combination of just u and v.

48. **(a)** [BB] Explain how one could use vectors to show that three points X, Y and Z are *collinear.* (Points are collinear if they lie on a line.)

    (b) Of the five points $A(-1,1,1)$, $B(2,2,1)$, $C(0,3,4)$, $D(2,7,10)$, $E(6,0,-5)$, which triples, if any, are collinear?

49. Show that the vector $\overrightarrow{AX}$ is a linear combination of $\overrightarrow{AB}$ and $\overrightarrow{AC}$ with the coefficients summing to 1.

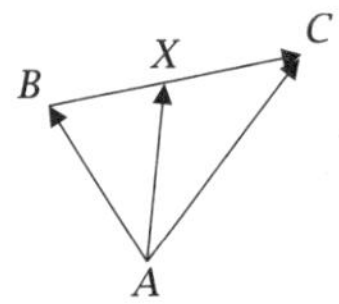

50. Given a triangle $A_0B_0C_0$, the first *Gunther* triangle is constructed as follows: A_1 lies on A_0B_0 and is such that $\overrightarrow{A_0A_1} = \frac{1}{3}\overrightarrow{A_0B_0}$; B_1 lies on B_0C_0 and is such that $\overrightarrow{B_0B_1} = \frac{1}{3}\overrightarrow{B_0C_0}$; C_1 lies on C_0A_0 and is such that $\overrightarrow{C_0C_1} = \frac{1}{3}\overrightarrow{C_0A_0}$. Do this again to obtain triangle $A_2B_2C_2$, the first Gunther triangle of $A_1B_1C_1$. (This is called the second Gunther triangle of $A_0B_0C_0$.) Use vectors to show that the lengths of corresponding sides of $\triangle A_2B_2C_2$ and $\triangle A_0B_0C_0$ are in the same proportions, hence that these two triangles are similar.

51. **(a)** [BB] If u and v are parallel vectors, show that some **nontrivial** linear combination of u and v is the zero vector. (*Nontrivial* means that 0u + 0v = 0 doesn't count!)

    (b) Show that if some nontrivial linear combination of vectors u and v is 0, then u and v are parallel.

1.2 Length and Direction

How long is an arrow that pictures the vector $\mathsf{v} = \begin{bmatrix} a \\ b \end{bmatrix}$?

Remembering the Theorem of Pythagoras and looking at the figure on the left of Figure 1.10, we see that the length is $\sqrt{a^2 + b^2}$. It is easily seen that this formula extends to 3-dimensional vectors—see the figure on the right. We use the notation $\|\mathsf{v}\|$ for the length of vector v.

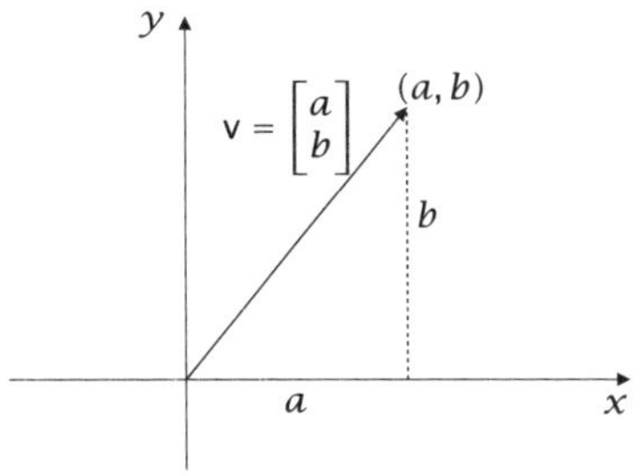

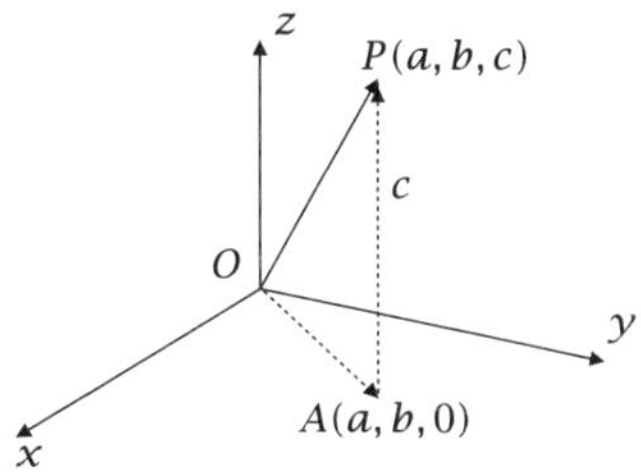

Figure 1.10: The length of v is $\|\mathsf{v}\| = \sqrt{a^2+b^2}$; $\left\|\overrightarrow{OP}\right\| = \sqrt{a^2+b^2+c^2}$.

1.2.1 Definition. For $n \geq 1$, the *length* of $\mathsf{x} = \begin{bmatrix} x_1 \\ x_2 \\ \vdots \\ x_n \end{bmatrix}$ is $\sqrt{x_1^2 + x_2^2 + \cdots + x_n^2}$.

Thus in R^2, $\left\| \begin{bmatrix} a \\ b \end{bmatrix} \right\| = \sqrt{a^2+b^2}$ and in R^3, $\left\| \begin{bmatrix} a \\ b \\ c \end{bmatrix} \right\| = \sqrt{a^2+b^2+c^2}$. These numbers are the lengths of the arrows that picture each vector so, in this respect, they are realistic. When $n > 3$, the definition of length is simply that, a definition, albeit one that seems sensible because it agrees with fact in the plane and in 3-space.

1.2.2 Examples.

- If $\mathsf{v} = \begin{bmatrix} 3 \\ -4 \end{bmatrix}$, then $\|\mathsf{v}\| = \sqrt{3^2 + (-4)^2} = 5$.
- If $\mathsf{v} = \begin{bmatrix} 1 \\ 0 \\ -1 \end{bmatrix}$, then $\|\mathsf{v}\| = \sqrt{2}$.
- $\left\| \begin{bmatrix} -1 \\ 1 \\ 1 \\ -1 \end{bmatrix} \right\| = \sqrt{4} = 2$.

:-)

READING CHECK 1. Only the zero vector has length 0. Why?

Suppose $\mathsf{v} = \begin{bmatrix} a \\ b \\ c \end{bmatrix}$ and k is a scalar. Then $k\mathsf{v} = \begin{bmatrix} ka \\ kb \\ kc \end{bmatrix}$, so

$$\|k\mathsf{v}\| = \sqrt{(ka)^2 + (kb)^2 + (kc)^2}$$
$$= \sqrt{k^2(a^2+b^2+c^2)} = \sqrt{k^2}\sqrt{a^2+b^2+c^2} = |k|\,\|\mathsf{v}\|,$$

an observation that extends to any Euclidean space.

1.2.3 If v is a vector in R^n and k is a scalar, $\|k\mathsf{v}\| = |k|\,\|\mathsf{v}\|$.

READING CHECK 2. If k is a real number, what is $\sqrt{k^2}$?

The fact expresssed in 1.2.3 is just an algebraic formulation of something we have noticed in R^2. The length of (an arrow for) $2\mathsf{v}$ is twice the length of v; the length of $-\frac{1}{2}\mathsf{v}$ is one half the length of v, and so on.

1.2.4 Definition. A *unit vector* is a vector of length 1.

1.2.5 Examples.

- Each of the standard basis vectors in R^n is a unit vector. Recall that these are the vectors $\mathsf{e}_1 = \begin{bmatrix} 1 \\ 0 \\ 0 \\ \vdots \\ 0 \end{bmatrix}$, $\mathsf{e}_2 = \begin{bmatrix} 0 \\ 1 \\ 0 \\ \vdots \\ 0 \end{bmatrix}$, ..., $\mathsf{e}_n = \begin{bmatrix} 0 \\ 0 \\ 0 \\ \vdots \\ 1 \end{bmatrix}$.
- $\begin{bmatrix} \frac{1}{\sqrt{2}} \\ \frac{1}{\sqrt{2}} \end{bmatrix}$ is a unit vector in R^2 and $\begin{bmatrix} \frac{1}{\sqrt{6}} \\ -\frac{2}{\sqrt{6}} \\ \frac{1}{\sqrt{6}} \end{bmatrix}$ is a unit vector in R^3. ☺

We can use 1.2.3 to make unit vectors. If v is any vector (except the zero vector) and we put $k = \frac{1}{\|\mathsf{v}\|}$, then $\|k\mathsf{v}\| = |k|\,\|\mathsf{v}\| = \frac{1}{\|\mathsf{v}\|}\|\mathsf{v}\| = 1$. Thus $\frac{1}{\|\mathsf{v}\|}\mathsf{v}$ is a unit vector, and it has the same direction as v since it is a **positive** scalar multiple of v.

1.2.6 For any v, the vector $\frac{1}{\|\mathsf{v}\|}\mathsf{v}$ is a unit vector in the direction of v.

1.2.7 Example. To find a unit vector with the same direction as $\mathsf{v} = \begin{bmatrix} 2 \\ -2 \\ 1 \end{bmatrix}$, we compute $k = \|\mathsf{v}\| = \sqrt{2^2 + (-2)^2 + 1^2} = \sqrt{9} = 3$. The desired vector is $\frac{1}{3}\mathsf{v} = \begin{bmatrix} \frac{2}{3} \\ -\frac{2}{3} \\ \frac{1}{3} \end{bmatrix}$. ☺

Dot Product and the Angle Between Vectors

Let $\mathsf{v} = \begin{bmatrix} x \\ y \end{bmatrix}$ be a two-dimensional vector. If $\|\mathsf{v}\| = 1$, then $x^2 + y^2 = 1$. This means that if v is pictured by an arrow that starts at the origin, then the arrowhead lies on the unit circle. Since any point on the unit circle has coordinates of the form $(\cos\alpha, \sin\alpha)$ (see Figure 1.11), it follows that

1.2.8 | Any two-dimensional unit vector is $\begin{bmatrix} \cos\alpha \\ \sin\alpha \end{bmatrix}$ for some angle α. |

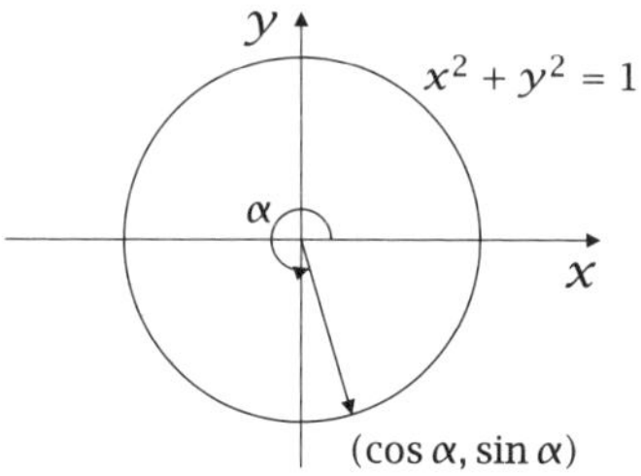

Figure 1.11: Any vector of length 1 has the form $\begin{bmatrix} \cos\alpha \\ \sin\alpha \end{bmatrix}$ for some angle α.

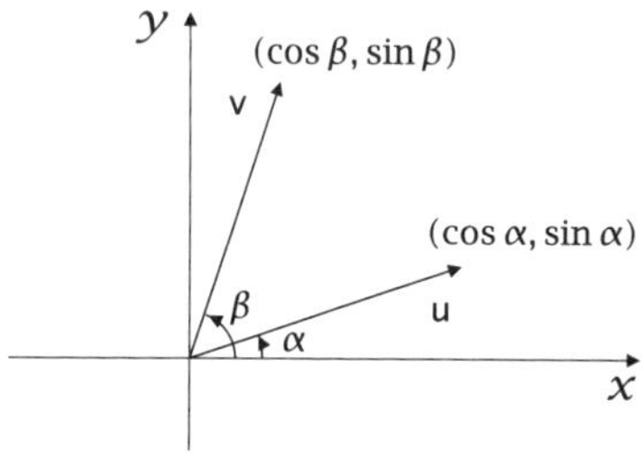

Figure 1.12: The angle between u and v is $\beta - \alpha$.

Suppose $\mathsf{u} = \begin{bmatrix} \cos\alpha \\ \sin\alpha \end{bmatrix}$ and $\mathsf{v} = \begin{bmatrix} \cos\beta \\ \sin\beta \end{bmatrix}$ each have length 1. For convenience, let's suppose that $0 < \alpha < \beta < \frac{\pi}{2}$. See Figure 1.12. The angle between u and v is $\beta - \alpha$ and the author is hoping that you remember one of the important "addition formulas" of trigonometry, this one:

$$\cos(\beta - \alpha) = \cos\beta\cos\alpha + \sin\beta\sin\alpha.$$

Look at the right hand side. This is the product of the first coordinates of u and v added to the product of the second coordinates of u and v. We have discovered that if θ is the angle between unit vectors $\mathsf{u} = \begin{bmatrix} a_1 \\ b_1 \end{bmatrix}$ and $\mathsf{v} = \begin{bmatrix} a_2 \\ b_2 \end{bmatrix}$, then $\cos\theta = a_1a_2 + b_1b_2$. The expression on the right has a special name: it's called the *dot product* of u and v.

1.2.9 Definition. The *dot product* of $\mathsf{u} = \begin{bmatrix} a_1 \\ a_2 \\ \vdots \\ a_n \end{bmatrix}$ and $\mathsf{v} = \begin{bmatrix} b_1 \\ b_2 \\ \vdots \\ b_n \end{bmatrix}$ is the number $a_1b_1 + a_2b_2 + \cdots + a_nb_n$. It is denoted $\mathsf{u} \cdot \mathsf{v}$.

$$\boxed{\mathsf{u} \cdot \mathsf{v} = a_1 b_1 + a_2 b_2 + \cdots + a_n b_n}$$

1.2.10 Examples. • If $\mathsf{u} = \begin{bmatrix} -1 \\ 2 \end{bmatrix}$ and $\mathsf{v} = \begin{bmatrix} 4 \\ -3 \end{bmatrix}$, then $\mathsf{u} \cdot \mathsf{v} = -1(4) + 2(-3) = -10$.

• If $\mathsf{u} = \begin{bmatrix} -1 \\ 0 \\ 2 \\ 3 \\ 5 \end{bmatrix}$ and $\mathsf{v} = \begin{bmatrix} 1 \\ 2 \\ 3 \\ -2 \\ -2 \end{bmatrix}$, then $\mathsf{u} \cdot \mathsf{v} = -1(1) + 0(2) + 2(3) + 3(-2) + 5(-2) = -11$. ☺

We were led to the idea of dot product because of our observation that if u and v are unit vectors, then u · v is the cosine of the angle between u and v.

1.2.11 Example. The vectors $\mathsf{u} = \begin{bmatrix} \frac{1}{\sqrt{2}} \\ \frac{1}{\sqrt{2}} \end{bmatrix}$ and $\mathsf{v} = \begin{bmatrix} -\frac{3}{5} \\ -\frac{4}{5} \end{bmatrix}$ are unit vectors, so the cosine of the angle θ between them is $\cos\theta = \mathsf{u} \cdot \mathsf{v} = -\frac{3}{5\sqrt{2}} - \frac{4}{5\sqrt{2}} = -\frac{7}{5\sqrt{2}} \approx -0.990$. So $\theta \approx \arccos(-0.990) \approx 3.000$ rads $\approx 171.87°$. ☺

The expression "angle between vectors" that we have been using is somewhat ambiguous as shown in the figure. Is this angle θ or $2\pi - \theta$? By general agreement, **the angle between vectors is an angle in the range $0 \le \theta \le \pi$.**

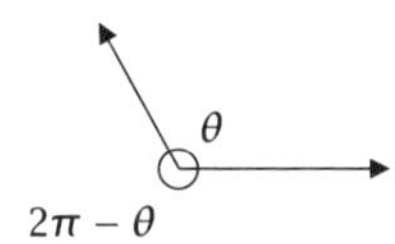

If θ is the angle between **unit** vectors u and v, then $\cos\theta = \mathsf{u} \cdot \mathsf{v}$, but what if u and v are not unit vectors? What is the angle between u and v in general? Notice that the angle in question has nothing to do with the length of the vectors. For example, this angle is the same as the angle between unit vectors that have the same direction as u and v. In 1.2.6, we showed that $\frac{1}{\|\mathsf{u}\|}\mathsf{u}$ and $\frac{1}{\|\mathsf{v}\|}\mathsf{v}$ are just such vectors. Using what we know about unit vectors,

$$\cos\theta = \frac{1}{\|\mathsf{u}\|}\mathsf{u} \cdot \frac{1}{\|\mathsf{v}\|}\mathsf{v} = \frac{1}{\|\mathsf{u}\|\,\|\mathsf{v}\|}\mathsf{u} \cdot \mathsf{v}. \tag{1}$$

in summary,

1.2.12
$$\boxed{\cos\theta = \frac{\mathsf{u} \cdot \mathsf{v}}{\|\mathsf{u}\|\,\|\mathsf{v}\|}.}$$

Written differently, this is

1.2.13
$$\boxed{\mathsf{u} \cdot \mathsf{v} = \|\mathsf{u}\|\,\|\mathsf{v}\|\cos\theta.}$$

Both of these formulas are important. We have attempted to give plausible arguments explaining their validity in R^2 where vectors can be pictured with arrows and the angle between arrows is a real angle. In higher dimensions, we take 1.2.12 simply as a **definition** of $\cos\theta$ yet to be justified—see later in this section where we discuss the "Cauchy-Schwarz" inequality.

1.2.14 Remark. The author confesses that he is getting a bit ahead of himself here. Proper justification for the assertion $\frac{1}{\|\mathsf{u}\|}\mathsf{u}\cdot\frac{1}{\|\mathsf{v}\|}\mathsf{v} = \frac{1}{\|\mathsf{u}\|\|\mathsf{v}\|}\mathsf{u}\cdot\mathsf{v}$ in equation (1) is actually a consequence of part 3 of Theorem 1.2.20, which is still to come. On the other hand, the author doesn't like making obvious things seem hard and hopes the reader feels the same way.

1.2.15 Example. If $\mathsf{u} = \begin{bmatrix}\sqrt{3}\\1\end{bmatrix}$ and $\mathsf{v} = \begin{bmatrix}1\\\sqrt{3}\end{bmatrix}$, then $\mathsf{u}\cdot\mathsf{v} = \sqrt{3}+\sqrt{3} = 2\sqrt{3}$. Since $\|\mathsf{u}\| = \sqrt{(\sqrt{3})^2+1^2} = \sqrt{4} = 2 = \|\mathsf{v}\|$, $\cos\theta = \frac{\mathsf{u}\cdot\mathsf{v}}{\|\mathsf{u}\|\|\mathsf{v}\|} = \frac{2\sqrt{3}}{4} = \frac{\sqrt{3}}{2}$, so $\theta = \frac{\pi}{6}$ radians $= 30°$. This is the actual angle between suitable arrows for u and v. ☺

1.2.16 Example. Suppose $\mathsf{u} = \begin{bmatrix}1\\2\\-3\\1\end{bmatrix}$ and $\mathsf{v} = \begin{bmatrix}-2\\1\\-1\\4\end{bmatrix}$. Then

$$\begin{aligned}\|\mathsf{u}\| &= \sqrt{1^2+2^2+(-3)^2+1^2} = \sqrt{15},\\ \|\mathsf{v}\| &= \sqrt{(-2)^2+1^2+(-1)^2+4^2} = \sqrt{22},\\ \mathsf{u}\cdot\mathsf{v} &= 1(-2)+2(1)+(-3)(-1)+1(4) = 7,\end{aligned}$$

so $\cos\theta = \dfrac{\mathsf{u}\cdot\mathsf{v}}{\|\mathsf{u}\|\,\|\mathsf{v}\|} = \dfrac{7}{\sqrt{15}\sqrt{22}} \approx .385$. Therefore $\theta \approx \arccos(.385) = 1.175$ radians $\approx 67.3°$. In contrast with the previous example, this is the angle between the four-dimensional vectors u and v **by definition**. ☺

1.2.17 Examples.

- If $\mathsf{u} = \begin{bmatrix}-4\\0\\2\\-2\end{bmatrix}$ and $\mathsf{v} = \begin{bmatrix}2\\0\\-1\\1\end{bmatrix}$, then $\cos\theta = \dfrac{\mathsf{u}\cdot\mathsf{v}}{\|\mathsf{u}\|\,\|\mathsf{v}\|} = \dfrac{-8-2-2}{\sqrt{24}\sqrt{6}} = \dfrac{-12}{\sqrt{144}} = -1$, so $\theta = \pi$, a fact consistent with the observation that $\mathsf{u} = -2\mathsf{v}$.

- If $\mathsf{u} = \begin{bmatrix}1\\2\\1\\1\end{bmatrix}$ and $\mathsf{v} = \begin{bmatrix}-1\\1\\3\\-4\end{bmatrix}$, then $\cos\theta = \dfrac{\mathsf{u}\cdot\mathsf{v}}{\|\mathsf{u}\|\,\|\mathsf{v}\|} = \dfrac{-1+2+3-4}{\|\mathsf{u}\|\,\|\mathsf{v}\|} = 0$. The angle between u and v is $\theta = \frac{\pi}{2}$. ☺

Vectors u and v are perpendicular—in linear algebra, the word "orthogonal" is preferred—if the angle between them is $\frac{\pi}{2}$ radians. Since $\cos\frac{\pi}{2} = 0$ and since a fraction is 0 if and only if its numerator is 0, equation 1.2.13 shows that orthogonality can be defined in terms of the dot product.

1.2.18 Definition. Vectors u and v are *orthogonal* if and only if $\mathsf{u}\cdot\mathsf{v} = 0$. More generally, vectors $\mathsf{x}_1, \mathsf{x}_2, \ldots, \mathsf{x}_n$ are orthogonal if and only if they are orthogonal pairwise: $\mathsf{x}_i \cdot \mathsf{x}_j = 0$ when $i \neq j$.

1.2.19 Examples.

- The vectors $\mathsf{u} = \begin{bmatrix} 3 \\ 4 \end{bmatrix}$ and $\mathsf{v} = \begin{bmatrix} -4 \\ 3 \end{bmatrix}$ are orthogonal.
- The vectors $\mathsf{u} = \begin{bmatrix} 3 \\ 2 \\ -1 \\ 4 \end{bmatrix}$ and $\mathsf{v} = \begin{bmatrix} 1 \\ -1 \\ 1 \\ 0 \end{bmatrix}$ are orthogonal.
- The zero vector is orthogonal to any vector because $\mathsf{u}\cdot\mathsf{0} = 0$ for any u.
- The standard basis vectors in R^n are orthogonal because if $i \neq j$, the 1s in e_i and e_j are in different positions:

$$\mathsf{e}_i \cdot \mathsf{e}_j = \begin{bmatrix} 0 \\ \vdots \\ 1 \\ \vdots \\ 0 \\ \vdots \\ 0 \end{bmatrix} \cdot \begin{bmatrix} 0 \\ \vdots \\ 0 \\ \vdots \\ 1 \\ \vdots \\ 0 \end{bmatrix} = 0.$$

:)

We summarize the basic properties of the dot product in a theorem, leaving proofs for the exercises.

1.2.20 Theorem (Properties of the Dot Product). *Let* u, v *and* w *be vectors and let* c *be a scalar.*

1. *(Commutativity)* $\mathsf{u}\cdot\mathsf{v} = \mathsf{v}\cdot\mathsf{u}$.
2. *(Distributivity)* $\mathsf{u}\cdot(\mathsf{v}+\mathsf{w}) = \mathsf{u}\cdot\mathsf{v} + \mathsf{u}\cdot\mathsf{w}$ *and* $(\mathsf{u}+\mathsf{v})\cdot\mathsf{w} = \mathsf{u}\cdot\mathsf{w} + \mathsf{v}\cdot\mathsf{w}$.
3. *(Scalar associativity)* $(c\mathsf{u})\cdot\mathsf{v} = c(\mathsf{u}\cdot\mathsf{v}) = \mathsf{u}\cdot(c\mathsf{v})$.
4. $\mathsf{u}\cdot\mathsf{u} = \|\mathsf{u}\|^2$.

As with Theorem 1.1.10, you will find yourself using this latest theorem without realizing it.

1.2.21 Example. Computing $(2\mathsf{u}+3\mathsf{v})\cdot(\mathsf{u}+4\mathsf{v})$ is very similar to the way we find the product $(2x+3y)(x+4y)=2x^2+11xy+12y^2$.

Consider $(2\mathsf{u}+3\mathsf{v})\cdot(\mathsf{u}+4\mathsf{v})$

$$
\begin{aligned}
&= (2\mathsf{u}+3\mathsf{v})\cdot\mathsf{u}+(2\mathsf{u}+3\mathsf{v})\cdot(4\mathsf{v}) && \text{distributivity}\\
&= (2\mathsf{u})\cdot\mathsf{u}+(3\mathsf{v})\cdot\mathsf{u}+(2\mathsf{u})\cdot(4\mathsf{v})+(3\mathsf{v})\cdot(4\mathsf{v}) && \text{distributivity again}\\
&= 2(\mathsf{u}\cdot\mathsf{u})+3(\mathsf{v}\cdot\mathsf{u})+2(4)(\mathsf{u}\cdot\mathsf{v})+3(4)(\mathsf{v}\cdot\mathsf{v}) && \text{scalar associativity}\\
&= 2(\mathsf{u}\cdot\mathsf{u})+3(\mathsf{u}\cdot\mathsf{v})+8(\mathsf{u}\cdot\mathsf{v})+12(\mathsf{v}\cdot\mathsf{v}) && \text{commutativity}\\
&= 2\,\|\mathsf{u}\|^2+11\mathsf{u}\cdot\mathsf{v}+12\,\|\mathsf{v}\|^2. && \mathsf{u}\cdot\mathsf{u}=\|\mathsf{u}\|^2
\end{aligned}
$$

:)

READING CHECK 3. Find $(3x-7y)(x+2y)$ and $(3\mathsf{u}-7\mathsf{v})\cdot(\mathsf{u}+2\mathsf{v})$.

The Cauchy-Schwarz Inequality[5]

Remember that we have defined the cosine of the angle θ between vectors in any Euclidean space by the formula

$$\cos\theta=\frac{\mathsf{u}\cdot\mathsf{v}}{\|\mathsf{u}\|\,\|\mathsf{v}\|},$$

but is this valid? Suppose the fraction on the right side is 1.3. There is no angle θ with this cosine! As it turns out, and we are very lucky (!), the right side is always between -1 and 1, so it is always a cosine. It's easy to believe this in R^2 if you have been convinced that the fraction here is truly the cosine between arrows for u and v. The problem is that in higher dimensions, where the idea of "angle" may seem dubious, the formula needs proof! The desired fact,

$$-1\le\frac{\mathsf{u}\cdot\mathsf{v}}{\|\mathsf{u}\|\,\|\mathsf{v}\|}\le 1,$$

can be written as a single inequality.

1.2.22 $$\boxed{|\mathsf{u}\cdot\mathsf{v}|\le\|\mathsf{u}\|\,\|\mathsf{v}\|}\qquad\textbf{Cauchy-Schwarz Inequality}$$

which, as noted, is known as the "Cauchy-Schwarz" inequality.

[5]With both a crater on the moon and a Paris street named after him, Augustin-Louis Cauchy (1789–1857) contributed to almost every area of mathematics. Hermann Amandus Schwarz (1843–1921), a student of Weierstrass, is best known for his work in conformal mappings.

1.2.23 Example. Suppose $u = \begin{bmatrix} -1 \\ 0 \\ 2 \\ 1 \end{bmatrix}$ and $v = \begin{bmatrix} 2 \\ -1 \\ 1 \\ -4 \end{bmatrix}$. Then $\|u\| = \sqrt{1+0+4+1} = \sqrt{6}$, $\|v\| = \sqrt{4+1+1+16} = \sqrt{22}$ and $u \cdot v = -2+0+2-4 = -4$. The Cauchy–Schwarz inequality says that $|u \cdot v| \le \|u\|\,\|v\|$. Here the left side is $|-4|$, the right side is $\sqrt{6}\sqrt{22}$ and certainly $4 \le \sqrt{132}$. ☺

To prove the Cauchy–Schwarz inequality, we begin with the observation that the inequality is certainly true if $u = 0$ because in this case $|u \cdot v| = 0$ and $\|u\|\,\|v\| = 0$. This leaves the case $u \ne 0$, and here we use something that students usually call a "trick." A better term is "insight," an idea that somebody once had and which led to something useful.

The "insight" is the fact that $(xu+v)\cdot(xu+v) \ge 0$ for any real number x because the dot product of any vector with itself, being a sum of squares, cannot be negative. Remembering that $(xu+v)\cdot(xu+v)$ expands just like $(xa+b)^2$—see Example 1.2.21—we obtain

$$(xu+v)\cdot(xu+v) = x^2(u \cdot u) + 2x(u \cdot v) + v \cdot v \ge 0. \tag{2}$$

Letting $a = u \cdot u$, $b = 2(u \cdot v)$ and $c = v \cdot v$, equation (2) says that $ax^2+bx+c \ge 0$ for all x. Also $a \ne 0$ because we are assuming $u \ne 0$. Thus $a > 0$ and the graph of $y = ax^2+bx+c$ is a parabola pointing upwards. Since $y \ge 0$ for all x, this parabola intersects the x-axis at most once, so the quadratic ax^2+bx+c is 0 for at most one x. See Figure 1.13. The roots of

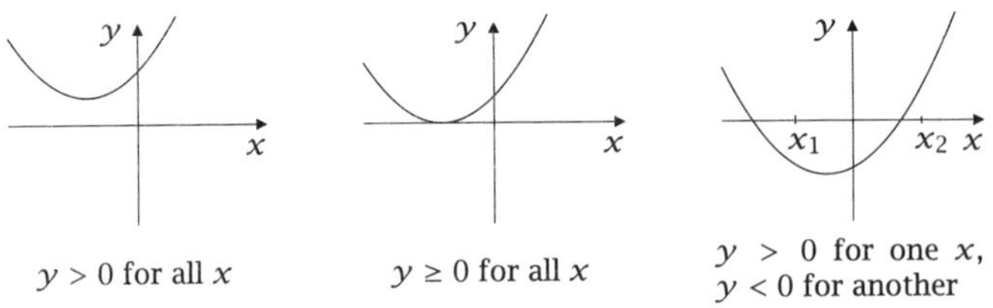

Figure 1.13: If $y = ax^2+bx+c \ge 0$, then $y = 0$ for at most one x.

$ax^2+bx+c = 0$ are given by the well-known formula

$$x = \frac{-b \pm \sqrt{b^2-4ac}}{2a}$$

and, since there is at most one root, the term under the root (known as the *discriminant*) cannot be positive (otherwise, we would have two roots). Thus

$$b^2 - 4ac \le 0,$$

$$b^2 \le 4ac,$$

$$4(u \cdot v)^2 \le 4(u \cdot u)(v \cdot v),$$

$$(u \cdot v)^2 \le (u \cdot u)(v \cdot v) = \|u\|^2\,\|v\|^2.$$

Taking the square root of each side and remembering that $\sqrt{k^2} = |k|$ gives the desired result $|\mathsf{u} \cdot \mathsf{v}| \le \|\mathsf{u}\| \, \|\mathsf{v}\|$.

More Inequalities

The Cauchy–Schwarz inequality leads to many other inequalities, some obvious applications and others more subtle.

1.2.24 **Problem.** Show that $|ab+cd| \le \sqrt{a^2+c^2}\sqrt{b^2+d^2}$ for any real numbers a, b, c and d.

Solution. This is a pretty obvious application of Cauchy–Schwarz. Does the left side resemble a dot product? Does the right side resemble the product of the lengths of vectors? Let $\mathsf{u} = \begin{bmatrix} a \\ c \end{bmatrix}$ and $\mathsf{v} = \begin{bmatrix} b \\ d \end{bmatrix}$. Then $\mathsf{u} \cdot \mathsf{v} = ab + cd$, $\|\mathsf{u}\| = \sqrt{a^2+c^2}$ and $\|\mathsf{v}\| = \sqrt{b^2+d^2}$. The given inequality is exactly the Cauchy–Schwarz inequality applied to these two vectors. ☝

READING CHECK 4. Verify the formula in Problem 1.2.24 with $a = 2$, $b = 3$, $c = 4$, $d = 5$.

READING CHECK 5. Show that $|ab + cd + ef| \le \sqrt{a^2+c^2+e^2}\sqrt{b^2+d^2+f^2}$ for any real numbers a, b, c, d, e, f.

Suppose a and b are nonnegative real numbers and we take $\mathsf{u} = \begin{bmatrix} \sqrt{a} \\ \sqrt{b} \end{bmatrix}$ and $\mathsf{v} = \begin{bmatrix} \sqrt{b} \\ \sqrt{a} \end{bmatrix}$. Then $\mathsf{u} \cdot \mathsf{v} = \sqrt{a}\sqrt{b} + \sqrt{a}\sqrt{b} = 2\sqrt{ab}$ while $\|\mathsf{u}\| = \|\mathsf{v}\| = \sqrt{a+b}$. The Cauchy–Schwarz inequality says that $|2\sqrt{ab}| \le \sqrt{a+b}\sqrt{a+b} = a+b$; that is,

1.2.25 $$\boxed{\sqrt{ab} \le \frac{a+b}{2}}$$ **Arithmetic/Geometric Mean Inequality**[6]

The expression $\frac{a+b}{2}$ on the right-hand side is, of course, just the average, or *arithmetic mean*, of a and b. The expression on the left, $\sqrt{ab}$, is known as the *geometric mean* of a and b, hence the name of this particular inequality. It says, for example, that $\sqrt{6} = \sqrt{3(2)} \le \frac{3+2}{2} = 2\frac{1}{2}$.

[6]The geometric mean appears in the work of Euclid. It is the number that "squares the rectangle," in the sense that if you want to build a square with area equal to that of a rectangle with sides of lengths a and b, then each side of the square must have length $\sqrt{ab}$.

A glance to the right shows why the next theorem is so-named.

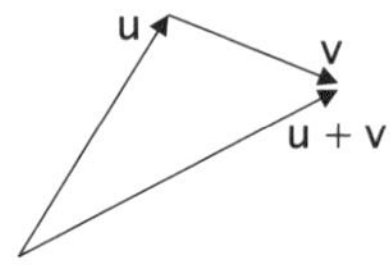

1.2.26 Theorem (The Triangle Inequality). $\|\mathsf{u}+\mathsf{v}\| \le \|\mathsf{u}\| + \|\mathsf{v}\|$ *for any vectors* u *and* v *in* R^n.

Proof. The proof follows directly from the fact that $\|\mathsf{x}\|^2 = \mathsf{x}\cdot\mathsf{x}$ for any vector x.

$$\begin{aligned}\|\mathsf{u}+\mathsf{v}\|^2 &= (\mathsf{u}+\mathsf{v})\cdot(\mathsf{u}+\mathsf{v}) = \mathsf{u}\cdot\mathsf{u} + 2\mathsf{u}\cdot\mathsf{v} + \mathsf{v}\cdot\mathsf{v}\\ &= \|\mathsf{u}\|^2 + 2\mathsf{u}\cdot\mathsf{v} + \|\mathsf{v}\|^2 \qquad (3)\\ &\le \|\mathsf{u}\|^2 + 2|\mathsf{u}\cdot\mathsf{v}| + \|\mathsf{v}\|^2.\end{aligned}$$

Applying the Cauchy-Schwarz inequality to the middle term in the last expression here, we get

$$\|\mathsf{u}\|^2 + 2|\mathsf{u}\cdot\mathsf{v}| + \|\mathsf{v}\|^2 \le \|\mathsf{u}\|^2 + 2\,\|\mathsf{u}\|\,\|\mathsf{v}\| + \|\mathsf{v}\|^2 = (\|\mathsf{u}\| + \|\mathsf{v}\|)^2.$$

We have shown that $\|\mathsf{u}+\mathsf{v}\|^2 \le (\|\mathsf{u}\| + \|\mathsf{v}\|)^2$, so the result follows by taking square roots. ■

At step (3), we had $\|\mathsf{u}+\mathsf{v}\|^2 = \|\mathsf{u}\|^2 + 2\mathsf{u}\cdot\mathsf{v} + \|\mathsf{v}\|^2$. Since vectors u and v are orthogonal if and only if $\mathsf{u}\cdot\mathsf{v} = 0$, we obtain the well-known Theorem of Pythagoras (in any Euclidean space R^n).

1.2.27 Theorem. (Pythagoras) *Vectors* u *and* v *are orthogonal if and only if* $\|\mathsf{u}+\mathsf{v}\|^2 = \|\mathsf{u}\|^2 + \|\mathsf{v}\|^2$.

The statement, "In a right angled triangle, the square of the hypotenuse is the sum of the squares of the other two sides," is well known. Did you realize the **converse** is true as well? If the square of one side of a triangle is the sum of the squares of the other two sides, then the triangle has a right angle.

Answers to Reading Checks

1. Let $\mathsf{x} = \begin{bmatrix} x_1 \\ \vdots \\ x_n \end{bmatrix}$ and suppose $\|\mathsf{x}\| = 0$. Then $x_1^2 + \cdots + x_n^2 = 0$. But the only way the sum of squares of real numbers is 0 is if all the numbers being squared are 0. So all $x_i = 0$. So $\mathsf{x} = 0$.

2. For any real number k, $\sqrt{k^2} = |k|$. Some people might think $\sqrt{k^2} = k$, but this would imply that $\sqrt{(-2)^2} = -2$. Any calculator, however, will tell you that $\sqrt{(-2)^2} = \sqrt{4} = +2 = |-2|$.

3. $(3x-7y)(x+2y) = 3x^2-xy-14y^2$; $(3\mathsf{u}-7\mathsf{v})\cdot(\mathsf{u}+2\mathsf{v}) = 3\,\|\mathsf{u}\|^2-\mathsf{u}\cdot\mathsf{v}-14\,\|\mathsf{v}\|^2$.

4. $|ab+cd| = 26$ while $\sqrt{a^2+c^2}\sqrt{b^2+d^2} = \sqrt{20}\sqrt{34} =\approx 26.08$. Close, but the inequality is not violated!

5. Apply the Cauchy-Schwarz inequality to the vectors $\mathsf{u} = \begin{bmatrix} a \\ c \\ e \end{bmatrix}$ and $\mathsf{v} = \begin{bmatrix} b \\ d \\ f \end{bmatrix}$.

True/False Questions

Decide, with as little calculation as possible, whether each of the following statements is true or false and, if you say "false," explain your answer. (Answers are at the back of the book.)

1. If c is a scalar and v is a vector, the length of $c\mathsf{v}$ is c times the length of v.

2. $\begin{bmatrix} 1 \\ 1 \\ 1 \end{bmatrix}$ is a unit vector.

3. For vectors u and v, $\|\mathsf{u}+\mathsf{v}\| = \|\mathsf{u}\| + \|\mathsf{v}\|$.

4. The vectors $\begin{bmatrix} 1 \\ 1 \\ 1 \\ 1 \\ 1 \end{bmatrix}$ and $\begin{bmatrix} 1 \\ 0 \\ -1 \\ -1 \\ 1 \end{bmatrix}$ are orthogonal.

5. The cosine of the angle between vectors u and v is $\mathsf{u}\cdot\mathsf{v}$.

6. For any vectors u and v, $\mathsf{u}\cdot\mathsf{v} \le \|\mathsf{u}\|\,\|\mathsf{v}\|$.

7. The Cauchy-Schwarz inequality says that for any vectors u and v, $\|\mathsf{u}\cdot\mathsf{v}\| \le \|\mathsf{u}\|\,\|\mathsf{v}\|$.

8. There exist unit vectors u and v with $\mathsf{u}\cdot\mathsf{v} = 2$.

9. If u and v are vectors that are not orthogonal, the sign of the dot product $\mathsf{u}\cdot\mathsf{v}$ (positive or negative) tells you whether or not the angle between u and v is acute $(0 < \theta < \frac{\pi}{2})$ or obtuse.

10. There exists a triangle with sides of lengths $1, 2, 4$.

EXERCISES

Answers to exercises marked [BB] can be found at the Back of the Book.

1. [BB] Find all vectors parallel to $\mathsf{v} = \begin{bmatrix} 1 \\ -2 \\ 1 \end{bmatrix}$ that have length 3.

2. Let $\mathsf{v} = \begin{bmatrix} -3 \\ 1 \\ 2 \end{bmatrix}$. Find all numbers k so that $\|k\mathsf{v}\| = 7$.

3. **(a)** [BB] Find a unit vector in the direction of $\mathsf{u} = \begin{bmatrix} 0 \\ 3 \\ 4 \end{bmatrix}$ and a vector of length 2 in the direction opposite to $\mathsf{v} = \begin{bmatrix} 1 \\ 2 \\ 2 \end{bmatrix}$.

   (b) Find a vector of length 3 in the direction of $2\mathsf{u} - \mathsf{v}$.

4. **(a)** Find a unit vector in the direction of $\mathsf{v} = \begin{bmatrix} 2 \\ -6 \\ -1 \end{bmatrix}$.

   (b) Find a vector of length 5 in the direction of v.

   (c) Find a vector of length 3 in the direction opposite that of v.

5. [BB] Let $\mathsf{u} = \begin{bmatrix} 3 \\ 4 \\ 0 \end{bmatrix}$, $\mathsf{v} = \begin{bmatrix} 2 \\ 1 \\ 2 \end{bmatrix}$ and $\mathsf{w} = \begin{bmatrix} 1 \\ 1 \\ 1 \end{bmatrix}$.

   (a) Find $\|\mathsf{u}\|$, $\|\mathsf{u} + \mathsf{v}\|$ and $\left\| \frac{\mathsf{w}}{\|\mathsf{w}\|} \right\|$.

   (b) Find a unit vector in the same direction as u.

   (c) Find a vector of length 4 in the direction opposite that of v.

6. **(a)** [BB] Let A and P be the points $A(1, 2)$, $P(4, 0)$. In geometrical terms, describe the set of points (x, y) in R^2 for which $\left\|\overrightarrow{AP} - \overrightarrow{AQ}\right\| = 1$.

   (b) Let A and P be the points $A(1, -1, 2)$, $P(4, 0, -1)$. Describe geometrically the set of all points (x, y, z) in R^3 for which $\left\|\overrightarrow{AP} - \overrightarrow{AQ}\right\| = 4$.

7. Find the angle between each of the following pairs of vectors. Give your answers in radians and in degrees. If it is necessary to approximate, give radians to two decimal places of accuracy and degrees to the nearest degree.

   (a) [BB] $\mathsf{u} = \begin{bmatrix} 3 \\ 4 \end{bmatrix}$, $\mathsf{v} = \begin{bmatrix} 4 \\ -3 \end{bmatrix}$

   (b) [BB] $\mathsf{u} = \begin{bmatrix} 3 \\ 4 \end{bmatrix}$, $\mathsf{v} = \begin{bmatrix} -1 \\ 1 \end{bmatrix}$

   (c) $\mathsf{u} = \begin{bmatrix} 6 \\ 8 \end{bmatrix}$, $\mathsf{v} = \begin{bmatrix} -3 \\ 3 \end{bmatrix}$

   (d) $\mathsf{u} = \begin{bmatrix} 2 \\ -1 \\ 1 \end{bmatrix}$ and $\mathsf{v} = \begin{bmatrix} 1 \\ 1 \\ 2 \end{bmatrix}$

(e) $u = \begin{bmatrix} 1 \\ 3 \\ 2 \end{bmatrix}$ and $v = \begin{bmatrix} -4 \\ -1 \\ 1 \end{bmatrix}$ **(f)** $u = \begin{bmatrix} 1 \\ -2 \\ 3 \end{bmatrix}$ and $v = \begin{bmatrix} -1 \\ 2 \\ 1 \end{bmatrix}$

(g) $u = \begin{bmatrix} 1 \\ -1 \\ 2 \end{bmatrix}$ and $v = \begin{bmatrix} -1 \\ 1 \\ 0 \end{bmatrix}$ **(h)** $u = \begin{bmatrix} 3 \\ 0 \\ 1 \\ -2 \end{bmatrix}$ and $v = \begin{bmatrix} 1 \\ 1 \\ 1 \\ 5 \end{bmatrix}$

(i) $u = \begin{bmatrix} 0 \\ 3 \\ 1 \\ -2 \end{bmatrix}$ and $v = \begin{bmatrix} -3 \\ 6 \\ 1 \\ 1 \end{bmatrix}$ **(j)** $u = \begin{bmatrix} 0 \\ 1 \\ 1 \\ 2 \\ 1 \end{bmatrix}$ and $v = \begin{bmatrix} 2 \\ -2 \\ 3 \\ 0 \\ 1 \end{bmatrix}$.

8. Find all numbers x (if any) such that the angle between $u = \begin{bmatrix} 1 \\ 1 \\ 0 \end{bmatrix}$ and $v = \begin{bmatrix} x \\ 1 \\ 1 \end{bmatrix}$ is 60°.

9. Let $u = \begin{bmatrix} 1 \\ 2 \\ 3 \\ -1 \end{bmatrix}$, $v = \begin{bmatrix} 0 \\ -1 \\ 1 \end{bmatrix}$, $x = \begin{bmatrix} 1 \\ 2 \\ 3 \end{bmatrix}$ and $y = \begin{bmatrix} 0 \\ -1 \\ 1 \\ 2 \end{bmatrix}$.

   Find each of the following, if possible.

   (a) [BB] $2x - 3y + u$ **(b)** [BB] $\|y\|$
   (c) $3v + 4x$ **(d)** $x \cdot v$
   (e) the cosine of the angle between u and y.

10. Find $u \cdot v$ in each of the following cases:

    (a) [BB] $\|u\| = 1$, $\|v\| = 3$ and the angle between u and v is 45°
    (b) $\|u\| = 4$, $\|v\| = 7$ and the angle between u and v is 2 radians.

11. **(a)** [BB] For vectors u and v, can $u \cdot v = -7$ if $\|u\| = 3$ and $\|v\| = 2$?
    (b) Can $u \cdot v = 9$ if $\|u\| = 5$ and $\|v\| = 2$?
    (c) Can $u \cdot v = 9$ if $\|u\| = 5$ and $\|v\| = 1$?

    Justify your answers.

12. Let $u = \begin{bmatrix} 2 \\ 5 \\ 3 \end{bmatrix}$, $v = \begin{bmatrix} -4 \\ 1 \\ 0 \end{bmatrix}$ and $w = \begin{bmatrix} -3 \\ 3 \\ 5 \end{bmatrix}$. Find $\|u\|$, $\|v\|$, $\|w\|$ and $\|u + v - w\|$.

13. Let u and v be vectors of lengths 3 and 5 respectively and suppose that $u \cdot v = 8$. Find the following:

    (a) [BB] $(u - v) \cdot (2u - 3v)$ **(b)** $(-3u + 4v) \cdot (2u + 5v)$
    (c) [BB] $\|u + v\|^2$ **(d)** $(2u + 3v) \cdot (8v - u)$
    (e) $(-4v + 5u) \cdot (-3v - 6u)$ **(f)** $(2u - 5v) \cdot (3u - 7v)$.

14. Let u, v and w be vectors of lengths 3, 2 and 6, respectively. Suppose $u \cdot v = 3$, $u \cdot w = 5$ and $v \cdot w = 2$. Find the following:

    (a) $(u - 2v) \cdot (3w + 2u)$ **(b)** $(3w - 2v) \cdot (5v - w)$
    (c) $(-3u + 4w) \cdot (6u - 2v)$ **(d)** $(8w - 12v) \cdot (37v + 24w)$
    (e) $\|u - v - w\|^2$.

15. [BB] Suppose u is a vector of length $\sqrt{6}$ and v is a unit vector orthogonal to $\mathsf{u}+\mathsf{v}$. Find $\|2\mathsf{u}-\mathsf{v}\|$.

16. Let $\mathsf{u}=\begin{bmatrix}1\\2\end{bmatrix}$ and $\mathsf{v}=\begin{bmatrix}1\\-1\end{bmatrix}$.

    (a) [BB] Find all numbers k such that $\mathsf{u}+k\mathsf{v}$ has length 3.

    (b) Find all numbers k such that $k\mathsf{u}-3\mathsf{v}$ is orthogonal to $\begin{bmatrix}-4\\5\end{bmatrix}$.

17. Let $\mathsf{u}=\begin{bmatrix}1\\0\\-1\end{bmatrix}$ and $\mathsf{v}=\begin{bmatrix}2\\1\\0\end{bmatrix}$.

    (a) Find a unit vector in the same direction as $3\mathsf{u}-\mathsf{v}$.

    (b) If possible, find a real number (or numbers) k so that $\mathsf{u}+k\mathsf{v}$ is orthogonal to $\mathsf{u}-k\mathsf{v}$.

18. Let $\mathsf{u}=\begin{bmatrix}1\\0\\-1\end{bmatrix}$ and $\mathsf{v}=\begin{bmatrix}2\\1\\0\end{bmatrix}$.

    (a) Find a real number k so that $\mathsf{u}+k\mathsf{v}$ is orthogonal to u, if such k exists.

    (b) Find a real number ℓ so that $\ell\mathsf{u}+\mathsf{v}$ is a unit vector, if such ℓ exists.

19. Let $\mathsf{u}=\begin{bmatrix}1\\1\\0\\-2\end{bmatrix}$ and $\mathsf{v}=\begin{bmatrix}-3\\0\\0\\1\end{bmatrix}$. Find all scalars c with the property that

    (a) [BB] $\|c\mathsf{u}\|=8$.

    (b) u is orthogonal to $\mathsf{u}+c\mathsf{v}$.

    (c) the angle between $c\mathsf{u}+\mathsf{v}$ and v is $60°$.

    (d) $c\mathsf{u}+\mathsf{v}$ and $\mathsf{u}+c\mathsf{v}$ are orthogonal.

    (e) $\frac{1}{2}\mathsf{u}-c\mathsf{v}$ is a unit vector.

20. Given vectors u and v that are not orthogonal, find all scalars k such that $\mathsf{u}+k\mathsf{v}$ is orthogonal to u.

21. [BB] If a nonzero vector u is orthogonal to another vector v and k is a scalar, show that $\mathsf{u}+k\mathsf{v}$ is not orthogonal to u.

22. Show that $A(8,1,-1)$, $B(3,0,2)$, $C(4,3,0)$ are the vertices of a right angled triangle. Where is the right angle?

23. [BB] Given the points $A(-1,0)$ and $B(2,3)$ in the plane, find all points C such that A, B and C are the vertices of a right angled triangle with right angle at A and AC of length 2.

24. Given the points $A(0,-2)$ and $B(1,1)$, find all points C so that A, B and C are the vertices of a right angled triangle with right angle at B and hypotenuse of length 10.

25. Find a point D such that $(1, 2)$, $(-3, -1)$, $(4, -2)$ and D are the vertices of a square and justify your answer.

26. [BB; 1, 5] Prove all parts of Theorem 1.2.20 for the two-dimensional vectors $\mathsf{u} = \begin{bmatrix} x \\ y \end{bmatrix}$, $\mathsf{v} = \begin{bmatrix} z \\ w \end{bmatrix}$ and $\mathsf{w} = \begin{bmatrix} r \\ s \end{bmatrix}$.

27. Use the formula $\mathsf{u} \cdot \mathsf{v} = \|\mathsf{u}\| \, \|\mathsf{v}\| \cos\theta$ to derive the *Law of Cosines*, that goes as follows: If $\triangle ABC$ has sides of lengths a, b, c as shown and $\theta = \angle BCA$, then $c^2 = a^2 + b^2 - 2ab\cos\theta$.

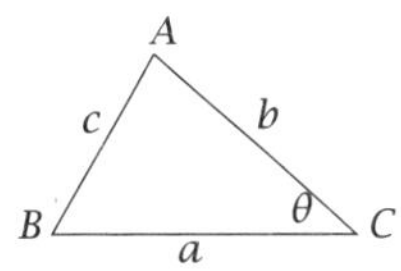

28. Given any $2n$ real numbers $x_1, x_2, \ldots, x_n$, $y_1, y_2, \ldots, y_n$, it is a fact that

$$|x_1y_1 + x_2y_2 + \cdots + x_ny_n| \le \sqrt{x_1^2 + x_2^2 + \cdots + x_n^2}\sqrt{y_1^2 + y_2^2 + \cdots + y_n^2}.$$

Why?

29. Verify the Cauchy-Schwarz inequality for each of the following pairs of vectors.

(a) [BB] $\mathsf{u} = \begin{bmatrix} -2 \\ 1 \end{bmatrix}$, $\mathsf{v} = \begin{bmatrix} 3 \\ 5 \end{bmatrix}$ **(b)** $\mathsf{u} = \begin{bmatrix} 1 \\ -2 \\ 3 \end{bmatrix}$, $\mathsf{v} = \begin{bmatrix} 2 \\ 0 \\ 1 \end{bmatrix}$

(c) $\mathsf{u} = \begin{bmatrix} 1 \\ 1 \\ -2 \end{bmatrix}$, $\mathsf{v} = \begin{bmatrix} 1 \\ -3 \\ -2 \end{bmatrix}$ **(d)** [BB] $\mathsf{u}\begin{bmatrix} 1 \\ 0 \\ -2 \end{bmatrix}$, $\mathsf{v} = \begin{bmatrix} -3 \\ 1 \\ 0 \end{bmatrix}$

(e) $\mathsf{u} = \begin{bmatrix} -1 \\ 0 \\ 1 \\ 2 \end{bmatrix}$, $\mathsf{v} = \begin{bmatrix} 3 \\ -1 \\ 1 \\ 0 \end{bmatrix}$ **(f)** $\mathsf{u} = \begin{bmatrix} -1 \\ -1 \\ 0 \\ 0 \\ 1 \end{bmatrix}$, $\mathsf{v} = \begin{bmatrix} 0 \\ 2 \\ 1 \\ 1 \\ 3 \end{bmatrix}$.

30. Verify the triangle inequality for each pair of vectors in Exercise 29.

31. Prove that $|a + b| \le |a| + |b|$ for any real numbers a and b.

32. [BB] Use the Cauchy-Schwarz inequality to prove that $(a\cos\theta + b\sin\theta)^2 \le a^2 + b^2$ for all $a, b \in \mathsf{R}$.

33. Prove that $\|\mathsf{u} + \mathsf{v}\|^2 + \|\mathsf{u} - \mathsf{v}\|^2 = 2\|\mathsf{u}\|^2 + 2\|\mathsf{v}\|^2$ and interpret this result geometrically when u and v lie in a plane.

34. **(a)** [BB] Use the Cauchy-Schwarz inequality to prove that $(3ac + bd)^2 \le (3a^2 + b^2)(3c^2 + d^2)$ for any real numbers a, b, c, d.

(b) Use part (a) to prove that $abcd \le \frac{1}{2}(a^2d^2 + b^2c^2)$.

35. Show that $\sqrt{\sin 2\theta} \le \frac{\sqrt{2}}{2}(\sin\theta + \cos\theta)$ for any θ, $0 \le \theta \le \frac{\pi}{2}$.

[Hint: Arithmetic/Geometric Mean inequality]

36. Do there exist points A, B and C such that $\overrightarrow{AB} = \begin{bmatrix} 7 \\ 4 \\ -1 \\ 0 \end{bmatrix}$, $\overrightarrow{AC} = \begin{bmatrix} 0 \\ 1 \\ -1 \\ 3 \end{bmatrix}$, $\overrightarrow{BC} = \begin{bmatrix} 2 \\ 1 \\ -1 \\ -2 \end{bmatrix}$? [Hint: Triangle inequality]

Critical Reading

37. [BB] Susie is asked to find the length of the vector $\mathsf{u} = \frac{4}{7}\begin{bmatrix} -2 \\ 1 \\ 5 \end{bmatrix}$. She proceeds like this: $\mathsf{u} = \begin{bmatrix} -\frac{8}{7} \\ \frac{4}{7} \\ \frac{20}{7} \end{bmatrix}$, so $\|\mathsf{u}\| = \sqrt{(-\frac{8}{7})^2 + (\frac{4}{7})^2 + (\frac{20}{7})^2} = \frac{\sqrt{480}}{7}$.

    Do you like this approach? Can you suggest a better one?

38. **(a)** i. Show that $\mathsf{w} = \begin{bmatrix} 8 \\ 5 \\ 4 \end{bmatrix}$ is orthogonal to $\mathsf{u} = \begin{bmatrix} 3 \\ -4 \\ -1 \end{bmatrix}$ and to $\mathsf{v} = \begin{bmatrix} 1 \\ 0 \\ -2 \end{bmatrix}$.

    ii. Show that w is orthogonal to $3\mathsf{u} - 2\mathsf{v}$.

    iii. Show that w is orthogonal to $a\mathsf{u} + b\mathsf{v}$ for any scalars a and b.

    (b) Generalize the results of part (a). Suppose a vector w is orthogonal to two vectors u and v. Show that w is orthogonal to every linear combination of u and v.

    (c) What might be a good use for the principle discovered in part (b)?

39. If u is a vector in R^n that is orthogonal to every vector in R^n, then u must be the zero vector. Why?

40. A rectangular box has sides of lengths in the ratio $1 : 2 : 3$. Find the angle between a diagonal to the box and one of its shortest sides.

41. Let $\mathcal{P}$ be a parallelogram with sides representing the vectors u and v as shown.

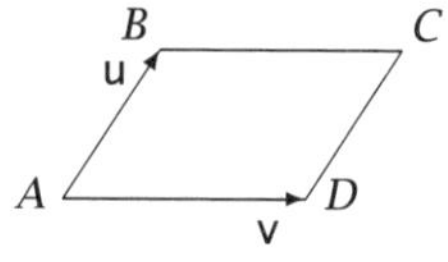

    (a) [BB] Express the diagonals $\overrightarrow{AC}$ and $\overrightarrow{DB}$ of $\mathcal{P}$ in terms of u and v.

    (b) Prove that the sum of the squares of the lengths of the diagonals equals the sum of the squares of the lengths of the four sides.

42. Let ABC be a triangle inscribed in a semicircle as shown. Use vectors to show that the angle at B is a right angle.

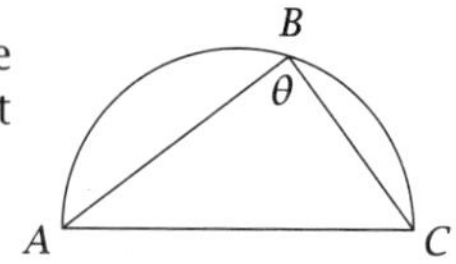

43. [BB] Suppose vectors u and v are pictured by arrows as shown. Find a formula for a vector whose arrow bisects angle AOB.

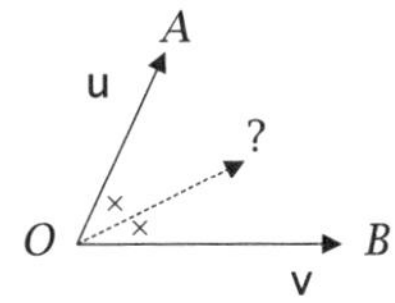

44. Let u and v be two vectors that are not parallel and let $w = u + v$. Explain why w bisects the angle between u and v.

45. Let u and v be nonzero vectors and set $w = \|v\|\, u + \|u\|\, v$. Show that w bisects the angle between u and v.

46. **(a)** Prove that the sum of the lengths of the diagonals of a rhombus is at least as great as the sum of the lengths of two opposite sides. Specifically, in the figure, explain why the sum of the lengths of AC and BD is greater than or equal to the sum of the lengths of AB and CD.

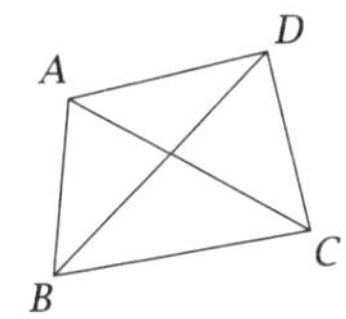

(b) There are 1000 blue dots on a page and 1000 red dots, all in different places. You are asked to pair each red dot with a blue dot by a straight line. Prove that you can do this (1000 lines) without any lines crossing each other. [Hint: Minimize the sum of the lengths.]

47. How many triangles have sides with integer lengths and perimeter 15? [Hint: First ask yourself what's the length of the longest side of such a triangle.]

1.3 Lines, Planes and the Cross Product

To university students who have taken some mathematics courses, "$x^2 + y^2 = 1$ is a circle" probably doesn't seem a strange thing to say, but it would certainly confuse a seven-year-old who thinks she knows what a circle is, but has no idea about equations. Of course, $x^2 + y^2 = 1$ is not a circle. It's an **equation** whose **graph** in the xy-plane happens to be a circle. What's the connection between an equation and its graph?

The graph of an equation is the set of all points whose coordinates satisfy that equation and only those points.

1.3.1 Example. A point (x, y) in the plane is on the line of slope 2 through the origin if and only if its y-coordinate is twice its x-coordinate. Thus $y = 2x$ is the equation of the line. ☺

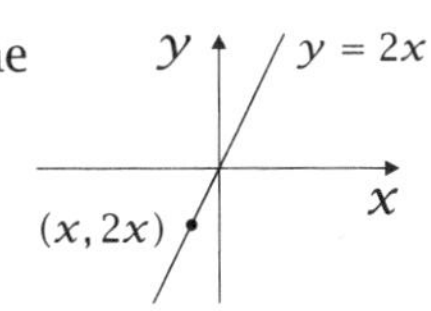

1.3.2 Example. A point (x, y) in the plane is on the circle with centre $(0, 0)$ and radius 1 if and only if its coordinates x and y satisfy $x^2 + y^2 = 1$. Thus the graph of $x^2 + y^2 = 1$ is a circle. ‿̈

What is the graph of $x^2 + y^2 = 1$ if we are working in 3-space? What points in 3-**space** satisfy this equation? Here are some examples (all with the same first two coordinates):

$$(1,0,0), (1,0,1), (1,0,2), (1,0,3), (1,0,4), \ldots, (1,0,-1), (1,0,-2), \ldots \quad (1)$$

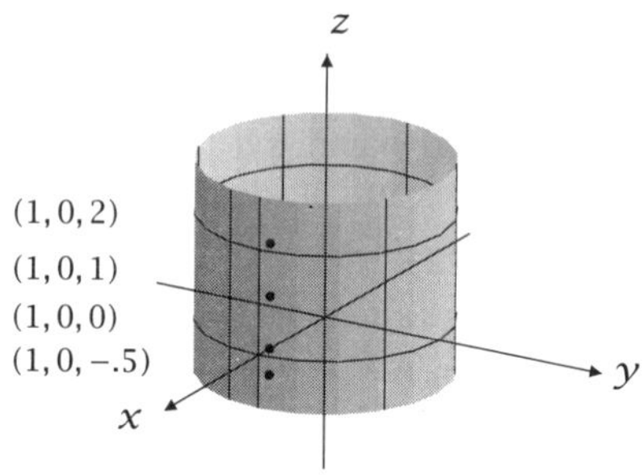

Figure 1.14: In 3-space, the graph of $x^2 + y^2 = 1$ is a cylinder.

In 3-space, all of these points lie on the graph of $x^2 + y^2 = 1$. In every case, the sum of the squares of the x- and y-coordinates is 1. Think of the floor as the xy-plane, identify a point near the middle of the floor as the origin and imagine the circle in the floor with its centre at that origin and radius 1. In 3-space, not only is this circle on the graph of $x^2 + y^2 = 1$, but so also is every point directly above or below this circle: $(1, 0, z)$ satisfies the equation for any z. In 3-space, the graph of $x^2 + y^2 = 1$ is a **cylinder**, specifically, that cylinder which slices through the xy-plane in the familiar unit circle with centre the origin. (See Figure 1.14.)

What is the graph of the equation $y = 2x$? A line, perhaps? True, if we are working in the plane: The set of points (x, y) with $y = 2x$ is a line of slope 2 through the origin. In 3-space, however, not only do the points on this line satisfy $y = 2x$, but so does every point $(x, 2x, z)$ directly above or below this line. In 3-space, the graph of $y = 2x$ is a **plane**, specifically, that plane perpendicular to the xy-plane that cuts it in the familiar "line $y = 2x$."

In the plane, lines are precisely the graphs of *linear equations*, equations of the form $ax + by = c$. The graph of such an equation in 3-space, however, is a plane perpendicular to the xy-plane. The graphs of linear equations in 3-space are **not** lines; in fact, they are planes.

The Equation of a Plane

We'd like to obtain the equation of a plane. This means finding an equation in variables x, y and z with the property that $P(x,y,z)$ is on the plane if and only if its coordinates x, y, z satisfy the equation. Just as a line in the xy-plane is determined by a slope m and a point on it, a plane in R^3 is determined by a vector perpendicular to it—such a vector is called a *normal*—and a point on it.

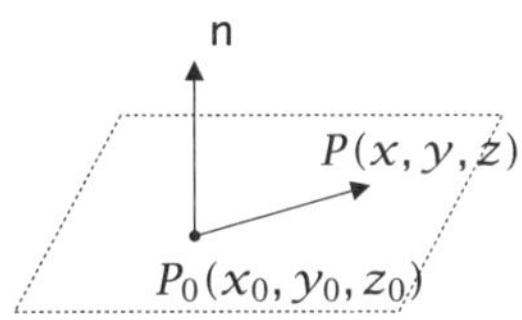

Figure 1.15: $P(x,y,z)$ is on the plane through $P_0(x_0,y_0,z_0)$ with normal n if and only if $\overrightarrow{P_0P}$ is orthogonal to n.

Let $\mathsf{n} = \begin{bmatrix} a \\ b \\ c \end{bmatrix}$ be a normal vector to a plane and let $P_0(x_0,y_0,z_0)$ be a point in the plane. See Figure 1.15. Then $P(x,y,z)$ will lie in the plane precisely if the vector represented by the arrow from P_0 to P is orthogonal to n; that is, if and only if

$$\begin{bmatrix} x - x_0 \\ y - y_0 \\ z - z_0 \end{bmatrix} \cdot \begin{bmatrix} a \\ b \\ c \end{bmatrix} = 0.$$

This occurs if and only if

$$a(x - x_0) + b(y - y_0) + c(z - z_0) = 0,$$

an equation that can be rewritten

$$ax + by + cz = ax_0 + by_0 + cz_0.$$

The right-hand side is a number d, since a, b, c are the given components of a normal n and x_0, y_0, z_0 are the given coordinates of point P_0. Thus the plane has equation

$$ax + by + cz = d. \tag{2}$$

Conversely, if $P_0(x_0,y_0,z_0)$ is any point whose coordinates satisfy the equation $ax + by + cz = d$—that is, $ax_0 + by_0 + cz_0 = d$—then $ax + by + cz = ax_0 + by_0 + cz_0$, which can be rewritten $a(x - x_0) + b(y - y_0) + c(z - z_0) = 0$. So the points P satisfying this equation are precisely those for which the vector $P_0P = \begin{bmatrix} x - x_0 \\ y - y_0 \\ z - z_0 \end{bmatrix}$ is orthogonal to $\mathsf{n} = \begin{bmatrix} a \\ b \\ c \end{bmatrix}$, and these are the points of a plane. We summarize.

1.3.3

> In 3-space, the equation $ax+by+cz=d$ defines a plane with normal $\begin{bmatrix} a \\ b \\ c \end{bmatrix}$.

1.3.4 Examples. • The equation $2x-3y+4z=0$ is the equation of a plane through the origin $(0,0,0)$ because $x=0, y=0, z=0$ satisfies the equation. A normal to the plane is $\begin{bmatrix} 2 \\ -3 \\ 4 \end{bmatrix}$.

- The equation $2x-3y+4z=10$ defines the plane with normal $\begin{bmatrix} 2 \\ -3 \\ 4 \end{bmatrix}$ that passes through the point $(5,0,0)$. This plane is parallel to the previous one since the two planes have the same normal.
- The equation $y=2x$ defines the plane through $(0,0,0)$ with normal $\begin{bmatrix} -2 \\ 1 \\ 0 \end{bmatrix}$ (the coefficients of x, y, z in the equation $-2x+y=0$).
- In 3-space, the xy-plane has the equation $z=0$ since $\mathsf{k}=\begin{bmatrix} 0 \\ 0 \\ 1 \end{bmatrix}$ is a normal and $(0,0,0)$ is a point on the plane.

1.3.5 Problem. Find an equation of the plane with normal $\mathsf{n}=\begin{bmatrix} 1 \\ -2 \\ 3 \end{bmatrix}$ and containing the point $(4,0,-1)$.

Solution. An equation has the form $1x-2y+3z=d$ and is satisfied by $x=4$, $y=0$, $z=-1$. Thus $1(4)-2(0)+3(-1)=d$, so $d=1$ and an equation is $x-2y+3z=1$. (We say "an" equation because $2x-4y+6z=2$ and $-3x+6y-9z=-3$ are also equations.)

1.3.6 Problem. Find an equation of the plane through $(4,-1,5)$ that is parallel to the plane with equation $2x-3y=6$.

Solution. Parallel planes have the same normals, so $\mathsf{n}=\begin{bmatrix} 2 \\ -3 \\ 0 \end{bmatrix}$ is a normal for the desired plane, which will have an equation of the form $2x-3y=d$. Since $(4,-1,5)$ is on the plane, $2(4)-3(-1)=d$. So $d=11$ and our plane has equation $2x-3y=11$.

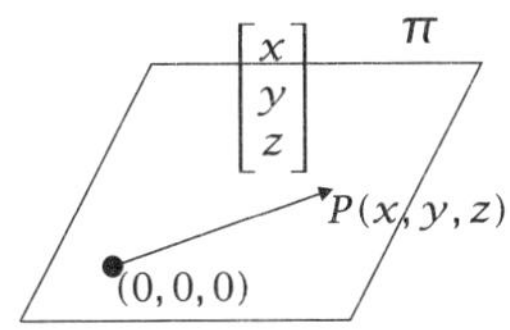

1.3.7 Remark. In this section, we have talked about **points** lying in planes. It is convenient to be able to talk about **vectors** lying in planes too.

As with points, when we say that a vector $\begin{bmatrix} x \\ y \\ z \end{bmatrix}$ lies in a plane, we mean that its components satisfy the equation of the plane, but also, by convention, that **the plane passes through the origin** so that the arrow from $(0, 0, 0)$ to $P(x, y, z)$ lies in the plane. So an equation of the plane that contains a given vector always looks like $ax + by + cz = 0$.

1.3.8 Example. The plane with equation $-x + 4y + 5z = 0$ contains the vectors $\begin{bmatrix} -8 \\ 3 \\ -4 \end{bmatrix}$ and $\begin{bmatrix} 9 \\ 1 \\ 1 \end{bmatrix}$. ⌣̈

1.3.9 Example. The vector $\begin{bmatrix} x \\ y \\ z \end{bmatrix}$ is in the plane π with equation $-x+4y+5z = 0$ if and only if $x = 4y+5z$. So π consists of all vectors of the form $\begin{bmatrix} 4y+5z \\ y \\ z \end{bmatrix} = \begin{bmatrix} 4y \\ y \\ 0 \end{bmatrix} + \begin{bmatrix} 5z \\ 0 \\ z \end{bmatrix} = y\begin{bmatrix} 4 \\ 1 \\ 0 \end{bmatrix} + z\begin{bmatrix} 5 \\ 0 \\ 1 \end{bmatrix}$, that is, the set of all linear combinations of $\begin{bmatrix} 4 \\ 1 \\ 0 \end{bmatrix}$ and $\begin{bmatrix} 5 \\ 0 \\ 1 \end{bmatrix}$. So π is spanned by these two vectors. Of course, π is spanned by many other pairs of nonparallel vectors too, in fact by any two vectors it contains, as we saw geometrically in Section 1.1. ⌣̈

READING CHECK 1. Find two vectors that span the plane with equation $-x + 4y + 5z = 0$ different from the two just given.

The Cross Product

The vectors $\mathsf{u} = \begin{bmatrix} -8 \\ 3 \\ -4 \end{bmatrix}$ and $\mathsf{v} = \begin{bmatrix} 9 \\ 1 \\ 1 \end{bmatrix}$ are not parallel, so they span a plane, that plane consisting of vectors of the form $a\mathsf{u} + b\mathsf{v} = a\begin{bmatrix} -8 \\ 3 \\ -4 \end{bmatrix} + b\begin{bmatrix} 9 \\ 1 \\ 1 \end{bmatrix} = \begin{bmatrix} -8a+9b \\ 3a+b \\ -4a+b \end{bmatrix}$. This is a perfectly good description of the plane in question, but we would like an equation too. How would we discover $-x+4y+5z = 0$—see

Example 1.3.8? Since the plane must pass through the origin, any equation has the form $ax+by+cz = 0$ and the coefficients a, b, c are the components of a normal to the plane. So the problem is to find a normal and there is a procedure for this. To describe what we have in mind, we first introduce the notion of 2×2 (read "two by two") **determinant**.

1.3.10 **Definition.** If a, b, c and d are four numbers, the 2×2 *determinant* $\begin{vmatrix} a & b \\ c & d \end{vmatrix}$ is the number $ad - bc$:

$$\boxed{\begin{vmatrix} a & b \\ c & d \end{vmatrix} = ad - bc.}$$

1.3.11 **Examples.**
- $\begin{vmatrix} 2 & 1 \\ -4 & 5 \end{vmatrix} = 2(5) - 1(-4) = 14;$
- $\begin{vmatrix} 3 & -1 \\ -6 & 2 \end{vmatrix} = 3(2) - (-1)(-6) = 0.$

1.3.12 **Definition.** The *cross product* of vectors $\mathsf{u} = \begin{bmatrix} u_1 \\ u_2 \\ u_3 \end{bmatrix}$ and $\mathsf{v} = \begin{bmatrix} v_1 \\ v_2 \\ v_3 \end{bmatrix}$ is a vector denoted $\mathsf{u} \times \mathsf{v}$ (read "u cross v") and defined by

$$\mathsf{u} \times \mathsf{v} = \begin{vmatrix} u_2 & u_3 \\ v_2 & v_3 \end{vmatrix} \mathsf{i} - \begin{vmatrix} u_1 & u_3 \\ v_1 & v_3 \end{vmatrix} \mathsf{j} + \begin{vmatrix} u_1 & u_2 \\ v_1 & v_2 \end{vmatrix} \mathsf{k}.$$

Expanding the three determinants in this definition, we have

$$\begin{aligned} \mathsf{u} \times \mathsf{v} &= (u_2v_3 - u_3v_2)\mathsf{i} - (u_1v_3 - u_3v_1)\mathsf{j} + (u_1v_2 - u_2v_1)\mathsf{k} \\ &= \begin{bmatrix} u_2v_3 - u_3v_2 \\ -(u_1v_3 - u_3v_1) \\ u_1v_2 - u_2v_1 \end{bmatrix}, \end{aligned}$$

although most people prefer to remember the coefficients of i, j and k in terms of determinants, using the following scheme:

1.3.13 $$\boxed{\mathsf{u} \times \mathsf{v} = \begin{vmatrix} \mathsf{i} & \mathsf{j} & \mathsf{k} \\ u_1 & u_2 & u_3 \\ v_1 & v_2 & v_3 \end{vmatrix}.}$$

Here's how this works.

To find the first component of $\mathsf{u} \times \mathsf{v}$, look at the array on the right side of 1.3.13, mentally remove the row and column in which i appears (the first row and the first column) and compute the determinant of the 2×2 array that remains,

$$\begin{vmatrix} u_2 & u_3 \\ v_2 & v_3 \end{vmatrix} = u_2v_3 - u_3v_2.$$

The third component is obtained by removing the row and column in which k appears (the first row and the third column) and computing the determinant of the two by two array that remains,

$$\begin{vmatrix} u_1 & u_2 \\ v_1 & v_2 \end{vmatrix} = u_1v_2 - u_2v_1.$$

The second component is obtained in a similar way, except for a sign change. Notice the minus sign highlighted by the arrow in formula 1.3.12.

1.3.14 Example. Let $\mathsf{u} = \begin{bmatrix} 1 \\ 3 \\ 2 \end{bmatrix}$ and $\mathsf{v} = \begin{bmatrix} 0 \\ 2 \\ -1 \end{bmatrix}$. Then

$$\mathsf{u} \times \mathsf{v} = \begin{vmatrix} \mathsf{i} & \mathsf{j} & \mathsf{k} \\ 1 & 3 & 2 \\ 0 & 2 & -1 \end{vmatrix} = \begin{vmatrix} 3 & 2 \\ 2 & -1 \end{vmatrix} \mathsf{i} - \begin{vmatrix} 1 & 2 \\ 0 & -1 \end{vmatrix} \mathsf{j} + \begin{vmatrix} 1 & 3 \\ 0 & 2 \end{vmatrix} \mathsf{k}$$

$$= -7\mathsf{i} - (-1)\mathsf{j} + 2\mathsf{k} = -7\mathsf{i} + \mathsf{j} + 2\mathsf{k} = \begin{bmatrix} -7 \\ 1 \\ 2 \end{bmatrix}.$$

Note that this vector is orthogonal to the two given vectors because

$$(\mathsf{u} \times \mathsf{v}) \cdot \mathsf{u} = \begin{bmatrix} -7 \\ 1 \\ 2 \end{bmatrix} \cdot \begin{bmatrix} 1 \\ 3 \\ 2 \end{bmatrix} = -7 + 3 + 4 = 0$$

and

$$(\mathsf{u} \times \mathsf{v}) \cdot \mathsf{v} = \begin{bmatrix} -7 \\ 1 \\ 2 \end{bmatrix} \cdot \begin{bmatrix} 0 \\ -2 \\ 1 \end{bmatrix} = -2 + 2 = 0.$$

Since $\mathsf{n} = \begin{bmatrix} -7 \\ 1 \\ 2 \end{bmatrix}$ is orthogonal to u and to v, it is orthogonal to any linear combination of u and v,

$$\mathsf{n} \cdot (a\mathsf{u} + b\mathsf{v}) = a(\mathsf{n} \cdot \mathsf{u}) + b(\mathsf{n} \cdot \mathsf{v}) = 0,$$

so it's a normal vector to the plane spanned by u and v. An equation for this plane is $7x - y - 2z = 0$. ☺

As we shall see, the cross product of two vectors is always normal to the plane spanned by the vectors. This observation makes it easy to check that what you think is the cross product of two vectors probably is. Check that the vector you found actually has dot product 0 with each of the two given vectors.

READING CHECK 2. Verify that $-x + 4y + 5z = 0$ is an equation for the plane spanned by $\mathsf{u} = \begin{bmatrix} -8 \\ 3 \\ -4 \end{bmatrix}$ and $\mathsf{v} = \begin{bmatrix} 9 \\ 1 \\ 1 \end{bmatrix}$ —see Example 1.3.8.

1.3.15 Problem. Find an equation of the plane through the points $A(1, 3, -2)$, $B(1, 1, 5)$ and $C(2, -2, 3)$.

Solution. A normal vector to the plane is orthogonal to both $\overrightarrow{AB} = \begin{bmatrix} 0 \\ -2 \\ 7 \end{bmatrix}$ and $\overrightarrow{AC} = \begin{bmatrix} 1 \\ -5 \\ 5 \end{bmatrix}$. So one normal is the cross product of $\overrightarrow{AB}$ and $\overrightarrow{AC}$:

$$\overrightarrow{AB} \times \overrightarrow{AC} = \begin{vmatrix} \mathsf{i} & \mathsf{j} & \mathsf{k} \\ 0 & -2 & 7 \\ 1 & -5 & 5 \end{vmatrix} = 25\mathsf{i} - (-7)\mathsf{j} + (-(-2))\mathsf{k} = 25\mathsf{i} + 7\mathsf{j} + 2\mathsf{k} = \begin{bmatrix} 25 \\ 7 \\ 2 \end{bmatrix}.$$

So the plane through A, B, C has an equation $25x + 7y + 2z = d$, and we can find d using the fact that $A(1, 3, -2)$ (or B or C) is on the plane. Thus $25(1)+7(3)+2(-2) = d = 42$. An equation of the plane is $25x+7y+2z = 42$.

READING CHECK 3. Check that the same value for d is obtained using the fact that the coordinates of $B(1, 1, 5)$ satisfy $25x + 7y + 2z = 42$. Do the same with $(2, -2, 3)$.

Before continuing, let's record some basic properties of this new concept, the cross product.

1.3.16 Theorem (Properties of the Cross Product). *If* $\mathsf{u}, \mathsf{v}, \mathsf{w}$ *are vectors in* 3*-space and* c *is a scalar,*

1. *(Anticommutativity)* $\mathsf{v} \times \mathsf{u} = -(\mathsf{u} \times \mathsf{v})$.
2. *(Distributivity)* $\mathsf{u} \times (\mathsf{v} + \mathsf{w}) = (\mathsf{u} \times \mathsf{v}) + (\mathsf{u} \times \mathsf{w})$.
3. *(Scalar associativity)* $(c\mathsf{u}) \times \mathsf{v} = c(\mathsf{u} \times \mathsf{v}) = \mathsf{u} \times (c\mathsf{v})$.
4. $\mathsf{u} \times 0 = 0 = 0 \times \mathsf{u}$.

5. $\mathsf{u} \times \mathsf{v}$ *is orthogonal to both* u *and* v *and hence orthogonal to any vector in the plane spanned by* u *and* v.

6. $\mathsf{u} \times \mathsf{u} = 0$.

7. $\mathsf{u} \times \mathsf{v} = 0$ *if and only if* u *is a scalar multiple of* v *or* v *is a scalar multiple of* u, *that is, if and only if* u *and* v *are parallel.*

8. $\|\mathsf{u} \times \mathsf{v}\| = \|\mathsf{u}\|\,\|\mathsf{v}\| \sin\theta$, *where* θ *is the angle between* u *and* v. *(Remember that we always assume* $0 \le \theta \le \pi$.*)*

9. $\|\mathsf{u} \times \mathsf{v}\|$ *is the area of the parallelogram having* u *and* v *as adjacent sides.*

That really is a minus sign in property 1. An operation $\star$ is *commutative* if $a \star b = b \star a$ for any a and b. Many operations are commutative, ordinary addition and multiplication of real numbers, for instance. The cross product is not commutative, however. It is what we call *anticommutative*: $\mathsf{v} \times \mathsf{u} = -(\mathsf{u} \times \mathsf{v})$.

1.3.17 Example. To illustrate, take u and v as in Example 1.3.14 and compute

$$\mathsf{v} \times \mathsf{u} = \begin{vmatrix} \mathsf{i} & \mathsf{j} & \mathsf{k} \\ 0 & 2 & -1 \\ 1 & 3 & 2 \end{vmatrix} = 7\mathsf{i} - 1\mathsf{j} + (-2)\mathsf{k} = \begin{bmatrix} 7 \\ -1 \\ -2 \end{bmatrix} = -(\mathsf{u} \times \mathsf{v}).$$

In the exercises, we ask you to verify the first five statements of Theorem 1.3.16. If we put $\mathsf{v} = \mathsf{u}$ in the first statement of Theorem 1.3.16, we obtain $\mathsf{u} \times \mathsf{u} = -(\mathsf{u} \times \mathsf{u})$, so $\mathsf{u} \times \mathsf{u} = 0$, which is statement 6. Half of statement 7 is then straightforward. If $\mathsf{v} = c\mathsf{u}$ is a scalar multiple of u, then $\mathsf{u} \times \mathsf{v} = \mathsf{u} \times (c\mathsf{v}) = c(\mathsf{u} \times \mathsf{v}) = c0 = 0$. The other direction, that $\mathsf{u} \times \mathsf{v} = 0$ implies u is a scalar multiple of v or v is a scalar multiple of u, we leave to the exercises (Exercise 27). The remarkable formula given in statement 8 follows from

$$\mathsf{u} \cdot \mathsf{v} = \|\mathsf{u}\|\,\|\mathsf{v}\| \cos\theta \tag{3}$$

and an identity attributed to the Italian mathematician Joseph-Louis Lagrange (1736–1813), namely,

$$\|\mathsf{u} \times \mathsf{v}\|^2 + (\mathsf{u} \cdot \mathsf{v})^2 = \|\mathsf{u}\|^2\,\|\mathsf{v}\|^2,$$

that we ask the reader to verify in the exercises.

1.3.18 Example. To illustrate statement 8, let $\mathsf{u} = \begin{bmatrix} -1 \\ 1 \\ 0 \end{bmatrix}$ and $\mathsf{v} = \begin{bmatrix} 1 \\ 1 \\ 1 \end{bmatrix}$. Then

$$\mathsf{u} \times \mathsf{v} = \begin{vmatrix} \mathsf{i} & \mathsf{j} & \mathsf{k} \\ -1 & 1 & 0 \\ 1 & 1 & 1 \end{vmatrix} = 1\mathsf{i} - (-1)\mathsf{j} + (-2)\mathsf{k} = \begin{bmatrix} 1 \\ 1 \\ -2 \end{bmatrix},$$

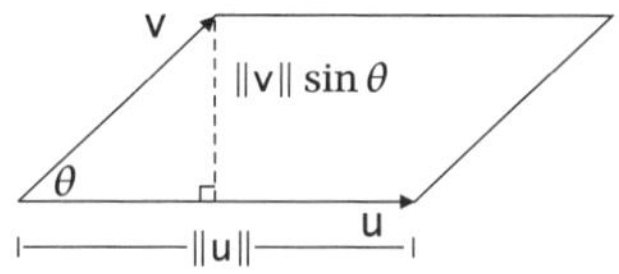

Figure 1.16: The area of the parallelogram is base times height: $\|\mathsf{u}\|\,(\|\mathsf{v}\|\sin\theta)$, and this is precisely $\|\mathsf{u}\times\mathsf{v}\|$.

so $\|\mathsf{u}\times\mathsf{v}\| = \sqrt{6}$. Since $\|\mathsf{u}\| = \sqrt{2}$ and $\|\mathsf{v}\| = \sqrt{3}$, statement 8 of Theorem 1.3.16 says

$$\sqrt{6} = \sqrt{2}\sqrt{3}\sin\theta,$$

so $\sin\theta = 1$ and $\theta = \frac{\pi}{2}$, which is indeed the case: the given vectors are orthogonal ($\mathsf{u}\cdot\mathsf{v} = 0$). ☺

READING CHECK 4. What is the angle between the vectors $\mathsf{u} = \begin{bmatrix}1\\2\\3\end{bmatrix}$ and $\mathsf{v} = \begin{bmatrix}-2\\1\\4\end{bmatrix}$? Give your answer in radians to two decimal places accuracy and in degrees to the nearest degree.

Figure 1.16 explains statement 9 of Theorem 1.3.16. The area of a parallelogram is the length of the base times the height. A parallelogram with sides u and v has base of length $\|\mathsf{u}\|$ and height $\|\mathsf{v}\|\sin\theta$. So the area is $\|\mathsf{u}\|\,\|\mathsf{v}\|\sin\theta = \|\mathsf{u}\times\mathsf{v}\|$.

1.3.19 Problem. What is the area of the parallelogram with adjacent sides $\mathsf{u} = \begin{bmatrix}-3\\4\\1\end{bmatrix}$ and $\mathsf{v} = \begin{bmatrix}1\\0\\-2\end{bmatrix}$?

Solution. We compute $\mathsf{u}\times\mathsf{v} = \begin{vmatrix}\mathsf{i} & \mathsf{j} & \mathsf{k}\\ -3 & 4 & 1\\ 1 & 0 & -2\end{vmatrix} = -8\mathsf{i} - 5\mathsf{j} - 4\mathsf{k} = \begin{bmatrix}-8\\-5\\-4\end{bmatrix}$ and conclude that the area is $\|\mathsf{u}\times\mathsf{v}\| = \sqrt{64+25+16} = \sqrt{105}$. ☝

READING CHECK 5. What is the area of the triangle two of whose sides are the vectors u and v of Problem 1.3.19?

One property that does not appear in Theorem 1.3.16 is associativity. An operation $\star$ is *associative* if $(a \star b) \star c = a \star (b \star c)$ for all a, b and c.

Ordinary addition and multiplication of real numbers are examples of associative operations. The cross product of vectors, however, is **not** associative. In general,

1.3.20
$$(\mathsf{u} \times \mathsf{v}) \times \mathsf{w} \neq \mathsf{u} \times (\mathsf{v} \times \mathsf{w}).$$

We illustrate.

1.3.21 Example. Let $\mathsf{u} = \mathsf{i}$, $\mathsf{v} = \mathsf{j}$ and $\mathsf{w} = \mathsf{k}$ be the standard basis vectors in R^3. Then

$$\mathsf{i} \times \mathsf{i} = \begin{vmatrix} \mathsf{i} & \mathsf{j} & \mathsf{k} \\ 1 & 0 & 0 \\ 1 & 0 & 0 \end{vmatrix} = 0\mathsf{i} - 0\mathsf{j} + 0\mathsf{k} = \begin{bmatrix} 0 \\ 0 \\ 0 \end{bmatrix}$$

and so $(\mathsf{i} \times \mathsf{i}) \times \mathsf{j} = \mathsf{0} \times \mathsf{j} = \mathsf{0}$. On the other hand, $\mathsf{i} \times \mathsf{j} = \begin{vmatrix} \mathsf{i} & \mathsf{j} & \mathsf{k} \\ 1 & 0 & 0 \\ 0 & 1 & 0 \end{vmatrix} =$ $0\mathsf{i} - 0\mathsf{j} + 1\mathsf{k} = \begin{bmatrix} 0 \\ 0 \\ 1 \end{bmatrix} = \mathsf{k}$, so that

$$\mathsf{i} \times (\mathsf{i} \times \mathsf{j}) = \mathsf{i} \times \mathsf{k} = \begin{vmatrix} \mathsf{i} & \mathsf{j} & \mathsf{k} \\ 1 & 0 & 0 \\ 0 & 0 & 1 \end{vmatrix} = 0\mathsf{i} - 1\mathsf{j} + 0\mathsf{k} = -\mathsf{j}.$$

Most certainly, $(\mathsf{i} \times \mathsf{i}) \times \mathsf{j} \neq \mathsf{i} \times (\mathsf{i} \times \mathsf{j})$.

The cross products of any two of i, j, k are quite easy to remember. First of all, the cross product of any of these with itself is 0; remember that $\mathsf{u} \times \mathsf{u} = \mathsf{0}$ for any u. To get cross products amongst i, j, k when the vectors are different, arrange these vectors clockwise in a circle, as at the right. The cross product of any two consecutive of vectors is the next; that is,

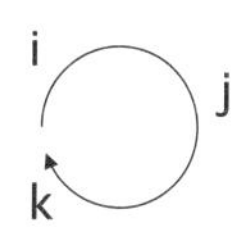

1.3.22
$$\mathsf{i} \times \mathsf{j} = \mathsf{k}, \quad \mathsf{j} \times \mathsf{k} = \mathsf{i}, \quad \mathsf{k} \times \mathsf{i} = \mathsf{j}.$$

Cross products in the reverse order are the additive inverses of these, by Theorem 1.3.16(1).

$$\mathsf{j} \times \mathsf{i} = -\mathsf{k}, \quad \mathsf{k} \times \mathsf{j} = -\mathsf{i}, \quad \mathsf{i} \times \mathsf{k} = -\mathsf{j}.$$

The cross product was introduced with the implication that it would give us a normal to the plane spanned by two given vectors. This is what Theorem 1.3.16, statement 5, says.

1.3.23 Example. Let $\mathsf{u} = \begin{bmatrix} 1 \\ -4 \\ 1 \end{bmatrix}$ and $\mathsf{v} = \begin{bmatrix} 2 \\ 3 \\ 0 \end{bmatrix}$. Then $\mathsf{u} \times \mathsf{v} = \begin{vmatrix} \mathsf{i} & \mathsf{j} & \mathsf{k} \\ 1 & -4 & 1 \\ 2 & 3 & 0 \end{vmatrix} =$ $-3\mathsf{i} - (-2)\mathsf{j} + 11\mathsf{k} = \begin{bmatrix} -3 \\ 2 \\ 11 \end{bmatrix}$. So $(\mathsf{u} \times \mathsf{v}) \cdot \mathsf{u} = -3(1) + 2(-4) + 11(1) = 0$ and $(\mathsf{u} \times \mathsf{v}) \cdot \mathsf{v} = -3(2) + 2(3) + 11(0) = 0$. Indeed, $\mathsf{u} \times \mathsf{v}$ is orthogonal to both u and v (and so it is orthogonal to the plane spanned by u and v—see Example 1.3.14).

1.3.24 Problem. Find an equation of the plane spanned by $\mathsf{u} = \begin{bmatrix} 1 \\ -4 \\ 1 \end{bmatrix}$ and $\mathsf{v} = \begin{bmatrix} 2 \\ 3 \\ 0 \end{bmatrix}$.

Solution. We just showed that $\mathsf{n} = \begin{bmatrix} -3 \\ 2 \\ 11 \end{bmatrix}$ is orthogonal to u and to v, so it is orthogonal to the plane spanned by u and v; in other words, it's a normal to the plane. The plane contains $(0,0,0)$, so an equation is $-3x + 2y + 11z = 0$.

1.3.25 The orientation of the cross product. Thinking of u and v as vectors represented by arrows in the xy-plane, there are two possibilities for an arrow orthogonal to both vectors: up and down. Which is correct depends on the relative positions of the vectors and is illustrated by the fact that $\mathsf{i} \times \mathsf{j} = \mathsf{k}$. If you are familiar with a screwdriver, you know that a screw goes into the wood if turned clockwise and comes out of the wood if turned counterclockwise: . The orientation of the cross product is determined in the same way depending on whether the turn from u to v (through that angle θ, $0 \le \theta \le \pi$) is counterclockwise or clockwise. (See Figure 1.17.)

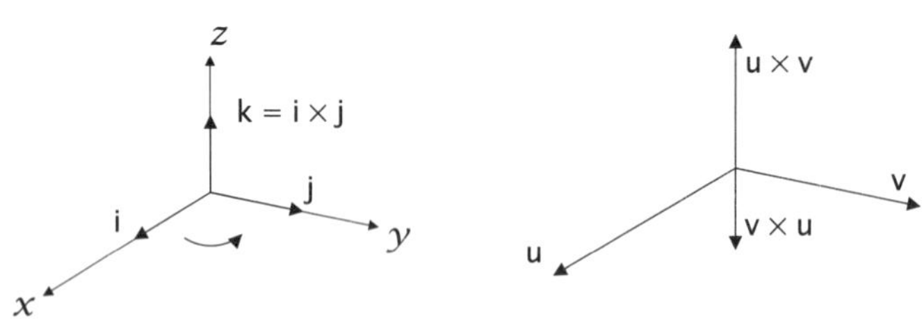

Figure 1.17: $\mathsf{u} \times \mathsf{v}$ points in the direction a screw would move if turned in the direction u to v.

Another way to determine the direction of the cross product uses the *right hand rule*—see Figure 1.18. Place the thumb, index finger and middle finger of your right hand at right angles to each other. If your thumb corresponds to u and your index finger to v, then your middle finger will point in the direction of $u \times v$.

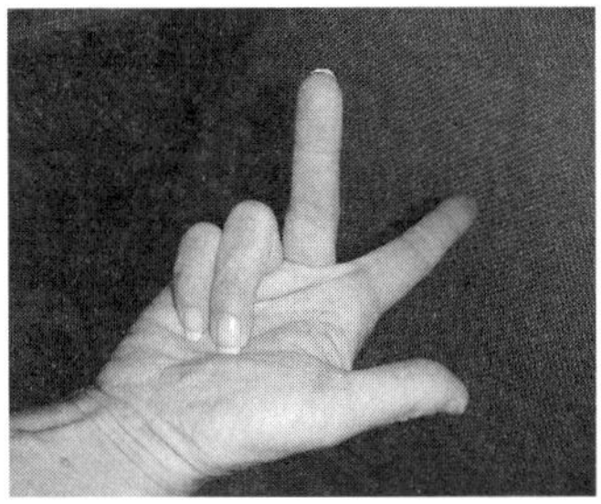

Figure 1.18: The right hand rule: If the thumb and first finger of your right hand represent u and v, respectively, then your middle finger points in the direction of $u \times v$.

The poor joke that follows is an attempt to remind you that whereas the dot product of vectors is a **scalar**, the cross product is a **vector**.

READING CHECK 6. What is the difference between the cross product and a mountaineer?

The Equation of a Line

We now turn our attention to the equation of a line in 3-space. What will such an equation look like? It won't be linear: equations like $ax + by + cz = d$ represent planes.

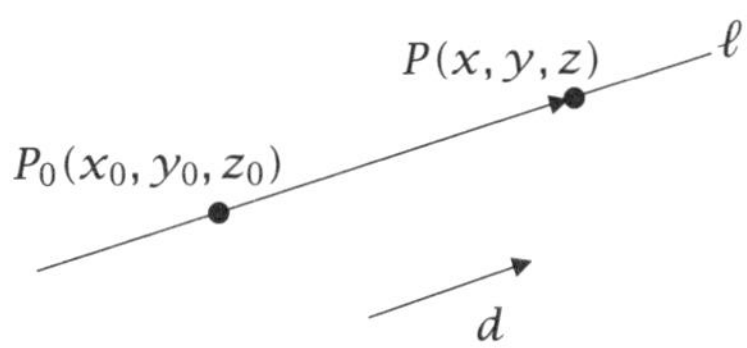

Figure 1.19: A line is determined by a direction d and a point P_0 on the line.

With reference to Figure 1.19, observe that a line ℓ is determined by a direction $d = \begin{bmatrix} a \\ b \\ c \end{bmatrix}$ and a point $P_0(x_0, y_0, z_0)$ on it. In fact, $P(x, y, z)$ is on ℓ if and only if the vector $\overrightarrow{P_0P}$ is a multiple of d; that is, if and only if

$$\begin{bmatrix} x - x_0 \\ y - y_0 \\ z - z_0 \end{bmatrix} = t\mathsf{d} = \begin{bmatrix} ta \\ tb \\ tc \end{bmatrix}$$

for some scalar t, and this occurs if and only if

$$\begin{bmatrix} x \\ y \\ z \end{bmatrix} = \begin{bmatrix} x_0 \\ y_0 \\ z_0 \end{bmatrix} + t \begin{bmatrix} a \\ b \\ c \end{bmatrix}$$

for some scalar t.

1.3.26 **Problem.** Find an equation of the line through $P(-1, 2, 5)$ in the direction of $\mathsf{d} = \begin{bmatrix} 2 \\ -3 \\ 4 \end{bmatrix}$.

Solution. We simply write it down: $\begin{bmatrix} x \\ y \\ z \end{bmatrix} = \begin{bmatrix} -1 \\ 2 \\ 5 \end{bmatrix} + t \begin{bmatrix} 2 \\ -3 \\ 4 \end{bmatrix}$.

1.3.27 **Problem.** Does the line in Problem 1.3.26 contain the point $P(-7, 11, -7)$? Does it pass through $Q(0, -1, 2)$?

Solution. The line contains P if and only if $\begin{bmatrix} -7 \\ 11 \\ -7 \end{bmatrix} = \begin{bmatrix} -1 \\ 2 \\ 5 \end{bmatrix} + t \begin{bmatrix} 2 \\ -3 \\ 4 \end{bmatrix}$ for some number t. We find that $t = -3$: the line indeed contains P. It passes through Q if and only if there is a number t such that $\begin{bmatrix} 0 \\ -1 \\ 2 \end{bmatrix} = \begin{bmatrix} -1 \\ 2 \\ 5 \end{bmatrix} + t \begin{bmatrix} 2 \\ -3 \\ 4 \end{bmatrix}$. Such t must satisfy three equations:

$$\begin{aligned} 0 &= -1 + 2t \\ -1 &= 2 - 3t \\ 2 &= 5 + 4t. \end{aligned}$$

There is no t; the line does not pass through Q.

1.3.28 **Problem.** Find an equation of the line through $P(-2, 8, 5)$ and $Q(4, -1, 1)$.

Solution. The line has direction $\mathsf{d} = \overrightarrow{PQ} = \begin{bmatrix} 6 \\ -9 \\ -4 \end{bmatrix}$ and so equation $\begin{bmatrix} x \\ y \\ z \end{bmatrix} = \begin{bmatrix} -2 \\ 8 \\ 5 \end{bmatrix} + t \begin{bmatrix} 6 \\ -9 \\ -4 \end{bmatrix}$. An equally correct answer is $\begin{bmatrix} x \\ y \\ z \end{bmatrix} = \begin{bmatrix} 4 \\ -1 \\ 1 \end{bmatrix} + t \begin{bmatrix} 6 \\ -9 \\ -4 \end{bmatrix}$.

READING CHECK 7. There are many possible answers to Problem 1.3.28. Find two equations for the line defined in Problem 1.3.28 other than the ones already given.

1.3.29 Problem. Find an equation of the line through $P(3,-1,2)$ that is parallel to the line with equation $\begin{bmatrix} x \\ y \\ z \end{bmatrix} = \begin{bmatrix} -3 \\ 1 \\ 6 \end{bmatrix} + t \begin{bmatrix} 4 \\ -1 \\ -7 \end{bmatrix}$.

Solution. The two lines must have the same direction. So any multiple of the given direction vector gives us a direction for the line we want: The most obvious $\mathsf{d} = \begin{bmatrix} 4 \\ -1 \\ -7 \end{bmatrix}$ gives equation $\begin{bmatrix} x \\ y \\ z \end{bmatrix} = \begin{bmatrix} 3 \\ -1 \\ 2 \end{bmatrix} + t \begin{bmatrix} 4 \\ -1 \\ -7 \end{bmatrix}$. ☝

1.3.30 Problem. Do the lines with equations

$$\begin{bmatrix} x \\ y \\ z \end{bmatrix} = \begin{bmatrix} 1 \\ 2 \\ 1 \end{bmatrix} + t \begin{bmatrix} -3 \\ 5 \\ 1 \end{bmatrix} \quad \text{and} \quad \begin{bmatrix} x \\ y \\ z \end{bmatrix} = \begin{bmatrix} -1 \\ 3 \\ 1 \end{bmatrix} + t \begin{bmatrix} 1 \\ -4 \\ -1 \end{bmatrix}$$

intersect? If so, where?

Solution. If the lines intersect, say at the point (x, y, z), then (x, y, z) is on both lines. So there is a t such that

$$\begin{bmatrix} x \\ y \\ z \end{bmatrix} = \begin{bmatrix} 1 \\ 2 \\ 1 \end{bmatrix} + t \begin{bmatrix} -3 \\ 5 \\ 1 \end{bmatrix}$$

and an s (**not necessarily the same as** t) such that

$$\begin{bmatrix} x \\ y \\ z \end{bmatrix} = \begin{bmatrix} -1 \\ 3 \\ 1 \end{bmatrix} + s \begin{bmatrix} 1 \\ -4 \\ -1 \end{bmatrix}.$$

The question then is whether there exist t and s so that

$$\begin{bmatrix} 1 \\ 2 \\ 1 \end{bmatrix} + t \begin{bmatrix} -3 \\ 5 \\ 1 \end{bmatrix} = \begin{bmatrix} -1 \\ 3 \\ 1 \end{bmatrix} + s \begin{bmatrix} 1 \\ -4 \\ -1 \end{bmatrix}.$$

We try to solve

$$\begin{aligned} 1 - 3t &= -1 + s \\ 2 + 5t &= 3 - 4s \\ 1 + t &= 1 - s, \end{aligned}$$

and find that $s = -1$, $t = 1$ is a solution. The lines intersect where

$$\begin{bmatrix} x \\ y \\ z \end{bmatrix} = \begin{bmatrix} 1 \\ 2 \\ 1 \end{bmatrix} + t\begin{bmatrix} -3 \\ 5 \\ 1 \end{bmatrix} = \begin{bmatrix} 1 \\ 2 \\ 1 \end{bmatrix} + 1\begin{bmatrix} -3 \\ 5 \\ 1 \end{bmatrix} = \begin{bmatrix} -2 \\ 7 \\ 2 \end{bmatrix};$$

that is, at the point $(-2, 7, 2)$.

1.3.31 **Problem.** Does the line with equation $\begin{bmatrix} x \\ y \\ z \end{bmatrix} = \begin{bmatrix} -1 \\ 7 \\ 0 \end{bmatrix} + t\begin{bmatrix} 1 \\ 2 \\ -1 \end{bmatrix}$ intersect the plane with equation $2x - 3y + 5z = 4$? If yes, find the point of intersection.

Solution. The direction of the line is $\mathsf{d} = \begin{bmatrix} 1 \\ 2 \\ -1 \end{bmatrix}$. A normal to the plane is $\mathsf{n} = \begin{bmatrix} 2 \\ -3 \\ 5 \end{bmatrix}$. Since $\mathsf{d} \cdot \mathsf{n} \neq 0$, d and n are not orthogonal, so the line is not parallel to the plane. The plane and the line must intersect. To find the point of intersection, substitute

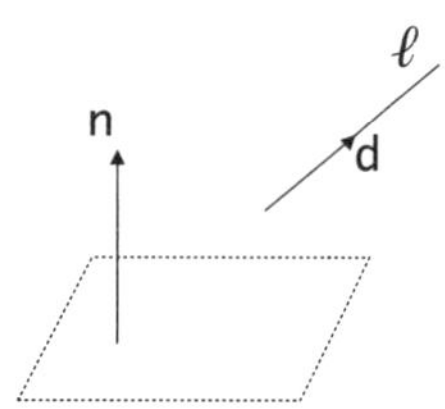

$x = -1 + t$, $y = 7 + 2t$, $z = -t$ in the equation for the plane:

$$\begin{aligned} 2(-1+t) - 3(7+2t) + 5(-t) &= 4 \\ -23 - 9t &= 4 \\ -9t &= 27 \\ t &= -3. \end{aligned}$$

So the point of intersection occurs where $x = -1 + t = -4$, $y = 7 + 2t = 1$, $z = -t = 3$; that is, $(-4, 1, 3)$.

Answers to Reading Checks

1. We gave two such vectors in Example 1.3.8; namely, $\begin{bmatrix} -8 \\ 3 \\ -4 \end{bmatrix}$ and $\begin{bmatrix} 9 \\ 1 \\ 1 \end{bmatrix}$. There are many other possibilities.

2. A normal to this plane is

$$\mathsf{u} \times \mathsf{v} = \begin{vmatrix} \mathsf{i} & \mathsf{j} & \mathsf{k} \\ -8 & 3 & -4 \\ 9 & 1 & 1 \end{vmatrix} = \begin{vmatrix} 3 & -4 \\ 1 & 1 \end{vmatrix}\mathsf{i} - \begin{vmatrix} -8 & -4 \\ 9 & 1 \end{vmatrix}\mathsf{j} + \begin{vmatrix} -8 & 3 \\ 9 & 1 \end{vmatrix}\mathsf{k}$$

$$= 7\mathsf{i} - 28\mathsf{j} - 35\mathsf{k} = \begin{bmatrix} 7 \\ -28 \\ -35 \end{bmatrix} = -7\begin{bmatrix} -1 \\ 4 \\ 5 \end{bmatrix}.$$

So $\begin{bmatrix} 7 \\ -28 \\ -35 \end{bmatrix}$ is a normal, as is its scalar multiple $\begin{bmatrix} -1 \\ 4 \\ 5 \end{bmatrix}$. The plane spanned by vectors always passes through $(0,0,0)$, so an equation is $-x + 4y + 5z = 0$, as in Example 1.3.8.

3. Substituting the coordinates of B, we obtain $25(1) + 7(1) + 2(5) = 42$. Substituting the coordinates of C, we obtain $25(2) + 7(-2) + 2(3) = 42$ too.

4. We have $\mathsf{u} \times \mathsf{v} = 5\begin{bmatrix} 1 \\ -2 \\ 1 \end{bmatrix}$, so $\|\mathsf{u} \times \mathsf{v}\| = 5\sqrt{6} = \|\mathsf{u}\|\,\|\mathsf{v}\| \sin\theta = \sqrt{14}\sqrt{21}\sin\theta$. Thus $\sin\theta = \frac{5}{7}$, so $\theta = \arcsin\frac{5}{7} \approx 0.80$ rads $\approx 46°$.

5. The triangle has an area of one-half the area of the parallelogram, $\frac{1}{2}\sqrt{105}$.

6. The cross product is not a "scaler."

7. We could replace the direction vector by any multiple of $\begin{bmatrix} 6 \\ -9 \\ -4 \end{bmatrix}$, so, for example, $\begin{bmatrix} x \\ y \\ z \end{bmatrix} = \begin{bmatrix} -2 \\ 8 \\ 5 \end{bmatrix} + t\begin{bmatrix} 12 \\ -18 \\ -8 \end{bmatrix}$ is also an equation of the line. We could also replace the point by any other point on the line. Given $\begin{bmatrix} x \\ y \\ z \end{bmatrix} = \begin{bmatrix} -2 \\ 8 \\ 5 \end{bmatrix} + t\begin{bmatrix} 6 \\ -9 \\ -4 \end{bmatrix}$, with $t = 2$ we see that $(10, -10, -3)$ is also on the line, so another equation is $\begin{bmatrix} x \\ y \\ z \end{bmatrix} = \begin{bmatrix} 10 \\ -10 \\ -3 \end{bmatrix} + t\begin{bmatrix} 6 \\ -9 \\ -4 \end{bmatrix}$.

True/False Questions

Decide, with as little calculation as possible, whether each of the following statements is true or false and, if you say "false," explain your answer. (Answers are at the back of the book.)

1. A line is an equation of the form $y = mx + b$.
2. If θ is the angle between vectors u and v, $|\mathsf{u} \times \mathsf{v}| = \|\mathsf{u}\|\,\|\mathsf{v}\| \sin\theta$.
3. If θ is the angle between vectors u and v, $\|\mathsf{u} \cdot \mathsf{v}\| = \|\mathsf{u}\|\,\|\mathsf{v}\| \cos\theta$.
4. If $\|\mathsf{u} \times \mathsf{v}\| = \|\mathsf{u}\|\,\|\mathsf{v}\|$, then u and v are orthogonal.
5. The point $(1,1,1)$ lies on the plane with equation $2x - 3y + 5z = 4$.
6. The line with equation $\begin{bmatrix} x \\ y \\ z \end{bmatrix} = \begin{bmatrix} -1 \\ 0 \\ 2 \end{bmatrix} + t\begin{bmatrix} 1 \\ 2 \\ 3 \end{bmatrix}$ is parallel to the plane with equation $x + 2y + 3z = 0$.

7. The planes with equations $-2x+3y-z=2$ and $4x-6y+2z=0$ are parallel.
8. If u and v are vectors and $\mathsf{u} \times \mathsf{v} = 0$, then $\mathsf{u} = \mathsf{v}$.
9. $\mathsf{i} \times (\mathsf{i} \times \mathsf{j}) = -\mathsf{j}$.
10. A line is parallel to a plane if the dot product of a direction vector of the line and a normal to the plane is 0.

EXERCISES

Answers to exercises marked [BB] can be found at the Back of the Book.

1. Express the planes with the given equations as the set of all linear combinations of two nonparallel vectors. (In each case, there are many possibilities.)

   (a) [BB] $3x-2y+z=0$ **(b)** $x-3y=0$
   (c) $2x-3y+4z=0$ **(d)** $2x+5y-z=0$
   (e) $3x-2y+4z=0$ **(f)** $4x-y+3z=0$.

2. Find an equation of each of the following planes.

   (a) [BB] parallel to the plane with equation $4x+2y-z=7$ and passing through the point $(1,2,3)$

   (b) parallel to the plane with equation $18x+6y-5z=0$ and passing through the point $(-1,1,7)$

   (c) parallel to the plane with equation $x-3y+7z=2$ and passing through the point $(0,4,-1)$.

3. In each of the following cases, find the cross product $\mathsf{u} \times \mathsf{v}$, verify that this vector is orthogonal to u and to v and that $\mathsf{v} \times \mathsf{u} = -(\mathsf{u} \times \mathsf{v})$.

   (a) [BB] $\mathsf{u} = \begin{bmatrix} 1 \\ -2 \\ 1 \end{bmatrix}, \mathsf{v} = \begin{bmatrix} 3 \\ 1 \\ -2 \end{bmatrix}$ **(b)** $\mathsf{u} = \begin{bmatrix} -3 \\ 1 \\ 0 \end{bmatrix}, \mathsf{v} = \begin{bmatrix} 4 \\ -1 \\ 5 \end{bmatrix}$

   (c) $\mathsf{u} = \begin{bmatrix} 1 \\ -3 \\ 0 \end{bmatrix}, \mathsf{v} = \begin{bmatrix} 2 \\ 1 \\ -5 \end{bmatrix}$ **(d)** $\mathsf{u} = \begin{bmatrix} 4 \\ -2 \\ -1 \end{bmatrix}, \mathsf{v} = \begin{bmatrix} 5 \\ 6 \\ -3 \end{bmatrix}$.

4. Compute $(\mathsf{u} \times \mathsf{v}) \times \mathsf{w}$ and $\mathsf{u} \times (\mathsf{v} \times \mathsf{w})$ for each of the following triples of vectors. Should these vectors be the same? Explain.

   (a) [BB] $\mathsf{u} = \begin{bmatrix} -2 \\ 1 \\ 1 \end{bmatrix}, \mathsf{v} = \begin{bmatrix} 0 \\ 1 \\ 3 \end{bmatrix}, \mathsf{w} = \begin{bmatrix} 4 \\ 0 \\ -3 \end{bmatrix}$

   (b) $\mathsf{u} = \begin{bmatrix} 1 \\ 2 \\ 3 \end{bmatrix}, \mathsf{v} = \begin{bmatrix} -2 \\ 3 \\ 0 \end{bmatrix}, \mathsf{w} = \begin{bmatrix} -1 \\ 1 \\ 1 \end{bmatrix}$ **(c)** $\mathsf{u} = \begin{bmatrix} 4 \\ 1 \\ 2 \end{bmatrix}, \mathsf{v} = \begin{bmatrix} -1 \\ 2 \\ 1 \end{bmatrix}, \mathsf{w} = \begin{bmatrix} 0 \\ 3 \\ 2 \end{bmatrix}$.

5. Do there exist vectors u and v such that $\|\mathsf{u} \times \mathsf{v}\| = 15$, $\|\mathsf{u}\| = 3$ and $\|\mathsf{v}\| = 4$? Explain.

6. [BB] Find two nonzero vectors orthogonal to both $u = \begin{bmatrix} 1 \\ 2 \\ 0 \end{bmatrix}$ and $v = \begin{bmatrix} 0 \\ 0 \\ 3 \end{bmatrix}$.

7. In each case, find a unit vector that is orthogonal to both vectors.

   (a) $\begin{bmatrix} 4 \\ 3 \\ -1 \end{bmatrix}$ and $\begin{bmatrix} 2 \\ 3 \\ 0 \end{bmatrix}$ **(b)** $\begin{bmatrix} -1 \\ 1 \\ 3 \end{bmatrix}$ and $\begin{bmatrix} 0 \\ -2 \\ 9 \end{bmatrix}$ **(c)** $\begin{bmatrix} 1 \\ 2 \\ 7 \end{bmatrix}$ and $\begin{bmatrix} -5 \\ 0 \\ 4 \end{bmatrix}$.

8. **(a)** Find a vector of length 5 orthogonal to both $u = \begin{bmatrix} 1 \\ 3 \\ -1 \end{bmatrix}$ and $v = \begin{bmatrix} 5 \\ 0 \\ 1 \end{bmatrix}$.

   (b) Find a vector of length 2 orthogonal to the plane with equation $3x - 2y + z = 9$.

9. Establish the first five statements of Theorem 1.3.16 for the vectors $u = \begin{bmatrix} u_1 \\ u_2 \\ u_3 \end{bmatrix}$, $v = \begin{bmatrix} v_1 \\ v_2 \\ v_3 \end{bmatrix}$ and $w = \begin{bmatrix} w_1 \\ w_2 \\ w_3 \end{bmatrix}$. Specifically, prove

   (a) [BB] $u \times v = -(v \times u)$;

   (b) $u \times (v + w) = (u \times v) + (u \times w)$;

   (c) $(cu) \times v = c(u \times v)$ and $u \times (cv) = c(u \times v)$, for any scalar c;

   (d) $u \times 0 = 0$ and $0 \times u = 0$;

   (e) $u \times v$ is orthogonal to u and to v.

10. In each case,

    i. verify Lagrange's identity: $\|u \times v\|^2 + (u \cdot v)^2 = \|u\|^2 \|v\|^2$;

    ii. find the sine and the cosine of the angle between u and v.

    (a) [BB] $u = \begin{bmatrix} 1 \\ 2 \\ 3 \end{bmatrix}$, $v = \begin{bmatrix} 0 \\ -1 \\ 1 \end{bmatrix}$ **(b)** $u = \begin{bmatrix} -2 \\ -4 \\ 1 \end{bmatrix}$, $v = \begin{bmatrix} 1 \\ 5 \\ -2 \end{bmatrix}$

    (c) $u = \begin{bmatrix} -1 \\ 2 \\ 1 \end{bmatrix}$, $v = \begin{bmatrix} -3 \\ 0 \\ 2 \end{bmatrix}$ **(d)** $u = \begin{bmatrix} 0 \\ 1 \\ -1 \end{bmatrix}$, $v = \begin{bmatrix} 2 \\ 1 \\ 2 \end{bmatrix}$.

11. **(a)** Establish Lagrange's identity $\|u \times v\|^2 + (u \cdot v)^2 = \|u\|^2 \|v\|^2$ in general, by taking $u = \begin{bmatrix} u_1 \\ u_2 \\ u_3 \end{bmatrix}$ and $v = \begin{bmatrix} v_1 \\ v_2 \\ v_3 \end{bmatrix}$.

    (b) Use Lagrange's identity to prove that $\|u \times v\| = \|u\| \|v\| \sin\theta$, where θ is the angle between u and v.

12. Let a and b denote arbitrary scalars. In each case, find $au + bv$. Then find an equation of the plane that consists of all such linear combinations of u and v.

(a) [BB] $u = \begin{bmatrix} 1 \\ 0 \\ 4 \end{bmatrix}, v = \begin{bmatrix} -1 \\ 2 \\ 0 \end{bmatrix}$ (b) $u = \begin{bmatrix} 1 \\ -3 \\ 0 \end{bmatrix}, v = \begin{bmatrix} 1 \\ 2 \\ 6 \end{bmatrix}$

(c) $u = \begin{bmatrix} 2 \\ 5 \\ 1 \end{bmatrix}, v = \begin{bmatrix} -1 \\ 1 \\ 1 \end{bmatrix}$ (d) $u = \begin{bmatrix} -3 \\ -1 \\ 1 \end{bmatrix}, v = \begin{bmatrix} 2 \\ -6 \\ -2 \end{bmatrix}$

(e) $u = \begin{bmatrix} -1 \\ 0 \\ 1 \end{bmatrix}, v = \begin{bmatrix} 2 \\ 5 \\ -3 \end{bmatrix}$ (f) $u = \begin{bmatrix} 3 \\ 2 \\ 1 \end{bmatrix}, v = \begin{bmatrix} 4 \\ -1 \\ -1 \end{bmatrix}$.

13. Find an equation of the plane containing A, B and C given

   (a) [BB] $A(2,1,3), B(3,-1,5), C(0,2,-4)$
   (b) $A(-1,2,1), B(0,1,1), C(7,-3,0)$
   (c) $A(1,2,3), B(-1,5,2), C(3,-1,1)$
   (d) $A(3,2,0), B(-1,0,4), C(0,-3,2)$.

14. Let $A(1,-1,-3), B(2,0,-3), C(2,-5,1), D(3,-4,1)$ be the indicated points.

   (a) [BB] Show that A, B, C, D are the vertices of a parallelogram and find the area of this figure.
   (b) [BB] Find the area of triangle ABC.
   (c) Find the area of triangle ACD.

15. Repeat Exercise 14 with $A(-1,2,1)$, $B(6,2,-3)$, $C(-4,6,-2)$ and $D(3,6,-6)$.

16. Find the area of the triangles in the xy-plane with each of the indicated vertices A, B, C.

   (a) [BB] $A(-2,3), B(4,4), C(5,-3)$ (b) $A(1,2), B(-3,0), C(2,-2)$
   (c) $A(0,6), B(4,-2), C(5,1)$.

17. Let $A(a_1,a_2), B(b_1,b_2), C(c_1,c_2)$ be three points in the plane. Find a formula for the area of $\triangle ABC$ in terms of the coordinates of A, B and C. (See Exercise 16.)

18. Find equations of the form $ax + by = c$ for the lines in the Euclidean plane with the following equations:

   (a) [BB] $\begin{bmatrix} x \\ y \end{bmatrix} = \begin{bmatrix} 1 \\ 2 \end{bmatrix} + t\begin{bmatrix} -1 \\ 1 \end{bmatrix}$ (b) $\begin{bmatrix} x \\ y \end{bmatrix} = \begin{bmatrix} -3 \\ 5 \end{bmatrix} + t\begin{bmatrix} 1 \\ 2 \end{bmatrix}$

   (c) $\begin{bmatrix} x \\ y \end{bmatrix} = \begin{bmatrix} 2 \\ -7 \end{bmatrix} + t\begin{bmatrix} 4 \\ -3 \end{bmatrix}$ (d) $\begin{bmatrix} x \\ y \end{bmatrix} = \begin{bmatrix} 4 \\ -1 \end{bmatrix} + t\begin{bmatrix} 2 \\ -2 \end{bmatrix}$.

19. Given $P(-1,1,2), Q(-3,0,4), R(3,2,1)$, find an equation of the line through P that is parallel to the line through Q and R.

20. (a) How do you know that the line ℓ with equation $\begin{bmatrix} x \\ y \\ z \end{bmatrix} = \begin{bmatrix} 1 \\ 1 \\ 2 \end{bmatrix} + t\begin{bmatrix} -1 \\ 1 \\ 2 \end{bmatrix}$ intersects the plane π with equation $2x + y - z = 7$?

(b) Find the point of intersection of π and ℓ.

(c) Find the angle between π and ℓ.

21. Find an equation of each of the following planes.

(a) [BB] containing $P(3,1,-1)$ and the line with equation
$$\begin{bmatrix} x \\ y \\ z \end{bmatrix} = \begin{bmatrix} 1 \\ 0 \\ 2 \end{bmatrix} + t\begin{bmatrix} -4 \\ 3 \\ 1 \end{bmatrix}$$

(b) containing $P(-1,2,5)$ and the line with equation $\begin{bmatrix} x \\ y \\ z \end{bmatrix} = \begin{bmatrix} 3 \\ -1 \\ 1 \end{bmatrix} + t\begin{bmatrix} -5 \\ 7 \\ 7 \end{bmatrix}$

(c) containing $P(2,-3,1)$ and the line with equation $\begin{bmatrix} x \\ y \\ z \end{bmatrix} = \begin{bmatrix} 0 \\ -3 \\ 4 \end{bmatrix} + t\begin{bmatrix} 1 \\ -3 \\ 2 \end{bmatrix}$

(d) containing $P(5,1,2)$ and the line with equation $\begin{bmatrix} x \\ y \\ z \end{bmatrix} = \begin{bmatrix} 1 \\ 4 \\ -2 \end{bmatrix} + t\begin{bmatrix} 1 \\ 0 \\ 1 \end{bmatrix}$.

22. [BB] The intersection of two nonparallel planes is a line.

(a) Find three points on the line that is the intersection of the planes with equations $2x + y + z = 5$ and $x - y + z = 1$.

(b) Find an equation for the line in part (a).

23. Find an equation of the line of intersection of

(a) [BB] the planes with equations $x + 2y + 6z = 5$ and $x - y - 3z = -1$;

(b) the planes with equations $x + y + z = 1$ and $2x - z = 0$;

(c) the planes with equations $-2x + 3y + 7z = 4$ and $3x - y + 5z = 1$.

24. (a) Find an equation of the line ℓ through $A(1,1,1)$ parallel to the vector $\mathsf{u} = \begin{bmatrix} -2 \\ 2 \\ 0 \end{bmatrix}$.

(b) Find the intersection point P of the line in (a) with the plane whose equation is $x - 3y + 2z = 1$.

25. In each case, determine, with a simple reason, whether the line and the plane with the given equations intersect. Find any points of intersection.

(a) [BB] $\begin{bmatrix} x \\ y \\ z \end{bmatrix} = \begin{bmatrix} 1 \\ -2 \\ 3 \end{bmatrix} + t\begin{bmatrix} 2 \\ 5 \\ -1 \end{bmatrix}$, $x - 3y + 2z = 4$

(b) $\begin{bmatrix} x \\ y \\ z \end{bmatrix} = \begin{bmatrix} -1 \\ 1 \\ 2 \end{bmatrix} + t\begin{bmatrix} 2 \\ 1 \\ -4 \end{bmatrix}$, $3x - 10y - z = 8$

(c) $\begin{bmatrix} x \\ y \\ z \end{bmatrix} = \begin{bmatrix} -1 \\ 0 \\ 4 \end{bmatrix} + t\begin{bmatrix} 2 \\ 2 \\ -7 \end{bmatrix}$, $3x - y + z = 7$

(d) $\begin{bmatrix} x \\ y \\ z \end{bmatrix} = \begin{bmatrix} 4 \\ -1 \\ 3 \end{bmatrix} + t\begin{bmatrix} 0 \\ -1 \\ 1 \end{bmatrix}$, $2x - 5y - 4z = 4$.

26. Let π be the plane with equation $2x-3y+5z = 1$, let $\mathsf{u} = \begin{bmatrix} 1 \\ -2 \\ 3 \end{bmatrix}$ and $\mathsf{v} = \begin{bmatrix} 2 \\ 0 \\ 1 \end{bmatrix}$.

(a) [BB] Find a vector orthogonal to π.

(b) Find a vector orthogonal to both u and v.

(c) Find a point in the plane π.

(d) Find an equation of the line through the point in (c) and parallel to u.

27. (a) [BB] Suppose a 2×2 determinant $\begin{vmatrix} a & b \\ c & d \end{vmatrix} = 0$. Show that one of the vectors $\begin{bmatrix} c \\ d \end{bmatrix}$, $\begin{bmatrix} a \\ b \end{bmatrix}$ is a scalar multiple of the other. There are four cases to consider, depending on whether or not $a = 0$ or $b = 0$.

(b) Suppose $\mathsf{u} = \begin{bmatrix} u_1 \\ u_2 \\ u_3 \end{bmatrix}$, $\mathsf{v} = \begin{bmatrix} v_1 \\ v_2 \\ v_3 \end{bmatrix}$ and $\mathsf{u} \times \mathsf{v} = 0$. Show that for some scalar k, either $\mathsf{v} = k\mathsf{u}$ or $\mathsf{u} = k\mathsf{v}$, thus completing the proof of Theorem 1.3.16(7). [Hint: If $\mathsf{u} = 0$, then $\mathsf{u} = 0\mathsf{v}$, the desired result. So assume $\mathsf{u} \neq 0$ and use the result of 27(a) to show that v is a multiple of u.]

28. Determine whether or not each pair of lines with the equations given below intersect. Find any points of intersection.

(a) [BB] $\begin{bmatrix} x \\ y \\ z \end{bmatrix} = \begin{bmatrix} 1 \\ 0 \\ -2 \end{bmatrix} + t\begin{bmatrix} -3 \\ 1 \\ 1 \end{bmatrix}$ and $\begin{bmatrix} x \\ y \\ z \end{bmatrix} = \begin{bmatrix} -4 \\ 1 \\ 1 \end{bmatrix} + t\begin{bmatrix} 11 \\ -3 \\ -5 \end{bmatrix}$

(b) $\begin{bmatrix} x \\ y \\ z \end{bmatrix} = \begin{bmatrix} -1 \\ 2 \\ -4 \end{bmatrix} + t\begin{bmatrix} 4 \\ 6 \\ 0 \end{bmatrix}$ and $\begin{bmatrix} x \\ y \\ z \end{bmatrix} = \begin{bmatrix} -3 \\ 1 \\ 6 \end{bmatrix} + t\begin{bmatrix} 0 \\ 1 \\ 1 \end{bmatrix}$

(c) $\begin{bmatrix} x \\ y \\ z \end{bmatrix} = \begin{bmatrix} -1 \\ 0 \\ 4 \end{bmatrix} + t\begin{bmatrix} 5 \\ 1 \\ 1 \end{bmatrix}$ and $\begin{bmatrix} x \\ y \\ z \end{bmatrix} = \begin{bmatrix} 0 \\ 1 \\ 7 \end{bmatrix} + t\begin{bmatrix} -2 \\ 0 \\ 1 \end{bmatrix}$

29. (a) Suppose u, v and w are three vectors and $\mathsf{w} = \mathsf{u}+\mathsf{v}$. Show that every linear combination of u, v, w is actually just a linear combination of u and v.

(b) Let $\mathsf{u} = \begin{bmatrix} 1 \\ 0 \\ -1 \end{bmatrix}$, $\mathsf{v} = \begin{bmatrix} 2 \\ 1 \\ 0 \end{bmatrix}$ and $\mathsf{w} = \begin{bmatrix} 3 \\ 1 \\ -1 \end{bmatrix}$. Show that the set of all linear combinations of these vectors is a plane and find an equation of this plane. [Hint: $\mathsf{w} = \mathsf{u} + \mathsf{v}$.]

Critical Reading

30. Is the dot product an associative operation? Does this question make sense?

31. Explain why any plane not containing $(0,0,0)$ has an equation of the form $ax + by + cz + 1 = 0$.

32. [BB] Let ℓ_1 and ℓ_2 be the lines with equations

$$\ell_1\colon \begin{bmatrix} x \\ y \\ z \end{bmatrix} = \begin{bmatrix} 2 \\ -1 \\ 3 \end{bmatrix} + t \begin{bmatrix} -1 \\ 2 \\ 1 \end{bmatrix}; \qquad \ell_2\colon \begin{bmatrix} x \\ y \\ z \end{bmatrix} = \begin{bmatrix} 3 \\ 1 \\ a \end{bmatrix} + t \begin{bmatrix} 1 \\ -1 \\ b \end{bmatrix}.$$

Given that these lines intersect at right angles, find the values of a and b.

33. Let π_1 and π_2 be the planes with equations $x - y + z = 3$ and $2x - 3y + z = 5$, respectively.

(a) Find an equation of the line of intersection of π_1 and π_2.

(b) For any scalars α and β, the equation $\alpha(x-y+z-3)+\beta(2x-3y+z-5) = 0$ is that of a plane. Why?

(c) Show that the plane in (b) contains the line of intersection of π_1 and π_2.

34. Show that the lines with equations

$$\begin{bmatrix} x \\ y \\ z \end{bmatrix} = \begin{bmatrix} -1 \\ 4 \\ 4 \end{bmatrix} + t \begin{bmatrix} -1 \\ 5 \\ 2 \end{bmatrix} \quad \text{and} \quad \begin{bmatrix} x \\ y \\ z \end{bmatrix} = \begin{bmatrix} 1 \\ -6 \\ 0 \end{bmatrix} + t \begin{bmatrix} 2 \\ -10 \\ -4 \end{bmatrix}$$

are the same.

35. **(a)** Find an equation of the plane that contains the points $A(1,0,-1)$, $B(0,2,3)$ and $C(-2,1,1)$.

(b) Find the area of triangle ABC, with A, B, C as in part (a).

(c) Find a point that is one unit away from the plane in part (a).

36. Find an equation of the plane each point of which is equidistant from the points $P(2,-1,3)$ and $Q(1,1,-1)$.

37. Let π be the plane through $P(5,4,8)$ that contains the line with equation $\begin{bmatrix} x \\ y \\ z \end{bmatrix} = \begin{bmatrix} 9 \\ 3 \\ 5 \end{bmatrix} + t \begin{bmatrix} 4 \\ 5 \\ -3 \end{bmatrix}$.

(a) Find an equation of π.

(b) Find the centres of two spheres of equal radii that touch π at P and also touch the unit sphere in R^3.

1.4 Projections

Projections onto Vectors

In this section, we discuss the notion of the **projection** of a vector onto a vector or a plane.[7] In each case, a picture really is worth a thousand words. In Figure 1.20, we show the projection of a vector u onto a vector v in two different situations according to whether the angle between u and v is acute or obtuse.

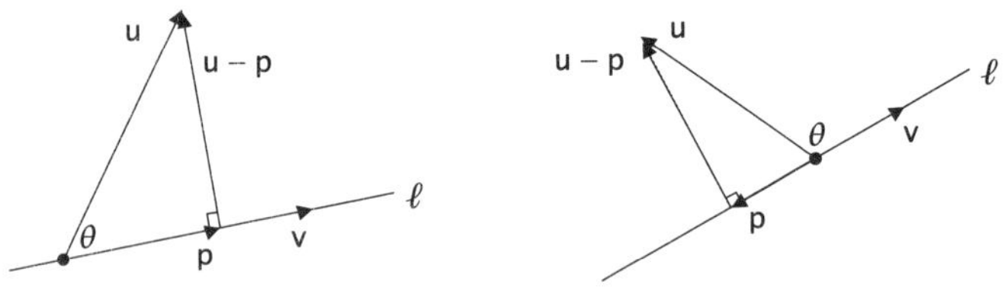

Figure 1.20: In both cases, $\mathsf{p} = \operatorname{proj}_{\mathsf{v}} \mathsf{u}$ is the projection of u on v.

1.4.1 Definition. The *projection* of a vector u onto a (nonzero) vector v is that multiple $\mathsf{p} = c\mathsf{v}$ of v with the property that $\mathsf{u} - \mathsf{p}$ is orthogonal to v. The projection of a vector u onto a line ℓ is the projection of u onto a direction vector for ℓ. We use the notation $\operatorname{proj}_{\mathsf{v}} \mathsf{u}$ and $\operatorname{proj}_{\ell} \mathsf{u}$, respectively.

It's easy to determine the scalar c in this definition. If $\mathsf{u} - \mathsf{p}$ is orthogonal to v,

$$0 = (\mathsf{u} - \mathsf{p}) \cdot \mathsf{v} = \mathsf{u} \cdot \mathsf{v} - \mathsf{p} \cdot \mathsf{v} = \mathsf{u} \cdot \mathsf{v} - (c\mathsf{v}) \cdot \mathsf{v} = \mathsf{u} \cdot \mathsf{v} - c(\mathsf{v} \cdot \mathsf{v}),$$

so $c(\mathsf{v} \cdot \mathsf{v}) = \mathsf{u} \cdot \mathsf{v}$. Assuming $\mathsf{v} \neq 0$, then $\mathsf{v} \cdot \mathsf{v} \neq 0$ and $c = \frac{\mathsf{u} \cdot \mathsf{v}}{\mathsf{v} \cdot \mathsf{v}}$.

1.4.2 Theorem. *The projection of a vector* u *onto a (nonzero) vector* v *is the vector*

$$\boxed{\operatorname{proj}_{\mathsf{v}} \mathsf{u} = \frac{\mathsf{u} \cdot \mathsf{v}}{\mathsf{v} \cdot \mathsf{v}} \mathsf{v}.}$$

The projection of u *onto a line* ℓ *with direction vector* v *is* $\operatorname{proj}_{\ell} \mathsf{u} = \operatorname{proj}_{\mathsf{v}} \mathsf{u} = \frac{\mathsf{u} \cdot \mathsf{v}}{\mathsf{v} \cdot \mathsf{v}} \mathsf{v}$.

READING CHECK 1. Show that the projection of a vector u onto a line ℓ does not depend on the direction vector for ℓ; that is, if both v and w are direction vectors for ℓ, then $\operatorname{proj}_{\mathsf{v}} \mathsf{u} = \operatorname{proj}_{\mathsf{w}} \mathsf{u}$.

---

[7]What we here call "projection," might more correctly be called "orthogonal projection." One can certainly talk about projections that are not orthogonal, but not in this book.

1.4.3 Problem. Let $\mathsf{u} = \begin{bmatrix} 1 \\ -3 \\ 4 \end{bmatrix}$ and $\mathsf{v} = \begin{bmatrix} -3 \\ 1 \\ 2 \end{bmatrix}$. Find the projection of u onto v.

Solution. We have $\frac{\mathsf{u}\cdot\mathsf{v}}{\mathsf{v}\cdot\mathsf{v}} = \frac{-3-3+8}{9+1+4} = \frac{2}{14} = \frac{1}{7}$ so $\operatorname{proj}_{\mathsf{v}} \mathsf{u} = \frac{1}{7}\mathsf{v} = \begin{bmatrix} -\frac{3}{7} \\ \frac{1}{7} \\ \frac{2}{7} \end{bmatrix}$.

To check that this is right, we compute $\mathsf{u} - \operatorname{proj}_{\mathsf{v}} \mathsf{u} = \begin{bmatrix} \frac{10}{7} \\ -\frac{22}{7} \\ \frac{26}{7} \end{bmatrix}$ and note that this is indeed orthogonal to v: the dot product of the two vectors is $-3(\frac{10}{7}) + \frac{-22}{7} + 2(\frac{26}{7}) = 0$.

READING CHECK 2. Let $\mathsf{u} = \begin{bmatrix} 1 \\ -3 \\ 4 \end{bmatrix}$ and let ℓ be the line with direction vector $\mathsf{v} = \begin{bmatrix} -3 \\ 1 \\ 2 \end{bmatrix}$. Find the projection of u onto ℓ.

It is often useful to be able to find two (nonzero) orthogonal vectors in a plane. (Trust me!) The next problem describes a procedure for doing this.

1.4.4 Problem. Find two (nonzero) orthogonal vectors e and f in the plane with equation $3x + y - z = 0$.

Solution. First, we find two **nonparallel** vectors u and v in the plane. Then we find the projection p of u onto v and take $\mathsf{e} = \mathsf{v}$ and $\mathsf{f} = \mathsf{u} - \mathsf{p}$ as our vectors. See Figure 1.20. For example, we might choose $\mathsf{u} = \begin{bmatrix} 1 \\ -3 \\ 0 \end{bmatrix}$ and $\mathsf{v} = \begin{bmatrix} 0 \\ 1 \\ 1 \end{bmatrix}$. (The components of each of these vectors satisfy $3x + y - z = 0$.) Then

$$\mathsf{p} = \operatorname{proj}_{\mathsf{v}} \mathsf{u} = \frac{\mathsf{u} \cdot \mathsf{v}}{\mathsf{v} \cdot \mathsf{v}} \mathsf{v} = \frac{-3}{2} \begin{bmatrix} 0 \\ 1 \\ 1 \end{bmatrix} \quad \text{and} \quad \mathsf{u} - \mathsf{p} = \begin{bmatrix} 1 \\ -3 \\ 0 \end{bmatrix} + \frac{3}{2} \begin{bmatrix} 0 \\ 1 \\ 1 \end{bmatrix} = \begin{bmatrix} 1 \\ -\frac{3}{2} \\ \frac{3}{2} \end{bmatrix},$$

so we can answer the problem with the vectors $\begin{bmatrix} 0 \\ 1 \\ 1 \end{bmatrix}$ and $\begin{bmatrix} 1 \\ -\frac{3}{2} \\ \frac{3}{2} \end{bmatrix}$ or, equally, with $\begin{bmatrix} 0 \\ 1 \\ 1 \end{bmatrix}$ and $\begin{bmatrix} 2 \\ -3 \\ 3 \end{bmatrix}$ (multiplying the second vector by 2 does not change its direction). We encourage the reader to verify that these vectors answer the question: They are orthogonal and each is in the given plane.

The Distance from a Point to a Line

1.4.5 Problem. Find the distance from the point $P(1,8,-5)$ to the line ℓ with equation $\begin{bmatrix} x \\ y \\ z \end{bmatrix} = \begin{bmatrix} -1 \\ 1 \\ 6 \end{bmatrix} + t \begin{bmatrix} 3 \\ 0 \\ -4 \end{bmatrix}$. Find also the point of ℓ closest to P.

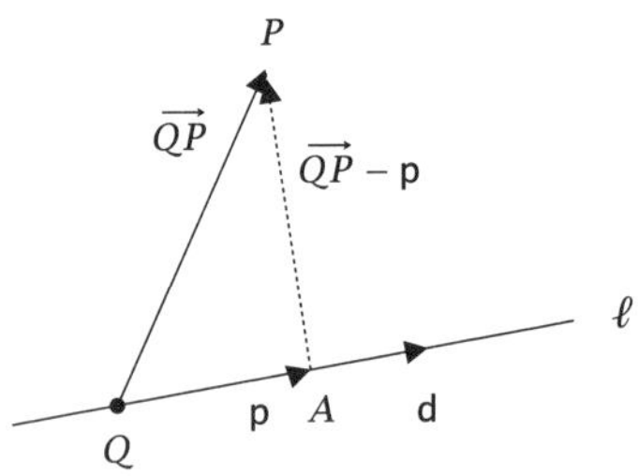

Solution. Let Q be any point on the line, say $Q(-1,1,6)$. The line has direction $\mathsf{d} = \begin{bmatrix} 3 \\ 0 \\ -4 \end{bmatrix}$ and the distance we want is the length of $\overrightarrow{QP} - \mathsf{p}$, where $\mathsf{p} = \operatorname{proj}_\ell \overrightarrow{QP}$. The sketch should make all this clear.

Now $\overrightarrow{QP} = \begin{bmatrix} 2 \\ 7 \\ -11 \end{bmatrix}$, so $\mathsf{p} = \operatorname{proj}_\ell \overrightarrow{QP} = \frac{\overrightarrow{QP} \cdot \mathsf{d}}{\mathsf{d} \cdot \mathsf{d}} \mathsf{d} = \frac{50}{25} \begin{bmatrix} 3 \\ 0 \\ -4 \end{bmatrix} = \begin{bmatrix} 6 \\ 0 \\ -8 \end{bmatrix}$. Thus $\overrightarrow{QP} - \mathsf{p} = \begin{bmatrix} 2 \\ 7 \\ -11 \end{bmatrix} - \begin{bmatrix} 6 \\ 0 \\ -8 \end{bmatrix} = \begin{bmatrix} -4 \\ 7 \\ -3 \end{bmatrix}$ and the required distance is $\left\| \overrightarrow{QP} - \mathsf{p} \right\| = \sqrt{74}$. As for the point of ℓ closest to P, call this $A(x,y,z)$ and note that $\overrightarrow{QA} = \mathsf{p}$. Thus $\begin{bmatrix} x+1 \\ y-1 \\ z-6 \end{bmatrix} = \begin{bmatrix} 6 \\ 0 \\ -8 \end{bmatrix}$, so A is $(5,1,-2)$.

The Distance from a Point to a Plane

1.4.6 Problem. Find the distance from the point $P(-1,2,1)$ to the plane π with equation $5x + y + 3z = 7$.

Solution. Let Q be any point in the plane; for instance, $Q(1,-1,1)$. The required distance is the length of the projection p of $\overrightarrow{PQ}$ onto n, a normal to the plane. We have $\overrightarrow{PQ} = \begin{bmatrix} 2 \\ -3 \\ 0 \end{bmatrix}$ and a normal $\mathsf{n} = \begin{bmatrix} 5 \\ 1 \\ 3 \end{bmatrix}$, so $\operatorname{proj}_{\mathsf{n}} \overrightarrow{PQ} = \frac{\overrightarrow{PQ} \cdot \mathsf{n}}{\mathsf{n} \cdot \mathsf{n}} \mathsf{n} = \frac{7}{35} \mathsf{n} = \frac{1}{5} \mathsf{n}$. The required distance is $\left\| \frac{1}{5} \mathsf{n} \right\| = \frac{1}{5} \|\mathsf{n}\| = \frac{1}{5}\sqrt{35}$.

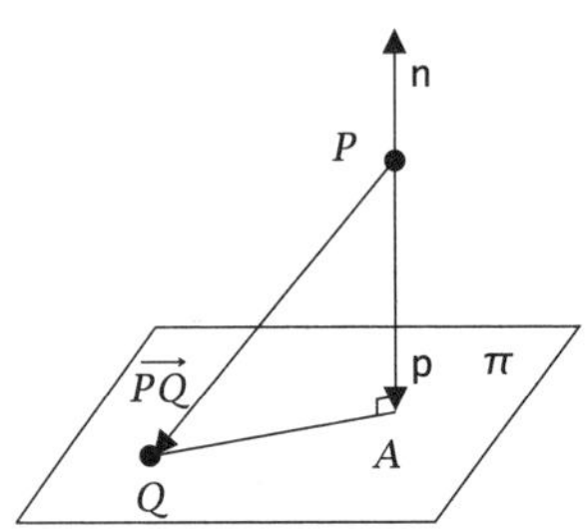

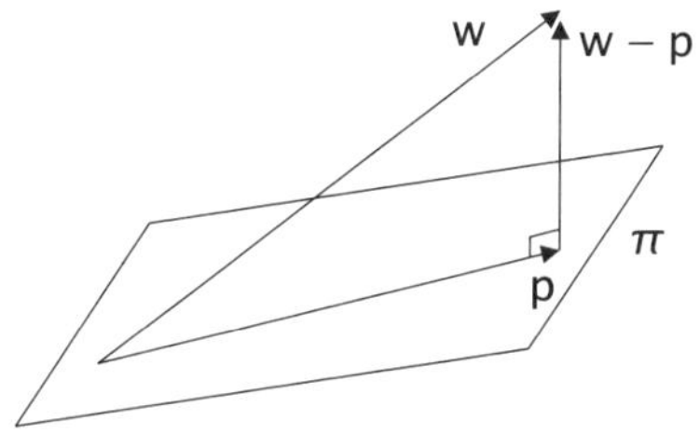

Figure 1.21: Vector $\mathsf{p} = \operatorname{proj}_\pi \mathsf{w}$ is the projection of w onto the plane π.

Projections onto Planes through the Origin

1.4.7 Definition. The *projection* of a vector w onto a plane π is a vector p in π with the property that $\mathsf{w} - \mathsf{p}$ is orthogonal to every vector in π. We denote this vector $\operatorname{proj}_\pi \mathsf{w}$.

See Figure 1.21.

1.4.8 Example. The projection of $\mathsf{w} = \begin{bmatrix} a \\ b \\ c \end{bmatrix}$ onto the xy-plane is the vector $\mathsf{p} = \begin{bmatrix} a \\ b \\ 0 \end{bmatrix}$ since this is a vector in the xy-plane and $\mathsf{w} - \mathsf{p} = \begin{bmatrix} 0 \\ 0 \\ c \end{bmatrix}$ is orthogonal to every vector $\begin{bmatrix} x \\ y \\ 0 \end{bmatrix}$ in the xy-plane. ☺

As just illustrated, it is easy to find the projection of a vector on any of the three coordinate planes—set the appropriate component equal to 0—but what about the projection of a vector onto some other plane? How can we find a vector orthogonal to every single vector in a plane? In the discussion that follows, we make use of the very useful idea that comes next.

1.4.9 Theorem. *Suppose a set U in R^n is spanned by vectors $\mathsf{u}_1, \mathsf{u}_2, \ldots, \mathsf{u}_k$. If a vector* x *is orthogonal to each of these vectors, then* x *is orthogonal to U, that is,* x *is orthogonal to every vector in U.*

Proof. If u is in U, then $\mathsf{u} = c_1\mathsf{u}_1 + c_2\mathsf{u}_2 + \cdots + c_k\mathsf{u}_k$ is a linear combination of $\mathsf{u}_1, \ldots, \mathsf{u}_k$. Now suppose that x is orthogonal to each of these vectors. Then $\mathsf{x} \cdot \mathsf{u}_1 = 0, \mathsf{x} \cdot \mathsf{u}_2 = 0, \ldots, \mathsf{x} \cdot \mathsf{u}_k = 0$. Using scalar associativity of the dot product and the fact that the dot product distributes over sums (see Theorem 1.2.20), $\mathsf{x} \cdot \mathsf{u} = c_1(\mathsf{x} \cdot \mathsf{u}_1) + c_2(\mathsf{x} \cdot \mathsf{u}_2) + \cdots + c_k(\mathsf{x} \cdot \mathsf{u}_k) = 0 + 0 + \cdots + 0 = 0$. Thus x is orthogonal to the arbitrary vector u in U, just as we claimed. ■

This theorem implies that if you want to find a vector orthogonal to a plane, it suffices to find a vector orthogonal to two vectors that span that plane, and for this there is a formula, a special case of the "Gram–Schmidt Algorithm" that we discuss later (in Section 6.3).

1.4.10 Theorem. *The projection of a vector* w *onto the plane spanned by* ***orthogonal*** *vectors* e *and* f *is*

$$\boxed{\operatorname{proj}_\pi \mathsf{w} = \frac{\mathsf{w}\cdot\mathsf{e}}{\mathsf{e}\cdot\mathsf{e}}\mathsf{e} + \frac{\mathsf{w}\cdot\mathsf{f}}{\mathsf{f}\cdot\mathsf{f}}\mathsf{f}.}$$

Note the similarity between the formula here and that given in Theorem 1.4.2 for the projection of w onto a vector.

Proof. Label the plane π. The projection $\mathsf{p} = \operatorname{proj}_\pi \mathsf{w}$ of w onto π is a vector in π, so it is a linear combination of e and f; that is, $\mathsf{p} = a\mathsf{e} + b\mathsf{f}$ for scalars a and b. Now $\mathsf{w} - \mathsf{p}$ is perpendicular to every vector in π and Theorem 1.4.9 says it suffices to ensure that $\mathsf{w} - \mathsf{p}$ is perpendicular to just e and f. The equations $(\mathsf{w}-\mathsf{p})\cdot\mathsf{e} = 0$ and $(\mathsf{w}-\mathsf{p})\cdot\mathsf{f} = 0$ give $\mathsf{w}\cdot\mathsf{e} = \mathsf{p}\cdot\mathsf{e}$ and $\mathsf{w}\cdot\mathsf{f} = \mathsf{p}\cdot\mathsf{f}$ and here is where the orthogonality of e and f becomes useful. We have

$$\mathsf{w}\cdot\mathsf{e} = (a\mathsf{e} + b\mathsf{f})\cdot\mathsf{e} = a(\mathsf{e}\cdot\mathsf{e}) + b(\mathsf{f}\cdot\mathsf{e}) = a(\mathsf{e}\cdot\mathsf{e}) \qquad (*)$$

because $\mathsf{f}\cdot\mathsf{e} = 0$. Assuming $\mathsf{e} \neq 0$ and hence $\mathsf{e}\cdot\mathsf{e} = \|\mathsf{e}\|^2 \neq 0$, equation (*) gives $a = \frac{\mathsf{w}\cdot\mathsf{e}}{\mathsf{e}\cdot\mathsf{e}}$. Similarly, $b = \frac{\mathsf{w}\cdot\mathsf{f}}{\mathsf{f}\cdot\mathsf{f}}$, so $\mathsf{p} = a\mathsf{e} + b\mathsf{f} = \frac{\mathsf{w}\cdot\mathsf{e}}{\mathsf{e}\cdot\mathsf{e}}\mathsf{e} + \frac{\mathsf{w}\cdot\mathsf{f}}{\mathsf{f}\cdot\mathsf{f}}\mathsf{f}$. ■

1.4.11 Problem. Find the projection of $\mathsf{w} = \begin{bmatrix}3\\1\\1\end{bmatrix}$ onto the plane π with equation $3x + y - z = 0$.

Solution. We note that $\mathsf{e} = \begin{bmatrix}0\\1\\1\end{bmatrix}$ and $\mathsf{f} = \begin{bmatrix}2\\-3\\3\end{bmatrix}$ are orthogonal vectors spanning π (see Problem 1.4.4) and simply write down the answer:

$$\begin{aligned}\operatorname{proj}_\pi \mathsf{w} &= \frac{\mathsf{w}\cdot\mathsf{e}}{\mathsf{e}\cdot\mathsf{e}}\mathsf{e} + \frac{\mathsf{w}\cdot\mathsf{f}}{\mathsf{f}\cdot\mathsf{f}}\mathsf{f}\\ &= \frac{2}{2}\begin{bmatrix}0\\1\\1\end{bmatrix} + \frac{6}{22}\begin{bmatrix}2\\-3\\3\end{bmatrix} = \begin{bmatrix}0\\1\\1\end{bmatrix} + \frac{3}{11}\begin{bmatrix}2\\-3\\3\end{bmatrix} = \begin{bmatrix}\frac{6}{11}\\ \frac{2}{11}\\ \frac{20}{11}\end{bmatrix}.\end{aligned}$$

Answers to Reading Checks

1. Suppose v and w are both direction vectors for the same line ℓ. Then $\mathsf{w} = c\mathsf{v}$ is a multiple of v so the projection of u onto w is $\frac{\mathsf{u}\cdot\mathsf{w}}{\mathsf{w}\cdot\mathsf{w}}\mathsf{w} = \frac{\mathsf{u}\cdot c\mathsf{v}}{c\mathsf{v}\cdot c\mathsf{v}}c\mathsf{v} = \frac{\mathsf{u}\cdot\mathsf{v}}{\mathsf{v}\cdot\mathsf{v}}\mathsf{v}$ (the cs cancel), which is the same as the projection of u onto v.

2. We just found this in Problem 1.4.3: $\operatorname{proj}_\ell \mathsf{u} = \operatorname{proj}_\mathsf{v} \mathsf{u} = \begin{bmatrix} -\frac{3}{7} \\ \frac{1}{7} \\ \frac{2}{7} \end{bmatrix}$.

True/False Questions

Decide, with as little calculation as possible, whether each of the following statements is true or false and, if you say "false," explain your answer. (Answers are at the back of the book.)

1. The projection of a vector u onto a nonzero vector v is $\dfrac{\mathsf{u}\cdot\mathsf{v}}{\mathsf{v}\cdot\mathsf{v}}\mathsf{u}$.

2. If nonzero vectors u and v are orthogonal, you cannot find the projection of u onto v.

3. With $\mathsf{i}, \mathsf{j}, \mathsf{k}$ the standard basis vectors of R^3, the projection of i onto j is k.

4. With $\mathsf{i}, \mathsf{j}, \mathsf{k}$ the standard basis vectors of R^3, the projection of i onto j is $\mathsf{0}$.

5. The projection of a vector on a line is the projection of that vector onto any direction vector of that line.

6. The projection of $\begin{bmatrix} 1 \\ 1 \\ 1 \end{bmatrix}$ onto the plane with equation $x - y + z = 0$ is $\begin{bmatrix} 1 \\ -1 \\ 1 \end{bmatrix}$.

7. The projection of a vector w onto the plane spanned by vectors e and f is given by the formula $\operatorname{proj}_\pi \mathsf{w} = \dfrac{\mathsf{w}\cdot\mathsf{e}}{\mathsf{e}\cdot\mathsf{e}}\mathsf{e} + \dfrac{\mathsf{w}\cdot\mathsf{f}}{\mathsf{f}\cdot\mathsf{f}}\mathsf{f}$.

8. The projection of $\begin{bmatrix} 1 \\ 1 \\ 1 \end{bmatrix}$ onto the xy-plane is $\begin{bmatrix} 1 \\ 1 \\ 0 \end{bmatrix}$.

9. The distance from $(-2, 3, 1)$ to the plane with equation $x - y + 5z = 0$ is 0.

10. The line whose vector equation is $\begin{bmatrix} x \\ y \\ z \end{bmatrix} = \begin{bmatrix} 0 \\ -3 \\ 5 \end{bmatrix} + t\begin{bmatrix} 2 \\ -4 \\ 1 \end{bmatrix}$ is parallel to the plane whose equation is $2x - 4y + z = 0$.

EXERCISES

Answers to exercises marked [BB] can be found at the Back of the Book.

1. [BB] When does the projection of u onto a nonzero vector v have a direction opposite that of v?

2. **(a)** [BB] What's the projection of u onto v? 

   (b) What's the projection of $\overrightarrow{AB}$ onto $\overrightarrow{AC}$? 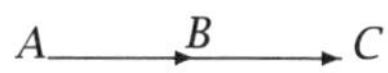

3. In each case, find the projection of u onto v and the projection of v on u.

   (a) [BB] $\mathsf{u} = \begin{bmatrix} 1 \\ 2 \\ -1 \end{bmatrix}$ and $\mathsf{v} = \begin{bmatrix} 3 \\ 1 \\ 1 \end{bmatrix}$ **(b)** $\mathsf{u} = \begin{bmatrix} 1 \\ 2 \\ 3 \end{bmatrix}$ and $\mathsf{v} = \begin{bmatrix} 4 \\ 5 \\ 6 \end{bmatrix}$

   (c) $\mathsf{u} = \begin{bmatrix} -1 \\ 0 \\ 1 \end{bmatrix}$ and $\mathsf{v} = \begin{bmatrix} 3 \\ 4 \\ -2 \end{bmatrix}$ **(d)** $\mathsf{u} = \begin{bmatrix} 3 \\ -4 \\ -2 \end{bmatrix}$ and $\mathsf{v} = \begin{bmatrix} 7 \\ 2 \\ 0 \end{bmatrix}$

   (e) $\mathsf{u} = \begin{bmatrix} 6 \\ -2 \\ 3 \end{bmatrix}$ and $\mathsf{v} = \begin{bmatrix} -5 \\ 2 \\ 4 \end{bmatrix}$ **(f)** $\mathsf{u} = \begin{bmatrix} 7 \\ 0 \\ 5 \end{bmatrix}$ and $\mathsf{v} = \begin{bmatrix} 3 \\ 1 \\ -4 \end{bmatrix}$.

4. In each case, find the distance from point P to line ℓ and the point of ℓ closest to P.

   (a) [BB] $P(-1,2,1)$; ℓ has equation $\begin{bmatrix} x \\ y \\ z \end{bmatrix} = \begin{bmatrix} 1 \\ 2 \\ 3 \end{bmatrix} + t\begin{bmatrix} 1 \\ 1 \\ 1 \end{bmatrix}$

   (b) $P(4,7,3)$; ℓ has equation $\begin{bmatrix} x \\ y \\ z \end{bmatrix} = \begin{bmatrix} 3 \\ 1 \\ 1 \end{bmatrix} + t\begin{bmatrix} 3 \\ -2 \\ 3 \end{bmatrix}$

   (c) $P(2,1,4)$; ℓ has equation $\begin{bmatrix} x \\ y \\ z \end{bmatrix} = \begin{bmatrix} -1 \\ 2 \\ 6 \end{bmatrix} + t\begin{bmatrix} 1 \\ -7 \\ 4 \end{bmatrix}$

   (d) $P(0,-1,1)$; ℓ has equation $\begin{bmatrix} x \\ y \\ z \end{bmatrix} = \begin{bmatrix} 1 \\ 0 \\ -1 \end{bmatrix} + t\begin{bmatrix} 4 \\ -1 \\ -5 \end{bmatrix}$.

5. [BB] In the text, we describe how to find the point A on a line that is closest to a given point P. With reference to the figure associated with Problem 1.4.6, describe a procedure for finding the point A on a plane closest to a given point P.

6. In each case,

   i. find the distance from point P to the plane π;
   ii. find the point of π closest to P.

   [See Exercise 5.]

   (a) [BB] P is $(2,3,0)$, π has equation $5x - y + z = 1$

(b) P is $(1, 2, 3)$ and π has equation $2x - y - z = 3$

(c) P is $(4, -1, 2)$, π has equation $3x - 2y + 4z = -2$

(d) P is $(1, 2, 1)$, π has equation $x - y + z = 0$

(e) P is $(-3, 0, 8)$, π has equation $2x - y - 7z = 1$

(f) P is $(1, -1, 1)$; π has equation $2x - y + z = 2$.

7. Find two nonzero orthogonal vectors that span each of the given planes.

(a) [BB] the plane spanned by $\mathsf{u} = \begin{bmatrix} -1 \\ 1 \\ 4 \end{bmatrix}$ and $\mathsf{v} = \begin{bmatrix} 2 \\ 0 \\ 1 \end{bmatrix}$

(b) the plane spanned by $\mathsf{u} = \begin{bmatrix} 1 \\ -2 \\ 3 \end{bmatrix}$ and $\mathsf{v} = \begin{bmatrix} 1 \\ 1 \\ 1 \end{bmatrix}$

(c) the plane spanned by $\mathsf{u} = \begin{bmatrix} -2 \\ 4 \\ 1 \end{bmatrix}$ and $\mathsf{v} = \begin{bmatrix} 3 \\ 0 \\ -1 \end{bmatrix}$

(d) [BB] the plane with equation $2x - 3y + 4z = 0$

(e) the plane with equation $x - y + 5z = 0$

(f) the plane with equation $3x + 2y - z = 0$.

8. Given that $\mathsf{u} = \begin{bmatrix} 1 \\ -1 \\ 0 \end{bmatrix}$ and $\mathsf{v} = \begin{bmatrix} 1 \\ 2 \\ -1 \end{bmatrix}$ are two vectors that span a plane π,

(a) find a nonzero vector w in π that is orthogonal to u;

(b) find the projection of $\mathsf{q} = \begin{bmatrix} 5 \\ 1 \\ -1 \end{bmatrix}$ on π and express this as a linear combination of u and w.

9. [BB] **(a)** Find two nonzero orthogonal vectors in the plane π with equation $2x - y + z = 0$.

(b) Use your answer to part (a) to find the projection of $\mathsf{w} = \begin{bmatrix} 1 \\ 1 \\ 1 \end{bmatrix}$ on π.

10. Repeat Exercise 9 taking π with equation $4x + y - 3z = 0$ and $\mathsf{w} = \begin{bmatrix} -1 \\ 1 \\ 2 \end{bmatrix}$.

11. Repeat Exercise 9 taking π with equation $8x - 3y - 2z = 0$ and $\mathsf{w} = \begin{bmatrix} -2 \\ 0 \\ 3 \end{bmatrix}$.

12. [BB] **(a)** What is the projection of $\mathsf{w} = \begin{bmatrix} 3 \\ 1 \\ 1 \end{bmatrix}$ on the plane π with equation $2x + y - z = 0$?

(b) What is the projection of $\mathsf{w} = \begin{bmatrix} x \\ y \\ z \end{bmatrix}$ on π?

13. Repeat both parts of Exercise 12 with π the plane with equation $2x - y + 3z = 0$ and $\mathsf{w} = \begin{bmatrix} 1 \\ -1 \\ 2 \end{bmatrix}$.

14. Find the projection of w on π in each of the following cases.

(a) [BB] $\mathsf{w} = \begin{bmatrix} -2 \\ 3 \\ 0 \end{bmatrix}$, π has equation $x + y - 2z = 0$

(b) $\mathsf{w} = \begin{bmatrix} 1 \\ 1 \\ 1 \end{bmatrix}$, π has equation $2x - y + 2z = 0$

(c) $\mathsf{w} = \begin{bmatrix} 3 \\ 1 \\ -4 \end{bmatrix}$, π has equation $3x - y + 2z = 0$

(d) $\mathsf{w} = \begin{bmatrix} x \\ y \\ z \end{bmatrix}$, π has equation $3x - 4y + 2z = 0$.

15. (a) [BB] Suppose a vector w is a linear combination of nonzero orthogonal vectors e and f. Prove that
$\mathsf{w} = \frac{\mathsf{w}\cdot\mathsf{e}}{\mathsf{e}\cdot\mathsf{e}}\mathsf{e} + \frac{\mathsf{w}\cdot\mathsf{f}}{\mathsf{f}\cdot\mathsf{f}}\mathsf{f}$.

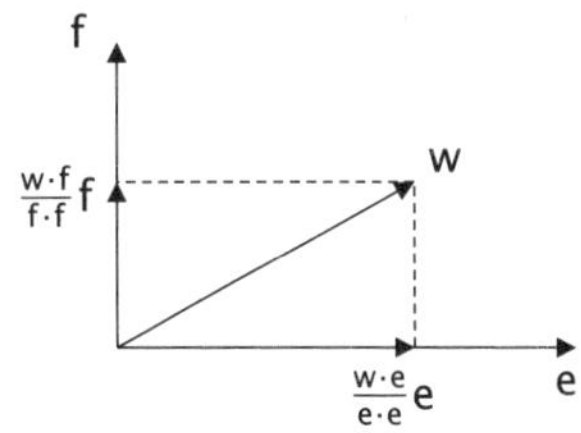

(b) The situation described in part (a) is pictured at the right. What happens if $\mathsf{e} = \mathsf{i}$, $\mathsf{f} = \mathsf{j}$ and $\mathsf{w} = \begin{bmatrix} a \\ b \end{bmatrix}$?

(c) Given that $\mathsf{w} = \begin{bmatrix} 13 \\ -20 \\ 15 \end{bmatrix}$ is in the span of the orthogonal vectors $\mathsf{e} = \begin{bmatrix} 1 \\ -2 \\ 3 \end{bmatrix}$ and $\mathsf{f} = \begin{bmatrix} -1 \\ 1 \\ 1 \end{bmatrix}$, write w as a linear combination of e and f.

16. In Problem 1.4.4, we showed how to find two nonzero orthogonal vectors in a plane. Write a short note explaining the method and why it works. (Your answer should include an explanation as to why the two vectors you find are in the plane.)

Critical Reading

17. (a) Show that the distance of the point $P(x_0, y_0, z_0)$ from the plane with equation $ax + by + cz = d$ is $\dfrac{|ax_0 + by_0 + cz_0 - d|}{\sqrt{a^2 + b^2 + c^2}}$.

[Yes, dear reader, there is a formula for the distance from a point to a plane. It was not mentioned in the text because the author wishes to emphasize the method, not the result.]

(b) Use the formula in part (a) to determine the distance of $P(1,2,3)$ from the plane with equation $x - 3y + 5z = 4$.

(c) Find an equation of the plane that contains the points $(1,1,0)$ and $(0,1,2)$ and is $\sqrt{6}$ units from the point $(3,-3,-2)$.

18. [BB] Don has a theory that to find the projection of a vector w on a plane π (through the origin), one could project w on a normal n to the plane, and take $\operatorname{proj}_\pi \mathsf{w} = \mathsf{w} - \operatorname{proj}_{\mathsf{n}} \mathsf{w}$.

(a) Is Don's theory correct? Explain by applying the definition of "projection on a plane."

(b) Illustrate Don's theory using the plane with equation $3x - 2y + z = 0$ and $\mathsf{w} = \begin{bmatrix} 1 \\ 1 \\ 1 \end{bmatrix}$.

19. Wendy has a theory that the projection of a vector w on the plane spanned by e and f should be the projection of w on the vector e + f. Let's test this theory with the vectors $\mathsf{w} = \begin{bmatrix} 1 \\ 3 \\ 2 \end{bmatrix}$, $\mathsf{e} = \begin{bmatrix} 2 \\ -1 \\ 1 \end{bmatrix}$ and $\mathsf{f} = \begin{bmatrix} 1 \\ 1 \\ -1 \end{bmatrix}$, noting that e and f are orthogonal.

(a) Find the projection p of w on the plane π spanned by e and f.

(b) Verify that w − p is orthogonal to e and to f.

(c) Find the projection p′ of w on the vector e + f.

(d) Is Wendy's theory correct?

(e) Is Wendy's vector in the right plane?

20. Let $\mathsf{u} = \begin{bmatrix} x \\ y \end{bmatrix}$ be a vector in R^2 and let v be the reflection of u in the line with equation $y = mx$. Find a formula for v. [Hint: Find a formula that involves u and the projection of u on the line.]

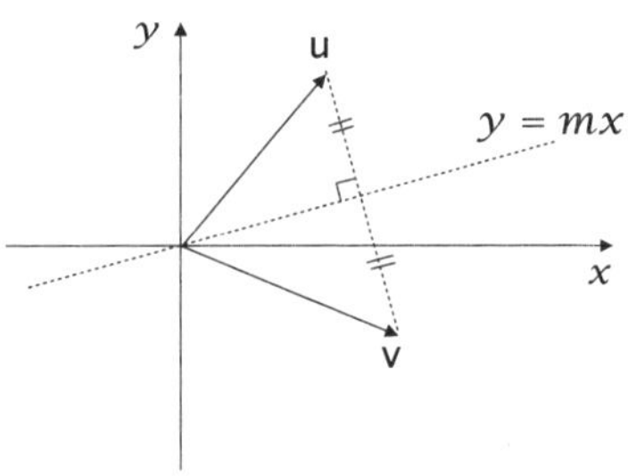

21. [BB] (The distance between skew lines) Let ℓ_1 and ℓ_2 be the lines with equations

$$\ell_1: \begin{bmatrix} x \\ y \\ z \end{bmatrix} = \begin{bmatrix} 3 \\ 0 \\ -1 \end{bmatrix} + t \begin{bmatrix} 4 \\ 3 \\ 1 \end{bmatrix}, \qquad \ell_2: \begin{bmatrix} x \\ y \\ z \end{bmatrix} = \begin{bmatrix} 2 \\ -1 \\ -3 \end{bmatrix} + t \begin{bmatrix} 2 \\ 6 \\ 7 \end{bmatrix}.$$

(a) Show that ℓ_1 and ℓ_2 are not parallel and that they do not intersect. (Such lines are called *skew.*)

(b) Find an equation of the plane that contains ℓ_1 and is parallel to ℓ_2.

(c) Use the result of (b) to find the (shortest) distance between ℓ_1 and ℓ_2.

22. **(a)** The lines with vector equations $\begin{bmatrix} x \\ y \\ z \end{bmatrix} = \begin{bmatrix} 0 \\ 1 \\ 4 \end{bmatrix} + t\begin{bmatrix} -2 \\ 1 \\ 2 \end{bmatrix}$ and $\begin{bmatrix} x \\ y \\ z \end{bmatrix} = \begin{bmatrix} 4 \\ -2 \\ 2 \end{bmatrix} + t\begin{bmatrix} 6 \\ -3 \\ -6 \end{bmatrix}$ are parallel. Why?

(b) Find the distance between the lines in (a).

23. Let ℓ_1 and ℓ_2 be the lines with the equations shown.

$$\ell_1\colon \begin{bmatrix} x \\ y \\ z \end{bmatrix} = \begin{bmatrix} -1 \\ 0 \\ 1 \end{bmatrix} + t\begin{bmatrix} 2 \\ 1 \\ -3 \end{bmatrix} \quad \text{and} \quad \ell_2\colon \begin{bmatrix} x \\ y \\ z \end{bmatrix} = \begin{bmatrix} 4 \\ 1 \\ -2 \end{bmatrix} + t\begin{bmatrix} 0 \\ 1 \\ -1 \end{bmatrix}.$$

(a) Show that ℓ_1 and ℓ_2 are not parallel and that they do not intersect.

(b) Find an equation of the plane that contains ℓ_1 and is parallel to ℓ_2.

(c) Use the result of (b) to find the (shortest) distance between ℓ_1 and ℓ_2.

24. Let ℓ_1 and ℓ_2 be the lines with the equations shown:

$$\ell_1\colon \begin{bmatrix} x \\ y \\ z \end{bmatrix} = \begin{bmatrix} 3 \\ 0 \\ 5 \end{bmatrix} + t\begin{bmatrix} 0 \\ -2 \\ 1 \end{bmatrix} \quad \text{and} \quad \ell_2\colon \begin{bmatrix} x \\ y \\ z \end{bmatrix} = \begin{bmatrix} -2 \\ -1 \\ -1 \end{bmatrix} + t\begin{bmatrix} 3 \\ 2 \\ -2 \end{bmatrix}.$$

(a) Show that ℓ_1 and ℓ_2 are not parallel and that they do not intersect.

(b) [BB] Find the shortest distance between ℓ_1 and ℓ_2.

(c) Find a point A on ℓ_1 and a point B on ℓ_2 so that the distance from A to B is the shortest distance between the lines.

25. [BB] Let ℓ be the line with equation $\begin{bmatrix} x \\ y \\ z \end{bmatrix} = \begin{bmatrix} 3 \\ 5 \\ -4 \end{bmatrix} + t\begin{bmatrix} 2 \\ 1 \\ 3 \end{bmatrix}$ and let π be the plane with equation $-2x + y + z = 3$.

(a) Explain why ℓ is parallel to π.

(b) Find a point P on ℓ and a point Q on π.

(c) Find the (shortest) distance from ℓ to π.

26. **(a)** Repeat Exercise 25 for π with equation $3x - 2y + z = 0$ and ℓ the line with equation $\begin{bmatrix} x \\ y \\ z \end{bmatrix} = \begin{bmatrix} 2 \\ 3 \\ 0 \end{bmatrix} + t\begin{bmatrix} 1 \\ 3 \\ 3 \end{bmatrix}$.

(b) Could you have predicted your answer to (a) with a little forethought?

27. Find an equation of the plane that is perpendicular to the plane with equation $2x - 4y + 2z = 9$ and that contains the line with equation $\begin{bmatrix} x \\ y \\ z \end{bmatrix} = \begin{bmatrix} -1 \\ 5 \\ 2 \end{bmatrix} + t\begin{bmatrix} 3 \\ 2 \\ -1 \end{bmatrix}$.

28. **(a)** Find an equation of a plane π that is perpendicular to the line with equation $\begin{bmatrix} x \\ y \\ z \end{bmatrix} = \begin{bmatrix} 1 \\ 0 \\ -1 \end{bmatrix} + t\begin{bmatrix} 2 \\ -1 \\ 2 \end{bmatrix}$ and 5 units away from the point $P(1,4,0)$. Is π unique? Explain.

    (b) Find a point in π that is 5 units away from P.

29. Describe in geometrical terms the set of all points one unit away from the plane with equation $2x + y + 2z = 5$. Find an equation for this set of points.

30. Think about the procedure outlined in Problem 1.4.6 for finding the distance from a point P to a plane π. Suppose two students choose different points Q_1, Q_2 in π. Explain why they will (should!) get the same answer for the distance.

31. **(a)** Suppose u and v are nonzero parallel vectors. Show that the projection of u on v is u.

    (b) Think about the procedure we outlined in Problem 1.4.5 for finding the distance from a point P to a line ℓ. Suppose two students select two different points Q_1 and Q_2 on ℓ. Explain why they will get the same answer for the distance.

1.5 Linear Dependence and Independence

This last section of our first chapter is very short, but very important.

Suppose we have some vectors $\mathsf{v}_1, \mathsf{v}_2, \ldots, \mathsf{v}_n$. It is easy to find a linear combination of these that equals the zero vector:

$$0\mathsf{v}_1 + 0\mathsf{v}_2 + \cdots + 0\mathsf{v}_n = \mathsf{0}.$$

This linear combination, with all coefficients 0, is called the *trivial* one. Vectors are *linearly independent* if this is the **only** way to get $\mathsf{0}$.

1.5.1 Definition. A set of vectors $\mathsf{v}_1, \mathsf{v}_2, \ldots, \mathsf{v}_n$ is *linearly independent* if

$$c_1\mathsf{v}_1 + c_2\mathsf{v}_2 + \cdots + c_n\mathsf{v}_n = \mathsf{0} \quad \text{implies that} \quad c_1 = 0, c_2 = 0, \ldots, c_n = 0.$$

Vectors are *linearly dependent* if they are not linearly independent; that is, there is a linear combination $c_1\mathsf{v}_1 + c_2\mathsf{v}_2 + \cdots + c_n\mathsf{v}_n = \mathbf{0}$ with **at least one** coefficient not 0.

1.5.2 Example. Are the vectors $\mathsf{v}_1 = \begin{bmatrix} 1 \\ 2 \end{bmatrix}$ and $\mathsf{v}_2 = \begin{bmatrix} -1 \\ 0 \end{bmatrix}$ linearly independent? Certainly

$$0\begin{bmatrix} 1 \\ 2 \end{bmatrix} + 0\begin{bmatrix} -1 \\ 0 \end{bmatrix} = \begin{bmatrix} 0 \\ 0 \end{bmatrix},$$

but this is not the point! The question is whether $0\begin{bmatrix} 1 \\ 2 \end{bmatrix} + 0\begin{bmatrix} -1 \\ 0 \end{bmatrix}$ is the **only** way to get $\mathbf{0}$? Does $c_1\mathsf{v}_1 + c_2\mathsf{v}_2 = \mathbf{0}$ force both c_1 and c_2 to be 0? We check. Suppose $c_1\mathsf{v}_1 + c_2\mathsf{v}_2 = \mathbf{0}$. This means

$$c_1\begin{bmatrix} 1 \\ 2 \end{bmatrix} + c_2\begin{bmatrix} -1 \\ 0 \end{bmatrix} = \begin{bmatrix} 0 \\ 0 \end{bmatrix}, \quad \text{which gives} \quad \begin{bmatrix} c_1 - c_2 \\ 2c_1 \end{bmatrix} = \begin{bmatrix} 0 \\ 0 \end{bmatrix}.$$

Equating components, we have two equations

$$\begin{aligned} c_1 - c_2 &= 0 \\ 2c_1 \qquad &= 0 \end{aligned}$$

which we must solve for the unknowns c_1 and c_2. This is not hard. The second equation gives $c_1 = 0$ and, with this knowledge, the first equation gives $c_2 = 0$ too. The vectors v_1 and v_2 are linearly independent because the assumption $c_1\mathsf{v}_1 + c_2\mathsf{v}_2 = \mathbf{0}$ implies that both coefficients are 0. :)

1.5.3 Problem. Are the vectors $\mathsf{v}_1 = \begin{bmatrix} -2 \\ 4 \end{bmatrix}$ and $\mathsf{v}_2 = \begin{bmatrix} 3 \\ -6 \end{bmatrix}$ linearly dependent or linearly independent?

Solution. We assume $c_1\mathsf{v}_1 + c_2\mathsf{v}_2 = \mathbf{0}$ and see if this forces $c_1 = c_2 = 0$. Thus we suppose

$$c_1\begin{bmatrix} -2 \\ 4 \end{bmatrix} + c_2\begin{bmatrix} 3 \\ -6 \end{bmatrix} = \begin{bmatrix} 0 \\ 0 \end{bmatrix}, \text{ which is } \begin{bmatrix} -2c_1 + 3c_2 \\ 4c_1 - 6c_2 \end{bmatrix} = \begin{bmatrix} 0 \\ 0 \end{bmatrix}.$$

Here there are many *nontrivial* solutions, that is, solutions with not all coefficients 0, $c_1 = 3$, $c_2 = 2$, for instance. Therefore, the vectors v_1 and v_2 are linearly dependent.

READING CHECK 1. Let $\mathsf{v}_1 = \begin{bmatrix} -2 \\ 4 \end{bmatrix}$ and $\mathsf{v}_2 = \begin{bmatrix} 3 \\ -6 \end{bmatrix}$ be as in Problem 1.5.3. Let $\mathsf{v}_3 = \begin{bmatrix} 7 \\ -2 \end{bmatrix}$. Are $\mathsf{v}_1, \mathsf{v}_2, \mathsf{v}_3$ linearly independent or linearly dependent?

1.5.4 Example. Let $\mathsf{v}_1 = \begin{bmatrix} 1 \\ -2 \\ 5 \end{bmatrix}$, $\mathsf{v}_2 = \begin{bmatrix} 0 \\ 5 \\ -7 \end{bmatrix}$ and $\mathsf{v}_3 = \begin{bmatrix} 2 \\ 1 \\ 3 \end{bmatrix}$. To determine whether these vectors are linearly independent or linearly dependent, we suppose $c_1\mathsf{v}_1 + c_2\mathsf{v}_2 + c_3\mathsf{v}_3 = \mathbf{0}$. This vector equation leads to the three equations involving the coefficients,

$$\begin{aligned} c_1 + 2c_3 &= 0 \\ -2c_1 + 5c_2 + c_3 &= 0 \\ 5c_1 - 7c_2 + 3c_3 &= 0. \end{aligned}$$

The first equation gives $c_1 = -2c_3$. Substituting this into the second and third equations gives

$$\begin{aligned} 5c_2 + 5c_3 &= 0 \\ -7c_2 - 7c_3 &= 0. \end{aligned}$$

So $c_3 = -c_2$, but there seems no way to get all the coefficients 0. In fact, $c_2 = 1$, $c_3 = -c_2 = -1$ and $c_1 = -2c_3 = 2$ gives a solution that is not all 0s: $2\mathsf{v}_1 + \mathsf{v}_2 - \mathsf{v}_3 = \mathbf{0}$. We conclude that the vectors are linearly dependent.

READING CHECK 2. Decide whether the vectors $\mathsf{v}_1 = \begin{bmatrix} 2 \\ 3 \\ 1 \\ 4 \end{bmatrix}$, $\mathsf{v}_2 = \begin{bmatrix} -1 \\ 1 \\ 2 \\ 3 \end{bmatrix}$, $\mathsf{v}_3 = \begin{bmatrix} 4 \\ 0 \\ -1 \\ 1 \end{bmatrix}$, $\mathsf{v}_4 = \begin{bmatrix} 4 \\ 6 \\ 2 \\ 8 \end{bmatrix}$ are linearly independent or linearly dependent. [Hint: Look at v_1 and v_4.]

In the exercises, we ask you to show that two vectors are linearly dependent if and only if one is a scalar multiple of the other, and that three vectors are linearly dependent if and only if one is a linear combination of the other two. These facts generalize.

1.5.5 Theorem. *A set of $n > 1$ vectors is linearly dependent if and only if one of the vectors is a linear combination of the others.*

Proof. ($\Longrightarrow$) Suppose vectors $\mathsf{v}_1, \mathsf{v}_2, \ldots, \mathsf{v}_n$ are linearly dependent. Then some nontrivial linear combination of them is the zero vector; that is, there are scalars $c_1, c_2, \ldots, c_n$, not all 0, such that

$$c_1\mathsf{v}_1 + c_2\mathsf{v}_2 + \cdots + c_n\mathsf{v}_n = \mathbf{0}. \tag{1}$$

Suppose $c_1 \neq 0$. Then we can rewrite equation (1)

$$c_1\mathsf{v}_1 = -c_2\mathsf{v}_2 - \cdots - c_n\mathsf{v}_n,$$

divide by c_1, and obtain

$$\mathsf{v}_1 = -\frac{c_2}{c_1}\mathsf{v}_2 - \cdots - \frac{c_n}{c_1}\mathsf{v}_n,$$

which says that v_1 is a linear combination of the other vectors. If $c_1 = 0$, but $c_2 \neq 0$, a similar argument would show that v_2 is a linear combination of the rest. In general, if we let c_k be the first nonzero coefficient, we can show that v_k is a linear combination of the others.

($\Leftarrow$) Suppose one of vectors $\mathsf{v}_1, \mathsf{v}_2, \ldots, \mathsf{v}_n$ is a linear combination of the others, say v_1. Thus, there are scalars $c_2, \ldots, c_n$ so that

$$\mathsf{v}_1 = c_2\mathsf{v}_2 + \cdots + c_n\mathsf{v}_n.$$

From this, we obtain

$$\underset{\uparrow}{1}\mathsf{v}_1 - c_2\mathsf{v}_2 - \cdots - c_n\mathsf{v}_n = \mathsf{0},$$

which expresses $\mathsf{0}$ as a **nontrivial** linear combination of all the vectors (the arrow points to one coefficient that is certainly not 0). If it is not v_1 that is a linear combination of the others, but another vector v_k, a similar argument would again lead to a representation of $\mathsf{0}$ as a nontrivial linear combination of the other vectors. ■

1.5.6 Example. The three vectors $\begin{bmatrix} 1 \\ 1 \\ 2 \end{bmatrix}, \begin{bmatrix} 0 \\ 3 \\ 4 \end{bmatrix}, \begin{bmatrix} 3 \\ 0 \\ 2 \end{bmatrix}$ are linearly dependent because the second vector is a linear combination of the other two:

$$\begin{bmatrix} 0 \\ 3 \\ 4 \end{bmatrix} = 3\begin{bmatrix} 1 \\ 1 \\ 2 \end{bmatrix} - \begin{bmatrix} 3 \\ 0 \\ 2 \end{bmatrix}.$$

Answers to Reading Checks

1. Since $3\mathsf{v}_1 + 2\mathsf{v}_2 = \mathsf{0}$, we have $3\mathsf{v}_1 + 2\mathsf{v}_2 + 0\mathsf{v}_3 = \mathsf{0}$. Since there is a linear combination of the three vectors equal to the zero vector without **all** coefficients 0, the vectors are linearly dependent.

2. Notice that $\mathsf{v}_4 = 2\mathsf{v}_1$. Thus $2\mathsf{v}_1 + 0\mathsf{v}_2 + 0\mathsf{v}_3 - 1\mathsf{v}_4$ is a nontrivial linear combination of the four vectors which equals $\mathsf{0}$ (*nontrivial* meaning not all coefficients 0). So the vectors are linearly dependent.

True/False Questions

Decide, with as little calculation as possible, whether each of the following statements is true or false and, if you say "false," explain your answer. (Answers are at the back of the book.)

1. Linear independence of vectors u and v means that $0\mathsf{u} + 0\mathsf{v} = \mathsf{0}$.

2. Linear independence of vectors u and v means that $a\mathsf{u} + b\mathsf{v} = 0$ where a and b are 0.

3. For any u and v, the vectors u, v, 0 are linearly dependent.

4. If $\mathsf{v}_1, \mathsf{v}_2, \ldots, \mathsf{v}_k$ are linearly dependent vectors, then there is an equation of the form $c_1\mathsf{v}_1 + c_2\mathsf{v}_2 + \cdots + c_k\mathsf{v}_k = 0$ with none of the scalars c_i equal to 0.

5. If vectors are linearly dependent, then one of them is a scalar multiple of another.

6. If one of the vectors $\mathsf{v}_1, \mathsf{v}_2, \ldots, \mathsf{v}_n$ is a scalar multiple of another, then the vectors are linearly dependent.

7. The equation $2\mathsf{v}_1 - 3\mathsf{v}_2 + \mathsf{v}_3 = 0$ implies that each of the three vectors $\mathsf{v}_1, \mathsf{v}_2, \mathsf{v}_3$ is a linear combination of the other two.

8. If u, v and w are vectors, v is not a scalar multiple of u, and w is not in the span of u and v, then u,v,w are linearly independent.

9. If u, v, w are linearly dependent, then $a\mathsf{u} + b\mathsf{v} + c\mathsf{w} = 0$ implies at least one of the coefficients a, b, c is not zero.

10. If $\mathsf{u}_1, \mathsf{u}_2, \ldots, \mathsf{u}_m$ are linearly independent, then $c_1\mathsf{u}_1 + c_2\mathsf{u}_2 + \cdots + c_m\mathsf{u}_m = 0$ with $c_1 = 0$, $c_2 = 0$, ..., $c_m = 0$.

EXERCISES

Answers to exercises marked [BB] can be found at the **Back of the Book.**

1. Determine whether each of the given sets of vectors is linearly independent or linearly dependent. If linearly dependent, give an explicit equation to support; that is, a nontrivial linear combination of the vectors that is 0.

(a) [BB] $\mathsf{v}_1 = \begin{bmatrix} -1 \\ 1 \end{bmatrix}, \mathsf{v}_2 = \begin{bmatrix} 3 \\ 0 \end{bmatrix}$

(b) $\mathsf{v}_1 = \begin{bmatrix} -3 \\ 5 \end{bmatrix}, \mathsf{v}_2 = \begin{bmatrix} 7 \\ 9 \end{bmatrix}$

(c) $\mathsf{v}_1 = \begin{bmatrix} -3 \\ 5 \end{bmatrix}, \mathsf{v}_2 = \begin{bmatrix} 7 \\ 9 \end{bmatrix}, \mathsf{v}_3 = \begin{bmatrix} 1 \\ 19 \end{bmatrix}$

(d) $\mathsf{v}_1 = \begin{bmatrix} 1 \\ 2 \end{bmatrix}, \mathsf{v}_2 = \begin{bmatrix} 3 \\ 4 \end{bmatrix}, \mathsf{v}_3 = \begin{bmatrix} 5 \\ 6 \end{bmatrix}.$

(e) $\mathsf{v}_1 = \begin{bmatrix} 1 \\ 0 \\ 0 \end{bmatrix}, \mathsf{v}_2 = \begin{bmatrix} -2 \\ 1 \\ 2 \end{bmatrix}, \mathsf{v}_3 = \begin{bmatrix} 3 \\ 2 \\ 5 \end{bmatrix}$

(f) [BB] $\mathsf{v}_1 = \begin{bmatrix} 1 \\ 0 \\ 0 \end{bmatrix}, \mathsf{v}_2 = \begin{bmatrix} -2 \\ 1 \\ 2 \end{bmatrix}, \mathsf{v}_3 = \begin{bmatrix} 3 \\ 2 \\ 4 \end{bmatrix}$

(g) $\mathsf{v}_1 = \begin{bmatrix} 1 \\ 0 \\ 0 \end{bmatrix}, \mathsf{v}_2 = \begin{bmatrix} -2 \\ 1 \\ 2 \end{bmatrix}, \mathsf{v}_3 = \begin{bmatrix} 3 \\ 2 \\ 4 \end{bmatrix}, \mathsf{v}_4 = \begin{bmatrix} -3 \\ 4 \\ 1 \end{bmatrix}$

(h) $\mathsf{v}_1 = \begin{bmatrix} -1 \\ 1 \\ 2 \end{bmatrix}, \mathsf{v}_2 = \begin{bmatrix} 0 \\ 1 \\ 4 \end{bmatrix}, \mathsf{v}_3 = \begin{bmatrix} 3 \\ 1 \\ 2 \end{bmatrix}, \mathsf{v}_4 = \begin{bmatrix} 2 \\ -2 \\ -4 \end{bmatrix}$

(i) $\mathsf{v}_1 = \begin{bmatrix} 1 \\ 1 \\ 2 \\ 1 \end{bmatrix}, \mathsf{v}_2 = \begin{bmatrix} 3 \\ 3 \\ 6 \\ 3 \end{bmatrix}, \mathsf{v}_3 = \begin{bmatrix} 2 \\ 3 \\ 4 \\ 1 \end{bmatrix}, \mathsf{v}_4 = \begin{bmatrix} 1 \\ 2 \\ 3 \\ 1 \end{bmatrix}$.

2. [BB] Suppose you were asked to prove that vectors $\mathsf{v}_1, \mathsf{v}_2, \dots, \mathsf{v}_k$ are linearly independent. What would be the first line of your proof? What would be the last line?

3. Suppose you are asked to prove that vectors u, v, w are linearly independent. What would be the first line of your proof? What would be the last line?

4. [BB] Suppose nonzero vectors e and f are orthogonal. Show that e and f are linearly independent. [Hint: Suppose $a\mathsf{e} + b\mathsf{f} = 0$ for scalars a and b. Then take the dot product with e.]

5. Suppose $\mathsf{v}_1, \mathsf{v}_2, \dots, \mathsf{v}_n$ are vectors and one of them is the zero vector. To be specific, suppose $\mathsf{v}_1 = 0$. Show that $\mathsf{v}_1, \mathsf{v}_2, \dots, \mathsf{v}_n$ are linearly dependent. (Thus the zero vector is never part of a linearly independent set.)

6. Suppose nonzero vectors u, v and w are orthogonal (meaning that any two of these are orthogonal). Show that u, v and w are linearly independent.

7. You are given that vectors $\mathsf{v}_1, \mathsf{v}_2, \dots, \mathsf{v}_n$ are linearly dependent. What do you know? Write your answer using mathematical notation and then entirely in English (no symbols).

8. [BB] Why have generations of linear algebra students refused to memorize the definitions of linear independence and linear dependence?

Critical Reading

9. [BB] Suppose $\mathsf{u}, \mathsf{v}, \mathsf{x}, \mathsf{y}$ are linearly independent vectors. Show that $\mathsf{u}, \mathsf{v}, \mathsf{x}$ are also linearly independent vectors.

10. Prove that two vectors are linearly dependent if and only if one is a scalar multiple of the other. This is a special case of Theorem 1.5.5. Do not answer by quoting the theorem.

11. Prove that three vectors are linearly dependent if and only if one of the vectors lies in the plane spanned by the other two. This is a special case of Theorem 1.5.5. Do not answer by quoting the theorem.

12. Suppose $\mathsf{u}_1, \mathsf{u}_2, \dots, \mathsf{u}_m$ are linearly independent vectors. Prove that $\mathsf{u}_2, \dots, \mathsf{u}_m$ are also linearly independent.

13. Let u, v and w be vectors such that $\mathsf{u} \neq 0$, v is not a scalar multiple of u, and w is not in the span of u and v. Show that u,v and w are linearly independent.

14. Suppose u, v and w are vectors with u, v linearly independent but $\mathsf{u}, \mathsf{v}, \mathsf{w}$ linearly dependent. Show that w is in the span of u and v.

15. Complete the following sentence and explain: "A single vector x is linearly dependent if and only if ..."

16. Suppose $\mathsf{u}, \mathsf{v}, \mathsf{x}$ are linearly dependent vectors. Show that $\mathsf{u}, \mathsf{v}, \mathsf{x}, \mathsf{y}$ are also linearly dependent vectors, for any additional vector y.

17. **(a)** Let $S = \{\mathsf{v}_1, \mathsf{v}_2, \dots, \mathsf{v}_k\}$ be a set of linearly dependent vectors and let $\mathcal{T} = \{\mathsf{v}_1, \dots, \mathsf{v}_k, \mathsf{v}_{k+1}, \dots, \mathsf{v}_n\}$ be a set of vectors containing S. Show that the vectors in $\mathcal{T}$ are also linearly dependent.

    (b) Is the result of (a) true if we replace the word "dependent" with "independent"?

CHAPTER KEY WORDS AND IDEAS: Here are some technical words and phrases that were used in this chapter. Do you know the meaning of each? If you're not sure, check the glossary or index at the back of the book.

2×2 determinant
n-dimensional vector
additive inverse (of a vector)
anticommutative
associative
Cauchy–Schwarz inequality
commutative
component (of a vector)
componentwise addition
dot product
Euclidean n-space
Euclidean plane
length of a vector
linear combination of vectors
linearly dependent vectors
linearly independent vectors
nontrivial linear combination
normal vector
orthogonal vectors
parallelogram rule
parallel vectors
plane spanned by vectors
projection of a vector on a plane
projection of a vector on a vector
scalar multiple (of a vector)
standard basis (vectors)
Theorem of Pythagoras
three-dimensional vector
two-dimensional vector
triangle inequality
trivial linear combination
unit vector
vectors are equal
vectors having opposite directions
vectors having the same direction
vectors span a plane
zero vector

Review Exercises for Chapter 1

1. **(a)** Determine whether or not $\begin{bmatrix} -1 \\ 3 \\ 2 \end{bmatrix}$ is a linear combination of $\begin{bmatrix} 2 \\ 0 \\ 1 \end{bmatrix}$ and $\begin{bmatrix} 0 \\ 2 \\ 4 \end{bmatrix}$.

   (b) Do $\begin{bmatrix} 2 \\ 0 \\ 1 \end{bmatrix}$ and $\begin{bmatrix} 0 \\ 2 \\ 4 \end{bmatrix}$ span R^3? Justify your answer.

2. If u and v are vectors and $\mathsf{w} = 2\mathsf{u} - 5\mathsf{v}$, show that any linear combination of u, v and w is in fact a linear combination of just u and v.

3. Suppose vectors u and v are each scalar multiples of a third vector w. Show that any linear combination of u and v is a scalar multiple of w.

4. Suppose u and v are vectors that are not parallel, a and b are scalars, and $a\mathsf{u} + b\mathsf{v} = \mathsf{0}$ is the zero vector. Prove that $a = b = 0$.

5. A *median* of a triangle is a line which joins a vertex to the midpoint of the opposite side. Prove that the three medians of a triangle are *concurrent*; that is, they meet at a common point.

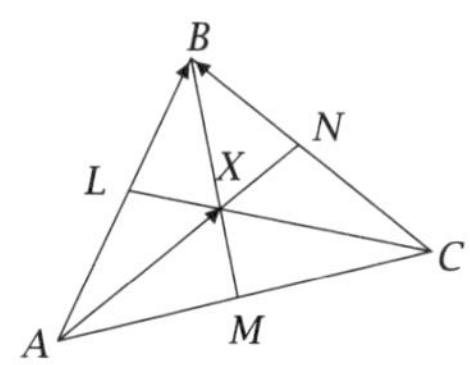

   [Hint: Let the triangle have vertices A, B, C, as shown, and let the midpoints of AB, AC, BC be L, M, N, respectively. Show that if X is on BM, then $\overrightarrow{AX}$ is a linear combination of $\overrightarrow{AB}$ and $\overrightarrow{AC}$. Do the same assuming X is also on CL and AN.]

6. The *right bisector* of a line segment AB is the line through the midpoint of AB that is perpendicular to AB.

   (a) Let A, B, C be the vertices of a triangle. Prove that the right bisectors of AB, BC and AC are concurrent.

   [Hint: Let P, Q, R be the midpoints of BC, AC, AB, respectively. Let X be the intersection of the right bisectors of AC and BC. It suffices to prove that $\overrightarrow{RX} \cdot \overrightarrow{AB} = 0$. Why?]

   (b) Explain how to find the centre of the circle through three noncollinear points.

7. Let $\mathsf{u} = \begin{bmatrix} -1 \\ a \\ b \\ 2 \\ 0 \end{bmatrix}$, $\mathsf{v} = \begin{bmatrix} c \\ 1 \\ 1 \\ a \\ -4 \end{bmatrix}$, $\mathsf{x} = \begin{bmatrix} 1 \\ 2 \\ 3 \\ 4 \\ 5 \end{bmatrix}$ and $\mathsf{y} = \begin{bmatrix} -1 \\ 1 \\ -2 \\ 4 \\ 8 \end{bmatrix}$.

   Find a, b and c, if possible, so that

   (a) $3\mathsf{u} - 2\mathsf{v} = \mathsf{y}$, **(b)** $2\mathsf{u} + \mathsf{x} - \mathsf{y} = \mathsf{0}$,

   (c) u and x are orthogonal, **(d)** $\|\mathsf{v}\| = \sqrt{c^2 + 27}$.

8. Let $\mathsf{u} = \begin{bmatrix} 1 \\ 2 \\ 3 \\ -1 \end{bmatrix}$, $\mathsf{v} = \begin{bmatrix} 0 \\ -1 \\ 1 \end{bmatrix}$, $\mathsf{x} = \begin{bmatrix} 1 \\ 2 \\ 3 \end{bmatrix}$ and $\mathsf{y} = \begin{bmatrix} 0 \\ -1 \\ 1 \\ 2 \end{bmatrix}$.

Find each of the following, if possible.

(a) $2x - 3y + u$ **(b)** $\|y\|$ **(c)** $3v + 4x$
(d) $x \cdot v$ **(e)** the cosine of the angle between u and y.

9. **(a)** Find a unit vector that has the same direction as $u = \begin{bmatrix} 1 \\ 0 \\ -1 \\ 1 \end{bmatrix}$.

   (b) Find a vector of length 3 that has the same direction as $u = \begin{bmatrix} 1 \\ 2 \\ 1 \\ 1 \\ 1 \end{bmatrix}$.

   (c) Find a vector of length 7 that has direction opposite that of $u = \begin{bmatrix} 0 \\ -3 \\ 1 \\ 1 \\ 2 \end{bmatrix}$.

   (d) Find a vector that has twice the length of $u = \begin{bmatrix} -1 \\ 2 \\ 0 \\ 2 \\ 7 \end{bmatrix}$.

   (e) Find a unit vector that has direction opposite that of $u = \begin{bmatrix} -1 \\ 1 \\ -1 \\ 4 \\ 1 \end{bmatrix}$.

10. Suppose u, v, x and y are unit vectors such that $v \cdot x = -\frac{1}{3}$, u and y are orthogonal, the angle between v and y is 45° and the angle between u and x is 60°. Find

    (a) $(u + v) \cdot (x - y)$ **(b)** $(2u - 3v) \cdot (x + 4y)$
    (c) $(3u + 4v) \cdot (2x - 5y)$ **(d)** $\|5u - y\|^2$.

11. **(a)** Find the exact value of the angle between $u = \begin{bmatrix} 2 \\ -1 \\ 1 \end{bmatrix}$ and $v = \begin{bmatrix} 1 \\ 1 \\ 2 \end{bmatrix}$.

    (b) Find a vector of length 3 with direction opposite v.

    (c) Find any number(s) k such that $u + kv$ is orthogonal to $u - kv$.

12. Find $u \times v$ given $u = \begin{bmatrix} -2 \\ -3 \\ 1 \end{bmatrix}$ and $v = \begin{bmatrix} 1 \\ -1 \\ 1 \end{bmatrix}$.

13. Let $u = \begin{bmatrix} 3 \\ 2 \\ 0 \end{bmatrix}$, $v = \begin{bmatrix} 1 \\ -2 \\ 3 \end{bmatrix}$ and $w = \begin{bmatrix} -1 \\ -2 \\ 3 \end{bmatrix}$.

    (a) Find the cosine of the angle between u and w.

    (b) Find a vector of length 3 that is perpendicular to both u and v.

    (c) Is it possible to express u as a linear combination of v and w? Explain.

(d) Find the area of the triangle two of whose sides are u and v.

14. (a) Find a unit vector in the direction opposite that of $\mathsf{v} = \begin{bmatrix} 2 \\ 0 \\ 1 \end{bmatrix}$.

(b) Find two nonzero orthogonal vectors in the plane spanned by $\mathsf{u} = \begin{bmatrix} -1 \\ 1 \\ 4 \end{bmatrix}$ and v.

(c) Find a real number k (if possible) so that $\mathsf{u} + k\mathsf{v}$ is orthogonal to u.

15. Do there exist vectors u and v with $\mathsf{u} \cdot \mathsf{v} = -13$, $\|\mathsf{u}\| = 5$ and $\|\mathsf{v}\| = 2$? Explain.

16. Given $\mathsf{u} \times \mathsf{v} = 0$, $\|\mathsf{u}\| = 2$ and $\|\mathsf{v}\| = 4$, find $\mathsf{u} \cdot \mathsf{v}$.

17. Does there exist a number k such that lines with equations

$$\begin{bmatrix} x \\ y \\ z \end{bmatrix} = \begin{bmatrix} 1 \\ -5 \\ 7 \end{bmatrix} + t\begin{bmatrix} 4 \\ k \\ -56 \end{bmatrix} \quad \text{and} \quad \begin{bmatrix} x \\ y \\ z \end{bmatrix} = \begin{bmatrix} -2 \\ 3 \\ 6 \end{bmatrix} + t\begin{bmatrix} 1 \\ 7 \\ -\frac{k}{2} \end{bmatrix}$$

are **(a)** parallel? **(b)** orthogonal? Explain.

18. Does there exist a number k such that the planes with equations $kx + y - 2z = 1$ and $-x + 3y + z = 2$ are **(a)** parallel? **(b)** orthogonal? Explain.

19. Given $\mathsf{u} = \begin{bmatrix} -t \\ t-1 \\ t \\ -t+1 \end{bmatrix}$ and $\mathsf{v} = \begin{bmatrix} 2t \\ t \\ t+2 \\ 3 \end{bmatrix}$, find t so that u and v are orthogonal.

20. Let $\mathsf{u} = \begin{bmatrix} -1 \\ 2 \\ 1 \end{bmatrix}$ and $\mathsf{v} = \begin{bmatrix} 2 \\ -3 \\ -5 \end{bmatrix}$.

(a) Find all numbers k, if any, so that $\|k\mathsf{u} - \mathsf{v}\| = 4$. Give k accurate to two decimal places.

(b) Find exact value(s) of k, if any, so that $k\mathsf{u} + 2\mathsf{v}$ is orthogonal to u.

(c) Find the angle between the vectors, giving your answer in radians to two decimal places and in degrees to the nearest degree.

21. (a) State the Cauchy-Schwarz inequality.

(b) Prove that $(a + b + c)^2 \le 3(a^2 + b^2 + c^2)$ for any real numbers a, b, c.

22. State the Theorem of Pythagoras and use it to determine whether or not the triangle with vertices $A(1, 0, 1, 0)$, $B(0, 1, 1, 1)$, $C(1, 1, 1, 0)$ has a right angle.

23. Given the points $A(0, -4)$ and $B(2, 2)$, find all points C so that A, B and C are the vertices of a right angled triangle with right angle at A and hypotenuse of length $\sqrt{50}$.

24. Suppose u and v are vectors in 3-space with $\|\mathsf{u}\| = 2$, $\|\mathsf{v}\| = 3$ and a $30°$ angle between them. Find the following:

(a) $\mathsf{u} \cdot \mathsf{v}$ **(b)** $\|\mathsf{u} + \mathsf{v}\|^2$
(c) the area of the parallelogram with sides u and v.

25. In each case, find an equation of the plane containing A, B and C.

(a) $A(-5, 6, 1)$, $B(-1, 0, 1)$, $C(4, 2, 9)$

(b) $A(1,2,3)$, $B(-1,0,1)$, $C(1,-2,1)$.

26. Let ℓ_1 be the line with equation $\begin{bmatrix} x \\ y \\ z \end{bmatrix} = \begin{bmatrix} 3 \\ 0 \\ 2 \end{bmatrix} + t\begin{bmatrix} -2 \\ 1 \\ 1 \end{bmatrix}$ and let ℓ_2 be the line with equation $\begin{bmatrix} x \\ y \\ z \end{bmatrix} = \begin{bmatrix} 1 \\ 1 \\ -4 \end{bmatrix} + t\begin{bmatrix} 1 \\ -3 \\ 1 \end{bmatrix}$.

(a) Show that ℓ_1 and ℓ_2 are not parallel.

(b) Show that ℓ_1 and ℓ_2 do not intersect.

(c) Find an equation for the plane containing ℓ_1 that intersects ℓ_2 at $(1,1,-4)$.

27. Let u, v and w be nonzero vectors.

(a) If $\mathsf{u}\cdot\mathsf{w} = \mathsf{v}\cdot\mathsf{w}$, does it follow that $\mathsf{u} = \mathsf{v}$?

(b) If $\mathsf{u}\times\mathsf{w} = \mathsf{v}\times\mathsf{w}$, does it follow that $\mathsf{u} = \mathsf{v}$?

28. Let ℓ_1 be the line with equation $\begin{bmatrix} x \\ y \\ z \end{bmatrix} = \begin{bmatrix} 2 \\ -3 \\ -4 \end{bmatrix} + t\begin{bmatrix} 1 \\ 2 \\ 3 \end{bmatrix}$ and let ℓ_2 be the line through $P(2,-2,5)$ in the direction $\mathsf{d} = \begin{bmatrix} 2 \\ 3 \\ -3 \end{bmatrix}$.

(a) Show that ℓ_1 and ℓ_2 intersect by finding the point of intersection.

(b) Give a simple reason why ℓ_1 and the plane with equation $3x - 4y + z = 18$ must intersect. (No calculation is required.)

(c) Find the point of intersection of ℓ_1 and the plane in part (a).

29. The lines ℓ_1 and ℓ_2 with equations

$$\ell_1\colon \begin{bmatrix} x \\ y \\ z \end{bmatrix} = \begin{bmatrix} -7 \\ 2 \\ 1 \end{bmatrix} + t\begin{bmatrix} -3 \\ -2 \\ 2 \end{bmatrix} \quad \text{and} \quad \ell_2\colon \begin{bmatrix} x \\ y \\ z \end{bmatrix} = \begin{bmatrix} 4 \\ 6 \\ 3 \end{bmatrix} + t\begin{bmatrix} -1 \\ 1 \\ -4 \end{bmatrix}$$

have a point in common.

(a) Find this point.

(b) Find an equation of the plane that contains both lines.

30. Let π be a plane through the origin. Suppose $\mathsf{u} = \begin{bmatrix} 1 \\ 0 \\ 1 \end{bmatrix}$ and $\mathsf{v} = \begin{bmatrix} 2 \\ -1 \\ 1 \end{bmatrix}$ are two vectors in this plane.

(a) Find a nonzero vector w in this plane which is orthogonal to u.

(b) Find the projection of $\mathsf{q} = \begin{bmatrix} 1 \\ -1 \\ 4 \end{bmatrix}$ on π and express this as a linear combination of u and w.

31. Find the point on the line through $P(0,0,0)$ and $Q(1,-1,7)$ that is closest to $\mathsf{w} = \begin{bmatrix} 2 \\ 4 \\ 5 \end{bmatrix}$.

32. In each case, find an equation of the plane spanned by u and v.

(a) $\mathsf{u} = \begin{bmatrix} -2 \\ 3 \\ 1 \end{bmatrix}$ and $\mathsf{v} = \begin{bmatrix} 1 \\ 0 \\ 1 \end{bmatrix}$ **(b)** $\mathsf{u} = \begin{bmatrix} 2 \\ 1 \\ 3 \end{bmatrix}$ and $\mathsf{v} = \begin{bmatrix} -1 \\ 0 \\ 4 \end{bmatrix}$

(c) $\mathsf{u} = \begin{bmatrix} -1 \\ 2 \\ 3 \end{bmatrix}$ and $\mathsf{v} = \begin{bmatrix} 2 \\ 0 \\ 5 \end{bmatrix}$ **(d)** $\mathsf{u} = \begin{bmatrix} 1 \\ -1 \\ 4 \end{bmatrix}$ and $\mathsf{v} = \begin{bmatrix} 1 \\ 1 \\ 2 \end{bmatrix}$.

33. **(a)** Find the equation of the plane containing the points $P(1,0,1)$, $Q(-1,5,3)$ and $R(3,3,0)$.

    (b) Write down an equation of the line through the points P and Q of part (a).

34. **(a)** Find the equation of the plane containing the points $A(1,-1,2)$, $B(2,1,1)$ and $C(-1,-1,2)$.

    (b) Does this plane contain the line with equation $\begin{bmatrix} x \\ y \\ z \end{bmatrix} = \begin{bmatrix} \frac{3}{2} \\ \frac{3}{2} \\ 0 \end{bmatrix} + t\begin{bmatrix} 3 \\ 2 \\ -1 \end{bmatrix}$? Explain.

35. Find the equation of the plane through $P(2,-1,0)$ that contains the line with equation $\begin{bmatrix} x \\ y \\ z \end{bmatrix} = \begin{bmatrix} 1 \\ -1 \\ 1 \end{bmatrix} + t\begin{bmatrix} 2 \\ 1 \\ 3 \end{bmatrix}$.

36. Given the points $P(0,3,1)$, $Q(4,-2,3)$ and $R(1,3,-2)$, find

    (a) the cosine of the angle between $\overrightarrow{PQ}$ and $\overrightarrow{PR}$,

    (b) the equation of the line through P and Q,

    (c) the equation of the plane through P, Q and R,

    (d) the area of $\triangle PQR$.

37. Let $\theta = \angle PQR$ for $P(-1,0,4,0)$, $Q(-2,2,7,1)$ and $R(-3,-2,5,0)$. Show that θ is acute (that is, $0 \le \theta < \frac{\pi}{2}$).

38. Find the equation of the plane π parallel to $\mathsf{v} = \begin{bmatrix} 1 \\ 1 \\ -1 \end{bmatrix}$ and containing the points $A(1,0,-1)$ and $B(3,1,1)$.

39. **(a)** Find the angle between $\mathsf{u} = \begin{bmatrix} 3 \\ 4 \\ 0 \end{bmatrix}$ and $\mathsf{v} = \begin{bmatrix} 0 \\ 0 \\ 3 \end{bmatrix}$.

    (b) Find a vector of length 7 that is perpendicular to both u and v.

    (c) Find the equation of the plane passing through $P(1,2,3)$ and containing vectors parallel to u and v.

40. **(a)** Find the equation of the line through $(-2,-2,0)$ parallel to $\mathsf{u} = \begin{bmatrix} 1 \\ 1 \\ 1 \end{bmatrix}$.

    (b) Find the point of intersection point of the line in (a) and the plane with equation is $2x + y - z = 1$.

41. Explain why the planes with equations $x - y + 3z = 2$ and $3x + y + z = 2$ intersect and find the equation of the line of intersection.

42. Find the equation of the line where the planes with equations $x - y + 2z = -5$ and $2x + 3y - z = 0$ intersect.

43. Find the equation of the line through $(1, 2, -1)$ that is parallel to the line with equation $\begin{bmatrix} x \\ y \\ z \end{bmatrix} = \begin{bmatrix} -1 \\ 0 \\ 3 \end{bmatrix} + t \begin{bmatrix} -3 \\ 4 \\ 1 \end{bmatrix}$.

44. Show that the line with equation $\begin{bmatrix} x \\ y \\ z \end{bmatrix} = \begin{bmatrix} 3 \\ 2 \\ 1 \end{bmatrix} + t \begin{bmatrix} 1 \\ 3 \\ 1 \end{bmatrix}$ lies in the plane with equation $2x - y + z = 5$.

45. Let P and Q be the points $P(2, 1, -1)$ and $Q(1, -3, 4)$ of Euclidean 3-space. Let π be the plane with equation $x + 4y - 5z = 53$.

    (a) Find an equation of the line passing through the points P and Q.

    (b) Is the line in (a) perpendicular to π? Explain.

    (c) Where does the line through P and Q intersect the plane?

46. Find two nonzero orthogonal unit vectors in R^3 each orthogonal to $\mathsf{u} = \begin{bmatrix} -1 \\ 6 \\ 1 \end{bmatrix}$.

47. Find two nonzero orthogonal vectors that span the planes with the given equations.

    (a) $x + y + z = 0$ **(b)** $2x - 3y + z = 0$ **(c)** $x - 3y + 7z = 0$

48. In this exercise, π is the plane with equation $3x - 2y + 2z = 0$.

    (a) Find two nonzero orthogonal vectors in plane π.

    (b) What is the projection of $\mathsf{w} = \begin{bmatrix} -1 \\ 1 \\ 2 \end{bmatrix}$ on π?

    (c) What is the projection of $\mathsf{w} = \begin{bmatrix} x \\ y \\ z \end{bmatrix}$ on π?

49. Find the projection of $\mathsf{w} = \begin{bmatrix} 1 \\ 1 \\ 1 \end{bmatrix}$ on the plane π with equation $x - 7y + 4z = 0$ by the following two methods: 1) projecting w on a normal to π and 2) the method described in Problem 1.4.11.

50. Find the distance from $P(1, 1, 1)$ to the plane through $(2, -1, 3)$ with normal $\begin{bmatrix} 3 \\ 0 \\ 1 \end{bmatrix}$.

51. Find the distance from the point $P(-3, 1, 12)$ to the plane with equation $2x - y - 4z = 8$. Find also the point in the plane that is closest to P.

52. Find the distance of $P(1, 1, 1)$ from the plane π with equation $x - 3y + 4z = 10$ and the point of π closest to P.

53. Let P be the point $(-1, 2, 1)$ and π the plane with equation $4x - 3y = 7$. Write down the equation of a line that is contained in π.

54. In each case, find the distance from P to line ℓ and the point of ℓ closest to P.

(a) $P(2,-5,1)$; ℓ has equation $\begin{bmatrix} x \\ y \\ z \end{bmatrix} = \begin{bmatrix} -1 \\ 1 \\ 2 \end{bmatrix} + t \begin{bmatrix} -4 \\ 3 \\ 5 \end{bmatrix}$

(b) $P(-1,2,1)$; ℓ has equation $\begin{bmatrix} x \\ y \\ z \end{bmatrix} = \begin{bmatrix} -3 \\ 3 \\ 1 \end{bmatrix} + t \begin{bmatrix} -2 \\ 1 \\ 5 \end{bmatrix}$

(c) $P(4,-4,17)$; ℓ has equation $\begin{bmatrix} x \\ y \\ z \end{bmatrix} = \begin{bmatrix} -1 \\ 2 \\ 0 \end{bmatrix} + t \begin{bmatrix} 4 \\ -1 \\ 8 \end{bmatrix}$.

55. **(a)** Find the equation of the line passing through the point $P(3,-4,5)$ and perpendicular to the plane with equation $2x - 3y + z = 12$.

    (b) Find the distance from $P(3,-4,5)$ to the plane in (a).

56. Let ℓ be the line with equation $\begin{bmatrix} x \\ y \\ z \end{bmatrix} = \begin{bmatrix} -1 \\ 2 \\ 3 \end{bmatrix} + t \begin{bmatrix} 4 \\ 0 \\ -2 \end{bmatrix}$ and let π be the plane with equation $2x - 3y + 4z = 2$.

    (a) Show that ℓ is parallel to π.

    (b) Find a point P on ℓ and a point Q on π.

    (c) Find the shortest distance from ℓ to π.

57. Suppose (x_1, y_1, z_1), (x_2, y_2, z_2), ..., (x_n, y_n, z_n) are n points that lie on the plane π with equation $ax + by + cz = d$. Let $\overline{x} = \frac{x_1 + x_2 + \cdots + x_n}{n}$ denote the average of $x_1, \ldots, x_n$. Similarly, let $\overline{y}$ and $\overline{z}$ denote the averages of the ys and zs, respectively. Show that the points $(x_1 - \overline{x}, y_1 - \overline{y}, z_1 - \overline{z})$, $(x_2 - \overline{x}, y_2 - \overline{y}, z_2 - \overline{z})$, ..., $(x_n - \overline{x}, y_n - \overline{y}, z_n - \overline{z})$ lie on a plane parallel to π passing through the origin.

58. Find a point Q on the line with equation $\begin{bmatrix} x \\ y \\ z \end{bmatrix} = \begin{bmatrix} 4 \\ 1 \\ 0 \end{bmatrix} + t \begin{bmatrix} 2 \\ 1 \\ -1 \end{bmatrix}$ that is $\sqrt{19}$ units distant from $P(3,-,3,-2)$. Is Q unique?

59. Let u and v be vectors. Prove that $\|\mathsf{u} - \mathsf{v}\|^2 = \|\mathsf{u}\|^2 + \|\mathsf{v}\|^2 - 2\mathsf{u} \cdot \mathsf{v}$.

60. **(a)** Given two lines, how would you determine whether or not they are in the same plane? Write a short note.

    (b) Let ℓ_1 and ℓ_2 be the lines with equations

    $$\ell_1: \begin{bmatrix} x \\ y \\ z \end{bmatrix} = \begin{bmatrix} 1 \\ 2 \\ -1 \end{bmatrix} + t \begin{bmatrix} 0 \\ 3 \\ -2 \end{bmatrix}; \qquad \ell_2: \begin{bmatrix} x \\ y \\ z \end{bmatrix} = \begin{bmatrix} 1 \\ -1 \\ 4 \end{bmatrix} + t \begin{bmatrix} 2 \\ 2 \\ 5 \end{bmatrix}.$$

    Do these lines lie in the same plane? If yes, find the equation of this plane.

    (c) Answer (b) for

    $$\ell_1: \begin{bmatrix} x \\ y \\ z \end{bmatrix} = \begin{bmatrix} 1 \\ 2 \\ -1 \end{bmatrix} + t \begin{bmatrix} 1 \\ 3 \\ -2 \end{bmatrix}; \qquad \ell_2: \begin{bmatrix} x \\ y \\ z \end{bmatrix} = \begin{bmatrix} 1 \\ -1 \\ 4 \end{bmatrix} + t \begin{bmatrix} -2 \\ -6 \\ 4 \end{bmatrix}.$$

61. In each case, determine whether the vectors are linearly independent or linearly dependent. If linearly dependent, give an explicit equation that proves this.

(a) $u = \begin{bmatrix} 1 \\ 2 \\ 3 \end{bmatrix}, v = \begin{bmatrix} 4 \\ 5 \\ 6 \end{bmatrix}$

(b) $u = \begin{bmatrix} 1 \\ 2 \\ 3 \end{bmatrix}, v = \begin{bmatrix} 4 \\ 5 \\ 6 \end{bmatrix}, w = \begin{bmatrix} 7 \\ 8 \\ 9 \end{bmatrix}$

(c) $v_1 = \begin{bmatrix} 1 \\ 0 \\ 0 \\ 0 \end{bmatrix}, v_2 = \begin{bmatrix} 1 \\ 1 \\ 0 \\ 0 \end{bmatrix}, v_3 = \begin{bmatrix} 1 \\ 1 \\ 1 \\ 0 \end{bmatrix}$ and $v_4 = \begin{bmatrix} 1 \\ 1 \\ 1 \\ 1 \end{bmatrix}$

(d) the standard basis vectors in R^n.

62. Suppose u, v and w are linearly dependent vectors in R^n. Prove that $\sqrt{2}u$, $-v$ and $\frac{1}{3}w$ are also linearly dependent.

63. If $u_1, u_2, \ldots, u_m$ are linearly independent vectors, prove that the same is true for $u_1 + u_2, u_3, \ldots, u_m$.

64. Suppose $\{v_1, v_2, \ldots, v_k\}$ is any set of vectors and we add the zero vector to this set. Is the set $\{0, v_1, v_2, \ldots, v_k\}$ linearly independent? linearly dependent? can't say?

65. Suppose vectors u and v are linearly independent. Show that the vectors $w = 2u + 3v$ and $z = u + v$ are also linearly independent.

66. Assume u, v, w, x are nonzero vectors such that u and v are linearly dependent while v, w and x are linearly independent.

    (a) Show that there exists a scalar $k \neq 0$ such that $u = kv$.

    (b) Show that $\{u, w, x\}$ is linearly independent.

Chapter 2

Matrices and Linear Equations

2.1 The Algebra of Matrices

What is a Matrix?

The heart of linear algebra is systems of linear equations. Here is such a system:

$$\begin{array}{rcrcrcr} 2x_1 & + & x_2 & - & x_3 & = & 2 \\ 3x_1 & - & 2x_2 & + & 6x_3 & = & -1 \\ x_1 & & & - & 5x_3 & = & 1. \end{array} \tag{1}$$

The goal is to find numbers x_1, x_2, x_3 that make each equation true. Notice that each row is a statement about the dot product of a certain vector with the vector $\mathsf{x} = \begin{bmatrix} x_1 \\ x_2 \\ x_3 \end{bmatrix}$ of unknowns. Think of the coefficients in each row as the components of a vector. Specifically, let

$$\mathsf{r}_1 = \begin{bmatrix} 2 \\ 1 \\ -1 \end{bmatrix}, \quad \mathsf{r}_2 = \begin{bmatrix} 3 \\ -2 \\ 6 \end{bmatrix}, \quad \mathsf{r}_3 = \begin{bmatrix} 1 \\ 0 \\ -5 \end{bmatrix}.$$

Then our equations are just the assertions that

$$\begin{array}{rcr} \mathsf{r}_1 \cdot \mathsf{x} & = & 2 \\ \mathsf{r}_2 \cdot \mathsf{x} & = & -1 \\ \mathsf{r}_3 \cdot \mathsf{x} & = & 1. \end{array}$$

As we shall see in this chapter, these equations can also be written

$$\begin{bmatrix} 2 & 1 & -1 \end{bmatrix}\begin{bmatrix} x_1 \\ x_2 \\ x_3 \end{bmatrix} = 2$$

$$\begin{bmatrix} 3 & -2 & 6 \end{bmatrix}\begin{bmatrix} x_1 \\ x_2 \\ x_3 \end{bmatrix} = -1$$

$$\begin{bmatrix} 1 & 0 & -5 \end{bmatrix}\begin{bmatrix} x_1 \\ x_2 \\ x_3 \end{bmatrix} = 1$$

and summarized in the single *matrix equation*

$$\begin{bmatrix} 2 & 1 & -1 \\ 3 & -2 & 6 \\ 1 & 0 & -5 \end{bmatrix}\begin{bmatrix} x_1 \\ x_2 \\ x_3 \end{bmatrix} = \begin{bmatrix} 2 \\ -1 \\ 1 \end{bmatrix},$$

which is of the simple form

$$A\mathsf{x} = \mathsf{b},$$

with

$$A = \begin{bmatrix} 2 & 1 & -1 \\ 3 & -2 & 6 \\ 1 & 0 & -5 \end{bmatrix}, \quad \mathsf{x} = \begin{bmatrix} x_1 \\ x_2 \\ x_3 \end{bmatrix} \quad \text{and} \quad \mathsf{b} = \begin{bmatrix} 2 \\ -1 \\ 1 \end{bmatrix}.$$

Look at the numbers in A. They are the coefficients of x_1, x_2 and x_3 in the equations of system (1). In the form $A\mathsf{x} = \mathsf{b}$, the system doesn't look much different from the single equation $ax = b$ with a and b given real numbers and x the unknown. We begin this chapter with a discussion of *matrices.*

2.1.1 Definition. An $m \times n$ *matrix* is a rectangular arrangement of mn numbers into m rows and n numbers, all surrounded by square brackets.

For example, $\begin{bmatrix} 2 & 1 & -1 \\ 3 & -2 & 6 \\ 1 & 0 & -5 \end{bmatrix}$ is a 3×3 ("three by three") matrix, $\begin{bmatrix} 1 & 2 & 3 \\ 4 & 5 & 6 \end{bmatrix}$ is a 2×3 matrix and $\begin{bmatrix} -4 \\ 0 \\ 5 \end{bmatrix}$ is a 3×1 matrix. An n-dimensional vector $\begin{bmatrix} x_1 \\ x_2 \\ \vdots \\ x_n \end{bmatrix}$ is an $n \times 1$ matrix called a *column matrix.* A $1 \times n$ matrix is called a *row matrix* because it's a row: $\begin{bmatrix} x_1 & x_2 & \dots & x_n \end{bmatrix}$. Ignoring the technicality that a vector should properly be enclosed in brackets, it is often convenient to think of an $m \times n$ matrix as n column vectors written side by side and enclosed in a single pair of brackets. We write

$$A = \begin{bmatrix} \mathsf{a}_1 & \mathsf{a}_2 & \dots & \mathsf{a}_n \\ \downarrow & \downarrow & & \downarrow \\ & & & \end{bmatrix}$$

to indicate that the columns of A are the vectors $\mathsf{a}_1, \mathsf{a}_2, \ldots, \mathsf{a}_n$. Again ignoring the problem of brackets, each row of a matrix is the *transpose* of a vector in the following sense.

2.1.2 Definition. The *transpose* of an $m \times n$ matrix A is the $n \times m$ matrix whose rows are the columns of A in the same order. The transpose of A is denoted A^T.

2.1.3 Examples. • If $A = \begin{bmatrix} 1 & 2 & 3 \\ 4 & 5 & 6 \\ 7 & 8 & 9 \end{bmatrix}$, then $A^T = \begin{bmatrix} 1 & 4 & 7 \\ 2 & 5 & 8 \\ 3 & 6 & 9 \end{bmatrix}$ has the three rows of A for its columns.

- If $A = \begin{bmatrix} -1 & 2 \\ 3 & 4 \\ 0 & -5 \end{bmatrix}$, then $A^T = \begin{bmatrix} -1 & 3 & 0 \\ 2 & 4 & -5 \end{bmatrix}$.

- $\begin{bmatrix} -1 & 0 & 2 \end{bmatrix}^T = \begin{bmatrix} -1 \\ 0 \\ 2 \end{bmatrix}$; $\begin{bmatrix} 0 \\ 2 \\ 3 \\ 4 \end{bmatrix}^T = \begin{bmatrix} 0 & 2 & 3 & 4 \end{bmatrix}$. ☺

The last bullet shows that the transpose of a row matrix is a vector and the transpose of a vector is a row matrix.

Sometimes, we wish to think of a matrix in terms of its rows, in which case we write

$$A = \begin{bmatrix} \mathsf{a}_1^T & \rightarrow \\ \mathsf{a}_2^T & \rightarrow \\ \vdots & \\ \mathsf{a}_m^T & \rightarrow \end{bmatrix}$$

to indicate that the rows of A are the transposes of the vectors $\mathsf{a}_1, \mathsf{a}_2, \ldots, \mathsf{a}_m$. For example, if

$$A = \begin{bmatrix} 1 & 2 & 3 & 4 \\ 8 & 7 & 6 & 5 \\ 0 & 1 & 1 & 0 \end{bmatrix}, \tag{2}$$

then the columns of A are the vectors

$$\begin{bmatrix} 1 \\ 8 \\ 0 \end{bmatrix}, \begin{bmatrix} 2 \\ 7 \\ 1 \end{bmatrix}, \begin{bmatrix} 3 \\ 6 \\ 1 \end{bmatrix}, \begin{bmatrix} 4 \\ 5 \\ 0 \end{bmatrix}$$

and the rows of A are

$$\begin{bmatrix} 1 \\ 2 \\ 3 \\ 4 \end{bmatrix}^T, \begin{bmatrix} 8 \\ 7 \\ 6 \\ 5 \end{bmatrix}^T, \begin{bmatrix} 0 \\ 1 \\ 1 \\ 0 \end{bmatrix}^T.$$

Occasionally, we may want to refer to the individual entries of a matrix. With reference to the matrix A in (2), we call the numbers 1, 2, 3, 4 in the first row the $(1,1)$, $(1,2)$, $(1,3)$ and $(1,4)$ *entries* of A and write

$$a_{11} = 1, \quad a_{12} = 2, \quad a_{13} = 3, \quad a_{14} = 4.$$

In general, the entry in row i and column j of a matrix called A is denoted a_{ij} and it is common to write $A = [a_{ij}]$. If the matrix is B, the (i,j) entry will be denoted b_{ij} and we will write $B = [b_{ij}]$.

2.1.4 Example. Let $A = \begin{bmatrix} 1 & 2 \\ 3 & 4 \\ 5 & 6 \end{bmatrix}$. Then $a_{31} = 5$, $a_{12} = 2$ and $a_{22} = 4$. ☺

2.1.5 Example. Let $A = [i - j]$ be a 2×3 matrix, that is, $a_{ij} = i - j$. The $(1,1)$ entry of A is $a_{11} = 1 - 1 = 0$, the $(1,2)$ entry is $a_{12} = 1 - 2 = -1$ and so on, so $A = \begin{bmatrix} 0 & -1 & -2 \\ 1 & 0 & -1 \end{bmatrix}$. ☺

2.1.6 Example. If $A = [a_{ij}]$ and $B = A^T$, then $b_{ij} = a_{ji}$. ☺

Equality of Matrices

2.1.7 Matrices A and B are *equal* if and only if they have the same number of rows, the same number of columns and the corresponding entries are equal.

2.1.8 Example. Let $A = \begin{bmatrix} -1 & x \\ 2y & -3 \end{bmatrix}$ and $B = \begin{bmatrix} a & -4 \\ 4 & a-b \end{bmatrix}$. If $A = B$, then $-1 = a$, $x = -4$, $2y = 4$ and $a - b = -3$. Thus $a = -1$, $b = a + 3 = 2$, $x = -4$ and $y = 2$. ☺

Addition of Matrices

If matrices $A = [a_{ij}]$ and $B = [b_{ij}]$ have the same *size* (meaning same number of columns and same number of rows), then $A + B = [a_{ij} + b_{ij}]$, that is, the (i,j) entry of $A + B$ is the sum of the (i,j) entries of A and of B.

2.1.9 Example. If $A = \begin{bmatrix} 1 & 2 & 3 \\ 4 & 5 & 6 \end{bmatrix}$, $B = \begin{bmatrix} -2 & 3 & 0 \\ 2 & 1 & -1 \end{bmatrix}$ and $C = \begin{bmatrix} 1 & 0 \\ 0 & -1 \end{bmatrix}$, then $A + B = \begin{bmatrix} -1 & 5 & 3 \\ 6 & 6 & 5 \end{bmatrix}$, while $A + C$ and $C + B$ are not defined. **You can only add matrices of the same size.** ☺

For any positive integers m and n, there is an $m \times n$ *zero matrix*, which we denote with bold face type $\mathbf{0}$ (or $\underline{0}$ in handwritten work). This is the matrix with all entries 0. It has the property that $A + \mathbf{0} = \mathbf{0} + A = A$ for any $m \times n$ matrix A.

2.1.10 Example. The 2×2 zero matrix is $\begin{bmatrix} 0 & 0 \\ 0 & 0 \end{bmatrix}$. This matrix works just like zero should, with respect to 2×2 matrices. For example,

$$\begin{bmatrix} -1 & 2 \\ 3 & 5 \end{bmatrix} + \begin{bmatrix} 0 & 0 \\ 0 & 0 \end{bmatrix} = \begin{bmatrix} -1 & 2 \\ 3 & 5 \end{bmatrix} = \begin{bmatrix} 0 & 0 \\ 0 & 0 \end{bmatrix} + \begin{bmatrix} -1 & 2 \\ 3 & 5 \end{bmatrix}.$$ ☺

2.1.11 Example. The 2×3 zero matrix is the matrix $\mathbf{0} = \begin{bmatrix} 0 & 0 & 0 \\ 0 & 0 & 0 \end{bmatrix}$. For any 2×3 matrix $A = \begin{bmatrix} a & c & e \\ b & d & f \end{bmatrix}$,

$$A + \mathbf{0} = \begin{bmatrix} a & c & e \\ b & d & f \end{bmatrix} + \begin{bmatrix} 0 & 0 & 0 \\ 0 & 0 & 0 \end{bmatrix} = \begin{bmatrix} a & c & e \\ b & d & f \end{bmatrix} = A$$

and, similarly, $\mathbf{0} + A = A$. ☺

Scalar Multiplication of Matrices

We have already said that the term "scalar" in linear algebra means "number," usually real, but sometimes complex.[1] Just as there is a notion of multiplication of vectors by scalars, so also is there a *scalar multiplication* of matrices.

If $A = [a_{ij}]$ and c is a scalar, then $cA = [ca_{ij}]$ is the matrix obtained from A by multiplying every entry by c.

2.1.12 Example. $$(-4)\begin{bmatrix} 1 & -1 \\ 2 & 0 \\ -3 & 5 \end{bmatrix} = \begin{bmatrix} -4 & 4 \\ -8 & 0 \\ 12 & -20 \end{bmatrix}.$$ ☺

Just as -5 is the "additive inverse" of 5 in the sense that $5 + (-5) = 0$, so any matrix has an additive inverse.

2.1.13 Definition. The *additive inverse* of a matrix A is the matrix $(-1)A$, denoted $-A$ ("minus A") and defined by $-A = (-1)A$.

[1] "Complex" here doesn't mean hard! See Appendix A for a brief introduction to complex numbers.

2.1.14 Example. If $A = \begin{bmatrix} -1 & 3 \\ -2 & 4 \end{bmatrix}$, then $-A = (-1)\begin{bmatrix} -1 & 3 \\ -2 & 4 \end{bmatrix} = \begin{bmatrix} 1 & -3 \\ 2 & -4 \end{bmatrix}$. ☺

Here are the basic properties of addition and scalar multiplication of matrices, which are identical to those of vector addition and scalar multiplication recorded in Theorem 1.1.10.

2.1.15 Theorem (Properties of Matrix Addition and Scalar Multiplication). *Let A, B and C be matrices of the same size and let c and d be scalars.*

1. *(Closure under addition) $A + B$ is a matrix.*
2. *(Commutativity of addition) $A + B = B + A$.*
3. *(Addition is associative) $(A + B) + C = A + (B + C)$.*
4. *(Zero) There is a zero matrix 0 with the property that $A + 0 = 0 + A = A$.*
5. *(Additive inverse) For every matrix A, there is a matrix $-A$ called "minus A" with the property that $A + (-A) = 0$ is the zero matrix.*
6. *(Closure under scalar multiplication) cA is a matrix.*
7. *(Scalar associativity) $c(dA) = (cd)A$.*
8. *(One) $1A = A$.*
9. *(Distributivity) $c(A + B) = cA + cB$ and $(c + d)A = cA + dA$.*

2.1.16 Problem. Solve the equation $2A + 4X = 3B$ for X in general and in the special case $A = \begin{bmatrix} 1 & 2 \\ 3 & 5 \\ -6 & -7 \end{bmatrix}$ and $B = \begin{bmatrix} -2 & 4 \\ 6 & 2 \\ -8 & 2 \end{bmatrix}$.

Solution. Just as in ordinary algebra, $4X = 3B - 2A$, so $X = \frac{1}{4}(3B - 2A)$. In the special case,

$$3B = \begin{bmatrix} -6 & 12 \\ 18 & 6 \\ -24 & 6 \end{bmatrix}, \quad 2A = \begin{bmatrix} 2 & 4 \\ 6 & 10 \\ -12 & -14 \end{bmatrix} \text{ and } 3B - 2A = \begin{bmatrix} -8 & 8 \\ 12 & -4 \\ -12 & 20 \end{bmatrix},$$

$$\text{so } X = \frac{1}{4}\begin{bmatrix} -8 & 8 \\ 12 & -4 \\ -12 & 20 \end{bmatrix} = \begin{bmatrix} -2 & 2 \\ 3 & -1 \\ -3 & 5 \end{bmatrix}.$$ 👍

Multiplying Matrices

We explain how to multiply matrices in three stages.[2] First we show how to multiply a row matrix by a column matrix, then how to multiply an arbitrary matrix by a column matrix, and finally how to multiply two matrices of "appropriate" sizes.

STAGE 1: Row times column. Suppose $A = [\quad]$ is a $1 \times n$ row matrix and $B = \begin{bmatrix} \\ \\ \end{bmatrix}$ is an $n \times 1$ column matrix. Then B and A^T can be regarded as vectors. The matrix product AB is defined to be the dot product of the vectors A^T and B (within square brackets):

$$AB = [A^T \cdot B]. \tag{3}$$

2.1.17 Examples. • $\begin{bmatrix} 2 & 4 & -3 \end{bmatrix} \begin{bmatrix} 3 \\ -1 \\ 1 \end{bmatrix} = \begin{bmatrix} 2(3) + 4(-1) - 3(1) \end{bmatrix} = \begin{bmatrix} -1 \end{bmatrix}$

• $\begin{bmatrix} 3 & -1 & 2 & 5 \end{bmatrix} \begin{bmatrix} 4 \\ -1 \\ -1 \\ 2 \end{bmatrix} = \begin{bmatrix} 12 + 1 - 2 + 10 \end{bmatrix} = \begin{bmatrix} 21 \end{bmatrix}.$:-)

It is useful to rephrase the fact expressed in (3). Suppose u and v are vectors. Then u^T is a row matrix and v is a column matrix, so $u^T v = [u \cdot v]$ (because $(u^T)^T = u$). Except for the square brackets, it says that the dot product of vectors is a matrix product:

2.1.18 The dot product of vectors is a matrix product: $u \cdot v = u^T v$.

2.1.19 Example. If $u = \begin{bmatrix} 2 \\ 3 \\ -1 \end{bmatrix}$ and $v = \begin{bmatrix} 1 \\ 4 \\ 2 \end{bmatrix}$, then $u \cdot v = 12 = u^T v = \begin{bmatrix} 2 & 3 & -1 \end{bmatrix} \begin{bmatrix} 1 \\ 4 \\ 2 \end{bmatrix}$. :-)

STAGE 2: Matrix times column. Knowing how to multiply a row by a column enables us to multiply a matrix with any number of rows by a column. The product is computed row by row.

[2]While the definition may seem strange at first, matrix multiplication comes about very naturally. See Section 5.2.

Suppose the rows of A are the $n \times 1$ vectors $\mathsf{a}_1^T, \mathsf{a}_2^T, \ldots, \mathsf{a}_m^T$ and b is a column in R^n. The product

$$A\mathsf{b} = \begin{bmatrix} \mathsf{a}_1^T & \rightarrow \\ \mathsf{a}_2^T & \rightarrow \\ \vdots & \\ \mathsf{a}_m^T & \rightarrow \end{bmatrix} \begin{bmatrix} \mathsf{b} \\ \downarrow \\ \\ \\ \end{bmatrix} = \begin{bmatrix} \mathsf{a}_1^T\mathsf{b} \\ \mathsf{a}_2^T\mathsf{b} \\ \vdots \\ \mathsf{a}_m^T\mathsf{b} \end{bmatrix} \tag{4}$$

is a column, the column whose entries are the products of the rows of A with b.

2.1.20 Example. If $A = \begin{bmatrix} 0 & 1 & 3 \\ -1 & 1 & 2 \\ 2 & 0 & -1 \\ 4 & 5 & -1 \end{bmatrix}$ and $B = \begin{bmatrix} -1 \\ 2 \\ 3 \end{bmatrix}$, then AB is a column whose first entry is the product of the first row of A with B,

$$\begin{bmatrix} 0 & 1 & 3 \end{bmatrix} \begin{bmatrix} -1 \\ 2 \\ 3 \end{bmatrix} = 11,$$

whose second entry is the product of the second row of A with B,

$$\begin{bmatrix} -1 & 1 & 2 \end{bmatrix} \begin{bmatrix} -1 \\ 2 \\ 3 \end{bmatrix} = 9,$$

whose third entry is the product of the third row of A with B,

$$\begin{bmatrix} 2 & 0 & -1 \end{bmatrix} \begin{bmatrix} -1 \\ 2 \\ 3 \end{bmatrix} = -5$$

and whose fourth entry is the product of the fourth row of A with B,

$$\begin{bmatrix} 4 & 5 & -1 \end{bmatrix} \begin{bmatrix} -1 \\ 2 \\ 3 \end{bmatrix} = 3.$$

Thus $AB = \begin{bmatrix} 0 & 1 & 3 \\ -1 & 1 & 2 \\ 2 & 0 & -1 \\ 4 & 5 & -1 \end{bmatrix} \begin{bmatrix} -1 \\ 2 \\ 3 \end{bmatrix} = \begin{bmatrix} 11 \\ 9 \\ -5 \\ 3 \end{bmatrix}$. ☺

Before showing how to multiply matrices in general, we note two important consequences of the definition of AB when $B = \mathsf{x}$ is a column.

2.1.21 Any system of equations has the form $A\mathsf{x} = \mathsf{b}$.

As noted at the start of this section, the system

$$\begin{aligned} 2x_1 + x_2 - x_3 &= 2 \\ 3x_1 - 2x_2 + 6x_3 &= -1 \\ x_1 \phantom{{}+2x_2} - 5x_3 &= 1 \end{aligned}$$

can be written as $A\mathsf{x} = \mathsf{b}$, with

$$A = \begin{bmatrix} 2 & 1 & -1 \\ 3 & -2 & 6 \\ 1 & 0 & -5 \end{bmatrix}, \quad \mathsf{x} = \begin{bmatrix} x_1 \\ x_2 \\ x_3 \end{bmatrix} \quad \text{and} \quad \mathsf{b} = \begin{bmatrix} 2 \\ -1 \\ 1 \end{bmatrix}.$$

2.1.22 $A\mathsf{e}_i$ **is column** i **of** A**.**

Recall that the standard basis vectors in R^n are the vectors

$$\mathsf{e}_1 = \begin{bmatrix} 1 \\ 0 \\ 0 \\ \vdots \\ 0 \end{bmatrix}, \quad \mathsf{e}_2 = \begin{bmatrix} 0 \\ 1 \\ 0 \\ \vdots \\ 0 \end{bmatrix}, \quad \ldots, \quad \mathsf{e}_n = \begin{bmatrix} 0 \\ 0 \\ \vdots \\ 0 \\ 1 \end{bmatrix}.$$

Suppose $A = \begin{bmatrix} \mathsf{a}_1^T & \rightarrow \\ \mathsf{a}_2^T & \rightarrow \\ \vdots & \\ \mathsf{a}_m^T & \rightarrow \end{bmatrix}$ is an $m \times n$ matrix with rows as indicated. Then

$$A\mathsf{e}_1 = \begin{bmatrix} \mathsf{a}_1^T & \rightarrow \\ \mathsf{a}_2^T & \rightarrow \\ \vdots & \\ \mathsf{a}_m^T & \rightarrow \end{bmatrix} \begin{bmatrix} 1 \\ 0 \\ \vdots \\ 0 \end{bmatrix} = \begin{bmatrix} \mathsf{a}_1^T\mathsf{e}_1 \\ \mathsf{a}_2^T\mathsf{e}_1 \\ \vdots \\ \mathsf{a}_m^T\mathsf{e}_1 \end{bmatrix}. \tag{5}$$

The first component, $\mathsf{a}_1^T\mathsf{e}_1$, is just the first component of a_1, $\mathsf{a}_2^T\mathsf{e}_1$ is the first component of a_2 and so on. The vector on the right of (5) is just the first column of A. Similarly, $A\mathsf{e}_2$ is the second column of A and so on.

For example, if $n = 2$ and $A = \begin{bmatrix} 1 & 4 \\ 2 & 5 \\ 3 & 6 \end{bmatrix}$, then

$$A\mathsf{e}_1 = \begin{bmatrix} 1 & 4 \\ 2 & 5 \\ 3 & 6 \end{bmatrix} \begin{bmatrix} 1 \\ 0 \end{bmatrix} = \begin{bmatrix} 1 \\ 2 \\ 3 \end{bmatrix}$$

is the first column of A and

$$A\mathsf{e}_2 = \begin{bmatrix} 1 & 4 \\ 2 & 5 \\ 3 & 6 \end{bmatrix} \begin{bmatrix} 0 \\ 1 \end{bmatrix} = \begin{bmatrix} 4 \\ 5 \\ 6 \end{bmatrix}$$

is the second column of A.

2.1.23 Problem. Suppose A and B are $m \times n$ matrices such that $A\mathsf{e}_i = B\mathsf{e}_i$ for each $i = 1, 2, \ldots, n$. Prove that $A = B$.

Solution. Using 2.1.22, the hypothesis $A\mathsf{e}_i = B\mathsf{e}_i$ says that each column of A equals the corresponding column of B, so $A = B$.

READING CHECK 1. Suppose A and B are $m \times n$ matrices and $A\mathsf{x} = B\mathsf{x}$ for all vectors x. Prove that $A = B$.

STAGE 3: Matrix times matrix. Now that we know how to multiply a matrix by a column, it is straightforward to explain how to find the product of any two matrices of "compatible" sizes. Suppose that A is an $m \times n$ matrix and B is an $n \times p$ matrix. (Notice the repeated n here. That's what we mean by *compatible*.) Let B have columns $\mathsf{b}_1, \ldots, \mathsf{b}_p$. Then

2.1.24
$$AB = A\begin{bmatrix} \mathsf{b}_1 & \mathsf{b}_2 & \ldots & \mathsf{b}_p \\ \downarrow & \downarrow & & \downarrow \\ & & & \end{bmatrix} = \begin{bmatrix} A\mathsf{b}_1 & A\mathsf{b}_2 & \ldots & A\mathsf{b}_p \\ \downarrow & \downarrow & & \downarrow \\ & & & \end{bmatrix}$$

is the matrix whose columns are the products $A\mathsf{b}_1, A\mathsf{b}_2, \ldots, A\mathsf{b}_p$ of A with the columns of B.

2.1.25 Example. Let $A = \begin{bmatrix} 2 & 4 & -1 \\ 3 & 5 & 1 \\ 0 & 4 & -2 \end{bmatrix}$ and $B = \begin{bmatrix} 2 & 4 & 0 & 1 & -1 \\ 3 & -2 & 0 & 2 & 4 \\ 1 & 2 & -6 & 2 & 1 \end{bmatrix}$. Then the first column of AB is the product of A and the first column of B,

$$A\begin{bmatrix} 2 \\ 3 \\ 1 \end{bmatrix} = \begin{bmatrix} 2 & 4 & -1 \\ 3 & 5 & 1 \\ 0 & 4 & -2 \end{bmatrix}\begin{bmatrix} 2 \\ 3 \\ 1 \end{bmatrix} = \begin{bmatrix} 15 \\ 22 \\ 10 \end{bmatrix},$$

the second column of AB is the product of A and the second column of B,

$$A\begin{bmatrix} 4 \\ -2 \\ 2 \end{bmatrix} = \begin{bmatrix} 2 & 4 & -1 \\ 3 & 5 & 1 \\ 0 & 4 & -2 \end{bmatrix}\begin{bmatrix} 4 \\ -2 \\ 2 \end{bmatrix} = \begin{bmatrix} -2 \\ 4 \\ -12 \end{bmatrix},$$

and so on: $AB = \begin{bmatrix} 2 & 4 & -1 \\ 3 & 5 & 1 \\ 0 & 4 & -2 \end{bmatrix}\begin{bmatrix} 2 & 4 & 0 & 1 & -1 \\ 3 & -2 & 0 & 2 & 4 \\ 1 & 2 & -6 & 2 & 1 \end{bmatrix} = \begin{bmatrix} 15 & -2 & 6 & 8 & 13 \\ 22 & 4 & -6 & 15 & 18 \\ 10 & -12 & 12 & 4 & 14 \end{bmatrix}$.

This is a good spot to observe that if

$$A = \begin{bmatrix} \mathsf{a}_1^T & \rightarrow \\ \mathsf{a}_2^T & \rightarrow \\ \vdots & \\ \mathsf{a}_m^T & \rightarrow \end{bmatrix} \quad \text{and} \quad B = \begin{bmatrix} \mathsf{b}_1 & \mathsf{b}_2 & \ldots & \mathsf{b}_p \\ \downarrow & \downarrow & & \downarrow \end{bmatrix},$$

then

$$AB = \begin{bmatrix} \mathsf{a}_1^T\mathsf{b}_1 & \mathsf{a}_1^T\mathsf{b}_2 & \dots & \mathsf{a}_1^T\mathsf{b}_p \\ \mathsf{a}_2^T\mathsf{b}_1 & \mathsf{a}_2^T\mathsf{b}_2 & \dots & \mathsf{a}_2^T\mathsf{b}_p \\ \vdots & \vdots & & \vdots \\ \mathsf{a}_m^T\mathsf{b}_1 & \mathsf{a}_m^T\mathsf{b}_2 & \dots & \mathsf{a}_m^T\mathsf{b}_p \end{bmatrix} = \begin{bmatrix} \mathsf{a}_1 \cdot \mathsf{b}_1 & \mathsf{a}_1 \cdot \mathsf{b}_2 & \dots & \mathsf{a}_1 \cdot \mathsf{b}_p \\ \mathsf{a}_2 \cdot \mathsf{b}_1 & \mathsf{a}_2 \cdot \mathsf{b}_2 & \dots & \mathsf{a}_2 \cdot \mathsf{b}_p \\ \vdots & \vdots & & \vdots \\ \mathsf{a}_m \cdot \mathsf{b}_1 & \mathsf{a}_m \cdot \mathsf{b}_2 & \dots & \mathsf{a}_m \cdot \mathsf{b}_p \end{bmatrix}$$

is a matrix of dot products.

2.1.26 The (i, j) entry of AB is the product of row i of A and column j of B.

In Example 2.1.25, for example, the $(3, 2)$ entry of AB is the product of the third row of A and the second column of B:

$$-12 = \begin{bmatrix} 0 & 4 & -2 \end{bmatrix} \begin{bmatrix} 4 \\ -2 \\ 2 \end{bmatrix}.$$

We have used the term *compatible sizes* a couple of times to emphasize that only certain pairs of matrices can be multiplied. The fact expressed in 2.1.26 makes it implicit that the number of entries in the ith row of A must match the number of entries in the jth column of B; that is, the number of columns of A must equal the number of rows of B. Thus, if A is $m \times n$ and we wish to form the product AB for some other matrix B, then B must be $n \times p$ for some p, and then AB will be $m \times p$. Students sometimes use the equation

$$\frac{m}{\not{n}} \frac{\not{n}}{p} = \frac{m}{p}$$

to help remember the sizes of matrices that can be multiplied and also the size of the product.

2.1.27 Example. Suppose $A = \begin{bmatrix} 1 & -2 \\ 3 & -4 \end{bmatrix}$ and $B = \begin{bmatrix} -1 & 0 & 1 \\ 3 & 4 & 2 \end{bmatrix}$.

Then $AB = \begin{bmatrix} -7 & -8 & -3 \\ -15 & -16 & -5 \end{bmatrix}$ while BA is not defined: B is $2 \times ③$ but A is $② \times 2$. (The circled numbers are different.) ☺

2.1.28 Example. If $A = \begin{bmatrix} 1 \\ 2 \\ 3 \\ 4 \end{bmatrix}$ and $B = \begin{bmatrix} -2 & 0 & 4 & 1 \end{bmatrix}$, then A is 4×1 and B is 1×4, so we can compute both AB and BA: AB will be 4×4; BA will be 1×1.

$$AB = \begin{bmatrix} -2 & 0 & 4 & 1 \\ -4 & 0 & 8 & 2 \\ -6 & 0 & 12 & 3 \\ -8 & 0 & 16 & 4 \end{bmatrix}, \quad BA = \begin{bmatrix} 14 \end{bmatrix}.$$

The next theorem summarizes the properties of matrix multiplication. Most parts are quite straightforward with the exception of associativity, a proof of which is deferred to Section 5.2.

2.1.29 Theorem (Properties of Matrix Multiplication). *Let A, B and C be matrices of appropriate sizes.*

1. *Matrix multiplication is associative: $(AB)C = A(BC)$.*
2. *Matrix multiplication is not commutative, in general: there exist matrices A and B for which $AB \neq BA$.*
3. *Matrix multiplication distributes over addition: $(A + B)C = AC + BC$ and $A(B + C) = AB + AC$.*
4. *(Scalar associativity) $c(AB) = (cA)B = A(cB)$ for any scalar c.*
5. *For any positive integer n, there is an $n \times n$* identity *matrix denoted I_n with the property that $AI_n = A$ and $I_m B = B$ whenever these products are defined.*

A discussion of the failure of commutativity is in order. Examples are easy to find. If A is a 3×4 matrix, for example, and B is 4×7, then AB is defined while BA is not, so there is no chance for $AB = BA$. Other times AB and BA are both defined but still $AB \neq BA$; we saw this in Example 2.1.28. Even square matrices of the same size usually fail to commute. For example,

$$\begin{bmatrix} 1 & 2 \\ 3 & 4 \end{bmatrix} \begin{bmatrix} 0 & 1 \\ 1 & 0 \end{bmatrix} = \begin{bmatrix} 2 & 1 \\ 4 & 3 \end{bmatrix}$$

whereas

$$\begin{bmatrix} 0 & 1 \\ 1 & 0 \end{bmatrix} \begin{bmatrix} 1 & 2 \\ 3 & 4 \end{bmatrix} = \begin{bmatrix} 3 & 4 \\ 1 & 2 \end{bmatrix}.$$

The fact that matrices do not commute, in general, means special care must be exercised with algebraic expressions involving matrices. For instance, the product $ABAB$ in general cannot be simplified. Note also that

$$\begin{aligned}(A+B)^2 &= (A+B)(A+B)\\ &= A(A+B)+B(A+B) = AA+AB+BA+BB = A^2+AB+BA+B^2\end{aligned}$$

which, in general, is *not* the same as $A^2 + 2AB + B^2$. On the other hand, there are certain special matrices which commute with all matrices of an appropriate size. For example, if $0 = \begin{bmatrix} 0 & 0 \\ 0 & 0 \end{bmatrix}$ denotes the 2×2 zero matrix, then $A0 = 0$ for any $m \times 2$ matrix A and $0B = 0$ for any $2 \times n$ matrix B.

The $n \times n$ Identity Matrix

An *algebraic system* is a pair $(A, \star)$ where A is a set and $\star$ is a binary operation on A; that is, whenever a and b are in A, there is defined an element $a \star b$. An element e in such a system is called an *identity* if $a \star e = e \star a = a$ for all a. The number 1, for example, is an identity for the real numbers (when the operation is multiplication) and 0 is an identity when the operation is addition.

The matrix $I = \begin{bmatrix} 1 & 0 \\ 0 & 1 \end{bmatrix}$ is called the 2×2 *identity* matrix because it is an identity element for 2×2 matrices under multiplication. Just like $a1 = 1a = a$ for any real number a, so $AI = IA = A$ for any 2×2 matrix A. For example,

$$\begin{bmatrix} 1 & 2 \\ 3 & -4 \end{bmatrix}\begin{bmatrix} 1 & 0 \\ 0 & 1 \end{bmatrix} = \begin{bmatrix} 1 & 2 \\ 3 & -4 \end{bmatrix} = \begin{bmatrix} 1 & 0 \\ 0 & 1 \end{bmatrix}\begin{bmatrix} 1 & 2 \\ 3 & -4 \end{bmatrix}.$$

In general, the $n \times n$ *identity matrix* is the matrix whose columns are the standard basis vectors $\mathsf{e}_1, \mathsf{e}_2, \ldots, \mathsf{e}_n$ in order. It is denoted I_n or just I when there is no chance of confusion.

2.1.30 Examples. $I_2 = \begin{bmatrix} 1 & 0 \\ 0 & 1 \end{bmatrix}, \quad I_3 = \begin{bmatrix} 1 & 0 & 0 \\ 0 & 1 & 0 \\ 0 & 0 & 1 \end{bmatrix}, \quad I_4 = \begin{bmatrix} 1 & 0 & 0 & 0 \\ 0 & 1 & 0 & 0 \\ 0 & 0 & 1 & 0 \\ 0 & 0 & 0 & 1 \end{bmatrix}.$ ☺

It is not hard to see that $AI_n = A$ for any $n \times n$ matrix A. The columns of I_n are $\mathsf{e}_1, \mathsf{e}_2, \ldots, \mathsf{e}_n$, so the columns of AI_n are $A\mathsf{e}_1, A\mathsf{e}_2, \ldots, A\mathsf{e}_n$, and these are just the columns of A, in order, by 2.1.22. It is almost as easy to see that $I_nA = A$, but we leave our reasoning for later. (See Problem 2.1.36.)

2.1.31 Problem. Show that $A(bB + cC) = b(AB) + c(AC)$ for any matrices A, B, C and scalars c, d whenever all products here are defined.

Solution. $A(bB + cC) = A(bB) + A(cC)$ by Theorem 2.1.29, part 3
$= b(AB) + c(AC)$ by part 4.

Linear Combinations of Matrices

We can combine the operations of addition and scalar multiplication and form *linear combinations* of matrices just as with vectors.

2.1.32 **Definition.** A *linear combination* of matrices $A_1, A_2, \dots, A_k$ (all of the same size) is a matrix of the form $c_1A_1 + c_2A_2 + \cdots + c_kA_k$, where $c_1, c_2, \dots, c_k$ are scalars.

The matrix $bB + cC$ that appeared in Problem 2.1.31, for example, is a linear combination of the matrices B and C. That problem was designed to illustrate a general property of matrix multiplication.

Suppose that A is an $m \times n$ matrix, that $B_1, B_2, \dots, B_k$ are $n \times p$ matrices, and that $c_1, c_2, \dots, c_k$ are scalars. Then

$$A(c_1B_1 + c_2B_2 + \cdots + c_kB_k) = c_1(AB_1) + c_2(AB_2) + \cdots + c_k(AB_k). \tag{6}$$

The matrix $c_1B_1 + c_2B_2 + \cdots + c_kB_k$ is a linear combination of $B_1, \dots, B_k$ so the left side of (6) is the product of A with a linear combination of matrices. The right side of (6) is a linear combination of the matrices $AB_1, \dots, AB_k$ with the same coefficients. We summarize (6) by saying "matrix multiplication preserves linear combinations." We have an important use in mind for this observation.

Let A be an $m \times n$ matrix and let $\mathsf{x} = \begin{bmatrix} x_1 \\ x_2 \\ \vdots \\ x_n \end{bmatrix}$ be a vector in R^n. Remember that $\mathsf{x} = x_1\mathsf{e}_1 + x_2\mathsf{e}_2 + \cdots + x_n\mathsf{e}_n$—see 1.1.32. Since matrix multiplication preserves linear combinations,

$$A\mathsf{x} = A(x_1\mathsf{e}_1 + x_2\mathsf{e}_2 + \cdots + x_n\mathsf{e}_n) = x_1A\mathsf{e}_1 + x_2A\mathsf{e}_2 + \cdots + x_nA\mathsf{e}_n.$$

In view of 2.1.22, the expression on the right is x_1 times the first column of A, plus x_2 times the second column of A and, finally, x_n times the last column of A.

In the author's opinion, the fact that comes next is the most important one in this book!

The product $A\mathsf{x}$ of a matrix A and a vector x is a linear combination of the columns of A, the coefficients being the components of x:

2.1.33
$$\begin{bmatrix} \mathsf{a}_1 & \mathsf{a}_2 & \dots & \mathsf{a}_n \\ \downarrow & \downarrow & & \downarrow \\ & & & \end{bmatrix}\begin{bmatrix} x_1 \\ x_2 \\ \vdots \\ x_n \end{bmatrix} = x_1\begin{bmatrix} \mathsf{a}_1 \\ \downarrow \\ \end{bmatrix} + x_2\begin{bmatrix} \mathsf{a}_2 \\ \downarrow \\ \end{bmatrix} + \cdots + x_n\begin{bmatrix} \mathsf{a}_n \\ \downarrow \\ \end{bmatrix}.$$

Conversely, any linear combination of vectors is $A\mathsf{x}$ for some matrix A and some vector x.

READING CHECK 2. The last statement says any linear combination of vectors is $A\mathsf{x}$. What is A? What is x?

READING CHECK 3. Complete the following sentence:

"A linear combination of the columns of a matrix A is ____________."

2.1.34 **Example.** Let $A = \begin{bmatrix} 1 & 2 & 3 \\ 4 & 5 & 6 \\ 7 & 8 & 9 \end{bmatrix}$ and $\mathsf{x} = \begin{bmatrix} 2 \\ -1 \\ 4 \end{bmatrix}$. Then

$$\begin{aligned} A\mathsf{x} &= \begin{bmatrix} 1 & 2 & 3 \\ 4 & 5 & 6 \\ 7 & 8 & 9 \end{bmatrix}\begin{bmatrix} 2 \\ -1 \\ 4 \end{bmatrix} \\ &= \begin{bmatrix} 12 \\ 27 \\ 42 \end{bmatrix} = 2\begin{bmatrix} 1 \\ 4 \\ 7 \end{bmatrix} - \begin{bmatrix} 2 \\ 5 \\ 8 \end{bmatrix} + 4\begin{bmatrix} 3 \\ 6 \\ 9 \end{bmatrix} \\ &= \boxed{2} \times \text{ column } 1 + \boxed{-1} \times \text{ column } 2 + \boxed{4} \times \text{ column } 3. \end{aligned}$$

This is a linear combination of the columns of A, the coefficients, $2, -1, 4$, being the components of x.

More generally, $A\begin{bmatrix} x_1 \\ x_2 \\ x_3 \end{bmatrix} = x_1\begin{bmatrix} 1 \\ 4 \\ 7 \end{bmatrix} + x_2\begin{bmatrix} 2 \\ 5 \\ 8 \end{bmatrix} + x_3\begin{bmatrix} 3 \\ 6 \\ 9 \end{bmatrix}$. ☺

2.1.35 **Example.** $\begin{bmatrix} 2 & -3 \\ 4 & 5 \\ 6 & 8 \end{bmatrix}\begin{bmatrix} 5 \\ 2 \end{bmatrix} = 5\begin{bmatrix} 2 \\ 4 \\ 6 \end{bmatrix} + 2\begin{bmatrix} -3 \\ 5 \\ 8 \end{bmatrix}$. ☺

2.1.36 **Problem.** Let A be an $m \times n$ matrix and $I = I_m$ the $m \times m$ identity matrix. Explain why $IA = A$.

Solution. Let the columns of A be $\mathsf{a}_1, \mathsf{a}_2, \ldots, \mathsf{a}_n$. Then

$$IA = I\begin{bmatrix} \mathsf{a}_1 & \mathsf{a}_2 & \ldots & \mathsf{a}_n \\ \downarrow & \downarrow & & \downarrow \end{bmatrix} = \begin{bmatrix} I\mathsf{a}_1 & I\mathsf{a}_2 & \ldots & I\mathsf{a}_n \\ \downarrow & \downarrow & & \downarrow \end{bmatrix}$$

(see (2.1.24)). The first column of IA is $I\mathsf{a}_1$, which is a linear combination of the columns of I with coefficients the components of a_1. The columns of I are the standard basis vectors $\mathsf{e}_1, \mathsf{e}_2, \ldots, \mathsf{e}_m$, so if $\mathsf{a}_1 = \begin{bmatrix} a_{11} \\ a_{21} \\ \vdots \\ a_{m1} \end{bmatrix}$, then the first column of IA is $a_{11}\mathsf{e}_1 + a_{21}\mathsf{e}_2 + \cdots + a_{m1}\mathsf{e}_m = \begin{bmatrix} a_{11} \\ a_{21} \\ \vdots \\ a_{m1} \end{bmatrix} = \mathsf{a}_1$. Similarly, column k of IA is a_k, so $IA = A$.

2.1.37 Problem. Suppose A is an $m \times n$ matrix with columns $\mathsf{a}_1, \mathsf{a}_2, \ldots, \mathsf{a}_n$ and D is the $n \times n$ *diagonal* matrix whose $(1,1)$, $(2,2)$, ..., (n,n) entries are $d_1, \ldots, d_n$, respectively; that is,

$$A = \begin{bmatrix} \mathsf{a}_1 & \mathsf{a}_2 & \ldots & \mathsf{a}_n \\ \downarrow & \downarrow & & \downarrow \end{bmatrix} \quad \text{and} \quad D = \begin{bmatrix} d_1 & 0 & 0 & \ldots & 0 \\ 0 & d_2 & 0 & \ldots & 0 \\ 0 & 0 & d_3 & & 0 \\ \vdots & \vdots & & \ddots & \\ 0 & 0 & \ldots & 0 & d_n \end{bmatrix}.$$

Describe the columns of AD.

Solution. The first column of AD is the product $A\begin{bmatrix} d_1 \\ 0 \\ \vdots \\ 0 \end{bmatrix}$ of A and the first column of D. This product is the linear combination of the columns of A whose coefficients are $d_1, 0, \ldots, 0$. So the product is just $d_1\mathsf{a}_1$, the product of d_1 and the first column of A. The second column of AD is the product $A\begin{bmatrix} 0 \\ d_2 \\ 0 \\ \vdots \\ 0 \end{bmatrix}$ of A and the second column of D. This product is the linear combination of the columns of A whose coefficients are $0, d_2, 0, \ldots, 0$, which is $d_2\mathsf{a}_2$. Similarly, the third column of AD is $d_3\mathsf{a}_3$. In general, column k of AD is $d_k\mathsf{a}_k$, d_k times the kth column of A.

$$AD = \begin{bmatrix} d_1\mathsf{a}_1 & d_2\mathsf{a}_2 & \ldots & d_n\mathsf{a}_n \\ \downarrow & \downarrow & & \downarrow \end{bmatrix}.$$

READING CHECK 4. Write down the product $\begin{bmatrix} 1 & 2 & 3 & 4 \\ 1 & 2 & 3 & 4 \\ 1 & 2 & 3 & 4 \\ 1 & 2 & 3 & 4 \end{bmatrix} \begin{bmatrix} 2 & 0 & 0 & 0 \\ 0 & 3 & 0 & 0 \\ 0 & 0 & 4 & 0 \\ 0 & 0 & 0 & 5 \end{bmatrix}$ without any calculation.

Block Multiplication

To *partition* a matrix A is to group the entries of A into smaller arrays so that A can be viewed as a smaller matrix whose entries are themselves matrices. For example,

$$\left[\begin{array}{c|cc} 1 & 2 & 3 \\ 4 & 5 & 6 \\ 7 & 8 & 9 \end{array}\right], \quad \left[\begin{array}{ccc} 1 & 2 & 3 \\ \hline 4 & 5 & 6 \\ \hline 7 & 8 & 9 \end{array}\right], \quad \left[\begin{array}{c|cc} 1 & 2 & 3 \\ \hline 4 & 5 & 6 \\ 7 & 8 & 9 \end{array}\right], \quad \left[\begin{array}{cc|c} 1 & 2 & 3 \\ 4 & 5 & 6 \\ \hline 7 & 8 & 9 \end{array}\right]$$

are four ways to partition the matrix $A = \begin{bmatrix} 1 & 2 & 3 \\ 4 & 5 & 6 \\ 7 & 8 & 9 \end{bmatrix}$. In the first case, we view $A = \begin{bmatrix} A_1 & A_2 \end{bmatrix}$ as a 1×2 matrix whose entries are the matrices $A_1 = \begin{bmatrix} 1 \\ 4 \\ 7 \end{bmatrix}$ and $A_2 = \begin{bmatrix} 2 & 3 \\ 5 & 6 \\ 8 & 9 \end{bmatrix}$. We refer to A_1 and A_2 as the *blocks* of the partitioned matrix. In the last situation, we view A as a 2×2 matrix with entries the blocks $\begin{bmatrix} 1 & 2 \\ 4 & 5 \end{bmatrix}$, $\begin{bmatrix} 3 \\ 6 \end{bmatrix}$, $\begin{bmatrix} 7 & 8 \end{bmatrix}$ and $\begin{bmatrix} 9 \end{bmatrix}$.

We have met partitioned matrices before. To define the product of matrices A and B, we viewed B as a list of columns—see (2.1.24). For example, let $A = \begin{bmatrix} 1 & 0 & -1 \\ 0 & -2 & 3 \\ -3 & 1 & 2 \\ 4 & -1 & -1 \end{bmatrix}$ and $B = \begin{bmatrix} 1 & 4 & 7 \\ 2 & 5 & 8 \\ 3 & 6 & 9 \end{bmatrix}$. Thinking of $B = \begin{bmatrix} \mathsf{b}_1 & \mathsf{b}_2 & \mathsf{b}_3 \end{bmatrix}$ as a 1×3 matrix with blocks the columns $\mathsf{b}_1 = \begin{bmatrix} 1 \\ 2 \\ 3 \end{bmatrix}$, $\mathsf{b}_2 = \begin{bmatrix} 4 \\ 5 \\ 6 \end{bmatrix}$, $\mathsf{b}_3 = \begin{bmatrix} 7 \\ 8 \\ 9 \end{bmatrix}$, then

$$AB = A\begin{bmatrix} \mathsf{b}_1 & \mathsf{b}_2 & \mathsf{b}_3 \end{bmatrix} = \begin{bmatrix} A\mathsf{b}_1 & A\mathsf{b}_2 & A\mathsf{b}_3 \end{bmatrix} = \left[\begin{array}{c|c|c} -2 & -2 & -2 \\ 5 & 8 & 2 \\ 5 & 5 & 5 \\ -1 & 5 & 11 \end{array}\right].$$

On the other hand, suppose we view $B = \begin{bmatrix} B_1 & B_2 \end{bmatrix}$ with blocks $B_1 = \begin{bmatrix} 1 \\ 2 \\ 3 \end{bmatrix}$ and

$B_2 = \begin{bmatrix} 4 & 7 \\ 5 & 8 \\ 6 & 9 \end{bmatrix}$. Then

$$AB = A\begin{bmatrix} B_1 & B_2 \end{bmatrix} = \begin{bmatrix} AB_1 & AB_2 \end{bmatrix} = \left[\begin{array}{c|cc} -2 & -2 & -2 \\ 5 & 8 & 2 \\ 5 & 5 & 5 \\ -1 & 5 & 11 \end{array}\right].$$

Then again, we might view $A = \begin{bmatrix} A_1 \\ A_2 \end{bmatrix}$ as a 2×1 matrix with $A_1 = \begin{bmatrix} 1 & 0 & -1 \\ 0 & -2 & 3 \end{bmatrix}$ and $A_2 = \begin{bmatrix} -3 & 1 & 2 \\ 4 & -1 & -1 \end{bmatrix}$. This time, we have

$$AB = \begin{bmatrix} A_1 \\ A_2 \end{bmatrix} \begin{bmatrix} B_1 & B_2 \end{bmatrix} = \begin{bmatrix} A_1B_1 & A_1B_2 \\ A_2B_1 & A_2B_2 \end{bmatrix} = \left[\begin{array}{c|cc} -2 & -2 & -2 \\ 5 & 8 & 2 \\ \hline 5 & 5 & 5 \\ -1 & 5 & 11 \end{array}\right].$$

Notice that we computed AB just as we would compute the product of a 2×1 matrix and a 1×2.

We have been demonstrating what is termed "block multiplication" of partitioned matrices. In each of three cases, we partitioned A or B or both into blocks and found the product AB as if the blocks were just numbers instead of matrices.

Let

$$A = \left[\begin{array}{c|cc} -1 & 1 & 1 \\ -2 & 1 & 2 \\ \hline 3 & -1 & -2 \end{array}\right] \quad \text{and} \quad B = \left[\begin{array}{cc|cc} 0 & 3 & 3 & 0 \\ \hline 1 & 1 & -2 & 2 \\ 1 & 0 & 2 & -2 \end{array}\right]$$

partitioned as shown; that is, we think of $A = \begin{bmatrix} A_1 & A_2 \\ A_3 & A_4 \end{bmatrix}$ and $B = \begin{bmatrix} B_1 & B_2 \\ B_3 & B_4 \end{bmatrix}$ as 2×2 matrices with

$$A_1 = \begin{bmatrix} -1 \\ -2 \end{bmatrix}, \quad A_2 = \begin{bmatrix} 1 & 1 \\ 1 & 2 \end{bmatrix}, \quad A_3 = \begin{bmatrix} 3 \end{bmatrix}, \quad A_4 = \begin{bmatrix} -1 & -2 \end{bmatrix}$$

and

$$B_1 = \begin{bmatrix} 0 & 3 \end{bmatrix}, \quad B_2 = \begin{bmatrix} 3 & 0 \end{bmatrix}, \quad B_3 = \begin{bmatrix} 1 & 1 \\ 1 & 0 \end{bmatrix}, \quad B_4 = \begin{bmatrix} -2 & 2 \\ 2 & -2 \end{bmatrix}.$$

The we compute AB as if multiplying 2×2 matrices:

$$AB = \begin{bmatrix} A_1 & A_2 \\ A_3 & A_4 \end{bmatrix} \begin{bmatrix} B_1 & B_2 \\ B_3 & B_4 \end{bmatrix} = \begin{bmatrix} A_1B_1 + A_2B_3 & A_1B_2 + A_2B_4 \\ A_3B_1 + A_4B_3 & A_3B_2 + A_4B_4 \end{bmatrix}$$

with

$$\begin{aligned}
A_1B_1 + A_2B_3 &= \begin{bmatrix} -1 \\ -2 \end{bmatrix}\begin{bmatrix} 0 & 3 \end{bmatrix} + \begin{bmatrix} 1 & 1 \\ 1 & 2 \end{bmatrix}\begin{bmatrix} 1 & 1 \\ 1 & 0 \end{bmatrix} \\
&= \begin{bmatrix} 0 & -3 \\ 0 & -6 \end{bmatrix} + \begin{bmatrix} 2 & 1 \\ 3 & 1 \end{bmatrix} = \begin{bmatrix} 2 & -2 \\ 3 & -5 \end{bmatrix} \\
A_1B_2 + A_2B_4 &= \begin{bmatrix} -1 \\ -2 \end{bmatrix}\begin{bmatrix} 3 & 0 \end{bmatrix} + \begin{bmatrix} 1 & 1 \\ 1 & 2 \end{bmatrix}\begin{bmatrix} -2 & 2 \\ 2 & -2 \end{bmatrix} \\
&= \begin{bmatrix} -3 & 0 \\ -6 & 0 \end{bmatrix} + \begin{bmatrix} 0 & 0 \\ 2 & -2 \end{bmatrix} = \begin{bmatrix} -3 & 0 \\ -4 & -2 \end{bmatrix} \\
A_3B_1 + A_4B_3 &= \begin{bmatrix} 3 \end{bmatrix}\begin{bmatrix} 0 & 3 \end{bmatrix} + \begin{bmatrix} -1 & -2 \end{bmatrix}\begin{bmatrix} 1 & 1 \\ 1 & 0 \end{bmatrix} \\
&= \begin{bmatrix} 0 & 9 \end{bmatrix} + \begin{bmatrix} -3 & -1 \end{bmatrix} = \begin{bmatrix} -3 & 8 \end{bmatrix} \\
A_3B_2 + A_4B_4 &= \begin{bmatrix} 3 \end{bmatrix}\begin{bmatrix} 3 & 0 \end{bmatrix} + \begin{bmatrix} 9 & 0 \end{bmatrix}\begin{bmatrix} -2 & 2 \end{bmatrix} = \begin{bmatrix} 7 & 2 \end{bmatrix}
\end{aligned}$$

giving

$$AB = \left[\begin{array}{cc|cc} 2 & -2 & -3 & 0 \\ 3 & -5 & -4 & -2 \\ \hline -3 & 8 & 7 & 2 \end{array}\right].$$

In general, if matrices A and B can be partitioned into blocks such that

- the matrices have compatible sizes $m \times n$, $n \times p$ when viewed as partitioned matrices, and
- the blocks have compatible sizes in the sense that each entry of AB, which is the sum of matrix products, can be determined

then AB can be found in the obvious way. This is called "block multiplication." To illustrate, in the example just completed, we could not have partitioned A and B like this,

$$A = \left[\begin{array}{c|cc} -1 & 1 & 1 \\ -2 & 1 & 2 \\ \hline 3 & -1 & -2 \end{array}\right] \quad \text{and} \quad B = \left[\begin{array}{cc|cc} 0 & 3 & 3 & 0 \\ 1 & 1 & -2 & 2 \\ 1 & 0 & 2 & -2 \end{array}\right],$$

since the product of a 2×2 matrix and a 1×2 matrix is not defined. Nor could we have partitioned A and B as

$$A = \left[\begin{array}{c|cc} -1 & 1 & 1 \\ -2 & 1 & 2 \\ \hline 3 & -1 & -2 \end{array}\right] \quad \text{and} \quad B = \left[\begin{array}{ccc|c} 0 & 3 & 3 & 0 \\ 1 & 1 & -2 & 2 \\ \hline 1 & 0 & 2 & -2 \end{array}\right].$$

While the partitioned matrices have compatible sizes—the product of two 2×2 matrices is defined—the 1×1 entry of the product, for example, is

$\begin{bmatrix} -1 \\ -2 \end{bmatrix} \begin{bmatrix} 0 & 3 & 3 \\ 1 & 1 & -2 \end{bmatrix} + \begin{bmatrix} 1 & 1 \\ 1 & 2 \end{bmatrix} \begin{bmatrix} 1 & 0 & 2 \end{bmatrix}$ which cannot be found: the block sizes are not compatible.

In conclusion, the reader should be aware that block multiplication is more important than it might appear. In practice today, some enormous matrices, with hundreds of thousands of rows and columns and billions of entries appear. Some matrices are too large for the memory of many computers, but the same computers can handle the smaller matrices that appear as blocks.

Answers to Reading Checks

1. The given condition implies, in particular, that $Ae_i = Be_i$ for all i. So $A = B$ follows from Problem 2.1.23.

2. A is the matrix whose columns are the given vectors and x is the vector of coefficients.

3. A linear combination of the columns of a matrix A is $A\mathsf{x}$ for some vector x, the vector whose components are the coefficients in the linear combination.

4. Using the result of Problem 2.1.37,

$$\begin{bmatrix} 1 & 2 & 3 & 4 \\ 1 & 2 & 3 & 4 \\ 1 & 2 & 3 & 4 \\ 1 & 2 & 3 & 4 \end{bmatrix} \begin{bmatrix} 2 & 0 & 0 & 0 \\ 0 & 3 & 0 & 0 \\ 0 & 0 & 4 & 0 \\ 0 & 0 & 0 & 5 \end{bmatrix} \begin{bmatrix} 2 & 6 & 12 & 20 \\ 2 & 6 & 12 & 20 \\ 2 & 6 & 12 & 20 \\ 2 & 6 & 12 & 20 \end{bmatrix}.$$

True/False Questions

Decide, with as little calculation as possible, whether each of the following statements is true or false and, if you say "false," explain your answer. (Answers are at the back of the book.)

1. If A and B are matrices such that both AB and BA are defined, then A and B are square.

2. If u and v are vectors, then $\mathsf{u} \cdot \mathsf{v}$ is the matrix product $\mathsf{u}^T\mathsf{v}$.

3. If A and B are matrices with $A \neq 0$ but $AB = 0$, then $B = 0$.

4. If A and B are matrices and $AB = 0$, then $AXB = 0$ for any matrix X (of appropriate size).

5. The $(2,1)$ entry of $\begin{bmatrix} 1 & -1 & 0 \\ 2 & 3 & 4 \end{bmatrix}$ is -1.

6. If $A = \begin{bmatrix} 1 & 2 & 3 & 4 \\ 5 & 6 & 7 & 8 \\ 9 & 10 & 11 & 12 \end{bmatrix}$ and $\mathsf{b} = \begin{bmatrix} 0 \\ 0 \\ 1 \\ 0 \end{bmatrix}$, then $A\mathsf{b} = \begin{bmatrix} 3 \\ 7 \\ 11 \end{bmatrix}$.

7. If $A = \begin{bmatrix} 1 & 2 & 3 & 4 \\ 5 & 6 & 7 & 8 \\ 9 & 10 & 11 & 12 \end{bmatrix}$ and $\mathsf{x} = \begin{bmatrix} -2 \\ 3 \\ 5 \\ -8 \end{bmatrix}$, then $A\mathsf{x} = -2\begin{bmatrix} 1 \\ 5 \\ 9 \end{bmatrix} + 3\begin{bmatrix} 2 \\ 6 \\ 10 \end{bmatrix} + 5\begin{bmatrix} 3 \\ 7 \\ 11 \end{bmatrix} - 8\begin{bmatrix} 4 \\ 8 \\ 12 \end{bmatrix}$.

8. The statement $(AB)C = A(BC)$ says that matrix multiplication is commutative.

9. If A is an $m \times n$ matrix and $A\mathsf{x} = 0$ for every vector x in R^n, then $A = 0$ is the zero matrix.

10. If x_0 is a solution to the system $A\mathsf{x} = \mathsf{b}$ with $\mathsf{b} \neq 0$, then $2\mathsf{x}_0$ is also a solution.

EXERCISES

Answers to exercises marked [BB] can be found at the **Back of the Book.**

1. Given $A = B$, find x and y, if possible.

(a) [BB] $A = \begin{bmatrix} 2x - 3y & -y \\ x - y & x + y \\ -x + y & x + 2y \end{bmatrix}$, $B = \begin{bmatrix} 8 & 2 \\ 3 & -1 \\ -3 & -3 \end{bmatrix}$

(b) $A = \begin{bmatrix} 5x - 4y & -x \\ 3x + 3y & -3x - 3y \\ \frac{1}{2}x + y & x + 2y \\ -x + y & x - 2y \end{bmatrix}$, $B = \begin{bmatrix} 8 & 4 \\ -33 & 33 \\ -9 & -18 \\ -3 & 10 \end{bmatrix}$

(c) $A = \begin{bmatrix} 6x - y & x + y \\ -x - y & 3x - y \\ x + y & x - y \end{bmatrix}$, $B = \begin{bmatrix} 10 & 4 \\ -4 & 4 \\ 6 & 0 \end{bmatrix}$.

2. Given $A = B$, find x, y, a and b, if possible.

(a) [BB] $A = \begin{bmatrix} 2a + 3x & 2 \\ -3 & y + b \end{bmatrix}$, $B = \begin{bmatrix} 0 & -2a - 2x \\ a + x & 2 \end{bmatrix}$

(b) $A = \begin{bmatrix} a - b & x + a \\ y & 2 \\ -x & x + y \end{bmatrix}$, $B = \begin{bmatrix} -2 & b \\ x - 2 & x - y \\ a - b & b - a \end{bmatrix}$

(c) $A = \begin{bmatrix} a - b & x + a \\ y & 2 \\ -x & x + y \end{bmatrix}$, $B = \begin{bmatrix} 4 & b \\ x - 2 & x - y \\ a - b & b - a \end{bmatrix}$

(d) $A = \begin{bmatrix} 2a - b & x + y & -7 \\ a - b + x & y - a & 3y - b \end{bmatrix}$, $B = \begin{bmatrix} x + 4 & 2a - x & x + 3y \\ -y - a & b - 7x & -x \end{bmatrix}$.

3. Find the 2×3 matrix A for which
   (a) [BB] $a_{ij} = j$ **(b)** $a_{ij} = 3i - 2j - 1$
   (c) $a_{ij} = ij$ **(d)** a_{ij} is the larger of i and j.

4. Write down the 4×4 matrix A whose (i, j) entry is
   (a) $i - 3$ **(b)** $2^i - 3j$ **(c)** i^{j-1} **(d)** $i^2 j - ij^2$.

5. Let $\mathsf{v} = \begin{bmatrix} 1 \\ 2 \\ -3 \end{bmatrix}$ and $\mathsf{w} = \begin{bmatrix} 0 \\ 4 \\ 5 \end{bmatrix}$. Let $A = \begin{bmatrix} \mathsf{v} & \mathsf{w} \\ \downarrow & \downarrow \end{bmatrix}$ be the 3×2 matrix whose columns are v and w and let $B = \begin{bmatrix} \mathsf{v}^T & \rightarrow \\ \mathsf{w}^T & \rightarrow \end{bmatrix}$ be the 2×3 matrix whose rows are v^T and w^T. Find a_{11}, a_{13}, a_{21}, b_{32}, b_{12} and b_{22} if possible.

6. [BB] What can you say about the size of a matrix P if there is a vector x such that $\mathsf{x} + P\mathsf{x}$ is defined?

7. If A is an $n \times n$ matrix and x is a vector in R^n, what is the size of $\mathsf{x}^T A \mathsf{x}$ and why?

8. Suppose A is a matrix and $A^2 = 0$. Comment on the size of A.

9. A matrix P is called *idempotent* if $P^2 = P$.

   (a) Explain why an idempotent matrix must be square.

   (b) Suppose P is idempotent and $Q = I - 2P$, where I denotes the identity matrix the size of P. What is Q^2?

   (c) Let $P = \begin{bmatrix} -5 & 15 & -15 \\ -4 & 11 & -10 \\ -2 & 5 & -4 \end{bmatrix}$. Determine whether or not P is idempotent.

10. Find $A + B$ and $2A^T - B$, where possible, in each of the following cases.

    (a) [BB] $A = \begin{bmatrix} 2 & 1 & 1 \\ -1 & -1 & 4 \end{bmatrix}$; $B = \begin{bmatrix} 2 & -3 & 4 \\ -3 & 1 & 2 \end{bmatrix}$

    (b) $A = \begin{bmatrix} 8 & 2 \\ -6 & -3 \end{bmatrix}$; $B = \begin{bmatrix} 7 & -4 \\ -3 & -1 \end{bmatrix}$

    (c) $A = \begin{bmatrix} 3 & -4 & 8 \\ -7 & 1 & 0 \\ 2 & 2 & 3 \end{bmatrix}$; $B = \begin{bmatrix} 0 & -2 & 0 \\ 4 & 3 & -6 \\ 5 & 7 & 1 \end{bmatrix}$

    (d) $A = \begin{bmatrix} -2 & 4 & -3 \\ 4 & -6 & 8 \end{bmatrix}$; $B = \begin{bmatrix} -2 & -4 & 3 \\ 5 & 6 & 8 \end{bmatrix}$.

11. Let $A = \begin{bmatrix} 3 & x \\ y & 4 \end{bmatrix}$, $B = \begin{bmatrix} z & 3 \\ 4 & -1 \end{bmatrix}$, $C = \begin{bmatrix} -1 & 6 \\ 4 & 14 \end{bmatrix}$ and $D = \begin{bmatrix} -3 & 9 \\ 15 & -1 \end{bmatrix}$.

    If possible, find x, y and z so that

    (a) [BB] $3A - 2B = C$ **(b)** $3A - 2B = C^T$

    (c) [BB] $AB = C$ **(d)** $AB = D$

    (e) $BA - D + C = \begin{bmatrix} 2 & 7 \\ 2 & 3 \end{bmatrix}$ **(f)** $(DB)^T = 2C$.

12. Let $A = \begin{bmatrix} -1 & 2 \\ 3 & 4 \end{bmatrix}$, $B = \begin{bmatrix} 0 & -3 & 4 \\ 4 & 1 & 2 \end{bmatrix}$, $C = \begin{bmatrix} 1 & 2 & 0 \\ -1 & 0 & 5 \end{bmatrix}$.
Compute AB, AC, $AB + AC$, $B + C$ and $A(B + C)$.

13. [BB] Let $A = \begin{bmatrix} 2 & 2 & 2 \end{bmatrix}$ and $B = \begin{bmatrix} 4 \\ 5 \\ 6 \end{bmatrix}$. If possible, find AB, BA, A^TB, AB^T, A^TB^T and B^TA^T.

14. Let $A = \begin{bmatrix} 1 & 2 \\ 3 & 4 \end{bmatrix}$ and $B = \begin{bmatrix} 0 & -1 \\ 5 & -2 \end{bmatrix}$. Compute $(2A + B)(A - 3B)$ and $2A^2 - 5AB - 3B^2$. Are these equal? What is the correct expansion of $(2A + B)(A - 3B)$?

15. Find AB, BA, A^TB^T, B^TA^T, $(AB)^T$ and $(BA)^T$ (if defined) in each of the following cases:

   (a) [BB] $A = \begin{bmatrix} 1 & 2 & -4 \\ -2 & 1 & 0 \end{bmatrix}$; $B = \begin{bmatrix} -3 & 1 \\ 2 & -2 \\ 0 & 1 \end{bmatrix}$

   (b) $A = \begin{bmatrix} 1 & 2 & -1 \\ -1 & 1 & 1 \\ 4 & 3 & -1 \end{bmatrix}$; $B = \begin{bmatrix} 2 & 1 \\ -1 & 4 \\ 2 & 3 \end{bmatrix}$

   (c) $A = \begin{bmatrix} 2 \\ -2 \\ 0 \\ 3 \end{bmatrix}$; $B = \begin{bmatrix} 1 & 2 & 3 & -1 \end{bmatrix}$

   (d) $A = \begin{bmatrix} -2 & 3 \\ 5 & 8 \end{bmatrix}$; $B = \begin{bmatrix} 3 & 0 \\ -5 & 3 \\ 7 & 4 \end{bmatrix}$.

16. Let $A = \begin{bmatrix} a & b & c \\ d & e & f \end{bmatrix}$ and $B = \begin{bmatrix} x & u \\ y & v \\ z & w \end{bmatrix}$.

   (a) [BB] What are the $(1, 1)$ and $(2, 2)$ entries of AB? These are called the *diagonal entries* of AB.

   (b) What are the diagonal entries of BA?

   (c) Verify that the sum of the diagonal entries of AB and BA are equal. [The sum of the diagonal entries of a matrix is called its *trace*.]

17. Let $A = \begin{bmatrix} 0 & -1 \\ 2 & 1 \end{bmatrix}$, $B = \begin{bmatrix} 1 & 0 \\ 3 & -1 \end{bmatrix}$ and $C = \begin{bmatrix} -2 & -3 \\ 0 & 5 \end{bmatrix}$.

   (a) Is C a linear combination of A and B?

   (b) Find a nontrivial linear combination of A, B and C that equals $\begin{bmatrix} 0 & 0 \\ 0 & 0 \end{bmatrix}$. (*Nontrivial* means that not all coefficients can be 0.)

18. Let $A = \begin{bmatrix} 1 & 0 & 1 & 1 \\ 2 & 1 & 0 & 1 \\ 1 & 2 & 1 & 0 \end{bmatrix}$, $\mathsf{x} = \begin{bmatrix} 1 \\ 2 \\ 1 \\ 1 \end{bmatrix}$ and $\mathsf{y} = \begin{bmatrix} 1 \\ 0 \\ 1 \end{bmatrix}$.

(a) Find $A\mathsf{x}$.

(b) Find $\mathsf{x}^T A^T$ by two different methods.

(c) Find $\mathsf{y}^T A\mathsf{x}$.

19. Find A^{10} with $A = \begin{bmatrix} 1 & 1 \\ 0 & 1 \end{bmatrix}$.

20. Find the matrix A, given $\left(\begin{bmatrix} 3 & 7 \\ 4 & 2 \\ -4 & 1 \end{bmatrix} - 2A\right)^T = \begin{bmatrix} 4 & 6 & 6 \\ 12 & 18 & 6 \end{bmatrix}$.

21. Find all matrices A and B with the property that $2A + 3B^T = \begin{bmatrix} 3 & -1 \\ 1 & 2 \end{bmatrix}$.

22. Write each of the following vectors in the form $A\mathsf{x}$, for a suitable matrix A and vector x. [This question tests your understanding of the very important fact expressed in 2.1.33.]

(a) [BB] $3\begin{bmatrix} -1 \\ 0 \end{bmatrix} + 4\begin{bmatrix} 2 \\ 1 \end{bmatrix} - \begin{bmatrix} 1 \\ 8 \end{bmatrix}$ (b) $5\begin{bmatrix} 1 \\ 3 \\ -5 \\ 7 \\ 9 \end{bmatrix} + \begin{bmatrix} -2 \\ 1 \\ 1 \\ 0 \\ 7 \end{bmatrix} - 3\begin{bmatrix} 2 \\ 4 \\ 6 \\ 0 \\ -1 \end{bmatrix}$

(c) $-2\begin{bmatrix} 1 \\ 1 \\ -2 \end{bmatrix} + 0\begin{bmatrix} 2 \\ 0 \\ 1 \end{bmatrix} + \begin{bmatrix} -3 \\ 1 \\ 1 \end{bmatrix} - \begin{bmatrix} 4 \\ 2 \\ 7 \end{bmatrix} + 3\begin{bmatrix} 6 \\ 5 \\ 4 \end{bmatrix}$

(d) $4\begin{bmatrix} 5 \\ 0 \\ 7 \\ 6 \end{bmatrix} - 2\begin{bmatrix} 2 \\ 3 \\ -1 \\ 7 \end{bmatrix} + \begin{bmatrix} 0 \\ 0 \\ 0 \\ 1 \end{bmatrix} - \begin{bmatrix} 5 \\ 1 \\ 4 \\ 5 \end{bmatrix}$.

23. In each case, find $A\mathsf{x}$ and express this vector as a linear combination of the columns of A.

(a) [BB] $A = \begin{bmatrix} 1 & 2 & 3 \\ -2 & 0 & 4 \end{bmatrix}$, $\mathsf{x} = \begin{bmatrix} -4 \\ 4 \\ 1 \end{bmatrix}$ (b) $A = \begin{bmatrix} 1 & 2 \\ 3 & 4 \\ 5 & 6 \end{bmatrix}$, $\mathsf{x} = \begin{bmatrix} 3 \\ -1 \end{bmatrix}$

(c) $A = \begin{bmatrix} 1 & -1 & 2 & 0 \\ 3 & 1 & 0 & -1 \\ 5 & 0 & 0 & -3 \end{bmatrix}$, $\mathsf{x} = \begin{bmatrix} 2 \\ 1 \\ -1 \\ 2 \end{bmatrix}$.

24. Express each of the following systems of equations in the form $A\mathsf{x} = \mathsf{b}$.

(a) [BB]
$$\begin{aligned} 2x - y &= 10 \\ 3x + 5y &= 7 \\ -x + y &= -3 \\ 2x - 5y &= 1 \end{aligned}$$

(b)
$$\begin{aligned} y - z &= 1 \\ 3x + 5y - 2z &= -9 \\ x + z &= 0 \end{aligned}$$

(c)
$$\begin{aligned} x - y + z - 2w &= 0 \\ 3x + 5y - 2z + 2w &= 1 \\ 3y + 4z - 7w &= -1 \end{aligned}$$

(d) $$\begin{aligned} x_1 + x_2 + x_3 + x_4 - x_5 &= 3 \\ 3x_1 - x_2 + 5x_3 - x_5 &= -2. \end{aligned}$$

25. [BB] The parabola with equation $y = ax^2 + bx + c$ passes through the points $(1,4)$ and $(2,8)$. Find an equation $A\mathsf{x} = \mathsf{b}$ whose solution is the vector $\begin{bmatrix} a \\ b \\ c \end{bmatrix}$.

26. It is conjectured that the points $(\frac{\pi}{3}, 2)$ and $(-\frac{\pi}{4}, 1)$ lie on a curve with equation of the form $y = a \sin x + b \cos x$. Assuming this is the case, find an equation $A\mathsf{x} = \mathsf{b}$ whose solution is $\mathsf{x} = \begin{bmatrix} a \\ b \end{bmatrix}$. [You are not being asked to find a and b.]

27. Express $\begin{bmatrix} 7 \\ -5 \\ 6 \end{bmatrix}$ as a linear combination of the columns of

$A = \begin{bmatrix} 3 & 3 & -7 & -3 \\ 6 & 1 & 5 & 0 \\ -5 & 2 & -6 & 9 \end{bmatrix}$. Then find a vector x so that $A\mathsf{x} = \begin{bmatrix} 7 \\ -5 \\ 6 \end{bmatrix}$.

28. Express $\begin{bmatrix} -1 \\ 14 \\ 2 \end{bmatrix}$ as a linear combination of the columns of $A = \begin{bmatrix} 1 & 2 & 0 \\ 0 & 3 & 1 \\ 0 & 0 & 1 \end{bmatrix}$.

Then find a vector x such that $A\mathsf{x} = \begin{bmatrix} -1 \\ 14 \\ 2 \end{bmatrix}$.

29. Suppose x_1, x_2 and x_3 are four-dimensional vectors and A is a 2×4 matrix such that $A\mathsf{x}_1 = \begin{bmatrix} -1 \\ 2 \end{bmatrix}$ and $A\mathsf{x}_2 = \begin{bmatrix} 10 \\ 12 \end{bmatrix}$. Find AX, given $X = \begin{bmatrix} \mathsf{x}_1 & \mathsf{x}_2 \\ \downarrow & \downarrow \end{bmatrix}$.

30. Let $A = \begin{bmatrix} 1 & 2 & 3 \\ 1 & 2 & 3 \\ 1 & 2 & 3 \end{bmatrix}$ and $D = \begin{bmatrix} -7 & 0 & 0 \\ 0 & 8 & 0 \\ 0 & 0 & 5 \end{bmatrix}$. Find AD without calculation and explain your reasoning.

31. Let $A = \begin{bmatrix} 1 & 2 & 3 \\ 4 & 5 & 6 \\ 7 & 8 & 9 \end{bmatrix}$ and $D = \begin{bmatrix} 2 & 0 & 0 \\ 0 & 3 & 0 \\ 0 & 0 & -1 \end{bmatrix}$. Find AD and DA without explicitly multiplying matrices. Explain your reasoning.

32. Suppose A, B, C are $n \times n$ matrices and A *commutes* with both B and C; that is, $AB = BA$ and $AC = CA$. Prove that A commutes with $B + C$ [BB] and also with BC.

33. Suppose a 2×2 matrix A commutes with $B = \begin{bmatrix} 0 & -1 \\ 1 & 0 \end{bmatrix}$. Discover what A must look like.

34. **(a)** A student is asked to prove that if A and B are matrices that commute, then $A^2B^2 = B^2A^2$. He presents the following argument.

$$\begin{aligned} A^2B^2 &= B^2A^2 \\ AABB &= BBAA \\ A(AB)B &= B(BA)A \\ A(BA)B &= B(AB)A \\ (AB)(AB) &= (BA)(BA) \\ ABAB &= ABAB. \end{aligned}$$

What is wrong with this "proof?"

(b) Prove that if A and B are matrices that commute, then $A^2B^2 = B^2A^2$.

35. [BB] Let A be a square matrix. Does the equation $(A-3I)(A+2I) = 0$ imply $A = 3I$ or $A = -2I$? Answer by considering the matrix $A = \begin{bmatrix} 3 & -1 \\ 0 & -2 \end{bmatrix}$.

36. **(a)** Suppose A and B are matrices such that $AB = 0$. Does this imply $A = 0$ or $B = 0$? If you say "yes," give a proof. If you say "no," give an example of two nonzero matrices A and B for which $AB = 0$.

(b) If A is a 2×2 matrix, $B = \begin{bmatrix} 1 & 2 \\ 0 & -1 \end{bmatrix}$ and $AB = 0$, show that $A = 0$. Does this result contradict your answer to part (a)?

(c) If X and Y are any 2×2 matrices and B is the matrix of part (b) and if $XB = YB$, show that $X = Y$.

37. Shown below is a matrix $A = \begin{bmatrix} 1 & 2 & 3 & 4 \\ 4 & 5 & 6 & 7 \\ 7 & 8 & 9 & 0 \\ 0 & 1 & 2 & 3 \end{bmatrix}$ partitioned in eight ways. In case (a), for instance, $A = \begin{bmatrix} A_1 \\ A_2 \end{bmatrix}$ is partitioned as a 2×1 matrix with blocks $A_1 = \begin{bmatrix} 1 & 2 & 3 & 4 \end{bmatrix}$ and $A_2 = \begin{bmatrix} 4 & 5 & 6 & 7 \\ 7 & 8 & 9 & 0 \\ 0 & 1 & 2 & 3 \end{bmatrix}$. Describe in a similar way the partitioning of A in the seven remaining cases.

(a) $A = \left[\begin{array}{cccc} 1 & 2 & 3 & 4 \\ \hline 4 & 5 & 6 & 7 \\ 7 & 8 & 9 & 0 \\ 0 & 1 & 2 & 3 \end{array}\right]$ **(b)** [BB] $A = \left[\begin{array}{cc|cc} 1 & 2 & 3 & 4 \\ 4 & 5 & 6 & 7 \\ 7 & 8 & 9 & 0 \\ 0 & 1 & 2 & 3 \end{array}\right]$

(c) $A = \left[\begin{array}{c|ccc} 1 & 2 & 3 & 4 \\ 4 & 5 & 6 & 7 \\ 7 & 8 & 9 & 0 \\ \hline 0 & 1 & 2 & 3 \end{array}\right]$ **(d)** [BB] $A = \left[\begin{array}{c|cc|c} 1 & 2 & 3 & 4 \\ 4 & 5 & 6 & 7 \\ \hline 7 & 8 & 9 & 0 \\ 0 & 1 & 2 & 3 \end{array}\right]$

(e) $A = \left[\begin{array}{ccc|c} 1 & 2 & 3 & 4 \\ \hline 4 & 5 & 6 & 7 \\ 7 & 8 & 9 & 0 \\ \hline 0 & 1 & 2 & 3 \end{array}\right]$ **(f)** $A = \left[\begin{array}{cc|cc} 1 & 2 & 3 & 4 \\ 4 & 5 & 6 & 7 \\ \hline 7 & 8 & 9 & 0 \\ 0 & 1 & 2 & 3 \end{array}\right]$

(g) $A = \left[\begin{array}{c|c|c|c} 1 & 2 & 3 & 4 \\ \hline 4 & 5 & 6 & 7 \\ 7 & 8 & 9 & 0 \\ 0 & 1 & 2 & 3 \end{array}\right]$ **(h)** $A = \left[\begin{array}{c|ccc} 1 & 2 & 3 & 4 \\ \hline 4 & 5 & 6 & 7 \\ \hline 7 & 8 & 9 & 0 \\ \hline 0 & 1 & 2 & 3 \end{array}\right]$.

38. In which of the eight cases described in Exercise 37 is it possible to find A^2? Find A^2 with the given partitioning when this is possible.

39. Shown below is a matrix $A = \begin{bmatrix} 1 & 2 \\ 0 & -2 \\ 4 & 4 \end{bmatrix}$ and a matrix $B = \begin{bmatrix} 2 & 3 & 2 & -2 \\ 1 & 0 & 2 & 4 \end{bmatrix}$ each partitioned in four ways.

(a) [BB] $A = \left[\begin{array}{c|c} 1 & 2 \\ 0 & -2 \\ 4 & 4 \end{array}\right]$ **i.** $B = \left[\begin{array}{cccc} 2 & 3 & 2 & -2 \\ \hline 1 & 0 & 2 & 4 \end{array}\right]$

(b) $A = \left[\begin{array}{cc} 1 & 2 \\ \hline 0 & -2 \\ \hline 4 & 4 \end{array}\right]$ **ii.** $B = \left[\begin{array}{c|c|cc} 2 & 3 & 2 & -2 \\ 1 & 0 & 2 & 4 \end{array}\right]$

(c) $A = \left[\begin{array}{cc} 1 & 2 \\ 0 & -2 \\ \hline 4 & 4 \end{array}\right]$ **iii.** $B = \left[\begin{array}{c|ccc} 2 & 3 & 2 & -2 \\ \hline 1 & 0 & 2 & 4 \end{array}\right]$

(d) $A = \left[\begin{array}{c|c} 1 & 2 \\ \hline 0 & -2 \\ 4 & 4 \end{array}\right]$ **iv.** $B = \left[\begin{array}{c|cc|c} 2 & 3 & 2 & -2 \\ \hline 1 & 0 & 2 & 4 \end{array}\right]$.

In which cases is the product AB **not** defined? Explain clearly.

40. Suppose that $A = \begin{bmatrix} A_1 & A_2 \\ A_3 & A_4 \end{bmatrix}$ and $B = \begin{bmatrix} B_1 & B_2 \\ B_3 & B_4 \end{bmatrix}$ are partitioned as 2×2 matrices and suppose that with this partitioning, the $(1,1)$ entry of AB is defined. Show that AB is defined.

41. If A and B are matrices of appropriate sizes and c is a scalar, then $c(AB) = (cA)B$—see part 4 of Theorem 2.1.29. This is more useful than it perhaps looks. Illustrate by computing the product XY with $X = \frac{1}{3}\begin{bmatrix} -3 & 5 & -2 \\ 4 & -21 & 16 \\ -1 & 14 & -8 \end{bmatrix}$ and $Y = \begin{bmatrix} 0 & 2 & 3 \\ 4 & 5 & 6 \\ 7 & 8 & 9 \end{bmatrix}$.

42. In part 7 of Theorem 2.1.15, *scalar associativity* is defined as the property $c(dA) = (cd)A$ where A is a matrix and c, d are scalars. Show that you understand what this means by illustrating with $A = \begin{bmatrix} -7 & 2 \\ 1 & 5 \end{bmatrix}$, $c = -2$ and $d = 9$.

Critical Reading

43. Let A be any $m \times n$ matrix. Let x be a vector in R^n and let y be a vector in R^m.

    (a) [BB] Explain the equation $\mathsf{y} \cdot (A\mathsf{x}) = \mathsf{y}^T A\mathsf{x}$.

    (b) Prove that $\|A\mathsf{x}\|^2 = \mathsf{x}^T A^T A\mathsf{x}$.

44. Your linear algebra professor wants to compute an assignment grade for each student in her class. She has a spreadsheet with the names of the 38 students in her class written down the side the numbers 1–8 written across the top, representing the eight assignments given during the semester, and cells completed with appropriate marks. Explain how matrix multiplication could be used to do the job

    (a) assuming the assignments are to be weighted equally,

    (b) assuming that assignments 4 and 8 are worth twice each of the others.

    (Each assignment mark is out of 100 and an assignment grade for the semester should be out of 100 as well.)

45. Let $A = \begin{bmatrix} \mathsf{a}_1 & \mathsf{a}_2 & \cdots & \mathsf{a}_n \\ \downarrow & \downarrow & & \downarrow \end{bmatrix}$ be an $m \times n$ matrix with columns as indicated. Let

$$B = \begin{bmatrix} b_{11} & b_{12} & \cdots & b_{1n} \\ 0 & b_{22} & \cdots & b_{2n} \\ \vdots & & \ddots & \vdots \\ 0 & 0 & & b_{nn} \end{bmatrix} \text{ and } C = \begin{bmatrix} c_{11} & 0 & \cdots & 0 \\ c_{12} & c_{22} & & 0 \\ \vdots & \vdots & \ddots & 0 \\ c_{n1} & c_{n2} & & c_{nn} \end{bmatrix}.$$

    (a) [BB] How are the columns of AB related to the columns of B?

    (b) How are the columns of AC related to the columns of A?

46. Suppose $A = \begin{bmatrix} \mathsf{a}_1 & \mathsf{a}_2 & \cdots & \mathsf{a}_n \\ \downarrow & \downarrow & & \downarrow \end{bmatrix}$ is a matrix and suppose that the entries in each row of A sum to 0. Show that $A\mathsf{v} = 0$, where $\mathsf{v} = \begin{bmatrix} 1 \\ 1 \\ \vdots \\ 1 \end{bmatrix}$.

47. Let $m \geq 1$ and $n \geq 1$ be positive integers. Let x be a nonzero vector in R^n and let y be any vector in R^m. Prove that there exists an $m \times n$ matrix A such that $A\mathsf{x} = \mathsf{y}$. [Hint: Never forget 2.1.33!]

2.2 Application: Generating Codes with Matrices

This section focuses on one area where matrices naturally come into play. When information is transmitted electronically, it is usual for words to be converted to numbers, which are more easily manipulated. One can imagine many ways in which this conversion might take place, for example, by assigning to "A" the number 1, to "B" the number 2, ..., to Z the number 26, and to other characters such as punctuation marks and spaces, numbers 27 and higher. In practice, numbers are usually expressed in "base 2" and hence appear as sequences of 0s and 1s. Specifically, abc is the base 2 representation of the number $a(2^2)+b(2)+c(2^0)$, just as abc is the base 10 representation of the number $a(10^2)+b(10)+c(10^0)$. For example, in base ten, 327 means $3(10^2)+2(10)+7(1)$. In base two, 011 means $0(2^2)+1(2)+1(1)=3$.

READING CHECK 1. Using four digits for each representation, write down the base 2 representations of all the integers from 0 to 10. The list begins 0000, 0001, 0010.

In base two, addition and multiplication are defined "modulo 2" ("mod 2," for short), that is, by these tables:

+	0	1
0	0	1
1	1	0

·	0	1
0	0	0
1	0	1

The only surprise here is that $1+1=0 \pmod 2$.

READING CHECK 2. What is $-1 \pmod 2$? (Think of what -1 means.)

A *word* is a sequence of 0s and 1s, such as 001, and a *code* is a set of words. In this book, we assume that all the words of a code have the same length and that the sum of code words is also a code word. (Such a code is said to be *linear.*) For example, a code might consist of all sequences of 0s and 1s of length three. There are eight such words, all written out below:

$$000, 001, 010, 011, 100, 101, 110, 111.$$

If we identify the word abc with the column vector $\begin{bmatrix} a \\ b \\ c \end{bmatrix}$, then the words of this code are just the eight possible vectors with three components, each a 0 or 1. Equivalently, the words are linear combinations of the columns of the 3×3 identity matrix $I=\begin{bmatrix} 1&0&0\\0&1&0\\0&0&1 \end{bmatrix}$. In this sense, I is called a *generator matrix* for the code because the words of the code are generated by (are linear combinations of) the columns. Here's a more interesting example of a generator matrix.

2.2.1 Example. (The Hamming (7, 4) Code[3]) Let $G = \begin{bmatrix} 1 & 0 & 0 & 0 \\ 0 & 1 & 0 & 0 \\ 0 & 0 & 1 & 0 \\ 0 & 0 & 0 & 1 \\ 0 & 1 & 1 & 1 \\ 1 & 0 & 1 & 1 \\ 1 & 1 & 0 & 1 \end{bmatrix}$ and let C be the code whose words are linear combinations of the columns of G, that is, vectors of the form $G\mathsf{x}$, $\mathsf{x} = \begin{bmatrix} x_1 \\ x_2 \\ x_3 \\ x_4 \end{bmatrix}$. Since $G\mathsf{x}+G\mathsf{y} = G(\mathsf{x}+\mathsf{y})$, the sum of words is a word, so C is indeed a (linear) code. The word 1010101 is in C because $\begin{bmatrix} 1 \\ 0 \\ 1 \\ 0 \\ 1 \\ 0 \\ 1 \end{bmatrix} = G \begin{bmatrix} 1 \\ 0 \\ 1 \\ 0 \end{bmatrix}$ is the sum of the first and third columns of G (remember that we add mod 2). On the other hand, 0101011 is not in C because it is not $G\mathsf{x}$ for any x. To see this, compute

$$G\begin{bmatrix} x_1 \\ x_2 \\ x_3 \\ x_4 \end{bmatrix} = \begin{bmatrix} x_1 \\ x_2 \\ x_3 \\ x_4 \\ x_2 + x_3 + x_4 \\ x_1 + x_3 + x_4 \\ x_1 + x_2 + x_4 \end{bmatrix},$$

and notice that if this equals $\begin{bmatrix} 0 \\ 1 \\ 0 \\ 1 \\ 0 \\ 1 \\ 1 \end{bmatrix}$, then $x_1 = 0$, $x_2 = 1$, $x_3 = 0$ and $x_4 = 1$. Examining last components, we would need $x_1 + x_2 + x_4 = 1$, which is not the case here because $x_1 + x_2 + x_4 = 0$. ☺

In real communication, over wires or through space, reliability of transmission is a problem. Words get garbled, 0s become 1s and 1s are changed to 0s. How can one be sure that the message received was the message sent? Such concerns make the construction of codes that can detect and correct errors an important (and lucrative!) activity. A common approach in the theory of "error-correcting codes" is to append to each code word being sent a sequence of *bits* (a sequence of one or more 0s and 1s), the purpose of which is to make errors obvious and, ideally, to make it possible to correct errors.

[3] After the American Richard Hamming (1915–1998)

Thus each transmitted word consists of a sequence of information bits, the message word itself, followed by a sequence of error correction bits.

Here's an example of what we have in mind. Suppose the message words we wish to transmit each have length four. We might attach to each message word a fifth bit which is the sum of the first four (mod 2). Thus, if the message word is 1010, we send 10100, the final 0 being the sum $1+0+1+0$ (mod 2) of the bits in the message word. If 11111 is received, an error is immediately detected because the sum of the first four digits is $0 \neq 1$ (mod 2). This simple attaching of a "parity check" digit makes any single error obvious, but not correctable, because the recipient has no idea which of the first four digits is wrong. Also, a single parity check digit can't detect two errors.

READING CHECK 3. Give an example to illustrate.

Each word in the Hamming code in Example 2.2.1 actually consists of four information bits followed by **three** error correction bits. For example, the actual message in 1010101 is 1010, the first four bits. The final 101 not only makes it possible to detect an error, but also to identify and correct it! We have essentially seen already how this comes about. A word of the Hamming code corresponds to a vector of the form

$$\begin{bmatrix} x_1 \\ x_2 \\ x_3 \\ x_4 \\ x_2+x_3+x_4 \\ x_1+x_3+x_4 \\ x_1+x_2+x_4 \end{bmatrix},$$

so, if the fifth bit of the transmitted sequence is not the sum of the second, third and fourth, or if the sixth bit is not the sum of the first, third and fourth, or if the last transmitted bit is not the sum of the first, second and fourth, the received word is not correct. Algebraically, $\begin{bmatrix} x_1 \\ x_2 \\ x_3 \\ x_4 \\ x_5 \\ x_6 \\ x_7 \end{bmatrix}$ is in C if and only if

$$\begin{aligned} x_5 &= x_2+x_3+x_4 \\ x_6 &= x_1+x_3+x_4 \\ x_7 &= x_1+x_2+x_4, \end{aligned} \tag{1}$$

equations which can be written

$$\begin{aligned} x_2+x_3+x_4+x_5 &= 0 \\ x_1+x_3+x_4+x_6 &= 0 \\ x_1+x_2+x_4+x_7 &= 0, \end{aligned} \tag{2}$$

using the fact that $-1 = 1 \pmod 2$. Let $H = \begin{bmatrix} 0 & 1 & 1 & 1 & 1 & 0 & 0 \\ 1 & 0 & 1 & 1 & 0 & 1 & 0 \\ 1 & 1 & 0 & 1 & 0 & 0 & 1 \end{bmatrix}$. Then $H\begin{bmatrix} x_1 \\ x_2 \\ x_3 \\ x_4 \\ x_5 \\ x_6 \\ x_7 \end{bmatrix} = \begin{bmatrix} x_2 + x_3 + x_4 + x_5 \\ x_1 + x_3 + x_4 + x_6 \\ x_1 + x_2 + x_4 + x_7 \end{bmatrix}$. So a vector $\mathsf{x} = \begin{bmatrix} x_1 \\ x_2 \\ x_3 \\ x_4 \\ x_5 \\ x_6 \\ x_7 \end{bmatrix}$ corresponds to a code word if and only if $H\mathsf{x} = \mathsf{0}$. The equations in (2) are called *parity check equations* and the matrix H a *parity check matrix.*

2.2.2 Definition. If C is a code with $n \times k$ generator matrix G, a *parity check matrix* for C is an $(n-k) \times n$ matrix H with the property that x is in C if and only if $H\mathsf{x} = \mathsf{0}$.

For example, in the Hamming code, the generator matrix G is 7×4 ($n = 7$, $k = 4$) and the parity check matrix H is 3×7.

Neither parity check equations nor a parity check matrix are unique, and some are more useful than others. One can show that the parity check equations in (2) are *equivalent* to

$$\begin{aligned} x_4 + x_5 + x_6 + x_7 &= 0 \\ x_2 + x_3 + x_6 + x_7 &= 0 \\ x_1 + x_3 + x_5 + x_7 &= 0 \end{aligned} \tag{3}$$

in the sense that a vector $\mathsf{x} = \begin{bmatrix} x_1 \\ x_2 \\ x_3 \\ x_4 \\ x_5 \\ x_6 \\ x_7 \end{bmatrix}$ satisfies equations (2) if and only if it satisfies equations (3). (Ways to discover this equivalence will be explained in Section 2.4.) The parity check matrix corresponding to system (3) is

$$H = \begin{bmatrix} 0 & 0 & 0 & 1 & 1 & 1 & 1 \\ 0 & 1 & 1 & 0 & 0 & 1 & 1 \\ 1 & 0 & 1 & 0 & 1 & 0 & 1 \end{bmatrix}. \tag{4}$$

Notice that the columns of H are the numbers $1, 2, 3, 4, 5, 6, 7$ written in base 2 in their natural order. Remember that if x is really a code word, then $H\mathsf{x} = \mathsf{0}$. Since $H\mathsf{x} = \begin{bmatrix} x_4 + x_5 + x_6 + x_7 \\ x_2 + x_3 + x_6 + x_7 \\ x_1 + x_3 + x_5 + x_7 \end{bmatrix}$, if x_1 is transmitted incorrectly (but all other bits are correct), then $H\mathsf{x} = \begin{bmatrix} 0 \\ 0 \\ 1 \end{bmatrix}$. Notice that the number 001 in base 2 is 1. If x_2 is in error (but all other bits are correct), then $H\mathsf{x} = \begin{bmatrix} 0 \\ 1 \\ 0 \end{bmatrix}$. The number 010 in base 2 is 2. If an error is made in x_3 (but all other bits are correct), then $H\mathsf{x} = \begin{bmatrix} 0 \\ 1 \\ 1 \end{bmatrix}$. The number 011 in base 2 is 3. Generally, if there is just one error in transmission, and that error is digit i, then $H\mathsf{x}$ is the number i in base 2. (It is not hard to understand why this is the case. See Exercise 7.) This particular parity check matrix not only tells us whether or not a transmitted message is correct, but it also allows us to make the correction. Clever? The $(7,4)$ Hamming code is an example of a "single error correcting code."

READING CHECK 4. Assuming the $(7,4)$ Hamming code is being used, if the word 0010101 is received, what was the message word?

Answers to Reading Checks

1. 0000, 0001, 0010, 0011, 0100, 0101, 0110, 0111, 1000, 1001, 1010.

2. -1 is the number that added to 1 gives 0. Working mod 2, that means 1: $-1 = 1 \pmod{2}$.

3. Suppose the message word we wish to send is 1010. Thus we transmit 10100. If there are errors in the second and fourth digits, say, then 11110 is received. The parity check digit, the final 0, is the sum of the first four, so the two errors would not be caught.

4. Since $H\begin{bmatrix} 0 \\ 0 \\ 1 \\ 0 \\ 1 \\ 0 \\ 1 \end{bmatrix} = \begin{bmatrix} 0 \\ 0 \\ 1 \end{bmatrix} \pmod{2}$, the received message 0010 is incorrect. Since 001 = 1 in base 2, the error is in the first digit. The message is 1010.

EXERCISES

*Answers to exercises marked [BB] can be found at the **Back of the Book**.*

1. Let C be the code with generator matrix $G = \begin{bmatrix} 1 & 0 & 0 \\ 0 & 1 & 0 \\ 0 & 0 & 1 \\ 1 & 0 & 1 \\ 1 & 1 & 1 \end{bmatrix}$.

Which of the following words are in C?

(a) [BB] 10011 **(b)** 10100 **(c)** 11101 **(d)** 11001 **(e)** 01100.

2. Let C be the code with generator matrix $G = \begin{bmatrix} 1 & 0 & 0 \\ 0 & 1 & 0 \\ 0 & 0 & 1 \\ 0 & 1 & 1 \\ 1 & 1 & 0 \end{bmatrix}$.

   Which of the following words are in C?

   (a)[BB] 00011 **(b)** 10101 **(c)** 01010 **(d)** 11010 **(e)** 01101.

3. [BB] Let G be the generator matrix of Exercise 1. Find a parity check matrix H.

4. Let G be the generator matrix of Exercise 2. Find a parity check matrix H. Assuming that the information in the code which G generates is carried in the first three bits, try to find an H that permits the correction of a single error in an information bit.

5. Assuming the $(7, 4)$ Hamming code is the code being used and the transmitted word contains at most one error, try to find the correct word assuming each of the following words is received.

   (a) [BB] 1010101 **(b)** 0101010 **(c)** [BB] 1011111
   (d) 1111111 **(e)** 1101101 **(f)** 0010001.

Critical Reading

6. When vectors consist of just 0s and 1s and we add mod 2, a linear combination of vectors is just a sum of vectors. Why?

7. Let $H = \begin{bmatrix} 0 & 0 & 0 & 1 & 1 & 1 & 1 \\ 0 & 1 & 1 & 0 & 0 & 1 & 1 \\ 1 & 0 & 1 & 0 & 1 & 0 & 1 \end{bmatrix}$ be the parity check matrix of (4) whose columns are the integers from 1 to 7 written in their natural order in base 2. Let $\mathsf{x} = \begin{bmatrix} x_1 \\ x_2 \\ x_3 \\ x_4 \\ x_5 \\ x_6 \\ x_7 \end{bmatrix}$ with $H\mathsf{x} = 0$. Suppose exactly one of $x_1, x_2, x_3, \ldots, x_7$ is transmitted incorrectly. Explain why $H\mathsf{x}$ identifies the wrong digit. [Hint: Since only 0s and 1s are being transmitted, if x_i is wrong, the digit sent is $x_i + 1$.]

8. Suppose a parity check matrix for a code has the property that every four columns are linearly independent. Show that the code is *doubly error correcting*, that is, it will correct errors in two digits.

2.3 The Inverse and Transpose of a Matrix

We have noted that the $n \times n$ identity matrix I behaves with respect to matrices the way the number 1 does with numbers. For any matrix A, $IA = A$ whenever IA is defined, just as $1a = a$ for any real number a. How far does this analogy extend? If a is a nonzero real number, there is another number b with $ab = 1$. The number $b = \frac{1}{a} = a^{-1}$ is called the (multiplicative) inverse of a. This idea motivates the idea of *inverse of a matrix.*

2.3.1 Definition. A matrix A is *invertible* or *has an inverse* if there is another matrix B such that $AB = I$ and $BA = I$. The matrix B is called the *inverse* of A and we write $B = A^{-1}$.

2.3.2 Remarks. 1. Since matrix multiplication is not commutative, we **never** write something like $\frac{1}{A}$ or $\frac{AB}{C}$, the latter being ambiguous: Would $\frac{AB}{C} = \frac{1}{C}AB$ or $AB\frac{1}{C}$?

2. It is not explicit in the definition, but the two Is in Definition 2.3.1 are the same. We shall see later, in Corollary 4.5.7, that if $AB = I_m$, the $m \times m$ identity matrix, and $BA = I_n$, then $m = n$. In particular, invertible matrices are always square.

3. (**Inverses are unique**) In the definition, we called B **the** inverse of A. This language suggests that a matrix can have at most one inverse. This is indeed the case, and here's why.

   Suppose A has inverses B and C. Then $AB = BA = I$ and $AC = CA = I$. Now evaluate the product BAC using the fact that matrix multiplication is associative. On the one hand, $BAC = (BA)C = IC = C$, while, on the other, $BAC = B(AC) = BI = B$. So $B = C$.

2.3.3 Example. Let $A = \begin{bmatrix} 1 & -2 \\ 2 & -3 \end{bmatrix}$ and $B = \begin{bmatrix} -3 & 2 \\ -2 & 1 \end{bmatrix}$. We compute

$$AB = \begin{bmatrix} 1 & -2 \\ 2 & -3 \end{bmatrix}\begin{bmatrix} -3 & 2 \\ -2 & 1 \end{bmatrix} = \begin{bmatrix} 1 & 0 \\ 0 & 1 \end{bmatrix} = I$$

and

$$BA = \begin{bmatrix} -3 & 2 \\ -2 & 1 \end{bmatrix}\begin{bmatrix} 1 & -2 \\ 2 & -3 \end{bmatrix} = \begin{bmatrix} 1 & 0 \\ 0 & 1 \end{bmatrix} = I.$$

Thus A is invertible and $B = A^{-1}$. Also B is invertible and $A = B^{-1}$. ☺

2.3.4 Example. Let $A = \begin{bmatrix} 1 & 2 & 3 \\ 4 & 5 & 6 \end{bmatrix}$ and $B = \begin{bmatrix} -\frac{2}{3} & -\frac{1}{3} \\ -\frac{2}{3} & \frac{5}{3} \\ 1 & -1 \end{bmatrix}$. The reader may check that $AB = I$. On the other hand, $BA = \begin{bmatrix} -2 & -3 & -4 \\ 1 & 7 & 8 \\ -3 & -3 & -3 \end{bmatrix} \neq I$. Thus A is not invertible.

The equations $AB = I$ and $BA = I$, which say that A is invertible with inverse B, say also that B is invertible with inverse A. We say that A and B are *inverses*: each is the inverse of the other, $A^{-1} = B$, $B^{-1} = A$. This implies that $A = B^{-1} = (A^{-1})^{-1}$.

2.3.5

> If A is invertible, so is A^{-1} and $(A^{-1})^{-1} = A$.

2.3.6 Example. Whereas "$a \neq 0$" in the real numbers implies that a is invertible, this is **not true** for matrices. As an example, we cite the nonzero matrix $A = \begin{bmatrix} 1 & 2 \\ 2 & 4 \end{bmatrix}$. For any $B = \begin{bmatrix} x & z \\ y & w \end{bmatrix}$, the second row of

$$AB = \begin{bmatrix} 1 & 2 \\ 2 & 4 \end{bmatrix} \begin{bmatrix} x & y \\ z & w \end{bmatrix} = \begin{bmatrix} x + 2z & y + 2w \\ 2x + 4z & 2y + 4w \end{bmatrix}$$

is twice the first, so AB can never be I. The matrix A does not have an inverse.

According to Definition 2.3.1, to show that B is the inverse of A, one must show that each product AB and BA is the identity matrix. Since matrix multiplication is not commutative, it seems logical that each of these products should be checked. Later (see Theorem 2.9.12), we shall show the following:

2.3.7

> If A is a square matrix and $AB = I$, then $BA = I$ too, so A and B are inverses.

This is a very useful fact that you are encouraged to start using right away.

2.3.8 Problem. Show that $A = \begin{bmatrix} 1 & 2 & 3 \\ -1 & 0 & 1 \\ 4 & 1 & 1 \end{bmatrix}$ and $B = \frac{1}{6}\begin{bmatrix} -1 & 1 & 2 \\ 5 & -11 & -4 \\ -1 & 7 & 2 \end{bmatrix}$ are inverses.

Solution. Since A is square, it is enough to show that $AB = I$. Since

$$AB = \frac{1}{6}\begin{bmatrix} 1 & 2 & 3 \\ -1 & 0 & 1 \\ 4 & 1 & 1 \end{bmatrix}\begin{bmatrix} -1 & 1 & 2 \\ 5 & -11 & -4 \\ -1 & 7 & 2 \end{bmatrix} = \frac{1}{6}\begin{bmatrix} 6 & 0 & 0 \\ 0 & 6 & 0 \\ 0 & 0 & 6 \end{bmatrix} = I,$$

A and B are inverses.

2.3.9 **Remark.** Notice how part 4 of Theorem 2.1.29 was used in the solution of this problem to move the $\frac{1}{6}$ to the front in the calculation of AB. For matrices X, Y and a scalar a, $X(aY) = a(XY)$.

It's easy to determine whether or not a 2×2 matrix is invertible and to find the inverse of an invertible 2×2 matrix.

2.3.10 **Proposition.** *The matrix* $A = \begin{bmatrix} a & b \\ c & d \end{bmatrix}$ *is invertible if and only if* $ad - bc \neq 0$. *If* $ad - bc \neq 0$, *then* $A^{-1} = \dfrac{1}{ad - bc}\begin{bmatrix} d & -b \\ -c & a \end{bmatrix}$.

Proof. We begin by observing that

$$A\begin{bmatrix} d & -b \\ -c & a \end{bmatrix} = (ad - bc)I. \tag{1}$$

The first statement is "if and only if" so it requires two arguments, one to prove "if" ($\Longleftarrow$) and the other to prove "only if" ($\Longrightarrow$).

($\Longleftarrow$) Suppose $ad - bc \neq 0$. Then we can divide each side of equation (1) by $ad - bc$ obtaining $A\left(\frac{1}{ad-bc}\begin{bmatrix} d & -b \\ -c & a \end{bmatrix}\right) = I$. Thus A is invertible and $A^{-1} = \frac{1}{ad-bc}\begin{bmatrix} d & -b \\ -c & a \end{bmatrix}$.

($\Longrightarrow$) Assume A is invertible. We wish to prove that $ad - bc \neq 0$. If this is not the case, then equation (1) says $A\begin{bmatrix} d & -b \\ -c & a \end{bmatrix} = 0$. Multiplying this equation on the left by A^{-1} then gives $\begin{bmatrix} d & -b \\ -c & a \end{bmatrix} = \begin{bmatrix} 0 & 0 \\ 0 & 0 \end{bmatrix}$, implying $d = -b = -c = a = 0$ and hence $A = 0$, contradicting the assumption that A is invertible. ■

It's not hard to remember the formula for the inverse of $A = \begin{bmatrix} a & b \\ c & d \end{bmatrix}$. The inverse is $\frac{1}{ad-bc}$ times $\begin{bmatrix} d & -b \\ -c & a \end{bmatrix}$. This matrix is obtained from A by switching the diagonal entries and changing the sign of the off-diagonal elements.

For example, if $A = \begin{bmatrix} 1 & -2 \\ 3 & 5 \end{bmatrix}$, then $ad - bc = 1(5) - (-2)(3) = 11$ and $A^{-1} = \frac{1}{11}\begin{bmatrix} 5 & 2 \\ -3 & 1 \end{bmatrix}$.

2.3.11 Remark. You may remember that we have called $ad-bc$ the *determinant* of $A = \begin{bmatrix} a & b \\ c & d \end{bmatrix}$. This is written either $\begin{vmatrix} a & b \\ c & d \end{vmatrix}$ or $\det A$. So Proposition 2.3.10 says a 2×2 matrix A is invertible if and only if $\det A \neq 0$, in which case A^{-1} is $\frac{1}{\det A}$ times a certain matrix. All this holds for larger matrices, as we shall see in Section 3.1.

2.3.12 Problem. Suppose A and B are $n \times n$ matrices, each of which is invertible. Show that AB is invertible by finding its inverse.

Solution. We need a matrix X such that $(AB)X = I$. What might X be? We remember that if a and b are nonzero real numbers, then $(ab)^{-1} = a^{-1}b^{-1}$, so we try $A^{-1}B^{-1}$. This does not seem to work, however, because in general $ABA^{-1}B^{-1}$ cannot be simplified (matrix multiplication is not commutative). On the other hand, if we let $X = B^{-1}A^{-1}$, then

$$(AB)X = (AB)(B^{-1}A^{-1}) = A(BB^{-1})A^{-1} = AIA^{-1} = AA^{-1} = I.$$

Since AB is square and $(AB)X = I$, AB is invertible with inverse X. ♘

We remember the result described in Problem 2.3.12 like this:

2.3.13 A product of invertible matrices is invertible with inverse the product of the inverses, order reversed: $(AB)^{-1} = B^{-1}A^{-1}$.

2.3.14 Example. Let $A = \begin{bmatrix} 1 & 3 \\ -2 & 5 \end{bmatrix}$ and $B = \begin{bmatrix} 2 & -1 \\ 1 & -1 \end{bmatrix}$. Then $AB = \begin{bmatrix} 5 & -4 \\ 1 & -3 \end{bmatrix}$ and $(AB)^{-1} = \frac{1}{11}\begin{bmatrix} 3 & -4 \\ 1 & -5 \end{bmatrix}$, by Proposition 2.3.10. Now $A^{-1} = \frac{1}{11}\begin{bmatrix} 5 & -3 \\ 2 & 1 \end{bmatrix}$ and $B^{-1} = \begin{bmatrix} 1 & -1 \\ 1 & -2 \end{bmatrix}$, so $B^{-1}A^{-1} = \frac{1}{11}\begin{bmatrix} 1 & -1 \\ 1 & -2 \end{bmatrix}\begin{bmatrix} 5 & -3 \\ 2 & 1 \end{bmatrix} = \frac{1}{11}\begin{bmatrix} 3 & -4 \\ 1 & -5 \end{bmatrix} = (AB)^{-1}$. ☺

READING CHECK 1. With A and B as in Example 2.3.14, verify that $A^{-1}B^{-1} \neq (AB)^{-1}$.

Matrix Equations

2.3.15 **Problem.** If A is an invertible $n \times n$ matrix and B is an $n \times t$ matrix, show that the equation $AX = B$ has a solution X and find it.

Solution. Multiplying $AX = B$ on the left by A^{-1} gives $A^{-1}AX = A^{-1}B$. Since $A^{-1}A = I$, this is $IX = A^{-1}B$, so $X = A^{-1}B$. It remains to verify that such X is a solution and it is, because $A(A^{-1}B) = IB = B$. 👍

The solution to $ax = b$ with a, x and b real numbers is $x = \frac{b}{a}$, provided $a \neq 0$ (that is, assuming a is invertible). Using exponents, this solution is just $x = a^{-1}b$. As just shown, the solution to $AX = B$ with A, X and B matrices is similar.

2.3.16 **Remark.** Since matrix multiplication is not commutative, the solution $X = A^{-1}B$ is not (generally) the same as $X = BA^{-1}$ because $A^{-1}B \neq BA^{-1}$ (in general). To turn AX into X, we must multiply by A^{-1} **on the left**, thus we must multiply B also **on the left**. This is quite different from ordinary algebra where, assuming $a \neq 0$, the solution to $ax = b$ can be written $x = a^{-1}b$ or $x = ba^{-1}$.

2.3.17 **Problem.** Solve the system of equations $\begin{aligned} x - 2y &= 5 \\ 2x - 3y &= 7. \end{aligned}$

Solution. The given system is $A\mathsf{x} = \mathsf{b}$, where $A = \begin{bmatrix} 1 & -2 \\ 2 & -3 \end{bmatrix}$ is the matrix of coefficients, $\mathsf{x} = \begin{bmatrix} x \\ y \end{bmatrix}$ and $\mathsf{b} = \begin{bmatrix} 5 \\ 7 \end{bmatrix}$. Since $\det A = 1 \neq 0$, A has an inverse, so

$$\mathsf{x} = A^{-1}\mathsf{b} = \begin{bmatrix} -3 & 2 \\ -2 & 1 \end{bmatrix}\begin{bmatrix} 5 \\ 7 \end{bmatrix} = \begin{bmatrix} -1 \\ -3 \end{bmatrix}.$$

The solution is $x = -1$, $y = -3$. 👍

READING CHECK 2. If A is an invertible $n \times n$ matrix and B is a $t \times n$ matrix, show that the equation $XA = B$ has a solution by finding it.

Algebra with matrices is not terribly different from ordinary algebra with numbers, but one must always be careful not to assume commutativity. For example, a matrix of the form $A^{-1}XA$ cannot usually be simplified.

2.3.18 **Problem.** Suppose A, B and X are invertible matrices and $AX^{-1}A^{-1} = B^{-1}$. What is X?

Solution. First, we multiply the given equation on the right by A, obtaining $AX^{-1}A^{-1}A = B^{-1}A$, that is, $AX^{-1} = B^{-1}A$. Now multiply this equation on the left by A^{-1} to get $X^{-1} = A^{-1}B^{-1}A$. Now $X = (X^{-1})^{-1} = (A^{-1}B^{-1}A)^{-1}$ and, using 2.3.13, this is $A^{-1}(B^{-1})^{-1}(A^{-1})^{-1} = A^{-1}BA$.

The Inverse of a Partitioned Matrix

The reader may remember that a matrix M has been *partitioned* if its entries have been grouped into smaller arrays so that we can view M as a matrix whose entries are themselves matrices—see Section 2.1. For example,

$$M = \left[\begin{array}{cc|cc} 1 & 2 & 0 & 0 \\ 2 & 3 & 0 & 0 \\ \hline 3 & 4 & 5 & 6 \\ 6 & 7 & 8 & 9 \end{array}\right]$$

partitions the matrix M so that we can view it as a 2×2 matrix $\begin{bmatrix} A & 0 \\ B & C \end{bmatrix}$ with "blocks" $A = \begin{bmatrix} 1 & 2 \\ 2 & 3 \end{bmatrix}$, $B = \begin{bmatrix} 3 & 4 \\ 6 & 7 \end{bmatrix}$, $C = \begin{bmatrix} 5 & 6 \\ 8 & 9 \end{bmatrix}$ and $0 = \begin{bmatrix} 0 & 0 \\ 0 & 0 \end{bmatrix}$. Suppose we want to find the inverse of A and make the assumption that A^{-1} will look like A. Thus, we seek a partitioned matrix $\begin{bmatrix} X & 0 \\ Y & Z \end{bmatrix}$ such that

$$\begin{bmatrix} A & 0 \\ B & C \end{bmatrix} \begin{bmatrix} X & 0 \\ Y & Z \end{bmatrix} = \left[\begin{array}{cc|cc} 1 & 0 & 0 & 0 \\ 0 & 1 & 0 & 0 \\ \hline 0 & 0 & 1 & 0 \\ 0 & 0 & 0 & 1 \end{array}\right].$$

Since partitioned matrices can be multiplied in the "normal" way if the block sizes are compatible, the product on the left is $\begin{bmatrix} AX & 0 \\ BX + CY & CZ \end{bmatrix}$, so we wish to find 2×2 matrices X, Y, Z so that

$$\begin{aligned} AX &= I_2 \\ BX + CY &= 0 \\ CZ &= I_2, \end{aligned}$$

$I_2 = \begin{bmatrix} 1 & 0 \\ 0 & 1 \end{bmatrix}$ denoting the 2×2 identity matrix. We see immediately that $X = A^{-1}$ and $Z = C^{-1}$. The second equation says $CY = -BX = -BA^{-1}$, so $Y = -C^{-1}BA^{-1}$. It is now straightforward to verify that

$$\begin{bmatrix} A & 0 \\ B & C \end{bmatrix}^{-1} = \begin{bmatrix} A^{-1} & 0 \\ -C^{-1}BA^{-1} & C^{-1} \end{bmatrix}.$$

In our example, $A = \begin{bmatrix} 1 & 2 \\ 2 & 3 \end{bmatrix}$, $B = \begin{bmatrix} 3 & 4 \\ 6 & 7 \end{bmatrix}$ and $C = \begin{bmatrix} 5 & 6 \\ 8 & 9 \end{bmatrix}$ so, making use of Proposition 2.3.10, we have

$$A^{-1} = \begin{bmatrix} -3 & 2 \\ 2 & -1 \end{bmatrix} \text{ and } C^{-1} = -\frac{1}{3}\begin{bmatrix} 9 & -6 \\ -8 & 5 \end{bmatrix}$$

so $C^{-1}BA^{-1} = \frac{1}{3}\begin{bmatrix} 15 & -12 \\ -12 & 9 \end{bmatrix} = \begin{bmatrix} 5 & -4 \\ -4 & 3 \end{bmatrix}$ and $M^{-1} = \frac{1}{9}\begin{bmatrix} -3 & 2 & 0 & 0 \\ 2 & -1 & 0 & 0 \\ 5 & -4 & -3 & 2 \\ -4 & 3 & \frac{8}{3} & -\frac{5}{3} \end{bmatrix}$.

Incidentally, while the three matrices A, B and C in this example were square and of the same size, not all these conditions are necessary. In fact, as the next theorem implies, it is necessary only for A and C to be square. (The proof, which is essentially given above, is left to the exercises.)

2.3.19 Theorem. *The partitioned matrix* $\begin{bmatrix} A & 0 \\ B & C \end{bmatrix}$ *is invertible if and only if the matrices A and C are invertible.*

Some Final Words about the Inverse. Except for the 2×2 case, our discussion of the inverse of a matrix has intentionally avoided mention of how to determine if a matrix is invertible and also how to find the inverse of an invertible matrix. There are various ways to determine whether or not a matrix has an inverse, one of which is described in Section 2.9 and another of which involves the concept of *determinant*, the subject of Chapter 3.

More on Transposes

We conclude this section with a summary of the most basic properties of the transpose of a matrix whose properties, in some ways, mimic those of the inverse.

2.3.20 Theorem. *Let A and B be matrices and let c be a scalar.*

1. $(A+B)^T = A^T + B^T$ *(assuming A and B have the same size).*
2. $(cA)^T = cA^T$.
3. $(A^T)^T = A$.
4. *If A is invertible, so is A^T and* $(A^T)^{-1} = (A^{-1})^T$.
5. $(AB)^T = B^T A^T$ *(assuming AB is defined): the transpose of a product is the product of the transposes, order reversed.*

READING CHECK 3. Property 3 here holds also for the inverse of a matrix: If A is invertible, then $(A^{-1})^{-1} = A$. Are there other properties of the transpose that hold for the inverse?

The last of these properties is the most interesting, and it should remind us of the formula for the inverse of the product of two invertible matrices. Just as $(AB)^{-1} = B^{-1}A^{-1}$, so also $(AB)^T = B^TA^T$.

Just as $(AB)^{-1}$ cannot possibly be $A^{-1}B^{-1}$ because $ABA^{-1}B^{-1}$ cannot usually be simplified, the "obvious" rule for the transpose of a product, $(AB)^T = A^TB^T$, cannot possibly work either. If A is 2×3 and B is 3×4, for example, then A^T is 3×2 and B^T is 4×3, so the product A^TB^T is not even defined. On the other hand, the product B^TA^T is defined and produces a 4×2 matrix, precisely the size of $(AB)^T$ because AB is 2×4. This argument does not **prove** that $(AB)^T = B^TA^T$, of course, but it gives us reason to hope. In a similar spirit, examples may also give hope.

2.3.21 Example. Suppose $A = \begin{bmatrix} 1 & 2 & 3 \\ -4 & 0 & -1 \end{bmatrix}$ and $B = \begin{bmatrix} 0 & 1 \\ 1 & -3 \\ 4 & 2 \end{bmatrix}$. Then

$$AB = \begin{bmatrix} 1 & 2 & 3 \\ -4 & 0 & -1 \end{bmatrix} \begin{bmatrix} 0 & 1 \\ 1 & -3 \\ 4 & 2 \end{bmatrix} = \begin{bmatrix} 14 & 1 \\ -4 & -6 \end{bmatrix}.$$

Now $A^T = \begin{bmatrix} 1 & -4 \\ 2 & 0 \\ 3 & -1 \end{bmatrix}$ and $B^T = \begin{bmatrix} 0 & 1 & 4 \\ 1 & -3 & 2 \end{bmatrix}$, so

$$B^TA^T = \begin{bmatrix} 0 & 1 & 4 \\ 1 & -3 & 2 \end{bmatrix} \begin{bmatrix} 1 & -4 \\ 2 & 0 \\ 3 & -1 \end{bmatrix} = \begin{bmatrix} 14 & -4 \\ 1 & -6 \end{bmatrix} = (AB)^T.$$ ☺

One example does not prove $(AB)^T = B^TA^T$ for all matrices A and B, but here is a general argument which does. Let $\mathsf{a}_1^T, \mathsf{a}_2^T, \ldots, \mathsf{a}_m^T$ be the rows of A and let $\mathsf{b}_1, \mathsf{b}_2, \ldots, \mathsf{b}_p$ be the columns of B, so that we picture A and B like this:

$$A = \begin{bmatrix} \mathsf{a}_1^T & \rightarrow \\ \mathsf{a}_2^T & \rightarrow \\ \vdots & \\ \mathsf{a}_m^T & \rightarrow \end{bmatrix}, \qquad B = \begin{bmatrix} \mathsf{b}_1 & \mathsf{b}_2 & \ldots & \mathsf{b}_p \\ \downarrow & \downarrow & & \downarrow \\ & & & \end{bmatrix}.$$

Then

$$AB = \begin{bmatrix} \mathsf{a}_1 \cdot \mathsf{b}_1 & \mathsf{a}_1 \cdot \mathsf{b}_2 & \ldots & \mathsf{a}_1 \cdot \mathsf{b}_p \\ \mathsf{a}_2 \cdot \mathsf{b}_1 & \mathsf{a}_2 \cdot \mathsf{b}_2 & \ldots & \mathsf{a}_2 \cdot \mathsf{b}_p \\ \vdots & \vdots & & \vdots \\ \mathsf{a}_m \cdot \mathsf{b}_1 & \mathsf{a}_m \cdot \mathsf{b}_2 & \ldots & \mathsf{a}_m \cdot \mathsf{b}_p \end{bmatrix},$$

so

$$(AB)^T = \begin{bmatrix} \mathsf{a}_1 \cdot \mathsf{b}_1 & \mathsf{a}_2 \cdot \mathsf{b}_1 & \dots & \mathsf{a}_m \cdot \mathsf{b}_1 \\ \mathsf{a}_1 \cdot \mathsf{b}_2 & \mathsf{a}_2 \cdot \mathsf{b}_2 & \dots & \mathsf{a}_m \cdot \mathsf{b}_2 \\ \vdots & \vdots & & \vdots \\ \mathsf{a}_1 \cdot \mathsf{b}_p & \mathsf{a}_2 \cdot \mathsf{b}_p & \dots & \mathsf{a}_m \cdot \mathsf{b}_p \end{bmatrix}. \tag{2}$$

Now

$$A^T = \begin{bmatrix} \mathsf{a}_1 & \mathsf{a}_2 & \dots & \mathsf{a}_m \\ \downarrow & \downarrow & & \downarrow \end{bmatrix} \text{ and } B^T = \begin{bmatrix} \mathsf{b}_1^T & \rightarrow \\ \mathsf{b}_2^T & \rightarrow \\ \vdots & \\ \mathsf{b}_p^T & \rightarrow \end{bmatrix}$$

so $B^T A^T$

$$= \begin{bmatrix} \mathsf{b}_1^T & \rightarrow \\ \mathsf{b}_2^T & \rightarrow \\ \vdots & \\ \mathsf{b}_p^T & \rightarrow \end{bmatrix} \begin{bmatrix} \mathsf{a}_1 & \mathsf{a}_2 & \dots & \mathsf{a}_m \\ \downarrow & \downarrow & & \downarrow \end{bmatrix} = \begin{bmatrix} \mathsf{b}_1 \cdot \mathsf{a}_1 & \mathsf{b}_1 \cdot \mathsf{a}_2 & \dots & \mathsf{b}_1 \cdot \mathsf{a}_m \\ \mathsf{b}_2 \cdot \mathsf{a}_1 & \mathsf{b}_2 \cdot \mathsf{a}_2 & \dots & \mathsf{b}_2 \cdot \mathsf{a}_m \\ \vdots & \vdots & & \vdots \\ \mathsf{b}_p \cdot \mathsf{a}_1 & \mathsf{b}_p \cdot \mathsf{a}_2 & \dots & \mathsf{b}_p \cdot \mathsf{a}_m \end{bmatrix}. \tag{3}$$

It remains only to observe that the matrices in (3) and (2) are the same because $\mathsf{a} \cdot \mathsf{b} = \mathsf{b} \cdot \mathsf{a}$ for any vectors a and b.

Answers to Reading Checks

1. $A^{-1}B^{-1} = \frac{1}{11}\begin{bmatrix} 5 & -3 \\ 2 & 1 \end{bmatrix}\begin{bmatrix} 1 & -1 \\ 1 & -2 \end{bmatrix} = \frac{1}{11}\begin{bmatrix} 2 & 1 \\ 3 & -4 \end{bmatrix}$

   whereas $(AB)^{-1} = \frac{1}{11}\begin{bmatrix} 3 & -4 \\ 1 & -5 \end{bmatrix}$, as shown in Example 2.3.14.

2. The solution is $X = BA^{-1}$ since $(BA^{-1})A = BI = B$.

3. Property 5 is the most obvious: $(AB)^{-1} = B^{-1}A^{-1}$.

True/False Questions

Decide, with as little calculation as possible, whether each of the following statements is true or false and, if you say "false," explain your answer. (Answers are at the back of the book.)

1. If A, B and C are matrices and $AC = BC$, then $A = B$.
2. Only square matrices are invertible.
3. Only square matrices have transposes.

4. If A and B are matrices and A is invertible, then $A^{-1}BA = B$.

5. If A, B and C are invertible matrices (of the same size), then $(ABC)^{-1} = C^{-1}B^{-1}A^{-1}$.

6. If A and B are invertible matrices and $XA = B$, then $X = A^{-1}B$.

7. Assuming A and B are invertible matrices, the solution to $AXB = I$ is $X = A^{-1}B^{-1}$.

8. If A is a 2×3 matrix and B is 3×5, then $(AB)^T$ is 5×2.

9. If A and B are matrices and both products A^TB^T and B^TA^T are defined, then A and B are square.

10. For matrices A and B, $(AB)^T = A^TB^T$.

EXERCISES

*Answers to exercises marked [BB] can be found at the **Back** of the **Book**.*

1. Determine whether each of the following pairs of matrices are inverses.

   (a) [BB] $A = \begin{bmatrix} 1 & 0 & 1 \\ 2 & 1 & 0 \\ 3 & -1 & 0 \end{bmatrix}, \quad B = \frac{1}{5}\begin{bmatrix} 0 & 1 & 1 \\ 0 & 3 & -2 \\ 5 & -1 & -1 \end{bmatrix}$

   (b) $A = \begin{bmatrix} 0 & 1 \\ 1 & 0 \end{bmatrix}, \quad B = A$

   (c) $A = \begin{bmatrix} 1 & 2 & 3 \\ 4 & 5 & 6 \end{bmatrix}, \quad B = \begin{bmatrix} -1 & 1 \\ 0 & -1 \\ \frac{2}{3} & \frac{1}{3} \end{bmatrix}$

   (d) $A = \begin{bmatrix} 1 & 0 & 0 \\ 2 & 1 & 0 \\ -1 & 2 & 1 \end{bmatrix}, \quad B = \begin{bmatrix} 1 & 0 & 0 \\ -2 & 1 & 1 \\ 5 & -1 & -1 \end{bmatrix}$

   (e) $A = \begin{bmatrix} 2 & 0 & -\frac{1}{2} \\ -1 & 0 & \frac{1}{2} \end{bmatrix}, \quad B = \begin{bmatrix} 1 & 1 \\ -1 & -2 \\ 2 & 4 \end{bmatrix}$

   (f) $A = \begin{bmatrix} 1 & 0 & 0 & 0 \\ 0 & 2 & 0 & 0 \\ 0 & 0 & 3 & 0 \\ 0 & 0 & 0 & 4 \end{bmatrix}, \quad B = \frac{1}{24}\begin{bmatrix} 24 & 0 & 0 & 0 \\ 0 & 12 & 0 & 0 \\ 0 & 0 & 8 & 0 \\ 0 & 0 & 0 & 6 \end{bmatrix}.$

2. Given $A^{-1} = \begin{bmatrix} 1 & 2 & 3 \\ 4 & 2 & 2 \\ 5 & 6 & -4 \end{bmatrix}$ and $\mathsf{b} = \begin{bmatrix} 2 \\ 4 \\ 6 \end{bmatrix}$, find a vector x such that $A\mathsf{x} = \mathsf{b}$.

3. Find A^{-1} and use this to solve each system of equations.

   (a) [BB] $A = \begin{bmatrix} -2 & 5 \\ 1 & -3 \end{bmatrix}; \quad \begin{aligned} -2x + 5y &= 3 \\ x - 3y &= 7. \end{aligned}$

(b) $A = \begin{bmatrix} 1 & 1 \\ 1 & -2 \end{bmatrix}$; $\begin{aligned} x + y &= -1 \\ x - 2y &= 4. \end{aligned}$ **(c)** $A = \begin{bmatrix} 3 & -1 \\ 1 & 2 \end{bmatrix}$; $\begin{aligned} 3x - y &= 0 \\ x + 2y &= 1. \end{aligned}$

4. Let $A = \begin{bmatrix} -1 & 2 & 2 \\ 3 & 0 & 5 \\ 2 & -1 & 0 \end{bmatrix}$. Use the fact that $A^{-1} = \frac{1}{9}\begin{bmatrix} 5 & -2 & 10 \\ 10 & -4 & 11 \\ -3 & 3 & -6 \end{bmatrix}$ to solve the system $\begin{aligned} -x + 2y + 2z &= 12 \\ 3x + 5z &= -1 \\ 2x - y &= -8. \end{aligned}$

5. Let $A = \begin{bmatrix} -1 & 5 & -6 \\ -1 & 4 & -8 \\ -2 & 10 & -13 \end{bmatrix}$. Use the fact that $A^{-1} = \begin{bmatrix} -28 & -5 & 16 \\ -3 & -1 & 2 \\ 2 & 0 & -1 \end{bmatrix}$ to solve the system $\begin{aligned} -28x - 5y + 16z &= -1 \\ -3x - y + 2z &= 2 \\ 2x - z &= -3. \end{aligned}$

6. [BB] Suppose that A is an invertible matrix and c is a nonzero scalar. Show that cA is invertible by finding its inverse.

7. If a matrix A is invertible, show that A^2 is also invertible. (Do **not** use the result of Problem 2.3.12.)

8. Suppose A and B are matrices with B invertible.

   (a) If $A = B^{-1}AB$, show that A and B commute.

   (b) Part (a) is called an *implication*; that is, a statement of the form "If $\mathcal{X}$, then $\mathcal{Y}$." What are $\mathcal{X}$ and $\mathcal{Y}$? The *converse* of "If $\mathcal{X}$, then $\mathcal{Y}$" is the implication "If $\mathcal{Y}$, then $\mathcal{X}$." What is the converse of the implication in part (a)? Is the converse true? Explain.

9. **(a)** [BB] Suppose A and B are square matrices that commute. Prove that $(AB)^T = A^T B^T$.

   (b) If A and B are square matrices and $(AB)^T = A^T B^T$, must A and B commute? Justify your answer with a proof if you say "yes" or an example that proves "no."

10. If A and B are invertible $n \times n$ matrices that commute, prove that B and A^{-1} commute.

11. [BB] Let $A = \begin{bmatrix} 1 & -7 & 1 \\ 2 & -9 & 1 \end{bmatrix}$ and $B = \begin{bmatrix} -1 & 3 \\ 0 & 1 \\ 2 & 4 \end{bmatrix}$.

    (a) Compute AB and BA. **(b)** Is A invertible? Explain.

12. Show that the matrix $A = \begin{bmatrix} 2 & 6 \\ 1 & 3 \end{bmatrix}$ is not invertible.

13. Let A, B and C be matrices with C invertible.

    (a) [BB] If $AC = BC$ then $A = B$. **(b)** If $CA = CB$, then $A = B$.

14. A matrix A is *diagonal* if its only nonzero entries occupy the diagonal positions $a_{11}, a_{22}, a_{33}, \ldots$.

(a) Show that the diagonal matrix $A = \begin{bmatrix} 2 & 0 & 0 \\ 0 & -3 & 0 \\ 0 & 0 & 5 \end{bmatrix}$ is invertible.

(b) Show that the diagonal matrix $A = \begin{bmatrix} 2 & 0 & 0 \\ 0 & 0 & 0 \\ 0 & 0 & 5 \end{bmatrix}$ is not invertible.

(c) Complete the following sentence:

"A square diagonal matrix with diagonal entries $d_1, d_2, \dots, d_n$ is invertible if and only if ______________. "

15. Let $A = \begin{bmatrix} 1 & 3 \\ 2 & 1 \end{bmatrix}$ and $B = \begin{bmatrix} 1 & 3 \\ 0 & 1 \end{bmatrix}$. Compute AB, $(AB)^T$, A^TB^T and B^TA^T. These calculations illustrate what property of the transpose?

16. [BB] Verify part 4 of Theorem 2.3.20 by showing that if A is an invertible matrix, then so is A^T, and $(A^T)^{-1} = (A^{-1})^T$.

17. Find a formula for $((AB)^T)^{-1}$ in terms of $(A^T)^{-1}$ and $(B^T)^{-1}$.

18. Let $\mathsf{x} = \begin{bmatrix} x_1 \\ x_2 \\ \vdots \\ x_n \end{bmatrix}$ be a vector in R^n and let $A = \begin{bmatrix} \mathsf{a}_1^T & \rightarrow \\ \mathsf{a}_2^T & \rightarrow \\ \vdots & \\ \mathsf{a}_n^T & \rightarrow \end{bmatrix}$ be an $n \times m$ matrix. Show that x^TA is a linear combination of the rows of A with coefficients the components of x. [Hint: 2.1.33 and Theorem 2.3.20.]

19. Suppose $A = \begin{bmatrix} \mathsf{a}_1^T & \rightarrow \\ \mathsf{a}_2^T & \rightarrow \\ \vdots & \\ \mathsf{a}_m^T & \rightarrow \end{bmatrix}$ is an $m \times n$ matrix with rows as indicated and suppose B is an $n \times t$ matrix. Show that $AB = \begin{bmatrix} \mathsf{a}_1^TB & \rightarrow \\ \mathsf{a}_2^TB & \rightarrow \\ \vdots & \\ \mathsf{a}_m^TB & \rightarrow \end{bmatrix}$.

[Hint: Theorem 2.3.20 and equation (2.1.24), p. 102.]

20. [BB] Suppose A and B are $n \times n$ matrices and $I - AB$ is invertible. Show that $I - BA$ is invertible by showing that $(I - BA)^{-1} = I + B(I - AB)^{-1}A$.

21. If A, B and $A + B$ are invertible $n \times n$ matrices, show that $A(A^{-1} + B^{-1})B(A + B)^{-1} = I$.

22. Suppose that B and $B - I$ are invertible matrices. Show that $I - B^{-1}$ is invertible by showing that $B(B - I)^{-1}(I - B^{-1}) = I$.

23. In each case, use the given information to find the matrix A.

(a) [BB] $(2A)^{-1} = \begin{bmatrix} 1 & 2 \\ 3 & 4 \end{bmatrix}^{-1}$

(b) $2A^{-1} = \begin{bmatrix} 1 & 2 \\ 3 & 4 \end{bmatrix}^{-1}$

(c) $\left(A \begin{bmatrix} 1 & -1 \\ 0 & 1 \end{bmatrix}\right)^{-1} = \begin{bmatrix} 2 & 3 \\ 1 & 2 \end{bmatrix}$

(d) $(A^T - 5I)^{-1} = \begin{bmatrix} 2 & 4 \\ 1 & 3 \end{bmatrix}$

(e) $\begin{bmatrix} 1 & 2 \\ 3 & 0 \end{bmatrix}^{-1} A \begin{bmatrix} 5 & 1 \\ -1 & 1 \end{bmatrix}^{-1} = \begin{bmatrix} -3 & 4 \\ 0 & 2 \end{bmatrix}$.

24. (a) [BB] Suppose A and B are invertible matrices and X is a matrix such that $AXB = A + B$. What is X?

(b) Suppose A, B, C and X are matrices, X and C invertible, such that $X^{-1}A = C - X^{-1}B$. What is X?

(c) Suppose A, B and X are invertible matrices such that $BAX = XABX$. What is X?

25. Find A given $\begin{bmatrix} 4 & 0 & 6 \\ 0 & -3 & 1 \\ 0 & 5 & -4 \end{bmatrix}^{-1} A \begin{bmatrix} 2 & 0 & 1 \\ 1 & 4 & -1 \\ 1 & 3 & 0 \end{bmatrix}^{-1} = \begin{bmatrix} 1 & 2 & -1 \\ -4 & 0 & 6 \\ -2 & 1 & 1 \end{bmatrix}$.

26. Prove Theorem 2.3.19 by showing that a partitioned matrix $\begin{bmatrix} A & 0 \\ B & C \end{bmatrix}$ has an inverse if and only if the matrices A and C have inverses.

27. Find the inverses of the matrices

(a) [BB] $\begin{bmatrix} -1 & 3 & 0 & 0 \\ 2 & -5 & 0 & 0 \\ 4 & 1 & 2 & 9 \\ -2 & 3 & 1 & 7 \end{bmatrix}$ (b) $\begin{bmatrix} -1 & 0 & 0 \\ 2 & 4 & 5 \\ 3 & 2 & 3 \end{bmatrix}$

(c) $\begin{bmatrix} 0 & -4 & 0 & 0 \\ 1 & 2 & 0 & 0 \\ -3 & 1 & 1 & 2 \\ 5 & 0 & -3 & 2 \end{bmatrix}$ (d) $\begin{bmatrix} 1 & 2 & 0 \\ -1 & 4 & 0 \\ 3 & 1 & 5 \end{bmatrix}$

(e) [BB] $\begin{bmatrix} 1 & 2 & 0 & 0 \\ -3 & -4 & 0 & 0 \\ 5 & 6 & 7 & 8 \\ -1 & 0 & 1 & 2 \end{bmatrix}$ (f) $\begin{bmatrix} 3 & -2 & 0 & 0 \\ 1 & 1 & 0 & 0 \\ 3 & -1 & -4 & 5 \\ 1 & 2 & 4 & -6 \end{bmatrix}$

(g) $\begin{bmatrix} 1 & 2 & 0 & 0 & 0 & 0 & 0 & 0 \\ -3 & -4 & 0 & 0 & 0 & 0 & 0 & 0 \\ 5 & 6 & 7 & 8 & 0 & 0 & 0 & 0 \\ -1 & 0 & 1 & 2 & 0 & 0 & 0 & 0 \\ 1 & 1 & 1 & 1 & 3 & -2 & 0 & 0 \\ 1 & 1 & 1 & 1 & 1 & 1 & 0 & 0 \\ 1 & 1 & 1 & 1 & 3 & -1 & -4 & 5 \\ 1 & 1 & 1 & 1 & 1 & 2 & 4 & -6 \end{bmatrix}$.

[Hint: Partition and view these matrices as 2×2.]

28. (a) Let $A = \begin{bmatrix} 1 & 0 & 0 & 0 \\ 2 & 1 & 0 & 0 \\ 4 & 4 & 1 & 0 \\ 8 & 12 & 6 & 1 \end{bmatrix}$ and $B = \begin{bmatrix} 1 & 0 & 0 & 0 \\ 3 & 1 & 0 & 0 \\ 9 & 6 & 1 & 0 \\ 27 & 27 & 9 & 1 \end{bmatrix}$. Find AB.

(b) Let $P(a) = \begin{bmatrix} 1 & 0 & 0 & 0 \\ a & 1 & 0 & 0 \\ a^2 & 2a & 1 & 0 \\ a^3 & 3a^2 & 3a & 1 \end{bmatrix}$ and $P(b) = \begin{bmatrix} 1 & 0 & 0 & 0 \\ b & 1 & 0 & 0 \\ b^2 & 2b & 1 & 0 \\ b^3 & 3b^2 & 3b & 1 \end{bmatrix}$.

Show that $P(a)P(b)$ is $P(x)$ for some x.

(c) The identity matrix has the form $P(a)$ for some a. Explain.

(d) Without any calculation, show that the matrix A in part (a) has an inverse by exhibiting the inverse.

[This exercise was suggested by an article that appeared in the *American Mathematical Monthly*, "Inverting the Pascal Matrix Plus One," by Rita Aggarwala and Michael P. Lamoureux, **109** (April 2002), no. 9.]

29. **(a)** Let A be an $n \times n$ matrix and let $\mathsf{v} = \begin{bmatrix} 1 \\ 1 \\ \vdots \\ 1 \end{bmatrix}$ be a column of n 1s. Show that the condition "each column of A sums to 1" is equivalent to $A^T\mathsf{v} = \mathsf{v}$.

    (b) Let A and B be $n \times n$ matrices. Suppose each of the columns of A sums to 1 and each of the columns of B sums to 1. Show that each of the columns of AB sums to 1.

Critical Reading

30. [BB] Suppose A and B are $n \times n$ matrices and AB is invertible. Show that A and B are each invertible.

31. Given matrices A and B with AB and B invertible, prove that A is invertible.

32. Suppose A, B and C are invertible matrices and X is a matrix such that $A = BXC$. Prove that X is invertible.

33. Show that the matrix $A = \begin{bmatrix} 1 & 3 & 3 \\ 3 & 1 & 3 \\ -3 & -3 & -5 \end{bmatrix}$ has an inverse of the form $aI + bA$ for scalars a and b.

34. **(a)** Suppose that B and $I - B^{-1}$ are invertible matrices and the matrix equation $(X - A)^{-1} = BX^{-1}$ has a solution X. Find X.

    (b) Show that $X = A(I - B^{-1})^{-1}$ is a solution to the matrix equation $(X - A)^{-1} = BX^{-1}$.

35. Suppose A, B and C are matrices with $A = BC$ and $B^2 = C^2 = I$. Explain why A is invertible.

36. If A is a matrix and $A^TA = 0$ is the zero matrix, show that $A = 0$. [Hint: What is the $(1,1)$ entry of A^TA?]

37. Let A be a 3×3 matrix such that $A\begin{bmatrix} 1 \\ 2 \\ 3 \end{bmatrix} = \begin{bmatrix} 0 \\ 0 \\ 0 \end{bmatrix}$. Can A be invertible? Explain.

38. Suppose A is an $n \times n$ matrix such that $A + A^2 = I$. Show that A is invertible.

2.4 Systems of Linear Equations

A *linear equation* in variables $x_1, x_2, \ldots$ is an equation such as $2x_1 - 3x_2 + x_3 - \sqrt{2}x_4 = 17$, where each variable appears by itself (and to the first power) and the coefficients are scalars. A set of one or more linear equations is called a *linear system*. *To solve* a linear system means to find values of the variables that make each equation true.

In this section, we describe a method for solving systems of linear equations. Known as "Gaussian Elimination," this procedure goes back to 200 BC and the Han Dynasty in China. It first appeared in Chapter 8 of a book entitled *Jiuzhang Suanshu* (*The Nine Chapters on the Mathematical Art*), but takes its name from Karl Friedrich Gauss who rediscovered the method in the early 1800s.[4]

Many people could find numbers x and y so that

$$\begin{aligned} x + y &= 7 \\ x - 2y &= -2 \end{aligned} \tag{1}$$

via some ad hoc process and, in fact, in previous sections, we have assumed you have this ability. Suppose you were faced with a system of eight equations in 17 variables, however. How would you go about finding a solution in this case?

Two linear systems are said to be *equivalent* if they have the same solutions. For example, system (1) is equivalent to

$$\begin{aligned} x + y &= 7 \\ 3y &= 9 \end{aligned} \tag{2}$$

because each system has the same solution, $x = 4$, $y = 3$.

Our procedure for solving a linear system involves transforming the given system to another system that is equivalent, but easier to solve. For example, system (2) makes it immediately clear that $y = 3$, so it is easier to solve than system (1).

There are three very useful ways to change a system of linear equations to an equivalent system. First, we can interchange any two of the equations (and hence write down the equations in any order we please). For example, the numbers x, y and z that satisfy

$$\begin{aligned} -2x + y + z &= -9 \\ x + y + z &= 0 \\ x + 3y &= -3 \end{aligned} \tag{3}$$

are the same as the numbers that satisfy

$$\begin{aligned} x + y + z &= 0 \\ -2x + y + z &= -9 \\ x + 3y &= -3, \end{aligned}$$

[4]The author is a ninth generation student of Gauss, through Gudermann, Weierstrass, Frobenius, Schur, Brauer, Bruck, Kleinfeld and Anderson.

(interchanging the first two equations of (3)), so the two systems are equivalent. Second, in any system of equations, we can multiply an equation by a number (different from 0) without changing the solution. For example,

$$x + 3y = -3 \quad \text{if and only if} \quad \tfrac{1}{3}x + y = -1.$$

Third, and this is the least obvious, in any system of equations, we may always add or subtract a multiple of one equation from another without changing the solution. Suppose, for instance, that we replace the first equation in system (3) by the first equation plus twice the third to get

$$(-2x + y + z) + 2(x + 3y) = -9 + 2(-3), \text{ that is, } 7y + z = -15$$

so that the system becomes

$$\begin{aligned} 7y + z &= -15 \\ x + y + z &= 0 \\ x + 3y &= -3. \end{aligned} \tag{4}$$

Any values of x, y and z that satisfy this new system (4) must also satisfy system (3) since the first equation in that system is a consequence of the equations in (4):

$$-2x + y + z = (7y + z) - 2(x + 3y) = -15 - 2(-3) = -9.$$

Conversely, any values of x, y and z that satisfy the original system also satisfy (4) since

$$7y + z = (-2x + y + z) + 2(x + 3y) = -9 + 2(-3) = -15.$$

Thus systems (3) and (4) are equivalent.

So we have described three ways in which a system of equations can be changed to an equivalent system:

1. $E \leftrightarrow E'$: interchange two equations;
2. $E \to cE$: multiply an equation by any number other than 0;
3. $E \to E - cE'$: replace an equation by that equation less a multiple of another.

By the way, it makes no difference whether we say "less" or "plus" in Statement 3 since, for example, adding $2E$ is the same as subtracting $-2E$, but we prefer the language of subtraction for reasons that will become clear later.

We illustrate how these principles can be applied to solve system (3). First, interchange the first two equations to get

$$\begin{aligned} x + y + z &= 0 \\ -2x + y + z &= -9 \\ x + 3y &= -3. \end{aligned}$$

Now replace the second equation by that equation plus twice the first—we denote this as $E2 \to E2 + 2(E1)$—to obtain

$$\begin{array}{rcr} x + y + z & = & 0 \\ 3y + 3z & = & -9 \\ x + 3y \quad & = & -3. \end{array} \tag{5}$$

We describe the transformation from system (3) to system (5) like this:

$$\begin{array}{rcr} -2x + y + z & = & -9 \\ x + y + z & = & 0 \\ x + 3y \quad & = & -3 \end{array} \xrightarrow{E1 \leftrightarrow E2} \begin{array}{rcr} x + y + z & = & 0 \\ -2x + y + z & = & -9 \\ x + 3y \quad & = & -3 \end{array}$$

$$\xrightarrow{E2 \to E2 + 2(E1)} \begin{array}{rcr} x + y + z & = & 0 \\ 3y + 3z & = & -9 \\ x + 3y \quad & = & -3. \end{array}$$

We continue with our solution, replacing the third equation with the third equation less the first:

$$\begin{array}{rcr} x + y + z & = & 0 \\ 3y + 3z & = & -9 \\ x + 3y \quad & = & -3 \end{array} \xrightarrow{E3 \to E3 - E1} \begin{array}{rcr} x + y + z & = & 0 \\ 3y + 3z & = & -9 \\ 2y - z & = & -3. \end{array}$$

Now we multiply the second equation by $\frac{1}{3}$ and replace the third equation with the third equation less twice the second:

$$\begin{array}{rcr} x + y + z & = & 0 \\ 3y + 3z & = & -9 \\ 2y - z & = & -3 \end{array} \xrightarrow{E2 \to \frac{1}{3}E2} \begin{array}{rcr} x + y + z & = & 0 \\ y + z & = & -3 \\ 2y - z & = & -3 \end{array}$$

$$\xrightarrow{E3 \to E3 - 2(E2)} \begin{array}{rcr} x + y + z & = & 0 \\ y + z & = & -3 \\ -3z & = & 3. \end{array}$$

We have reached our goal, which was to transform the original system to one which is *upper triangular*, like this: ◹ . At this point, we can obtain a solution in a straightforward manner using a technique called *back substitution*, which means solving the equations from the bottom up. The third equation gives $z = -1$. The second says $y + z = -3$, so $y = -3 - z = -2$. The first says $x + y + z = 0$, so $x = -y - z = 3$.

It is easy to check that $x = 3$, $y = -2$, $z = -1$ satisfies the original system (3), so we have our solution. **We strongly advise you always to check solutions!**

Did you get tired of writing the x, the y, the z and the $=$ in the solution just illustrated? Only the coefficients change from one stage to another, so it is only the coefficients we really have to write down. We solve system (3) again, by different means.

First we write the system in matrix form, like this:

$$\begin{bmatrix} -2 & 1 & 1 \\ 1 & 1 & 1 \\ 1 & 3 & 0 \end{bmatrix} \begin{bmatrix} x \\ y \\ z \end{bmatrix} = \begin{bmatrix} -9 \\ 0 \\ 3 \end{bmatrix}.$$

This is $A\mathsf{x} = \mathsf{b}$ with $A = \begin{bmatrix} -2 & 1 & 1 \\ 1 & 1 & 1 \\ 1 & 3 & 0 \end{bmatrix}$, the *matrix of coefficients*, $\mathsf{x} = \begin{bmatrix} x \\ y \\ z \end{bmatrix}$ and $\mathsf{b} = \begin{bmatrix} -9 \\ 0 \\ -3 \end{bmatrix}$. Next, we write down the *augmented matrix*

$$[A \mid \mathsf{b}] = \left[\begin{array}{ccc|c} -2 & 1 & 1 & -9 \\ 1 & 1 & 1 & 0 \\ 1 & 3 & 0 & -3 \end{array}\right],$$

which consists of A followed by a vertical line and then b. Just keeping track of coefficients, with augmented matrices, the solution to system (3) now looks like this:

$$\left[\begin{array}{ccc|c} -2 & 1 & 1 & -9 \\ 1 & 1 & 1 & 0 \\ 1 & 3 & 0 & -3 \end{array}\right] \xrightarrow{R1 \leftrightarrow R2} \left[\begin{array}{ccc|c} 1 & 1 & 1 & 0 \\ -2 & 1 & 1 & -9 \\ 1 & 3 & 0 & -3 \end{array}\right]$$

$$\xrightarrow[R3 \to R3 - R1]{R2 \to R2 + 2(R1)} \left[\begin{array}{ccc|c} 1 & 1 & 1 & 0 \\ 0 & 3 & 3 & -9 \\ 0 & 2 & -1 & -3 \end{array}\right] \xrightarrow{R2 \to \frac{1}{3}R2} \left[\begin{array}{ccc|c} 1 & 1 & 1 & 0 \\ 0 & 1 & 1 & -3 \\ 0 & 2 & -1 & -3 \end{array}\right] \qquad (6)$$

$$\xrightarrow{R3 \to R3 - 2(R2)} \left[\begin{array}{ccc|c} 1 & 1 & 1 & 0 \\ 0 & 1 & 1 & -3 \\ 0 & 0 & -3 & 3 \end{array}\right]$$

Here we have used R for **row** rather than **E** for *equation* to describe the changes to each system, writing, for example, $R3 \to R3 - 2(R2)$ at the last step, rather than $E3 \to E3 - 2(E2)$ as we did before. The three basic operations that we have been performing on equations now become operations on rows, called the *elementary row operations.*

2.4.1 The Elementary Row Operations

1. $R \leftrightarrow R'$: Interchange two rows;
2. $R \to cR$: Multiply a row by a scalar $c \neq 0$;
3. $R \to R - cR'$: Replace a row by that row less a multiple of another row.

We could equally well have stated the third elementary row operation as

3. replace a row by that row **plus** a multiple of another,

since $R + cR' = R - (-c)R'$, but, as we said before, there are advantages to using the language of subtraction. In particular, when doing so, the multiple has a special name. It's called a *multiplier*, $-c$ in our example, a word that will be important to us later.

2.4.2 Definitions. The *(main) diagonal* of an $m \times n$ matrix A is the set of elements $a_{11}, a_{22}, \ldots$. A matrix is *diagonal* if its only nonzero entries lie on the main diagonal, *upper triangular* if all entries below (and to the left) of the main diagonal are 0, and *lower triangular* if all entries above (and to the right of) the main diagonal are 0.

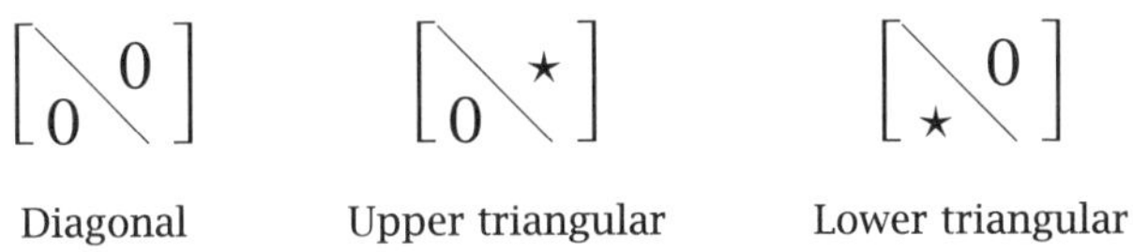

Figure 2.1: Schematic representations of a diagonal, upper triangular and lower triangular matrix.

2.4.3 Examples. The main diagonal of $\begin{bmatrix} 1 & 2 & 3 \\ 4 & 5 & 6 \\ 7 & 8 & 9 \end{bmatrix}$ is $1, 5, 9$, and the diagonal of $\begin{bmatrix} 7 & 9 & 3 \\ 6 & 5 & 4 \end{bmatrix}$ is $7, 5$. (Sometimes we omit "main.") The matrix $\begin{bmatrix} -3 & 4 & 1 & 1 \\ 0 & 8 & 2 & -3 \end{bmatrix}$ is upper triangular while the matrix $\begin{bmatrix} 1 & 0 \\ -2 & 1 \end{bmatrix}$ is lower triangular. The matrix $\begin{bmatrix} 1 & 0 & 0 \\ 0 & 0 & 0 \\ 0 & 0 & -2 \end{bmatrix}$ is diagonal. ☺

To solve a system of equations, the goal is to transform a matrix into a special kind of upper triangular matrix called **row echelon**.

2.4.4 Definition. A *row echelon matrix* is an upper triangular matrix with the following properties.

1. If there are any rows consisting entirely of 0s, these are at the bottom.
2. The pivots step from left to right as you read down the matrix. ("Pivot" means the first nonzero entry in a nonzero row).
3. All the entries in a column below a pivot are 0.

A *row echelon form* of a matrix A is a row echelon matrix U to which A can be moved via elementary row operations.

The concept of "pivot" applies to any matrix.

2.4.5 Definition. A *pivot* of a row echelon matrix U is the first nonzero number in a nonzero row. A column containing a pivot is called a *pivot column* of U. If U is a row echelon form of a matrix A, the pivots of A are the pivots of U and the pivot columns of A are the columns of A that become pivot columns of U.

2.4.6 Examples. • The matrix $U = \begin{bmatrix} 0 & 0 & \textcircled{2} & 1 & 5 \\ 0 & 0 & 0 & 0 & \textcircled{1} \\ 0 & 0 & 0 & 0 & 0 \end{bmatrix}$ is a row echelon matrix. The pivots are the circled entries. The pivot columns are columns three and five.

- $U = \begin{bmatrix} 0 & 0 & 2 & 1 & 5 \\ 0 & 0 & 0 & 0 & 1 \\ 0 & 0 & 0 & 0 & 3 \end{bmatrix}$ is an upper triangular matrix that is not row echelon. If row echelon, the 1 in the $(2,5)$ position would be a pivot, but then the number below it should be 0, not 3.

- $U = \begin{bmatrix} \textcircled{1} & 1 & 0 & 0 & 1 \\ 0 & 0 & \textcircled{-2} & -3 & 4 \\ 0 & \textcircled{-2} & 0 & -3 & -14 \\ 0 & 0 & 0 & \textcircled{6} & 9 \end{bmatrix}$ is also not row echelon because the pivots (the circled numbers) do not step left to right as you read down the matrix.

- The matrix $U = \begin{bmatrix} \textcircled{7} & 1 & 3 & 4 & 5 \\ 0 & 0 & \textcircled{-2} & 3 & 1 \\ 0 & 0 & 0 & 0 & \textcircled{4} \end{bmatrix}$ is a row echelon matrix. The pivots are circled. Since these are in columns one, three and five, the pivot columns are columns one, three and five.

- $U = \begin{bmatrix} 1 & -1 & -1 \\ 0 & 4 & 1 \\ 0 & 0 & 0 \end{bmatrix}$ is a row echelon form of $A = \begin{bmatrix} -2 & 2 & 2 \\ 1 & -1 & -1 \\ 1 & 3 & 0 \end{bmatrix}$ as the following sequence of elementary row operations shows:

$$A \to \begin{bmatrix} 1 & -1 & -1 \\ -2 & 2 & 2 \\ 1 & 3 & 0 \end{bmatrix} \to \begin{bmatrix} 1 & -1 & -1 \\ 0 & 0 & 0 \\ 0 & 4 & 1 \end{bmatrix} \to \begin{bmatrix} \textcircled{1} & -1 & -1 \\ 0 & \textcircled{4} & 1 \\ 0 & 0 & 0 \end{bmatrix}.$$

So the pivots of A are 1 and 4 (the pivots of U) and the pivot columns of A are its first two columns. Here is a different sequence of elementary row operations that begins by multiplying the first row of A by $\frac{1}{2}$:

$$A \to \begin{bmatrix} -1 & 1 & 1 \\ 1 & -1 & -1 \\ 1 & 3 & 0 \end{bmatrix} \to \begin{bmatrix} -1 & 1 & 1 \\ 0 & 0 & 0 \\ 0 & 4 & 1 \end{bmatrix} \to \begin{bmatrix} \textcircled{-1} & 1 & 1 \\ 0 & \textcircled{4} & 1 \\ 0 & 0 & 0 \end{bmatrix}$$

This is a different row echelon matrix from the previous (so we see that row echelon form is not unique), the pivots of A have changed—they are now -1 and 4—but the pivot columns remain columns one and two. This example shows that the pivots of a matrix are not unique, but the pivot columns are: the pivot columns of a matrix do not depend on a particular row echelon form of that matrix. :-)

A row echelon matrix takes its name from the French word "échelon" meaning "step." When a matrix A has been transformed by elementary row operations to a row echelon matrix (we say that A has been moved to *row echelon form*), the path formed by the pivots resembles a staircase.

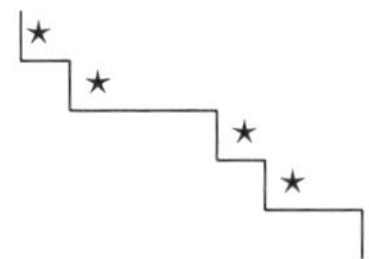

2.4.7 Examples. Here are some more matrices that are not in row echelon form:

$$\begin{bmatrix} 2 & 0 & 7 \\ 0 & 0 & 1 \\ 0 & 0 & 1 \end{bmatrix}, \quad \begin{bmatrix} 6 & 0 & 3 & 2 \\ 0 & 0 & 5 & 4 \\ 0 & 1 & 0 & 2 \end{bmatrix}, \quad \begin{bmatrix} 3 & 0 & -1 & 5 \\ 0 & 0 & 0 & 0 \\ 0 & -4 & 2 & -3 \end{bmatrix}.$$:-)

READING CHECK 1. Explain why each of these matrices is not row echelon.

2.4.8 Examples. • Let $A = \begin{bmatrix} 3 & -2 & 1 & -4 \\ 6 & -4 & 8 & -7 \\ 9 & -6 & 9 & -12 \end{bmatrix}$. Then

$$A \xrightarrow[R3 \to R3 - 3(R1)]{R2 \to R2 - 2(R1)} \begin{bmatrix} 3 & -2 & 1 & -4 \\ 0 & 0 & 6 & 1 \\ 0 & 0 & 6 & 0 \end{bmatrix} \xrightarrow{R3 \to R3 - 6(R2)} \begin{bmatrix} 3 & -2 & 1 & -4 \\ 0 & 0 & 6 & 1 \\ 0 & 0 & 0 & -1 \end{bmatrix} = U,$$

so one possible set of pivots of A is $\{3, 6, -1\}$. The pivot columns of A are one, three and four.

• Let $A = \begin{bmatrix} -2 & 3 & 4 \\ 1 & 0 & 1 \\ 4 & -5 & 6 \end{bmatrix}$. Then

$$A \xrightarrow{R1 \leftrightarrow R2} \begin{bmatrix} 1 & 0 & 1 \\ -2 & 3 & 4 \\ 4 & -5 & 6 \end{bmatrix} \xrightarrow[R3 \to R3 - 4(R1)]{R2 \to R2 + 2(R1)} \begin{bmatrix} 1 & 0 & 1 \\ 0 & 3 & 6 \\ 0 & -5 & 2 \end{bmatrix}$$

$$\xrightarrow{R3 \to R3 + \frac{5}{3}(R2)} \begin{bmatrix} 1 & 0 & 1 \\ 0 & 3 & 6 \\ 0 & 0 & 12 \end{bmatrix} = U,$$

so 1, 3 and 12 are pivots of A and the pivot columns of A are columns one, two and three. Perhaps a more natural way to move to row echelon form is this:

$$A \to \begin{bmatrix} 1 & 0 & 1 \\ -2 & 3 & 4 \\ 4 & -5 & 6 \end{bmatrix} \to \begin{bmatrix} 1 & 0 & 1 \\ 0 & 3 & 6 \\ 0 & -5 & 2 \end{bmatrix}$$
$$\to \begin{bmatrix} 1 & 0 & 1 \\ 0 & 1 & 2 \\ 0 & -5 & 2 \end{bmatrix} \to \begin{bmatrix} 1 & 0 & 1 \\ 0 & 1 & 2 \\ 0 & 0 & 12 \end{bmatrix}.$$

This is another row echelon form, the pivots have changed (they are now 1, 1, 12), but the pivot columns have not. We are reminded again that row echelon form is not unique and the pivots of a matrix are not unique, but the pivot columns are unique. Here is yet another reduction of A to a row echelon matrix.

$$A \to \begin{bmatrix} -2 & 3 & 4 \\ 0 & \frac{3}{2} & 3 \\ 0 & 1 & 14 \end{bmatrix} \to \begin{bmatrix} -2 & 3 & 4 \\ 0 & \frac{3}{2} & 3 \\ 0 & 0 & 12 \end{bmatrix}.$$

Now the pivots are $-2, \frac{3}{2}, 12$. :-)

READING CHECK 2. Find a set of pivots and identify the pivot columns of $A = \begin{bmatrix} -3 & 6 & 4 \\ 2 & -4 & 0 \end{bmatrix}$.

2.4.9 Remark. Since a matrix can be transformed to different row echelon matrices by different sequences of elementary row operations, we refer to **a** row echelon form, not **the** row echelon form of a matrix and to **a** set of pivots, not **the** pivots. Remember, however, that while the pivots of a matrix are not unique, the number of pivots and the columns in which they appear (the pivot columns) are indeed unique. Uniqueness of the number of pivots follows because this number is the "dimension" of the row space—see Chapter 4. Uniqueness of the pivot columns is a consequence of uniqueness of the "reduced row echelon form." For this, Appendix A of *Linear Algebra and its Applications* by David C. Lay, Addison-Wesley (2000).

2.4.10 Remark. It was long the custom, and some authors still require, that the leading nonzero entry in a nonzero row of a row echelon matrix be 1. A row echelon matrix in which all pivots are 1 and all column entries above as well as below a pivot are 0 is said to be in *reduced row echelon form*, a concept to which we return briefly in Section 2.9. Converting pivots to 1s makes a lot of sense when doing calculations by hand, but is much less important today with the ready availability of computing power. Having said this, in the rest of this section, we will often divide each row by a first nonzero entry, thus making that first entry a 1. Putting 0s below a 1 by hand is much easier (and less susceptible to errors) than putting 0s below $\frac{2}{3}$, for instance.

2.4.11 Example. Suppose we wish to solve the system

$$\begin{aligned} 2x - y + z &= -7 \\ x + y + z &= -2 \\ 3x + y + z &= 0 \end{aligned}$$

without the use of a computer. The steps of Gaussian elimination we might employ are these:

$$[A \mid \mathsf{b}] = \left[\begin{array}{rrr|r} 2 & -1 & 1 & -7 \\ 1 & 1 & 1 & -2 \\ 3 & 1 & 1 & 0 \end{array}\right] \xrightarrow{R1 \leftrightarrow R2} \left[\begin{array}{rrr|r} 1 & 1 & 1 & -2 \\ 2 & -1 & 1 & -7 \\ 3 & 1 & 1 & 0 \end{array}\right]$$

$$\xrightarrow[R3 \to R3 - 3(R1)]{R2 \to R2 - 2(R1)} \left[\begin{array}{rrr|r} 1 & 1 & 1 & -2 \\ 0 & -3 & -1 & -3 \\ 0 & -2 & -2 & 6 \end{array}\right] \xrightarrow{R2 \leftrightarrow R3} \left[\begin{array}{rrr|r} 1 & 1 & 1 & -2 \\ 0 & -2 & -2 & 6 \\ 0 & -3 & -1 & -3 \end{array}\right]$$

$$\xrightarrow{R2 \to (-\frac{1}{2})R2} \left[\begin{array}{rrr|r} 1 & 1 & 1 & -2 \\ 0 & 1 & 1 & -3 \\ 0 & -3 & -1 & -3 \end{array}\right] \xrightarrow{R3 \to R3 + 3(R2)} \left[\begin{array}{rrr|r} 1 & 1 & 1 & -2 \\ 0 & 1 & 1 & -3 \\ 0 & 0 & 2 & -12 \end{array}\right]$$

$$\xrightarrow{R3 \to \frac{1}{2}(R3)} \left[\begin{array}{rrr|r} 1 & 1 & 1 & -2 \\ 0 & 1 & 1 & -3 \\ 0 & 0 & 1 & -6 \end{array}\right].$$

The final matrix is in row echelon form and, by back substitution, we quickly read off the solution: $z = -6$, $y = -3 - z = 3$, $x = -2 - y - z = 1$.

We went to a lot of effort to make all pivots 1. Look how much more efficiently the process would proceed if we didn't so insist:

$$[A \mid \mathsf{b}] = \left[\begin{array}{rrr|r} 2 & -1 & 1 & -7 \\ 1 & 1 & 1 & -2 \\ 3 & 1 & 1 & 0 \end{array}\right] \xrightarrow[R3 \to R3 - \frac{3}{2}(R1)]{R2 \to R2 - \frac{1}{2}(R1)} \left[\begin{array}{rrr|r} 2 & -1 & 1 & -7 \\ 0 & \frac{3}{2} & \frac{1}{2} & \frac{3}{2} \\ 0 & \frac{5}{2} & -\frac{1}{2} & \frac{21}{2} \end{array}\right]$$

$$\xrightarrow{R3 \to R3 - \frac{5}{3}(R2)} \left[\begin{array}{rrr|r} 2 & -1 & 1 & -7 \\ 0 & \frac{3}{2} & \frac{1}{2} & \frac{3}{2} \\ 0 & 0 & -\frac{4}{3} & 8 \end{array}\right]$$

(two steps instead of six). The second elementary row operation—multiply a row by a nonzero scalar—is not very important. It is useful when doing calculations by hand, but otherwise not particularly. ☺

The process by which we use the elementary row operations to transform a matrix to row echelon form, and which we have illustrated repeatedly throughout this section, is called *Gaussian elimination* after the German mathematician Karl Friedrich Gauss (1777–1855), who first used this process to solve systems of 17 linear equations in order to determine the orbit of a new planet.

2.4.12 Gaussian Elimination. To move a matrix to a row echelon matrix, follow these steps.

1. If the matrix is all 0, do nothing. Otherwise, change the order of the rows, if necessary, to ensure that the first nonzero entry in the first row is a leftmost nonzero entry in the matrix.
2. Use the second elementary row operation, if you like, to get a more desirable first nonzero entry. This becomes the first pivot.
3. Use the third elementary row operation to get 0s in the rest of the column below this pivot.
4. Repeat steps 1, 2 and 3 on the submatrix immediately to the right of and below the pivot.
5. Repeat step 4 until you reach the last nonzero row. The first nonzero entry in this row is the last pivot.

If all this looks complicated, it's only because the procedure has to be written to cover all possible situations, and there are many possible situations. Here are some examples we hope will convince you that Gausssian elimination is not tough at all.

2.4.13 Example. We use Gaussian elimination to move $A = \begin{bmatrix} -1 & 2 & 3 \\ 4 & 1 & 0 \\ -2 & 5 & 8 \end{bmatrix}$ to a row echelon matrix. The -1 in the $(1,1)$-position is just what we need. It's our first pivot and we use this and the third elementary operation to put 0s in column one below the pivot, like this.

$$A \to \begin{bmatrix} \textcircled{-1} & 2 & 3 \\ 0 & 9 & 12 \\ 0 & 1 & 2 \end{bmatrix} \tag{7}$$

Now we have two choices. Working by hand, we might interchange rows two and three in order to get a 1 in the $(2,2)$-position:

$$\begin{bmatrix} -1 & 2 & 3 \\ 0 & 9 & 12 \\ 0 & 1 & 2 \end{bmatrix} \to \begin{bmatrix} -1 & 2 & 3 \\ 0 & \textcircled{1} & 2 \\ 0 & 9 & 12 \end{bmatrix}$$

and use this 1 (the second pivot) and the third elementary row operation to put a 0 where the 9 is:

$$\begin{bmatrix} -1 & 2 & 3 \\ 0 & 1 & 2 \\ 0 & 9 & 12 \end{bmatrix} \to \begin{bmatrix} -1 & 2 & 3 \\ 0 & 1 & 2 \\ 0 & 0 & -6 \end{bmatrix}.$$

Or, returning to (7), if we do not interchange rows two and three, we could leave the second row as it is (or divide by 9) and use the $(2,2)$-entry to put a 0 below. So, either

$$\begin{bmatrix} -1 & 2 & 3 \\ 0 & 9 & 12 \\ 0 & 1 & 2 \end{bmatrix} \to \begin{bmatrix} -1 & 2 & 3 \\ 0 & 9 & 12 \\ 0 & 0 & 2 - \frac{12}{9} \end{bmatrix} = \begin{bmatrix} -1 & 2 & 3 \\ 0 & 9 & 12 \\ 0 & 0 & \frac{2}{3} \end{bmatrix}$$

or

$$\begin{bmatrix} -1 & 2 & 3 \\ 0 & 9 & 12 \\ 0 & 1 & 2 \end{bmatrix} \to \begin{bmatrix} -1 & 2 & 3 \\ 0 & 1 & \frac{4}{3} \\ 0 & 1 & 2 \end{bmatrix} \to \begin{bmatrix} -1 & 2 & 3 \\ 0 & 1 & \frac{4}{3} \\ 0 & 0 & \frac{2}{3} \end{bmatrix}.$$

There are many ways to skin a cat! ☺

2.4.14 Example. Sometimes we have to fiddle to make a desired entry nonzero. Suppose $A = \begin{bmatrix} -2 & 1 & 0 & 1 \\ 3 & -3 & 1 & 1 \\ 1 & -1 & 1 & 5 \end{bmatrix}$. Here's a start at Gaussian elimination:

$$A \to \begin{bmatrix} 1 & -1 & 1 & 5 \\ 3 & -3 & 1 & 1 \\ -2 & 1 & 0 & 1 \end{bmatrix}$$

(interchanging rows one and three because it is easier to work by hand with a 1 in the top left corner). Now we make the rest of column one all 0. Thus we begin

$$A \to \begin{bmatrix} 1 & -1 & 1 & 5 \\ 3 & -3 & 1 & 1 \\ -2 & 1 & 0 & 1 \end{bmatrix} \to \begin{bmatrix} 1 & -1 & 1 & 5 \\ 0 & 0 & -2 & -14 \\ 0 & -1 & 2 & 11 \end{bmatrix}.$$

To work on column two, we would like the $(2,2)$ entry to be nonzero. Here then, we "fiddle," interchanging rows two and three and continuing, like this

$$\to \begin{bmatrix} 1 & -1 & 1 & 5 \\ 0 & -1 & 2 & 11 \\ 0 & 0 & -2 & -14 \end{bmatrix}.$$

Column two is fine and—Hey! this matrix is row echelon—so we are finished. $\ddot{\smile}$

2.4.15 Example. It's not always possible to "fiddle." Here's the start of Gaussian elimination:

$$\begin{bmatrix} 1 & -2 & -1 & 1 \\ -3 & 6 & 5 & 1 \\ -1 & 2 & 4 & 3 \\ 1 & -2 & 1 & 1 \end{bmatrix} \to \begin{bmatrix} 1 & -2 & -1 & 1 \\ 0 & 0 & 2 & 4 \\ 0 & 0 & 3 & 4 \\ 0 & 0 & 2 & 0 \end{bmatrix}.$$

There is no way to make the $(2, 2)$ entry a 0 (we are now working entirely with the 3×3 submatrix to the right of and below the $(1, 1)$ entry). The Gaussian elimination process requires us to move to column three. Here is one way to achieve row echelon form:

$$\to \begin{bmatrix} 1 & -2 & -1 & 1 \\ 0 & 0 & 1 & 2 \\ 0 & 0 & 3 & 4 \\ 0 & 0 & 2 & 0 \end{bmatrix} \to \begin{bmatrix} 1 & -2 & -1 & 1 \\ 0 & 0 & 1 & 2 \\ 0 & 0 & 0 & -2 \\ 0 & 0 & 0 & -4 \end{bmatrix} \to \begin{bmatrix} 1 & -2 & -1 & 1 \\ 0 & 0 & 1 & 2 \\ 0 & 0 & 0 & -2 \\ 0 & 0 & 0 & 0 \end{bmatrix}. \qquad \ddot{\smile}$$

The purpose of Gaussian elimination is to simplify the system $A\mathsf{x} = \mathsf{b}$ by moving the augmented matrix $[A \mid \mathsf{b}]$ to a row echelon matrix $[U \mid \mathsf{c}]$. The systems $A\mathsf{x} = \mathsf{b}$ and $U\mathsf{x} = \mathsf{c}$ are equivalent—they have the same solutions—and it is easy to solve $U\mathsf{x} = \mathsf{c}$ by back substitution. So we have a procedure for solving $A\mathsf{x} = \mathsf{b}$.

2.4.16 To Solve a System of Linear Equations

1. Write the system in the form $A\mathsf{x} = \mathsf{b}$, where A is the matrix of coefficients, x is the vector of unknowns and b is the vector whose components are the constants to the right of the equals signs.

2. Add a vertical line and b to the right of A to form the *augmented matrix* $[A \mid \mathsf{b}]$.

3. Use Gaussian elimination to change $[A \mid \mathsf{b}]$ to a row echelon matrix $[U \mid \mathsf{c}]$.

4. Write down the equations that correspond to $[U \mid \mathsf{c}]$ and solve by back substitution.

The system (3) we considered earlier in this section had $x = 3$, $y = -2$, $z = -1$ for the solution. The solution was "unique," just one. There are other possibilities.

2.4.17 Problem. Solve the system
$$\begin{aligned} x - y + z &= 2 \\ 2x + y - 6z &= 7 \\ 6x - 10z &= 5. \end{aligned}$$

Solution. This is $A\mathsf{x} = \mathsf{b}$ with $A = \begin{bmatrix} 1 & -1 & 1 \\ 2 & 1 & -6 \\ 6 & 0 & -10 \end{bmatrix}$, $\mathsf{x} = \begin{bmatrix} x \\ y \\ z \end{bmatrix}$ and $\mathsf{b} = \begin{bmatrix} 2 \\ 7 \\ 5 \end{bmatrix}$. Here is one way to move $[A \mid \mathsf{b}]$ to row echelon form via Gaussian elimination:

$$[A \mid \mathsf{b}] = \left[\begin{array}{rrr|r} 1 & -1 & 1 & 2 \\ 2 & 1 & -6 & 7 \\ 6 & 0 & -10 & 5 \end{array}\right]$$

$$\xrightarrow[R3 \to R3 - 6(R1)]{R2 \to R2 - 2(R1)} \left[\begin{array}{rrr|r} 1 & -1 & 1 & 2 \\ 0 & 3 & -8 & 3 \\ 0 & 6 & -16 & -7 \end{array}\right] \xrightarrow{R3 \to R3 - 2(R2)} \left[\begin{array}{rrr|r} 1 & -1 & 1 & 2 \\ 0 & 3 & -8 & 3 \\ 0 & 0 & 0 & -13 \end{array}\right].$$

The equation that corresponds to the last row is $0x + 0y + 0z = -13$, that is, $0 = -13$. Since this is obviously wrong, the system of equations here has no solution. A system like this, with no solution, is called *inconsistent.*

2.4.18 Problem. Solve the system
$$\begin{aligned} 2x - 3y + 6z &= 14 \\ x + y - 2z &= -3. \end{aligned}$$

Solution. This is $A\mathsf{x} = \mathsf{b}$ with

$$A = \begin{bmatrix} 2 & -3 & 6 \\ 1 & 1 & -2 \end{bmatrix}, \quad \mathsf{x} = \begin{bmatrix} x \\ y \\ z \end{bmatrix} \quad \text{and} \quad \mathsf{b} = \begin{bmatrix} 14 \\ -3 \end{bmatrix}. \tag{8}$$

We form the augmented matrix

$$[A \mid \mathsf{b}] = \left[\begin{array}{rrr|r} 2 & -3 & 6 & 14 \\ 1 & 1 & -2 & -3 \end{array}\right].$$

While there is no need for initial row interchanges, it is easier to work with a 1 rather than a 2 in the $(1,1)$ position, so we interchange rows one and two and then apply the third elementary row operation to get a 0 below the pivot.

$$\left[\begin{array}{rrr|r} 2 & -3 & 6 & 14 \\ 1 & 1 & -2 & -3 \end{array}\right] \xrightarrow{R1 \leftrightarrow R2} \left[\begin{array}{rrr|r} 1 & 1 & -2 & -3 \\ 2 & -3 & 6 & 14 \end{array}\right]$$

$$\xrightarrow{R2 \to R2 - 2(R1)} \left[\begin{array}{rrr|r} 1 & 1 & -2 & -3 \\ 0 & -5 & 10 & 20 \end{array}\right].$$

This is row echelon form. We can stop here or divide the last row by -5 obtaining

$$\left[\begin{array}{rrr|r} 1 & 1 & -2 & -3 \\ 0 & 1 & -2 & -4 \end{array}\right].$$

The equations corresponding to this matrix are

$$\begin{aligned} x + y - 2z &= -3 \\ y - 2z &= -4. \end{aligned}$$

We begin back substitution with the last equation, rewriting as $y = -4 + 2z$. The first equation says $x = -3 - y + 2z = -3 - (-4 + 2z) + 2z = 1$. We have conditions on y and x, but there is no restriction on z. To emphasize this fact, we introduce the *parameter* t, write $z = t$ and the solution like this:

$$\begin{aligned} x &= 1 \\ y &= -4 + 2t \\ z &= t. \end{aligned}$$

The variable z is called *free* because, as the equation $z = t$ tries to show, z can be any number.[5] The system of equations in this problem has infinitely many solutions, one for each value of t. Writing the solution as a single vector, we have

$$\begin{bmatrix} x \\ y \\ z \end{bmatrix} = \begin{bmatrix} 1 \\ -4 + 2t \\ t \end{bmatrix} = \begin{bmatrix} 1 \\ -4 \\ 0 \end{bmatrix} + t \begin{bmatrix} 0 \\ 2 \\ 1 \end{bmatrix}.$$

Such an equation should look familiar. It is the equation of a line, the one through $(1, -4, 0)$ with direction $\mathsf{d} = \begin{bmatrix} 0 \\ 2 \\ 1 \end{bmatrix}$. Geometrically, this is what we expect. Each of the equations in the system we have solved describes a plane in 3-space. When we solve the system, we are finding the points that lie on both planes. Since the planes are not parallel, the points that lie on both form a line. 👍

READING CHECK 3. How do you know that the planes described by the equations in Problem 2.4.18 are not parallel?

Each solution to the linear system in Problem 2.4.18 is the sum of the vector $\begin{bmatrix} 1 \\ -4 \\ 0 \end{bmatrix}$ and a scalar multiple of $\begin{bmatrix} 0 \\ 2 \\ 1 \end{bmatrix}$. The vector $\begin{bmatrix} 1 \\ -4 \\ 0 \end{bmatrix}$ is a solution called a *particular solution*. You should check that the vector $\begin{bmatrix} 0 \\ 2 \\ 1 \end{bmatrix}$ is a solution to the system $A\mathsf{x} = 0$. It is characteristic of linear systems that whenever there is more than one solution, each solution is the sum of a particular solution and the solution to $A\mathsf{x} = 0$. We shall have more to say about this in Section 2.6.

[5]Free variables are those associated with columns that are not pivot columns, that is, the *nonpivot columns*. More explanation soon!

READING CHECK 4. Verify that $\begin{bmatrix} 1 \\ -4 \\ 0 \end{bmatrix}$ is a solution to the system $A\mathsf{x} = \mathsf{b}$ in Problem 2.4.18 and that $\begin{bmatrix} 0 \\ 2 \\ 1 \end{bmatrix}$ is a solution to $A\mathsf{x} = \begin{bmatrix} 0 \\ 0 \end{bmatrix}$.

2.4.19 Problem. Solve the system
$$\begin{aligned} 3y - 12z - 7w &= -15 \\ y - 4z \qquad &= 2 \\ -2x + 4y \qquad + 5w &= -1. \end{aligned}$$

Solution. This is $A\mathsf{x} = \mathsf{b}$ with

$$A = \begin{bmatrix} 0 & 3 & -12 & -7 \\ 0 & 1 & -4 & 0 \\ -2 & 4 & 0 & 5 \end{bmatrix}, \quad \mathsf{x} = \begin{bmatrix} x \\ y \\ z \\ w \end{bmatrix} \text{ and } \mathsf{b} = \begin{bmatrix} -15 \\ 2 \\ -1 \end{bmatrix}.$$

We form the augmented matrix and immediately interchange rows one and three in order to ensure that the $(1,1)$ entry is different from 0:

$$\left[\begin{array}{cccc|c} 0 & 3 & -12 & -7 & -15 \\ 0 & 1 & -4 & 0 & 2 \\ -2 & 4 & 0 & 5 & -1 \end{array}\right] \xrightarrow{R1 \leftrightarrow R3} \left[\begin{array}{cccc|c} \textcircled{-2} & 4 & 0 & 5 & -1 \\ 0 & 1 & -4 & 0 & 2 \\ 0 & 3 & -12 & -7 & -15 \end{array}\right].$$

The first pivot is -2. Now we continue with the submatrix immediately to the right of and below this pivot. The second pivot is the 1 in the $(2,2)$ position. We use the third elementary row operation to put a 0 below it

$$\xrightarrow{R3 \to R3-3(R2)} \left[\begin{array}{cccc|c} \textcircled{-2} & 4 & 0 & 5 & -1 \\ 0 & \textcircled{1} & -4 & 0 & 2 \\ 0 & 0 & 0 & -7 & -21 \end{array}\right],$$

obtaining a row echelon matrix. Back substitution proceeds like this:

$$\begin{aligned} -7w &= -21 && \text{so } w = 3, \\ y - 4z &= 2, && \text{which we write as } y = 2 + 4z. \end{aligned}$$

This suggests that z is free, so we set $z = t$ and obtain $y = 2 + 4t$. Finally,

$$-2x + 4y + 5w = -1,$$

so $-2x = -1 - 4y - 5w = -1 - 4(2 + 4t) - 15 = -24 - 16t$ and $x = 12 + 8t$. Our solution vector is

$$\mathsf{x} = \begin{bmatrix} 12 + 8t \\ 2 + 4t \\ t \\ 3 \end{bmatrix} = \begin{bmatrix} 12 \\ 2 \\ 0 \\ 3 \end{bmatrix} + t \begin{bmatrix} 8 \\ 4 \\ 1 \\ 0 \end{bmatrix}.$$

As before, the vector $\begin{bmatrix} 12 \\ 2 \\ 0 \\ 3 \end{bmatrix}$ is a particular solution to $A\mathsf{x} = \mathsf{b}$ and $\begin{bmatrix} 8 \\ 4 \\ 1 \\ 0 \end{bmatrix}$ is a solution to $A\mathsf{x} = \mathsf{0}$.

In general, there are three possible outcomes when we attempt to solve a system of linear equations.

2.4.20 A system of linear equations may have

1. a unique solution,
2. no solution, or
3. infinitely many solutions.

The solution to system (3) was unique. In Problem 2.4.17, we met an inconsistent system (no solution) while Problems 2.4.18 and 2.4.19 presented us with systems that have infinitely many solutions. Interestingly, a linear system can't have just three solutions or five solutions or ten solutions, because if a solution is not unique, there is a free variable that can be chosen in infinitely many ways. But we are getting a bit ahead of ourselves!

READING CHECK 5. How many solutions does the system $\begin{aligned} x^2 + y^2 &= 8 \\ x - y &= 0 \end{aligned}$ have? (Note that this system is *not* linear!)

Free Variables Correspond to Nonpivot Columns

The pivot columns of a matrix are important for a variety of reasons. The variables that correspond to nonpivot columns are free; those that correspond to pivot columns are not. For example, look at the pivot in column three of

$$\left[\begin{array}{cccc|c} 1 & 1 & 3 & 4 & 5 \\ 0 & 0 & 1 & 3 & 1 \\ 0 & 0 & 0 & 1 & -3 \end{array}\right]. \tag{9}$$

The second equation reads $x_3 + 3x_4 = 1$. Thus $x_3 = 1 - 3x_4$ is not free; it is determined by x_4. On the other hand, a variable that corresponds to a column that is not a pivot column is always free. The equations that correspond to the rows of the matrix in (9) are

$$\begin{aligned} x_1 + x_2 + 3x_3 + 4x_4 &= 5 \\ x_3 + 3x_4 &= 1 \\ x_4 &= -3. \end{aligned}$$

The variables x_1, x_3 and x_4 are not free since

$$\begin{aligned} x_4 &= -3 \\ x_3 &= 1 - 3x_4 \\ x_1 &= 5 - x_2 - 3x_3 - 4x_4, \end{aligned}$$

but there is no equation of the form $x_2 = *$, so x_2 is free. To summarize,

2.4.21 Free variables are those that correspond to nonpivot columns.

READING CHECK 6. Suppose a system of equations in the variables x_1, x_2, x_3, x_4 leads to the row echelon matrix $\left[\begin{array}{cccc|c} 0 & 1 & 2 & 7 & 5 \\ 0 & 0 & 0 & 1 & 1 \\ 0 & 0 & 0 & 0 & 0 \end{array}\right]$. What are the free variables?

2.4.22 **Problem.** Solve $\begin{array}{rcrcrcrcr} 2x_1 & - & 2x_2 & - & x_3 & + & 4x_4 & = & 9 \\ -x_1 & + & x_2 & + & 2x_3 & + & x_4 & = & -3. \end{array}$

Solution. The given system is $A\mathsf{x} = \mathsf{b}$ where

$$A = \begin{bmatrix} 2 & -2 & -1 & 4 \\ -1 & 1 & 2 & 1 \end{bmatrix}, \quad \mathsf{x} = \begin{bmatrix} x_1 \\ x_2 \\ x_3 \\ x_4 \end{bmatrix} \quad \text{and} \quad \mathsf{b} = \begin{bmatrix} 9 \\ -3 \end{bmatrix}.$$

The augmented matrix is

$$[A \mid \mathsf{b}] = \left[\begin{array}{cccc|c} 2 & -2 & -1 & 4 & 9 \\ -1 & 1 & 2 & 1 & -3 \end{array}\right].$$

Gaussian elimination begins

$$\left[\begin{array}{cccc|c} 2 & -2 & -1 & 4 & 9 \\ -1 & 1 & 2 & 1 & -3 \end{array}\right] \xrightarrow{R1 \leftrightarrow R2} \left[\begin{array}{cccc|c} \textcircled{-1} & 1 & 2 & 1 & -3 \\ 2 & -2 & -1 & 4 & 9 \end{array}\right]$$

(the first pivot is the -1 in the $(1,1)$ position), and continues

$$\xrightarrow{R2 \to R2 + 2(R1)} \left[\begin{array}{cccc|c} \textcircled{-1} & 1 & 2 & 1 & -3 \\ 0 & 0 & \textcircled{3} & 6 & 3 \end{array}\right]$$

(using the third elementary row operation to put a 0 below the first pivot). We have reached row echelon form. The second pivot is 3, the first nonzero entry of the last row. The pivot columns are columns one and three. The nonpivot columns are columns two and four. The corresponding variables are free and we so signify by setting $x_2 = t$ and $x_4 = s$. The equations corresponding to the rows of the row echelon matrix are

$$\begin{aligned} -x_1 + x_2 + 2x_3 + x_4 &= -3 \\ 3x_3 + 6x_4 &= 3. \end{aligned}$$

We solve these by back substitution (that is, from the bottom up).

The second equation says $3x_3 = 3 - 6x_4 = 3 - 6s$, so $x_3 = 1 - 2s$.
The first equation says
$$\begin{aligned} -x_1 &= -x_2 - 2x_3 - x_4 - 3 \\ &= -t - 2(1-2s) - s - 3 \\ &= -t + 3s - 5, \quad \text{so } x_1 = t - 3s + 5. \end{aligned}$$

The solution is

$$\begin{aligned} x_1 &= t - 3s + 5 \\ x_2 &= t \\ x_3 &= 1 - 2s \\ x_4 &= s \end{aligned}$$

which, in vector form, is

$$\mathsf{x} = \begin{bmatrix} x_1 \\ x_2 \\ x_3 \\ x_4 \end{bmatrix} = \begin{bmatrix} t - 3s + 5 \\ t \\ 1 - 2s \\ s \end{bmatrix} = \begin{bmatrix} 5 \\ 0 \\ 1 \\ 0 \end{bmatrix} + t \begin{bmatrix} 1 \\ 1 \\ 0 \\ 0 \end{bmatrix} + s \begin{bmatrix} -3 \\ 0 \\ -2 \\ 1 \end{bmatrix}.$$

You should check that $\begin{bmatrix} 5 \\ 0 \\ 1 \\ 0 \end{bmatrix}$ is a particular solution to $A\mathsf{x} = \mathsf{b}$ while $t\begin{bmatrix} 1 \\ 1 \\ 0 \\ 0 \end{bmatrix} + s\begin{bmatrix} -3 \\ 0 \\ -2 \\ 1 \end{bmatrix}$ is the solution to $A\mathsf{x} = \mathsf{0}$.

2.4.23 Problem. Is $\begin{bmatrix} 1 \\ -6 \end{bmatrix}$ a linear combination of $\begin{bmatrix} 1 \\ -4 \end{bmatrix}$, $\begin{bmatrix} 2 \\ 0 \end{bmatrix}$ and $\begin{bmatrix} 3 \\ -1 \end{bmatrix}$?

Solution. The question asks if there exist scalars a, b, c so that

$$\begin{bmatrix} 1 \\ -6 \end{bmatrix} = a\begin{bmatrix} 1 \\ -4 \end{bmatrix} + b\begin{bmatrix} 2 \\ 0 \end{bmatrix} + c\begin{bmatrix} 3 \\ -1 \end{bmatrix} = \begin{bmatrix} 1 & 2 & 3 \\ -4 & 0 & -1 \end{bmatrix}\begin{bmatrix} a \\ b \\ c \end{bmatrix}$$

(using 2.1.33). So we want to know whether there exists a solution to $A\mathsf{x} = \mathsf{b}$, with $\mathsf{x} = \begin{bmatrix} a \\ b \\ c \end{bmatrix}$ and $\mathsf{b} = \begin{bmatrix} 1 \\ -6 \end{bmatrix}$. We apply Gaussian elimination to the augmented matrix as follows:

$$[A \mid \mathsf{b}] = \left[\begin{array}{ccc|c} 1 & 2 & 3 & 1 \\ -4 & 0 & -1 & -6 \end{array}\right] \to \left[\begin{array}{ccc|c} 1 & 2 & 3 & 1 \\ 0 & 8 & 11 & -2 \end{array}\right].$$

This is row echelon form. There is one nonpivot column, column three. Thus the third variable, c, is free. There are infinitely many solutions. Yes, the vector $\begin{bmatrix} 1 \\ -6 \end{bmatrix}$ is a linear combination of the other three.

READING CHECK 7. Find specific numbers a, b and c so that $\begin{bmatrix} 1 \\ -6 \end{bmatrix} = a\begin{bmatrix} 1 \\ -4 \end{bmatrix} + b\begin{bmatrix} 2 \\ 0 \end{bmatrix} + c\begin{bmatrix} 3 \\ -1 \end{bmatrix}$.

Do you, dear reader, feel more comfortable with numbers than letters? Why is that? Here are a couple of worked problems we hope will alleviate fears. You work with letters just as if they were numbers.

2.4.24 Problem. Show that the system $\begin{aligned} x + y + z &= a \\ x + 2y - z &= b \\ 2x - 2y + z &= c \end{aligned}$ has a solution for any values of a, b and c.

Solution. We apply Gaussian elimination to the augmented matrix.

$$\left[\begin{array}{ccc|c} 1 & 1 & 1 & a \\ 1 & 2 & -1 & b \\ 2 & -2 & 1 & c \end{array}\right] \to \left[\begin{array}{ccc|c} 1 & 1 & 1 & a \\ 0 & 1 & -2 & b-a \\ 0 & -4 & -1 & c-2a \end{array}\right]$$
$$\to \left[\begin{array}{ccc|c} 1 & 1 & 1 & a \\ 0 & 1 & -2 & b-a \\ 0 & 0 & -9 & (c-2a)+4(b-a) \end{array}\right].$$

The equation which corresponds to the last row is $-9z = -6a + 4b + c$. For any values of a, b, c, we can solve for z. The equation corresponding to the second row says $y = 2z + b - a$, so knowing z gives us y. Finally we determine x from the first row.

2.4.25 Problem. Find conditions on a and b under which the system

$$\begin{aligned} x + y + z &= -2 \\ x + 2y - z &= 1 \\ 2x + ay + bz &= 2 \end{aligned}$$

has

i. no solution,

ii. infinitely many solutions,

iii. a unique solution.

Solution. We apply Gaussian elimination to the augmented matrix (exactly as if the letters were numbers).

$$\left[\begin{array}{ccc|c} 1 & 1 & 1 & -2 \\ 1 & 2 & -1 & 1 \\ 2 & a & b & 2 \end{array}\right] \to \left[\begin{array}{ccc|c} 1 & 1 & 1 & -2 \\ 0 & 1 & -2 & 3 \\ 0 & a-2 & b-2 & 6 \end{array}\right]$$
$$\to \left[\begin{array}{ccc|c} 1 & 1 & 1 & -2 \\ 0 & 1 & -2 & 3 \\ 0 & 0 & (b-2)+2(a-2) & 6-3(a-2) \end{array}\right].$$

The equation corresponding to the last row reads $2a + b - 6 = -3a + 12$.

i. If $2a + b - 6 = 0$ while $-3a + 12 \neq 0$, there is no solution. Thus there is no solution if $a \neq 4$ and $2a + b - 6 = 0$.

ii. If $2a + b - 6 = 0$ and $-3a + 12 = 0$, the equation corresponding to the last row reads $0 = 0$. The variable z is free and there are infinitely many solutions. So there are infinitely many solutions in the case $a = 4$, $b = -2$.

iii. If $2a + b - 6 \neq 0$, we can divide the third equation by this number and determine z (uniquely). Then, continuing with back substitution, we can determine y and x. There is a unique solution if $2a + b - 6 \neq 0$.

Answers to Reading Checks

1. The 1 in the $(2,3)$ position of $\begin{bmatrix} 2 & 0 & 7 \\ 0 & 0 & 1 \\ 0 & 0 & 1 \end{bmatrix}$ is a pivot, so the number below it should be a 0.

   The leading nonzero entries in $\begin{bmatrix} 6 & 0 & 3 & 2 \\ 0 & 0 & 5 & 4 \\ 0 & 1 & 0 & 2 \end{bmatrix}$ are 6, 5 and 1, but these do not step from left to right as you read down the matrix.

   The zero row of $\begin{bmatrix} 3 & 0 & -1 & 5 \\ 0 & 0 & 0 & 0 \\ 0 & -4 & 2 & -3 \end{bmatrix}$ is not at the bottom of the matrix.

2. Since $A \to \begin{bmatrix} -3 & 6 & 4 \\ 0 & 0 & \frac{8}{3} \end{bmatrix} = U$, the pivots are -3 and $\frac{8}{3}$, and the pivot columns are columns one and three.

3. The two normals, $\begin{bmatrix} 2 \\ -3 \\ 6 \end{bmatrix}$ and $\begin{bmatrix} 1 \\ 1 \\ -2 \end{bmatrix}$, are not parallel.

4. $A\begin{bmatrix}1\\-4\\0\end{bmatrix} = \begin{bmatrix}2 & -3 & 6\\1 & 1 & -2\end{bmatrix}\begin{bmatrix}1\\-4\\0\end{bmatrix} = \begin{bmatrix}14\\-3\end{bmatrix} = \mathsf{b}$;

$A\begin{bmatrix}0\\2\\1\end{bmatrix} = \begin{bmatrix}2 & -3 & 6\\1 & 1 & -2\end{bmatrix}\begin{bmatrix}0\\2\\1\end{bmatrix} = \begin{bmatrix}0\\0\end{bmatrix}$.

5. Setting $x = y$ in the first equation, we obtain $2x^2 = 8$, so $x^2 = 4$. The system has *two* solutions, $x = 2$, $y = 2$ and $x = -2$, $y = -2$.

6. There are pivots in columns two and four. The nonpivot columns are columns one and three, so the corresponding variables x_1 and x_3 are free.

7. Continuing Problem 2.4.23, we solve the system whose augmented matrix has row echelon form $\left[\begin{array}{ccc|c}1 & 2 & 3 & 1\\0 & 8 & 11 & -2\end{array}\right]$. Thus $c = t$ is free, $8b = -2 - 11c$, so $b = -\frac{1}{4} - \frac{11}{8}t$ and $a + 2b + 3c = 1$, so that $a = 1 - 2b - 3c = 1 - 2(-\frac{1}{4} - \frac{11}{8}t) - 3t = \frac{3}{2} - \frac{1}{4}t$. One particular solution is obtained with $t = 0$, giving $a = \frac{3}{2}$, $b = -\frac{1}{4}$, $c = 0$. Checking our answer, we have $\begin{bmatrix}1\\-6\end{bmatrix} = \frac{3}{2}\begin{bmatrix}1\\-4\end{bmatrix} - \frac{1}{4}\begin{bmatrix}2\\0\end{bmatrix} + 0\begin{bmatrix}3\\-1\end{bmatrix}$.

True/False Questions

Decide, with as little calculation as possible, whether each of the following statements is true or false and, if you say "false," explain your answer. (Answers are at the back of the book.)

1. $\pi x_1 + \sqrt{3}x_2 + 5\sqrt{x_3} = \ln 7$ is a linear equation.

2. The systems $\begin{array}{rcrcrcr} x & - & y & - & z & = & 0\\ -4x & + & y & + & 2z & = & -2\\ 2x & + & y & - & z & = & 3\end{array}$ and $\begin{array}{rcrcrcr} x & - & y & - & z & = & 0\\ -4x & + & y & + & 2z & = & -2\\ & & 3y & + & z & = & 3\end{array}$ are equivalent.

3. It is possible for a system of linear equations to have exactly two (different) solutions.

4. The matrix $\begin{bmatrix}0 & 4 & 3 & -4 & 1\\0 & 0 & 0 & 5 & 1\\0 & 0 & 0 & 0 & -1\end{bmatrix}$ is in row echelon form.

5. If U is a row echelon form of A, the systems $A\mathsf{x} = \mathsf{b}$ and $U\mathsf{x} = \mathsf{b}$ have the same solutions.

6. If Gaussian elimination on an augmented matrix leads to $\left[\begin{array}{cccc|c}0 & 1 & 3 & -4 & 1\\0 & 0 & 0 & 1 & 1\\0 & 0 & 0 & 0 & 1\end{array}\right]$, the system has a unique solution.

7. In the solution to a system of linear equations, free variables are those that correspond to columns of row echelon form that contain the pivots.

8. If a linear system $A\mathsf{x} = \mathsf{b}$ has a solution, then b is a linear combination of the columns of A.

9. The pivots of the matrix $\begin{bmatrix} 4 & 0 & 2 \\ 0 & 7 & 1 \\ 0 & 0 & 0 \end{bmatrix}$ are 4, 7 and 0.

10. If every column of a matrix A is a pivot column, then the system $A\mathsf{x} = \mathsf{b}$ has a solution for every vector b.

EXERCISES

Answers to exercises marked [BB] can be found at the ***Back of the Book.***

1. Which of the following matrices are **not** in row echelon form? Give a reason in each case.

   (a) [BB] $\begin{bmatrix} 0 & 1 \\ 1 & 0 \end{bmatrix}$ **(b)** $\begin{bmatrix} 1 & 2 & 3 \\ 0 & 1 & 0 \\ 0 & 0 & 1 \end{bmatrix}$ **(c)** $\begin{bmatrix} 1 & 5 \\ 0 & 0 \\ 0 & 1 \\ 0 & 0 \\ 0 & 0 \end{bmatrix}$ **(d)** $\begin{bmatrix} 1 & 2 & 3 & 4 \\ 0 & 0 & -1 & 10 \\ 0 & 0 & 0 & 0 \end{bmatrix}$.

2. In each case, specify the value (or values) of a for which the matrix is **not** in row echelon form. Explain.

   (a) [BB] $\begin{bmatrix} 1 & 2 & 3 & 1 \\ 0 & a-2 & 0 & a-2 \\ 0 & 0 & a & 1 \end{bmatrix}$ **(b)** $\begin{bmatrix} 3 & 4 & 5 \\ a^2-9 & 1 & -2 \\ 0 & a+1 & 0 \end{bmatrix}$

   (c) $\begin{bmatrix} a & -2 & 1 \\ 0 & a-2 & 3 \\ 0 & a+2 & 4 \end{bmatrix}$ **(d)** $\begin{bmatrix} 1 & a & -1 & 1 \\ 0 & -3a & 1 & -2 \\ 0 & 0 & a-1 & a+2 \end{bmatrix}$.

3. Reduce each of the following matrices to row echelon form using only the third elementary row operation. In each case, identify the pivots and the pivot columns.

   (a) [BB] $\begin{bmatrix} 1 & -1 & -2 \\ 2 & -3 & -5 \\ -1 & 4 & 5 \end{bmatrix}$ **(b)** $\begin{bmatrix} 2 & 1 & 4 & 2 \\ 3 & 0 & 2 & 1 \\ 5 & 2 & 3 & 5 \end{bmatrix}$

   (c) $\begin{bmatrix} 1 & 4 & 5 & 2 \\ 3 & 13 & 20 & 8 \\ -2 & -10 & -16 & -4 \\ 1 & 10 & 38 & 28 \end{bmatrix}$ **(d)** $\begin{bmatrix} 1 & 2 & 1 & 3 & 1 \\ -1 & -1 & 2 & -1 & 4 \\ 2 & 8 & 15 & 11 & 26 \\ 0 & 2 & 8 & -2 & 21 \end{bmatrix}$.

4. Identify the pivot columns of each of the following matrices.

   (a) [BB] $\begin{bmatrix} 3 & -1 & 5 & 3 & 6 & -3 \\ 7 & -7 & 1 & 5 & 4 & 2 \\ -4 & 6 & 4 & -2 & 4 & 10 \\ 16 & -10 & 16 & 14 & 22 & -10 \\ -13 & 9 & -11 & -11 & -16 & 12 \end{bmatrix}$ **(b)** $\begin{bmatrix} 6 & -4 & 0 & -5 \\ 3 & -7 & -2 & 2 \\ -9 & 5 & 83 & 12 \\ -3 & 5 & 0 & 2 \\ -1 & 1 & 4 & -2 \\ 5 & 7 & 0 & 0 \end{bmatrix}$.

5. Find two different sets of pivots of each of the following matrices.

(a) $\begin{bmatrix} 2 & -1 & 1 & 0 & -4 \\ 0 & 1 & 3 & 6 & -4 \\ 1 & 2 & 1 & 1 & -7 \end{bmatrix}$ **(b)** $\begin{bmatrix} -3 & 3 & 1 & -10 & 9 \\ 2 & -2 & -2 & 7 & -4 \\ -1 & 1 & 3 & -4 & 1 \end{bmatrix}$.

6. In our discussion of the Hamming code in Section 2.2, we said that the two systems of parity check equations

$$\begin{aligned} x_2 + x_3 + x_4 + x_5 &= 0 \\ x_1 + x_3 + x_4 + x_6 &= 0 \\ x_1 + x_2 + x_4 + x_7 &= 0 \end{aligned}$$

and

$$\begin{aligned} x_4 + x_5 + x_6 + x_7 &= 0 \\ x_2 + x_3 + x_6 + x_7 &= 0 \\ x_1 + x_3 + x_5 + x_7 &= 0 \end{aligned}$$

are equivalent. Why is this so?

7. In each case, write the system in the form $A\mathsf{x} = \mathsf{b}$. What is A? What is x? What is b? Solve the system, expressing your answer as a vector or a linear combination of vectors as appropriate.

(a) [BB] $\begin{aligned} -2x_1 + x_2 + 5x_3 &= -10 \\ -8x_1 + 7x_2 + 19x_3 &= -42 \end{aligned}$ **(b)** $\begin{aligned} x_1 - 2x_2 + 3x_3 &= 1 \\ 2x_1 - 2x_2 + 9x_3 - 4x_4 &= -7 \\ -2x_1 + 10x_2 + 2x_3 - 5x_4 &= -9 \end{aligned}$

(c) $\begin{aligned} x_1 - 2x_2 + 3x_3 &= 2 \\ -x_1 + 2x_2 - x_3 &= -4 \\ 6x_1 - 10x_2 + 5x_3 &= 25 \\ 13x_1 - 19x_2 + 12x_3 &= 53 \end{aligned}$

(d) $\begin{aligned} -x_1 + x_2 + 2x_3 + x_4 + 3x_5 &= 10 \\ x_1 - x_2 + x_3 + 3x_4 - 4x_5 &= -15 \\ -8x_1 + 8x_2 + 7x_3 - 4x_4 + 36x_5 &= 131. \end{aligned}$

8. Find a value or values of t for which the system

$$\begin{aligned} x + y &= 2 \\ x - y &= 0 \\ 3x - y &= t \end{aligned}$$

has a solution. Find any solution(s) that may exist.

9. [BB] In solving the system $A\mathsf{x} = \mathsf{b}$, Gaussian elimination on the augmented matrix $[A \mid \mathsf{b}]$ led to the row echelon matrix $\left[\begin{array}{cccc|c} 1 & -2 & 3 & -1 & 5 \end{array}\right]$. The variables were x_1, x_2, x_3, x_4.

(a) Circle the pivots. Identify the free variables.

(b) Write the solution as a vector or as a linear combination of vectors, as appropriate.

10. In solving the system $A\mathsf{x} = \mathsf{b}$, Gaussian elimination on the augmented matrix $[A \mid \mathsf{b}]$ led to the row echelon matrix $\left[\begin{array}{ccc|c} 1 & 2 & 3 & 0 \\ 0 & 1 & -1 & 0 \end{array}\right]$. The variables were x_1, x_2, x_3.

(a) Circle the pivots. Identify the free variables.

(b) Express the solution as a vector or as a linear combination of vectors, as appropriate.

11. In solving the system $A\mathsf{x} = \mathsf{b}$, Gaussian elimination on the augmented matrix $[A \mid \mathsf{b}]$ led to the row echelon matrix $\left[\begin{array}{cccc|c} 1 & -1 & 0 & 1 & 1 \\ 0 & 0 & 1 & 2 & 3 \\ 0 & 0 & 0 & 0 & 0 \end{array}\right]$. The variables were x_1, x_2, x_3, x_4.

(a) Circle the pivots. Identify the free variables.

(b) Express the solution as a vector or as a linear combination of vectors, as appropriate.

12. Shown below are some matrices in row echelon form. Each represents the final row echelon matrix after Gaussian elimination was applied to an augmented matrix $[A \mid \mathsf{b}]$ in an attempt to solve $A\mathsf{x} = \mathsf{b}$. Find the solution (in vector form) of $A\mathsf{x} = \mathsf{b}$. In each case, if there is a solution, state whether this is unique or whether there are infinitely many solutions. In each case, $\mathsf{x} = \begin{bmatrix} x_1 \\ x_2 \\ \vdots \end{bmatrix}$.

(a) [BB] $\left[\begin{array}{cccc|c} 1 & 0 & 0 & 3 & 2 \\ 0 & 1 & 1 & 2 & 3 \\ 0 & 0 & 0 & 1 & \frac{1}{3} \\ 0 & 0 & 0 & 0 & 0 \end{array}\right]$ **(b)** $\left[\begin{array}{ccccc|c} 1 & 6 & 0 & 3 & 0 & 0 \\ 0 & 0 & 1 & -4 & 0 & 5 \\ 0 & 0 & 0 & 0 & 1 & 7 \end{array}\right]$

(c) $\left[\begin{array}{cccc|c} 0 & 3 & 1 & 2 & 1 \\ 0 & 0 & 0 & 0 & 4 \\ 0 & 0 & 0 & 0 & 0 \end{array}\right]$ **(d)** $\left[\begin{array}{cccccc|c} 1 & 4 & 0 & 2 & 3 & 0 & 2 \\ 0 & 0 & 1 & 1 & 0 & 5 & -3 \\ 0 & 0 & 0 & 0 & 1 & -4 & 0 \\ 0 & 0 & 0 & 0 & 0 & 1 & -7 \end{array}\right]$.

13. Solve each of the following systems of linear equations by Gaussian elimination and back substitution. Express your answers as vectors or as linear combinations of vectors if appropriate.

(a) [BB] $\begin{aligned} 2x - y + 2z &= -4 \\ 3x + 2y &= 1 \\ x + 3y - 6z &= 5 \end{aligned}$ **(b)** $\begin{aligned} x + 3z &= 6 \\ 3x + 4y - z &= 4 \\ 2x + 5y - 4z &= -3 \end{aligned}$

(c) [BB] $x - y + 2z = 4$ **(d)** $\begin{aligned} 2x + 3y - 2z &= 4 \\ -4x + z &= 5 \\ 2x + y - z &= 1 \end{aligned}$

(e) [BB] $\begin{aligned} 2x - y + z &= 2 \\ 3x + y - 6z &= -9 \\ -x + 2y - 5z &= -4 \end{aligned}$ **(f)** $\begin{aligned} 2x - y + z &= 3 \\ 4x - y - 2z &= 7 \end{aligned}$

(g) $\begin{aligned} 2x_1 + 4x_2 - 2x_3 &= 2 \\ 4x_1 + 9x_2 - 3x_3 &= 8 \\ -2x_1 - 3x_2 + 7x_3 &= 10 \end{aligned}$ **(h)** $\begin{aligned} x + y + 7z &= 2 \\ 2x - 4y + 14z &= -1 \\ 5x + 11y - 7z &= 8 \\ 2x + 5y - 4z &= -3 \end{aligned}$

(i) [BB]
$$\begin{aligned} -6x_1 + 8x_2 - 5x_3 - 5x_4 &= -11 \\ -6x_1 + 7x_2 - 10x_3 - 8x_4 &= -9 \\ -8x_1 + 10x_2 - 10x_3 - 9x_4 &= -13 \end{aligned}$$

(j)
$$\begin{aligned} -x_1 + x_2 + x_3 + 2x_4 &= 4 \\ 2x_3 - x_4 &= -7 \\ 3x_1 - 3x_2 - 7x_3 - 4x_4 &= 2 \end{aligned}$$

(k)
$$\begin{aligned} 3x_1 + x_2 &= -10 \\ x_1 + 3x_2 + x_3 &= 0 \\ x_2 + 3x_3 + x_4 &= 0 \\ x_3 + 3x_4 &= 10 \end{aligned}$$

(l)
$$\begin{aligned} -x_1 + x_3 + 3x_4 &= 4 \\ x_1 + x_2 - 2x_3 - 3x_4 &= 5 \\ -2x_1 - 3x_2 + 5x_3 + 3x_4 &= -15 \\ x_2 - x_3 - 6x_4 &= 18 \end{aligned}$$

(m)
$$\begin{aligned} x - y + z - w &= 0 \\ 2x - 2z + 3w &= 11 \\ 5x - 2y + z - w &= 6 \\ -x + y + w &= 0 \end{aligned}$$

(n)
$$\begin{aligned} -x_1 + 2x_2 + 3x_3 + 5x_4 - x_5 &= 0 \\ -2x_1 + 5x_2 + 10x_3 + 13x_4 - 4x_5 &= -5 \\ -3x_1 + 7x_2 + 13x_3 + 19x_4 - 11x_5 &= 1 \\ -x_1 + 4x_2 + 11x_3 + 11x_4 - 5x_5 &= -10 \end{aligned}$$

(o)
$$\begin{aligned} x_1 - 2x_2 + x_3 + x_4 + 4x_5 + 3x_6 &= 3 \\ -x_1 + x_2 + x_3 + 4x_6 &= 7 \\ -3x_1 + 5x_2 - x_3 - 2x_4 - 9x_5 &= 3 \\ -2x_1 + 4x_2 - 2x_3 - 2x_4 - 9x_5 - 4x_6 &= -4 \\ -5x_1 + 9x_2 - 3x_3 - 4x_4 - 16x_5 - 8x_6 &= -5. \end{aligned}$$

14. [BB] Determine whether $\begin{bmatrix} 1 \\ 6 \\ -4 \end{bmatrix}$ is a linear combination of the columns of $A = \begin{bmatrix} 2 & 3 & 4 \\ 4 & 7 & 5 \\ 6 & -1 & 9 \end{bmatrix}$.

15. Determine whether each of the given vectors is a linear combination of the columns of $A = \begin{bmatrix} 2 & -1 \\ -1 & 4 \\ 5 & 9 \end{bmatrix}$.

(a) [BB] $\begin{bmatrix} 8 \\ -11 \\ -3 \end{bmatrix}$ (b) $\begin{bmatrix} 0 \\ -1 \\ 3 \end{bmatrix}$ (c) $\begin{bmatrix} 0 \\ 0 \\ 0 \end{bmatrix}$ (d) $\begin{bmatrix} -17 \\ 33 \\ 38 \end{bmatrix}$.

16. Determine whether the given vector v is a linear combination of the other vectors $\mathsf{v}_1, \mathsf{v}_2, \ldots$ in each of the following cases.

(a) [BB] $\mathsf{v} = \begin{bmatrix} 1 \\ 3 \\ 2 \end{bmatrix}$; $\mathsf{v}_1 = \begin{bmatrix} 0 \\ 1 \\ 1 \end{bmatrix}, \mathsf{v}_2 = \begin{bmatrix} 1 \\ 1 \\ 0 \end{bmatrix}, \mathsf{v}_3 = \begin{bmatrix} 1 \\ 0 \\ 2 \end{bmatrix}$

(b) $\mathsf{v} = \begin{bmatrix} 1 \\ 2 \\ 3 \\ 4 \\ 5 \end{bmatrix}$; $\mathsf{v}_1 = \begin{bmatrix} 5 \\ 1 \\ 2 \\ 3 \\ 4 \end{bmatrix}, \mathsf{v}_2 = \begin{bmatrix} 4 \\ 5 \\ 1 \\ 2 \\ 3 \end{bmatrix}, \mathsf{v}_3 = \begin{bmatrix} 2 \\ 3 \\ 4 \\ 5 \\ 1 \end{bmatrix}, \mathsf{v}_4 = \begin{bmatrix} -1 \\ -2 \\ -3 \\ -4 \\ -5 \end{bmatrix}, \mathsf{v}_5 = \begin{bmatrix} 0 \\ 1 \\ 2 \\ 3 \\ -1 \end{bmatrix}$

(c) $\mathsf{v} = \begin{bmatrix} 1 \\ 2 \\ 3 \\ 4 \end{bmatrix}$; $\mathsf{v}_1 = \begin{bmatrix} -2 \\ 1 \\ 0 \\ 1 \end{bmatrix}, \mathsf{v}_2 = \begin{bmatrix} 3 \\ 4 \\ 0 \\ -3 \end{bmatrix}, \mathsf{v}_3 = \begin{bmatrix} 6 \\ 5 \\ 0 \\ 4 \end{bmatrix}$

(d) $\mathsf{v} = \begin{bmatrix} 1 \\ -2 \\ 3 \\ 0 \end{bmatrix}$; $\mathsf{v}_1 = \begin{bmatrix} 2 \\ -3 \\ 1 \\ 0 \end{bmatrix}, \mathsf{v}_2 = \begin{bmatrix} 4 \\ -2 \\ 1 \\ 0 \end{bmatrix}, \mathsf{v}_3 = \begin{bmatrix} 0 \\ 1 \\ 1 \\ 1 \end{bmatrix}, \mathsf{v}_4 = \begin{bmatrix} 1 \\ -1 \\ 1 \\ -1 \end{bmatrix}$.

17. [BB] Let $\mathsf{v}_1 = \begin{bmatrix} 2 \\ 0 \\ 2 \end{bmatrix}, \mathsf{v}_2 = \begin{bmatrix} 1 \\ 1 \\ 1 \end{bmatrix}, \mathsf{v}_3 = -\begin{bmatrix} -1 \\ 1 \\ -1 \end{bmatrix}$ and $\mathsf{v}_4 = \begin{bmatrix} 0 \\ 2 \\ 0 \end{bmatrix}$.

Is every vector in R^3 a linear combination of $\mathsf{v}_1, \mathsf{v}_2, \mathsf{v}_3$ and v_4?

18. Is every vector in R^4 a linear combination of the given vectors?

(a) $\mathsf{v}_1 = \begin{bmatrix} 1 \\ 1 \\ 1 \\ 1 \end{bmatrix}, \mathsf{v}_2 = \begin{bmatrix} 1 \\ 1 \\ 1 \\ 0 \end{bmatrix}, \mathsf{v}_3 = \begin{bmatrix} 1 \\ 1 \\ 0 \\ 0 \end{bmatrix}, \mathsf{v}_4 = \begin{bmatrix} 1 \\ 0 \\ 0 \\ 0 \end{bmatrix}$

(b) $\mathsf{v}_1 = \begin{bmatrix} 1 \\ 2 \\ 3 \\ 4 \end{bmatrix}, \mathsf{v}_2 = \begin{bmatrix} 5 \\ 6 \\ 7 \\ 8 \end{bmatrix}, \mathsf{v}_3 = \begin{bmatrix} 9 \\ 10 \\ 11 \\ 12 \end{bmatrix}, \mathsf{v}_4 = \begin{bmatrix} 13 \\ 14 \\ 15 \\ 16 \end{bmatrix}$.

19. [BB] Let $A = \begin{bmatrix} 2 & 5 \\ 1 & 3 \end{bmatrix}$ and $\mathsf{b} = \begin{bmatrix} b_1 \\ b_2 \end{bmatrix}$.

(a) Solve the system $A\mathsf{x} = \mathsf{b}$ for $\mathsf{x} = \begin{bmatrix} x_1 \\ x_2 \end{bmatrix}$.

(b) Write b as a linear combination of the columns of A.

20. Repeat Exercise 19 with $A = \begin{bmatrix} 1 & -1 & -1 \\ 2 & -1 & -3 \\ 1 & 0 & -3 \end{bmatrix}$, $\mathsf{x} = \begin{bmatrix} x_1 \\ x_2 \\ x_3 \end{bmatrix}$ and $\mathsf{b} = \begin{bmatrix} b_1 \\ b_2 \\ b_3 \end{bmatrix}$.

21. [BB] The planes with equations $x + 2z = 5$ and $2x + y = 2$ intersect in a line.

(a) Why?

(b) Find an equation of the line of intersection.

22. Answer Exercise 21 for the planes with equations $x + 2y + 6z = 5$ and $x - y - 3z = -1$.

23. [BB] A circle in the Cartesian plane has an equation of the form $x^2 + y^2 + ax + by + c = 0$. Write down a linear system whose solution would give an equation for the circle that passes through the points $(10, 7)$, $(-6, -1)$ and $(-4, -7)$.

24. Suppose we want to find a quadratic polynomial $p(x) = a + bx + cx^2$ whose graph passes through the points $(-2, 3)$, $(0, -11)$, $(5, 24)$.

    (a) Write down a system of equations whose solution is a, b and c.

    (b) Find the desired polynomial.

25. Repeat Exercise 24 using the points $(3, 0)$, $(-1, 4)$, $(0, 6)$.

26. Find a polynomial of degree two whose graph passes through the points $(-1, 1)$, $(0, 1)$, $(1, 4)$. Graph your polynomial showing clearly the given points.

27. Find a polynomial of degree three whose graph passes through the points $(-2, 1)$, $(0, 3)$, $(3, -2)$, $(4, 5)$. Graph this polynomial showing clearly the given points.

28. Find values of a, b and c such that the graph of the function $f(x) = a2^x + b2^{2x} + c2^{3x}$ passes through the points $(-1, 2)$, $(0, 0)$ and $(1, -2)$.

29. [BB] Find a condition or conditions on k so that the system

$$\begin{aligned} x + y + 2z &= 1 \\ 3x - y - 2z &= 3 \\ -x + 3y + 6z &= k \end{aligned}$$

has (i) no solution; (ii) a unique solution; (iii) infinitely many solutions.

30. For which value or values of a will the system

$$\begin{aligned} x + 2y - 3z &= 4 \\ 3x - y + 5z &= 2 \\ 4x + y + (a^2 - 14)z &= a + 2 \end{aligned}$$

have (i) no solution; (ii) exactly one solution; (iii) infinitely many solutions? Find any solutions that may exist.

31. [BB] Consider the system $\begin{aligned} 5x + 2y &= a \\ -15x - 6y &= b. \end{aligned}$

    (a) Under what conditions on a and b, if any, does this system fail to have a solution?

    (b) Under what conditions, if any, does the system have a unique solution? Find any unique solution that may exist.

    (c) Under what conditions are there infinitely many solutions? Find these solutions if and when they exist.

32. Find conditions on a and b which guarantee that the system

$$\begin{aligned} x - 2y &= 4 \\ 2x + az &= 5 \\ 3x - 4y + 5z &= b \end{aligned}$$

has (i) no solution; (ii) a unique solution; (iii) infinitely many solutions.

33. Find conditions on a, b and c (if any) such that the system

$$\begin{aligned} x \qquad + z &= -1 \\ 2x - y \qquad &= 2 \\ y + 2z &= -4 \\ ax + by + cz &= 3 \end{aligned}$$

has (i) a unique solution; (ii) no solution; (iii) infinitely many solutions.

34. Find necessary and sufficient conditions on a, b and c in order that the system

$$\begin{aligned} x + y - z &= a \\ 2x - 3y + 5z &= b \\ 5x \qquad + 2z &= c \end{aligned}$$

should have a solution. When there is a solution, is this unique? Explain.

35. [BB] **(a)** Find a condition on a, b and c that is both necessary and sufficient for the system

$$\begin{aligned} x - 2y &= a \\ -5x + 3y &= b \\ 3x + y &= c \end{aligned}$$

to have a solution.

(b) Could the system in part (a) have infinitely many solutions? Explain.

36. Answer Exercise 35 for the system $\begin{aligned} x - 5y &= a \\ -2x + 6y &= b \\ 3x + 1y &= c \end{aligned}$.

37. Find a condition on a, b and c which implies that the system

$$\begin{aligned} 7x_1 + x_2 - 4x_3 + 5x_4 &= a \\ x_1 + x_2 + 2x_3 - x_4 &= b \\ 3x_1 + x_2 \qquad + x_4 &= c \end{aligned}$$

does **not** have a solution.

Critical Reading

38. In each case, determine whether or not the given points are collinear, that is, lie on a line. If they are, find the equation of the line; otherwise, find the equation of the plane they determine.

(a) $(1, 2, 4), (-1, 0, 6), (5, -4, -10)$

(b) $(-1, -2, 0), (9, 4, 2), (4, 1, 1)$

(c) $(-3, -2, 1), (0, 5, 2), (-1, 1, 1)$

(d) $(1, 2, -1), (3, 6, 0), (5, 0, -4)$

(e) $(-4,1,-15), (2,5,-7), (5,7,-3)$.

39. In each case, determine whether or not the given points are *coplanar*, that is, lie in a plane. When the answer is "yes," give an equation of a plane on which the points all lie.

(a) $(1,4,7), (-2,1,1), (5,0,-3)$

(b) [BB] $(-2,1,1), (2,2,-2), (4,7,-1), (-10,11,-9)$

(c) $(-1,0,2), (1,1,3), (-3,-1,1), (3,2,4)$

(d) $(5,1,3), (10,3,5), (20,3,20), (-15,2,0), (5,0,25)$

(e) $(1,0,1), (-1,-1,2), (0,16,7), (2,-10,3), (4,0,-1)$.

40. Find the equation of the line of intersection of the planes with equations $x - y + z = 3$ and $2x - 3y + z = 5$.

41. In the solution of $A\mathsf{x} = \mathsf{b}$, elementary row operations on the augmented matrix $[A \mid \mathsf{b}]$ produce the matrix $\left[\begin{array}{cccc|c} 1 & 2 & 3 & 4 & 5 \\ 0 & -3 & 4 & 0 & -1 \\ 0 & 0 & 2 & 4 & 0 \\ 0 & 0 & -1 & -2 & 0 \end{array}\right]$. Does $A\mathsf{x} = \mathsf{b}$ have a unique solution, infinitely many solutions, no solution?

42. Suppose A is an invertible matrix. Explain why the system $A\mathsf{x} = \mathsf{0}$ has only the trivial solution, $\mathsf{x} = \mathsf{0}$.

2.5 Application: Electric Circuits

In this brief section, we illustrate one situation where systems of linear equations naturally arise. We consider electrical networks comprised of batteries and resistors connected with wires. Each battery has a positive terminal and a negative terminal causing current to flow around the circuit out of the positive terminal and into the negative terminal. We are all familiar with nine volt batteries and 1.5 volt AA batteries. "Voltage" is a measure of the power of a battery. Each resistor, as the name implies, consumes voltage (one refers to a "voltage drop") and affects the current in the circuit. According to Ohm's Law,[6] the connection between voltage E, current I (measured in amperes, amps for short) and resistance R (measured in Ohms) is given by this formula:

$$\begin{array}{ccccc} V & = & I & & R \\ \text{volts} & = & \text{amps} & \times & \text{Ohms.} \end{array}$$

[6]after Georg Ohm (1789-1854)

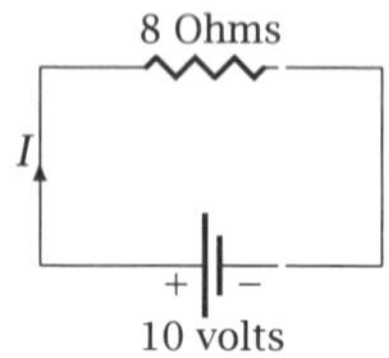

In a simple circuit such as the one shown, Ohm's Law gives us two pieces of information. It says that the size of the current flowing around the circuit is $I = \frac{E}{R} = \frac{10}{8} = 1.25$ amps. It also gives us the size of the voltage drop at the resistor: $V = IR = 1.25(8) = 10$ volts. (It is standard to use the symbol 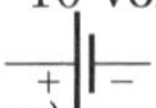 for battery and  for resistor.)

Figure 2.2

A basic principle due to Gustav Kirchhoff (1824–1887) says that the sum of the voltage drops at the resistors on a circuit must equal the total voltage provided by the batteries. For the circuit on the left of Figure 2.2, the voltage drops are $4I$ at one resistor and $6I$ at the other for a total of $10I$. The total voltage provided by the batteries is $3 + 12 = 15$. Kirchhoff's Law says $10I = 15$, so the current around the circuit is $I = \frac{15}{10} = 1.5$ amps. In this circuit (the one on the left) , the current flows counterclockwise around the circuit, from the positive terminal to the negative terminal in each battery. The situation is different for the circuit on the right where current leaves the positive terminal of the top battery, but enters the positive terminal of the lower battery. For this reason, the lower battery contributes -3 volts to the total voltage of the circuit, which is then $12 - 3 = 9$ volts. This time, we get $10I = 9$, so $I = 0.9$ amps. Since batteries contribute "signed" voltages to the total voltage in a circuit, Kirchhoff's first law is stated like this.

2.5.1 **Kirchhoff's Circuit Law:** The sum of the voltage drops around a circuit equals the algebraic sum of the voltages provided by the batteries on the circuit.

"Algebraic" means taking the direction of the current into account.

The reason for introducing electrical circuits in a chapter devoted to systems of linear equations becomes evident when we consider circuits that contain two or more "subcircuits." The circuit depicted in Figure 2.3 contains two subcircuits which are joined at A and B. The points A and B are called *nodes*, these being places where one subcircuit meets another. At A, there are two entering currents of I_2 and I_3 amps and a single departing current of I_1 amps. At B, the entering current of I_1 amps splits into currents of I_2 and I_3 amps. A second law is useful.

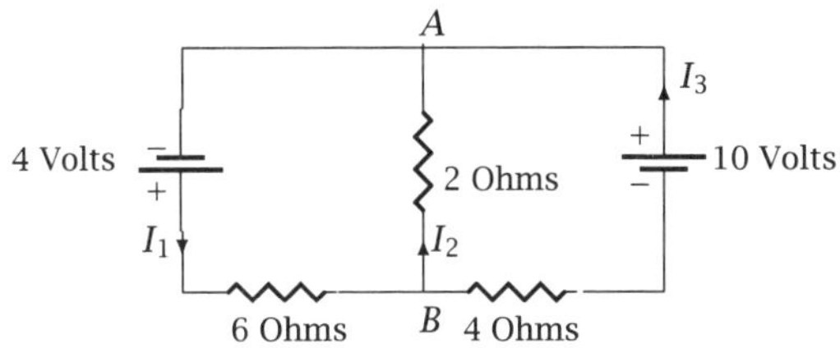

Figure 2.3

2.5.2 Kirchhoff's Node Law: The sum of the currents flowing into any node is the sum of the currents flowing out of that node.

So in Figure 2.3,

$$\begin{aligned} \text{At } A: &\quad I_2 + I_3 = I_1 \\ \text{At } B: &\quad I_1 = I_2 + I_3. \end{aligned}$$

These equations both say $I_1 - I_2 - I_3 = 0$. In addition, Kirchhoff's Circuit Law applied to the subcircuit on the left gives

$$6I_1 + 2I_2 = 4,$$

since there are voltage drops of $6I_1$ and $2I_2$ at the two resistors on this circuit ($V = IR$) and the total voltage supplied by the battery is 4 volts. We must be careful with the subcircuit on the right. Do we wish to follow this circuit clockwise or counterclockwise? Suppose we follow it counterclockwise, in the direction of the arrow for I_3. Then there is a $4I_3$ voltage drop at the 4 Ohm resistor and a $-2I_2$ voltage drop at the 2 Ohm resistor because the current passing through this resistor, in the counterclockwise direction, is $-I_2$. Kirchhoff's Circuit Law gives

$$-2I_2 + 4I_3 = 10.$$

To find the three currents, we must solve the system

$$\begin{aligned} I_1 - I_2 - I_3 &= 0 \\ 6I_1 + 2I_2 \phantom{{}- 4I_3} &= 4 \\ -2I_2 + 4I_3 &= 10. \end{aligned} \tag{1}$$

Here is one instance of Gaussian elimination applied to the augmented matrix of coefficients:

$$\left[\begin{array}{rrr|r} 1 & -1 & -1 & 0 \\ 6 & 2 & 0 & 4 \\ 0 & -2 & 4 & 10 \end{array}\right] \to \left[\begin{array}{rrr|r} 1 & -1 & -1 & 0 \\ 0 & 8 & 6 & 4 \\ 0 & -1 & 2 & 5 \end{array}\right]$$

$$\to \left[\begin{array}{rrr|r} 1 & -1 & -1 & 0 \\ 0 & 1 & -2 & -5 \\ 0 & 4 & 3 & 2 \end{array}\right] \to \left[\begin{array}{rrr|r} 1 & -1 & -1 & 0 \\ 0 & 1 & -2 & -5 \\ 0 & 0 & 11 & 22 \end{array}\right].$$

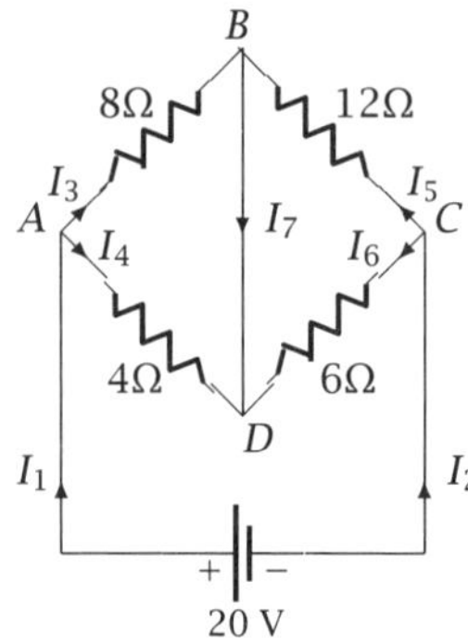

Figure 2.4: A Wheatstone Bridge Circuit

Back substitution gives $11I_3 = 22$, so $I_3 = 2$ and $I_2 - 2I_3 = -5$. We obtain $I_2 = -5 + 2I_3 = -1$ and $I_1 = I_2 + I_3 = 1$. Don't worry about the negative I_2. This simply means we put the arrow on the middle wire in the wrong direction. The current is actually flowing from A to B.

READING CHECK 1. Suppose we follow the subcircuit on the right of Figure 2.3 in the clockwise direction, that is, in the direction indicated by the arrow for I_2. Verify that our answer does not change.

2.5.3 Remark. The observant student may point out that there is a third circuit in Figure 2.3, the one that goes around the outer edge. What does Kirchhoff's Circuit Law say about this? The sum of the voltage drops is $6I_1 + 4I_3$ and the total voltage supplied by the two batteries is 14, so we get $6I_1 + 4I_3 = 14$. Since this is satisfied by $I_1 = 1$, $I_3 = 2$, it is implied by the three equations we used in (1). This should always be the case.

READING CHECK 2. Reverse the terminals of the 10 volt battery in Figure 2.3, draw the circuit and find the currents.

2.5.4 Example. We present in Figure 2.4 what is known in electrical engineering as a "Wheatstone Bridge Circuit." If the ratio of the resistances to the left of the "bridge" BD is the same as the corresponding ratio on the right, there should be no current through the bridge. Thus, testing that $I_7 = 0$ provides a way to be sure that the ratios are the same. Let's see if this is the case.

There are four nodes, A, B, C, D where Kirchhoff's Node Law gives, respectively, these equations:

$$\begin{aligned} I_1 &= I_3 + I_4 \\ I_7 &= I_3 + I_5 \\ I_2 &= I_5 + I_6 \\ I_4 + I_6 + I_7 &= 0. \end{aligned}$$

There are three subcircuits. Kirchhoff's Law applied to the one at the bottom says $4I_4 - 6I_6 = 20$. Applied to the two triangular subcircuits at the top, the law gives $8I_3 - 4I_4 = 0$ and $6I_6 - 12I_5 = 0$. To find the individual currents, we solve the system

$$\begin{array}{rcrcrcrcrcrcrcr}
I_1 & & & - & I_3 & - & I_4 & & & & & & & = & 0 \\
 & & & & I_3 & & & + & I_5 & & & - & I_7 & = & 0 \\
 & & I_2 & & & & & - & I_5 & - & I_6 & & & = & 0 \\
 & & & & & & I_4 & & & + & I_6 & + & I_7 & = & 0 \\
 & & & & & & 4I_4 & & & - & 6I_6 & & & = & 20 \\
 & & & & 8I_3 & - & 4I_4 & & & & & & & = & 0 \\
 & & & & & & & - & 12I_5 & + & 6I_6 & & & = & 0.
\end{array}$$

Gaussian elimination proceeds

$$\left[\begin{array}{rrrrrrr|r}
1 & 0 & -1 & -1 & 0 & 0 & 0 & 0 \\
0 & 0 & 1 & 0 & 1 & 0 & -1 & 0 \\
0 & 1 & 0 & 0 & -1 & -1 & 0 & 0 \\
0 & 0 & 0 & 1 & 0 & 1 & 1 & 0 \\
0 & 0 & 0 & 4 & 0 & -6 & 0 & 20 \\
0 & 0 & 8 & -4 & 0 & 0 & 0 & 0 \\
0 & 0 & 0 & 0 & -12 & 6 & 0 & 0
\end{array}\right] \to \left[\begin{array}{rrrrrrr|r}
1 & 0 & -1 & -1 & 0 & 0 & 0 & 0 \\
0 & 1 & 0 & 0 & -1 & -1 & 0 & 0 \\
0 & 0 & 1 & 0 & 1 & 0 & -1 & 0 \\
0 & 0 & 0 & 1 & 0 & 1 & 1 & 0 \\
0 & 0 & 0 & 4 & 0 & -6 & 0 & 20 \\
0 & 0 & 8 & -4 & 0 & 0 & 0 & 0 \\
0 & 0 & 0 & 0 & -12 & 6 & 0 & 0
\end{array}\right]$$

$$\to \left[\begin{array}{rrrrrrr|r}
1 & 0 & -1 & -1 & 0 & 0 & 0 & 0 \\
0 & 1 & 0 & 0 & -1 & -1 & 0 & 0 \\
0 & 0 & 1 & 0 & 1 & 0 & -1 & 0 \\
0 & 0 & 0 & 1 & 0 & 1 & 1 & 0 \\
0 & 0 & 0 & 4 & 0 & -6 & 0 & 20 \\
0 & 0 & 0 & -4 & -8 & 0 & 8 & 0 \\
0 & 0 & 0 & 0 & -12 & 6 & 0 & 0
\end{array}\right] \to \left[\begin{array}{rrrrrrr|r}
1 & 0 & -1 & -1 & 0 & 0 & 0 & 0 \\
0 & 1 & 0 & 0 & -1 & -1 & 0 & 0 \\
0 & 0 & 1 & 0 & 1 & 0 & -1 & 0 \\
0 & 0 & 0 & 1 & 0 & 1 & 1 & 0 \\
0 & 0 & 0 & 0 & 0 & -10 & -4 & 20 \\
0 & 0 & 0 & 0 & -8 & 4 & 12 & 0 \\
0 & 0 & 0 & 0 & -12 & 6 & 0 & 0
\end{array}\right]$$

$$\to \left[\begin{array}{rrrrrrr|r}
1 & 0 & -1 & -1 & 0 & 0 & 0 & 0 \\
0 & 1 & 0 & 0 & -1 & -1 & 0 & 0 \\
0 & 0 & 1 & 0 & 1 & 0 & -1 & 0 \\
0 & 0 & 0 & 1 & 0 & 1 & 1 & 0 \\
0 & 0 & 0 & 0 & -8 & 4 & 12 & 0 \\
0 & 0 & 0 & 0 & 0 & -10 & -4 & 20 \\
0 & 0 & 0 & 0 & -12 & 6 & 0 & 0
\end{array}\right] \to \left[\begin{array}{rrrrrrr|r}
1 & 0 & -1 & -1 & 0 & 0 & 0 & 0 \\
0 & 1 & 0 & 0 & -1 & -1 & 0 & 0 \\
0 & 0 & 1 & 0 & 1 & 0 & -1 & 0 \\
0 & 0 & 0 & 1 & 0 & 1 & 1 & 0 \\
0 & 0 & 0 & 0 & 2 & -1 & -3 & 0 \\
0 & 0 & 0 & 0 & 0 & -10 & -4 & 20 \\
0 & 0 & 0 & 0 & -12 & 6 & 0 & 0
\end{array}\right]$$

$$\to \left[\begin{array}{rrrrrrr|r}
1 & 0 & -1 & -1 & 0 & 0 & 0 & 0 \\
0 & 1 & 0 & 0 & -1 & -1 & 0 & 0 \\
0 & 0 & 1 & 0 & 1 & 0 & -1 & 0 \\
0 & 0 & 0 & 1 & 0 & 1 & 1 & 0 \\
0 & 0 & 0 & 0 & 2 & -1 & -3 & 0 \\
0 & 0 & 0 & 0 & 0 & -10 & -4 & 20 \\
0 & 0 & 0 & 0 & 0 & 0 & -18 & 0
\end{array}\right].$$

Back substitution now says that $I_7 = 0$ (oh boy!), $-10I_6 - 4I_7 = 20$, so $-10I_6 = 20$ and $I_6 = -2$ (the arrow on CD should be reversed). Furthermore, $2I_5 - I_6 - 3I_7 = 0$, so $2I_5 = I_6$ and $I_5 = -1$; $I_4 + I_6 + I_7 = 0$, so $I_4 = -I_6 = 2$; $I_3 + I_5 - I_7 = 0$, so $I_3 = -I_5 = 1$; $I_2 - I_5 - I_6 = 0$ and, finally, $I_2 = I_5 + I_6 = -3$

and $I_1 = I_3 + I_4 = 3$. Evidence that our solution is probably correct can be obtained by checking the entire circuit, where Kirchhoff's Circuit Law says that $8I_3 - 12I_5 = 20$, in agreement with our results. ☺

Answers to Reading Checks

1. Kirchhoff's Node Law gives us the same equation as before, as does Kirchhoff's Circuit Law for the left subcircuit. For the right subcircuit, the total voltage is -10 because the current flow in the clockwise direction is into the positive terminal of the 10 volt battery. The voltage drops are $2I_2$ (at the 2 Ohm resistor) and $-4I_3$ at the 4 Ohm resistor, so Kirchhoff's Circuit Law applied to the right subcircuit gives $2I_2 - 4I_3 = -10$, the equation we had before with change of signs. The solution is the same as before.

2. 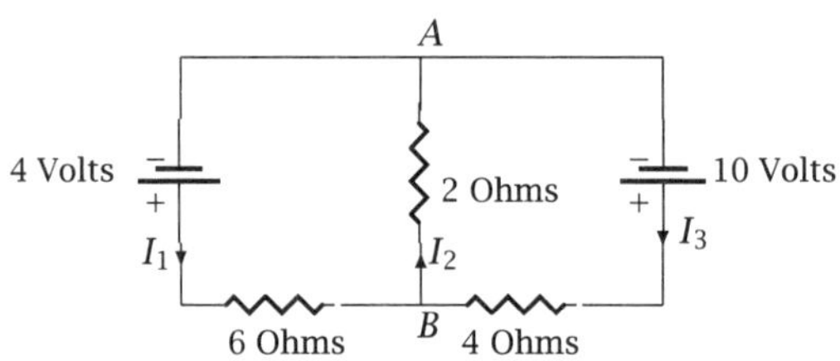

At node A and at node B, Kirchhoff's Node Law says $I_2 = I_1 + I_3$. Kirchhoff's Circuit Law on the left subcircuit is $6I_1 + 2I_2 = 4$, as before. In the right subcircuit followed in the direction of I_3 (and I_2), the voltage drops are $2I_2$ at the 2 Ohm resistor and $4I_3$ at the 4 Ohm resistor. The voltage is 10, so Kirchhoff's Circuit Law applied to the right subcircuit gives $2I_2 + 4I_3 = 10$. The system to solve is

$$\begin{aligned} I_1 - I_2 + I_3 &= 0 \\ 6I_1 + 2I_2 \qquad &= 4 \\ 2I_2 + 4I_3 &= 10. \end{aligned}$$

Gaussian elimination applied to the augmented matrix of coefficients proceeds

$$\left[\begin{array}{rrr|r} 1 & -1 & 1 & 0 \\ 6 & 2 & 0 & 4 \\ 0 & 2 & 4 & 10 \end{array}\right] \to \left[\begin{array}{rrr|r} 1 & -1 & 1 & 0 \\ 0 & 8 & -6 & 4 \\ 0 & 1 & 2 & 5 \end{array}\right]$$
$$\to \left[\begin{array}{rrr|r} 1 & -1 & 1 & 0 \\ 0 & 1 & 2 & 5 \\ 0 & 4 & -3 & 2 \end{array}\right] \to \left[\begin{array}{rrr|r} 1 & -1 & 1 & 0 \\ 0 & 1 & 2 & 5 \\ 0 & 0 & -11 & -18 \end{array}\right].$$

so $I_3 = \frac{18}{11}$ amps, $I_2 = 5 - 2I_3 = \frac{19}{11}$ amps and $I_1 = I_2 - I_3 = \frac{1}{11}$ amps.

EXERCISES

Answers to exercises marked [BB] can be found at the Back of the Book.

1. Find all indicated currents in each circuit. The standard symbol for Ohms is the Greek symbol Ω, "Omega."

(a) [BB]

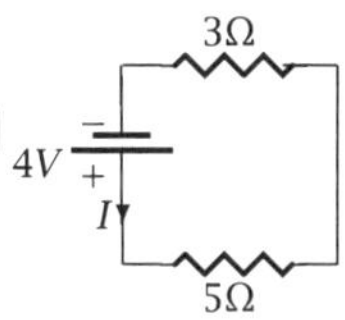

(b)

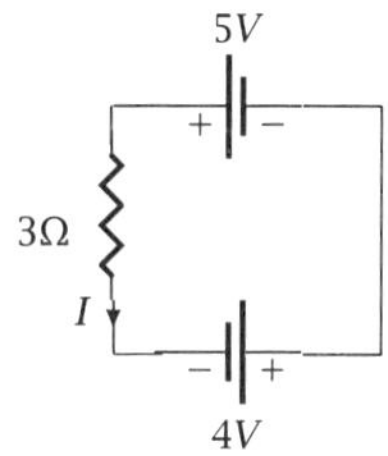

(c)

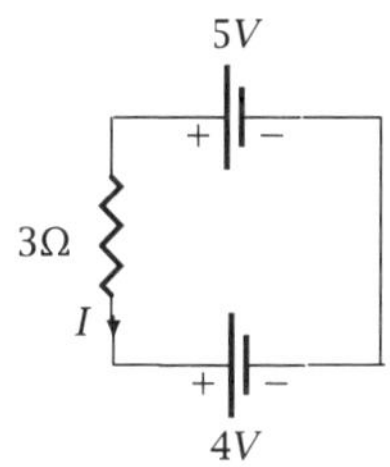

(d)

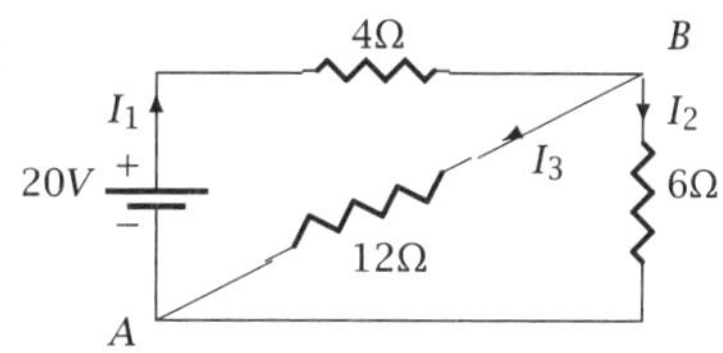

(e) [BB]

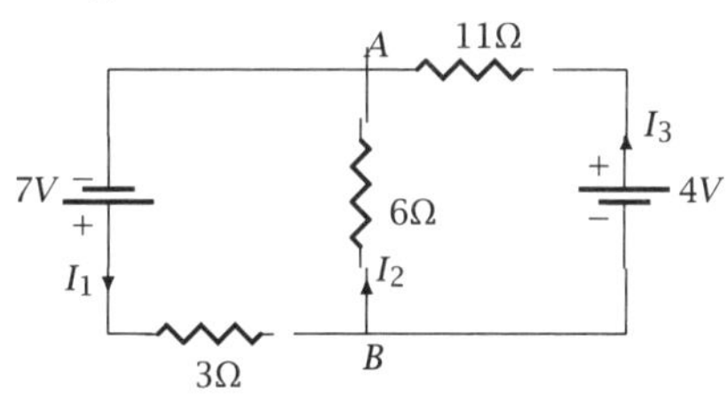

(f)

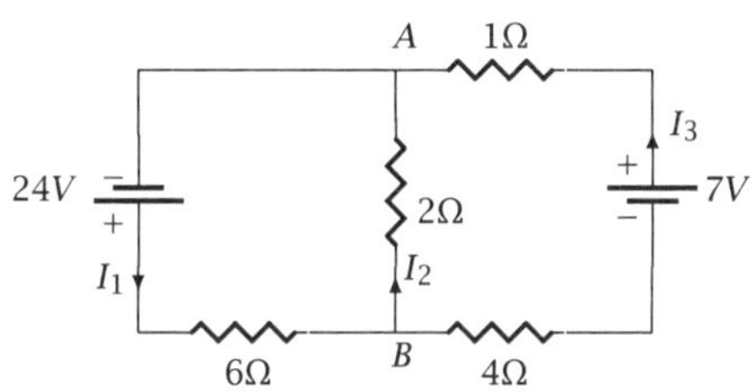

(g)

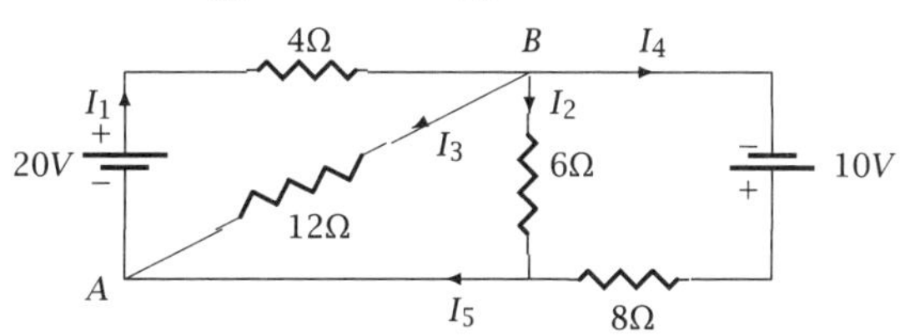

(h)

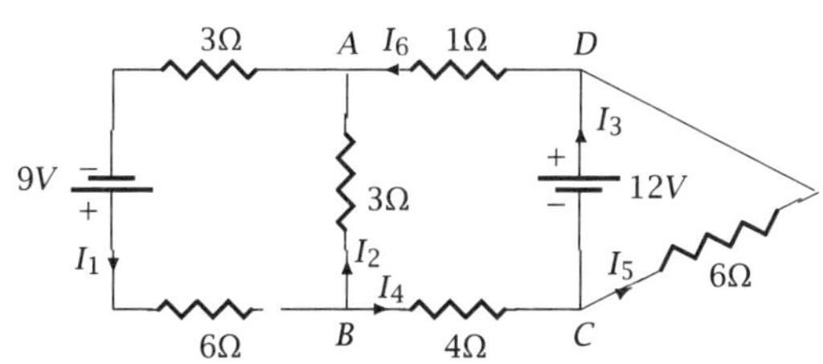

(i)

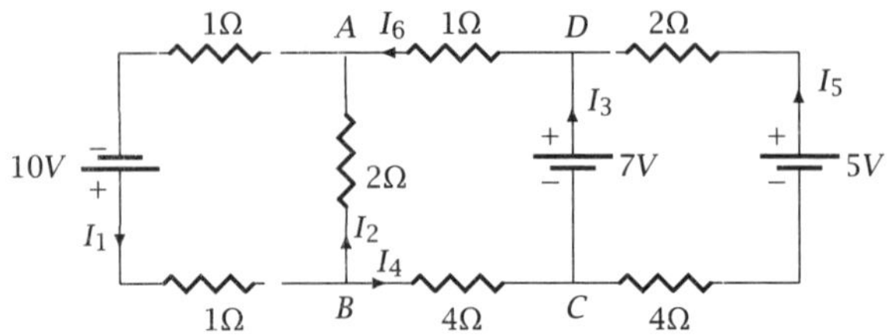

2. **(a)** Find all indicated currents in the circuit shown.

   (b) Is it possible to change the resistance along the top right wire (that is to say, along BC) so as to make the current $I_4 = 0$?

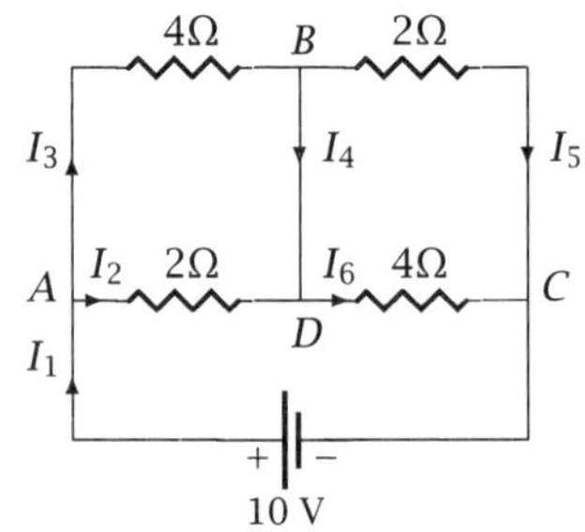

3. Look at the Wheatstone Bridge Circuit in Figure 2.4 and consider the possibility that the resistance along AB is unknown and a current of $I_7 = \frac{5}{6}$ is detected along BD. What is the resistance in AB?

4. Two resistors are said to be in *series* in a circuit if they are positioned as on the left. They are in *parallel* if they are appear as on the right. In each case, we wish to replace the two resistors by one resistor that will provide the same resistance in a simpler circuit. In each circuit, determine the "effective" single resistance of the two resistors.

(a) [BB]

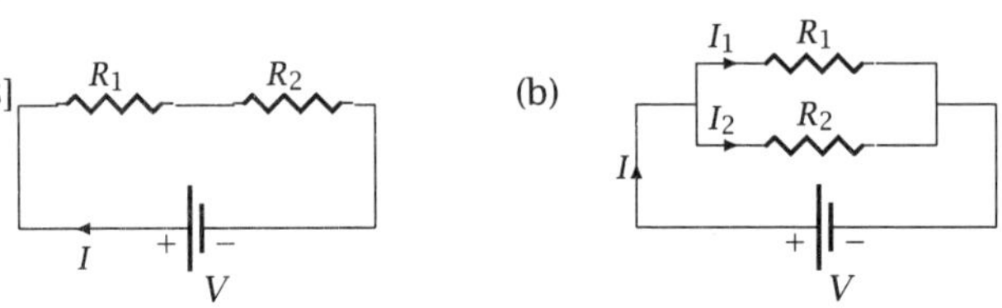

5. Assuming V and all four resistances are positive, show that the current I_7 in the Wheatstone Bridge is 0 if and only if $\frac{a}{b} = \frac{c}{d}$.

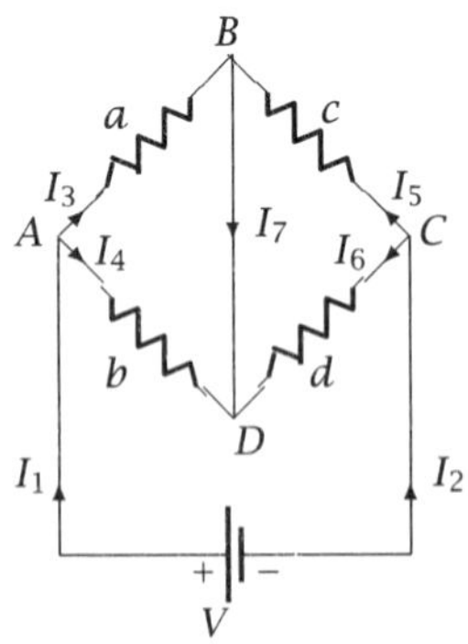

2.6 Homogeneous Systems; More on Linear Independence

In this section, we consider systems of linear equations of the form $A\mathsf{x} = \mathsf{0}$, in which the vector to the right of the = sign is the zero vector. Astute readers might quickly observe that solving $A\mathsf{x} = \mathsf{0}$ is equivalent to determining whether or not the columns of the matrix A are linearly independent.

READING CHECK 1. Why?

We call a system of the type $A\mathsf{x} = \mathsf{0}$ *homogeneous*. For example, the linear system

$$\begin{aligned} x_1 + 2x_2 + x_3 &= 0 \\ x_1 + 3x_2 &= 0 \\ 2x_1 + x_2 + 5x_3 &= 0 \end{aligned} \tag{1}$$

is homogeneous. Unlike linear systems in general, a homogeneous system always has a solution, namely $\mathsf{x} = \mathsf{0}$. We call this the *trivial* solution. Our primary interest in a homogeneous system, therefore, is whether or not there is a *nontrivial* solution, that is, a solution $\mathsf{x} \neq \mathsf{0}$.

When we speak of the homogeneous system *corresponding to* the system $A\mathsf{x} = \mathsf{b}$, we mean the system $A\mathsf{x} = \mathsf{0}$, with the same matrix of coefficients. For example, system (1) is the homogeneous system that corresponds to

$$\begin{aligned} x_1 + 2x_2 + x_3 &= 2 \\ x_1 + 3x_2 &= -1 \\ 2x_1 + x_2 + 5x_3 &= 13. \end{aligned} \tag{2}$$

When solving a homogeneous system of equations by Gaussian elimination, it is not necessary to augment the matrix of coefficients with a column of 0s because the elementary row operations will not change these 0s. Thus, to solve the homogeneous system (1), we apply Gaussian elimination directly to the matrix of coefficients proceeding like this:

$$\begin{bmatrix} 1 & 2 & 1 \\ 1 & 3 & 0 \\ 2 & 1 & 5 \end{bmatrix} \to \begin{bmatrix} 1 & 2 & 1 \\ 0 & 1 & -1 \\ 0 & -3 & 3 \end{bmatrix} \to \begin{bmatrix} 1 & 2 & 1 \\ 0 & 1 & -1 \\ 0 & 0 & 0 \end{bmatrix}.$$

The last matrix is in row echelon form. There are two pivot columns. Column three is the only nonpivot column and we set the corresponding variable $x_3 = t$ to indicate that x_3 is free. Each row of the row echelon matrix determines an equation the right side of which is 0. The equations are

$$\begin{aligned} x_1 + 2x_2 + x_3 &= 0 \\ x_2 - x_3 &= 0. \end{aligned}$$

Using back substitution, we see that $x_2 = x_3 = t$ and $x_1 = -2x_2 - x_3 = -3t$. In vector form, the solution to the homogeneous system (1) is

$$\mathsf{x} = \begin{bmatrix} x_1 \\ x_2 \\ x_3 \end{bmatrix} = \begin{bmatrix} -3t \\ t \\ t \end{bmatrix} = t \begin{bmatrix} -3 \\ 1 \\ 1 \end{bmatrix}. \tag{3}$$

Interestingly, this solution is closely related to the solution of the *nonhomogeneous system* (2). To solve (2), we employ the same sequence of steps in Gaussian elimination as we did before, except this time we start with the augmented matrix and proceed

$$\left[\begin{array}{ccc|c} 1 & 2 & 1 & 2 \\ 1 & 3 & 0 & -1 \\ 2 & 1 & 5 & 13 \end{array}\right] \to \left[\begin{array}{ccc|c} 1 & 2 & 1 & 2 \\ 0 & 1 & -1 & -3 \\ 0 & -3 & 3 & 9 \end{array}\right] \to \left[\begin{array}{ccc|c} 1 & 2 & 1 & 2 \\ 0 & 1 & -1 & -3 \\ 0 & 0 & 0 & 0 \end{array}\right].$$

As before, $x_3 = t$ is the lone free variable, but this time back substitution gives $x_2 = -3 + x_2 = -3 + t$, $x_1 = 2 - 2x_2 - x_3 = 2 - 3t$, so the solution

$$\mathsf{x} = \begin{bmatrix} x_1 \\ x_2 \\ x_3 \end{bmatrix} = \begin{bmatrix} 2-3t \\ -3+t \\ t \end{bmatrix} = \begin{bmatrix} 2 \\ -3 \\ 0 \end{bmatrix} + t\begin{bmatrix} -3 \\ 1 \\ 1 \end{bmatrix}. \tag{4}$$

Part of this solution should seem familiar. The vector $\mathsf{x}_h = t\begin{bmatrix} -3 \\ 1 \\ 1 \end{bmatrix}$ is the solution to the homogeneous system (1) presented at the start of this section. The first vector in (4), $\mathsf{x}_p = \begin{bmatrix} 2 \\ -3 \\ 0 \end{bmatrix}$, is a *particular solution* to system (2); we get $\mathsf{x} = \begin{bmatrix} 2 \\ -3 \\ 0 \end{bmatrix}$ when $t = 0$. The general solution is $\mathsf{x}_p + \mathsf{x}_h$, the sum of x_p and a solution to the corresponding homogeneous system. This is one instance of a general pattern.

2.6.1 Theorem. *Suppose* x_p *is a particular solution to the system* $A\mathsf{x} = \mathsf{b}$ *and* x *is any other solution. Then* $\mathsf{x} = \mathsf{x}_p + \mathsf{x}_h$ *is the sum of* x_p *and a solution* x_h *to the homogeneous system* $A\mathsf{x} = \mathsf{0}$.

Proof. Since both x_p and x are solutions to $A\mathsf{x} = \mathsf{b}$, we have both $A\mathsf{x}_p = \mathsf{b}$ and $A\mathsf{x} = \mathsf{b}$. Thus $A(\mathsf{x} - \mathsf{x}_p) = \mathsf{0}$, so $\mathsf{x} - \mathsf{x}_p = \mathsf{x}_h$ is a solution to $A\mathsf{x} = \mathsf{0}$. So $\mathsf{x} = \mathsf{x}_p + \mathsf{x}_h$. ■

2.6.2 Example. In Problem 2.4.18, we found that the solution to

$$\begin{aligned} 2x - 3y + 6z &= 14 \\ x + y - 2z &= -3, \end{aligned}$$

is $\begin{bmatrix} x \\ y \\ z \end{bmatrix} = \begin{bmatrix} 1 \\ 2z-4 \\ z \end{bmatrix} = \begin{bmatrix} 1 \\ -4 \\ 0 \end{bmatrix} + t\begin{bmatrix} 0 \\ 2 \\ 1 \end{bmatrix}$. We noted that the vector $\begin{bmatrix} 1 \\ -4 \\ 0 \end{bmatrix}$ is one particular solution to the given system and that the other vector, $\begin{bmatrix} 0 \\ 2 \\ 1 \end{bmatrix}$, is a solution to the corresponding homogeneous system. Here $\mathsf{x}_p = \begin{bmatrix} 1 \\ -4 \\ 0 \end{bmatrix}$ and $\mathsf{x}_h = t\begin{bmatrix} 0 \\ 2 \\ 1 \end{bmatrix}$. (See also Problem 2.4.22.)

2.6.3 **Problem.** Given that the solution to

$$\begin{aligned} x_1 + 2x_2 + 3x_3 - 5x_4 + x_5 &= -2 \\ -3x_1 - 6x_2 - 7x_3 + 12x_4 + 6x_5 &= 0 \\ x_1 + 2x_2 + 2x_3 - 5x_4 + x_5 &= 7 \end{aligned}$$

is $\mathsf{x} = \begin{bmatrix} 5 \\ 0 \\ -9 \\ -4 \\ 0 \end{bmatrix} + t\begin{bmatrix} -2 \\ 1 \\ 0 \\ 0 \\ 0 \end{bmatrix} + s\begin{bmatrix} 14 \\ 0 \\ 0 \\ 3 \\ 1 \end{bmatrix}$, find a particular solution to the given system and the solution to the corresponding homogeneous system.

Solution. Setting $t = s = 0$, a particular solution is $\mathsf{x}_p = \begin{bmatrix} 5 \\ 0 \\ -9 \\ -4 \\ 0 \end{bmatrix}$. The general solution to the corresponding homogeneous system is $\mathsf{x}_h = t\begin{bmatrix} -2 \\ 1 \\ 0 \\ 0 \\ 0 \end{bmatrix} + s\begin{bmatrix} 14 \\ 0 \\ 0 \\ 3 \\ 1 \end{bmatrix}$.

More on Linear Independence

Recall that vectors $\mathsf{x}_1, \mathsf{x}_2, \ldots, \mathsf{x}_n$ are linearly independent if the condition

$$c_1\mathsf{x}_1 + c_2\mathsf{x}_2 + \cdots + c_n\mathsf{x}_n = \mathsf{0} \tag{5}$$

implies that $c_1 = 0$, $c_2 = 0$, ..., $c_n = 0$. Vectors that are not linearly independent are linearly dependent, that is, there exists an equation of the form (5) where **not all** the coefficients are 0. The left side of (5) is $A\mathsf{c}$, where

$A = \begin{bmatrix} \mathsf{x}_1 & \mathsf{x}_2 & \ldots & \mathsf{x}_n \\ \downarrow & \downarrow & & \downarrow \\ & & & \end{bmatrix}$ and $\mathsf{c} = \begin{bmatrix} c_1 \\ c_2 \\ \vdots \\ c_n \end{bmatrix}$. Thus

2.6.4

> $\mathsf{x}_1, \ldots, \mathsf{x}_n$ are linearly independent if and only if $A\mathsf{c} = \mathsf{0}$ implies $\mathsf{c} = \mathsf{0}$ where $A = \begin{bmatrix} \mathsf{x}_1 & \mathsf{x}_2 & \ldots & \mathsf{x}_n \\ \downarrow & \downarrow & & \downarrow \\ & & & \end{bmatrix}$.

2.6.5 Example. The standard basis vectors of R^n are linearly independent. To see why, we suppose that some linear combination of these is the zero vector, say

$$c_1\mathsf{e}_1 + c_2\mathsf{e}_2 + \cdots + c_n\mathsf{e}_n = \mathsf{0}.$$

This is $A\mathsf{c} = \mathsf{0}$ where $\mathsf{c} = \begin{bmatrix} c_1 \\ c_2 \\ \vdots \\ c_n \end{bmatrix}$ and $A = \begin{bmatrix} \mathsf{e}_1 & \mathsf{e}_2 & \cdots & \mathsf{e}_n \\ \downarrow & \downarrow & & \downarrow \end{bmatrix} = I$, the $n \times n$ identity matrix. Thus $A\mathsf{c} = I\mathsf{c} = \mathsf{0}$ certainly implies $\mathsf{c} = \mathsf{0}$, so the vectors are linearly independent. ☺

2.6.6 Problem. Decide whether the vectors $\mathsf{x}_1 = \begin{bmatrix} 2 \\ 3 \\ 1 \\ 4 \end{bmatrix}$, $\mathsf{x}_2 = \begin{bmatrix} -1 \\ 1 \\ 2 \\ 3 \end{bmatrix}$ and $\mathsf{x}_3 = \begin{bmatrix} 4 \\ 0 \\ -1 \\ 1 \end{bmatrix}$ are linearly independent or linearly dependent.

Solution. Suppose $c_1\mathsf{x}_1 + c_2\mathsf{x}_2 + c_3\mathsf{x}_3 = \mathsf{0}$. This is the homogeneous system $A\mathsf{c} = \mathsf{0}$ with $A = \begin{bmatrix} 2 & -1 & 4 \\ 3 & 1 & 0 \\ 1 & 2 & -1 \\ 4 & 3 & 1 \end{bmatrix}$ and $\mathsf{c} = \begin{bmatrix} c_1 \\ c_2 \\ c_3 \end{bmatrix}$. Using Gaussian elimination, we obtain

$$\begin{bmatrix} 2 & -1 & 4 \\ 3 & 1 & 0 \\ 1 & 2 & -1 \\ 4 & 3 & 1 \end{bmatrix} \to \begin{bmatrix} 1 & 2 & -1 \\ 0 & -5 & 6 \\ 0 & -5 & 3 \\ 0 & -5 & 5 \end{bmatrix} \to \begin{bmatrix} 1 & 2 & -1 \\ 0 & 1 & -1 \\ 0 & 0 & 1 \\ 0 & 0 & -2 \end{bmatrix} \to \begin{bmatrix} 1 & 2 & -1 \\ 0 & 1 & -1 \\ 0 & 0 & 1 \\ 0 & 0 & 0 \end{bmatrix}.$$

Back substitution gives $c_3 = 0$, then $c_2 - c_3 = 0$, so $c_2 = 0$ and, finally, $c_1 + 2c_2 - c_3 = 0$, so $c_1 = 0$. Thus $\mathsf{c} = \mathsf{0}$, so the vectors are linearly independent. 👍

READING CHECK 2. Decide whether or not the vectors $\mathsf{x}_1 = \begin{bmatrix} 1 \\ 1 \\ 2 \end{bmatrix}$, $\mathsf{x}_2 = \begin{bmatrix} 2 \\ 3 \\ 1 \end{bmatrix}$, $\mathsf{x}_3 = \begin{bmatrix} 1 \\ 0 \\ 5 \end{bmatrix}$, $\mathsf{x}_4 = \begin{bmatrix} 1 \\ 1 \\ 1 \end{bmatrix}$ are linearly dependent or linearly independent. If the vectors are linearly dependent, exhibit a specific nontrivial linear combination of the vectors which is 0. [Hint: The first few pages of this section.]

Answers to Reading Checks

1. $A\mathsf{x}$ is a linear combination of the columns of A.

2. Suppose $c_1\mathsf{x}_1 + c_2\mathsf{x}_2 + c_3\mathsf{x}_3 + c_4\mathsf{x}_4 = 0$. Then $A\mathsf{c} = 0$ where $A = \begin{bmatrix} 1 & 2 & 1 & 1 \\ 1 & 3 & 0 & 1 \\ 2 & 1 & 5 & 1 \end{bmatrix}$. The submatrix A_1, the first three columns of A, is the matrix of coefficients of system (1). As shown in the text, $A_1\mathsf{x} = 0$ has solution $\mathsf{x} = t\begin{bmatrix} -3 \\ 1 \\ 1 \end{bmatrix}$. So the first three columns of A, which are $\mathsf{x}_1, \mathsf{x}_2, \mathsf{x}_3$, are linearly dependent: specifically, $-3\mathsf{x}_1 + \mathsf{x}_2 + \mathsf{x}_3 = 0$. So $-3\mathsf{x}_1 + \mathsf{x}_2 + \mathsf{x}_3 + 0\mathsf{x}_4 = 0$, showing that the given four vectors are also linearly dependent.

True/False Questions

Decide, with as little calculation as possible, whether each of the following statements is true or false and, if you say "false," explain your answer. (Answers are at the back of the book.)

1. The system $\begin{aligned} -2x_1 \qquad\quad - x_3 &= 0 \\ 5x_2 + 6x_3 &= 0 \\ 7x_1 + 5x_2 - 4x_3 &= 0 \end{aligned}$ is homogeneous.

2. If a 4×4 matrix A has row echelon form $U = \begin{bmatrix} 1 & 2 & 3 & 4 \\ 0 & 1 & 5 & -1 \\ 0 & 0 & 1 & 7 \\ 0 & 0 & 0 & 0 \end{bmatrix}$, the system $A\mathsf{x} = 0$ has only the trivial solution.

3. If u and v are solutions to $A\mathsf{x} = \mathsf{b}$, then $\mathsf{u} - \mathsf{v}$ is a solution to the corresponding homogeneous system.

4. If vectors u, v and w are linearly dependent and A is the matrix whose columns are these vectors, then the homogeneous system $A\mathsf{x} = 0$ has a nontrivial solution.

5. If the columns of a matrix A are linearly dependent, there is a nonzero vector x satisfying $A\mathsf{x} = 0$.

EXERCISES

Answers to exercises marked [BB] can be found at the **Back** *of the* **Book***.*

1. Solve each of the following systems of linear equations by Gaussian elimination and back substitution. Write your answers as vectors or as linear combinations of vectors if appropriate.

(a) [BB] $\begin{aligned} x_1 + x_2 + x_3 &= 0 \\ x_1 + 3x_2 \qquad &= 0 \\ 2x_1 - x_2 - x_3 &= 0 \end{aligned}$

(b) $\begin{aligned} x - 2y + z &= 0 \\ 3x - 7y + 2z &= 0 \end{aligned}$

(c)
$$\begin{aligned} 2x_1 - 7x_2 + x_3 + x_4 &= 0 \\ x_1 - 2x_2 + x_3 &= 0 \\ 3x_1 + 6x_2 + 7x_3 - 4x_4 &= 0 \end{aligned}$$

(d)
$$\begin{aligned} x_1 - 2x_2 + x_3 + 3x_4 - x_5 &= 0 \\ 4x_1 - 7x_2 + 5x_3 + 16x_4 - 2x_5 &= 0 \\ -3x_1 + 4x_2 - 4x_3 - 18x_4 + 6x_5 &= 0. \end{aligned}$$

2. Find the value or values of a for which the system
$$\begin{aligned} ax + y + z &= 0 \\ x + y - z &= 0 \\ x + y + az &= 0 \end{aligned}$$
has nontrivial solutions and find all such solutions.

3. [BB] **(a)** Express the zero vector $\mathsf{0} = \begin{bmatrix} 0 \\ 0 \\ 0 \end{bmatrix}$ as a nontrivial linear combination of the columns of $A = \begin{bmatrix} 1 & -1 & 1 \\ 0 & 1 & 1 \\ 1 & 0 & 2 \end{bmatrix}$.

   (b) Use part (a) to show that A is not invertible.

4. [BB] Find conditions on a, b and c (if any) such that the system
$$\begin{aligned} x + y \quad\quad &= 0 \\ y + z &= 0 \\ x \quad - z &= 0 \\ ax + by + cz &= 0 \end{aligned}$$
has (i) a unique solution; (ii) no solution; (iii) an infinite number of solutions.

5. A homogeneous system of four linear equations in the six unknowns x_1, x_2, x_3, x_4, x_5, x_6 has coefficient matrix A whose row echelon form is
$$\begin{bmatrix} 1 & 2 & 0 & 3 & 1 & 1 \\ 0 & 0 & 0 & 1 & -2 & 0 \\ 0 & 0 & 0 & 0 & 0 & 1 \\ 0 & 0 & 0 & 0 & 0 & 0 \end{bmatrix}.$$
Solve the system.

6. Write the solution to each of the following linear systems in the form $\mathsf{x} = \mathsf{x}_p + \mathsf{x}_h$ where x_p is a particular solution and x_h is a solution of the corresponding homogeneous system.

   (a) [BB]
$$\begin{aligned} x_1 + 5x_2 + 7x_3 &= -2 \\ 9x_3 &= 3 \end{aligned}$$

   (b)
$$\begin{aligned} x_1 + 2x_2 + 3x_3 + 4x_4 &= 1 \\ 2x_1 + 4x_2 + 8x_3 + 10x_4 &= 6 \\ 3x_1 + 7x_2 + 11x_3 + 14x_4 &= 7 \end{aligned}$$

   (c)
$$\begin{aligned} 3x_1 - x_2 + x_3 + 2x_4 &= 4 \\ -4x_1 + x_2 + 2x_3 + 7x_4 &= 3 \\ x_1 - 2x_2 + 3x_3 + 7x_4 &= 1 \\ 3x_2 + x_3 + 6x_4 &= 18 \end{aligned}$$

(d)
$$\begin{aligned} 3x_1 + 6x_2 \qquad\quad - 2x_4 + 7x_5 &= -1 \\ -2x_1 - 4x_2 + x_3 + 4x_4 - 9x_5 &= 0 \\ x_1 + 2x_2 \qquad\quad - x_4 + 2x_5 &= 4 \end{aligned}$$

(e)
$$\begin{aligned} -x_1 \qquad\qquad\qquad - x_4 + 2x_5 &= -2 \\ 3x_1 + x_2 + 2x_3 + 4x_4 - x_5 &= 1 \\ 4x_1 - x_2 + x_3 \qquad\quad + 2x_5 &= 1 \end{aligned}$$

(f)
$$\begin{aligned} x_1 + 2x_2 - x_3 \qquad + 3x_5 + x_6 &= 0 \\ -2x_1 - 4x_2 + 3x_3 - 3x_4 - 6x_5 - 2x_6 &= 2 \\ 3x_1 + 6x_2 - x_3 - 6x_4 + 11x_5 + 7x_6 &= 2 \\ x_1 + 2x_2 - 2x_3 + 3x_4 \qquad + 2x_6 &= -1. \end{aligned}$$

7. [BB] **(a)** Write the system

$$\begin{aligned} x_1 - 2x_2 + 3x_3 + x_4 + 3x_5 + 4x_6 &= -1 \\ -3x_1 + 6x_2 - 8x_3 + 2x_4 - 11x_5 - 15x_6 &= 2 \\ x_1 - 2x_2 + 2x_3 - 4x_4 + 6x_5 + 9x_6 &= 3 \\ -2x_1 + 4x_2 - 6x_3 - 2x_4 - 6x_5 - 7x_6 &= 1 \end{aligned}$$

in the form $A\mathsf{x} = \mathsf{b}$.

(b) Solve the system expressing your answer as a vector $\mathsf{x} = \mathsf{x}_p + \mathsf{x}_h$, where x_p is a particular solution and x_h is a solution to the corresponding homogeneous system.

(c) Express $\begin{bmatrix} -1 \\ 2 \\ 3 \\ 1 \end{bmatrix}$ as a linear combination of the columns of A.

8. Repeat Exercise 7 for the system

$$\begin{aligned} x_1 - x_2 + 2x_3 \qquad\quad + 3x_5 - 5x_6 &= -1 \\ -3x_1 + 4x_2 - 2x_3 + x_4 - 7x_5 + 15x_6 &= 5 \\ 2x_1 - 2x_2 + 4x_3 + x_4 + 5x_5 - 9x_6 &= -2 \\ x_1 - 2x_2 - 2x_3 - 2x_4 + 2x_5 - 5x_6 &= -2 \end{aligned}$$

using the vector $\begin{bmatrix} -1 \\ 5 \\ -2 \\ -2 \end{bmatrix}$ in part (c).

9. Determine whether each of the given sets of vectors is linearly independent or linearly dependent. In the case of linear dependence, write down a nontrivial linear combination of the vectors that equals the zero vector.

(a) [BB] $\mathsf{v}_1 = \begin{bmatrix} 2 \\ 0 \\ 2 \end{bmatrix}, \mathsf{v}_2 = \begin{bmatrix} 1 \\ 1 \\ 1 \end{bmatrix}, \mathsf{v}_3 = \begin{bmatrix} -1 \\ 1 \\ -1 \end{bmatrix}, \mathsf{v}_4 = \begin{bmatrix} 0 \\ 2 \\ 0 \end{bmatrix}$

(b) $\mathsf{v}_1 = \begin{bmatrix} 1 \\ 2 \\ 3 \end{bmatrix}, \mathsf{v}_2 = \begin{bmatrix} -3 \\ 0 \\ 1 \end{bmatrix}, \mathsf{v}_3 = \begin{bmatrix} 4 \\ 2 \\ 2 \end{bmatrix}, \mathsf{v}_4 = \begin{bmatrix} 1 \\ 1 \\ 13 \end{bmatrix}$

(c) $\mathsf{v}_1 = \begin{bmatrix} 1 \\ 1 \\ 2 \\ 1 \end{bmatrix}, \mathsf{v}_2 = \begin{bmatrix} 3 \\ 3 \\ 3 \\ 1 \end{bmatrix}, \mathsf{v}_3 = \begin{bmatrix} 2 \\ 5 \\ 10 \\ 4 \end{bmatrix}, \mathsf{v}_4 = \begin{bmatrix} 1 \\ 2 \\ 3 \\ 1 \end{bmatrix}$

(d) $\mathsf{v}_1 = \begin{bmatrix} -1 \\ 5 \\ -1 \\ 3 \\ 1 \end{bmatrix}, \mathsf{v}_2 = \begin{bmatrix} -2 \\ 9 \\ 1 \\ 6 \\ 1 \end{bmatrix}, \mathsf{v}_3 = \begin{bmatrix} 3 \\ -11 \\ -1 \\ -9 \\ -3 \end{bmatrix}$

(e) $\mathsf{v}_1 = \begin{bmatrix} 1 \\ -1 \\ -1 \\ -4 \end{bmatrix}, \mathsf{v}_2 = \begin{bmatrix} -1 \\ 2 \\ -1 \\ 1 \end{bmatrix}, \mathsf{v}_3 = \begin{bmatrix} 2 \\ 4 \\ -2 \\ 6 \end{bmatrix}, \mathsf{v}_4 = \begin{bmatrix} 3 \\ 5 \\ -1 \\ 7 \end{bmatrix}$

(f) $\mathsf{v}_1 = \begin{bmatrix} 4 \\ 5 \\ 1 \\ 7 \end{bmatrix}, \mathsf{v}_2 = \begin{bmatrix} 8 \\ 6 \\ 10 \\ 8 \end{bmatrix}, \mathsf{v}_3 = \begin{bmatrix} 4 \\ 2 \\ -3 \\ 4 \end{bmatrix}$

(g) $\mathsf{v}_1 = \begin{bmatrix} -2 \\ 1 \\ 6 \\ 1 \\ 1 \end{bmatrix}, \mathsf{v}_2 = \begin{bmatrix} 0 \\ -1 \\ 2 \\ 9 \\ 7 \end{bmatrix}, \mathsf{v}_3 = \begin{bmatrix} 1 \\ 3 \\ 4 \\ 0 \\ -1 \end{bmatrix}, \mathsf{v}_4 = \begin{bmatrix} 1 \\ 1 \\ 1 \\ 2 \\ 1 \end{bmatrix}.$

10. Find a condition on k which is both *necessary and sufficient* for the vectors $\begin{bmatrix} 1 \\ -1 \\ 0 \end{bmatrix}, \begin{bmatrix} k \\ 1 \\ 0 \end{bmatrix}$ and $\begin{bmatrix} 0 \\ 2 \\ 3 \end{bmatrix}$ to be linearly independent; that is, complete the following sentence:

    "The vectors $\begin{bmatrix} 1 \\ -1 \\ 0 \end{bmatrix}, \begin{bmatrix} k \\ 1 \\ 0 \end{bmatrix}$ and $\begin{bmatrix} 0 \\ 2 \\ 3 \end{bmatrix}$ are linearly independent if and only if __________."

Critical Reading

11. Complete the following sentence: The columns of a matrix T are linearly independent if and only if ________ .

12. Suppose the homogeneous system $A\mathsf{x} = 0$ has a nontrivial solution x_0. Explain why the system has infinitely many nontrivial solutions.

13. Suppose the quadratic polynomial $ax^2 + bx + c$ is 0 when evaluated at three different values of x. Show that $a = b = c = 0$.

14. Suppose A and B are matrices with linearly independent columns. Prove that AB also has linearly independent columns (assuming this matrix product is defined).

2.7 Elementary Matrices and LU Factorization

Elementary Matrices

The three elementary row operations on a matrix are

1. $R \leftrightarrow R'$: the interchange of two rows,
2. $R \rightarrow cR$: multiplication of a row by a nonzero scalar, and
3. $R \rightarrow R - CR'$: replacement of a row by that row less a multiple of another.

2.7.1 Definition. An *elementary matrix* is a square matrix obtained from the identity matrix by a single elementary row operation.

2.7.2 Examples. • $E_1 = \begin{bmatrix} 0 & 1 & 0 \\ 1 & 0 & 0 \\ 0 & 0 & 1 \end{bmatrix}$ is an elementary matrix, obtained from the 3×3 identity matrix I by interchanging rows one and two, an operation we denote $R1 \leftrightarrow R2$.

- The matrix $E_2 = \begin{bmatrix} 1 & 0 & 0 \\ 0 & 3 & 0 \\ 0 & 0 & 1 \end{bmatrix}$ is elementary; it is obtained from I by multiplying row two by 3, an operation we denote $R2 \rightarrow 3(R2)$.

- The matrix $E_3 = \begin{bmatrix} 1 & 0 & 0 \\ 0 & 1 & 0 \\ -4 & 0 & 1 \end{bmatrix}$ is elementary; it is obtained from I by replacing row three by row three less 4 times row one, an operation denoted $R3 \rightarrow R3 - 4(R1)$. ☺

The next observation is very useful.[7]

2.7.3 | If A is a matrix and E is an elementary matrix with EA defined, then EA is that matrix obtained by applying to A the elementary operation that produced E.

To illustrate, let $A = \begin{bmatrix} a & d \\ b & e \\ c & f \end{bmatrix}$ and leave E_1, E_2 and E_3 as above. The matrix E_1A is

$$E_1A = \begin{bmatrix} 0 & 1 & 0 \\ 1 & 0 & 0 \\ 0 & 0 & 1 \end{bmatrix} \begin{bmatrix} a & d \\ b & e \\ c & f \end{bmatrix} = \begin{bmatrix} b & e \\ a & d \\ c & f \end{bmatrix},$$

[7]For some reason, this always reminds the author of the golden rule. An elementary matrix E does unto others what was done unto the identity when E was created.

which is just A with rows one and two interchanged. The matrix E_2A is

$$E_2A = \begin{bmatrix} 1 & 0 & 0 \\ 0 & 3 & 0 \\ 0 & 0 & 1 \end{bmatrix} \begin{bmatrix} a & d \\ b & e \\ c & f \end{bmatrix} = \begin{bmatrix} a & d \\ 3b & 3e \\ c & f \end{bmatrix},$$

which is A with its second row multiplied by 3, and

$$E_3A = \begin{bmatrix} 1 & 0 & 0 \\ 0 & 1 & 0 \\ -4 & 0 & 1 \end{bmatrix} \begin{bmatrix} a & d \\ b & e \\ c & f \end{bmatrix} = \begin{bmatrix} a & d \\ b & e \\ -4a+c & -4d+f \end{bmatrix},$$

which is A, but with row three replaced by row three less 4 times row one.

The inverse of an elementary matrix is another elementary matrix, easily determined.

2.7.4 If E is an elementary matrix, then E has an inverse that is also elementary: E^{-1} is the elementary matrix that reverses (or undoes) what E does.

We illustrate.

2.7.5 Example. $E_1 = \begin{bmatrix} 0 & 1 & 0 \\ 1 & 0 & 0 \\ 0 & 0 & 1 \end{bmatrix}$ is the elementary matrix that interchanges rows one and two of a matrix. To undo such an interchange, we interchange rows one and two again. According to 2.7.4 we should have $E_1^{-1} = E_1$. Since E_1 is square, to prove that E_1 is the inverse of E_1 it is enough to show that $E_1E_1 = I$. This follows from 2.7.3 since E_1E_1 is E_1 with its first two rows interchanged. Alternatively, we can compute

$$E_1E_1 = \begin{bmatrix} 0 & 1 & 0 \\ 1 & 0 & 0 \\ 0 & 0 & 1 \end{bmatrix} \begin{bmatrix} 0 & 1 & 0 \\ 1 & 0 & 0 \\ 0 & 0 & 1 \end{bmatrix} = \begin{bmatrix} 1 & 0 & 0 \\ 0 & 1 & 0 \\ 0 & 0 & 1 \end{bmatrix} = I.$$

2.7.6 Example. To multiply row two of a matrix by 3, multiply (on the left) by the elementary matrix $E_2 = \begin{bmatrix} 1 & 0 & 0 \\ 0 & 3 & 0 \\ 0 & 0 & 1 \end{bmatrix}$. To undo the effect of this operation, multiply row two by $\frac{1}{3}$. According to 2.7.4, E_2^{-1} should be $E = \begin{bmatrix} 1 & 0 & 0 \\ 0 & \frac{1}{3} & 0 \\ 0 & 0 & 1 \end{bmatrix}$. We can compute E_2E using 2.7.3: E_2E is E with row two multiplied by 3. Alternatively, we can compute

$$E_2E = \begin{bmatrix} 1 & 0 & 0 \\ 0 & 3 & 0 \\ 0 & 0 & 1 \end{bmatrix} \begin{bmatrix} 1 & 0 & 0 \\ 0 & \frac{1}{3} & 0 \\ 0 & 0 & 1 \end{bmatrix} = \begin{bmatrix} 1 & 0 & 0 \\ 0 & 1 & 0 \\ 0 & 0 & 1 \end{bmatrix} = I.$$

2.7.7 Example. Multiplying a matrix by the elementary matrix $E_3 = \begin{bmatrix} 1 & 0 & 0 \\ 0 & 1 & 0 \\ -4 & 0 & 1 \end{bmatrix}$ replaces row three of that matrix by row three less 4 times row one. To undo this operation, we should replace row three by row three plus 4 times row one. Apparently then, the inverse of E_3 is $E = \begin{bmatrix} 1 & 0 & 0 \\ 0 & 1 & 0 \\ 4 & 0 & 1 \end{bmatrix}$. Since E_3E is B with row three replaced by that row less 4 times row one, $E_3E = I$. Alternatively, we can compute directly

$$E_3E = \begin{bmatrix} 1 & 0 & 0 \\ 0 & 1 & 0 \\ -4 & 0 & 1 \end{bmatrix} \begin{bmatrix} 1 & 0 & 0 \\ 0 & 1 & 0 \\ 4 & 0 & 1 \end{bmatrix} = \begin{bmatrix} 1 & 0 & 0 \\ 0 & 1 & 0 \\ 0 & 0 & 1 \end{bmatrix} = I.$$

LU Factorization

What can one do with these ideas about elementary matrices?

2.7.8 Example. We can move $A = \begin{bmatrix} 2 & 4 & 1 & 1 \\ -6 & 0 & 0 & 5 \end{bmatrix}$ to row echelon form in one step using $R2 \to R2 + 3(R1)$:

$$A = \begin{bmatrix} 2 & 4 & 1 & 1 \\ -6 & 0 & 0 & 5 \end{bmatrix} \to \begin{bmatrix} 2 & 4 & 1 & 1 \\ 0 & 12 & 3 & 8 \end{bmatrix} = U.$$

As we have seen, the elementary row operation $R2 \to R2 + 3(R1)$ can be accomplished upon multiplying A on the left by $E = \begin{bmatrix} 1 & 0 \\ 3 & 1 \end{bmatrix}$. Thus $U = EA$ and, since E is invertible, $A = E^{-1}U$. The inverse of E is the elementary matrix which "undoes" E: $E^{-1} = \begin{bmatrix} 1 & 0 \\ -3 & 1 \end{bmatrix}$. Setting $L = E^{-1}$, we have a factorization

$$\underset{A}{\begin{bmatrix} 2 & 4 & 1 & 1 \\ -6 & 0 & 0 & 5 \end{bmatrix}} = \underset{L}{\begin{bmatrix} 1 & 0 \\ -3 & 1 \end{bmatrix}} \underset{U}{\begin{bmatrix} 2 & 4 & 1 & 1 \\ 0 & 12 & 3 & 8 \end{bmatrix}} \tag{1}$$

of A as the product of a lower triangular matrix L and a row echelon (and hence upper triangular) matrix U. This kind of factorization is the subject of this section—LU factorization, L for lower and U for upper.

2.7.9 Definition. An *LU factorization* of a matrix A is a factorization $A = LU$ where L is a (necessarily) square lower triangular matrix with 1s on the diagonal and U is a row echelon form of A.

READING CHECK 1. If A is an $m \times n$ matrix and we factor $A = LU$ with U a row echelon form of A, the matrix L must be square. Why?

2.7.10 Example. We find an LU factorization of the matrix $A = \begin{bmatrix} -1 & 1 & -2 & 1 \\ 2 & 1 & 7 & 0 \\ -3 & 0 & 1 & 5 \end{bmatrix}$ by reducing A to a row echelon matrix, making note of the elementary matrices that effect the row operations we use in Gaussian elimination. To begin we use $R2 \to R2 + 2(R1)$, which can be achieved multiplying A (on the left) by a certain 3×3 elementary matrix (3×3 because A is 3×4). The elementary matrix we need is $E_1 = \begin{bmatrix} 1 & 0 & 0 \\ 2 & 1 & 0 \\ 0 & 0 & 1 \end{bmatrix}$, and

$$E_1 A = \begin{bmatrix} 1 & 0 & 0 \\ 2 & 1 & 0 \\ 0 & 0 & 1 \end{bmatrix} \begin{bmatrix} -1 & 1 & -2 & 1 \\ 2 & 1 & 7 & 0 \\ -3 & 0 & 1 & 5 \end{bmatrix} = \begin{bmatrix} -1 & 1 & -2 & 1 \\ 0 & 3 & 3 & 2 \\ -3 & 0 & 1 & 5 \end{bmatrix}.$$

Now we use $R3 \to R3 - 3(R1)$ which is achieved upon multiplication by $E_2 = \begin{bmatrix} 1 & 0 & 0 \\ 0 & 1 & 0 \\ -3 & 1 & 0 \end{bmatrix}$, giving

$$E_2(E_1 A) = \begin{bmatrix} 1 & 0 & 0 \\ 0 & 1 & 0 \\ -3 & 1 & 0 \end{bmatrix} \begin{bmatrix} -1 & 1 & -2 & 1 \\ 0 & 3 & 3 & 2 \\ -3 & 0 & 1 & 5 \end{bmatrix} = \begin{bmatrix} -1 & 1 & -2 & 1 \\ 0 & 3 & 3 & 2 \\ 0 & -3 & 7 & 2 \end{bmatrix}.$$

Finally, we use $R3 \to R3 + R2$, so we multiply by $E_3 = \begin{bmatrix} 1 & 0 & 0 \\ 0 & 1 & 0 \\ 0 & 1 & 1 \end{bmatrix}$ and get

$$E_3(E_2 E_1 A) = \begin{bmatrix} 1 & 0 & 0 \\ 0 & 1 & 0 \\ 0 & 1 & 1 \end{bmatrix} \begin{bmatrix} -1 & 1 & -2 & 1 \\ 0 & 3 & 3 & 2 \\ 0 & -3 & 7 & 2 \end{bmatrix} = \begin{bmatrix} -1 & 1 & -2 & 1 \\ 0 & 3 & 3 & 2 \\ 0 & 0 & 10 & 4 \end{bmatrix} = U.$$

So $(E_3 E_2 E_1)A = U$ and, since each of E_3, E_2 and E_1 is invertible, so is their product and

$$\begin{aligned} (E_3 E_2 E_1)^{-1} &= E_1^{-1} E_2^{-1} E_3^{-1} \\ &= \begin{bmatrix} 1 & 0 & 0 \\ -2 & 1 & 0 \\ 0 & 0 & 1 \end{bmatrix} \begin{bmatrix} 1 & 0 & 0 \\ 0 & 1 & 0 \\ 3 & 0 & 1 \end{bmatrix} \begin{bmatrix} 1 & 0 & 0 \\ 0 & 1 & 0 \\ 0 & -1 & 0 \end{bmatrix} = \begin{bmatrix} 1 & 0 & 0 \\ -2 & 1 & 0 \\ 3 & -1 & 1 \end{bmatrix}. \end{aligned}$$

Finally $(E_3 E_2 E_1)A = U$ implies

$$A = (E_3 E_2 E_1)^{-1} U = \begin{bmatrix} 1 & 0 & 0 \\ -2 & 1 & 0 \\ 3 & -1 & 1 \end{bmatrix} \begin{bmatrix} -1 & 1 & -2 & 1 \\ 0 & 3 & 3 & 2 \\ 0 & 0 & 10 & 4 \end{bmatrix},$$

an equation the student should verify.

2.7.11 Problem. Find an LU factorization of $A = \begin{bmatrix} 1 & 2 \\ 4 & -6 \\ -7 & -7 \\ 1 & 0 \end{bmatrix}$.

Solution. We reduce A to row echelon form via Gaussian elimination, keeping track of the elementary matrices we use.

The operation $R2 \to R2 - 4(R1)$ is effected multiplying on the left by $E_1 = \begin{bmatrix} 1 & 0 & 0 & 0 \\ -4 & 1 & 0 & 0 \\ 0 & 0 & 1 & 0 \\ 0 & 0 & 0 & 1 \end{bmatrix}$, giving $E_1A = \begin{bmatrix} 1 & 2 \\ 0 & -14 \\ -7 & -7 \\ 1 & 0 \end{bmatrix}$. The operation $R_3 \to R_3 + 7R_1$ is effected multiplying by $E_2 = \begin{bmatrix} 1 & 0 & 0 & 0 \\ 0 & 1 & 0 & 0 \\ 7 & 0 & 1 & 0 \\ 0 & 0 & 0 & 1 \end{bmatrix}$, giving $E_2(E_1A) = \begin{bmatrix} 1 & 2 \\ 0 & -14 \\ 0 & 7 \\ 1 & 0 \end{bmatrix}$.

Next we apply $R_4 \to R_4 - R_1$, which is effected by $E_3 = \begin{bmatrix} 1 & 0 & 0 & 0 \\ 0 & 1 & 0 & 0 \\ 0 & 0 & 1 & 0 \\ -1 & 0 & 0 & 1 \end{bmatrix}$, giving $E_3(E_2E_1A) = \begin{bmatrix} 1 & 2 \\ 0 & -14 \\ 0 & 7 \\ 0 & -2 \end{bmatrix}$. The operation $R_3 \to R_3 + \frac{1}{2}R_2$ requires $E_4 = \begin{bmatrix} 1 & 0 & 0 & 0 \\ 0 & 1 & 0 & 0 \\ 0 & \frac{1}{2} & 1 & 0 \\ 0 & 0 & 0 & 1 \end{bmatrix}$ and gives $E_4(E_3E_2E_1A) = \begin{bmatrix} 1 & 2 \\ 0 & -14 \\ 0 & 0 \\ 0 & -2 \end{bmatrix}$. The operation $R_4 \to R_4 - \frac{1}{7}R_2$ requires $E_5 = \begin{bmatrix} 1 & 0 & 0 & 0 \\ 0 & 1 & 0 & 0 \\ 0 & 0 & 1 & 0 \\ 0 & \frac{1}{7} & 0 & 1 \end{bmatrix}$ and gives, finally, $E_5(E_4E_3E_2E_1A) = \begin{bmatrix} 1 & 2 \\ 0 & -14 \\ 0 & 0 \\ 0 & 0 \end{bmatrix} = U$, so $A = (E_5E_4E_3E_2E_1)^{-1}U = E_1^{-1}E_2^{-1}E_3^{-1}E_4^{-1}E_5^{-1}U$, which is $A = LU$ with $L = E_1^{-1}E_2^{-1}E_3^{-1}E_4^{-1}E_5^{-1}$

$$= \begin{bmatrix} 1 & 0 & 0 & 0 \\ 4 & 1 & 0 & 0 \\ 0 & 0 & 1 & 0 \\ 0 & 0 & 0 & 1 \end{bmatrix} \begin{bmatrix} 1 & 0 & 0 & 0 \\ 0 & 1 & 0 & 0 \\ -7 & 0 & 1 & 0 \\ 0 & 0 & 0 & 1 \end{bmatrix} \begin{bmatrix} 1 & 0 & 0 & 0 \\ 0 & 1 & 0 & 0 \\ 0 & 0 & 1 & 0 \\ 1 & 0 & 0 & 1 \end{bmatrix}$$

$$\cdot \begin{bmatrix} 1 & 0 & 0 & 0 \\ 0 & 1 & 0 & 0 \\ 0 & -\frac{1}{2} & 1 & 0 \\ 0 & 0 & 0 & 1 \end{bmatrix} \begin{bmatrix} 1 & 0 & 0 & 0 \\ 0 & 1 & 0 & 0 \\ 0 & 0 & 1 & 0 \\ 0 & \frac{1}{7} & 0 & 1 \end{bmatrix} = \begin{bmatrix} 1 & 0 & 0 & 0 \\ 4 & 1 & 0 & 0 \\ -7 & -\frac{1}{2} & 1 & 0 \\ 1 & \frac{1}{7} & 0 & 1 \end{bmatrix}.$$

So we have $A = \begin{bmatrix} 1 & 2 \\ 4 & -6 \\ -7 & -7 \\ 1 & 0 \end{bmatrix} = \begin{bmatrix} 1 & 0 & 0 & 0 \\ 4 & 1 & 0 & 0 \\ -7 & -\frac{1}{2} & 1 & 0 \\ 1 & \frac{1}{7} & 0 & 1 \end{bmatrix} \begin{bmatrix} 1 & 2 \\ 0 & -14 \\ 0 & 0 \\ 0 & 0 \end{bmatrix} = LU$, an equation which should be checked (of course).

Preceding examples illustrate the following theorem.

2.7.12 Theorem. *If a matrix A can be transformed to a row echelon matrix from the top down using only elementary operations of the type $R_j \mapsto R_j - cR_i$ with $i < j$, then A has an LU factorization.*

The argument requires a couple of ideas. The third elementary row operation $R_j \mapsto R_j - cR_i$ with $i < j$ means you can only replace a row with that row less a multiple of a **previous** row and corresponds to multiplication by an elementary matrix that is **lower triangular**. Since the inverse of an elementary matrix is an elementary matrix of the same type, the inverse of each elementary matrix used to reduce A to row echelon form is a lower triangular matrix. Since the product of lower triangular matrices is lower triangular, the first matrix in the factorization is lower triangular.

Note that row interchanges are not allowed because the elementary matrix corresponding to an interchange of rows is not lower triangular. Consider, for example, $\begin{bmatrix} 0 & 1 \\ 1 & 0 \end{bmatrix}$. The reason for avoiding the second elementary row operation $R_i \to cR_i$ will become clear soon.

We'll explain later what "top down" in Theorem 2.7.12 is all about —see Remark 2.7.15.

2.7.13 Problem. Does $A = \begin{bmatrix} -2 & 6 & 0 & 5 \\ 1 & -3 & 3 & 1 \\ -1 & 0 & 1 & 9 \end{bmatrix}$ have an LU factorization?

Solution. We try to bring A to row echelon form using only the third elementary row operation $R_j \to R_j - cR_i$ with $i < j$. We begin

$$A = \begin{bmatrix} -2 & 6 & 0 & 5 \\ 1 & -3 & 3 & 1 \\ -1 & 0 & 1 & 9 \end{bmatrix} \to \begin{bmatrix} -2 & 6 & 0 & 5 \\ 0 & 0 & 3 & \frac{7}{2} \\ 0 & -3 & 1 & \frac{13}{2} \end{bmatrix}$$

Now we need either a row interchange, or the operation $R2 \mapsto R2 + R3$ which is not allowed, so we conclude that this matrix does **not** have an LU factorization.

L without effort

Our primary purpose in introducing elementary matrices was theoretical. They show why Theorem 2.7.12 is true: the matrix L is lower triangular because it is the product of lower triangular (elementary) matrices. In practice, this is **not** how L is determined, however. If you use only the third elementary row operation in the Gaussian elimination then L can be found with ease.

Consider again Problem 2.7.11. There we found the LU factorization

$$\begin{bmatrix} 1 & 2 \\ 4 & -6 \\ -7 & -7 \\ 1 & 0 \end{bmatrix} = \begin{bmatrix} 1 & 0 & 0 & 0 \\ 4 & 1 & 0 & 0 \\ -7 & -\frac{1}{2} & 1 & 0 \\ 1 & \frac{1}{7} & 0 & 1 \end{bmatrix} \begin{bmatrix} 1 & 2 \\ 0 & -14 \\ 0 & 0 \\ 0 & 0 \end{bmatrix},$$

using only the third elementary row operation, like this:

$$A = \begin{bmatrix} 1 & 2 \\ 4 & -6 \\ -7 & -7 \\ 1 & 0 \end{bmatrix} \overset{R2 \to R2-4(R1)}{\longrightarrow} \begin{bmatrix} 1 & 2 \\ 0 & -14 \\ -7 & -7 \\ 1 & 0 \end{bmatrix}$$

$$\overset{R3 \to R3-(-7)(R1)}{\longrightarrow} \begin{bmatrix} 1 & 2 \\ 0 & -14 \\ 0 & 7 \\ 1 & 0 \end{bmatrix} \overset{R4 \to R4-R1}{\longrightarrow} \begin{bmatrix} 1 & 2 \\ 0 & -14 \\ 0 & 7 \\ 0 & -2 \end{bmatrix}$$

$$\overset{R3 \to R3-(-\frac{1}{2})(R2)}{\longrightarrow} \begin{bmatrix} 1 & 2 \\ 0 & -14 \\ 0 & 0 \\ 0 & -2 \end{bmatrix} \overset{R4 \to R4-\frac{1}{7})(R2)}{\longrightarrow} \begin{bmatrix} 1 & 2 \\ 0 & -14 \\ 0 & 0 \\ 0 & 0 \end{bmatrix} = U.$$

We found $L = \begin{bmatrix} 1 & 0 & 0 & 0 \\ 4 & 1 & 0 & 0 \\ -7 & -\frac{1}{2} & 1 & 0 \\ 1 & \frac{1}{7} & 0 & 1 \end{bmatrix}$ as the product of certain elementary matrices. We could, however, have simply written L down without any calculation whatsoever! Remember our insistence on using the language of subtraction to describe the third elementary row operation.

Have a close look at L. It's square and lower triangular with diagonal entries all 1. That's easy to remember. What about the entries below the diagonal, the 4, the -7, the $-\frac{1}{2}$, the 1, the $\frac{1}{7}$? These are just the **multipliers** that were needed in the Gaussian elimination, **since we used the language of subtraction**. Our operations were

$$\begin{aligned} R2 &\to R2-\mathbf{4}(R1), \\ R3 &\to R3-(\mathbf{-7})(R1), \\ R4 &\to R4-\mathbf{1}(R1), \\ R3 &\to R3-(-\tfrac{\mathbf{1}}{\mathbf{2}})(R2), \\ R4 &\to R4-\tfrac{\mathbf{1}}{\mathbf{7}}(R2), \end{aligned}$$

hence L has the multipliers

4 in row two, column one,
-7 in row three, column one,
1 in row row, column one,
$-\frac{1}{2}$ in row three, column two and
$\frac{1}{7}$ in row four, column two.

Notice that the multipliers in L correspond to the positions where they achieve 0s.

2.7.14

> To find an LU factorization of a matrix A, try to reduce A to row echelon form U from the top down using only the third elementary row operation in the form $R_j \mapsto R_j - cR_i$ with $i < j$.

2.7.15 Remark. "Top down" here and in Theorem 2.7.12 means that once a pivot has been identified, the third elementary row operation must put 0s below that pivot in order starting with the entry immediately below that pivot. Suppose $A = \begin{bmatrix} 1 & 2 & 3 \\ 1 & 1 & 1 \\ 2 & 2 & 3 \end{bmatrix}$ and we apply the following sequence of elementary row operations:

$$A = \begin{bmatrix} 1 & 2 & 3 \\ 1 & 1 & 1 \\ 2 & 2 & 3 \end{bmatrix} \xrightarrow{R3 \to R3 - 2R2} \begin{bmatrix} 1 & 2 & 3 \\ 1 & 1 & 1 \\ 0 & 0 & 1 \end{bmatrix} \xrightarrow{R2 \to R2 - R1} \begin{bmatrix} 1 & 2 & 3 \\ 0 & -1 & -2 \\ 0 & 0 & 1 \end{bmatrix} = U.$$

The 0s below the first pivot were not produced from the top down. The matrix recording the multipliers is $L = \begin{bmatrix} 1 & 0 & 0 \\ 1 & 1 & 0 \\ 2 & 0 & 1 \end{bmatrix}$, but $A \neq LU$. For this observation and example, the author is indebted to an anonymous reviewer of this manuscript whom we thank. See also Exercise 25 for another example of where failure to move top down causes problems.

2.7.16 Example. We move $A = \begin{bmatrix} 1 & 2 & 3 \\ 5 & 14 & 7 \\ 9 & 10 & 0 \\ 0 & 2 & 3 \end{bmatrix}$ to row echelon form from the top down.

$$A \xrightarrow{R2 \to R2 - 5(R1)} \begin{bmatrix} 1 & 2 & 3 \\ 0 & 4 & -8 \\ 9 & 10 & 0 \\ 0 & 2 & 3 \end{bmatrix} \xrightarrow{R3 \to R3 - 9(R1)} \begin{bmatrix} 1 & 2 & 3 \\ 0 & 4 & -8 \\ 0 & -8 & -27 \\ 0 & 2 & 3 \end{bmatrix}$$

$$\xrightarrow{R3\to R3-(-2)(R2)} \begin{bmatrix} 1 & 2 & 3 \\ 0 & 4 & -8 \\ 0 & 0 & -43 \\ 0 & 2 & 3 \end{bmatrix} \xrightarrow{R4\to R4-\frac{1}{2}(R2)} \begin{bmatrix} 1 & 2 & 3 \\ 0 & 4 & -8 \\ 0 & 0 & -43 \\ 0 & 0 & 7 \end{bmatrix}$$

$$\xrightarrow{R4\to R4-(-\frac{7}{43})(R3)} \begin{bmatrix} 1 & 2 & 3 \\ 0 & 4 & -8 \\ 0 & 0 & -43 \\ 0 & 0 & 0 \end{bmatrix} = U.$$

We conclude that $A = LU$ with a lower triangular L that we can just write down: put 1s on the main diagonal, 0s above, and the multipliers below.

$$L = \begin{bmatrix} 1 & 0 & 0 & 0 \\ 5 & 1 & 0 & 0 \\ 9 & -2 & 1 & 0 \\ 0 & \frac{1}{2} & -\frac{7}{43} & 1 \end{bmatrix}.$$

:)

READING CHECK 2. The matrix $A = \begin{bmatrix} 1 & 2 & 3 \\ 1 & 1 & 1 \\ 2 & 2 & 3 \end{bmatrix}$ in 2.7.15 does in fact have an LU factorization. Find it.

Why LU?

The author's favourite high school math teacher used to say frequently that "any method is a good method:" Any solution to a problem is better than no solution. Nowadays, the adage is qualified. Suppose one has two solutions to the same problem. When implemented on a computer, one solution takes a month and the other a tenth of a second. In an age when time is money, there is no doubt as to which solution is preferable.

The importance of LU factorization lies in the fact that it increases the speed with which we solve a linear system. Suppose we want to solve the system $A\mathsf{x} = \mathsf{b}$ and we know that $A = LU$. Then our system is $LU\mathsf{x} = \mathsf{b}$, which we think of as $L(U\mathsf{x}) = \mathsf{b}$, or $L\mathsf{y} = \mathsf{b}$ with $\mathsf{y} = U\mathsf{x}$. The idea is to solve $L\mathsf{y} = \mathsf{b}$ for y and then $U\mathsf{x} = \mathsf{y}$ for x. The single system $A\mathsf{x} = \mathsf{y}$ is replaced by the two systems $L\mathsf{y} = \mathsf{b}$ and $U\mathsf{x} = \mathsf{y}$ each of which is triangular and therefore solved much more quickly, as we shall soon show. First we demonstrate.

2.7.17 Example. Suppose we want to solve the system

$$\begin{aligned} -3x_1 \qquad\quad + x_3 + \quad x_4 &= 2 \\ -12x_1 + 2x_2 + 2x_3 + 11x_4 &= -2 \\ 9x_1 + 4x_2 + x_3 + 7x_4 &= -10 \\ -3x_1 - 2x_2 + 5x_3 - 6x_4 &= 14, \end{aligned}$$

which is $A\mathsf{x} = \mathsf{b}$ with $A = \begin{bmatrix} -3 & 0 & 1 & 1 \\ -12 & 2 & 2 & 11 \\ 9 & 4 & 1 & 7 \\ -3 & -2 & 5 & -6 \end{bmatrix}$, $\mathsf{x} = \begin{bmatrix} x_1 \\ x_2 \\ x_3 \\ x_4 \end{bmatrix}$ and $\mathsf{b} = \begin{bmatrix} 2 \\ -2 \\ -10 \\ 14 \end{bmatrix}$.
Suppose we know $A = LU$ with

$$L = \begin{bmatrix} 1 & 0 & 0 & 0 \\ 4 & 1 & 0 & 0 \\ -3 & 2 & 1 & 0 \\ 1 & -1 & \frac{1}{4} & 1 \end{bmatrix} \text{ and } U = \begin{bmatrix} -3 & 0 & 1 & 1 \\ 0 & 2 & -2 & 7 \\ 0 & 0 & 8 & -4 \\ 0 & 0 & 0 & 1 \end{bmatrix}.$$

To solve $A\mathsf{x} = \mathsf{b}$, which is $L(U\mathsf{x}) = \mathsf{b}$, we first solve $L\mathsf{y} = \mathsf{b}$ for $\mathsf{y} = U\mathsf{x} = \begin{bmatrix} y_1 \\ y_2 \\ y_3 \\ y_4 \end{bmatrix}$.

This is the system

$$\begin{aligned} y_1 &&&&= 2 \\ 4y_1 + y_2 &&&&= -2 \\ -3y_1 + 2y_2 + y_3 &&&&= -10 \\ y_1 - y_2 + \tfrac{1}{4}y_3 + y_4 &&&&= 14. \end{aligned}$$

By **forward substitution** we have

$y_1 = 2$,

$4y_1 + y_2 = -2$, so $y_2 = -2 - 8 = -10$,

$-3y_1 + 2y_2 + y_3 = -10$, so $y_3 = -10 + 6 + 20 = 16$,

$y_1 - y_2 + \frac{1}{4}y_3 + y_4 = 14$, so $y_4 = 14 - 2 - 10 - 4 = -2$.

So $\mathsf{y} = \begin{bmatrix} 2 \\ -10 \\ 16 \\ -2 \end{bmatrix}$. Now we solve $U\mathsf{x} = \mathsf{y}$, which is

$$\begin{aligned} -3x_1 \qquad + x_3 + x_4 &= 2 \\ 2x_2 - 2x_3 + 7x_4 &= -10 \\ 8x_3 - 4x_4 &= 16 \\ x_4 &= -2. \end{aligned}$$

By back substitution,

$x_4 = -2$,

$8x_3 - 4x_4 = 16$, so $8x_3 = 16 - 8 = 8$ and $x_3 = -1$,

$2x_2 - 2x_3 + 7x_4 = -10$, so $2x_2 = -10 + 2 + 14 = 6$ and $x_2 = 3$,

$-3x_1 + x_3 + x_4 = 2$, so $-3x_1 = 2 - 1 + 2 = 3$ and $x_1 = -1$.

We get $\mathsf{x} = \begin{bmatrix} -1 \\ 3 \\ 1 \\ -2 \end{bmatrix}$ and check that this is correct by verifying that $A\mathsf{x} = \mathsf{b}$:

$$\begin{bmatrix} -3 & 0 & 1 & 1 \\ -12 & 2 & 2 & 11 \\ 9 & 4 & 1 & 7 \\ -3 & -2 & 5 & -6 \end{bmatrix} \begin{bmatrix} -1 \\ 3 \\ 1 \\ -2 \end{bmatrix} = \begin{bmatrix} 2 \\ -2 \\ -10 \\ 14 \end{bmatrix}.$$

:)

2.7.18 Given $A = LU$, an LU factorization of A, solve $A\mathsf{x} = \mathsf{b}$ by first solving $L\mathsf{y} = \mathsf{b}$ for y and then $U\mathsf{x} = \mathsf{y}$ for x.

Now we make precise our suggestion that solving two triangular systems is "faster" than solving an arbitrary system. Assuming that speed is proportional to the total number of arithmetical operations (multiplications and subtractions) involved, we determine this number in two situations.

Think carefully about how we solve the general system $A\mathsf{x} = \mathsf{b}$ with A an $n \times n$ matrix. First we apply Gaussian elimination. Assuming the $(1, 1)$ entry of A is not 0, the first step requires us to put a 0 in the first position of every row below the first. This requires $n - 1$ row operations of the sort $R \to R - cR'$. In each of the $n - 1$ rows below the first, the first entry becomes 0, but there remain $n - 1$ numbers which get changed with a multiplication and a subtraction. This gives a total of $2(n - 1)(n - 1) = 2(n - 1)^2$ basic arithmetic operations to put 0s below the $(1, 1)$ entry.

Similarly, to put 0s below the $(2, 2)$ entry requires $2(n - 2)^2$ operations, and so on. In all, moving an $n \times n$ matrix to row echelon form requires

$$\begin{aligned} 2(n-1)^2 + 2(n-2)^2 + \cdots + 2(1)^2 + 0 \\ &= 2[1^2 + 2^2 + \cdots + (n-1)^2] \\ &= 2\frac{(n-1)n(2n-1)}{6} = \frac{(n-1)n(2n-1)}{3}, \end{aligned}$$

roughly n^3, operations. Here, we have assumed the reader is familiar with the formula

$$1^2 + 2^2 + 3^2 + \cdots + k^2 = \frac{k(k+1)(2k+1)}{6}$$

for the sum of the squares of the first k positive integers. In what comes next, we assume that the reader also knows that the sum of the first k odd positive integers is k^2, that is,

$$1 + 3 + 5 + \cdots + (2k - 1) = k^2.$$

Having moved our matrix to row echelon form, we next count the number of operations involved in back substitution. The last equation, of the form $bx_n = a$, gives x_n with at most one division. The second last equation, of the form $ax_{n-1} + bx_n = c$, gives x_{n-1} with at most one multiplication, one subtraction and one division—three operations. The third equation from the bottom, of the form $ax_{n-2} + bx_{n-1} + cx_n = d$, gives x_{n-2} with at most two multiplications, two subtractions and one division—five operations in all. The equation fourth from the bottom, of the form $ax_{n-3} + bx_{n-2} + cx_{n-1} + dx_n = e$, gives x_{n-3} with at most three multiplications, three subtractions, and one division—seven operations total. The first equation, of the form $a_1x_1 + a_2x_2 + \cdots + a_nx_n = b$, gives x_1 after $n-1$ multiplications, $n-1$ subtractions and one division, $2(n-1) + 1 = 2n - 1$ operations altogether. In total, the number of operations involved with back substitution is

$$1 + 3 + 5 + \cdots + (2n - 1) = n^2,$$

a number which is so much smaller than n^3 that $n^3 + n^2 \approx n^3$ when n is large. We conclude that the solution of $A\mathsf{x} = \mathsf{b}$ in general requires roughly n^3 operations.

On the other hand, the solution of an upper triangular system, $U\mathsf{x} = \mathsf{y}$, involves just n^2 operations because only back substitution is required, and the same goes for the solution of a lower triangular system $L\mathsf{y} = \mathsf{b}$. Factoring $A = LU$ reduces the "work" involved in solving $A\mathsf{x} = \mathsf{b}$ from n^3 to $2n^2$. This difference is substantial for large n. For instance, if $n = 1000$, then $n^2 = 10^6$ and $n^3 = 10^9$. On an electronic device performing 10^8 operations per second (this is very slow!), the difference is 0.02 seconds versus 10 seconds.

PLU Factorization; Row Interchanges

We conclude this section by discussing a possibility hitherto avoided. To find an LU factorization, we have said that we should try to move A to a row echelon matrix without interchanging rows. Such a goal may not always be attainable. For example, if we apply Gaussian elimination to $A = \begin{bmatrix} 2 & 4 & 2 \\ 1 & 2 & 3 \\ 3 & 2 & 1 \end{bmatrix}$, trying to avoid row interchanges, we begin

$$A \to \begin{bmatrix} 2 & 4 & 2 \\ 0 & 0 & 2 \\ 0 & -4 & -2 \end{bmatrix},$$

but now are forced to interchange rows two and three. To avoid this problem, we change A at the outset by interchanging its second and third rows. This gives a matrix A' that we can reduce to row echelon form without row interchanges:

$$A' = \begin{bmatrix} 2 & 4 & 2 \\ 3 & 2 & 1 \\ 1 & 2 & 3 \end{bmatrix} \to \begin{bmatrix} 2 & 4 & 2 \\ 0 & -4 & -2 \\ 0 & 0 & 2 \end{bmatrix} = U.$$

So

$$A' = LU = \begin{bmatrix} 1 & 0 & 0 \\ \frac{3}{2} & 1 & 0 \\ \frac{1}{2} & 0 & 1 \end{bmatrix} \begin{bmatrix} 2 & 4 & 2 \\ 0 & -4 & -2 \\ 0 & 0 & 2 \end{bmatrix}. \tag{2}$$

Of course, this is a factorization of A', not of A, but there is a nice connection between A' and A, a connection we know. The matrix A' is obtained from A by interchanging rows two and three, so $A' = PA$ where $P = \begin{bmatrix} 1 & 0 & 0 \\ 0 & 0 & 1 \\ 0 & 1 & 0 \end{bmatrix}$ is the elementary matrix obtained from the 3×3 identity matrix by interchanging rows two and three. Thus (2) is an equation of the form $PA = LU$, and this gives $A = P^{-1}LU$. Now the inverse of the elementary matrix P is another elementary matrix—in this case, $P^{-1} = P$—so we obtain the factorization $A = PLU$.

In general, Gaussian elimination may involve several interchanges of rows so that the initial matrix A' is A with its rows quite scrambled ("permuted" is the correct term). In this case, $A' = P'A$ where P' is the matrix obtained from the identity matrix by reordering rows just as the rows of A were reordered to give A'. Such a matrix is called a "permutation matrix."

2.7.19 Definition. A *permutation matrix* is a matrix whose rows are the rows of the identity matrix in some order.

2.7.20 Example. $P_1 = \begin{bmatrix} 0 & 1 \\ 1 & 0 \end{bmatrix}$, $P_2 = \begin{bmatrix} 0 & 1 & 0 \\ 0 & 0 & 1 \\ 1 & 0 & 0 \end{bmatrix}$ and $P_3 = \begin{bmatrix} 0 & 0 & 1 & 0 \\ 0 & 1 & 0 & 0 \\ 0 & 0 & 0 & 1 \\ 1 & 0 & 0 & 0 \end{bmatrix}$ are all permutation matrices. Multiplying a matrix A on the left by a permutation matrix changes the rows of A in a predictable way. The rows of P_1 are the rows of the identity in the order 2, 1 so whenever P_1A is defined, P_1A is A with its rows arranged in the same order, 2, 1. For example, $P_1 \begin{bmatrix} 1 & 2 \\ 3 & 4 \end{bmatrix} = \begin{bmatrix} 3 & 4 \\ 1 & 2 \end{bmatrix}$. The rows of P_2 are those of the identity matrix in the order 2, 3, 1 so, whenever P_2A is defined, this matrix is the matrix A with its rows appearing in the order 2, 3, 1. For example, $P_2 \begin{bmatrix} 1 & 2 \\ 3 & 4 \\ 5 & 6 \end{bmatrix} = \begin{bmatrix} 3 & 4 \\ 5 & 6 \\ 1 & 2 \end{bmatrix}$. ☺

READING CHECK 3. Find P_3A with $A = \begin{bmatrix} 1 & 2 & 3 & 4 \\ 5 & 6 & 7 & 8 \\ 9 & 10 & 11 & 12 \\ 13 & 14 & 15 & 16 \end{bmatrix}$.

It is easy to find the inverse of a permutation matrix.

2.7.21 Proposition. *If P is a permutation matrix, then $P^{-1} = P^T$.*

Proof. The (i, j) entry of PP^T is the dot product of row i of P and column j of P^T. This is the dot product of row i of P and row j of P, which is the dot product of two of the standard basis vectors $\mathsf{e}_1, \mathsf{e}_2, \ldots, \mathsf{e}_n$ of R^n. The dot product $\mathsf{e}_i \cdot \mathsf{e}_i = 1$ for any i, so each diagonal entry of PP^T is 1. On the other hand, the dot product $\mathsf{e}_i \cdot \mathsf{e}_j = 0$ when $i \neq j$, so each off-diagonal entry of PP^T is 0. These observations show that $PP^T = I$, the $n \times n$ identity matrix. Since permutation matrices are square, this is sufficient to guarantee that P^T is the inverse of P. ■

READING CHECK 4. Find the inverse of the matrix $\begin{bmatrix} 0 & 1 & 0 \\ 0 & 0 & 1 \\ 1 & 0 & 0 \end{bmatrix}$ without any calculation. Explain your reasoning.

In general, we cannot expect to factor a given matrix A in the form $A = LU$, but we can always achieve $P'A = LU$, so that $A = PLU$, with $P = (P')^{-1} = (P')^T$.

2.7.22 **Problem.** If possible, find an LU factorization of $A = \begin{bmatrix} 0 & 1 & 1 & 0 \\ 0 & -1 & -1 & 4 \\ -2 & 3 & 2 & 1 \\ -6 & 4 & 2 & -8 \end{bmatrix}$; otherwise, find a PLU factorization.

Solution. The first step in moving A to a row echelon matrix requires a row interchange. There is no LU factorization, but there is a PLU factorization. After experimenting, we discover that if we write the rows of A in the order $3, 1, 4, 2$, then we can carry the new matrix to row echelon form without row interchanges. So we start with

$$PA = \begin{bmatrix} -2 & 3 & 2 & 1 \\ 0 & 1 & 1 & 0 \\ -6 & 4 & 2 & -8 \\ 0 & -1 & -1 & 4 \end{bmatrix},$$

where $P = \begin{bmatrix} 0 & 0 & 1 & 0 \\ 1 & 0 & 0 & 0 \\ 0 & 0 & 0 & 1 \\ 0 & 1 & 0 & 0 \end{bmatrix}$ is the permutation matrix formed from the identity by reordering the rows of the identity as we wish to reorder the rows of A, in the order $3, 1, 4, 2$. Gaussian elimination, using only the third elementary row operation, yields $PA = \begin{bmatrix} -2 & 3 & 2 & 1 \\ 0 & 1 & 1 & 0 \\ -6 & 4 & 2 & -8 \\ 0 & -1 & -1 & 4 \end{bmatrix}$

$$\to \begin{bmatrix} -2 & 3 & 2 & 1 \\ 0 & 1 & 1 & 0 \\ 0 & -5 & -4 & -11 \\ 0 & -1 & -1 & 4 \end{bmatrix} \to \begin{bmatrix} -2 & 3 & 2 & 1 \\ 0 & 1 & 1 & 0 \\ 0 & 0 & 1 & -11 \\ 0 & 0 & 0 & 4 \end{bmatrix} = U.$$

So $PA = LU$ with $L = \begin{bmatrix} 1 & 0 & 0 & 0 \\ 0 & 1 & 0 & 0 \\ 3 & -5 & 1 & 0 \\ 0 & -1 & 0 & 1 \end{bmatrix}$

and $A = P^{-1}LU$ with $P^{-1} = P^T = \begin{bmatrix} 0 & 1 & 0 & 0 \\ 0 & 0 & 0 & 1 \\ 1 & 0 & 0 & 0 \\ 0 & 0 & 1 & 0 \end{bmatrix}$.

2.7.23 A Final Remark. In this section, we defined a permutation matrix to be a matrix whose rows are the rows of the identity matrix, in some order. Equivalently, a permutation matrix is one whose **columns** are the columns of the identity matrix, in some order. A square matrix has rows the standard basis vectors in some order if and only if its columns are the standard basis vectors in some order. See Exercise 40.

Answers to Reading Checks

1. U is also $m \times n$, so L is $? \times m$ and, since LU is $m \times n$, L must be $m \times m$.

2. We reduce A to row echelon form from the top down using only the third elementary operation: $A \to \begin{bmatrix} 1 & 2 & 3 \\ 0 & -1 & -2 \\ 0 & -2 & -3 \end{bmatrix} \to \begin{bmatrix} 1 & 2 & 3 \\ 0 & -1 & -2 \\ 0 & 0 & 1 \end{bmatrix} = U$ with $L = \begin{bmatrix} 1 & 0 & 0 \\ 1 & 1 & 0 \\ 2 & 2 & 1 \end{bmatrix}$, the matrix of multipliers.

3. The rows of P_3 are the rows of the identity matrix in the order 3, 2, 4, 1, so P_3A is A with its rows appearing in the same order: $P_1A = \begin{bmatrix} 9 & 10 & 11 & 12 \\ 5 & 6 & 7 & 8 \\ 13 & 14 & 15 & 16 \\ 1 & 2 & 3 & 4 \end{bmatrix}$.

4. The given matrix is a permutation matrix, so its inverse is its transpose:

$$\begin{bmatrix} 0 & 1 & 0 \\ 0 & 0 & 1 \\ 1 & 0 & 0 \end{bmatrix}^{-1} = \begin{bmatrix} 0 & 0 & 1 \\ 1 & 0 & 0 \\ 0 & 1 & 0 \end{bmatrix}.$$

True/False Questions

Decide, with as little calculation as possible, whether each of the following statements is true or false and, if you say "false," explain your answer. (Answers are at the back of the book.)

1. The matrix $\begin{bmatrix} 1 & 0 & 0 \\ 0 & 2 & 1 \\ 0 & 0 & 1 \end{bmatrix}$ is elementary.

2. If $E = \begin{bmatrix} 1 & 0 & 0 \\ 0 & 1 & -2 \\ 0 & 0 & 1 \end{bmatrix}$, then $E^{-1} = \begin{bmatrix} 1 & 0 & 0 \\ 0 & 1 & 2 \\ 0 & 0 & 1 \end{bmatrix}$.

3. If $E = \begin{bmatrix} 1 & 0 & -3 \\ 0 & 1 & 0 \\ 0 & 0 & 1 \end{bmatrix}$ and A is a $3 \times n$ matrix, rows two and three of EA are the same as rows two and three of A.

4. The product of elementary matrices is lower triangular.

5. Any matrix that can be transformed to row echelon form without row interchanges has an LU factorization.

6. If $A = LU$ is an LU factorization of A, we can solve $A\mathsf{x} = \mathsf{b}$ by first solving $U\mathsf{y} = \mathsf{b}$ for y and then $L\mathsf{x} = \mathsf{y}$ for x.

7. $1 + 3 + 5 + \cdots + 99 = 2500$.

8. The inverse of $\begin{bmatrix} 0 & 1 & 0 & 0 & 0 \\ 0 & 0 & 0 & 1 & 0 \\ 1 & 0 & 0 & 0 & 0 \\ 0 & 0 & 0 & 0 & 1 \\ 0 & 0 & 1 & 0 & 0 \end{bmatrix}$ is $\begin{bmatrix} 0 & 0 & 1 & 0 & 0 \\ 1 & 0 & 0 & 0 & 0 \\ 0 & 0 & 0 & 0 & 1 \\ 0 & 1 & 0 & 0 & 0 \\ 0 & 0 & 0 & 1 & 0 \end{bmatrix}$. (No calculation!)

9. Any matrix can be factored PLU.

10. The pivots of an invertible matrix A are the nonzero diagonal entries of U in an LU or a PLU factorization.

EXERCISES

Answers to exercises marked [BB] can be found at the Back of the Book.

1. [BB] Let $E = \begin{bmatrix} 1 & 0 & 0 \\ 0 & 1 & 4 \\ 0 & 0 & 1 \end{bmatrix}$ and let A be an arbitrary $3 \times n$ matrix, $n \geq 1$.

   (a) Explain the connection between A and EA.

   (b) What is the name for a matrix such as E?

   (c) Is E invertible? If so, what is its inverse?

2. Suppose A is a $3 \times n$ matrix. In each case, write down the elementary matrix E for which EA is A modified as specified.

   (a) [BB] third row replaced by third row less the second;

   (b) twice the third row added to the first;

   (c) first and third rows interchanged;

   (d) second row divided by -2.

3. Let $A = \begin{bmatrix} 1 & 2 & 3 \\ 4 & 5 & 6 \\ 6 & 8 & 9 \end{bmatrix}$. Exhibit an elementary matrix E so that

(a) [BB] $EA = \begin{bmatrix} 1 & 2 & 3 \\ 4 & 5 & 6 \\ 0 & -4 & -9 \end{bmatrix}$ **(b)** $EA = \begin{bmatrix} 1 & 2 & 3 \\ 0 & -3 & -6 \\ 6 & 8 & 9 \end{bmatrix}$

(c) $EA = \begin{bmatrix} 1 & 2 & 3 \\ 4 & 5 & 6 \\ 0 & \frac{1}{2} & 0 \end{bmatrix}$ **(d)** $EA = \begin{bmatrix} 1 & 2 & 3 \\ 6 & 8 & 9 \\ 4 & 5 & 6 \end{bmatrix}$

4. [BB] Let $E = \begin{bmatrix} -2 & 0 \\ 0 & 1 \end{bmatrix}$ and $F = \begin{bmatrix} 1 & 0 \\ 3 & 0 \end{bmatrix}$. Suppose A is a 2×5 matrix. Describe EA and FA.

5. Let $A = \begin{bmatrix} 2 & 1 & 0 \\ 4 & -2 & 1 \\ 5 & 0 & 2 \\ -3 & 4 & 1 \end{bmatrix}$. Find a matrix E so that EA has a 0 in the $(2,1)$ position and a matrix F so that FA has a 0 in the $(4,1)$ position.

6. [BB] Let E be the 3×3 elementary matrix that adds 4 times row one to row two and F the 3×3 elementary matrix that subtracts 3 times row two from row three.

(a) Find E, F, EF and FE without calculation.

(b) Write down the inverses of the matrices in part (a) without calculation.

7. Let E be the 4×4 elementary matrix that subtracts 5 times row two from row four and F the 4×4 elementary matrix that adds one half row three to row one.

(a) Find E, F, EF and FE without calculation.

(b) Write down the inverses of the matrices in part (a) without calculation.

8. Let E be a 4×4 matrix such that for any $4 \times n$ matrix $A = \begin{bmatrix} \mathsf{a}_1^T & \rightarrow \\ \mathsf{a}_2^T & \rightarrow \\ \mathsf{a}_3^T & \rightarrow \\ \mathsf{a}_4^T & \rightarrow \end{bmatrix}$ with rows $\mathsf{a}_1^T, \mathsf{a}_2^T, \mathsf{a}_3^T, \mathsf{a}_4^T$, $EA = \begin{bmatrix} \mathsf{a}_1^T + 2\mathsf{a}_3^T & \rightarrow \\ \mathsf{a}_2^T & \rightarrow \\ \mathsf{a}_3^T & \rightarrow \\ \mathsf{a}_4^T & \rightarrow \end{bmatrix}$. What is E? What is E^{-1}?

9. [BB] Let $A = \begin{bmatrix} -3 & 3 & 6 \\ 2 & 5 & 10 \\ 0 & 1 & 4 \end{bmatrix}$. Find an LU factorization of A using the elementary row operations to move A to a row echelon matrix keeping track of the elementary matrices that correspond to the row operations you use. Express L as the product of elementary matrices.

10. Answer Exercise 9 again with $A = \begin{bmatrix} 2 & 4 & 6 & 18 \\ 4 & 5 & 6 & 24 \\ 3 & 1 & -2 & 4 \end{bmatrix}$.

11. In each case, you are given a row echelon form U of a matrix A. Find a matrix M, the product of elementary matrices, with $U = MA$.

(a) [BB] $A = \begin{bmatrix} -3 & 6 \\ 8 & -5 \end{bmatrix}$; $U = \begin{bmatrix} -3 & 6 \\ 0 & 11 \end{bmatrix}$ (b) $A = \begin{bmatrix} 3 & 2 & 1 \\ 6 & 5 & 4 \\ 9 & 8 & 7 \end{bmatrix}$; $U = \begin{bmatrix} 3 & 2 & 1 \\ 0 & 1 & 2 \\ 0 & 0 & 0 \end{bmatrix}$

(c) $A = \begin{bmatrix} -1 & 1 & 3 \\ 2 & 6 & 8 \\ 3 & 2 & 6 \end{bmatrix}$; $U = \begin{bmatrix} -1 & 1 & 3 \\ 0 & 1 & 3 \\ 0 & 0 & -10 \end{bmatrix}$

(d) $A = \begin{bmatrix} 4 & -1 & 0 & 1 \\ 2 & 1 & 1 & 4 \\ -6 & 0 & 5 & 2 \end{bmatrix}$; $U = \begin{bmatrix} 2 & 1 & 1 & 4 \\ 0 & -3 & -2 & -7 \\ 0 & 0 & 6 & 7 \end{bmatrix}$.

12. In each case, express L as the product of elementary matrices (without calculation) and use this factorization to find L^{-1}.

(a) [BB] $L = \begin{bmatrix} 1 & 0 & 0 \\ -2 & 1 & 0 \\ 3 & 5 & 1 \end{bmatrix}$ (b) $L = \begin{bmatrix} 1 & 0 & 0 \\ 4 & 1 & 0 \\ -1 & 0 & 1 \end{bmatrix}$

(c) $L = \begin{bmatrix} 1 & 0 & 0 \\ a & 1 & 0 \\ b & c & 1 \end{bmatrix}$ (d) $L = \begin{bmatrix} 1 & 0 & 0 & 0 \\ -3 & 1 & 0 & 0 \\ 4 & 2 & 1 & 0 \\ -1 & -5 & 7 & 1 \end{bmatrix}$.

13. In each case, express U as the product of elementary matrices (without calculation) and use this factorization to find U^{-1}.

(a) [BB] $U = \begin{bmatrix} 1 & 2 & 0 \\ 0 & 1 & -5 \\ 0 & 0 & 1 \end{bmatrix}$ (b) $U = \begin{bmatrix} 1 & -1 & 9 \\ 0 & 1 & 4 \\ 0 & 0 & 1 \end{bmatrix}$

(c) $U = \begin{bmatrix} 1 & a & b \\ 0 & 1 & c \\ 0 & 0 & 1 \end{bmatrix}$ (d) $U = \begin{bmatrix} 1 & -2 & -4 & 3 \\ 0 & 1 & -7 & -5 \\ 0 & 0 & 1 & 4 \\ 0 & 0 & 0 & 1 \end{bmatrix}$.

14. [BB] Find an LU factorization of $A = \begin{bmatrix} -1 & 1 & -2 \\ 2 & 1 & 7 \end{bmatrix}$ and express L as the product of elementary matrices.

15. Answer Exercise 14 again with $A = \begin{bmatrix} 2 & -6 & -2 & 4 \\ -1 & 3 & 3 & 2 \\ -1 & 3 & 7 & 10 \end{bmatrix}$.

16. [BB] Let $A = \begin{bmatrix} 2 & 1 & 0 \\ 0 & 4 & 2 \\ 6 & 3 & 5 \end{bmatrix}$ and $\mathsf{v} = \begin{bmatrix} -2 \\ 3 \\ 5 \end{bmatrix}$.

(a) Find an elementary matrix E and a row echelon matrix U with $EA = U$.

(b) Find a lower triangular matrix L such that $A = LU$.

(c) Express $A\mathsf{v}$ as a linear combination of the columns of A.

17. Write down the inverse of each of the following matrices without calculation. Explain your reasoning.

(a) [BB] $A = \begin{bmatrix} 1 & 3 \\ 0 & 1 \end{bmatrix}$ **(b)** $A = \begin{bmatrix} 1 & 0 \\ -5 & 1 \end{bmatrix}$

(c) $E = \begin{bmatrix} a & 0 \\ 0 & 1 \end{bmatrix}$ (assume $a \neq 0$) **(d)** $E = \begin{bmatrix} 1 & 0 & 0 \\ 0 & 1 & a \\ 0 & 0 & 1 \end{bmatrix}$

(e) [BB] $A = \begin{bmatrix} 1 & 0 & 0 \\ 0 & 1 & -2 \\ 0 & 0 & 1 \end{bmatrix}$ **(f)** $A = \begin{bmatrix} 1 & 4 & 0 \\ 0 & 1 & 0 \\ 0 & 0 & 1 \end{bmatrix}$

(g) $A = \begin{bmatrix} 0 & 0 & 0 & 1 \\ 0 & 1 & 0 & 0 \\ 0 & 0 & 1 & 0 \\ 1 & 0 & 0 & 0 \end{bmatrix}$ **(h)** $A = \begin{bmatrix} 1 & 0 & 0 \\ 0 & \frac{1}{2} & 0 \\ 0 & 0 & 1 \end{bmatrix}$.

18. [BB] What number x will force an interchange of rows when Gaussian elimination is applied to $\begin{bmatrix} 2 & 5 & 1 \\ 4 & x & 1 \\ 0 & 1 & -1 \end{bmatrix}$?

19. To move A to row echelon form, Gaussian elimination is applied to the matrix $\begin{bmatrix} -1 & 0 & 1 & 1 \\ 3 & 0 & x & -4 \\ 4 & 1 & 7 & 8 \end{bmatrix}$. For any x, an interchange of rows is necessary. Why?

20. **(a)** [BB] Show that the lower triangular matrix $L = \begin{bmatrix} 1 & 0 \\ a & 1 \end{bmatrix}$ is invertible by finding a matrix M with $LM = I$, the 2×2 identity matrix.

(b) Show that the lower triangular matrix $L = \begin{bmatrix} 1 & 0 & 0 \\ a & 1 & 0 \\ b & c & 1 \end{bmatrix}$ is invertible.

21. [BB] Let $E = \begin{bmatrix} 1 & 0 & 0 \\ -1 & 1 & 0 \\ 2 & 0 & 1 \end{bmatrix}$, $D = \begin{bmatrix} 4 & 0 & 0 \\ 0 & 1 & 0 \\ 0 & 0 & -1 \end{bmatrix}$ and $P = \begin{bmatrix} 0 & 0 & 1 \\ 1 & 0 & 0 \\ 0 & 1 & 0 \end{bmatrix}$, and let A be any $3 \times n$ matrix.

(a) How are the rows of EA, of DA and of PA related to the rows of A?

(b) Find P^{-1}.

22. Answer Exercise 21 again with $E = \begin{bmatrix} 1 & 0 & 3 \\ 0 & 1 & 0 \\ 0 & -2 & 1 \end{bmatrix}$, $D = \begin{bmatrix} 1 & 0 & 0 \\ 0 & -1 & 0 \\ 0 & 0 & 2 \end{bmatrix}$ and $P = \begin{bmatrix} 1 & 0 & 0 \\ 0 & 0 & 1 \\ 0 & 1 & 0 \end{bmatrix}$.

23. Let $A = \begin{bmatrix} 3 & 4 & 8 \\ 6 & 2 & 9 \\ 0 & 0 & -3 \end{bmatrix}$ and $\mathsf{v} = \begin{bmatrix} 4 \\ -7 \\ 2 \end{bmatrix}$.

(a) Find an elementary matrix E and a row echelon matrix U with $EA = U$.

(b) Find a lower triangular matrix L such that $A = LU$.

(c) Express $A\mathsf{v}$ as a linear combination of the columns of A.

24. [BB] Ian says $\begin{bmatrix} 0 & 1 \\ 1 & 2 \end{bmatrix}$ has no LU factorization but Lynn disagrees. Lynn adds the second row to the first, then subtracts the first from the second, like this

$$\begin{bmatrix} 0 & 1 \\ 1 & 2 \end{bmatrix} \to \begin{bmatrix} 1 & 3 \\ 1 & 2 \end{bmatrix} \to \begin{bmatrix} 1 & 3 \\ 0 & -1 \end{bmatrix} = U.$$

She gets L as the product of certain elementary matrices. Who is right? Explain.

25. To find an LU factorization of $A = \begin{bmatrix} 1 & -3 & -2 \\ 2 & -6 & -2 \\ 1 & -2 & 5 \end{bmatrix}$, a logical start is

$A \to \begin{bmatrix} 1 & -3 & -2 \\ 0 & 0 & 2 \\ 0 & 1 & 7 \end{bmatrix}$. Knowing she must avoid row interchanges, a clever student continues like this:

$$\begin{bmatrix} 1 & -3 & -2 \\ 0 & 0 & 2 \\ 0 & 1 & 7 \end{bmatrix} \xrightarrow{R2 \to R2+R3} \begin{bmatrix} 1 & -3 & -2 \\ 0 & 1 & 9 \\ 0 & 1 & 7 \end{bmatrix} \xrightarrow{R3 \to R3-R2} \begin{bmatrix} 1 & -3 & -2 \\ 0 & 1 & 9 \\ 0 & 0 & -2 \end{bmatrix}$$

and writes $A = LU$ with $U = \begin{bmatrix} 1 & -3 & -2 \\ 0 & 1 & 9 \\ 0 & 0 & -2 \end{bmatrix}$ and $L = \begin{bmatrix} 1 & 0 & 0 \\ 2 & 1 & 0 \\ 1 & -1 & 1 \end{bmatrix}$.

(a) Show that this is wrong.

(b) What was the problem?

26. Let $A = \begin{bmatrix} 0 & 1 \\ 1 & 0 \end{bmatrix}$. Show that the equation $A = LU$ with L lower triangular and U upper triangular leads to a contradiction.

27. Repeat Exercise 26 with $A = \begin{bmatrix} 0 & 1 \\ 2 & 3 \end{bmatrix}$.

28. In each case, either find an LU factorization or explain why there is no LU factorization.

(a) [BB] $\begin{bmatrix} 2 & 4 \\ -1 & 6 \end{bmatrix}$

(b) $\begin{bmatrix} 2 & 1 \\ 6 & 8 \end{bmatrix}$

(c) $\begin{bmatrix} 2 & -4 & 1 \\ 1 & 3 & -1 \end{bmatrix}$

(d) $\begin{bmatrix} 5 & 0 & 15 \\ 0 & 1 & 2 \\ 4 & -3 & 8 \\ -2 & 2 & -1 \end{bmatrix}$

(e) $\begin{bmatrix} 1 & 2 & 3 \\ 2 & 3 & 4 \\ 3 & 4 & 5 \end{bmatrix}$

(f) [BB] $\begin{bmatrix} 2 & -6 & 5 \\ -4 & 12 & -9 \\ 2 & -9 & 8 \end{bmatrix}$

(g) $\begin{bmatrix} 1 & -1 & 1 \\ 1 & -1 & 1 \\ 1 & -1 & 1 \end{bmatrix}$ (h) $\begin{bmatrix} 1 & 3 & 8 \\ 2 & 5 & 21 \\ 1 & 7 & -5 \end{bmatrix}$

(i) $\begin{bmatrix} -3 & 8 & 10 & 1 \\ 3 & -2 & -1 & 0 \\ -3 & 4 & 4 & 0 \\ 6 & -4 & -2 & 1 \\ 9 & 0 & 1 & 0 \end{bmatrix}$ (j) $\begin{bmatrix} -2 & 4 & 6 & -8 \\ 1 & 2 & -8 & 2 \\ -4 & 0 & -5 & 3 \\ 2 & 1 & -4 & 1 \end{bmatrix}$

(k) $\begin{bmatrix} 1 & 2 & 3 & 4 \\ 5 & 6 & 7 & 8 \\ 9 & 10 & 0 & 1 \\ 0 & -2 & 3 & 4 \\ -1 & 6 & 4 & 9 \end{bmatrix}$ (l) $\begin{bmatrix} 1 & -1 & -2 & 1 & 1 \\ 2 & -4 & -5 & 6 & 0 \\ -1 & -5 & 2 & 12 & -2 \end{bmatrix}$.

29. [BB] **(a)** Use the LU factorization $\begin{bmatrix} 2 & 1 \\ 6 & 8 \end{bmatrix} = \begin{bmatrix} 2 & 0 \\ 6 & 5 \end{bmatrix} \begin{bmatrix} 1 & \frac{1}{2} \\ 0 & 1 \end{bmatrix}$ to solve $A\begin{bmatrix} x \\ y \end{bmatrix} = \begin{bmatrix} -2 \\ 9 \end{bmatrix}$ with $A = \begin{bmatrix} 2 & 1 \\ 6 & 8 \end{bmatrix}$.

(b) Express $\begin{bmatrix} -2 \\ 9 \end{bmatrix}$ as a linear combination of the columns of A.

30. Use the factorization $\begin{bmatrix} -2 & -6 \\ 1 & 15 \end{bmatrix} = \begin{bmatrix} -2 & 0 \\ 1 & 1 \end{bmatrix} \begin{bmatrix} 1 & 3 \\ 0 & 12 \end{bmatrix}$ to solve $A\begin{bmatrix} x_1 \\ x_2 \end{bmatrix} = \begin{bmatrix} -3 \\ 4 \end{bmatrix}$ with $A = \begin{bmatrix} -2 & -6 \\ 1 & 15 \end{bmatrix}$.

31. **(a)** Use the factorization $A = \begin{bmatrix} 1 & 6 & 7 \\ 5 & 25 & 36 \\ -2 & -27 & -4 \end{bmatrix} = \begin{bmatrix} 1 & 0 & 0 \\ 5 & 1 & 0 \\ -2 & 3 & 1 \end{bmatrix} \begin{bmatrix} 1 & 6 & 7 \\ 0 & -5 & 1 \\ 0 & 0 & 7 \end{bmatrix}$ to solve the linear system $A\begin{bmatrix} x_1 \\ x_2 \\ x_3 \end{bmatrix} = \mathsf{b}$ with $\mathsf{b} = \begin{bmatrix} 1 \\ 2 \\ 3 \end{bmatrix}$.

(b) Express b as a linear combination of the columns of A.

32. **(a)** Use the factorization

$$A = \begin{bmatrix} 1 & 4 & 3 & 4 \\ 3 & 10 & 10 & 17 \\ -4 & -20 & -14 & -1 \\ 7 & 38 & 28 & -6 \end{bmatrix} = \begin{bmatrix} 1 & 0 & 0 & 0 \\ 3 & 1 & 0 & 0 \\ -4 & 2 & 1 & 0 \\ 7 & -5 & -3 & 1 \end{bmatrix} \begin{bmatrix} 1 & 4 & 3 & 4 \\ 0 & -2 & 1 & 5 \\ 0 & 0 & -4 & 5 \\ 0 & 0 & 0 & 6 \end{bmatrix}$$

to solve the linear system $A\mathsf{x} = \mathsf{b}$ for $\mathsf{x} = \begin{bmatrix} x_1 \\ x_2 \\ x_3 \\ x_4 \end{bmatrix}$ with $\mathsf{b} = \begin{bmatrix} 3 \\ 4 \\ 2 \\ -2 \end{bmatrix}$.

(b) Express b as a linear combination of the columns of A.

33. Use the factorization $\begin{bmatrix} 1 & 2 \\ 4 & -6 \\ -7 & -7 \\ 1 & 0 \end{bmatrix} = \begin{bmatrix} 1 & 0 & 0 & 0 \\ 4 & 1 & 0 & 0 \\ -7 & -\frac{1}{2} & 1 & 0 \\ 1 & \frac{1}{7} & 0 & 1 \end{bmatrix} \begin{bmatrix} 1 & 2 \\ 0 & -14 \\ 0 & 0 \\ 0 & 0 \end{bmatrix}$ to solve

the system $\begin{aligned} x_1 + 2x_2 &= 0 \\ 4x_1 - 6x_2 &= -14 \\ -7x_1 - 7x_2 &= 7 \\ x_1 \qquad &= -2. \end{aligned}$

34. Use the factorization $\begin{bmatrix} -2 & 1 & 3 & 4 \\ 4 & 0 & -9 & -7 \\ -6 & 10 & 4 & 15 \end{bmatrix} = \begin{bmatrix} 1 & 0 & 0 \\ -2 & 1 & 0 \\ 3 & \frac{7}{2} & 1 \end{bmatrix} \begin{bmatrix} -2 & 1 & 3 & 4 \\ 0 & 2 & -3 & 1 \\ 0 & 0 & \frac{11}{2} & -\frac{1}{2} \end{bmatrix}$

to solve the system $\begin{aligned} -2x_1 + x_2 + 3x_3 + 4x_4 &= -1 \\ 4x_1 \qquad - 9x_3 - 7x_4 &= 2 \\ -6x_1 + 10x_2 + 4x_3 + 15x_4 &= 3. \end{aligned}$

35. Find a PLU factorization of each of the following matrices A.

(a) [BB] $A = \begin{bmatrix} 0 & 3 & 1 \\ 1 & 2 & 1 \end{bmatrix}$ **(b)** $A = \begin{bmatrix} 0 & 2 & 4 \\ 1 & 1 & 0 \\ 3 & 4 & 5 \end{bmatrix}$ **(c)** $A = \begin{bmatrix} 2 & -1 & 2 \\ 6 & -3 & 1 \\ 4 & -2 & 4 \\ -2 & 1 & 8 \end{bmatrix}$

(d) $A = \begin{bmatrix} -2 & 1 & 0 & -3 \\ 2 & -1 & 1 & 5 \\ -4 & 18 & 4 & -2 \end{bmatrix}$ **(e)** $A = \begin{bmatrix} 0 & 3 & 0 & 2 \\ 1 & 2 & 1 & -1 \\ 1 & 2 & 1 & 1 \\ 0 & -4 & 2 & 0 \end{bmatrix}$ **(f)** $A = \begin{bmatrix} 1 & -2 & 1 \\ 0 & 1 & 2 \\ 0 & 4 & 8 \\ 3 & -6 & 3 \\ 1 & 2 & 5 \end{bmatrix}$.

36. Suppose A is an $m \times n$ matrix that can be transformed by Gaussian elimination to a row echelon matrix U without the use of row interchanges. How does this lead to an LU factorization of A? Explain clearly how the lower triangular matrix L arises.

37. Explain how an LU factorization of a matrix A can be used to obtain an LU factorization of A^T.

Critical Reading

38. Is the product of elementary matrices invertible? Explain.

39. Suppose A is an invertible matrix and its first two rows are interchanged to give a matrix B. Is B invertible? Explain.

40. If the rows on an $n \times n$ matrix P are the (transposes of the) standard basis vectors $\mathsf{e}_1, \ldots, \mathsf{e}_n$ in some order, show that the same holds true for the columns. Is the order the same? [Hint: What is PP^T?]

41. Prove that the product of two $n \times n$ permutation matrices is a permutation matrix. [Hint: Let P_1 and P_2 be permutation matrices and examine the columns of P_1P_2.]

42. Suppose P is the 5×5 permutation matrix whose columns are the standard basis vectors of R^5 in the order $\mathsf{e}_1, \mathsf{e}_5, \mathsf{e}_3, \mathsf{e}_4, \mathsf{e}_2$. Let A be a 5×5 matrix with columns $\mathsf{c}_1, \mathsf{c}_2, \mathsf{c}_3, \mathsf{c}_4, \mathsf{c}_5$ and rows $\mathsf{r}_1, \mathsf{r}_2, \mathsf{r}_3, \mathsf{r}_4, \mathsf{r}_5$.

(a) Describe the matrix AP.

(b) Describe the matrix P^TA.

43. The system $A\mathsf{x} = \mathsf{b}$ has been solved by applying Gaussian elimination to the augmented matrix $[A \mid \mathsf{b}]$, thereby reducing A to a row echelon matrix U and the augmented matrix $[A \mid \mathsf{b}]$ to $[U \mid \mathsf{c}]$ for some vector c. In the elimination, only the third elementary operation was used. Now suppose we wish to solve $A\mathsf{x} = \mathsf{b}$ by factoring $A = LU$, solving $L\mathsf{y} = \mathsf{b}$ for y and then $U\mathsf{x} = \mathsf{y}$ for x. Gerard claims $\mathsf{y} = \mathsf{c}$. Is he right?

Guess an answer by examining $A\mathsf{x} = \mathsf{b}$, with $A = \begin{bmatrix} 1 & 0 & 1 \\ 2 & -2 & 3 \\ -45 & 4 & -2 \end{bmatrix}$, $\mathsf{x} = \begin{bmatrix} x_1 \\ x_2 \\ x_3 \end{bmatrix}$ and $\mathsf{b} = \begin{bmatrix} 2 \\ 8 \\ -16 \end{bmatrix}$. Then explain what is happening in general.

44. Suppose a matrix A has two (different) row echelon forms U_1 and U_2. Find two (different) LU factorizations of U_2.

2.8 LDU Factorization

In Example 2.7.16, with $A = \begin{bmatrix} 1 & 2 & 3 \\ 5 & 14 & 7 \\ 9 & 10 & 0 \\ 0 & 2 & 3 \end{bmatrix}$, we found $A = LU'$ with

$$L = \begin{bmatrix} 1 & 0 & 0 & 0 \\ 5 & 1 & 0 & 0 \\ 9 & -2 & 1 & 0 \\ 0 & \frac{1}{2} & -\frac{7}{43} & 1 \end{bmatrix} \quad \text{and} \quad U' = \begin{bmatrix} 1 & 2 & 3 \\ 0 & 4 & -8 \\ 0 & 0 & -43 \\ 0 & 0 & 0 \end{bmatrix}.$$

Factoring the diagonal entries of U'—the 1, 4 and −43—from the rows of U, gives a factorization $U' = DU$ with

$$D = \begin{bmatrix} 1 & 0 & 0 & 0 \\ 0 & 4 & 0 & 0 \\ 0 & 0 & -43 & 0 \\ 0 & 0 & 0 & 1 \end{bmatrix} \quad \text{and} \quad U = \begin{bmatrix} 1 & 2 & 3 \\ 0 & 1 & -2 \\ 0 & 0 & 1 \\ 0 & 0 & 0 \end{bmatrix}$$

and hence an **LDU factorization** of A:

$$\begin{bmatrix} 1 & 2 & 3 \\ 5 & 14 & 7 \\ 9 & 10 & 0 \\ 0 & 2 & 3 \end{bmatrix} = \begin{bmatrix} 1 & 0 & 0 & 0 \\ 5 & 1 & 0 & 0 \\ 9 & -2 & 1 & 0 \\ 0 & \frac{1}{2} & -\frac{7}{43} & 1 \end{bmatrix} \begin{bmatrix} 1 & 0 & 0 & 0 \\ 0 & 4 & 0 & 0 \\ 0 & 0 & -43 & 0 \\ 0 & 0 & 0 & 1 \end{bmatrix} \begin{bmatrix} 1 & 2 & 3 \\ 0 & 1 & -2 \\ 0 & 0 & 1 \\ 0 & 0 & 0 \end{bmatrix}.$$

2.8.1 **Definition.** An *LDU factorization* of a matrix A is a representation of A as the product LDU of a (necessarily square) lower triangular matrix L with 1s on the diagonal, a (square) diagonal matrix D and a row echelon matrix U (the same size as A) with 1s on the diagonal.

2.8.2

> To find an LDU factorization of a matrix A, find an LU factorization $A = LU'$, if possible—see 2.7.14. If none of the diagonal entries of U' is 0, factor these from the rows of U' to write $U' = DU$ with D diagonal and U upper triangular with 1s on the diagonal.

2.8.3 **Problem.** Find an LDU factorization of $A = \begin{bmatrix} -1 & 2 & 0 & -2 \\ 2 & 4 & 6 & 8 \\ -3 & 2 & 4 & 1 \end{bmatrix}$.

Solution. We bring A to upper triangular form using only the third elementary row operation:

$$A \to \begin{bmatrix} -1 & 2 & 0 & -2 \\ 0 & 8 & 6 & 4 \\ 0 & -4 & 4 & 7 \end{bmatrix} \to \begin{bmatrix} -1 & 2 & 0 & -2 \\ 0 & 8 & 6 & 4 \\ 0 & 0 & 7 & 9 \end{bmatrix} = U'.$$

This implies $A = LU'$ with $L = \begin{bmatrix} 1 & 0 & 0 \\ -2 & 1 & 0 \\ 3 & -\frac{1}{2} & 1 \end{bmatrix}$. Now we write $U' = DU$ by factoring the diagonal entries of U' from its rows and obtain the required LDU factorization of A:

$$\begin{bmatrix} -1 & 2 & 0 & -2 \\ 2 & 4 & 6 & 8 \\ -3 & 2 & 4 & 1 \end{bmatrix} = \begin{bmatrix} 1 & 0 & 0 \\ -2 & 1 & 0 \\ 3 & -\frac{1}{2} & 1 \end{bmatrix} \begin{bmatrix} -1 & 0 & 0 \\ 0 & 8 & 0 \\ 0 & 0 & 7 \end{bmatrix} \begin{bmatrix} 1 & -2 & 0 & 2 \\ 0 & 1 & \frac{3}{4} & \frac{1}{2} \\ 0 & 0 & 1 & \frac{9}{7} \end{bmatrix}.$$

READING CHECK 1. In 2.8.2, why did we say "if none of the diagonal entries of U' is 0?" Suppose there is a $U' = DU$ factorization of $U' = \begin{bmatrix} 2 & 4 & 6 \\ 0 & 0 & 9 \\ 0 & 0 & 0 \end{bmatrix}$. What goes wrong?

Uniqueness Questions

In general, neither LU nor LDU factorizations are unique. The fact that

$$A = \begin{bmatrix} 1 & 2 & 3 \\ -3 & -6 & -9 \\ 5 & 10 & 15 \end{bmatrix} = \begin{bmatrix} 1 & 0 & 0 \\ -3 & 1 & 0 \\ 5 & a & 1 \end{bmatrix} \begin{bmatrix} 1 & 2 & 3 \\ 0 & 0 & 0 \\ 0 & 0 & 0 \end{bmatrix}$$

for any a shows that A can be factored $A = LU$ in many ways. The fact that

$$B = \begin{bmatrix} 0 & 0 \\ 0 & 1 \end{bmatrix} = \begin{bmatrix} 1 & 0 \\ a & 1 \end{bmatrix} \begin{bmatrix} 0 & 0 \\ 0 & 1 \end{bmatrix} \begin{bmatrix} 1 & b \\ 0 & 1 \end{bmatrix}$$

shows that B can be written in the form LDU in many ways.

On the other hand, if A is invertible, then there is just one way to factor A in the forms $A = LU$ and $A = LDU$. To explain why, we first list a few facts about triangular matrices that we invite the reader to investigate with examples. (The proofs are not hard, but too ugly for the author's taste to write down here.)

1. A square triangular matrix with nonzero entries on the diagonal is invertible. [See 3.2.1 in Section 3.1.]

2. The product of lower triangular matrices is lower triangular and the product of upper triangular matrices is upper triangular. In each case, the diagonal entries of the product are the products of the diagonal entries of the two matrices. For example,

$$\begin{bmatrix} 2 & 4 & 5 \\ 0 & -1 & 6 \\ 0 & 0 & 3 \end{bmatrix} \begin{bmatrix} -4 & 1 & 2 \\ 0 & 7 & 8 \\ 0 & 0 & -3 \end{bmatrix} = \begin{bmatrix} -8 & \star & \star \\ 0 & -7 & \star \\ 0 & 0 & -9 \end{bmatrix}.$$

   In particular, the product of two lower triangular matrices with 1s on the diagonal is a lower triangular matrix with 1s on the diagonal, and the same holds true for upper triangular matrices.

3. If a lower triangular matrix is invertible, its inverse is lower triangular. If an upper triangular matrix is invertible, its inverse is upper triangular.

Now suppose we have two LU factorizations of an invertible matrix A,

$$A = LU = L_1U_1 \tag{1}$$

where L and L_1 are (necessarily square) lower triangular matrices with 1s on the diagonal, and U and U_1 are row echelon and hence upper triangular matrices. The matrices L and L_1 are invertible, hence so are U and U_1 because each is the product of invertible matrices, $U = L^{-1}A$ and $U_1 = L_1^{-1}A$. From equation (1), we obtain

$$L^{-1}L_1 = UU_1^{-1},$$

which is interesting because the matrix on the left is lower triangular while the matrix on the right is upper triangular. This can happen only if each matrix is diagonal.

READING CHECK 2. Why?

Since the diagonal entries of $L^{-1}L_1$ are 1s, the diagonal matrix in question must be the identity. So $L^{-1}L_1 = UU_1^{-1} = I$, hence $L_1 = L$ and $U_1 = U$: the factorization $A = LU$ is unique.

An LDU factorization is often unique as well. We leave the proof of the following theorem to the exercises (Exercise 13).

2.8.4 Theorem. *If an invertible matrix A can be factored $A = LDU$ and also $A = L_1D_1U_1$ with L and L_1 lower triangular matrices with 1s on the diagonal, U and U_1 row echelon matrices with 1s on the diagonal, and D and D_1 diagonal matrices, then $L = L_1$, $U = U_1$ and $D = D_1$.*

Symmetric Matrices

We describe a cool application of Theorem 2.8.4.

2.8.5 Definition. A matrix A is *symmetric* if it equals its transpose: $A^T = A$.

The name symmetric is derived from the fact that the entries of a symmetric matrix are symmetric with respect to the main diagonal. The $(2,3)$ entry equals the $(3,2)$ entry, the $(1,4)$ entry equals the $(4,1)$ entry, the (i,j) entry equals the (j,i) entry for any i and j.

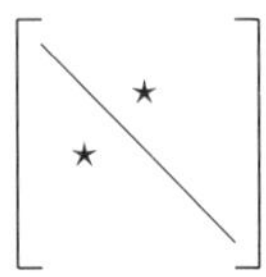

For example, $\begin{bmatrix} 1 & 2 \\ 2 & 4 \end{bmatrix}$ is a symmetric matrix. Its first column equals its first row and its second column equals its second row, so $A^T = A$. The matrix $\begin{bmatrix} 1 & 2 & 3 \\ 2 & 7 & 0 \\ 3 & 0 & 5 \end{bmatrix}$ is also symmetric.

READING CHECK 3. A symmetric matrix must be square. Why?

It is easy to make symmetric matrices. Take any matrix A and let $S = AA^T$. This is symmetric because it equals its transpose: $S^T = (AA^T)^T = (A^T)^T A^T = AA^T = S$. (Remember that $(AB)^T = B^T A^T$—Theorem 2.3.20.)

2.8.6 Example. If $A = \begin{bmatrix} -1 & 0 \\ 1 & 3 \\ 2 & -2 \end{bmatrix}$, then $A^T = \begin{bmatrix} -1 & 1 & 2 \\ 0 & 3 & -2 \end{bmatrix}$

and $AA^T = \begin{bmatrix} 1 & -1 & -2 \\ -1 & 10 & -4 \\ -2 & -4 & 8 \end{bmatrix}$ is symmetric.

Now suppose we have a factorization $A = LDU$ of a symmetric matrix A. Remember that both L and U have 1s on the diagonal. Then $LDU = A = A^T = (LDU)^T = U^T D^T L^T$. The transpose of the upper triangular matrix U is lower triangular with 1s on the diagonal, the transpose of the diagonal matrix D is D itself, and the transpose of the lower triangular matrix L is row echelon with 1s on the diagonal, so $LDU = U^T D^T L^T$ gives two LDU factorizations of A. Provided A is invertible, uniqueness of the LDU factorization implies

$L = U^T$ and $U = L^T$, so $LDU = U^T DU$. It will never be easier to find L!

2.8.7 Remark. In fact, the assumption in this argument that A is invertible is not necessary. In the exercises, we ask you to show that if any symmetric matrix has an LDU factorization, then it can also be factored $A = U^T DU$. (See Exercise 18.)

2.8.8 Problem. Find an LDU factorization of $A = \begin{bmatrix} 1 & 2 & -1 \\ 2 & 3 & 0 \\ -1 & 0 & 5 \end{bmatrix}$.

Solution. We reduce A to an upper triangular matrix using only the third elementary row operation,

$$A \to \begin{bmatrix} 1 & 2 & -1 \\ 0 & -1 & 2 \\ 0 & 2 & 4 \end{bmatrix} \to \begin{bmatrix} 1 & 2 & -1 \\ 0 & -1 & 2 \\ 0 & 0 & 8 \end{bmatrix} = U',$$

and factor $U' = DU$ with $D = \begin{bmatrix} 1 & 0 & 0 \\ 0 & -1 & 0 \\ 0 & 0 & 8 \end{bmatrix}$ and $U = \begin{bmatrix} 1 & 2 & -1 \\ 0 & 1 & -2 \\ 0 & 0 & 1 \end{bmatrix}$. Since A is symmetric, we have $A = LDU$ with $L = U^T = \begin{bmatrix} 1 & 0 & 0 \\ 2 & 1 & 0 \\ -1 & -2 & 1 \end{bmatrix}$:

$$\begin{bmatrix} 1 & 2 & -1 \\ 2 & 3 & 0 \\ -1 & 0 & 5 \end{bmatrix} = \begin{bmatrix} 1 & 0 & 0 \\ 2 & 1 & 0 \\ -1 & -2 & 1 \end{bmatrix} \begin{bmatrix} 1 & 0 & 0 \\ 0 & -1 & 0 \\ 0 & 0 & 8 \end{bmatrix} \begin{bmatrix} 1 & 2 & -1 \\ 0 & 1 & -2 \\ 0 & 0 & 1 \end{bmatrix}.$$ ☝

READING CHECK 4. Find an LDU factorization of $A = \begin{bmatrix} 1 & 2 & 3 \\ 2 & 7 & 0 \\ 3 & 0 & 5 \end{bmatrix}$.

PLDU Factorization

Suppose one or more row interchanges are needed in the Gaussian elimination process when a matrix A is carried to an upper triangular matrix U'. In Section 2.7, we observed that while such A will not have an LU factorization, it can be factored $A = PLU'$, with P a permutation matrix. If U' can be factored $U' = DU$, then we obtain $A = PLDU$.

2.8.9 Example. Let $A = \begin{bmatrix} 2 & 4 & 2 \\ 1 & 2 & 3 \\ 3 & 2 & 1 \end{bmatrix}$. With $P' = \begin{bmatrix} 1 & 0 & 0 \\ 0 & 0 & 1 \\ 0 & 1 & 0 \end{bmatrix}$, we have $PA = A' = \begin{bmatrix} 2 & 4 & 2 \\ 3 & 2 & 1 \\ 1 & 2 & 3 \end{bmatrix}$ and $A' = LU' = \begin{bmatrix} 1 & 0 & 0 \\ \frac{3}{2} & 1 & 0 \\ \frac{1}{2} & 0 & 1 \end{bmatrix} \begin{bmatrix} 2 & 4 & 2 \\ 0 & -4 & -2 \\ 0 & 0 & 2 \end{bmatrix}$. Factoring $U' = DU$ gives $A' = LDU = \begin{bmatrix} 1 & 0 & 0 \\ \frac{3}{2} & 1 & 0 \\ \frac{1}{2} & 0 & 1 \end{bmatrix} \begin{bmatrix} 2 & 0 & 0 \\ 0 & -4 & 0 \\ 0 & 0 & 2 \end{bmatrix} \begin{bmatrix} 1 & 2 & 1 \\ 0 & 1 & \frac{1}{2} \\ 0 & 0 & 1 \end{bmatrix}$. Now $PA = A'$ implies $A = P^{-1}A'$. The inverse of the elementary matrix P is P itself, so $A = PA' = PLDU$. ☺

2.8.10 Problem. Find an LDU factorization of $A = \begin{bmatrix} 0 & 1 & 1 & 0 \\ 0 & -1 & -1 & 4 \\ -2 & 3 & 2 & 1 \\ -6 & 4 & 2 & -8 \end{bmatrix}$ if this is possible. Otherwise find a PLDU factorization.

Solution. In Problem 2.7.22, we noted that A does not have an LU factorization, but $A = PLU'$ with $P = \begin{bmatrix} 0 & 0 & 1 & 0 \\ 0 & 0 & 0 & 1 \\ 1 & 0 & 0 & 0 \\ 0 & 1 & 0 & 0 \end{bmatrix}$, $L = \begin{bmatrix} 1 & 0 & 0 & 0 \\ 3 & 1 & 0 & 0 \\ 0 & -\frac{1}{5} & 1 & 0 \\ 0 & \frac{1}{5} & -1 & 1 \end{bmatrix}$ and $U' = \begin{bmatrix} -2 & 3 & 2 & 1 \\ 0 & -5 & -4 & -11 \\ 0 & 0 & \frac{1}{5} & -\frac{11}{5} \\ 0 & 0 & 0 & 4 \end{bmatrix}$. Factoring $U' = DU$ with $D = \begin{bmatrix} -2 & 0 & 0 & 0 \\ 0 & -5 & 0 & 0 \\ 0 & 0 & \frac{1}{5} & 0 \\ 0 & 0 & 0 & 4 \end{bmatrix}$ and $U = \begin{bmatrix} 1 & -\frac{3}{2} & -1 & -\frac{1}{2} \\ 0 & 1 & \frac{4}{5} & \frac{11}{5} \\ 0 & 0 & 1 & -11 \\ 0 & 0 & 0 & 1 \end{bmatrix}$ gives a factorization $A = PLDU$. ☝

Answers to Reading Checks

1. Let $D = \begin{bmatrix} a & 0 & 0 \\ 0 & b & 0 \\ 0 & 0 & c \end{bmatrix}$ and $U = \begin{bmatrix} 1 & x & y \\ 0 & 1 & z \\ 0 & 0 & 1 \end{bmatrix}$. Then $DU = \begin{bmatrix} a & ax & ay \\ 0 & b & bz \\ 0 & 0 & c \end{bmatrix}$. Look at the second row. If $DU = U' = \begin{bmatrix} 2 & 4 & 6 \\ 0 & 0 & 9 \\ 0 & 0 & 0 \end{bmatrix}$, then $b = 0$, so the second row of DU is all zero. So $DU \neq U'$.
2. A lower triangular matrix has 0s above the diagonal and an upper triangular

matrix has 0s below the diagonal. Therefore, if a matrix is both lower and upper triangular, all entries off the diagonal (above and below) are 0. This is what we mean by a *diagonal* matrix.

3. If A is $m \times n$, then A^T is $n \times m$. If A is to equal A^T, these two matrices must have the same size, so $m = n$.

4. Gaussian elimination proceeds

$$A \to \begin{bmatrix} 1 & 2 & 3 \\ 0 & 3 & -6 \\ 0 & -6 & -4 \end{bmatrix} \to \begin{bmatrix} 1 & 2 & 3 \\ 0 & 3 & -6 \\ 0 & 0 & -16 \end{bmatrix} = U'.$$

We have $U' = DU$ with $D = \begin{bmatrix} 1 & 0 & 0 \\ 0 & 3 & 0 \\ 0 & 0 & -16 \end{bmatrix}$ and $U = \begin{bmatrix} 1 & 2 & 3 \\ 0 & 1 & -2 \\ 0 & 0 & 1 \end{bmatrix}$, so $A =$ LDU with $L = U^T = \begin{bmatrix} 1 & 0 & 0 \\ 2 & 1 & 0 \\ 3 & -2 & 1 \end{bmatrix}$.

True/False Questions

Decide, with as little calculation as possible, whether each of the following statements is true or false and, if you say "false," explain your answer. (Answers are at the back of the book.)

1. If $A = LU$ is an LU factorization of a matrix, it is understood that L and U have 1s on the diagonal.
2. If $A = LDU$ is an LDU factorization of a matrix, it is understood that L and U have 1s on the diagonal.
3. The matrix $\begin{bmatrix} 0 & 0 \\ 0 & 0 \end{bmatrix}$ has an LDU factorization.
4. An LDU factorization of a matrix is unique.
5. The product of symmetric matrices is symmetric.

EXERCISES

Answers to exercises marked [BB] can be found at the **Back** *of the* **Book.**

1. Find an LDU factorization of each of the following matrices. In each case, find L without calculation, just by keeping track of multipliers.

   (a) [BB] $\begin{bmatrix} -2 & 1 & 6 \\ 4 & 0 & 8 \\ 6 & 3 & -10 \end{bmatrix}$ **(b)** $\begin{bmatrix} 1 & -1 & -2 \\ 2 & -3 & -5 \\ -1 & 3 & 5 \end{bmatrix}$

(c) $\begin{bmatrix} 2 & 8 & -5 \\ 2 & 12 & -8 \\ 8 & 40 & -29 \\ 6 & 20 & -6 \end{bmatrix}$ (d) $\begin{bmatrix} 1 & 0 & -1 & 5 & -1 \\ -1 & 4 & 0 & -3 & 2 \\ 0 & 2 & -3 & 7 & 6 \\ 2 & -1 & -2 & 12 & 3 \end{bmatrix}$.

2. Find an LDU factorization of

   (a) [BB] the matrix $A = \begin{bmatrix} -3 & 3 & 6 \\ 2 & 5 & 10 \\ 0 & 1 & 4 \end{bmatrix}$ in Exercise 9, Section 2.7,

   (b) the matrix $A = \begin{bmatrix} 2 & 4 & 6 & 18 \\ 4 & 5 & 6 & 24 \\ 3 & 1 & -2 & 4 \end{bmatrix}$ in Exercise 10, Section 2.7.

3. [BB; (a), (f)] Find an LDU factorization of all the matrices of Exercise 28, Section 2.7 for which an LDU factorization exists.

4. Show that $\begin{bmatrix} 1 & 3 & -1 \\ 2 & 6 & 1 \end{bmatrix}$ does not have an LDU factorization. Why does this fact not contradict Theorem 2.8.4?

5. Suppose A and B are symmetric matrices.

   (a) [BB] Is AB symmetric (in general)?

   (b) What about A^2, ABA and $A - B$?

6. Suppose that A and P are $n \times n$ matrices and A is symmetric. Prove that $P^T AP$ is symmetric.

7. [BB] A matrix A is *skew-symmetric* if $A^T = -A$. Show that if A and B are $n \times n$ skew-symmetric matrices, then so is $A + B$.

8. Let A be an $n \times n$ matrix.

   (a) Show that $A - A^T$ is skew-symmetric. (The concept of skew-symmetric matrix is defined in Exercise 7.)

   (b) Show that $A + A^T$ is symmetric.

   (c) Show that $A = S + K$ can be written as the sum of a symmetric matrix S and a skew-symmetric matrix K.

9. Find an LDU factorization of each of the following symmetric matrices:

   (a) [BB] $A = \begin{bmatrix} 2 & 5 \\ 5 & 9 \end{bmatrix}$ **(b)** $A = \begin{bmatrix} 1 & 2 & 0 \\ 2 & 6 & 4 \\ 0 & 4 & 11 \end{bmatrix}$

   (c) $A = \begin{bmatrix} 2 & -1 & 0 \\ -1 & 2 & -1 \\ 0 & -1 & 2 \end{bmatrix}$ **(d)** $A = \begin{bmatrix} -2 & 1 & 4 \\ 1 & 0 & 1 \\ 4 & 1 & 3 \end{bmatrix}$

(e) $A = \begin{bmatrix} 1 & -4 & 2 \\ -4 & 1 & -2 \\ 2 & -2 & 2 \end{bmatrix}$ **(f)** $A = \begin{bmatrix} 5 & 4 & -4 \\ 4 & 5 & 4 \\ -4 & 4 & 5 \end{bmatrix}$

(g) $A = \begin{bmatrix} 1 & 2 & 2 & 1 \\ 2 & 7 & -2 & -1 \\ 2 & -2 & 3 & 8 \\ 1 & -1 & 8 & 4 \end{bmatrix}$ **(h)** $\begin{bmatrix} 7 & 5 & 0 & 3 & 0 \\ 5 & 4 & -2 & 1 & -2 \\ 0 & -2 & 1 & 0 & 1 \\ 3 & 1 & 0 & 1 & 0 \\ 0 & -2 & 1 & 0 & 1 \end{bmatrix}$.

10. Find an LU or LDU factorization of $A = \begin{bmatrix} 0 & 1 & 1 & -1 \\ 1 & 2 & 3 & 1 \\ 2 & 7 & 9 & 3 \\ -3 & 1 & 1 & 2 \end{bmatrix}$, if possible. If not possible, find a factorization $A = PLU$ and discuss the possibility of a factorization $A = PLDU$.

Critical Reading

11. If $A = LU$ and A is invertible, then U is invertible too. Why?

12. Suppose $A = DU$ is the product of a diagonal matrix D and an upper triangular matrix U. Show that there exists a lower triangular matrix L such that LA is a symmetric matrix.

13. Prove Theorem 2.8.4.

14. **(a)** [BB] We have seen in this section how to write an upper triangular matrix U as the product of a diagonal matrix D and an upper triangular matrix U_1 with 1s on its diagonal. Show how to express a **lower** triangular matrix L as the product L_1D_1 of a diagonal matrix and a triangular matrix L_1 with 1s on its diagonal. Illustrate using $L = \begin{bmatrix} 2 & 0 & 0 \\ 4 & -3 & 0 \\ 6 & 7 & 4 \end{bmatrix}$.

    The point of this question was to reveal a quick way to factor a lower triangular matrix $L = L_1D$ where L_1 has 1s on the diagonal. Explain. [Hint: transpose.]

    (b) Find an LDU decomposition of $A = \begin{bmatrix} 2 & 0 & 0 \\ 4 & -3 & 0 \\ 6 & 7 & 4 \end{bmatrix} \begin{bmatrix} -1 & 4 & 8 \\ 0 & 3 & 9 \\ 0 & 0 & 5 \end{bmatrix}$ without calculation. (Check the definition of LDU factorization.)

15. If A is a symmetric $n \times n$ matrix and x is a vector in R^n, show that $\mathsf{x}^T A^2 \mathsf{x} = \|A\mathsf{x}\|^2$.

16. If A is a matrix such that $A^T = A^T A$, show that A is symmetric and that $A = A^2$.

17. If A is a symmetric $n \times n$ matrix and x, y are vectors in R^n, prove that $A\mathsf{x} \cdot \mathsf{y} = \mathsf{x} \cdot A\mathsf{y}$. [Hint: 2.1.18.]

18. **(a)** If A is a symmetric matrix and $A = LDU$ is an LDU factorization, show that $A = U^T DU$.

    (b) If A is an invertible symmetric matrix and $A = LDU$, then $L = U^T$.

    (c) Do you detect the subtle difference between (a) and (b)?

2.9 More on the Inverse of a Matrix

In Section 2.3, we introduced the concept of the inverse of a matrix. A matrix A *has an inverse* or is *invertible* if there is another matrix B, called the inverse of A, such that $AB = I$ and $BA = I$. We write $B = A^{-1}$ for the inverse of A just as 5^{-1} denotes the (multiplicative) inverse of the number 5. An invertible matrix is necessarily square and, as we shall see, if A is square, it is only necessary to verify that $AB = I$ (or $BA = I$) to establish that $B = A^{-1}$ (Theorem 2.9.12). This idea has been used previously, and it is the very best way to determine whether two square matrices A and B are inverses; compute AB or BA and see if you get the identity.

2.9.1 Problem. Determine whether or not $A = \begin{bmatrix} -3 & 2 \\ -2 & 1 \end{bmatrix}$ and $B = \begin{bmatrix} 1 & -2 \\ 2 & -3 \end{bmatrix}$ are inverses.

Solution. We compute $AB = \begin{bmatrix} -3 & 2 \\ -2 & 1 \end{bmatrix} \begin{bmatrix} 1 & -2 \\ 2 & -3 \end{bmatrix} = \begin{bmatrix} 1 & 0 \\ 0 & 1 \end{bmatrix} = I$ and, since the matrices are square, conclude that A and B are inverses. ♤

2.9.2 Problem. Let $A = \begin{bmatrix} 1 & 2 & 3 \\ 4 & 5 & 6 \end{bmatrix}$ and $B = \begin{bmatrix} -1 & 1 \\ 0 & -1 \\ \frac{2}{3} & \frac{1}{3} \end{bmatrix}$. Then

$$AB = \begin{bmatrix} 1 & 2 & 3 \\ 4 & 5 & 6 \end{bmatrix} \begin{bmatrix} -1 & 1 \\ 0 & -1 \\ \frac{2}{3} & \frac{1}{3} \end{bmatrix} = \begin{bmatrix} 1 & 0 \\ 0 & 1 \end{bmatrix}.$$

Is A invertible?

Solution. No, A is not invertible because it is not square. Alternatively, you may check that $BA \neq I$. ♤

READING CHECK 1. Are $A = \begin{bmatrix} 2 & 7 & 1 \\ 0 & -3 & 1 \\ 1 & 1 & 0 \end{bmatrix}$ and $B = \begin{bmatrix} 3 & -5 & 0 \\ -1 & 2 & 1 \\ 2 & -4 & -7 \end{bmatrix}$ inverses?

A Method for Finding the Inverse

Given two matrices A and B, it is easy to determine if they are inverses: they must be square and AB must be the identity matrix.

Suppose we are given just one matrix and we wish to know whether or not it has an inverse. If it does, we'd also like to find this inverse. We answered this question for 2×2 matrices in Proposition 2.3.10. Here we study the general situation.

Let A be an $n \times n$ matrix. If A is invertible, there is a matrix B with

$$AB = I = \begin{bmatrix} 1 & 0 & \dots & 0 \\ 0 & 1 & \dots & 0 \\ \vdots & \vdots & \ddots & \vdots \\ 0 & 0 & \dots & 1 \end{bmatrix} = \begin{bmatrix} \mathsf{e}_1 & \mathsf{e}_2 & \dots & \mathsf{e}_n \\ \downarrow & \downarrow & & \downarrow \\ & & & \end{bmatrix}.$$

Let $B = \begin{bmatrix} \mathsf{b}_1 & \mathsf{b}_2 & \dots & \mathsf{b}_n \\ \downarrow & \downarrow & & \downarrow \\ & & & \end{bmatrix}$ have columns $\mathsf{b}_1, \mathsf{b}_2, \dots, \mathsf{b}_n$. Then

$$AB = \begin{bmatrix} A\mathsf{b}_1 & A\mathsf{b}_2 & \dots & A\mathsf{b}_n \\ \downarrow & \downarrow & & \downarrow \\ & & & \end{bmatrix}.$$

We want $AB = I$, so we require

$$A\mathsf{b}_1 = \begin{bmatrix} 1 \\ 0 \\ 0 \\ \vdots \\ 0 \end{bmatrix} = \mathsf{e}_1, \quad A\mathsf{b}_2 = \begin{bmatrix} 0 \\ 1 \\ 0 \\ \vdots \\ 0 \end{bmatrix} = \mathsf{e}_2, \quad \dots, \quad A\mathsf{b}_n = \begin{bmatrix} 0 \\ 0 \\ \vdots \\ 0 \\ 1 \end{bmatrix} = \mathsf{e}_n.$$

So we must solve the n linear systems $A\mathsf{x} = \mathsf{e}_i$, $i = 1, 2, \dots, n$.

2.9.3 Example. Suppose $A = \begin{bmatrix} 1 & 4 \\ 2 & 7 \end{bmatrix}$. We seek the matrix $B = \begin{bmatrix} \mathsf{b}_1 & \mathsf{b}_2 \\ \downarrow & \downarrow \end{bmatrix}$ with $AB = I$. This requires solving two systems of equations. First we solve $A\mathsf{x} = \mathsf{e}_1$ for b_1. Gaussian elimination proceeds

$$\left[\begin{array}{cc|c} 1 & 4 & 1 \\ 2 & 7 & 0 \end{array}\right] \to \left[\begin{array}{cc|c} 1 & 4 & 1 \\ 0 & -1 & -2 \end{array}\right] \to \left[\begin{array}{cc|c} 1 & 4 & 1 \\ 0 & \textcircled{1} & 2 \end{array}\right]$$

so, if $\mathsf{b}_1 = \begin{bmatrix} x \\ y \end{bmatrix}$, we have $y = 2$, $x + 4y = 1$, so $x = 1 - 4y = -7$. The first column of B is $\mathsf{b}_1 = \begin{bmatrix} -7 \\ 2 \end{bmatrix}$. This solution can also be obtained by continuing the Gaussian elimination, using the circled pivot to get a 0 **above** it:

$$\left[\begin{array}{cc|c} 1 & 4 & 1 \\ 0 & \textcircled{1} & 2 \end{array}\right] \to \left[\begin{array}{cc|c} 1 & 0 & -7 \\ 0 & 1 & 2 \end{array}\right],$$

thus $x = -7$, $y = 2$ comes directly and the desired column b_1 appears as the final column of the augmented matrix. Solving the second system this way,

$$\left[\begin{array}{cc|c} 1 & 4 & 0 \\ 2 & 7 & 1 \end{array}\right] \to \left[\begin{array}{cc|c} 1 & 4 & 0 \\ 0 & -1 & 1 \end{array}\right] \to \left[\begin{array}{cc|c} 1 & 4 & 0 \\ 0 & 1 & -1 \end{array}\right] \to \left[\begin{array}{cc|c} 1 & 0 & 4 \\ 0 & 1 & -1 \end{array}\right],$$

gives $\mathsf{b}_2 = \begin{bmatrix} 4 \\ -1 \end{bmatrix}$, so $B = A^{-1} = \begin{bmatrix} -7 & 4 \\ 2 & -1 \end{bmatrix}$. ☺

In the two systems of linear equations just solved, the sequences of row operations used to bring each matrix to row echelon form were the same because the coefficient matrix was the same in each case. It follows that both systems can be solved at the same time, like this:

$$\left[\begin{array}{cc|cc} 1 & 4 & 1 & 0 \\ 2 & 7 & 0 & 1 \end{array}\right] \to \left[\begin{array}{cc|cc} 1 & 4 & 1 & 0 \\ 0 & -1 & -2 & 1 \end{array}\right]$$
$$\to \left[\begin{array}{cc|cc} 1 & 4 & 1 & 0 \\ 0 & 1 & 2 & -1 \end{array}\right] \to \left[\begin{array}{cc|cc} 1 & 0 & -7 & 4 \\ 0 & 1 & 2 & -1 \end{array}\right].$$

Notice that the identity matrix appears to the left of the vertical line at the last step and A^{-1} appears to the right. This example suggests a general procedure for determining whether or not a matrix has an inverse and for finding the inverse if such exists.

2.9.4 If A can be moved to the identity matrix I by a sequence of elementary row operations, then A is invertible and, after moving $[A \mid I]$ to $[I \mid B]$, the matrix $B = A^{-1}$; otherwise, A is not invertible.

2.9.5 Example. Suppose $A = \begin{bmatrix} 1 & 2 & 0 \\ 3 & -1 & 2 \\ -2 & 3 & -2 \end{bmatrix}$. Gaussian elimination applied to

$[A \mid I]$ proceeds

$$\left[\begin{array}{rrr|rrr} 1 & 2 & 0 & 1 & 0 & 0 \\ 3 & -1 & 2 & 0 & 1 & 0 \\ -2 & 3 & -2 & 0 & 0 & 1 \end{array}\right]$$

$$\to \left[\begin{array}{rrr|rrr} 1 & 2 & 0 & 1 & 0 & 0 \\ 0 & -7 & 2 & -3 & 1 & 0 \\ 0 & 7 & -2 & 2 & 0 & 1 \end{array}\right] \to \left[\begin{array}{rrr|rrr} 1 & 2 & 0 & 1 & 0 & 0 \\ 0 & -7 & 2 & -3 & 1 & 0 \\ 0 & 0 & 0 & -1 & 1 & 1 \end{array}\right].$$

There is no point in continuing. The three initial 0s in the third row can't be changed. We cannot make the left 3×3 block the identity matrix. The given matrix A has no inverse. :-)

2.9.6 Example. Suppose $A = \begin{bmatrix} 2 & 7 & 0 \\ 1 & 4 & -1 \\ 1 & 3 & 0 \end{bmatrix}$.

Gaussian elimination proceeds $[A \mid I] = \left[\begin{array}{rrr|rrr} 2 & 7 & 0 & 1 & 0 & 0 \\ 1 & 4 & -1 & 0 & 1 & 0 \\ 1 & 3 & 0 & 0 & 0 & 1 \end{array}\right]$

$$\to \left[\begin{array}{rrr|rrr} 1 & 4 & -1 & 0 & 1 & 0 \\ 2 & 7 & 0 & 1 & 0 & 0 \\ 1 & 3 & 0 & 0 & 0 & 1 \end{array}\right] \to \left[\begin{array}{rrr|rrr} 1 & 4 & -1 & 0 & 1 & 0 \\ 0 & -1 & 2 & 1 & -2 & 0 \\ 0 & -1 & 1 & 0 & -1 & 1 \end{array}\right]$$

$$\to \left[\begin{array}{rrr|rrr} 1 & 4 & -1 & 0 & 1 & 0 \\ 0 & 1 & -2 & -1 & 2 & 0 \\ 0 & -1 & 1 & 0 & -1 & 1 \end{array}\right] \to \left[\begin{array}{rrr|rrr} 1 & 0 & 7 & 4 & -7 & 0 \\ 0 & 1 & -2 & -1 & 2 & 0 \\ 0 & 0 & -1 & -1 & 1 & 1 \end{array}\right]$$

$$\to \left[\begin{array}{rrr|rrr} 1 & 0 & 7 & 4 & -7 & 0 \\ 0 & 1 & -2 & -1 & 2 & 0 \\ 0 & 0 & 1 & 1 & -1 & -1 \end{array}\right] \to \left[\begin{array}{rrr|rrr} 1 & 0 & 0 & -3 & 0 & 7 \\ 0 & 1 & 0 & 1 & 0 & -2 \\ 0 & 0 & 1 & 1 & -1 & -1 \end{array}\right] = [I \mid B].$$

We have moved $[A \mid I]$ to $[I \mid B]$, so $B = \begin{bmatrix} -3 & 0 & 7 \\ 1 & 0 & -2 \\ 1 & -1 & -1 \end{bmatrix} = A^{-1}$. :-)

READING CHECK 2. Let $A = \begin{bmatrix} -1 & 0 & 2 \\ 1 & 1 & -1 \\ 3 & -1 & 0 \end{bmatrix}$ and $B = \frac{1}{7}\begin{bmatrix} 1 & 2 & 2 \\ 3 & 6 & -1 \\ 4 & 1 & 1 \end{bmatrix}$.

Are A and B inverses?

Suppose a and b are real numbers and we want to solve the equation $ax = b$. If $a \neq 0$, then $x = \frac{b}{a}$, a solution that can also be written $x = a^{-1}b$. Similarly, suppose we want to solve the system of linear equations $A\mathsf{x} = \mathsf{b}$. If A is invertible, then $\mathsf{x} = A^{-1}\mathsf{b}$ because $A\mathsf{x} = A(A^{-1}\mathsf{b}) = I\mathsf{b} = \mathsf{b}$. Moreover, this is the only solution to $A\mathsf{x} = \mathsf{b}$—the solution is unique—because if $A\mathsf{x}_1 = A\mathsf{x}_2 =$

b, multiplying on the left by A^{-1} would give $A^{-1}A\mathsf{x}_1 = A^{-1}A\mathsf{x}_2$, so $I\mathsf{x}_1 = I\mathsf{x}_2$ and $\mathsf{x}_1 = \mathsf{x}_2$.

2.9.7 Example. The system $\begin{aligned} 2x_1 + 7x_2 \quad &= -2 \\ x_1 + 4x_2 - x_3 &= 4 \\ x_1 + 3x_2 \quad &= 5 \end{aligned}$ is $A\mathsf{x} = \mathsf{b}$ with $A = \begin{bmatrix} 2 & 7 & 0 \\ 1 & 4 & -1 \\ 1 & 3 & 0 \end{bmatrix}$, the matrix of Example 2.9.6, $\mathsf{x} = \begin{bmatrix} x_1 \\ x_2 \\ x_3 \end{bmatrix}$ and $\mathsf{b} = \begin{bmatrix} -2 \\ 4 \\ 5 \end{bmatrix}$. Since A is invertible, the unique solution is

$$\mathsf{x} = A^{-1}\mathsf{b} = \begin{bmatrix} -3 & 0 & 7 \\ 1 & 0 & -2 \\ 1 & -1 & -1 \end{bmatrix} \begin{bmatrix} -2 \\ 4 \\ 5 \end{bmatrix} = \begin{bmatrix} 41 \\ -12 \\ -11 \end{bmatrix}.$$

We check that this is the right answer by verifying that $A\mathsf{x} = \mathsf{b}$:

$$A\mathsf{x} = \begin{bmatrix} 2 & 7 & 0 \\ 1 & 4 & -1 \\ 1 & 3 & 0 \end{bmatrix} \begin{bmatrix} 41 \\ -12 \\ -11 \end{bmatrix} = \begin{bmatrix} -2 \\ 4 \\ 5 \end{bmatrix} = \mathsf{b}.$$

2.9.8 Problem. Express the vector $\mathsf{b} = \begin{bmatrix} -2 \\ 4 \\ 5 \end{bmatrix}$ as a linear combination of the columns of the matrix A in Example 2.9.7.

Solution. We just saw that $A \begin{bmatrix} 41 \\ -12 \\ -11 \end{bmatrix} = \mathsf{b}$. Since $A\mathsf{x}$ is a linear combination of the columns of A with coefficients the components of x (recall 2.1.33 and never forget), we have $\mathsf{b} = 41 \begin{bmatrix} 2 \\ 1 \\ 1 \end{bmatrix} - 12 \begin{bmatrix} 7 \\ 4 \\ 3 \end{bmatrix} - 11 \begin{bmatrix} 0 \\ -1 \\ 0 \end{bmatrix}$.

2.9.9 Remark. Whether working by hand or with a computer, we virtually never solve $A\mathsf{x} = \mathsf{b}$ by computing A^{-1}. By hand, we much prefer Gaussian elimination and back substitution on the augmented matrix $[A \mid \mathsf{b}]$. In the first place, this is easier. Secondly, at the outset, there is no reason at all to suppose that A^{-1} actually exists, so trying to find A^{-1} in order to solve $A\mathsf{x} = \mathsf{b}$ is potentially a waste of time. The real value to the approach illustrated in Example 2.9.7 is its contribution to the theory of linear equations, which we discuss in more detail in Section 4.1. For now, we simply remark that

> If A has an inverse, then $A\mathsf{x} = \mathsf{b}$ has the unique solution $\mathsf{x} = A^{-1}\mathsf{b}$.

We continue this section with some theorems about invertible matrices.

2.9.10 **Theorem.** *A matrix is invertible if and only if it can be written as the product of elementary matrices.*

Proof. ($\Longleftarrow$) Suppose A is the product of elementary matrices. An elementary matrix is invertible and the product of invertible matrices is invertible, so A is invertible.

($\Longrightarrow$) Suppose A is invertible. Then we can transform A to the identity matrix by a sequence of elementary row operations. The Gaussian elimination process looks like this:

$$A \to E_1A \to E_2(E_1A) \to \cdots \to (E_kE_{k-1}\cdots E_2E_1)A = I,$$

where $E_1, E_2, \ldots, E_k$ are elementary matrices. So $A = (E_kE_{k-1}\cdots E_2E_1)^{-1} = E_1^{-1}E_2^{-1}\cdots E_{k-1}^{-1}E_k^{-1}$. Since the inverse of an elementary matrix is elementary, $A = E_1^{-1}E_2^{-1}\cdots E_k^{-1}$ expresses A as the product of elementary matrices. ■

2.9.11 **Problem.** Express $A = \begin{bmatrix} 2 & 3 \\ 1 & 1 \end{bmatrix}$ as the product of elementary matrices.

Solution. We move A to the identity matrix by a sequence of elementary row operations.

$$A = \begin{bmatrix} 2 & 3 \\ 1 & 1 \end{bmatrix} \to \begin{bmatrix} 1 & 1 \\ 2 & 3 \end{bmatrix} = E_1A$$
$$\to \begin{bmatrix} 1 & 1 \\ 0 & 1 \end{bmatrix} = E_2E_1A \to \begin{bmatrix} 1 & 0 \\ 0 & 1 \end{bmatrix} = E_3E_2E_1A = I$$

with $E_1 = \begin{bmatrix} 0 & 1 \\ 1 & 0 \end{bmatrix}$, $E_2 = \begin{bmatrix} 1 & 0 \\ -2 & 1 \end{bmatrix}$ and $E_3 = \begin{bmatrix} 1 & -1 \\ 0 & 1 \end{bmatrix}$.
So $(E_3E_2E_1)A = I$ and $A = (E_3E_2E_1)^{-1} = E_1^{-1}E_2^{-1}E_3^{-1}$; that is,

$$\begin{bmatrix} 2 & 3 \\ 1 & 1 \end{bmatrix} = \begin{bmatrix} 0 & 1 \\ 1 & 0 \end{bmatrix}\begin{bmatrix} 1 & 0 \\ 2 & 1 \end{bmatrix}\begin{bmatrix} 1 & 1 \\ 0 & 1 \end{bmatrix}.$$

We now justify a property of square matrices we have used on a number of occasions.

2.9.12 **Theorem.** *If A and B are square $n \times n$ matrices and $AB = I$ (the $n \times n$ identity matrix), then $BA = I$, so A is invertible.*

Proof. If $AB = I$, then $B\mathsf{x} = 0$ has the unique solution $\mathsf{x} = 0$ because $B\mathsf{x} = 0$ implies $AB\mathsf{x} = A0$, so $\mathsf{x} = I\mathsf{x} = 0$. Thus, when we try to solve $B\mathsf{x} = 0$ by reducing B to row echelon form, there are no free variables (free variables imply infinitely many solutions), every column contains a pivot, so every row contains a pivot. It follows that we can transform B to I with elementary row operations. As in Theorem 2.9.10, this means that $EB = I$ for some product E of elementary matrices. Since E is invertible, $B = E^{-1}$, $AB = I$ reads $AE^{-1} = I$, so $A = E$, $BA = E^{-1}E = I$ and A is invertible. ■

READING CHECK 3. The hypothesis in Theorem 2.9.12 that both A and B be square is redundant. It is sufficient, for example, that only B be square. Why?

Recall that vectors $\mathsf{x}_1, \mathsf{x}_2, \ldots, \mathsf{x}_n$ are *linearly independent* if and only if

$$c_1\mathsf{x}_1 + c_2\mathsf{x}_2 + \cdots + c_n\mathsf{x}_n = 0 \quad \text{implies} \quad c_1 = 0, c_2 = 0, \ldots, c_n = 0.$$

Since $c_1\mathsf{x}_1 + c_2\mathsf{x}_2 + \cdots + c_n\mathsf{x}_n = A\mathsf{c}$, where $A = \begin{bmatrix} \mathsf{x}_1 & \mathsf{x}_2 & \cdots & \mathsf{x}_n \\ \downarrow & \downarrow & & \downarrow \end{bmatrix}$ and $\mathsf{c} = \begin{bmatrix} c_1 \\ c_2 \\ \vdots \\ c_n \end{bmatrix}$, the condition for linear independence can be restated as $A\mathsf{c} = 0$ implies $\mathsf{c} = 0$. We exploit this idea in the next theorem.

2.9.13 Theorem. *A square matrix is invertible if and only if its columns are linearly independent.*

Proof. ($\Longrightarrow$) Assume that A is an invertible matrix. If $A\mathsf{x} = 0$, multiplying on the left by A^{-1} gives $\mathsf{x} = 0$. We have just seen that this implies that the columns of A are linearly independent.

($\Longleftarrow$) Assume the columns of a square matrix A are linearly independent. Then the homogeneous system $A\mathsf{x} = 0$ has the unique solution $\mathsf{x} = 0$. This means that every column of A is a pivot column because nonpivot columns imply free variables and infinitely many solutions. Since A is square, there is a pivot in every row of A, so A can be reduced to the identity matrix. Thus $EA = I$ for some product E of elementary matrices, so $A = E^{-1}$ is the product elementary matrices and hence invertible by Theorem 2.9.10. ■

Reduced Row Echelon Form

In this section, we have used the Gaussian elimination process to move a matrix not just to row echelon form, but to **reduced** row echelon form.

2.9.14 Definition. A matrix is in *reduced row echelon form* if

1. it is in row echelon form;
2. each pivot is a 1;
3. every entry above (and below) a pivot is 0.

2.9.15 Examples. The following matrices are all in reduced row echelon form:

$$\begin{bmatrix} 1 & 0 \\ 0 & 1 \end{bmatrix}, \quad \begin{bmatrix} 1 & 0 & -2 \\ 0 & 1 & 1 \end{bmatrix}, \quad \begin{bmatrix} 1 & 0 & 0 & 0 \\ 0 & 0 & 1 & 0 \\ 0 & 0 & 0 & 1 \end{bmatrix} \quad \text{and} \quad \begin{bmatrix} 1 & 0 & 0 & 0 & 0 & 4 \\ 0 & 1 & 0 & 5 & 0 & -2 \\ 0 & 0 & 0 & 0 & 1 & 3 \end{bmatrix}.$$

READING CHECK 4. Find the reduced row echelon form of $A = \begin{bmatrix} 1 & 2 & 0 \\ 3 & -1 & 2 \\ -2 & 3 & -2 \end{bmatrix}$, the matrix of Example 2.9.5.

2.9.16 Remark. Unlike row echelon form, *reduced* row echelon form is unique, so we refer to **the** reduced row echelon form of a matrix. See Appendix A of *Linear Algebra and its Applications* by David C. Lay, Addison-Wesley (2000) for a proof.

To bring a matrix to reduced row echelon form, use Gaussian elimination in the usual way, sweeping across the columns of the matrix from left to right, but making sure all pivots are 1, and then, each time we get such a 1, using this to get 0s in the rest of the corresponding column, above as well as below the 1.

2.9.17 Example. $\begin{bmatrix} \textcircled{1} & 2 & -5 \\ 3 & 1 & 5 \\ -2 & 3 & 4 \end{bmatrix} \to \begin{bmatrix} 1 & 2 & -5 \\ 0 & -5 & 20 \\ 0 & 7 & -6 \end{bmatrix} \to \begin{bmatrix} 1 & 2 & -5 \\ 0 & \textcircled{1} & -4 \\ 0 & 7 & -6 \end{bmatrix}$

$$\to \begin{bmatrix} 1 & 0 & 3 \\ 0 & 1 & -4 \\ 0 & 0 & 22 \end{bmatrix} \to \begin{bmatrix} 1 & 0 & 3 \\ 0 & 1 & -4 \\ 0 & 0 & \textcircled{1} \end{bmatrix} \to \begin{bmatrix} 1 & 0 & 0 \\ 0 & 1 & 0 \\ 0 & 0 & 1 \end{bmatrix}.$$

Thus the reduced row echelon form of $\begin{bmatrix} 1 & 2 & -5 \\ 3 & 1 & 5 \\ -2 & 3 & 4 \end{bmatrix}$ is $\begin{bmatrix} 1 & 0 & 0 \\ 0 & 1 & 0 \\ 0 & 0 & 1 \end{bmatrix}$.

Our interest in reduced row echelon form derives from the method we have described for finding the inverse of a matrix, when there is an inverse. If the reduced row echelon form of A is the identity matrix, then A has an inverse. See Example 2.9.6.

Answers to Reading Checks

1. $AB = \begin{bmatrix} 1 & 0 & 0 \\ 5 & -10 & -10 \\ 2 & -3 & 1 \end{bmatrix} \neq I$, so A and B are not inverses.

2. This is easy to answer. Since A is square and $AB = I$, A and B are inverses. [The author hopes very much that you did not attempt to find the inverse of A.]

3. Suppose A is $r \times s$. Since we can form the product AB with B a square $n \times n$ matrix, we must have $s = n$. Thus AB is $r \times n$. Since AB is an identity matrix, $r = n$ and A is $r \times s$ which is $n \times n$.

4. In Example 2.9.5, Gaussian elimination moved A to $\begin{bmatrix} 1 & 2 & 0 \\ 0 & -7 & 2 \\ 0 & 0 & 0 \end{bmatrix}$. Continuing, $\begin{bmatrix} 1 & 2 & 0 \\ 0 & -7 & 2 \\ 0 & 0 & 0 \end{bmatrix} \to \begin{bmatrix} 1 & 2 & 0 \\ 0 & 1 & -\frac{2}{7} \\ 0 & 0 & 0 \end{bmatrix} \to \begin{bmatrix} 1 & 0 & \frac{4}{7} \\ 0 & 1 & -\frac{2}{7} \\ 0 & 0 & 0 \end{bmatrix}$. The reduced row echelon form of $A = \begin{bmatrix} 1 & 2 & 0 \\ 3 & -1 & 2 \\ -2 & 3 & -2 \end{bmatrix}$ is $\begin{bmatrix} 1 & 0 & \frac{4}{7} \\ 0 & 1 & -\frac{2}{7} \\ 0 & 0 & 0 \end{bmatrix}$.

True/False Questions

Decide, with as little calculation as possible, whether each of the following statements is true or false and, if you say "false," explain your answer. (Answers are at the back of the book.)

1. The matrix $\begin{bmatrix} 1 & 0 & 4 \\ 0 & 0 & 9 \\ 0 & 0 & 0 \end{bmatrix}$ is in row echelon form.
2. The matrix $\begin{bmatrix} 1 & 0 & 4 \\ 0 & 0 & 9 \\ 0 & 0 & 0 \end{bmatrix}$ is in reduced row echelon form.
3. A row echelon form for a matrix is unique.
4. Reduced row echelon form is unique.
5. If A and B are elementary matrices, so is AB.
6. If A and B are invertible matrices, so is AB.
7. If AB is an invertible matrix, then so is A.
8. If E and F are elementary matrices, EF is invertible.
9. The inverse of the invertible matrix ABC is $A^{-1}B^{-1}C^{-1}$.
10. If A and B are matrices with $AB = I$, then $BA = I$.

EXERCISES

Answers to exercises marked [BB] can be found at the Back of the Book.

1. Determine whether or not each of the following pairs of matrices are inverses:

   (a) [BB] $A = \begin{bmatrix} -1 & 2 \\ -7 & 10 \end{bmatrix}$, $B = \frac{1}{4}\begin{bmatrix} 10 & -2 \\ 7 & -1 \end{bmatrix}$

   (b) $A = \begin{bmatrix} 1 & 2 \\ 3 & 4 \end{bmatrix}$, $B = \frac{1}{2}\begin{bmatrix} -4 & 2 & 1 \\ 3 & -1 & 0 \end{bmatrix}$

   (c) $A = \begin{bmatrix} 0 & 2 & 3 \\ 4 & 5 & 6 \\ 7 & 8 & 9 \end{bmatrix}$, $B = \frac{1}{3}\begin{bmatrix} -3 & 6 & -3 \\ 6 & -21 & 12 \\ -3 & 14 & -8 \end{bmatrix}$

   (d) $A = \begin{bmatrix} -1 & 0 & 1 \\ 2 & 4 & -1 \\ 0 & 1 & 1 \end{bmatrix}$, $B = \begin{bmatrix} -5 & -1 & 4 \\ 2 & 1 & -1 \\ -2 & -1 & 4 \end{bmatrix}$

   (e) $A = \begin{bmatrix} 0 & 4 & 3 & 5 \\ 1 & 7 & 2 & -1 \\ 5 & 3 & -1 & 0 \\ -2 & 1 & 7 & 1 \end{bmatrix}$, $B = \begin{bmatrix} 1 & -3 & 4 & 2 \\ 5 & -1 & 6 & 1 \\ -7 & 5 & 2 & 3 \\ 0 & -2 & 5 & 1 \end{bmatrix}$

   (f) $A = \frac{1}{2}\begin{bmatrix} 1 & -2 & 2 & 1 \\ 2 & 1 & 1 & -2 \\ 1 & 0 & 1 & 3 \\ 1 & -1 & 1 & 1 \end{bmatrix}$, $B = \frac{1}{4}\begin{bmatrix} -7 & 1 & -2 & 15 \\ 2 & 2 & 4 & -10 \\ 10 & 2 & 4 & -18 \\ -1 & -1 & 2 & 1 \end{bmatrix}$.

2. Find the reduced row echelon form of each of the following matrices:

   (a) [BB] $\begin{bmatrix} 1 & -1 & 2 \\ -2 & 0 & 1 \\ -3 & 3 & -5 \\ 2 & -1 & 4 \end{bmatrix}$ (b) $\begin{bmatrix} 1 & 1 & 5 & 17 \\ -1 & 0 & -2 & -8 \\ 1 & 5 & 18 & 56 \\ 0 & 3 & 8 & 24 \end{bmatrix}$

   (c) $\begin{bmatrix} 1 & 0 & 2 & 2 & 0 \\ 4 & 1 & 12 & 8 & 4 \\ 0 & -1 & -4 & 1 & -2 \\ 1 & 2 & 10 & 6 & 14 \end{bmatrix}$ (d) $\begin{bmatrix} 0 & -1 & 2 & 1 & 2 & 1 & -1 \\ 0 & 1 & -2 & 2 & 7 & 2 & 4 \\ 0 & -2 & 4 & 3 & 7 & 1 & 0 \\ 0 & 3 & -6 & 1 & 6 & 4 & 1 \end{bmatrix}$.

3. Determine whether or not each of the following matrices has an inverse. Find the inverse whenever this exists.

   (a) [BB] $\begin{bmatrix} 3 & 1 \\ -6 & -3 \end{bmatrix}$ (b) $\begin{bmatrix} 1 & 2 \\ -3 & 4 \end{bmatrix}$ (c) $\begin{bmatrix} -2 & 2 \\ 1 & 1 \end{bmatrix}$

   (d) [BB] $\begin{bmatrix} 2 & 4 & 2 \\ 1 & 2 & 3 \\ 3 & 2 & 1 \end{bmatrix}$ (e) $\begin{bmatrix} -2 & 3 & 4 \\ 1 & 0 & 1 \\ -4 & 3 & 2 \end{bmatrix}$ (f) $\begin{bmatrix} -\frac{1}{4} & \frac{3}{4} & 1 \\ \frac{3}{4} & -\frac{5}{4} & -1 \\ \frac{1}{4} & \frac{1}{4} & 0 \end{bmatrix}$

   (g) $\begin{bmatrix} 1 & -2 & 1 & 3 \\ 1 & 4 & 0 & 2 \\ -3 & 1 & 1 & 5 \\ -4 & 1 & 0 & 1 \end{bmatrix}$ (h) $\begin{bmatrix} 0 & -1 & 1 & 1 \\ 2 & 2 & -2 & 2 \\ 3 & 1 & 1 & -1 \\ 0 & 1 & 4 & -15 \end{bmatrix}$.

4. [BB] **(a)** Write the system $\begin{aligned} x_1 - 2x_2 + 2x_3 &= 3 \\ 2x_1 + x_2 + x_3 &= 0 \\ x_1 \qquad + x_3 &= -2 \end{aligned}$ in the form $A\mathbf{x} = \mathbf{b}$.

   (b) Solve the system by finding A^{-1}.

   (c) Express $\begin{bmatrix} 3 \\ 0 \\ -2 \end{bmatrix}$ as a linear combination of the vectors $\begin{bmatrix} 1 \\ 2 \\ 1 \end{bmatrix}, \begin{bmatrix} -2 \\ 1 \\ 0 \end{bmatrix}, \begin{bmatrix} 2 \\ 1 \\ 1 \end{bmatrix}$.

5. **(a)** Write the system $\begin{aligned} x_2 - x_3 &= 8 \\ x_1 + 2x_2 + x_3 &= 5 \\ x_1 \qquad + x_3 &= -7 \end{aligned}$ in the form $A\mathbf{x} = \mathbf{b}$.

   (b) Solve the system by finding A^{-1}.

   (c) Express $\begin{bmatrix} 8 \\ 5 \\ -7 \end{bmatrix}$ as a linear combination of the vectors $\begin{bmatrix} 0 \\ 1 \\ 1 \end{bmatrix}, \begin{bmatrix} 1 \\ 2 \\ 0 \end{bmatrix}, \begin{bmatrix} -1 \\ 1 \\ 1 \end{bmatrix}$.

6. **(a)** Write the system $\begin{aligned} x_1 + 2x_2 - 4x_3 &= 4 \\ -x_1 + 3x_2 + 2x_3 &= -2 \\ 5x_1 - 2x_2 - x_3 &= 0 \end{aligned}$ in the form $A\mathbf{x} = \mathbf{b}$.

   (b) Solve the system by finding A^{-1}.

   (c) Express $\begin{bmatrix} 4 \\ -2 \\ 0 \end{bmatrix}$ as a linear combination of the vectors $\begin{bmatrix} 1 \\ -1 \\ 5 \end{bmatrix}, \begin{bmatrix} 2 \\ 3 \\ -2 \end{bmatrix}, \begin{bmatrix} -4 \\ 2 \\ -1 \end{bmatrix}$.

7. [BB] Given $\begin{bmatrix} 1 & 2 \\ 3 & 0 \end{bmatrix}^{-1} A \begin{bmatrix} 5 & 1 \\ -1 & 1 \end{bmatrix} = \begin{bmatrix} 1 & 0 \\ 0 & 1 \end{bmatrix}$, find the matrix A.

8. Find a matrix X satisfying $\begin{bmatrix} 4 & 7 & 8 \\ 2 & -3 & 1 \\ 0 & 5 & -3 \end{bmatrix}^{-1} X = \begin{bmatrix} 1 & 2 & 3 \\ -4 & 0 & 6 \\ -2 & 1 & 1 \end{bmatrix} \begin{bmatrix} 2 & 0 & 0 \\ 1 & 4 & -1 \\ 1 & 3 & 0 \end{bmatrix}^{-1}$.

9. Let $A = \begin{bmatrix} 1 & 1 \\ 2 & 4 \end{bmatrix}$ and $C = \begin{bmatrix} 5 & 3 \\ 2 & 2 \end{bmatrix}$. Given that $ABC^{-1} = I$, the identity matrix, find the matrix B.

10. Express each of the following invertible matrices as the product of elementary matrices:

    (a) [BB] $\begin{bmatrix} 0 & -1 \\ 2 & 1 \end{bmatrix}$ **(b)** $\begin{bmatrix} 1 & 4 \\ 1 & -2 \end{bmatrix}$ **(c)** $\begin{bmatrix} -1 & 2 \\ 3 & 1 \end{bmatrix}$

    (d) $\begin{bmatrix} 0 & -2 & 1 \\ 0 & 1 & 0 \\ 1 & -5 & 2 \end{bmatrix}$ **(e)** $\begin{bmatrix} 0 & -1 & -2 \\ 0 & 2 & 6 \\ 1 & -1 & 3 \end{bmatrix}$ **(f)** $\begin{bmatrix} 1 & -1 & 1 \\ 0 & 2 & -1 \\ 2 & 1 & 0 \end{bmatrix}$

    (g) $\begin{bmatrix} 3 & 0 & 3 & -11 \\ 5 & 1 & 5 & -20 \\ 1 & 0 & 1 & -4 \\ -9 & -2 & -10 & 40 \end{bmatrix}$ **(h)** $\begin{bmatrix} 0 & -2 & 0 & 0 & 14 \\ -4 & 0 & 0 & 12 & 1 \\ 0 & 0 & 1 & 0 & 0 \\ 0 & 0 & 2 & 1 & 0 \\ 1 & 0 & 0 & -3 & 0 \end{bmatrix}$.

11. Use Theorem 2.9.13 to determine whether or not each of the given matrices is invertible.

(a) [BB] $\begin{bmatrix} -1 & 2 \\ 3 & -7 \end{bmatrix}$ **(b)** $\begin{bmatrix} -1 & 2 \\ 3 & -6 \end{bmatrix}$ **(c)** $\begin{bmatrix} 1 & 2 & 3 \\ 4 & 5 & 6 \\ 7 & 8 & 9 \end{bmatrix}$

(d) $\begin{bmatrix} 1 & 2 & 3 \\ 4 & 5 & 6 \\ 7 & 8 & 0 \end{bmatrix}$ **(e)** $\begin{bmatrix} 2 & 1 & 8 & 3 \\ -1 & 1 & 12 & 2 \\ 0 & 5 & 0 & 1 \\ 3 & 0 & -9 & 3 \end{bmatrix}$ **(f)** $\begin{bmatrix} 2 & 1 & 8 & 3 \\ -1 & 1 & 12 & 2 \\ 0 & 5 & 0 & 1 \\ 3 & 0 & -9 & 0 \end{bmatrix}$.

Critical Reading

12. **(a)** [BB] Given two $n \times n$ matrices X and Y, how do you determine whether or not $Y = X^{-1}$?

(b) Given $A = \begin{bmatrix} 0 & 2 & 3 \\ 4 & 5 & 6 \\ 7 & 8 & 9 \end{bmatrix}$ and $B = \frac{1}{3}\begin{bmatrix} 3 & 6 & -3 \\ 6 & -21 & 12 \\ -3 & 14 & -8 \end{bmatrix}$, is $B = A^{-1}$? Explain.

For the next two parts, let A be an $n \times n$ matrix and let I denote the $n \times n$ identity matrix.

(c) Suppose $A^3 = 0$. Verify that $(I - A)^{-1} = I + A + A^2$.

(d) Use part (c) to find the inverse of $B = \begin{bmatrix} 1 & 2 & -1 \\ 0 & 1 & 3 \\ 0 & 0 & 1 \end{bmatrix}$.

13. Suppose A is a square matrix with linearly independent columns and $BA = 0$. What can you say about B and why?

14. [BB] Is the factorization of an invertible matrix as the product of elementary matrices unique? Explain.

15. Find the inverse of $A = \begin{bmatrix} 1 & a & b & c \\ 0 & 1 & x & y \\ 0 & 0 & 1 & z \\ 0 & 0 & 0 & 1 \end{bmatrix}$.

16. Suppose A and B are invertible matrices. Show that $B = EA$ where E is the product of elementary matrices.

CHAPTER KEY WORDS AND IDEAS: Here are some technical words and phrases that were used in this chapter. Do you know the meaning of each? If you're not sure, check the glossary or index at the back of the book.

additive inverse of a matrix
associative
augmented matrix
back substitution
block multiplication
commutative
commute (matrices do not)
elementary matrix
elementary row operations
equal matrices
equivalent (linear systems)
forward substitution
Gaussian elimination
has an inverse
homogeneous system
identity matrix
inconsistent system
inverse of a matrix
invertible matrix
linear combination of matrices
linear equation
linearly dependent vectors
linearly independent vectors
lower triangular matrix
LDU factorization
LU factorization
main diagonal of a matrix
matrix
matrix of coefficients
multiplier
parameter
particular solution
partitioned matrix
permutation matrix
pivot
reduced row echelon form
row echelon form
standard basis vectors
symmetric matrix
system of linear equations
transpose
unique solution
upper triangular matrix
zero matrix

Review Exercises for Chapter 2

1. Find x and y so that

(a) $\begin{bmatrix} -x & 5 \\ 0 & y \end{bmatrix} + \begin{bmatrix} 7y & -1 \\ -2y & 2 \end{bmatrix} = \begin{bmatrix} 1 & 4 \\ 6 & -1 \end{bmatrix}$

(b) $\begin{bmatrix} -1 & x & y \\ 2x & 3 & 4 \end{bmatrix}^T = \begin{bmatrix} -1 & 8 \\ 4 & 3 \\ 9 & 4 \end{bmatrix}$.

2. Write down the 3×2 matrix A with $a_{ij} = 2ij - \cos \frac{\pi j}{3}$.

3. Find each of the following products:

(a) $\begin{bmatrix} 1 & 3 & 0 & 4 \\ 2 & 1 & 1 & 0 \end{bmatrix} \begin{bmatrix} 1 & 0 \\ 0 & 1 \\ 1 & 0 \\ 0 & 1 \end{bmatrix}$ **(b)** $\begin{bmatrix} 1 & 3 & 2 \\ 0 & 2 & 1 \\ 1 & 3 & 1 \end{bmatrix} \begin{bmatrix} 1 & 2 \\ 0 & 1 \\ 1 & 1 \end{bmatrix}$ **(c)** $\begin{bmatrix} 1 \\ 2 \\ 3 \\ 4 \end{bmatrix} \begin{bmatrix} 1 \\ 2 \\ 3 \\ 4 \end{bmatrix}^T$

(d) $\begin{bmatrix} 1 \\ 2 \\ 3 \\ 4 \end{bmatrix}^T \begin{bmatrix} 1 \\ 2 \\ 3 \\ 4 \end{bmatrix}$ **(e)** $\begin{bmatrix} x & y & z \end{bmatrix} \begin{bmatrix} a & d & e \\ d & b & f \\ e & f & c \end{bmatrix} \begin{bmatrix} x \\ y \\ z \end{bmatrix}$.

4. Find A^2 and A^3 with $A = \begin{bmatrix} 1 & 2 & 3 \\ 2 & 1 & 0 \\ 1 & 0 & 1 \end{bmatrix}$.

5. Given $A = \begin{bmatrix} 2 & 1 \\ 1 & 1 \\ 3 & -5 \end{bmatrix}$, $B = \begin{bmatrix} -1 & 3 \\ 2 & 4 \end{bmatrix}$ and $C = \begin{bmatrix} 1 & 2 \\ 8 & -5 \\ 0 & 2 \end{bmatrix}$, if possible, find **(a)** $AB - C$; **(b)** B^{-1}; **(c)** a matrix X such that $XA = X$.

6. **(a)** Given $A = \begin{bmatrix} 1 & 2 \\ 3 & 5 \end{bmatrix}$, $B = \begin{bmatrix} -1 & 0 \\ 2 & 4 \end{bmatrix}$, $C = \begin{bmatrix} -1 & -5 \\ 1 & 4 \end{bmatrix}$ and $XA - B = C$, find X.

   (b) Given A, B, C as in part (a) and $AX + 2B = C$, find X.

7. Given $A = \begin{bmatrix} 1 & 2 & 3 \\ 4 & 5 & 6 \end{bmatrix}$, $B = \begin{bmatrix} -1 & 2 \\ 1 & 0 \\ 1 & 3 \end{bmatrix}$ and $C = \begin{bmatrix} -1 & 0 & 1 \\ 2 & 1 & 0 \end{bmatrix}$, find each of the following, if possible. If not possible, explain why not.
   (a) $A + 2C$; **(b)** $(AB)^T$; **(c)** $B^T A^T$; **(d)** $AC + B$.

8. Let $A = \begin{bmatrix} 1 & 2 \\ -1 & 11 \end{bmatrix}$ and $B = \begin{bmatrix} -2 & 3 \\ -2 & 1 \end{bmatrix}$.

   (a) Find AB, BA, A^2, B^2, $A - B$, $(A - B)^2$, $A^2 - 2AB + B^2$ and $A^2 - AB - BA + B^2$.

   (b) In general, for arbitrary matrices A and B, is it the case that $(A - B)^2 = A^2 - 2AB + B^2$? If not, find a correct expansion of $(A - B)^2$.

9. Complete the following sentence:
   "A linear combination of the columns of a matrix A is ____________."

10. Suppose $A = \begin{bmatrix} \mathsf{x}_1 & \mathsf{x}_2 & \mathsf{x}_3 & \mathsf{x}_4 \\ \downarrow & \downarrow & \downarrow & \downarrow \end{bmatrix}$ and $B = \begin{bmatrix} -2 & 0 & -1 & 0 \\ 0 & 0 & 0 & 0 \\ 0 & 1 & 0 & 0 \\ 0 & 0 & 3 & 4 \end{bmatrix}$.

    Find AB and explain.

11. Let A be an $n \times n$ matrix and let x, y, z be vectors in R^n. Suppose $A\mathsf{x} = 3\mathsf{y}$, $A\mathsf{y} = \mathsf{x} - 2\mathsf{z}$, and $A\mathsf{z} = \mathsf{x} + \mathsf{y} + \mathsf{z}$. Let $B = \begin{bmatrix} \mathsf{x} & \mathsf{y} & \mathsf{z} \\ \downarrow & \downarrow & \downarrow \end{bmatrix}$ be the $n \times 3$ matrix whose columns are x, y, z.

    (a) What is AB?

    (b) Find a 3×3 matrix X such that $AB = BX$.

12. Use elementary row operations to move the matrix $\begin{bmatrix} 0 & -1 & 2 & 1 & 2 & 1 & -1 \\ 0 & 1 & -2 & 2 & 7 & 2 & 4 \\ 0 & -2 & 4 & 3 & 7 & 1 & 0 \\ 0 & 3 & -6 & 1 & 6 & 4 & 1 \end{bmatrix}$ to row echelon form. Identify the pivots and the pivot columns.

13. Let $A = \begin{bmatrix} 3 & 1 & -2 \\ 2 & -2 & 0 \\ -1 & 1 & 2 \end{bmatrix}$ and $B = \begin{bmatrix} 1 & 1 & 1 \\ 1 & -1 & 1 \\ 0 & 1 & 2 \end{bmatrix}$. Use the fact that $AB = 4I$ to solve the system $\begin{aligned} x + y + z &= 2 \\ x - y + z &= 0 \\ y + 2z &= -1. \end{aligned}$

14. (a) Solve the system $\begin{aligned} x - 2y + 3z &= 2 \\ 2x - 5y + &= -2. \end{aligned}$

    (b) Observe that the point $P(1,1,1)$ is on each of the planes defined by the equations in part (a). Use this fact and the notion of cross product to find the equation of the line of intersection of the two planes.

    (c) Show that the answers to (a) and (b) describe the same line.

15. Find the equation of the line where the planes with equations $x - 3y + 5z = 2$ and $5x + y - 7z = 6$ intersect.

16. (a) Find a row echelon form for $A = \begin{bmatrix} 1 & -1 & -1 & 2 \\ 0 & 0 & 5 & 10 \\ 2 & -1 & 1 & 9 \end{bmatrix}$ without interchanging any rows.

    (b) Find the reduced row echelon form of A.

17. Solve each of the following systems of equations, in each case expressing any solution as a vector or a linear combination of vectors.

    (a) $\begin{aligned} x + 3y &= -1 \\ 4x - 3y &= 7 \\ x - 2y &= 4 \end{aligned}$ (b) $\begin{aligned} x + 2y - z &= 1 \\ 2x + 2y + 4z &= 1 \\ x + 3y - 3z &= 1 \end{aligned}$

    (c) $\begin{aligned} -2x + y + z &= 0 \\ y - 3z &= 0 \\ -4x + 3y - z &= 0 \end{aligned}$ (d) $\begin{aligned} x + 3y - z &= -1 \\ 2x + y - 2z &= 4 \\ 3x + 4y - z &= 2 \end{aligned}$

    (e) $\begin{aligned} x_1 + 2x_2 - x_3 + 2x_4 &= 4 \\ x_2 - x_3 + 4x_4 - x_5 &= -1 \\ 2x_1 + x_2 + x_3 - 2x_4 + 3x_5 &= 5 \end{aligned}$

    (f) $\begin{aligned} x_1 + x_2 + x_3 + x_4 &= 4 \\ 2x_1 - x_2 + 5x_3 + x_4 &= -1 \\ 4x_1 + 3x_2 + x_3 - x_4 &= 3 \\ 5x_1 + 6x_2 - 4x_3 - 4x_4 &= 3 \end{aligned}$

    (g) $\begin{aligned} x_1 + x_2 + x_3 + x_4 &= 10 \\ x_1 - x_2 - x_3 + 2x_4 &= 4 \\ 2x_1 + x_2 + 3x_3 - x_4 &= 9 \\ 5x_2 + 7x_3 - 6x_4 &= 6 \end{aligned}$

    (h) $\begin{aligned} x_1 + 2x_2 + 3x_3 + 5x_5 &= 6 \\ x_3 + 3x_5 &= 7 \\ x_1 + 2x_2 + 2x_3 + 2x_4 + 2x_5 &= 3 \\ 2x_1 + 4x_2 + 6x_3 + 2x_4 + 10x_5 &= 16. \end{aligned}$

18. **(a)** Determine whether or not $\begin{bmatrix} -1 \\ 0 \\ 6 \\ 3 \end{bmatrix}$ is a linear combination of $\mathsf{u} = \begin{bmatrix} -1 \\ 0 \\ 2 \\ 1 \end{bmatrix}$, $\mathsf{v} = \begin{bmatrix} 0 \\ 1 \\ 3 \\ 2 \end{bmatrix}$ and $\mathsf{w} = \begin{bmatrix} -1 \\ 1 \\ 0 \\ 1 \end{bmatrix}$.

    (b) Do u, v and w span R^4? Justify your answer.

19. Determine whether or not $\mathsf{w} = \begin{bmatrix} 10 \\ 6 \\ -4 \end{bmatrix}$ is in the plane spanned by $\mathsf{u} = \begin{bmatrix} 1 \\ 2 \\ 1 \end{bmatrix}$ and $\mathsf{v} = \begin{bmatrix} 2 \\ 1 \\ -4 \end{bmatrix}$.

20. Express the zero vector as a nontrivial linear combination of $\mathsf{v}_1 = \begin{bmatrix} -4 \\ 2 \\ 0 \\ -3 \end{bmatrix}$, $\mathsf{v}_2 = \begin{bmatrix} 1 \\ 3 \\ 4 \\ 1 \end{bmatrix}$, $\mathsf{v}_3 = \begin{bmatrix} -2 \\ 8 \\ 8 \\ 3 \end{bmatrix}$ and $\mathsf{v}_4 = \begin{bmatrix} 3 \\ -5 \\ -4 \\ -3 \end{bmatrix}$.
    What does this say about the linear dependence/independence of $\mathsf{v}_1, \ldots, \mathsf{v}_4$?

21. Determine whether each of the following sets of vectors is linearly independent or linearly dependent. In the case of linear dependence, express the zero vector explicitly as a nontrivial linear combination of the given vectors.

    (a) $\mathsf{v}_1 = \begin{bmatrix} 1 \\ 0 \\ 1 \end{bmatrix}$, $\mathsf{v}_2 = \begin{bmatrix} 1 \\ 2 \\ -1 \end{bmatrix}$, $\mathsf{v}_3 = \begin{bmatrix} 2 \\ 1 \\ -4 \end{bmatrix}$ **(b)** $\mathsf{v}_1 = \begin{bmatrix} 1 \\ 2 \\ 3 \end{bmatrix}$, $\mathsf{v}_2 = \begin{bmatrix} 2 \\ 1 \\ -1 \end{bmatrix}$, $\mathsf{v}_3 = \begin{bmatrix} -1 \\ 4 \\ 11 \end{bmatrix}$

    (c) $\mathsf{v}_1 = \begin{bmatrix} -2 \\ 3 \\ 3 \\ 1 \end{bmatrix}$, $\mathsf{v}_2 = \begin{bmatrix} 3 \\ 4 \\ 0 \\ 7 \end{bmatrix}$, $\mathsf{v}_3 = \begin{bmatrix} 2 \\ 1 \\ 1 \\ 3 \end{bmatrix}$

    (d) $\mathsf{v}_1 = \begin{bmatrix} 1 \\ 1 \\ 2 \\ 7 \\ 8 \\ -2 \end{bmatrix}$, $\mathsf{v}_2 = \begin{bmatrix} -2 \\ 3 \\ 1 \\ 6 \\ 4 \\ -11 \end{bmatrix}$, $\mathsf{v}_3 = \begin{bmatrix} 1 \\ 2 \\ 3 \\ 11 \\ 12 \\ -5 \end{bmatrix}$

    (e) $\mathsf{v}_1 = \begin{bmatrix} -4 \\ 2 \\ 0 \\ -3 \end{bmatrix}$, $\mathsf{v}_2 = \begin{bmatrix} 1 \\ 3 \\ 4 \\ 1 \end{bmatrix}$, $\mathsf{v}_3 = \begin{bmatrix} -2 \\ 8 \\ 8 \\ 3 \end{bmatrix}$, $\mathsf{v}_4 = \begin{bmatrix} 3 \\ -5 \\ -4 \\ -3 \end{bmatrix}$.

22. If A is an $n \times n$ matrix and x is a vector in R^n, what is the size of the matrix $\mathsf{x}^T A \mathsf{x}$? Explain.

23. Suppose A is a matrix such that $A^2 + A = I$, the identity matrix.

(a) A must be square. Why?

(b) A must be invertible. Why?

24. Let $A = \begin{bmatrix} 4 & -1 \\ 1 & 5 \end{bmatrix}$. Does there exist a **nonzero** vector $\mathsf{x} = \begin{bmatrix} x \\ y \end{bmatrix}$ such that $A\mathsf{x} = 2\mathsf{x}$? Explain your answer.

25. (a) Solve the system $\begin{aligned} x_2 + x_3 - x_4 &= 1 \\ 2x_1 \quad + 3x_3 + x_4 &= 2 \end{aligned}$ expressing the solution as the sum of a particular solution x_p to the given system and a solution x_h of the corresponding homogeneous system. Identify x_p and x_h.

(b) Find numbers a, b, c, d so that $\begin{bmatrix} 1 \\ 2 \end{bmatrix} = a\begin{bmatrix} 0 \\ 2 \end{bmatrix} + b\begin{bmatrix} 1 \\ 0 \end{bmatrix} + c\begin{bmatrix} 1 \\ 3 \end{bmatrix} + d\begin{bmatrix} -1 \\ 1 \end{bmatrix}$ is a linear combination of the columns of A. (There are many possibilities.)

26. Find a value or values of t (if any) for which the system $\begin{aligned} x + y &= 1 \\ tx + y &= t \\ (1+t)x + 2y &= 3 \end{aligned}$ has a solution and find any solution that may exist.

27. Find all value(s) of a such that the linear system

$$\begin{aligned} x_1 + ax_2 - x_3 &= 1 \\ -2x_1 - ax_2 + 3x_3 &= -4 \\ -x_1 - ax_2 + ax_3 &= a+1 \end{aligned}$$

has (i) no solution; (ii) a unique solution; (iii) infinitely many solutions.

28. Find conditions on a and b that guarantee that the system $\begin{aligned} x - 3y &= a \\ -2x + 6y &= b \end{aligned}$ has (i) a unique solution; (ii) no solution; (iii) infinitely many solutions.

29. Explain why the system $\begin{aligned} 3x - 4y &= 2 \\ mx + 5y &= -1 \end{aligned}$ has a unique solution for any m.

30. (a) Write the system $\begin{aligned} y + z &= a \\ -x - y + 4z &= b \\ 2x \quad - 3z &= c \end{aligned}$ in the form $A\mathsf{x} = \mathsf{b}$.

(b) Under what condition or conditions on a, b and c does the system have a solution?

(c) If and when the system has a solution, is it unique?

31. Answer all parts of Exercise 30 for the system $\begin{aligned} y + 2z &= a \\ -x - y + 4z &= b \\ 2x \quad - 12z &= c. \end{aligned}$

32. How do the solutions of $\begin{aligned} ax \quad + z &= 1 \\ x + y \quad &= a \\ ay + z &= 1. \end{aligned}$ depend on the value of a?

33. For which a, if any, is $\begin{aligned} 2x_2 + 3x_3 + x_4 &= 11 \\ x_1 + 4x_2 + 6x_3 + 2x_4 &= 8 \\ 2x_1 + 3x_2 + 4x_3 + 2x_4 &= 3 \\ -5x_1 + 3x_2 + 5x_3 + x_4 &= a \end{aligned}$ inconsistent?

34. Consider the system

$$\begin{aligned} x + 2y + kz + \quad\quad w &= 0 \\ 2x + 3y - 2z + (1-k)w &= k \\ x + 2y - z + (2k+1)w &= k+1 \\ kx + y + z + (k^2-2)w &= 3. \end{aligned}$$

(a) Find a value or values of k so that this system has no solution.

(b) Find a value or values of k so that this system has infinitely many solutions.

(c) For each value of k found in (b), find all solutions (in vector form).

35. Consider this linear system:

$$\begin{aligned} -3x_1 + 8x_2 - 5x_3 + 19x_4 - 9x_5 - 7x_6 &= 5 \\ x_1 - 3x_2 + 2x_3 - 9x_4 + 5x_5 - 2x_6 &= -3 \\ -x_1 + 2x_2 - x_3 + 2x_4 \quad\quad - 8x_6 &= 0 \\ -5x_1 + 13x_2 - 8x_3 + 29x_4 - 13x_5 - 15x_6 &= 3. \end{aligned}$$

(a) Write this in the form $A\mathsf{x} = \mathsf{b}$.

(b) Solve the system by reducing the augmented matrix of coefficients to a matrix in row echelon form. Express your answer in the form $\mathsf{x} = \mathsf{x}_p + \mathsf{x}_h$, where x_p is a particular solution vector and x_h is the general solution to the homogeneous system $A\mathsf{x} = \mathsf{0}$.

(c) Express $\begin{bmatrix} 5 \\ -3 \\ 0 \\ 3 \end{bmatrix}$ as a linear combination of the columns of A.

36. **(a)** Let $P = \begin{bmatrix} 0 & 0 & 1 & 0 \\ 1 & 0 & 0 & 0 \\ 0 & 0 & 0 & 1 \\ 0 & 1 & 0 & 0 \end{bmatrix}$ and let A be a $4 \times n$ matrix. What kind of a matrix is P? Explain the connection between PA and A.

(b) Let P be an $n \times n$ permutation matrix and let J be the $n \times n$ matrix consisting entirely of 1s. Find PJ, JP^T and PJP^T.

(c) Suppose A and B are matrices such that $AB = B$. Does this imply $A = I$? Explain.

37. In each case, find the inverse if such exists.

(a) $\begin{bmatrix} 1 & 2 & 1 \\ 3 & 1 & 1 \\ 1 & 1 & 1 \end{bmatrix}$ **(b)** $\begin{bmatrix} 1 & 1 & 2 \\ 3 & 2 & 3 \\ 4 & 2 & 1 \end{bmatrix}$

(c) $\begin{bmatrix} 2 & -1 & 3 & 3 \\ -8 & 4 & -6 & -5 \\ 0 & 1 & -1 & 2 \\ -3 & 2 & -2 & 0 \end{bmatrix}$ **(d)** $\begin{bmatrix} 0 & 1 & 0 & 0 & 0 \\ 0 & 0 & 0 & 1 & 0 \\ 1 & 0 & 0 & 0 & 0 \\ 0 & 0 & 0 & 0 & 1 \\ 0 & 0 & 1 & 0 & 0 \end{bmatrix}$.

38. Find the inverse of $\begin{bmatrix} 1 & 3 & 1 \\ -2 & -5 & -2 \\ 3 & 14 & 4 \end{bmatrix}$ and use this to solve the system

$$\begin{aligned} x_1 + 3x_2 + x_3 &= 1 \\ -2x_1 - 5x_2 - 2x_3 &= 1 \\ 3x_1 + 14x_2 + 4x_3 &= 1. \end{aligned}$$

39. **(a)** Suppose A, B, C and $M = \begin{bmatrix} A & B \\ 0 & C \end{bmatrix}$ are square matrices with A and C invertible. Show that M is invertible by finding its inverse (as a 2×2 partitioned matrix).

    (b) Find the inverses of (i) $\begin{bmatrix} -1 & 3 & 2 & -5 \\ -4 & 10 & 2 & 9 \\ 0 & 0 & 1 & -3 \\ 0 & 0 & -2 & 3 \end{bmatrix}$ and (ii) $\begin{bmatrix} 2 & -4 & 3 & -1 \\ 1 & -1 & -5 & 1 \\ 0 & 0 & -9 & 5 \\ 0 & 0 & -3 & 2 \end{bmatrix}$.

40. Suppose A, B, C and D are square matrices of the same size with A, C and $C^{-1} + DA^{-1}B$ all invertible. Show that $A + BCD$ is invertible by proving that $(A + BCD)^{-1} = A^{-1} - A^{-1}B(C^{-1} + DA^{-1}B)^{-1}DA^{-1}$.
    [Hint: Let $X = (C^{-1} + DA^{-1}B)^{-1}$.]

41. Is the inverse of a symmetric matrix symmetric? Explain.

42. **(a)** What is meant by an *elementary matrix*?

    (b) Given that E is an elementary matrix such that $E\begin{bmatrix} 1 & 4 \\ 5 & 3 \end{bmatrix} = \begin{bmatrix} 1 & 4 \\ 0 & -17 \end{bmatrix}$, write down E and E^{-1} without any calculation.

43. Write a short note, in good clear English, explaining why every invertible matrix is the product of elementary matrices. Illustrate your answer with reference to the matrix $A = \begin{bmatrix} 1 & -2 \\ 3 & 1 \end{bmatrix}$.

44. Express $A = \begin{bmatrix} 1 & 2 \\ 3 & 4 \end{bmatrix}$ as the product of elementary matrices.

45. Find the inverse of each of the following matrices simply by writing it down. No calculation is necessary, but explain your reasoning.

    (a) $\begin{bmatrix} 0 & 1 \\ 1 & 0 \end{bmatrix}$ **(b)** $\begin{bmatrix} 0 & 0 & 1 \\ 1 & 0 & 0 \\ 0 & 1 & 0 \end{bmatrix}$

    (c) $\begin{bmatrix} 0 & 0 & 1 \\ 0 & 1 & 0 \\ 1 & 0 & 0 \end{bmatrix}$ **(d)** $\begin{bmatrix} 0 & 1 & 0 & 0 \\ 0 & 0 & 1 & 0 \\ 0 & 0 & 0 & 1 \\ 1 & 0 & 0 & 0 \end{bmatrix}$.

46. **(a)** Reduce $A = \begin{bmatrix} 1 & 1 \\ 0 & 3 \\ -2 & 1 \end{bmatrix}$ to row echelon form.

    (b) Show that the equation $A\mathsf{x} = \mathsf{b}$ has a unique solution for any $\mathsf{b} = \begin{bmatrix} b_1 \\ b_2 \\ b_3 \end{bmatrix}$ provided $2b_1 - b_2 + b_3 = 0$.

    (c) Show that there is a unique matrix B such that $AB = \begin{bmatrix} 1 & -1 & -2 \\ 1 & 2 & -3 \\ -1 & 4 & 1 \end{bmatrix}$.

47. Find an LU factorization of $\begin{bmatrix} 2 & -4 & 6 & -12 \\ -3 & 9 & -3 & 36 \\ 10 & -25 & 41 & -41 \end{bmatrix}$ using only the third elementary row operation.

48. Use the factorization $\begin{bmatrix} 1 & -2 & 3 \\ -5 & 0 & -2 \\ 3 & 0 & 1 \end{bmatrix} = \begin{bmatrix} 1 & 0 & 0 \\ -5 & 1 & 0 \\ 3 & -\frac{3}{5} & 1 \end{bmatrix} \begin{bmatrix} 1 & -2 & 3 \\ 0 & -10 & 13 \\ 0 & 0 & -\frac{1}{5} \end{bmatrix}$ to solve
$$\begin{aligned} x - 2y + 3z &= 1 \\ -5x \qquad - 2z &= 2 \\ 3x \qquad + z &= 3. \end{aligned}$$

49. **(a)** Find an LU factorization of $A = \begin{bmatrix} 2 & 4 & 6 & 8 \\ 0 & -1 & 2 & 5 \\ 1 & 3 & 6 & 9 \end{bmatrix}$.

(b) Use your factorization of A to solve the system $A\begin{bmatrix} x_1 \\ x_2 \\ x_3 \\ x_4 \end{bmatrix} = \begin{bmatrix} 4 \\ 3 \\ -6 \end{bmatrix}$.

50. **(a)** Find an LU factorization of $A = \begin{bmatrix} 1 & 2 & 3 \\ 4 & -6 & -6 \\ -7 & -7 & 9 \end{bmatrix}$ and express L as the product of elementary matrices.

(b) Use the factorization in (a) to solve $A\mathsf{x} = \mathsf{b}$ with $\mathsf{b} = \begin{bmatrix} 0 \\ 1 \\ 0 \end{bmatrix}$.

51. Find an LDU factorization of $A = \begin{bmatrix} -2 & 4 \\ 0 & 5 \\ 3 & 1 \end{bmatrix}$ or explain why there is no such factorization.

52. Find an LDU factorization of $A = \begin{bmatrix} -1 & -1 & 0 & -4 \\ 2 & -1 & -6 & 8 \\ -1 & -1 & 2 & 6 \\ -3 & -6 & -6 & -8 \\ 0 & 6 & 14 & 10 \end{bmatrix}$.

53. Find an LDU factorization of the symmetric matrix $A = \begin{bmatrix} 5 & 2 & 2 \\ 2 & 2 & -4 \\ 2 & -4 & 2 \end{bmatrix}$.

54. Factor $A = PLDU$ with $A = \begin{bmatrix} 0 & 2 & 1 & 3 & 1 \\ 3 & 1 & 2 & 1 & 1 \\ 2 & 0 & -1 & 1 & 1 \\ 4 & 1 & 2 & 2 & 1 \end{bmatrix}$.

55. **(a)** Show that any square matrix A can be factored $A = BS$ with S symmetric. [Hint: Start with an LDU or a PLDU factorization.]

(b) Factor $A = BS$ as in part (a) with $A = \begin{bmatrix} -1 & 2 & 4 \\ 4 & 0 & 2 \\ -2 & 5 & 10 \end{bmatrix}$.

Chapter 3

Determinants, Eigenvalues, Eigenvectors

3.1 The Determinant of a Matrix

In Section 1.3, when we introduced the notion of cross product, we defined the determinant of a 2×2 matrix and showed that this number was closely connected to the invertibility of the matrix: Specifically, $A = \begin{bmatrix} a & b \\ c & d \end{bmatrix}$ is invertible if and only if $\det A = 0$—see Proposition 2.3.10. In this section, we define the determinant of an $n \times n$ matrix A[1] using the notation $\det A$ or $|A|$ for the determinant of A and show that in general A is invertible if and only if $\det A \neq 0$. The determinant of the 1×1 matrix $[a]$ is the number a. If A is the 2×2 matrix $\begin{bmatrix} a & b \\ c & d \end{bmatrix}$, then $\det A = ad - bc$. The determinant of the general $n \times n$ matrix is defined "inductively."

3.1.1 **Definitions.** With $n \geq 2$, the (i, j) *minor* of an $n \times n$ matrix A, denoted m_{ij}, is the determinant of the $(n-1) \times (n-1)$ matrix obtained from A by deleting row i and column j. The (i, j) *cofactor* of A is $(-1)^{i+j} m_{ij}$ and denoted c_{ij}.

This may seem like a mouthful, but as with many new ideas, the concepts of "minor" and "cofactor" will soon seem second nature. Let's look at an example.

[1]Only square matrices have determinants.

3.1.2 Example. The $(1,1)$ minor of $A = \begin{bmatrix} 1 & 2 & 3 \\ 0 & -2 & 2 \\ 3 & 7 & -4 \end{bmatrix}$ is the determinant of

$$\begin{bmatrix} -2 & 2 \\ 7 & -4 \end{bmatrix},$$

which is what remains when we remove row one and column one of A. So $m_{11} = -2(-4) - 2(7) = -6$, and the $(1,1)$ cofactor is $c_{11} = (-1)^{1+1}m_{11} = m_{11} = -6$. The $(2,3)$ minor of A is the determinant of

$$\begin{bmatrix} 1 & 2 \\ 3 & 7 \end{bmatrix},$$

which is what remains after removing row two and column three of A. Thus $m_{23} = 7 - 6 = 1$ and $c_{23} = (-1)^{2+3}m_{23} = -m_{23} = -1$.

Continuing in this way, we can find a minor and a cofactor corresponding to every position (i,j) of the matrix A. The *matrix of minors* of A is

$$M = \begin{bmatrix} -6 & -6 & 6 \\ -29 & -13 & 1 \\ 10 & 2 & -2 \end{bmatrix}$$

and the *matrix of cofactors* is

$$C = \begin{bmatrix} -6 & 6 & 6 \\ 29 & -13 & -1 \\ 10 & -2 & -2 \end{bmatrix}.$$

Since the only powers of -1 are $+1$ and -1, the factor $(-1)^{i+j}$ that appears in the definition of cofactor is ± 1. Thus any cofactor is either plus or minus the corresponding minor. Whether it is plus or minus is most easily remembered by the following pattern, which shows the values of $(-1)^{i+j}$ for all positive integers i and j.

$$\begin{matrix} + & - & + & - & \cdots \\ - & + & - & + & \cdots \\ + & - & + & - & \cdots \\ - & + & - & + & \cdots \\ \vdots & \vdots & & & \end{matrix} \qquad (1)$$

Each "+" means "multiply by $+1$," that is, leave the minor unchanged. Each "−" means "multiply by -1," that is, change the sign of the minor.

3.1.3 Example. Let $A = \begin{bmatrix} -1 & 0 & 2 \\ 3 & -7 & 8 \\ -4 & 4 & -5 \end{bmatrix}$. The matrix of minors is

$M = \begin{bmatrix} 3 & 17 & -16 \\ -8 & 13 & -4 \\ 14 & -14 & 7 \end{bmatrix}$. To obtain the matrix of cofactors of A, we modify the entries of M according to the pattern given in (1). Thus, we leave the

$(1,1)$ entry of M alone, but multiply the $(1,2)$ entry by -1. We leave the $(1,3)$ entry alone, but multiply the $(2,3)$ entry by -1, and so forth. The matrix of cofactors is $C = \begin{bmatrix} 3 & -17 & -16 \\ 8 & 13 & 4 \\ 14 & 14 & 7 \end{bmatrix}$. ☺

As hard as this may be to believe, the **transpose of the matrix of cofactors**, also known as the *adjoint*, is an extraordinary object!

3.1.4 Definition. The *adjoint* of a matrix A, denoted $\operatorname{adj}(A)$, is the transpose of its matrix of cofactors.

For example, the matrix of minors of $A = \begin{bmatrix} a & b \\ c & d \end{bmatrix}$ is $M = \begin{bmatrix} d & c \\ b & a \end{bmatrix}$, the matrix of cofactors is $C = \begin{bmatrix} d & -c \\ -b & a \end{bmatrix}$ and the adjoint is

$$\operatorname{adj}(A) = C^T = \begin{bmatrix} d & -b \\ -c & a \end{bmatrix}.$$

The reader might check that the product of A and its adjoint is $A\operatorname{adj}(A) = (ad - bc)I = (\det A)I$—see Proposition 2.3.10.

3.1.5 Example. Continuing Example 3.1.3 where $A = \begin{bmatrix} -1 & 0 & 2 \\ 3 & -7 & 8 \\ -4 & 4 & -5 \end{bmatrix}$, the matrix of cofactors was $C = \begin{bmatrix} 3 & -17 & -16 \\ 8 & 13 & 4 \\ 14 & 14 & 7 \end{bmatrix}$, so $\operatorname{adj}(A) = C^T = \begin{bmatrix} 3 & 8 & 14 \\ -17 & 13 & 14 \\ -16 & 4 & 7 \end{bmatrix}$. The product of A and its adjoint is

$$\begin{bmatrix} -1 & 0 & 2 \\ 3 & -7 & 8 \\ -4 & 4 & -5 \end{bmatrix} \begin{bmatrix} 3 & 8 & 14 \\ -17 & 13 & 14 \\ -16 & 4 & 7 \end{bmatrix} = \begin{bmatrix} -35 & 0 & 0 \\ 0 & -35 & 0 \\ 0 & 0 & -35 \end{bmatrix} = -35I,$$

a scalar multiple of the identity matrix, as with 2×2 matrices. ☺

READING CHECK 1. If A is 2×2, why is $A(\operatorname{adj} A)$ a scalar multiple of I?

3.1.6 Example. The matrix of cofactors of the matrix $A = \begin{bmatrix} 1 & 2 & 3 \\ 0 & -2 & 2 \\ 3 & 7 & -4 \end{bmatrix}$ is $C = \begin{bmatrix} -6 & 6 & 6 \\ 29 & -13 & -1 \\ 10 & -2 & -2 \end{bmatrix}$—see Example 3.1.2—so $\operatorname{adj}(A) = C^T = \begin{bmatrix} -6 & 29 & 10 \\ 6 & -13 & -2 \\ 6 & -1 & -2 \end{bmatrix}$.

The product of A and its adjoint is

$$\begin{bmatrix} 1 & 2 & 3 \\ 0 & -2 & 2 \\ 3 & 7 & -4 \end{bmatrix} \begin{bmatrix} -6 & 29 & 10 \\ 6 & -13 & -2 \\ 6 & -1 & -2 \end{bmatrix} = 24 \begin{bmatrix} 1 & 0 & 0 \\ 0 & 1 & 0 \\ 0 & 0 & 1 \end{bmatrix} = 24I,$$

again a scalar multiple of the identity matrix. ☺

It's a fact, which we shall not justify, that

3.1.7 A square matrix A commutes with its adjoint, and the product $A\,\text{adj}(A) = \text{adj}(A)A = kI$ is a scalar multiple of the identity matrix.

(See W. Keith Nicholson, *Linear Algebra with Applications*, Fourth edition, McGraw-Hill Ryerson, Toronto, 2003, for a proof of this fact and other facts that may appear in this chapter without proof.)

The number k that appears in 3.1.7 is called the *determinant* of A. Thus, the determinant of a matrix A is defined by the equation

3.1.8 $$A\,\text{adj}(A) = \text{adj}(A)A = (\det A)I.$$

3.1.9 Examples. The determinant of the matrix A in Example 3.1.6 is 24—since $A\,\text{adj}(A) = 24I$—and the determinant of the matrix in Example 3.1.5 is -35. ☺

Fortunately, it isn't necessary to compute the entire matrix of cofactors if we are only interested in finding the determinant of a matrix. Suppose $A = [a_{ij}]$ is the general 3×3 matrix,

$$A = \begin{bmatrix} a_{11} & a_{12} & a_{13} \\ a_{21} & a_{22} & a_{23} \\ a_{31} & a_{32} & a_{33} \end{bmatrix}.$$

The matrix of cofactors is

$$C = \begin{bmatrix} c_{11} & c_{12} & c_{13} \\ c_{21} & c_{22} & c_{23} \\ c_{31} & c_{32} & c_{33} \end{bmatrix},$$

so

$$A\,\text{adj}(A) = AC^T$$
$$= \begin{bmatrix} a_{11} & a_{12} & a_{13} \\ a_{21} & a_{22} & a_{23} \\ a_{31} & a_{32} & a_{33} \end{bmatrix} \begin{bmatrix} c_{11} & c_{21} & c_{31} \\ c_{12} & c_{22} & c_{32} \\ c_{13} & c_{23} & c_{33} \end{bmatrix} = \begin{bmatrix} \det A & 0 & 0 \\ 0 & \det A & 0 \\ 0 & 0 & \det A \end{bmatrix}.$$

The $(1,1)$, the $(2,2)$ and the $(3,3)$ entries of AC^T all equal $\det A$:

$$\begin{aligned}a_{11}c_{11} + a_{12}c_{12} + a_{13}c_{13} &= \det A\\ a_{21}c_{21} + a_{22}c_{22} + a_{23}c_{23} &= \det A\\ a_{31}c_{31} + a_{32}c_{32} + a_{33}c_{33} &= \det A.\end{aligned}$$

These calculations show that the determinant of a 3×3 matrix A is the dot product of any row of A with the corresponding row of cofactors. This holds in general. Moreover, we shall see in the next section that the determinant of a matrix is the determinant of its transpose (3.2.24), so what goes for rows goes for columns too.

3.1.10 Laplace Expansion: The determinant of a matrix A is the dot product of any row (or column) of A and the corresponding row (or column) of cofactors.

READING CHECK 2. Show that the dot product of any **column** of a 3×3 matrix A with the corresponding column of cofactors is again $\det A$. [Hint: Use $C^T A = (\det A)I$.]

As noted, this method of finding a determinant via a dot product of a row or column with the corresponding cofactors is called the "Laplace Expansion" after the mathematician and astronomer Pierre-Simon Laplace (1749–1827).

Have another look at Example 3.1.5. The dot product of the first column of A and the first column of C is -35:

$$\begin{bmatrix}-1\\3\\-4\end{bmatrix} \cdot \begin{bmatrix}3\\8\\14\end{bmatrix} = -35 = \det A.$$

The dot product of the second row of A and the second row of C is -35:[2]

$$\begin{bmatrix}3\\-7\\8\end{bmatrix} \cdot \begin{bmatrix}8\\13\\4\end{bmatrix} = -35 = \det A.$$

3.1.11 Example. Suppose we want to find the determinant of the matrix $A = \begin{bmatrix}-1 & 0 & 2\\ 3 & -7 & 8\\ -4 & 4 & -5\end{bmatrix}$ that we met in Example 3.1.5. Take any row or column of A and compute the dot product with the corresponding row of cofactors. Given this degree of choice, we should take advantage of the 0 in the $(1,2)$

[2]Strictly speaking, we shouldn't be talking about the dot product of rows since we have only defined the dot product of vectors and, in this book, vectors are columns. On the other hand, we think the reader will know what we mean.

position and choose either the first row of A or second column of A. If we choose to *expand by cofactors of the first row*, we get

$$-1\begin{vmatrix} -7 & 8 \\ 4 & -5 \end{vmatrix} + 2\begin{vmatrix} 3 & -7 \\ -4 & 4 \end{vmatrix} = -1(3) + 2(-16) = -35$$

whereas, if we *expand by cofactors of the second column*, we get

$$-7\begin{vmatrix} -1 & 2 \\ -4 & -5 \end{vmatrix} \ominus 4\begin{vmatrix} -1 & 2 \\ 3 & 8 \end{vmatrix} = -7(13) - 4(-14) = -35.$$

Notice the circled minus sign here. We must change the sign of the 4 in the $(3,2)$ position of A because the corresponding cofactor requires multiplication of the minor by -1. ☺

3.1.12 Problem. Find the determinant of the matrix $A = \begin{bmatrix} -1 & -2 & 0 \\ 0 & 1 & 5 \\ 2 & 3 & -4 \end{bmatrix}$.

Solution. A Laplace expansion along the first row gives

$$\det A = -1\begin{vmatrix} 1 & 5 \\ 3 & -4 \end{vmatrix} - (-2)\begin{vmatrix} 0 & 5 \\ 2 & -4 \end{vmatrix} = -(-19) + 2(-10) = -1.$$

READING CHECK 3. Use a Laplace expansion down the first column to again confirm $\det A = -1$.

3.1.13 Problem. Find the determinant of $A = \begin{bmatrix} -1 & 2 & -3 & 4 \\ 0 & -2 & 3 & 0 \\ 1 & 2 & 3 & -5 \\ -2 & 9 & -1 & 2 \end{bmatrix}$.

Solution. The easiest thing to do here is to use a Laplace expansion along the second row (to take advantage of two 0s). We get

$$\det A = -2\begin{vmatrix} -1 & -3 & 4 \\ 1 & 3 & -5 \\ -2 & -1 & 2 \end{vmatrix} - 3\begin{vmatrix} -1 & 2 & 4 \\ 1 & 2 & -5 \\ -2 & 9 & 2 \end{vmatrix}.$$

Now we must find two 3×3 determinants. With a Laplace expansion along the first row, the first determinant is

$$\begin{vmatrix} -1 & -3 & 4 \\ 1 & 3 & -5 \\ -2 & -1 & 2 \end{vmatrix} = -1\begin{vmatrix} 3 & -5 \\ -1 & 2 \end{vmatrix} + 3\begin{vmatrix} 1 & -5 \\ -2 & 2 \end{vmatrix} + 4\begin{vmatrix} 1 & 3 \\ -2 & -1 \end{vmatrix}$$
$$= -1(1) + 3(-8) + 4(5) = -5$$

and with a Laplace expansion along the first row, the second determinant is

$$\begin{vmatrix} -1 & 2 & 4 \\ 1 & 2 & -5 \\ -2 & 9 & 2 \end{vmatrix} = -1\begin{vmatrix} 2 & -5 \\ 9 & 2 \end{vmatrix} - 2\begin{vmatrix} 1 & -5 \\ -2 & 2 \end{vmatrix} + 4\begin{vmatrix} 1 & 2 \\ -2 & 9 \end{vmatrix}$$
$$= -1(49) - 2(-8) + 4(13) = 19.$$

Thus, $\det A = -2(-5) - 3(19) = -47$. ☝

We said at the start of this section that if the determinant of a matrix is not zero, then that matrix is invertible. We now see why.

3.1.14 Theorem. *Let A be a square matrix with $\det A \neq 0$. Then A is invertible and $A^{-1} = \frac{1}{\det A}\operatorname{adj}(A)$.*

Proof. Since $A\operatorname{adj}(A) = (\det A)I$ and $\det A \neq 0$, we can divide by $\det A$ to obtain $A\left(\frac{1}{\det A}\operatorname{adj}(A)\right) = I$. This says that $\frac{1}{\det A}\operatorname{adj}(A)$ is the inverse of A. ■

3.1.15 Example. In Example 3.1.5, with $A = \begin{bmatrix} -1 & 0 & 2 \\ 3 & -7 & 8 \\ -4 & 4 & -5 \end{bmatrix}$, we had $A\operatorname{adj}(A) = -35I$. So A is invertible and

$$A^{-1} = -\frac{1}{35}\operatorname{adj}(A) = -\frac{1}{35}\begin{bmatrix} 3 & 8 & 14 \\ -17 & 13 & 14 \\ -16 & 4 & 7 \end{bmatrix}.$$ ☺

The converse of Theorem 3.1.14 is also true; that is, if A is invertible, then $\det A \neq 0$. We will have more to say about this in Section 3.2. For now, we content ourselves with an example.

3.1.16 Example. The matrix of cofactors of the matrix $A = \begin{bmatrix} -1 & 2 & 2 \\ 3 & 0 & -1 \\ 1 & 4 & 3 \end{bmatrix}$ is $C = \begin{bmatrix} 4 & -10 & 12 \\ 2 & -5 & 6 \\ -2 & 5 & -6 \end{bmatrix}$, so $\operatorname{adj}(A) = C^T = \begin{bmatrix} 4 & 2 & -2 \\ -10 & -5 & 5 \\ 12 & 6 & -6 \end{bmatrix}$ and

$$A\operatorname{adj}(A) = \begin{bmatrix} -1 & 2 & 2 \\ 3 & 0 & -1 \\ 1 & 4 & 3 \end{bmatrix}\begin{bmatrix} 4 & 2 & -2 \\ -10 & -5 & 5 \\ 12 & 6 & -6 \end{bmatrix} = \begin{bmatrix} 0 & 0 & 0 \\ 0 & 0 & 0 \\ 0 & 0 & 0 \end{bmatrix}$$

is the zero matrix. Thus $\det A = 0$. If A were invertible, we could multiply the equation $A\operatorname{adj}(A) = 0$ by A^{-1} to get $\operatorname{adj}(A) = 0$, which is not true. ☺

Answers to Reading Checks

1. $A(\operatorname{adj} A) = (\det A)I$

2. Since $C^T A = \begin{bmatrix} c_{11} & c_{21} & c_{31} \\ c_{12} & c_{22} & c_{32} \\ c_{13} & c_{23} & c_{33} \end{bmatrix} \begin{bmatrix} a_{11} & a_{12} & a_{13} \\ a_{21} & a_{22} & a_{23} \\ a_{31} & a_{32} & a_{33} \end{bmatrix} = (\det A)I$, the $(1,1)$, the $(2,2)$ and the $(3,3)$ entries of this product are all equal to $\det A$. These numbers are $c_{11}a_{11} + c_{21}a_{21} + c_{31}a_{31}$, $c_{12}a_{12} + c_{22}a_{22} + c_{32}a_{32}$ and $c_{13}a_{13} + c_{23}a_{23} + c_{33}a_{33}$, these being the dot products of the columns of A with the corresponding cofactors.

3. A Laplace expansion down the third column gives $\begin{vmatrix} -1 & -2 & 0 \\ 0 & 1 & 5 \\ 2 & 3 & -4 \end{vmatrix}$

$$= 0 \begin{vmatrix} \star & \star \\ \star & \star \end{vmatrix} - 5 \begin{vmatrix} -1 & -2 \\ 2 & 3 \end{vmatrix} - 4 \begin{vmatrix} -1 & -2 \\ 0 & 1 \end{vmatrix} = -5(1) - 4(-1) = -1.$$

True/False Questions

Decide, with as little calculation as possible, whether each of the following statements is true or false and, if you say "false," explain your answer. (Answers are at the back of the book.)

1. The $(1,2)$ minor of $\begin{bmatrix} -6 & 7 \\ 2 & 1 \end{bmatrix}$ is $m_{12} = -2$.

2. The $(1,2)$ cofactor of $\begin{bmatrix} -6 & 7 \\ 2 & 1 \end{bmatrix}$ is $c_{12} = -2$.

3. The product $A(\operatorname{adj} A)$ of a matrix and its adjoint is a scalar multiple of the identity matrix.

4. If I is an identity matrix, then $\det(-2I) = -2$.

5. The determinant of $\begin{bmatrix} 1 & 2 & 3 \\ 0 & 0 & 1 \\ 4 & 5 & 6 \end{bmatrix}$ is -3.

6. There exists a 3×2 matrix with determinant 0.

7. If A and B are square matrices of the same size, $\det(A + B) = \det A + \det B$.

8. If the matrix of cofactors of a matrix A is the zero matrix, then $A = 0$ too.

9. If an $n \times n$ matrix A has nonzero determinant, the system $A\mathsf{x} = \mathsf{b}$ can be solved for any vector b in R^n.

10. One important use of the determinant is to give a test for the invertibility of a matrix.

EXERCISES

Answers to exercises marked [BB] can be found at the Back of the Book.

1. For each of the following matrices,

   i. find the matrix M of minors, find the matrix C of cofactors and compute the products AC^T and C^TA;

   ii. use part i to find $\det A$ and explain;

   iii. use the result of part i to determine if A is invertible and find A^{-1} when an inverse exists.

   (a) [BB] $A = \begin{bmatrix} 2 & -4 \\ -7 & 9 \end{bmatrix}$ **(b)** $A = \begin{bmatrix} 5 & 3 \\ 15 & 9 \end{bmatrix}$

   (c) [BB] $A = \begin{bmatrix} 2 & 3 & 4 \\ 0 & -1 & 3 \\ 4 & 7 & 5 \end{bmatrix}$ **(d)** $A = \begin{bmatrix} 3 & -1 & 4 \\ 6 & 2 & 0 \\ -4 & 5 & 1 \end{bmatrix}$

   (e) $A = \begin{bmatrix} 3 & 5 & 2 \\ 4 & 8 & 9 \\ -1 & 2 & 5 \end{bmatrix}$ **(f)** $A = \begin{bmatrix} -6 & 8 & 9 \\ 7 & -4 & 0 \\ 11 & 2 & -5 \end{bmatrix}$.

2. Explain why $\begin{vmatrix} \cos\theta & -\sin\theta \\ \sin\theta & \cos\theta \end{vmatrix}$ is independent of θ.

3. **(a)** [BB] Find the determinant of $A = \begin{bmatrix} 1 & -1 & 2 \\ 3 & 1 & 1 \\ 2 & -1 & 3 \end{bmatrix}$ with a Laplace expansion along the third row.

   (b) Find the determinant of A with a Laplace expansion down the second column.

4. **(a)** Find the determinant of $A = \begin{bmatrix} -4 & 2 & 1 \\ 0 & -3 & 5 \\ 6 & 4 & -1 \end{bmatrix}$ with a Laplace expansion down the first column.

   (b) Find the determinant of A with a Laplace expansion along the third row.

5. [BB] **(a)** Find the matrix of minors of $A = \begin{bmatrix} 2 & 0 & -3 \\ 4 & 1 & 7 \\ -3 & 1 & 5 \end{bmatrix}$.

   (b) Find C, the matrix of cofactors of A.

   (c) Find the dot product of the second row of A and the third row of C. Could this result have been anticipated? Explain.

   (d) Find the dot product of the first column of A and the second column of C. Could this result have been anticipated? Explain.

   (e) Find the dot product of the second column of A and the second column of C. What is the significance of this number?

6. (a) Find the matrix of minors of $A = \begin{bmatrix} -7 & 1 & 5 \\ 4 & -3 & 1 \\ 2 & 0 & 6 \end{bmatrix}$.

   (b) Find C, the cofactor matrix of A.

   (c) Find the dot product of the third row of A and the first row of C. Could this result have been anticipated? Explain.

   (d) Find the dot product of the second column of A and the first column of C. Could this result have been anticipated? Explain.

   (e) Find the dot product of the third row of A and the third row of C. What is the significance of this number?

7. [BB] Let $A = \begin{bmatrix} 1 & 0 & 1 \\ 2 & 1 & 1 \\ 1 & 2 & 1 \end{bmatrix}$ and suppose $C = \begin{bmatrix} -1 & -1 & c_{13} \\ c_{21} & 0 & -2 \\ -1 & c_{32} & 1 \end{bmatrix}$ is the matrix of cofactors of A.

   (a) Find the values of c_{21}, c_{13} and c_{32}.

   (b) Find $\det A$.

   (c) Is A invertible? Explain without further calculation.

   (d) Find A^{-1}, if this exists.

8. The matrix of cofactors of $A = \begin{bmatrix} -1 & 2 & 3 \\ 0 & 4 & 4 \\ -2 & 8 & 10 \end{bmatrix}$ is $C = \begin{bmatrix} 8 & c_{12} & 8 \\ 4 & -4 & c_{23} \\ c_{31} & 4 & -4 \end{bmatrix}$.

   (a) Find the values of c_{12}, c_{23} and c_{31}.

   (b) Find $\det A$.

   (c) Is A invertible? Explain without further calculation.

   (d) Find A^{-1}, if this exists.

9. For what values of x are the given matrices not invertible? (Such matrices are called *singular*.)

   (a) [BB] $\begin{bmatrix} x & 1 \\ -2 & x+3 \end{bmatrix}$ (b) $\begin{bmatrix} -1 & x \\ 3 & x \end{bmatrix}$

   (c) $\begin{bmatrix} 1 & 2 & x \\ 0 & -3 & 0 \\ 4 & x & 7 \end{bmatrix}$ (d) $\begin{bmatrix} -1 & x & x \\ 2 & 0 & x \\ 4 & 4 & -2 \end{bmatrix}$.

10. [BB] Professor G. was making up a homework problem for his class when he spilled a cup of coffee over his work. (This happens to Prof. G. all too frequently.) All that remained legible was this:

$$A = \begin{bmatrix} -1 & 2 & * \\ 0 & 3 & * \\ 2 & -2 & * \end{bmatrix}; \quad \text{matrix of cofactors} = C = \begin{bmatrix} 19 & 10 & * \\ -14 & -11 & * \\ -2 & 5 & * \end{bmatrix}.$$

   (a) Find all missing entries. (b) Find $\det A$. (c) Find A^{-1}.

11. Answer Exercise 10 with $A = \begin{bmatrix} 2 & * & -3 \\ -4 & * & 2 \\ 1 & * & 0 \end{bmatrix}$ and $C = \begin{bmatrix} -10 & * & -21 \\ -15 & * & -6 \\ 11 & * & 18 \end{bmatrix}$.

12. The *triple product* of vectors u, v, w is defined as $\mathsf{u} \cdot (\mathsf{v} \times \mathsf{w})$, the dot product of u and the cross product of v and w.

    (a) Show that $|\mathsf{u} \cdot (\mathsf{v} \times \mathsf{w})|$ (the absolute value of the triple product) is the volume of the *parallelepiped* with sides u, v, w. (A parallelepiped is like a cube, but its sides are parallelograms that are not necessarily squares.)

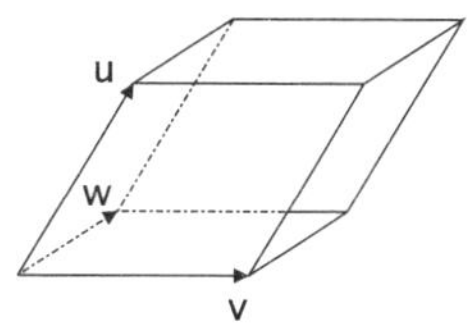

    (b) Show that $\mathsf{u} \cdot (\mathsf{v} \times \mathsf{w})$ is the determinant of the 3×3 matrix $\begin{bmatrix} \mathsf{u} & \longrightarrow \\ \mathsf{v} & \longrightarrow \\ \mathsf{w} & \longrightarrow \end{bmatrix}$ whose rows are the vectors u, v, w in this order.

    (c) What is an easy way to determine whether three vectors in R^3 lie in a plane? (Such vectors are called *coplanar.*)

Critical Reading

13. Let $A = \begin{bmatrix} 1 & 2 \\ 3 & 4 \end{bmatrix}$. Write down some matrices that commute with A and explain. (At least one of your matrices should have a connection with this section.)

14. [BB] The matrix of cofactors of a matrix A is $\begin{bmatrix} -1 & 7 \\ 2 & 4 \end{bmatrix}$. Find A.

15. The 2×2 matrix A has cofactor matrix $C = \begin{bmatrix} 3 & -4 \\ 2 & 1 \end{bmatrix}$. Find A.

16. Find A, given $\det A$ and the cofactor matrix C.

    (a) [BB] $\det A = 5$, $C = \begin{bmatrix} 3 & 1 \\ -2 & 1 \end{bmatrix}$ (b) $\det A = -24$, $C = \begin{bmatrix} -1 & 2 \\ 6 & 3 \end{bmatrix}$

    (c) $\det A = 11$, $C = \begin{bmatrix} 1 & 5 & 4 \\ 2 & -1 & -3 \\ 10 & -5 & -4 \end{bmatrix}$ (d) $\det A = 100$, $C = \begin{bmatrix} 2 & 34 & 6 \\ 8 & 36 & 24 \\ 13 & -29 & -11 \end{bmatrix}$.

17. Find a formula for the inverse of the adjoint of a matrix A, when this inverse exists.

18. Suppose the matrix $T = \begin{bmatrix} a & b & c \\ p & q & r \\ x & y & z \end{bmatrix}$ has cofactor matrix $\begin{bmatrix} A & B & C \\ P & Q & R \\ X & Y & Z \end{bmatrix}$.

    (a) Find P, Q and R.

    (b) What is $aA + pP + xX$? Explain.

    (c) What is $aB + pQ + xY$? Explain.

3.2 Properties of Determinants

Now that we know what a determinant is, we turn to a discussion of the many properties that this function possesses. We begin with an observation about the determinant of a triangular matrix.

3.2.1 Property 1. The determinant of a triangular matrix is the product of its diagonal entries. The determinant of a diagonal matrix is the product of its diagonal entries.

3.2.2 Examples.

- $\det \begin{bmatrix} a & 0 \\ b & c \end{bmatrix} = ac$
- $\det \begin{bmatrix} a & b & c \\ 0 & d & e \\ 0 & 0 & f \end{bmatrix} = a \det \begin{bmatrix} d & e \\ 0 & f \end{bmatrix} = adf$, using a Laplace expansion down the first column
- $\det \begin{bmatrix} a & 0 & 0 & 0 \\ b & c & 0 & 0 \\ d & e & f & 0 \\ g & h & i & j \end{bmatrix} = a \det \begin{bmatrix} c & 0 & 0 \\ e & f & 0 \\ h & i & j \end{bmatrix} = acfj$, using the result for 3×3 matrices already established.

  At this point, the student might see why Property 1 holds in general. If it holds for matrices of a certain size, then it holds for matrices of the next size because if $A = \begin{bmatrix} a_{11} & 0 & \cdots & 0 \\ a_{21} & & & \\ \vdots & & B & \\ a_{k1} & & & \end{bmatrix}$, then (with a Laplace expansion along the first row) $\det A = a_{11} \det B$ so, if $\det B$ is the product of its diagonal entries, so is $\det A$. This is an argument by "mathematical induction" with which some readers may be familiar.

- $\det \begin{bmatrix} -2 & 0 & 0 & 0 \\ 0 & 3 & 0 & 0 \\ 0 & 0 & 7 & 0 \\ 0 & 0 & 0 & 5 \end{bmatrix} = (-2)(3)(7)(5) = -210.$ ☺

READING CHECK 1. What is $\det I$, where I denotes the 100×100 identity matrix? What is $\det(-I)$, where I denotes the 101×101 identity matrix?

Readers of this book should be familiar with the word "function." A function from a set A to a set B is a procedure that turns elements of A into elements of B. If the function is named f, then we write $f\colon A \to B$ to express the fact that f is a function from A to B.

The author likes to think of a function as a hopper, like that pictured in Figure 3.1. You toss an element a from A into the hopper and an element b from B comes out. The element b is usually written $f(a)$. The squaring

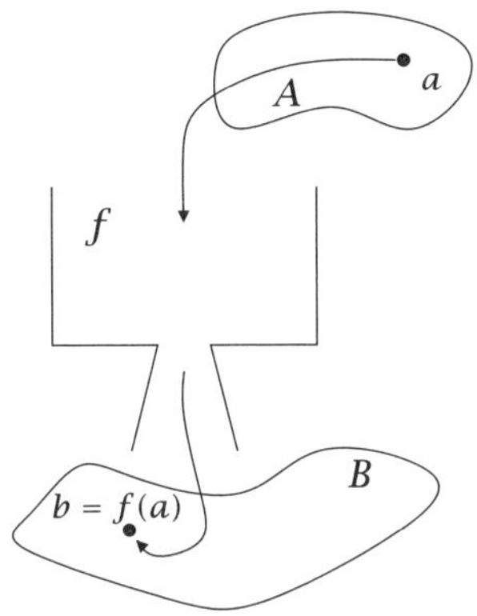

Figure 3.1: How to think of a function

function is the function $f\colon \mathsf{R} \to \mathsf{R}$ defined by $f(a) = a^2$ for a in R, the real numbers. "Addition" is a function from $A = \{(x,y) \mid x,y \in \mathsf{R}\}$ (the set of ordered pairs of real numbers) that turns the pair $a = (x,y)$ into the number $b = x + y$. The determinant is a function from the set A of $n \times n$ matrices to the set B of real numbers.

Most readers have probably seen the term "linear function" applied to functions described with rules like $f(x) = ax+b$. The term makes sense because the graphs of such functions are straight lines. In linear algebra, the concept of "linearity" is a little different and more restrictive.

3.2.3 Definition. A *linear function* from R^n to R^m is a function f satisfying

1. $f(\mathsf{x} + \mathsf{y}) = f(\mathsf{x}) + f(\mathsf{y})$ for all vectors x and y, and
2. $f(c\mathsf{x}) = cf(\mathsf{x})$ for all scalars c and all vectors x.

3.2.4 Examples. • The function $f\colon \mathsf{R} \to \mathsf{R}$ defined by $f(x) = 3x$ is linear since, for any real numbers x and y and any scalar c,

$f(x+y) = 3(x+y) = 3x + 3y = f(x) + f(y)$
and $f(cx) = 3(cx) = c(3x) = cf(x)$.

- The function $f\colon \mathsf{R}^2 \to \mathsf{R}$ defined by $f\left(\begin{bmatrix} x_1 \\ x_2 \end{bmatrix}\right) = \begin{vmatrix} x_1 & x_2 \\ 3 & 2 \end{vmatrix} = 2x_1 - 3x_2$ is linear since, for any vectors $\mathsf{x} = \begin{bmatrix} x_1 \\ x_2 \end{bmatrix}$ and $\mathsf{y} = \begin{bmatrix} y_1 \\ y_2 \end{bmatrix}$, and for any scalar c,

$$\begin{aligned} f(\mathsf{x}+\mathsf{y}) &= f\left(\begin{bmatrix} x_1+y_1 \\ x_2+y_2 \end{bmatrix}\right) = 2(x_1+y_1) - 3(x_2+y_2) \\ &= (2x_1 - 3x_2) + (2y_1 - 3y_2) = f(\mathsf{x}) + f(\mathsf{y}), \text{ and} \\ f(c\mathsf{x}) &= f\left(\begin{bmatrix} cx_1 \\ cx_2 \end{bmatrix}\right) = 2(cx_1) - 3(cx_2) = c(2x_1 - 3x_2) = cf(\mathsf{x}). \end{aligned}$$

- The function $f\colon \mathsf{R} \to \mathsf{R}^3$ defined by $f(x) = \begin{bmatrix} x \\ 2x \\ 3x \end{bmatrix}$ is linear since, for any real numbers x and y, and for any scalar c,

$$\begin{aligned} f(x+y) &= \begin{bmatrix} x+y \\ 2(x+y) \\ 3(x+y) \end{bmatrix} = \begin{bmatrix} x \\ 2x \\ 3x \end{bmatrix} + \begin{bmatrix} y \\ 2y \\ 3y \end{bmatrix} = f(x) + f(y), \text{ and} \\ f(cx) &= \begin{bmatrix} cx \\ 2(cx) \\ 3(cx) \end{bmatrix} = c\begin{bmatrix} x \\ 2x \\ 3x \end{bmatrix} = cf(x). \end{aligned}$$

- The dot product is linear "in each variable" meaning that for a fixed vector u, the function $f\colon \mathsf{R}^n \to \mathsf{R}$ defined by $f(\mathsf{x}) = \mathsf{x} \cdot \mathsf{u}$ is linear because

  - $(\mathsf{x}_1 + \mathsf{x}_2) \cdot \mathsf{u} = \mathsf{x}_1 \cdot \mathsf{u} + \mathsf{x}_2 \cdot \mathsf{u}$ and
  - $(c\mathsf{x}) \cdot \mathsf{u} = c(\mathsf{x} \cdot \mathsf{u})$

  and, for a fixed vector v, the function $g\colon \mathsf{R}^n \to \mathsf{R}$ defined by $g(\mathsf{x}) = \mathsf{v} \cdot \mathsf{x}$ is linear because

  - $\mathsf{v} \cdot (\mathsf{x}_1 + \mathsf{x}_2) = \mathsf{v} \cdot \mathsf{x}_1 + \mathsf{v} \cdot \mathsf{x}_2$ and
  - $\mathsf{v} \cdot (c\mathsf{x}) = c(\mathsf{v} \cdot \mathsf{x})$

  for all vectors $\mathsf{x}, \mathsf{x}_1, \mathsf{x}_2$ in R^n and any scalar c.

- The cross product is linear in each variable because for fixed vectors u, v in R^3, the functions $f, g\colon \mathsf{R}^3 \to \mathsf{R}^3$ defined by $f(\mathsf{x}) = \mathsf{x} \times \mathsf{v}$ and $g(\mathsf{x}) = \mathsf{u} \times \mathsf{x}$ are each linear. ☺

READING CHECK 2. Write down the two properties implied by the statement "the cross product is linear in its first variable."

3.2.5 Examples. • The function $g\colon \mathsf{R} \to \mathsf{R}$ defined by $g(x) = 3x + 1$ is not linear (within the context of linear algebra) since, for example, $g(x+y) = 3(x+y)+1$, while $g(x)+g(y) = (3x+1)+(3y+1)$. (One could also point out that $g(cx) \neq cg(x)$, in general.)

- The sine function $\mathsf{R} \to \mathsf{R}$ is not linear: $\sin(\alpha+\beta) \neq \sin\alpha + \sin\beta$. (Also, $\sin c\alpha \neq c\sin\alpha$.)

- The determinant is certainly not a linear function: $\det(A+B) \neq \det A + \det B$. For example, if $A = \begin{bmatrix} 1 & 2 \\ 3 & 4 \end{bmatrix}$ and $B = \begin{bmatrix} -1 & -2 \\ -3 & -4 \end{bmatrix}$, then $A + B = \begin{bmatrix} 0 & 0 \\ 0 & 0 \end{bmatrix}$, so $\det(A+B) = 0$, while $\det A + \det B = -2 + (-2) = -4$. ☺

While the determinant is not linear, all is not lost! Suppose we fix every row of an $n \times n$ matrix except the first one. So we have a matrix $A = \begin{bmatrix} ??? & \rightarrow & \\ \mathsf{a}_2 & \rightarrow & \\ \mathsf{a}_3 & \rightarrow & \\ \vdots & & \\ \mathsf{a}_n & \rightarrow & \end{bmatrix}$ just waiting for its first row. Define $f\colon \mathsf{R}^n \rightarrow \mathsf{R}$ by the rule

$f(\mathsf{x}) = \det \begin{bmatrix} \mathsf{x} & \rightarrow & \\ \mathsf{a}_2 & \rightarrow & \\ \mathsf{a}_3 & \rightarrow & \\ \vdots & & \\ \mathsf{a}_n & \rightarrow & \end{bmatrix}$, the determinant of A with x in the first row. It is this sort of function we have in mind when we state our next property.

3.2.6 Property 2. The determinant is linear in each row.

For example, the function $f\colon \mathsf{R}^2 \rightarrow \mathsf{R}$ defined by $f\left(\begin{bmatrix} x_1 \\ x_2 \end{bmatrix}\right) = \det \begin{bmatrix} x_1 & x_2 \\ 3 & 2 \end{bmatrix}$ is linear— see Examples 3.2.4.

We introduce some notation which will be useful in this section.

Suppose $A = \begin{bmatrix} \mathsf{a}_1 & \rightarrow & \\ \mathsf{a}_2 & \rightarrow & \\ \vdots & & \\ \mathsf{a}_n & \rightarrow & \end{bmatrix}$ is an $n \times n$ matrix with rows $\mathsf{a}_1, \ldots, \mathsf{a}_n$. It will be useful to write $\det(\mathsf{a}_1, \ldots, \mathsf{a}_n)$ for the determinant of A. For example, with this notation, the statement "The determinant is linear in each row" means that for any vectors x and y and for any scalar c,

$$\det(\mathsf{a}_1, \ldots, \mathsf{x} + \mathsf{y}, \ldots, \mathsf{a}_n) = \det(\mathsf{a}_1, \ldots, \mathsf{x}, \ldots, \mathsf{a}_n) + \det(\mathsf{a}_1, \ldots, \mathsf{y}, \ldots, \mathsf{a}_n)$$

and

$$\det(\mathsf{a}_1, \ldots, c\mathsf{x}, \ldots, \mathsf{a}_n) = c \det(\mathsf{a}_1, \ldots \mathsf{x}, \ldots, \mathsf{a}_n).$$

The first property says that if one row of a matrix is the sum of two rows x and y, then the determinant of the new matrix is the sum of the determinants of the matrices with rows x and y, respectively. For instance,

$$\det\begin{bmatrix} 7 & 8 \\ 1 & 5 \end{bmatrix} = \det\begin{bmatrix} [3, & 2] \\ +[4, & 6] \\ 1 & 5 \end{bmatrix}$$

$$= \det\begin{bmatrix} 3 & 2 \\ 1 & 5 \end{bmatrix} + \det\begin{bmatrix} 4 & 6 \\ 1 & 5 \end{bmatrix} = 13 + 14 = 27.$$

The second property says that if one row of A contains a factor of c, then this c can be pulled outside. For instance,

$$\det\begin{bmatrix} -1 & 5 \\ 3 & 6 \end{bmatrix} = 3\det\begin{bmatrix} -1 & 5 \\ 1 & 2 \end{bmatrix} = 3(-7) = -21.$$

Note that the second property does **not** say that $\det(cA) = c\det A$, which is not true. In fact,

$$\det\begin{bmatrix} cx & cy \\ cz & cw \end{bmatrix} = c\det\begin{bmatrix} x & y \\ cz & cw \end{bmatrix} = c^2\begin{bmatrix} x & y \\ z & w \end{bmatrix}$$

factoring c first from the first row and then from the second.

3.2.7 Property 3. If A is an $n \times n$ matrix and c is a scalar, $\det cA = c^n \det A$.

3.2.8 Example. If A is a $5{\times}5$ matrix and $\det A = 3$, then $\det(-2A) = (-2)^5 \det A = -32(3) = -96$. ☺

3.2.9 Property 4. The determinant of a matrix with a row of 0s is 0.

This is quite easy to see—if one row of a matrix A consists entirely of 0s, a Laplace expansion along that row shows quickly that the determinant is 0. It is also a consequence of linearity: Let x be your favourite vector (all 1s perhaps?), then $\mathsf{0} = 0\mathsf{x}$, so $\det(\mathsf{a}_1, \dots, \mathsf{0}, \dots, \mathsf{a}_n) = \det(\mathsf{a}_1, \dots, 0\mathsf{x}, \dots, \mathsf{a}_n) = 0\det(\mathsf{a}_1, \dots, \mathsf{x}, \dots, \mathsf{a}_n) = 0$.

3.2.10 Property 5. The determinant of a matrix with two equal rows is 0.

This is clear for 2×2 matrices because $\det\begin{bmatrix} a & b \\ a & b \end{bmatrix} = ab - ab = 0$ and an argument by mathematical induction finishes the job in general. To see this, suppose an $n \times n$ matrix, $n \geq 3$, has two equal rows. Using a Laplace expansion along any row different from the two equal rows produces a linear combination of determinants, each determinant that of an $(n-1) \times (n-1)$ matrix with two equal rows. So, if each such determinant is 0, so is the $n \times n$ determinant.

3.2.11 Property 6. If one row of a matrix is a scalar multiple of another, the determinant is 0.

To see this, let A be a matrix with two rows of the form x and $c\mathsf{x}$. Using linearity to factor c from row $c\mathsf{x}$ gives a matrix with two equal rows,

$$\det(\mathsf{a}_1,\ldots,\mathsf{x},\ldots,c\mathsf{x},\ldots,\mathsf{a}_n) = c\det(\mathsf{a}_1,\ldots,\mathsf{x},\ldots,\mathsf{x},\ldots,\mathsf{a}_n),$$

which is 0 by Property 5.

3.2.12 Property 7. If two rows of a matrix are interchanged, the determinant changes sign.

This follows quickly from linearity in each row and Property 5. We wish to show that if x and y are two rows of a matrix, say rows i and j, then $\det(\mathsf{a}_1,\ldots,\mathsf{y},\ldots,\mathsf{x},\ldots,\mathsf{a}_n) = -\det(\mathsf{a}_1,\ldots,\mathsf{x},\ldots,\mathsf{y},\ldots,\mathsf{a}_n)$. To see this, make a matrix with $\mathsf{x}+\mathsf{y}$ in each of rows i and j. Since the matrix has two equal rows, its determinant is 0:

$$\det(\mathsf{a}_1,\ldots,\mathsf{x}+\mathsf{y},\ldots,\mathsf{x}+\mathsf{y},\ldots,\mathsf{a}_n) = 0.$$

Linearity in row i says that this determinant is

$$\det(\mathsf{a}_1,\ldots,\mathsf{x},\ldots,\mathsf{x}+\mathsf{y},\ldots,\mathsf{a}_n) + \det(\mathsf{a}_1,\ldots,\mathsf{y},\ldots,\mathsf{x}+\mathsf{y},\ldots,\mathsf{a}_n)$$

and linearity in row j allows us to write each of these determinants as the sum of two more determinants. Altogether, we get

$$\det(\mathsf{a}_1,\ldots,\mathsf{x},\ldots,\mathsf{x},\ldots,\mathsf{a}_n) + \det(\mathsf{a}_1,\ldots,\mathsf{x},\ldots,\mathsf{y},\ldots,\mathsf{a}_n)$$
$$+ \det(\mathsf{a}_1,\ldots,\mathsf{y},\ldots,\mathsf{x},\ldots,\mathsf{a}_n) + \det(\mathsf{a}_1,\ldots,\mathsf{y},\ldots,\mathsf{y},\ldots,\mathsf{a}_n) = 0.$$

The first and last of these four determinants here are 0 because they have two equal rows, so

$$\det(\mathsf{a}_1,\ldots,\mathsf{x},\ldots,\mathsf{y},\ldots,\mathsf{a}_n) + \det(\mathsf{a}_1,\ldots,\mathsf{y},\ldots,\mathsf{x},\ldots,\mathsf{a}_n) = 0$$

which gives what we wanted. Here is a little application of this idea.

3.2.13 Proposition. *The determinant of a permutation matrix is* ± 1.

Proof. A permutation matrix is obtained by a sequence of row interchanges of the identity matrix. The determinant of the identity matrix is 1 and each interchange of rows changes the sign of the determinant. ■

3.2.14 Example. The matrix $P = \begin{bmatrix} 0&1&0&0\\0&0&1&0\\0&0&0&1\\1&0&0&0 \end{bmatrix}$ is a permutation matrix whose rows are the rows of I, the 4×4 identity matrix, in order 2, 3, 4, 1. We obtain P from I by successive interchanges of rows:

$$I = \begin{bmatrix} 1&0&0&0\\0&1&0&0\\0&0&1&0\\0&0&0&1 \end{bmatrix} \to P_1 = \begin{bmatrix} 0&1&0&0\\1&0&0&0\\0&0&1&0\\0&0&0&1 \end{bmatrix} \to P_2 = \begin{bmatrix} 0&1&0&0\\0&0&1&0\\1&0&0&0\\0&0&0&1 \end{bmatrix} \to P = \begin{bmatrix} 0&1&0&0\\0&0&1&0\\0&0&0&1\\1&0&0&0 \end{bmatrix},$$

so $\det P = -\det P_2 = -(-\det P_1) = (-1)(-1)(-1)\det I = -1$. :)

3.2.15 Property 8. The determinant is *multiplicative*: $\det AB = \det A \det B$.

For example, if $A = \begin{bmatrix} -1 & 2 \\ 4 & 5 \end{bmatrix}$ and $B = \begin{bmatrix} 0 & 3 \\ 4 & -2 \end{bmatrix}$, then $AB = \begin{bmatrix} 8 & -7 \\ 20 & 2 \end{bmatrix}$, so $\det AB = 156$, which is $(-13)(-12) = \det A \det B$.

One proof of Property 8 which the author finds particularly instructive involves three preliminary steps.

Step 1. $\det AB = (\det A)(\det B)$ if A is singular (that is, not invertible).

It remains to get the same result when A is invertible.

Step 2. If E is an elementary matrix, $\det(EB) = (\det E)(\det B)$.

Step 3. If $E_1, E_2, \ldots, E_k$ are elementary matrices, then $\det(E_1E_2\cdots E_kB) = (\det E_1)(\det E_2)\cdots(\det E_k)(\det B)$.

Taking $B = I$ here, we get $\det(E_1E_2\cdots E_k) = (\det E_1)(\det E_2)\cdots(\det E_k)$. Now the fact that an invertible matrix is the product of elementary matrices (Theorem 2.9.10) completes the argument. We leave the details to an exercise (Exercise 37).

3.2.16 Property 9. If A is invertible, then $\det A \neq 0$ and $\det(A^{-1}) = \dfrac{1}{\det A}$.

This follows directly from Property 8. Since $AA^{-1} = I$, the identity matrix, $1 = \det I = \det AA^{-1} = (\det A)(\det A^{-1})$. In particular, $\det A \neq 0$ and $\det A^{-1} = \dfrac{1}{\det A}$.

Here's a nice application of this fact.

3.2.17 Proposition. *If P is a permutation matrix,* $\det P^{-1} = \det P$.

Proof. As noted in 3.2.13, $\det P = \pm 1$. Thus

$$\det P^{-1} = \frac{1}{\det P} = \begin{cases} +1 & \text{if } \det P = 1 \\ -1 & \text{if } \det P = -1 \end{cases}$$

and we notice that $\det P^{-1} = \det P$ in either case. ■

Theorem 3.1.14 said that if a matrix has a nonzero determinant, then it is invertible. Property 9 says that the "converse" is also true: if A is invertible, then $\det A \neq 0$. Thus the determinant gives a test for invertibility.

3.2.18 Theorem. *A square matrix A is invertible if and only if* $\det A \neq 0$.

Under what conditions is a real number a invertible? Surely, a is invertible if and only if $a \neq 0$. For matrices, A is invertible if and only if the number $\det A \neq 0$. The analogy is beautiful!

3.2.19 Example. $A = \begin{bmatrix} -1 & 2 \\ 5 & 9 \end{bmatrix}$ is invertible because $\det A = -19 \neq 0$.

3.2.20 Problem. Show that the matrix $A = \begin{bmatrix} 1 & 0 & 3 \\ -4 & 1 & 6 \\ -2 & 1 & 4 \end{bmatrix}$ is invertible.

Solution. With a Laplace expansion down the second column, we have

$$\det A = \begin{vmatrix} 1 & 3 \\ -2 & 4 \end{vmatrix} - \begin{vmatrix} 1 & 3 \\ -4 & 6 \end{vmatrix} = 10 - 18 = -8 \neq 0,$$

so A is invertible by Theorem 3.2.18.

3.2.21 Problem. Explain why the vectors $\begin{bmatrix} 1 \\ -4 \\ -2 \end{bmatrix}, \begin{bmatrix} 0 \\ 1 \\ 1 \end{bmatrix}, \begin{bmatrix} 3 \\ 6 \\ 4 \end{bmatrix}$ are linearly independent.

Solution. A matrix is invertible if and only if its columns are linearly independent. (This was Theorem 2.9.13.) The matrix which has the given vectors as columns is the matrix A of Problem 3.2.20. Since $\det A \neq 0$, the given vectors are linearly independent.

There is a commonly used term for "not invertible."

3.2.22 Definition. A square matrix is *singular* if it is not invertible.

So Theorem 3.2.18 says that A is singular if and only if $\det A = 0$.

3.2.23 Problem. For what values of c is the matrix $A = \begin{bmatrix} 1 & 0 & -c \\ -1 & 3 & 1 \\ 0 & 2c & -4 \end{bmatrix}$ singular?

Solution. With a Laplace expansion along the first row, we have

$$\begin{aligned} \det A &= \begin{vmatrix} 3 & 1 \\ 2c & -4 \end{vmatrix} - c \begin{vmatrix} -1 & 3 \\ 0 & 2c \end{vmatrix} = (-12 - 2c) - c(-2c) \\ &= 2c^2 - 2c - 12 = 2(c^2 - c - 6) = 2(c-3)(c+2). \end{aligned}$$

The matrix is singular if and only if $\det A = 0$, and this is the case if and only if $c = 3$ or $c = -2$.

3.2.24 Property 10. $\det A^T = \det A$.

Here's an argument. If P is a permutation matrix, then $P^T = P^{-1}$—see Proposition 2.7.21—and $\det P^{-1} = \det P$ by Proposition 3.2.17, so $\det P^T = \det P$, the desired result for permutation matrices. Now remember that any matrix A can be factored $A = PLU$ as the product of a permutation matrix P (perhaps $P = I$), a lower triangular matrix L with 1s on the diagonal, and a row echelon matrix U—see Section 2.7. Thus $\det L = 1$ and $\det A = \det PLU = (\det P)(\det L)(\det U) = (\det P)(\det U)$. On the other hand, $A = PLU$ implies $A^T = (PLU)^T = U^T L^T P^T$. Now $\det P^T = \det P$ and $\det L^T = 1$ because L^T is triangular with 1s on the diagonal. Also U^T is triangular with the same diagonal as U, so $\det U^T = \det U$. Thus $\det A^T = \det U^T \det L^T \det P^T = (\det U)(\det P) = (\det P)(\det U) = \det A$.

3.2.25 Problem. Suppose A and B are $n \times n$ matrices with $\det A = 2$ and $\det B = 5$. What is $\det A^3 B^{-1} A^T$?

Solution. By Property 8, $\det A^3 = \det AAA = (\det A)(\det A)(\det A) = (\det A)^3 = 2^3 = 8$. By Property 9, $\det B^{-1} = \frac{1}{\det B} = \frac{1}{5}$. By Property 10, $\det A^T = \det A = 2$. A final application of Property 8 gives $\det A^3 B^{-1} A^T = 8(\frac{1}{5})(2) = \frac{16}{5}$. ♘

READING CHECK 3. With A and B as above, find $\det(-A^3 B^{-1} A^T)$, or is this possible?

Property 10 is especially important because it implies that any property of the determinant concerning rows holds equally for columns. We get two facts for the price of one! For example, suppose we interchange columns one and two of the matrix $\begin{bmatrix} a & d & g \\ b & e & h \\ c & f & i \end{bmatrix}$. We have

$$\begin{vmatrix} d & a & g \\ e & b & h \\ f & c & i \end{vmatrix} = \begin{vmatrix} d & e & f \\ a & b & c \\ g & h & i \end{vmatrix} \qquad \text{since } \det A = \det A^T$$

$$= -\begin{vmatrix} a & b & c \\ d & e & f \\ g & h & i \end{vmatrix} \qquad \text{interchanging rows one and two}$$

$$= -\begin{vmatrix} a & d & g \\ b & e & h \\ c & f & i \end{vmatrix} \qquad \text{since } \det A = \det A^T.$$

So, as with rows, interchanging two columns of a matrix changes the sign of the determinant.

Consider the effect of multiplying a column of a matrix by a scalar. For example,

$$\begin{vmatrix} ka & d & g \\ kb & e & h \\ kc & f & i \end{vmatrix} = \begin{vmatrix} ka & kb & kc \\ d & e & f \\ g & h & i \end{vmatrix} \qquad \text{since } \det A = \det A^T$$

$$= k \begin{vmatrix} a & b & c \\ d & e & f \\ g & h & i \end{vmatrix} \qquad \text{factoring } k \text{ from the first row}$$

$$= k \begin{vmatrix} a & d & g \\ b & e & h \\ c & f & i \end{vmatrix} \qquad \text{since } \det A = \det A^T,$$

and so, if B is obtained from A by multiplying a column by k, $\det B = k \det A$. In a similar manner, using the fact $\det A = \det A^T$, it is straightforward to show that all the properties of determinants stated for rows are also true for columns.

Evaluating a Determinant with Elementary Row Operations

There is another way to find the determinant of a matrix which, especially for large matrices, is much simpler than the cofactor method. Remember the three elementary row operations:

1. Interchange two rows.
2. Multiply a row by any nonzero scalar.
3. Replace a row by that row less a multiple of another row.

Earlier in this section, we showed the effect of the first two of these operations on the determinant.

- The determinant changes sign if two rows are interchanged.
- If a certain row of a matrix contains the factor c, then c factors out of the determinant (part of "linearity in each row").

What happens to the determinant when we apply the third elementary row operation to a matrix? Suppose A has rows $\mathsf{a}_1, \mathsf{a}_2, \dots, \mathsf{a}_n$ and we apply the operation $R_j \to R_j - cR_i$ to get a new matrix B. Since the determinant is linear in row j,

$$\begin{aligned} \det B &= \det(\mathsf{a}_1, \dots, \mathsf{a}_i, \dots, \mathsf{a}_j - c\mathsf{a}_i, \dots, \mathsf{a}_n) \\ &= \det(\mathsf{a}_1, \dots, \mathsf{a}_i, \dots, \mathsf{a}_j, \dots \mathsf{a}_n) + \det(\mathsf{a}_1, \dots, \mathsf{a}_i, \dots, -c\mathsf{a}_i, \dots, \mathsf{a}_n). \end{aligned}$$

The second determinant is 0 because one row is a scalar multiple of another, so $\det B = \det(\mathsf{a}_1, \dots, \mathsf{a}_i, \dots, \mathsf{a}_j, \dots \mathsf{a}_n) = \det A$. The determinant has not changed! We summarize.

Matrix operation	Effect on determinant
interchange two rows	multiply by -1
factor $c \neq 0$ from a row	factor c from the determinant
replace a row by that row less a multiple of another	no change

This suggests a method for finding the determinant of a matrix.

3.2.26 To find the determinant of a matrix A, reduce A to an upper triangular matrix using the elementary row operations, making note of the effect of each operation on the determinant. Then use the fact that the determinant of a triangular matrix is the product of its diagonal entries.

3.2.27 Example. $\begin{vmatrix} 1 & -1 & 4 \\ 2 & 1 & -1 \\ -2 & 0 & 3 \end{vmatrix} = \begin{vmatrix} 1 & -1 & 4 \\ 0 & 3 & -9 \\ -2 & 0 & 3 \end{vmatrix}$ $\quad R2 \to R2 - 2(R1)$

$= \begin{vmatrix} 1 & -1 & 4 \\ 0 & 3 & -9 \\ 0 & -2 & 11 \end{vmatrix}$ $\quad R3 \to R3 + 2(R1)$

$= 3\begin{vmatrix} 1 & -1 & 4 \\ 0 & 1 & -3 \\ 0 & -2 & 11 \end{vmatrix}$ $\quad$ factoring a 3 from row two

$= 3\begin{vmatrix} 1 & -1 & 4 \\ 0 & 1 & -3 \\ 0 & 0 & 5 \end{vmatrix}$ $\quad R3 \to R3 + 2(R2)$

$= 3(5) = 15$ $\quad$ the determinant of a triangular matrix is the product of its diagonal entries.

Sometimes, it is not necessary to reduce to triangular because we can see what the determinant must be before achieving triangular.

3.2.28 Example. $\begin{vmatrix} 1 & 2 & 3 \\ 4 & 5 & 6 \\ 7 & 8 & 9 \end{vmatrix} = \begin{vmatrix} 1 & 2 & 3 \\ 0 & -3 & -6 \\ 7 & 8 & 9 \end{vmatrix}$ $\quad R2 \to R2 - 4(R1)$

$= \begin{vmatrix} 1 & 2 & 3 \\ 0 & -3 & -6 \\ 0 & -6 & -12 \end{vmatrix}$ $\quad R3 \to R3 - 7(R1)$

$= 0$ $\quad$ since row three is a multiple of row two.

3.2.29 Example. Let $A = \begin{bmatrix} 1 & 0 & 3 & 1 \\ 2 & 2 & 6 & 0 \\ 4 & 0 & -3 & 1 \\ 4 & 1 & 12 & 1 \end{bmatrix}$. Using the 1 in the $(1,1)$ position to put 0s below it (via the third elementary row operation which does not change the determinant), we have

$$\det A = \begin{vmatrix} 1 & 0 & 3 & 1 \\ 0 & 2 & 0 & -2 \\ 0 & 0 & -15 & -3 \\ 0 & 1 & 0 & -3 \end{vmatrix}.$$

Factoring 2 from row two and -3 from row three and then using $R4 \to R4 - R2$, we obtain

$$\det A = 2(-3) \begin{vmatrix} 1 & 0 & 3 & 1 \\ 0 & 1 & 0 & -1 \\ 0 & 0 & 5 & 1 \\ 0 & 1 & 0 & -3 \end{vmatrix} = -6 \begin{vmatrix} 1 & 0 & 3 & 1 \\ 0 & 1 & 0 & -1 \\ 0 & 0 & 5 & 1 \\ 0 & 0 & 0 & -2 \end{vmatrix} = 60$$

because the determinant of a triangular matrix is the product of its diagonal entries. ☺

3.2.30 Problem. Suppose $A = \begin{bmatrix} a & b & c \\ d & e & f \\ g & h & i \end{bmatrix}$ is a 3×3 matrix with $\det A = 2$.

Find the determinant of $\begin{bmatrix} -g & 3d+5a & 7d \\ -h & 3e+5b & 7e \\ -i & 3f+5c & 7f \end{bmatrix}$.

Solution. We use the fact that $\det X = \det X^T$ and then use elementary row operations to move the new matrix towards A.

$$\begin{vmatrix} -g & 3d+5a & 7d \\ -h & 3e+5b & 7e \\ -i & 3f+5c & 7f \end{vmatrix} = \begin{vmatrix} -g & -h & -i \\ 3d+5a & 3e+5b & 3f+5c \\ 7d & 7e & 7f \end{vmatrix}$$

$$= -7 \begin{vmatrix} g & h & i \\ 3d+5a & 3e+5b & 3f+5c \\ d & e & f \end{vmatrix} \quad \text{factoring } -1 \text{ from row one and 7 from row three}$$

$$= -7 \begin{vmatrix} g & h & i \\ 5a & 5b & 5c \\ d & e & f \end{vmatrix} \quad R2 \to R2 - 3(R3)$$

$$= -35 \begin{vmatrix} g & h & i \\ a & b & c \\ d & e & f \end{vmatrix} \quad \text{factoring 5 from row two}$$

$$= -35 \begin{vmatrix} a & b & c \\ g & h & i \\ d & e & f \end{vmatrix} \quad R1 \longleftrightarrow R2$$

$$= -35 \begin{vmatrix} a & b & c \\ d & e & f \\ g & h & i \end{vmatrix} \qquad R2 \longleftrightarrow R3$$

The desired determinant is $-35 \det A = -70$.

As we have said, the fact that $\det A^T = \det A$ means all properties of the determinant that are true for rows are true for columns. While the author's strong preference is to use row operations as far as possible, it is nonetheless true that it sometimes helps a lot to take advantage of column as well as row properties.

3.2.31 Example. $\begin{vmatrix} 3 & -1 & 2 \\ 2 & 4 & -9 \\ -3 & 5 & 6 \end{vmatrix} = -\begin{vmatrix} -1 & 3 & 2 \\ 4 & 2 & -9 \\ 5 & -3 & 6 \end{vmatrix}$ interchanging columns one and two

$= -\begin{vmatrix} -1 & 3 & 2 \\ 0 & 14 & -1 \\ 0 & 12 & 16 \end{vmatrix}$ using the -1 in the $(1,1)$ position to put 0s below it, via the third elementary row operation

$= -(-1)\begin{vmatrix} -1 & 2 & 3 \\ 0 & -1 & 14 \\ 0 & 16 & 12 \end{vmatrix}$ interchanging columns two and three

$= \begin{vmatrix} -1 & 2 & 3 \\ 0 & -1 & 14 \\ 0 & 0 & 236 \end{vmatrix}$ using the -1 in the $(2,2)$ position to get a 0 below, with the third elementary row operation

$= 236$ since the determinant of a triangular matrix is the product of the diagonal entries.

Concluding Remarks

This section and the previous have introduced a number of ways to compute determinants. Which is best depends on the matrix and who is doing the calculation. The author's preference is often to use a computer algebra package like *Maple*™. Figure 3.2 shows how to enter a matrix, which Maple echos, and then how easy it is to find the determinant.

If no computing tools are available, the author finds 2×2 determinants in his head, 3×3 determinants by the method of cofactors (because 2×2 determinants are mental exercises), and higher order determinants by reducing to triangular form, usually exclusively by row operations. There are always exceptions to these general rules, of course, depending on the nature of the matrix.

```
> A := Matrix([[-1,2,0],[4,5,-3],[-2,1,6]]);
```

$$A := \begin{bmatrix} -1 & 2 & 0 \\ 4 & 5 & -3 \\ -2 & 1 & 6 \end{bmatrix}$$

```
> Determinant(A);
```

$$-69$$

Figure 3.2: How to enter a matrix and find its determinant with *Maple*™.

3.2.32 Examples.

- $\begin{vmatrix} 2 & 3 \\ -1 & 7 \end{vmatrix} = 17$

- $\begin{vmatrix} 1 & -2 & 3 \\ 4 & 5 & 6 \\ 9 & 8 & 7 \end{vmatrix} = 1(-13)-(-2)(-26)+3(-13) = -13-52-39 = -104,$ using a Laplace expansion along the first row.

- $\begin{vmatrix} 6 & -4 & 0 & -5 \\ 3 & -7 & -2 & 2 \\ -9 & 5 & -5 & 12 \\ -3 & 5 & 1 & 0 \end{vmatrix} = -\begin{vmatrix} 3 & -7 & -2 & 2 \\ 6 & -4 & 0 & -5 \\ -9 & 5 & -5 & 12 \\ -3 & 5 & 1 & 0 \end{vmatrix}$ interchanging rows one and two

$= -\begin{vmatrix} 3 & -7 & -2 & 2 \\ 0 & 10 & 4 & -9 \\ 0 & -16 & -11 & 18 \\ 0 & -2 & -1 & 2 \end{vmatrix}$ using the 3 in the $(1,1)$ position to put 0s in the rest of column one

$= \begin{vmatrix} 3 & -7 & -2 & 2 \\ 0 & -2 & -1 & 2 \\ 0 & -16 & -11 & 18 \\ 0 & 10 & 4 & -9 \end{vmatrix}$ interchanging rows two and four

$= \begin{vmatrix} 3 & -7 & -2 & 2 \\ 0 & -2 & -1 & 2 \\ 0 & 0 & -3 & 2 \\ 0 & 0 & -1 & 1 \end{vmatrix}$ using the -2 in the $(2,2)$ position to put 0s below it

$= -\begin{vmatrix} 3 & -7 & -2 & 2 \\ 0 & -2 & -1 & 2 \\ 0 & 0 & -1 & 1 \\ 0 & 0 & -3 & 2 \end{vmatrix}$ interchanging rows three and four

$= -\begin{vmatrix} 3 & -7 & -2 & 2 \\ 0 & -2 & -1 & 2 \\ 0 & 0 & -1 & 1 \\ 0 & 0 & 0 & -1 \end{vmatrix}$ using the -1 in the $(3,3)$ position to put a 0 below it

$= -3(-2)(-1)(-1) = 6$ because the determinant of a triangular matrix is the product of its diagonal entries.

- $\begin{vmatrix} 1 & 2 & 3 & 4 & 5 \\ 6 & 7 & 8 & 9 & 10 \\ 1 & 2 & 3 & 4 & 5 \\ 10 & 9 & 8 & 7 & 6 \\ 5 & 4 & 3 & 2 & 1 \end{vmatrix} = 0$ because there are two equal rows. No calculation required here!

- The three 0s in column three quickly turn the next 4×4 into a 3×3 determinant.

$$\begin{vmatrix} 2 & -3 & 0 & 7 \\ 9 & 2 & 0 & -3 \\ -5 & 6 & -4 & 8 \\ -7 & 0 & 0 & 2 \end{vmatrix} = -4 \begin{vmatrix} 2 & -3 & 7 \\ 9 & 2 & -3 \\ -7 & 0 & 2 \end{vmatrix}$$

$$= -4\left[-(-3)\begin{vmatrix} 9 & -3 \\ -7 & 2 \end{vmatrix} + 2\begin{vmatrix} 2 & 7 \\ -7 & 2 \end{vmatrix}\right] \quad \text{with a Laplace expansion down the second column}$$

$$= -4[3(-3) + 2(53)] = -4[-9 + 106] = -4(97) = -388.$$

:)

Answers to Reading Checks

1. Any identity matrix is diagonal with 1s on the diagonal. Its determinant is $+1$. The additive inverse of the identity has -1s on the diagonal, so the 101×101 matrix which is $-I$ has determinant $(-1)^{101} = -1$.

2. For a fixed vector v in R^3,

   i. $(\mathsf{x}_1 + \mathsf{x}_2) \times \mathsf{v} = (\mathsf{x}_1 \times \mathsf{v}) + (\mathsf{x}_2 \times \mathsf{v})$ and

   ii. $(c\mathsf{x}) \times \mathsf{v} = c(\mathsf{x} \times \mathsf{v})$

   for all vectors $\mathsf{x}, \mathsf{x}_1, \mathsf{x}_2$ in R^3 and all scalars c.

3. Using 3.2.7, $\det(-A^3B^{-1}A^T) = \det((-1)A^3B^{-1}A^T) = (-1)^n \det(A^3B^{-1}A^T)$, assuming A and B are $n \times n$, and, using the result of Problem 3.2.25, this is $(-1)^n \frac{16}{5}$. We cannot say more unless we know whether n is even or odd. If n is even, the answer is $+\frac{16}{5}$; if n is odd, it's $-\frac{16}{5}$.

True/False Questions

Decide, with as little calculation as possible, whether each of the following statements is true or false and, if you say "false," explain your answer. (Answers are at the back of the book.)

1. A triangular matrix with no 0s on the diagonal is invertible.

2. The determinant of $\begin{bmatrix} 1 & -2 & 4 & 3 \\ 7 & 8 & -3 & 0 \\ 1 & -2 & 4 & 3 \end{bmatrix}$ is 0.

3. $\det(-A) = -\det A$ for any square matrix A.

4. $\det \frac{1}{5}\begin{bmatrix} 4 & 3 \\ 1 & 2 \end{bmatrix} = 1$.

5. A matrix is singular if its determinant is 0.

6. If A is 3×3, $\det A = 5$ and C is the matrix of cofactors of A, then $\det C^T = 25$.

7. An elementary row operation does not change the determinant.

8. If A is a square matrix and $\det A = 0$, then A must have a row of 0s.

9. If A and B are square matrices of the same size, then $\det(A+B) = \det A + \det B$.

10. If A and B are square matrices of the same size, then $\det AB = \det BA$.

EXERCISES

Answers to exercises marked [BB] can be found at the **Back of the Book**.

1. [BB] What does it mean when we say that the cross product is "linear in its second variable?"

2. If x is a real number, the *floor* of x, denoted $\lfloor x \rfloor$, is the greatest integer less than or equal to x. The function $\lfloor \ \rfloor \colon \mathsf{R} \to \mathsf{Z}$ is called the *floor function* (sometimes the *greatest integer function*). Formally, if n is the integer such that $n \le x < n+1$, then $\lfloor x \rfloor = n$.

   (a) Determine whether or not $\lfloor x + y \rfloor = \lfloor x \rfloor + \lfloor y \rfloor$.

   (b) Determine whether or not $\lfloor cx \rfloor = c\lfloor x \rfloor$ for a scalar c.

   (c) Is the floor function linear? Explain.

3. Given $\begin{vmatrix} -1 & 7 & 3 & 4 \\ 0 & -5 & -3 & 2 \\ 6 & 1 & -1 & 3 \\ 2 & 2 & 0 & 1 \end{vmatrix} = 118$, find

   (a) [BB] $\begin{vmatrix} -1 & 7 & 3 & 4 \\ 0 & -5 & -3 & 2 \\ 6 & 1 & -1 & 3 \\ -1 & 7 & 3 & 4 \end{vmatrix}$ **(b)** $\begin{vmatrix} -1 & 7 & 3 & 4 \\ 0 & -5 & -3 & 2 \\ 12 & 2 & -2 & 6 \\ 2 & 2 & 0 & 1 \end{vmatrix}$

   (c) $\begin{vmatrix} 8 & 7 & 3 & 4 \\ 4 & -5 & -3 & 2 \\ 6 & 1 & -1 & 3 \\ 2 & 2 & 0 & 1 \end{vmatrix}$ **(d)** $\begin{vmatrix} -1 & 7 & 3 & 4 \\ 6 & 1 & -1 & 3 \\ 0 & -5 & -3 & 2 \\ 2 & 2 & 0 & 1 \end{vmatrix}$

4. Let $A = \begin{bmatrix} 1 & 2 & 3 & 4 \\ 0 & -1 & 1 & 2 \\ 1 & 2 & 3 & 4 \\ 8 & -2 & 5 & 5 \end{bmatrix}$. Find $\det A$ by inspection and explain.

5. [BB] Let A and B be 5×5 matrices with $\det(-3A) = 4$ and $\det B^{-1} = 2$. Find $\det A$, $\det B$ and $\det AB$.

6. If A and B are 5×5 matrices, $\det A = 4$ and $\det B = \frac{1}{2}$, find $\det[B^{-1}A^2(-B)^T]$.

7. [BB] Let $A = \begin{bmatrix} 1 & 7 & 0 & 17 & 9 \\ 0 & 2 & -35 & 10 & 15 \\ 0 & 0 & 3 & -5 & 11 \\ 0 & 0 & 0 & 2 & 77 \\ 0 & 0 & 0 & 0 & 5 \end{bmatrix}$. Find $\det A$, $\det A^{-1}$ and $\det A^2$.

8. Suppose A and B are 3×3 matrices with $\det A = 2$ and $\det B = -5$. Let $X = A^T B^{-1} A^3(-B)$. Find $\det X$ and $\det(-X)$.

9. Let A and B be 3×3 matrices such that $\det A = 10$ and $\det B = 12$. Find $\det AB$, $\det A^4$, $\det 2B$, $\det(AB)^T$, $\det A^{-1}$ and $\det(A^{-1}B^{-1}AB)$.

10. A 6×6 matrix A has determinant 5. Find $\det 2A$, $\det(-A)$, $\det A^{-1}$ and $\det A^2$.

11. [BB] Given that A is a 4×4 matrix and $\det A = 2$, find $\det(15A^{-1} - 6 \operatorname{adj} A)$.

12. Let A be a (square) matrix satisfying $\det A^2 + 3 \det A^T = -2$. Find $\det A$ and $\det A^{-1}$.

13. If P is a matrix, $P \neq I$, and $P^2 = P$, then P is square. Why? Find $\det P$.

14. Let $B = \begin{bmatrix} 1 & 2 & 1 \\ 3 & -2 & 0 \\ -1 & 4 & 1 \end{bmatrix}$.

    (a) Find $\det B$, $\det \frac{1}{3}B$ and $\det B^{-1}$.

    (b) Suppose A is a matrix whose inverse is B. Find the cofactor matrix of A.

15. Suppose A is a 4×4 matrix and P is an invertible 4×4 matrix such that

$$PAP^{-1} = D = \begin{bmatrix} 1 & 0 & 0 & 0 \\ 0 & -1 & 0 & 0 \\ 0 & 0 & -1 & 0 \\ 0 & 0 & 0 & -1 \end{bmatrix}.$$

    (a) Find $\det A$. **(b)** Find A^2. **(c)** Find A^{101}.

16. [BB] **(a)** Find the determinant of $A = \begin{bmatrix} -1 & -1 & 1 & 0 \\ 2 & 1 & 1 & 3 \\ 0 & 1 & 1 & 2 \\ 1 & 3 & -1 & 2 \end{bmatrix}$ with a Laplace expansion down the first column.

    (b) Find the determinant of A by reducing it to an upper triangular matrix.

17. (a) Find the determinant of $A = \begin{bmatrix} 5 & 15 & 10 & 0 \\ 4 & 11 & 12 & 2 \\ 7 & 25 & 0 & -2 \\ -3 & -14 & 17 & 23 \end{bmatrix}$ with a Laplace expansion down the fourth column.

(b) Find the determinant of A by a Laplace expansion along the third row.

(c) Find the determinant of A by reducing it to an upper triangular matrix.

(d) Is A invertible? If yes, what is $\det A^{-1}$? If no, why not?

18. Determine whether each of the given sets of vectors is linearly independent or linearly dependent.

(a) [BB] $\begin{bmatrix} 1 \\ -2 \end{bmatrix}, \begin{bmatrix} 3 \\ 4 \end{bmatrix}$ (b) $\begin{bmatrix} -1 \\ 2 \\ 1 \end{bmatrix}, \begin{bmatrix} -1 \\ 1 \\ 4 \end{bmatrix}, \begin{bmatrix} 1 \\ -3 \\ 2 \end{bmatrix}$

(c) $\begin{bmatrix} 1 \\ 2 \\ 3 \end{bmatrix}, \begin{bmatrix} 4 \\ 5 \\ 6 \end{bmatrix}, \begin{bmatrix} 7 \\ 8 \\ 9 \end{bmatrix}$ (d) $\begin{bmatrix} -1 \\ 0 \\ 2 \\ 3 \end{bmatrix}, \begin{bmatrix} 0 \\ 1 \\ 1 \\ 6 \end{bmatrix}, \begin{bmatrix} 2 \\ -4 \\ 0 \\ 0 \end{bmatrix}, \begin{bmatrix} -3 \\ 1 \\ 0 \\ 1 \end{bmatrix}$.

19. For what values of a are the vectors $\begin{bmatrix} a \\ 0 \\ 0 \end{bmatrix}, \begin{bmatrix} -1 \\ 2a+1 \\ 0 \end{bmatrix}, \begin{bmatrix} 2 \\ 3 \\ a^2 - 2a \end{bmatrix}$ linearly dependent?

20. Find the determinants of each of the following matrices by reducing to upper triangular matrices.

(a) [BB] $\begin{bmatrix} -2 & 1 & 2 \\ 1 & 3 & 6 \\ -4 & 5 & 9 \end{bmatrix}$ (b) $\begin{bmatrix} -3 & 1 & 1 \\ 0 & -7 & 1 \\ 3 & 2 & 1 \end{bmatrix}$

(c) $\begin{bmatrix} 8 & 1 & -2 \\ -6 & 1 & 1 \\ 4 & -2 & 3 \end{bmatrix}$ (d) [BB] $\begin{bmatrix} -3 & 0 & 1 & 1 \\ 3 & 1 & 2 & 2 \\ -6 & -2 & -4 & 2 \\ 1 & -1 & 0 & -1 \end{bmatrix}$

(e) $\begin{bmatrix} 4 & -1 & 3 & -1 \\ 0 & 1 & 2 & 2 \\ 3 & 1 & 0 & 2 \\ 1 & 2 & -1 & 1 \end{bmatrix}$ (f) $\begin{bmatrix} -2 & 1 & 1 & 2 \\ 1 & -2 & 1 & 2 \\ 4 & -3 & 5 & 1 \\ 0 & -2 & 2 & 3 \end{bmatrix}$

(g) $\begin{bmatrix} 1 & 0 & 0 & 1 & 0 \\ 0 & 2 & 4 & 0 & -8 \\ -3 & 0 & 1 & 0 & 5 \\ 4 & 1 & 0 & 1 & -2 \\ -1 & 0 & 1 & 0 & 1 \end{bmatrix}$ (h) $\begin{bmatrix} 1 & 0 & 0 & 0 & 5 \\ 0 & 0 & 1 & 0 & 1 \\ 2 & 5 & 0 & 0 & 0 \\ 0 & 0 & 1 & -4 & 0 \\ 0 & 1 & 0 & 1 & 0 \end{bmatrix}$

(i) $\begin{bmatrix} -3 & 0 & 1 & 1 & 0 \\ 3 & 1 & 2 & -2 & 2 \\ -6 & -2 & -4 & 1 & 2 \\ 1 & 0 & -4 & 4 & 5 \\ -1 & 0 & 0 & 0 & -1 \end{bmatrix}$ (j) $\begin{bmatrix} 0 & 6 & 42 & -18 & 12 \\ 1 & 3 & 4 & 5 & 1 \\ 2 & 11 & 40 & -20 & 6 \\ -3 & -12 & -27 & 25 & 7 \\ 0 & 2 & 11 & -14 & 29 \end{bmatrix}$.

21. Let $A = \begin{bmatrix} 1 & 2 & 3 \\ 1 & 1 & 1 \\ 1 & 0 & 1 \end{bmatrix}$ and $B = \begin{bmatrix} 2 & 0 & 1 \\ 1 & 1 & 0 \\ 0 & 1 & 1 \end{bmatrix}$. Find the matrices AB, A^TB, AB^T, A^TB^T and explain why they all have the same determinant. What is this number?

22. Let $A = \begin{bmatrix} a & b \\ c & d \end{bmatrix}$ and $B = \begin{bmatrix} a+c & 2c \\ b+d & 2d \end{bmatrix}$. If $\det A = 2$, find $\det(A^2B^TA^{-1})$.

23. [BB] If P is a permutation matrix and

$$PA = \begin{bmatrix} 1 & 6 & 7 \\ 5 & 25 & 36 \\ -2 & -27 & -4 \end{bmatrix} = \begin{bmatrix} 1 & 0 & 0 \\ 5 & 1 & 0 \\ -2 & 3 & 1 \end{bmatrix} \begin{bmatrix} 1 & 6 & 7 \\ 0 & -5 & 1 \\ 0 & 0 & 7 \end{bmatrix} = LU,$$

what are the possible values for $\det A$?

24. **(a)** [BB] Find the value or values of a that make $P(a) = \begin{bmatrix} 1 & 0 & 0 \\ a & 2 & 0 \\ 1 & 2a & a^2+a \end{bmatrix}$ singular.

(b) For which value or values of a are the columns of $P(a)$ linearly independent?

[Hint: Theorem 2.9.13.]

25. If A is $n \times n$, n is odd and $A^T = -A$, then A is singular. Why?

26. For what value(s) of k is $A = \begin{bmatrix} 1 & 2 & 3 \\ 3 & 4 & k \\ k & 6 & 7 \end{bmatrix}$ invertible?

27. Find the following determinants by methods of your choice.

(a) [BB] $\begin{vmatrix} 6 & -3 \\ 1 & -4 \end{vmatrix}$

(b) $\begin{vmatrix} -1 & 2 & 0 \\ 2 & -7 & -8 \\ 3 & 1 & 5 \end{vmatrix}$

(c) [BB] $\begin{vmatrix} 5 & 3 & 8 \\ -4 & 1 & 4 \\ -2 & 3 & 6 \end{vmatrix}$

(d) $\begin{vmatrix} 4 & 28 & 16 & 12 \\ 12 & 81 & 27 & 18 \\ 4 & 31 & 36 & 28 \\ 1 & 15 & 56 & 41 \end{vmatrix}$

(e) [BB] $\begin{vmatrix} -1 & 2 & 3 & 4 \\ 3 & -9 & 2 & 1 \\ 0 & -5 & 7 & 6 \\ 2 & -4 & -6 & -8 \end{vmatrix}$

(f) $\begin{vmatrix} 1 & 0 & 0 & 0 & 1 \\ 0 & 0 & 1 & 0 & 1 \\ 1 & 1 & 0 & 0 & 0 \\ 0 & 0 & 1 & 1 & 0 \\ 0 & 1 & 0 & 1 & 0 \end{vmatrix}$

(g) $\begin{vmatrix} 1 & -1 & 2 & 0 & -2 \\ 0 & 1 & 0 & 4 & 1 \\ 1 & 1 & 3 & 0 & 0 \\ 0 & 0 & 0 & 3 & -1 \\ 0 & 0 & 0 & 1 & 1 \end{vmatrix}$

(h) $\begin{vmatrix} 12 & 48 & 72 & 36 \\ 3 & 17 & 28 & -26 \\ 2 & 11 & 11 & 6 \\ -2 & -10 & -21 & 31 \end{vmatrix}$.

28. [BB] Find $\begin{vmatrix} a & -g & 2d \\ b & -h & 2e \\ c & -i & 2f \end{vmatrix}$ given $\begin{vmatrix} a & b & c \\ d & e & f \\ g & h & i \end{vmatrix} = -3$.

29. Let $A = \begin{bmatrix} a & b & c \\ p & q & r \\ u & v & w \end{bmatrix}$ and suppose $\det A = 5$. Find $\begin{vmatrix} 2p & -a+u & 3u \\ 2q & -b+v & 3v \\ 2r & -c+w & 3w \end{vmatrix}$.

30. [BB] Given that $\det \begin{bmatrix} a & b & c \\ p & q & r \\ x & y & z \end{bmatrix} = 4$, find $\det \begin{bmatrix} a+2x & b+2y & c+2z \\ x+p & y+q & z+r \\ 3p & 3q & 3r \end{bmatrix}$.

31. Let $A = \begin{bmatrix} a & b & c \\ p & q & r \\ x & y & z \end{bmatrix}$ and $B = \begin{bmatrix} 2x & a+2p & p-3x \\ 2y & b+2q & q-3y \\ 2z & c+2r & r-3z \end{bmatrix}$. Given $\det A = 3$, find $\det(-2B^{-1})$.

32. Given that the determinant $\begin{vmatrix} a & b & c \\ u & v & w \\ x & y & z \end{vmatrix} = 5$, find $\begin{vmatrix} x & y & z \\ a+u & b+v & c+w \\ a-u & b-v & c-w \end{vmatrix}$.

33. Let $A = \begin{bmatrix} a+h+g & h+b+f & g+f+c \\ 2a+3h+4g & 2h+3b+4f & 2g+3f+4c \\ 3a+2h+5g & 3h+2b+5f & 3g+2f+5c \end{bmatrix}$ and $B = \begin{bmatrix} a & h & g \\ h & b & f \\ g & f & c \end{bmatrix}$.
Show that $\det A = 4 \det B$.

34. **(a)** Let a, b, c be three real numbers and let $A = \begin{bmatrix} 1 & 1 & 1 \\ a & b & c \\ a^2 & b^2 & c^2 \end{bmatrix}$.

Show that $\det A = (a-b)(a-c)(b-c)$.

(b) Guess the value of $\det \begin{bmatrix} 1 & 1 & 1 & 1 \\ a & b & c & d \\ a^2 & b^2 & c^2 & d^2 \\ a^3 & b^3 & c^3 & d^3 \end{bmatrix}$, with a, b, c, d real numbers. [No proof required.]

35. Show that $\begin{vmatrix} 1 & a & p & q \\ x & 1 & b & r \\ x^2 & x & 1 & c \\ x^3 & x^2 & x & 1 \end{vmatrix} = (1-ax)(1-bx)(1-cx)$ for any choice of p, q and r.

36. [BB] Show that $\det \begin{bmatrix} 1 & x & x^2 & x^3 \\ x & x^2 & x^3 & 1 \\ x^2 & x^3 & 1 & x \\ x^3 & 1 & x & x^2 \end{bmatrix} = (x^4-1)^3$.

37. Show that $\det AB = (\det A)(\det B)$ as suggested in the text by establishing each of the following statements.

(a) $\det(AB) = (\det A)(\det B)$ if A is not invertible.

It remains to obtain the same result when A is invertible.

(b) $\det(EB) = (\det E)(\det B)$ for any elementary matrix E.

(c) $\det(E_1 E_2 \cdots E_k B) = (\det E_1)(\det E_2) \cdots (\det E_k)(\det B)$ for any matrix B and any elementary matrices $E_1, E_2, \ldots, E_k$.

(d) $\det AB = (\det A)(\det B)$ for any invertible matrix A.

38. State at least three conditions on the rows of a matrix A that guarantee that A is singular.

39. Explain why a square triangular matrix with no zero entries on its diagonal must be invertible.

Critical Reading

40. Let E be the $n \times n$ elementary matrix obtained by interchanging rows i and j of the $n \times n$ identity matrix I. Let A be any $n \times n$ matrix. Find $\det(EA)$.

41. If A is invertible, $\det A$ is the product of the pivots of A. Why?

42. [BB] Suppose A is a square matrix. Explain how an LDU factorization of A would help finding the determinant of A.

Illustrate with $A = \begin{bmatrix} 2 & -1 & 4 & 1 \\ 1 & 1 & -10 & -2 \\ 4 & 0 & -7 & 6 \\ 6 & -3 & 0 & 1 \end{bmatrix}$.

43. Suppose that A is a square matrix and the sum of the entries in each column is 0. Explain why $\det A = 0$.

44. [BB] Suppose each column of a 3×3 matrix A is a linear combination of two vectors u and v. Find an argument involving the determinant that shows that A is singular.

45. Let $A = \begin{bmatrix} 3 & 1 & 1 & \cdots & 1 \\ 1 & 3 & 1 & \cdots & 1 \\ 1 & 1 & 3 & & 1 \\ \vdots & \vdots & & \ddots & \\ 1 & 1 & \cdots & 1 & 3 \end{bmatrix}$ be $n \times n$.

   (a) Find $\det A$. [Hint: Investigate small values of n and look for a way to apply elementary row operations.]

   (b) Does there exist a matrix B, with integer entries, such that $A = BB^T$ when $n = 40$? Explain.

46. [BB] Suppose A is an $n \times n$ invertible matrix and $A^{-1} = \frac{1}{2}(I - A)$, where I is the $n \times n$ identity matrix. Show that $\det A$ is a power of 2.

47. If A is an $n \times n$ matrix with every entry an odd integer, explain why $\det A$ must be divisible by 2^{n-1}.

48. Let A be a 3×3 matrix with columns $\mathsf{a}_1, \mathsf{a}_2, \mathsf{a}_3$, respectively. Let B be the matrix with columns $\mathsf{a}_1 + \alpha \mathsf{a}_2$, $\mathsf{a}_2 + \beta \mathsf{a}_1$, a_3, respectively, for scalars α and β. Show that $\det B = (1 - \alpha\beta) \det A$. [Use the notation $\det(\mathsf{a}_1, \mathsf{a}_2, \mathsf{a}_3)$ for $\det A$.]

49. Let $A = \begin{bmatrix} 4x & -x-y & 2y+z \\ 4x+2y-z & -x-3y+z & 2x+4y+z \\ x+y & -x-y & 3x+y+z \end{bmatrix}$ and $B = \begin{bmatrix} 1 & 1 & 1 \\ 2 & 3 & 1 \\ 1 & 1 & 0 \end{bmatrix}$.

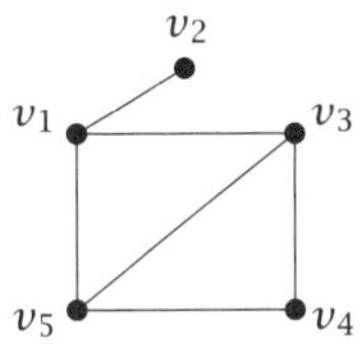

$$A = \begin{bmatrix} 0 & 1 & 1 & 0 & 1 \\ 1 & 0 & 0 & 0 & 0 \\ 1 & 0 & 0 & 1 & 1 \\ 0 & 0 & 1 & 0 & 1 \\ 1 & 0 & 1 & 1 & 0 \end{bmatrix}$$

Figure 3.3: A graph and its adjacency matrix

(a) Find AB.

(b) Express $\det A$ as the product of factors that are linear in x, y and z.

50. Let $A = \begin{bmatrix} a_1 & a_2 & a_3 \\ b_1 & b_2 & b_3 \\ c_1 & c_2 & c_3 \end{bmatrix}$ and $B = \begin{bmatrix} a_1 & a_2 & a_3 & 0 \\ a_1b_1 + a_2c_1 & a_1b_2 + a_2c_2 & a_1b_3 + a_2c_3 & a_3 \\ b_1b_1 + b_2c_1 & b_1b_2 + b_2c_2 & b_1b_3 + b_2c_3 & b_3 \\ c_1b_1 + c_2c_1 & c_1b_2 + c_2c_2 & c_1b_3 + c_2c_3 & c_3 \end{bmatrix}$.
Prove that $\det B = (\det A)^2$. [Hint: $\det XY = \det X \det Y$.]

51. Let $A = \begin{bmatrix} B & D \\ 0 & C \end{bmatrix}$ with B and D square matrices. Prove that $\det A = \det B \det C$.

3.3 Application: Graphs

A *graph* is a pair $(\mathcal{V}, \mathcal{E})$ of sets with $\mathcal{V}$ nonempty and each element of $\mathcal{E}$ a set $\{u, v\}$ of two distinct elements of $\mathcal{V}$. The elements of $\mathcal{V}$ are called *vertices* and the elements of $\mathcal{E}$ are called *edges*. The edge $\{u, v\}$ is usually denoted just uv or vu. Edge vu is the same as uv because the sets $\{u, v\}$ and $\{v, u\}$ are the same.

One nice thing about graphs is that they can easily be pictured (if they are sufficiently small). Figure 3.3 shows a picture of a graph $\mathcal{G}$ with vertex set $\mathcal{V} = \{v_1, v_2, v_3, v_4, v_5\}$ and edge set $\mathcal{E} = \{v_1v_2, v_1v_3, v_1v_5, v_3v_4, v_4v_5, v_3v_5\}$. The five vertices are represented by solid dots, the edges by lines: there is a line between vertices v_i and v_j if v_iv_j is an edge.

Graphs arise in numerous contexts. In an obvious way, they can be used to illustrate a map of towns and connecting roads or a network of computers. Less obvious perhaps are applications in biochemistry concerned with the recovery of RNA and DNA chains from fragmentary data, to the design of facilities such as hospitals, where it is important that certain rooms be close to others (an operating room and intensive care unit, for example), and to examination scheduling, where the goal is to minimize conflicts (that is, two exams scheduled for the same time, with one or more students having to write both). For more detail about these and other applications of graph theory, we refer the reader to the text *Discrete Mathematics with Graph Theory* by E. G. Goodaire and M. M. Parmenter, 4th edition, Pearson/Prentice-Hall, 2006.

The vertices in Figure 3.3 might correspond to subjects—math, English, chemistry, physics, biology—with an edge between subjects signifying that exams in these subjects should not be scheduled at the same time. The reader might wish to convince herself that three examination periods are required for the problem pictured in the figure.

Why talk about graphs in a linear algebra book? In practice, graphs are often very large (lots of vertices) and it is only feasible to analyze a graph theoretical problem with the assistance of a high-speed computer. Since computers are blind, you can't give a computer a picture; instead, it is common to give a computer the *adjacency matrix* of the graph, this being a matrix whose entries consist only of 0s and 1s, with a 1 in position (i, j) if v_iv_j is an edge and otherwise a 0. The adjacency matrix of the graph in Figure 3.3 appears next to its picture.

READING CHECK 1. The adjacency matrix of a graph is symmetric. Why?

If A is the adjacency matrix of a graph, the powers $A, A^2, A^3, \ldots$ record the number of *walks* between pairs of vertices, where a walk is a sequence of vertices with an edge between each consecutive pair. The *length* of a walk is the number of edges in the walk, which is one less than the number of vertices. In the graph of Figure 3.3, v_1v_3 is a walk of length one from v_1 to v_3, $v_1v_3v_5v_4$ is a walk of length three from v_1 to v_4, and $v_1v_2v_1$ is a walk of length two from v_1 to v_1.

The (i, j) entry of the adjacency matrix A is the number of walks of length **one** from v_i to v_j since there exists such a walk if and only if there is an edge between v_i and v_j. Interestingly, the (i, j) entry of A^2 is the number of walks of length **two** from v_i to v_j, and the (i, j) entry of A^3 is the number of walks of length **three** from v_i to v_j. The general situation is summarized in the next proposition.

3.3.1 Proposition. *Let A be the adjacency matrix of a graph with vertices labelled $v_1, v_2, \ldots, v_n$. For any integer $k \geq 1$, the (i, j) entry of A^k is the number of walks of length k from v_i to v_j.*

This proposition can be proved directly using a technique called "mathematical induction," with which we do not presume all our readers to be familiar, so we omit the proof, contenting ourselves with illustrations here and the proof of two specific cases in the exercises. (See Exercise 3.)

In Figure 3.4, we reproduce the graph of Figure 3.3 and show the powers A^2 and A^3 of its adjacency matrix A.

The $(1, 1)$ entry of A^2 is 3, corresponding to the fact that there are three walks of length 2 from v_1 to v_1, namely, $v_1v_2v_1$, $v_1v_3v_1$ and $v_1v_5v_1$.

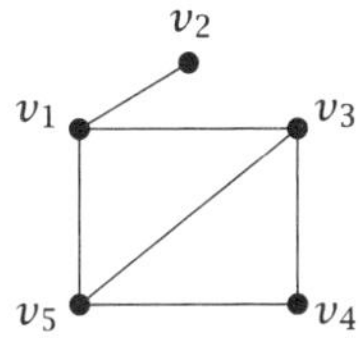

$$A^2 = \begin{bmatrix} 3 & 0 & 1 & 2 & 1 \\ 0 & 1 & 1 & 0 & 1 \\ 1 & 1 & 3 & 1 & 2 \\ 2 & 0 & 1 & 2 & 1 \\ 1 & 1 & 2 & 1 & 3 \end{bmatrix}, \quad A^3 = \begin{bmatrix} 2 & 3 & 6 & 2 & 6 \\ 3 & 0 & 1 & 2 & 1 \\ 6 & 1 & 4 & 5 & 5 \\ 2 & 2 & 5 & 2 & 5 \\ 6 & 1 & 5 & 5 & 4 \end{bmatrix}$$

Figure 3.4: The (i, j) entry of A^k is the number of walks of length k from v_i to v_j.

The $(3, 5)$ entry is 2, corresponding to the fact that there are two walks of length 2 from v_3 to v_5, namely, $v_3v_1v_5$ and $v_3v_4v_5$.

The $(1, 3)$ entry of A^3 is 6, corresponding to the six walks of length 3 from v_1 to v_3:

$$v_1v_2v_1v_3,\ v_1v_3v_1v_3,\ v_1v_5v_1v_3,\ v_1v_5v_4v_3,\ v_1v_3v_4v_3,\ v_1v_3v_5v_3.$$

READING CHECK 2. Explain why the $(2, 4)$ entry of A^3 is 2 and why the $(4, 3)$ entry of A^3 is 5.

Spanning Trees

A *cycle* in a graph is a walk from a vertex v back to v that passes through vertices that are all different (except the first and the last). In the graph shown to the right, $v_2v_3v_6v_2$ and $v_1v_5v_4v_6v_2v_1$ are cycles, but $v_1v_6v_4v_3v_6v_1$ is not.

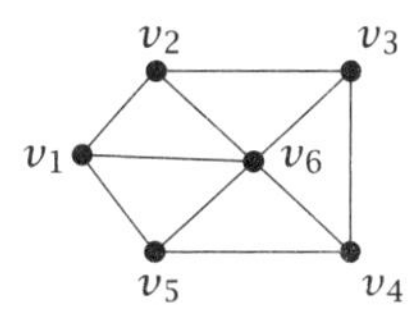

Figure 3.5

A *tree* is a connected graph that contains no cycles.[3] Several trees are illustrated in Figure 3.6. A *spanning tree* in a connected graph $\mathcal{G}$ is a tree

Figure 3.6: These are trees (of the graphical variety)

that contains every vertex of $\mathcal{G}$. A graph and three of its spanning trees are shown in Figure 3.7. A graph and all (eight of) its spanning trees are shown in Figure 3.8.

[3]"Connected" means just what you'd expect. There's a walk from any vertex to any other vertex along a sequence of edges.

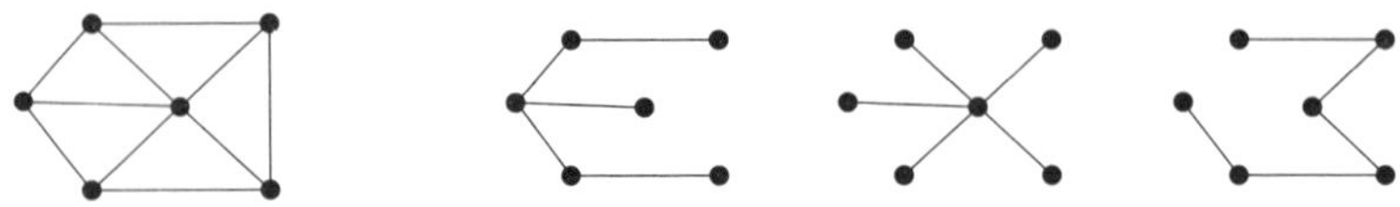

Figure 3.7: A graph and three of its spanning trees.

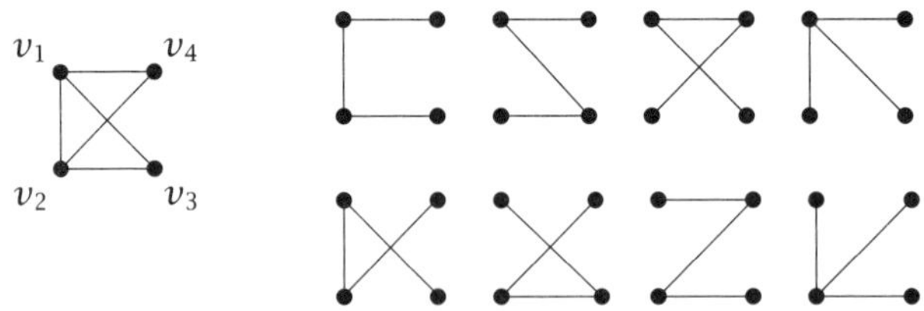

Figure 3.8: A graph and its eight spanning trees.

What do all these interesting pictures have to do with linear algebra? Well, it is of interest to know how many spanning trees a given graph has. Are we sure, for instance, that the graph in Figure 3.8 has just the spanning trees shown? Are there others we have missed?

In 1847, the German physicist Gustav Kirchhoff (1824–1887) found something quite remarkable.

3.3.2 Theorem (Kirchhoff).[4] *Let M be the matrix obtained from the adjacency matrix of a connected graph $\mathcal{G}$ by changing all 1s to -1s and each diagonal 0 to the degree of the corresponding vertex. Then all the cofactors of M are equal, and this common number is the number of spanning trees in $\mathcal{G}$.*

3.3.3 Example. The graph in Figure 3.8, labelled as shown, has adjacency matrix $A = \begin{bmatrix} 0 & 1 & 1 & 1 \\ 1 & 0 & 1 & 1 \\ 1 & 1 & 0 & 0 \\ 1 & 1 & 0 & 0 \end{bmatrix}$, so the matrix specified by Kirchhoff's Theorem is $M = \begin{bmatrix} 3 & -1 & -1 & -1 \\ -1 & 3 & -1 & -1 \\ -1 & -1 & 2 & 0 \\ -1 & -1 & 0 & 2 \end{bmatrix}$. Using a Laplace expansion along the first row, the $(1,1)$

[4]Also known as the "Matrix Tree Theorem."

cofactor of M is

$$\det\begin{bmatrix} 3 & -1 & -1 \\ -1 & 2 & 0 \\ -1 & 0 & 2 \end{bmatrix} = 3\begin{vmatrix} 2 & 0 \\ 0 & 2 \end{vmatrix} - (-1)\begin{vmatrix} -1 & 0 \\ -1 & 2 \end{vmatrix} + (-1)\begin{vmatrix} -1 & 2 \\ -1 & 0 \end{vmatrix}$$

$$= 3(4) + 1(-2) + (-1)2 = 8.$$

A Laplace expansion down the third column shows that the $(2,3)$ cofactor of M is

$$-\det\begin{bmatrix} 3 & -1 & -1 \\ -1 & -1 & 0 \\ -1 & -1 & 2 \end{bmatrix} = -\left[(-1)\begin{vmatrix} -1 & -1 \\ -1 & -1 \end{vmatrix} + 2\begin{vmatrix} 3 & -1 \\ -1 & -1 \end{vmatrix}\right]$$

$$= -[(-1)(0) + 2(-3-1)] = -(-8) = 8.$$

It is no coincidence that each of these cofactors is the same. This is guaranteed by Kirchhoff's Theorem. ☺

3.3.4 Problem. In Figure 3.7, we showed a graph $\mathcal{G}$ and three of its spanning trees. How many spanning trees does $\mathcal{G}$ have in all?

Solution. With vertices labelled as shown, the adjacency matrix of $\mathcal{G}$ is

$$A = \begin{bmatrix} 0 & 1 & 0 & 0 & 1 & 1 \\ 1 & 0 & 1 & 0 & 0 & 1 \\ 0 & 1 & 0 & 1 & 0 & 1 \\ 0 & 0 & 1 & 0 & 1 & 1 \\ 1 & 0 & 0 & 1 & 0 & 1 \\ 1 & 1 & 1 & 1 & 1 & 0 \end{bmatrix},$$

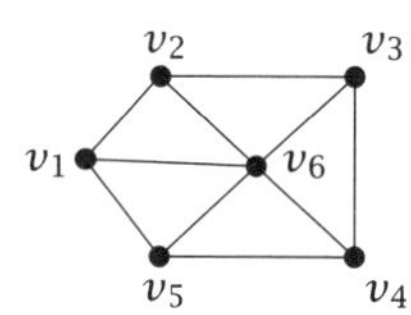

so the matrix specified by Kirchhoff's Theorem is

$$M = \begin{bmatrix} 3 & -1 & 0 & 0 & -1 & -1 \\ -1 & 3 & -1 & 0 & 0 & -1 \\ 0 & -1 & 3 & -1 & 0 & -1 \\ 0 & 0 & -1 & 3 & -1 & -1 \\ -1 & 0 & 0 & -1 & 3 & -1 \\ -1 & -1 & -1 & -1 & -1 & 5 \end{bmatrix}.$$

The $(1,1)$ cofactor is $\det\begin{bmatrix} 3 & -1 & 0 & 0 & -1 \\ -1 & 3 & -1 & 0 & -1 \\ 0 & -1 & 3 & -1 & -1 \\ 0 & 0 & -1 & 3 & -1 \\ 1 & -1 & -1 & -1 & 5 \end{bmatrix} = 121$, so the graph has 121 spanning trees. 👍

Isomorphism

The concept of "isomorphism" recurs throughout many fields of mathematics. Objects are isomorphic if they differ only in appearance rather than in some fundamental way.

In Figure 3.9, we show the pictures of two graphs that are fundamentally different—they have different numbers of edges. On the other hand, the graphs pictured in Figure 3.10 are the same, technically, *isomorphic.*

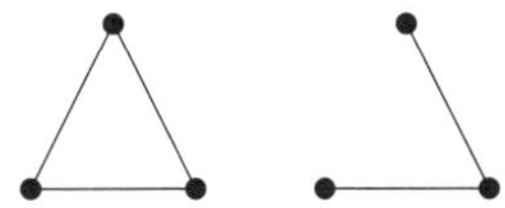

Figure 3.9: The graphs pictured here are not isomorphic.

3.3.5 Definition. Graphs $\mathcal{G}_1$ and $\mathcal{G}_2$ are *isomorphic* if their vertices can be labelled with the same symbols in such a way that uv is an edge in $\mathcal{G}_1$ if and only if uv is an edge in $\mathcal{G}_2$.

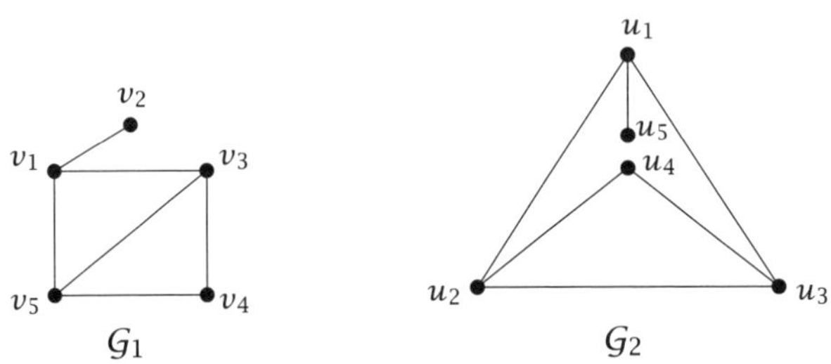

Figure 3.10: Two pictures of the same graph.

For instance, consider this relabelling of vertices of the graph $\mathcal{G}_1$, pictured on the left in Figure 3.10.

$$\begin{aligned} v_1 &\to u_1 \\ v_2 &\to u_5 \\ v_3 &\to u_2 \\ v_4 &\to u_4 \\ v_5 &\to u_3 \end{aligned}$$

The edges in each graph are now the same—u_1u_5, u_1u_2, u_1u_3, u_2u_4, u_2u_3, u_3u_4—so the graphs are isomorphic. Equivalently, the vertices of $\mathcal{G}_1$ have been rearranged so that the adjacency matrices of the two graphs are the same. Before rearranging, $\mathcal{G}_1$ and $\mathcal{G}_2$ had the adjacency matrices

$$A_1 = \begin{bmatrix} 0 & 1 & 1 & 0 & 1 \\ 1 & 0 & 0 & 0 & 0 \\ 1 & 0 & 0 & 1 & 1 \\ 0 & 0 & 1 & 0 & 1 \\ 1 & 0 & 1 & 1 & 0 \end{bmatrix} \quad \text{and} \quad A_2 = \begin{bmatrix} 0 & 1 & 1 & 0 & 1 \\ 1 & 0 & 1 & 1 & 0 \\ 1 & 1 & 0 & 1 & 0 \\ 0 & 1 & 1 & 0 & 0 \\ 1 & 0 & 0 & 0 & 0 \end{bmatrix},$$

respectively. Rearranging the vertices of $\mathcal{G}_1$ has the effect of rearranging the columns and the rows of A_1 in the order 15243, so that the rearranged matrix becomes A_2. Rearranging the columns of A_1 in order 15243 can be effected by multiplying this matrix on the right by the permutation matrix

$$P = \begin{bmatrix} \mathsf{e}_1 & \mathsf{e}_5 & \mathsf{e}_2 & \mathsf{e}_4 & \mathsf{e}_3 \\ \downarrow & \downarrow & \downarrow & \downarrow & \downarrow \\ & & & & \end{bmatrix} = \begin{bmatrix} 1 & 0 & 0 & 0 & 0 \\ 0 & 0 & 1 & 0 & 0 \\ 0 & 0 & 0 & 0 & 1 \\ 0 & 0 & 0 & 1 & 0 \\ 0 & 1 & 0 & 0 & 0 \end{bmatrix}$$

whose columns are the standard basis vectors in order 15243 because

$$A_1P = \begin{bmatrix} A_1\mathsf{e}_1 & A_1\mathsf{e}_5 & A_1\mathsf{e}_2 & A_1\mathsf{e}_4 & A_1\mathsf{e}_3 \\ \downarrow & \downarrow & \downarrow & \downarrow & \downarrow \\ & & & & \end{bmatrix}$$

and $A_1\mathsf{e}_i$ is column i of A_1. Rearranging the rows of A_1 in the order 15243 can be accomplished by multiplying A_1 on the left by P^T, the transpose of P. To see why, write $A_1 = \begin{bmatrix} \mathsf{a}_1 & \rightarrow \\ \mathsf{a}_2 & \rightarrow \\ \mathsf{a}_3 & \rightarrow \\ \mathsf{a}_4 & \rightarrow \\ \mathsf{a}_5 & \rightarrow \end{bmatrix}$ and note that

$$A_1^TP = \begin{bmatrix} A_1^T\mathsf{e}_1 & A_1^T\mathsf{e}_5 & A_1^T\mathsf{e}_2 & A_1^T\mathsf{e}_4 & A_1^T\mathsf{e}_3 \\ \downarrow & \downarrow & \downarrow & \downarrow & \downarrow \\ & & & & \end{bmatrix} = \begin{bmatrix} \mathsf{a}_1 & \mathsf{a}_5 & \mathsf{a}_2 & \mathsf{a}_4 & \mathsf{a}_3 \\ \downarrow & \downarrow & \downarrow & \downarrow & \downarrow \\ & & & & \end{bmatrix}.$$

Thus the rows of the transpose, which is P^TA_1, are $\mathsf{a}_1, \mathsf{a}_5, \mathsf{a}_2, \mathsf{a}_4, \mathsf{a}_3$. These arguments support a general fact.

3.3.6 Theorem. *Let graphs $\mathcal{G}_1$ and $\mathcal{G}_2$ have adjacency matrices A_1 and A_2, respectively. Then $\mathcal{G}_1$ is isomorphic to $\mathcal{G}_2$ if and only if there exists a permutation matrix P with the property that $P^TA_1P = A_2$.*

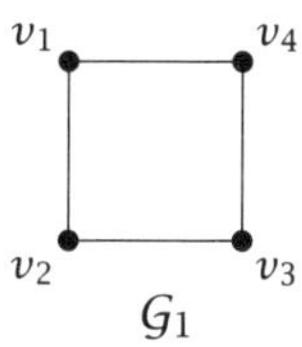

$$\begin{bmatrix} 0 & 1 & 0 & 1 \\ 1 & 0 & 1 & 0 \\ 0 & 1 & 0 & 1 \\ 1 & 0 & 1 & 0 \end{bmatrix}$$

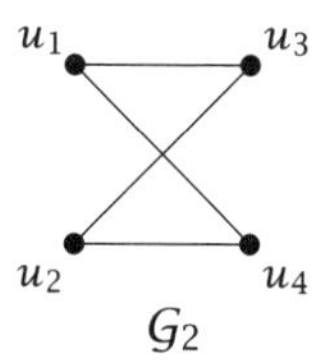

$$\begin{bmatrix} 0 & 0 & 1 & 1 \\ 0 & 0 & 1 & 1 \\ 1 & 1 & 0 & 0 \\ 1 & 1 & 0 & 0 \end{bmatrix}$$

Figure 3.11: The pictures of two isomorphic graphs and their adjacency matrices.

3.3.7 Example. The pictures of two isomorphic graphs, along with their adjacency matrices, are shown in Figure 3.11. If we relabel the vertices of $\mathcal{G}_1$,

$$\begin{aligned} v_1 &\to u_1 \\ v_2 &\to u_4 \\ v_3 &\to u_2 \\ v_4 &\to u_3, \end{aligned}$$

and form the permutation matrix $P = \begin{bmatrix} 1 & 0 & 0 & 0 \\ 0 & 0 & 0 & 1 \\ 0 & 1 & 0 & 0 \\ 0 & 0 & 1 & 0 \end{bmatrix}$ whose columns are the standard basis vectors in R^4 in the order 1423, then $P^T A_1 P = A_2$.

3.3.8 Problem. Let $A_1 = \begin{bmatrix} 0 & 1 & 0 & 1 & 0 \\ 1 & 0 & 1 & 0 & 1 \\ 0 & 1 & 0 & 1 & 0 \\ 1 & 0 & 1 & 0 & 1 \\ 0 & 1 & 0 & 1 & 0 \end{bmatrix}$ and $A_2 = \begin{bmatrix} 0 & 1 & 0 & 1 & 1 \\ 1 & 0 & 1 & 0 & 0 \\ 0 & 1 & 0 & 1 & 0 \\ 1 & 0 & 1 & 0 & 1 \\ 1 & 0 & 0 & 1 & 0 \end{bmatrix}$. Is there a permutation matrix P such that $P^T A_1 P = A_2$?

Solution. The matrices are the adjacency matrices of the graphs pictured in Figure 3.12. The graphs are not isomorphic since, for example, $\mathcal{G}_2$ contains

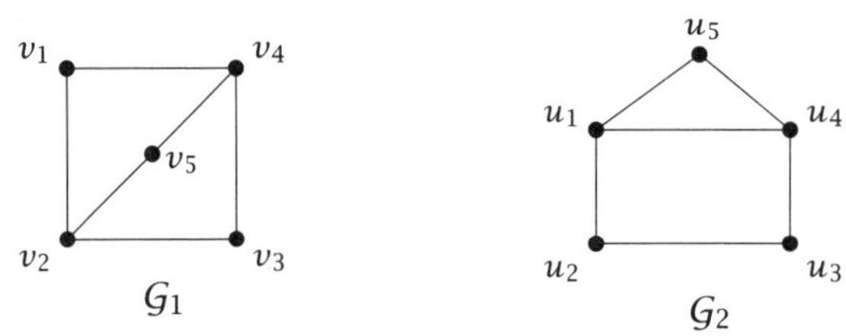

Figure 3.12

a *triangle*—three vertices u_1, u_4, u_5, each of which is joined by an edge—whereas $\mathcal{G}_1$ has no triangles. By Theorem 3.3.6, there is no permutation matrix P with $P^T A_1 P = A_2$.

Answers to Reading Checks

1. If there is an edge between v_i and v_j, there is an edge between v_j and v_i. So $a_{ij} = 1$ if and only if $a_{ji} = 1$.

2. There are two walks of length 3 from v_2 to v_4: $v_2v_1v_5v_4$ and $v_2v_1v_3v_4$. There are five walks of length 3 from v_4 to v_3: $v_4v_3v_4v_3$, $v_4v_5v_4v_3$, $v_4v_3v_1v_3$, $v_4v_3v_5v_3$, $v_4v_5v_1v_3$.

EXERCISES

Answers to exercises marked [BB] can be found at the Back of the Book.

1. [BB] If A is the adjacency matrix of a graph, the (i, i) entry of A^2 is the number of edges that meet at vertex v_i. Why?

2. Let A be the adjacency matrix of the graph shown in Figure 3.3. Find the $(1, 2)$ [BB], $(2, 5)$ and $(5, 4)$ $(3, 3)$ entries of A^4 without calculating this matrix. Explain your answers.

3. Let A be the adjacency matrix of a graph with vertices $v_1, v_2, \ldots$.

   (a) [BB] Explain why the (i, j) entry of A^2 is the number of walks of length two from v_i to v_j.

   (b) Use the result of (a) to explain why the (i, j) entry of A^3 is the number of walks of length three from v_i to v_j.

4. The "Petersen" graph shown on the left below is of interest in graph theory for many reasons.

   (a) What is the adjacency matrix of the Petersen graph, as shown?

   (b) Show that the graph on the right is isomorphic to the Petersen graph by labelling its vertices so that its adjacency matrix is the matrix of part (a).

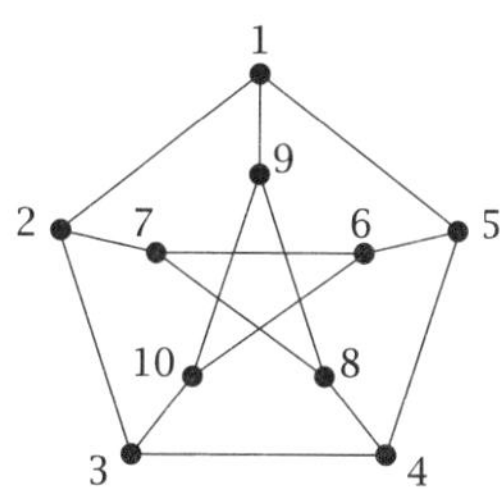

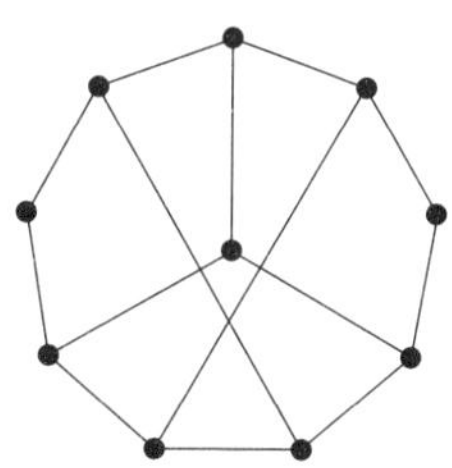

5. [BB] Let $\mathcal{T}$ be a tree and let u and v be vertices of $\mathcal{T}$ that are not joined by an edge. Adding edge uv to $\mathcal{T}$ must produce a cycle. Why?

6. There is a procedure for obtaining all the spanning trees of a graph that we describe with reference to the graph in Figure 3.5. First of all, delete edges one at a time until the graph that remains is a spanning tree, say edges v_6v_2, v_6v_3, v_6v_4, v_6v_5 and v_3v_4 in this order. This should produce the first spanning tree shown in Figure 3.7. Now replace edge v_3v_4.

   (a) [BB] Draw the current graph.

   (b) Your graph should contain exactly one cycle. What is this cycle? How many edges does it contain?

   (c) Delete edges of the cycle one at a time. In each case, you should be left with a spanning tree. Draw pictures of all such spanning trees.

7. Use Kirchhoff's Theorem to determine the number of spanning trees in each of the graphs shown.

(a) [BB] 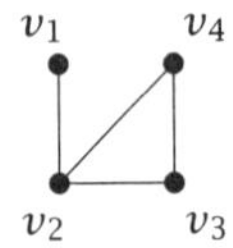

(b) 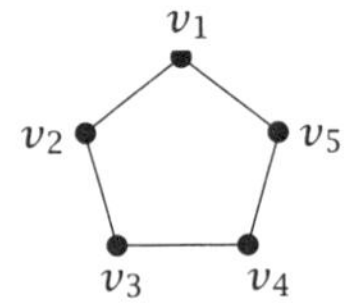

(c) 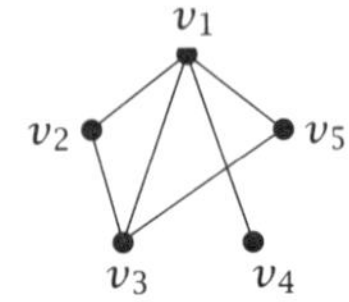

8. The *complete graph* on n vertices is the graph with n vertices and all possible edges. It is denoted $\mathcal{K}_n$.

   (a) Draw pictures of $\mathcal{K}_2$, $\mathcal{K}_3$, $\mathcal{K}_4$ [BB] and $\mathcal{K}_4$.

   (b) How many spanning trees has $\mathcal{K}_n$?

9. Let A be the adjacency matrix of the graph shown in Figure 3.8 and let M be the matrix specified by Kirchhoff's Theorem. Find the $(1,4)$, the $(3,3)$ and the $(4,3)$ cofactors of M and comment on your findings.

10. Find the adjacency matrices A_1, A_2 of each pair of graphs shown below. If there is a permutation matrix P such that $P^T A_1 P = A_2$, write down P and explain how you found it.

    (a) [BB] $\mathcal{G}_1$: 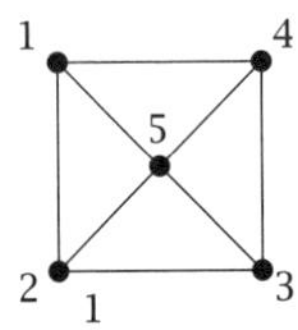

    $\mathcal{G}_2$: 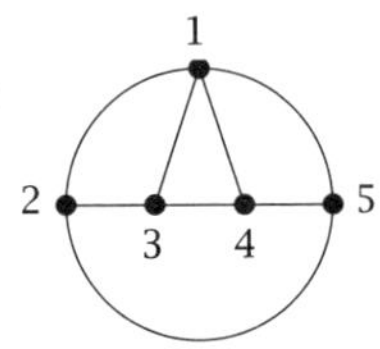

    (b) $\mathcal{G}_1$: 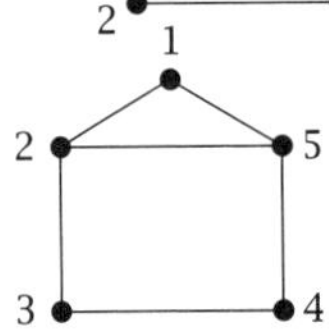

    $\mathcal{G}_2$: 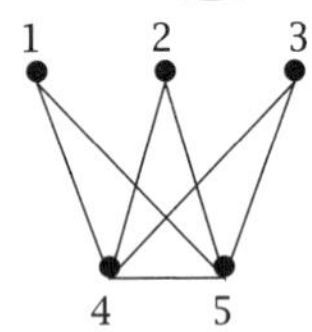

    (c) $\mathcal{G}_1$: 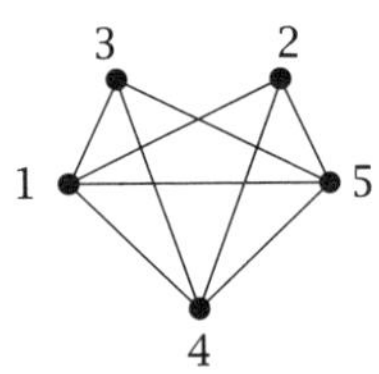

    $\mathcal{G}_2$: 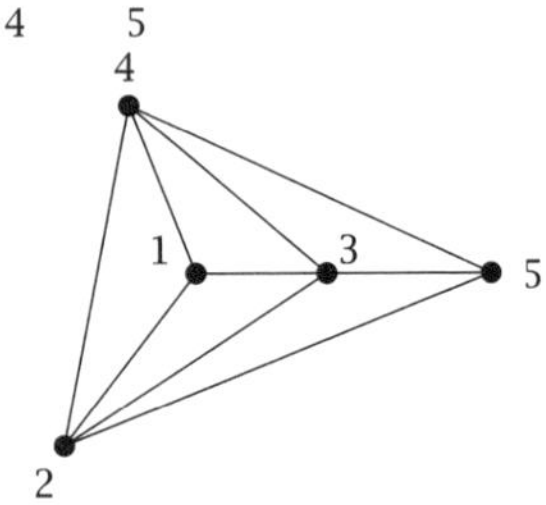

    (d) $\mathcal{G}_1$: 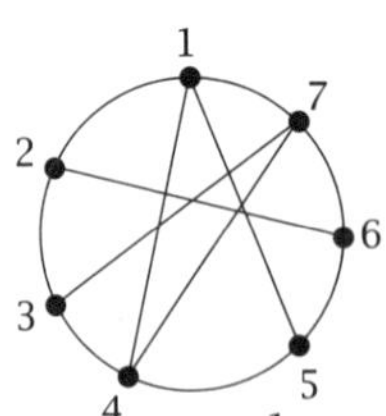

    $\mathcal{G}_2$: 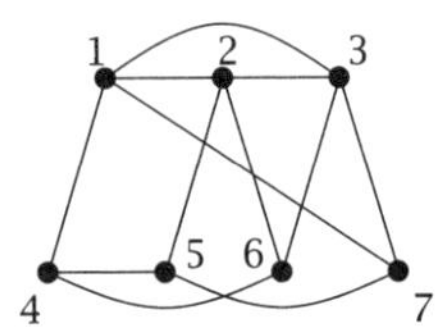

    (e) [BB] $\mathcal{G}_1$: 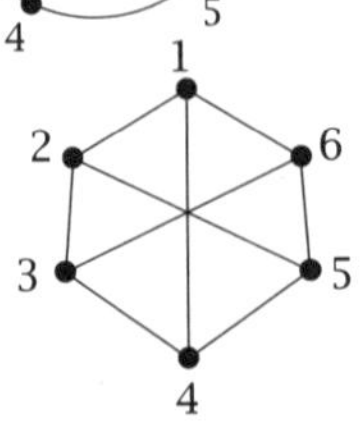

    $\mathcal{G}_2$: 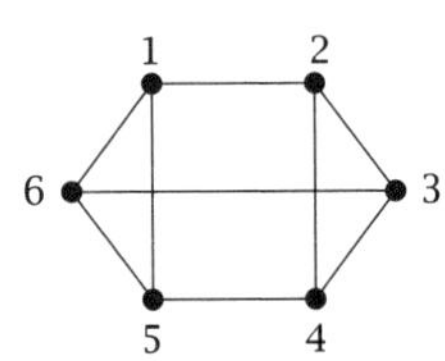

(f) $\mathcal{G}_1$:

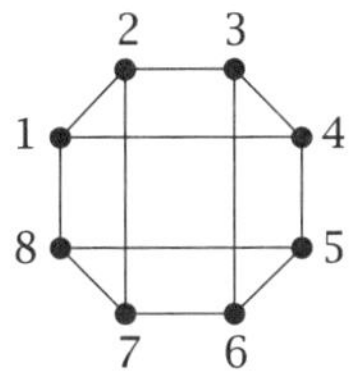

$\mathcal{G}_2$:

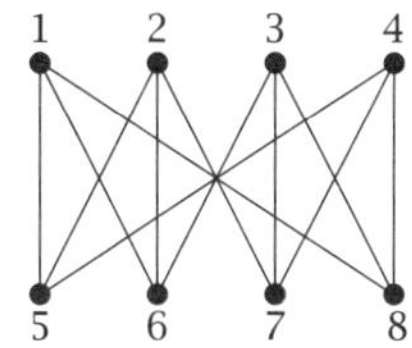

11. In each case, decide whether or not there exists a permutation matrix P with $A_2 = P^T A_1 P$. Explain your answers.

(a) [BB] $A_1 = \begin{bmatrix} 0 & 1 & 0 & 1 \\ 1 & 0 & 1 & 1 \\ 0 & 1 & 0 & 1 \\ 1 & 1 & 1 & 0 \end{bmatrix}$; $A_2 = \begin{bmatrix} 0 & 1 & 1 & 1 \\ 1 & 0 & 1 & 1 \\ 1 & 1 & 0 & 1 \\ 1 & 1 & 1 & 0 \end{bmatrix}$

(b) $A_1 = \begin{bmatrix} 0 & 1 & 0 & 0 & 1 \\ 1 & 0 & 1 & 0 & 0 \\ 0 & 1 & 0 & 1 & 0 \\ 0 & 0 & 1 & 0 & 1 \\ 1 & 0 & 1 & 1 & 0 \end{bmatrix}$; $A_2 = \begin{bmatrix} 0 & 1 & 0 & 0 & 1 \\ 1 & 0 & 1 & 0 & 1 \\ 0 & 1 & 0 & 1 & 0 \\ 0 & 0 & 1 & 0 & 1 \\ 1 & 1 & 0 & 1 & 0 \end{bmatrix}$

(c) $A_1 = \begin{bmatrix} 0 & 0 & 0 & 0 & 1 \\ 0 & 0 & 1 & 0 & 1 \\ 0 & 1 & 0 & 1 & 1 \\ 0 & 0 & 1 & 0 & 1 \\ 1 & 1 & 1 & 1 & 0 \end{bmatrix}$; $A_2 = \begin{bmatrix} 0 & 0 & 1 & 1 & 0 \\ 0 & 0 & 1 & 1 & 0 \\ 1 & 1 & 0 & 1 & 1 \\ 1 & 1 & 1 & 0 & 0 \\ 0 & 0 & 1 & 0 & 0 \end{bmatrix}$

(d) $A_1 = \begin{bmatrix} 0 & 1 & 0 & 1 & 0 & 1 \\ 1 & 0 & 1 & 1 & 0 & 0 \\ 0 & 1 & 0 & 1 & 1 & 0 \\ 1 & 0 & 1 & 0 & 1 & 0 \\ 0 & 0 & 1 & 1 & 0 & 1 \\ 1 & 1 & 0 & 0 & 1 & 0 \end{bmatrix}$; $A_2 = \begin{bmatrix} 0 & 0 & 0 & 1 & 1 & 1 \\ 0 & 0 & 0 & 1 & 1 & 1 \\ 0 & 0 & 0 & 1 & 1 & 1 \\ 1 & 1 & 1 & 0 & 0 & 0 \\ 1 & 1 & 1 & 0 & 0 & 0 \\ 1 & 1 & 1 & 0 & 0 & 0 \end{bmatrix}$

(e) $A_1 = \begin{bmatrix} 0 & 1 & 1 & 1 & 1 & 1 \\ 1 & 0 & 1 & 0 & 0 & 0 \\ 1 & 1 & 0 & 1 & 0 & 0 \\ 1 & 0 & 1 & 0 & 1 & 0 \\ 1 & 0 & 0 & 1 & 0 & 1 \\ 1 & 0 & 0 & 0 & 1 & 0 \end{bmatrix}$; $A_2 = \begin{bmatrix} 0 & 1 & 1 & 0 & 0 & 1 \\ 1 & 0 & 1 & 1 & 0 & 0 \\ 1 & 1 & 0 & 1 & 1 & 1 \\ 0 & 1 & 1 & 0 & 1 & 0 \\ 0 & 0 & 1 & 1 & 0 & 0 \\ 1 & 0 & 1 & 0 & 0 & 0 \end{bmatrix}$.

3.4 Eigenvalues and Eigenvectors

A common topic on the syllabus of a first year calculus course is "exponential growth and decay." Bacteria grow at a rate proportional to the size of the culture. Radioactive materials decay at a rate proportional to their mass. A cup of coffee cools at a rate proportional to the difference between the temperature of the coffee and the temperature of the surroundings. Each of these scenarios corresponds to an equation of the form $y'(t) = ky(t)$, k a constant, the solution to which is $y(t) = y_0 e^{kt}$.

Let's consider a 2-dimensional version of this problem. Suppose $\mathsf{v}(t) = \begin{bmatrix} v_1(t) \\ v_2(t) \end{bmatrix}$ is a vector and we know that $\mathsf{v}'(t) = A\mathsf{v}(t)$ where A is a 2×2 matrix of constants. Guided by previous experience with growth and decay problems, we might expect a solution of the form $\mathsf{v}(t) = e^{\lambda t}\mathsf{v}_0$. The Greek letter λ, pronounced "lambda," is preferred over k as the constant in this situation and the vector $\mathsf{v}_0 = \begin{bmatrix} v_1(0) \\ v_2(0) \end{bmatrix}$ gives the initial values of $v_1(t)$ and $v_2(t)$. If $\mathsf{v}(t) = e^{\lambda t}\mathsf{v}_0$ is indeed a solution to $\mathsf{v}'(t) = A\mathsf{v}(t)$, then $\lambda e^{\lambda t}\mathsf{v}_0 = Ae^{\lambda t}\mathsf{v}_0$. Since $e^{\lambda t} \neq 0$, we can divide by this factor and get $A\mathsf{v}_0 = \lambda\mathsf{v}_0$, a famous equation that we study in this section.

Are you impressed with Google's searching capability? Considering that there are at least 25 billion pages on the web and probably only about 10000 different words used on all those pages put together, we would expect a given word to appear on 2.5 million web pages on average. So when you search "butler, England, Halifax," how does Google find four million pages in a fifth of a second with the one you wanted at or very near the top? The answer lies in a procedure called the "PageRank Algorithm" that ranks pages according to their importance.

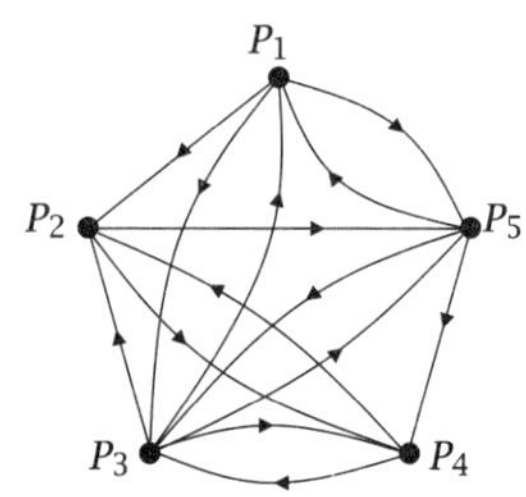

Figure 3.13: The arrows correspond to web pages and an arrow from P_i to P_j indicates a link from page P_i to page P_j.

Here's a very simple example that shows how this algorithm works. In Figure 3.13, we see what is called a *directed graph* consisting of five *vertices* (the dots) and some arrows between pairs of vertices. Each vertex corresponds to a web page and an arrow from P_i to P_j indicates that there is a link from page P_i to page P_j. The algorithm produces a vector $\mathsf{x} = \begin{bmatrix} x_1 \\ x_2 \\ x_3 \\ x_4 \\ x_5 \end{bmatrix}$ with x_i a measure of the importance of page P_i, the presumption being that the importance of a web page is proportional to the number of links from other pages to that particular web page. We assume that a given page shares its importance equally amongst the pages to which it points, so if there are n arrows pointing from P_i to other vertices of which P_j is one, P_j acquires $\frac{1}{n}$ of the importance of P_i.

Let's return to our example. The importance of P_1 derives from the importance of the two pages, P_3 and P_5 which have links to it. There are four arrows out of P_3 and three out of P_5, so P_1 acquires one quarter the importance of P_3 and one third the importance of P_5: $x_1 = \frac{1}{4}x_3 + \frac{1}{3}x_5$. Similarly, page P_2 derives its importance from P_1, P_3 and P_4. The three arrows out of P_1 mean that the importance of P_1 is split three ways, one third going to P_2. This reasoning yields $x_2 = \frac{1}{3}x_1 + \frac{1}{4}x_3 + \frac{1}{2}x_4$ and, finally, to this system of equations:

$$\begin{aligned}
\tfrac{1}{4}x_3 + \tfrac{1}{3}x_5 &= x_1 \\
\tfrac{1}{3}x_1 + \tfrac{1}{4}x_3 + \tfrac{1}{2}x_4 &= x_2 \\
\tfrac{1}{3}x_1 + \tfrac{1}{2}x_4 + \tfrac{1}{3}x_5 &= x_3 \\
\tfrac{1}{2}x_2 + \tfrac{1}{4}x_3 + \tfrac{1}{3}x_5 &= x_4 \\
\tfrac{1}{3}x_1 + \tfrac{1}{2}x_2 + \tfrac{1}{4}x_3 &= x_5.
\end{aligned}$$

But this is just $A\mathsf{x} = \mathsf{x}$, with $A = \begin{bmatrix} 0 & 0 & \frac{1}{4} & 0 & \frac{1}{3} \\ \frac{1}{3} & 0 & \frac{1}{4} & \frac{1}{2} & 0 \\ \frac{1}{3} & 0 & 0 & \frac{1}{2} & \frac{1}{3} \\ 0 & \frac{1}{2} & \frac{1}{4} & 0 & \frac{1}{3} \\ \frac{1}{3} & \frac{1}{2} & \frac{1}{4} & 0 & 0 \end{bmatrix}$. Again we meet the equation $A\mathsf{x} = \lambda\mathsf{x}$, this time with $\lambda = 1$. The solution, by the way, is $\mathsf{x} = \begin{bmatrix} 0.125 \\ 0.214 \\ 0.226 \\ 0.231 \\ 0.205 \end{bmatrix}$, so page P_4 is the most important with P_3 a close second. Did you guess?

By the way, the matrix A in this example is easy to write down. The first column corresponds to page P_1 which shares its importance equally amongst pages P_2, P_3 and P_5, each getting $\frac{1}{3}$ the importance of P_1, and so on for the other columns.

READING CHECK 1. What matrix A corresponds to the graph in Figure 3.14? Determine the relative importance of the vertices.

This section revolves around the equation

$$A\mathsf{x} = \lambda\mathsf{x},$$

where A is a (necessarily square) matrix and x is a vector.

3.4.1 Definitions. An *eigenvalue* of a matrix A is a scalar λ with the property that $A\mathsf{x} = \lambda\mathsf{x}$ for some nonzero vector x. The vector x is called an *eigenvector of*

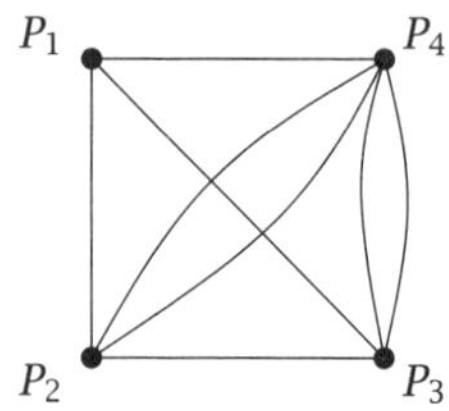

Figure 3.14: The arrows correspond to web pages and an arrow from P_i to P_j indicates a link from page P_i to page P_j.

A corresponding to λ. The set of all vectors x satisfying $A\mathsf{x} = \lambda\mathsf{x}$ is called the *eigenspace of A corresponding to λ.*

3.4.2 Remark. The equation $A\mathsf{x} = \lambda\mathsf{x}$ is not very interesting if we allow $\mathsf{x} = 0$ since, in this case, it holds for any scalar λ. This is why eigenvectors are required to be **nonzero** vectors. On the other hand, the zero vector belongs to the eigenspace of λ because $A0 = \lambda 0$, so the eigenspace for λ consists of all eigenvectors for λ, together with the zero vector.

3.4.3 Example. Let $A = \begin{bmatrix} 1 & 4 \\ 2 & 3 \end{bmatrix}$ and $\mathsf{x} = \begin{bmatrix} 1 \\ 1 \end{bmatrix}$. Then x is an eigenvector of A corresponding to $\lambda = 5$ because

$$A\mathsf{x} = \begin{bmatrix} 1 & 4 \\ 2 & 3 \end{bmatrix}\begin{bmatrix} 1 \\ 1 \end{bmatrix} = \begin{bmatrix} 5 \\ 5 \end{bmatrix} = 5\mathsf{x}.$$

The eigenspace of A corresponding to $\lambda = 5$ is the set of solutions to $A\mathsf{x} = 5\mathsf{x}$, or equally, rewriting $A\mathsf{x} = 5\mathsf{x}$ as $(A - 5I)\mathsf{x} = 0$, the set of solutions to the homogeneous system with coefficient matrix $A - 5I$. Gaussian elimination proceeds $A - 5I = \begin{bmatrix} -4 & 4 \\ 2 & -2 \end{bmatrix} \to \begin{bmatrix} 1 & -1 \\ 0 & 0 \end{bmatrix}$. If $\mathsf{x} = \begin{bmatrix} x_1 \\ x_2 \end{bmatrix}$ is an eigenvector, then $x_2 = t$ is free and $x_1 - x_2 = 0$, so $x_1 = x_2 = t$ and $\mathsf{x} = \begin{bmatrix} t \\ t \end{bmatrix} = t\begin{bmatrix} 1 \\ 1 \end{bmatrix}$. The eigenspace corresponding to $\lambda = 5$ is the set of all multiples of $\begin{bmatrix} 1 \\ 1 \end{bmatrix}$. ☺

READING CHECK 2. In Example 3.4.3, we rewrote $A\mathsf{x} = 5\mathsf{x}$ as $(A - 5I)\mathsf{x} = 0$. Why didn't we write $(A - 5)\mathsf{x} = 0$? Why $5I$?

Suppose x is an eigenvector of A corresponding to λ; thus $A\mathsf{x} = \lambda\mathsf{x}$. Any vector on the line through $(0,0)$ with direction x is of the form $c\mathsf{x}$ and $A(c\mathsf{x}) = c(A\mathsf{x}) = c(\lambda\mathsf{x}) = (c\lambda)\mathsf{x}$. So multiplication by A moves $c\mathsf{x}$ to $(c\lambda)\mathsf{x}$, which is another multiple of x.

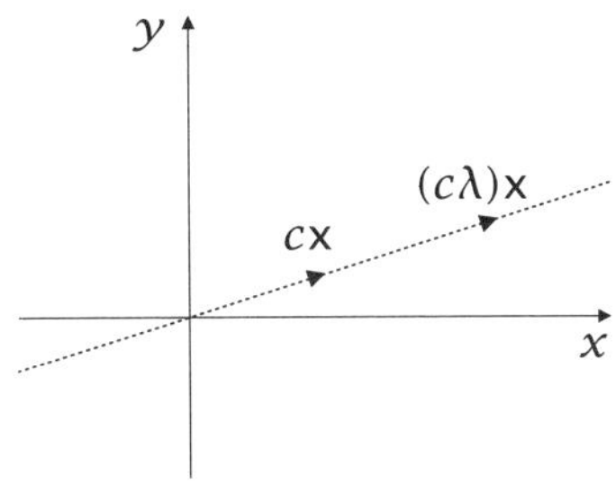

This shows that vectors on the line determined by x are moved to other vectors on this same line. The line itself is fixed ("invariant") under multiplication by A.

3.4.4 | Eigenvectors of a matrix A correspond to lines that are invariant under multiplication by A. |

Be careful here. We aren't saying here that every vector on an invariant line is fixed, just that every vector on the line is moved to another vector on the line. The line, as a whole, is fixed. In Example 3.4.3, for instance, the line consisting of scalar multiples of $\begin{bmatrix} 1 \\ 1 \end{bmatrix}$ (the line with equation $y = x$) is invariant under multiplication by A. If x is on this line, however, x is not fixed: $A\mathsf{x} = 5\mathsf{x}$.

3.4.5 Example. The scalar $\lambda = -3$ is an eigenvalue of $A = \begin{bmatrix} 5 & 8 & 16 \\ 4 & 1 & 8 \\ -4 & -4 & -11 \end{bmatrix}$ and $\mathsf{x} = \begin{bmatrix} 3 \\ -1 \\ -1 \end{bmatrix}$ is a corresponding eigenvector because

$$A\mathsf{x} = \begin{bmatrix} 5 & 8 & 16 \\ 4 & 1 & 8 \\ -4 & -4 & -11 \end{bmatrix}\begin{bmatrix} 3 \\ -1 \\ -1 \end{bmatrix} = \begin{bmatrix} -9 \\ 3 \\ 3 \end{bmatrix} = -3\mathsf{x}.$$

To find the eigenspace corresponding to $\lambda = -3$, we must find all the vectors x that satisfy $A\mathsf{x} = -3\mathsf{x}$. Rewriting this equation as $(A + 3I)\mathsf{x} = 0$, with I the 3×3 identity matrix, the desired eigenspace is the set of solutions to the homogeneous system with coefficient matrix $A + 3I$ which turns out to be all linear combinations of $\begin{bmatrix} -1 \\ 1 \\ 0 \end{bmatrix}$ and $\begin{bmatrix} -2 \\ 0 \\ 1 \end{bmatrix}$, a plane in Euclidean 3-space—see 1.1.27. ☺

READING CHECK 3. What is the equation of this plane?

3.4.6 Example. Multiplication by the matrix $A = \begin{bmatrix} 1 & 0 \\ 0 & -1 \end{bmatrix}$ transforms $\begin{bmatrix} x \\ y \end{bmatrix}$ to $\begin{bmatrix} x \\ -y \end{bmatrix}$: $A\begin{bmatrix} x \\ y \end{bmatrix} = \begin{bmatrix} 1 & 0 \\ 0 & -1 \end{bmatrix}\begin{bmatrix} x \\ y \end{bmatrix} = \begin{bmatrix} x \\ -y \end{bmatrix}$.

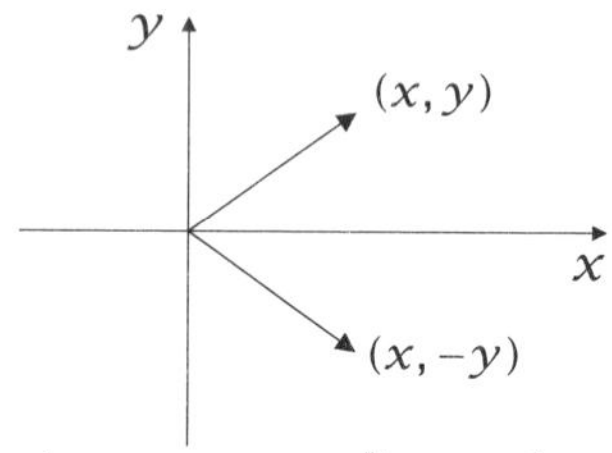

So multiplication by A reflects vectors in the x-axis. Reflection in a line fixes that line, in this case, every vector on the line because

$$A\begin{bmatrix} x \\ 0 \end{bmatrix} = \begin{bmatrix} 1 & 0 \\ 0 & -1 \end{bmatrix}\begin{bmatrix} x \\ 0 \end{bmatrix} = \begin{bmatrix} x \\ 0 \end{bmatrix}.$$

Thus every vector on the x-axis is an eigenvector corresponding to $\lambda = 1$. The reflection in question also fixes the y-axis because every vector on this line is moved to its additive inverse (which is still on the line):

$$A\begin{bmatrix} 0 \\ y \end{bmatrix} = \begin{bmatrix} 1 & 0 \\ 0 & -1 \end{bmatrix}\begin{bmatrix} 0 \\ y \end{bmatrix} = \begin{bmatrix} 0 \\ -y \end{bmatrix} = -\begin{bmatrix} 0 \\ y \end{bmatrix}.$$

Every vector on the y-axis is an eigenvector corresponding to $\lambda = -1$. The matrix A has two eigenspaces, the two axes. Both these lines are fixed under multiplication by A. ☺

3.4.7 Example. Multiplication by $A = \begin{bmatrix} 0 & -1 \\ 1 & 0 \end{bmatrix}$ moves $\begin{bmatrix} x \\ y \end{bmatrix}$ to $\begin{bmatrix} -y \\ x \end{bmatrix}$ and it can be shown that this transformation rotates vectors counter-clockwise through an angle of 90°. (See Section 5.1 and especially Example 5.1.16.) A rotation leaves no lines fixed, so A has no eigenvectors. Geometrically this is obvious, but verification algebraically is straightforward too.

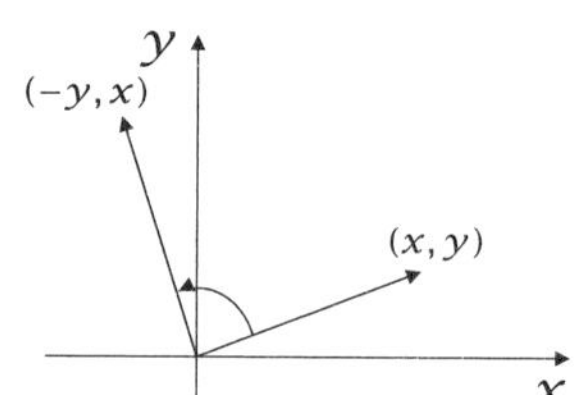

If $A\mathsf{x} = \lambda\mathsf{x}$, then $A(A\mathsf{x}) = A(\lambda\mathsf{x}) = \lambda(A\mathsf{x}) = \lambda(\lambda\mathsf{x}) = \lambda^2\mathsf{x}$, so $\lambda^2\mathsf{x} = -\mathsf{x}$ (because $A^2 = -I$). Since $\mathsf{x} \neq 0$, we obtain $\lambda^2 = -1$. This cannot be the case for a real number λ. We conclude that A has no **real** eigenvalues, hence no fixed lines. ☺

How to Find Eigenvalues and Eigenvectors

Examples 3.4.3 and 3.4.5 illustrate a general method for finding the eigenvalues and eigenspaces of a matrix. The key idea is that if $A\mathsf{x} = \lambda\mathsf{x}$ with $\mathsf{x} \neq 0$, then $A\mathsf{x} - \lambda\mathsf{x} = 0$, so $(A - \lambda I)\mathsf{x} = 0$. Since $\mathsf{x} \neq 0$, the matrix $A - \lambda I$ cannot be invertible. Hence

3.4.8
$$\boxed{\det(A - \lambda I) = 0.}$$

The determinant on the left is a polynomial in the variable λ called the *characteristic polynomial* of A. Its roots are the eigenvalues of A.

3.4.9

To find eigenvalues and eigenvectors of a (square) matrix A:

1. Compute the characteristic polynomial $\det(A - \lambda I)$.
2. Set $\det(A - \lambda I) = 0$. The solutions of this equation are the eigenvalues.
3. For each eigenvalue λ, solve the homogeneous system $(A - \lambda I)\mathsf{x} = 0$. The solution is the eigenspace corresponding to λ.

3.4.10 Example. The characteristic polynomial of the matrix $A = \begin{bmatrix} 1 & 4 \\ 2 & 3 \end{bmatrix}$ is

$$\det(A - \lambda I) = \begin{vmatrix} 1-\lambda & 4 \\ 2 & 3-\lambda \end{vmatrix} = (1-\lambda)(3-\lambda) - 8 = \lambda^2 - 4\lambda - 5\,, \quad (1)$$

a polynomial of degree 2 in λ. It factors, $\lambda^2 - 4\lambda - 5 = (\lambda - 5)(\lambda + 1)$, and its roots, $\lambda = 5$, $\lambda = -1$, are the eigenvalues of A. We found the eigenspace corresponding to $\lambda = 5$ in Example 3.4.3. To find the eigenspace corresponding to $\lambda = -1$, we solve the homogeneous system $(A + 1I)\mathsf{x} = 0$. The coefficient matrix is $A - \lambda I$ with $\lambda = -1$, which is $\begin{bmatrix} 2 & 4 \\ 2 & 4 \end{bmatrix}$—see (1). The eigenspace is the set of multiples of $\begin{bmatrix} -2 \\ 1 \end{bmatrix}$ which is, geometrically, another line through $(0,0)$. The critical reader will verify our calculations by checking that $A\begin{bmatrix} -2 \\ 1 \end{bmatrix} = (-1)\begin{bmatrix} -2 \\ 1 \end{bmatrix}$. ☺

3.4.11 Understanding matrix multiplication via eigenvectors

When we multiply a vector x by a matrix A, the vector x moves to the vector $A\mathsf{x}$. Knowing the eigenvalues and eigenvectors of a matrix A can help us understand the nature of the transformation from x to $A\mathsf{x}$. In Example 3.4.10, for instance, the matrix A has eigenvalues -1 and 5 corresponding to eigenspaces $t\begin{bmatrix} -2 \\ 1 \end{bmatrix}$ and $t\begin{bmatrix} 1 \\ 1 \end{bmatrix}$, respectively. If x is a multiple of $\begin{bmatrix} -2 \\ 1 \end{bmatrix}$, then $A\mathsf{x} = -\mathsf{x}$, while if x is a multiple of $\begin{bmatrix} 1 \\ 1 \end{bmatrix}$, then $A\mathsf{x} = 5\mathsf{x}$. What does multiplication by A do to a vector v that is not a multiple of $\mathsf{x}_1 = \begin{bmatrix} -2 \\ 1 \end{bmatrix}$ or

$\mathsf{x}_2 = \begin{bmatrix} 1 \\ 1 \end{bmatrix}$? We construct a parallelogram with diagonal v and sides on the lines with directions x_1 and x_2. See Figure 3.15. Thus $\mathsf{v} = a\mathsf{x}_1 + b\mathsf{x}_2$ is a linear combination of x_1 and x_2. Since matrix multiplication preserves linear combinations,

$$\begin{aligned} A\mathsf{v} &= A(a\mathsf{x}_1 + b\mathsf{x}_2) \\ &= a(A\mathsf{x}_1) + b(A\mathsf{x}_2) = a(-\mathsf{x}_1) + b(5\mathsf{x}_2) = -a\mathsf{x}_1 + 5b\mathsf{x}_2, \end{aligned}$$

so $A\mathsf{v}$ is the diagonal of the parallelogram with sides $-a\mathsf{x}_1$ and $5b\mathsf{x}_2$.

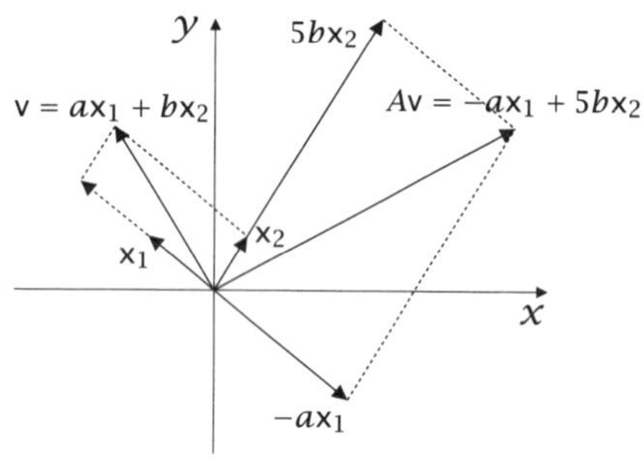

Figure 3.15

3.4.12 Problem. Find all eigenvalues of $A = \begin{bmatrix} 5 & 8 & 16 \\ 4 & 1 & 8 \\ -4 & -4 & -11 \end{bmatrix}$ and the corresponding eigenspaces.

Solution. The characteristic polynomial of A is the determinant of

$$A - \lambda I = \begin{bmatrix} 5-\lambda & 8 & 16 \\ 4 & 1-\lambda & 8 \\ -4 & -4 & -11-\lambda \end{bmatrix}. \tag{2}$$

Using a Laplace expansion along the first row,

$$\begin{aligned} \det(A - \lambda I) &= (5-\lambda)\begin{vmatrix} 1-\lambda & 8 \\ -4 & -11-\lambda \end{vmatrix} - 8\begin{vmatrix} 4 & 8 \\ -4 & -11-\lambda \end{vmatrix} \\ &\quad + 16\begin{vmatrix} 4 & 1-\lambda \\ -4 & -4 \end{vmatrix} \\ &= (5-\lambda)[(1-\lambda)(-11-\lambda) + 32] - 8[4(-11-\lambda) + 32] \\ &\quad + 16[-16 + 4(1-\lambda)] \\ &= (5-\lambda)(21 + 10\lambda + \lambda^2) - 8(-12 - 4\lambda) + 16(-12 - 4\lambda) \\ &= -\lambda^3 - 5\lambda^2 - 3\lambda + 9 = -(\lambda^3 + 5\lambda^2 + 3\lambda - 9). \end{aligned}$$

In Example 3.4.5, we noted that $\lambda = -3$ is an eigenvalue. Thus $\lambda = -3$ is one root of this polynomial and we can find the others by long division—see Figure 3.16. We obtain

$$\begin{aligned}\lambda^3 + 5\lambda^2 + 3\lambda - 9 &= (\lambda+3)(\lambda^2 + 2\lambda - 3)\\ &= (\lambda+3)(\lambda+3)(\lambda-1) = (\lambda+3)^2(\lambda-1).\end{aligned}$$

So the eigenvalues are $\lambda = 1$ and $\lambda = -3$. We found the eigenspace corresponding to $\lambda = -3$ in Example 3.4.5. The eigenspace corresponding to $\lambda = 1$ is the set of solutions to $(A - \lambda I)\mathsf{x} = 0$ with $\lambda = 1$. The coefficient matrix of this homogeneous system is just the matrix in (2) with $\lambda = 1$, namely, $\begin{bmatrix} 4 & 8 & 16\\ 4 & 0 & 8\\ -4 & -4 & -12\end{bmatrix}$. The solution, the eigenspace corresponding to $\lambda = 1$, is the set of multiples of $\begin{bmatrix}-2\\-1\\1\end{bmatrix}$, the line in 3-space through $(0,0,0)$ with direction $\begin{bmatrix}-2\\-1\\1\end{bmatrix}$.

$$\begin{array}{r|rrrrrrr}
\multicolumn{1}{r}{} & \lambda^2 & + & 2\lambda & - & 3 & & \\ \cline{2-8}
\lambda + 3 & \lambda^3 & + & 5\lambda^2 & + & 3\lambda & - & 9\\
\multicolumn{1}{r}{} & \lambda^3 & + & 3\lambda^2 & & & & \\ \cline{2-4}
\multicolumn{1}{r}{} & & & 2\lambda^2 & + & 3\lambda & - & 9\\
\multicolumn{1}{r}{} & & & 2\lambda^2 & + & 6\lambda & & \\ \cline{4-6}
\multicolumn{1}{r}{} & & & & & -3\lambda & - & 9\\
\multicolumn{1}{r}{} & & & & & -3\lambda & - & 9\\ \cline{6-8}
\multicolumn{1}{r}{} & & & & & & & 0
\end{array}$$

Figure 3.16: Long division of polynomials

3.4.13 Remark. Finding the eigenvalues and eigenvectors of a matrix often involves a lot of calculations. Without the assistance of a computer or sophisticated calculator, errors can creep into the workings of even the most careful person. Fortunately, there are ways to make us confident that we are on the right track as we move along. In Problem 3.4.12, for instance, we found that the characteristic polynomial was $-(\lambda+3)^2(\lambda-1)$ and concluded that $\lambda = 1$ was an eigenvalue of A. If this really is the case, then the homogeneous system $(A - \lambda I)\mathsf{x} = 0$ must have nonzero solutions when $\lambda = 1$. It was reassuring to find that $A - I$ had row echelon form $\begin{bmatrix}1 & 2 & 4\\ 0 & 1 & 1\\ 0 & 0 & 0\end{bmatrix}$, guaranteeing nonzero solutions.

A Final Thought

Most of the matrices we have presented in this section have (real) eigenvalues, but not all. The matrix $A = \begin{bmatrix} 0 & 1 \\ -1 & 0 \end{bmatrix}$ of Example 3.4.7 has characteristic polynomial

$$\det(A - \lambda I) = \begin{vmatrix} -\lambda & 1 \\ -1 & -\lambda \end{vmatrix} = \lambda^2 + 1.$$

Since this polynomial has no **real** roots, A has no **real** eigenvalues. In Chapter 7, we extend our discussion of eigenvalues and eigenvectors to the complex numbers.

Answers to Reading Checks

1. $A = \begin{bmatrix} 0 & 0 & 0 & \frac{1}{3} \\ \frac{1}{2} & 0 & \frac{1}{2} & \frac{1}{3} \\ \frac{1}{2} & 0 & 0 & \frac{1}{3} \\ 0 & 1 & \frac{1}{2} & 0 \end{bmatrix}$ and $\mathsf{x} = \frac{1}{31}\begin{bmatrix} 4 \\ 9 \\ 6 \\ 12 \end{bmatrix} \approx \begin{bmatrix} 0.129 \\ 0.290 \\ 0.194 \\ 0.387 \end{bmatrix}$, so P_4 is the most important, followed by P_2, P_3 and P_1.

2. If A is $n \times n$ with $n > 1$, $A - 5$ has no meaning. You can't subtract a number from a matrix!

3. We seek an equation for the plane which consists of all vectors of the form $\begin{bmatrix} -t-2s \\ t \\ s \end{bmatrix}$. Let $x = -t - 2s$, $y = t$, $z = s$. Then $x = -y - 2z$, so an equation is $x + y + 2z = 0$.

True/False Questions

Decide, with as little calculation as possible, whether each of the following statements is true or false and, if you say "false," explain your answer. (Answers are at the back of the book.)

1. Eigenvalues are important to Google.
2. The eigenvectors of a matrix A which correspond to a scalar λ are the solutions of the homogeneous system $(A - \lambda I)\mathsf{x} = 0$.
3. If x is an eigenvector of a matrix A, then so is every vector on the line through x.
4. The characteristic polynomial of a 4×4 matrix is a polynomial of degree 4.
5. Every 2×2 matrix has at least one real eigenvector.

6. If 0 is an eigenvalue of a matrix A, then A cannot be invertible.
7. If all the eigenvalues of a matrix are 0, then A is the zero matrix.
8. If 1 is the only eigenvalue of a 2×2 matrix A, then $A = I$.
9. If 3 is an eigenvalue of matrix A, then 9 is an eigenvalue of A^2.
10. If 3 is an eigenvalue of a matrix, so is 6.

EXERCISES

Answers to exercises marked [BB] can be found at the Back of the Book.

1. [BB] Is $\mathsf{x} = \begin{bmatrix} 2 \\ 4 \\ 0 \end{bmatrix}$ and eigenvector of $A = \begin{bmatrix} 2 & 1 & 1 \\ -2 & 5 & 0 \\ 2 & -1 & 4 \end{bmatrix}$? If yes, state the corresponding eigenvalue.

2. Determine whether or not the given vectors are eigenvectors of the given matrix. State the eigenvalue that corresponds to any eigenvector.

   (a) $A = \begin{bmatrix} 1 & 2 \\ 2 & 4 \end{bmatrix}$; $\mathsf{v}_1 = \begin{bmatrix} 1 \\ 2 \end{bmatrix}$, $\mathsf{v}_2 = \begin{bmatrix} -2 \\ 1 \end{bmatrix}$, $\mathsf{v}_3 = \begin{bmatrix} 1 \\ 1 \end{bmatrix}$, $\mathsf{v}_4 = \begin{bmatrix} 1 \\ -2 \end{bmatrix}$, $\mathsf{v}_5 = \begin{bmatrix} 0 \\ 0 \end{bmatrix}$

   (b) $A = \begin{bmatrix} 5 & -7 & 7 \\ 4 & -3 & 4 \\ 4 & -1 & 2 \end{bmatrix}$; $\mathsf{v}_1 = \begin{bmatrix} 0 \\ 0 \\ 0 \end{bmatrix}$, $\mathsf{v}_2 = \begin{bmatrix} 1 \\ 0 \\ 0 \end{bmatrix}$, $\mathsf{v}_3 = \begin{bmatrix} 0 \\ 2 \\ 2 \end{bmatrix}$, $\mathsf{v}_4 = \begin{bmatrix} 1 \\ 0 \\ -1 \end{bmatrix}$, $\mathsf{v}_5 = \begin{bmatrix} 3 \\ 3 \\ 3 \end{bmatrix}$

   (c) $A = \begin{bmatrix} 5 & 8 & 16 \\ 4 & 1 & 8 \\ -4 & -4 & -11 \end{bmatrix}$; $\mathsf{v}_1 = \begin{bmatrix} 3 \\ 4 \\ -1 \end{bmatrix}$, $\mathsf{v}_2 = \begin{bmatrix} -1 \\ 1 \\ 0 \end{bmatrix}$, $\mathsf{v}_3 = \begin{bmatrix} -1 \\ 1 \\ -1 \end{bmatrix}$, $\mathsf{v}_4 = \begin{bmatrix} -2 \\ 0 \\ 1 \end{bmatrix}$,
   $\mathsf{v}_5 = \begin{bmatrix} 2 \\ 1 \\ -1 \end{bmatrix}$

   (d) $A = \begin{bmatrix} 1 & -6 & 5 & -10 \\ -1 & 6 & -5 & 10 \\ 5 & -6 & 1 & -2 \\ -2 & 0 & 2 & -4 \end{bmatrix}$; $\mathsf{v}_1 = \begin{bmatrix} 1 \\ -1 \\ -1 \\ 1 \end{bmatrix}$, $\mathsf{v}_2 = \begin{bmatrix} 1 \\ 1 \\ 1 \\ 0 \end{bmatrix}$, $\mathsf{v}_3 = \begin{bmatrix} 0 \\ 0 \\ 0 \\ 0 \end{bmatrix}$, $\mathsf{v}_4 = \begin{bmatrix} 1 \\ -1 \\ 1 \\ 0 \end{bmatrix}$,
   $\mathsf{v}_5 = \begin{bmatrix} -1 \\ -1 \\ 1 \\ 1 \end{bmatrix}$.

3. Determine whether or not each of the given scalars is an eigenvalue of the given matrix.

   (a) [BB] $A = \begin{bmatrix} 1 & 2 \\ 2 & 4 \end{bmatrix}$; $-1, 0, 1, 3, 5$ **(b)** $A = \begin{bmatrix} 5 & -2 \\ 4 & -1 \end{bmatrix}$; $-2, -1, 1, 3, 4$

   (c) $A = \begin{bmatrix} -4 & -2 \\ 3 & 1 \end{bmatrix}$; $0, \sqrt{3}, \frac{-3+\sqrt{5}}{2}, -1, 4$

(d) [BB] $A = \begin{bmatrix} 5 & -7 & 7 \\ 4 & -3 & 4 \\ 4 & -1 & 2 \end{bmatrix}$; 1, 2, 4, 5, 6

(e) $A = \begin{bmatrix} 3 & 4 & 2 \\ 2 & -2 & 6 \\ 0 & -7 & 5 \end{bmatrix}$; 2, 1, 3, 6, −2

(f) $A = \begin{bmatrix} 2 & 1 & -1 \\ -2 & -1 & 2 \\ 2 & 1 & 0 \end{bmatrix}$; 2, −1, 3, 0, 4.

4. What are the eigenvalues of $\begin{bmatrix} 2 & 0 & 0 \\ 0 & 2 & 0 \\ 0 & 0 & 2 \end{bmatrix}$? What are the corresponding eigenspaces?

5. [BB] **(a)** Show that $\begin{bmatrix} 2 \\ 3 \end{bmatrix}$ is an eigenvector of $A = \begin{bmatrix} 1 & 2 \\ 3 & 2 \end{bmatrix}$ and state the corresponding eigenvalue.

   (b) What is the characteristic polynomial of A?

   (c) Find all the eigenvalues of A.

6. **(a)** Show that $\begin{bmatrix} -3 \\ 3 \end{bmatrix}$ is an eigenvector of $A = \begin{bmatrix} 5 & 1 \\ -1 & 3 \end{bmatrix}$ and state the corresponding eigenvalue.

   (b) What is the characteristic polynomial of A?

   (c) What are the eigenvalues of A?

7. Find the characteristic polynomial, the (real) eigenvalues and the corresponding eigenspaces of each of the following matrices A.

   (a) [BB] $\begin{bmatrix} 5 & 8 \\ 4 & 1 \end{bmatrix}$ **(b)** $\begin{bmatrix} -1 & 3 \\ 2 & 0 \end{bmatrix}$ **(c)** $\begin{bmatrix} 1 & 2 \\ 3 & 2 \end{bmatrix}$

   (d) [BB] $\begin{bmatrix} 1 & -2 & 3 \\ 2 & 6 & -6 \\ 1 & 2 & -1 \end{bmatrix}$ **(e)** $\begin{bmatrix} 0 & 1 & 1 \\ -2 & 3 & 0 \\ 2 & -1 & 2 \end{bmatrix}$ **(f)** $\begin{bmatrix} 4 & 2 & 2 \\ 4 & 2 & -4 \\ -2 & 0 & 4 \end{bmatrix}$

   (g) $\begin{bmatrix} 1 & 6 & 0 & 0 \\ 5 & 2 & 0 & 0 \\ 0 & 0 & 3 & -2 \\ 0 & 0 & 1 & 0 \end{bmatrix}$ **(h)** $\begin{bmatrix} 3 & 4 & -4 & -4 \\ 4 & 3 & -4 & -4 \\ 0 & 4 & -1 & -4 \\ 4 & 0 & -4 & -1 \end{bmatrix}$.

8. Let $A = \begin{bmatrix} 0 & 1 \\ 1 & 0 \end{bmatrix}$, let P be the point (x, y) in the Cartesian plane, and let v be the vector $\overrightarrow{OP}$. Let Q be the point such that $A\mathsf{v} = \overrightarrow{OQ}$.

   (a) Find the coordinates of Q.

   (b) Show that the line ℓ with equation $y = x$ is the right bisector of the line PQ joining P to Q; that is, PQ is perpendicular to ℓ and the lines intersect at the midpoint of PQ.

   (c) Describe the action of multiplication of A in geometrical terms.

(d) Find the eigenvalues and the corresponding eigenspaces of A without calculation.

9. Multiplication by each of the following matrices is reflection in a line. Find the equation of this line. Use the fact that the matrix is a reflection to find all eigenvectors and eigenvalues.

(a) [BB] $A = \begin{bmatrix} -1 & 0 \\ 0 & 1 \end{bmatrix}$ **(b)** $A = \begin{bmatrix} -\frac{3}{5} & \frac{4}{5} \\ \frac{4}{5} & \frac{3}{5} \end{bmatrix}$

(c) $A = \begin{bmatrix} \frac{7}{25} & -\frac{24}{25} \\ -\frac{24}{25} & -\frac{7}{25} \end{bmatrix}$ **(d)** $A = \begin{bmatrix} \frac{4}{5} & \frac{3}{5} \\ \frac{3}{5} & -\frac{4}{5} \end{bmatrix}$.

10. [BB] Let A be an $n \times n$ matrix and let U be the eigenspace of A corresponding to a scalar λ.

(a) Show that U is closed under scalar multiplication; that is, if x is in U and a is a scalar, then $a\mathsf{x}$ is also in U.

(b) Show that U is closed under addition; that is, if x and y are in U, then so is $\mathsf{x} + \mathsf{y}$.

[Together, these two properties show that any eigenspace of a matrix is a *subspace* of R^n. See Section 4.2.]

11. Let A be a 3×3 matrix, the entries in each row adding to 0.

(a) Show that $\begin{bmatrix} 1 \\ 1 \\ 1 \end{bmatrix}$ is an eigenvector of A and find the corresponding eigenvalue.

(b) Explain why $\det A = 0$.

(c) **Without calculation**, explain why $\det \begin{bmatrix} 2 & -1 & -1 \\ -3 & 1 & 2 \\ 0 & -2 & 2 \end{bmatrix} = 0$.

12. Suppose the components of each row of an $n \times n$ matrix A add to 17. Prove that 17 is an eigenvalue of A.

13. [BB] Suppose $\begin{bmatrix} 2 \\ 2 \end{bmatrix}$ is an eigenvector of a matrix A corresponding to eigenvalue $\lambda = 6$. Find the sum of the columns of A.

14. Suppose A is a 3×3 matrix and $\mathsf{x} = \begin{bmatrix} -2 \\ -2 \\ -2 \end{bmatrix}$ is an eigenvector corresponding to $\lambda = 5$. Find the sum of the columns of A.

15. [BB] Suppose A, B and P are $n \times n$ matrices with P invertible and $PA = BP$. Show that every eigenvalue of A is an eigenvalue of B.

16. Suppose x is an eigenvector of an invertible matrix A corresponding to λ.

(a) Show that $\lambda \neq 0$.

(b) Show that x is also an eigenvector of A^{-1}. What is the corresponding eigenvalue?

Critical Reading

17. Suppose A is a square matrix. Show that A and A^T have the same eigenvalues.

18. Suppose x is an eigenvector of a matrix A corresponding to an eigenvalue λ.

    (a) Could x be unique? That is, might x be the only eigenvector associated with λ?

    (b) Could λ be unique? That is, might λ be the only eigenvalue associated with x?

19. Suppose $\begin{bmatrix} 1 \\ 2 \\ 3 \end{bmatrix}$ is an eigenvector of a matrix A with corresponding eigenvalue -3 and $\begin{bmatrix} -1 \\ 0 \\ 5 \end{bmatrix}$ is an eigenvector with corresponding eigenvalue 5. Find the matrix product $A \begin{bmatrix} -1 & 1 \\ 0 & 2 \\ 5 & 3 \end{bmatrix}$.

20. Let v be an eigenvector of a matrix A with λ the corresponding eigenvalue.

    (a) [BB] Is v an eigenvector of $5A$? If so, what is the corresponding eigenvalue?

    (b) Is 5v an eigenvector of $5A$? If so, what is the corresponding eigenvalue?

    (c) Is 5v an eigenvector of A? If so, what is the corresponding eigenvalue?

    Explain your answers.

21. Suppose v is an eigenvector of the matrix A corresponding to λ and an eigenvector of the matrix B corresponding to μ. Show that v is also an eigenvector of AB. What is the corresponding eigenvalue?

22. Given a matrix A, let λ be an eigenvalue of A^TA. Show that $\lambda \geq 0$.

23. If 0 is an eigenvalue of a matrix A, then A cannot be invertible. Why not?

24. Let A be a square matrix with characteristic polynomial $f(\lambda)$ that factors $f(\lambda) = (\lambda_1 - \lambda)(\lambda_2 - \lambda) \cdots (\lambda_n - \lambda)$ into the product of linear polynomials over R. Explain why $\det A$ is the product of the eigenvalues of A.

3.5 Similarity and Diagonalization

3.5.1 Definition. Matrices A and B are *similar* if there exists an invertible matrix P with $P^{-1}AP = B$.

As the wording implies, this definition is symmetric: if $P^{-1}AP = B$ for some invertible P, then $Q^{-1}BQ = A$ for some invertible Q, so "A and B are similar," "A is similar to B" and "B is similar to A" all mean the same thing.

READING CHECK 1. If $B = P^{-1}AP$ for some P, show that $A = Q^{-1}AQ$ for some invertible Q.

3.5.2 Example. Matrices $A = \begin{bmatrix} 1 & 2 \\ 3 & 4 \end{bmatrix}$ and $B = \begin{bmatrix} -55 & -97 \\ 34 & 60 \end{bmatrix}$ are similar. To see why, let $P = \begin{bmatrix} 3 & 5 \\ 1 & 2 \end{bmatrix}$, note that $P^{-1} = \begin{bmatrix} 2 & -5 \\ -1 & 3 \end{bmatrix}$ and simply compute:

$$P^{-1}AP = \begin{bmatrix} 2 & -5 \\ -1 & 3 \end{bmatrix}\begin{bmatrix} 1 & 2 \\ 3 & 4 \end{bmatrix}\begin{bmatrix} 3 & 5 \\ 1 & 2 \end{bmatrix} = \begin{bmatrix} -55 & -97 \\ 34 & 60 \end{bmatrix} = B.$$ ☺

As Example 3.5.2 shows, "similar" matrices need not look very similar. On the other hand, they are similar in many ways.

- Similar matrices have the same determinant since
$$\det(P^{-1}AP) = \det(P^{-1})\det A \det P = \frac{1}{\det P}\det A \det P = \det A.$$
- The matrices $A - \lambda I$ and $P^{-1}AP - \lambda I$ have the same determinant because they too are similar: $P^{-1}AP - \lambda I = P^{-1}(A - \lambda I)P$. These determinants are the characeristic polynomials of A and $P^{-1}AP$. So similar matrices have the same characteristic polynomial and hence the same eigenvalues.
- It is not a coincidence that the sum of the diagonal entries of the matrices A and B in Example 3.5.2 is 5 in each case. Similar matrices even have the same trace. The *trace* of a matrix is the sum of its diagonal entries.

In this section, we focus attention on diagonal matrices. Remember that a diagonal matrix is a matrix like $\begin{bmatrix} 2 & 0 \\ 0 & 7 \end{bmatrix}$ whose only nonzero entries lie on the (main) diagonal. Calculations with diagonal matrices are much easier than with arbitrary matrices. For example, compare

$$\begin{bmatrix} 2 & 0 \\ 0 & -5 \end{bmatrix}\begin{bmatrix} -17 & 0 \\ 0 & 8 \end{bmatrix} = \begin{bmatrix} -34 & 0 \\ 0 & -40 \end{bmatrix}$$

with

$$\begin{bmatrix} 9 & -7 \\ 6 & 25 \end{bmatrix}\begin{bmatrix} 12 & 11 \\ -18 & 8 \end{bmatrix} = \begin{bmatrix} ? & ? \\ ? & ? \end{bmatrix}.$$

Finding a power of an arbitrary matrix can seem impossible compared with the ease of finding a power of a diagonal matrix. Compare

$$\begin{bmatrix} 2 & 0 \\ 0 & -5 \end{bmatrix}^{100} = \begin{bmatrix} 2^{100} & 0 \\ 0 & 5^{100} \end{bmatrix}$$

with

$$\begin{bmatrix} 3 & -4 \\ 5 & 2 \end{bmatrix}^{100} = ??\,.$$

The next best thing to "diagonal" is "similar to diagonal" (we also say "diagonalizable"). We investigate when this is the case.

3.5.3 Suppose A is an $n \times n$ matrix with n eigenvalues $\lambda_1, \dots, \lambda_n$ corresponding to eigenvectors $\mathrm{x}_1, \dots, \mathrm{x}_n$, respectively. Let $P = \begin{bmatrix} \mathrm{x}_1 & \mathrm{x}_2 & \cdots & \mathrm{x}_n \\ \downarrow & \downarrow & & \downarrow \end{bmatrix}$ be the matrix with these eigenvectors as columns and compute

$$AP = A \begin{bmatrix} \mathrm{x}_1 & \mathrm{x}_2 & \cdots & \mathrm{x}_n \\ \downarrow & \downarrow & & \downarrow \\ & & & \end{bmatrix} = \begin{bmatrix} A\mathrm{x}_1 & A\mathrm{x}_2 & \cdots & A\mathrm{x}_n \\ \downarrow & \downarrow & & \downarrow \\ & & & \end{bmatrix}.$$

Since $A\mathrm{x}_1 = \lambda_1\mathrm{x}_1$, $A\mathrm{x}_2 = \lambda_2\mathrm{x}_2$, and so on, we obtain

$$AP = \begin{bmatrix} \lambda_1\mathrm{x}_1 & \lambda_2\mathrm{x}_2 & \cdots & \lambda_n\mathrm{x}_n \\ \downarrow & \downarrow & & \downarrow \\ & & & \end{bmatrix}.$$

As in Problem 2.1.37, this matrix is the same as

$$\begin{bmatrix} \mathrm{x}_1 & \mathrm{x}_2 & \cdots & \mathrm{x}_n \\ \downarrow & \downarrow & & \downarrow \\ & & & \\ & & & \end{bmatrix} \begin{bmatrix} \lambda_1 & 0 & \cdots & 0 \\ 0 & \lambda_2 & \cdots & 0 \\ \vdots & & \ddots & \\ 0 & \cdots & & \lambda_n \end{bmatrix},$$

which is the product PD of P and the diagonal matrix $D = \begin{bmatrix} \lambda_1 & 0 & \cdots & 0 \\ 0 & \lambda_2 & \cdots & 0 \\ \vdots & & \ddots & \\ 0 & \cdots & & \lambda_n \end{bmatrix}$.

All this shows that $AP = PD$. If P happens to be invertible, we can multiply on the left by P^{-1} and obtain $P^{-1}AP = D$: A is similar to D.

3.5.4 **Definition.** A matrix A is *diagonalizable* if it is similar to a diagonal matrix—that is, if there exists an invertible matrix P and a diagonal matrix D such that $P^{-1}AP = D$.

If A is diagonalizable, say $P^{-1}AP = D$, then $AP = PD$ and it follows, as above, that the columns of P are eigenvectors of A corresponding to eigenvalues that are precisely the diagonal entries of D, in order.

3.5.5

> $P^{-1}AP = D$ if and only if the columns of P are eigenvectors of A and the diagonal entries of D are eigenvalues that correspond in order to the columns of P.

3.5.6 Example. Let $A = \begin{bmatrix} 1 & 4 \\ 2 & 3 \end{bmatrix}$ be the matrix considered first in Example 3.4.3 and again in 3.4.10. There are two eigenvalues, $\lambda = -1$ and $\lambda = 5$, with corresponding eigenvectors $\begin{bmatrix} -2 \\ 1 \end{bmatrix}$ and $\begin{bmatrix} 1 \\ 1 \end{bmatrix}$, respectively. If we put these into the columns of a matrix $P = \begin{bmatrix} -2 & 1 \\ 1 & 1 \end{bmatrix}$, we have

$$AP = \begin{bmatrix} 1 & 4 \\ 2 & 3 \end{bmatrix}\begin{bmatrix} -2 & 1 \\ 1 & 1 \end{bmatrix} = \begin{bmatrix} 2 & 5 \\ -1 & 5 \end{bmatrix} = \begin{bmatrix} -2 & 1 \\ 1 & 1 \end{bmatrix}\begin{bmatrix} -1 & 0 \\ 0 & 5 \end{bmatrix} = PD$$

with $D = \begin{bmatrix} -1 & 0 \\ 0 & 5 \end{bmatrix}$. Since P is invertible, $P^{-1}AP = D$. Notice that the first column of P is an eigenvector of A corresponding to $\lambda = -1$ and the second column of P is an eigenvector corresponding to $\lambda = 5$. If we change the order of the columns of P and use $Q = \begin{bmatrix} 1 & -2 \\ 1 & 1 \end{bmatrix}$, we find $Q^{-1}AQ = \begin{bmatrix} 5 & 0 \\ 0 & -1 \end{bmatrix}$. The order of the columns of Q and the diagonal entries of D must match.

☺

3.5.7 Remarks. 1. As just noted, the *diagonalizing matrix* P is far from unique. You can get a variety of Ds by rearranging the columns of P. And even for a fixed D, there are many Ps that work. For example, you can always replace a column of P by a nonzero multiple of this column. Throughout Section 3.4, we noted if x is an eigenvector corresponding to λ, then so is $c\mathsf{x}$ for any $c \neq 0$. See also Exercise 10 of Section 3.4.

2. Not all matrices are diagonalizable. Suppose we try to find the eigenvalues of $A = \begin{bmatrix} 0 & 1 \\ 0 & 0 \end{bmatrix}$. We have

$$\det(A - \lambda I) = \begin{vmatrix} -\lambda & 1 \\ 0 & -\lambda \end{vmatrix} = \lambda^2,$$

so $\lambda = 0$ is the only eigenvalue. If A were diagonalizable, the diagonal matrix D, having eigenvalues of A on its diagonal, would be the zero matrix. Since the equation $P^{-1}AP = 0$ implies $A = 0$, A cannot be diagonalizable.

READING CHECK 2. Alternatively, show that the only eigenvectors of $A = \begin{bmatrix} 0 & 1 \\ 0 & 0 \end{bmatrix}$ are (nonzero) multiples of $\begin{bmatrix} 1 \\ 0 \end{bmatrix}$. Conclude again that A is not diagonalizable.

3.5.8 **Remark.** The diagonalizability of a matrix is very much connected to the **field of scalars** over which we are working. Over the real numbers, $A = \begin{bmatrix} 0 & -1 \\ 1 & 0 \end{bmatrix}$ is not diagonalizable because its characteristic polynomial,

$$\det(A - \lambda I) = \begin{vmatrix} -\lambda & -1 \\ 1 & -\lambda \end{vmatrix} = \lambda^2 + 1,$$

has no real roots. Over the complex numbers, however, this polynomial has the roots $\pm i$, so A has two eigenvalues and, in fact, it's diagonalizable (via a matrix P with entries that are not all real). Until now, all scalars have been assumed real, so we do not push this issue further at the moment. The subject of complex matrices will be discussed in earnest in Chapter 7.

We have seen that there is an equation of the type $P^{-1}AP = D$ if and only if the columns of P are eigenvectors of A. We also know that a matrix is invertible if and only if its columns are linearly independent—this was Theorem 2.9.13. These are the two key ingredients in the proof of a fundamental theorem characterizing diagonalizable matrices.

3.5.9 **Theorem.** *A square $n \times n$ matrix is diagonalizable if and only if it has n linearly independent eigenvectors.*

Proof. ($\Longrightarrow$) Assume A is a diagonalizable matrix. Then there is an invertible matrix P such that $P^{-1}AP = D$ is a diagonal matrix. Since P is invertible, its columns are linearly independent (by Theorem 2.9.13). By 3.5.5, the columns of P are eigenvectors of A, so A has n linearly independent eigenvectors.

($\Longleftarrow$) Assume A has n linearly independent eigenvectors. Theorem 2.9.13 implies that the matrix P whose columns are these eigenvectors is invertible. As shown in paragraph 3.5.3, we have an equation $AP = PD$ where D is the diagonal matrix whose entries are the eigenvalues of A. Since P is invertible, we obtain $P^{-1}AP = D$ and conclude that A is diagonalizable. ■

To *diagonalize* a matrix A means to find an invertible matrix P such that $P^{-1}AP = D$ is diagonal. Theorem 3.5.9 (and its proof) imply a procedure for diagonalization.

3.5.10

Diagonalization Algorithm: Given an $n \times n$ matrix A,
1. find the eigenvalues of A;
2. find (if possible) n linearly independent eigenvectors $\mathsf{x}_1, \ldots, \mathsf{x}_n$;
3. let P be the matrix whose columns are $\mathsf{x}_1, \ldots, \mathsf{x}_n$;
4. then P is invertible and $P^{-1}AP = D$ is diagonal, the diagonal entries being the eigenvalues of A in the order corresponding to the columns of P.

3.5.11 Example. Let $A = \begin{bmatrix} 1 & 4 \\ 2 & 3 \end{bmatrix}$ as in Example 3.4.10. The characteristic polynomial of A is $\lambda^2 - 4\lambda - 5 = (\lambda - 5)(\lambda + 1)$. There are two eigenvalues, $\lambda = 5$ and $\lambda = -1$. An eigenvector for $\lambda = 5$ is $\mathsf{x}_1 = \begin{bmatrix} 1 \\ 1 \end{bmatrix}$, while $\mathsf{x}_2 = \begin{bmatrix} -2 \\ 1 \end{bmatrix}$ is an eigenvector for $\lambda = -1$. With $P = \begin{bmatrix} \mathsf{x}_1 & \mathsf{x}_2 \\ \downarrow & \downarrow \end{bmatrix} = \begin{bmatrix} 1 & -2 \\ 1 & 1 \end{bmatrix}$, we have

$$\begin{aligned} AP &= \begin{bmatrix} A\mathsf{x}_1 & A\mathsf{x}_2 \\ \downarrow & \downarrow \end{bmatrix} \\ &= \begin{bmatrix} 5\mathsf{x}_1 & -\mathsf{x}_2 \\ \downarrow & \downarrow \end{bmatrix} = \begin{bmatrix} 5 & 2 \\ 5 & -1 \end{bmatrix} = \begin{bmatrix} 1 & -2 \\ 1 & 1 \end{bmatrix}\begin{bmatrix} 5 & 0 \\ 0 & -1 \end{bmatrix} = PD, \end{aligned}$$

with $D = \begin{bmatrix} 5 & 0 \\ 0 & -1 \end{bmatrix}$. The columns of P are linearly independent, so P is invertible and $P^{-1}AP = D = \begin{bmatrix} 5 & 0 \\ 0 & -1 \end{bmatrix}$, the diagonal matrix whose entries are the eigenvalues of A in the order determined by the columns of P. ⌣̈

3.5.12 Example. Let $A = \begin{bmatrix} 5 & 8 & 16 \\ 4 & 1 & 8 \\ -4 & -4 & -11 \end{bmatrix}$ as in Problem 3.4.12. The eigenvalues of A are $\lambda = -3$ and $\lambda = 1$. For $\lambda = 1$, $\mathsf{x}_1 = \begin{bmatrix} -2 \\ -1 \\ 1 \end{bmatrix}$ is an eigenvector. For $\lambda = -3$, $\mathsf{x}_2 = \begin{bmatrix} -1 \\ 1 \\ 0 \end{bmatrix}$ and $\mathsf{x}_3 = \begin{bmatrix} -2 \\ 0 \\ 1 \end{bmatrix}$ are eigenvectors—see Example 3.4.5. We leave it to the reader to verify that x_1, x_2 and x_3 are linearly independent. So the matrix $P = \begin{bmatrix} -2 & -1 & -2 \\ -1 & 1 & 0 \\ 1 & 0 & 1 \end{bmatrix}$, whose columns are $\mathsf{x}_1, \mathsf{x}_2, \mathsf{x}_3$, is invertible and $P^{-1}AP = D = \begin{bmatrix} 1 & 0 & 0 \\ 0 & -3 & 0 \\ 0 & 0 & -3 \end{bmatrix}$, the diagonal matrix whose diagonal entries are the eigenvalues of A in the order determined by the columns of P. Since the first column of P is an eigenvector corresponding to $\lambda = 1$, the first diagonal entry of D is 1; since the last two columns of P are eigenvectors corresponding to $\lambda = -3$, the last two diagonal entries of D are -3. ⌣̈

READING CHECK 3. Take A as in Example 3.5.12, but $P = \begin{bmatrix} -2 & -2 & -1 \\ 0 & -1 & 1 \\ 1 & 1 & 0 \end{bmatrix}$.

Find $P^{-1}AP$.

3.5.13 Example. Let $A = \begin{bmatrix} 2 & 1 & -12 \\ 0 & 1 & 11 \\ 1 & 1 & 4 \end{bmatrix}$. Using a Laplace expansion down the first column, the characteristic polynomial of A is

$$\begin{vmatrix} 2-\lambda & 1 & -12 \\ 0 & 1-\lambda & 11 \\ 1 & 1 & 4-\lambda \end{vmatrix} = (2-\lambda)[(1-\lambda)(4-\lambda)-11] + [11+12(1-\lambda)]$$

$$= -\lambda^3 + 7\lambda^2 - 15\lambda + 9 = -(\lambda-1)(\lambda-3)^2.$$

The eigenvalues of A are $\lambda = 1$ and $\lambda = 3$. To find the eigenspace of A corresponding to $\lambda = 1$, we must solve $(A - \lambda I)\mathsf{x} = \mathsf{0}$ with $\lambda = 1$. The eigenspace consists of multiples of $\begin{bmatrix} -1 \\ 1 \\ 0 \end{bmatrix}$. Similarly, the eigenspace corresponding to $\lambda = 3$ consists of multiples of $\begin{bmatrix} -\frac{13}{2} \\ \frac{11}{2} \\ 1 \end{bmatrix}$. Since three linearly independent eigenvectors do not exist, A is not diagonalizable. ☺

There is one instance when we are assured that an $n \times n$ matrix has n linearly independent eigenvectors and is therefore diagonalizable.

3.5.14 Theorem. *Eigenvectors corresponding to different eigenvalues are linearly independent. Hence an $n \times n$ matrix with n different eigenvalues is diagonalizable.*

Proof. The second statement is an immediate consequence of the first and Theorem 3.5.9. To validate the first sentence, suppose we have a matrix A and a set of eigenvectors of A that correspond to different eigenvalues. We prove these are linearly independent with a *proof by contradiction*; that is, we assume the vectors are linearly dependent and derive a consequence that is false. So assume the vectors are linearly dependent. Amongst all the equations establishing linear dependence, choose one that is shortest; that is, an equation of the form

$$c_1\mathsf{x}_1 + c_2\mathsf{x}_2 + \cdots + c_k\mathsf{x}_k = \mathsf{0}, \tag{1}$$

not all coefficients 0, with the property that there is no such equation involving fewer than k vectors. This means, in particular, that none of the coefficients c_i is 0 and, moreover, that k must be more than 1: if $k = 1$, our equation would read $c_1\mathsf{x}_1 = \mathsf{0}$ which cannot be because the eigenvector $\mathsf{x}_1 \neq \mathsf{0}$ and $c_1 \neq 0$.

Assume that λ_i is the eigenvalue corresponding to x_i and multiply both sides of (1) by A. Since $A\mathsf{x}_i = \lambda_i\mathsf{x}_i$, we obtain

$$c_1\lambda_1\mathsf{x}_1 + c_2\lambda_2\mathsf{x}_2 + \cdots + c_k\lambda_k\mathsf{x}_k = \mathsf{0}.$$

Next multiply (1) by λ_1,

$$c_1\lambda_1\mathsf{x}_1 + c_2\lambda_1\mathsf{x}_2 + \cdots + c_k\lambda_1\mathsf{x}_k = \mathsf{0},$$

and then subtract this from the previous equation. We get

$$c_2(\lambda_2 - \lambda_1)\mathsf{x}_2 + \cdots + c_k(\lambda_k - \lambda_1)\mathsf{x}_k = \mathsf{0}.$$

This is an equation involving just $k-1$ vectors and none of the coefficients is 0 since $\lambda_i \neq \lambda_1$ for $i \neq 1$ and $c_i \neq 0$. This fact contradicts the assumption that there is no such equation involving fewer than k vectors and our proof is complete. ■

3.5.15 Example. The 2×2 matrix A in Example 3.5.11 has different eigenvalues 5 and -1, so it is diagonalizable. End of story. There is no need to find eigenspaces or the invertible matrix P. ☺

3.5.16 Problem. Show that $A = \begin{bmatrix} 1 & 2 \\ 1 & 1 \end{bmatrix}$ is diagonalizable.

Solution. The characteristic polynomial of A is

$$\begin{vmatrix} 1-\lambda & 2 \\ 1 & 1-\lambda \end{vmatrix} = (1-\lambda)^2 - 2.$$

The roots satisfy $(1-\lambda)^2 = 2$, so $1 - \lambda = \pm\sqrt{2}$ and $\lambda = 1 \pm \sqrt{2}$. These are different, so A is diagonalizable and similar to $\begin{bmatrix} 1+\sqrt{2} & 0 \\ 0 & 1-\sqrt{2} \end{bmatrix}$.

3.5.17 Remark. Theorem 3.5.14 doesn't explain why the matrix in Example 3.5.12 is diagonalizable. This 3×3 matrix has just two eigenvalues, $\lambda = -3$ and $\lambda = 1$. For $\lambda = 1$, we found an eigenvector x_1. For $\lambda = -3$, we found two eigenvectors x_2 and x_3. Theorem 3.5.14 guarantees that x_1 and x_2 are linearly independent since they correspond to different eigenvalues and, for the same reason, it says that x_1 and x_3 are linearly independent. It doesn't say, however, that the **three** vectors $\mathsf{x}_1, \mathsf{x}_2, \mathsf{x}_3$ are linearly independent. This is true however, and not difficult to see. See Exercise 30.

Answers to Reading Checks

1. Solving $B = P^{-1}AP$ for A gives $A = PBP^{-1}$, which is $Q^{-1}BQ$ with $Q = P^{-1}$.
2. As shown in the text, the only eigenvalue of A is 0. To find the corresponding eigenvectors, we reduce A to row echelon form. In this case, row echelon form

is the original matrix $A = \begin{bmatrix} 0 & 1 \\ 0 & 0 \end{bmatrix}$. We have $x_1 = t$ free and $x_2 = 0$, giving $\mathsf{x} = \begin{bmatrix} x_1 \\ x_2 \end{bmatrix} = \begin{bmatrix} t \\ 0 \end{bmatrix} = t\begin{bmatrix} 1 \\ 0 \end{bmatrix}$ as desired. If A is similar to a diagonal matrix, the (invertible) matrix P has columns that are eigenvectors of A. Since all eigenvectors of A are multiples of $\begin{bmatrix} 1 \\ 0 \end{bmatrix}$, A does not have two linearly independent eigenvectors, so no invertible P exists.

3. The columns of P are eigenvectors of A with eigenvalues -3, 1, -3, respectively, so $P^{-1}AP = \begin{bmatrix} -3 & 0 & 0 \\ 0 & 1 & 0 \\ 0 & 0 & -3 \end{bmatrix}$.

True/False Questions

Decide, with as little calculation as possible, whether each of the following statements is true or false and, if you say "false," explain your answer. (Answers are at the back of the book.)

1. The only matrix similar to $A = \begin{bmatrix} 4 & 0 & 0 \\ 0 & 4 & 0 \\ 0 & 0 & 4 \end{bmatrix}$ is A itself.

2. The characteristic polynomial of $\begin{bmatrix} 1 & 2 & 3 \\ 0 & 4 & 5 \\ 0 & 0 & 6 \end{bmatrix}$ is $(1-\lambda)(4-\lambda)(6-\lambda)$.

3. If $A = \begin{bmatrix} -1 & 0 \\ 0 & 2 \end{bmatrix}$, then $\lambda = +1$ is an eigenvalue of A^{10}.

4. The eigenvalues of $\begin{bmatrix} 2 & 3 \\ 0 & -1 \end{bmatrix}$ are -2 and $+1$.

5. The matrices $\begin{bmatrix} 2 & 3 \\ 1 & -1 \end{bmatrix}$ and $\begin{bmatrix} 2 & 1 \\ 3 & 1 \end{bmatrix}$ are similar.

6. The matrix $A = \begin{bmatrix} -1 & 0 \\ 0 & 5 \end{bmatrix}$ is diagonalizable.

7. If a 2×2 matrix has just one eigenvalue, it is not diagonalizable.

8. An $n \times n$ matrix with n different eigenvalues is diagonalizable.

9. If an $n \times n$ matrix is diagonalizable, it must have n different eigenvalues.

10. The eigenvalues of a symmetric matrix are the entries on its diagonal.

EXERCISES

Answers to exercises marked [BB] can be found at the **Back** *of the* **Book.**

1. [BB] **(a)** Suppose a matrix A is similar to the identity matrix. What can you say about A? Explain.

   (b) Can $\begin{bmatrix} 1 & 2 \\ 0 & 1 \end{bmatrix}$ be similar to the 2×2 identity matrix? Explain.

   (c) Is $A = \begin{bmatrix} 3 & 2 \\ -2 & -1 \end{bmatrix}$ diagonalizable? Explain.

2. **(a)** Show that similar matrices have the same characteristic polynomial.

   (b) Are the matrices $A = \begin{bmatrix} 2 & 1 \\ 1 & -1 \end{bmatrix}$ and $B = \begin{bmatrix} 3 & 0 \\ 1 & -1 \end{bmatrix}$ similar? Explain.

   (c) State the "converse" of part (a) and show that this is false by considering the matrices $A = \begin{bmatrix} 1 & 0 \\ 0 & 1 \end{bmatrix}$ and $B = \begin{bmatrix} 1 & 1 \\ 0 & 1 \end{bmatrix}$.

3. In each case, the matrices A and B are similar. Use this fact to find the determinant and the characteristic polynomial of B.

   (a) [BB] $A = \begin{bmatrix} 1 & 2 \\ 3 & 4 \end{bmatrix}$, $B = \begin{bmatrix} -3 & -1 \\ 22 & 8 \end{bmatrix}$ **(b)** $A = \begin{bmatrix} -4 & 1 \\ 0 & 2 \end{bmatrix}$, $B = \begin{bmatrix} -24 & -8 \\ 65 & 22 \end{bmatrix}$

   (c) $A = \begin{bmatrix} 1 & 0 & 0 \\ 2 & 1 & 1 \\ 0 & -3 & 2 \end{bmatrix}$, $B = \begin{bmatrix} -1 & -3 & -5 \\ 1 & -1 & -2 \\ 2 & 3 & 6 \end{bmatrix}$

   (d) $A = \begin{bmatrix} -3 & 1 & 1 \\ 2 & 0 & 4 \\ 5 & 1 & -2 \end{bmatrix}$, $B = \frac{1}{9}\begin{bmatrix} -37 & 7 & 29 \\ 10 & 20 & -2 \\ 5 & 37 & -28 \end{bmatrix}$

   (e) $A = \begin{bmatrix} 2 & -4 & 7 \\ 1 & 3 & 5 \\ -1 & -1 & 3 \end{bmatrix}$, $B = \begin{bmatrix} \frac{11}{2} & \frac{2\sqrt{6}}{3} & -\frac{19\sqrt{3}}{6} \\ -\frac{\sqrt{6}}{6} & 3 & \frac{3\sqrt{2}}{2} \\ \frac{19\sqrt{3}}{6} & 2\sqrt{2} & -\frac{1}{2} \end{bmatrix}$.

4. Given that the eigenvalues of a 3×3 matrix A are -1 and 2, what are the possibilities for the characteristic polynomial of A?

5. [BB] Give an easy reason that explains why $A = \begin{bmatrix} 5 & 3 \\ 2 & 4 \end{bmatrix}$ is diagonalizable. To how many matrices is A similar? (You are not expected to find eigenvectors.)

6. Answer Exercise 5 for the matrix $A = \begin{bmatrix} 5 & -2 & 6 \\ 0 & -1 & 9 \\ 0 & 0 & 3 \end{bmatrix}$.

7. [BB] What is an easy way to see that the matrix $A = \begin{bmatrix} 2 & -16 & -2 \\ 0 & 5 & 0 \\ 2 & -8 & -3 \end{bmatrix}$ is diagonalizable? Find a diagonal matrix to which A is similar.

8. Answer Exercise 7 with $A = \begin{bmatrix} -1 & 2 & 2 \\ 2 & 2 & 2 \\ -3 & -6 & -6 \end{bmatrix}$.

9. [BB] **(a)** Show that $\lambda = 2$ is an eigenvalue of $A = \begin{bmatrix} 1 & 2 \\ 2 & -2 \end{bmatrix}$ and find the corresponding eigenspace.

   (b) Given that the eigenvalues of A are 2 and -3, what is the characteristic polynomial of A?

   (c) Why is A diagonalizable?

   (d) Given that $\begin{bmatrix} -1 \\ 2 \end{bmatrix}$ is an eigenvector of A corresponding to $\lambda = -3$, find a matrix P such that $P^{-1}AP = \begin{bmatrix} -3 & 0 \\ 0 & 2 \end{bmatrix}$.

   (e) Let $Q = \begin{bmatrix} -2 & -6 \\ 4 & -3 \end{bmatrix}$. Find $Q^{-1}AQ$ without calculation.

10. Given that the eigenvalues of $A = \begin{bmatrix} -2 & 3 & 1 & 0 \\ -1 & 2 & 1 & 0 \\ -4 & 3 & 3 & 1 \\ 0 & 0 & 0 & 2 \end{bmatrix}$ are -1 and 2, determine (with reasons) whether or not A is diagonalizable.

11. [BB] **(a)** Explain how the equation $P^{-1}AP = D$ with D a diagonal matrix makes it (relatively) easy to find powers of A.

    (b) Find A^{10} if $A = \begin{bmatrix} 1 & 2 \\ 0 & 3 \end{bmatrix}$.

12. **(a)** Find an invertible matrix P such that $P^{-1}\begin{bmatrix} 0 & 1 & 0 \\ 0 & 0 & 1 \\ 2 & 1 & -2 \end{bmatrix}P = \begin{bmatrix} -1 & 0 & 0 \\ 0 & -2 & 0 \\ 0 & 0 & 1 \end{bmatrix}$.

    (b) Find A^{20}. [Hint: Exercise 11.]

13. **(a)** Show that $\lambda = -2$ is an eigenvalue of $A = \begin{bmatrix} 5 & -7 & 7 \\ 4 & -3 & 4 \\ 4 & -1 & 2 \end{bmatrix}$ and find the corresponding eigenspace.

    (b) Given that the eigenvalues of A are 1, -2 and 5, what is the characteristic polynomial of A?

    (c) Given that $\begin{bmatrix} 1 \\ 1 \\ 1 \end{bmatrix}$ is an eigenvector of A corresponding to $\lambda = 5$ and $\begin{bmatrix} 0 \\ 1 \\ 1 \end{bmatrix}$ is an eigenvector of A corresponding to $\lambda = 1$, find a matrix P such that $P^{-1}AP = \begin{bmatrix} 1 & 0 & 0 \\ 0 & -2 & 0 \\ 0 & 0 & 5 \end{bmatrix}$.

    (d) Let $Q = \begin{bmatrix} -1 & 1 & 0 \\ 0 & 1 & 1 \\ 1 & 1 & 1 \end{bmatrix}$. Find $Q^{-1}AQ$ without calculation.

(e) Find A^{100}.

14. [BB] (a) What is the characteristic polynomial of $A = \begin{bmatrix} 4 & 2 & 2 \\ -5 & -3 & -2 \\ 5 & 5 & 4 \end{bmatrix}$?

(b) Why is A diagonalizable?

(c) Find a matrix P such that $P^{-1}AP = \begin{bmatrix} 4 & 0 & 0 \\ 0 & -1 & 0 \\ 0 & 0 & 2 \end{bmatrix}$.

15. Let $A = \begin{bmatrix} 3 & 1 & 1 \\ -4 & -2 & -5 \\ 2 & 2 & 5 \end{bmatrix}$.

(a) What is the characteristic polynomial of A?

(b) Why is A diagonalizable?

(c) Find a matrix P such that $P^{-1}AP = \begin{bmatrix} 1 & 0 & 0 \\ 0 & 2 & 0 \\ 0 & 0 & 3 \end{bmatrix}$.

16. Determine whether or not each of the matrices A given below is diagonalizable. If it is, find an invertible matrix P and a diagonal matrix D such that $P^{-1}AP = D$. If it isn't, explain why.

(a) [BB] $A = \begin{bmatrix} 1 & 0 \\ 2 & 3 \end{bmatrix}$

(b) $A = \begin{bmatrix} 2 & 1 \\ -1 & 0 \end{bmatrix}$

(c) $A = \begin{bmatrix} 5 & -2 \\ 1 & 2 \end{bmatrix}$

(d) [BB] $A = \begin{bmatrix} -1 & 3 & 0 \\ 0 & 2 & 0 \\ 2 & 1 & -1 \end{bmatrix}$

(e) $A = \begin{bmatrix} 8 & 9 & -9 \\ 0 & 2 & 0 \\ 4 & 6 & -4 \end{bmatrix}$

(f) $A = \begin{bmatrix} 4 & -3 & 3 \\ 0 & 1 & 0 \\ -6 & 6 & -5 \end{bmatrix}$

(g) $A = \begin{bmatrix} 22 & 0 & 9 \\ -9 & -5 & -3 \\ 9 & 0 & -2 \end{bmatrix}$

(h) [BB] $A = \begin{bmatrix} 2 & 1 & 1 \\ 0 & 1 & 0 \\ 1 & -1 & 2 \end{bmatrix}$

(i) $A = \begin{bmatrix} 3 & 0 & 6 \\ 0 & -3 & 0 \\ 5 & 0 & 2 \end{bmatrix}$

(j) $A = \begin{bmatrix} 1 & 6 & 0 & 0 \\ 5 & 2 & 0 & 0 \\ 0 & 0 & 3 & -2 \\ 0 & 0 & 1 & 0 \end{bmatrix}$.

17. [BB] (a) Let $A = \begin{bmatrix} 3 & -2 \\ 1 & 0 \end{bmatrix}$. Find an invertible matrix P and a diagonal matrix D such that $P^{-1}AP = D$.

(b) Find a matrix B such that $B^2 = A$. [Hint: First, find a matrix D_1 such that $D = D_1^2$.]

18. If A is a diagonalizable matrix with all eigenvalues nonnegative, show that A has a square root; that is, $A = B^2$ for some matrix B. [Hint: Exercise 17.]

19. (a) Let $A = \begin{bmatrix} 10 & -5 & 7 \\ -5 & 22 & -5 \\ 7 & -5 & 10 \end{bmatrix}$. Find an invertible matrix P and a diagonal matrix D such that $P^{-1}AP = D$.

    (b) Find a matrix B such that $B^2 = A$. [Hint: Exercise 17.]

20. [BB] One of the most important and useful theorems in linear algebra says that a real symmetric matrix A is always diagonalizable—in fact, *orthogonally diagonalizable*; that is, there is a matrix P **with orthogonal columns** such that $P^{-1}AP$ is a diagonal matrix. (See Corollary 7.2.12.) Illustrate this fact using the matrix $A = \begin{bmatrix} 103 & -96 \\ -96 & 47 \end{bmatrix}$.

21. Answer Exercise 20 again using each of the following symmetric matrices.

    (a) $A = \begin{bmatrix} 3 & -5 \\ -5 & 3 \end{bmatrix}$ (b) $A = \begin{bmatrix} 2 & 1 & 0 \\ 1 & 2 & 0 \\ 0 & 0 & 1 \end{bmatrix}$

    (c) $A = \begin{bmatrix} 2 & -2 & -1 \\ -2 & 2 & 1 \\ -1 & 1 & 5 \end{bmatrix}$ (d) $A = \begin{bmatrix} 1 & -2 & -1 \\ -2 & 4 & 2 \\ -1 & 2 & 1 \end{bmatrix}$.

22. Show that the notion of matrix similarity is

    (a) *reflexive*; that is, any square matrix A is similar to itself;

    (b) *symmetric*; that is, if $B = P^{-1}AP$ for some invertible matrix P, then $A = Q^{-1}BQ$ for some invertible matrix Q;

    (c) *transitive*; that is, if matrices A and B are similar, and matrices B and C are similar, then the matrices A and C are similar.

    [Together, these three properties show that matrix similarity is an *equivalence relation.*]

Critical Reading

23. (a) If P is invertible, show that P^T is also invertible and find its inverse.

    (b) If A is diagonalizable, explain why A^T must be diagonalizable too.

24. [BB] Let $A = \begin{bmatrix} 2 & 1 & -12 \\ 0 & 1 & 11 \\ 1 & 1 & 4 \end{bmatrix}$ be the matrix of Example 3.5.13 and let $D_1 = \begin{bmatrix} 1 & 0 & 0 \\ 0 & 3 & 0 \\ 0 & 0 & 3 \end{bmatrix}$, $D_2 = \begin{bmatrix} 3 & 0 & 0 \\ 0 & 1 & 0 \\ 0 & 0 & 3 \end{bmatrix}$ and $D_3 = \begin{bmatrix} 1 & 0 & 0 \\ 0 & 3 & 0 \\ 0 & 0 & 1 \end{bmatrix}$.

    (a) Find matrices P_1, P_2, P_3 such that $AP_1 = P_1D_1$, $AP_2 = P_2D_2$ and $AP_3 = P_3D_3$.

    (b) Are any of the matrices P_1, P_2, P_3 invertible? Explain.

(c) Does there exist an invertible matrix P such that $AP = PD$ for some diagonal matrix D? Explain.

25. Answer Exercise 24 again with $A = \begin{bmatrix} -58 & 20 & 70 \\ -45 & 42 & 35 \\ 0 & 0 & 32 \end{bmatrix}$, $D_1 = \begin{bmatrix} -48 & 0 & 0 \\ 0 & -48 & 0 \\ 0 & 0 & 32 \end{bmatrix}$,
$D_2 = \begin{bmatrix} 32 & 0 & 0 \\ 0 & -48 & 0 \\ 0 & 0 & 32 \end{bmatrix}$ and $D_3 = \begin{bmatrix} 32 & 0 & 0 \\ 0 & 32 & 0 \\ 0 & 0 & 32 \end{bmatrix}$.

26. (a) Find the characteristic polynomial of $A = \begin{bmatrix} a & b \\ c & d \end{bmatrix}$.

(b) Identify the constant term of the characteristic polynomial.

(c) If $\det A = 1$ and the characteristic polynomial of A factors as the product of polynomials of degree 1, show that each eigenvalue of A is the reciprocal of the other.

27. If A an $n \times n$ diagonalizable matrix, show that A^{-1} is also diagonalizable.

28. Suppose A and B are matrices, one invertible, and $AB = BA$.

(a) Show that AB and BA are similar.

(b) Explain why AB and BA have the same eigenvalues. (This is true without the invertibility assumption, but the author is unaware of an elementary reason.)

29. The *converse* of an implication "if $\mathcal{A}$ then $\mathcal{B}$" is the implication "if $\mathcal{B}$, then $\mathcal{A}$." Write down the converse of Theorem 3.5.14 and describe, with a proof or a counterexample, whether or not the converse is true.

30. Suppose x and y are linearly independent eigenvectors of a matrix A corresponding to an eigenvalue λ and that z is an eigenvector corresponding to μ. Assume $\lambda \neq \mu$. Show that the three vectors x, y, z are linearly independent. (See 3.5.17.) [Hint: Use the fact that $a\mathsf{x} + b\mathsf{y}$ is either $\mathsf{0}$ or an eigenvector of A corresponding to λ. See Exercise 10 of Section 3.4.]

31. Let $A = \begin{bmatrix} -11 & 18 \\ -6 & 10 \end{bmatrix}$. Show that $A \neq B^2$ for any matrix B. [Hint: Show that A is diagonalizable and then examine the solution to Exercise 17.]

32. The instructor of a linear algebra course preparing a homework assignment for her class wants to produce a 3×3 matrix with eigenvalues $1, 2, 3$ with corresponding eigenvectors x_1, x_2, x_3, respectively. How should she proceed if

(a) [BB] $\mathsf{x}_1 = \begin{bmatrix} 1 \\ 2 \\ 3 \end{bmatrix}$, $\mathsf{x}_2 = \begin{bmatrix} 4 \\ 5 \\ 6 \end{bmatrix}$ and $\mathsf{x}_3 = \begin{bmatrix} 5 \\ 7 \\ 9 \end{bmatrix}$?

(b) $\mathsf{x}_1 = \begin{bmatrix} 1 \\ 2 \\ 3 \end{bmatrix}$, $\mathsf{x}_2 = \begin{bmatrix} 4 \\ 5 \\ 6 \end{bmatrix}$ and $\mathsf{x}_3 = \begin{bmatrix} 1 \\ 0 \\ 0 \end{bmatrix}$?

33. Explain why an $n \times n$ matrix with n different nonzero eigenvalues must be invertible.

3.6 Application: Linear Recurrence Relations

Here is a sequence of numbers defined "recursively:"

$$a_0 = 1, \quad a_1 = 2 \quad \text{and, for } n \geq 1, \quad a_{n+1} = -a_n + 2a_{n-1}. \tag{1}$$

The third equation here

$$a_{n+1} = -a_n + 2a_{n-1} \tag{2}$$

tells us how to find the terms after a_0 and a_1. To get a_2, we set $n = 1$ in (2) and find $a_2 = -a_1 + 2a_0 = 0$. We get a_3 by setting $n = 2$ in (2), and so on. Thus

$$\begin{aligned} a_3 &= -a_2 + 2a_1 = -0 + 2(2) = 4 \\ a_4 &= -a_3 + 2a_2 = -4 + 2(0) = -4 \\ a_5 &= -a_4 + 2a_3 = 4 + 2(4) = 12. \end{aligned}$$

Recursive means that each number in the sequence, after the first two, is not defined explicitly, but in terms of previous numbers in the sequence. We don't know a_{20}, for example, until we know a_{18} and a_{19}; we don't know a_{18} without knowing a_{16} and a_{17}, so it seems as if we can't determine a_{20} until we have found all preceding terms. Actually, this is not correct. As we shall soon see, linear algebra can be used to find an explicit formula for a_n.

To begin, notice that

$$\begin{bmatrix} a_{n+1} \\ a_n \end{bmatrix} = \begin{bmatrix} -1 & 2 \\ 1 & 0 \end{bmatrix} \begin{bmatrix} a_n \\ a_{n-1} \end{bmatrix},$$

for $n \geq 1$, and this is $\mathsf{v}_n = A\mathsf{v}_{n-1}$ with $\mathsf{v}_n = \begin{bmatrix} a_{n+1} \\ a_n \end{bmatrix}$ and $A = \begin{bmatrix} -1 & 2 \\ 1 & 0 \end{bmatrix}$. Thus $\mathsf{v}_1 = A\mathsf{v}_0$, $\mathsf{v}_2 = A\mathsf{v}_1 = A^2\mathsf{v}_0$, $\mathsf{v}_3 = A\mathsf{v}_2 = A^3\mathsf{v}_0$ and, in general, $\mathsf{v}_n = A^n\mathsf{v}_0 = A^n\begin{bmatrix} 2 \\ 1 \end{bmatrix}$. Our job then is to find the matrix A^n, a rather unpleasant thought: $A^2 = AA, A^3 = AAA, A^4 = AAAA, \ldots$. The task is much simpler, however, if we can diagonalize!

The characteristic polynomial of A is

$$\det(A - \lambda I) = \begin{vmatrix} -1-\lambda & 2 \\ 1 & -\lambda \end{vmatrix} = \lambda^2 + \lambda - 2 = (\lambda + 2)(\lambda - 1).$$

Since the 2×2 matrix A has two distinct eigenvalues, it is diagonalizable—see Theorem 3.5.14. Thus there exists an invertible matrix P with the property that $P^{-1}AP$ is the diagonal matrix $D = \begin{bmatrix} -2 & 0 \\ 0 & 1 \end{bmatrix}$, whose diagonal entries are the eigenvalues of A. Recall from Section 3.5 that the columns of P are eigenvectors of A corresponding to the eigenvalues -2 and 1, respectively.

To find the eigenspace for $\lambda = -2$, we solve the homogeneous system $(A - \lambda I)\mathsf{x} = \mathsf{0}$ for $\mathsf{x} = \begin{bmatrix} x_1 \\ x_2 \end{bmatrix}$ with $\lambda = -2$. We let the reader confirm that the solution is $\mathsf{x} = \begin{bmatrix} -2t \\ t \end{bmatrix} = t\begin{bmatrix} -2 \\ 1 \end{bmatrix}$. Similarly, the eigenspace for $\lambda = 1$ consists of vectors of the form $\mathsf{x} = \begin{bmatrix} t \\ t \end{bmatrix} = t\begin{bmatrix} 1 \\ 1 \end{bmatrix}$. Placing eigenvectors in the columns (in order corresponding to $-2, 1$ respectively), a suitable $P = \begin{bmatrix} -2 & 1 \\ 1 & 1 \end{bmatrix}$, which gives $P^{-1}AP = D$. This implies $A = PDP^{-1}$.

Now notice that

$$\begin{aligned} A^2 &= AA = PDP^{-1}PDP^{-1} = PD^2P^{-1}, \\ A^3 &= A^2A = PD^2P^{-1}PDP^{-1} = PD^3P^{-1}, \\ A^4 &= PD^4P^{-1} \end{aligned}$$

and, in general,

$$\begin{aligned} A^n = PD^nP^{-1} &= \frac{1}{3}\begin{bmatrix} -2 & 1 \\ 1 & 1 \end{bmatrix}\begin{bmatrix} (-2)^n & 0 \\ 0 & 1 \end{bmatrix}\begin{bmatrix} -1 & 1 \\ 1 & 2 \end{bmatrix} \\ &= \frac{1}{3}\begin{bmatrix} -2 & 1 \\ 1 & 1 \end{bmatrix}\begin{bmatrix} -(-2)^n & (-2)^n \\ 1 & 2 \end{bmatrix} \\ &= \frac{1}{3}\begin{bmatrix} -(-2)^{n+1}+1 & (-2)^{n+1}+2 \\ -(-2)^n+1 & (-2)^n+2 \end{bmatrix}. \end{aligned}$$

Thus

$$\begin{aligned} \mathsf{v}_n = \begin{bmatrix} a_{n+1} \\ a_n \end{bmatrix} = A^n\begin{bmatrix} 2 \\ 1 \end{bmatrix} &= \frac{1}{3}\begin{bmatrix} -(-2)^{n+1}+1 & (-2)^{n+1}+2 \\ -(-2)^n+1 & (-2)^n+2 \end{bmatrix}\begin{bmatrix} 2 \\ 1 \end{bmatrix} \\ &= \frac{1}{3}\begin{bmatrix} (-2)^{n+2}+2+(-2)^{n+1}+2 \\ (-2)^{n+1}+2+(-2)^n+2 \end{bmatrix}. \end{aligned}$$

The second component of this vector gives a formula for a_n:

$$a_n = \frac{1}{3}[4-(-2)^n] = \frac{4}{3} - \frac{1}{3}(-2)^n.$$

We check that

$$a_2 = \tfrac{4}{3} - \tfrac{4}{3} = 0, \quad a_3 = \tfrac{4}{3} + \tfrac{8}{3} = 4, \quad a_4 = \tfrac{4}{3} - \tfrac{16}{3} = -4, \quad a_5 = \tfrac{4}{3} + \tfrac{32}{3} = 12,$$

in agreement with what we found earlier.

The Fibonacci Sequence

Leonardo Fibonacci (≈1180–1228), one of the brightest mathematicians of the middle ages, is credited with posing a problem about rabbits that goes more or less like this.

> Suppose that newborn rabbits start producing offspring by the end of their second month of life, after which they produce a pair a month, one male and one female. Starting with one pair of rabbits on January 1, and assuming that rabbits never die, how many pairs will be alive at the end of the year (and available for New Year's dinner)?

At the beginning, on January 1, only the initial pair is alive, and the same is true at the start of February. By March 1, however, the initial pair of rabbits has produced its first pair of offspring, so now two pairs of rabbits are alive. On April 1, the initial pair has produced one more pair and the second pair hasn't started reproducing yet, so three pairs are alive. On May 1, there are five pairs of rabbits. Do you see why?

At the beginning of a month, the number of pairs alive is the number alive at the start of the previous month plus the number alive two months ago (since the pairs of this last group have started to produce offspring, one pair per pair). The numbers of pairs of rabbits alive at the start of each month are the terms of the famous *Fibonacci sequence* $1, 1, 2, 3, 5, 8, \ldots$, each number after the first two being the sum of the previous two.

The Fibonacci sequence is best defined recursively like this:

$$f_0 = 1, \quad f_1 = 1 \quad \text{and, for } n \geq 1, \quad f_{n+1} = f_n + f_{n-1},$$

which can be written

$$\begin{bmatrix} f_{n+1} \\ f_n \end{bmatrix} = \begin{bmatrix} 1 & 1 \\ 1 & 0 \end{bmatrix} \begin{bmatrix} f_n \\ f_{n-1} \end{bmatrix}.$$

This is $\mathsf{v}_n = A\mathsf{v}_{n-1}$, with $\mathsf{v}_n = \begin{bmatrix} f_{n+1} \\ f_n \end{bmatrix}$ and $A = \begin{bmatrix} 1 & 1 \\ 1 & 0 \end{bmatrix}$. As before, this implies $\mathsf{v}_n = A^n \mathsf{v}_0$ with $\mathsf{v}_0 = \begin{bmatrix} 1 \\ 1 \end{bmatrix}$. Our job is to find A^n.

The characteristic polynomial of A is

$$\det(A - \lambda I) = \begin{vmatrix} 1-\lambda & 1 \\ 1 & -\lambda \end{vmatrix} = \lambda^2 - \lambda - 1,$$

whose roots are $\lambda_1 = \frac{1+\sqrt{5}}{2}$ and $\lambda_2 = \frac{1-\sqrt{5}}{2}$. Since the eigenvalues are different, A is diagonalizable. In fact, $P^{-1}AP = D = \begin{bmatrix} \lambda_1 & 0 \\ 0 & \lambda_2 \end{bmatrix}$ with P the matrix whose columns are eigenvectors corresponding to λ_1 and λ_2, respectively.

It will pay dividends now to work in some generality. Let λ be either λ_1 or λ_2 and remember that $\lambda^2 - \lambda - 1 = 0$. Thus $\lambda(\lambda - 1) = 1$, so $\lambda \neq 0$ and

$$\lambda - 1 = \frac{1}{\lambda}, \tag{3}$$

a fact that will prove very useful.

To find the eigenspace for λ, we solve the homogeneous system $(A-\lambda I)\mathsf{x} = 0$. Gaussian elimination proceeds

$$\begin{bmatrix} 1-\lambda & 1 \\ 1 & -\lambda \end{bmatrix} \to \begin{bmatrix} 1 & \frac{1}{1-\lambda} \\ 1 & -\lambda \end{bmatrix} = \begin{bmatrix} 1 & -\lambda \\ 1 & -\lambda \end{bmatrix} \to \begin{bmatrix} 1 & -\lambda \\ 0 & 0 \end{bmatrix}.$$

With $\mathsf{x} = \begin{bmatrix} x_1 \\ x_2 \end{bmatrix}$, $x_2 = t$ is free and $x_1 = \lambda x_2 = \lambda t$, giving $\mathsf{x} = \begin{bmatrix} \lambda t \\ t \end{bmatrix} = t\begin{bmatrix} \lambda \\ 1 \end{bmatrix}$.

It follows that $P = \begin{bmatrix} \lambda_1 & \lambda_2 \\ 1 & 1 \end{bmatrix}$, so $P^{-1} = \frac{1}{\sqrt{5}}\begin{bmatrix} 1 & -\lambda_2 \\ -1 & \lambda_1 \end{bmatrix}$ (using Proposition 2.3.10 and $\det P = \lambda_1 - \lambda_2 = \sqrt{5}$). Now

$$\begin{bmatrix} f_{n+1} \\ f_n \end{bmatrix} = \mathsf{v}_n = A^n \mathsf{v}_0 = PD^nP^{-1}\mathsf{v}_0 = PD^nP^{-1}\begin{bmatrix} 1 \\ 1 \end{bmatrix} = \frac{1}{\sqrt{5}}PD^n\begin{bmatrix} 1-\lambda_2 \\ -1+\lambda_1 \end{bmatrix}.$$

Since $PD^n = \begin{bmatrix} \lambda_1 & \lambda_2 \\ 1 & 1 \end{bmatrix}\begin{bmatrix} \lambda_1^n & 0 \\ 0 & \lambda_2^n \end{bmatrix} = \begin{bmatrix} \lambda_1^{n+1} & \lambda_2^{n+1} \\ \lambda_1^n & \lambda_2^n \end{bmatrix}$, we have

$$\begin{aligned} \begin{bmatrix} f_{n+1} \\ f_n \end{bmatrix} &= \frac{1}{\sqrt{5}}\begin{bmatrix} \lambda_1^{n+1} & \lambda_2^{n+1} \\ \lambda_1^n & \lambda_2^n \end{bmatrix}\begin{bmatrix} 1-\lambda_2 \\ -1+\lambda_1 \end{bmatrix} \\ &= \frac{1}{\sqrt{5}}\left[(1-\lambda_2)\begin{bmatrix} \lambda_1^{n+1} \\ \lambda_1^n \end{bmatrix} + (-1+\lambda_1)\begin{bmatrix} \lambda_2^{n+1} \\ \lambda_2^n \end{bmatrix}\right]. \end{aligned}$$

(Never forget 2.1.33!)

Since $\lambda - 1 = \frac{1}{\lambda}$ for each λ, by (3), the second component of this vector is

$$\begin{aligned} f_n &= \frac{1}{\sqrt{5}}\left[(1-\lambda_2)\lambda_1^n + (-1+\lambda_1)\lambda_2^n\right] = \frac{1}{\sqrt{5}}\left(-\frac{\lambda_1^n}{\lambda_2} + \frac{\lambda_2^n}{\lambda_1}\right) \\ &= \frac{1}{\sqrt{5}}\left(\frac{-\lambda_1^{n+1} + \lambda_2^{n+1}}{\lambda_1\lambda_2}\right). \end{aligned}$$

Now $\lambda_1\lambda_2 = \frac{1+\sqrt{5}}{2}\,\frac{1-\sqrt{5}}{2} = \frac{1-5}{4} = -1$, so $\frac{-\lambda_1^{n+1} + \lambda_2^{n+1}}{\lambda_1\lambda_2} = \lambda_1^{n+1} - \lambda_2^{n+1}$ and

$$f_n = \frac{1}{\sqrt{5}}\lambda_1^{n+1} - \frac{1}{\sqrt{5}}\lambda_2^{n+1} = \frac{1}{\sqrt{5}}\left(\frac{1+\sqrt{5}}{2}\right)^{n+1} - \frac{1}{\sqrt{5}}\left(\frac{1-\sqrt{5}}{2}\right)^{n+1}. \tag{4}$$

It is almost incredible that this formula should give integer values (see READING CHECK 1), but keeping this in mind, we can get a better formula.

Since $|\frac{1-\sqrt{5}}{2}| < 1$, $\left|\frac{1}{\sqrt{5}}\left(\frac{1-\sqrt{5}}{2}\right)^{n+1}\right| < \frac{1}{\sqrt{5}} < \frac{1}{2}$ and then, from (4), we discover

$$\left|f_n - \frac{1}{\sqrt{5}}\left(\frac{1+\sqrt{5}}{2}\right)^{n+1}\right| = \left|\frac{1}{\sqrt{5}}\left(\frac{1-\sqrt{5}}{2}\right)^{n+1}\right| < \frac{1}{2}.$$

This inequality says that the integer f_n is at a distance less than $\frac{1}{2}$ from $x = \frac{1}{\sqrt{5}}(\frac{1+\sqrt{5}}{2})^{n+1}$. See Figure 3.17. Since there is at most one such integer, we discover that f_n is the integer closest to $\frac{1}{\sqrt{5}}(\frac{1+\sqrt{5}}{2})^{n+1}$.

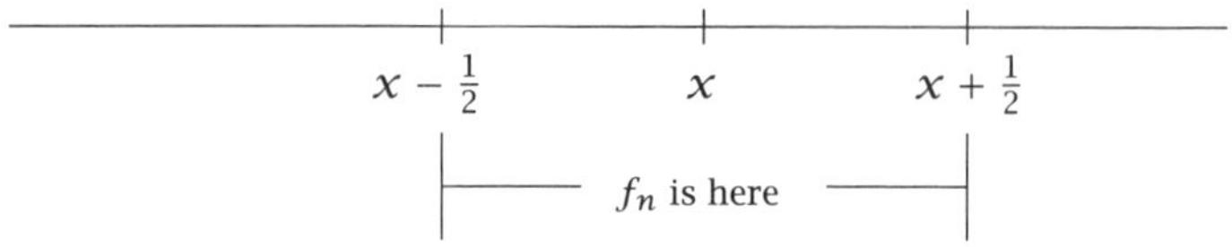

Figure 3.17: x is the integer closest to f_n

For example, since $\frac{1}{\sqrt{5}}(\frac{1+\sqrt{5}}{2})^{13} \approx 232.999$, we know that $f_{12} = 233$. (New Year's dinner looks promising.)

READING CHECK 1. Why does the expression on the right of formula in (4) give integer values?

Answers to Reading Checks

1. The expression gives the values of the Fibonacci sequence, which are all integers.

EXERCISES

Answers to exercises marked [BB] can be found at the **Back of the Book.**

1. [BB] Consider the sequence of numbers defined recursively by

$$a_0 = 0, a_1 = 1 \text{ and, for } n \geq 1, a_{n+1} = a_n + 2a_{n-1}.$$

   (a) List the first eight terms of this sequence.

   (b) Find a general formula for a_n.

2. Answer Exercise 1 again for the sequence defined recursively by $a_{n+1} = -4a_n + 5a_{n-1}$, $n \geq 1$. Assume $a_0 = 1$ and $a_1 = -1$.

3. Answer Exercise 1 again for the sequence defined recursively by $a_{n+1} = 2a_n + 3a_{n-1}$, $n \geq 1$. Assume $a_0 = -3$ and $a_1 = 4$.

4. For each of the following recursively defined sequences, find a formula for a_n.

   (a) [BB] $a_0 = -4$, $a_1 = 0$ and, for $n \geq 1$, $a_{n+1} = 3a_n - 2a_{n-1}$

   (b) $a_0 = -1$, $a_1 = 5$ and, for $n \geq 1$, $a_{n+1} = 4a_n - 3a_{n-1}$

   (c) $a_1 = 1$, $a_2 = 2$ and, for $n \geq 1$, $a_{n+1} = 3a_n + 4a_{n-1}$

   (d) $a_1 = -8$, $a_2 = 1$ and, for $n \geq 1$, $a_{n+1} = a_n + 12a_{n-1}$.

5. The eigenvalues of $A = \begin{bmatrix} -1 & 2 & 2 \\ 2 & 2 & 2 \\ -3 & -6 & -6 \end{bmatrix}$ are 0, -2 and -3.

   (a) Find the eigenspaces of A.

   (b) Find A^4.

   (c) Find A^n for an integer $n \geq 1$.

   (d) Solve $\mathsf{v}_{n+1} = A\mathsf{v}_n$ with $\mathsf{v}_0 = \begin{bmatrix} 1 \\ 0 \\ 1 \end{bmatrix}$.

6. **(a)** Obtain a formula for A^n with $A = \begin{bmatrix} -1 & 3 \\ 2 & 0 \end{bmatrix}$ and n a positive integer.

   (b) The sequence of vectors $\mathsf{v}_n = \begin{bmatrix} x_n \\ y_n \end{bmatrix}$, $n = 0, 1, 2, \ldots$, is defined by

   $\mathsf{v}_0 = \begin{bmatrix} 1 \\ -1 \end{bmatrix}$ and $\mathsf{v}_{n+1} = A\mathsf{v}_n$ for $n \geq 0$.

   Obtain formulas for x_n and y_n and show that $\frac{x_n}{y_n} \to -\frac{3}{2}$ as $n \to \infty$.

7. Let $f_0, f_1, f_2, \ldots$ be the terms of the Fibonacci sequence.

   Find the values of $\frac{1}{\sqrt{5}}(\frac{1+\sqrt{5}}{2})^{n+1}$ for $n = 0, 1, \ldots, 9$ (to three decimal place accuracy) and compare with the values of $f_0, f_1, \ldots, f_9$. What do you observe?

8. Let λ_1 and λ_2 be the roots of the polynomial $\lambda^2 - r\lambda - s$ and assume that $\lambda_1 \neq \lambda_2$. Show that the solution of the recurrence relation $a_n = ra_{n-1} + sa_{n-2}$, $n \geq 2$ has the form $a_n = c_1\lambda_1^n + c_2\lambda_2^n$ for some constants c_1 and c_2.

3.7 Application: Markov Chains

Three thousand members of the International Order of Raccoons arrive in town for a convention. They go out to dinner the night before the convention begins, three thousand strong. Fifteen hundred go to local Italian restaurants, one thousand to Chinese restaurants and five hundred go Thai. Not everyone is pleased with their choice. The first night of the convention, for instance, of those who ate Italian the previous night, just 70% decided to eat Italian again, 20% deciding to try Chinese and 10% Thai cuisine. The table below shows how the Raccoons changed their choices of cuisine.

		Last Night		
		I	C	T
	I	.7	.1	.1
Tonight	C	.2	.8	.2
	T	.1	.1	.7

Suppose we set $x_0 = 1500$, $y_0 = 1000$, $z_0 = 500$, the numbers eating Italian, Chinese and Thai, respectively, the night before the convention and let x_k, y_k and z_k denote the numbers of Raccoons eating Italian, Chinese and Thai, respectively, on the kth night of the convention. The table shows that the numbers eating Italian, Chinese and Thai on the first night of the convention are

$$\begin{aligned} x_1 &= 0.7x_0 + 0.1y_0 + 0.1z_0 = 0.7(1500) + 0.1(1000) + 0.1(500) = 1200 \\ y_1 &= 0.2x_0 + 0.8y_0 + 0.2z_0 = 0.2(1500) + 0.8(1000) + 0.2(500) = 1200 \\ z_1 &= 0.1x_0 + 0.1y_0 + 0.7z_0 = 0.1(1500) + 0.1(1000) + 0.7(500) = 600. \end{aligned}$$

These numbers, 1200, 1200, 600, are the components of $A\mathsf{v}_0$ where $A = \begin{bmatrix} 0.7 & 0.1 & 0.1 \\ 0.2 & 0.8 & 0.2 \\ 0.1 & 0.1 & 0.7 \end{bmatrix}$ and $\mathsf{v}_0 = \begin{bmatrix} x_0 \\ y_0 \\ z_0 \end{bmatrix} = \begin{bmatrix} 1500 \\ 1000 \\ 500 \end{bmatrix}$:

$$A\mathsf{v}_0 = \begin{bmatrix} 0.7 & 0.1 & 0.1 \\ 0.2 & 0.8 & 0.2 \\ 0.1 & 0.1 & 0.7 \end{bmatrix} \begin{bmatrix} 1500 \\ 1000 \\ 500 \end{bmatrix} = \begin{bmatrix} 1200 \\ 1200 \\ 600 \end{bmatrix}.$$

The matrix A is called a *transition matrix* because it describes the transition of dining preferences from one night to the next. Let $\mathsf{v}_k = \begin{bmatrix} x_k \\ y_k \\ z_k \end{bmatrix}$. Then $\mathsf{v}_1 = A\mathsf{v}_0$. If the trend continues, on the second night of the convention, the numbers of Raccoons dining Italian, Chinese and Thai, respectively, will be

$$\begin{aligned} x_2 &= 0.7x_1 + 0.1y_1 + 0.1z_1 = 1020 \\ y_2 &= 0.2x_1 + 0.8y_1 + 0.2z_1 = 1320 \\ z_2 &= 0.1x_1 + 0.1y_1 + 0.7z_1 = 660, \end{aligned}$$

these being the components of $\mathsf{v}_2 = A\mathsf{v}_1$:

$$A\mathsf{v}_1 = \begin{bmatrix} 0.7 & 0.1 & 0.1 \\ 0.2 & 0.8 & 0.2 \\ 0.1 & 0.1 & 0.7 \end{bmatrix} \begin{bmatrix} 1200 \\ 1200 \\ 600 \end{bmatrix} = \begin{bmatrix} 1020 \\ 1320 \\ 660 \end{bmatrix} = \mathsf{v}_2.$$

On the third night, assuming the trend continues, the numbers will be the components of $\mathsf{v}_3 = A\mathsf{v}_2 = \begin{bmatrix} 912 \\ 1392 \\ 696 \end{bmatrix}$. In Table 3.1, we show the numbers eating Chinese, Italian and Thai over the first ten nights of the convention. The numbers are striking. Despite the overwhelming popularity of Italian cuisine at the beginning, and the relatively few people who initially preferred

	Day k					
k	0	1	2	3	4	5
x_k (Chinese)	1500	1200	1020	912	847	808
y_k (Italian)	1000	1200	1320	1392	1435	1461
z_k (Thai)	500	600	660	696	718	731

	Day k				
k	6	7	8	9	10
x_k (Chinese)	785	771	763	758	756
y_k (Italian)	1477	1786	1492	1495	1497
z_k (Thai)	738	743	746	748	749

Table 3.1

Thai, after ten days (Raccoon conventions are long), most people are eating Italian while the numbers going Chinese and Thai are almost the same and about half the number eating Italian.

In this section, we continue our study of the recurrence relation $\mathsf{v}_{n+1} = A\mathsf{v}_n$, but with a very special class of matrices.

3.7.1 Definition. A *Markov matrix*[5] is a matrix $A = [a_{ij}]$ with each entry $a_{ij} \geq 0$ and the entries in each column summing to 1. A *Markov chain* is a sequence of vectors $\mathsf{v}_0, \mathsf{v}_1, \mathsf{v}_2, \ldots$, where $\mathsf{v}_{k+1} = A\mathsf{v}_k$ for $k \geq 0$ and the matrix A is Markov. The components of v_k are called the *states* of the Markov chain at step k.

In the example above, the states correspond to Italian, Chinese and Thai.

3.7.2 Examples. The matrix $\begin{bmatrix} 0.7 & 0.1 & 0.1 \\ 0.2 & 0.8 & 0.2 \\ 0.1 & 0.1 & 0.7 \end{bmatrix}$ which appears in our restaurant example is Markov, as is $\begin{bmatrix} 0 & \frac{2}{3} \\ 1 & \frac{1}{3} \end{bmatrix}$, but not $\begin{bmatrix} \frac{1}{2} & \frac{3}{4} \\ \frac{1}{2} & \frac{1}{3} \end{bmatrix}$ and not $\begin{bmatrix} \frac{5}{4} & \frac{1}{3} \\ -\frac{1}{4} & \frac{2}{3} \end{bmatrix}$. ☺

Markov matrices have some remarkable properties. In order to describe some of these, we first discuss (square) matrices each of whose **rows** sums to 0.

[5]After the Russian probabilist Andrei Markov (1856–1922).

Suppose then that $B = \begin{bmatrix} \mathsf{b}_1 & \mathsf{b}_2 & \cdots & \mathsf{b}_n \\ \downarrow & \downarrow & & \downarrow \end{bmatrix}$ is such a matrix. Let $\mathsf{u} = \begin{bmatrix} 1 \\ \vdots \\ 1 \end{bmatrix}$ be a column of n 1s. Remembering that $B\mathsf{x}$ is a linear combination of the columns of B with coefficients the components of the vector x (the oft-used 2.1.33), we see that

$$B\mathsf{u} = 1\mathsf{b}_1 + 1\mathsf{b}_2 + \cdots + 1\mathsf{b}_n = \mathsf{b}_1 + \mathsf{b}_2 + \cdots + \mathsf{b}_n$$

is just the sum of the columns of B. This sum is 0 because each of the rows of B sums to 0. So $B\mathsf{u} = 0$; thus, B is not invertible and $\det B = 0$.

Suppose B is a square matrix each of whose **columns** sums to 0. Then each of the rows of B^T sums to 0, so $0 = \det B^T = \det B$. Whether the rows or the columns of a matrix sum to 0, the determinant of the matrix must be 0.

Let A be a Markov matrix. Then each column of $B = A - I$ sums to 0, so $\det(A-I) = 0$, showing that 1 is an eigenvalue of A. Moreover, if A is *regular* in the sense that some power of A has all entries positive, then the eigenspace corresponding to $\lambda = 1$ is a line, that is, it consists of scalar multiples of a single vector. (We justify this statement later in Theorem 3.7.9.)

3.7.3 Definition. A Markov matrix A is *regular* if, for some positive integer n, all entries of A^n are positive.

3.7.4 Examples. The matrix $\begin{bmatrix} 0.7 & 0.1 & 0.1 \\ 0.2 & 0.8 & 0.2 \\ 0.1 & 0.1 & 0.7 \end{bmatrix}$ is regular because it is Markov and all its entries are positive. The reader should check that the eigenspace corresponding to $\lambda = 1$ consists of multiples of $\begin{bmatrix} 1 \\ 2 \\ 1 \end{bmatrix}$. The matrix $A = \begin{bmatrix} 0 & \frac{2}{3} \\ 1 & \frac{1}{3} \end{bmatrix}$ is regular because it is Markov and all the entries of $A^2 = \begin{bmatrix} \frac{2}{3} & \frac{2}{9} \\ \frac{1}{3} & \frac{7}{9} \end{bmatrix}$ are positive. The eigenspace corresponding to $\lambda = 1$ consists of multiples of $\begin{bmatrix} 2 \\ 3 \end{bmatrix}$. On the other hand, while $A = \begin{bmatrix} 0 & 1 \\ 1 & 0 \end{bmatrix}$ is Markov, it is not regular because the only powers of A are A and I so no power will have all entries positive. ☺

Remember that we are studying the recurrence relation $\mathsf{v}_{n+1} = A\mathsf{v}_n$ with A Markov. We have $\mathsf{v}_1 = A\mathsf{v}_0$, $\mathsf{v}_2 = A\mathsf{v}_1 = A^2\mathsf{v}_0$, $\mathsf{v}_3 = A\mathsf{v}_2 = A^3\mathsf{v}_0$ and, in general, $\mathsf{v}_k = A^k\mathsf{v}_0$ for any $k \geq 1$. To see if our guesses about the eventual dining tastes of our conventioneers are really true, we have to discover what happens to v_k, and hence to A^k, as k gets bigger and bigger.

The columns of A^k are $A^k\mathsf{e}_1$, $A^k\mathsf{e}_2$, ..., $A^k\mathsf{e}_n$ where, as always, $\mathsf{e}_1, \mathsf{e}_2, \ldots, \mathsf{e}_n$ denote the standard basis vectors of R^n. It turns out that as k increases, each of these vectors $A^k\mathsf{e}_i$ "converges" to one and the same vector v_∞—called the *steady state* vector—and the sum of the components of v_∞ is 1. (It is not hard to show that the components of v_∞ sum to 1—see statement (2) below and subsequent remarks—but it's much harder to show that all the vectors $A^k\mathsf{e}_i$ converge to the same vector—see Theorem 3.7.10.)

Summarizing, as k gets larger and larger, A^k converges, that is, gets closer and closer, to the matrix $B = \begin{bmatrix} \mathsf{v}_\infty & \mathsf{v}_\infty & \cdots & \mathsf{v}_\infty \\ \downarrow & \downarrow & & \downarrow \\ & & & \end{bmatrix}$. For example, we shall show that the powers of $A = \begin{bmatrix} 0.7 & 0.1 & 0.1 \\ 0.2 & 0.8 & 0.2 \\ 0.1 & 0.1 & 0.7 \end{bmatrix}$ converge to $B = \begin{bmatrix} 0.25 & 0.25 & 0.25 \\ 0.50 & 0.50 & 0.50 \\ 0.25 & 0.25 & 0.25 \end{bmatrix}$. Thus, in the long run, the numbers of Raccoons eating Italian, Chinese and Thai, respectively, are the components of $B\mathsf{v}_0 = \begin{bmatrix} 750 \\ 1500 \\ 750 \end{bmatrix}$, in agreement with our earlier suspicions.

It remains only to explain how to find v_∞, this wonderful "steady state vector." This is the vector in each column of B, in particular, it's the first column, $B\mathsf{e}_1$: $\mathsf{v}_\infty = B\mathsf{e}_1$. Remember too that B is the matrix to which A^k converges as k gets bigger and bigger. Since A^k and $A^{k+1} = A(A^k)$ both converge to B, it follows that $B = AB$,

$$\begin{array}{ccc} A^{k+1} & = & A(A^k), \\ \downarrow & & \downarrow \\ B & & B \end{array}$$

so $A\mathsf{v}_\infty = A(B\mathsf{e}_1) = (AB)\mathsf{e}_1 = B\mathsf{e}_1 = \mathsf{v}_\infty$. Thus v_∞ is an eigenvector of A corresponding to the eigenvalue 1. Now the eigenspace of A corresponding to 1 is a line, so it consists of all multiples $t\mathsf{v}_\infty$ of v_∞. The sum of the components of $t\mathsf{v}_\infty$ is t (because the sum of the components of v_∞ is 1). There is just one vector in the eigenspace for which the sum $t = 1$, and this is v_∞. So we have a way to find this vector.

3.7.5 v_∞ is the unique eigenvector corresponding to $\lambda = 1$ whose components sum to 1.

For example, the eigenspace of $A = \begin{bmatrix} 0.7 & 0.1 & 0.1 \\ 0.2 & 0.8 & 0.2 \\ 0.1 & 0.1 & 0.7 \end{bmatrix}$ corresponding to $\lambda = 1$ is multiples of $\begin{bmatrix} 1 \\ 2 \\ 1 \end{bmatrix}$, so $\mathsf{v}_\infty = t\begin{bmatrix} 1 \\ 2 \\ 1 \end{bmatrix}$ for some t that we can determine from the fact that the components of v_∞ sum to 1. The components of $t\begin{bmatrix} 1 \\ 2 \\ 1 \end{bmatrix}$ sum to $4t$, so $4t = 1$, $t = \frac{1}{4}$ and $\mathsf{v}_\infty = \begin{bmatrix} \frac{1}{4} \\ \frac{1}{2} \\ \frac{1}{4} \end{bmatrix}$.

READING CHECK 1. If an eigenspace of a matrix A is a line, that is, it consists of all vectors of the form $t\mathsf{v}$ for some v, and if the components of v do not sum to 0, then the eigenspace always contains a vector whose components sum to 1. There is just one such vector. What is it?

3.7.6 Example. $A = \begin{bmatrix} 0 & \frac{2}{3} \\ 1 & \frac{1}{3} \end{bmatrix}$ is a regular Markov matrix. The eigenspace corresponding to $\lambda = 1$ consists of multiples of $\begin{bmatrix} 2 \\ 3 \end{bmatrix}$. The steady state vector is the only vector in this eigenspace whose components sum to 1, so it has the form $t\begin{bmatrix} 2 \\ 3 \end{bmatrix}$ where $2t + 3t = 1$. Thus $t = \frac{1}{5}$ and the steady state vector is $\mathsf{v}_\infty = \frac{1}{5}\begin{bmatrix} 2 \\ 3 \end{bmatrix}$. ☺

3.7.7 Problem. Three political parties P_1, P_2, P_3 are running for office in a forthcoming election. The transition matrix $A = \begin{bmatrix} \frac{1}{2} & \frac{1}{3} & \frac{1}{6} \\ \frac{1}{4} & \frac{1}{3} & \frac{1}{3} \\ \frac{1}{4} & \frac{1}{3} & \frac{1}{2} \end{bmatrix}$ shows how voter preferences are changing from week to week during the election campaign; specifically, the (i, j) entry of A shows the fraction of the voting population whose support changes from party j to party i each week. With $j = 3$, and $i = 1, 2, 3$, for instance, we see that each week $\frac{1}{6}$ of the supporters of P_3 switch to party P_1, $\frac{1}{3}$ switch to P_2 and $\frac{1}{2}$ maintain their allegiance to party P_3.

(a) Find the steady state vector.

(b) To what matrix does A^k converge as k increases?

(c) Assuming there are initially $3\frac{1}{2}$ million voters, one million of whom support P_1, 500,000 of whom support P_2 and two million who support P_3, how many voters will eventually support each party?

Solution. (a) The given matrix is Markov (columns sum to 1) and regular (all entries are positive), so there is a steady state vector which is the unique vector in the eigenspace of 1 whose components sum to 1. The eigenspace corresponding to $\lambda = 1$ consists of vectors of the form $t\begin{bmatrix} \frac{8}{9} \\ \frac{5}{6} \\ 1 \end{bmatrix}$. The sum of the components of this vector is $\frac{8}{9}t+\frac{5}{6}t+t = \frac{49}{18}t$. The vector whose components add to 1 is $\mathsf{v}_\infty = \frac{18}{49}\begin{bmatrix} \frac{8}{9} \\ \frac{5}{6} \\ 1 \end{bmatrix} = \frac{1}{49}\begin{bmatrix} 16 \\ 15 \\ 18 \end{bmatrix}$.

(b) A^k converges to $B = \frac{1}{49}\begin{bmatrix} 16 & 16 & 16 \\ 15 & 15 & 15 \\ 18 & 18 & 18 \end{bmatrix}$, each of whose columns is v_∞.

(c) The final numbers of supporters of each party are the components of

$$B\begin{bmatrix} 1{,}000{,}000 \\ 500{,}000 \\ 2{,}000{,}000 \end{bmatrix} = \frac{1}{7}\begin{bmatrix} 8{,}000{,}000 \\ 7{,}500{,}000 \\ 9{,}000{,}000 \end{bmatrix} \approx 10^6\begin{bmatrix} 1.14 \\ 1.07 \\ 1.29 \end{bmatrix},$$

so eventually, about 1.14×10^6 people will support party P_1, 1.07×10^6 will support P_2 and 1.29×10^6 will support P_3.

READING CHECK 2. In Problem 3.7.7, the fractions of voters eventually supporting each party are, approximately, $\frac{1.14}{3.5}$, $\frac{1.07}{3.5}$ and $\frac{1.29}{3.5}$. These numbers, roughly, 0.326, 0.306 and 0.369, are the components of the steady state vector: $\frac{16}{49}, \frac{15}{49}, \frac{18}{49}$. Show that this is always the case: Namely, if $\mathsf{v}_0 = \begin{bmatrix} x_0 \\ y_0 \\ z_0 \end{bmatrix}$ and $\mathsf{v}_\infty = \begin{bmatrix} p_1 \\ p_2 \\ p_3 \end{bmatrix}$, then, as $k \to \infty$, the vectors v_k approach the vector $\begin{bmatrix} p_1(x_0+y_0+z_0) \\ p_2(x_0+y_0+z_0) \\ p_3(x_0+y_0+z_0) \end{bmatrix}$ whose components, as fractions of the total voting population $x_0 + y_0 + z_0$, are precisely the components of v_∞. Note that this long term behaviour is entirely independent of the initial proportions. No matter how the electorate felt initially, it will eventually support the parties in proportions that are the components of the steady state vector.

Theory

In the last part of this section, we justify most of the claims we have made about Markov and regular Markov matrices. This is material to which the interested reader may wish to return because some of our arguments rely on ideas discussed in later chapters. Moreover, we warn that the theorems and proofs that appear here are at a higher level than most other theorems in this book.

3.7.8 Theorem. *If A is a Markov matrix, then 1 is an eigenvalue of A and all other eigenvalues λ satisfy $|\lambda| \leq 1$.*

Proof. We have already shown that 1 is an eigenvalue, so it remains to prove that $|\lambda| \leq 1$ for any eigenvalue λ. Our proof uses the fact that

$$|a_1 + a_2 + \cdots + a_n| \leq |a_1| + |a_2| + \cdots + |a_n|$$

for any n real numbers $a_1, a_2, a_3, \ldots, a_n$, an easy extension of the triangle equality that we met in Section 1.2. Moreover, if $|a_1 + a_2 + \cdots + a_n| = |a_1| + |a_2| + \cdots + |a_n|$, then the numbers a_i all have the same sign (all are nonnegative or all nonpositive). (See Exercise 11.)

The matrices $A = \begin{bmatrix} \mathsf{a}_1 & \mathsf{a}_2 & \ldots & \mathsf{a}_n \\ \downarrow & \downarrow & & \downarrow \\ & & & \end{bmatrix}$ and A^T have the same characteristic polynomials because $A^T - \lambda I = (A - \lambda I)^T$, so they have the same eigenvalues. In particular, λ is an eigenvalue of $A^T = \begin{bmatrix} \mathsf{a}_1^T & \rightarrow \\ \mathsf{a}_2^T & \rightarrow \\ \vdots & \\ \mathsf{a}_n^T & \rightarrow \end{bmatrix}$, so $A^T\mathsf{x} = \lambda\mathsf{x}$ for some nonzero vector $\mathsf{x} = \begin{bmatrix} x_1 \\ x_2 \\ \vdots \\ x_n \end{bmatrix}$. Let $|x_k|$ be the largest of $|x_1|, |x_2|, \ldots, |x_n|$ and compare the absolute values of the kth components of $\lambda\mathsf{x}$ and $A^T\mathsf{x}$. We have

$$\begin{aligned} |\lambda x_k| &= |\mathsf{a}_k \cdot \mathsf{x}| = |a_{1k}x_1 + a_{2k}x_2 + \cdots + a_{nk}x_n| \\ &\leq |a_{1k}x_1| + |a_{2k}x_2| + \cdots + |a_{nk}x_n| \qquad \text{(triangle inequality)} \\ &= |a_{1k}||x_1| + |a_{2k}||x_2| + \cdots + |a_{nk}||x_n| \qquad (1) \\ &\leq |a_{1k}||x_k| + |a_{2k}||x_k| + \cdots + |a_{nk}||x_k| \\ &= (a_{1k} + a_{2k} + \cdots + a_{nk})|x_k| = |x_k| \end{aligned}$$

because each $a_{ik} \geq 0$ and these numbers sum to 1. Dividing by the nonzero $|x_k|$ gives $|\lambda| \leq 1$ as desired. ∎

The proof just given suggests another theorem.

3.7.9 Theorem. *Let λ be an eigenvalue of a regular Markov matrix A and suppose $|\lambda| = 1$. Then $\lambda = 1$ and the corresponding eigenspace is a line: it consists of scalar multiples of a single vector.*

Proof. Our proof uses several ideas, including some major results from Chapters 4 and 6. Theorem 6.4.17 says that if B is an $n \times n$ matrix, then $(\text{col sp}\, B)^\perp = \text{null sp}\, B^T$. Here col sp B denotes the "column space" of B (the span of the columns of B—see Section 4.1) and null sp B^T denotes the "null space" of B^T, which is the set of solutions to the homogeneous system $B^T\mathsf{x} = 0$. It follows that the dimension of null sp B^T is the dimension of $(\text{col sp}\, B)^\perp$, the orthogonal complement of col sp B, and, by a fundamental result about orthogonal complements (Theorem 6.4.26), this is $n - \dim \text{col sp} B$. One of the most basic results in linear algebra (Theorem 4.4.17) says that the sum of the dimensions of the null space and column space of an $n \times n$ matrix is n, so $\dim \text{null sp} B^T = n - \dim \text{col sp} B = \dim \text{null sp} B$.

We want to prove that the eigenspace of A corresponding to $\lambda = 1$ is a line, that is, that the null space of $B = A - I$ has dimension 1, and we have just observed that this can be established by proving that $\dim \text{null sp} B^T = \dim \text{null sp}(A^T - I) = 1$, that is, that the eigenspace of A^T corresponding to $\lambda = 1$ is a line. We now prove the theorem.

First assume that all the entries of the Markov matrix A are positive. Let λ be an eigenvalue of A, and hence of A^T too, with $|\lambda| = 1$ and suppose $A^T\mathsf{x} = \lambda\mathsf{x}$ as in the proof of Theorem 3.7.8. We wish to show that $\lambda = 1$ and that the eigenspace of A^T corresponding to λ is a line. By Theorem 3.7.8, $|\lambda| \leq 1$. If $|\lambda| = 1$, then the two inequalities in (1) must be equalities. The second of these is

$$|a_{1k}||x_1| + |a_{2k}||x_2| + \cdots + |a_{nk}||x_n| \\ = |a_{1k}||x_k| + |a_{2k}||x_k| + \cdots + |a_{nk}||x_k|.$$

Since each $a_{ik} \geq 0$, this says

$$a_{1k}(|x_k| - |x_1|) + a_{2k}(|x_k| - |x_1|) + \cdots + a_{nk}(|x_k| - |x_n|) = 0.$$

Each term on the left is nonnegative because $|x_k| \geq |x_i|$ by maximality of $|x_k|$, so the only way the sum can be 0 is if each term is 0, that is, $|x_i| = |x_k|$ for all i.

One of our preliminary remarks gives that all $a_{ik}x_i$ have the same sign. Each $a_{ik} > 0$ by assumption, so all x_i have the same sign. Since $|x_i| = |x_k|$, we get $x_i = x_k$ for all i, which says that $\mathsf{x} = \begin{bmatrix} x_k \\ x_k \\ \vdots \\ x_k \end{bmatrix} = x_k\mathsf{u}$, with $\mathsf{u} = \begin{bmatrix} 1 \\ 1 \\ \vdots \\ 1 \end{bmatrix}$. This shows that the eigenspace of A^T corresponding to λ is a line, the desired result. Now u is in this eigenspace, so $A^T\mathsf{u} = \lambda\mathsf{u}$. But $A^T\mathsf{u} = \mathsf{u}$ because each row of A^T sums to 1, so $\lambda = 1$. This completes the proof in the case that the entries of A itself are positive.

For an arbitrary regular matrix, we know only that the entries of some power A^m of A are positive. Now A^m is a regular Markov matrix (see Exercise 9), so, by what we have already seen, the eigenspace of A^m corresponding to the eigenvalue 1 is a line, consisting of multiples of some vector x_0. Let λ be an eigenvalue of A with $|\lambda| = 1$ and let x be an eigenvector of A corresponding to λ. Thus $A\mathsf{x} = \lambda\mathsf{x}$, which implies $A^m\mathsf{x} = \lambda^m\mathsf{x}$. Since $|\lambda^m| = 1$, by what we have already seen, we must have $\lambda^m = 1$. So x, being an eigenvector of A^m corresponding to $\lambda^m = 1$, is a multiple of x_0. But x was any eigenvector of A corresponding to λ. It follows that the eigenspace of A corresponding to λ is multiples of x_0. To see that $\lambda = 1$, we note that A^{m+1} is Markov and regular (Exercise 9). Just as we showed $\lambda^m = 1$ for the regular Markov matrix

A^m, we have $\lambda^{m+1} = 1$ for the regular Markov matrix A^{m+1}. So $\lambda^{m+1} = \lambda^m$, $\lambda^m(\lambda - 1) = 0$, implying $\lambda = 1$ ($\lambda \neq 0$ because $|\lambda| = 1$). ∎

In our discussion of the properties of a Markov matrix, we stated without proof that the components of the steady state vector v_∞ sum to 1. This fact requires only an appeal to the result 2.1.33 that we emphasize throughout this text.

Let $A = \begin{bmatrix} a_1 & a_2 & \dots & a_n \\ \downarrow & \downarrow & & \downarrow \\ & & & \end{bmatrix}$ be a Markov matrix, let $x = \begin{bmatrix} x_1 \\ x_2 \\ \vdots \\ x_n \end{bmatrix}$ and suppose that $Ax = y = \begin{bmatrix} y_1 \\ y_2 \\ \vdots \\ y_n \end{bmatrix}$. Then $\begin{bmatrix} y_1 \\ \vdots \\ y_n \end{bmatrix} = x_1 a_1 + x_2 a_2 + \cdots + x_n a_n$, so the sum of the components of y is

$$x_1 \times \text{ sum of components of } a_1 \\ + x_2 \times \text{ sum of components of } a_2 \\ + \cdots + x_n \times \text{ sum of components of } a_n.$$

Since the components of each a_i sum to 1, we get

$$y_1 + y_2 + \cdots + y_n = x_1 + x_2 + \cdots + x_n$$

and conclude that

$$Ax = y \quad \text{implies} \quad \begin{matrix}\text{the sum of the components of } x \text{ is the} \\ \text{sum of the components of } y.\end{matrix} \tag{2}$$

In particular, if e_i is one of the standard basis vectors of R^n, the components of e_i sum to 1, so the components of Ae_i sum to 1 as well, as do the components of $A^2e_i = A(Ae_i)$, A^3e_i, A^4e_i and, in general, A^ke_i for any positive integer k. Since A^ke_i converges to the vector v_∞, the sum of the components of v_∞ is 1 too.

The final property of a Markov matrix upon which we have relied follows.

3.7.10 Theorem. *Let A be an $n \times n$ Markov matrix and v_0 a vector. Let $v_1 = Av_0$, $v_2 = Av_1 = A^2v_0$, $v_3 = Av_2 = A^3v_0$, and $v_{k+1} = Av_k = A^{k+1}v_0$ in general. Then, as k grows larger and larger, the vectors v_k converge to a vector v_∞ that depends only on the sum of the components of v_0, not on v_0 itself. In particular, the vectors $A^ke_1, A^ke_2, \dots, A^ke_n$ converge to the same vector.*

Proof. We prove this only in the case that A is diagonalizable, that is, A has n linearly independent eigenvectors $x_1, x_2, \dots, x_n$—see Theorem 3.5.9. Let $\lambda_1, \lambda_2, \dots, \lambda_n$ be corresponding eigenvalues, respectively. The matrix

$$P = \begin{bmatrix} \mathsf{x}_1 & \mathsf{x}_2 & \cdots & \mathsf{x}_n \\ \downarrow & \downarrow & & \downarrow \\ & & & \end{bmatrix} \text{ is invertible and } P^{-1}AP = D = \begin{bmatrix} \lambda_1 & 0 & \cdots & 0 \\ 0 & \lambda_2 & & 0 \\ \vdots & & \ddots & \\ 0 & & & \lambda_n \end{bmatrix}$$

(see 3.5.3). Thus $A = PDP^{-1}$ and $A^k = PD^kP^{-1}$ for any positive integer k. Thus $\mathsf{v}_k = A^k\mathsf{v}_0 = PD^kP^{-1}\mathsf{v}_0$

$$= \begin{bmatrix} \mathsf{x}_1 & \mathsf{x}_2 & \cdots & \mathsf{x}_n \\ \downarrow & \downarrow & & \downarrow \\ & & & \end{bmatrix} \begin{bmatrix} \lambda_1^k & & & \\ & \lambda_2^k & & \\ & & \ddots & \\ & & & \lambda_n^k \end{bmatrix} \underbrace{P^{-1}\mathsf{v}_0}_{\mathsf{t}}$$

$$= \begin{bmatrix} \lambda_1^k\mathsf{x}_1 & \lambda_2^k\mathsf{x}_2 & \cdots & \lambda_n^k\mathsf{x}_n \\ \downarrow & \downarrow & & \downarrow \\ & & & \end{bmatrix} \begin{bmatrix} t_1 \\ t_2 \\ \vdots \\ t_n \end{bmatrix} = t_1\lambda_1^k\mathsf{x}_1 + t_2\lambda_2^k\mathsf{x}_2 + \cdots + t_n\lambda_n^k\mathsf{x}_n,$$

where, as indicated, $\mathsf{t} = \begin{bmatrix} t_1 \\ \vdots \\ t_n \end{bmatrix} = P^{-1}\mathsf{v}_0$. (We have now used 2.1.33 on several occasions!) Summarizing,

$$\mathsf{v}_k = t_1\lambda_1^k\mathsf{x}_1 + t_2\lambda_2^k\mathsf{x}_2 + \cdots + t_n\lambda_n^k\mathsf{x}_n, \tag{3}$$

an equation that can be used to see what happens to the vectors v_k in the long run, as k increases. By Theorem 3.7.8, 1 is an eigenvalue of A and by Theorem 3.7.9 all other eigenvalues have absolute value less than 1. Order the eigenvalues so that $\lambda_1 = 1$. Then $|\lambda_i| < 1$ for $i > 1$, so $\lambda_i^k \to 0$ as $k \to \infty$ and

$$\mathsf{v}_k \to t_1\mathsf{x}_1. \tag{4}$$

As recorded in (2), the sum of the components of each v_k is the sum of the components of v_0 and the same is true of the limit vector $t_1\mathsf{x}_1$. Had we started with any of the standard basis vectors, for instance, we would have that the sum of the components of $t_1\mathsf{x}_1$ is 1. Let r be the sum of the components of x_1. Thus $r \neq 0$.

Suppose the starting vector v_0 has components that sum to c. It remains only to argue that the limit vector, $t_1\mathsf{x}_1$ depends only on c. By (2), the components of each vector v_k sum to c, hence the same is true of the limit vector, so $c = t_1r$. Since $r \neq 0$, $t_1 = \frac{c}{r}$, so the vectors v_k converge to $\frac{c}{r}\mathsf{x}_1$, a vector depending only on c and the eigenvector x_1. This completes the proof. ■

Answers to Reading Checks

1. Any vector in the eigenspace is of the form $t\mathsf{v}$ for some scalar t. If the sum of the components of v is r, the sum of the components of $t\mathsf{x}$ is tr. There is just one t with $tr = 1$, namely, $t = \frac{1}{r}$, so the unique vector is $\frac{1}{r}\mathsf{v}$.

2. The vectors $\mathsf{v}_k = A^k\mathsf{v}_0$ approach $B\mathsf{v}_0$ as $k \to \infty$. Now $B = \begin{bmatrix} \mathsf{v}_\infty & \mathsf{v}_\infty & \mathsf{v}_\infty \\ \downarrow & \downarrow & \downarrow \end{bmatrix} =$ $\begin{bmatrix} p_1 & p_1 & p_1 \\ p_2 & p_2 & p_2 \\ p_3 & p_3 & p_3 \end{bmatrix}$, so $\mathsf{v}_k \to B\mathsf{v}_\infty = \begin{bmatrix} p_1(x_0 + y_0 + z_0) \\ p_2(x_0 + y_0 + z_0) \\ p_3(x_0 + y_0 + z_0 \end{bmatrix}$.

True/False Questions

Decide, with as little calculation as possible, whether each of the following statements is true or false and, if you say "false," explain your answer. (Answers are at the back of the book.)

1. The matrix $\begin{bmatrix} .333 & .505 & .250 \\ .333 & 0 & .350 \\ .333 & .495 & .400 \end{bmatrix}$ is Markov.

2. If A is an $n \times n$ Markov matrix and $\mathsf{u} = \begin{bmatrix} 2 \\ 2 \\ \vdots \\ 2 \end{bmatrix}$ is the vector of n 2s, then $A^T\mathsf{u} = \mathsf{u}$.

3. The determinant of a Markov matrix is 0.

4. The matrix $\begin{bmatrix} 1 & 0 \\ 0 & -1 \end{bmatrix}$ is regular.

5. $\lambda = 1$ is an eigenvalue of $\begin{bmatrix} .25 & .25 & .25 \\ .50 & .50 & .50 \\ .25 & .25 & .25 \end{bmatrix}$.

6. A Markov matrix has a unique eigenvector corresponding to $\lambda = 1$.

7. A regular Markov matrix has just one eigenvector corresponding to 1 whose components have sum 1.

8. It is possible for $\begin{bmatrix} 1 \\ 0 \end{bmatrix}$ and $\begin{bmatrix} 0 \\ 1 \end{bmatrix}$ to each be eigenvectors of a regular Markov matrix corresponding to $\lambda = 1$.

9. If A is a regular Markov matrix, then the matrices $A, A^2, A^3, \ldots$ converge to a matrix all of whose columns are the same.

10. If A is a regular Markov matrix, then the two sequences $A\mathsf{e}_1, A^2\mathsf{e}_1, A^3\mathsf{e}_1, \ldots$ and $A\mathsf{e}_2, A^2\mathsf{e}_2, A^3\mathsf{e}_2, \ldots$ converge to the same vector.

EXERCISES

Answers to exercises marked [BB] can be found at the Back of the Book.

1. [BB] Let $A = \begin{bmatrix} \frac{1}{3} & \frac{1}{4} \\ \frac{2}{3} & \frac{3}{4} \end{bmatrix}$.

   (a) Find a matrix P such that $P^{-1}AP$ is diagonal.

   (b) Let $k \geq 1$ be an integer. Find A^k.

   (c) Use the result of part (b) to determine the matrix B to which A^k converges as k gets large.

   (d) Could B have been determined without computing A^k? Explain.

2. In each case, show that A is regular and find the steady state vector.

   (a) [BB] $A = \begin{bmatrix} 0 & \frac{1}{2} & 0 & 0 \\ 0 & 0 & 0 & \frac{1}{2} \\ 1 & \frac{1}{2} & \frac{1}{2} & \frac{1}{2} \\ 0 & 0 & \frac{1}{2} & 0 \end{bmatrix}$ **(b)** $A = \begin{bmatrix} \frac{2}{3} & \frac{2}{3} & \frac{1}{3} \\ 0 & 0 & \frac{1}{3} \\ \frac{1}{3} & \frac{1}{3} & \frac{1}{3} \end{bmatrix}$.

3. The powers of $A = \begin{bmatrix} \frac{1}{2} & \frac{1}{4} & \frac{1}{4} \\ \frac{1}{4} & \frac{1}{2} & \frac{1}{4} \\ \frac{1}{4} & \frac{1}{4} & \frac{1}{2} \end{bmatrix}$ converge to a unique matrix B. Why? Find B.

4. [BB] After the first day of lectures one semester, one tenth of Dr. G's linear algebra students transferred to Dr. L's section and one fifth moved to Dr. P's section. After his first class, poor Dr. L lost half his students to Dr. G and a quarter of them to Dr. P. Meanwhile, Dr. P wasn't doing so well. He found that three tenths of his class had moved to Dr. G while another tenth had moved to Dr. L. If this movement occurs lecture after lecture, determine the eventual relative proportions of students in each section.

5. Each year, 20% of the mathematics majors at a certain university become physics majors, 10% of the chemistry majors turn to mathematics and 10% to physics, while 10% of physics majors become mathematics majors. Assuming each of these subjects has 100 students at the outset, how will these 300 eventually be distributed? (Life is so sweet at this university, that students choose to stay indefinitely.)

6. [BB] Each year the populations of British Columbia, Ontario and Newfoundland migrate as follows:

   - one quarter of those in British Columbia and one quarter of those in Newfoundland move to Ontario;
   - one sixth of those in Ontario move to British Columbia and one third of those in Ontario move to Newfoundland.

   If the total population of these three provinces is ten million, what is the eventual long term distribution of the population. Does the answer depend on the initial distribution? Explain.

7. Three companies A, B and C simultaneously introduce new tablets. Every year, A loses 30% of its market to B, B loses 40% of its market to C and C loses 20% of its market to A. Find the percentage share of the market each company will eventually have.

8. Suppose that each year $1/1000$ of the population of Canada that resides outside the province of Newfoundland and Labrador moves into the province while $3/100$ of those inside move out. Assume that these numbers remain the same year after year and there is no other population movement. Assume an initial population of 500,000 for Newfoundland and Labrador and 30,000,000 for the rest of Canada.

   (a) What will be the respective populations after one year? after two years? after three years? after 10 years?

   (b) What will the respective populations be in the long run? (Carry at least 8 decimal digits in your calculations.)

Critical Reading

9. **(a)** Let A be an $n \times n$ matrix and let $\mathsf{u} = \begin{bmatrix} 1 \\ 1 \\ \vdots \\ 1 \end{bmatrix}$ be a column of n 1s. Show that each column of A has sum 1 if and only if $A^T\mathsf{u} = \mathsf{u}$.

   (b) Show that the product of Markov matrices is Markov.

   (c) Show that if R is a regular Markov matrix and M is Markov, then RM is a regular Markov matrix. Thus, if A is a regular Markov matrix, so is A^m for any $m \geq 1$.

10. If A is a Markov matrix and x is an eigenvector of A corresponding to $\lambda \neq 1$, then the components of x sum to 0. Why?

11. Let $a_1, a_2, \ldots, a_n$ be n real numbers.

   (a) Show that $|a_1 + a_2 + \cdots + a_n| \leq |a_1| + |a_2| + \cdots + |a_n|$.

   (b) If $|a_1 + a_2 + \cdots + a_n| = |a_1| + |a_2| + \cdots + |a_n|$, show that all $a_i \geq 0$ or all $a_i \leq 0$.

12. Consider a Markov chain with transition matrix $A = \begin{bmatrix} \frac{5}{8} & \frac{1}{8} \\ \frac{3}{8} & \frac{7}{8} \end{bmatrix}$. Assume that the initial state vector is $\mathsf{v}_0 = \begin{bmatrix} 1 \\ 0 \end{bmatrix}$.

   (a) Find the state vector, v_3, after 3 transitions.

   (b) Find the state vector, v_k, after k transitions.

   (c) Find v_∞, the steady state vector.

(d) What can you say about v_k when k is large?

13. A mouse is put into a maze of compartments. The compartments are connected by tunnels as shown on the diagram. Some tunnels are two-way and others are one-way. The direction of one-way tunnels is indicated by arrows.

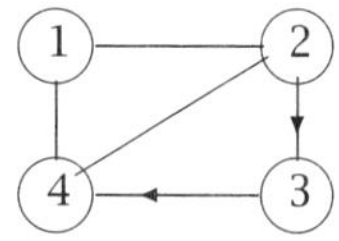

Assume that after visiting any compartment, the mouse always attempts to leave by choosing a tunnel at random so that all available tunnel entries are equally likely to be chosen.

(a) Write down the transition matrix A and show that it is regular.

(b) What proportion of time does the mouse spend visiting each compartment if we observe it moving in the maze for a long time?

CHAPTER KEY WORDS AND IDEAS: Here are some technical words and phrases that were used in this chapter. Do you know the meaning of each? If you're not sure, check the glossary or index at the back of the book.

characteristic polynomial	Markov
cofactor	matrix minor
commute (as in, matrices do not)	recursive
determinant	regular Markov
diagonalizable	sequence
eigenvalue	similar matrices
eigenvector	singular matrix
Fibonacci sequence	steady state
linear function	

Review Exercises for Chapter 3

1. True or false: If A is an $n \times n$ matrix, $n \geq 2$, then $\det(\frac{1}{\det A} \det A) = 1$.

2. Given 4×4 matrices X, Y, Z with Y invertible, express $\det(-2X^TY^{-1}Z^2)$ in terms of $\det X$, $\det Y$, and $\det Z$.

3. An $n \times n$ matrix A has the property that $A^T = -A$. Suppose $\det A \neq 0$. Show that n is even.

4. (a) Show that $A = \begin{bmatrix} 9 & -\frac{3}{2} & -5 \\ -5 & 1 & 3 \\ -2 & \frac{1}{2} & 1 \end{bmatrix}$ is invertible with inverse $B = \begin{bmatrix} 1 & 2 & -1 \\ 2 & 2 & 4 \\ 1 & 3 & -3 \end{bmatrix}$.

(b) Find $\det A$, $\det B$, and $\det 2A$.

5. An invertible matrix A satisfies $A^{-1} = A$. Show that $\det A = \pm 1$.

6. If A is an $n \times n$ matrix, $A^2 = A$ and $\det A \neq 1$, explain why A is not invertible.

7. (a) Find C, the matrix of cofactors of $A = \begin{bmatrix} 1 & 3 & 5 \\ 2 & 1 & 1 \\ 3 & 4 & 2 \end{bmatrix}$.

   (b) Find AC^T. (c) What is $\det A$? (d) What is A^{-1}?

   (e) Use your answer to (d) to solve the system $A\mathsf{x} = \mathsf{b}$, where $\mathsf{b} = \begin{bmatrix} 1 \\ 0 \\ 0 \end{bmatrix}$.

8. Find the matrix of cofactors of $A = \begin{bmatrix} -1 & 3 & 2 \\ 0 & -2 & 1 \\ 1 & 0 & -2 \end{bmatrix}$ and use this to find $\det A$.

   Why is A invertible? What is A^{-1}?

9. (a) Let $A = \begin{bmatrix} 1 & 3 & 5 \\ 2 & 4 & 6 \\ 3 & 5 & 9 \end{bmatrix}$. Find $\det A$ with Laplace expansions along the first row and down the third column.

   (b) Let $B = \begin{bmatrix} 6 & -4 & -3 \\ 2 & 8 & 5 \\ 8 & 4 & 0 \end{bmatrix}$. Find $\det B$ with Laplace expansions along the second row and down the second column.

   (c) Are either of the matrices A and B singular? Explain.

10. Find the determinant of $A = \begin{bmatrix} 3 & -2 & 1 & 1 \\ 0 & 0 & 0 & -5 \\ 0 & 4 & -3 & 7 \\ 0 & 0 & -2 & 9 \end{bmatrix}$.

11. Find the determinant of each of the following matrices by reducing to triangular.

    (a) $\begin{bmatrix} 0 & 2 & 1 & 1 \\ 1 & 2 & -1 & -2 \\ -1 & -4 & 3 & 1 \\ 1 & 6 & 4 & 2 \end{bmatrix}$ (b) $\begin{bmatrix} 2 & 1 & 1 & 1 \\ 0 & 3 & 0 & 1 \\ 1 & 0 & 2 & 1 \\ 0 & 0 & 0 & 1 \end{bmatrix}$

    (c) $\begin{bmatrix} -3 & 0 & 1 & 1 \\ 3 & 1 & 2 & 2 \\ -6 & -2 & -4 & 2 \\ 1 & -1 & 0 & -1 \end{bmatrix}$ (d) $\begin{bmatrix} -2 & 1 & 2 & 1 \\ 0 & 2 & 0 & 3 \\ 1 & 3 & -1 & 0 \\ 1 & 2 & 1 & 1 \end{bmatrix}$.

12. Given $A = \begin{bmatrix} 1 & 2 & -3 \\ 0 & 4 & 9 \\ 0 & 0 & -3 \end{bmatrix}$ and $B = \begin{bmatrix} -2 & 1 & 3 \\ 1 & -4 & 0 \\ 5 & 1 & 2 \end{bmatrix}$, find $\det(2A^{-1}B^3A^T)$.

13. (a) Find the determinant of $A = \begin{bmatrix} 2 & 1 & 1 & 1 \\ 1 & 2 & -1 & 2 \\ 1 & -1 & 2 & 1 \\ 1 & 3 & 3 & 2 \end{bmatrix}$.

    (b) Find $\det(A^{-1}(3A^T))$.

14. Let A and B be $6{\times}6$ matrices with $\det A = \frac{1}{4}$ and $\det B = -2$. Let $C = -ABA^TB^T$.

(a) Find $\det C$.

(b) Matrix C is invertible. Why?

(c) Find $\det C^{-1}$.

15. Show that $\begin{bmatrix} 1 & 8 & 2 & 4 \\ 3 & 0 & 1 & -2 \\ 1 & 7 & 4 & 1 \\ -3 & 2 & 1 & 3 \end{bmatrix}$ and $\begin{bmatrix} 2 & 8 & 1 & 4 \\ 4 & 7 & 1 & 1 \\ 1 & 0 & 3 & -2 \\ 1 & 2 & -3 & 3 \end{bmatrix}$ have the same determinant, without finding the determinant of either matrix.

16. Show that the line in the plane through the points $P_1(x_1, y_1)$ and $P_2(x_2, y_2)$ has equation $\begin{vmatrix} x & y & 1 \\ x_1 & y_1 & 1 \\ x_2 & y_2 & 1 \end{vmatrix} = 0$.

17. Find $\begin{vmatrix} a & b & c & 0 \\ b & a & d & 0 \\ c & d & 0 & a \\ d & c & 0 & b \end{vmatrix}$.

18. Show that the matrix $A = \begin{bmatrix} 1+a & 2+b & 3+c \\ a & b & c \\ 2 & 4 & 6 \end{bmatrix}$ is never invertible, no matter the values of a, b and c.

19. For which values of x is the matrix $A = \begin{bmatrix} 1 & -1 & x \\ 2 & 1 & x^2 \\ 4 & -1 & x^3 \end{bmatrix}$ singular?

20. Let $A = \begin{bmatrix} a & b & c \\ u & v & w \\ x & y & z \end{bmatrix}$. Given $\det A = 2$, find

(a) $\begin{vmatrix} 5x & 5y & 5z \\ u-a & v-b & w-c \\ u+a & v+b & w+c \end{vmatrix}$; **(b)** $\det(A^{-1}A^TA^2)$; **(c)** $\det(-5A)$.

21. Suppose A is a 4×4 with row echelon form $U = \begin{bmatrix} 1 & 2 & 3 & 4 \\ 0 & 1 & 5 & -1 \\ 0 & 0 & 1 & 7 \\ 0 & 0 & 0 & 1 \end{bmatrix}$. For each statement, either prove true or provide a counterexample to show false.

(a) $\det A = \det U$ **(b)** $\det A \neq 0$.

22. If A and B are $n \times n$ matrices, n odd, and $AB = -BA$, show that either A or B is not invertible.

23. Suppose $A = \begin{bmatrix} 50 & -1 & \cdots & -1 \\ -1 & 50 & \cdots & -1 \\ -1 & -1 & \cdots & -1 \\ \vdots & \vdots & \ddots & \vdots \\ -1 & -1 & \cdots & -1 \\ -1 & -1 & \cdots & 50 \end{bmatrix}$ is 50×50. Find $\det A$.

24. **(a)** Find the determinant of $\begin{bmatrix} 1+x_1 & x_2 & x_3 & \dots & x_n \\ x_1 & 1+x_2 & x_3 & \dots & x_n \\ x_1 & x_2 & 1+x_3 & \dots x_n & \\ \vdots & \vdots & \vdots & & \vdots \\ x_1 & x_2 & x_3 & \dots & 1+x_n \end{bmatrix}$.

(b) Find the determinant of the $n \times n$ matrix $\begin{bmatrix} 0 & 1 & 1 & \dots & 1 \\ 1 & 0 & 1 & \dots & 1 \\ 1 & 1 & 0 & \dots & 1 \\ \vdots & \vdots & \vdots & & \vdots \\ 1 & 1 & 1 & \dots & 0 \end{bmatrix}$.

25. Suppose one row of an $n \times n$ matrix A is a linear combination of some other rows. Show that $\det A = 0$. [Hint: Since the determinant only changes sign when rows are interchanged, you may assume that the row in question is the last row. Also, by allowing the possibility that some coefficients are 0, you may assume that the last row is a linear combination of **all** the remaining rows. Use the notation $\det(\mathsf{a}_1, \mathsf{a}_2, \dots, \mathsf{a}_n)$ for the determinant of a matrix whose rows are (the transposes of) $\mathsf{a}_1, \mathsf{a}_2, \dots, \mathsf{a}_n$.]

26. Show that $\mathsf{v} = \begin{bmatrix} 2 \\ 3 \end{bmatrix}$ is an eigenvector of $A = \begin{bmatrix} 1 & 2 \\ 3 & 2 \end{bmatrix}$. Find a corresponding eigenvalue.

27. Let $A = \begin{bmatrix} 2 & 1 & 1 \\ -2 & 5 & 0 \\ 2 & -1 & 4 \end{bmatrix}$ and $\mathsf{v} = \begin{bmatrix} 2 \\ 4 \\ 0 \end{bmatrix}$. Is v an eigenvector of A? If not, explain; if so, what is the corresponding eigenvalue?

28. **(a)** What is the characteristic polynomial of $A = \begin{bmatrix} 4 & 8 \\ 6 & 2 \end{bmatrix}$?

    (b) Find the eigenvalues of A and the corresponding eigenspaces.

    (c) Exhibit an invertible matrix P such that $P^{-1}AP$ is a diagonal matrix D whose smallest diagonal entry is in position $(1, 1)$.

29. Find the eigenvalues and the corresponding eigenspaces.

    (a) $A = \begin{bmatrix} 1 & 3 \\ 3 & 1 \end{bmatrix}$ **(b)** $A = \begin{bmatrix} 1 & 2 \\ 4 & 3 \end{bmatrix}$.

30. **(a)** If λ is an eigenvalue of an invertible matrix A and x is a corresponding eigenvector, then $\lambda \neq 0$. Why not?

    (b) Show that x is also an eigenvector for A^{-1}. What is the corresponding eigenvalue?

31. Suppose A is a matrix that satisfies $A^9 = 0$. Find all eigenvalues of A.

32. **(a)** Find all eigenvalues of $A = \begin{bmatrix} 6 & -1 & -3 \\ 8 & 0 & -6 \\ 8 & -1 & -5 \end{bmatrix}$ and the corresponding eigenspaces.

    (b) Is A diagonalizable? Explain and, if the answer is "yes," find an invertible matrix P such that $P^{-1}AP$ is diagonal.

33. Repeat Exercise 32 with $A = \begin{bmatrix} 7 & 1 & 2 \\ -1 & 7 & 0 \\ 1 & -1 & 6 \end{bmatrix}$.

34. Without calculation, explain why $\begin{bmatrix} 3 & -2 & 0 \\ -2 & 3 & 0 \\ 0 & 0 & 5 \end{bmatrix}$ is a diagonalizable matrix.

35. Given that $\mathsf{x}_1 = \begin{bmatrix} -1 \\ 0 \\ 1 \end{bmatrix}$ and $\mathsf{x}_2 = \begin{bmatrix} 2 \\ 1 \\ 0 \end{bmatrix}$ are eigenvectors of $A = \begin{bmatrix} -1 & -14 & 7 \\ -7 & 6 & -7 \\ 7 & -14 & -1 \end{bmatrix}$ and that 20 is an eigenvalue, find an invertible matrix P and a diagonal matrix D such that $P^{-1}AP = D$.

36. **(a)** What does it mean to say that matrices A and B are *similar*?

    (b) Prove that similar matrices have the same determinant.

37. **(a)** Find all the (real) eigenvalues of $A = \begin{bmatrix} 3 & -2 & 0 \\ 0 & 3 & 0 \\ 0 & 0 & 5 \end{bmatrix}$ and the corresponding eigenspaces.

    (b) Is A similar to a diagonal matrix?

38. **(a)** Find the characteristic polynomial of $A = \begin{bmatrix} -2 & -4 \\ 3 & 6 \end{bmatrix}$ and use this to explain why A is similar to a diagonal matrix.

    (b) Find a diagonal matrix D and an invertible matrix P such that $P^{-1}AP = D$.

    (c) Find a diagonal matrix D and an invertible matrix S such that $S^{-1}DS = A$.

39. Suppose λ is an eigenvalue of a matrix A and x is a corresponding eigenvector. If $B = P^{-1}AP$ for some invertible matrix P, show that λ is also an eigenvalue of B. Find a corresponding eigenvector.

40. If A and B are similar matrices, show that A^2 and B^2 are also similar.

41. Consider the sequence of numbers defined recursively by

$$a_0 = -3, a_1 = -2 \text{ and, for } n \geq 1,\ a_{n+1} = 5a_n - 6a_{n-1}.$$

    (a) List the first eight terms of this sequence.

    (b) Find a general formula for a_n.

42. Let $A = \begin{bmatrix} .2 & .1 \\ .8 & .9 \end{bmatrix}$. Without any calculation, explain why there exists a nonzero vector x satisfying $A\mathsf{x} = \mathsf{x}$.

43. Consider a Markov chain $\mathsf{v}_0, \mathsf{v}_1, \mathsf{v}_2, \ldots$ with transition matrix $A = \begin{bmatrix} 0 & \frac{1}{2} & \frac{1}{2} \\ \frac{1}{2} & \frac{1}{4} & \frac{1}{2} \\ \frac{1}{2} & \frac{1}{4} & 0 \end{bmatrix}$. The vectors v_k will converge to what vector as $k \to \infty$? Justify your answer.

44. The transition matrix $A = \begin{bmatrix} .7 & .3 & .2 \\ .2 & .5 & .2 \\ .1 & .2 & .6 \end{bmatrix}$ shows the movement of people between cities, suburbs and the rural countryside. Column one, for instance, shows that each year 70% of the people who live in cities remain there, while 20% move to the suburbs and 10% move to the countryside. If these numbers remain the same year after year, what will be the eventual proportions of the population living in the city, the suburbs and the country?

Chapter 4

Vector Spaces

4.1 The Theory of Linear Equations

As we know, any linear system of equations can be written in the form $A\mathsf{x} = \mathsf{b}$, where A is the matrix that records the coefficients of the variables, x is the vector of unknowns and b is the vector whose components are the constants appearing to the right of the $=$ signs. For example, the system

$$\begin{aligned} -2x_1 + x_2 + x_3 &= -9 \\ x_1 + x_2 + x_3 &= 0 \\ x_1 + 3x_2 &= -3 \end{aligned}$$

is $A\mathsf{x} = \mathsf{b}$ with

$$A = \begin{bmatrix} -2 & 1 & 1 \\ 1 & 1 & 1 \\ 1 & 3 & 0 \end{bmatrix}, \quad \mathsf{x} = \begin{bmatrix} x_1 \\ x_2 \\ x_3 \end{bmatrix} \quad \text{and } \mathsf{b} = \begin{bmatrix} -9 \\ 0 \\ -3 \end{bmatrix}.$$

In Section 2.4, we saw that a linear system may have no solution, or just one solution, or infinitely many solutions. Thinking geometrically for a moment, it is not hard to see how each of these possibilities occurs. Suppose we are solving two equations in three unknowns, x, y, z. In 3-space, we know that the set of points (x, y, z) that satisfy an equation of the form $ax + by + cz = d$ is a plane. So, solving two such equations means we are trying to find the points that lie in two planes. If the planes are parallel and not the same—think of the floor and ceiling of a room—there is no intersection; the system of two equations has no solution. If the planes are not parallel, they intersect in a line—think of a wall and the ceiling of a room—and the system has infinitely many solutions. Except in special cases, the solution of three equations in three unknowns is unique; the first two equations determine a line, and the unique solution is the point where this line punctures the plane defined by the third equation.

One of the goals of this section is to make it possible to determine in advance what sort of solution, if any, to expect from a given system. In so doing,

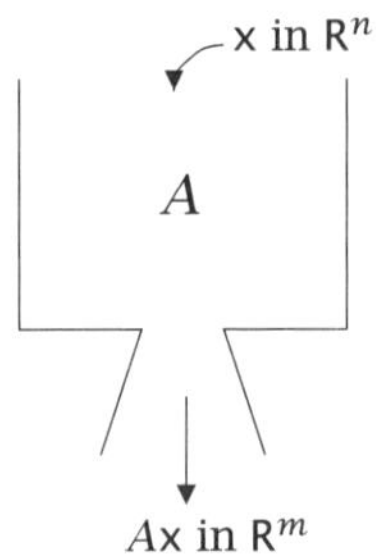

Figure 4.1: Multiplication by an $m \times n$ matrix A is a function $\mathsf{R}^n \to \mathsf{R}^m$.

we get a lot of mileage from the fact that $A\mathsf{x}$ is a linear combination of the columns of the matrix A. Remember the important fact highlighted in 2.1.33.

The Null Space of a Matrix

Suppose A is an $m \times n$ matrix, x and b are vectors, and $A\mathsf{x} = \mathsf{b}$. In what Euclidean spaces do x and b live? For the product of A and x to be defined, the vector x must be $n \times 1$, so x is in R^n and b, the product of the $m \times n$ matrix A and the $n \times 1$ vector x, is $m \times 1$, hence in R^m.

As illustrated in Figure 4.1, these observations show that we can view multiplication by an $m \times n$ matrix as a function that sends vectors in R^n to vectors in R^m, written $A \colon \mathsf{R}^n \to \mathsf{R}^m$.

4.1.1 **Definition.** The *null space* of a matrix A is the set of all vectors x that satisfy $A\mathsf{x} = \mathsf{0}$, that is, the set of all solutions to the homogeneous system $A\mathsf{x} = \mathsf{0}$. We use the notation null sp A.

If A is $m \times n$, the null space is a set of vectors in R^n. This set always contains the zero vector because $A\mathsf{x} = \mathsf{0}$ always has the trivial solution $\mathsf{x} = \mathsf{0}$. Sometimes this is the only solution.

4.1.2 **Example.** Let $A = \begin{bmatrix} 1 & 5 \\ 0 & 1 \end{bmatrix}$. If $A\begin{bmatrix} x \\ y \end{bmatrix} = \begin{bmatrix} 0 \\ 0 \end{bmatrix}$, then $\begin{bmatrix} x + 5y \\ y \end{bmatrix} = \begin{bmatrix} 0 \\ 0 \end{bmatrix}$, so $y = x = 0$. If $A\mathsf{x} = \mathsf{0}$, then $\mathsf{x} = \mathsf{0}$. The null space of A is just the zero vector. In fact, we can draw the same conclusion with A any invertible matrix since if $A\mathsf{x} = \mathsf{0}$ with A invertible, then $A^{-1}A\mathsf{x} = \mathsf{0}$, so $\mathsf{x} = \mathsf{0}$. ☺

4.1.3 Example. The null space of $A = \begin{bmatrix} 1 & -3 & -2 & 4 \\ 1 & -3 & 1 & 1 \\ 0 & 0 & 1 & -1 \end{bmatrix}$ is the set of all $\mathsf{x} = \begin{bmatrix} x_1 \\ x_2 \\ x_3 \\ x_4 \end{bmatrix}$ such that $A\mathsf{x} = 0$. Using Gaussian elimination to move A to row echelon form, we have

$$A \to \begin{bmatrix} 1 & -3 & -2 & 4 \\ 0 & 0 & 3 & -3 \\ 0 & 0 & 1 & -1 \end{bmatrix} \to \begin{bmatrix} 1 & -3 & -2 & 4 \\ 0 & 0 & 1 & -1 \\ 0 & 0 & 0 & 0 \end{bmatrix},$$

so $x_2 = t$ and $x_4 = s$ are free variables, $x_3 = x_4 = s$ and $x_1 = 3x_2 + 2x_3 - 4x_4 = 3t + 2s - 4s = 3t - 2s$. The null space consists of all vectors of the form $\begin{bmatrix} x_1 \\ x_2 \\ x_3 \\ x_4 \end{bmatrix} = \begin{bmatrix} 3t-2s \\ t \\ s \\ s \end{bmatrix} = t\begin{bmatrix} 3 \\ 1 \\ 0 \\ 0 \end{bmatrix} + s\begin{bmatrix} -2 \\ 0 \\ 1 \\ 1 \end{bmatrix}$. ☺

Since the null space of a matrix always contains the zero vector, the interesting question is whether there are **nonzero** vectors in the null space. Are there "nontrivial" solutions to $A\mathsf{x} = 0$? With just a little thinking, we could have answered this question for the matrix in Example 4.1.3 without any calculation! Consider the fact that A is 3×4, so A can have at most three pivots (at most per row). But there are four columns, so there has to be a nonpivot column, hence at least one free variable in the solution of $A\mathsf{x} = 0$. It is free variables that give nontrivial solutions.

This idea generalizes.

4.1.4 Theorem. *If A is an $m \times n$ matrix and $m < n$, then A has a nonzero null space; there exist vectors $\mathsf{x} \neq 0$ with $A\mathsf{x} = 0$.*

[Remark. In some books, this theorem is stated differently, like this: A homogeneous system of linear equations with more unknowns than equations has a nontrivial solution.]

Proof. The matrix A has at most m pivots, at most one per row. Since $m < n$, there is at least one nonpivot column. So there is at least one free variable in the solution of $A\mathsf{x} = 0$, hence there are an infinite number of solutions; in particular, there are nontrivial solutions. ■

Theorem 4.1.4 has an obvious corollary.

4.1.5 Corollary. *If $m < n$, any n vectors in R^m are linearly dependent.*

Proof. Let the vectors be $\mathsf{x}_1, \mathsf{x}_2, \ldots, \mathsf{x}_n$ and put them into the columns of an $m \times n$ matrix $A = \begin{bmatrix} \mathsf{x}_1 & \mathsf{x}_2 & \cdots & \mathsf{x}_n \\ \downarrow & \downarrow & & \downarrow \end{bmatrix}$. Suppose $c_1\mathsf{x}_1 + c_2\mathsf{x}_2 + \cdots + c_n\mathsf{x} = 0$.

Then $A\begin{bmatrix} c_1 \\ c_2 \\ \vdots \\ c_n \end{bmatrix} = 0$. The homogeneous system $A\mathsf{x} = 0$ has an $m \times n$ coefficient matrix with $m < n$. So there are nontrivial solutions $\mathsf{c} = \begin{bmatrix} c_1 \\ c_2 \\ \vdots \\ c_n \end{bmatrix}$. The columns are linearly dependent. ■

4.1.6 Problem. Suppose $\mathsf{x}_1 = \begin{bmatrix} 1 \\ 2 \\ 5 \end{bmatrix}$, $\mathsf{x}_2 = \begin{bmatrix} 1 \\ 0 \\ 0 \end{bmatrix}$, $\mathsf{x}_3 = \begin{bmatrix} -2 \\ 1 \\ 3 \end{bmatrix}$ and $\mathsf{x}_4 = \begin{bmatrix} 1 \\ 1 \\ 1 \end{bmatrix}$. Are these vectors linearly dependent or linearly independent?

Solution. Any four vectors in R^3 are linearly dependent, by the corollary.

The Column Space of a Matrix

4.1.7 Definition. The *column space* of a matrix A is the span of the columns of A, that is, the set of all linear combinations of the columns. It is denoted col sp A.

Since A is $m \times n$, the columns of A are vectors in R^m, so col sp A is a collection of vectors in R^m. Contrast this with the null space, which consists of vectors in R^n. These concepts are illustrated in Figure 4.2.

Before continuing, we note that

4.1.8

A vector b is in the column space of A if and only if $\mathsf{b} = A\mathsf{x}$ for some vector x—that is, if and only if the linear system $A\mathsf{x} = \mathsf{b}$ has a solution.

READING CHECK 1. Why?

For example, if A is an $n \times n$ invertible matrix, then col sp $A = \mathsf{R}^n$ because $\mathsf{b} = A(A^{-1}\mathsf{b})$ for any $b \in \mathsf{R}^n$, so $A\mathsf{x} = \mathsf{b}$ has the solution $\mathsf{x} = A^{-1}\mathsf{b}$.

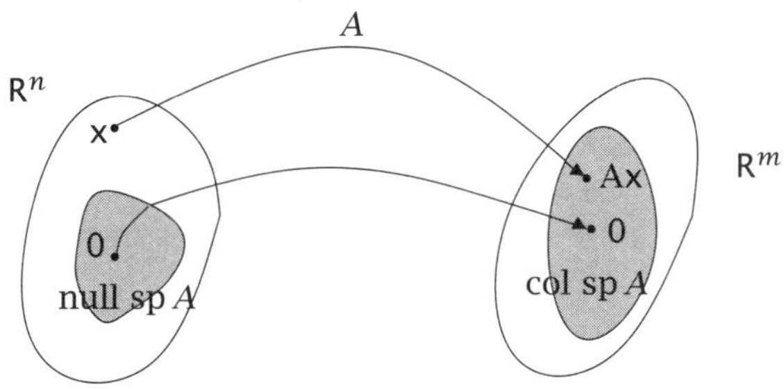

Figure 4.2: If A is an $m \times n$ matrix, the null space is in R^n and the column space is in R^m.

4.1.9 Example. The column space of $A = \begin{bmatrix} 1 & 2 & 3 \\ 4 & 5 & 6 \end{bmatrix}$ is a collection of vectors in R^2. Notice that the submatrix $B = \begin{bmatrix} 1 & 2 \\ 4 & 5 \end{bmatrix}$ is invertible and so, as noted, its column space is R^2. Thus every vector in R^2 is a linear combination of the first two columns of A and hence certainly of all three columns of A (use 0 for the coefficient of the third column $\begin{bmatrix} 3 \\ 6 \end{bmatrix}$). Again col sp $A = \mathsf{R}^2$. :-)

4.1.10 Example. The column space of the 3×4 matrix $A = \begin{bmatrix} 1 & -3 & -2 & 4 \\ 1 & -3 & 1 & 1 \\ 0 & 0 & 1 & -1 \end{bmatrix}$ is a collection of vectors in R^3. Label the columns $\mathsf{a}_1, \mathsf{a}_2, \mathsf{a}_3, \mathsf{a}_4$, in order, and notice that $\mathsf{a}_2 = -3\mathsf{a}_1$ and $\mathsf{a}_3 + \mathsf{a}_4 = \begin{bmatrix} 2 \\ 2 \\ 0 \end{bmatrix} = 2\mathsf{a}_1$. A vector in the column space looks like

$$\begin{aligned} c_1\mathsf{a}_1 + c_2\mathsf{a}_2 + c_3\mathsf{a}_3 + c_4\mathsf{a}_4 &= c_1\mathsf{a}_1 + c_2(-3\mathsf{a}_1) + c_3\mathsf{a}_3 + c_4(2\mathsf{a}_1 - \mathsf{a}_3) \\ &= (c_1 - 3c_2 + 2c_4)\mathsf{a}_1 + (c_3 - c_4)\mathsf{a}_3, \end{aligned}$$

so the column space of A is a plane, the span of a_1 and a_3. :-)

READING CHECK 2. What is the equation of this plane?

4.1.11 Example. Is $\mathsf{b} = \begin{bmatrix} 1 \\ 2 \\ 2 \end{bmatrix}$ in the column space of $A = \begin{bmatrix} 1 & 1 & -5 & -1 \\ -1 & 8 & 11 & 4 \\ 2 & 5 & -8 & -1 \end{bmatrix}$?

The question asks if the system $A\mathsf{x} = \mathsf{b}$ has a solution $\mathsf{x} = \begin{bmatrix} x_1 \\ x_2 \\ x_3 \\ x_4 \end{bmatrix}$. The augmented matrix is $[A \mid \mathsf{b}] = \left[\begin{array}{rrrr|r} 1 & 1 & -5 & -1 & 1 \\ -1 & 8 & 11 & 4 & 2 \\ 2 & 5 & -8 & -1 & 2 \end{array}\right]$. Gaussian elimination leads to $\left[\begin{array}{rrrr|r} 1 & 1 & -5 & -1 & 1 \\ 0 & 3 & 2 & 1 & 0 \\ 0 & 0 & 0 & 0 & 3 \end{array}\right]$. The equation corresponding to the last row is $0 = 3$. This absurdity tells us there are no solutions. The system is *inconsistent*; the vector $\mathsf{b} = \begin{bmatrix} 1 \\ 2 \\ 2 \end{bmatrix}$ is **not** in the column space of A.

Having determined that the column space of A is not R^3, one might still be interested in discovering just what it is. For this, we might proceed by arguing that $\mathsf{b} = \begin{bmatrix} b_1 \\ b_2 \\ b_3 \end{bmatrix}$ is in the column space if and only if $A\mathsf{x} = \mathsf{b}$ has a solution. The augmented matrix in this more general situation is

$$\left[\begin{array}{rrrr|l} 1 & 1 & -5 & -1 & b_1 \\ -1 & 8 & 11 & 4 & b_2 \\ 2 & 5 & -8 & -1 & b_3 \end{array}\right], \tag{1}$$

and one possible application of Gaussian elimination is this:

$$\to \left[\begin{array}{rrrr|c} 1 & 1 & -5 & -1 & b_1 \\ 0 & 9 & 6 & 3 & b_2 + b_1 \\ 0 & 3 & 2 & 1 & b_3 - 2b_1 \end{array}\right] \to \left[\begin{array}{rrrr|c} 1 & 1 & -5 & -1 & b_1 \\ 0 & 3 & 2 & 1 & b_3 - 2b_1 \\ 0 & 9 & 6 & 3 & b_2 + b_1 \end{array}\right]$$

$$\to \left[\begin{array}{rrrr|c} 1 & 1 & -5 & -1 & b_1 \\ 0 & 3 & 2 & 1 & b_3 - 2b_1 \\ 0 & 0 & 0 & 0 & (b_2 + b_1) - 3(b_3 - 2b_1) \end{array}\right].$$

The last line corresponds to the equation $0 = (b_2 + b_1) - 3(b_3 - 2b_1)$, which says that if the system has a solution then $7b_1 + b_2 - 3b_3 = 0$. On the other hand, if $7b_1 + b_2 - 3b_3 = 0$, the system with augmented matrix (1) has a solution (in fact, infinitely many solutions), for in this case, the final augmented matrix is

$$\left[\begin{array}{rrrr|c} 1 & 0 & -2 & -1 & b_1 \\ 0 & 1 & 1 & 1 & b_3 - 2b_1 \\ 0 & 0 & 0 & 0 & 0 \end{array}\right],$$

the variables x_3 and x_4 are free and there are infinitely many solutions. We conclude that the vector $\mathsf{b} = \begin{bmatrix} b_1 \\ b_2 \\ b_3 \end{bmatrix}$ is in the column space of A if and only if $7b_1 + b_2 - 3b_3 = 0$. So the column space of A is the plane with equation $7x + y - 3z = 0$. In particular, this gives us another way to see why $\mathsf{b} = \begin{bmatrix} 1 \\ 2 \\ 2 \end{bmatrix}$ is not in the column space: Its components do not satisfy the equation $7x + y - 3z = 0$. ☺

Application: Generator and Parity Check Matrices

In Section 2.1, we defined a code C as a set of sequences of 0s and 1s, all of the same length, with the property that the sum of code words is again a code word. We showed how the words of C could be identified with the set of linear combinations of the columns of a generator matrix G, all arithmetic performed mod 2—see Section 2.2. For example, if G is $\begin{bmatrix} 1 & 0 \\ 0 & 1 \\ 1 & 0 \\ 1 & 1 \end{bmatrix}$, then

$$C = \{G\mathsf{x} \mid \mathsf{x} = \begin{bmatrix} x_1 \\ x_2 \end{bmatrix}, x_i = 0, 1\} = \{0000, 1011, 0101, 1110\}.$$

Since $G\begin{bmatrix} x_1 \\ x_2 \end{bmatrix} = \begin{bmatrix} x_1 \\ x_2 \\ x_1 \\ x_1 + x_2 \end{bmatrix}$, the vector $\begin{bmatrix} x_1 \\ x_2 \\ x_3 \\ x_4 \end{bmatrix}$ is a code word if and only if $x_3 = x_1$ and $x_4 = x_1 + x_2$. Modulo 2, $+a = -a$, so these equations become

$$\begin{aligned} x_1 + x_3 &= 0 \\ x_1 + x_2 + x_4 &= 0. \end{aligned}$$

These are called "parity check equations" and the matrix of coefficients $H = \begin{bmatrix} 1 & 0 & 1 & 0 \\ 1 & 1 & 0 & 1 \end{bmatrix}$ is the "parity check matrix." The term "parity check" derives from the fact that H can be used to check whether or not a vector is a code word: $\mathsf{x} = \begin{bmatrix} x_1 \\ x_2 \\ x_3 \\ x_4 \end{bmatrix}$ is a code word if and only if $H\mathsf{x} = \mathsf{0}$. Thus a code can be described equally as the column space of its generator matrix or the null space of its parity check matrix.

The Rank of a Matrix

Before continuing, it would be wise for the reader to review the concept of pivot discussed in Section 2.4. See especially Definition 2.4.5 and Remark 2.4.9. While the pivots of a matrix are not unique, the **number** of pivots is, so the next definition makes sense.

4.1.12 Definition. The *rank* of a matrix is the number of pivots.

4.1.13 Example. The matrix $\begin{bmatrix} 4 & 0 & 2 \\ 0 & 1 & 3 \end{bmatrix}$ has rank 2. It is already in row echelon form and there are two pivots. ☺

4.1.14 Example. The matrix $\begin{bmatrix} 1 & 2 & 3 \\ 2 & 4 & 6 \end{bmatrix}$ has rank 1 since a row echelon form is $\begin{bmatrix} 1 & 2 & 3 \\ 0 & 0 & 0 \end{bmatrix}$. There is just one pivot. ☺

Suppose A is an $m \times n$ matrix. As you might guess, the particular cases $\operatorname{rank} A = m$ and $\operatorname{rank} A = n$ are special and, as we shall see, related to the nature of the solutions to the linear system $A\mathsf{x} = \mathsf{b}$. In what follows, we use the fact that a linear system is *inconsistent* (meaning no solutions) if and only if the Gaussian elimination process applied to the augmented matrix $[A \mid \mathsf{b}]$ leads to a matrix containing a row of the form

$$0 \quad 0 \quad 0 \quad \cdots \quad 0 \mid \star$$

with $\star \neq 0$, since such a row corresponds to the equation $0 = \star$, which is not true.

The condition rank A = m is equivalent to the statement that there is a pivot in every row. So there are no zero rows in a row echelon form of A; there is no possibility that the system $A\mathsf{x} = \mathsf{b}$ is inconsistent. Thus $\operatorname{rank} A = m$ is equivalent to the statement that for any b in R^m, the system $A\mathsf{x} = \mathsf{b}$ has at least one solution. Here's an example.

4.1.15 Example. The 3×4 matrix $A = \begin{bmatrix} 1 & 2 & 3 & 4 \\ 5 & 6 & 7 & 8 \\ 0 & 12 & 11 & 10 \end{bmatrix}$ has row echelon form $U = \begin{bmatrix} 1 & 2 & 3 & 4 \\ 0 & 1 & 2 & 3 \\ 0 & 0 & 1 & 2 \end{bmatrix}$, so $\operatorname{rank} A = m = 3$, the number of rows. This should imply we can solve $A\mathsf{x} = \mathsf{b}$ for any $\mathsf{b} = \begin{bmatrix} b_1 \\ b_2 \\ b_3 \end{bmatrix}$. Using Gaussian elimination,

$$\left[\begin{array}{cccc|c} 1 & 2 & 3 & 4 & b_1 \\ 5 & 6 & 7 & 8 & b_2 \\ 0 & 12 & 11 & 10 & b_3 \end{array}\right] \to \left[\begin{array}{cccc|c} 1 & 2 & 3 & 4 & b_1 \\ 0 & -4 & -8 & -12 & b_2 - 5b_1 \\ 0 & 12 & 11 & 10 & b_3 \end{array}\right]$$

$$\rightarrow \left[\begin{array}{cccc|c} 1 & 2 & 3 & 4 & b_1 \\ 0 & -4 & -8 & -12 & b_2 - 5b_1 \\ 0 & 0 & -13 & -26 & b_3 + 3(b_2 - 5b_1) \end{array}\right]$$

so $\mathsf{x}_4 = t$ is free and there are infinitely many solutions. ⍨

The condition rank A = n is equivalent to every column being a pivot column; that is, no free variables arise in the solution to $A\mathsf{x} = \mathsf{b}$. Thus rank $A = n$ is equivalent to the statement that for any b in R^m, the system $A\mathsf{x} = \mathsf{b}$ has at most one solution. Here's an example.

4.1.16 Example. A row echelon form of $A = \begin{bmatrix} 1 & 2 \\ 3 & 4 \\ 5 & 6 \end{bmatrix}$ is $U = \begin{bmatrix} 1 & 2 \\ 0 & 1 \\ 0 & 0 \end{bmatrix}$, so rank $A = n = 2$, the number of columns of A. This should imply at most one solution to $A\mathsf{x} = \mathsf{b} = \begin{bmatrix} b_1 \\ b_2 \\ b_3 \end{bmatrix}$. What happens when we try to solve this equation? Here is one instance of Gaussian elimination:

$$\left[\begin{array}{cc|c} 1 & 2 & b_1 \\ 3 & 4 & b_2 \\ 5 & 6 & b_3 \end{array}\right] \rightarrow \left[\begin{array}{cc|c} 1 & 2 & b_1 \\ 0 & -2 & b_2 - 3b_1 \\ 0 & -4 & b_3 - 5b_1 \end{array}\right]$$

$$\rightarrow \left[\begin{array}{cc|c} 1 & 2 & b_1 \\ 0 & -2 & b_2 - 3b_1 \\ 0 & 0 & b_3 - 5b_1 - 2(b_2 - 3b_1) \end{array}\right].$$

The equation corresponding to the last row reads $0 = b_1 - 2b_2 + b_3$. If the components of b do not satisfy this equation, there is no solution. If they do, there is a unique solution because back substitution would give $\mathsf{x}_2 = -\frac{1}{2}(b_2 - 3b_1)$ and then $x_1 = b_1 - 2x_2 = -2b_1 + b_2$. ⍨

Our discussion has included proofs of the first two parts of the next theorem. Look carefully at the equivalence of parts iii and iv of part 3. The equivalence is quite remarkable. If A is square and $A\mathsf{x} = \mathsf{b}$ has a unique solution when $\mathsf{b} = \mathsf{0}$, then it has a unique solution for any b whatsoever.

4.1.17 Theorem. *Let A be an $m \times n$ matrix.*

1. rank $A = m$ *if and only if, for any vector* b *in* R^m, $A\mathsf{x} = \mathsf{b}$ *has at least one solution.*

2. rank $A = n$ *if and only if, for any vector* b *in* R^m, $A\mathsf{x} = \mathsf{b}$ *has at most one solution.*

3. *If $m = n$, the following statements are equivalent:*

   i. rank $A = n$.

   ii. *For any vector* b *in* R^n, *the system* $A\mathsf{x} = \mathsf{b}$ *has at most one solution.*

   iii. *The homogeneous system* $A\mathsf{x} = 0$ *has just the trivial solution,* $\mathsf{x} = 0$.

   iv. *For any vector* b *in* R^n, $A\mathsf{x} = \mathsf{b}$ *has a unique solution.*

   v. *For any vector* b *in* R^n, $A\mathsf{x} = \mathsf{b}$ *has at least one solution.*

   vi. *A is invertible.*

   vii. $\det A \neq 0$.

Proof. We have already given proofs of parts 1 and 2 and illustrated with examples. For part 3, we first note that the equivalence of vi and vii was Theorem 3.2.18. The rest of the theorem will be proved by establishing the sequence of implications

$$\text{i} \implies \text{ii} \implies \text{iii} \implies \text{iv} \implies \text{v} \implies \text{vi} \implies \text{i}.$$

(i $\implies$ ii) is part 2 of the theorem.

(ii $\implies$ iii) follows from the fact that $A\mathsf{x} = 0$ always has the solution $\mathsf{x} = 0$.

(iii $\implies$ iv) Assume $A\mathsf{x} = 0$ has just the trivial solution. Then there are no free variables; every column is a pivot column. So rank $A = n = m$. By parts 1 and 2, we see that for any b in R^m, the system $A\mathsf{x} = \mathsf{b}$ has at most one solution and at least one solution, so it has a unique solution.

(iv $\implies$ v) If a system has a unique solution, it certainly has at least one solution.

(v $\implies$ vi) Assume v, that $A\mathsf{x} = \mathsf{b}$ has at least one solution for any b. Let $\mathsf{e}_1, \mathsf{e}_2, \dots, \mathsf{e}_n$ be the standard basis vectors of R^n. By assumption, we can solve the systems

$$\begin{aligned} A\mathsf{x} &= \mathsf{e}_1 \quad \text{for } \mathsf{x}_1, \\ A\mathsf{x} &= \mathsf{e}_2 \quad \text{for } \mathsf{x}_2, \\ &\vdots \\ A\mathsf{x} &= \mathsf{e}_n \quad \text{for } \mathsf{x}_n. \end{aligned}$$

So $\mathsf{x}_1, \mathsf{x}_2, \dots, \mathsf{x}_n$ are vectors satisfying $A\mathsf{x}_1 = \mathsf{e}_1$, $A\mathsf{x}_2 = \mathsf{e}_2$, ..., $A\mathsf{x}_n = \mathsf{e}_n$.

Let $B = \begin{bmatrix} \mathsf{x}_1 & \mathsf{x}_2 & \cdots & \mathsf{x}_n \\ \downarrow & \downarrow & & \downarrow \end{bmatrix}$ be the matrix whose columns are x_1, x_2, ..., x_n.

Then

$$\begin{aligned} AB &= A \begin{bmatrix} \mathsf{x}_1 & \mathsf{x}_2 & \cdots & \mathsf{x}_n \\ \downarrow & \downarrow & & \downarrow \end{bmatrix} \\ &= \begin{bmatrix} A\mathsf{x}_1 & A\mathsf{x}_2 & \cdots & A\mathsf{x}_n \\ \downarrow & \downarrow & & \downarrow \end{bmatrix} = \begin{bmatrix} \mathsf{e}_1 & \mathsf{e}_2 & \cdots & \mathsf{e}_n \\ \downarrow & \downarrow & & \downarrow \end{bmatrix} = I. \end{aligned}$$

Thus B is a *right inverse* of A. Since A is square, Theorem 2.9.12 says that A is invertible, which is v. (Another way to see this is to note that $AB = I$ with A and B square implies $\det A \det B = 1$, so $\det A \neq 0$.)

(vi $\implies$ i) Assume that A is invertible. Then, for any b in R^n, the system $A\mathsf{x} = \mathsf{b}$ has the solution $\mathsf{x} = A^{-1}\mathsf{b}$, so $\operatorname{rank} A = n$ by part 1. ■

Answers to Reading Checks

1. The answer is 2.1.33: A linear combination of the columns of A is a vector of the form $A\mathsf{x}$.

2. The plane is the set of all vectors of the form $a\mathsf{a}_1 + b\mathsf{a}_2 = a\begin{bmatrix}1\\2\\0\end{bmatrix} + b\begin{bmatrix}-3\\0\\4\end{bmatrix} = \begin{bmatrix}a-3b\\2a\\4b\end{bmatrix}$. Let $x = a - 3b$, $y = 2a$ and $z = 4b$. Then $a = \frac{1}{2}y$ and $b = \frac{1}{4}z$ so $x = \frac{1}{2}y - \frac{3}{4}z$. The equation is $4x - 2y + 3z = 0$.

   [Alternatively, use the fact that $\mathsf{a}_1 \times \mathsf{a}_2$ is a normal.]

True/False Questions

Decide, with as little calculation as possible, whether each of the following statements is true or false and, if you say "false," explain your answer. (Answers are at the back of the book.)

1. Left multiplication by a $k \times \ell$ matrix defines a function from $\mathsf{R}^k \to \mathsf{R}^\ell$.

2. The column space of $\begin{bmatrix}-1&0&4&1\\0&0&0&0\\0&2&5&1\end{bmatrix}$ is the xz-plane.

3. The matrix $A = \begin{bmatrix}1&2&3&4\\-7&15&9&18\\2&-6&8&-1\end{bmatrix}$ has a nontrivial null space.

4. If A and B are $n \times n$ matrices and $AB = cI$ for some $c \neq 0$, then $\operatorname{rank} A = \operatorname{rank} B$.

5. If $A\mathsf{x} = 2\mathsf{x}$ for some matrix A and vector x, then x is in the column space of A.

6. If A is a 7×5 matrix of rank 5, the linear system $A\mathsf{x} = \mathsf{b}$ has at least one solution for any b.

7. The columns of a matrix A are linearly independent if and only if the null space of A consists only of the zero vector.

8. If U is a row echelon form of the matrix A, null sp A = null sp U.

9. If U is a row echelon form of the matrix A, col sp A = col sp U.

10. An $n \times n$ singular matrix has rank less than n.

EXERCISES

*Answers to exercises marked [BB] can be found at the **Back of the Book.***

1. [BB] Suppose A is a 7×9 matrix with null space in R^n and column space in R^m. What are m and n?

2. Is $\begin{bmatrix} 3 \\ 2 \\ 1 \end{bmatrix}$ in the null space of $A = \begin{bmatrix} -1 & 1 & 1 \\ 1 & 1 & -5 \\ 1 & 2 & 3 \end{bmatrix}$? What about $\begin{bmatrix} -1 \\ 0 \\ 1 \end{bmatrix}$?

3. Express the null space of each of the following matrices as the span of certain vectors. Also give the rank of each matrix.

   (a) [BB] $\begin{bmatrix} 1 & -2 & 1 & 1 \\ 3 & 0 & 2 & -2 \\ 0 & 4 & -1 & -1 \\ 5 & 0 & 3 & -1 \end{bmatrix}$ **(b)** $\begin{bmatrix} 0 & 0 & 2 & 3 & 1 \\ 1 & -2 & 2 & 1 & -2 \\ 1 & -2 & 6 & 7 & 0 \end{bmatrix}$

   (c) $\begin{bmatrix} 1 & 1 & 1 & -1 \\ -1 & 1 & -1 & 1 \\ 2 & -1 & 1 & 3 \end{bmatrix}$ **(d)** $\begin{bmatrix} 1 & 2 & 3 & 4 \\ 2 & 4 & 8 & 10 \\ 3 & 6 & 11 & 14 \end{bmatrix}$.

4. [BB] The column space of $\begin{bmatrix} 2 & -1 & 1 \\ 1 & 2 & 5 \end{bmatrix}$ is R^2. Why?

5. In each case, determine whether or not the given vector is in the column space of the matrix. If it is, express b as a linear combination of the columns of A.

   (a) [BB] $\mathsf{b} = \begin{bmatrix} 1 \\ 3 \end{bmatrix}$, $A = \begin{bmatrix} 2 & 0 \\ 2 & 3 \end{bmatrix}$ **(b)** $\mathsf{b} = \begin{bmatrix} 5 \\ 3 \\ 4 \end{bmatrix}$, $A = \begin{bmatrix} 2 & 2 & -4 \\ 0 & 3 & 5 \\ 1 & 2 & -1 \end{bmatrix}$,

   (c) $\mathsf{b} = \begin{bmatrix} -6 \\ 7 \\ 3 \end{bmatrix}$, $A = \begin{bmatrix} 1 & 1 & 1 \\ 0 & 1 & 1 \\ 0 & 0 & 1 \end{bmatrix}$ **(d)** $\mathsf{b} = \begin{bmatrix} -4 \\ -6 \\ 1 \\ 0 \end{bmatrix}$, $A = \begin{bmatrix} 1 & 2 \\ -5 & 3 \\ -4 & 9 \\ 2 & 4 \end{bmatrix}$.

6. **(a)** [BB] Is $\mathsf{b} = \begin{bmatrix} 1 \\ 1 \\ 1 \\ 1 \end{bmatrix}$ in the column space of $A = \begin{bmatrix} 2 & -1 & 4 \\ 0 & 2 & 3 \\ -1 & -5 & 1 \\ 1 & -3 & 2 \end{bmatrix}$?

   If it is, express this vector as a linear combination of the columns of A. If it isn't, explain why.

(b) Same as (a) for $b = \begin{bmatrix} 1 \\ 1 \\ -1 \\ 0 \end{bmatrix}$.

(c) Find conditions on b_1, b_2, b_3, b_4 that are necessary and sufficient for $b = \begin{bmatrix} b_1 \\ b_2 \\ b_3 \\ b_4 \end{bmatrix}$ to belong to the column space of A.

7. [BB] Let $A = \begin{bmatrix} 6 & -12 & -5 & 16 & -2 & -53 \\ -3 & 6 & 3 & -9 & 1 & 29 \\ -4 & 8 & 3 & -10 & 1 & 33 \end{bmatrix}$, $u = \begin{bmatrix} 3 \\ 2 \\ -1 \\ 0 \\ 1 \\ 0 \end{bmatrix}$ and $b = \begin{bmatrix} -53 \\ 29 \\ 33 \end{bmatrix}$.

(a) Is u in the null space of A? Explain.

(b) Is b in the column space of A? If it is, express b as a specific linear combination of the columns of A (the coefficients should be numbers). If it isn't, explain why not.

(c) Find the null space of A.

(d) Find the column space of A.

(e) Find rank A.

8. Answer Exercise 7 again with $A = \begin{bmatrix} 2 & -8 & 7 & -4 & 3 \\ 6 & -24 & 24 & -7 & 12 \\ 10 & -40 & 41 & -8 & 27 \end{bmatrix}$, $u = \begin{bmatrix} 1 \\ -2 \\ -2 \\ 1 \\ 0 \end{bmatrix}$ and $b = \begin{bmatrix} -4 \\ 5 \\ 4 \end{bmatrix}$.

9. Let $A = \begin{bmatrix} 3 & 1 & 7 & 2 & 1 \\ 2 & -4 & 14 & -2 & 0 \\ 5 & 11 & -7 & 10 & 3 \\ 2 & 5 & -4 & -3 & 1 \end{bmatrix}$, $u = \begin{bmatrix} 1 \\ 2 \\ 5 \\ 9 \end{bmatrix}$, $v = \begin{bmatrix} 13 \\ -10 \\ 59 \\ 39 \end{bmatrix}$ and $b = \begin{bmatrix} b_1 \\ b_2 \\ b_3 \\ b_4 \end{bmatrix}$.

(a) Find the general solution of $Ax = u$ and of $Ax = v$ [BB], if a solution exists.

(b) Find a condition on b_1, b_2, b_3, b_4 that is necessary and sufficient for $Ax = b$ to have a solution; that is, complete the sentence

$Ax = b$ has a solution if and only if ________.

(c) The column space of A is a subset of R^n for what n?

10. The column space of each of the following matrices is a plane in R^3. Find the equation of each such plane. Also find the null space of each matrix.

(a) [BB] $A = \begin{bmatrix} 5 & 0 & -3 \\ 4 & 3 & 6 \\ -2 & 1 & 4 \end{bmatrix}$

(b) $A = \begin{bmatrix} 1 & 6 & 5 & 7 & -8 \\ -5 & -4 & 1 & -9 & 14 \\ 0 & 2 & 2 & 2 & -2 \end{bmatrix}$

(c) $A = \begin{bmatrix} 1 & 7 & 4 & 3 & 8 \\ 8 & 59 & 35 & 21 & 69 \\ -3 & -24 & -15 & -6 & -29 \end{bmatrix}$

(d) $A = \begin{bmatrix} 2 & -5 & 4 & 3 \\ 6 & -14 & 15 & 7 \\ 4 & -14 & -4 & 14 \end{bmatrix}$.

11. Find the rank of each of the following matrices.

(a) [BB] $\begin{bmatrix} 1 & 2 \\ 4 & 8 \end{bmatrix}$ (b) $\begin{bmatrix} 1 & 2 & 3 \\ 4 & 5 & 6 \\ 7 & 8 & 9 \end{bmatrix}$

(c) $\begin{bmatrix} 1 & 2 & 3 & 5 \\ 1 & 0 & 1 & 3 \\ 2 & 3 & 5 & 9 \end{bmatrix}$ (d) $\begin{bmatrix} -1 & 2 & 3 & 4 & 0 \\ 0 & 1 & 1 & 2 & 1 \\ -3 & 7 & 10 & 14 & 1 \end{bmatrix}$.

12. [BB] Find the column space and the null space of $A = \begin{bmatrix} 1 & 0 & 0 \\ 0 & 1 & 2 \\ 0 & 0 & 1 \end{bmatrix}$ without any calculation.

13. How does the rank of $\begin{bmatrix} 1 & 2 & 3 \\ a & 4 & 6 \end{bmatrix}$ depend on a?

14. [BB] Fill in the missing entries of $A = \begin{bmatrix} 1 & 2 & 3 & 4 \\ 2 & \star & \star & \star \\ 3 & \star & \star & \star \end{bmatrix}$ so that A has rank 1.

15. What can you say about a matrix of rank 0? Explain.

16. [BB] **(a)** Find the null space of $A = \begin{bmatrix} 2 & 3 & 4 \\ 4 & 7 & 5 \\ 0 & -1 & 3 \end{bmatrix}$.

(b) What is the rank of A?

(c) Solve $A\begin{bmatrix} x \\ y \\ z \end{bmatrix} = \begin{bmatrix} 1 \\ 6 \\ -4 \end{bmatrix}$.

(d) Express $\begin{bmatrix} 1 \\ 6 \\ -4 \end{bmatrix}$ as a linear combination of the columns of A.

(e) Find the column space of A and describe this in geometrical terms.

17. **(a)** Find the null space of $A = \begin{bmatrix} 1 & 0 & 3 \\ 3 & 1 & 16 \\ 5 & 2 & 29 \end{bmatrix}$.

(b) What is the rank of A?

(c) Solve $A\begin{bmatrix} x \\ y \\ z \end{bmatrix} = \begin{bmatrix} 2 \\ 3 \\ 4 \end{bmatrix}$.

(d) Express $\begin{bmatrix} 2 \\ 3 \\ 4 \end{bmatrix}$ as a linear combination of the columns of A.

(e) Find the column space of A and describe this in geometrical terms.

18. **(a)** Find the null space of $A = \begin{bmatrix} 1 & 2 & 3 & 2 \\ -1 & -2 & -2 & 1 \\ 2 & 4 & 8 & 12 \end{bmatrix}$.

(b) Solve $A\mathsf{x} = \begin{bmatrix} -1 \\ 2 \\ 4 \end{bmatrix}$.

(c) What is the rank of A and why?

(d) Find the column space of A and describe this in geometrical terms.

19. (a) [BB] If x is in the null space of a matrix A, show that x is in the null space of BA for any matrix B (of appropriate size).

(b) If $A = BC$ and x is in the column space of A, show that x is in the column space of B.

20. [BB] Give an example of a 3×3 matrix whose column space contains $\mathsf{i} = \begin{bmatrix} 1 \\ 0 \\ 0 \end{bmatrix}$ and $\mathsf{j} = \begin{bmatrix} 0 \\ 1 \\ 0 \end{bmatrix}$, but not $\mathsf{k} = \begin{bmatrix} 0 \\ 0 \\ 1 \end{bmatrix}$.

21. (a) [BB] If x is in the null space of AB, is $B\mathsf{x}$ in the null space of A? Why?

(b) If A is invertible and x is in the null space of AB, is x in the null space of B? Why?

22. (a) Complete the following sentence:

"An $n \times n$ matrix A is singular if and only if the matrix equation $A\mathsf{x} = \mathsf{0}$ has ________ solution."

(b) Use your answer to (a) to explain whether the following sentence is true or false.

"If A is a 3×3 matrix such that $A\mathsf{x} = \mathsf{x}$ for some nonzero column vector $\mathsf{x} = \begin{bmatrix} x_1 \\ x_2 \\ x_3 \end{bmatrix}$, then $I - A$ must have an inverse." (Here, I denotes the 3×3 identity matrix.)

23. Let $A = \begin{bmatrix} 2 & -1 & -1 & 1 & 2 \\ -4 & 2 & 2 & 0 & -3 \\ -2 & 2 & 5 & 4 & 0 \\ -6 & 3 & 3 & 1 & -4 \end{bmatrix}$.

(a) Find an equation of the form $A = PLU$, where L is lower triangular with 1s on the diagonal, U is upper triangular and P is a permutation matrix—see Section 2.7.

(b) What is the rank of A and why?

(c) Find the null space of A.

(d) Find a condition on b_1, b_2, b_3, b_4 that is necessary and sufficient for $A\mathsf{x} = \begin{bmatrix} b_1 \\ b_2 \\ b_3 \\ b_4 \end{bmatrix}$ to have a solution.

(e) What is the column space of A and why?

24. [BB] **(a)** Find the rank of $A = \begin{bmatrix} 2 & 1 & 5 & 4 \\ 1 & -1 & 1 & -1 \\ 1 & 0 & 2 & 1 \end{bmatrix}$.

(b) For which values of c does the system

$$\begin{aligned} 2x_1 + x_2 + 5x_3 + 4x_4 &= c^2 + 2 \\ x_1 - x_2 + x_3 - x_4 &= 4c + 5 \\ x_1 \qquad + 2x_3 + x_4 &= c + 3 \end{aligned}$$

have a solution? Find all solutions.

25. **(a)** Let x be a real number and $A = \begin{bmatrix} 1 & x & x^2 \\ x & x^2 & 1 \\ x^2 & 1 & x \end{bmatrix}$. Explain how the rank of A depends on x.

(b) Find a necessary and sufficient condition on c which guarantees that the system

$$\begin{aligned} x + cy + c^2z &= b_1 \\ cx + c^2y + z &= b_2 \\ c^2x + y + cz &= b_3 \end{aligned}$$

has a solution for all scalars b_1, b_2, b_3. Explain your answer.

(c) Determine whether or not the system

$$\begin{aligned} x + cy + c^2z &= 1 \\ cx + c^2y + z &= 1 \\ c^2x + y + cz &= 1 \end{aligned}$$

has a solution. Does your answer depend on c? Explain.

26. Let $m \geq 1$ and $n \geq 1$ be positive integers. Let x be a nonzero vector in R^n and let y be any vector in R^m. Prove that there exists an $m \times n$ matrix A such that $A\mathsf{x} = \mathsf{y}$. [Hint: Never forget 2.1.33!]

27. Prove Corollary 4.1.5 without appealing to Theorem 4.1.4.

Critical Reading

28. [BB] Suppose A and B are matrices and $A = ABA$. Show that A and AB have the same column space.

29. The eigenspace of a matrix A corresponding to the eigenvalue λ is the null space of $A - \lambda I$. Explain.

30. Can a 1×3 matrix A have scalar multiples of $\begin{bmatrix} 1 \\ 0 \\ 1 \end{bmatrix}$ as its null space? Explain.

31. Suppose A is an $n \times n$ invertible matrix. What's the column space of A? Why?

32. If A is a square matrix whose column space is contained in its null space, what can you say about A^2? Why?

33. Suppose A is an $m \times n$ matrix and its null space is R^n. What can you say about A and why?

34. Prove that an $n \times n$ matrix A is invertible if and only if the homogeneous system $A\mathsf{x} = 0$ has only the trivial solution.

35. **(a)** Suppose the system $A\mathsf{x} = 0$ has only the trivial solution. If $A\mathsf{x} = \mathsf{b}$ has a solution, show that this solution is unique.

    (b) Suppose A is an $m \times n$ matrix such that $A\mathsf{x} = 0$ has only the trivial solution. Suppose also that $A\mathsf{x} = \mathsf{b}$ has no solution for some b. What can you conclude about the relative sizes of m and n and why?

    (c) Give an example of a matrix A and a vector b such that $A\mathsf{x} = 0$ has only the trivial solution and $A\mathsf{x} = \mathsf{b}$ has **no** solution.

36. **(a)** Suppose A is an $n \times n$ matrix and $\mathsf{x}^T A\mathsf{x} > 0$ for every nonzero $\mathsf{x} \in \mathsf{R}^n$. Show that all eigenvalues of A are positive.

    (b) Let B be an $m \times n$ matrix of rank n. Show that the eigenvalues of $B^T B$ are positive.

37. **(a)** Let $m \geq 1$ and $n \geq 1$ be positive integers. Let x be a nonzero vector in R^n and let y be any vector in R^m. Prove that there exists an $m \times n$ matrix A such that $A\mathsf{x} = \mathsf{y}$. [Hint: Never forget 2.1.33!]

    (b) Suppose A and B are given matrices with the property that for every matrix C (of appropriate size), there exist matrices X and Y such that $AX - YB = C$. Prove that either null sp $B = \{0\}$ or the column space of A is as large as possible. [Hint: Give a proof **by contradiction**; that is, assume that null sp $B \neq \{0\}$ and let v be a nonzero vector in this null space. Also assume there is a vector w not in the column space of A. By part (a), there is a matrix C such that $C\mathsf{v} = \mathsf{w}$. Now derive something that is wrong (a contradiction).]

4.2 Subspaces

Remember that *Euclidean n-space* refers to the set

$$\mathsf{R}^n = \left\{ \begin{bmatrix} x_1 \\ x_2 \\ \vdots \\ x_n \end{bmatrix} \mid x_1, x_2, \ldots, x_n \in \mathsf{R} \right\}.$$

This is an example, and probably the most important example, of a "finite dimensional vector space." The definition of "vector space" involves the concept of a "field," which is beyond the scope of this text. Roughly speaking, a field is a system of numbers with a 0 and a 1 that can be added and multiplied subject to certain rules and in which division is possible by every nonzero number, as in R, the field of real numbers.

4.2.1 Definition. A *vector space* is a set V of objects called vectors and a field F of elements called scalars. There are notions of vector addition and multiplication of vectors by scalars that obey certain rules.

For any vectors u, v and w and any scalars c and d,

1. (Closure under addition) $\mathsf{u} + \mathsf{v}$ is in V.
2. (Commutativity of addition) $\mathsf{u} + \mathsf{v} = \mathsf{v} + \mathsf{u}$.
3. (Associativity of addition) $(\mathsf{u} + \mathsf{v}) + \mathsf{w} = \mathsf{u} + (\mathsf{v} + \mathsf{w})$.
4. (Zero) There is a vector denoted 0, called the *zero vector*, with the property that $0 + \mathsf{u} = \mathsf{u}$.
5. (Additive Inverses) There exists a vector called the *additive inverse* of u and denoted $-\mathsf{u}$ (*minus* u) with the property that $\mathsf{u} + (-\mathsf{u}) = 0$.
6. (Closure under scalar multiplication) $c\mathsf{u}$ is in V.
7. (Scalar associativity) $c(d\mathsf{u}) = (cd)\mathsf{u}$.
8. (One) $1\mathsf{u} = \mathsf{u}$.
9. (Distributivity) $c(\mathsf{u} + \mathsf{v}) = c\mathsf{u} + c\mathsf{v}$ and $(c + d)\mathsf{u} = c\mathsf{u} + d\mathsf{u}$.

With $V = \mathsf{R}^2$ and $F = \mathsf{R}$, these are exactly the properties of two-dimensional vectors listed in Theorem 1.1.10 and observed later to hold in any Euclidean space. They are exactly the properties of matrix addition and scalar multiplication recorded in Theorem 2.1.15 and not so difficult to remember if they are grouped. The first five properties deal only with addition and the next three just with scalar multiplication; the last (distributivity) is the only property that involves both addition and scalar multiplication.

4.2.2 Remark. Throughout this book, unless stated to the contrary, the field F will always be the real numbers, so "scalar" will always mean "real number." Exceptions appear in Chapter 7 where F is sometimes the set of complex numbers and in those parts of this book where we discuss applications to coding, where $F = \{0, 1\}$ is the set of two elements 0 and 1 with addition and multiplication defined "mod 2"—see Section 2.2. Though we assume for the most part in this book that scalars are real, in fact, most results hold for any field.

4.2.3 Examples. 1. Euclidean n-space, R^n, or more generally F^n, for any field F. For example, if $F = \{0, 1\}$, F^3 is the set of all column vectors $\begin{bmatrix} a \\ b \\ c \end{bmatrix}$ where a, b, c are each either 0 or 1. There are just eight vectors in F^3.

$$\begin{bmatrix} 0 \\ 0 \\ 0 \end{bmatrix}, \begin{bmatrix} 0 \\ 0 \\ 1 \end{bmatrix}, \begin{bmatrix} 0 \\ 1 \\ 0 \end{bmatrix}, \begin{bmatrix} 0 \\ 1 \\ 1 \end{bmatrix}, \begin{bmatrix} 1 \\ 0 \\ 0 \end{bmatrix}, \begin{bmatrix} 1 \\ 0 \\ 1 \end{bmatrix}, \begin{bmatrix} 1 \\ 1 \\ 0 \end{bmatrix}, \begin{bmatrix} 1 \\ 1 \\ 1 \end{bmatrix}.$$

READING CHECK 1. What is $\begin{bmatrix} 1 \\ 0 \\ 1 \end{bmatrix} + \begin{bmatrix} 1 \\ 1 \\ 1 \end{bmatrix}$ in F^3, $F = \{0, 1\}$?

2. The set $M_{mn}(\mathsf{R})$ of $m \times n$ matrices over R, where m and n are positive integers. (We write $M_n(\mathsf{R})$ rather than $M_{nn}(\mathsf{R})$ for square $n \times n$ matrices.) In this vector space, "vectors" are matrices and the zero "vector" is the $m \times n$ matrix all of whose entries are 0. If $A = [a_{ij}]$, then $-A = [-a_{ij}]$ is the matrix whose entries are the negatives of those of A. Notice that $m \times 1$ matrices are precisely the column vectors of R^m: $M_{m1}(\mathsf{R}) = \mathsf{R}^m$.

3. Another special case of the previous example occurs with $m = 1$ where the matrices of $M_{1n}(\mathsf{R})$ are row matrices. If a and b are row matrices, say
$$\mathsf{a} = [a_1\ a_2\ \cdots\ a_n] \quad \text{and} \quad \mathsf{b} = [b_1\ b_2\ \cdots\ b_n],$$
then
$$\mathsf{a} + \mathsf{b} = [a_1 + b_1\ a_2 + b_2\ \cdots\ a_n + b_n]$$
and, for any scalar c,
$$c\mathsf{a} = [ca_1\ ca_2\ \cdots\ ca_n].$$
In this case, the zero vector is $\mathsf{0} = [0\ 0\ \cdots\ 0]$ and the additive inverse of a is $-\mathsf{a} = [-a_1\ -a_2\ \cdots\ -a_n]$.

4. The set $\mathsf{R}[x]$ of polynomials over R in a variable x with usual addition and scalar multiplication; that is, if
$$f(x) = a_0 + a_1x + a_2x^2 + \cdots + a_nx^n$$
and
$$g(x) = b_0 + b_1x + b_2x^2 + \cdots + b_mx^m,$$
then addition and scalar multiplication by c are defined by
$$\begin{aligned} f(x) + g(x) &= (a_0 + b_0) + (a_1 + b_1)x + (a_2 + b_2)x^2 + \cdots \\ cf(x) &= ca_0 + ca_1x + ca_2x^2 + \cdots + ca_nx^n. \end{aligned}$$

In this vector space, "vectors" are polynomials and the zero vector is the polynomial $0 + 0x + 0x^2 + \cdots$, all of whose coefficients are zero. If $f(x) = a_0 + a_1x + a_2x^2 + \cdots + a_nx^n$, then

$$-f(x) = -a_0 - a_1x - a_2x^2 - \cdots - a_nx^n.$$

5. The set of all functions $\mathsf{R} \to \mathsf{R}$ with "usual" addition and scalar multiplication; that is, $f + g \colon \mathsf{R} \to \mathsf{R}$ is defined by

$$(f+g)(x) = f(x) + g(x), \quad \text{for any real number } x$$

and, for a scalar c, $cf \colon \mathsf{R} \to \mathsf{R}$ is defined by

$$(cf)(x) = c[f(x)], \quad \text{for any real number } x.$$

6. The set $C(\mathsf{R})$ of continuous functions $\mathsf{R} \to \mathsf{R}$ with usual addition and scalar multiplication.

7. The set $\mathcal{D}(\mathsf{R})$ of differentiable functions $\mathsf{R} \to \mathsf{R}$ with usual addition and scalar multiplication. ☺

There are many other examples, but those that are "finite dimensional" (see Section 4.3) are essentially R^n.[1] For this reason, in this book, the word "vector" almost always means "column vector." Sometimes, it may mean "row vector," that is, a $1 \times n$ matrix as described in Example 4.2.3(3) above.

4.2.4 Definition. A *subspace* of a vector space V is a subset U of V that is itself a vector space under the operations of addition and scalar multiplication which it inherits from V.

The definition suggests that verifying a certain subset is a subspace will take a lot of work, but such is not the case! To show that a certain set of vectors forms a subspace, it is sufficient to check just three things, not nine. The proof of Lemma 4.2.6 below requires a couple of facts that may seem "obvious," but actually have to be justified. If $\mathsf{x} = \begin{bmatrix} x_1 \\ \vdots \\ x_n \end{bmatrix}$ is a vector in R^n, it is obvious that $0\mathsf{x} = \mathsf{0}$ because $0\mathsf{x} = \begin{bmatrix} 0x_1 \\ \vdots \\ 0x_n \end{bmatrix} = \begin{bmatrix} 0 \\ \vdots \\ 0 \end{bmatrix}$. Equally, $(-1)\mathsf{x} = \begin{bmatrix} -x_1 \\ -x_2 \\ \vdots \\ -x_n \end{bmatrix}$ is the additive inverse $-\mathsf{x}$ of x because $(-1)\mathsf{x} + \mathsf{x} = \begin{bmatrix} -x_1 \\ -x_2 \\ \vdots \\ -x_n \end{bmatrix} + \begin{bmatrix} x_1 \\ x_2 \\ \vdots \\ x_n \end{bmatrix} = \begin{bmatrix} 0 \\ 0 \\ \vdots \\ 0 \end{bmatrix}$. In a different vector space, however, why should $0\mathsf{v} = \mathsf{0}$ or $(-1)\mathsf{v} = -\mathsf{v}$?. Neither of these properties is amongst the nine axioms of a vector space.

[1]See the discussion of coordinate vectors in Chapter 5 and especially 5.4.8.

4.2.5 Lemma. *If V is a vector space over a field F, v is a vector in V and 0 is the zero element of F, then $0\mathsf{v} = \mathsf{0}$ is the zero vector and $(-1)\mathsf{v} = -\mathsf{v}$ is the additive inverse of v.*

Proof. The element 0 in a field behaves like it should; in particular, $0 = 0+0$. Thus $0\mathsf{v} = (0+0)\mathsf{v} = 0\mathsf{v}+0\mathsf{v}$, using one of the distributive axioms of a vector space. Now $0\mathsf{v}$ is a vector (by closure under scalar multiplication), so it has an additive inverse $-0\mathsf{v}$. Adding this to each side of $0\mathsf{v} = 0\mathsf{v} + 0\mathsf{v}$, we get $\mathsf{0} = (0\mathsf{v} + 0\mathsf{v}) + (-0\mathsf{v}) = 0\mathsf{v} + [0\mathsf{v} + (-0\mathsf{v})]$ (using associativity of addition) $= 0\mathsf{v} + \mathsf{0} = 0\mathsf{v}$ (by the defining property of the zero vector $\mathsf{0}$). So, indeed, $0\mathsf{v} = \mathsf{0}$.

We leave the proof of $(-1)\mathsf{v} = -\mathsf{v}$ to the exercises. See Exercise 9. ■

4.2.6 Lemma. *A subset U of a vector space V is a subspace if and only if it is*

1) *nonempty,*

2) *closed under addition—if u and v are in U, then $\mathsf{u}+\mathsf{v}$ is also in U—and*

3) *closed under scalar multiplication—if u is in U and c is a scalar, then $c\mathsf{u}$ is also in U.*

Proof. ($\Longrightarrow$) Assume U is a subspace of V. Then U is nonempty because it contains the zero vector, and it is closed under addition and scalar multiplication since these are two of the nine axioms for a vector space (see Definition 4.2.1).

($\Longleftarrow$) Conversely, suppose U is a nonempty subset of a vector space V that is closed under addition and scalar multiplication. We must show that U is a subspace, hence that U satisfies the nine axioms of Definition 4.2.1. We are assuming U satisfies axioms 1 and 6. Axioms 2, 3, 7, 8 and 9 hold in U because they hold in V. This leaves axioms 4 and 5.

First we show that U contains the zero vector. Since U is not empty, there is some u in U. Using Lemma 4.2.5, we get $\mathsf{0} = 0\mathsf{u}$ and this is in U because U is closed under scalar multiplication. Finally, we must show that U contains the additive inverse of each of its vectors. This is equally straightforward since if u is in U, so is $-\mathsf{u} = (-1)\mathsf{u}$, because U is closed under scalar multiplication. ■

4.2.7 Remark. Every vector space V contains at least two subspaces, namely V itself and (at the other extreme) the one element set $\{\mathsf{0}\}$ consisting of just the zero vector. Neither V nor $\{\mathsf{0}\}$ is empty; both are closed under addition and scalar multiplication.

READING CHECK 2. Prove that $\left\{\mathsf{0}\right\}$ is closed under addition and scalar multiplication.

4.2.8 Problem. Show that $U = \left\{ \begin{bmatrix} x \\ y \\ z \end{bmatrix} \mid 2x - y + 3z = 0 \right\}$ is a subspace of R^3.

Solution. We use Lemma 4.2.6. The set U is nonempty because it contains $\mathsf{0} = \begin{bmatrix} 0 \\ 0 \\ 0 \end{bmatrix}$; indeed, $2(0) - 0 + 3(0) = 0$.

Let $\mathsf{u}_1 = \begin{bmatrix} x_1 \\ y_1 \\ z_1 \end{bmatrix}$ and $\mathsf{u}_2 = \begin{bmatrix} x_2 \\ y_2 \\ z_2 \end{bmatrix}$ belong to U. Then $2x_1 - y_1 + 3z_1 = 0$ and $2x_2 - y_2 + 3z_2 = 0$, so $\mathsf{u}_1 + \mathsf{u}_2 = \begin{bmatrix} x_1 + x_2 \\ y_1 + y_2 \\ z_1 + z_2 \end{bmatrix}$ is in U (so U is closed under addition) because

$$2(x_1 + x_2) - (y_1 + y_2) + 3(z_1 + z_2) = (2x_1 - y_1 + 3z_1) + (2x_2 - y_2 + 3z_2) = 0 + 0 = 0.$$

Finally, if $\mathsf{u} = \begin{bmatrix} x \\ y \\ z \end{bmatrix}$ is in U and c is a scalar, then $c\mathsf{u} = \begin{bmatrix} cx \\ cy \\ cz \end{bmatrix}$ is in U (so U is closed under scalar multiplication) because $2(cx) - (cy) + 3(cz) = c(2x - y + 3z) = c(0) = 0$. 👍

READING CHECK 3. The set $U = \left\{ \begin{bmatrix} x \\ y \\ z \end{bmatrix} \mid 2x - y + 3z = 1 \right\}$ is not a subspace of R^3. Why not?

4.2.9 Example. The set $U = \left\{ \begin{bmatrix} x \\ x - y \\ y \end{bmatrix} \mid x, y \in \mathsf{R} \right\}$ of vectors in R^3 whose middle component is the first component less the third is a subspace of R^3. This set is nonempty because it contains $\mathsf{0} = \begin{bmatrix} 0 \\ 0 - 0 \\ 0 \end{bmatrix}$. It is closed under addition because if $\mathsf{u}_1 = \begin{bmatrix} x_1 \\ x_1 - y_1 \\ y_1 \end{bmatrix}$ and $\mathsf{u}_2 = \begin{bmatrix} x_2 \\ x_2 - y_2 \\ y_2 \end{bmatrix}$ are in U, so is their sum, because $\mathsf{u}_1 + \mathsf{u}_2 = \begin{bmatrix} x_1 + x_2 \\ (x_1 + x_2) - (y_1 + y_2) \\ y_1 + y_2 \end{bmatrix}$. It is closed under scalar multiplication because if $\mathsf{u} = \begin{bmatrix} x \\ x - y \\ y \end{bmatrix}$ is in U and c is a scalar, then $c\mathsf{u}$ is in U because $c\mathsf{u} = \begin{bmatrix} cx \\ cx - cy \\ cy \end{bmatrix}$. ☺

4.2.10 Example. Let n be a fixed positive integer. In Section 2.1, we defined a "word" as a sequence of n 0s and 1s and a "code" as a set C of words with the property that the sum of two words in C is again in C. Making the obvious connection between words and vectors (think of 0101 as the vector $\begin{bmatrix}0\\1\\0\\1\end{bmatrix}$, for instance) a code is a set of vectors closed under addition. In this context, a scalar is an element of the field $F = \{0, 1\}$, so if v is a vector and α is a scalar then $\alpha\mathsf{v} = 0$ or $\alpha\mathsf{v} = \mathsf{v}$ because $\alpha = 0$ or $\alpha = 1$. Thus a code is closed under scalar multiplication as well as addition, so it is a subspace of the vector space F^n. ☺

4.2.11 Example. Let $V = \mathsf{R}[x]$ be the vector space of polynomials over R and let $U = \{a + bx + cx^2 \mid a, b, c \in \mathsf{R}\}$ be the set of all polynomials of degree at most 2. Certainly U is not empty: the polynomial $1 - x$ is in U, for instance. The sum of two polynomials in U is also in U, as is the scalar multiple of any polynomial in U. Thus U is a subspace of V. ☺

4.2.12 Example.[2] Let $V = C(\mathsf{R})$ be the vector space of continuous functions $\mathsf{R} \to \mathsf{R}$ and let $U = \mathcal{D}(\mathsf{R})$ be the set of differentiable functions from R to R. In Examples 4.2.3, we noted that both U and V are vector spaces under usual addition and scalar multiplication of functions. Since any differentiable function is continuous, U is a subset of V. Thus U is a subspace of V. ☺

4.2.13 Example. The set U of all symmetric[3] $n \times n$ matrices is a subspace of the vector space $V = M_n(\mathsf{R})$ of all $n \times n$ matrices over R. This set is not empty because, for example, the $n \times n$ matrix of all 0s is symmetric. It is closed under addition because if A and B are symmetric, then $(A+B)^T = A^T + B^T = A + B$, so $A + B$ is also symmetric. It is closed under scalar multiplication because if A is symmetric and c is a scalar, $(cA)^T = cA^T = cA$, so cA is symmetric as well. ☺

4.2.14 Example. The set $U_1 = \left\{\begin{bmatrix}x\\y\end{bmatrix} \in \mathsf{R}^2 \mid x = 0 \text{ or } y = 0\right\}$ is not a subspace of R^2. While it is closed under scalar multiplication, it is **not** closed under addition. For example, $\mathsf{u} = \begin{bmatrix}1\\0\end{bmatrix}$ is in U_1 and $\mathsf{v} = \begin{bmatrix}0\\1\end{bmatrix}$ is in U_1, but $\mathsf{u}+\mathsf{v} = \begin{bmatrix}1\\1\end{bmatrix}$ is not in U_1. The set $U_2 = \left\{\begin{bmatrix}x\\y\end{bmatrix} \in \mathsf{R}^2 \mid x \text{ and } y \text{ both integers}\right\}$ is not a subspace of R^2. It's closed under addition, but not scalar multiplication: for example, $\mathsf{u} = \begin{bmatrix}1\\2\end{bmatrix}$ is in U_2, but $\frac{1}{2}\mathsf{u} = \begin{bmatrix}\frac{1}{2}\\1\end{bmatrix}$ is not. ☺

READING CHECK 4. In Example 4.2.14, why is U_1 closed under scalar multiplication? Why is U_2 closed under addition?

[2]This example requires some knowledge of first-year calculus.
[3]Remember that A is symmetric means $A^T = A$.

4.2.15 Example. (With thanks to Michael Doob) The set of vectors in the plane which can be pictured by arrows from the origin to a point in the first or second quadrant is closed under addition, but not scalar multiplication while those vectors pictured by arrows from the origin to a point in the first or third quadrants is closed under scalar multiplication but not addition. (See Exercise 2.) ⍨

4.2.16 Problem. Matthew says R^2 is not a subspace of R^3 because vectors in R^2 have two components while vectors in R^3 have three. Jayne, on the other hand, says this is just a minor irritation: "Everybody knows the xy-plane is part of 3-space." Who's right?

Solution. Technically, Matthew is correct, but Jayne has given the mature answer. Writing the vector $\begin{bmatrix} x \\ y \end{bmatrix}$ as $\begin{bmatrix} x \\ y \\ 0 \end{bmatrix}$ allows us to imagine R^2 as part of R^3. With this understanding, R^2 is indeed a subspace of R^3 and, indeed, R^m is a subspace of R^n if $m \leq n$.

It is good to be a bit flexible about what you think of as "n-space." Suppose two people are asked to shade the region of the plane where $x \geq 0$, $y \geq 0$ and $x \leq 3 - 3y$. Chelsea goes to the front blackboard and produces a picture like the one shown, while Johnny draws more or less the same picture on the blackboard on the side wall. Has each student drawn a picture in R^2?

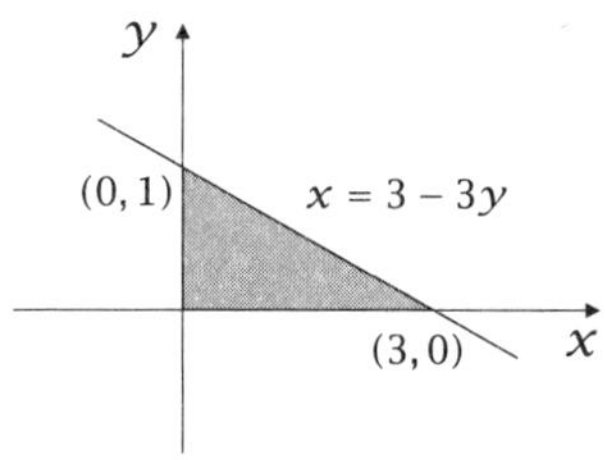

Most of us would say yes, while acknowledging that the planes in which the pictures were drawn were not the same. Are both planes "essentially" R^2? Surely yes. Does each picture make the point that R^2 is a subspace of R^3? Surely, yes.

4.2.17 Example (The null space of a matrix). In Section 4.1, we defined the *null space* of an $m \times n$ matrix A as the set of solutions to the homogeneous system $A\mathsf{x} = \mathsf{0}$:

$$\text{null sp}\, A = \{\mathsf{x} \in \mathsf{R}^n \mid A\mathsf{x} = \mathsf{0}\}.$$

The properties of matrix multiplication allow us to show that the null space of A is a subspace of R^n. Since $\mathsf{0} = A\mathsf{0}$, the zero vector of R^n is in null sp A, so null sp A is not empty. If x_1 and x_2 are in null sp A, so is their sum, since $A(\mathsf{x}_1 + \mathsf{x}_2) = A\mathsf{x}_1 + A\mathsf{x}_2 = \mathsf{0} + \mathsf{0} = \mathsf{0}$. Finally, if x is in null sp A and c is a scalar, $A(c\mathsf{x}) = c(A\mathsf{x}) = c\mathsf{0} = \mathsf{0}$, which shows that $c\mathsf{x}$ is in null sp A too. Since null sp A is not empty and closed under both addition and scalar multiplication, it is a subspace of R^n. ⍨

The Span of Vectors

Recall that the "span" of a set of vectors is the set of all linear combinations of these vectors. (See 1.1.24.) We will often use the notation $\operatorname{sp}\{\mathsf{u}_1, \mathsf{u}_2, \ldots, \mathsf{u}_k\}$ for the span of vectors $\mathsf{u}_1, \mathsf{u}_2, \ldots, \mathsf{u}_k$. For example, the span of the three standard basis vectors $\mathsf{i} = \begin{bmatrix} 1 \\ 0 \\ 0 \end{bmatrix}$, $\mathsf{j} = \begin{bmatrix} 0 \\ 1 \\ 0 \end{bmatrix}$ and $\mathsf{k} = \begin{bmatrix} 0 \\ 0 \\ 1 \end{bmatrix}$ is R^3, as is the span of the four vectors i, j, k, $\begin{bmatrix} 4 \\ 5 \\ 8 \end{bmatrix}$. Any vector in R^3 is a linear combination of these four vectors:

$$\begin{bmatrix} x \\ y \\ z \end{bmatrix} = x\begin{bmatrix} 1 \\ 0 \\ 0 \end{bmatrix} + y\begin{bmatrix} 0 \\ 1 \\ 0 \end{bmatrix} + z\begin{bmatrix} 0 \\ 0 \\ 1 \end{bmatrix} + 0\begin{bmatrix} 4 \\ 5 \\ 8 \end{bmatrix}.$$

4.2.18 Example. Another spanning set for R^3 is $\left\{\begin{bmatrix} 1 \\ 2 \\ 3 \end{bmatrix}, \begin{bmatrix} -1 \\ 3 \\ 5 \end{bmatrix}, \begin{bmatrix} 7 \\ 10 \\ -1 \end{bmatrix}\right\}$. To see why, we have to show that any $\begin{bmatrix} x \\ y \\ z \end{bmatrix}$ can be written

$$\begin{bmatrix} x \\ y \\ z \end{bmatrix} = a\begin{bmatrix} 1 \\ 2 \\ 3 \end{bmatrix} + b\begin{bmatrix} -1 \\ 3 \\ 5 \end{bmatrix} + c\begin{bmatrix} 7 \\ 10 \\ -1 \end{bmatrix}$$

for a certain choice of scalars a, b, c. In matrix form, this is

$$\begin{bmatrix} x \\ y \\ z \end{bmatrix} = \begin{bmatrix} 1 & -1 & 7 \\ 2 & 3 & 10 \\ 3 & 5 & -1 \end{bmatrix}\begin{bmatrix} a \\ b \\ c \end{bmatrix}. \tag{1}$$

It is easy to verify that the coefficient matrix here has rank 3, so the system has a (unique) solution for any x, y, z, by part 3 of Theorem 4.1.17. ☺

Any plane in R^3 which contains $\begin{bmatrix} 0 \\ 0 \\ 0 \end{bmatrix}$ is a subspace of R^3 because such a plane is the span of any two nonparallel vectors in it: the span of any set of vectors in a vector space V is always a subspace of V.

4.2.19 Proposition. *Let $\mathsf{u}_1, \mathsf{u}_2, \ldots, \mathsf{u}_k$ be vectors in a vector space V and let $U = \operatorname{sp}\{\mathsf{u}_1, \mathsf{u}_2, \ldots, \mathsf{u}_k\}$ be the span of these vectors. Then U is a subspace of V.*

Proof. Almost always, to show that an alleged subspace is not empty, we show that it contains $\mathsf{0}$.[4] Since $\mathsf{0} = 0\mathsf{u}_1 + 0\mathsf{u}_2 + \cdots + 0\mathsf{u}_k$ is a linear combi-

[4] If we can't do that, we are in deep trouble!

nation of $u_1, \ldots, u_k$, the set U defined here contains the zero vector, so it is not empty. To show that U is closed under addition, we take two vectors in U, say

$$u = c_1 u_1 + c_2 u_2 + \cdots + c_k u_k$$

and

$$v = d_1 u_1 + d_2 u_2 + \cdots + d_k u_k$$

and show that the sum $u + v$ is in U. This is the case because $u + v = (c_1 + d_1)u_1 + (c_2 + d_2)u_2 + \cdots + (c_k + d_k)u_k$ is again a linear combination of $u_1, \ldots, u_k$. To show that U is closed under scalar multiplication, we take u as above and a scalar c and show that cu is in U. This is the case because

$$cu = c(c_1 u_1 + c_2 u_2 + \cdots + c_k u_k) = (cc_1)u_1 + (cc_2)u_2 + \cdots + (cc_k)u_k$$

is again a linear combination of $u_1, \ldots, u_k$. ■

The Row Space of a Matrix

In Section 4.1, we defined the column space of an $m \times n$ matrix A as the span of the columns of A. So

4.2.20 The column space of an $m \times n$ matrix is a subspace of R^m.

In a similar fashion, we define the *row space* of a matrix A to be the span of its rows, thought of as row vectors as in Example 4.2.3(3). This is denoted row sp A. Since spanning sets are always subspaces,

4.2.21 The row space of an $m \times n$ matrix A is a subspace of $M_{1n}(\mathsf{R})$, a vector space like R^n, but where vectors are written as rows $\begin{bmatrix} x_1 & x_2 & \cdots & x_n \end{bmatrix}$.

Remember that the three elementary row operations on a matrix A are

1. $R_i \leftrightarrow R_j$: the interchange of two rows,
2. $R_i \to cR_i$: the multiplication of a row by a nonzero scalar, and
3. $R_i \to R_i - cR_j$: the replacement of a row by that row less a multiple of another row.

We wish to show that these operations do not change the row space of a matrix and illustrate in the case of a $3 \times n$ matrix

$$A = \begin{bmatrix} u_1 & \rightarrow & \\ u_2 & \rightarrow & \\ u_3 & \rightarrow & \end{bmatrix}$$

with rows labelled $\mathbf{u}_1$, $\mathbf{u}_2$, $\mathbf{u}_3$. Thus the row space of A is the subspace spanned by $\mathbf{u}_1, \mathbf{u}_2, \mathbf{u}_3$: row sp $A = \operatorname{sp}\{\mathbf{u}_1, \mathbf{u}_2, \mathbf{u}_3\}$.

Suppose we interchange the first two rows of A to get a matrix $B = \begin{bmatrix} \mathbf{u}_2 & \rightarrow \\ \mathbf{u}_1 & \rightarrow \\ \mathbf{u}_3 & \rightarrow \end{bmatrix}$. The row space of B is $\operatorname{sp}\{\mathbf{u}_2, \mathbf{u}_1, \mathbf{u}_3\}$, which is the same as row sp A: any linear combination of $\mathbf{u}_2, \mathbf{u}_1, \mathbf{u}_3$ is a linear combination of $\mathbf{u}_1, \mathbf{u}_2, \mathbf{u}_3$ and conversely. This argument shows that **the first elementary row operation does not change the row space.**

Suppose $B = \begin{bmatrix} \mathbf{u}_1 & \rightarrow \\ \mathbf{u}_2 & \rightarrow \\ 2\mathbf{u}_3 & \rightarrow \end{bmatrix}$ is obtained from A by multiplying row three by 2. The row space of B is $\operatorname{sp}\{\mathbf{u}_1, \mathbf{u}_2, 2\mathbf{u}_3\}$, which we assert is the same as row sp A. The fact that

$$a\mathbf{u}_1 + b\mathbf{u}_2 + c(2\mathbf{u}_3) = a\mathbf{u}_1 + b\mathbf{u}_2 + (2c)\mathbf{u}_3$$

shows that row sp $B \subseteq$ row sp A,[5] and the fact that

$$a\mathbf{u}_1 + b\mathbf{u}_2 + c\mathbf{u}_3 = a\mathbf{u}_1 + b\mathbf{u}_2 + \frac{c}{2}(2\mathbf{u}_3)$$

shows that row sp $A \subseteq$ row sp B, so the row spaces of A and B are the same.[6] **The second elementary row operation does not change the row space.**

Finally, suppose $B = \begin{bmatrix} \mathbf{u}_1 & \rightarrow \\ \mathbf{u}_2 + 4\mathbf{u}_1 & \rightarrow \\ \mathbf{u}_3 & \rightarrow \end{bmatrix}$ is obtained from A via the operation $R2 \rightarrow R2 + 4(R1)$. The row space of B is $\operatorname{sp}\{\mathbf{u}_1, \mathbf{u}_2 + 4\mathbf{u}_1, \mathbf{u}_3\}$ and, again, we can show that this is the row space of A. The fact that

$$a\mathbf{u}_1 + b(\mathbf{u}_2 + 4\mathbf{u}_1) + c\mathbf{u}_3 = (a + 4b)\mathbf{u}_1 + b\mathbf{u}_2 + c\mathbf{u}_3$$

shows that row sp $B \subseteq$ row sp A, and the equation

$$a\mathbf{u}_1 + b\mathbf{u}_2 + c\mathbf{u}_3 = (a - 4b)\mathbf{u}_1 + b(\mathbf{u}_2 + 4\mathbf{u}_1) + c\mathbf{u}_3$$

shows that row sp $A \subseteq$ row sp B. So the row spaces of A and B are the same. **The third elementary row operation does not change the row space.**

Generalizing in an obvious way the arguments just presented for a $3 \times n$ matrix to any matrix, we see that

4.2.22 The elementary row operations do not change the row space of a matrix. So the row space of a matrix is the row space of any of its row echelon forms.

[5] The notation $X \subseteq Y$ means "Set X is contained in set Y."

[6] To show sets X and Y are equal, we almost always show that $X \subseteq Y$ and $Y \subseteq X$.

4.2.23 Example. The row space of $A = \begin{bmatrix} 1 & 3 & 1 & -1 & 1 \\ -1 & -1 & 3 & 1 & 3 \\ 1 & 4 & 3 & -1 & 3 \end{bmatrix}$ is the span of its rows $\begin{bmatrix} 1 & 3 & 1 & -1 & 1 \end{bmatrix}$, $\begin{bmatrix} -1 & -1 & 3 & 1 & 3 \end{bmatrix}$ and $\begin{bmatrix} 1 & 4 & 3 & -1 & 3 \end{bmatrix}$. Here is one possible application of Gaussian elimination:

$$A \to \begin{bmatrix} 1 & 3 & 1 & -1 & 1 \\ 0 & 2 & 4 & 0 & 4 \\ 0 & 1 & 2 & 0 & 2 \end{bmatrix} \to \begin{bmatrix} 1 & 3 & 1 & -1 & 1 \\ 0 & 1 & 2 & 0 & 2 \\ 0 & 0 & 0 & 0 & 0 \end{bmatrix}.$$

From this, we see that the row space of A is the span of $\begin{bmatrix} 1 & 3 & 1 & -1 & 1 \end{bmatrix}$ and $\begin{bmatrix} 0 & 1 & 2 & 0 & 2 \end{bmatrix}$. :-)

Answers to Reading Checks

1. $\begin{bmatrix} 1 \\ 0 \\ 1 \end{bmatrix} + \begin{bmatrix} 1 \\ 1 \\ 1 \end{bmatrix} = \begin{bmatrix} 0 \\ 1 \\ 0 \end{bmatrix}$. (Remember that $1 + 1 = 0$ in $F = \{0, 1\}$.)

2. The set $U = \{\mathbf{0}\}$ is closed under addition because $\mathbf{0} + \mathbf{0} = \mathbf{0}$ is in U and closed under scalar multiplication because, for any scalar c, $c\mathbf{0} = \mathbf{0}$ is in U.

3. U does not contain the zero vector.

4. Let $\mathsf{u} = \begin{bmatrix} x \\ y \end{bmatrix}$ belong to U_1 and let c be a scalar. We know that $x = 0$ or $y = 0$. If $x = 0$, then $c\mathsf{x} = \begin{bmatrix} 0 \\ cy \end{bmatrix}$ is in U because the first component of this vector is 0. Similarly, if $y = 0$, then $c\mathsf{x} = \begin{bmatrix} cx \\ 0 \end{bmatrix}$ is in U_1.

   U_2 is closed under addition because if $\mathsf{u}_1 = \begin{bmatrix} x_1 \\ y_1 \end{bmatrix}$ and $\mathsf{u}_2 = \begin{bmatrix} x_2 \\ y_2 \end{bmatrix}$ are both in U_2, then the numbers x_1, y_1, x_2, y_2 are integers. The sum of integers is an integer, so $\mathsf{u}_1 + \mathsf{u}_2 = \begin{bmatrix} x_1 + x_2 \\ y_1 + y_2 \end{bmatrix}$ is in U_2.

True/False Questions

Decide, with as little calculation as possible, whether each of the following statements is true or false and, if you say "false," explain your answer. (Answers are at the back of the book.)

1. The line with equation $x + y = 1$ is a subspace of R^2.

2. The set of vectors in R^3 of the form $\begin{bmatrix} x \\ x \\ x \end{bmatrix}$ is closed under scalar multiplication.

3. The vector $\begin{bmatrix} 0 \\ 1 \end{bmatrix}$ is in the column space of $\begin{bmatrix} -1 & 2 & 1 & 5 \\ 0 & 1 & 7 & -9 \end{bmatrix}$.

4. Given an $n \times n$ matrix A, the set U of vectors x satisfying $A\mathsf{x} = 0$ is a subspace of R^n.

5. Given an $n \times n$ matrix A, the set U of vectors x satisfying $A\mathsf{x} = 2\mathsf{x}$ is a subspace of R^n.

6. The matrices $\begin{bmatrix} 1 & 2 & 3 \\ -1 & -2 & -3 \\ 2 & 5 & 6 \end{bmatrix}$ and $\begin{bmatrix} 1 & 2 & 3 \\ 0 & 1 & 0 \\ 0 & 0 & 0 \end{bmatrix}$ have the same row space.

7. The matrices $\begin{bmatrix} 1 & 2 & 3 \\ -1 & -2 & -3 \\ 2 & 5 & 6 \end{bmatrix}$ and $\begin{bmatrix} 1 & 2 & 3 \\ 0 & 1 & 0 \\ 0 & 0 & 0 \end{bmatrix}$ have the same column space.

8. If S is a set of linearly independent vectors that spans a subspace U, then the zero vector is not in U.

9. If a set $\{\mathsf{v}_1, \mathsf{v}_2, \ldots, \mathsf{v}_k\}$ of vectors is linearly dependent, then each vector in this set is a linear combination of the others.

10. If $\mathsf{v}_1, \mathsf{v}_2, \ldots, \mathsf{v}_k$ are vectors and one vector v_i is a scalar multiple of another v_j, then the vectors are linearly dependent.

EXERCISES

Answers to exercises marked [BB] can be found at the **Back** *of the* **Book**.

1. Determine whether or not each of the following sets is closed under the usual addition and scalar multiplication of vectors in R^3.

   (a) [BB] $U = \left\{ \begin{bmatrix} x \\ y \\ 2x - 3y \end{bmatrix} \mid x, y \in \mathsf{R} \right\}$ **(b)** $U = \left\{ \begin{bmatrix} x \\ y \\ z \end{bmatrix} \mid x + 4y - 5z = 0 \right\}$

   (c) $U = \left\{ \begin{bmatrix} x \\ y \\ z \end{bmatrix} \mid x > 0 \right\}$ **(d)** $U = \left\{ \begin{bmatrix} x \\ y \\ z \end{bmatrix} \mid xyz = 0 \right\}$.

2. Let $U_1 = \left\{ \begin{bmatrix} x \\ y \end{bmatrix} \mid y \geq 0 \right\}$ be the set of vectors which can be pictured by arrows in the plane that go from the origin to a point in the first or second quadrant. Let $U_2 = \left\{ \begin{bmatrix} x \\ y \end{bmatrix} \mid xy \geq 0 \right\}$ be the set of vectors that can be pictured by arrows that go from the origin to a point in the first or third quadrant. Explain why U_1 is closed under addition but not scalar multiplication while U_2 is closed under scalar multiplication but not addition.

3. Determine whether or not each of the following sets U is a subspace of the indicated vector space V. Justify your answers.

   (a) [BB] $V = \mathsf{R}^3$, $U = \left\{ \begin{bmatrix} a \\ a - 1 \\ c \end{bmatrix} \mid a, c \in \mathsf{R} \right\}$ **(b)** $V = \mathsf{R}^3$, $U = \left\{ \begin{bmatrix} a \\ a \\ c \end{bmatrix} \mid a, c \in \mathsf{R} \right\}$

(c) $V = \mathsf{R}^3$, $U = \left\{ \begin{bmatrix} x \\ y \\ x+y+1 \end{bmatrix} \mid x, y \in \mathsf{R} \right\}$

(d) $V = \mathsf{R}^4$, $U = \left\{ \begin{bmatrix} x \\ y \\ z \\ x+y-z \end{bmatrix} \mid x, y, z \in \mathsf{R} \right\}$

(e) $V = \mathsf{R}^3$, $U = \left\{ \begin{bmatrix} x \\ y \\ z \end{bmatrix} \mid x, y, z \text{ are integers} \right\}$

(f) V is the vector space of polynomials over R in the variable x and U is the subset of polynomials of degree 2.

(g) V is the vector space of polynomials over R of degree at most 2 and U is the subset of polynomials of the form $a + bx + ax^2$.

(h) V is the vector space of all $n \times n$ matrices and U is the set of all skew-symmetric matrices. [A matrix A is *skew-symmetric* if $A^T = -A$.]

4. Let A be an $n \times n$ matrix and let λ be a scalar. In Section 3.4, the eigenspace of A corresponding to λ was defined as the set $U = \{\mathsf{x} \in \mathsf{R}^n \mid A\mathsf{x} = \lambda\mathsf{x}\}$.

   (a) In the case that $\lambda = 0$, U has another name. What is it?

   (b) Show that U is a subspace of R^n.

5. [BB] Let U be a subspace of a vector space V and let u_1 and u_2 be vectors in U. Use the definition of "subspace" to explain why the vector $\mathsf{u}_1 - \mathsf{u}_2$ is also in U.

6. In each of the following cases, determine whether the vector v is in the subspace spanned by the other vectors.

   (a) [BB] $\mathsf{v} = \begin{bmatrix} 1 \\ 3 \\ -1 \\ 1 \end{bmatrix}$; $\mathsf{v}_1 = \begin{bmatrix} 1 \\ -1 \\ 2 \\ 1 \end{bmatrix}$, $\mathsf{v}_2 = \begin{bmatrix} 2 \\ 1 \\ 1 \\ 0 \end{bmatrix}$

   (b) $\mathsf{v} = \begin{bmatrix} 1 \\ -7 \\ 8 \\ 5 \end{bmatrix}$; $\mathsf{v}_1 = \begin{bmatrix} 1 \\ -1 \\ 2 \\ 1 \end{bmatrix}$, $\mathsf{v}_2 = \begin{bmatrix} 2 \\ 1 \\ 1 \\ 0 \end{bmatrix}$

   (c) $\mathsf{v} = \begin{bmatrix} 1 \\ -1 \\ 1 \\ 9 \end{bmatrix}$; $\mathsf{v}_1 = \begin{bmatrix} 0 \\ -1 \\ 1 \\ 2 \end{bmatrix}$, $\mathsf{v}_2 = \begin{bmatrix} 3 \\ 0 \\ 0 \\ -1 \end{bmatrix}$, $\mathsf{v}_3 = \begin{bmatrix} 2 \\ 1 \\ -1 \\ 1 \end{bmatrix}$

   (d) $\mathsf{v} = \begin{bmatrix} 2 \\ 0 \\ 1 \\ -3 \end{bmatrix}$; $\mathsf{v}_1 = \begin{bmatrix} 0 \\ -1 \\ 1 \\ 2 \end{bmatrix}$, $\mathsf{v}_2 = \begin{bmatrix} 3 \\ 0 \\ 0 \\ -1 \end{bmatrix}$, $\mathsf{v}_3 = \begin{bmatrix} 2 \\ 1 \\ -1 \\ 1 \end{bmatrix}$

7. (a) Do the vectors $\mathsf{u}_1 = \begin{bmatrix} 2 \\ 1 \\ -3 \end{bmatrix}$ and $\mathsf{u}_2 = \begin{bmatrix} 6 \\ 0 \\ 1 \end{bmatrix}$ span R^3? If the answer is yes, provide a proof; otherwise, identify the subspace spanned by the given vectors.

(b) Answer (a) for the vectors u_1, u_2, u_3, with u_1 and u_2 as before and $\mathsf{u}_3 = \begin{bmatrix} 1 \\ 1 \\ 1 \end{bmatrix}$.

(c) Answer (a) for the vectors u_1, u_2, u_3, with u_1 and u_2 as before and $\mathsf{u}_3 = \begin{bmatrix} 2 \\ -2 \\ 7 \end{bmatrix}$. [Hint: $\mathsf{u}_3 = \mathsf{u}_2 - 2\mathsf{u}_1$.]

8. Determine whether the vectors $\mathsf{v}_1 = \begin{bmatrix} 1 \\ 0 \\ 0 \\ 1 \end{bmatrix}$, $\mathsf{v}_2 = \begin{bmatrix} 1 \\ 0 \\ 0 \\ -1 \end{bmatrix}$, $\mathsf{v}_3 = \begin{bmatrix} 0 \\ 1 \\ 1 \\ 0 \end{bmatrix}$, $\mathsf{v}_4 = \begin{bmatrix} 0 \\ 1 \\ -1 \\ 0 \end{bmatrix}$ span R^4. If the answer is yes, supply a proof; otherwise, identity the subspace spanned by the given vectors.

9. Complete the proof of Lemma 4.2.5 by showing that $(-1)\mathsf{v}$ is the additive inverse of v for any vector v in any vector space. [Hint: The proof uses the first part of the lemma.]

Critical Reading

10. [BB] Suppose U and W are subspaces of a vector space V. Show that the *union* of U and W is a subspace of V if and only if one of U, W contains the other. [Hint: In the part of the proof in which you assume $U \cup W$ is a subspace, suppose neither U nor W is contained in the other and derive a contradiction.]

    The union of U and W is denoted $U \cup W$ and defined, symbolically, by $U \cup W = \{\mathsf{v} \in V \mid \mathsf{v} \in U \text{ or } \mathsf{v} \in W\}$.

11. Suppose A is an $m \times n$ matrix and B is an $n \times t$ matrix. Suppose the columns of A span R^m and the columns of B span R^n. Show that the columns of AB span R^m.

4.3 Basis and Dimension

Linear independence in a vector space is defined just as it is in R^n, and most results stated and proved in previous sections about linearly independent sets in R^n carry over verbatim to any vector space.

4.3.1 **Definition.** Vectors $\mathsf{v}_1, \mathsf{v}_2, \ldots, \mathsf{v}_n$ are *linearly independent* if and only if

$$c_1\mathsf{v}_1 + c_2\mathsf{v}_2 + \cdots + c_n\mathsf{v}_n = 0 \text{ implies } c_1 = 0, c_2 = 0, \ldots, c_n = 0.$$

Vectors are *linearly dependent* if they are not linearly independent; that is, there exist scalars $c_1, c_2, \ldots, c_n$, not all 0, with $c_1\mathsf{v}_1 + c_2\mathsf{v}_2 + \cdots + c_n\mathsf{v}_n = 0$.

In Theorem 1.5.5, we proved that vectors are linearly dependent if and only if one of them is a linear combination of the others. It is sometimes useful to have available a stronger version of this statement. In the exercises, we ask you to show that if $\mathsf{v}_1, \mathsf{v}_2, \ldots, \mathsf{v}_n$ are linearly dependent vectors none of which is $\mathbf{0}$, then one of these is a linear combination of those vectors that **precede** it in the list. See Exercise 2.

4.3.2 Example. The matrix $U = \begin{bmatrix} \textcircled{3} & -1 & 4 & 1 & 2 & 8 \\ 0 & \textcircled{-2} & 0 & 5 & 7 & 5 \\ 0 & 0 & 0 & \textcircled{8} & 3 & -2 \\ 0 & 0 & 0 & 0 & 0 & \textcircled{9} \\ 0 & 0 & 0 & 0 & 0 & 0 \end{bmatrix}$ is in row echelon form. The pivots, 3, −2, 8 and 9 are in columns one, two, four and six. If these columns are linearly dependent, then one of them is a linear combination of columns that precede it. But this is not possible because of the nature of row echelon form. For example, column two is not a multiple of column one because any multiple of column one has second component 0. Column four is not a linear combination of columns one and two because any such linear combination has third component 0. For a similar reason, column six is not a linear combination of columns one, two and four. The pivot columns of U are linearly independent.

The nonzero rows of U (starting with the last nonzero row and moving upwards), namely

$$\begin{array}{l} [0 \quad 0 \quad 0 \quad 0 \quad 0 \quad 9], \\ [0 \quad 0 \quad 0 \quad 8 \quad 3 \quad -2], \\ [0 \quad -2 \quad 0 \quad 5 \quad 7 \quad 5], \\ [3 \quad -1 \quad 4 \quad 1 \quad 2 \quad 8] \end{array}$$

are also linearly independent because none is a linear combination of the rows that precede it. ☺

A general proof of the next theorem would follow very closely the arguments just presented in a special case.

4.3.3 Theorem. *Let U be a matrix in row echelon form.*

1. *The pivot columns of U are linearly independent.*
2. *The nonzero rows of U are linearly independent.*

4.3.4 Problem. Find a linearly independent set of vectors that spans the null space of $A = \begin{bmatrix} 1 & 2 & -1 & 0 \\ 1 & 3 & -2 & 0 \\ 1 & -1 & 1 & -1 \\ 5 & 4 & 0 & -1 \\ 1 & 8 & -3 & 4 \\ 0 & 7 & -2 & 5 \end{bmatrix}$, the row space of A and the column space of A.

Solution. Null Space. Row echelon form is $\begin{bmatrix} 1 & 2 & -1 & 0 \\ 0 & 1 & -1 & 0 \\ 0 & 0 & 1 & 1 \\ 0 & 0 & 0 & 0 \\ 0 & 0 & 0 & 0 \\ 0 & 0 & 0 & 0 \end{bmatrix}$. If $\mathsf{x} = \begin{bmatrix} x_1 \\ x_2 \\ x_3 \\ x_4 \end{bmatrix}$ is in the null space of A, $x_4 = t$ is free, $x_3 = -x_4 = -t$, $x_2 = x_3 = -t$, and $x_1 = -2x_2 + x_3 = 2t - t = t$, so $\mathsf{x} = \begin{bmatrix} t \\ -t \\ -t \\ t \end{bmatrix} = t \begin{bmatrix} 1 \\ -1 \\ -1 \\ 1 \end{bmatrix}$. Thus the vector $\begin{bmatrix} 1 \\ -1 \\ -1 \\ 1 \end{bmatrix}$ spans the null space (and any single nonzero vector always comprises a linearly independent set).

Row Space. The row space of A is the row space of any row echelon form of A because the elementary row operations do not change the row space. Thus the nonzero vectors in a row echelon form span the row space and they are linearly independent, so the row space of A is spanned by the linearly independent vectors $\begin{bmatrix} 1 & 2 & -1 & 0 \end{bmatrix}$, $\begin{bmatrix} 0 & 1 & -1 & 0 \end{bmatrix}$, and $\begin{bmatrix} 0 & 0 & 1 & 1 \end{bmatrix}$.

Column space. The column space is spanned by the rows of

$$A^T = \begin{bmatrix} 1 & 1 & 1 & 5 & 1 & 0 \\ 0 & 1 & -3 & -6 & 6 & 7 \\ 0 & 0 & 1 & 1 & -4 & -5 \\ 0 & 0 & 0 & 0 & 0 & 0 \end{bmatrix}.$$

Row echelon form of A^T is $\begin{bmatrix} 1 & 1 & 1 & 5 & 1 & 0 \\ 0 & 1 & -3 & -6 & 6 & 7 \\ 0 & 0 & 1 & 1 & -4 & -5 \\ 0 & 0 & 0 & 0 & 0 & 0 \end{bmatrix} = U$. The nonzero rows of U are linearly independent and their transposes span the column space of A: $\begin{bmatrix} 1 \\ 1 \\ 1 \\ 5 \\ 1 \\ 0 \end{bmatrix}$, $\begin{bmatrix} 0 \\ 1 \\ -3 \\ -6 \\ 6 \\ 7 \end{bmatrix}$ and $\begin{bmatrix} 0 \\ 0 \\ 1 \\ 1 \\ -4 \\ -5 \end{bmatrix}$. ☝

Basis

Suppose a vector space V is spanned by a finite set S of linearly independent vectors. In this section, we prove that the number of vectors in S, denoted $|S|$, is an "invariant" of V: this number does not depend on S, only on V.[7]

> Any two finite linearly independent sets of vectors that span a vector space contain the same number of vectors.

[7]For any finite set X, $|X|$ means the number of elements in X.

This fact, the most fundamental in linear algebra, hinges on a theorem which the author likes to remember as $|\mathcal{L}| \le |S|$. The proof will go smoothly if we remember three things.

First, if A is a matrix and x is a vector, then $A\mathsf{x}$ is a linear combination of the columns of A with coefficients the components of x. By now, I hope the reader is very tired of being reminded of this central idea, which first appeared as 2.1.33. Second, if A is an $m \times n$ matrix and $B = \begin{bmatrix} \mathsf{b}_1 & \mathsf{b}_2 & \cdots & \mathsf{b}_p \\ \downarrow & \downarrow & \cdots & \downarrow \\ & & & \end{bmatrix}$ is a matrix with columns $\mathsf{b}_1, \mathsf{b}_2, \ldots, \mathsf{b}_p$, then the matrix product AB has columns $A\mathsf{b}_1, A\mathsf{b}_2, \ldots, A\mathsf{b}_p$:

$$A \begin{bmatrix} \mathsf{b}_1 & \mathsf{b}_2 & \cdots & \mathsf{b}_p \\ \downarrow & \downarrow & \cdots & \downarrow \\ & & & \end{bmatrix} = \begin{bmatrix} A\mathsf{b}_1 & A\mathsf{b}_2 & \cdots & A\mathsf{b}_p \\ \downarrow & \downarrow & \cdots & \downarrow \\ & & & \end{bmatrix}.$$

Third, if A is an $m \times n$ matrix and $m < n$, then A has a nontrivial null space; that is, there is a nonzero vector x satisfying $A\mathsf{x} = \mathsf{0}$—see Theorem 4.1.4.

While most of the proofs of propositions and theorems given in this chapter are valid in any vector space—R^n or polynomials or matrices or functions—the proof of the next theorem assumes V is a subspace of R^n, for some n. (On the other hand, the concept of "coordinate" vector—see Definition 5.4.2 and Remark 5.4.8—makes any finite dimensional vector space look like R^n, so it is not hard to modify our proof so that it works generally.)

4.3.5 Theorem. ($|\mathcal{L}| \le |S|$) *Let V be a vector space that is a subspace of some Euclidean space. Let $\mathcal{L}$ be a finite set of linearly independent vectors in V and S a finite set of vectors that spans V. Then the number of vectors in $\mathcal{L}$ is less than or equal to the number of vectors in S, a fact we record symbolically by $|\mathcal{L}| \le |S|$.*

Proof. Suppose $\mathcal{L} = \{\mathsf{u}_1, \mathsf{u}_2, \ldots, \mathsf{u}_n\}$ and $S = \{\mathsf{w}_1, \mathsf{w}_2, \ldots, \mathsf{w}_m\}$. We wish to prove that $n \le m$.

Each of the vectors $\mathsf{u}_1, \mathsf{u}_2, \ldots, \mathsf{u}_n$ is a linear combination of $\mathsf{w}_1, \mathsf{w}_2, \ldots, \mathsf{w}_m$, in particular, u_1 is. So there exist scalars $a_{11}, a_{21}, \ldots, a_{m1}$ such that

$$\mathsf{u}_1 = a_{11}\mathsf{w}_1 + a_{21}\mathsf{w}_2 + \cdots + a_{m1}\mathsf{w}_m.$$

This is just the matrix equation

$$\mathsf{u}_1 = \begin{bmatrix} \mathsf{w}_1 & \mathsf{w}_2 & \cdots & \mathsf{w}_m \\ \downarrow & \downarrow & & \downarrow \\ & & & \end{bmatrix} \begin{bmatrix} a_{11} \\ a_{21} \\ \vdots \\ a_{m1} \end{bmatrix} = W \begin{bmatrix} a_{11} \\ a_{21} \\ \vdots \\ a_{m1} \end{bmatrix}$$

with $W = \begin{bmatrix} \mathsf{w}_1 & \mathsf{w}_2 & \cdots & \mathsf{w}_m \\ \downarrow & \downarrow & & \downarrow \\ & & & \end{bmatrix}$. Similarly, there exist scalars a_{12}, a_{22}, ..., a_{m2}

such that

$$u_2 = a_{12}w_1 + a_{22}w_2 + \cdots + a_{m2}w_m,$$

that is, $u_2 = W \begin{bmatrix} a_{12} \\ a_{22} \\ \vdots \\ a_{m2} \end{bmatrix}$, and there are similar equations for $u_3, \ldots, u_n$. Letting

$$A = \begin{bmatrix} a_{11} & a_{12} & \cdots & a_{1n} \\ a_{21} & a_{22} & \cdots & a_{2n} \\ \vdots & \vdots & & \vdots \\ a_{m1} & a_{n2} & \cdots & a_{mn} \end{bmatrix} \quad \text{and} \quad U = \begin{bmatrix} u_1 & u_2 & \cdots & u_n \\ \downarrow & \downarrow & & \downarrow \end{bmatrix},$$

we have $U = WA$. Suppose $m < n$. Since A is $m \times n$, the null space of A is nontrivial, so there is a nonzero vector x with $Ax = 0$. Thus $Ux = W(Ax) = 0$, contradicting the fact that the columns of U are linearly independent. Our assumption $m < n$ must be wrong, so $n \leq m$ as desired. ■

4.3.6 Example. R^m has a spanning set of m vectors, namely, the standard basis vectors, so linearly independent sets in R^m contain at most m vectors: $|\mathcal{L}| \leq |S| = m$. This fact was observed earlier in Corollary 4.1.5. By the same token, the standard basis vectors in R^m are linearly independent, so sets that span R^m contain at least m vectors: $m = |\mathcal{L}| \leq |S|$. $\ddot{\smile}$

4.3.7

Any linearly independent set of vectors in R^m contains at most m vectors. Any spanning set for R^m contains at least m vectors.

4.3.8 Example. Let π be the plane in R^3 with equation $2x - y - z = 0$. Then π is a subspace spanned by the two vectors $\begin{bmatrix} 1 \\ 2 \\ 0 \end{bmatrix}$ and $\begin{bmatrix} 1 \\ 0 \\ 2 \end{bmatrix}$, so linearly independent sets in π contain at most two vectors; three or more vectors in π are linearly dependent. $\ddot{\smile}$

4.3.9 Example. The vectors $u_1 = \begin{bmatrix} 1 \\ 2 \\ 3 \\ 4 \end{bmatrix}$, $u_2 = \begin{bmatrix} 5 \\ 6 \\ 7 \\ 8 \end{bmatrix}$, $u_3 = \begin{bmatrix} 9 \\ 10 \\ 11 \\ 12 \end{bmatrix}$ do not span R^4. At least four vectors are needed to span R^4. $\ddot{\smile}$

4.3.10 Definition. A *basis* of a vector space V is a linearly independent set of vectors that spans V. (The plural of "basis" is "bases.")

4.3.11 Example. The standard basis vectors $e_1, e_2, \ldots, e_n$ in R^n form a basis for R^n. $\ddot{\smile}$

4.3.12 Example. If $\begin{bmatrix} x \\ y \\ z \end{bmatrix}$ is in the plane π with equation $2x - y + z = 0$, then $\begin{bmatrix} x \\ y \\ z \end{bmatrix} = \begin{bmatrix} x \\ y \\ -2x + y \end{bmatrix} = x\begin{bmatrix} 1 \\ 0 \\ -2 \end{bmatrix} + y\begin{bmatrix} 0 \\ 1 \\ 1 \end{bmatrix}$. So the vectors $\mathsf{u}_1 = \begin{bmatrix} 1 \\ 0 \\ -2 \end{bmatrix}$ and $\mathsf{u}_2 = \begin{bmatrix} 0 \\ 1 \\ 1 \end{bmatrix}$ span π. Since they are also linearly independent, they comprise a basis for π. ☺

4.3.13 Example. The six matrices $\begin{bmatrix} 1 & 0 & 0 \\ 0 & 0 & 0 \end{bmatrix}$, $\begin{bmatrix} 0 & 1 & 0 \\ 0 & 0 & 0 \end{bmatrix}$, $\begin{bmatrix} 0 & 0 & 1 \\ 0 & 0 & 0 \end{bmatrix}$, $\begin{bmatrix} 0 & 0 & 0 \\ 1 & 0 & 0 \end{bmatrix}$, $\begin{bmatrix} 0 & 0 & 0 \\ 0 & 1 & 0 \end{bmatrix}$, $\begin{bmatrix} 0 & 0 & 0 \\ 0 & 0 & 1 \end{bmatrix}$ are linearly independent and they also span the vector space $M_{23}(\mathsf{R})$ of 2×3 matrices over R—the argument is almost identical to that which shows $\mathsf{e}_1, \mathsf{e}_2, \ldots, \mathsf{e}_n$ are linearly independent vectors that span R^n—so they are a basis for $M_{23}(\mathsf{R})$. ☺

Suppose $\mathcal{V}$ is a basis of n vectors for some vector space V and $\mathcal{U}$ is another basis for V. Then $\mathcal{V}$ is a spanning set and $\mathcal{U}$ is a linearly independent set, so $|\mathcal{U}| \le n$, by Theorem 4.3.5. On the other hand, $\mathcal{U}$ is a spanning set and $\mathcal{V}$ is a linearly independent set, so $n \le |\mathcal{U}|$. We conclude that $|\mathcal{U}| = n$ and state this fact as a theorem.

4.3.14 Theorem (Invariance of Dimension). *If a vector space V has a basis of $n \ge 1$ vectors, then any basis of V contains n vectors.*

We call the number n in this theorem the *dimension* of V.

4.3.15 Definitions. The *dimension* of a vector space $V \ne \{0\}$ is the number of elements in a basis of V, if such a number n exists. If it does, we write $n = \dim V$, the dimension of V. The dimension of the vector space $V = \{0\}$ is 0, by general agreement. A vector space V is *finite dimensional* if $V = \{0\}$ or $\dim V = n$ for some positive integer n. A vector space that is not finite dimensional is *infinite dimensional.*

4.3.16 Examples. $\dim \mathsf{R}^n = n$, $\dim M_{23}(\mathsf{R}) = 6$. ☺

4.3.17 Example. Let U be a line through the origin in Euclidean 3-space, that is, $U = \mathsf{R}\mathsf{u} = \{c\mathsf{u} \mid c \in \mathsf{R}\}$ for some nonzero vector u in R^3. Since $\mathsf{u} \ne 0$, $\{\mathsf{u}\}$ is a linearly independent set and since this set also spans the line, it is a basis for U. It follows that any basis for U will contain exactly **one** vector, so $\dim U = 1$. ☺

4.3.18 Remark. Our examples have all been vector spaces of finite dimension. Although all the vector spaces of interest to us in this book are finite dimensional, the reader should be aware that there are many important infinite dimensional vector spaces as well, for example, the set of all polynomials over R in a variable x, with usual addition and scalar multiplication (see Exercise 11) and the set of all continuous functions $\mathsf{R} \to \mathsf{R}$.

True/False Questions

Decide, with as little calculation as possible, whether each of the following statements is true or false and, if you say "false," explain your answer. (Answers are at the back of the book.)

1. The vectors $\mathsf{v}_1 = \begin{bmatrix} -1 \\ 2 \\ 3 \\ 4 \end{bmatrix}, \mathsf{v}_2 = \begin{bmatrix} 0 \\ 0 \\ -1 \\ 2 \end{bmatrix}, \mathsf{v}_3 = \begin{bmatrix} 2 \\ 1 \\ 3 \\ 5 \end{bmatrix}$ span R^4.

2. The vectors $\begin{bmatrix} 1 \\ 0 \\ 0 \\ 0 \end{bmatrix}, \begin{bmatrix} 0 \\ 1 \\ 0 \\ 0 \end{bmatrix}, \begin{bmatrix} 0 \\ 0 \\ 1 \\ 0 \end{bmatrix}, \begin{bmatrix} 0 \\ 0 \\ 0 \\ 1 \end{bmatrix}, \begin{bmatrix} 1 \\ 1 \\ 1 \\ 1 \end{bmatrix}$ span R^4.

3. The vectors $\mathsf{v}_1 = \begin{bmatrix} 1 \\ 2 \\ 3 \end{bmatrix}, \mathsf{v}_2 = \begin{bmatrix} -1 \\ -3 \\ 7 \end{bmatrix}, \mathsf{v}_3 = \begin{bmatrix} 0 \\ -1 \\ 1 \end{bmatrix}, \mathsf{v}_4 = \begin{bmatrix} 2 \\ 3 \\ -7 \end{bmatrix}$ are linearly independent.

4. Any set of vectors of R^n that spans R^n contains at least n vectors.

5. A finite dimensional vector space is a vector space which has only a finite number of vectors.

6. The vectors $\begin{bmatrix} 1 \\ 0 \\ 0 \end{bmatrix}, \begin{bmatrix} 0 \\ 1 \\ 0 \end{bmatrix}$ and $\begin{bmatrix} 0 \\ 0 \\ 1 \end{bmatrix}$ constitute a basis for the column space of $\begin{bmatrix} 1 & 3 & 2 & 0 & 0 \\ 0 & 0 & 1 & 0 & 2 \\ 0 & 0 & 0 & 1 & 4 \end{bmatrix}$.

7. A basis for the row space of a matrix can be determined by reducing it to row echelon form.

8. If A is an $m \times n$ matrix and $n > m$, the columns of A are linearly dependent.

9. A vector space of dimension 5 has precisely five subspaces.

10. A vector space of dimension 5 has subspaces of just six possible dimensions.

EXERCISES

Answers to exercises marked [BB] can be found at the Back of the Book.

1. In each case, the given vectors are linearly dependent. Give an elementary reason why this is the case, then express one of the vectors as a linear combination of the vectors preceding it.

   (a) [BB] $\mathsf{v}_1 = \begin{bmatrix} 1 \\ 2 \end{bmatrix}, \mathsf{v}_2 = \begin{bmatrix} -3 \\ 4 \end{bmatrix}, \mathsf{v}_3 = \begin{bmatrix} 5 \\ 0 \end{bmatrix}$

   (b) $\mathsf{v}_1 = \begin{bmatrix} -2 \\ 1 \end{bmatrix}, \mathsf{v}_2 = \begin{bmatrix} 4 \\ -2 \end{bmatrix}, \mathsf{v}_3 = \begin{bmatrix} 8 \\ 7 \end{bmatrix}, \mathsf{v}_4 = \begin{bmatrix} -3 \\ 5 \end{bmatrix}$

   (c) $\mathsf{v}_1 = \begin{bmatrix} 0 \\ -2 \\ 1 \end{bmatrix}, \mathsf{v}_2 = \begin{bmatrix} 1 \\ 1 \\ 1 \end{bmatrix}, \mathsf{v}_3 = \begin{bmatrix} 7 \\ 11 \\ 5 \end{bmatrix}, \mathsf{v}_4 = \begin{bmatrix} 2 \\ 4 \\ 1 \end{bmatrix}$

   (d) $\mathsf{v}_1 = \begin{bmatrix} 1 \\ 2 \\ 3 \\ 4 \end{bmatrix}, \mathsf{v}_2 = \begin{bmatrix} -1 \\ 1 \\ 7 \\ -3 \end{bmatrix}, \mathsf{v}_3 = \begin{bmatrix} 0 \\ -1 \\ 2 \\ 0 \end{bmatrix}, \mathsf{v}_4 = \begin{bmatrix} 1 \\ 8 \\ 23 \\ 6 \end{bmatrix}, \mathsf{v}_5 = \begin{bmatrix} -1 \\ 1 \\ 6 \\ 0 \end{bmatrix}$

   (e) $\mathsf{v}_1 = \begin{bmatrix} 0 \\ -1 \\ 1 \\ 2 \end{bmatrix}, \mathsf{v}_2 = \begin{bmatrix} 3 \\ 0 \\ -1 \\ 1 \end{bmatrix}, \mathsf{v}_3 = \begin{bmatrix} -3 \\ -7 \\ 8 \\ 13 \end{bmatrix}, \mathsf{v}_4 = \begin{bmatrix} -3 \\ -2 \\ 3 \\ 3 \end{bmatrix}, \mathsf{v}_5 = \begin{bmatrix} 0 \\ -5 \\ 5 \\ 1 \end{bmatrix}$.

2. **(a)** Sharpen Theorem 1.5.5 by proving that if $n > 1$ and $\mathsf{v}_1, \mathsf{v}_2, \ldots, \mathsf{v}_n$ are linearly dependent vectors none of which is the zero vector, then some vector is a linear combination of vectors that precede it in the list. [Hint: Examine the proof of Theorem 1.5.5.]

   (b) Show that the result of part (a) does not hold without the assumption that none of the vectors is $\mathbf{0}$.

3. For each matrix, find a linearly independent set of vectors that spans

   i. the null space, ii. the row space, iii. the column space.

   (a) [BB] $A = \begin{bmatrix} -1 & 1 & 0 & 2 \\ 3 & 1 & 1 & 10 \\ 2 & 4 & -1 & 0 \end{bmatrix}$ **(b)** $A = \begin{bmatrix} 0 & 1 & 3 \\ -1 & 3 & 8 \\ -13 & 2 & -7 \\ 1 & 1 & 4 \\ 4 & -2 & -2 \end{bmatrix}$

   (c) $A = \begin{bmatrix} 1 & 2 & -1 & 0 \\ 1 & 3 & -2 & 0 \\ 1 & -1 & 1 & -1 \\ 5 & 4 & 0 & -1 \\ 1 & 8 & -3 & 4 \\ 0 & 7 & -2 & 5 \end{bmatrix}$ **(d)** $A = \begin{bmatrix} 1 & 2 & -1 & 0 & 1 & 3 \\ -3 & -10 & -3 & 3 & 8 & -18 \\ 3 & 6 & -3 & 2 & -7 & 11 \\ 1 & 4 & 2 & -2 & -2 & 7 \\ 2 & 2 & -5 & 3 & 0 & 3 \end{bmatrix}$.

4. Decide whether each of the following sets of vectors is a basis for the identified vector space.

   (a) [BB] $\left\{ \begin{bmatrix} 1 \\ 4 \\ 7 \end{bmatrix}, \begin{bmatrix} 2 \\ 5 \\ 8 \end{bmatrix}, \begin{bmatrix} 3 \\ 6 \\ 9 \end{bmatrix} \right\}$ for R^3 **(b)** $\left\{ \begin{bmatrix} 1 \\ 4 \\ 7 \end{bmatrix}, \begin{bmatrix} 2 \\ 5 \\ 8 \end{bmatrix} \right\}$ for R^3

(c) $\left\{\begin{bmatrix}3\\6\\3\\-6\end{bmatrix}, \begin{bmatrix}0\\-1\\-1\\0\end{bmatrix}, \begin{bmatrix}0\\-8\\-12\\-4\end{bmatrix}, \begin{bmatrix}1\\0\\-1\\2\end{bmatrix}\right\}$ for R^4

(d) $\left\{\begin{bmatrix}1\\8\\-9\\16\end{bmatrix}, \begin{bmatrix}-1\\-6\\10\\-15\end{bmatrix}, \begin{bmatrix}-3\\6\\11\\-14\end{bmatrix}, \begin{bmatrix}4\\5\\-12\\13\end{bmatrix}\right\}$ for R^4

(e) $\left\{\begin{bmatrix}1\\8\\9\\16\end{bmatrix}, \begin{bmatrix}2\\7\\10\\15\end{bmatrix}, \begin{bmatrix}3\\6\\11\\14\end{bmatrix}, \begin{bmatrix}4\\5\\12\\13\end{bmatrix}\right\}$ for R^4

(f) $\left\{\begin{bmatrix}-1\\2\\0\\1\end{bmatrix}, \begin{bmatrix}0\\0\\1\\1\end{bmatrix}, \begin{bmatrix}-3\\4\\0\\2\end{bmatrix}, \begin{bmatrix}0\\-5\\6\\14\end{bmatrix}, \begin{bmatrix}-3\\2\\4\\1\end{bmatrix}\right\}$ for R^4.

5. [BB] Let $\mathsf{v}_1 = \begin{bmatrix}1\\2\end{bmatrix}$, $\mathcal{L} = \{\mathsf{v}_1\}$ and $\mathsf{v}_2 = \begin{bmatrix}a\\b\end{bmatrix}$. Find necessary and sufficient conditions on a and b in order that $\mathcal{L} \cup \{\mathsf{v}_2\}$ should be a basis for R^2, then complete this sentence: "$\mathcal{L} \cup \{\mathsf{v}_2\}$ is a basis of R^2 if and only if ..."

6. Let $\mathsf{v}_1 = \begin{bmatrix}1\\0\\-1\end{bmatrix}$, $\mathsf{v}_2 = \begin{bmatrix}0\\1\\2\end{bmatrix}$ and $\mathcal{L} = \{\mathsf{v}_1, \mathsf{v}_2\}$.

   (a) Find necessary and sufficient conditions on a, b, c for $\mathcal{L} \cup \{\mathsf{v}_3\}$ **not** to be a basis of R^3, where $\mathsf{v}_3 = \begin{bmatrix}a\\b\\c\end{bmatrix}$. So you are being asked to complete this sentence: "$\mathcal{L} \cup \{\mathsf{v}_3\}$ is not a basis of R^3 if and only if ...."

   (b) Explain how you would extend $\mathcal{L}$ to a basis of R^3.

7. Let $\mathsf{v}_1 = \begin{bmatrix}0\\0\\1\\-1\end{bmatrix}$, $\mathsf{v}_2 = \begin{bmatrix}1\\2\\-1\\-3\end{bmatrix}$, $\mathsf{v}_3 = \begin{bmatrix}-1\\-2\\0\\1\end{bmatrix}$ and $\mathcal{L} = \{\mathsf{v}_1, \mathsf{v}_2, \mathsf{v}_3\}$.

   (a) Find necessary and sufficient conditions on a, b, c, d for $\mathcal{L} \cup \{\mathsf{v}_4\}$ **not** to be a basis of R^4, where $\mathsf{v}_4 = \begin{bmatrix}a\\b\\c\\d\end{bmatrix}$. So you are asked to complete the sentence,

   "$\mathcal{L} \cup \{\mathsf{v}_4\}$ is not a basis of $\mathsf{R}^3 \iff$ ...."

   (b) Explain how you would extend $\mathcal{L}$ to a basis of R^3.

8. [BB] Let W be the subspace of R^4 consisting of all vectors $\mathsf{x} = \begin{bmatrix}x_1\\x_2\\x_3\\x_4\end{bmatrix}$ that satisfy

$$\begin{aligned} x_1 + 2x_2 + 3x_3 + 5x_4 &= 0 \\ x_1 + x_3 + 3x_4 &= 0 \\ 2x_1 + 3x_2 + 5x_3 + 9x_4 &= 0. \end{aligned}$$

Find a basis for and the dimension of W.

9. Let V be a finite dimensional vector space. Suppose U is a subspace of V and $\dim U = \dim V$. Prove that $U = V$.

Critical Reading

10. Given four nonzero vectors v_1, v_2, v_3 and v_4 in a vector space V, suppose that $\{\mathsf{v}_1, \mathsf{v}_2, \mathsf{v}_3\}$ and $\{\mathsf{v}_2, \mathsf{v}_3, \mathsf{v}_4\}$ are linearly dependent sets of vectors, but $\{\mathsf{v}_1, \mathsf{v}_2, \mathsf{v}_4\}$ is a linearly independent set. Prove that v_2 is a multiple of v_3.

11. Let V be the vector space of all polynomials in the variable x with real coefficients. Prove that V does not have finite dimension.

12. [BB] Given that two vectors u and v span R^n, what can you say about n and why?

13. Let $\mathsf{w}, \mathsf{v}_1, \mathsf{v}_2, \mathsf{v}_3$ be vectors. State a condition under which $\operatorname{sp}\{\mathsf{v}_1, \mathsf{v}_2, \mathsf{v}_3\} = \operatorname{sp}\{\mathsf{w}, \mathsf{v}_1, \mathsf{v}_2, \mathsf{v}_3\}$.

14. [BB] **(a)** Suppose you wanted to exhibit three nonzero vectors in R^3 that span a plane. What would you do?

    (b) Suppose you were offered \$100 to produce three linearly independent vectors in R^3 that span a plane. What would you do?

15. Let V be a finite dimensional vector space. A set $S = \{\mathsf{v}_1, \mathsf{v}_2, \dots, \mathsf{v}_k\}$ of vectors is *maximal linearly independent* if it is a linearly independent set, but $S \cup \{\mathsf{v}\}$ is linearly **dependent** for any $\mathsf{v} \notin S$. Prove that a maximal linearly independent set of vectors in V is a basis for V.

16. Let V be a finite dimensional vector space. A set $S = \{\mathsf{v}_1, \mathsf{v}_2, \dots, \mathsf{v}_k\}$ of vectors that spans V is said to be a *minimal spanning set* if no subset formed by removing one or more vectors of S spans V. Prove that a minimal spanning set of vectors in V is a basis for V.

17. Suppose $\{\mathsf{u}_1, \mathsf{u}_2, \dots, \mathsf{u}_r\}$ and $\{\mathsf{w}_1, \mathsf{w}_2, \dots, \mathsf{w}_s\}$ are bases for vector spaces U and W respectively. Let V be the set of all ordered pairs (u, w) of vectors, u in U, w in W, with the agreement that $(\mathsf{u}_1, \mathsf{w}_1) = (\mathsf{u}_2, \mathsf{w}_2)$ if and only if $\mathsf{u}_1 = \mathsf{u}_2$ and $\mathsf{w}_1 = \mathsf{w}_2$. Then V is a vector space with addition and scalar multiplication defined by $(\mathsf{u}_1, \mathsf{w}_1) + (\mathsf{u}_2, \mathsf{w}_2) = (\mathsf{u}_1 + \mathsf{u}_2, \mathsf{w}_1 + \mathsf{w}_2)$ and $c(\mathsf{u}, \mathsf{w}) = (c\mathsf{u}, c\mathsf{w})$ for vectors $\mathsf{u}, \mathsf{u}_1, \mathsf{u}_2$ in U, $\mathsf{w}, \mathsf{w}_1, \mathsf{w}_2$ in W, and a scalar c (the proof is routine), but what is the dimension of V, and why?

4.4 Finite Dimensional Vector Spaces

In this section, we discuss further properties of linear independence, spanning sets and rank.

Linear Independence

Let V be a vector space of dimension n and let $\mathcal{V}$ be a basis for V. Then $\mathcal{V}$ contains n vectors. Let $\mathcal{L}$ be a linearly independent set of vectors. While Theorem 4.3.5 was proven just for subspaces of Euclidean space, we have pointed out that it holds in general, so $|\mathcal{L}| \leq |\mathcal{V}| = n$. This gives us information about the size of linearly independent and spanning sets.

4.4.1 If $\dim V = n$, any set of linearly independent vectors contains at most n vectors.

4.4.2 **Lemma.** *Suppose V is a finite dimensional vector space. Let v be a vector in V and let $\mathcal{L}$ be a set of linearly independent vectors in V. If v is not in the span of $\mathcal{L}$, then the set $\mathcal{L} \cup \{\mathsf{v}\}$ obtained by adding v to $\mathcal{L}$ is still linearly independent.*[8]

Proof. Let $\mathcal{L} = \{\mathsf{v}_1, \mathsf{v}_2, \ldots, \mathsf{v}_k\}$. We wish to show that

$$\mathcal{L} \cup \{\mathsf{v}\} = \{\mathsf{v}, \mathsf{v}_1, \mathsf{v}_2, \ldots, \mathsf{v}_k\}$$

is linearly independent. So suppose

$$c\mathsf{v} + c_1\mathsf{v}_1 + c_2\mathsf{v}_2 + \cdots + c_k\mathsf{v}_k = 0 \qquad (1)$$

for scalars $c, c_1, c_2, \ldots, c_k$. If $c \neq 0$, we can divide by c and rewrite (1) as

$$\mathsf{v} = -\frac{c_1}{c}\mathsf{v}_1 - \frac{c_2}{c}\mathsf{v}_2 - \cdots - \frac{c_k}{c}\mathsf{v}_k.$$

This expresses v as a linear combination of $\mathsf{v}_1, \ldots, \mathsf{v}_k$, which says that v is in the span of $\mathcal{L}$. This is not true, so $c = 0$ and (1) now reads

$$c_1\mathsf{v}_1 + c_2\mathsf{v}_2 + \cdots + c_k\mathsf{v}_k = 0.$$

Since the vectors in $\mathcal{L}$ are linearly independent, all the coefficients $c_i = 0$, so all the coefficients in (1) are 0. The vectors $\mathsf{v}, \mathsf{v}_1, \mathsf{v}_2, \ldots, \mathsf{v}_k$ are linearly independent, as claimed. ■

4.4.3 **Corollary.** *If V is a vector space of dimension n, then any set of n linearly independent vectors must span V and hence form a basis.*

Proof. Suppose $\mathcal{L}$ is a set of n linearly independent vectors. If $\mathcal{L}$ does not span V, there is a vector v not in the span of $\mathcal{L}$. The lemma says that $\mathcal{L} \cup \{\mathsf{v}\}$ is linearly independent, but there are $n+1$ vectors in this set and $\dim V = n$, so this cannot happen. ■

[8]The expression "$\mathcal{L} \cup \{\mathsf{v}\}$" is read "$\mathcal{L}$ *union* v"; it means the set consisting of all the vectors in $\mathcal{L}$ together with v.

4.4.4 Corollary. *If V is a finite dimensional vector space, any set of linearly independent vectors can be extended to a basis of V; that is, V has a basis containing the given vectors.*

Proof. Suppose V has dimension n and $\mathcal{L}$ is a set of linearly independent vectors in V. If $\mathcal{L}$ contains n vectors, then it is a basis containing $\mathcal{L}$ and we have the desired result. Otherwise, there are fewer than n vectors in $\mathcal{L}$, so $\mathcal{L}$ does not span V. Thus there exists a vector v not in the span of $\mathcal{L}$ and, by Lemma 4.4.2, $\mathcal{L} \cup \{\mathsf{v}\}$ is a linearly independent set. If this set contains n vectors, we have the result. Otherwise, we continue to add vectors until we get a set of n linearly independent vectors, which is a basis containing $\mathcal{L}$. ■

4.4.5 Problem. Show that $\begin{bmatrix}1\\2\\3\\0\end{bmatrix}, \begin{bmatrix}4\\6\\2\\1\end{bmatrix}, \begin{bmatrix}9\\10\\11\\12\end{bmatrix}, \begin{bmatrix}8\\-4\\23\\-2\end{bmatrix}$ comprise a basis for R^4.

Solution. Since $\dim \mathsf{R}^4 = 4$, Corollary 4.4.3 says it is sufficient to show that these four vectors are linearly independent. Let $A = \begin{bmatrix}1&4&9&8\\2&6&10&-4\\3&2&11&23\\0&1&12&-2\end{bmatrix}$ be the matrix whose columns are the given vectors. The columns are linearly independent if and only if $A\mathsf{x} = \mathsf{0}$ has only the trivial solution $\mathsf{x} = \mathsf{0}$. A row echelon form of A is $\begin{bmatrix}1&4&9&8\\0&-2&-8&-20\\0&0&24&99\\0&0&0&-45\end{bmatrix}$. Every column is a pivot column, there are no free variables, the system $A\mathsf{x} = \mathsf{0}$ has only the trivial solution. So the given four vectors are linearly independent in a vector space of dimension 4 and hence form a basis.

4.4.6 Example. Let $\mathsf{v}_1 = \begin{bmatrix}1\\2\\3\end{bmatrix}$ and $\mathsf{v}_2 = \begin{bmatrix}-1\\0\\1\end{bmatrix}$. The set $\mathcal{L} = \{\mathsf{v}_1, \mathsf{v}_2\}$ is linearly independent, so it can be extended to a basis of R^3, consisting necessarily of three vectors. Thus there exists a third vector v_3, which, together with v_1 and v_2, forms a linearly independent set (and hence a basis for R^3). In fact, there are many many possibilities, for example, $\mathsf{v}_3 = \begin{bmatrix}1\\0\\0\end{bmatrix}$. The reader should verify that $c_1\mathsf{v}_1 + c_2\mathsf{v}_2 + c_3\mathsf{v}_3 = \mathsf{0}$ implies $c_1 = c_2 = c_3 = 0$.

4.4.7 Example. The vectors $\mathsf{v}_1 = \begin{bmatrix}1\\0\\1\\1\end{bmatrix}$ and $\mathsf{v}_2 = \begin{bmatrix}0\\-1\\2\\1\end{bmatrix}$ are linearly independent in R^4, so the set $\{\mathsf{v}_1, \mathsf{v}_2\}$ can be extended to a basis. There are numerous ways to do this, for instance, by checking that $\{\mathsf{v}_1, \mathsf{v}_2, \mathsf{e}_1, \mathsf{e}_2\}$ is a basis of R^4.

We continue with some comments about spanning sets.

Spanning Sets

To begin, we return to the opening remarks of this chapter and note another consequence of $|\mathcal{L}| \leq |S|$.

4.4.8 If $\dim V = n$, any set that spans V contains at least n vectors.

4.4.9 **Lemma.** *Let V be a vector space and let S be a finite spanning set for V containing a certain vector v. If v is a linear combination of the vectors in $S \setminus \{\mathsf{v}\}$, then v is redundant, not needed; V is spanned by $S \setminus \{\mathsf{v}\}$.*[9]

Proof. We are given that S is a set of the form $S = \{\mathsf{v}, \mathsf{v}_1, \mathsf{v}_2, \dots, \mathsf{v}_k\}$ and that

$$\mathsf{v} = a_1\mathsf{v}_1 + a_2\mathsf{v}_2 + \cdots + a_k\mathsf{v}_k \tag{2}$$

is a linear combination of $\mathsf{v}_1, \dots, \mathsf{v}_k$. Every vector in V is a linear combination of the vectors in S and we are asked to show that v is redundant, that every vector in V is a linear combination of just $\mathsf{v}_1, \dots, \mathsf{v}_k$. Since S is a spanning set, a vector in V has the form

$$c\mathsf{v} + c_1\mathsf{v}_1 + c_2\mathsf{v}_2 + \cdots + c_k\mathsf{v}_k. \tag{3}$$

In this, if we replace v with the expression on the right of (2), we get a linear combination of just $\mathsf{v}_1, \mathsf{v}_2, \dots, \mathsf{v}_k$, so the linear combination in (3) becomes a linear combination of the vectors in $S \setminus \{\mathsf{v}\}$. This is exactly what we wanted! ■

4.4.10 **Corollary.** *If V is a vector space of dimension $n \geq 1$, any n vectors that span V must be linearly independent and hence form a basis.*

Proof. Suppose S is a set of n vectors that spans V. If these vectors are not linearly independent, one of them, call it v, is a linear combination of the others—remember Theorem 1.5.5. The Lemma says that $S \setminus \{\mathsf{v}\}$ still spans V. But there are only $n - 1$ vectors in this set, contradicting $\dim V = n$. ■

4.4.11 **Corollary.** *If a vector space V has dimension $n \geq 1$, then any spanning set of vectors contains a basis.*

[9] The expression $S \setminus \{\mathsf{v}\}$ is read "S minus v". As you might guess, it means the set obtained from S by removing v.

Proof. Suppose S is a set of vectors that spans V. Then S contains at least n vectors. If it contains exactly n vectors, it is a basis by the previous corollary. Otherwise it contains more than n vectors. These must be linearly dependent, so some vector in S, call it v, is a linear combination of the other vectors. Lemma 4.4.9 says that this vector is redundant: it can be removed from S and the smaller set $S \setminus \{\mathsf{v}\}$ still spans V. Continue in this way to remove vectors until a set of n spanning vectors is reached. Such a set is a basis contained in S. ■

4.4.12 **Problem.** Use Corollary 4.4.10 to give a different solution to Problem 4.4.5.

Solution. Since $\dim \mathsf{R}^4 = 4$, Corollary 4.4.10 says it is sufficient to show that these vectors span R^4. Make these vectors the columns of a matrix A and reduce A^T to row echelon form.

$$A^T = \begin{bmatrix} 1 & 2 & 3 & 0 \\ 4 & 6 & 2 & 1 \\ 9 & 10 & 11 & 12 \\ 8 & -4 & 23 & -2 \end{bmatrix} \to \begin{bmatrix} 1 & 2 & 3 & 0 \\ 0 & -2 & -10 & 1 \\ 0 & 0 & 3 & 1 \\ 0 & 0 & 0 & -45 \end{bmatrix} = U.$$

The rows of U are linearly independent and there are four of them, so the row space of U has dimension 4. The elementary row operations do not change the row space of a matrix, so the row space of A^T has dimension 4, so the transposes of these rows, namely the given vectors, span R^4.

4.4.13 **Problem.** Let $\mathsf{u}_1 = \begin{bmatrix} -1 \\ 2 \\ 1 \end{bmatrix}$, $\mathsf{u}_2 = \begin{bmatrix} -1 \\ 8 \\ 2 \end{bmatrix}$, $\mathsf{u}_3 = \begin{bmatrix} 3 \\ 0 \\ -2 \end{bmatrix}$, $\mathsf{u}_4 = \begin{bmatrix} 2 \\ -1 \\ -3 \end{bmatrix}$, let $S = \{\mathsf{u}_1, \mathsf{u}_2, \mathsf{u}_3, \mathsf{u}_4\}$ and let V be the subspace of R^3 spanned by S. Find a basis for V contained in S.

Solution. It is helpful first to determine the dimension of V and we do this by finding the row space of the matrix $A = \begin{bmatrix} -1 & 2 & 1 \\ -1 & 8 & 2 \\ 3 & 0 & -2 \\ 2 & -1 & -3 \end{bmatrix}$ whose rows are the transposes of the four given vectors. Gaussian elimination proceeds

$$A \to \begin{bmatrix} -1 & 2 & 1 \\ 0 & 6 & 1 \\ 0 & 6 & 1 \\ 0 & 3 & -1 \end{bmatrix} \to \begin{bmatrix} -1 & 2 & 1 \\ 0 & 3 & -1 \\ 0 & 0 & 3 \\ 0 & 0 & 0 \end{bmatrix} = U.$$

The nonzero rows of the row echelon matrix U are linearly independent—see Theorem 4.3.3—and they span the row space of A, so V has dimension 3. We can answer our question by finding three linearly independent vectors within S. Are $\mathsf{u}_1, \mathsf{u}_2, \mathsf{u}_3$ linearly independent? We solve $A\mathsf{x} = \mathsf{0}$ with $A = \begin{bmatrix} -1 & -1 & 3 \\ 2 & 8 & 0 \\ 1 & 2 & -2 \end{bmatrix}$, the matrix whose columns are these three vectors. Row echelon form is $\begin{bmatrix} 1 & 1 & -3 \\ 0 & 1 & 1 \\ 0 & 0 & 0 \end{bmatrix}$. There are nontrivial solutions, so the vectors $\mathsf{u}_1, \mathsf{u}_2, \mathsf{u}_3$ are linearly dependent.

What about u_1, u_2, u_4? Are these linearly independent? We solve $A\mathsf{x} = 0$ with $A = \begin{bmatrix} -1 & -1 & 2 \\ 2 & 8 & 1 \\ 1 & 2 & -3 \end{bmatrix}$. Row echelon form is $\begin{bmatrix} 1 & 1 & -2 \\ 0 & 1 & -1 \\ 0 & 0 & 1 \end{bmatrix}$. The only solution to $A\mathsf{x} = \mathsf{0}$ is $\mathsf{x} = \mathsf{0}$. Vectors $\mathsf{u}_1, \mathsf{u}_2, \mathsf{u}_4$ are linearly independent and hence a basis for V. We have shown that $\operatorname{sp} S = V = \mathsf{R}^3$ and that S contains the basis $\{\mathsf{u}_1, \mathsf{u}_2, \mathsf{u}_4\}$.

We continue by applying some of our results to matrices and, in particular, to learn more about the concept of "rank."

More about Rank

We begin with a seemingly "obvious" fact about the dimension of a subspace. Let V be a vector space of dimension n and let U be a subspace of V. If $\mathsf{u}_1, \mathsf{u}_2, \ldots, \mathsf{u}_k$ are vectors in U that are linearly independent in U, then they are certainly linearly independent as vectors in V. Thus $k \leq n$; that is, any set of linearly independent vectors in U consists of at most n vectors (see Exercise 15).

4.4.14 Proposition. *If U is a subspace of a finite dimensional vector space V, then U is also finite dimensional and* $\dim U \leq \dim V$.

For instance, since $\dim \mathsf{R}^3 = 3$, subspaces of R^3 have dimensions just 0, 1, 2 and 3. The only subspace of dimension 0 is the one element vector space $\{\mathsf{0}\}$; the only subspace of dimension 3 is R^3 itself—see Exercise 9. Subspaces of dimension 1 are lines through $\mathsf{0}$ while those of dimension 2 are planes that contain $\mathsf{0}$.

Recall that the rank of a matrix is its number of pivot columns. A related concept is "nullity."

4.4.15 Definition. The *nullity* of a matrix A is the dimension of its null space.

4.4.16 Example. Suppose $A = \begin{bmatrix} 1 & 2 & -1 & 1 & 3 \\ -3 & -6 & 4 & -13 & -17 \\ 4 & 8 & -6 & 25 & 29 \\ -1 & -2 & 0 & 10 & 6 \end{bmatrix}$. The nullity of A is the dimension of the null space, which is the set of solutions to $A\mathsf{x} = \mathsf{0}$. So we solve this homogeneous system and find $\mathsf{x} = \begin{bmatrix} -2t-4s \\ t \\ -2s \\ -s \\ s \end{bmatrix} = t \begin{bmatrix} -2 \\ 1 \\ 0 \\ 0 \\ 0 \end{bmatrix} + s \begin{bmatrix} -4 \\ 0 \\ -2 \\ -1 \\ 1 \end{bmatrix}$.

Every vector in the null space is a linear combination of $\mathsf{u} = \begin{bmatrix} -2 \\ 1 \\ 0 \\ 0 \\ 0 \end{bmatrix}$ and $\mathsf{v} = \begin{bmatrix} -4 \\ 0 \\ -2 \\ -1 \\ 1 \end{bmatrix}$, so these vectors span the null space. Moreover, u and v are linearly independent since

$$t\mathsf{u} + s\mathsf{v} = t \begin{bmatrix} -2 \\ 1 \\ 0 \\ 0 \\ 0 \end{bmatrix} + s \begin{bmatrix} -4 \\ 0 \\ -2 \\ -1 \\ 1 \end{bmatrix} = \begin{bmatrix} \star \\ t \\ \star \\ \star \\ s \end{bmatrix} = \begin{bmatrix} 0 \\ 0 \\ 0 \\ 0 \\ 0 \end{bmatrix}$$

certainly implies $t = s = 0$. (As on other occasions, we use $\star$ for entries that don't matter.) So u and v comprise a basis for the null space. The null space has dimension 2, the number of free variables in the solution to $A\mathsf{x} = 0$, which is the number of nonpivot columns of A. ☺

In this example, the matrix A has three pivot columns, so $\operatorname{rank} A = 3$. It is no coincidence that $\operatorname{rank} A + \operatorname{nullity} A = 5$, the number of columns of A, since this number is obviously the sum of the number of pivot columns and the number of nonpivot columns.

4.4.17 Theorem. *For any $m \times n$ matrix A, $\operatorname{rank} A + \operatorname{nullity} A = n$, the number of columns of A.*

4.4.18 Examples.

- Let $A = \begin{bmatrix} -3 & 6 \\ 2 & 5 \end{bmatrix}$. Row echelon form is $\begin{bmatrix} 1 & -2 \\ 0 & 9 \end{bmatrix}$, there are no free variables in the solution of $A\mathsf{x} = 0$, $\operatorname{nullity} A = 0$. There are two pivot columns, so $\operatorname{rank} A = 2$ and $\operatorname{rank} A + \operatorname{nullity} A = 2 + 0 = 2 = n$, the number of columns of A.
- Let $A = \begin{bmatrix} 1 & 2 & 3 \\ 4 & 5 & 6 \\ 7 & 8 & 9 \end{bmatrix}$. Row echelon form is $\begin{bmatrix} 1 & 2 & 3 \\ 0 & 1 & 2 \\ 0 & 0 & 0 \end{bmatrix}$, there is one free variable in the solution of $A\mathsf{x} = 0$, $\operatorname{nullity} A = 1$. There are two pivots, so $\operatorname{rank} A = 2$ and $\operatorname{rank} A + \operatorname{nullity} A = 2 + 1 = 3 = n$, the number of columns of A. ☺

Row Rank = Column Rank = Rank

In Section 4.2, we showed that the elementary row operations do not change the row space of a matrix. We also noted that the (nonzero) rows of a row echelon matrix are linearly independent, as are the pivot columns.

Suppose that Gaussian elimination moves a matrix A to a row echelon matrix U. The row space of A is the row space of U. Thus the nonzero rows of U span the row space of A and they are also linearly independent, so they are a basis for the row space. So the dimension of the row space is the number of nonzero rows, a number called the *row rank* of A. Notice that this number is also just the number of pivots of A, so the row rank of a matrix is the same as its rank.

4.4.19 Example. Let $A = \begin{bmatrix} 1 & 2 & -1 & 1 & 3 \\ -3 & -6 & 4 & -13 & -17 \\ 4 & 8 & -6 & 25 & 29 \\ -1 & -2 & 0 & 10 & 6 \end{bmatrix}$. A row echelon form is $U = \begin{bmatrix} 1 & 2 & -1 & 1 & 3 \\ 0 & 0 & 1 & -10 & -8 \\ 0 & 0 & 0 & 1 & 1 \\ 0 & 0 & 0 & 0 & 0 \end{bmatrix}$—see Example 4.4.16. The nonzero rows of U, namely,

$$\begin{bmatrix} 1 & 2 & -1 & 1 & 3 \end{bmatrix}, \quad \begin{bmatrix} 0 & 0 & 1 & -10 & -8 \end{bmatrix} \text{ and } \begin{bmatrix} 0 & 0 & 0 & 1 & 1 \end{bmatrix},$$

form a basis for the row space of A, so $\dim \text{row sp} A = 3 = \text{row rank } A = \text{rank} A$. ⌣̈

Now let's think about the column space of a matrix. We wish to show that the dimension of the column space, which is called the *column rank*, is also the number of pivot columns. This fact will tell us that

$$\boxed{\text{row rank } A = \text{column rank } A = \text{rank } A}$$

for any matrix A.

We can always find the column rank by reducing A^T to row echelon form and counting nonzero rows (as we have done several times previously), but there is a much easier way to find the column rank that we now present.

Let's think first about the special case where the matrix in question is an $m \times n$ matrix U in row echelon form. For example, suppose U is the matrix just considered in Example 4.4.19. There are three pivot columns—columns one, three and four—and these are linearly independent (by Theorem 4.3.3). We claim that these also span the column space, so that the column space has dimension 3. The point is that even though the columns of U are vectors in R^4, the last components are all 0, so we can think of them as belonging to R^3. Thus columns one, three and four of U are linearly independent vectors in a vector space of dimension 3. Corollary 4.4.3 says that these columns form a basis for the column space of U.

READING CHECK 1. What is the most obvious basis for the subspace of R^4 that consists of all vectors $\begin{bmatrix} x \\ y \\ z \\ 0 \end{bmatrix}$ with fourth component 0?

We argue similarly for any $m \times n$ row echelon matrix U. Suppose there are r pivot columns. These are linearly independent. Since U has m rows and r pivots, U has $m - r$ zero rows, so the columns of U are vectors in R^m whose last $m - r$ components are 0.

$$U = \begin{bmatrix} \star & \cdots & \star \\ \star & & \star \\ \vdots & & \vdots \\ \star & & \star \\ 0 & \cdots & 0 \\ \vdots & & \vdots \\ 0 & \cdots & 0 \end{bmatrix} \begin{matrix} \left.\vphantom{\begin{matrix} \star \\ \star \\ \vdots \\ \star \end{matrix}}\right\} r \text{ rows containing the pivots} \\ \left.\vphantom{\begin{matrix} 0 \\ \vdots \\ 0 \end{matrix}}\right\} m - r \text{ zero rows} \end{matrix}$$

Effectively, the columns of U are vectors in R^r. Thus, the pivot columns of U are r linearly independent vectors in a vector space of dimension r, so they comprise a basis for the column space. The dimension of the column space is r, the rank of A.

4.4.20 If U is in row echelon form, the pivot columns of U comprise a basis for the column space of U. The column rank of U is the rank of U.

Now let's think about the general situation where A is any $m \times n$ matrix. Let U be a row echelon form of A. The column spaces of A and U are usually quite different. For example, if $A = \begin{bmatrix} 1 & 2 \\ 2 & 4 \end{bmatrix}$, one elementary row operation carries A to row echelon form: $A = \begin{bmatrix} 1 & 2 \\ 2 & 4 \end{bmatrix} \to \begin{bmatrix} 1 & 2 \\ 0 & 0 \end{bmatrix} = U$. The column space of A is the set of scalar multiples of $\begin{bmatrix} 1 \\ 2 \end{bmatrix}$, but the column space of U is the set of scalar multiples of $\begin{bmatrix} 1 \\ 0 \end{bmatrix}$, a totally different vector space. Notice, however, that the column spaces of A and U each have basis column one, the same pivot column. We proceed to show that the pivot columns of A (which are in the same positions as the pivot columns of U, a row echelon form) are a basis for the column space of A.

Suppose that the pivot columns of U are columns $j_1, j_2, \ldots, j_r$. We have seen that these columns form a basis for the column space of U. We are going to show that columns $j_1, j_2, \ldots, j_r$ of A are a basis for the column space of A. Our argument depends on work in Section 2.7, where we saw that U can be written $U = MA$ with M invertible (M is the product of elementary matrices).

Suppose $A = \begin{bmatrix} \mathsf{a}_1 & \mathsf{a}_2 & \cdots & \mathsf{a}_n \\ \downarrow & \downarrow & \cdots & \downarrow \end{bmatrix}$ has columns $\mathsf{a}_1, \mathsf{a}_2, \ldots, \mathsf{a}_n$. We wish to show that the pivot columns $\mathsf{a}_{j_1}, \mathsf{a}_{j_2}, \ldots, \mathsf{a}_{j_r}$ are a basis for the column space of A. Now $U = MA = \begin{bmatrix} M\mathsf{a}_1 & M\mathsf{a}_2 & \cdots & M\mathsf{a}_n \\ \downarrow & \downarrow & \cdots & \downarrow \end{bmatrix}$, so the pivot columns of U are $M\mathsf{a}_{j_1}, M\mathsf{a}_{j_2}, \ldots, M\mathsf{a}_{j_r}$. We show that the pivot columns of A are linearly independent with the "classic" proof of linear independence, appealing directly to the definition.

Thus, we suppose

$$c_1\mathsf{a}_{j_1} + c_2\mathsf{a}_{j_2} + \cdots + c_r\mathsf{a}_{j_r} = 0$$

and multiply this equation by M. This gives

$$M(c_1\mathsf{a}_{j_1} + c_2\mathsf{a}_{j_2} + \cdots + c_r\mathsf{a}_{j_r}) = 0,$$

so

$$c_1(M\mathsf{a}_{j_1}) + c_2(M\mathsf{a}_{j_2}) + \cdots + c_r(M\mathsf{a}_{j_r}) = 0.$$

Now $M\mathsf{a}_{j_1}, M\mathsf{a}_{j_2}, \ldots, M\mathsf{a}_{j_r}$ are the pivot columns of U and these are linearly independent. Thus $c_1 = 0$, $c_2 = 0$, ..., $c_r = 0$. This proves that $\mathsf{a}_{j_1}, \ldots, \mathsf{a}_{j_r}$ are linearly independent. It remains to prove that these vectors also span the column space of A.

To do this, let y be any vector in the column space of A. We wish to show that y is a linear combination of $\mathsf{a}_{j_1}, \ldots, \mathsf{a}_{j_r}$. We know $\mathsf{y} = A\mathsf{x}$ for some vector x. Multiplying by M gives $M\mathsf{y} = MA\mathsf{x} = U\mathsf{x}$, a vector in the column space of U. The pivot columns of U—$M\mathsf{a}_{j_1}, M\mathsf{a}_{j_2}, \ldots, M\mathsf{a}_{j_r}$—span the column space of U, so there are scalars $c_1, \ldots, c_r$ so that

$$M\mathsf{y} = c_1M\mathsf{a}_{j_1} + c_2M\mathsf{a}_{j_2} + \cdots + c_rM\mathsf{a}_{j_r}.$$

The right side is $M(c_1\mathsf{a}_{j_1} + c_2\mathsf{a}_{j_2} + \cdots + c_r\mathsf{a}_{j_r})$, so

$$M\mathsf{y} = M(c_1\mathsf{a}_{j_1} + c_2\mathsf{a}_{j_2} + \cdots + c_r\mathsf{a}_{j_r}).$$

Since M is invertible, we can multiply by M^{-1} and obtain $\mathsf{y} = c_1\mathsf{a}_{j_1} + c_2\mathsf{a}_{j_2} + \cdots + c_r\mathsf{a}_{j_r}$. So y is in the span of $\mathsf{a}_{j_1}, \ldots, \mathsf{a}_{j_r}$, just what we wanted. We have proven that the pivot columns of A are linearly independent and span the column space of A, so they are a basis. There are r of them, so the column rank of A is r, the rank of A.

4.4.21 The pivot columns of any matrix A form a basis for the column space of A. The column rank of A is rank A.

4.4.22 Example. The matrix $A = \begin{bmatrix} 1 & 2 & -1 & 1 & 3 \\ -3 & -6 & 4 & -13 & -17 \\ 4 & 8 & -6 & 25 & 29 \\ -1 & -2 & 0 & 10 & 6 \end{bmatrix}$ of Example 4.4.16 has row echelon form $U = \begin{bmatrix} 1 & 2 & -1 & 1 & 3 \\ 0 & 0 & 1 & -10 & -8 \\ 0 & 0 & 0 & 1 & 1 \\ 0 & 0 & 0 & 0 & 0 \end{bmatrix}$. The pivot columns are columns one, three and four. Thus columns one, three and four of A, that is,

$$\begin{bmatrix} 1 \\ -3 \\ 4 \\ -1 \end{bmatrix}, \quad \begin{bmatrix} -1 \\ 4 \\ -6 \\ 0 \end{bmatrix} \quad \text{and} \quad \begin{bmatrix} 1 \\ -13 \\ 25 \\ 10 \end{bmatrix},$$

comprise a basis for the column space of A.

The column rank of a matrix A is the dimension of its column space. We have just seen that this number is r, the number of pivot columns of A. This number r is the rank of A and we saw earlier that this was also the row rank. This establishes what we've been after.

4.4.23 Theorem. *For any matrix A, row rank A = column rank A = rank A.*

That the row rank and the column rank of a matrix are equal is quite an astonishing fact when you think about it. If A is 15×28, for example, there seems to be no particular reason why the row space, a subspace of R^{28}, and the column space, a subspace of R^{15}, should have the same dimension, but they do!

4.4.24 Example. Let $A = \begin{bmatrix} 1 & 2 & 3 & -9 & -3 & 6 \\ -4 & -8 & 0 & 0 & -12 & 12 \\ 3 & 6 & 5 & -15 & -1 & 18 \\ -2 & -4 & 1 & -3 & -8 & 4 \end{bmatrix}$. We leave it to the reader to verify that a row echelon form of A is $U = \begin{bmatrix} 1 & 2 & 3 & -9 & -3 & 6 \\ 0 & 0 & 1 & -3 & -2 & 3 \\ 0 & 0 & 0 & 0 & 0 & 1 \\ 0 & 0 & 0 & 0 & 0 & 0 \end{bmatrix}$. There are three pivots, so the row rank of A is 3 and the nonzero rows of U comprise a basis for the row space of A, that is, the rows

$$\begin{bmatrix} 1 & 2 & 3 & -9 & -3 & 6 \end{bmatrix}, \quad \begin{bmatrix} 0 & 0 & 1 & -3 & -2 & -3 \end{bmatrix}, \text{ and } \begin{bmatrix} 0 & 0 & 0 & 0 & 0 & 1 \end{bmatrix}.$$

The pivot columns of A are columns one, three and six. Thus the first, third and sixth columns of A, $\begin{bmatrix} 1 \\ -4 \\ 3 \\ -2 \end{bmatrix}, \begin{bmatrix} 3 \\ 0 \\ 5 \\ 1 \end{bmatrix}$ and $\begin{bmatrix} 6 \\ 12 \\ 18 \\ 4 \end{bmatrix}$, are a basis for the column space of A. The column rank is 3, the same as the row rank.

Here's one place where equality of row and column rank is useful.

4.4.25 **Theorem.** *If A is an $m \times n$ matrix, then* $\operatorname{rank} A \le m$ *and* $\operatorname{rank} A \le n$.

Proof. The rank of A is the dimension of the row space of A. The row space of A is a subspace of the n-dimensional vector space $M_{1n}(\mathsf{R})$ of $1 \times n$ matrices which is just like R^n, but vectors are written as rows. By Proposition 4.4.14, $\operatorname{rank} A = \dim \operatorname{row} \operatorname{sp} A \le n$. The rank of A is also the dimension of the column space of A. The column space of A is a subspace of R^m so, by Proposition 4.4.14, the dimension of the column space of A is at most m; that is, $\operatorname{rank} A = \dim \operatorname{col} \operatorname{sp} A \le m$. ■

Because of Theorem 4.4.23, when we talk about the rank of a matrix A, we can think in terms of column rank or row rank, whichever we please, because these numbers are the same. Here is one cool consequence.

Suppose we have a matrix A. The rank of the transpose A^T is its column rank, the dimension of its column space. This column space is the row space of A, whose dimension is the row rank of A.

$$\operatorname{rank} A^T = \text{column rank } A^T = \text{row rank } A = \operatorname{rank} A.$$

We come upon a remarkable fact.

4.4.26 **Theorem.** *For any matrix A,* $\operatorname{rank} A = \operatorname{rank} A^T$.

We close this section with a second surprising consequence of "row rank equals column rank." If A is a 5×20 matrix, say, then $A^T A$ is 20×20 and one might expect that this matrix could have a rank as big as 20. This cannot happen, however.

4.4.27 **Theorem.** *For any matrix A,* $\operatorname{rank} A = \operatorname{rank} A^T A = \operatorname{rank} AA^T$.

Proof. First we show that $\operatorname{rank} A = \operatorname{rank} A^T A$. Suppose A is $m \times n$. Then both A and $A^T A$ have n columns, so rank + nullity $= n$ in each case (by Theorem 4.4.17). It follows that we can show A and $A^T A$ have the same rank by showing they have the same nullity. Thus we examine the null space of each matrix.

Suppose x is in the null space of A. Thus $A\mathsf{x} = 0$, so certainly $A^T A\mathsf{x} = 0$. This observation shows that

$$\text{null sp } A \subseteq \text{null sp } A^T A.$$[10]

On the other hand, if x is in the null space of $A^T A$, then $A^T A\mathsf{x} = 0$, so $\mathsf{x}^T A^T A\mathsf{x} = 0$; that is, $(A\mathsf{x})^T(A\mathsf{x}) = 0$. Remember that $\mathsf{u}^T\mathsf{u} = \mathsf{u} \cdot \mathsf{u} = \|\mathsf{u}\|^2$ for

[10]The symbol $\subseteq$ means "is contained in", or "is a subset of."

any vector u. Here, we have $\|A\mathsf{x}\|^2 = 0$, so $A\mathsf{x} = 0$. This shows that

$$\text{null sp}\, A^TA \subseteq \text{null sp}\, A$$

and gives null sp A = null sp A^TA. In particular, the dimensions of these two spaces are the same: nullity A = nullity A^TA, as desired.

The second equality of the theorem—rank A = rank AA^T—follows by applying what we have just shown to the matrix $B = A^T$, namely, rank B = rank B^TB. The theorem now follows from $B^TB = AA^T$ and rank B = rank A. ■

READING CHECK 2. If A is 5×20, what is the biggest possible rank of A^TA?

4.4.28 **Theorem (Matrix multiplication cannot increase rank).** *Suppose A is an $m \times n$ matrix, B is an $n \times p$ matrix and C is an $r \times m$ matrix. Then*

(a) rank $AB \le$ rank A, *and*

(b) rank $CA \le$ rank A.

Proof. (a) If y is in the column space of AB, then $\mathsf{y} = (AB)\mathsf{x}$ for some vector x. Since $(AB)\mathsf{x} = A(B\mathsf{x})$ is in the column space of A, we have col sp $AB \subseteq$ col sp A. Thus rank AB = dim col sp $AB \le$ dim col sp A = rank A.

(b) We have rank $CA = \text{rank}(CA)^T = \text{rank}\, A^TC^T \le \text{rank}\, A^T$ by part (a). Since rank A^T = rank A, we have the result. ■

Answers to Reading Checks

1. $\mathsf{e}_1 = \begin{bmatrix} 1 \\ 0 \\ 0 \\ 0 \end{bmatrix}, \mathsf{e}_2 = \begin{bmatrix} 0 \\ 1 \\ 0 \\ 0 \end{bmatrix}, \mathsf{e}_3 = \begin{bmatrix} 0 \\ 0 \\ 1 \\ 0 \end{bmatrix}.$

2. Theorem 4.4.27 says that rank A^TA = rank $A \le 5$.

True/False Questions

Decide, with as little calculation as possible, whether each of the following statements is true or false and, if you say "false," explain your answer. (Answers are at the back of the book.)

1. Given that the rows of $A = \begin{bmatrix} 1 & 1 & 1 \\ 1 & 2 & 3 \\ 1 & 3 & 6 \end{bmatrix}$ are linearly independent, the columns of A must be linearly independent too.

2. If the columns of an $n \times n$ matrix span R^n, the columns must be linearly independent.

3. A basis for the column space of a matrix can be determined by reducing it to row echelon form.

4. Given that $U = \begin{bmatrix} 1 & 2 & 3 & 4 \\ 0 & 1 & 2 & 3 \\ 0 & 0 & 0 & 0 \end{bmatrix}$ is a row echelon form of $A = \begin{bmatrix} 1 & 2 & 3 & 4 \\ 5 & 6 & 7 & 8 \\ 9 & 10 & 11 & 12 \end{bmatrix}$, the vectors $\begin{bmatrix} 1 \\ 0 \\ 0 \end{bmatrix}$ and $\begin{bmatrix} 2 \\ 1 \\ 0 \end{bmatrix}$ constitute a basis for the column space of A.

5. If matrix A can be changed to B by the elementary row operations, then the column spaces of A and B are the same.

6. If each column of a matrix is a multiple of the first, then each row of the matrix is a multiple of the first.

7. If each column of a matrix is a multiple of the first, then each row of the matrix is a multiple of some fixed row.

8. There exists a 5×8 matrix of rank 6.

9. If A is a 4×3 matrix, the maximum value for the rank of AA^T is 3.

10. If $A = \begin{bmatrix} 1 & 2 & 3 \\ 2 & 4 & 6 \\ 17 & 8 & 92 \end{bmatrix}$ and $B = \begin{bmatrix} 1 & 2 & 3 \\ 0 & 7 & 5 \\ 0 & 0 & 4 \end{bmatrix}$, then $\operatorname{rank} AB < \operatorname{rank} B$.

EXERCISES

Answers to exercises marked [BB] can be found at the ***Back of the Book.***

1. [BB] Let U be the span of the vectors $\mathsf{v}_1 = \begin{bmatrix} -1 \\ 0 \\ 1 \\ 1 \end{bmatrix}$, $\mathsf{v}_2 = \begin{bmatrix} 0 \\ 1 \\ 1 \\ 0 \end{bmatrix}$, $\mathsf{v}_3 = \begin{bmatrix} 2 \\ 1 \\ 1 \\ -1 \end{bmatrix}$, $\mathsf{v}_4 = \begin{bmatrix} -4 \\ -4 \\ -6 \\ 1 \end{bmatrix}$ and $\mathsf{v}_5 = \begin{bmatrix} 4 \\ 4 \\ -2 \\ -5 \end{bmatrix}$. Let A be the matrix with these vectors as columns. Use the principle stated in 4.4.21 to find a basis for U.

2. Repeat Exercise 1 with $\mathsf{v}_1 = \begin{bmatrix} -1 \\ 0 \\ 1 \\ 1 \end{bmatrix}$, $\mathsf{v}_2 = \begin{bmatrix} 0 \\ 1 \\ 1 \\ 0 \end{bmatrix}$, $\mathsf{v}_3 = \begin{bmatrix} 2 \\ 1 \\ 1 \\ -1 \end{bmatrix}$, $\mathsf{v}_4 = \begin{bmatrix} -4 \\ -4 \\ -6 \\ 1 \end{bmatrix}$ and $\mathsf{v}_5 = \begin{bmatrix} 4 \\ 4 \\ -1 \\ 2 \end{bmatrix}$.

3. Repeat Exercise 1 with $v_1 = \begin{bmatrix} 2 \\ 8 \\ 9 \\ -1 \end{bmatrix}$, $v_2 = \begin{bmatrix} 4 \\ 24 \\ 27 \\ -3 \end{bmatrix}$, $v_3 = \begin{bmatrix} 1 \\ 0 \\ 0 \\ 0 \end{bmatrix}$, $v_4 = \begin{bmatrix} 3 \\ -6 \\ 11 \\ -2 \end{bmatrix}$ and $v_5 = \begin{bmatrix} 4 \\ 5 \\ 2 \\ 6 \end{bmatrix}$.

4. In each of the following cases, you are given a set S of vectors. Let $V = \operatorname{sp} S$ be the span of S. Find a basis for V contained in S.

   (a) [BB] $S = \left\{ \begin{bmatrix} 1 \\ 2 \\ 3 \end{bmatrix}, \begin{bmatrix} 0 \\ 1 \\ 0 \end{bmatrix}, \begin{bmatrix} 4 \\ 5 \\ 6 \end{bmatrix}, \begin{bmatrix} 1 \\ 0 \\ 0 \end{bmatrix}, \begin{bmatrix} 0 \\ 0 \\ 1 \end{bmatrix} \right\}$

   (b) $S = \left\{ \begin{bmatrix} 1 \\ 1 \\ -1 \end{bmatrix}, \begin{bmatrix} 1 \\ 0 \\ -3 \end{bmatrix}, \begin{bmatrix} 0 \\ 1 \\ 2 \end{bmatrix}, \begin{bmatrix} 2 \\ 4 \\ 2 \end{bmatrix} \right\}$

   (c) [BB] $S = \left\{ \begin{bmatrix} -1 \\ 0 \\ 1 \\ 1 \end{bmatrix}, \begin{bmatrix} 0 \\ 1 \\ 1 \\ 0 \end{bmatrix}, \begin{bmatrix} 2 \\ 1 \\ 1 \\ -1 \end{bmatrix}, \begin{bmatrix} -4 \\ -4 \\ -6 \\ 1 \end{bmatrix}, \begin{bmatrix} 4 \\ 4 \\ -2 \\ -5 \end{bmatrix} \right\}$

   (d) $S = \left\{ \begin{bmatrix} 1 \\ 1 \\ -1 \end{bmatrix}, \begin{bmatrix} 1 \\ 1 \\ 11 \end{bmatrix}, \begin{bmatrix} 1 \\ 1 \\ 2 \end{bmatrix}, \begin{bmatrix} 2 \\ 4 \\ 1 \end{bmatrix} \right\}$

   (e) $S = \left\{ \begin{bmatrix} 1 \\ 1 \\ 1 \\ 1 \end{bmatrix}, \begin{bmatrix} -3 \\ -1 \\ 1 \\ -1 \end{bmatrix}, \begin{bmatrix} 3 \\ 2 \\ 1 \\ 2 \end{bmatrix}, \begin{bmatrix} 0 \\ -4 \\ -3 \\ 1 \end{bmatrix} \right\}$

   (f) $S = \left\{ \begin{bmatrix} 1 \\ 8 \\ 9 \\ 16 \end{bmatrix}, \begin{bmatrix} 2 \\ 7 \\ 10 \\ 15 \end{bmatrix}, \begin{bmatrix} 3 \\ 6 \\ 11 \\ 14 \end{bmatrix}, \begin{bmatrix} 4 \\ 5 \\ 12 \\ 13 \end{bmatrix} \right\}$.

5. [BB] Let $S = \left\{ \begin{bmatrix} 1 \\ -1 \\ 3 \\ 2 \end{bmatrix}, \begin{bmatrix} 2 \\ 1 \\ 1 \\ 3 \end{bmatrix}, \begin{bmatrix} 1 \\ 5 \\ -7 \\ 0 \end{bmatrix}, \begin{bmatrix} 4 \\ -1 \\ 7 \\ 7 \end{bmatrix} \right\}$ and let $V = \operatorname{sp} S$.

   (a) Find a basis for and the dimension of V.

   (b) Find all subsets of S that are bases for V. Explain.

6. Answer Exercise 5 again with $S = \left\{ \begin{bmatrix} -1 \\ 0 \\ 2 \\ 1 \end{bmatrix}, \begin{bmatrix} 0 \\ 0 \\ 1 \\ 1 \end{bmatrix}, \begin{bmatrix} 1 \\ 1 \\ 5 \\ 2 \end{bmatrix}, \begin{bmatrix} 3 \\ 1 \\ 1 \\ 0 \end{bmatrix} \right\}$.

7. **(a)** [BB] If V is a vector space of dimension n, a set of n vectors is linearly independent if and only if it spans. Why?

   (b) If A is an $n \times n$ matrix, Theorem 4.1.17 says that $A\mathsf{x} = \mathsf{0}$ has only the solution $\mathsf{x} = \mathsf{0}$ if and only if $A\mathsf{x} = \mathsf{b}$ has a unique solution for any b. Explain the connection with part (a).

8. [BB] This exercise is intended to help you come to grips with the important fact labelled 4.4.21. Please do not use the fact to answer any part of this question.

   (a) Find a row echelon form U of the matrix $A = \begin{bmatrix} 1 & -1 & 0 & 2 & 3 & 1 \\ 3 & -2 & 4 & 4 & 11 & 6 \\ 5 & -4 & 4 & 9 & 12 & 9 \\ 2 & -1 & 4 & 2 & 8 & 5 \end{bmatrix}$.

   (b) Find a basis for the row space of U and a basis for the row space of A. Explain your answer.

   (c) What is the row rank of U? What is the row rank of A? Explain.

   (d) What are the pivot columns of U? Use the definition to show that these columns are linearly independent.

   (e) Let the columns of A be $\mathsf{a}_1, \mathsf{a}_2, \mathsf{a}_3, \ldots$ in order. Which are the pivot columns? Use the definition to show that these columns are also linearly independent.

   (f) Show that each of the "other" columns of A (those that are not pivot columns) is in the span of the pivot columns. (Be explicit.) Thus, the pivot columns of A span the column space of A and hence form a basis.

   (g) What is the dimension of the column space of A? (This shows that row rank = column rank for this matrix.)

   (h) Why is $\operatorname{rank} A^T = \operatorname{rank} A$?

   (i) What is the nullity of A? Why?

9. Answer Exercise 8 again using the matrix $A = \begin{bmatrix} 1 & -3 & 2 & 4 & 5 \\ 0 & 1 & 1 & 4 & 3 \\ 2 & -9 & 1 & -4 & 2 \\ -1 & 6 & 1 & 8 & 4 \end{bmatrix}$.

10. [BB] Let $A = \begin{bmatrix} -1 & 0 & -1 & 1 \\ 0 & 1 & -1 & 2 \\ 1 & 1 & 0 & 1 \\ 2 & -1 & 3 & 1 \\ 0 & -1 & 1 & 0 \end{bmatrix}$.

    (a) Find a basis for the row space of A. What's $\dim \operatorname{row\,sp} A$?

    (b) Exhibit some columns of A that form a basis for the column space. What's $\dim \operatorname{col\,sp} A$?

    (c) Find a basis of the null space of A. What is the nullity of A?

11. Answer Exercise 10 again with $A = \begin{bmatrix} 11 & 8 & -2 & 3 \\ 2 & 3 & -1 & 2 \\ 7 & -15 & 7 & -17 \\ 4 & -11 & 5 & -12 \end{bmatrix}$.

12. Suppose A and U are $m \times n$ matrices and E is an invertible $m \times m$ matrix such that $U = EA$.

    (a) If the columns of U are linearly independent, show that the columns of A are linearly independent [BB], and conversely.

    (b) If the columns of U span R^m, show that the columns of A span R^m, and [BB] conversely.

13. Let the vectors $v_1, v_2, \dots, v_n$ be a basis of R^n and let A be an $n \times n$ matrix.

   (a) [BB] Prove that the $n \times n$ matrix B whose columns are $v_1, v_2, \dots, v_n$ is invertible.

   (b) Suppose $\{Av_1, Av_2, \dots, Av_n\}$ is a basis of R^n. Prove that A is invertible.

   (c) Establish the converse of (b). If A is invertible, then the set $\{Av_1, \dots, Av_n\}$ is a basis for R^n.

14. For each of the following matrices A,

   i. find a basis for the null space and the nullity;

   ii. find a basis for the row space and the row rank;

   iii. find a basis for the column space and the column rank;

   iv. verify that rank A + nullity A is the number of columns of A.

   (a) [BB] $A = \begin{bmatrix} 1 & 1 \\ 1 & -1 \\ 2 & 0 \\ 1 & 2 \end{bmatrix}$ **(b)** $A = \begin{bmatrix} 1 & 2 & 3 \\ 3 & 5 & 7 \\ 1 & 1 & 1 \end{bmatrix}$

   (c) $A = \begin{bmatrix} 1 & 2 & 3 & -1 \\ 2 & 4 & 6 & 4 \end{bmatrix}$ **(d)** [BB] $A = \begin{bmatrix} 1 & 0 & 1 & 1 \\ 1 & 1 & 2 & 0 \\ 0 & 1 & 1 & -1 \end{bmatrix}$

   (e) $A = \begin{bmatrix} 1 & 1 & -2 & 0 \\ 3 & -1 & 2 & -4 \\ -1 & 2 & -4 & 3 \end{bmatrix}$ **(f)** $A = \begin{bmatrix} 2 & 3 & 4 \\ 0 & 7 & 10 \\ -1 & 2 & 3 \\ 4 & -1 & -2 \end{bmatrix}$

   (g) $A = \begin{bmatrix} 1 & 0 & 3 & 0 & 1 \\ 2 & 5 & -1 & 5 & -4 \\ -3 & 2 & 5 & 2 & 3 \\ 2 & -5 & 13 & -5 & 8 \end{bmatrix}$

   (h) $A = \begin{bmatrix} 1 & -3 & 1 & -1 & 0 & -1 & 2 \\ -1 & 3 & 0 & 3 & 1 & 3 & -7 \\ 2 & -6 & 3 & 0 & -1 & 2 & 7 \\ -1 & 3 & 1 & 5 & 2 & 6 & -3 \\ 2 & -6 & 5 & 4 & 7 & 0 & -27 \end{bmatrix}.$

15. Suppose A is a 7×12 matrix. What are the sizes of A^TA and AA^T? What are the possible values for the rank of each matrix? Explain your answer.

16. Let A be an $m \times n$ matrix with linearly independent columns.

   (a) Why are the columns of A^TA linearly independent?

   (b) Is A^TA invertible? Explain.

   (c) Is AA^T invertible? Explain.

17. Suppose $v_1, \dots, v_r$ span a vector space V. Write a short note explaining why some subset of these vectors forms a basis. [Hint: Review the proof of Corollary 4.4.11.]

Critical Reading

18. Prove that multiplication by an **invertible** matrix preserves rank; that is, if A is invertible and B is any matrix for which AB is defined, then $\operatorname{rank} AB = \operatorname{rank} B$.

19. Suppose B is an $m \times n$ matrix, C is an $n \times m$ matrix and $m > n$.

    (a) Can BC be invertible?

    (b) What is $\det BC$?

20. Suppose A is an $m \times n$ matrix with linearly independent rows and linearly independent columns. Prove that $m = n$.

21. Suppose $\{u_1, u_2, \dots, u_r\}$ and $\{w_1, w_2, \dots, w_s\}$ are linearly independent sets of vectors in a vector space V, each of which can be extended to a basis of V with the same set of vectors $\{x_1, x_2, \dots, x_t\}$. Explain why $r = s$.

22. Suppose A and B are matrices, B has rank 1 and AB is defined. Explain why $\operatorname{rank} AB \le 1$.

23. [BB] If every row of a nonzero 7×7 matrix A is a multiple of the first, explain why every column of A is a multiple of one of its columns.

24. Show that the system $Ax = b$ has a solution if and only if the rank of A is the rank of the augmented matrix $[A \mid b]$.

25. **(a)** [BB] Let x and y be nonzero vectors in R^n, thought of as column matrices. What is the size of the matrix xy^T? Show that this matrix has rank 1.

    (b) Suppose A is an $m \times n$ matrix of rank 1. Show that there are column matrices x and y such that $A = xy^T$. [Hint: Think about the columns of A.]
    Remark. Note the difference between x^Ty and xy^T. The 1×1 matrix x^Ty is a scalar, the dot product of x and y, also known as an *inner product*. The matrix xy^T (which is not a scalar unless $n = 1$) is called the *outer product* of x and y.

    (c) Let A be the 2×2 matrix with columns the vectors $a_1 = \begin{bmatrix} a_{11} \\ a_{21} \end{bmatrix}$ and $a_2 = \begin{bmatrix} a_{12} \\ a_{22} \end{bmatrix}$. Let B be the 2×2 matrix with rows the transposes of $b_1 = \begin{bmatrix} b_{11} \\ b_{12} \end{bmatrix}$ and $b_2 = \begin{bmatrix} b_{21} \\ b_{22} \end{bmatrix}$. Show that $AB = a_1 b_1^T + a_2 b_2^T$ is the sum of outer products of the columns of A and the rows of B.

26. Let $A = CR$ where C is a nonzero $m \times 1$ (column) matrix and R is a nonzero $1 \times n$ (row) matrix.

    (a) Show that C is a basis for $\operatorname{col\,sp} A$.

    (b) Find $\operatorname{nullity} A$ (and explain).

    (c) Show that $\operatorname{null\,sp} A = \operatorname{null\,sp} R$.

27. If matrices B and AB have the same rank, prove that they have the same null spaces. [Hint: First prove that $\operatorname{null\,sp} B$ is contained in $\operatorname{null\,sp} AB$.]

28. If A is an $n \times n$ matrix of rank r and $A^2 = 0$, explain why $r \le \frac{n}{2}$.

29. Suppose an $n \times n$ matrix A satisfies $A^2 = A$. Show that $\operatorname{rank}(I-A) = n - \operatorname{rank} A$, where I denotes the $n \times n$ identity matrix.

30. Let A be an $(n-k) \times k$ matrix, I_k the $k \times k$ identity matrix, I_{n-k} the $(n-k) \times (n-k)$ identity matrix, $G = \begin{bmatrix} I_k \\ A \end{bmatrix}$ and $H = [-A \; I_{n-k}]$.

    (a) It is a fact that $HG = 0$ is the zero matrix. Convince yourself of this in the case that $n = 5$ and $k = 3$.

    (b) Prove that the column space of G is contained in the null space of H.

    (c) Prove that the column space of G **is** the null space of H. [Hint: What is rank G? What is rank H?]

    (This exercise provides an easy way to obtain a parity check matrix H from a generator matrix G for a code. See Section 2.1.)

4.5 One-sided Inverses

4.5.1 Definition. A *right inverse* of an $m \times n$ matrix A is an $n \times m$ matrix B such that $AB = I_m$, the $m \times m$ identity matrix. A *left inverse* of A is an $n \times m$ matrix C such that $CA = I_n$, the $n \times n$ identity matrix.

4.5.2 Example. Let $A = \begin{bmatrix} 1 & 2 & 3 \\ 4 & 5 & 6 \end{bmatrix}$ and $B = \begin{bmatrix} -\frac{2}{3} & -\frac{1}{3} \\ -\frac{2}{3} & \frac{5}{3} \\ 1 & -1 \end{bmatrix}$. The reader may check that $AB = I$. Thus B is a right inverse of A and A is a left inverse of B. Since the matrix $C = \begin{bmatrix} -\frac{5}{3} & \frac{8}{3} \\ \frac{4}{3} & -\frac{13}{3} \\ 0 & 2 \end{bmatrix}$ is also a right inverse for A, it is apparent that the right inverse of a matrix is not necessarily unique. (In fact, the right or left inverse of a matrix which is not square is never unique. See Exercise 11.)

READING CHECK 1. If a matrix A has both a right inverse X and a left inverse Y, prove that $X = Y$. [Hint: Compute YAX in two different ways.]

In Theorem 4.1.17, we gave one condition equivalent to "rank $A = m$" for an $m \times n$ matrix A. Here are some more.

4.5.3 Theorem. *Let A be an $m \times n$ matrix. Then the following statements are equivalent.*

1. *A has a right inverse.*
2. *The columns of A span R^m.*

3. rank $A = m$.

4. *The rows of A are linearly independent.*

Proof. (1 $\implies$ 2) There exists a matrix B with $AB = I$. Now given any y in R^m, $\mathsf{y} = (AB)\mathsf{y} = A(B\mathsf{y})$ has the form $A\mathsf{x}$, so it is in the column space of A.

(2 $\implies$ 3) Assume the columns of A span R^m. Then the column space of A is R^m, so the column rank is m. Therefore rank $A = m$.

(3 $\implies$ 4) The row space of A has dimension m and is spanned by the m rows, so these are linearly independent by Corollary 4.4.10.

(4 $\implies$ 1) The row space of A is spanned by m linearly independent rows, so the rows are a basis for the row space. The dimension of the row space is m, so rank $A = m$. In particular, the column rank is m, so the column space is R^m. This means that we can solve the equation $A\mathsf{x} = \mathsf{y}$ for any y in R^m. Solving this equation with $\mathsf{y} = \mathsf{e}_1$, then $\mathsf{y} = \mathsf{e}_2$ and finally $\mathsf{y} = \mathsf{e}_m$, gives us m vectors $\mathsf{x}_1, \mathsf{x}_2, \ldots, \mathsf{x}_m$ with $A\mathsf{x}_i = \mathsf{e}_i$. Putting these vectors into the columns of a matrix $B = \begin{bmatrix} \mathsf{x}_1 & \mathsf{x}_2 & \cdots & \mathsf{x}_m \\ \downarrow & \downarrow & & \downarrow \\ & & & \end{bmatrix}$, we have $AB = I$, so B is a right inverse of A. ■

4.5.4 Example. Let's look again at the matrix $A = \begin{bmatrix} 1 & 2 & 3 \\ 4 & 5 & 6 \end{bmatrix}$ that appeared in Example 4.5.2. Since the rows of A are linearly independent, A has a right inverse by Theorem 4.5.3. Let $B = \begin{bmatrix} x_1 & x_2 \\ y_1 & y_2 \\ z_1 & z_2 \end{bmatrix}$ denote such a right inverse. The matrix equation $AB = \begin{bmatrix} 1 & 0 \\ 0 & 1 \end{bmatrix}$ leads to the two systems

$$A\begin{bmatrix} x_1 \\ y_1 \\ z_1 \end{bmatrix} = \begin{bmatrix} 1 \\ 0 \end{bmatrix} \quad \text{and} \quad A\begin{bmatrix} x_2 \\ y_2 \\ z_2 \end{bmatrix} = \begin{bmatrix} 0 \\ 1 \end{bmatrix},$$

which we can solve simultaneously:

$$\left[\begin{array}{ccc|cc} 1 & 2 & 3 & 1 & 0 \\ 4 & 5 & 6 & 0 & 1 \end{array}\right] \to \left[\begin{array}{ccc|cc} 1 & 2 & 3 & 1 & 0 \\ 0 & -3 & -6 & -4 & 1 \end{array}\right] \to \left[\begin{array}{ccc|cc} 1 & 2 & 3 & 1 & 0 \\ 0 & 1 & 2 & \frac{4}{3} & -\frac{1}{3} \end{array}\right].$$

In the first system, $z_1 = t$ is free,

$$y_1 = \tfrac{4}{3} - 2z_1 = \tfrac{4}{3} - 2t,$$

and

$$x_1 = 1 - 2y_1 - 3z_1 = 1 - 2(\tfrac{4}{3} - 2t) - 3t = -\tfrac{5}{3} + t$$

so that $\begin{bmatrix} x_1 \\ y_1 \\ z_1 \end{bmatrix} = \begin{bmatrix} -\frac{5}{3} + 2t \\ \frac{4}{3} - 2t \\ t \end{bmatrix}$. In the second system, $z_2 = s$ is free,

$$\begin{aligned} y_2 &= -\tfrac{1}{3} - 2z_2 = -\tfrac{1}{3} - 2s, \\ x_2 &= -2y_2 - 3z_2 = -2(-\tfrac{1}{3} - 2s) - 3s = \tfrac{2}{3} + s \end{aligned}$$

so that $\begin{bmatrix} x_2 \\ y_2 \\ z_2 \end{bmatrix} = \begin{bmatrix} \frac{2}{3} + s \\ -\frac{1}{3} - 2s \\ s \end{bmatrix}$.

The most general right inverse for A is $\begin{bmatrix} t - \frac{5}{3} & s + \frac{2}{3} \\ \frac{4}{3} - 2t & -\frac{1}{3} - 2s \\ t & s \end{bmatrix}$. :)

READING CHECK 2. Show that $\frac{1}{18}\begin{bmatrix} -17 & 8 \\ -2 & 2 \\ 13 & -4 \end{bmatrix}$ is a particular case of the "general" right inverse just found.

In Theorem 4.1.17, we gave a condition equivalent to "rank $A = n$" for an $m \times n$ matrix A. Here are some more.

4.5.5 Theorem. *Let A be an $m \times n$ matrix. Then the following statements are equivalent.*

1. *A has a left inverse.*
2. *The columns of A are linearly independent.*
3. rank $A = n$.
4. *The rows of A span R^n.*

Proof. This theorem follows more or less directly from Theorem 4.5.3 and the facts $(AB)^T = B^T A^T$ and rank A = rank A^T.

($1 \implies 2$) Given a matrix B with $BA = I$, we have $(BA)^T = I^T = I$, so $A^T B^T = I$. Thus A^T has a right inverse. Using Theorem 4.5.3, we know that the rows of A^T are linearly independent, but these are the columns of A.

($2 \implies 3$) Assume the columns of A are linearly independent. Then the rows of A^T are linearly independent, so rank $A^T = n$, the number of rows of A^T, using Theorem 4.5.3. Therefore rank A = rank $A^T = n$.

($3 \implies 4$) Given rank $A = n$, we have rank $A^T = n$, the number of rows of A^T. Theorem 4.5.3 says that R^n is spanned by the columns of A^T, which are the rows of A.

($4 \implies 1$) Given that the rows of A span R^n, it follows that the columns of A^T span R^n so A^T has a right inverse. But $A^TB = I$ implies $B^TA = I$, so A has a left inverse. ■

4.5.6 Example. Let $A = \begin{bmatrix} 1 & 4 \\ 2 & 5 \\ 3 & 6 \end{bmatrix}$. Since rank $A = 2$, the number of columns, A has a left inverse C by Theorem 4.5.5. From $CA = I$, we get $A^TC^T = I$, so C^T is a right inverse of A^T. In Example 4.5.4, we found that the most general right inverse is $\begin{bmatrix} -\frac{5}{3} + t & \frac{2}{3} + s \\ \frac{4}{3} - 2t & -\frac{1}{3} - 2s \\ t & s \end{bmatrix}$. Setting this equal to C^T, we get $C = \begin{bmatrix} -\frac{5}{3} + t & \frac{4}{3} - 2t & t \\ \frac{2}{3} + s & -\frac{1}{3} - 2s & s \end{bmatrix}$, which is the most general left inverse of A. :-)

Theorems 4.5.3 and 4.5.5 have several interesting consequences.

In Section 2.3, we said that a square $n \times n$ matrix A is *invertible* if there is another matrix B such that $AB = BA = I$, that is, if A has both a left and a right inverse and these are the same. We said that only square matrices are invertible, but didn't explain why at the time.

4.5.7 Corollary. *If a matrix has both a right and a left inverse, then the matrix is square and hence invertible.*

Proof. Suppose A is an $m \times n$ matrix with both a right inverse and a left inverse. Since A has a right inverse, rank $A = m$ by Theorem 4.5.3. Since A has a left inverse, rank $A = n$ by Theorem 4.5.5. Thus $m = n$. Finally, we have matrices B and C with $AB = CA = I_n$, so $B = (CA)B = C(AB) = C$. So A is square with the same inverse on each side, thus invertible. ■

4.5.8 Corollary. *If A is a square matrix, then A has a right inverse if and only if A has a left inverse, these are necessarily equal, so A is invertible.*

Proof. Suppose A is $n \times n$. Then A has a right inverse if and only if rank $A = n$ by Theorem 4.5.3. By Theorem 4.5.5, this occurs if and only if A has a left inverse. By READING CHECK 1 or the last sentence of the proof of Corollary 4.5.7, these one-sided inverses are equal. ■

4.5.9 Corollary. *An $n \times n$ matrix A is invertible if and only if it has rank n.*

Proof. ($\implies$) Assume A is invertible. In particular, A has a left inverse, so rank $A = n$ by Theorem 4.5.5.

($\impliedby$) Assume rank $A = n$. Then A has a right inverse by Theorem 4.5.3 and a left inverse by Theorem 4.5.5, so it's invertible by Corollary 4.5.8. ■

4.5.10 Corollary. *Let A be an $m \times n$ matrix. If the rows of A are linearly independent, then AA^T is invertible. If the columns of A are linearly independent, then A^TA is invertible.*

Proof. We use the fact that $\operatorname{rank} A = \operatorname{rank} AA^T = \operatorname{rank} A^TA$—see Theorem 4.4.27. If the rows of A are linearly independent, $\operatorname{rank} A = m = \operatorname{rank} AA^T$. Since AA^T is $m \times m$, AA^T is invertible. If the columns of A are linearly independent, $\operatorname{rank} A = n = \operatorname{rank} A^TA$. Since A^TA is $n \times n$, A^TA is invertible. ■

In Section 2.3, we said without proof that if a matrix A is square and $AB = I$, then $BA = I$. Here's the proof.

In Theorem 4.1.17, we listed a number of conditions each equivalent to the statement that an $n \times n$ matrix has rank n. Here are some more.

4.5.11 Corollary. *Let A be an $n \times n$ matrix. Then the following statements are equivalent.*

1. *The rows of A are linearly independent.*
2. *The rows of A span* R^n.
3. *The columns of A are linearly independent.*
4. *The columns of A span* R^n.
5. *A is invertible.*
6. $\operatorname{rank} A = n$.

Proof. Statements 1, 4 and 6 are equivalent by Theorem 4.5.3. Statements 2, 3 and 6 are equivalent by Theorem 4.5.5. Corollary 4.5.9 give the equivalence of 5 and 6, hence the theorem. ■

Answers to Reading Checks

1. Following the hint, we have $(YA)X = IX = X$ and $Y(AX) = YI = Y$. Since $(YA)X = Y(AX)$ (matrix multiplication is associative), $X = Y$.
2. Set $t = \frac{13}{18}$ and $s = -\frac{4}{18}$ in the general right inverse.

True/False Questions

Decide, with as little calculation as possible, whether each of the following statements is true or false and, if you say "false," explain your answer. (Answers are at the back of the book.)

1. If A and B are matrices and $\operatorname{rank} BA = \operatorname{rank} A$, then B is invertible.

2. If the equation $A\mathsf{x} = \mathsf{b}$ has a solution for any vector b, the matrix A has a right inverse.

3. If the rows of an $m \times n$ matrix A are linearly independent, the columns of A span R^m.

4. If the rows of an $m \times n$ matrix A span R^n, the columns of A are linearly independent.

5. If a matrix A has a left inverse, the system $A\mathsf{x} = \mathsf{b}$ has a unique solution.

6. The matrix $A = \begin{bmatrix} 1 & -1 & 2 \\ -1 & 2 & 0 \end{bmatrix}$ has a right inverse.

7. The matrix $A = \begin{bmatrix} 1 & 3 & 7 \\ 9 & -10 & 8 \end{bmatrix}$ has a left inverse.

8. If A is $m \times n$, B is $n \times t$ and $AB = cI$ for some $c \neq 0$, then $\operatorname{rank} A = \operatorname{rank} B$.

9. If a matrix has a right inverse but not a left inverse then it cannot be square.

10. If the columns of a matrix A are linearly independent, then $A^T A$ is invertible.

EXERCISES

*Answers to exercises marked [BB] can be found at the **Back of the Book**.*

1. [BB] The matrix $A = \begin{bmatrix} 1 & 1 & -2 & 0 \\ 3 & -1 & 2 & -4 \\ -1 & 2 & -4 & 3 \end{bmatrix}$ has neither a right nor a left inverse. Why?

2. The matrix $A = \begin{bmatrix} 1 & 2 & 1 \\ 3 & 7 & 4 \\ 2 & 5 & 3 \\ 1 & 3 & 2 \end{bmatrix}$ has neither a right nor a left inverse. Why?

3. Answer these questions for each matrix.

i. Does A have a one-sided inverse? Which side? Why?
ii. Does A has a left inverse and a right inverse? Explain.
iii. Find the most general one-sided inverse of A.
iv. Give a specific example of a one-sided inverse.

(a) [BB] $A = \begin{bmatrix} -1 & 2 & 1 \\ 3 & 1 & 0 \end{bmatrix}$ **(b)** [BB] $A = \begin{bmatrix} 1 & 1 \\ -1 & -2 \\ 2 & 4 \end{bmatrix}$ **(c)** $A = \begin{bmatrix} 1 & 3 \\ 2 & 4 \\ 6 & 22 \end{bmatrix}$

(d) $A = \begin{bmatrix} 1 & 2 & 3 \\ -2 & 1 & 4 \end{bmatrix}$ **(e)** $A = \begin{bmatrix} 1 & 3 & -2 \\ 3 & 1 & 2 \end{bmatrix}$ **(f)** $A = \begin{bmatrix} 1 & 3 \\ 3 & 1 \\ -2 & 2 \end{bmatrix}$

(g) $A = \begin{bmatrix} 1 & 0 & 3 & 4 \\ -1 & 1 & 4 & 0 \end{bmatrix}$ **(h)** $A = \begin{bmatrix} 3 & 12 & -6 & 0 \\ 3 & 13 & -3 & -1 \\ -5 & -18 & 16 & -6 \end{bmatrix}$

4. Let A be an $m \times n$ matrix, B an invertible $n \times n$ matrix and C an invertible $m \times m$ matrix.

   (a) [BB] Explain why rank AB = rank A.

   (b) Explain why rank CA = rank A.

5. Let A be an $m \times n$ and B an $n \times t$ matrix.

   (a) [BB] If A and B each have a left inverse, show that AB also has a left inverse.

   (b) True or false: If A and B have linearly independent columns, so does AB. Justify your answer.

6. Suppose A is an $m \times n$ matrix and B is an $n \times t$ matrix.

   (a) If A and B each have a right inverse, show that AB also has a right inverse.

   (b) True or false: If the columns of A span R^m and the columns of B span R^n, then the columns of AB span R^m. Justify your answer.

7. Is the converse of each statement of Corollary 4.5.10 true? To be specific:

   (a) [BB] If A is $m \times n$ and A^TA is invertible, need the columns of A be linearly independent?

   (b) If A is $m \times n$ and AA^T is invertible, need the rows of A be linearly independent?

   Justify your answers.

8. **(a)** [BB] Suppose an $m \times n$ matrix A has a unique right inverse. Prove that $m = n$ and A is invertible. [Hint: If $A\mathsf{x} = 0$, then $\mathsf{x} = 0$.]

   (b) Suppose an $m \times n$ matrix A has a unique left inverse. Prove that $m = n$ and A is invertible. (Use 8(a) if you wish.)

Critical Reading

9. [BB] Let U be a row echelon matrix without zero rows. Does U have a one-sided inverse? Explain.

10. Suppose U is a row echelon matrix that has a pivot in every column. Does U have a one-sided inverse? Explain.

11. Suppose A is an $m \times n$ matrix.

    (a) If $m < n$ and A has one right inverse, show that A has infinitely many right inverses. [Hint: First argue that null sp $A \neq \{0\}$.]

    (b) If $n < m$ and A has one left inverse, show that A has infinitely many left inverses. [Hint: There is an easy way to see this using the result of part (a).]

12. Give an example of a matrix A with the property that $A\mathsf{x} = \mathsf{b}$ does not have a solution for all b, but it does have a unique solution for some b. Explain your answer.

13. Give an example of a matrix A for which the system $A\mathsf{x} = \mathsf{b}$ can be solved for any b, yet not always uniquely.

14. Does there exist a matrix A with the property that $A\mathsf{x} = \mathsf{b}$ has a solution for any b, sometimes a unique solution, sometimes not? Explain.

15. Let x_1, x_2, x_3, x_4 be real numbers and let

$$B = \begin{bmatrix} 4 & x_1 + x_2 + x_3 + x_4 \\ x_1 + x_2 + x_3 + x_4 & x_1^2 + x_2^2 + x_3^2 + x_4^2 \end{bmatrix}.$$

Show that $\det B = 0$ if and only if $x_1 = x_2 = x_3 = x_4$.

[Hint: Let $A = \begin{bmatrix} 1 & x_1 \\ 1 & x_2 \\ 1 & x_3 \\ 1 & x_4 \end{bmatrix}$ and compute $A^T A$.]

CHAPTER KEY WORDS AND IDEAS: Here are some technical words and phrases that were used in this chapter. Do you know the meaning of each? If you're not sure, check the glossary or index at the back of the book.

associative (vector addition is)
basis
closed under addition
closed under scalar multiplication
column rank
column space
commutative (vector addition is)
dimension
extend vectors to a basis
finite dimensional vector space
homogeneous linear system
inconsistent linear system
infinite dimensional vector space
left inverse of a matrix
linear combination of vectors
linearly dependent vectors
linearly independent vectors
nontrivial linear combination
null space
nullity
subspace spanned by vectors
rank
right inverse of a matrix
row rank
row space
spanning set
span a subspace
span of vectors
standard basis vectors
subspace
trivial linear combination
trivial solution
vector space

Review Exercises for Chapter 4

1. Complete the following sentence:

   "An $n \times n$ matrix A has an inverse if and only if the homogeneous system $A\mathsf{x} = 0$ has ________ ."

2. Let $A = \begin{bmatrix} -2 & -1 & 3 & 1 \\ 1 & 1 & -2 & 1 \\ 1 & 0 & -2 & 0 \\ 0 & 1 & -2 & 1 \\ 1 & 1 & 3 & -1 \end{bmatrix}$, $\mathsf{b}_1 = \begin{bmatrix} 0 \\ 1 \\ 0 \\ -1 \\ 4 \end{bmatrix}$ and $\mathsf{b}_2 = \begin{bmatrix} 1 \\ 0 \\ 1 \\ 0 \\ 1 \end{bmatrix}$.

   (a) Is b_1 in the column space of A? If it is, express this vector as a linear combination of the columns of A. If it isn't, explain why not.

   (b) Same as (a) for b_2.

   (c) Find a condition on $b_1, \ldots, b_5$ that is necessary and sufficient for $\mathsf{b} = \begin{bmatrix} b_1 \\ b_2 \\ b_3 \\ b_4 \\ b_5 \end{bmatrix}$ to belong to the column space of A; that is, complete the sentence

   $\mathsf{b} = \begin{bmatrix} b_1 \\ b_2 \\ b_3 \\ b_4 \\ b_5 \end{bmatrix}$ is in the column space of A if and only if ________.

3. Find the rank, the null space and the nullity of each matrix. In each case, verify that the sum of the rank and the nullity is what it should be.

   (a) $A = \begin{bmatrix} 2 & 3 & 1 & 6 \\ 4 & 5 & 2 & 12 \\ 3 & 4 & 2 & 9 \end{bmatrix}$ **(b)** $A = \begin{bmatrix} 1 & 2 & 3 & 0 & 1 \\ 2 & 1 & 0 & 1 & 0 \\ -1 & 2 & 2 & 2 & 1 \\ 2 & 2 & -2 & 0 & 0 \\ 4 & 1 & 2 & 3 & 0 \end{bmatrix}$.

4. Suppose A and B are matrices and $AB = 0$ is the zero matrix.

   (a) Explain why the column space of B is contained in the null space of A.

   (b) Prove that $\operatorname{rank} A + \operatorname{rank} B \le n$, the number of columns of A.

5. Suppose the $n \times n$ matrix A has rank $n - r$. Let $\mathsf{x}_1, \mathsf{x}_2, \ldots, \mathsf{x}_r$ be linearly independent vectors satisfying $A\mathsf{x} = 0$. Suppose x is a vector and $A\mathsf{x} = 0$. Explain why x must be a linear combination of $\mathsf{x}_1, \mathsf{x}_2, \ldots, \mathsf{x}_r$.

6. Suppose A is an $m \times n$ matrix and x is a vector in R^n. Suppose $\mathsf{y}^T A\mathsf{x} = 0$ for all vectors y in R^m. Explain why x must be in the null space of A.

7. Let $\mathsf{v} = \begin{bmatrix} 1 \\ 2 \\ 7 \end{bmatrix}$, $\mathsf{v}_1 = \begin{bmatrix} 1 \\ 5 \\ 1 \end{bmatrix}$ and $\mathsf{v}_2 = \begin{bmatrix} 1 \\ 6 \\ -1 \end{bmatrix}$.

   (a) Is v a linear combination of v_1 and v_2? If yes, show how; if no, explain why not.

   (b) Is v in the span of $\{\mathsf{v}_1, \mathsf{v}_2\}$?

   (c) Is $\{\mathsf{v}, \mathsf{v}_1, \mathsf{v}_2\}$ a basis for R^3?

(d) Extend $\{v_1, v_2\}$ to a basis of R^3.

8. Let $v_1 = \begin{bmatrix} 7 \\ -8 \\ -4 \\ 3 \end{bmatrix}, v_2 = \begin{bmatrix} -1 \\ 2 \\ 2 \\ 1 \end{bmatrix}, v_3 = \begin{bmatrix} 2 \\ -1 \\ 1 \\ 3 \end{bmatrix}, v_4 = \begin{bmatrix} 3 \\ -3 \\ -1 \\ 2 \end{bmatrix}, v_5 = \begin{bmatrix} 5 \\ -4 \\ 0 \\ 5 \end{bmatrix}$.

(a) Are these vectors linearly independent or linearly dependent?

(b) Do v_1, v_4 and v_5 span R^4?

(c) Do v_1 and v_3 span a two-dimensional subspace of R^4?

(d) What is the dimension of the subspace of R^4 spanned by the five given vectors?

Most of the answers are very short.

9. Let $v_1 = \begin{bmatrix} -1 \\ 0 \\ 2 \\ 1 \end{bmatrix}, v_2 = \begin{bmatrix} 3 \\ 1 \\ 2 \\ 0 \end{bmatrix}, v_3 = \begin{bmatrix} -5 \\ 2 \\ 5 \\ -1 \end{bmatrix}, v_4 = \begin{bmatrix} -1 \\ -4 \\ -2 \\ 5 \end{bmatrix}$ and $v_5 = \begin{bmatrix} 5 \\ 1 \\ 0 \\ 2 \end{bmatrix}$.

Some of the following questions can be answered **very** briefly.

(a) Do v_1, v_2 and v_3 span R^4?

(b) Are $v_1, \dots, v_5$ linearly independent?

Now let $A = \begin{bmatrix} v_1 & v_2 & v_3 & v_4 & v_5 \\ \downarrow & \downarrow & \downarrow & \downarrow & \downarrow \\ & & & & \end{bmatrix}$ be the matrix whose columns are the five given vectors.

(c) Find a basis for the row space of A.

(d) Find some columns of A which are a basis for the column space of A.

(e) What is the dimension of the subspace of R^4 spanned by $\{v_1, v_2, v_3, v_4\}$?

(f) What is the rank of A? What is the nullity of A? State a theorem which connects these numbers.

10. Let $U = \left\{ \begin{bmatrix} x+y \\ x-2y \\ 3x+y \end{bmatrix} \mid x, y \in R \right\}$.

Show that U is a subspace of R^3 and find a basis for U.

11. Let U be the subspace of R^4 spanned by $u_1 = \begin{bmatrix} 1 \\ 0 \\ 1 \\ 0 \end{bmatrix}, u_2 = \begin{bmatrix} 0 \\ 1 \\ 2 \\ -1 \end{bmatrix}, u_3 = \begin{bmatrix} 1 \\ -1 \\ 2 \\ -2 \end{bmatrix}$.

(a) Are u_1, u_2, u_3 linearly independent or linearly dependent? Explain.

(b) Is $\begin{bmatrix} 1 \\ 0 \\ 0 \\ 0 \end{bmatrix}$ a linear combination of u_1, u_2 and u_3?

(c) Do u_1, u_2, u_3 span R^4?

(d) What is $\dim U$?

12. (a) Find a basis for the null space of $A = \begin{bmatrix} 1 & -5 & 2 & -4 \\ 3 & -9 & 12 & 0 \\ -1 & 0 & -7 & -6 \end{bmatrix}$.

(b) Find a basis for the row space of A.

(c) Find a basis for the column space of A.

(d) Describe the column space of A in geometric terms.

(e) Decide whether or not AA^T is invertible without any calculation.

13. (a) Find a basis for the null space of $A = \begin{bmatrix} 2 & 1 & -1 & 0 & 1 \\ -1 & -1 & 2 & -3 & 1 \\ 1 & 1 & -2 & 0 & -1 \\ 0 & 0 & 1 & 1 & 1 \end{bmatrix}$.

(b) Find a basis for the column space of A.

(c) Verify that rank A + nullity $A = n$. What is n?

(d) Are the columns of A linearly independent?

(e) Find the dimension of the subspace of R^5 spanned by the rows of A.

14. Show that $U = \left\{ \begin{bmatrix} a \\ b \\ c \end{bmatrix} \mid a + 2b - 3c = 0 \right\}$ is a subspace of R^3 and find a basis for U.

15. (a) Show that $U = \left\{ \begin{bmatrix} a \\ b \\ c \\ d \end{bmatrix} \mid a + b + c + d = 0 \right\}$ is a subspace of R^4.

(b) Find a basis for U. What is the dimension of U?

16. Let $\mathsf{v}_1 = \begin{bmatrix} 1 \\ 1 \\ 1 \\ 1 \end{bmatrix}$ and $\mathsf{v}_2 = \begin{bmatrix} 1 \\ 0 \\ 1 \\ 0 \end{bmatrix}$. Extend $\mathsf{v}_1, \mathsf{v}_2$ to a basis of R^4.

17. Let $\mathsf{v}_1 = \begin{bmatrix} 1 \\ 2 \\ -1 \\ 1 \end{bmatrix}$, $\mathsf{v}_2 = \begin{bmatrix} -2 \\ -4 \\ 2 \\ -2 \end{bmatrix}$, $\mathsf{v}_3 = \begin{bmatrix} 1 \\ 2 \\ -4 \\ 0 \end{bmatrix}$, $\mathsf{v}_4 = \begin{bmatrix} 2 \\ 4 \\ 1 \\ 3 \end{bmatrix}$ and let V be the subspace of R^4 spanned by these vectors.

(a) Find a basis $\mathcal{B}_0$ for V.

(b) Extend $\mathcal{B}_0$ to a basis $\mathcal{B}$ of R^4.

(c) Express $\begin{bmatrix} 1 \\ 2 \\ 3 \\ 5 \end{bmatrix}$ as a linear combination of the vectors of $\mathcal{B}$.

18. Let $\mathsf{v}_1 = \begin{bmatrix} 2 \\ 4 \\ 6 \end{bmatrix}$, $\mathsf{v}_2 = \begin{bmatrix} -1 \\ 0 \\ 1 \end{bmatrix}$, $\mathsf{v}_3 = \begin{bmatrix} 1 \\ 4 \\ 7 \end{bmatrix}$, $\mathsf{v}_4 = \begin{bmatrix} 1 \\ 1 \\ 1 \end{bmatrix}$ and $\mathsf{v}_5 = \begin{bmatrix} 0 \\ 1 \\ 0 \end{bmatrix}$.
Find all possible subsets of $\{\mathsf{v}_1, \mathsf{v}_2, \mathsf{v}_3, \mathsf{v}_4, \mathsf{v}_5\}$ that form bases of R^3.

19. Let A be an $m \times n$ matrix of rank n. Prove that A^TA is invertible.

20. Use the fact that $\begin{bmatrix} 4 & 0 & -2 \\ 3 & 1 & -1 \\ 1 & 3 & 1 \end{bmatrix}\begin{bmatrix} 1 \\ -1 \\ 2 \end{bmatrix} = \begin{bmatrix} 0 \\ 0 \\ 0 \end{bmatrix}$ to show that there exist scalars a, b, c for which the system $\begin{aligned} 4x \quad\quad - 2z &= a \\ 3x + y - z &= b \\ x + 3y + z &= c \end{aligned}$ has no solution.

21. Let $S = \{\mathsf{v}_1, \mathsf{v}_2, \ldots, \mathsf{v}_m\}$ be a set of m vectors in R^n. Determine, if possible and with reasons, whether or not S is linearly independent or spans R^n in each of the following cases:
(a) $m = n$, **(b)** $m > n$, **(c)** $m < n$.

22. Suppose A is a 13×21 matrix. What are the possible values for nullity A and why?

23. If A is an $m \times n$ matrix with linearly independent rows and $m \neq n$, exactly one of the matrices AA^T, A^TA is invertible. Which one and why? Can you say anything about the relative sizes of m and n?

24. Suppose A is an $m \times n$ matrix and B is an $n \times p$ matrix.
    (a) Show that col sp $AB \subseteq$ col sp A.
    (b) Why is rank $AB \leq$ rank A?
    (c) If A is 7×3 and B is 3×7, is it possible that $AB = I$? Explain.

25. **(a)** The matrix $A = \begin{bmatrix} 1 & 2 & 3 \\ 0 & 1 & -1 \end{bmatrix}$ has a right inverse. Why? Find the most general right inverse.
    (b) The matrix $A = \begin{bmatrix} 1 & 0 \\ 2 & 1 \\ 3 & -1 \end{bmatrix}$ has a left inverse. Why? Find the most general left inverse.

26. The matrix $A = \begin{bmatrix} -2 & 1 & 3 & 0 \\ 1 & -\frac{1}{2} & -1 & 1 \end{bmatrix}$ has a right inverse. Why? Find the most general right inverse of A.

27. The matrix $A = \begin{bmatrix} 1 & 3 \\ 3 & 11 \\ -2 & 2 \end{bmatrix}$ has a left inverse.
    (a) Why?
    (b) Does it also have a right inverse?
    (c) Find the most general left inverse.

28. Write down as many statements as you can that are equivalent to "matrix A is invertible."

29. Let A be an $m \times n$ matrix. The linear system $A\mathsf{x} = \mathsf{b}$ has a solution for any b if and only if the column space of A is R^m. Explain why.

30. Explain why a homogeneous system of equations with more unknowns than equations always has a nontrivial solution.

31. Suppose A is a square matrix and the homogeneous system $A\mathsf{x} = \mathsf{0}$ has only the solution $\mathsf{x} = \mathsf{0}$. Explain why $A\mathsf{x} = \mathsf{b}$ has a solution for any b.

32. Let A be an $m \times n$ matrix whose columns span R^m. Explain how you know that $A\mathsf{x} = \mathsf{b}$ can be solved for any b in R^m.

33. Let A be an $m \times n$ matrix. Suppose the system $A\mathsf{x} = \mathsf{b}$ has at least one solution for any $\mathsf{b} \in \mathsf{R}^m$. Explain why the columns of A must span R^m.

Chapter 5

Linear Transformations

5.1 Fundamentals

5.1.1 Definition. A *linear transformation* is a function $T\colon V \to W$ from one vector space V to another W that satisfies

1. $T(\mathsf{u}+\mathsf{v}) = T(\mathsf{u}) + T(\mathsf{v})$ and
2. $T(c\mathsf{v}) = cT(\mathsf{v})$

for all vectors u, v in V and all scalars c.

5.1.2 Examples.

- The function $T\colon \mathsf{R}^1 \to \mathsf{R}^1$ defined by $T(x) = 2x$ is a linear transformation because $T(x+y) = 2(x+y) = 2x+2y = T(x)+T(y)$ and $T(cx) = 2(cx) = c(2x) = cT(x)$ for all vectors x, y in R^1 (= R) and all scalars c.

- The function sin: $\mathsf{R} \to \mathsf{R}$ does **not** define a linear transformation: it satisfies neither of the required properties. In general, $\sin(x+y) \neq \sin x + \sin y$ and $\sin cx \neq c \sin x$.

READING CHECK 1. Do you know formulas for $\sin(x+y)$ and $\sin 2x$?

- Let b be a fixed vector in R^2. Then the function $T\colon \mathsf{R}^2 \to \mathsf{R}^2$ defined by $T(\mathsf{v}) = \mathsf{v}\cdot\mathsf{b}$ is linear since for any vectors u, v, and for any scalar c, $T(\mathsf{u}+\mathsf{v}) = (\mathsf{u}+\mathsf{v})\cdot\mathsf{b} = \mathsf{u}\cdot\mathsf{b}+\mathsf{v}\cdot\mathsf{b} = T(\mathsf{u})+T(\mathsf{v})$ and $T(c\mathsf{v}) = (c\mathsf{v})\cdot\mathsf{b} = c(\mathsf{v}\cdot\mathsf{b}) = cT(\mathsf{v})$. (See Theorem 1.2.20.)

- The function $T\colon \mathsf{R}^3 \to \mathsf{R}$ defined by $T(\mathsf{x}) = \det\begin{bmatrix} x_1 & x_2 & x_3 \\ 1 & 2 & 3 \\ 4 & 5 & 6 \end{bmatrix}$ for $\mathsf{x} = \begin{bmatrix} x_1 \\ x_2 \\ x_3 \end{bmatrix}$ is a linear transformation; this is exactly what is meant when we say that the determinant is "linear in its first row." See Section 3.2.

- The differentiation operator, $D\colon \mathsf{R}[x] \to \mathsf{R}[x]$, defined by $D(f(x)) = f'(x)$ is linear. Any student of calculus will know that $D(f(x)+g(x)) = D(f(x)) + D(g(x))$ and $D(cf(x)) = cD(f(x))$ for any scalar c. (These rules may have been learned as $(f+g)' = f' + g'$ and $(cf)' = cf'$.)

:-)

5.1.3 Problem. The function $T\colon \mathsf{R}^3 \to \mathsf{R}^2$ defined by $T\left(\begin{bmatrix} x_1 \\ x_2 \\ x_3 \end{bmatrix}\right) = \begin{bmatrix} x_1 - x_3 \\ x_1 + 2x_2 \end{bmatrix}$ is a linear transformation. Explain.

Solution. Let $\mathsf{x} = \begin{bmatrix} x_1 \\ x_2 \\ x_3 \end{bmatrix}$ and $\mathsf{y} = \begin{bmatrix} y_1 \\ y_2 \\ y_3 \end{bmatrix}$ be vectors in R^3. Then $\mathsf{x} + \mathsf{y} = \begin{bmatrix} x_1 + y_1 \\ x_2 + y_2 \\ x_3 + y_3 \end{bmatrix}$, so $T(\mathsf{x}+\mathsf{y}) = \begin{bmatrix} (x_1 + y_1) - (x_3 + y_3) \\ (x_1 + y_1) + 2(x_2 + y_2) \end{bmatrix}$

$$= \begin{bmatrix} x_1 - x_3 + y_1 - y_3 \\ x_1 + 2x_2 + y_1 + 2y_2 \end{bmatrix} = \begin{bmatrix} x_1 - x_3 \\ x_1 + 2x_2 \end{bmatrix} + \begin{bmatrix} y_1 - y_3 \\ y_1 + 2y_2 \end{bmatrix} = T(\mathsf{x}) + T(\mathsf{y}).$$

Also, for any scalar c, $c\mathsf{x} = \begin{bmatrix} cx_1 \\ cx_2 \\ cx_3 \end{bmatrix}$, so

$$T(c\mathsf{x}) = \begin{bmatrix} cx_1 - cx_3 \\ cx_1 + 2cx_2 \end{bmatrix} = c\begin{bmatrix} x_1 - x_3 \\ x_1 + 2x_2 \end{bmatrix} = cT(\mathsf{x}).$$

5.1.4 Examples. Other examples of linear transformations include

- projection onto the xy-plane, that is, the function $T\colon \mathsf{R}^3 \to \mathsf{R}^3$ defined by $T\left(\begin{bmatrix} x \\ y \\ z \end{bmatrix}\right) = \begin{bmatrix} x \\ y \\ 0 \end{bmatrix}$ (see the leftmost picture in Figure 5.1), and

- rotation in the plane through the angle θ, that is, the function $T\colon \mathsf{R}^2 \to \mathsf{R}^2$ where the arrow representing $T(\mathsf{v})$ is the arrow for v rotated through θ (see the rightmost picture in Figure 5.1).

To see that the projection described here is a linear transformation, let $\mathsf{x} = \begin{bmatrix} x_1 \\ x_2 \\ x_3 \end{bmatrix}$ and $\mathsf{y} = \begin{bmatrix} y_1 \\ y_2 \\ y_3 \end{bmatrix}$ be vectors in R^3 and let c be a scalar. Then $\mathsf{x} + \mathsf{y} =$

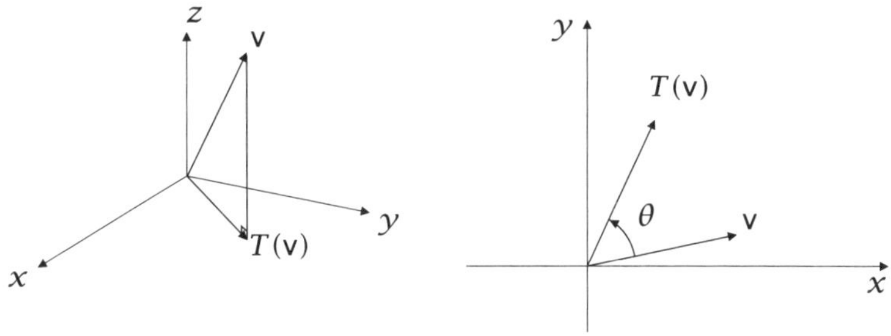

Figure 5.1: A projection (on the left) and a rotation

$\begin{bmatrix} x_1+y_1 \\ x_2+y_2 \\ x_3+y_3 \end{bmatrix}$ and $c\mathsf{x} = \begin{bmatrix} cx_1 \\ cx_2 \\ cx_3 \end{bmatrix}$. So $T(\mathsf{x}+\mathsf{y}) = \begin{bmatrix} x_1+y_1 \\ x_2+y_2 \\ 0 \end{bmatrix} = \begin{bmatrix} x_1 \\ x_2 \\ 0 \end{bmatrix} + \begin{bmatrix} y_1 \\ y_2 \\ 0 \end{bmatrix} =$ $T(\mathsf{x}) + T(\mathsf{y})$ and $T(c\mathsf{x}) = \begin{bmatrix} cx_1 \\ cx_2 \\ 0 \end{bmatrix} = c\begin{bmatrix} x_1 \\ x_2 \\ 0 \end{bmatrix} = cT(\mathsf{x})$.

To see that rotation in the plane through the angle θ is a linear transformation, it is probably easiest to draw pictures such as those shown in Figure 5.2. The sum $\mathsf{u}+\mathsf{v}$ can be pictured by the diagonal of the parallelogram with sides u and v. The statement $T(\mathsf{u}+\mathsf{v}) = T(\mathsf{u}) + T(\mathsf{v})$ expresses the fact that the rotated vector $T(\mathsf{u}+\mathsf{v})$ is the diagonal of the parallelogram with sides $T(\mathsf{u})$ and $T(\mathsf{v})$. Similarly, the statement $T(c\mathsf{u}) = cT(\mathsf{u})$ expresses the fact that the rotated vector $T(c\mathsf{u})$ is c times the vector $T(\mathsf{u})$.

The most important and, as we shall see, essentially the only example of a linear transformation (on a finite dimensional vector space) is "left multiplication by a matrix."

Left Multiplication by a Matrix

An $m \times n$ matrix A defines a function $T_A\colon \mathsf{R}^n \to \mathsf{R}^m$ by $T_A(\mathsf{x}) = A\mathsf{x}$. This function is a linear transformation because

$$T_A(\mathsf{x}+\mathsf{y}) = A(\mathsf{x}+\mathsf{y}) = A\mathsf{x} + A\mathsf{y} = T_A(\mathsf{x}) + T_A(\mathsf{y})$$

and

$$T_A(c\mathsf{x}) = A(c\mathsf{x}) = c(A\mathsf{x}) = cT_A(\mathsf{x})$$

for all vectors x, y in R^n and all scalars c.

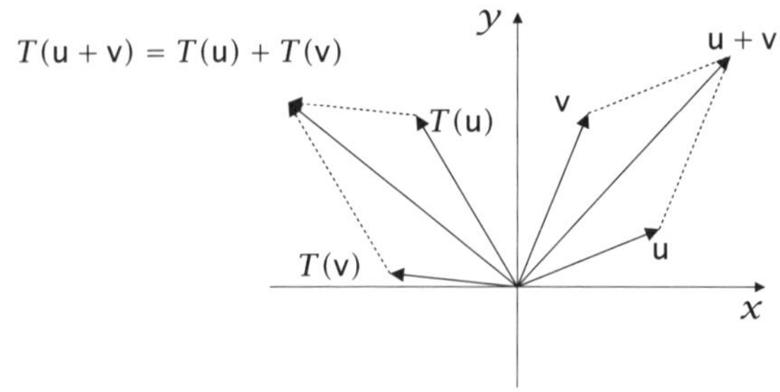

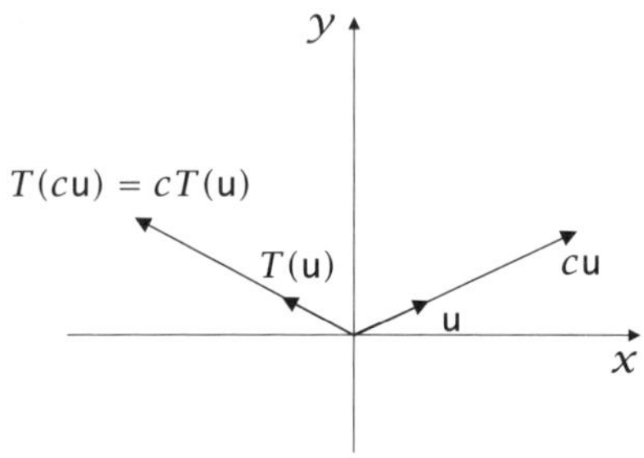

Figure 5.2: A rotation is a linear transformation.

Properties of Linear Transformations

Recall that a vector space is a set on which there are defined operations called addition and scalar multiplication (subject to a number of conditions). A linear transformation T is just a function that "preserves" these two operations in the sense that T applied to a sum $u+v$ is the sum $T(u)+T(v)$ and T applied to a scalar multiple cv is the same scalar multiple of $T(v)$. In fact, a linear transformation preserves more than just addition and scalar multiples.

5.1.5 Property 1. A linear transformation T preserves the zero vector: $T(0) = 0$.

To see why this is so, suppose $T\colon V \to W$ is a linear transformation. Take any vector v in V. Then $0v = 0$ is the zero vector (of V) (see Lemma 4.2.5), so $T(0) = T(0v) = 0T(v)$, since T preserves scalar multiples. Since $0T(v)$ is the zero vector (of W) we have $T(0) = 0$.

5.1.6 Problem. Explain why the function $T\colon \mathsf{R}^4 \to \mathsf{R}^3$ defined by $T\left(\begin{bmatrix} x_1 \\ x_2 \\ x_3 \\ x_4 \end{bmatrix}\right) =$

$\begin{bmatrix} x_1 + 2x_4 \\ 3x_1 + x_3 - 5x_4 \\ 2 \end{bmatrix}$ is not linear.

Solution. In R^4, $0 = \begin{bmatrix} 0 \\ 0 \\ 0 \\ 0 \end{bmatrix}$, but $T(0) = \begin{bmatrix} 0 \\ 0 \\ 2 \end{bmatrix}$ is not the zero vector of R^3.

5.1.7 Property 2. A linear transformation T preserves additive inverses: $T(-\mathsf{v}) = -T(\mathsf{v})$.

Again, suppose that $T\colon V \to W$ is a linear transformation and let v be a vector in the vector space V. Remember that the additive inverse of v is defined by the condition $\mathsf{v} + (-\mathsf{v}) = 0$. Applying T, we have $0 = T(0) = T[\mathsf{v} + (-\mathsf{v})] = T(\mathsf{v}) + T(-\mathsf{v})$. Thus $T(-\mathsf{v})$ is the additive inverse of $T(\mathsf{v})$: $T(-\mathsf{v}) = -T(\mathsf{v})$.

5.1.8 Property 3. A linear transformation preserves linear combinations:

$$T(c_1\mathsf{v}_1 + c_2\mathsf{v}_2 + \cdots + c_k\mathsf{v}_k) = c_1T(\mathsf{v}_1) + c_2T(\mathsf{v}_2) + \cdots + c_kT(\mathsf{v}_k)$$

for any vectors $\mathsf{v}_1, \ldots, \mathsf{v}_k$ and scalars $c_1, \ldots, c_k$.

This follows directly from the two defining properties of a linear transformation.

$$T(c_1\mathsf{v}_1 + c_2\mathsf{v}_2 + \cdots + c_k\mathsf{v}_k) = T(c_1\mathsf{v}_1) + T(c_2\mathsf{v}_2) + \cdots + T(c_k\mathsf{v}_k)$$

because T preserves sums, and then

$$T(c_1\mathsf{v}_1) + T(c_2\mathsf{v}_2) + \cdots + T(c_k\mathsf{v}_k) = c_1T(\mathsf{v}_1) + c_2T(\mathsf{v}_2) + \cdots + c_kT(\mathsf{v}_k)$$

because T preserves scalar multiples.

That a linear transformation T preserves linear combinations is extremely important. It says, for instance, that if $\mathcal{V} = \{\mathsf{v}_1, \mathsf{v}_2, \ldots, \mathsf{v}_k\}$ is a set of vectors that spans a vector space V, and if you know $T(\mathsf{v}_1), T(\mathsf{v}_2), \ldots, T(\mathsf{v}_k)$, then you know $T(\mathsf{v})$ for any v, because any v is a linear combination of $\mathsf{v}_1, \mathsf{v}_2, \ldots, \mathsf{v}_k$ and T preserves linear combinations!

5.1.9 Proposition. *A linear transformation $T\colon V \to W$ is determined by its effect on any spanning set for V, in particular, on any basis for V.*

5.1.10 Example. Let $T\colon \mathsf{R}^2 \to \mathsf{R}^2$ be a linear transformation satisfying $T(\mathsf{i}) = \begin{bmatrix} 2 \\ -1 \end{bmatrix}$ and $T(\mathsf{j}) = \begin{bmatrix} 4 \\ 7 \end{bmatrix}$. Any vector $\mathsf{v} = \begin{bmatrix} x \\ y \end{bmatrix}$ in R^2 can be written $\mathsf{v} = x\mathsf{i} + y\mathsf{j}$, so

$T(\mathsf{v}) = xT(\mathsf{i}) + yT(\mathsf{j}) = x\begin{bmatrix} 2 \\ -1 \end{bmatrix} + y\begin{bmatrix} 4 \\ 7 \end{bmatrix} = \begin{bmatrix} 2x + 4y \\ -x + 7y \end{bmatrix}$. Knowing just the two vectors $T(\mathsf{i})$ and $T(\mathsf{j})$ tells us $T(\mathsf{v})$ for any v.

5.1.11 Problem. If $f\colon \mathsf{R} \to \mathsf{R}$ is a function such that $f(1) = 2$, what can you say about $f(x)$? Would your answer change if you knew that f was a linear transformation?

Solution. In general, you can say nothing about f. Think, for example, of the functions defined by $f(x) = x^2+1$, $f(x) = 2\cos(x-1)$ and $f(x) = 2^x$. Each of these satisfies $f(1) = 2$. On the other hand, if f is a linear transformation and $f(1) = 2$, then $f(x) = f(x \cdot 1) = xf(1)$ because x is a scalar and f preserves scalar multiples. Thus $f(x) = 2x$.

5.1.12 Problem. Suppose $T\colon \mathsf{R}^2 \to \mathsf{R}^3$ is a linear transformation such that $T(\begin{bmatrix} 1 \\ 1 \end{bmatrix}) = \begin{bmatrix} -1 \\ 2 \\ 1 \end{bmatrix}$ and $T(\begin{bmatrix} -3 \\ 2 \end{bmatrix}) = \begin{bmatrix} 0 \\ 4 \\ -3 \end{bmatrix}$. Why does this information allow us to determine $T(\begin{bmatrix} x \\ y \end{bmatrix})$ for any vector $\begin{bmatrix} x \\ y \end{bmatrix}$ in R^2? Find $T(\begin{bmatrix} x \\ y \end{bmatrix})$.

Solution. The vectors $\begin{bmatrix} 1 \\ 1 \end{bmatrix}$ and $\begin{bmatrix} -3 \\ 2 \end{bmatrix}$ are linearly independent and so span the vector space R^2. This fact is enough to tell us $T(\begin{bmatrix} x \\ y \end{bmatrix})$ for any $\begin{bmatrix} x \\ y \end{bmatrix}$. Specifically, find real numbers a and b such that

$$\begin{bmatrix} x \\ y \end{bmatrix} = a\begin{bmatrix} 1 \\ 1 \end{bmatrix} + b\begin{bmatrix} -3 \\ 2 \end{bmatrix}.$$

We obtain $a = \frac{2}{5}x + \frac{3}{5}y$ and $b = -\frac{1}{5}x + \frac{1}{5}y$, so

$$\begin{aligned} T(\begin{bmatrix} x \\ y \end{bmatrix}) &= T\left(a\begin{bmatrix} 1 \\ 1 \end{bmatrix} + b\begin{bmatrix} -3 \\ 2 \end{bmatrix}\right) \\ &= aT\left(\begin{bmatrix} 1 \\ 1 \end{bmatrix}\right) + bT\left(\begin{bmatrix} -3 \\ 2 \end{bmatrix}\right) \\ &= (\frac{2}{5}x + \frac{3}{5}y)\begin{bmatrix} -1 \\ 2 \\ 1 \end{bmatrix} + (-\frac{1}{5}x + \frac{1}{5}y)\begin{bmatrix} 0 \\ 4 \\ -3 \end{bmatrix} = \begin{bmatrix} -\frac{2}{5}x - \frac{3}{5}y \\ 2y \\ x \end{bmatrix}. \end{aligned}$$

We have made the point that left multiplication by an $m \times n$ matrix is a linear transformation $\mathsf{R}^n \to \mathsf{R}^m$. Now we show that this is effectively the only linear transformation from R^n to R^m.

5.1.13 Theorem. *Let $T\colon \mathsf{R}^n \to \mathsf{R}^m$ be a linear transformation.*
Let $A = \begin{bmatrix} T(\mathsf{e}_1) & T(\mathsf{e}_2) & \cdots & T(\mathsf{e}_n) \\ \downarrow & \downarrow & & \downarrow \end{bmatrix}$ *be the $m \times n$ matrix whose columns are the values of T applied to the standard basis vectors $\mathsf{e}_1, \ldots, \mathsf{e}_n$ of R^n. Then $T = T_A$ is left multiplication by A, that is, $T(\mathsf{x}) = A\mathsf{x}$ for any x in R^n.*

Proof. Let $\mathsf{x} = \begin{bmatrix} x_1 \\ x_2 \\ \vdots \\ x_n \end{bmatrix}$ be a vector in R^n. Then $\mathsf{x} = x_1\mathsf{e}_1 + x_2\mathsf{e}_2 + \cdots + x_n\mathsf{e}_n$.
Since T preserves linear combinations,

$$T(\mathsf{x}) = x_1 T(\mathsf{e}_1) + x_2 T(\mathsf{e}_2) + \cdots + x_n T(\mathsf{e}_n) = A\mathsf{x},$$

so T is left multiplication by A. ■

Theorem 5.1.13 establishes a correspondence between linear transformations $\mathsf{R}^n \to \mathsf{R}^m$ and $m \times n$ matrices.

$$\begin{array}{ccc} \text{Linear transformations} & \longleftrightarrow & \text{Matrices} \\ T & \longrightarrow & \begin{bmatrix} T(\mathsf{e}_1) & T(\mathsf{e}_2) & \cdots & T(\mathsf{e}_n) \\ \downarrow & \downarrow & & \downarrow \end{bmatrix} \\ T_A & \longleftarrow & A \end{array}$$

5.1.14 Example. Let $T\colon \mathsf{R}^2 \to \mathsf{R}^2$ be the linear transformation defined by $T\left(\begin{bmatrix} x \\ y \end{bmatrix}\right) = \begin{bmatrix} 3x - y \\ 4x + 5y \end{bmatrix}$. Theorem 5.1.13 says that T is left multiplication by the matrix $A = \begin{bmatrix} T(\mathsf{e}_1) & T(\mathsf{e}_2) \\ \downarrow & \downarrow \end{bmatrix} = \begin{bmatrix} 3 & -1 \\ 4 & 5 \end{bmatrix}$. This is most certainly the case because for any $\mathsf{x} = \begin{bmatrix} x \\ y \end{bmatrix}$ in R^2, $A\mathsf{x} = \begin{bmatrix} 3 & -1 \\ 4 & 5 \end{bmatrix}\begin{bmatrix} x \\ y \end{bmatrix} = \begin{bmatrix} 3x - y \\ 4x + 5y \end{bmatrix} = T(\mathsf{x})$. ☺

5.1.15 Problem. Define $T\colon \mathsf{R}^3 \to \mathsf{R}^4$ by $T\left(\begin{bmatrix} x_1 \\ x_2 \\ x_3 \end{bmatrix}\right) = \begin{bmatrix} x_1 + 2x_2 + x_3 \\ 2x_1 - 4x_3 \\ -x_1 + x_2 - 2x_3 \\ 3x_1 - 2x_2 + 5x_3 \end{bmatrix}$.

Find a matrix A such that T is left multiplication by A.

Solution. Let $\mathsf{e}_1 = \begin{bmatrix} 1 \\ 0 \\ 0 \end{bmatrix}$, $\mathsf{e}_2 = \begin{bmatrix} 0 \\ 1 \\ 0 \end{bmatrix}$, $\mathsf{e}_3 = \begin{bmatrix} 0 \\ 0 \\ 1 \end{bmatrix}$ be the standard basis vectors for R^3 and let

$$A = \begin{bmatrix} T(\mathsf{e}_1) & T(\mathsf{e}_2) & T(\mathsf{e}_3) \\ \downarrow & \downarrow & \downarrow \\ & & \end{bmatrix} = \begin{bmatrix} 1 & 2 & 1 \\ 2 & 0 & -4 \\ -1 & 1 & -2 \\ 3 & -2 & 5 \end{bmatrix}.$$

Then

$$A\mathsf{x} = \begin{bmatrix} 1 & 2 & 1 \\ 2 & 0 & -4 \\ -1 & 1 & -2 \\ 3 & -2 & 5 \end{bmatrix} \begin{bmatrix} x_1 \\ x_2 \\ x_3 \end{bmatrix} = \begin{bmatrix} x_1 + 2x_2 + x_3 \\ 2x_1 - 4x_3 \\ -x_1 + x_2 - 2x_3 \\ 3x_1 - 2x_2 + 5x_3 \end{bmatrix} = T(\mathsf{x}),$$

so T is left multiplication by A.

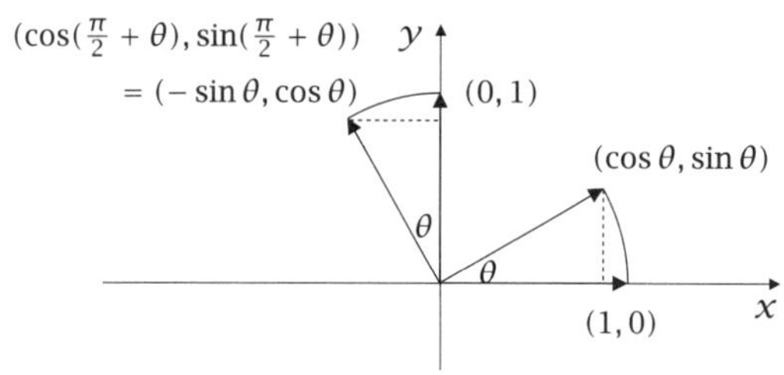

Figure 5.3

5.1.16 Example. In 5.1.4, we presented rotation in the plane as an example of a linear transformation and gave a geometrical argument making this statement plausible. With reference to Figure 5.3, we see that $T(\mathsf{e}_1) = T\left(\begin{bmatrix} 1 \\ 0 \end{bmatrix}\right) = \begin{bmatrix} \cos\theta \\ \sin\theta \end{bmatrix}$ and $T(\mathsf{e}_2) = T\left(\begin{bmatrix} 0 \\ 1 \end{bmatrix}\right) = \begin{bmatrix} -\sin\theta \\ \cos\theta \end{bmatrix}$. Thus T is left multiplication by the matrix $A = \begin{bmatrix} \cos\theta & -\sin\theta \\ \sin\theta & \cos\theta \end{bmatrix}$. For any $\begin{bmatrix} x \\ y \end{bmatrix}$ in the plane, the rotated vector is

$$T\left(\begin{bmatrix} x \\ y \end{bmatrix}\right) = A\begin{bmatrix} x \\ y \end{bmatrix} = \begin{bmatrix} \cos\theta & -\sin\theta \\ \sin\theta & \cos\theta \end{bmatrix}\begin{bmatrix} x \\ y \end{bmatrix} = \begin{bmatrix} x\cos\theta - y\sin\theta \\ x\sin\theta + y\cos\theta \end{bmatrix},$$

a fact perhaps not so easily seen using purely geometrical reasoning. For example, rotation through $\theta = 60°$ moves the vector $\begin{bmatrix} x \\ y \end{bmatrix}$ to $\frac{1}{2}\begin{bmatrix} x - \sqrt{3}y \\ \sqrt{3}x + y \end{bmatrix}$.

5.1.17 Example. Let $\mathsf{v} = \begin{bmatrix} -3 \\ 1 \end{bmatrix}$ and define $T\colon \mathsf{R}^2 \to \mathsf{R}^2$ by $T(\mathsf{u}) = \operatorname{proj}_{\mathsf{v}} \mathsf{u} = \frac{\mathsf{u}\cdot\mathsf{v}}{\mathsf{v}\cdot\mathsf{v}}\,\mathsf{v}$. The proof that T is a linear transformation is a straightforward exercise—see Exercise 12—so T is left multiplication by the matrix A defined in Theorem 5.1.13. Since

$$T(\mathsf{e}_1) = T\left(\begin{bmatrix} 1 \\ 0 \end{bmatrix}\right) = \frac{\mathsf{e}_1 \cdot \mathsf{v}}{\mathsf{v} \cdot \mathsf{v}}\mathsf{v} = \frac{-3}{10}\mathsf{v} = \begin{bmatrix} \frac{9}{10} \\ -\frac{3}{10} \end{bmatrix}$$

$$\text{and } T(\mathsf{e}_2) = T\left(\begin{bmatrix} 0 \\ 1 \end{bmatrix}\right) = \frac{1}{10}\mathsf{v} = \begin{bmatrix} -\frac{3}{10} \\ \frac{1}{10} \end{bmatrix},$$

$A = \frac{1}{10}\begin{bmatrix} 9 & -3 \\ -3 & 1 \end{bmatrix}$. By inspection, A is symmetric ($A^T = A$). Less obvious is the fact that T is also *idempotent* ($A^2 = A$). These features are characteristic of projection matrices and will be discussed more fully in Chapter 6. ☺

5.1.18 Example. The function $T\colon \mathsf{R}^3 \to \mathsf{R}^3$ defined by $T\left(\begin{bmatrix} x \\ y \\ z \end{bmatrix}\right) = \begin{bmatrix} x \\ y \\ 0 \end{bmatrix}$, which projects vectors (orthogonally) onto the xy-plane, is a linear transformation—see 5.1.4. Since $T(\mathsf{e}_1) = T\left(\begin{bmatrix} 1 \\ 0 \\ 0 \end{bmatrix}\right) = \begin{bmatrix} 1 \\ 0 \\ 0 \end{bmatrix}$, $T(\mathsf{e}_2) = T\left(\begin{bmatrix} 0 \\ 1 \\ 0 \end{bmatrix}\right) = \begin{bmatrix} 0 \\ 1 \\ 0 \end{bmatrix}$ and $T(\mathsf{e}_3) = T\left(\begin{bmatrix} 0 \\ 0 \\ 1 \end{bmatrix}\right) = \begin{bmatrix} 0 \\ 0 \\ 0 \end{bmatrix}$, T is left multiplication by the matrix $A = \begin{bmatrix} 1 & 0 & 0 \\ 0 & 1 & 0 \\ 0 & 0 & 0 \end{bmatrix}$:

$$A\begin{bmatrix} x \\ y \\ z \end{bmatrix} = \begin{bmatrix} 1 & 0 & 0 \\ 0 & 1 & 0 \\ 0 & 0 & 0 \end{bmatrix}\begin{bmatrix} x \\ y \\ z \end{bmatrix} = \begin{bmatrix} x \\ y \\ 0 \end{bmatrix} = T\left(\begin{bmatrix} x \\ y \\ z \end{bmatrix}\right).$$ ☺

5.1.19 Problem. Let $T\colon \mathsf{R}^3 \to \mathsf{R}^3$ be the function that projects vectors onto the plane with equation $x + y - z = 0$. (To see that T is a linear transformation is not difficult—see Exercise 12.) Find a matrix A such that T is left multiplication by A.

Solution. Recall that the projection of a vector w on the plane spanned by orthogonal vectors f_1 and f_2 is $T(\mathsf{w}) = \frac{\mathsf{w}\cdot\mathsf{f}_1}{\mathsf{f}_1\cdot\mathsf{f}_1}\mathsf{f}_1 + \frac{\mathsf{w}\cdot\mathsf{f}_2}{\mathsf{f}_2\cdot\mathsf{f}_2}\mathsf{f}_2$. In this problem, the given plane—call it π—is spanned by the orthogonal vectors $\mathsf{f}_1 = \begin{bmatrix} 1 \\ 0 \\ 1 \end{bmatrix}$ and $\mathsf{f}_2 = \begin{bmatrix} -1 \\ 2 \\ 1 \end{bmatrix}$. The projection of $\mathsf{e}_1 = \begin{bmatrix} 1 \\ 0 \\ 0 \end{bmatrix}$ on π is

$$T(\mathsf{e}_1) = \frac{\mathsf{e}_1 \cdot \mathsf{f}_1}{\mathsf{f}_1 \cdot \mathsf{f}_1}\mathsf{f}_1 + \frac{\mathsf{e}_1 \cdot \mathsf{f}_2}{\mathsf{f}_2 \cdot \mathsf{f}_2}\mathsf{f}_2 = \frac{1}{2}\begin{bmatrix} 1 \\ 0 \\ 1 \end{bmatrix} + \frac{-1}{6}\begin{bmatrix} -1 \\ 2 \\ 1 \end{bmatrix} = \begin{bmatrix} \frac{2}{3} \\ -\frac{1}{3} \\ \frac{1}{3} \end{bmatrix},$$

the projection of $\mathsf{e}_2 = \begin{bmatrix} 0 \\ 1 \\ 0 \end{bmatrix}$ on π is

$$T(\mathsf{e}_2) = \frac{\mathsf{e}_2 \cdot \mathsf{f}_1}{\mathsf{f}_1 \cdot \mathsf{f}_1}\mathsf{f}_1 + \frac{\mathsf{e}_2 \cdot \mathsf{f}_2}{\mathsf{f}_2 \cdot \mathsf{f}_2}\mathsf{f}_2 = 0 + \frac{2}{6}\begin{bmatrix} -1 \\ 2 \\ 1 \end{bmatrix} = \begin{bmatrix} -\frac{1}{3} \\ \frac{2}{3} \\ \frac{1}{3} \end{bmatrix},$$

and the projection of $\mathsf{e}_3 = \begin{bmatrix} 0 \\ 0 \\ 1 \end{bmatrix}$ on π is

$$T(\mathsf{e}_3) = \frac{\mathsf{e}_3 \cdot \mathsf{f}_1}{\mathsf{f}_1 \cdot \mathsf{f}_1}\mathsf{f}_1 + \frac{\mathsf{e}_3 \cdot \mathsf{f}_2}{\mathsf{f}_2 \cdot \mathsf{f}_2}\mathsf{f}_2 = \frac{1}{2}\begin{bmatrix} 1 \\ 0 \\ 1 \end{bmatrix} + \frac{1}{6}\begin{bmatrix} -1 \\ 2 \\ 1 \end{bmatrix} = \begin{bmatrix} \frac{1}{3} \\ \frac{1}{3} \\ \frac{2}{3} \end{bmatrix}.$$

Thus T is left multiplication by $A = \begin{bmatrix} \frac{2}{3} & -\frac{1}{3} & \frac{1}{3} \\ -\frac{1}{3} & \frac{2}{3} & \frac{1}{3} \\ \frac{1}{3} & \frac{1}{3} & \frac{2}{3} \end{bmatrix}$. Again, it is clear that A is symmetric. It is true also that A is idempotent. ♢

Answers to Reading Checks

1. $\sin(x + y) = \sin x \cos y + \cos x \sin y$; $\sin 2x = 2 \sin x \cos x$.

True/False Questions

Decide, with as little calculation as possible, whether each of the following statements is true or false and, if you say "false," explain your answer. (Answers are at the back of the book.)

1. The square root function $\sqrt{\ }: \mathsf{R} \to \mathsf{R}$ is a linear transformation.

2. The function $T\colon \mathsf{R}^3 \to \mathsf{R}^4$ defined by $T(\begin{bmatrix} x \\ y \\ z \end{bmatrix}) = \begin{bmatrix} x - y \\ y - z \\ z - x + 2 \\ x + y + z \end{bmatrix}$ is a linear transformation.

3. If $f\colon \mathsf{R} \to \mathsf{R}$ is a linear transformation such that $f(1) = 5$, then $f(x) = 5x$ for any x.

4. If $f\colon \mathsf{R} \to \mathsf{R}$ is a linear transformation such that $f(2) = 10$, then $f(x) = 5x$ for any x.

5. The matrix that rotates vectors in the plane through 45° is $\begin{bmatrix} \frac{\sqrt{2}}{2} & -\frac{\sqrt{2}}{2} \\ \frac{\sqrt{2}}{2} & \frac{\sqrt{2}}{2} \end{bmatrix}$.

6. The matrix $P = \begin{bmatrix} 1 & 0 & 0 \\ 0 & 1 & 0 \\ 0 & 0 & 0 \end{bmatrix}$ projects vectors in R^3 orthogonally onto the xy-plane.

7. The matrix $P = \begin{bmatrix} 1 & 1 \\ 1 & -1 \end{bmatrix}$ projects vectors in R^2 orthogonally onto the line with equation $y = x$.

8. The matrix $P = \begin{bmatrix} 1 & 1 \\ 1 & 1 \end{bmatrix}$ projects vectors in R^2 orthogonally onto the line with equation $y = x$.

9. If $T\colon \mathsf{R}^2 \to \mathsf{R}^2$ is a linear transformation such that $T(\begin{bmatrix} 1 \\ 2 \end{bmatrix}) = \begin{bmatrix} 0 \\ 1 \end{bmatrix}$ and $T(\begin{bmatrix} 2 \\ 1 \end{bmatrix}) = \begin{bmatrix} 1 \\ 0 \end{bmatrix}$, then $T(\begin{bmatrix} 3 \\ 3 \end{bmatrix}) = \begin{bmatrix} 1 \\ 1 \end{bmatrix}$.

10. A linear transformation is a function that is closed under addition and scalar multiplication.

EXERCISES

Answers to exercises marked [BB] can be found at the Back of the Book.

1. Determine whether or not each of the following functions is a linear transformation. Explain.

(a) [BB] $T\colon \mathsf{R}^3 \to \mathsf{R}^3$ is defined by $T(\begin{bmatrix} x \\ y \\ z \end{bmatrix}) = \begin{bmatrix} x + 2y + 3z \\ 1 \\ y - x - z \end{bmatrix}$

(b) [BB] $T\colon \mathsf{R}^3 \to \mathsf{R}^2$ is defined by $T(\begin{bmatrix} x_1 \\ x_2 \\ x_3 \end{bmatrix}) = \begin{bmatrix} x_1 + x_2 \\ x_2 + x_3 \end{bmatrix}$

(c) $T\colon \mathsf{R}^2 \to \mathsf{R}^2$ rotates vectors 45° about the point $(1, 0)$

(d) $T\colon \mathsf{R}^2 \to \mathsf{R}^3$ is defined by $T(\begin{bmatrix} x \\ y \end{bmatrix}) = \begin{bmatrix} x + y \\ xy \\ y \end{bmatrix}$

(e) $T\colon \mathsf{R}^2 \to \mathsf{R}^2$ is defined by $T(\begin{bmatrix} x \\ y \end{bmatrix}) = \begin{bmatrix} 2x + y \\ y \end{bmatrix}$.

2. Suppose $f\colon \mathsf{R} \to \mathsf{R}$ is a linear transformation such that $f(5) = 20$. Find $f(x)$ for $x \in \mathsf{R}$.

3. [BB] Let $T\colon \mathsf{R}^2 \to \mathsf{R}^3$ be a linear transformation such that $T\left(\begin{bmatrix}1\\0\end{bmatrix}\right) = \begin{bmatrix}3\\-1\\1\end{bmatrix}$ and $T\left(\begin{bmatrix}0\\1\end{bmatrix}\right) = \begin{bmatrix}4\\5\\-2\end{bmatrix}$. Find $T\left(\begin{bmatrix}x\\y\end{bmatrix}\right)$.

4. Suppose $T\colon \mathsf{R}^3 \to \mathsf{R}^2$ is a linear transformation such that $T\left(\begin{bmatrix}1\\0\\0\end{bmatrix}\right) = \begin{bmatrix}3\\1\end{bmatrix}$, $T\left(\begin{bmatrix}0\\1\\0\end{bmatrix}\right) = \begin{bmatrix}1\\-2\end{bmatrix}$ and $T\left(\begin{bmatrix}0\\0\\1\end{bmatrix}\right) = \begin{bmatrix}-1\\0\end{bmatrix}$. Find $T\left(\begin{bmatrix}x_1\\x_2\\x_3\end{bmatrix}\right)$.

5. [BB] Suppose $T\colon \mathsf{R}^2 \to \mathsf{R}^3$ is a linear transformation such that $T\left(\begin{bmatrix}1\\-1\end{bmatrix}\right) = \begin{bmatrix}1\\2\\3\end{bmatrix}$ and $T\left(\begin{bmatrix}-2\\3\end{bmatrix}\right) = \begin{bmatrix}-2\\-4\\7\end{bmatrix}$. Find $T\left(\begin{bmatrix}x\\y\end{bmatrix}\right)$.

6. Suppose $T\colon \mathsf{R}^3 \to \mathsf{R}^2$ is a linear transformation such that $T\left(\begin{bmatrix}1\\0\\1\end{bmatrix}\right) = \begin{bmatrix}-2\\3\end{bmatrix}$, $T\left(\begin{bmatrix}-1\\2\\1\end{bmatrix}\right) = \begin{bmatrix}1\\1\end{bmatrix}$ and $T\left(\begin{bmatrix}0\\2\\3\end{bmatrix}\right) = \begin{bmatrix}-2\\5\end{bmatrix}$. Find $T\left(\begin{bmatrix}x\\y\\z\end{bmatrix}\right)$.

7. Suppose $T\colon \mathsf{R}^3 \to \mathsf{R}^4$ is a linear transformation such that $T\left(\begin{bmatrix}-1\\1\\1\end{bmatrix}\right) = \begin{bmatrix}0\\1\\2\\3\end{bmatrix}$, $T\left(\begin{bmatrix}-2\\1\\0\end{bmatrix}\right) = \begin{bmatrix}-4\\-5\\1\\1\end{bmatrix}$ and $T\left(\begin{bmatrix}1\\0\\-1\end{bmatrix}\right) = \begin{bmatrix}2\\2\\-3\\1\end{bmatrix}$. Find $T\left(\begin{bmatrix}x\\y\\z\end{bmatrix}\right)$.

8. Does there exist a linear transformation

   (a) [BB] $T\colon \mathsf{R}^3 \to \mathsf{R}^2$ with $T\left(\begin{bmatrix}1\\0\\-2\end{bmatrix}\right) = \begin{bmatrix}1\\2\\3\end{bmatrix}$ and $T\left(\begin{bmatrix}2\\0\\-4\end{bmatrix}\right) = \begin{bmatrix}2\\3\\9\end{bmatrix}$?

   (b) $T\colon \mathsf{R}^3 \to \mathsf{R}^3$ with $T\left(\begin{bmatrix}2\\3\\1\end{bmatrix}\right) = \begin{bmatrix}1\\1\\1\end{bmatrix}$, $T\left(\begin{bmatrix}1\\0\\9\end{bmatrix}\right) = \begin{bmatrix}1\\2\\3\end{bmatrix}$ and $T\left(\begin{bmatrix}3\\3\\10\end{bmatrix}\right) = \begin{bmatrix}0\\1\\-1\end{bmatrix}$?

   (c) $T\colon \mathsf{R}^2 \to \mathsf{R}^2$ with $T\left(\begin{bmatrix}1\\0\end{bmatrix}\right) = \begin{bmatrix}1\\2\end{bmatrix}$ and $T\left(\begin{bmatrix}0\\1\end{bmatrix}\right) = \begin{bmatrix}-1\\3\end{bmatrix}$?

(d) $T\colon \mathsf{R}^2 \to \mathsf{R}^2$ with $T\left(\begin{bmatrix}1\\1\end{bmatrix}\right) = \begin{bmatrix}1\\2\end{bmatrix}$ and $T\left(\begin{bmatrix}-1\\1\end{bmatrix}\right) = \begin{bmatrix}-1\\4\end{bmatrix}$?

Explain your answers.

9. For each of the following linear transformations T, find a matrix A such that $T(\mathsf{x}) = A\mathsf{x}$.

(a) [BB] $T\colon \mathsf{R}^2 \to \mathsf{R}^3$, $T(\mathsf{x}) = \begin{bmatrix}3x_1\\x_1 - x_2\\-x_1 + 5x_2\end{bmatrix}$ for $\mathsf{x} = \begin{bmatrix}x_1\\x_2\end{bmatrix}$

(b) $T\colon \mathsf{R}^3 \to \mathsf{R}^2$, $T(\mathsf{x}) = \begin{bmatrix}x_1 - 2x_2 + x_3\\x_1 + 5x_2 - x_3\end{bmatrix}$ for $\mathsf{x} = \begin{bmatrix}x_1\\x_2\\x_3\end{bmatrix}$

(c) $T\colon \mathsf{R}^2 \to \mathsf{R}^4$, $T(\mathsf{x}) = \begin{bmatrix}3x_1 + 2x_2\\x_1 - x_2\\x_2\\7x_1\end{bmatrix}$ for $\mathsf{x} = \begin{bmatrix}x_1\\x_2\end{bmatrix}$

(d) $T\colon \mathsf{R}^5 \to \mathsf{R}^3$, $T(\mathsf{x}) = \begin{bmatrix}x_1 - x_2 + x_5\\2x_3 - x_4 + 5x_5\\-3x_1 + x_2 - 3x_3 + 2x_4 - x_5\end{bmatrix}$ for $\mathsf{x} = \begin{bmatrix}x_1\\x_2\\x_3\\x_4\\x_5\end{bmatrix}$

(e) $T\colon \mathsf{R}^3 \to \mathsf{R}^4$, $T(\mathsf{x}) = \begin{bmatrix}2x_1 - 3x_2 + 7x_3\\8x_1 + x_2 - x_3\\-5x_2 + 6x_3\\-3x_1 - 2x_2 + 9x_3\end{bmatrix}$ for $\mathsf{x} = \begin{bmatrix}x_1\\x_2\\x_3\end{bmatrix}$

(f) $T\colon \mathsf{R}^4 \to \mathsf{R}^3$, $T(\mathsf{x}) = \begin{bmatrix}-x_1 + x_2 + x_3 - 4x_4\\x_1 + 2x_2 + 5x_3 + x_4\\x_1 + x_3 + 3x_4\end{bmatrix}$ for $\mathsf{x} = \begin{bmatrix}x_1\\x_2\\x_3\\x_4\end{bmatrix}$.

10. Find a matrix A such that rotation in R^2 is left multiplication by A given

(a) [BB] $\theta = 30°$ (b) $\theta = \pi/2$ (c) $\theta = -\pi/3$

(d) $\theta = 3\pi/4$ (e) $\theta = -240°$ (f) $\theta = 23\pi/6$.

11. A vector v in the plane is the *reflection* of vector u in the line with equation $y = mx$ if this line is the right bisector of the line that joins the heads of u and v. Find the matrix that reflects vectors in the lines with each of these equations.

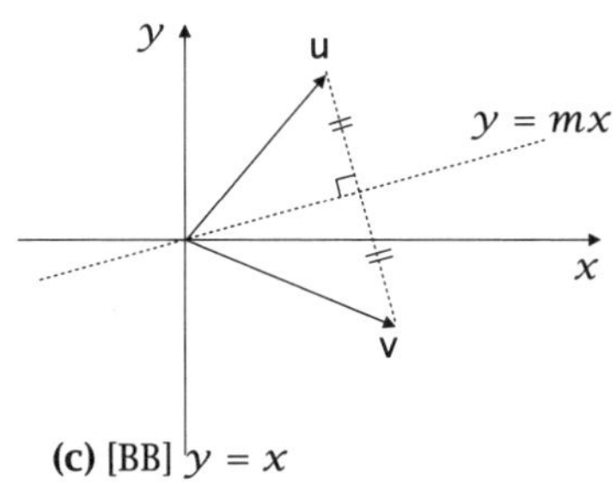

(a) [BB] $y = 0$ (b) $x = 0$ (c) [BB] $y = x$

(d) $y = -x$ (e) $y = \frac{1}{\sqrt{3}}x$ (f) $y = \frac{1}{2}x$.

12. (a) With v a vector in R^3, define $T\colon \mathsf{R}^3 \to \mathsf{R}^3$ by $T(\mathsf{u}) = \operatorname{proj}_{\mathsf{v}} \mathsf{u}$, the projection of u on v. (See Section 1.4.) Prove that T is a linear transformation.

(b) Let $T\colon \mathsf{R}^3 \to \mathsf{R}^3$ be the function that projects vectors onto a plane π. Prove that T is a linear transformation.

[Hint: As in Problem 5.1.19, $T(\mathsf{w}) = T_1(\mathsf{w}) + T_2(\mathsf{w})$ where $T_1(\mathsf{w})$ and $T_2(\mathsf{w})$ have expressions as in part (a). So apply the result of part (a).]

13. Let $T\colon \mathsf{R}^3 \to \mathsf{R}^3$ be projection onto the vector v. Find a matrix A such that $T(\mathsf{x}) = A\mathsf{x}$ in each case.

(a) [BB] $\mathsf{v} = \begin{bmatrix} 1 \\ 2 \\ -1 \end{bmatrix}$ **(b)** $\mathsf{v} = \begin{bmatrix} 3 \\ -5 \\ 4 \end{bmatrix}$ **(c)** $\mathsf{v} = \begin{bmatrix} \sqrt{3} \\ 2 \\ \sqrt{5} \end{bmatrix}$

(d) $\mathsf{v} = \begin{bmatrix} 6 \\ 2 \\ 3 \end{bmatrix}$ **(e)** $\mathsf{v} = \begin{bmatrix} 0 \\ -2 \\ 9 \end{bmatrix}$.

14. Let $T\colon \mathsf{R}^3 \to \mathsf{R}^3$ be the linear transformation which projects vectors onto the plane π. Find a matrix A such that T is left multiplication by A when π has each of these equations.

(a) [BB] $3x - y + 2z = 0$ **(b)** $x + y - 2z = 0$
(c) $2x - 3y + 4z = 0$ **(d)** $3x - 4y + 2z = 0$.

15. Find a matrix A such that the linear transformation T is left multiplication by A in each of the following cases.

(a) [BB] $T\colon \mathsf{R}^2 \to \mathsf{R}^2$, $T\left(\begin{bmatrix} -3 \\ -2 \end{bmatrix}\right) = \begin{bmatrix} 3 \\ 4 \end{bmatrix}$, $T\left(\begin{bmatrix} 2 \\ 1 \end{bmatrix}\right) = \begin{bmatrix} 2 \\ 5 \end{bmatrix}$

(b) $T\colon \mathsf{R}^3 \to \mathsf{R}^3$, $T\left(\begin{bmatrix} 2 \\ 1 \\ 1 \end{bmatrix}\right) = \begin{bmatrix} 1 \\ -1 \\ 0 \end{bmatrix}$, $T\left(\begin{bmatrix} 7 \\ 4 \\ 3 \end{bmatrix}\right) = \begin{bmatrix} 2 \\ -1 \\ 3 \end{bmatrix}$, $T\left(\begin{bmatrix} 0 \\ -1 \\ 0 \end{bmatrix}\right) = \begin{bmatrix} -1 \\ -1 \\ 2 \end{bmatrix}$

(c) $T\colon \mathsf{R}^2 \to \mathsf{R}^3$, $T\left(\begin{bmatrix} 1 \\ -3 \end{bmatrix}\right) = \begin{bmatrix} -1 \\ 0 \\ -1 \end{bmatrix}$, $T\left(\begin{bmatrix} 2 \\ -5 \end{bmatrix}\right) = \begin{bmatrix} -1 \\ -1 \\ 0 \end{bmatrix}$.

Critical Reading

16. Suppose π_1 and π_2 are planes through the origin in R^3. Prove that there exists a linear transformation $T\colon \mathsf{R}^3 \to \mathsf{R}^3$ such that $T(\pi_1) = \pi_2$; that is, any vector y in π_2 is $T(\mathsf{x})$ for some x in π_1.

17. Let $T\colon \mathsf{R}^2 \to \mathsf{R}^2$ be the linear transformation which reflects vectors (pictured as arrows starting at the origin) in the line with equation $y = mx$. Find a matrix A such that T is left multiplication by A. See the sketch accompanying Exercise 11.

18. [BB] Suppose $T\colon \mathsf{R}^2 \to \mathsf{R}^2$ rotates vectors through an angle θ and $S\colon \mathsf{R}^2 \to \mathsf{R}^2$ rotates vectors through $-\theta$. If T is left multiplication by the matrix A and S is left multiplication by B, what is the connection between A and B?

19. The *identity* on a vector space V is the function $\text{id}: V \to V$ defined by $\text{id}(\mathsf{v}) = \mathsf{v}$ for any vector v.

    (a) Show that the identity is a linear transformation.

    (b) In the case that $V = \mathsf{R}^n$, what is the matrix of the identity?

20. Show that any linear transformation $T: \mathsf{R}^2 \to \mathsf{R}^2$ has the form $T\left(\begin{bmatrix} x \\ y \end{bmatrix}\right) = \begin{bmatrix} ax+by \\ cx+dy \end{bmatrix}$ for certain constants a, b, c, d.

21. **(a)** Let V be a vector space of dimension n, let k be a positive integer, $k \le n$, and let $\mathsf{v}_1, \mathsf{v}_2, \ldots, \mathsf{v}_k$ be k linearly independent vectors. Let $\mathsf{w}_1, \mathsf{w}_2, \ldots, \mathsf{w}_k$ be any k vectors in a vector space W. Prove that there exists a linear transformation $T: V \to W$ such that $T(\mathsf{v}_i) = \mathsf{w}_i$ for $i = 1, 2, \ldots, k$. [Hint: Corollary 4.4.4 and 5.1.9.]

    (b) If x is a nonzero vector in R^n and y is any vector in R^m, prove that there exists an $m \times n$ matrix A such that $A\mathsf{x} = \mathsf{y}$.

5.2 Matrix Multiplication Revisited

Given matrices $A = \begin{bmatrix} 1 & 2 \\ 3 & 4 \end{bmatrix}$ and $B = \begin{bmatrix} 2 & 3 \\ 4 & 5 \end{bmatrix}$, why not define $AB = \begin{bmatrix} 2 & 6 \\ 12 & 20 \end{bmatrix}$ as the matrix whose entries are the products of the corresponding entries of A and B? After all, we add matrices entry-by-entry. Why should multiplication be different? We provide the answer in this short section.

In a calculus course, the student will surely have met the notion of function *composition*. If f and g are functions from R to R, the *composition* $f \circ g$ of f and g is defined by

$$(f \circ g)(x) = f(g(x)).$$

For example, if $f(x) = x^2$ and $g(x) = 3x - 1$, then

$$(f \circ g)(x) = f(g(x)) = f(3x-1) = (3x-1)^2$$

while

$$(g \circ f)(x) = g(f(x)) = g(x^2) = 3x^2 - 1.$$

This example makes the point that composition is not a commutative operation: in general, $f \circ g \neq g \circ f$. On the other hand, composition of functions is an associative operation and this is very straightforward to see.

5.2.1 **Theorem.** *Composition of functions is an associative operation: $(f \circ g) \circ h = f \circ (g \circ h)$ whenever f, g and h are functions for which both sides of this equation make sense.*

Proof. Suppose A, B, C and D are sets and $h\colon A \to B$, $g\colon B \to C$ and $f\colon C \to D$ are functions. For any a in A,

$$[(f \circ g) \circ h](a) = (f \circ g)[h(a)] = f(g(h(a))),$$

which is precisely the same as

$$[f \circ (g \circ h)](a) = f[(g \circ h)(a)] = f(g(h(a))).$$ □

Composition of Linear Transformations

Now let V, W and U be vector spaces and let $T\colon V \to W$ and $S\colon W \to U$ be linear transformations. Then the *composition* $S \circ T$ of S and T makes sense, although we usually suppress the $\circ$ and denote this function simply ST. Thus ST is the function $V \to U$ defined by $(ST)(\mathsf{v}) = S(T(\mathsf{v}))$ for v in V.

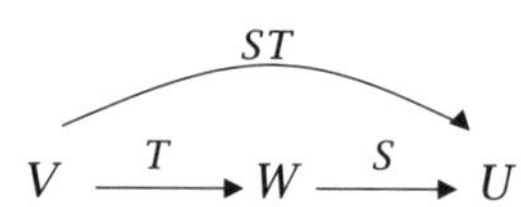

Of importance is the fact that the composition of S and T is not just a function, but also a linear transformation. To see why, let v_1 and v_2 be vectors in V. Then

$$\begin{aligned} (ST)(\mathsf{v}_1 + \mathsf{v}_2) &= S(T(\mathsf{v}_1 + \mathsf{v}_2)) && \text{definition of } ST \\ &= S(T(\mathsf{v}_1) + T(\mathsf{v}_2)) && \text{since } T \text{ is linear} \\ &= S(T(\mathsf{v}_1)) + S(T(\mathsf{v}_2)) && \text{since } S \text{ is linear} \\ &= (ST)(\mathsf{v}_1) + (ST)(\mathsf{v}_2) && \text{definition of } ST \end{aligned}$$

and, for any vector v in V and any scalar c,

$$\begin{aligned} (ST)(c\mathsf{v}) &= S(T(c\mathsf{v})) && \text{definition of } ST \\ &= S(c(T(\mathsf{v}))) && \text{since } T \text{ is linear} \\ &= cS(T(\mathsf{v})) && \text{since } S \text{ is linear} \\ &= c(ST)(\mathsf{v}) && \text{definition of } ST. \end{aligned}$$

When the vector spaces are Euclidean, we saw in Section 5.1 that every linear transformation is multiplication by a matrix. In particular, when V, W and U are Euclidean, ST is multiplication by a matrix and it is reasonable to ask how this matrix is related to the matrices for S and for T.

Let $V = \mathsf{R}^n$, $W = \mathsf{R}^m$ and $U = \mathsf{R}^p$. Then $S\colon \mathsf{R}^m \to \mathsf{R}^p$ is left multiplication by a $p \times m$ matrix A and $T\colon \mathsf{R}^n \to \mathsf{R}^m$ is left multiplication by an $m \times n$ matrix B.

$$\mathsf{R}^n \xrightarrow[B]{T} \mathsf{R}^m \xrightarrow[A]{S} \mathsf{R}^p$$

Let $\mathsf{e}_1, \ldots, \mathsf{e}_n$ be the standard basis vectors for R^n. By Theorem 5.1.13, the first column of the matrix for ST is $(ST)(\mathsf{e}_1) = S(T(\mathsf{e}_1))$. Now $T(\mathsf{e}_1) = \mathsf{b}_1$ is the first column of B, the matrix for T. So the first column of the matrix for ST is

$$(ST)(\mathsf{e}_1) = S(T(\mathsf{e}_1)) = S(\mathsf{b}_1) = A\mathsf{b}_1$$

because S is left multiplication by A. The second column of ST is $(ST)(\mathsf{e}_2) = S(T(\mathsf{e}_2)) = S(\mathsf{b}_2) = A\mathsf{b}_2$, where b_2 denotes the second column of B, and so on. In general, column k of the matrix for ST is $A\mathsf{b}_k$, where b_k is the kth column of B. Thus, the matrix for ST is

$$\begin{bmatrix} A\mathsf{b}_1 & A\mathsf{b}_2 & \cdots & A\mathsf{b}_n \\ \downarrow & \downarrow & & \downarrow \\ & & & \end{bmatrix} = A \begin{bmatrix} \mathsf{b}_1 & \mathsf{b}_2 & \cdots & \mathsf{b}_n \\ \downarrow & \downarrow & & \downarrow \\ & & & \end{bmatrix} = AB,$$

which is the product of the matrices for S and T.

5.2.2 Theorem. *Multiplication of matrices corresponds to composition of linear transformations. If S is left multiplication by matrix A and T is left multiplication by matrix B, then ST is left multiplication by the matrix AB.*

(See Theorem 5.4.13 for a more general version of this theorem.)

5.2.3 Example. Suppose $T\colon \mathsf{R}^3 \to \mathsf{R}^2$ and $S\colon \mathsf{R}^2 \to \mathsf{R}^3$ are defined by

$$T\left(\begin{bmatrix} x \\ y \\ z \end{bmatrix}\right) = \begin{bmatrix} 2x - y \\ y + z \end{bmatrix} \quad \text{and} \quad S\left(\begin{bmatrix} x \\ y \end{bmatrix}\right) = \begin{bmatrix} x + y \\ x - y \\ 3x + 2y \end{bmatrix}.$$

Then $ST\colon \mathsf{R}^3 \to \mathsf{R}^3$ is defined by

$$ST\left(\begin{bmatrix} x \\ y \\ z \end{bmatrix}\right) = S\left(T\left(\begin{bmatrix} x \\ y \\ z \end{bmatrix}\right)\right) = S\left(\begin{bmatrix} 2x - y \\ y + z \end{bmatrix}\right)$$
$$= \begin{bmatrix} (2x - y) + (y + z) \\ (2x - y) - (y + z) \\ 3(2x - y) + 2(y + z) \end{bmatrix} = \begin{bmatrix} 2x + z \\ 2x - 2y - z \\ 6x - y + 2z \end{bmatrix}.$$

By Theorem 5.1.13, the matrices A, B and C for S, T and ST respectively are

$$A = \begin{bmatrix} 1 & 1 \\ 1 & -1 \\ 3 & 2 \end{bmatrix}, \quad B = \begin{bmatrix} 2 & -1 & 0 \\ 0 & 1 & 1 \end{bmatrix}, \quad C = \begin{bmatrix} 2 & 0 & 1 \\ 2 & -2 & -1 \\ 6 & -1 & 2 \end{bmatrix}.$$

In agreement with Theorem 5.2.2, we observe that $C = AB$:

$$\begin{bmatrix} 2 & 0 & 1 \\ 2 & -2 & -1 \\ 6 & -1 & 2 \end{bmatrix} = \begin{bmatrix} 1 & 1 \\ 1 & -1 \\ 3 & 2 \end{bmatrix} \begin{bmatrix} 2 & -1 & 0 \\ 0 & 1 & 1 \end{bmatrix}.$$

The matrix for ST is the product of the matrices for S and T. ☺

READING CHECK 1. With S and T as in Example 5.2.3, the function TS is a linear transformation $\mathsf{R}^2 \to \mathsf{R}^2$. Find $TS(\begin{bmatrix} x \\ y \end{bmatrix})$ and from this the matrix of TS. Then verify that this is the product BA of the matrices of T and S.

5.2.4 Example. Let S and T be the linear transformations $\mathsf{R}^2 \to \mathsf{R}^2$ that rotate vectors through angles α and β, respectively. In Example 5.1.16, we saw that S is left multiplication by $A = \begin{bmatrix} \cos\alpha & -\sin\alpha \\ \sin\alpha & \cos\alpha \end{bmatrix}$ and T is left multiplication by $B = \begin{bmatrix} \cos\beta & -\sin\beta \\ \sin\beta & \cos\beta \end{bmatrix}$. The composition ST is rotation through $\alpha + \beta$, which is left multiplication by matrix $C = \begin{bmatrix} \cos(\alpha+\beta) & -\sin(\alpha+\beta) \\ \sin(\alpha+\beta) & \cos(\alpha+\beta) \end{bmatrix}$. On the other hand, the matrix of ST is AB; thus $AB = C$; that is,

$$\begin{bmatrix} \cos\alpha & -\sin\alpha \\ \sin\alpha & \cos\alpha \end{bmatrix} \begin{bmatrix} \cos\beta & -\sin\beta \\ \sin\beta & \cos\beta \end{bmatrix} = \begin{bmatrix} \cos(\alpha+\beta) & -\sin(\alpha+\beta) \\ \sin(\alpha+\beta) & \cos(\alpha+\beta) \end{bmatrix}.$$

Multiplying the matrices on the left gives

$$\begin{bmatrix} \cos\alpha\cos\beta - \sin\alpha\sin\beta & -\cos\alpha\sin\beta - \sin\alpha\cos\beta \\ \sin\alpha\cos\beta + \cos\alpha\sin\beta & -\sin\alpha\sin\beta + \cos\alpha\cos\beta \end{bmatrix}$$
$$= \begin{bmatrix} \cos(\alpha+\beta) & -\sin(\alpha+\beta) \\ \sin(\alpha+\beta) & \cos(\alpha+\beta) \end{bmatrix},$$

and equating the $(1,1)$ and $(2,1)$ entries establishes two familiar *addition rules* of trigonometry:

$$\begin{aligned} \cos(\alpha+\beta) &= \cos\alpha\cos\beta - \sin\alpha\sin\beta \\ \sin(\alpha+\beta) &= \sin\alpha\cos\beta + \cos\alpha\sin\beta. \end{aligned}$$

Matrix Multiplication is Associative

When you first encountered matrix multiplication, you may have thought it a pretty strange operation. In this section, however, we have seen that it is perfectly natural. It is defined precisely so that the product of matrices corresponds to the composition of the corresponding linear transformations.

Since $f \circ g$ is rarely the same as $g \circ f$, we gain a deeper understanding as to why matrices rarely commute. Since composition of functions is an associative operation, we get an especially pleasing explanation as to why matrix multiplication is associative. We make this explanation explicit.

Let A, B and C be matrices corresponding to the linear transformations S, T and U, respectively (that is, S is left multiplication by A, T is left multiplication by B, and U is left multiplication by C). If the composition $(ST)U = S(TU)$ is defined, then so are the products $(AB)C$ and $A(BC)$. Moreover, since $(AB)C$ is the matrix for $(ST)U$ and $A(BC)$ is the matrix for $S(TU)$, we have $(AB)C = A(BC)$.

Answers to Reading Checks

1. Since $TS\left(\begin{bmatrix} x \\ y \end{bmatrix}\right) = T\left(\begin{bmatrix} x+y \\ x-y \\ 3x+2y \end{bmatrix}\right) = \begin{bmatrix} 2(x+y)-(x-y) \\ (x-y)+(3x+2y) \end{bmatrix} = \begin{bmatrix} x+3y \\ 4x+y \end{bmatrix}$, the matrix for TS is $\begin{bmatrix} 1 & 3 \\ 4 & 1 \end{bmatrix} = \begin{bmatrix} 2 & -1 & 0 \\ 0 & 1 & 1 \end{bmatrix}\begin{bmatrix} 1 & 1 \\ 1 & -1 \\ 3 & 2 \end{bmatrix} = BA$, the product of the matrix of T and the matrix of S.

True/False Questions

Decide, with as little calculation as possible, whether each of the following statements is true or false and, if you say "false," explain your answer. (Answers are at the back of the book.)

1. If $A = \begin{bmatrix} 7 & 0 \\ 3 & 4 \end{bmatrix}$ is the matrix of a linear transformation T, then there exists a vector x in R^2 with $T(\mathsf{x}) = 4\mathsf{x}$.
2. If $T\colon \mathsf{R}^3 \to \mathsf{R}^5$ and $S\colon \mathsf{R}^7 \to \mathsf{R}^3$ are linear transformations, then TS is a linear transformation $\mathsf{R}^7 \to \mathsf{R}^5$.
3. If $T\colon \mathsf{R}^3 \to \mathsf{R}^5$ and $S\colon \mathsf{R}^7 \to \mathsf{R}^3$ are linear transformations, then ST is a linear transformation $\mathsf{R}^7 \to \mathsf{R}^5$.
4. If $f, g\colon \mathsf{R} \to \mathsf{R}$ are defined by $f(x) = 1$ and $g(x) = x$ for all real numbers x, then $(f \circ g)(x) = x$ for all x.
5. If $f, g\colon \mathsf{R} \to \mathsf{R}$ are defined by $f(x) = 1$ and $g(x) = x$ for all real numbers x, then $(g \circ f)(x) = x$ for all x.
6. Composition of functions is a commutative operation.
7. Composition of functions is an associative operation.
8. If T is the linear transformation of the plane that rotates vectors through the angle θ and A is the matrix of T, then the matrix that rotates angles through 2θ is A^2.

9. If $A = \begin{bmatrix} -1 & 0 \\ 0 & -1 \end{bmatrix}$ is the matrix of a linear transformation T, then for any $\mathsf{x} \in \mathsf{R}^2$, $T(T\mathsf{x}) = \mathsf{x}$.

10. If $T\colon \mathsf{R}^2 \to \mathsf{R}^2$ has matrix $A = \begin{bmatrix} 1 & 2 \\ 0 & 3 \end{bmatrix}$ and $S\colon \mathsf{R}^2 \to \mathsf{R}^2$ has matrix $B = \begin{bmatrix} 0 & 3 \\ 2 & 1 \end{bmatrix}$, then TS has matrix $\begin{bmatrix} 4 & 5 \\ 6 & 3 \end{bmatrix}$.

EXERCISES

*Answers to exercises marked [BB] can be found at the **Back of the Book**.*

1. Let S and T be linear transformations. In each of the following situations,

   i. find the vector $ST(\mathsf{x})$;

   ii. use your answer to i. to find the matrix of ST;

   iii. verify that the matrix for ST is the product of the matrices for S and T.

   (a) [BB] $S\colon \mathsf{R}^3 \to \mathsf{R}^4$ is defined by $S\left(\begin{bmatrix} x \\ y \\ z \end{bmatrix}\right) = \begin{bmatrix} x+y-z \\ y-z \\ x+z \\ y \end{bmatrix}$;

   $T\colon \mathsf{R}^2 \to \mathsf{R}^3$ is defined by $T\left(\begin{bmatrix} x \\ y \end{bmatrix}\right) = \begin{bmatrix} 2x+y \\ x-3y \\ 4x \end{bmatrix}$; $\mathsf{x} = \begin{bmatrix} x \\ y \end{bmatrix}$.

   (b) $S\colon \mathsf{R}^3 \to \mathsf{R}^2$ is defined by $S\left(\begin{bmatrix} x \\ y \\ z \end{bmatrix}\right) = \begin{bmatrix} -x+y \\ 2x+y-z \end{bmatrix}$;

   $T\colon \mathsf{R}^2 \to \mathsf{R}^3$ is defined by $T\left(\begin{bmatrix} x \\ y \end{bmatrix}\right) = \begin{bmatrix} 4x+y \\ 2x \\ x-3y \end{bmatrix}$; $\mathsf{x} = \begin{bmatrix} x \\ y \end{bmatrix}$.

   (c) $S\colon \mathsf{R}^2 \to \mathsf{R}^4$ is defined by $S\left(\begin{bmatrix} x \\ y \end{bmatrix}\right) = \begin{bmatrix} 3x-2y \\ x+y \\ -2x+7y \\ -x+3y \end{bmatrix}$;

   $T\colon \mathsf{R}^3 \to \mathsf{R}^2$ is defined by $T\left(\begin{bmatrix} x \\ y \\ z \end{bmatrix}\right) = \begin{bmatrix} x+2y-z \\ 2x-3y+z \end{bmatrix}$; $\mathsf{x} = \begin{bmatrix} x \\ y \\ z \end{bmatrix}$.

2. Let $S\colon \mathsf{R}^2 \to \mathsf{R}^3$ and $T\colon \mathsf{R}^3 \to \mathsf{R}^4$ be the linear transformations defined by
$S\left(\begin{bmatrix} x \\ y \end{bmatrix}\right) = \begin{bmatrix} 2x+3y \\ y \\ -x \end{bmatrix}$ and $T\left(\begin{bmatrix} x \\ y \\ z \end{bmatrix}\right) = \begin{bmatrix} x-y \\ y-z \\ z-x \\ x+y+z \end{bmatrix}$.

   (a) One of ST, TS is defined. Which one?

   (b) Find x and the vector $ST(\mathsf{x})$ or $TS(\mathsf{x})$, as the case may be.

3. [BB] Define $S, T\colon \mathsf{R}^2 \to \mathsf{R}^2$ by $S\left(\begin{bmatrix} x \\ y \end{bmatrix}\right) = \begin{bmatrix} x-y \\ -x+2y \end{bmatrix}$ and $T\left(\begin{bmatrix} x \\ y \end{bmatrix}\right) = \begin{bmatrix} y \\ x \end{bmatrix}$. Suppose S is multiplication by A, T is multiplication by B, ST is multiplication by C, and TS is multiplication by D.

   (a) Find $ST\left(\begin{bmatrix} x \\ y \end{bmatrix}\right)$ and $TS\left(\begin{bmatrix} x \\ y \end{bmatrix}\right)$ using the definition of function composition.

   (b) Find A, B, C and D.

   (c) Verify that $AB = C$ and $BA = D$.

4. Answer Exercise 3 again with $S, T\colon \mathsf{R}^2 \to \mathsf{R}^2$ defined by $S\left(\begin{bmatrix} x \\ y \end{bmatrix}\right) = \begin{bmatrix} 2x-3y \\ x+5y \end{bmatrix}$ and $T\left(\begin{bmatrix} x \\ y \end{bmatrix}\right) = \begin{bmatrix} -y \\ x+7y \end{bmatrix}$.

5. [BB] Given that $S\colon \mathsf{R}^3 \to \mathsf{R}^2$ and $T\colon \mathsf{R}^2 \to \mathsf{R}^2$ are linear transformations such that $T\left(\begin{bmatrix} x \\ y \end{bmatrix}\right) = \begin{bmatrix} x-2y \\ 4x \end{bmatrix}$ and $TS\left(\begin{bmatrix} x \\ y \\ z \end{bmatrix}\right) = \begin{bmatrix} -5x-4y-z \\ 12x-4z \end{bmatrix}$, find the matrix of S and $S\left(\begin{bmatrix} x \\ y \\ z \end{bmatrix}\right)$.

6. Answer Exercise 5 again with $S\colon \mathsf{R}^3 \to \mathsf{R}^2$, $T\colon \mathsf{R}^2 \to \mathsf{R}^2$, $T\left(\begin{bmatrix} x \\ y \end{bmatrix}\right) = \begin{bmatrix} x-2y \\ 3x-5y \end{bmatrix}$ and $TS\left(\begin{bmatrix} x \\ y \\ z \end{bmatrix}\right) = \begin{bmatrix} x-2y-z \\ 3x-z \end{bmatrix}$.

7. [BB] Define $S, T, U\colon \mathsf{R}^2 \to \mathsf{R}^2$ by

$$S\left(\begin{bmatrix} x \\ y \end{bmatrix}\right) = \begin{bmatrix} y \\ x \end{bmatrix}, \quad T\left(\begin{bmatrix} x \\ y \end{bmatrix}\right) = \begin{bmatrix} x+y \\ x-y \end{bmatrix} \quad \text{and} \quad U\left(\begin{bmatrix} x \\ y \end{bmatrix}\right) = \begin{bmatrix} 2x \\ 3x \end{bmatrix}.$$

   (a) Find $ST\left(\begin{bmatrix} x \\ y \end{bmatrix}\right)$ and $TU\left(\begin{bmatrix} x \\ y \end{bmatrix}\right)$.

   (b) Verify that $[(ST)U]\left(\begin{bmatrix} x \\ y \end{bmatrix}\right) = [S(TU)]\left(\begin{bmatrix} x \\ y \end{bmatrix}\right)$.

8. Define $S\colon \mathsf{R}^4 \to \mathsf{R}^2$, $T\colon \mathsf{R}^3 \to \mathsf{R}^4$ and $U\colon \mathsf{R}^2 \to \mathsf{R}^3$ by

$$S\left(\begin{bmatrix} x_1 \\ x_2 \\ x_3 \\ x_4 \end{bmatrix}\right) = \begin{bmatrix} x_1+x_2-x_3-x_4 \\ -x_1+2x_2+x_3-4x_4 \end{bmatrix}, \quad T\left(\begin{bmatrix} x_1 \\ x_2 \\ x_3 \end{bmatrix}\right) = \begin{bmatrix} 2x_1-x_2+x_3 \\ x_2-4x_3 \\ x_1-x_2 \\ x_1+x_2+x_3 \end{bmatrix},$$

$$U\left(\begin{bmatrix} x_1 \\ x_2 \end{bmatrix}\right) = \begin{bmatrix} 2x_2 \\ 3x_2 \\ 4x_1 \end{bmatrix}.$$

   (a) Find $ST\left(\begin{bmatrix} x_1 \\ x_2 \\ x_3 \end{bmatrix}\right)$ and $TU\left(\begin{bmatrix} x_1 \\ x_2 \end{bmatrix}\right)$.

(b) Find $[(ST)U]\left(\begin{bmatrix} x_1 \\ x_2 \end{bmatrix}\right)$ and $[S(TU)]\left(\begin{bmatrix} x_1 \\ x_2 \end{bmatrix}\right)$.

9. Define $S, T, U\colon \mathsf{R}^3 \to \mathsf{R}^3$ by

$$S\left(\begin{bmatrix} x \\ y \\ z \end{bmatrix}\right) = \begin{bmatrix} y \\ z \\ x \end{bmatrix}, \quad T\left(\begin{bmatrix} x \\ y \\ z \end{bmatrix}\right) = \begin{bmatrix} x-y \\ y-z \\ z-x \end{bmatrix} \quad \text{and} \quad U\left(\begin{bmatrix} x \\ y \\ z \end{bmatrix}\right) = \begin{bmatrix} -x \\ y \\ -z \end{bmatrix}.$$

(a) Find $ST\left(\begin{bmatrix} x \\ y \\ z \end{bmatrix}\right)$ and $TU\left(\begin{bmatrix} x \\ y \\ z \end{bmatrix}\right)$.

(b) Verify that $[(ST)U]\left(\begin{bmatrix} x \\ y \\ z \end{bmatrix}\right) = [S(TU)]\left(\begin{bmatrix} x \\ y \\ z \end{bmatrix}\right)$.

(c) Find the matrix for ST and the matrix for U, and use these to find the matrix for $(ST)U$ (all bases standard). If your work so far is correct, this should be the matrix corresponding to part of your answer to (b).

(d) Find the matrix for S and the matrix for TU and use these to find the matrix for $S(TU)$. If your work so far is correct, this should be the matrix corresponding to another part of your answer to (b) and it should also be the matrix you found in (c).

10. Use the approach of Example 5.2.4 to derive the trigonometric identities

$$\cos(\alpha - \beta) = \cos\alpha\cos\beta + \sin\alpha\sin\beta$$
$$\sin(\alpha - \beta) = \sin\alpha\cos\beta - \cos\alpha\sin\beta.$$

Critical Reading

11. Use the methods of this section to find formulas for $\sin 4\theta$ and $\cos 4\theta$ in terms of $\sin\theta$ and $\cos\theta$.

12. Suppose V is a vector space and $T\colon V \to V$ is a linear transformation. Suppose v is a vector in V such that $T(\mathsf{v}) \neq 0$, but $T^2(\mathsf{v}) = \mathsf{0}$ [T^2 means TT, that is $T \circ T$]. Prove that v and $T(\mathsf{v})$ are linearly independent.

5.3 Application: Computer Graphics

Do you play games on your computer? Do you watch television? Go to the movies? However you spend your leisure hours, it is hard to escape the reality that computer graphics are all around us. Perhaps you do not realize that linear algebra is at the heart of all the animated films of today. In this section, we explain how matrices can be used to move and transform objects. We consider only objects that are defined by points in the plane with straight lines joining certain pairs of points, such as the boat shown in the top left of Figure 5.4.

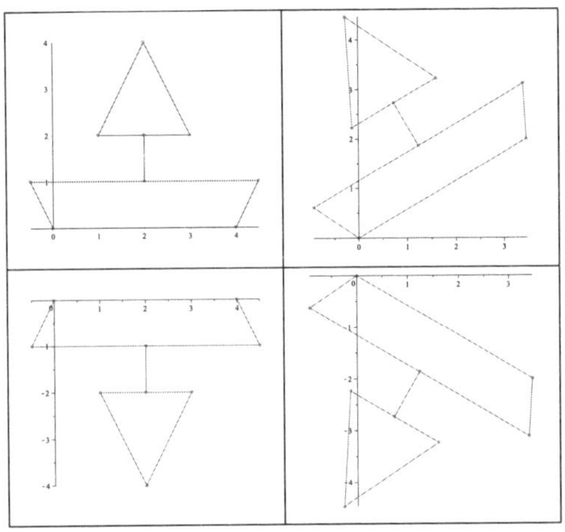

Figure 5.4: A boat rotated, reflected, then rotated and reflected.

Such an object can be moved with the help of a matrix A by moving each point $P(x, y)$ to the point P' whose coordinates are the components of $A\begin{bmatrix} x \\ y \end{bmatrix}$, and then joining P' and Q' with a line if and only if P and Q were joined with a line in the original figure. The matrix A, for example, might be a rotation matrix $\begin{bmatrix} \cos\theta & -\sin\theta \\ \sin\theta & \cos\theta \end{bmatrix}$. To the right, we show the effects of applying such a matrix to a line PQ.

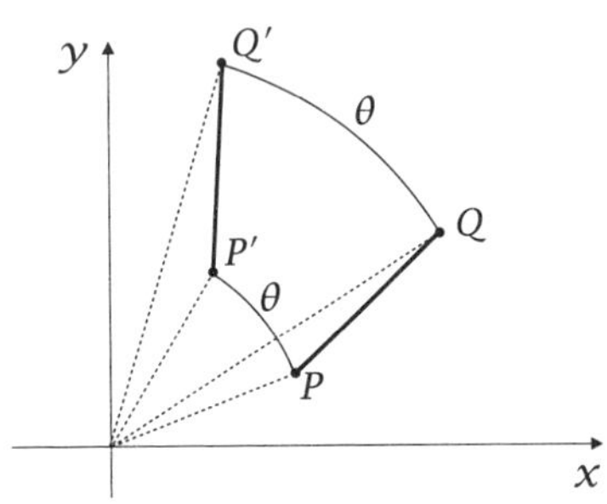

There are many other ways to move and change figures in the plane and, from what we learned in Section 5.1, it is not hard to write down matrices that accomplish these changes. Remember that the columns of the matrix that effect the linear transformation T are $T(\mathbf{e}_1)$ and $T(\mathbf{e}_2)$. For example, reflection in the x-axis leaves fixed $\mathbf{e}_1 = \begin{bmatrix} 1 \\ 0 \end{bmatrix}$, but moves $\mathbf{e}_2 = \begin{bmatrix} 0 \\ 1 \end{bmatrix}$ to the vector $\begin{bmatrix} 0 \\ -1 \end{bmatrix}$, so the matrix that performs this reflection is $\begin{bmatrix} 1 & 0 \\ 0 & -1 \end{bmatrix}$. In general, reflection in the line with equation $y = mx$ is effected by multiplying by $\begin{bmatrix} \frac{1-m^2}{1+m^2} & \frac{2m}{m^2+1} \\ \frac{2m}{m^2+1} & \frac{m^2-1}{m^2+1} \end{bmatrix}$ —see Exercise 17 of Section 5.1. In Table 5.1, we show the matrices that accomplish other transformations of figures. Rotations and reflections are probably known to most readers.

Scaling. A transformation that moves the point (x, y) to (kx, y) is called a *scaling* in the x-direction with factor k. If it moves (x, y) to (x, ky), it's a scaling in the y-direction with factor k.

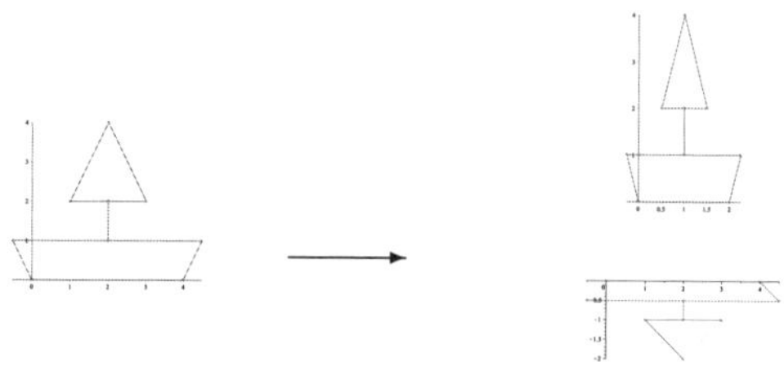

x-direction $k = \frac{1}{2}$

y-direction $k = -\frac{1}{2}$

Shears. A *shear* in the x-direction with factor k moves the point (x, y) to $(x + ky, y)$.

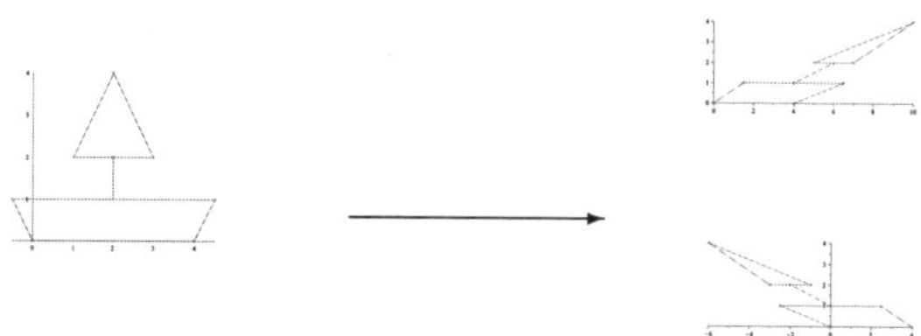

$k = 2$

$k = -2$

A shear in the y-direction with factor k moves the point (x, y) to $(x, y + kx)$.

$k = 1.5$

$k = -1.5$

Transformation	Matrix
Rotation through angle θ	$\begin{bmatrix} \cos\theta & -\sin\theta \\ \sin\theta & \cos\theta \end{bmatrix}$
Reflection in the x-axis	$\begin{bmatrix} 1 & 0 \\ 0 & -1 \end{bmatrix}$
Reflection in the y-axis	$\begin{bmatrix} -1 & 0 \\ 0 & 1 \end{bmatrix}$
Reflection in $y = x$	$\begin{bmatrix} 0 & 1 \\ 1 & 0 \end{bmatrix}$
Reflection in $y = mx$	$\begin{bmatrix} \frac{1-m^2}{1+m^2} & \frac{2m}{m^2+1} \\ \frac{2m}{m^2+1} & \frac{m^2-1}{m^2+1} \end{bmatrix}$
Shear in the x-direction by factor k	$\begin{bmatrix} 1 & k \\ 0 & 1 \end{bmatrix}$
Shear in the y-direction by factor k	$\begin{bmatrix} 1 & 0 \\ k & 1 \end{bmatrix}$
Scaling in the x-direction by factor k	$\begin{bmatrix} k & 0 \\ 0 & 1 \end{bmatrix}$
Scaling in the y-direction by factor k	$\begin{bmatrix} 1 & 0 \\ 0 & k \end{bmatrix}$
Dilation/contraction by k	$\begin{bmatrix} k & 0 \\ 0 & k \end{bmatrix}$

Table 5.1: Some ways to transform figures in the plane and their associated matrices

Dilations/Contractions. A transformation that moves (x, y) to (kx, ky) is a *dilation* if $k > 1$ and a *contraction* if $0 < k < 1$.

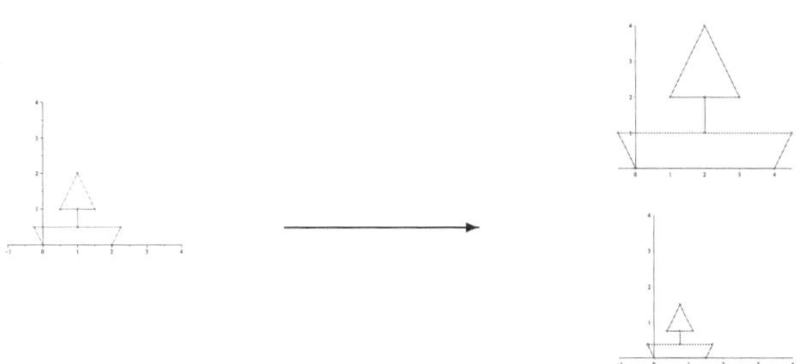

$k = 2$

$k = \frac{3}{4}$

Homogeneous Coordinates

The one obvious transformation of objects in the plane not mentioned so far is a *translation*: a map of the form $(x, y) \mapsto (x + h, y + k)$ for fixed h and k.

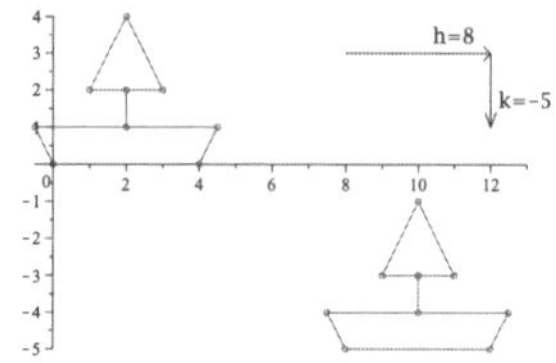

The reason, of course, is that a translation does not fix the origin, so it is not a linear transformation, hence not left multiplication by a matrix, or is it? Look what happens if we multiply the vector $\begin{bmatrix} x \\ y \\ 1 \end{bmatrix}$ by the matrix $\begin{bmatrix} 1 & 0 & h \\ 0 & 1 & k \\ 0 & 0 & 1 \end{bmatrix}$:

$$\begin{bmatrix} 1 & 0 & h \\ 0 & 1 & k \\ 0 & 0 & 1 \end{bmatrix} \begin{bmatrix} x \\ y \\ 1 \end{bmatrix} = \begin{bmatrix} x + h \\ y + k \\ 1 \end{bmatrix}.$$

So translations can be effected by changing $\begin{bmatrix} x \\ y \end{bmatrix}$ to $\begin{bmatrix} x \\ y \\ 1 \end{bmatrix}$ and multiplying by a certain 3×3 matrix. In fact, all the transformations previously discussed in this section can be effected in a similar way, by replacing the 2×2 matrix A with the 3×3 partitioned matrix $\begin{bmatrix} A & 0 \\ 0 & 1 \end{bmatrix}$ and changing the coordinates of the point $P(x, y)$ to what are called its *homogeneous coordinates* $(x, y, 1)$. For example, the matrix $\begin{bmatrix} 1 & 0 \\ 0 & -1 \end{bmatrix}$ that reflects vectors in the x-axis is, in practice, replaced by the 3×3 matrix $\begin{bmatrix} 1 & 0 & 0 \\ 0 & -1 & 0 \\ 0 & 0 & 1 \end{bmatrix}$, and the effect of this reflection on (x, y) is noted by observing the first two components of

$$\begin{bmatrix} 1 & 0 & 0 \\ 0 & -1 & 0 \\ 0 & 0 & 1 \end{bmatrix} \begin{bmatrix} x \\ y \\ 1 \end{bmatrix} = \begin{bmatrix} x \\ -y \\ 1 \end{bmatrix}.$$

Similarly, the rotation matrix $\begin{bmatrix} \cos\theta & -\sin\theta \\ \sin\theta & \cos\theta \end{bmatrix}$ becomes $\begin{bmatrix} \cos\theta & -\sin\theta & 0 \\ \sin\theta & \cos\theta & 0 \\ 0 & 0 & 1 \end{bmatrix}$ and the effect of this rotation on (x, y) is seen from the first two components of

$$\begin{bmatrix} \cos\theta & -\sin\theta & 0 \\ \sin\theta & \cos\theta & 0 \\ 0 & 0 & 1 \end{bmatrix} \begin{bmatrix} x \\ y \\ 1 \end{bmatrix} = \begin{bmatrix} x\cos\theta - y\sin\theta \\ x\sin\theta + y\cos\theta \\ 1 \end{bmatrix}.$$

The reason for using homogeneous coordinates and converting **all** transformation matrices to 3×3 matrices is so that we can compose. For example, the matrix that rotates through 30° and then translates in the direction $(8,-5)$ is

$$\begin{bmatrix} 1 & 0 & 8 \\ 0 & 1 & -5 \\ 0 & 0 & 1 \end{bmatrix} \begin{bmatrix} \cos\frac{\pi}{6} & -\sin\frac{\pi}{6} & 0 \\ \sin\frac{\pi}{6} & \cos\frac{\pi}{6} & 0 \\ 0 & 0 & 1 \end{bmatrix} = \begin{bmatrix} \frac{\sqrt{3}}{2} & -\frac{1}{2} & 8 \\ \frac{1}{2} & \frac{\sqrt{3}}{2} & -5 \\ 0 & 0 & 1 \end{bmatrix}.$$

Since

$$\begin{bmatrix} \frac{\sqrt{3}}{2} & -\frac{1}{2} & 8 \\ \frac{1}{2} & \frac{\sqrt{3}}{2} & -5 \\ 0 & 0 & 1 \end{bmatrix} \begin{bmatrix} x \\ y \\ 1 \end{bmatrix} = \begin{bmatrix} \frac{\sqrt{3}}{2}x - \frac{1}{2}y + 8 \\ \frac{1}{2}x + \frac{\sqrt{3}}{2}y - 5 \\ 1 \end{bmatrix},$$

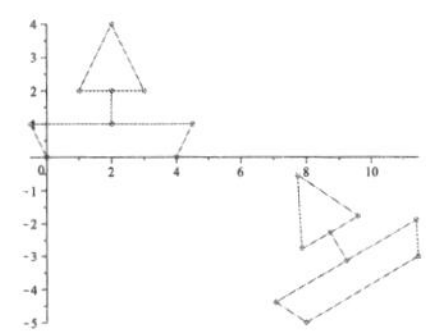

the transformation moves (x,y) to $(\frac{\sqrt{3}}{2}x - \frac{1}{2}y + 8, \frac{1}{2}x + \frac{\sqrt{3}}{2}y - 5)$.

EXERCISES

Answers to exercises marked [BB] can be found at the Back of the Book.

1. In each case, write down the 3×3 matrix that effects the transformation of R^2 described, all coordinates homogeneous.

   (a) [BB] Reflection in the line with equation $y = x$

   (b) Scale in the x-direction by the factor r and in the y-direction by the factor s (in each of the two possible orders)

   (c) Shear in the x-direction by the factor r and in the y-direction by the factor s (in each of the two possible orders)

   (d) Translation in the direction $(4,2)$, that is, four units right and two units up

   (e) Rotation in the clockwise direction through the angle $\theta = \frac{\pi}{4}$

   (f) Reflection in the x-axis, followed by translation in the direction $(-2,1)$

   (g) Counterclockwise rotation through $\frac{\pi}{3}$, then reflection in the y-axis, then translatation in the direction $(3,4)$.

2. The composition of reflection in the x-axis and then reflection in the y-axis is a single transformation, one of those discussed in this section. What is it? Is the product of reflections a reflection?

3. [BB] We have seen that scaling in the x-direction can be achieved upon multiplying by $\begin{bmatrix} k & 0 \\ 0 & 1 \end{bmatrix}$ and in the y-direction upon multiplying by $\begin{bmatrix} 1 & 0 \\ 0 & k \end{bmatrix}$. Suppose you want to scale in the x-direction by the factor k_1 and in the y-direction by the factor k_2. Show that this can be accomplished with just one matrix.

4. **(a)** Explain why multiplication by any elementary matrix corresponds to a reflection in the line with equation $y = x$, or a scaling, or a shear.

(b) Explain why multiplication by an invertible matrix is the product of transformations of the kind described in part (a).

5. **(a)** Show that multiplication by the matrix $\begin{bmatrix} 2 & 4 \\ 1 & 3 \end{bmatrix}$ maps the line with equation $y = 2x$ onto a line. What line?

   (b) Show that multiplication by the matrix in (a) maps the line with equation $y = 2x - 5$ onto a line. What line?

   (c) Show that multiplication by the matrix of part (a) maps any line to a line. (Consider first the case of vertical lines and then lines with equations of the form $y = mx + k$.)

6. **(a)** Show that multiplication by the matrix $A = \begin{bmatrix} a & b \\ c & d \end{bmatrix}$ maps any vertical line to another vertical line.

   (b) Show that multiplication by an invertible matrix $A = \begin{bmatrix} a & b \\ c & d \end{bmatrix}$ maps a line to a line. In particular, each of the transformations discussed in this section map lines to lines.

5.4 The Matrices of a Linear Transformation

In Section 5.1, we defined the term "linear transformation," discussed some elementary properties of linear transformations and showed that any linear transformation $\mathsf{R}^n \to \mathsf{R}^m$ is multiplication by an $m \times n$ matrix. We propose here to investigate linear transformations more deeply and to show that the matrix of a linear transformation is far from unique—a linear transformation has many matrices associated with it.

First, we need the idea of "coordinate vector" and, for this, an elementary result about linearly independent sets.

5.4.1 **Theorem. (Uniqueness of Coefficients)** *Suppose* $\{\mathsf{v}_1, \mathsf{v}_2, \ldots, \mathsf{v}_n\}$ *is a basis for a vector space* V *and* v *is a vector in* V. *Then* v *can be written uniquely in the form* $\mathsf{v} = c_1\mathsf{v}_1 + c_2\mathsf{v}_2 + \cdots + c_n\mathsf{v}_n$; *that is to say, the coefficients* $c_1, c_2, \ldots, c_n$ *are unique.*

Proof. The vector v certainly has a representation of the form $\mathsf{v} = c_1\mathsf{v}_1 + c_2\mathsf{v}_2 + \cdots + c_n\mathsf{v}_n$ because the vectors $\mathsf{v}_1, \mathsf{v}_2, \ldots, \mathsf{v}_n$ span V. The import of the theorem is that there is just one possibility for the scalars $c_1, c_2, \ldots, c_n$. To see this, suppose that v can also be written

$$\mathsf{v} = d_1\mathsf{v}_1 + d_2\mathsf{v}_2 + \cdots + d_n\mathsf{v}_n.$$

Subtracting the two representations of v gives

$$(c_1 - d_1)\mathsf{v}_1 + (c_2 - d_2)\mathsf{v}_2 + \cdots + (c_n - d_n)\mathsf{v}_n = 0.$$

Since the vectors $\mathsf{v}_1, \mathsf{v}_2, \ldots, \mathsf{v}_n$ are linearly independent, all the coefficients here are 0. Thus $c_1 = d_1$, $c_2 = d_2$, ..., $c_n = d_n$. ■

5.4.2 Definition. Let V be a vector space with basis $\mathcal{V} = \{\mathsf{v}_1, \mathsf{v}_2, \ldots, \mathsf{v}_n\}$ and let x be a vector in V. Write $\mathsf{x} = x_1\mathsf{v}_1 + x_2\mathsf{v}_2 + \cdots + x_n\mathsf{v}_n$. The *coordinates of* x *relative to* $\mathcal{V}$ are the unique coefficients $x_1, x_2, \ldots, x_n$ and the *coordinate vector of* x *relative to* $\mathcal{V}$ is the vector $\begin{bmatrix} x_1 \\ x_2 \\ \vdots \\ x_n \end{bmatrix}$. We use the notation $\mathsf{x}_\mathcal{V}$ for this vector. Thus

$$\mathsf{x}_\mathcal{V} = \begin{bmatrix} x_1 \\ x_2 \\ \vdots \\ x_n \end{bmatrix} \quad \text{if and only if} \quad \mathsf{x} = x_1\mathsf{v}_1 + x_2\mathsf{v}_2 + \cdots + x_n\mathsf{v}_n.$$

5.4.3 Example. If $\mathsf{x} = \begin{bmatrix} x \\ y \end{bmatrix}$ is a vector in R^2, the coordinates of x relative to the standard basis $\mathcal{E} = \{\mathsf{e}_1, \mathsf{e}_2\}$ are x and y because $\mathsf{x} = x\mathsf{e}_1 + y\mathsf{e}_2$. Thus $\mathsf{x}_\mathcal{E} = \begin{bmatrix} x \\ y \end{bmatrix}$. More generally, if $\mathcal{E}$ is the standard basis of R^n and $\mathsf{x} = \begin{bmatrix} x_1 \\ x_2 \\ \vdots \\ x_n \end{bmatrix}$, then $\mathsf{x}_\mathcal{E} = \mathsf{x} = \begin{bmatrix} x_1 \\ x_2 \\ \vdots \\ x_n \end{bmatrix}$. ☺

5.4.4 Example. Let $\mathsf{v}_1 = \begin{bmatrix} -1 \\ 2 \end{bmatrix}$ and $\mathsf{v}_2 = \begin{bmatrix} 2 \\ -3 \end{bmatrix}$. Then $\mathcal{V} = \{\mathsf{v}_1, \mathsf{v}_2\}$ is a basis for R^2 and

$$\begin{bmatrix} x \\ y \end{bmatrix} = (3x + 2y)\mathsf{v}_1 + (2x + y)\mathsf{v}_2, \qquad (1)$$

so the coordinates of $\mathsf{x} = \begin{bmatrix} x \\ y \end{bmatrix}$ relative to $\mathcal{V}$ are $3x + 2y$ and $2x + y$ and $\mathsf{x}_\mathcal{V} = \begin{bmatrix} 3x + 2y \\ 2x + y \end{bmatrix}$. Note the difference between the vector $\mathsf{x} = \begin{bmatrix} x \\ y \end{bmatrix}$ and its **coordinate vector** $\mathsf{x}_\mathcal{V} = \begin{bmatrix} 3x + 2y \\ 2x + y \end{bmatrix}$ relative to the basis $\mathcal{V}$. ☺

READING CHECK 1. How would you find the coefficients $3x+2y$ and $2x+y$ in equation (1) if they weren't handed to you on a platter?

5.4.5 Example. Let $\mathsf{v}_1 = \begin{bmatrix}1\\0\\0\end{bmatrix}$, $\mathsf{v}_2 = \begin{bmatrix}1\\1\\0\end{bmatrix}$ and $\mathsf{v}_3 = \begin{bmatrix}1\\1\\1\end{bmatrix}$. The set $\mathcal{V} = \{\mathsf{v}_1, \mathsf{v}_2, \mathsf{v}_3\}$ is a basis for R^3. The vector $\mathsf{x} = \begin{bmatrix}x\\y\\z\end{bmatrix}$ can be written uniquely

$$\begin{bmatrix}x\\y\\z\end{bmatrix} = (x-y)\mathsf{v}_1 + (y-z)\mathsf{v}_2 + z\mathsf{v}_3,$$

so the coordinates of x relative to $\mathcal{V}$ are $x-y$, $y-z$ and z: $\mathsf{x}_{\mathcal{V}} = \begin{bmatrix}x-y\\y-z\\z\end{bmatrix}$. ⌣̈

5.4.6 Example. Let $V = M_2(\mathsf{R})$ be the vector space of 2×2 matrices over R. Let

$$E_{11} = \begin{bmatrix}1&0\\0&0\end{bmatrix}, \quad E_{12} = \begin{bmatrix}0&1\\0&0\end{bmatrix}, \quad E_{21} = \begin{bmatrix}0&0\\1&0\end{bmatrix}, \quad E_{22} = \begin{bmatrix}0&0\\0&1\end{bmatrix}.$$

Then $\mathcal{E} = \{E_{11}, E_{12}, E_{21}, E_{22}\}$ is a basis for V and relative to this basis, $\mathsf{x} = \begin{bmatrix}a&b\\c&d\end{bmatrix}$ has coordinate vector $\begin{bmatrix}a\\b\\c\\d\end{bmatrix}$: $\mathsf{x}_{\mathcal{E}} = \begin{bmatrix}a\\b\\c\\d\end{bmatrix}$. ⌣̈

Before continuing, we record a fact that will soon prove useful.

5.4.7 Lemma. *If V is a vector space of dimension n with basis $\mathcal{V}$, then the function $V \to \mathsf{R}^n$ defined by $\mathsf{x} \mapsto \mathsf{x}_{\mathcal{V}}$, which assigns to each vector in V its coordinate vector relative to $\mathcal{V}$, is a linear transformation.*

Proof. Call the given function T. Thus $T(\mathsf{x}) = \mathsf{x}_{\mathcal{V}}$ for $\mathsf{x} \in V$. Let

$$\mathsf{x} = x_1\mathsf{v}_1 + x_2\mathsf{v}_2 + \cdots + x_n\mathsf{v}_n$$

and

$$\mathsf{y} = y_1\mathsf{v}_1 + y_2\mathsf{v}_2 + \cdots + y_n\mathsf{v}_n$$

be vectors in V. The associated coordinate vectors are

$$\mathsf{x}_{\mathcal{V}} = \begin{bmatrix} x_1 \\ x_2 \\ \vdots \\ x_n \end{bmatrix} \quad \text{and} \quad \mathsf{y}_{\mathcal{V}} = \begin{bmatrix} y_1 \\ y_2 \\ \vdots \\ y_n \end{bmatrix}.$$

Since $\mathsf{x} + \mathsf{y} = (x_1 + y_1)\mathsf{v}_1 + (x_2 + y_2)\mathsf{v}_2 + \cdots + (x_n + y_n)\mathsf{v}_n$, the coordinate vector of $\mathsf{x} + \mathsf{y}$ relative to $\mathcal{V}$ is

$$T(\mathsf{x} + \mathsf{y}) = \begin{bmatrix} x_1 + y_1 \\ x_2 + y_2 \\ \vdots \\ x_n + y_n \end{bmatrix} = \begin{bmatrix} x_1 \\ x_2 \\ \vdots \\ x_n \end{bmatrix} + \begin{bmatrix} y_1 \\ y_2 \\ \vdots \\ y_n \end{bmatrix} = T(\mathsf{x}) + T(\mathsf{y}).$$

Thus T preserves vector addition. Let c be a scalar. Since

$$c\mathsf{x} = (cx_1)\mathsf{v}_1 + (cx_2)\mathsf{v}_2 + \cdots + (cx_n)\mathsf{v}_n,$$

the coordinate vector of $c\mathsf{x}$ is $\begin{bmatrix} cx_1 \\ cx_2 \\ \vdots \\ cx_n \end{bmatrix} = c\begin{bmatrix} x_1 \\ x_2 \\ \vdots \\ x_n \end{bmatrix} = cT(\mathsf{x})$. Thus T also preserves scalar multiples. ■

5.4.8 Remark. The lemma says that any finite dimensional vector space V is "essentially" R^n, because vectors x in V can be replaced by their coordinate vectors $\mathsf{x}_{\mathcal{V}}$ relative to a basis $\mathcal{V}$. This is the reason we have chosen so frequently in this text to make "vector space" synonymous with "Euclidean n-space."

To illustrate, the set $V = \{a + bx + cx^2 \mid a, b, c \in \mathsf{R}\}$ of polynomials with real coefficients and of degree at most 2 is a three-dimensional vector space with basis $\{1, x, x^2\}$ relative to which $f(x) = a + bx + cx^2$ has coordinate vector $\begin{bmatrix} a \\ b \\ c \end{bmatrix}$. The linear transformation T of Lemma 5.4.7 is $f(x) \mapsto \begin{bmatrix} a \\ b \\ c \end{bmatrix}$. Thus working with V is essentially working in R^3. The set $M_2(\mathsf{R}) = \{ \begin{bmatrix} a & b \\ c & d \end{bmatrix} \mid a, b, c, d \in \mathsf{R}\}$ of 2×2 matrices over R is a vector space of dimension 4 with basis $\{E_{11}, E_{12}, E_{21}, E_{22}\}$, where

$$E_{11} = \begin{bmatrix} 1 & 0 \\ 0 & 0 \end{bmatrix}, \quad E_{12} = \begin{bmatrix} 0 & 1 \\ 0 & 0 \end{bmatrix}, \quad E_{21} = \begin{bmatrix} 0 & 0 \\ 1 & 0 \end{bmatrix}, \quad E_{22} = \begin{bmatrix} 0 & 0 \\ 0 & 1 \end{bmatrix},$$

relative to which $\begin{bmatrix} a & b \\ c & d \end{bmatrix}$ has coordinate vector $\begin{bmatrix} a \\ b \\ c \\ d \end{bmatrix}$. The linear transformation T of Lemma 5.4.7 is $\begin{bmatrix} a & b \\ c & d \end{bmatrix} \mapsto \begin{bmatrix} a \\ b \\ c \\ d \end{bmatrix}$. So when we are working in $M_2(\mathsf{R})$, we are effectively working in R^4.

The Matrix of T Relative to $\mathcal{V}$ and $\mathcal{W}$

Suppose we have a linear transformation $T\colon V \to W$, where

- V is a vector space with basis $\mathcal{V} = \{\mathsf{v}_1, \mathsf{v}_2, \ldots, \mathsf{v}_n\}$,
- W is a vector space with basis $\mathcal{W}$.

Let x be a vector in V. We are going to explore the connection between the coordinate vector $\mathsf{x}_{\mathcal{V}}$ of x relative to $\mathcal{V}$ and the coordinate vector $[T(\mathsf{x})]_{\mathcal{W}}$ of $T(\mathsf{x})$ relative to $\mathcal{W}$.

To begin, let the coordinate vector of x relative to $\mathcal{V}$ be $\mathsf{x}_{\mathcal{V}} = \begin{bmatrix} x_1 \\ x_2 \\ \vdots \\ x_n \end{bmatrix}$. Thus $\mathsf{x} = x_1\mathsf{v}_1 + x_2\mathsf{v}_2 + \cdots + x_n\mathsf{v}_n$. Since T preserves linear combinations,

$$T(\mathsf{x}) = x_1 T(\mathsf{v}_1) + x_2 T(\mathsf{v}_2) + \cdots + x_n T(\mathsf{v}_n).$$

By Lemma 5.4.7, the function $\mathsf{w} \mapsto \mathsf{w}_{\mathcal{W}}$ is a linear transformation, so it too preserves linear combinations. Thus, the coordinate vector of $T(\mathsf{x})$ relative to $\mathcal{W}$ is

$$\begin{aligned} [T(\mathsf{x})]_{\mathcal{W}} &= x_1[T(\mathsf{v}_1)]_{\mathcal{W}} + x_2[T(\mathsf{v}_2)]_{\mathcal{W}} + \cdots + x_n[T(\mathsf{v}_n)]_{\mathcal{W}} \\ &= \begin{bmatrix} [T(\mathsf{v}_1)]_{\mathcal{W}} & [T(\mathsf{v}_2)]_{\mathcal{W}} & \cdots & [T(\mathsf{v}_n)]_{\mathcal{W}} \\ \downarrow & \downarrow & & \downarrow \\ & & & \end{bmatrix} \begin{bmatrix} x_1 \\ x_2 \\ \vdots \\ x_n \end{bmatrix} = A \begin{bmatrix} x_1 \\ x_2 \\ \vdots \\ x_n \end{bmatrix} = A\mathsf{x}_{\mathcal{V}}, \end{aligned} \tag{2}$$

where

$$A = \begin{bmatrix} [T(\mathsf{v}_1)]_{\mathcal{W}} & [T(\mathsf{v}_2)]_{\mathcal{W}} & \cdots & [T(\mathsf{v}_n)]_{\mathcal{W}} \\ \downarrow & \downarrow & & \downarrow \\ & & & \end{bmatrix}$$

is the matrix whose columns are the coordinate vectors of $T(\mathsf{v}_1)$, $T(\mathsf{v}_2)$, $\ldots, T(\mathsf{v}_n)$ relative to $\mathcal{W}$. So T is effectively left multiplication by a matrix in the sense that

$$[T(\mathsf{x})]_{\mathcal{W}} = A\mathsf{x}_{\mathcal{V}}. \tag{3}$$

We use the notation $M_{\mathcal{W}\leftarrow\mathcal{V}}(T)$ for the matrix A in equation (3) and call this the *matrix of T relative to $\mathcal{V}$ and $\mathcal{W}$*.

5.4.9 Definition. Let $\mathcal{V} = \{\mathsf{v}_1, \mathsf{v}_2, \dots, \mathsf{v}_n\}$ and $\mathcal{W}$ be bases for vector spaces V and W. Let $T\colon V \to W$ be a linear transformation. Then the *matrix of T relative to $\mathcal{V}$ and $\mathcal{W}$* is the matrix

$$M_{\mathcal{W}\leftarrow\mathcal{V}}(T) = \begin{bmatrix} [T(\mathsf{v}_1)]_{\mathcal{W}} & [T(\mathsf{v}_2)]_{\mathcal{W}} & \cdots & [T(\mathsf{v}_n)]_{\mathcal{W}} \\ \downarrow & \downarrow & & \downarrow \end{bmatrix}$$

whose columns are the coordinate vectors of $T(\mathsf{v}_1), T(\mathsf{v}_2), \dots, T(\mathsf{v}_n)$ relative to $\mathcal{W}$. This matrix has the property that

5.4.10
$$\boxed{[T(\mathsf{x})]_{\mathcal{W}} = M_{\mathcal{W}\leftarrow\mathcal{V}}(T)\mathsf{x}_{\mathcal{V}}}$$

for any vector x in V.

The author hopes the notation is helpful. In plain English, equation (5.4.10) says that if you want the coordinates of $T(\mathsf{x})$ relative to $\mathcal{W}$, you take the coordinate vector of x relative to $\mathcal{V}$ and multiply on the left by the matrix $M_{\mathcal{W}\leftarrow\mathcal{V}}(T)$.

5.4.11 Example. Let $T\colon \mathsf{R}^n \to \mathsf{R}^m$ be a linear transformation. Let $\mathcal{E}$ and $\mathcal{F}$ be the standard bases for R^n and R^m respectively. Thus, for any x in R^n, $\mathsf{x}_{\mathcal{E}} = \mathsf{x}$ and $[T(\mathsf{x})]_{\mathcal{F}} = T(x)$. The matrix $M_{\mathcal{F}\leftarrow\mathcal{E}}(T)$

$$= \begin{bmatrix} [T(\mathsf{e}_1)]_{\mathcal{F}} & [T(\mathsf{e}_2)]_{\mathcal{F}} & \cdots & [T(\mathsf{e}_n)]_{\mathcal{F}} \\ \downarrow & \downarrow & & \downarrow \end{bmatrix} = \begin{bmatrix} T(\mathsf{e}_1) & T(\mathsf{e}_2) & \cdots & T(\mathsf{e}_n) \\ \downarrow & \downarrow & & \downarrow \end{bmatrix}$$

is the matrix defined in Theorem 5.1.13. Equation (5.4.10) simply reiterates that T is left multiplication by A. ☺

5.4.12 Example. Let $V = W = \mathsf{R}^2$ and let $T\colon V \to V$ be the linear transformation defined by $T(\begin{bmatrix} x \\ y \end{bmatrix}) = \begin{bmatrix} y \\ x \end{bmatrix}$. The set $\mathcal{V} = \{\mathsf{v}_1, \mathsf{v}_2\}$, with $\mathsf{v}_1 = \begin{bmatrix} -1 \\ 2 \end{bmatrix}$ and $\mathsf{v}_2 = \begin{bmatrix} 2 \\ -3 \end{bmatrix}$, is a basis for V. Let $\mathcal{E} = \{\mathsf{e}_1, \mathsf{e}_2\}$ be the standard basis for V. We find the four matrices $M_{\mathcal{E}\leftarrow\mathcal{E}}(T)$, $M_{\mathcal{E}\leftarrow\mathcal{V}}(T)$, $M_{\mathcal{V}\leftarrow\mathcal{E}}(T)$ and $M_{\mathcal{V}\leftarrow\mathcal{V}}(T)$ beginning with $M_{\mathcal{E}\leftarrow\mathcal{E}}(T)$, the matrix of T relative to $\mathcal{E}$ and $\mathcal{E}$.

The columns of this matrix are the coordinate vectors of $T(\mathsf{e}_1)$ and $T(\mathsf{e}_2)$ relative to $\mathcal{E}$. We have

$$T(\mathsf{e}_1) = T\left(\begin{bmatrix}1\\0\end{bmatrix}\right) = \begin{bmatrix}0\\1\end{bmatrix} \quad \text{and} \quad T(\mathsf{e}_2) = T\left(\begin{bmatrix}0\\1\end{bmatrix}\right) = \begin{bmatrix}1\\0\end{bmatrix},$$

and we have observed that the coordinate vector of a vector in R^2 relative to the standard basis $\mathcal{E}$ is the vector itself—see Example 5.4.3. Thus

$$M_{\mathcal{E}\leftarrow\mathcal{E}}(T) = \begin{bmatrix} 0 & 1 \\ 1 & 0 \end{bmatrix}$$

just as in Section 5.1 where, implicitly, all matrices of linear transformations were formed with respect to standard bases. Equation (5.4.10) states the obvious fact:

$$[T(\mathsf{x})]_{\mathcal{E}} = \begin{bmatrix}y\\x\end{bmatrix} = \begin{bmatrix} 0 & 1 \\ 1 & 0 \end{bmatrix}\begin{bmatrix}x\\y\end{bmatrix} = M_{\mathcal{E}\leftarrow\mathcal{E}}(T)\mathsf{x}_{\mathcal{E}}.$$

Now let's find $M_{\mathcal{E}\leftarrow\mathcal{V}}(T)$, the matrix of T relative to $\mathcal{V}$ and $\mathcal{E}$. The columns of this matrix are the coordinate vectors of $T(\mathsf{v}_1)$ and $T(\mathsf{v}_2)$ relative to $\mathcal{E}$. We find

$$T(\mathsf{v}_1) = T\left(\begin{bmatrix}-1\\2\end{bmatrix}\right) = \begin{bmatrix}2\\-1\end{bmatrix} = 2\mathsf{e}_1 + (-1)\mathsf{e}_2, \quad \text{so that} \quad [T(\mathsf{v}_1)]_{\mathcal{E}} = \begin{bmatrix}2\\-1\end{bmatrix};$$

$$T(\mathsf{v}_2) = T\left(\begin{bmatrix}2\\-3\end{bmatrix}\right) = \begin{bmatrix}-3\\2\end{bmatrix} = -3\mathsf{e}_1 + 2\mathsf{e}_2, \quad \text{so that} \quad [T(\mathsf{v}_2)]_{\mathcal{E}} = \begin{bmatrix}-3\\2\end{bmatrix}.$$

Thus $M_{\mathcal{E}\leftarrow\mathcal{V}}(T) = \begin{bmatrix} 2 & -3 \\ -1 & 2 \end{bmatrix}$. Let's be sure we understand the sense in which T is multiplication by this matrix.

Using 5.4.4, we know that the coordinate vector of $\mathsf{x} = \begin{bmatrix}x\\y\end{bmatrix}$ relative to $\mathcal{V}$ is $\mathsf{x}_{\mathcal{V}} = \begin{bmatrix}3x+2y\\2x+y\end{bmatrix}$. Thus

$$\begin{aligned} M_{\mathcal{E}\leftarrow\mathcal{V}}(T)\mathsf{x}_{\mathcal{V}} &= \begin{bmatrix} 2 & -3 \\ -1 & 2 \end{bmatrix}\begin{bmatrix}3x+2y\\2x+y\end{bmatrix} \\ &= \begin{bmatrix}2(3x+2y)-3(2x+y)\\-(3x+2y)+2(2x+y)\end{bmatrix} = \begin{bmatrix}y\\x\end{bmatrix} = [T(\mathsf{x})]_{\mathcal{E}}, \end{aligned}$$

the coordinate vector $T(\mathsf{x})$ relative to $\mathcal{E}$.

Next, let us find $M_{\mathcal{V}\leftarrow\mathcal{E}}(T)$, the matrix of T relative to $\mathcal{E}$ and $\mathcal{V}$. The columns of T are the coordinate vectors of $T(\mathsf{e}_1)$ and $T(\mathsf{e}_2)$ relative to $\mathcal{V}$. Again using what we discovered in Example 5.4.4, the coordinates of $\begin{bmatrix}1\\0\end{bmatrix}$ relative to $\mathcal{V}$

are 3 and 2, and the coordinates of $\begin{bmatrix} 0 \\ 1 \end{bmatrix}$ relative to $\mathcal{V}$ are 2 and 1. We have

$$T(\mathsf{e}_1) = \begin{bmatrix} 0 \\ 1 \end{bmatrix} \quad \text{so that} \quad [T(\mathsf{e}_1)]_{\mathcal{V}} = \begin{bmatrix} 2 \\ 1 \end{bmatrix};$$

$$T(\mathsf{e}_2) = \begin{bmatrix} 1 \\ 0 \end{bmatrix} \quad \text{so that} \quad [T(\mathsf{e}_2)]_{\mathcal{V}} = \begin{bmatrix} 3 \\ 2 \end{bmatrix}.$$

Thus $M_{\mathcal{V}\leftarrow\mathcal{E}}(T) = \begin{bmatrix} 2 & 3 \\ 1 & 2 \end{bmatrix}$. Again, we emphasize the sense in which T is multiplication by this matrix:

$$M_{\mathcal{V}\leftarrow\mathcal{E}}(T)\mathsf{x}_{\mathcal{E}} = \begin{bmatrix} 2 & 3 \\ 1 & 2 \end{bmatrix}\begin{bmatrix} x \\ y \end{bmatrix} = \begin{bmatrix} 2x+3y \\ x+2y \end{bmatrix} = [T(\mathsf{x})]_{\mathcal{V}},$$

the coordinate vector $[T(\mathsf{x})]_{\mathcal{V}}$ of $T(\mathsf{x})$ relative to $\mathcal{V}$—see Example 5.4.4.

Finally, we find $M_{\mathcal{V}\leftarrow\mathcal{V}}(T)$, the matrix of T relative to $\mathcal{V}$ and $\mathcal{V}$. The columns of T are the coordinate vectors of $T(\mathsf{v}_1)$ and $T(\mathsf{v}_2)$ relative to $\mathcal{V}$. Again Example 5.4.4 proves useful. We have

$$T(\mathsf{v}_1) = \begin{bmatrix} 2 \\ -1 \end{bmatrix} = 4\mathsf{v}_1 + 3\mathsf{v}_2, \qquad \text{so that} \quad [T(\mathsf{v}_1)]_{\mathcal{V}} = \begin{bmatrix} 4 \\ 3 \end{bmatrix}$$

and

$$T(\mathsf{v}_2) = \begin{bmatrix} -3 \\ 2 \end{bmatrix} = -5\mathsf{v}_1 - 4\mathsf{v}_2, \quad \text{so that} \quad [T(\mathsf{v}_2)]_{\mathcal{V}} = \begin{bmatrix} -5 \\ -4 \end{bmatrix}.$$

Thus $M_{\mathcal{V}\leftarrow\mathcal{V}}(T) = \begin{bmatrix} 4 & -5 \\ 3 & -4 \end{bmatrix}$. Since the coordinate vector of $\mathsf{x} = \begin{bmatrix} x \\ y \end{bmatrix}$ relative to $\mathcal{V}$ is $\mathsf{x}_{\mathcal{V}} = \begin{bmatrix} 3x+2y \\ 2x+y \end{bmatrix}$, the linear transformation T is left multiplication by $M_{\mathcal{V}\leftarrow\mathcal{V}}(T)$ in the sense that

$$\begin{aligned} M_{\mathcal{V}\leftarrow\mathcal{V}}(T)\mathsf{x}_{\mathcal{V}} &= \begin{bmatrix} 4 & -5 \\ 3 & -4 \end{bmatrix}\begin{bmatrix} 3x+2y \\ 2x+y \end{bmatrix} \\ &= \begin{bmatrix} 4(3x+2y) - 5(2x+y) \\ 3(3x+2y) - 4(2x+y) \end{bmatrix} = \begin{bmatrix} 2x+3y \\ x+2y \end{bmatrix} = [T(\mathsf{x})]_{\mathcal{V}}, \end{aligned}$$

the coordinate vector $T(\mathsf{x}) = \begin{bmatrix} y \\ x \end{bmatrix}$ relative to $\mathcal{V}$. ☺

Matrix Multiplication is Composition of Linear Transformations

In Section 5.2, we showed that matrix multiplication corresponds to composition of linear transformations—see Theorem 5.2.2. At that point, all matrices were implicitly formed with respect to standard bases. Here is the more general statement, whose proof is left to the exercises.

5.4.13 Theorem. *[Matrix multiplication corresponds to composition of linear transformations] Let $T\colon V \to W$ and $S\colon W \to U$ be linear transformations. Let $\mathcal{V}$, $\mathcal{W}$ and $\mathcal{U}$ be bases for V, W and U respectively. Then*

$$\boxed{M_{\mathcal{U}\leftarrow\mathcal{V}}(ST) = M_{\mathcal{U}\leftarrow\mathcal{W}}(S)M_{\mathcal{W}\leftarrow\mathcal{V}}(T).}$$

5.4.14 Example. Let $V = \mathsf{R}^2$, $W = \mathsf{R}^3$ and $U = \mathsf{R}^4$. Let $T\colon V \to W$ and $S\colon W \to U$ be the linear transformations defined by

$$T\left(\begin{bmatrix} x_1 \\ x_2 \end{bmatrix}\right) = \begin{bmatrix} 2x_1 - 3x_2 \\ x_2 \\ x_1 + x_2 \end{bmatrix} \quad \text{and} \quad S\left(\begin{bmatrix} x_1 \\ x_2 \\ x_3 \end{bmatrix}\right) = \begin{bmatrix} x_1 - x_2 - x_3 \\ 2x_2 + x_3 \\ x_1 - x_3 \\ 3x_1 + x_2 - 2x_3 \end{bmatrix}.$$

Let V, W, U have, respectively, the bases

$$\mathcal{V} = \{\mathsf{v}_1, \mathsf{v}_2\}, \quad \text{where } \mathsf{v}_1 = \begin{bmatrix} 1 \\ 1 \end{bmatrix}, \mathsf{v}_2 = \begin{bmatrix} -1 \\ 1 \end{bmatrix};$$

$$\mathcal{W} = \{\mathsf{w}_1, \mathsf{w}_2, \mathsf{w}_3\}, \quad \text{where } \mathsf{w}_1 = \begin{bmatrix} 1 \\ 0 \\ 0 \end{bmatrix}, \mathsf{w}_2 = \begin{bmatrix} 1 \\ 1 \\ 0 \end{bmatrix}, \mathsf{w}_3 = \begin{bmatrix} 1 \\ 1 \\ 1 \end{bmatrix};$$

$$\mathcal{U} = \{\mathsf{e}_1, \mathsf{e}_2, \mathsf{e}_3, \mathsf{e}_4\} = \mathcal{E}, \quad \text{the standard basis of } \mathsf{R}^4.$$

First, we find $M_{\mathcal{W}\leftarrow\mathcal{V}}(T)$, the matrix of T relative to $\mathcal{V}$ and $\mathcal{W}$. In view of Definition 5.4.9, to do so we must find the coordinate vectors of $T(\mathsf{v}_1)$ and $T(\mathsf{v}_2)$ relative to $\mathcal{W}$. (These vectors are the columns of the desired matrix.) In Example 5.4.5, we showed that the coordinate vector of $\mathsf{x} = \begin{bmatrix} x_1 \\ x_2 \\ x_3 \end{bmatrix}$ relative to $\mathcal{W}$ is $\begin{bmatrix} x_1 - x_2 \\ x_2 - x_3 \\ x_3 \end{bmatrix}$, so

$$T(\mathsf{v}_1) = \begin{bmatrix} -1 \\ 1 \\ 2 \end{bmatrix} \quad \text{has coordinate vector} \quad [T(\mathsf{v}_1)]_{\mathcal{W}} = \begin{bmatrix} -2 \\ -1 \\ 2 \end{bmatrix},$$

$$T(\mathsf{v}_2) = \begin{bmatrix} -5 \\ 1 \\ 0 \end{bmatrix} \quad \text{has coordinate vector} \quad [T(\mathsf{v}_2)]_{\mathcal{W}} = \begin{bmatrix} -6 \\ 1 \\ 0 \end{bmatrix},$$

and $M_{\mathcal{W}\leftarrow\mathcal{V}}(T) = \begin{bmatrix} -2 & -6 \\ -1 & 1 \\ 2 & 0 \end{bmatrix}$. Next, we find $M_{\mathcal{E}\leftarrow\mathcal{W}}(S)$, the matrix of S relative to $\mathcal{W}$ and $\mathcal{E}$. We have

$$S(\mathsf{w}_1) = \begin{bmatrix} 1 \\ 0 \\ 1 \\ 3 \end{bmatrix} = [S(\mathsf{w}_1)]_{\mathcal{E}}, \quad S(\mathsf{w}_2) = \begin{bmatrix} 0 \\ 2 \\ 1 \\ 4 \end{bmatrix} = [S(\mathsf{w}_2)]_{\mathcal{E}},$$

$$\text{and} \quad S(\mathsf{w}_3) = \begin{bmatrix} -1 \\ 3 \\ 0 \\ 2 \end{bmatrix} = [S(\mathsf{w}_3)]_{\mathcal{E}},$$

so $M_{\mathcal{E}\leftarrow\mathcal{W}}(S) = \begin{bmatrix} 1 & 0 & -1 \\ 0 & 2 & 3 \\ 1 & 1 & 0 \\ 3 & 4 & 2 \end{bmatrix}$. The linear transformation $ST\colon V \to U$ is defined by $ST(\mathsf{v}) = S(T(\mathsf{v}))$, so

$$ST(\mathsf{v}_1) = S\left(\begin{bmatrix} -1 \\ 1 \\ 2 \end{bmatrix}\right) = \begin{bmatrix} -4 \\ 4 \\ -3 \\ -6 \end{bmatrix} = [ST(\mathsf{v}_1)]_{\mathcal{E}}$$

and

$$ST(\mathsf{v}_2) = S\left(\begin{bmatrix} -5 \\ 1 \\ 0 \end{bmatrix}\right) = \begin{bmatrix} -6 \\ 2 \\ -5 \\ -14 \end{bmatrix} = [ST(\mathsf{v}_2)]_{\mathcal{E}}.$$

Thus the matrix of ST relative to $\mathcal{V}$ and $\mathcal{E}$ is $M_{\mathcal{E}\leftarrow\mathcal{V}}(ST) = \begin{bmatrix} -4 & -6 \\ 4 & 2 \\ -3 & -5 \\ -6 & -14 \end{bmatrix}$. We ask the reader to check that $M_{\mathcal{E}\leftarrow\mathcal{W}}(S)M_{\mathcal{W}\leftarrow\mathcal{V}}(T) = M_{\mathcal{E}\leftarrow\mathcal{V}}(ST)$:

$$\begin{bmatrix} 1 & 0 & -1 \\ 0 & 2 & 3 \\ 1 & 1 & 0 \\ 3 & 4 & 2 \end{bmatrix} \begin{bmatrix} -2 & -6 \\ -1 & 1 \\ 2 & 0 \end{bmatrix} = \begin{bmatrix} -4 & -6 \\ 4 & 2 \\ -3 & -5 \\ -6 & -14 \end{bmatrix},$$

as asserted in Theorem 5.4.13. ☺

Answers to Reading Checks

1. We wish to find the scalars a and b such that $\mathsf{x} = a\mathsf{v}_1 + b\mathsf{v}_2 = a\begin{bmatrix}-1\\2\end{bmatrix} + b\begin{bmatrix}2\\-3\end{bmatrix}$. This is the matrix equation $\begin{bmatrix}x\\y\end{bmatrix} = \begin{bmatrix}-1 & 2\\2 & -3\end{bmatrix}\begin{bmatrix}a\\b\end{bmatrix}$ whose solution is $a = 3x + 2y$, $b = 2x + y$.

True/False Questions

Decide, with as little calculation as possible, whether each of the following statements is true or false and, if you say "false," explain your answer. (Answers are at the back of the book.)

1. The coordinate vector of $\begin{bmatrix}-2\\3\end{bmatrix}$ relative to the standard basis of R^2 is $\begin{bmatrix}-2\\3\end{bmatrix}$.
2. The coordinate vector of $\begin{bmatrix}-2\\3\end{bmatrix}$ relative to the basis $\left\{\begin{bmatrix}0\\1\end{bmatrix}, \begin{bmatrix}1\\0\end{bmatrix}\right\}$ is $\begin{bmatrix}-2\\3\end{bmatrix}$.
3. The coordinate vector of $\begin{bmatrix}-2\\3\end{bmatrix}$ relative to the basis $\left\{\begin{bmatrix}-2\\3\end{bmatrix}, \begin{bmatrix}7\\5\end{bmatrix}\right\}$ is $\begin{bmatrix}1\\0\end{bmatrix}$.
4. Let $\mathsf{v}_1 = \begin{bmatrix}2\\5\end{bmatrix}$ and $\mathsf{v}_2 = \begin{bmatrix}-5\\1\end{bmatrix}$ and let $\mathcal{V} = \{\mathsf{v}_1, \mathsf{v}_2\}$. The coordinate vector of v_1 relative to $\mathcal{V}$ is $\begin{bmatrix}1\\0\end{bmatrix}$.
5. Let $\mathsf{v}_1 = \begin{bmatrix}2\\5\end{bmatrix}$ and $\mathsf{v}_2 = \begin{bmatrix}5\\1\end{bmatrix}$ and let $\mathcal{V} = \{\mathsf{v}_1, \mathsf{v}_2\}$. The coordinate vector of $\begin{bmatrix}3\\-4\end{bmatrix}$ relative to $\mathcal{V}$ is $\begin{bmatrix}-1\\1\end{bmatrix}$.
6. If V is a vector space with basis $\mathcal{V} = \{\mathsf{v}_1, \mathsf{v}_2\}$ and id: $V \to V$ denotes the identity on V, then $M_{\mathcal{V} \leftarrow \mathcal{V}}(\mathrm{id})$ is $\begin{bmatrix}1 & 0\\0 & 1\end{bmatrix}$.

   In the final four questions, $\mathsf{v}_1 = \begin{bmatrix}1\\0\end{bmatrix}$, $\mathsf{v}_2 = \begin{bmatrix}1\\1\end{bmatrix}$ and $\mathcal{V} = \{\mathsf{v}_1, \mathsf{v}_2\}$. The standard basis of R^2 is $\mathcal{E}$ and id denotes the identity linear transformation on R^2; that is, $\mathrm{id}(\mathsf{v}) = \mathsf{v}$ for any vector v.

7. The matrix $M_{\mathcal{E} \leftarrow \mathcal{E}}(\mathrm{id})$ is $\begin{bmatrix}1 & 0\\0 & 1\end{bmatrix}$.
8. The matrix $M_{\mathcal{E} \leftarrow \mathcal{V}}(\mathrm{id})$ is $\begin{bmatrix}1 & 0\\0 & 1\end{bmatrix}$.
9. The matrix $M_{\mathcal{V} \leftarrow \mathcal{E}}(\mathrm{id})$ is $\begin{bmatrix}1 & 0\\-1 & 1\end{bmatrix}$.

10. The matrix $M_{\mathcal{V}\leftarrow\mathcal{V}}(\mathrm{id})$ is $\begin{bmatrix} 1 & 0 \\ 0 & 1 \end{bmatrix}$.

EXERCISES

Answers to exercises marked [BB] can be found at the Back of the Book.

1. [BB] Let $\mathcal{V} = \{\mathsf{v}_1, \mathsf{v}_2\}$ be the basis of R^2, where $\mathsf{v}_1 = \begin{bmatrix} 2 \\ 5 \end{bmatrix}$ and $\mathsf{v}_2 = \begin{bmatrix} -3 \\ 1 \end{bmatrix}$. Find the coordinates of $\mathsf{x} = \begin{bmatrix} x_1 \\ x_2 \end{bmatrix}$ relative to $\mathcal{V}$.

2. Let $\mathcal{V} = \{\mathsf{v}_1, \mathsf{v}_2, \mathsf{v}_3\}$ be the basis of R^3 where $\mathsf{v}_1 = \begin{bmatrix} -1 \\ 1 \\ 1 \end{bmatrix}$, $\mathsf{v}_2 = \begin{bmatrix} 1 \\ 0 \\ 1 \end{bmatrix}$ and $\mathsf{v}_3 = \begin{bmatrix} 0 \\ 2 \\ -1 \end{bmatrix}$. Find the coordinates of $\mathsf{x} = \begin{bmatrix} x_1 \\ x_2 \\ x_3 \end{bmatrix}$ relative to $\mathcal{V}$.

3. Let $\mathcal{V} = \{\mathsf{v}_1, \mathsf{v}_2, \mathsf{v}_3\}$ be the basis of R^3 where $\mathsf{v}_1 = \begin{bmatrix} 2 \\ 3 \\ -1 \end{bmatrix}$, $\mathsf{v}_2 = \begin{bmatrix} -1 \\ -2 \\ 4 \end{bmatrix}$ and $\mathsf{v}_3 = \begin{bmatrix} 0 \\ 1 \\ -1 \end{bmatrix}$. Find the coordinate vector $\mathsf{x}_{\mathcal{V}}$ of $\mathsf{x} = \begin{bmatrix} x_1 \\ x_2 \\ x_3 \end{bmatrix}$ relative to $\mathcal{V}$.

4. [BB] Let $\mathsf{v}_1 = \begin{bmatrix} 1 \\ 0 \end{bmatrix}$, $\mathsf{v}_2 = \begin{bmatrix} 1 \\ 1 \end{bmatrix}$ and $\mathcal{V} = \{\mathsf{v}_1, \mathsf{v}_2\}$. Let $\mathcal{E}$ denote the standard basis of R^2. Let $A = \begin{bmatrix} 1 & 2 \\ 3 & 4 \end{bmatrix}$ be a matrix of a linear transformation $T \colon \mathsf{R}^2 \to \mathsf{R}^2$. Find $T\left(\begin{bmatrix} x_1 \\ x_2 \end{bmatrix}\right)$ in each of the following cases:

   (a) $A = M_{\mathcal{E}\leftarrow\mathcal{E}}(T)$ **(b)** $A = M_{\mathcal{E}\leftarrow\mathcal{V}}(T)$
   (c) $A = M_{\mathcal{V}\leftarrow\mathcal{E}}(T)$ **(d)** $A = M_{\mathcal{V}\leftarrow\mathcal{V}}(T)$.

5. Answer all parts of Exercise 4 with $\mathsf{v}_1 = \begin{bmatrix} -1 \\ 2 \end{bmatrix}$, $\mathsf{v}_2 = \begin{bmatrix} 2 \\ 1 \end{bmatrix}$ and $A = \begin{bmatrix} -2 & 0 \\ 1 & -3 \end{bmatrix}$.

6. [BB] Define $T \colon \mathsf{R}^2 \to \mathsf{R}^2$ by $T\left(\begin{bmatrix} x \\ y \end{bmatrix}\right) = \begin{bmatrix} x + 4y \\ 2x + 7y \end{bmatrix}$. Let $\mathcal{V} = \{\mathsf{v}_1, \mathsf{v}_2\}$, where $\mathsf{v}_1 = \begin{bmatrix} 1 \\ -1 \end{bmatrix}$ and $\mathsf{v}_2 = \begin{bmatrix} -2 \\ 1 \end{bmatrix}$. Let $\mathcal{E} = \{\mathsf{e}_1, \mathsf{e}_2\}$ denote the standard basis for V.

   (a) Find $M_{\mathcal{E}\leftarrow\mathcal{E}}(T)$, the matrix of T relative to $\mathcal{E}$ and $\mathcal{E}$.

   (b) Find $M_{\mathcal{E}\leftarrow\mathcal{V}}(T)$, the matrix of T relative to $\mathcal{V}$ and $\mathcal{E}$.

   (c) Find $M_{\mathcal{V}\leftarrow\mathcal{E}}(T)$, the matrix of T relative to $\mathcal{E}$ and $\mathcal{V}$.

   (d) Find $M_{\mathcal{V}\leftarrow\mathcal{V}}(T)$, the matrix of T relative to $\mathcal{V}$ and $\mathcal{V}$.

7. Let $\mathsf{v}_1 = \begin{bmatrix}1\\0\end{bmatrix}$, $\mathsf{v}_2 = \begin{bmatrix}1\\1\end{bmatrix}$ and $\mathcal{V} = \{\mathsf{v}_1, \mathsf{v}_2\}$. Let $\mathsf{w}_1 = \begin{bmatrix}1\\0\\1\end{bmatrix}$, $\mathsf{w}_2 = \begin{bmatrix}0\\1\\-1\end{bmatrix}$, $\mathsf{w}_3 = \begin{bmatrix}1\\1\\0\end{bmatrix}$ and $\mathcal{W} = \{\mathsf{w}_1, \mathsf{w}_2, \mathsf{w}_3\}$. Let $\mathcal{E} = \{\mathsf{e}_1, \mathsf{e}_2\}$ and $\mathcal{F} = \{\mathsf{f}_1, \mathsf{f}_2, \mathsf{f}_3\}$ denote the standard bases of R^2 and R^3, respectively. Let $A = \begin{bmatrix}0 & -1\\2 & 1\\-4 & 1\end{bmatrix}$ be a matrix of a linear transformation $T\colon \mathsf{R}^2 \to \mathsf{R}^3$. Find $T\left(\begin{bmatrix}x_1\\x_2\end{bmatrix}\right)$ in each of the following cases.

(a) $A = M_{\mathcal{F}\leftarrow\mathcal{E}}(T)$ **(b)** $A = M_{\mathcal{W}\leftarrow\mathcal{E}}(T)$
(c) $A = M_{\mathcal{F}\leftarrow\mathcal{V}}(T)$ **(d)** $A = M_{\mathcal{W}\leftarrow\mathcal{V}}(T)$.

8. Let $\mathsf{v}_1 = \begin{bmatrix}-1\\1\\1\end{bmatrix}$, $\mathsf{v}_2 = \begin{bmatrix}2\\0\\1\end{bmatrix}$, $\mathsf{v}_3 = \begin{bmatrix}1\\1\\3\end{bmatrix}$ and $\mathcal{V} = \{\mathsf{v}_1, \mathsf{v}_2, \mathsf{v}_3\}$. Let $\mathsf{w}_1 = \begin{bmatrix}1\\1\end{bmatrix}$, $\mathsf{w}_2 = \begin{bmatrix}-2\\4\end{bmatrix}$ and $\mathcal{W} = \{\mathsf{w}_1, \mathsf{w}_2\}$. Let $\mathcal{E} = \{\mathsf{e}_1, \mathsf{e}_2, \mathsf{e}_3\}$ and $\mathcal{F} = \{\mathsf{f}_1, \mathsf{f}_2\}$ denote the standard bases of R^3 and R^2, respectively. Let $T\colon \mathsf{R}^3 \to \mathsf{R}^2$ be the linear transformation defined by $T\left(\begin{bmatrix}x\\y\\z\end{bmatrix}\right) = \begin{bmatrix}2x - y + z\\-x + y + 4z\end{bmatrix}$.

(a) Find $M_{\mathcal{F}\leftarrow\mathcal{E}}(T)$, the matrix of T relative to $\mathcal{E}$ and $\mathcal{F}$.

(b) Find $M_{\mathcal{F}\leftarrow\mathcal{V}}(T)$, the matrix of T relative to $\mathcal{V}$ and $\mathcal{F}$.

(c) Find $M_{\mathcal{W}\leftarrow\mathcal{E}}(T)$, the matrix of T relative to $\mathcal{E}$ and $\mathcal{W}$.

(d) Find $M_{\mathcal{W}\leftarrow\mathcal{V}}(T)$, the matrix of T relative to $\mathcal{V}$ and $\mathcal{W}$.

9. Let $T\colon \mathsf{R}^3 \to \mathsf{R}^2$ and $S\colon \mathsf{R}^2 \to \mathsf{R}^4$ be the linear transformations defined by

$$T\left(\begin{bmatrix}x\\y\\z\end{bmatrix}\right) = \begin{bmatrix}x - y\\y + 2z\end{bmatrix} \quad \text{and} \quad S\left(\begin{bmatrix}x\\y\end{bmatrix}\right) = \begin{bmatrix}2x\\3y\\-x\\x + y\end{bmatrix}.$$

Let R^3, R^2 and R^4 have, respectively, the bases

$$\mathcal{V} = \{\mathsf{v}_1, \mathsf{v}_2, \mathsf{v}_3\}, \quad \text{where } \mathsf{v}_1 = \begin{bmatrix}1\\0\\0\end{bmatrix}, \mathsf{v}_2 = \begin{bmatrix}1\\1\\0\end{bmatrix}, \mathsf{v}_3 = \begin{bmatrix}1\\1\\1\end{bmatrix};$$

$$\mathcal{W} = \{\mathsf{w}_1, \mathsf{w}_2\}, \quad \text{where } \mathsf{w}_1 = \begin{bmatrix}-1\\1\end{bmatrix}, \mathsf{w}_2 = \begin{bmatrix}0\\2\end{bmatrix};$$

$$\mathcal{E} = \{\mathsf{e}_1, \mathsf{e}_2, \mathsf{e}_3, \mathsf{e}_4\}, \quad \text{the standard basis of } \mathsf{R}^4.$$

(a) Find $M_{\mathcal{W}\leftarrow\mathcal{V}}(T)$, the matrix of T relative to $\mathcal{V}$ and $\mathcal{W}$.

(b) Find $M_{\mathcal{E}\leftarrow\mathcal{W}}(S)$, the matrix of S relative to $\mathcal{W}$ and $\mathcal{E}$.

(c) Find $ST\left(\begin{bmatrix} x \\ y \\ z \end{bmatrix}\right)$ using the definition of function composition.

(d) Use part (c) to find $M_{\mathcal{E}\leftarrow\mathcal{V}}(ST)$, the matrix of ST relative to $\mathcal{V}$ and $\mathcal{E}$.

(e) Verify that $M_{\mathcal{E}\leftarrow\mathcal{V}}(ST) = M_{\mathcal{E}\leftarrow\mathcal{W}}(S)M_{\mathcal{W}\leftarrow\mathcal{V}}(T)$.

10. Let $T\colon \mathsf{R}^2 \to \mathsf{R}^5$ and $S\colon \mathsf{R}^5 \to \mathsf{R}^3$ be the linear transformations defined by $T\left(\begin{bmatrix} x \\ y \end{bmatrix}\right) = \begin{bmatrix} y \\ x \\ y \\ x+y \\ x-y \end{bmatrix}$ and $S\left(\begin{bmatrix} x_1 \\ x_2 \\ x_3 \\ x_4 \\ x_5 \end{bmatrix}\right) = \begin{bmatrix} 2x_1 - x_3 + x_5 \\ x_2 + x_4 \\ x_1 \end{bmatrix}$. Let $\mathcal{E}$ denote the standard basis for R^5. Here are bases $\mathcal{V}$ and $\mathcal{W}$ for R^2 and R^3 respectively:

$$\mathcal{V} = \{\mathsf{v}_1, \mathsf{v}_2\}, \qquad \mathsf{v}_1 = \begin{bmatrix} 1 \\ -2 \end{bmatrix}, \mathsf{v}_2 = \begin{bmatrix} -3 \\ 4 \end{bmatrix}$$

$$\mathcal{W} = \{\mathsf{w}_1, \mathsf{w}_2, \mathsf{w}_3\}, \qquad \mathsf{w}_1 = \begin{bmatrix} 1 \\ 0 \\ 1 \end{bmatrix}, \mathsf{w}_2 = \begin{bmatrix} -1 \\ 1 \\ 1 \end{bmatrix}, \mathsf{w}_3 = \begin{bmatrix} 0 \\ 0 \\ 1 \end{bmatrix}.$$

(a) Find $M_{\mathcal{E}\leftarrow\mathcal{V}}(T)$, the matrix of T relative to $\mathcal{V}$ and $\mathcal{E}$.

(b) Find $M_{\mathcal{W}\leftarrow\mathcal{E}}(S)$, the matrix of S relative to $\mathcal{E}$ and $\mathcal{W}$.

(c) Find $(ST)\left(\begin{bmatrix} x \\ y \end{bmatrix}\right)$ using the definition of function composition.

(d) Use your answer to (c) to find $M_{\mathcal{W}\leftarrow\mathcal{V}}(ST)$, the matrix of ST relative to $\mathcal{V}$ and $\mathcal{W}$.

(e) Verify that $M_{\mathcal{W}\leftarrow\mathcal{V}}(ST) = M_{\mathcal{W}\leftarrow\mathcal{E}}(S)M_{\mathcal{E}\leftarrow\mathcal{V}}(T)$.

11. Prove Theorem 5.4.13. [Hint: Let $A = M_{\mathcal{U}\leftarrow\mathcal{W}}(S)$, $B = M_{\mathcal{W}\leftarrow\mathcal{V}}(T)$ and $C = M_{\mathcal{U}\leftarrow\mathcal{V}}(ST)$. Use the definition of "matrix of a linear transformation" (5.4.9) to show that the first column of C is the first column of AB. (We assume the reader will see how to apply your argument to the remaining columns.)]

5.5 Changing Coordinates

In Section 5.4, we introduced the concept of "coordinates" and experienced the need to convert the coordinates of a vector from one basis to another. In this section, we describe a quick way to accomplish this task, with a formula!

Suppose $T\colon V \to W$ is a linear transformation from a vector space V to a vector space W. Once bases $\mathcal{V}$ and $\mathcal{W}$ have been specified for V and W, respectively, T is left multiplication by a matrix we have denoted $M_{\mathcal{W}\leftarrow\mathcal{V}}(T)$ in the sense that

$$[T(\mathsf{x})]_{\mathcal{W}} = M_{\mathcal{W}\leftarrow\mathcal{V}}(T)\mathsf{x}_{\mathcal{V}}.$$

Remember that $\mathsf{x}_{\mathcal{V}}$ is the coordinate vector of x relative to $\mathcal{V}$ and $[T(\mathsf{x})]_{\mathcal{W}}$ is the coordinate vector of $T(\mathsf{x})$ relative to $\mathcal{W}$.

We begin this section by examining the special case where $V = W$ and T is a special function $V \to V$ called the "identity."

The Identity and its Matrices

5.5.1 **Definition.** Let V be a vector space. The *identity* on V is the function id defined by $\mathrm{id}(\mathsf{v}) = \mathsf{v}$ for any v in V.

In fact, there is an identity function on any set. On R, for example, the identity is the familiar function f defined by $f(x) = x$, whose graph is a straight line through $(0,0)$ making a $45°$ angle with the x-axis.

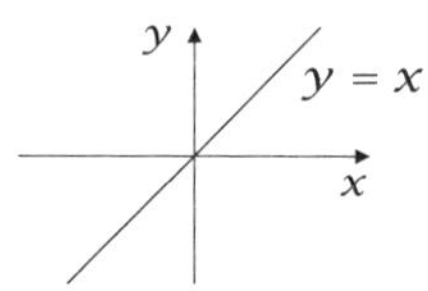

It was an exercise in Section 5.1 to show that the identity on a vector space V is a linear transformation (Exercise 19) and so, once bases are specified, the identity will be associated with a matrix.

Let $\mathcal{V} = \{\mathsf{v}_1, \mathsf{v}_2, \ldots, \mathsf{v}_n\}$ be a basis for V and let $P = M_{\mathcal{V}\leftarrow\mathcal{V}}(\mathrm{id})$ be the matrix of the identity $V \to V$ relative to $\mathcal{V}$ and $\mathcal{V}$. By Definition 5.4.9, the first column of P is the coordinate vector of $\mathrm{id}(\mathsf{v}_1)$ relative to $\mathcal{V}$. Since $\mathrm{id}(\mathsf{v}_1) = \mathsf{v}_1 = 1\mathsf{v}_1 + 0\mathsf{v}_2 + \cdots + 0\mathsf{v}_n$, this vector is $\begin{bmatrix}1\\0\\\vdots\\0\end{bmatrix}$. The second column of P is the coordinate vector of $\mathrm{id}(\mathsf{v}_2)$ relative to $\mathcal{V}$. Since $\mathrm{id}(\mathsf{v}_2) = \mathsf{v}_2 = 0\mathsf{v}_1 + 1\mathsf{v}_2 + 0\mathsf{v}_3 + \cdots + 0\mathsf{v}_n$, this vector is $\begin{bmatrix}0\\1\\0\\\vdots\\0\end{bmatrix}$, and so on. We obtain $P = \begin{bmatrix}1 & 0 & \cdots & 0\\ 0 & 1 & & 0\\ \vdots & & \ddots & \\ 0 & 0 & \cdots & 1\end{bmatrix} = I_n$, the $n \times n$ identity matrix.

5.5.2

> The matrix $P = M_{\mathcal{V}\leftarrow\mathcal{V}}(\mathrm{id})$ of the identity $V \to V$ relative to a single basis for V is the identity matrix.

READING CHECK 1. Let $V = \mathsf{R}^2$. Let $\mathcal{E}$ denote the standard basis of V and let $\mathcal{V} = \{\mathsf{v}_1, \mathsf{v}_2\}$ be the basis with $\mathsf{v}_1 = \begin{bmatrix}-1\\2\end{bmatrix}$ and $\mathsf{v}_2 = \begin{bmatrix}2\\-3\end{bmatrix}$. Find the matrix $M_{\mathcal{E}\leftarrow\mathcal{E}}(\mathrm{id})$ of the identity $V \to V$ relative to $\mathcal{E}$ and $\mathcal{E}$ and the matrix $M_{\mathcal{V}\leftarrow\mathcal{V}}(\mathrm{id})$ of the identity $V \to V$ relative to $\mathcal{V}$ and $\mathcal{V}$.

As seen in Section 5.4, a linear transformation has many different matrices associated with it. In particular, the identity matrix I is just one of many matrices associated with the identity linear transformation.

5.5.3 Proposition. *Any invertible matrix is a matrix of the identity.*

Proof. Let $P = \begin{bmatrix} \mathsf{v}_1 & \mathsf{v}_2 & \cdots & \mathsf{v}_n \\ \downarrow & \downarrow & & \downarrow \\ & & & \end{bmatrix}$ be an invertible $n \times n$ matrix. The columns of P form a basis $\mathcal{V} = \{\mathsf{v}_1, \mathsf{v}_2, \ldots, \mathsf{v}_n\}$ of R^n. (See Corollary 4.5.11, for example.) We claim that $P = M_{\mathcal{E} \leftarrow \mathcal{V}}(\mathrm{id})$ is the matrix of the identity linear transformation relative to $\mathcal{V}$ and $\mathcal{E}$, the standard basis of R^n. To see this, recall that the first column of $M_{\mathcal{E} \leftarrow \mathcal{V}}(\mathrm{id})$ is the coordinate vector of $\mathrm{id}(\mathsf{v}_1)$ relative to $\mathcal{E}$. But $\mathrm{id}(\mathsf{v}_1) = \mathsf{v}_1$, whose coordinate vector relative to $\mathcal{E}$ is v_1 itself (see Example 5.4.3). Similarly, column two of the matrix is v_2, and so on. The matrix of the identity in this situation is the given matrix P. ■

5.5.4 Example. Let $P = \begin{bmatrix} -3 & 0 \\ 1 & -7 \end{bmatrix}$. Let $\mathsf{v}_1 = \begin{bmatrix} -3 \\ 0 \end{bmatrix}$ be the first column of P and let $\mathsf{v}_2 = \begin{bmatrix} 0 \\ -7 \end{bmatrix}$ be the second column. Then $\mathcal{V} = \{\mathsf{v}_1, \mathsf{v}_2\}$ is a basis for R^2 and $P = M_{\mathcal{E} \leftarrow \mathcal{V}}(\mathrm{id})$ is the matrix of the identity from $\mathcal{V}$ to $\mathcal{E}$, the standard basis of R^2.

Remember the definition of the matrix $A = M_{\mathcal{W} \leftarrow \mathcal{V}}(T)$ of a linear transformation T relative to bases $\mathcal{V}$ and $\mathcal{W}$,

$$M_{\mathcal{W} \leftarrow \mathcal{V}}(T) = \begin{bmatrix} [T(\mathsf{v}_1)]_{\mathcal{W}} & [T(\mathsf{v}_2)]_{\mathcal{W}} & \cdots & [T(\mathsf{v}_n)]_{\mathcal{W}} \\ \downarrow & \downarrow & & \downarrow \\ & & & \end{bmatrix}, \tag{1}$$

and the sense in which T is left multiplication by A—see 5.4.10:

$$[T(\mathsf{x})]_{\mathcal{W}} = M_{\mathcal{W} \leftarrow \mathcal{V}}(T)\mathsf{x}_{\mathcal{V}}. \tag{2}$$

Consider the special case that $W = V$ and $T = \mathrm{id}$ is the identity linear transformation $V \to V$. Letting $\mathcal{V}$ and $\mathcal{V}'$ denote two different bases for V, we have, for any vector x,

$$\mathsf{x}_{\mathcal{V}'} = M_{\mathcal{V}' \leftarrow \mathcal{V}}(\mathrm{id})\mathsf{x}_{\mathcal{V}}.$$

This equation says that $\mathsf{x}_{\mathcal{V}'}$, which is the coordinate vector of x relative to $\mathcal{V}'$, can be obtained from its coordinate vector $\mathsf{x}_{\mathcal{V}}$ relative to $\mathcal{V}$, simply by multiplying $\mathsf{x}_{\mathcal{V}}$ by the matrix $M_{\mathcal{V}' \leftarrow \mathcal{V}}(\mathrm{id})$. This matrix is called the *change of coordinates matrix* $\mathcal{V}$ to $\mathcal{V}'$ and we choose to denote this $P_{\mathcal{V}' \leftarrow \mathcal{V}}$.

5.5.5 Definition. Given bases $\mathcal{V}$ and $\mathcal{V}'$ for a vector space V, the *change of coordinates matrix* from $\mathcal{V}$ to $\mathcal{V}'$ is denoted $P_{\mathcal{V}'\leftarrow\mathcal{V}}$ and defined by

$$P_{\mathcal{V}'\leftarrow\mathcal{V}} = M_{\mathcal{V}'\leftarrow\mathcal{V}}(\mathrm{id}).$$

This matrix is named appropriately: to change coordinates from $\mathcal{V}$ to $\mathcal{V}'$, multiply by the change of coordinates matrix:

5.5.6

$$\boxed{\mathsf{x}_{\mathcal{V}'} = P_{\mathcal{V}'\leftarrow\mathcal{V}}\mathsf{x}_{\mathcal{V}}.}$$

If $\mathcal{V} = \{\mathsf{v}_1, \mathsf{v}_2, \ldots, \mathsf{v}_n\}$, with reference to (1) with $T = \mathrm{id}$, we note that $P_{\mathcal{V}'\leftarrow\mathcal{V}}$ is the matrix whose columns are the coordinate vectors of $\mathsf{v}_1, \mathsf{v}_2, \ldots, \mathsf{v}_n$ relative to $\mathcal{V}'$:

$$P_{\mathcal{V}'\leftarrow\mathcal{V}} = \begin{bmatrix} [\mathsf{v}_1]_{\mathcal{V}'} & [\mathsf{v}_2]_{\mathcal{V}'} & \cdots & [\mathsf{v}_n]_{\mathcal{V}'} \\ \downarrow & \downarrow & & \downarrow \\ & & & \end{bmatrix}. \qquad (3)$$

Sometimes, the matrix $P_{\mathcal{V}'\leftarrow\mathcal{V}}$ seems hard to compute, while the matrix $P_{\mathcal{V}\leftarrow\mathcal{V}'}$ needed to change coordinates in the other direction $\mathcal{V}' \to \mathcal{V}$ can be found easily. In this case, we can exploit a nice connection between these two matrices. Remember that matrix multiplication corresponds to composition of linear transformations (Theorem 5.4.13):

$$M_{\mathcal{U}\leftarrow\mathcal{V}}(ST) = M_{\mathcal{U}\leftarrow\mathcal{W}}(S)M_{\mathcal{W}\leftarrow\mathcal{V}}(T).$$

Since the composition $(\mathrm{id})(\mathrm{id}) = \mathrm{id}$ is again the identity, this equation implies $M_{\mathcal{V}\leftarrow\mathcal{V}'}(\mathrm{id})M_{\mathcal{V}'\leftarrow\mathcal{V}}(\mathrm{id}) = M_{\mathcal{V}\leftarrow\mathcal{V}}(\mathrm{id}) = I$, the identity matrix—see 5.5.2—that is,

$$P_{\mathcal{V}\leftarrow\mathcal{V}'}P_{\mathcal{V}'\leftarrow\mathcal{V}} = I,$$

which says

5.5.7

$$\boxed{P_{\mathcal{V}'\leftarrow\mathcal{V}} = P^{-1}_{\mathcal{V}\leftarrow\mathcal{V}'}.}$$

5.5.8 Example. Let $V = \mathsf{R}^2$. Let $\mathcal{E} = \{\mathsf{e}_1, \mathsf{e}_2\}$ be the standard basis of V and let $\mathcal{V} = \{\mathsf{v}_1, \mathsf{v}_2\}$ be the basis with $\mathsf{v}_1 = \begin{bmatrix} -1 \\ 2 \end{bmatrix}$ and $\mathsf{v}_2 = \begin{bmatrix} 2 \\ -3 \end{bmatrix}$. The change of coordinates matrix $\mathcal{V}$ to $\mathcal{E}$ is $P = P_{\mathcal{E}\leftarrow\mathcal{V}} = M_{\mathcal{E}\leftarrow\mathcal{V}}(\mathrm{id})$. Since $[\mathsf{v}_1]_{\mathcal{E}} = \mathsf{v}_1$ and $[\mathsf{v}_2]_{\mathcal{E}} = \mathsf{v}_2$, $P = \begin{bmatrix} -1 & 2 \\ 2 & -3 \end{bmatrix}$. It is often the matrix of base change $\mathcal{E}$ to $\mathcal{V}$ in which we are interested, however, and this can be computed with the help of 5.5.7:

$$P_{\mathcal{V}\leftarrow\mathcal{E}} = P^{-1}_{\mathcal{E}\leftarrow\mathcal{V}} = \begin{bmatrix} -1 & 2 \\ 2 & -3 \end{bmatrix}^{-1} = \begin{bmatrix} 3 & 2 \\ 2 & 1 \end{bmatrix}.$$

Remember the purpose of this matrix: the coordinate vector of $\mathsf{x} = \begin{bmatrix} x_1 \\ x_2 \end{bmatrix}$ relative to $\mathcal{V}$ is

$$\mathsf{x}_{\mathcal{V}} = P_{\mathcal{V} \leftarrow \mathcal{E}} \mathsf{x}_{\mathcal{E}} = \begin{bmatrix} 3 & 2 \\ 2 & 1 \end{bmatrix} \begin{bmatrix} x_1 \\ x_2 \end{bmatrix} = \begin{bmatrix} 3x_1 + 2x_2 \\ 2x_1 + x_2 \end{bmatrix},$$

in agreement with Example 5.4.4. :)

5.5.9 Example. Let $\mathsf{v}_1 = \begin{bmatrix} 1 \\ 0 \\ 0 \\ 0 \end{bmatrix}$, $\mathsf{v}_2 = \begin{bmatrix} 1 \\ 0 \\ 0 \\ 1 \end{bmatrix}$, $\mathsf{v}_3 = \begin{bmatrix} 0 \\ 1 \\ 1 \\ 0 \end{bmatrix}$ and $\mathsf{v}_4 = \begin{bmatrix} 1 \\ 1 \\ 0 \\ 1 \end{bmatrix}$. These four vectors form a basis $\mathcal{V}$ of R^4, so we might for the coordinates of a vector $\mathsf{x} = \begin{bmatrix} x_1 \\ x_2 \\ x_3 \\ x_4 \end{bmatrix}$ in R^4 relative to $\mathcal{V}$. For this, we need the matrix $P_{\mathcal{V} \leftarrow \mathcal{E}}$, where $\mathcal{E}$ is the standard basis for R^4. The matrix $P_{\mathcal{E} \leftarrow \mathcal{V}}$ in the other direction we can simply write down:

$$P_{\mathcal{E} \leftarrow \mathcal{V}} = \begin{bmatrix} 1 & 1 & 0 & 1 \\ 0 & 0 & 1 & 1 \\ 0 & 0 & 1 & 0 \\ 0 & 1 & 0 & 1 \end{bmatrix}.$$

By 5.5.7,

$$P_{\mathcal{V} \leftarrow \mathcal{E}} = P_{\mathcal{E} \leftarrow \mathcal{V}}^{-1} = \begin{bmatrix} 1 & 1 & 0 & 1 \\ 0 & 0 & 1 & 1 \\ 0 & 0 & 1 & 0 \\ 0 & 1 & 0 & 1 \end{bmatrix}^{-1} = \begin{bmatrix} 1 & 0 & 0 & -1 \\ 0 & -1 & 1 & 1 \\ 0 & 0 & 1 & 0 \\ 0 & 1 & -1 & 0 \end{bmatrix}.$$

The coordinates of x relative to $\mathcal{V}$ are the components of the vector

$$P_{\mathcal{V} \leftarrow \mathcal{E}} \mathsf{x}_{\mathcal{E}} = P_{\mathcal{V} \leftarrow \mathcal{E}} \begin{bmatrix} x_1 \\ x_2 \\ x_3 \\ x_4 \end{bmatrix} = \begin{bmatrix} 1 & 0 & 0 & -1 \\ 0 & -1 & 1 & 1 \\ 0 & 0 & 1 & 0 \\ 0 & 1 & -1 & 0 \end{bmatrix} \begin{bmatrix} x_1 \\ x_2 \\ x_3 \\ x_4 \end{bmatrix} = \begin{bmatrix} x_1 - x_4 \\ -x_2 + x_3 + x_4 \\ x_3 \\ x_2 - x_3 \end{bmatrix},$$

namely, the numbers $x_1 - x_4$, $-x_2 + x_3 + x_4$, x_3 and $x_2 - x_3$. Let's check that the vector with these components is indeed x.

$$\begin{aligned}
&(x_1 - x_4)\mathsf{v}_1 + (-x_2 + x_3 + x_4)\mathsf{v}_2 + x_3\mathsf{v}_3 + (x_2 - x_3)\mathsf{v}_4 \\
&= (x_1 - x_4)\begin{bmatrix}1\\0\\0\\0\end{bmatrix} + (-x_2 + x_3 + x_4)\begin{bmatrix}1\\0\\0\\1\end{bmatrix} + x_3\begin{bmatrix}0\\1\\1\\0\end{bmatrix} + (x_2 - x_3)\begin{bmatrix}1\\1\\0\\1\end{bmatrix} \\
&= \begin{bmatrix}x_1\\x_2\\x_3\\x_4\end{bmatrix} = \mathsf{x}.
\end{aligned}$$

:-)

5.5.10 Example. Let $T\colon \mathsf{R}^2 \to \mathsf{R}^2$ be the linear transformation defined by $T\left(\begin{bmatrix}x_1\\x_2\end{bmatrix}\right) = \begin{bmatrix}2x_1 + 5x_2\\4x_1 - 3x_2\end{bmatrix}$. Let $\mathcal{V}$ and $\mathcal{W}$ be the bases of R^2 defined by

$$\mathcal{V} = \{\mathsf{v}_1, \mathsf{v}_2\}, \quad \text{where } \mathsf{v}_1 = \begin{bmatrix}-1\\2\end{bmatrix}, \mathsf{v}_2 = \begin{bmatrix}2\\-3\end{bmatrix}$$

and

$$\mathcal{W} = \{\mathsf{w}_1, \mathsf{w}_2\}, \quad \text{where } \mathsf{w}_1 = \begin{bmatrix}1\\1\end{bmatrix}, \mathsf{w}_2 = \begin{bmatrix}-1\\1\end{bmatrix}.$$

The matrix $M_{\mathcal{E}\leftarrow\mathcal{E}}(T)$ of T relative to the standard basis of R^2 is easy to write down:

$$M_{\mathcal{E}\leftarrow\mathcal{E}}(T) = \begin{bmatrix}2 & 5\\4 & -3\end{bmatrix},$$

but what is the matrix $M_{\mathcal{W}\leftarrow\mathcal{V}}(T)$ of T relative to the bases $\mathcal{V}$ and $\mathcal{W}$? With the aid of Theorem 5.4.13 and a picture, the answer is not hard to find.

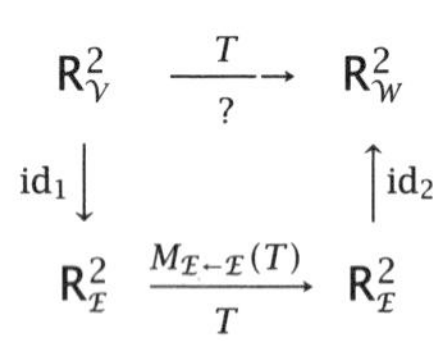

Figure 5.5

This picture, known as a *commutative diagram*, shows that the linear transformation T (at the top) is the composition $(\mathrm{id}_2)T(\mathrm{id}_1)$:

$$T(\mathsf{x}) = \mathrm{id}_2\left(\left(T(\mathrm{id}_1(x))\right)\right).$$

(There are two identities in this diagram which for the moment we distinguish by the labels id_1 and id_2.) By Theorem 5.4.13—matrix multiplication corresponds to composition of linear transformations—the desired matrix, denoted ? in the figure, is

$$M_{\mathcal{W}\leftarrow\mathcal{V}}(T) = M_{\mathcal{W}\leftarrow\mathcal{E}}(\text{id})M_{\mathcal{E}\leftarrow\mathcal{E}}(T)M_{\mathcal{E}\leftarrow\mathcal{V}}(\text{id}) = P_{\mathcal{W}\leftarrow\mathcal{E}}M_{\mathcal{E}\leftarrow\mathcal{E}}(T)P_{\mathcal{E}\leftarrow\mathcal{V}}.$$

Now $P_{\mathcal{E}\leftarrow\mathcal{V}} = \begin{bmatrix} -1 & 2 \\ 2 & -3 \end{bmatrix}$ and $P_{\mathcal{E}\leftarrow\mathcal{W}} = \begin{bmatrix} 1 & -1 \\ 1 & 1 \end{bmatrix}$, so $P_{\mathcal{W}\leftarrow\mathcal{E}} = [P_{\mathcal{E}\leftarrow\mathcal{W}}]^{-1} = \begin{bmatrix} \frac{1}{2} & \frac{1}{2} \\ -\frac{1}{2} & \frac{1}{2} \end{bmatrix}$. Thus

$$M_{\mathcal{W}\leftarrow\mathcal{V}}(T) = \begin{bmatrix} \frac{1}{2} & \frac{1}{2} \\ -\frac{1}{2} & \frac{1}{2} \end{bmatrix}\begin{bmatrix} 2 & 5 \\ 4 & -3 \end{bmatrix}\begin{bmatrix} -1 & 2 \\ 2 & -3 \end{bmatrix} = \begin{bmatrix} -1 & 3 \\ -9 & 14 \end{bmatrix}.$$

In what sense is T "left multiplication" by this matrix? The answer is given in formula (2): $[T(\mathsf{x})]_{\mathcal{W}} = M_{\mathcal{W}\leftarrow\mathcal{V}}(T)\mathsf{x}_{\mathcal{V}}$.

Take a vector $\mathsf{x} = \begin{bmatrix} x_1 \\ x_2 \end{bmatrix}$ in V. Its coordinate vector relative to $\mathcal{V}$ is

$$\mathsf{x}_{\mathcal{V}} = P_{\mathcal{V}\leftarrow\mathcal{E}}\begin{bmatrix} x_1 \\ x_2 \end{bmatrix} = \begin{bmatrix} 3x_1 + 2x_2 \\ 2x_1 + x_2 \end{bmatrix},$$

and multiplication by $M_{\mathcal{W}\leftarrow\mathcal{V}}(T)$ gives

$$\begin{bmatrix} -1 & 3 \\ -9 & 14 \end{bmatrix}\begin{bmatrix} 3x_1 + 2x_2 \\ 2x_1 + x_2 \end{bmatrix} = \begin{bmatrix} 3x_1 + x_2 \\ x_1 - 4x_2 \end{bmatrix},$$

which should be $[T(\mathsf{x})]_{\mathcal{W}}$, the coordinate vector of $T(\mathsf{x})$ relative to $\mathcal{W}$, and it is because

$$\begin{aligned}(3x_1 + x_2)\mathsf{w}_1 + (x_1 - 4x_2)\mathsf{w}_2 &= (3x_1 + x_2)\begin{bmatrix} 1 \\ 1 \end{bmatrix} + (x_1 - 4x_2)\begin{bmatrix} -1 \\ 1 \end{bmatrix} \\ &= \begin{bmatrix} 2x_1 + 5x_2 \\ 4x_1 - 3x_2 \end{bmatrix} = T(\mathsf{x}).\end{aligned}$$

:)

Similarity of Matrices

Suppose $T\colon V \to V$ is a linear transformation of a vector space V and $\mathcal{V}$ is a basis for V. Then T has a matrix $A = M_{\mathcal{V}\leftarrow\mathcal{V}}(T)$ relative to $\mathcal{V}$. If $\mathcal{V}'$ is another basis for V, then T also has a matrix $B = M_{\mathcal{V}'\leftarrow\mathcal{V}'}(T)$ relative to $\mathcal{V}'$. We might expect a connection between A and B, and there is. Figure 5.6 gives the clue. The linear transformation $T\colon V \to V$ (at the top of the figure) is the same as the composition $(\text{id})T(\text{id})$ of three linear transformations, just as in Figure 5.5. Using Theorem 5.4.13,

$$B = M_{\mathcal{V}'\leftarrow\mathcal{V}'}(T) = M_{\mathcal{V}'\leftarrow\mathcal{V}}(\text{id})M_{\mathcal{V}\leftarrow\mathcal{V}}(T)M_{\mathcal{V}\leftarrow\mathcal{V}'}(\text{id}) = P_{\mathcal{V}'\leftarrow\mathcal{V}}AP_{\mathcal{V}\leftarrow\mathcal{V}'} = P^{-1}AP,$$

$$
\begin{array}{ccc}
V_{\mathcal{V}} & \xrightarrow[A]{T} & V_{\mathcal{V}} \\
\text{id}\Big\uparrow P & & P^{-1}\Big\downarrow \text{id} \\
V_{\mathcal{V}'} & \xrightarrow[T]{B} & V_{\mathcal{V}'}
\end{array}
$$

Figure 5.6: One linear transformation T and two of its matrices A and $B = P^{-1}AP$.

where $P = P_{\mathcal{V} \leftarrow \mathcal{V}'}$ (and so $P_{\mathcal{V}' \leftarrow \mathcal{V}} = P^{-1}$ by 5.5.7). The matrices A and B are **similar**, a concept first introduced in Section 3.5.

Conversely, suppose A and B are similar $n \times n$ matrices and $B = P^{-1}AP$. The columns $\mathsf{v}_1, \mathsf{v}_2, \ldots, \mathsf{v}_n$ of P form a basis $\mathcal{V}$ for R^n and $P = M_{\mathcal{E} \leftarrow \mathcal{V}}(\text{id})$ is the change of coordinates matrix from $\mathcal{V}$ to $\mathcal{E}$, the standard basis of R^n. By 5.5.7, $P^{-1} = M_{\mathcal{V} \leftarrow \mathcal{E}}(\text{id})$ is the change of coordinates matrix from $\mathcal{E}$ to $\mathcal{V}$. Let $T\colon \mathsf{R}^n \to \mathsf{R}^n$ be the linear transformation that is left multiplication by A; that is, $A = M_{\mathcal{E} \leftarrow \mathcal{E}}(T)$. As above,

$$B = P^{-1}AP = M_{\mathcal{V} \leftarrow \mathcal{E}}(\text{id}) M_{\mathcal{E} \leftarrow \mathcal{E}}(T) M_{\mathcal{E} \leftarrow \mathcal{V}}(\text{id}) = M_{\mathcal{V} \leftarrow \mathcal{V}}(T)$$

is the matrix of T relative to $\mathcal{V}$ and $\mathcal{V}$. Thus A and B are each matrices for the same linear transformation. We have established this theorem.

5.5.11 Theorem. *Matrices are similar if and only if they are matrices of the same linear transformation $V \to V$, in each case with respect to a single basis for the vector space V.*

5.5.12 Example. Let $T\colon \mathsf{R}^3 \to \mathsf{R}^3$ be the linear transformation defined by $T\left(\begin{bmatrix} x_1 \\ x_2 \\ x_3 \end{bmatrix}\right) = \begin{bmatrix} -x_1 + x_2 + x_3 \\ x_3 \\ 2x_1 - 3x_2 - x_3 \end{bmatrix}$. The matrix of T relative to $\mathcal{E}$ and $\mathcal{E}$, the standard basis of R^3, is $A = M_{\mathcal{E} \leftarrow \mathcal{E}}(T) = \begin{bmatrix} -1 & 1 & 1 \\ 0 & 0 & 1 \\ 2 & -3 & -1 \end{bmatrix}$. Let $\mathsf{v}_1 = \begin{bmatrix} 1 \\ -1 \\ 1 \end{bmatrix}$, $\mathsf{v}_2 = \begin{bmatrix} 0 \\ 2 \\ 1 \end{bmatrix}$, $\mathsf{v}_3 = \begin{bmatrix} 1 \\ 0 \\ 1 \end{bmatrix}$ and $\mathcal{V} = \{\mathsf{v}_1, \mathsf{v}_2, \mathsf{v}_3\}$. The matrix of T relative to $\mathcal{V}$ and $\mathcal{V}$ is

$$B = M_{\mathcal{V} \leftarrow \mathcal{V}}(T) = P_{\mathcal{V} \leftarrow \mathcal{E}} M_{\mathcal{E} \leftarrow \mathcal{E}}(T) P_{\mathcal{E} \leftarrow \mathcal{V}} = P^{-1}AP$$

where $P = P_{\mathcal{E} \leftarrow \mathcal{V}} = \begin{bmatrix} 1 & 0 & 1 \\ -1 & 2 & 0 \\ 1 & 1 & 1 \end{bmatrix}$. Since $P_{\mathcal{V} \leftarrow \mathcal{E}} = P^{-1} = \begin{bmatrix} -2 & -1 & 2 \\ -1 & 0 & 1 \\ 3 & 1 & -2 \end{bmatrix}$, we have

$$B = \begin{bmatrix} -2 & -1 & 2 \\ -1 & 0 & 1 \\ 3 & 1 & -2 \end{bmatrix} \begin{bmatrix} -1 & 1 & 1 \\ 0 & 0 & 1 \\ 2 & -3 & -1 \end{bmatrix} \begin{bmatrix} 1 & 0 & 1 \\ -1 & 2 & 0 \\ 1 & 1 & 1 \end{bmatrix} = \begin{bmatrix} 9 & -21 & 1 \\ 5 & -10 & 1 \\ -10 & 24 & -1 \end{bmatrix}.$$

:-)

READING CHECK 2. Since B is similar to A in this example and since A is a matrix for T, Theorem 5.5.11 says that B is also a matrix for T.

(a) In what sense is B a matrix for T?

(b) Show that $[T(\mathsf{x})]_{\mathcal{V}} = B\mathsf{x}_{\mathcal{V}}$ for any $\mathsf{x} = \begin{bmatrix} x_1 \\ x_2 \\ x_3 \end{bmatrix}$ in R^3.

Answers to Reading Checks

1. As described in 5.5.2, the answer in each case is the 2×2 identity matrix $\begin{bmatrix} 1 & 0 \\ 0 & 1 \end{bmatrix}$. The reader is urged, however, to carry out the reasoning, without simply quoting 5.5.2.

2. **(a)** $[T(\mathsf{x})]_{\mathcal{V}} = B\mathsf{x}_{\mathcal{V}}$ for any vector $\mathsf{x} = \begin{bmatrix} x_1 \\ x_2 \\ x_3 \end{bmatrix}$.

   (b) First, since $P_{\mathcal{V} \leftarrow \mathcal{E}} = \begin{bmatrix} -2 & -1 & 2 \\ -1 & 0 & 1 \\ 3 & 1 & -2 \end{bmatrix}$ as in the example, we have

$$\mathsf{x}_{\mathcal{V}} = P_{\mathcal{V} \leftarrow \mathcal{E}} \mathsf{x}_{\mathcal{E}} = \begin{bmatrix} -2 & -1 & 2 \\ -1 & 0 & 1 \\ 3 & 1 & -2 \end{bmatrix} \begin{bmatrix} x_1 \\ x_2 \\ x_3 \end{bmatrix} = \begin{bmatrix} -2x_1 - x_2 + 2x_3 \\ -x_1 + x_3 \\ 3x_1 + x_2 - 2x_3 \end{bmatrix},$$

$$\text{so } B\mathsf{x}_{\mathcal{V}} = \begin{bmatrix} 9 & -21 & 1 \\ 5 & -10 & 1 \\ -10 & 24 & -1 \end{bmatrix} \begin{bmatrix} -2x_1 - x_2 + 2x_3 \\ -x_1 + x_3 \\ 3x_1 + x_2 - 2x_3 \end{bmatrix} = \begin{bmatrix} 6x_1 - 8x_2 - 5x_3 \\ 3x_1 - 4x_2 - 2x_3 \\ -7x_1 + 9x_2 + 6x_3 \end{bmatrix}.$$

   Next, we show that this is $[T(\mathsf{x})]_{\mathcal{V}}$. Since $T(\mathsf{x}) = \begin{bmatrix} -x_1 + x_2 + x_3 \\ x_3 \\ 2x_1 - 3x_2 - x_3 \end{bmatrix} =$

$[T(\mathbf{x})]_{\mathcal{E}}$,

$$[(T(\mathbf{x})]_{\mathcal{V}} = P_{\mathcal{V}\leftarrow\mathcal{E}}[T(\mathbf{x})]_{\mathcal{E}} = \begin{bmatrix} -2 & -1 & 2 \\ -1 & 0 & 1 \\ 3 & 1 & -2 \end{bmatrix} \begin{bmatrix} -x_1 + x_2 + x_3 \\ x_3 \\ 2x_1 - 3x_2 - x_3 \end{bmatrix}$$

$$= \begin{bmatrix} 6x_1 - 8x_2 - 5x_3 \\ 3x_1 - 4x_2 - 2x_3 \\ -7x_1 + 9x_2 + 6x_3 \end{bmatrix} = B\mathbf{x}_{\mathcal{V}}$$

as desired.

True/False Questions

Decide, with as little calculation as possible, whether each of the following statements is true or false and, if you say "false," explain your answer. (Answers are at the back of the book.)

1. The matrix of the identity transformation $\mathsf{R}^n \to \mathsf{R}^n$ is the identity matrix.

2. The matrix $\begin{bmatrix} 1 & 2 \\ 3 & 4 \end{bmatrix}$ is a matrix for the identity $\mathsf{R}^2 \to \mathsf{R}^2$.

3. If $P = \begin{bmatrix} 0 & 1 \\ 1 & 0 \end{bmatrix}$ is the matrix that changes coordinates $\mathcal{V} \to \mathcal{V}'$, then P also changes coordinates $\mathcal{V}' \to \mathcal{V}$.

4. If $\mathcal{V} = \{\mathbf{v}_1, \mathbf{v}_2\}$ and $\mathcal{W} = \{\mathbf{w}_1, \mathbf{w}_2\}$ are bases for R^2 and $P_{\mathcal{W}\leftarrow\mathcal{V}}\begin{bmatrix} a \\ b \end{bmatrix} = \begin{bmatrix} c \\ d \end{bmatrix}$, then $a\mathbf{v}_1 + b\mathbf{v}_2 = c\mathbf{w}_1 + d\mathbf{w}_2$.

5. Let V be a vector space with bases $\mathcal{V}$ and $\mathcal{V}'$. The change of coordinates matrix from $\mathcal{V}$ to $\mathcal{V}'$ is the matrix of the identity linear transformation $V_{\mathcal{V}} \to V_{\mathcal{V}'}$.

6. Let $\mathbf{v}_1 = \begin{bmatrix} 1 \\ 2 \end{bmatrix}$, $\mathbf{v}_2 = \begin{bmatrix} -2 \\ 3 \end{bmatrix}$, $\mathbf{v}_3 = \begin{bmatrix} 1 \\ 0 \end{bmatrix}$ and $\mathbf{v}_4 = \begin{bmatrix} 0 \\ 1 \end{bmatrix}$. Let $\mathcal{V} = \{\mathbf{v}_1, \mathbf{v}_2\}$ and $\mathcal{V}' = \{\mathbf{v}_3, \mathbf{v}_4\}$. Then the change of coordinates matrix $\mathcal{V} \to \mathcal{V}'$ is $\begin{bmatrix} 1 & -2 \\ 2 & 3 \end{bmatrix}$.

7. Let $\mathbf{v}_1 = \begin{bmatrix} 1 \\ 2 \end{bmatrix}$, $\mathbf{v}_2 = \begin{bmatrix} -2 \\ 3 \end{bmatrix}$, $\mathbf{v}_3 = \begin{bmatrix} 0 \\ 1 \end{bmatrix}$ and $\mathbf{v}_4 = \begin{bmatrix} 1 \\ 0 \end{bmatrix}$. Let $\mathcal{V} = \{\mathbf{v}_3, \mathbf{v}_4\}$ and $\mathcal{V}' = \{\mathbf{v}_1, \mathbf{v}_2\}$. Then the change of coordinates matrix $\mathcal{V} \to \mathcal{V}'$ is $\begin{bmatrix} 2 & 3 \\ 1 & -2 \end{bmatrix}$.

8. Let $\mathcal{V} = \left\{\begin{bmatrix} 3 \\ -1 \end{bmatrix}, \begin{bmatrix} 2 \\ 0 \end{bmatrix}\right\}$ and suppose $M_{\mathcal{V}\leftarrow\mathcal{E}}(T) = \begin{bmatrix} 1 & 2 \\ 3 & 4 \end{bmatrix}$ for some linear transformation $T\colon \mathsf{R}^2 \to \mathsf{R}^2$. Then $[T(\mathbf{x})]_{\mathcal{V}} = \begin{bmatrix} x + 2y \\ 3x + 4y \end{bmatrix}$ for $\mathbf{x} = \begin{bmatrix} x \\ y \end{bmatrix}$.

9. If $P_{\mathcal{V}\leftarrow\mathcal{E}} = \begin{bmatrix} -1 & 0 \\ 0 & 2 \end{bmatrix}$ and $M_{\mathcal{E}\leftarrow\mathcal{E}}(T) = \begin{bmatrix} 6 & 7 \\ 8 & 9 \end{bmatrix}$, then $M_{\mathcal{V}\leftarrow\mathcal{E}}(T) = \begin{bmatrix} -6 & -7 \\ 16 & 18 \end{bmatrix}$.

10. Let $\mathcal{V}$ and $\mathcal{V}'$ be bases for R^2 and let $T\colon \mathsf{R}^2 \to \mathsf{R}^2$ be a linear transformation. If $A = M_{\mathcal{V}\leftarrow\mathcal{V}}(T)$ and $B = M_{\mathcal{V}'\leftarrow\mathcal{V}'}(T)$, then A and B have the same determinant.

EXERCISES

Answers to exercises marked [BB] can be found at the Back of the Book.

1. Let $\mathsf{v}_1 = \begin{bmatrix} -1 \\ 1 \end{bmatrix}$, $\mathsf{v}_2 = \begin{bmatrix} 2 \\ 0 \end{bmatrix}$ and $\mathcal{V} = \{\mathsf{v}_1, \mathsf{v}_2\}$. Let $\mathcal{E}$ denote the standard basis of R^2 and $\mathrm{id}\colon \mathsf{R}^2 \to \mathsf{R}^2$ the identity. Find each of the following matrices.

   (a) [BB] $M_{\mathcal{E}\leftarrow\mathcal{E}}(\mathrm{id})$ **(b)** $M_{\mathcal{V}\leftarrow\mathcal{E}}(\mathrm{id})$ **(c)** [BB] $M_{\mathcal{E}\leftarrow\mathcal{V}}(\mathrm{id})$ **(d)** $M_{\mathcal{V}\leftarrow\mathcal{V}}(\mathrm{id})$.

2. [BB] Let $\mathcal{E} = \{\mathsf{e}_1, \mathsf{e}_2, \mathsf{e}_3\}$ be the standard basis for R^3 and let $\mathcal{V} = \{\mathsf{v}_1, \mathsf{v}_2, \mathsf{v}_3\}$ be the basis with $\mathsf{v}_1 = \begin{bmatrix} 1 \\ 1 \\ 0 \end{bmatrix}$, $\mathsf{v}_2 = \begin{bmatrix} 0 \\ 0 \\ 1 \end{bmatrix}$ and $\mathsf{v}_3 = \begin{bmatrix} 3 \\ 0 \\ 0 \end{bmatrix}$.

   (a) Find the change of coordinates matrix $P_{\mathcal{V}\leftarrow\mathcal{E}}$ from $\mathcal{E}$ to $\mathcal{V}$.

   (b) What are the coordinates of $\mathsf{x} = \begin{bmatrix} 2 \\ 1 \\ -1 \end{bmatrix}$ relative to $\mathcal{V}$?

3. Let $\mathcal{E} = \{\mathsf{e}_1, \mathsf{e}_2, \mathsf{e}_3, \mathsf{e}_4\}$ be the standard basis for R^4 and let $\mathcal{V}$ be the basis $\{\mathsf{v}_1, \mathsf{v}_2, \mathsf{v}_3, \mathsf{v}_4\}$ where

$$\mathsf{v}_1 = \begin{bmatrix} 1 \\ 1 \\ 0 \\ 0 \end{bmatrix}, \quad \mathsf{v}_2 = \begin{bmatrix} 1 \\ 0 \\ 1 \\ 0 \end{bmatrix}, \quad \mathsf{v}_3 = \begin{bmatrix} 1 \\ 0 \\ 0 \\ 1 \end{bmatrix}, \quad \mathsf{v}_4 = \begin{bmatrix} 0 \\ 1 \\ 1 \\ 0 \end{bmatrix}.$$

   (a) Find the change of coordinates matrix $P_{\mathcal{V}\leftarrow\mathcal{E}}$ from $\mathcal{E} \to \mathcal{V}$.

   (b) What are the coordinates of $\mathsf{x} = \begin{bmatrix} 1 \\ 3 \\ 2 \\ 4 \end{bmatrix}$ relative to $\mathcal{V}$?

4. Let $\mathsf{v}_1 = \begin{bmatrix} 2 \\ 2 \end{bmatrix}$, $\mathsf{v}_2 = \begin{bmatrix} 4 \\ 0 \end{bmatrix}$, $\mathsf{w}_1 = \begin{bmatrix} 1 \\ 3 \end{bmatrix}$, $\mathsf{w}_2 = \begin{bmatrix} -1 \\ -1 \end{bmatrix}$ and $\mathsf{x} = \begin{bmatrix} 3 \\ 6 \end{bmatrix}$. Let $\mathcal{V} = \{\mathsf{v}_1, \mathsf{v}_2\}$ and $\mathcal{W} = \{\mathsf{w}_1, \mathsf{w}_2\}$.

   (a) Find the change of coordinates matrices from $\mathcal{V}$ to $\mathcal{W}$ and from $\mathcal{W}$ to $\mathcal{V}$.

   (b) Find the coordinate vectors of x with respect to $\mathcal{V}$ and with respect to $\mathcal{W}$.

5. Let $T\colon \mathsf{R}^3 \to \mathsf{R}^3$ be the linear transformation defined by the rule $T(\begin{bmatrix} x \\ y \\ z \end{bmatrix}) = \begin{bmatrix} 2x + y \\ x + y - z \\ -z \end{bmatrix}$. Find $M_{\mathcal{V}\leftarrow\mathcal{V}}(T)$, the matrix of T relative to the basis $\mathcal{V}$ (in

each instance), where $\mathcal{V} = \{v_1, v_2, v_3\}$, $v_1 = \begin{bmatrix} 1 \\ 0 \\ 0 \end{bmatrix}$, $v_2 = \begin{bmatrix} 1 \\ 1 \\ 0 \end{bmatrix}$, $v_3 = \begin{bmatrix} 1 \\ 1 \\ 1 \end{bmatrix}$.

6. [BB] Let B be the matrix found in Example 5.5.12 and consider the equation $B\begin{bmatrix} 3 \\ 1 \\ -2 \end{bmatrix} = \begin{bmatrix} 4 \\ 3 \\ -4 \end{bmatrix}$. This says that if $x_{\mathcal{V}} = \begin{bmatrix} 3 \\ 1 \\ -2 \end{bmatrix}$ and $y_{\mathcal{V}} = \begin{bmatrix} 4 \\ 3 \\ -4 \end{bmatrix}$, then $T(x) = y$. Verify that this is correct.

7. Let $\mathcal{E}$ denote the standard basis of R^2 and let $\mathcal{V} = \{v_1, v_2\}$ and $\mathcal{V}' = \{v_1', v_2'\}$ be the bases with

$$v_1 = \begin{bmatrix} -1 \\ 1 \end{bmatrix}, v_2 = \begin{bmatrix} 4 \\ -5 \end{bmatrix}, v_1' = \begin{bmatrix} 3 \\ 1 \end{bmatrix}, v_2' = \begin{bmatrix} 1 \\ 2 \end{bmatrix}.$$

Let $T\colon R^2 \to R^2$ be a linear transformation and let $A = \begin{bmatrix} -109 & 494 \\ -24 & 109 \end{bmatrix}$.

Let $u = \begin{bmatrix} 1 \\ 0 \end{bmatrix}$, $v = \begin{bmatrix} -2 \\ 1 \end{bmatrix}$ and $w = \begin{bmatrix} 3 \\ 3 \end{bmatrix}$.

(a) [BB] Suppose $A = M_{\mathcal{E}\leftarrow\mathcal{E}}(T)$ is the matrix of T relative to $\mathcal{E}$ and $\mathcal{E}$. Find $T(u)$, $T(v)$ and $T(w)$.

(b) [BB] Suppose $A = M_{\mathcal{E}\leftarrow\mathcal{V}}(T)$ is the matrix of T relative to $\mathcal{E}$ and $\mathcal{V}$. Find $T(u)$, $T(v)$ and $T(w)$.

(c) Suppose $A = M_{\mathcal{V}\leftarrow\mathcal{E}}(T)$ is the matrix of T relative to $\mathcal{V}$ and $\mathcal{E}$. Find $T(u)$, $T(v)$ and $T(w)$.

(d) Suppose $A = M_{\mathcal{V}\leftarrow\mathcal{V}}(T)$ is the matrix of T relative to $\mathcal{V}$ and $\mathcal{V}$. Find $T(u)$, $T(v)$ and $T(w)$.

(e) Suppose $A = M_{\mathcal{V}\leftarrow\mathcal{V}}(T)$. Find $B = M_{\mathcal{E}\leftarrow\mathcal{E}}(T)$.

(f) Suppose $A = M_{\mathcal{V}\leftarrow\mathcal{V}}(T)$. Find $B = M_{\mathcal{V}'\leftarrow\mathcal{V}'}(T)$.

8. Let $\mathcal{E} = \{e_1, e_2, e_3, e_4\}$ be the standard basis for R^4 and let $\mathcal{V}$ be the basis $\{v_1, v_2, v_3, v_4\}$ where

$$v_1 = \begin{bmatrix} 1 \\ 1 \\ 0 \\ 0 \end{bmatrix}, \quad v_2 = \begin{bmatrix} 1 \\ 0 \\ 1 \\ 0 \end{bmatrix}, \quad v_3 = \begin{bmatrix} 0 \\ 0 \\ 1 \\ 1 \end{bmatrix}, \quad v_4 = \begin{bmatrix} 1 \\ 0 \\ 0 \\ 1 \end{bmatrix}.$$

(a) Find the change of coordinates matrix $P_{\mathcal{V}\leftarrow\mathcal{E}}$ from $\mathcal{E}$ to $\mathcal{V}$ and use this to find the coordinates of $x = \begin{bmatrix} 1 \\ 2 \\ -1 \\ 0 \end{bmatrix}$ relative to $\mathcal{V}$.

(b) Let $T\colon R^4 \to R^4$ be the linear transformation whose matrix relative to $\mathcal{E}$ and $\mathcal{E}$, the standard basis of R^4, is $A = M_{\mathcal{E}\leftarrow\mathcal{E}}(T) = \begin{bmatrix} -1 & 2 & 3 & 0 \\ 5 & 0 & 1 & 1 \\ 2 & -3 & 0 & -2 \\ 0 & 4 & 1 & 1 \end{bmatrix}$.
Find the matrix $B = M_{\mathcal{V}\leftarrow\mathcal{V}}(T)$ of T relative to $\mathcal{V}$ and $\mathcal{V}$.

(c) Consider the equation $B\begin{bmatrix}0\\2\\0\\-4\end{bmatrix} = \begin{bmatrix}-12\\13\\-9\\7\end{bmatrix}$. This says that if $\mathsf{x}_{\mathcal{V}} = \begin{bmatrix}0\\2\\0\\-4\end{bmatrix}$ and $\mathsf{y}_{\mathcal{V}} = \begin{bmatrix}-12\\13\\-9\\7\end{bmatrix}$, then $T(\mathsf{x}) = \mathsf{y}$. Verify that this is correct.

9. Let $T\colon \mathsf{R}^5 \to \mathsf{R}^2$ be the linear transformation defined by

$T\left(\begin{bmatrix}x_1\\x_2\\x_3\\x_4\\x_5\end{bmatrix}\right) = \begin{bmatrix}x_1 - x_2 + x_3 - x_4 + x_5\\ 2x_1 + 3x_3 + x_4\end{bmatrix}$. Let $\mathcal{E}$ and $\mathcal{F}$ be the standard bases for $V = \mathsf{R}^5$ and $W = \mathsf{R}^3$, respectively. Let $\mathcal{V} = \{\mathsf{v}_1, \mathsf{v}_2, \mathsf{v}_3, \mathsf{v}_4, \mathsf{v}_5\}$ with

$$\mathsf{v}_1 = \begin{bmatrix}1\\0\\0\\0\\0\end{bmatrix}, \mathsf{v}_2 = \begin{bmatrix}1\\1\\0\\0\\0\end{bmatrix}, \mathsf{v}_3 = \begin{bmatrix}1\\1\\1\\0\\0\end{bmatrix}, \mathsf{v}_4 = \begin{bmatrix}1\\1\\1\\1\\0\end{bmatrix}, \mathsf{v}_5 = \begin{bmatrix}1\\1\\1\\1\\1\end{bmatrix}.$$

Let $\mathsf{w}_1 = \begin{bmatrix}-1\\3\end{bmatrix}$, $\mathsf{w}_2 = \begin{bmatrix}2\\-5\end{bmatrix}$ and $\mathcal{W} = \{\mathsf{w}_1, \mathsf{w}_2\}$.

(a) Find the matrix $A = M_{\mathcal{F}\leftarrow\mathcal{E}}(T)$ of T relative to $\mathcal{E}$ and $\mathcal{F}$.

(b) Find the matrix $B = M_{\mathcal{W}\leftarrow\mathcal{V}}(T)$ of T relative to $\mathcal{V}$ and $\mathcal{W}$.

(c) Explain what is meant when we say that "T is left multiplication by B?"

(d) Illustrate your answer to (c).

10. Define $T\colon \mathsf{R}^3 \to \mathsf{R}^3$ by $T\left[\begin{bmatrix}x\\y\\z\end{bmatrix}\right] = \begin{bmatrix}2x+3y\\-2x+y+z\\x-y-z\end{bmatrix}$.

(a) [BB] Let $\mathcal{V} = \{\begin{bmatrix}1\\0\\1\end{bmatrix}, \begin{bmatrix}0\\1\\1\end{bmatrix}, \begin{bmatrix}1\\1\\0\end{bmatrix}\}$. Find $M_{\mathcal{V}\leftarrow\mathcal{V}}(T)$, the matrix of T relative to $\mathcal{V}$ and $\mathcal{V}$.

(b) Let $\mathcal{V}' = \{\begin{bmatrix}0\\0\\1\end{bmatrix}, \begin{bmatrix}0\\1\\1\end{bmatrix}, \begin{bmatrix}1\\1\\1\end{bmatrix}\}$. Find $M_{\mathcal{V}'\leftarrow\mathcal{V}'}(T)$, the matrix of T relative to $\mathcal{V}'$ and $\mathcal{V}'$.

(c) True or false: The matrices found in parts (a) and (b) are similar. Explain.

11. Let $\mathcal{V} = \{\mathsf{v}_1, \mathsf{v}_2, \mathsf{v}_3\}$ with $\mathsf{v}_1 = \begin{bmatrix}1\\2\\-1\end{bmatrix}$, $\mathsf{v}_2 = \begin{bmatrix}0\\1\\1\end{bmatrix}$ and $\mathsf{v}_3 = \begin{bmatrix}-3\\0\\6\end{bmatrix}$.

(a) [BB] Suppose u, v and w are vectors in R^3 with $\mathsf{u}_{\mathcal{V}} = \begin{bmatrix}1\\2\\3\end{bmatrix}$, $\mathsf{v}_{\mathcal{V}} = \begin{bmatrix}0\\1\\0\end{bmatrix}$ and $\mathsf{w}_{\mathcal{V}} = \begin{bmatrix}-1\\0\\1\end{bmatrix}$. What are u, v and w?

(b) If $\mathsf{u} = \begin{bmatrix} 1 \\ 2 \\ 3 \end{bmatrix}$, $\mathsf{v} = \begin{bmatrix} 0 \\ 1 \\ 0 \end{bmatrix}$ and $\mathsf{w} = \begin{bmatrix} -1 \\ 0 \\ 1 \end{bmatrix}$, find $\mathsf{u}_{\mathcal{V}}$, $\mathsf{v}_{\mathcal{V}}$ and $\mathsf{w}_{\mathcal{V}}$.

12. Let $V = \mathsf{R}^3$, $W = \mathsf{R}^2$ and $U = \mathsf{R}^3$. Let $T\colon V \to W$ and $S\colon W \to U$ be the linear transformations defined by

$$T\left(\begin{bmatrix} x_1 \\ x_2 \\ x_3 \end{bmatrix}\right) = \begin{bmatrix} 2x_1 - x_2 + x_3 \\ x_2 + x_3 \end{bmatrix} \quad \text{and} \quad S\left(\begin{bmatrix} x_1 \\ x_2 \end{bmatrix}\right) = \begin{bmatrix} x_1 + 2x_2 \\ x_1 - x_2 \\ 3x_1 \end{bmatrix}.$$

Let V, W, U have, respectively, the bases

$$\mathcal{V} = \{\mathsf{v}_1, \mathsf{v}_2, \mathsf{v}_3\}, \quad \text{where } \mathsf{v}_1 = \begin{bmatrix} -1 \\ 1 \\ 1 \end{bmatrix}, \mathsf{v}_2 = \begin{bmatrix} 1 \\ 0 \\ 1 \end{bmatrix}, \mathsf{v}_3 = \begin{bmatrix} 1 \\ 1 \\ 1 \end{bmatrix};$$

$$\mathcal{W} = \{\mathsf{w}_1, \mathsf{w}_2\}, \quad \text{where } \mathsf{w}_1 = \begin{bmatrix} 1 \\ 2 \end{bmatrix}, \mathsf{w}_2 = \begin{bmatrix} 2 \\ 1 \end{bmatrix};$$

$$\mathcal{U} = \{\mathsf{u}_1, \mathsf{u}_2, \mathsf{u}_3\}, \quad \text{where } \mathsf{u}_1 = \begin{bmatrix} 1 \\ 0 \\ -1 \end{bmatrix}, \mathsf{u}_2 = \begin{bmatrix} 0 \\ 0 \\ 1 \end{bmatrix}, \mathsf{u}_3 = \begin{bmatrix} -1 \\ 1 \\ 1 \end{bmatrix}.$$

(a) Find $A = M_{\mathcal{U} \leftarrow \mathcal{W}}(S)$ and $B = M_{\mathcal{W} \leftarrow \mathcal{V}}(T)$.

(b) Find $ST\left(\begin{bmatrix} x_1 \\ x_2 \\ x_3 \end{bmatrix}\right)$.

(c) Find $C = M_{\mathcal{U} \leftarrow \mathcal{V}}(ST)$ and verify that $AB = C$.

Use coordinate change matrices wherever possible.

13. Let $\mathcal{E} = \{\mathsf{e}_1, \mathsf{e}_2, \mathsf{e}_3\}$ be the standard basis of R^3, and let $\mathcal{V} = \{\mathsf{v}_1, \mathsf{v}_2, \mathsf{v}_3\}$ and $\mathcal{W} = \{\mathsf{w}_1, \mathsf{w}_2, \mathsf{w}_3\}$ be other bases, where $\mathsf{v}_1 = \begin{bmatrix} 1 \\ 1 \\ 1 \end{bmatrix}$, $\mathsf{v}_2 = \begin{bmatrix} 1 \\ -1 \\ 0 \end{bmatrix}$, $\mathsf{v}_3 = \begin{bmatrix} -1 \\ 0 \\ 1 \end{bmatrix}$, and $\mathsf{w}_1 = \mathsf{e}_3$, $\mathsf{w}_2 = \mathsf{e}_2$, $\mathsf{w}_3 = \mathsf{e}_1$.

(a) [BB] Find the change of coordinates matrix $\mathcal{V} \to \mathcal{E}$, the change of coordinates matrix $\mathcal{E} \to \mathcal{W}$ and the change of coordinates matrix $\mathcal{V} \to \mathcal{W}$, all straight from the definition. (See also (3).)

(b) Verify that $P_{\mathcal{W} \leftarrow \mathcal{V}} = P_{\mathcal{W} \leftarrow \mathcal{E}} P_{\mathcal{E} \leftarrow \mathcal{V}}$.

(c) Express $\mathsf{v} = \begin{bmatrix} a \\ b \\ c \end{bmatrix}$ as a linear combination of $\mathsf{v}_1, \mathsf{v}_2, \mathsf{v}_3$.

14. Let $T\colon \mathsf{R}^2 \to \mathsf{R}^2$ be the linear transformation whose matrix relative to the standard basis $\mathcal{E} = \{\mathsf{e}_1, \mathsf{e}_2\}$ of R^2 is $A = \begin{bmatrix} 3 & -1 \\ 4 & 0 \end{bmatrix}$. Let $\mathsf{v}_1 = \begin{bmatrix} -2 \\ 3 \end{bmatrix}$, $\mathsf{v}_2 = \begin{bmatrix} -1 \\ 1 \end{bmatrix}$, and $\mathcal{V} = \{\mathsf{v}_1, \mathsf{v}_2\}$.

(a) For a vector $\mathsf{x} = \begin{bmatrix} x_1 \\ x_2 \end{bmatrix}$ in R^2, what is $T(\mathsf{x})$?

(b) What is the matrix $B = M_{\mathcal{V} \leftarrow \mathcal{V}}(T)$?

(c) Verify that $[T(\mathsf{x})]_{\mathcal{V}} = B\mathsf{x}_{\mathcal{V}}$ for any $\mathsf{x} = \begin{bmatrix} x_1 \\ x_2 \end{bmatrix}$ in R^2.

15. Let $\mathsf{v}_1 = \begin{bmatrix} -2 \\ 3 \end{bmatrix}$, $\mathsf{v}_2 = \begin{bmatrix} -1 \\ 1 \end{bmatrix}$, and $\mathcal{V} = \{\mathsf{v}_1, \mathsf{v}_2\}$. Let $T\colon \mathsf{R}^2 \to \mathsf{R}^2$ be the linear transformation whose matrix relative to $\mathcal{V}$ is $A = \begin{bmatrix} 3 & -1 \\ 4 & 0 \end{bmatrix}$. For a vector $\mathsf{x} = \begin{bmatrix} x_1 \\ x_2 \end{bmatrix}$ in R^2, what is $T(\mathsf{x})$?

16. Let $T\colon \mathsf{R}^2 \to \mathsf{R}^2$ be defined by $T\left(\begin{bmatrix} x \\ y \end{bmatrix}\right) = \begin{bmatrix} 2x - y \\ 3y \end{bmatrix}$. Let $\mathcal{E}$ denote the standard basis of R^2 and let $\mathcal{V} = \{\mathsf{v}_1, \mathsf{v}_2\}$ be the basis with $\mathsf{v}_1 = \begin{bmatrix} 1 \\ 1 \end{bmatrix}$, $\mathsf{v}_2 = \begin{bmatrix} -2 \\ 5 \end{bmatrix}$.

(a) [BB] Find $A = M_{\mathcal{E}\leftarrow\mathcal{E}}(T)$.

(b) [BB] Find $B = M_{\mathcal{V}\leftarrow\mathcal{V}}(T)$ **from first principles**, that is, without the use of coordinate change matrices.

(c) [BB] Verify that A and B are similar.

(d) Let $\mathcal{W} = \{\mathsf{w}_1, \mathsf{w}_2\}$ be the basis with $\mathsf{w}_1 = \begin{bmatrix} -3 \\ 2 \end{bmatrix}$, $\mathsf{w}_2 = \begin{bmatrix} 4 \\ -1 \end{bmatrix}$. Find $C = M_{\mathcal{W}\leftarrow\mathcal{W}}(T)$, again from first principles.

(e) Verify that A and C are similar.

(f) Verify that B and C are similar. [Hint: This is not hard if you use previous parts.]

17. Answer Exercise 16 again given $T\colon \mathsf{R}^3 \to \mathsf{R}^3$ is $T\left(\begin{bmatrix} x \\ y \\ z \end{bmatrix}\right) = \begin{bmatrix} x - y + z \\ 2x + y \\ 3y - z \end{bmatrix}$ and $\mathcal{V} = \{\mathsf{v}_1, \mathsf{v}_2, \mathsf{v}_3\}$ with $\mathsf{v}_1 = \begin{bmatrix} 1 \\ 1 \\ 0 \end{bmatrix}$, $\mathsf{v}_2 = \begin{bmatrix} 1 \\ 0 \\ 1 \end{bmatrix}$, $\mathsf{v}_3 = \begin{bmatrix} 0 \\ 1 \\ 1 \end{bmatrix}$.

18. Let $T\colon \mathsf{R}^4 \to \mathsf{R}^4$ be the linear transformation whose matrix relative to $\mathcal{E}$ and $\mathcal{E}$, the standard basis of R^4, is $A = M_{\mathcal{E}\leftarrow\mathcal{E}}(T) = \begin{bmatrix} -1 & 2 & 3 & 0 \\ 5 & 0 & 1 & 1 \\ 2 & -3 & 0 & -2 \\ 0 & 4 & 1 & 1 \end{bmatrix}$. Let $\mathsf{v}_1 = \begin{bmatrix} 1 \\ 0 \\ 0 \\ 0 \end{bmatrix}$, $\mathsf{v}_2 = \begin{bmatrix} 1 \\ 0 \\ 0 \\ 1 \end{bmatrix}$, $\mathsf{v}_3 = \begin{bmatrix} 0 \\ 1 \\ 1 \\ 0 \end{bmatrix}$, $\mathsf{v}_4 = \begin{bmatrix} 1 \\ 1 \\ 0 \\ 1 \end{bmatrix}$ and let $\mathcal{V}$ be the basis $\{\mathsf{v}_1, \mathsf{v}_2, \mathsf{v}_3, \mathsf{v}_4\}$.

(a) For a vector $\mathsf{x} = \begin{bmatrix} x_1 \\ x_2 \\ x_3 \\ x_4 \end{bmatrix}$ in R^4, what is $T(\mathsf{x})$?

(b) Find the matrix $B = M_{\mathcal{V}\leftarrow\mathcal{V}}(T)$ of T relative to $\mathcal{V}$ and $\mathcal{V}$.

(c) There is a special word which describes the relationship between A and B. What is this word?

(d) Check that $[T(\mathsf{x})]_{\mathcal{V}} = B\mathsf{x}_{\mathcal{V}}$.

(e) Consider the equation $B\begin{bmatrix} 2 \\ -2 \\ -3 \\ 1 \end{bmatrix} = \begin{bmatrix} -2 \\ -3 \\ 10 \\ -9 \end{bmatrix}$ which says that if $\mathsf{x}_{\mathcal{V}} = \begin{bmatrix} 2 \\ -2 \\ -3 \\ 1 \end{bmatrix}$ and $\mathsf{y}_{\mathcal{V}} = \begin{bmatrix} -2 \\ -3 \\ 10 \\ -9 \end{bmatrix}$, then $T(\mathsf{x}) = \mathsf{y}$. Use part (a) to show that this is correct.

19. [BB] Let $\mathsf{v}_1 = \begin{bmatrix} 0 \\ 1 \end{bmatrix}$, $\mathsf{v}_2 = \begin{bmatrix} 2 \\ 4 \end{bmatrix}$ and $\mathcal{V} = \{\mathsf{v}_1, \mathsf{v}_2\}$. Let $\mathsf{w}_1 = \begin{bmatrix} 2 \\ -3 \end{bmatrix}$, $\mathsf{w}_2 = \begin{bmatrix} -2 \\ 5 \end{bmatrix}$ and $\mathcal{W} = \{\mathsf{w}_1, \mathsf{w}_2\}$. If the matrix of a linear transformation $T\colon \mathsf{R}^2 \to \mathsf{R}^2$ is $A = M_{\mathcal{V} \leftarrow \mathcal{V}}(T) = \begin{bmatrix} 3 & 2 \\ 2 & 1 \end{bmatrix}$ relative to the basis $\mathcal{V}$, find the matrix $B = M_{\mathcal{W} \leftarrow \mathcal{W}}(T)$ of T relative to the basis $\mathcal{W}$.

20. Answer Exercise 19 again with $\mathsf{v}_1 = \begin{bmatrix} -2 \\ 1 \end{bmatrix}$, $\mathsf{v}_2 = \begin{bmatrix} 1 \\ 1 \end{bmatrix}$, $\mathsf{w}_1 = \begin{bmatrix} -1 \\ 4 \end{bmatrix}$, $\mathsf{w}_2 = \begin{bmatrix} 0 \\ 1 \end{bmatrix}$ and $A = \begin{bmatrix} 1 & 2 \\ -1 & 4 \end{bmatrix}$.

21. [BB] Let $\mathsf{v}_1 = \begin{bmatrix} 1 \\ 0 \\ -1 \end{bmatrix}$, $\mathsf{v}_2 = \begin{bmatrix} 1 \\ 1 \\ 0 \end{bmatrix}$, $\mathsf{v}_3 = \begin{bmatrix} -2 \\ 1 \\ 1 \end{bmatrix}$ and $\mathcal{V} = \{\mathsf{v}_1, \mathsf{v}_2, \mathsf{v}_3\}$. Let $\mathsf{w}_1 = \begin{bmatrix} 1 \\ 1 \\ 1 \end{bmatrix}$, $\mathsf{w}_2 = \begin{bmatrix} 1 \\ 0 \\ 3 \end{bmatrix}$, $\mathsf{w}_3 = \begin{bmatrix} 1 \\ -1 \\ 2 \end{bmatrix}$ and $\mathcal{W} = \{\mathsf{w}_1, \mathsf{w}_2, \mathsf{w}_3\}$. If the matrix of a linear transformation $T\colon \mathsf{R}^3 \to \mathsf{R}^3$ is $A = M_{\mathcal{V} \leftarrow \mathcal{V}}(T) = \begin{bmatrix} 1 & 2 & 0 \\ 0 & -1 & 0 \\ 0 & 1 & 1 \end{bmatrix}$ relative to the basis $\mathcal{V}$, find the matrix $B = M_{\mathcal{W} \leftarrow \mathcal{W}}(T)$ of T relative to the basis $\mathcal{W}$.

22. Answer Exercise 21 again with $\mathsf{v}_1 = \begin{bmatrix} 1 \\ -2 \\ 1 \end{bmatrix}$, $\mathsf{v}_2 = \begin{bmatrix} 1 \\ 0 \\ 0 \end{bmatrix}$, $\mathsf{v}_3 = \begin{bmatrix} 0 \\ -1 \\ 1 \end{bmatrix}$, $\mathsf{w}_1 = \begin{bmatrix} 0 \\ 1 \\ 2 \end{bmatrix}$, $\mathsf{w}_2 = \begin{bmatrix} 1 \\ 2 \\ 0 \end{bmatrix}$, $\mathsf{w}_3 = \begin{bmatrix} -2 \\ 0 \\ 1 \end{bmatrix}$ and $A = \begin{bmatrix} -1 & 0 & -1 \\ 0 & 1 & -1 \\ 3 & -2 & 1 \end{bmatrix}$.

23. Let A be a 2×2 matrix, $A \neq 0$, with $A^2 = 0$. Let v be a vector in R^2 such that $A\mathsf{v} \neq 0$.

    (a) Prove that $\{\mathsf{v}, A\mathsf{v}\}$ is a basis for R^2.

    (b) Prove that A is similar to $\begin{bmatrix} 0 & 0 \\ 1 & 0 \end{bmatrix}$.

    (c) Prove that A is similar to $\begin{bmatrix} 0 & 1 \\ 0 & 0 \end{bmatrix}$.

24. "The matrix $P = \begin{bmatrix} 1 & -3 \\ 4 & 2 \end{bmatrix}$ is a matrix of the identity transformation $\mathsf{R}^2 \to \mathsf{R}^2$." Explain.

25. In a sentence or two, describe the meaning of the formula $[T(\mathsf{x})]_{\mathcal{W}} = M_{\mathcal{W} \leftarrow \mathcal{V}}(T)\mathsf{x}_{\mathcal{V}}$.

26. In a sentence or two, express the meaning of the formula $P_{\mathcal{V}' \leftarrow \mathcal{V}} = P^{-1}_{\mathcal{V} \leftarrow \mathcal{V}'}$.

Critical Reading

27. Let $\mathcal{V}$ and $\mathcal{V}'$ be bases for a vector space V and let $T \colon V \to V$ be a linear transformation. Find the product of the matrices $M_{\mathcal{V}' \leftarrow \mathcal{V}}(T)$ and $M_{\mathcal{V} \leftarrow \mathcal{V}'}(T)$ (in both orders).

28. Let $\mathcal{V} = \{\mathsf{v}_1, \mathsf{v}_2, \mathsf{v}_3\}$ be a basis for R^3 and let $A = M_{\mathcal{V} \leftarrow \mathcal{V}}(T)$ be the matrix of a linear transformation $T \colon \mathsf{R}^3 \to \mathsf{R}^3$ relative to $\mathcal{V}$ and $\mathcal{V}$. Suppose we change the order of the vectors in $\mathcal{V}$ to obtain the basis $\mathcal{W} = \{\mathsf{v}_2, \mathsf{v}_3, \mathsf{v}_1\}$. Let $B = M_{\mathcal{W} \leftarrow \mathcal{W}}(T)$ be the matrix of T relative to $\mathcal{W}$ and $\mathcal{W}$. What is the relationship between A and B? For example, if $A = \begin{bmatrix} 1 & 2 & 3 \\ 4 & 5 & 6 \\ 7 & 8 & 9 \end{bmatrix}$, what is B? Explain.

29. Suppose A and B are $n \times n$ matrices that are similar. Show that A and B are the matrices of the same linear transformation.

CHAPTER KEY WORDS AND IDEAS: Here are some technical words and phrases that were used in this chapter. Do you know the meaning of each? If you're not sure, check the glossary or index at the back of the book.

associative
change of coordinates
commutative
composition
    of linear transformations
coordinate vector
coordinates of a vector
identity linear transformation
linear transformation
linear transformations preserve
- scalar multiplication
- the zero vector
- linear combinations

matrix
    of a linear transformation
    change of coordinates
similar matrices

Review Exercises for Chapter 5

1. Let $T\colon \mathsf{R}^2 \to \mathsf{R}^2$ be a linear transformation such that $T\left(\begin{bmatrix}1\\0\end{bmatrix}\right) = \begin{bmatrix}2\\-1\end{bmatrix}$ and $T\left(\begin{bmatrix}1\\1\end{bmatrix}\right) = \begin{bmatrix}4\\4\end{bmatrix}$.

   (a) Find $T\left(\begin{bmatrix}0\\1\end{bmatrix}\right)$.

   (b) Find a matrix A such that $T = T_A$ is multiplication by A.

2. Suppose $T\colon \mathsf{R}^2 \to \mathsf{R}^4$ is a linear transformation with the property that $T\left(\begin{bmatrix}3\\-1\end{bmatrix}\right) = \begin{bmatrix}2\\-1\\1\\0\end{bmatrix}$ and $T\left(\begin{bmatrix}-1\\2\end{bmatrix}\right) = \begin{bmatrix}4\\0\\1\\-3\end{bmatrix}$. Find $T\left(\begin{bmatrix}x\\y\end{bmatrix}\right)$.

3. Let $T\colon \mathsf{R}^3 \to \mathsf{R}^2$ be a linear transformation such that $T\left(\begin{bmatrix}3\\-5\\0\end{bmatrix}\right) = \begin{bmatrix}2\\0\end{bmatrix}$ and $T\left(\begin{bmatrix}1\\-1\\2\end{bmatrix}\right) = \begin{bmatrix}1\\-1\end{bmatrix}$. Find $T\left(\begin{bmatrix}1\\3\\14\end{bmatrix}\right)$, if possible.

4. Find a matrix A such that rotation in R^2 through $\theta = \frac{11\pi}{6}$ is left multiplication by A.

5. Let $T\colon \mathsf{R}^3 \to \mathsf{R}^3$ be projection onto the vector $\mathsf{v} = \begin{bmatrix}1\\1\\1\end{bmatrix}$. Find a matrix A such that $T(\mathsf{x}) = A\mathsf{x}$.

6. Let $T\colon \mathsf{R}^3 \to \mathsf{R}^3$ be the linear transformation which projects vectors onto the plane π with equation $x - y + 5z = 0$. Find a matrix A such that T is left multiplication by A.

7. Define linear transformations $S\colon \mathsf{R}^2 \to \mathsf{R}^3$, $T\colon \mathsf{R}^4 \to \mathsf{R}^2$ and $U\colon \mathsf{R}^1 \to \mathsf{R}^4$ by

$$S\left(\begin{bmatrix}x_1\\x_2\end{bmatrix}\right) = \begin{bmatrix}x_1 - x_2\\2x_1\\3x_1 + x_2\end{bmatrix},\quad T\left(\begin{bmatrix}x_1\\x_2\\x_3\\x_4\end{bmatrix}\right) = \begin{bmatrix}x_1 - x_2 - x_3 + x_4\\3x_1 + x_3\end{bmatrix},\quad U(x) = \begin{bmatrix}-x\\0\\3x\\x\end{bmatrix}.$$

   (a) Find $ST\left(\begin{bmatrix}x_1\\x_2\\x_3\\x_4\end{bmatrix}\right)$ and $TU(x)$ using the definition of function composition.

   (b) Find $[(ST)U](x)$ and $[S(TU)](x)$ using the definition of function composition and comment on your answers.

8. Let $T\colon \mathsf{R}^3 \to \mathsf{R}^3$ be the linear transformation defined by $T\left(\begin{bmatrix}x\\y\\z\end{bmatrix}\right) = \begin{bmatrix}x-y\\x-z\\x\end{bmatrix}$. Find the matrix of T relative to the basis $\mathcal{V} = \{\mathsf{v}_1, \mathsf{v}_2, \mathsf{v}_3\}$ (in each instance),

where $\mathsf{v}_1 = \begin{bmatrix} 1 \\ 1 \\ 1 \end{bmatrix}$, $\mathsf{v}_2 = \begin{bmatrix} 1 \\ 1 \\ 0 \end{bmatrix}$, $\mathsf{v}_3 = \begin{bmatrix} 1 \\ 0 \\ 0 \end{bmatrix}$.

9. Let a and b be real numbers with $a^2 + b^2 = 1$. Let $\mathsf{u} = \begin{bmatrix} a \\ b \end{bmatrix}$, $\mathsf{v} = \begin{bmatrix} -b \\ a \end{bmatrix}$ and $\mathcal{V} = \{\mathsf{u}, \mathsf{v}\}$. Let $T\colon \mathsf{R}^2 \to \mathsf{R}^2$ be the matrix that reflects vectors in the line $\mathsf{x} = t\mathsf{u}$.

   (a) Find the matrix of T relative to $\mathcal{V}$.

   (b) Find the matrix of T relative to the standard basis $\mathcal{E} = \{\mathsf{e}_1, \mathsf{e}_2\}$.

   (c) Find the matrix relative to $\mathcal{E}$ that reflects vectors in the line with equation $y = ax$.

10. Let $\mathsf{v}_1 = \begin{bmatrix} 1 \\ 2 \\ 0 \\ 1 \end{bmatrix}$, $\mathsf{v}_2 = \begin{bmatrix} -1 \\ 1 \\ 3 \\ 0 \end{bmatrix}$, $\mathsf{v}_3 = \begin{bmatrix} 0 \\ 0 \\ 1 \\ 0 \end{bmatrix}$, $\mathsf{v}_4 = \begin{bmatrix} 0 \\ 1 \\ 2 \\ 3 \end{bmatrix}$ and $\mathcal{V} = \{\mathsf{v}_1, \mathsf{v}_2, \mathsf{v}_3, \mathsf{v}_4\}$. Let $\mathsf{w}_1 = \begin{bmatrix} 3 \\ 4 \end{bmatrix}$, $\mathsf{w}_2 = \begin{bmatrix} -1 \\ 5 \end{bmatrix}$ and $\mathcal{W} = \{\mathsf{w}_1, \mathsf{w}_2\}$. Define $T\colon \mathsf{R}^4 \to \mathsf{R}^2$ by $T(\mathsf{v}_1) = \mathsf{w}_1$, $T(\mathsf{v}_2) = \mathsf{w}_1$, $T(\mathsf{v}_3) = \mathsf{w}_2$, $T(\mathsf{v}_4) = \mathsf{w}_2$.

    (a) Find the matrix $M_{\mathcal{W}\leftarrow\mathcal{V}}(T)$ of T relative to $\mathcal{V}$ and $\mathcal{W}$.

    (b) Find the matrix $M_{\mathcal{E}\leftarrow\mathcal{V}}(T)$ of T relative to $\mathcal{V}$ and the standard basis of R^2.

    (c) Find the matrix $M_{\mathcal{E}\leftarrow\mathcal{E}}(T)$ of T relative to the standard bases of R^4 and R^2.

11. Let $\mathsf{v}_1 = \begin{bmatrix} 1 \\ 1 \\ 1 \end{bmatrix}$, $\mathsf{v}_2 = \begin{bmatrix} 4 \\ 2 \\ 1 \end{bmatrix}$, $\mathsf{v}_3 = \begin{bmatrix} 3 \\ 2 \\ 1 \end{bmatrix}$. Define $T\colon \mathsf{R}^3 \to \mathsf{R}^4$ by $T(\mathsf{v}_1) = \begin{bmatrix} 5 \\ 7 \\ -3 \\ -1 \end{bmatrix}$, $T(\mathsf{v}_2) = \begin{bmatrix} -3 \\ 0 \\ 2 \\ -2 \end{bmatrix}$, $T(\mathsf{v}_3) = \begin{bmatrix} -1 \\ 7 \\ 1 \\ -5 \end{bmatrix}$.

    (a) Show that $\mathcal{V} = \{\mathsf{v}_1, \mathsf{v}_2, \mathsf{v}_3\}$ is a basis for R^3.

    (b) Find $M_{\mathcal{E}\leftarrow\mathcal{V}}(T)$, the matrix of T relative to $\mathcal{V}$ and the standard basis $\mathcal{E}$ of R^4.

    (c) Find the matrix of T relative to the standard bases of R^3 and R^4.

12. Let $\mathcal{V} = \{\mathsf{v}_1, \mathsf{v}_2, \mathsf{v}_3\}$ be the basis of R^3 where $\mathsf{v}_1 = \begin{bmatrix} 1 \\ 0 \\ 1 \end{bmatrix}$, $\mathsf{v}_2 = \begin{bmatrix} -1 \\ 2 \\ 2 \end{bmatrix}$, $\mathsf{v}_3 = \begin{bmatrix} 0 \\ 1 \\ 2 \end{bmatrix}$ and $\mathcal{W} = \{\mathsf{w}_1, \mathsf{w}_2\}$ be the basis of R^2 with $\mathsf{w}_1 = \begin{bmatrix} 3 \\ 1 \end{bmatrix}$, $\mathsf{w}_2 = \begin{bmatrix} 5 \\ 2 \end{bmatrix}$.

    (a) Find the matrices of basis change $\mathcal{E} \to \mathcal{V}$ and $\mathcal{V} \to \mathcal{E}$, where $\mathcal{E}$ denotes the standard basis of R^3.

    (b) Find the coordinate vectors of $\begin{bmatrix} 1 \\ 0 \\ 0 \end{bmatrix}$ and $\begin{bmatrix} 1 \\ 0 \\ 1 \end{bmatrix}$ relative to $\mathcal{V}$.

Let $T\colon \mathsf{R}^2 \to \mathsf{R}^3$ be a linear transformation such that $T(\begin{bmatrix} 3 \\ 1 \end{bmatrix}) = \begin{bmatrix} 1 \\ 0 \\ 0 \end{bmatrix}$ and $T(\begin{bmatrix} 5 \\ 2 \end{bmatrix}) = \begin{bmatrix} 1 \\ 0 \\ 1 \end{bmatrix}$.

(c) Find $T(\begin{bmatrix} 0 \\ 1 \end{bmatrix})$ and $T(\begin{bmatrix} 1 \\ 0 \end{bmatrix})$.

(d) Find the matrix of T relative to the standard bases of R^2 and R^3.

(e) Find the matrix of T relative to $\mathcal{V}$ and $\mathcal{W}$.

13. Let $V = \mathsf{R}^2$. Let $\mathcal{E}$ denote the standard basis of V and let $\mathcal{V}$ be the basis $\{\mathsf{v}_1, \mathsf{v}_2\}$, where $\mathsf{v}_1 = \begin{bmatrix} 1 \\ -1 \end{bmatrix}$ and $\mathsf{v}_2 = \begin{bmatrix} 2 \\ 1 \end{bmatrix}$.

(a) Find the change of coordinate matrices $P_{\mathcal{V} \leftarrow \mathcal{E}}$ and $P_{\mathcal{E} \leftarrow \mathcal{V}}$.

(b) Write $\mathsf{x} = \begin{bmatrix} -3 \\ 4 \end{bmatrix}$ as a linear combination of v_1 and v_2.

(c) What is $\mathsf{x}_{\mathcal{V}}$, the coordinate vector of x relative to $\mathcal{V}$?

(d) If the coordinate vector $\mathsf{x}_{\mathcal{V}}$ of x relative to $\mathcal{V}$ is $\begin{bmatrix} -3 \\ 4 \end{bmatrix}$, what is x?

Now let $W = \mathsf{R}^3$. Let $\mathcal{F}$ denote the standard basis of W and let $\mathcal{W}$ be the basis $\{\mathsf{w}_1, \mathsf{w}_2, \mathsf{w}_3\}$, where $\mathsf{w}_1 = \begin{bmatrix} 1 \\ 0 \\ -1 \end{bmatrix}$, $\mathsf{w}_2 = \begin{bmatrix} 2 \\ 1 \\ -1 \end{bmatrix}$, $\mathsf{w}_3 = \begin{bmatrix} 0 \\ -1 \\ 2 \end{bmatrix}$.

Let $T\colon V \to W$ be the linear transformation whose matrix relative to $\mathcal{V}$ and $\mathcal{W}$ is $M_{\mathcal{W} \leftarrow \mathcal{V}}(T) = \begin{bmatrix} -1 & 4 \\ 0 & 3 \\ 4 & -7 \end{bmatrix}$.

(e) What is the matrix $M_{\mathcal{F} \leftarrow \mathcal{E}}(T)$ of T relative to $\mathcal{E}$ and $\mathcal{F}$?

(f) What is $T(\begin{bmatrix} x \\ y \end{bmatrix})$?

14. Let $V = \mathsf{R}^3$. Let $\mathsf{v}_1 = \begin{bmatrix} -1 \\ 2 \\ 1 \end{bmatrix}$, $\mathsf{v}_2 = \begin{bmatrix} 2 \\ 1 \\ 1 \end{bmatrix}$, $\mathsf{v}_3 = \begin{bmatrix} 1 \\ 1 \\ 0 \end{bmatrix}$, $\mathsf{v}'_1 = \begin{bmatrix} 0 \\ 1 \\ 1 \end{bmatrix}$, $\mathsf{v}'_2 = \begin{bmatrix} -1 \\ 3 \\ 0 \end{bmatrix}$ and $\mathsf{v}'_3 = \begin{bmatrix} 2 \\ -2 \\ 1 \end{bmatrix}$. Let $\mathcal{V}$ and $\mathcal{V}'$ be the bases of V which are $\{\mathsf{v}_1, \mathsf{v}_2, \mathsf{v}_3\}$ and $\{\mathsf{v}'_1, \mathsf{v}'_2, \mathsf{v}'_3\}$, respectively. Let $T\colon V \to V$ be the linear transformation whose matrix relative to $\mathcal{V}$ and $\mathcal{V}$ is $A = M_{\mathcal{V} \leftarrow \mathcal{V}}(T) = \begin{bmatrix} -1 & 2 & 3 \\ 0 & 1 & -1 \\ 1 & 4 & 0 \end{bmatrix}$.

(a) Find $P_{\mathcal{V}' \leftarrow \mathcal{V}}$, the change of coordinates matrix $\mathcal{V}$ to $\mathcal{V}'$.

(b) Let $B = M_{\mathcal{V}' \leftarrow \mathcal{V}'}(T)$ be the matrix of T relative to $\mathcal{V}'$ and $\mathcal{V}'$. Find B.

(c) There is a name for a pair of matrices like A and B. What is this name?

✍ **15.** "Any invertible matrix is the matrix of the identity." Explain.

Chapter 6

Orthogonality

6.1 Projection Matrices

Orthogonality: A Quick Review

We begin this chapter with a quick review of the basic terminology associated with the concept of "orthogonality."

If $\mathsf{x} = \begin{bmatrix} x_1 \\ x_2 \\ \vdots \\ x_n \end{bmatrix}$ and $\mathsf{y} = \begin{bmatrix} y_1 \\ y_2 \\ \vdots \\ y_n \end{bmatrix}$ are vectors in R^n, the dot product of x and y is the number $x_1y_1 + x_2y_2 + \cdots + x_ny_n$, which is also the matrix product $\mathsf{x}^T\mathsf{y}$, thinking of x and y as $n \times 1$ matrices.

1. $\mathsf{x} \cdot \mathsf{y} = x_1y_1 + x_2y_2 + \cdots + x_ny_n = \mathsf{x}^T\mathsf{y}$.

   Vectors x and y are *orthogonal* if and only if $\mathsf{x} \cdot \mathsf{y} = 0$. In R^2 or R^3, two vectors are orthogonal if and only if arrows that picture these vectors are perpendicular, but in higher dimensions "orthogonality" is just a definition.

   The *length* of vector x is a number denoted $\|\mathsf{x}\|$ and defined like this.

2. $\|\mathsf{x}\| = \sqrt{x_1^2 + x_2^2 + \cdots + x_n^2} = \sqrt{\mathsf{x} \cdot \mathsf{x}}$.

   In R^2 or R^3, the length of x is the actual length of an arrow that pictures x, but "length" is simply a definition in R^n when $n > 3$.

   Since a sum of squares of real numbers is 0 if and only if each number is 0,

3. $\|\mathsf{x}\| = 0$ if and only if $\mathsf{x} = \mathsf{0}$.

   If a vector x is orthogonal to all other vectors, in particular, it is orthogonal to itself, that is, $\mathsf{x} \cdot \mathsf{x} = 0$. As we have just seen, this means $\mathsf{x} = \mathsf{0}$.

4. $\mathsf{x} \cdot \mathsf{y} = 0$ for all y if and only if $\mathsf{x} = \mathsf{0}$.

Finally, since $\|\mathsf{x}+\mathsf{y}\|^2 = (\mathsf{x}+\mathsf{y})\cdot(\mathsf{x}+\mathsf{y}) = \mathsf{x}\cdot\mathsf{x}+2\mathsf{x}\cdot\mathsf{y}+\mathsf{y}\cdot\mathsf{y} = \|\mathsf{x}\|^2+2\mathsf{x}\cdot\mathsf{y}+\|\mathsf{y}\|^2$, we see that $\|\mathsf{x}+\mathsf{y}\|^2 = \|\mathsf{x}\|^2 + \|\mathsf{y}\|^2$ if and only if $\mathsf{x}\cdot\mathsf{y} = 0$.

6.1.1 Theorem. (Pythagoras) *Vectors* x *and* y *are orthogonal if and only if* $\|\mathsf{x}+\mathsf{y}\|^2 = \|\mathsf{x}\|^2 + \|\mathsf{y}\|^2$.

The "Best Solution" to $A\mathsf{x} = \mathsf{b}$

The primary focus of this section is the linear system $A\mathsf{x} = \mathsf{b}$, where A is an $m \times n$ matrix and $m > n$. In this situation, the system is "overdetermined." There are more equations than unknowns. Generally, such a system has no solution. Consider, for example,

$$\begin{aligned} x_1 + x_2 &= 2 \\ x_1 - x_2 &= 0 \\ 2x_1 + x_2 &= -4. \end{aligned} \tag{1}$$

The first two equations already give $x_1 = x_2 = 1$, so we would have to be awfully lucky for these values to satisfy more equations. Here, the third equation $2x_1 + x_2 = -4$ is not satisfied by $x_1 = x_2 = 1$ (we weren't lucky), so the system has no solution.

Suppose your boss doesn't like this answer. "I told you to find a solution," she yells. Is there a pair of numbers x_1, x_2 that might satisfy her, providing a "best solution" to (1) in some sense?

Our system is $A\mathsf{x} = \mathsf{b}$ with

$$A = \begin{bmatrix} 1 & 1 \\ 1 & -1 \\ 2 & 1 \end{bmatrix}, \quad \mathsf{x} = \begin{bmatrix} x_1 \\ x_2 \end{bmatrix} \quad \text{and} \quad \mathsf{b} = \begin{bmatrix} 2 \\ 0 \\ -4 \end{bmatrix}.$$

For any x, $A\mathsf{x}$ is in the column space of A, so $A\mathsf{x} = \mathsf{b}$ has no solution precisely because b is not in the column space of A. We'd like $\mathsf{b} - A\mathsf{x}$ to be 0, equivalently $\|\mathsf{b} - A\mathsf{x}\| = 0$. This can't be done, but we could at least try to make $\|\mathsf{b} - A\mathsf{x}\|$ as small as possible.

If there is a unique vector x that minimizes the length of $\mathsf{b} - A\mathsf{x}$, we call this the "best solution to $A\mathsf{x} = \mathsf{b}$ in the sense of least squares" and denote this vector x^+. This expression derives from the fact that the length of a vector is the square root of a sum of squares, so minimizing $\|\mathsf{b} - A\mathsf{x}\|$ is making a sum of squares least. As Figure 6.1 suggests, $\|\mathsf{b} - A\mathsf{x}\|$ is smallest when $A\mathsf{x}^+$ is the (orthogonal) *projection* of b onto the column space of A; that is, the vector $\mathsf{b} - A\mathsf{x}^+$ is orthogonal to the column space of A. (This is

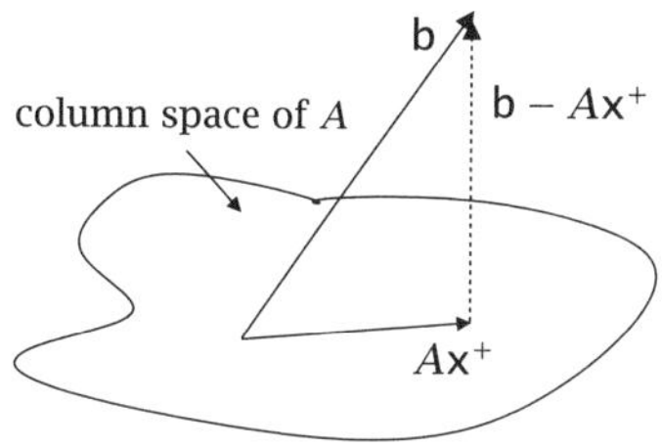

Figure 6.1: The best solution to $Ax = b$ is the vector x^+ with Ax^+ the projection of b on the column space of A.

perhaps "obvious" geometrically, but you should also think about a formal proof—see Exercise 25.)

We are going to find a formula for x^+ and, in so doing, we'll use most of the facts about orthogonality summarized at the start of this section.

We want $b - Ax^+$ to be orthogonal to every vector in the column space of A. A vector in the column space of A has the form Ay, so we want $(Ay) \cdot (b - Ax^+) = 0$ for all y. Now

$$\begin{aligned}(Ay) \cdot (b - Ax^+) &= (Ay)^T(b - Ax^+) \\ &= y^T A^T (b - Ax^+) \\ &= y^T(A^T b - A^T Ax^+) = y \cdot (A^T b - A^T Ax^+)\end{aligned}$$

and this is to be 0 for all y. Since the only vector orthogonal to every y is the zero vector, we must have $A^T b - A^T Ax^+ = 0$; otherwise put,

$$A^T Ax^+ = A^T b.$$

This is a system of equations with coefficient matrix $A^T A$. It will have a unique solution x^+ if the $n \times n$ matrix $A^T A$ is invertible, in which case the solution is

6.1.2
$$\boxed{x^+ = (A^T A)^{-1} A^T b.}$$

One condition for the invertibility of $A^T A$ is linear independence of the columns of A—see Corollary 4.5.10—so we obtain the following theorem.

6.1.3 Theorem. *If A has linearly independent columns, the best solution in the sense of least squares to the linear system $Ax = b$ is $x^+ = (A^T A)^{-1} A^T b$. This is the unique vector that minimizes $\|b - Ax\|$.*

6.1.4 Example. Let's return to the system of equations (1) that appeared earlier. There we had $A = \begin{bmatrix} 1 & 1 \\ 1 & -1 \\ 2 & 1 \end{bmatrix}$ and $\mathsf{b} = \begin{bmatrix} 2 \\ 0 \\ -4 \end{bmatrix}$. The columns of A are linearly independent, so

$$A^T A = \begin{bmatrix} 1 & 1 & 2 \\ 1 & -1 & 1 \end{bmatrix} \begin{bmatrix} 1 & 1 \\ 1 & -1 \\ 2 & 1 \end{bmatrix} = \begin{bmatrix} 6 & 2 \\ 2 & 3 \end{bmatrix}$$

is invertible. In fact, $(A^T A)^{-1} = \frac{1}{14} \begin{bmatrix} 3 & -2 \\ -2 & 6 \end{bmatrix}$, so

$$\mathsf{x}^+ = (A^T A)^{-1} A^T \mathsf{b} = \frac{1}{14} \begin{bmatrix} 3 & -2 \\ -2 & 6 \end{bmatrix} \begin{bmatrix} 1 & 1 & 2 \\ 1 & -1 & 1 \end{bmatrix} \begin{bmatrix} 2 \\ 0 \\ -4 \end{bmatrix} = \begin{bmatrix} -1 \\ 0 \end{bmatrix}.$$

Let's make sure we understand exactly what this means. If $\mathsf{x} = \begin{bmatrix} x_1 \\ x_2 \end{bmatrix}$, then

$$\mathsf{b} - A\mathsf{x} = \begin{bmatrix} 2 - x_1 - x_2 \\ -x_1 + x_2 \\ -4 - 2x_1 - x_2 \end{bmatrix}$$

is the vector we would like to be 0, so that each of the original equations is satisfied. It has length

$$\ell = \sqrt{(2 - x_1 - x_2)^2 + (-x_1 + x_2)^2 + (-4 - 2x_1 - x_2)^2},$$

which is smallest when $\mathsf{x} = \mathsf{x}^+ = \begin{bmatrix} -1 \\ 0 \end{bmatrix}$, that is, when $\mathsf{b} - A\mathsf{x}^+ = \begin{bmatrix} 3 \\ 1 \\ -2 \end{bmatrix}$. This has length $\ell_{\min} = \sqrt{9 + 1 + 4} = \sqrt{14}$, and we have proven that $\ell_{\min} \le \ell$ for any x_1, x_2. For example, with $x_1 = x_2 = 1$—the exact solution to the first two equations in (1)—we have $\ell = \sqrt{0 + 0 + 49} = 7 \ge \sqrt{14} = \ell_{\min}$. ☺

The author has often seen students start a proof by writing down the very thing they are asked to prove, assuming it is true and then doing some calculations that support this fact. The following little problem illustrates why this is not a good approach to a mathematics problem.

6.1.5 Problem. Suppose we wish to solve the system $A\mathsf{x} = \mathsf{b}$ where A is an $m \times n$ matrix with linearly independent columns and b is a given vector in R^m.

Let's begin by assuming $A\mathsf{x} = \mathsf{b}$. Multiplying by A^T gives $A^T A\mathsf{x} = A^T\mathsf{b}$. Since A has linearly independent columns, $A^T A$ is invertible, so $\mathsf{x} = (A^T A)^{-1} A^T \mathsf{b}$. We have discovered that the system $A\mathsf{x} = \mathsf{b}$ can always be solved exactly. Is this right? Explain.

Solution. The answer is no! Certainly $Ax = b$ cannot always be solved. The system we have been studying, system (1), has no solution even though $A = \begin{bmatrix} 1 & 1 \\ 1 & -1 \\ 2 & 1 \end{bmatrix}$ has linearly independent columns. The system

$$\begin{aligned} x_1 &= 1 \\ x_2 &= 2 \\ x_2 &= 3 \end{aligned}$$

has no solution even though this is $Ax = b$ with the coefficient matrix $A = \begin{bmatrix} 1 & 0 \\ 0 & 1 \\ 0 & 1 \end{bmatrix}$ having linearly independent columns. The problem with our reasoning is one of **logic**. By starting with the equation $Ax = b$, we assumed that there was a solution x. All we have done is to prove that $x = (A^TA)^{-1}A^Tb$, **if there is a solution**.

READING CHECK 1. Suppose the linear system $Ax = 0$ has only the trivial solution. What is the matter with the following argument, which purports to show that $Ax = b$ has a unique solution for any b?

> "If $Ax_1 = b$ and $Ax_2 = b$, then $A(x_1 - x_2) = 0$, so, by assumption, $x_1 - x_2 = 0$, so $x_1 = x_2$."

Now let's have another look at Figure 6.1 and the calculations leading to the formula for x^+ given in Theorem 6.1.3. The vector Ax^+ has the property that $b - Ax^+$ is orthogonal to the column space of A. We call Ax^+ the (orthogonal) *projection* of b on the column space and note that

$$Ax^+ = A(A^TA)^{-1}A^Tb = Pb,$$

with

6.1.6 $$\boxed{P = A(A^TA)^{-1}A^T.}$$

We now give a little argument showing that P and A have the same column spaces. First, since $Pb = A[(A^TA)^{-1}A^Tb]$ is of the form Ax, the column space of P is contained in the column space of A. Second, since $A = PA$, any $Ax = P(Ax)$ has the form Pb, so the column space of A is contained in the column space of P.

Indeed, A and P have the same column space, so $b - Pb = b - Ax^+$ is orthogonal to the column space of P. The matrix P is a *projection matrix* in the following sense.

6.1.7 **Definition.** A matrix P is a *projection matrix* if and only if $b - Pb$ is orthogonal to the column space of P for all vectors b. (See Figure 6.2.)

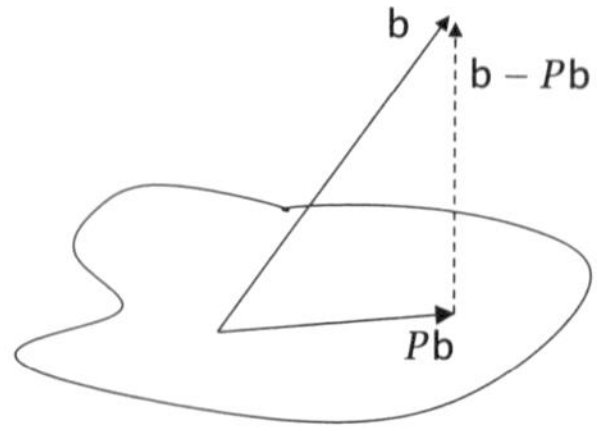

Figure 6.2: The action of a projection matrix: $P\mathsf{b}$ is the projection of b on the column space of P if $\mathsf{b} - P\mathsf{b}$ is orthogonal to the column space.

For example, we have seen that if A has linearly independent columns, then $P = A(A^TA)^{-1}A^T$ is a projection matrix. With $A = \begin{bmatrix} 1 & 1 \\ 1 & -1 \\ 2 & 1 \end{bmatrix}$, as above,

$$P = A(A^TA)^{-1}A^T = \frac{1}{14}\begin{bmatrix} 1 & 1 \\ 1 & -1 \\ 2 & 1 \end{bmatrix}\begin{bmatrix} 3 & -2 \\ -2 & 6 \end{bmatrix}\begin{bmatrix} 1 & 1 & 2 \\ 1 & -1 & 1 \end{bmatrix}$$
$$= \frac{1}{14}\begin{bmatrix} 5 & -3 & 6 \\ -3 & 13 & 2 \\ 6 & 2 & 10 \end{bmatrix}.$$

READING CHECK 2. We have argued that any P of the form $A(A^TA)^{-1}A^T$ has the same column space as A. Show that this is true in the specific case $A = \begin{bmatrix} 1 & 1 \\ 1 & -1 \\ 2 & 1 \end{bmatrix}$, $P = \frac{1}{14}\begin{bmatrix} 5 & -3 & 6 \\ -3 & 13 & 2 \\ 6 & 2 & 10 \end{bmatrix}$.

READING CHECK 3. If b is in the column space of a projection matrix P, then $P\mathsf{b} = \mathsf{b}$. Why?

We have established the following theorem.

6.1.8 Theorem. *For any matrix A with linearly independent columns, the matrix $P = A(A^TA)^{-1}A^T$ is a projection matrix. It projects vectors (orthogonally) onto the column space of A, which is the column space of P.*

6.1.9 Example. Let $A = \begin{bmatrix} 1 & 1 \\ 1 & -1 \\ 2 & 1 \end{bmatrix}$, $P == \frac{1}{14}\begin{bmatrix} 5 & -3 & 6 \\ -3 & 13 & 2 \\ 6 & 2 & 10 \end{bmatrix}$ and $\mathsf{b} = \begin{bmatrix} 2 \\ 0 \\ -4 \end{bmatrix}$, as at the start of this section. Then $P\mathsf{b} = \begin{bmatrix} -1 \\ -1 \\ -2 \end{bmatrix}$ is in the column space of P

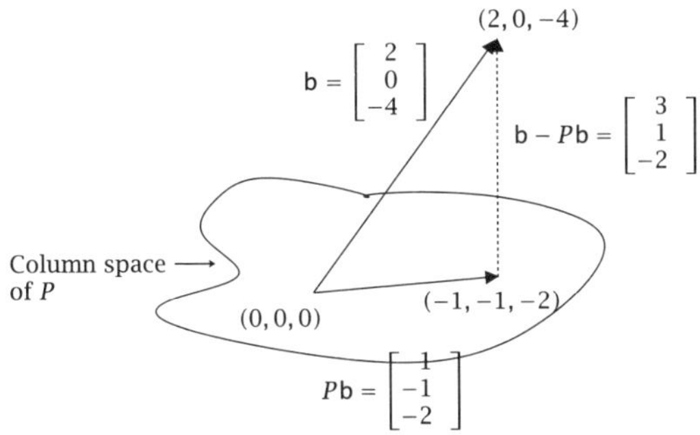

Figure 6.3

(which is the column space of A) and $\mathsf{b} - P\mathsf{b} = \begin{bmatrix} 2 \\ 0 \\ -4 \end{bmatrix} - \begin{bmatrix} -1 \\ -1 \\ -2 \end{bmatrix} = \begin{bmatrix} 3 \\ 1 \\ -2 \end{bmatrix}$ is orthogonal to the column space—see Figure 6.3. We can check by taking the dot product with each column of A:

$$\begin{bmatrix} 3 \\ 1 \\ -2 \end{bmatrix} \cdot \begin{bmatrix} 1 \\ 1 \\ 2 \end{bmatrix} = 3 + 1 - 4 = 0 \quad \text{and} \quad \begin{bmatrix} 3 \\ 1 \\ -2 \end{bmatrix} \cdot \begin{bmatrix} 1 \\ -1 \\ 1 \end{bmatrix} = 3 - 1 - 2 = 0.$$

Knowing the matrix that projects vectors onto the column space of A makes it simple to find the "best solution" to $A\mathsf{x} = \mathsf{b}$. Since $A\mathsf{x}^+ = P\mathsf{b}$ is the projection of b on the column space of A, x^+ is a solution to $A\mathsf{x} = P\mathsf{b}$ and it is easy to see that this solution is unique—see Exercise 34. So we have this result.

6.1.10 Theorem. *Assuming A has linearly independent columns, the "best solution" to $A\mathsf{x} = \mathsf{b}$ is the unique solution to $A\mathsf{x} = P\mathsf{b}$, where $P = A(A^TA)^{-1}A^T$.*

6.1.11 Example. With $A = \begin{bmatrix} 1 & 1 \\ 1 & -1 \\ 2 & 1 \end{bmatrix}$, $\mathsf{b} = \begin{bmatrix} 2 \\ 0 \\ -4 \end{bmatrix}$ and $P = \frac{1}{14}\begin{bmatrix} 5 & -3 & 6 \\ -3 & 3 & 2 \\ 6 & 2 & 10 \end{bmatrix}$, we have $P\mathsf{b} = \begin{bmatrix} -1 \\ -1 \\ -2 \end{bmatrix}$. The best solution to $A\mathsf{x} = \mathsf{b}$ is the exact solution to $A\mathsf{x} = P\mathsf{b}$. One instance of Gaussian elimination applied to the augmented matrix $[A \mid P\mathsf{b}]$ is

$$\left[\begin{array}{cc|c} 1 & 1 & -1 \\ 1 & -1 & -1 \\ 2 & 1 & -2 \end{array}\right] \to \left[\begin{array}{cc|c} 1 & 1 & -1 \\ 0 & -2 & 0 \\ 0 & -1 & 0 \end{array}\right] \to \left[\begin{array}{cc|c} 1 & 1 & -1 \\ 0 & 1 & 0 \\ 0 & 0 & 0 \end{array}\right],$$

so $x_2 = 0$, $x_1 = -1 - x_2 = -1$ and $\mathsf{x}^+ = \begin{bmatrix} -1 \\ 0 \end{bmatrix}$ as before.

How to Recognize a Projection Matrix

There is an easy way to identify a projection matrix.

6.1.12 Theorem. *A matrix P is a projection matrix if and only if P is* symmetric *($P^T = P$) and* idempotent *($P^2 = P$).*

Proof. Suppose first that P is a matrix satisfying $P^T = P$ and $P^2 = P$. To show that P is a projection matrix, we must show that $\mathsf{b} - P\mathsf{b}$ is orthogonal to the column space of P for all b. This is so because a vector in the column space has the form $P\mathsf{y}$ and

$$\begin{aligned}(P\mathsf{y}) \cdot (\mathsf{b} - P\mathsf{b}) &= \mathsf{y}^T P^T (\mathsf{b} - P\mathsf{b}) \\ &= \mathsf{y}^T P(\mathsf{b} - P\mathsf{b}) = \mathsf{y}^T (P\mathsf{b} - P^2\mathsf{b}) = \mathsf{y}^T (P\mathsf{b} - P\mathsf{b}) = 0.\end{aligned}$$

Conversely, suppose P is a projection matrix. Thus $\mathsf{b} - P\mathsf{b}$ is orthogonal to $P\mathsf{y}$ for any vector y. So, for all y and for all b,

$$\begin{aligned}0 &= (P\mathsf{y}) \cdot (\mathsf{b} - P\mathsf{b}) \\ &= (P\mathsf{y})^T (\mathsf{b} - P\mathsf{b}) \quad \longleftarrow \\ &= \mathsf{y}^T P^T (\mathsf{b} - P\mathsf{b}) \\ &= \mathsf{y}^T (P^T\mathsf{b} - P^T P\mathsf{b}) \\ &= \mathsf{y} \cdot (P^T\mathsf{b} - P^T P\mathsf{b}).\end{aligned}$$

The only vector orthogonal to all vectors y is the zero vector, so $P^T\mathsf{b} - P^T P\mathsf{b} = 0$, giving $P^T P\mathsf{b} = P^T\mathsf{b}$ for all b. So $P^T P = P^T$. (See READING CHECK 5.) Any matrix of the form $X^T X$ is symmetric, so P^T is symmetric; that is, $P^T = (P^T)^T = P$. So $P = P^T$ is symmetric too and $P^T P = P^T$ then reads $P^2 = P$. ■

READING CHECK 4. Explain the line marked with the arrow in the proof of Theorem 6.1.12. How did we move from $(P\mathsf{y}) \cdot (\mathsf{b} - P\mathsf{b})$ to $(P\mathsf{y})^T (\mathsf{b} - P\mathsf{b})$?

READING CHECK 5. In the proof of Theorem 6.1.12, we asserted that $P^T P\mathsf{b} = P^T\mathsf{b}$ for all b implies $P^T P = P^T$. Explain why $A\mathsf{x} = B\mathsf{x}$ for all x implies that $A = B$.

Projections onto Subspaces

Dimension 1. If U is a subspace of R^n of dimension 1, spanned by a vector u, say, and if v is some other vector in R^n, the "projection of v on U" is the vector

$$\operatorname{proj}_U \mathsf{v} = \frac{\mathsf{v} \cdot \mathsf{u}}{\mathsf{u} \cdot \mathsf{u}}\, \mathsf{u}.$$

See Section 1.4. This is a scalar multiple of u, but it's also $P\mathsf{v}$ for a certain projection matrix P. To find P, we begin by writing $\frac{\mathsf{v}\cdot\mathsf{u}}{\mathsf{u}\cdot\mathsf{u}}\,\mathsf{u} = \frac{1}{\mathsf{u}\cdot\mathsf{u}}\,(\mathsf{v}\cdot\mathsf{u})\mathsf{u}$. Noting that a scalar multiple $c\mathsf{u}$ is the matrix product $\mathsf{u}[c]$ of u and the 1×1 matrix $[c]$, we have

$$\begin{aligned}
(\mathsf{v}\cdot\mathsf{u})\mathsf{u} &= (\mathsf{u}\cdot\mathsf{v})\mathsf{u} && \text{because the dot product is commutative}\\
&= \mathsf{u}[\mathsf{u}\cdot\mathsf{v}] && \text{because } c\mathsf{u}=\mathsf{u}[c] \qquad (2)\\
&= \mathsf{u}(\mathsf{u}^T\mathsf{v}) && \text{(see the answer to READING CHECK 4)}\\
&= (\mathsf{u}\mathsf{u}^T)\mathsf{v} && \text{because matrix multiplication is associative.}
\end{aligned}$$

All this gives $\operatorname{proj}_U \mathsf{v} = \dfrac{1}{\mathsf{u}\cdot\mathsf{u}}(\mathsf{u}\mathsf{u}^T)\mathsf{v} = P\mathsf{v}$ with $P = \dfrac{\mathsf{u}\mathsf{u}^T}{\mathsf{u}\cdot\mathsf{u}}$.

READING CHECK 6. Use Theorem 6.1.12 to show that $P = \dfrac{\mathsf{u}\mathsf{u}^T}{\mathsf{u}\cdot\mathsf{u}}$ is a projection matrix.

6.1.13 Example. Suppose we wish to project vectors in the Euclidean plane onto the subspace spanned by the vector $\mathsf{u} = \begin{bmatrix}1\\2\end{bmatrix}$. Since $\mathsf{u}\mathsf{u}^T = \begin{bmatrix}1\\2\end{bmatrix}\begin{bmatrix}1 & 2\end{bmatrix} = \begin{bmatrix}1 & 2\\2 & 4\end{bmatrix}$, the projection matrix we need is $P = \frac{\mathsf{u}\mathsf{u}^T}{\mathsf{u}\cdot\mathsf{u}} = \frac{1}{5}\begin{bmatrix}1 & 2\\2 & 4\end{bmatrix}$, so the projection of $\begin{bmatrix}x\\y\end{bmatrix}$ on U is $P\begin{bmatrix}x\\y\end{bmatrix} = \frac{1}{5}\begin{bmatrix}x+2y\\2x+4y\end{bmatrix}$. ☺

Dimension 2. The projection of a vector b on the plane π spanned by orthogonal vectors e and f is

$$\operatorname{proj}_\pi \mathsf{b} = \frac{\mathsf{b}\cdot\mathsf{e}}{\mathsf{e}\cdot\mathsf{e}}\,\mathsf{e} + \frac{\mathsf{b}\cdot\mathsf{f}}{\mathsf{f}\cdot\mathsf{f}}\,\mathsf{f}.$$

(See Theorem 1.4.10.)

As in the one-dimensional case, this vector is also of the form $P\mathsf{b}$ with P a projection matrix. With the same manipulations as before, we deduce that $(\mathsf{b}\cdot\mathsf{e})\mathsf{e} = (\mathsf{e}\mathsf{e}^T)\mathsf{b}$ and $(\mathsf{b}\cdot\mathsf{f})\mathsf{f} = (\mathsf{f}\mathsf{f}^T)\mathsf{b}$. Thus

$$\operatorname{proj}_\pi \mathsf{b} = \frac{\mathsf{e}\mathsf{e}^T}{\mathsf{e}\cdot\mathsf{e}}\,\mathsf{b} + \frac{\mathsf{f}\mathsf{f}^T}{\mathsf{f}\cdot\mathsf{f}}\,\mathsf{b} = P\mathsf{b},$$

with $P = \dfrac{\mathsf{e}\mathsf{e}^T}{\mathsf{e}\cdot\mathsf{e}} + \dfrac{\mathsf{f}\mathsf{f}^T}{\mathsf{f}\cdot\mathsf{f}}$. [In Exercise 12, we ask you to verify that this matrix is both symmetric and idempotent and hence a projection matrix.]

6.1.14 Example. Suppose we want to project vectors in R^3 onto the plane π with equation $3x + y - 2z = 0$. We find two orthogonal vectors $\mathsf{e} = \begin{bmatrix}0\\2\\1\end{bmatrix}$ and

$\mathsf{f} = \begin{bmatrix} 5 \\ -3 \\ 6 \end{bmatrix}$ in π and compute the matrix

$$P = \frac{\mathsf{ee}^T}{\mathsf{e} \cdot \mathsf{e}} + \frac{\mathsf{ff}^T}{\mathsf{f} \cdot \mathsf{f}} = \frac{1}{5}\mathsf{ee}^T + \frac{1}{70}\mathsf{ff}^T.$$

Now

$$\mathsf{ee}^T = \begin{bmatrix} 0 \\ 2 \\ 1 \end{bmatrix} \begin{bmatrix} 0 & 2 & 1 \end{bmatrix} = \begin{bmatrix} 0 & 0 & 0 \\ 0 & 4 & 2 \\ 0 & 2 & 1 \end{bmatrix}$$

and

$$\mathsf{ff}^T = \begin{bmatrix} 5 \\ -3 \\ 6 \end{bmatrix} \begin{bmatrix} 5 & -3 & 6 \end{bmatrix} = \begin{bmatrix} 25 & -15 & 30 \\ -15 & 9 & -18 \\ 30 & -18 & 36 \end{bmatrix},$$

so

$$P = \frac{1}{5}\begin{bmatrix} 0 & 0 & 0 \\ 0 & 4 & 2 \\ 0 & 2 & 1 \end{bmatrix} + \frac{1}{70}\begin{bmatrix} 25 & -15 & 30 \\ -15 & 9 & -18 \\ 30 & -18 & 36 \end{bmatrix} = \frac{1}{14}\begin{bmatrix} 5 & -3 & 6 \\ -3 & 13 & 2 \\ 6 & 2 & 10 \end{bmatrix},$$

a matrix we have met before and which we know very well projects vectors onto π. ☺

Dimension k. Having seen how to project vectors onto lines and planes, it is not hard to imagine projections onto subspaces of any dimension. First, we should be precise as to what this means.

6.1.15 Definition. If U is a subspace of some Euclidean space V and b is in V, the (orthogonal) *projection of* b *on* U is the vector $\mathsf{p} = \operatorname{proj}_U \mathsf{b}$ in U with the property that $\mathsf{b} - \mathsf{p}$ is orthogonal to U, that is, orthogonal to every vector in U.

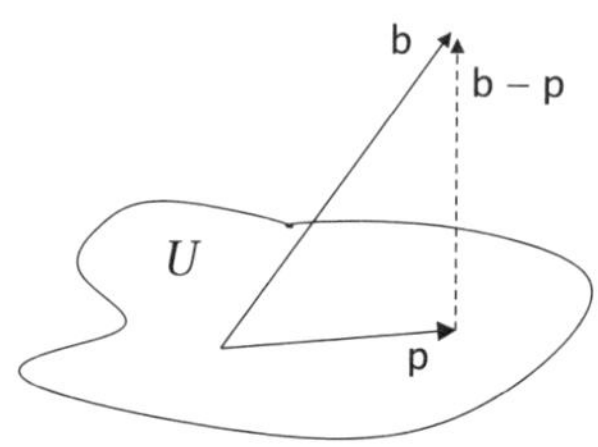

Uniqueness of the projection is a consequence of the next theorem, which gives an explicit formula for $\operatorname{proj}_U \mathsf{b}$.

6.1.16 Theorem. *Let U be a subspace of some Euclidean space and let $\mathsf{f}_1, \mathsf{f}_2, \ldots, \mathsf{f}_k$ be a basis of U consisting of pairwise orthogonal vectors.*[1] *For any vector b, the projection of b on U is*

$$\operatorname{proj}_U \mathsf{b} = \frac{\mathsf{b} \cdot \mathsf{f}_1}{\mathsf{f}_1 \cdot \mathsf{f}_1}\mathsf{f}_1 + \frac{\mathsf{b} \cdot \mathsf{f}_2}{\mathsf{f}_2 \cdot \mathsf{f}_2}\mathsf{f}_2 + \cdots + \frac{\mathsf{b} \cdot \mathsf{f}_k}{\mathsf{f}_k \cdot \mathsf{f}_k}\mathsf{f}_k. \qquad (3)$$

[1] "Pairwise" means that the dot product of any two is 0.

This vector can be written $P\mathsf{b}$ with $P = \dfrac{\mathsf{f}_1\mathsf{f}_1^T}{\mathsf{f}_1 \cdot \mathsf{f}_1} + \dfrac{\mathsf{f}_2\mathsf{f}_2^T}{\mathsf{f}_2 \cdot \mathsf{f}_2} + \cdots + \dfrac{\mathsf{f}_k\mathsf{f}_k^T}{\mathsf{f}_k \cdot \mathsf{f}_k}$, *the projection matrix onto U (= col sp P).*

Proof. Let $\mathsf{p} = \text{proj}_U \mathsf{b}$. Since p is in U, which has basis $\{\mathsf{f}_1, \mathsf{f}_2, \ldots, \mathsf{f}_k\}$, there are scalars $c_1, c_2, \ldots, c_k$ such that

$$\mathsf{p} = c_1\mathsf{f}_1 + c_2\mathsf{f}_2 + \cdots + c_k\mathsf{f}_k. \tag{4}$$

Since $\mathsf{b} - \mathsf{p}$ is orthogonal to U, it is orthogonal to each f_i, so $(\mathsf{b} - \mathsf{p}) \cdot \mathsf{f}_i = 0$, that is, $\mathsf{b} \cdot \mathsf{f}_i = \mathsf{p} \cdot \mathsf{f}_i$, for each i. Now $\mathsf{p} \cdot \mathsf{f}_i = c_i(\mathsf{f}_i \cdot \mathsf{f}_i)$ because $\mathsf{f}_i \cdot \mathsf{f}_j = 0$ if $i \neq j$. So $\mathsf{b} \cdot \mathsf{f}_i = c_i(\mathsf{f}_i \cdot \mathsf{f}_i)$ and, since $\mathsf{f}_i \neq 0$, $c_i = \frac{\mathsf{b} \cdot \mathsf{f}_i}{\mathsf{f}_i \cdot \mathsf{f}_i}$. Thus the equation for p given in (4) is the equation appearing in (3). This establishes the first part of the theorem.

The formula for P follows from $(\mathsf{b} \cdot \mathsf{f})\mathsf{f} = (\mathsf{f}^T\mathsf{f})\mathsf{b}$, which we have observed several times already in this section. To show that P is symmetric and idempotent follows from an easy generalization of the argument with $k = 2$—see Exercise 12.

To show that U = col sp P, we observe first that col sp $P \subseteq U$ because $P\mathsf{b}$ is a linear combination of $\mathsf{f}_1, \ldots, \mathsf{f}_k$, which is a basis for U. On the other hand, if b is in U, then $P\mathsf{b}$ is in col sp P and hence in U too, so $\mathsf{b} - P\mathsf{b}$ is in U. It is also orthogonal to U and hence to itself, so it is 0. Thus $\mathsf{b} = P\mathsf{b}$ is in col sp P. So $U \subseteq$ col sp P proving that these subspaces are the same. ■

6.1.17 Example. The vectors $\mathsf{f}_1 = \begin{bmatrix} 1 \\ 0 \\ -1 \\ 1 \end{bmatrix}$, $\mathsf{f}_2 = \begin{bmatrix} 1 \\ 1 \\ 1 \\ 0 \end{bmatrix}$ and $\mathsf{f}_3 = \begin{bmatrix} -1 \\ 0 \\ 1 \\ 2 \end{bmatrix}$ are pairwise orthogonal in R^4. If U is the span of these vectors and we wish to project onto U, we need the matrix

$$P = \frac{\mathsf{f}_1\mathsf{f}_1^T}{\mathsf{f}_1 \cdot \mathsf{f}_1} + \frac{\mathsf{f}_2\mathsf{f}_2^T}{\mathsf{f}_2 \cdot \mathsf{f}_2} + \frac{\mathsf{f}_3\mathsf{f}_3^T}{\mathsf{f}_3 \cdot \mathsf{f}_3} = \tfrac{1}{3}\mathsf{f}_1\mathsf{f}_1^T + \tfrac{1}{3}\mathsf{f}_2\mathsf{f}_2^T + \tfrac{1}{6}\mathsf{f}_3\mathsf{f}_3^T.$$

Since

$$\mathsf{f}_1\mathsf{f}_1^T = \begin{bmatrix} 1 \\ 0 \\ -1 \\ 1 \end{bmatrix} \begin{bmatrix} 1 & 0 & -1 & 1 \end{bmatrix} = \begin{bmatrix} 1 & 0 & -1 & 1 \\ 0 & 0 & 0 & 0 \\ -1 & 0 & 1 & -1 \\ 1 & 0 & -1 & 1 \end{bmatrix},$$

$$\mathsf{f}_2\mathsf{f}_2^T = \begin{bmatrix} 1 \\ 1 \\ 1 \\ 0 \end{bmatrix} \begin{bmatrix} 1 & 1 & 1 & 0 \end{bmatrix} = \begin{bmatrix} 1 & 1 & 1 & 0 \\ 1 & 1 & 1 & 0 \\ 1 & 1 & 1 & 0 \\ 0 & 0 & 0 & 0 \end{bmatrix}$$

and $$\mathsf{f}_3\mathsf{f}_3^T = \begin{bmatrix} -1 \\ 0 \\ 1 \\ 2 \end{bmatrix} \begin{bmatrix} -1 & 0 & 1 & 2 \end{bmatrix} = \begin{bmatrix} 1 & 0 & -1 & -2 \\ 0 & 0 & 0 & 0 \\ -1 & 0 & 1 & 2 \\ -2 & 0 & 2 & 4 \end{bmatrix},$$

the desired matrix P is

$$\frac{1}{3}\begin{bmatrix} 1 & 0 & -1 & 1 \\ 0 & 0 & 0 & 0 \\ -1 & 0 & 1 & -1 \\ 1 & 0 & -1 & 1 \end{bmatrix} + \frac{1}{3}\begin{bmatrix} 1 & 1 & 1 & 0 \\ 1 & 1 & 1 & 0 \\ 1 & 1 & 1 & 0 \\ 0 & 0 & 0 & 0 \end{bmatrix} + \frac{1}{6}\begin{bmatrix} 1 & 0 & -1 & -2 \\ 0 & 0 & 0 & 0 \\ -1 & 0 & 1 & 2 \\ -2 & 0 & 2 & 4 \end{bmatrix}$$
$$= \frac{1}{6}\begin{bmatrix} 5 & 2 & -1 & 0 \\ 2 & 2 & 2 & 0 \\ -1 & 2 & 5 & 0 \\ 0 & 0 & 0 & 6 \end{bmatrix}.$$

:-)

READING CHECK 7. The vectors $\mathsf{u} = \begin{bmatrix} 1 \\ 1 \\ 0 \end{bmatrix}$ and $\mathsf{v} = \begin{bmatrix} 1 \\ 0 \\ -1 \end{bmatrix}$ span the plane U with equation $x + y + z = 0$. So, to project a vector b on U, we multiply by $P = \dfrac{\mathsf{uu}^T}{\mathsf{u}\cdot\mathsf{u}} + \dfrac{\mathsf{vv}^T}{\mathsf{v}\cdot\mathsf{v}} = \frac{1}{2}\begin{bmatrix} 1 & 1 & 0 \\ 1 & 1 & 0 \\ 0 & 0 & 0 \end{bmatrix} + \frac{1}{2}\begin{bmatrix} 1 & 0 & -1 \\ 0 & 0 & 0 \\ -1 & 0 & 1 \end{bmatrix} = \frac{1}{2}\begin{bmatrix} 2 & 1 & -1 \\ 1 & 1 & 0 \\ -1 & 0 & 1 \end{bmatrix}$. Is this correct?

Answers to Reading Checks

1. The given argument is correct **if there is a solution**. The matrix A in Problem 6.1.5 has the property that $A\mathsf{x} = 0$ has only the trivial solution $\mathsf{x} = 0$, but it is not true that $A\mathsf{x} = \mathsf{b}$ has a unique solution for any b.

2. A normal to the plane which is the column space of A is $\begin{bmatrix} 1 \\ 1 \\ 2 \end{bmatrix} \times \begin{bmatrix} 1 \\ -1 \\ 1 \end{bmatrix} = \begin{bmatrix} 3 \\ 1 \\ -2 \end{bmatrix}$, so this plane has equation $3x + y - 2z = 0$. Notice that each column of P satisfies this equation, so the column space of P is contained in the column space of A. Since the column spaces have the same dimension, they are the same.

3. By definition of "projection matrix," $\mathsf{b} - P\mathsf{b}$ is orthogonal to the column space of P. If b is in the column space, so is $\mathsf{b} - P\mathsf{b}$, so this vector is orthogonal to itself and hence 0.

4. For any vectors u and v, the dot product $\mathsf{u}\cdot\mathsf{v}$ is the **matrix product** $\mathsf{u}^T\mathsf{v}$, as observed at the start of this section.

5. The hypothesis $A\mathsf{x} = B\mathsf{x}$ for any x says, in particular, that $A\mathsf{e}_i = B\mathsf{e}_i$ for any i, so each column of A equals the corresponding column of B—see 2.1.22.

6. We show that P is symmetric and idempotent and for this use basic properties of the transpose operator, specifically $(cA)^T = cA^T$, $(AB)^T = B^TA^T$ and $(A^T)^T = A$. First, $P^T = (\frac{1}{\mathsf{v}\cdot\mathsf{v}}\mathsf{v}\mathsf{v}^T)^T = (\frac{1}{\mathsf{v}\cdot\mathsf{v}})(\mathsf{v}\mathsf{v}^T)^T = \frac{1}{\mathsf{v}\cdot\mathsf{v}}(\mathsf{v}^T)^T\mathsf{v}^T = \frac{1}{\mathsf{v}\cdot\mathsf{v}}\mathsf{v}\mathsf{v}^T = P$. Now $P^2 = \dfrac{\mathsf{v}\mathsf{v}^T\mathsf{v}\mathsf{v}^T}{(\mathsf{v}\cdot\mathsf{v})(\mathsf{v}\cdot\mathsf{v})}$ and $\mathsf{v}^T\mathsf{v} = \mathsf{v}\cdot\mathsf{v}$, so the numerator is $\mathsf{v}(\mathsf{v}\cdot\mathsf{v})\mathsf{v}^T$. Cancelling $\mathsf{v}\cdot\mathsf{v}$ from numerator and denominator gives $P^2 = \dfrac{\mathsf{v}\mathsf{v}^T}{\mathsf{v}\cdot\mathsf{v}} = P$.

7. This is not correct. The reader has misread Theorem 6.1.16. The vectors in the formula for P must be orthogonal. Here, u and v are not.

True/False Questions

Decide, with as little calculation as possible, whether each of the following statements is true or false and, if you say "false," explain your answer. (Answers are at the back of the book.)

1. The length of $\begin{bmatrix}1\\1\\1\\1\end{bmatrix}$ is 4.

2. If x is a vector and $\mathsf{x}\cdot\mathsf{x} = 0$, then $\mathsf{x} = 0$.

3. If u and v are vectors, $\mathsf{u}\cdot\mathsf{v} = \mathsf{u}\mathsf{v}^T$.

4. For any matrix A, the "best solution" to $A\mathsf{x} = \mathsf{b}$ in the sense of least squares is $\mathsf{x}^+ = (A^TA)^{-1}A^T\mathsf{b}$.

5. The "best solution" to $A\mathsf{x} = \mathsf{b}$ is a solution to $A\mathsf{x} = \mathsf{b}$.

6. A projection matrix is square.

7. The only invertible projection matrix is the identity.

8. For any unit vector u, the matrix $\mathsf{u}\mathsf{u}^T$ is a projection matrix.

9. If vectors u and v span a plane π, then the matrix that projects vectors onto π is $P = \dfrac{\mathsf{u}\mathsf{u}^T}{\mathsf{u}\cdot\mathsf{u}} + \dfrac{\mathsf{v}\mathsf{v}^T}{\mathsf{v}\cdot\mathsf{v}}$.

10. If b is a vector in a subspace U, the projection of b on U is b.

EXERCISES

Answers to exercises marked [BB] can be found at the Back of the Book.

1. [BB] In this section, a lot of attention was paid to the matrix $A(A^TA)^{-1}A^T$. What's the problem with the calculation $A(A^TA)^{-1}A^T = A[A^{-1}(A^T)^{-1}]A^T = (AA^{-1})(A^T)^{-1}A^T = I$? Was this whole section about the identity matrix?

2. Let x be a vector in R^n.

   (a) Show that $\mathsf{x}^T\mathsf{x} = 0$ implies $\mathsf{x} = 0$.

   (b) Show that $\mathsf{x}\mathsf{x}^T = 0$ implies $\mathsf{x} = 0$.

3. [BB] **(a)** Find the best solution x^+ in the sense of least squares to $\begin{aligned} x + y &= 1 \\ x - y &= 2 \\ 2x + y &= 5. \end{aligned}$

   (b) What does "best solution in the sense of least squares" mean? Explain with reference to part (a).

4. Find the best solution in the sense of least squares to each of the following systems of linear equations.

   (a) [BB] $\begin{aligned} 2x_1 - x_2 &= 1 \\ x_1 - x_2 &= 0 \\ 2x_1 + x_2 &= -1 \\ x_1 + 3x_2 &= 3 \end{aligned}$ **(b)** $\begin{aligned} x_1 + 2x_2 &= 1 \\ x_1 - x_2 &= -1 \\ 3x_1 - x_2 &= 0 \\ -4x_1 + x_2 &= -3 \end{aligned}$

   (c) $\begin{aligned} x_1 \quad\quad - 2x_3 &= 0 \\ x_1 + x_2 - x_3 &= 2 \\ - x_2 - x_3 &= 3 \\ 2x_1 + 3x_2 + x_3 &= -4 \end{aligned}$ **(d)** $\begin{aligned} x_2 + 2x_3 &= 1 \\ x_1 \quad - x_3 &= -1 \\ x_1 + x_2 + x_3 &= 0 \\ -x_1 + x_2 + 3x_3 &= 1 \\ -x_1 + x_2 \quad &= 2. \end{aligned}$

5. Which of the following are projection matrices? Explain your answers.

   (a) [BB] $P = \frac{1}{10}\begin{bmatrix} 2 & 5 \\ 5 & 10 \end{bmatrix}$ **(b)** $P = \frac{1}{3}\begin{bmatrix} 2 & 1 & 1 \\ 1 & 2 & -1 \\ 1 & -1 & 2 \end{bmatrix}$

   (c) $\frac{1}{25}\begin{bmatrix} 1 & 2 & -3 \\ 2 & 4 & 5 \\ -1 & 5 & 10 \end{bmatrix}$ **(d)** [BB] $\frac{1}{6}\begin{bmatrix} 5 & -2 & -1 \\ -2 & 2 & -2 \\ -1 & -2 & 5 \end{bmatrix}$

   (e) $\frac{1}{3}\begin{bmatrix} 1 & 0 & 1 & 1 \\ 0 & 1 & 1 & -1 \\ 1 & 1 & 2 & 0 \\ 1 & -1 & 0 & 2 \end{bmatrix}$ **(f)** $\begin{bmatrix} 2 & -1 & 1 & 1 \\ 1 & 0 & 1 & 1 \\ 1 & -1 & 2 & 1 \\ -2 & 2 & -2 & -1 \end{bmatrix}$.

6. Find the matrix P that projects vectors in R^4 onto the column space of each matrix.

   (a) [BB] $A = \begin{bmatrix} 1 & 1 \\ 2 & 1 \\ -2 & 1 \\ 0 & 1 \end{bmatrix}$ **(b)** $A = \begin{bmatrix} -2 & 1 \\ -1 & 1 \\ 0 & 1 \\ 1 & -1 \end{bmatrix}$

(c) $A = \begin{bmatrix} 1 & 1 & -1 \\ 1 & 2 & 0 \\ 1 & -3 & 1 \\ 2 & -1 & 1 \end{bmatrix}$ (d) $A = \begin{bmatrix} 1 & 0 & 1 & 0 \\ 1 & 1 & 1 & 0 \\ -1 & 0 & 3 & 1 \\ -1 & 1 & -1 & 0 \end{bmatrix}$.

7. [BB] **(a)** Find a condition on b_1, b_2, b_3, b_4 that is necessary and sufficient for $\mathsf{b} = \begin{bmatrix} b_1 \\ b_2 \\ b_3 \\ b_4 \end{bmatrix}$ to belong to the column space of $A = \begin{bmatrix} 1 & -2 & 0 \\ -1 & 0 & 1 \\ 0 & 0 & 1 \\ 1 & -1 & 1 \end{bmatrix}$.

   (b) Prove that the columns of A are linearly independent.

   (c) Without calculation, explain why the matrix A^TA must be invertible.

   For the rest of this exercise, let $\mathsf{b} = \mathsf{e}_1 = \begin{bmatrix} 1 \\ 0 \\ 0 \\ 0 \end{bmatrix}$.

   (d) Without calculation, explain why $A\mathsf{x} = \mathsf{b}$ has no solution.

   (e) Find the best solution x^+ in the sense of least squares to the system $A\mathsf{x} = \mathsf{b}$.

   (f) What is the significance of your answer to 7(e)?

   (g) Find the matrix P that projects vectors in R^4 onto the column space of A.

   (h) Find P^2 **without calculation**, and explain.

   (i) Solve the system $A\mathsf{x} = P\mathsf{b}$ and comment on your answer.

8. If the column space of an $m \times n$ matrix A is R^m, what is the projection matrix onto the column space of A? Explain.

9. Let A be any $m \times n$ matrix. Prove that A^TA and A have the same null space. [Hint: $\|A\mathsf{x}\|^2 = \mathsf{x}^TA^TA\mathsf{x}$. See Exercise 43 of Section 2.1.]

10. [BB] Prove that any linear combination of projection matrices is symmetric.

11. Let P be a projection matrix. Show that $I - P$ is also a projection matrix.

12. Let e and f be orthogonal vectors. Show that the matrix $P = \frac{\mathsf{e}\mathsf{e}^T}{\mathsf{e}\cdot\mathsf{e}} + \frac{\mathsf{f}\mathsf{f}^T}{\mathsf{f}\cdot\mathsf{f}}$ is symmetric and satisfies $P^2 = P$, and hence is a projection matrix. [Hint: See the solution to READING CHECK 6.]

13. [BB] If A is $m \times n$ and $A^TA = I$, show that $P = AA^T$ is a projection matrix.

14. Let Q be a matrix whose columns are pairwise orthogonal unit vectors $\mathsf{q}_1, \mathsf{q}_2, \ldots, \mathsf{q}_n$.

   (a) Show that QQ^T is a projection matrix. [Hint: Show that $Q^TQ = I$ and use Exercise 13.]

   (b) Show that $Q^TQ = \mathsf{q}_1^T\mathsf{q}_1 + \mathsf{q}_2^T\mathsf{q}_2 + \cdots + \mathsf{q}_n^T\mathsf{q}_n$. [Hint: Apply the left side to a vector x and use the fact that $(\mathsf{v}\cdot\mathsf{u})\mathsf{v}$ is the matrix product $\mathsf{v}(\mathsf{v}^T\mathsf{u})$.]

15. Let u and v be orthogonal vectors of length 1. Let $P = \mathsf{u}\mathsf{u}^T + \mathsf{v}\mathsf{v}^T$.

   (a) Prove that P is a projection matrix. [Use Theorem 6.1.12.]

(b) Identify the space onto which P project vectors. Say something more useful than "the column space of P."

[Hint: A vector is in the column space of P if and only if it has the form $P\mathsf{x}$ for some vector x. Look at the calculations preceding Example 6.1.13.]

(c) Compute P with $\mathsf{u} = \begin{bmatrix} 1 \\ 0 \\ 0 \end{bmatrix}$ and $\mathsf{v} = \begin{bmatrix} 0 \\ 1 \\ 0 \end{bmatrix}$. Is this the matrix you would expect? Explain.

16. For each of the following matrices A and vectors b,

i. use formula (6.1.2) to find the best (least squares) solution x^+ of the system $A\mathsf{x} = \mathsf{b}$;

ii. compute the vector $A\mathsf{x}^+$ and explain its significance;

iii. find the matrix P that projects vectors onto the column space of A;

iv. solve $A\mathsf{x} = P\mathsf{b}$ by Gaussian elimination and compare your answer with that found in part i.

(a) [BB] $A = \begin{bmatrix} 1 & -1 \\ 0 & 2 \\ 1 & 1 \\ 1 & -3 \end{bmatrix}$, $\mathsf{b} = \begin{bmatrix} 1 \\ 2 \\ -1 \\ 3 \end{bmatrix}$ **(b)** $A = \begin{bmatrix} 1 & -1 \\ 2 & 1 \\ 1 & -3 \end{bmatrix}$, $\mathsf{b} = \begin{bmatrix} 1 \\ -1 \\ 1 \end{bmatrix}$

(c) $A = \begin{bmatrix} 1 & 0 & 1 \\ 1 & 1 & 1 \\ 2 & 0 & 0 \\ 1 & 1 & -1 \end{bmatrix}$, $\mathsf{b} = \begin{bmatrix} -1 \\ 0 \\ 0 \\ 3 \end{bmatrix}$ **(d)** $A = \begin{bmatrix} 1 & 1 & 0 \\ 1 & 0 & 2 \\ -1 & 1 & -1 \\ 0 & 0 & -2 \end{bmatrix}$, $\mathsf{b} = \begin{bmatrix} 0 \\ 2 \\ -2 \\ 0 \end{bmatrix}$.

17. [BB] Find the matrix that projects vectors in R^3 onto the plane spanned by $\begin{bmatrix} 1 \\ 0 \\ 1 \end{bmatrix}$ and $\begin{bmatrix} 1 \\ 1 \\ 0 \end{bmatrix}$.

18. **(a)** Find the matrix that projects vectors in R^3 onto the plane with equation $x + 3y - z = 0$.

(b) Find the projection of $\mathsf{b} = \begin{bmatrix} 1 \\ 1 \\ 1 \end{bmatrix}$ onto the plane in (a) and verify that $\mathsf{b} - P\mathsf{b}$ is orthogonal to every vector in the plane.

19. [BB] Let U be the set of vectors in R^4 of the form $\begin{bmatrix} 2s - t \\ s \\ t \\ -5s + 2t \end{bmatrix}$. Find the matrix that projects vectors in R^4 onto U.

20. **(a)** [BB] Find three vectors that span $U = \left\{ \begin{bmatrix} x_1 \\ x_2 \\ x_3 \\ x_4 \end{bmatrix} \in \mathsf{R}^4 \mid x_1 - x_2 + x_4 = 0 \right\}$.

(b) Find a matrix that projects vectors in R^4 onto U.

21. Find the matrix P that projects vectors onto each of the following spaces.

(a) [BB] the subspace of R^3 spanned by $\mathsf{v} = \begin{bmatrix} 1 \\ -3 \\ 5 \end{bmatrix}$

(b) the subspace of R^3 spanned by $\mathsf{v} = \begin{bmatrix} -1 \\ 2 \\ 4 \end{bmatrix}$

(c) [BB] the plane spanned by the vectors $\mathsf{e} = \begin{bmatrix} -1 \\ 2 \\ 1 \end{bmatrix}$ and $\mathsf{f} = \begin{bmatrix} 3 \\ 1 \\ 1 \end{bmatrix}$

(d) the plane spanned by the vectors $\mathsf{e} = \begin{bmatrix} 0 \\ 2 \\ 1 \end{bmatrix}$ and $\mathsf{f} = \begin{bmatrix} 5 \\ 3 \\ -6 \end{bmatrix}$

(e) [BB] the plane in R^3 with equation $2x - y + 5z = 0$

(f) the plane in R^3 with equation $x + 3y + z = 0$

(g) the subspace U of R^4 spanned by the vectors $\mathsf{e} = \begin{bmatrix} 1 \\ -2 \\ 3 \\ 4 \end{bmatrix}$ and $\mathsf{f} = \begin{bmatrix} 0 \\ 1 \\ 2 \\ -1 \end{bmatrix}$

(h) the subspace U of R^4 spanned by $\mathsf{f}_1 = \begin{bmatrix} 1 \\ 0 \\ 0 \\ 1 \end{bmatrix}$, $\mathsf{f}_2 = \begin{bmatrix} -1 \\ 0 \\ 2 \\ 1 \end{bmatrix}$, $\mathsf{f}_3 = \begin{bmatrix} -1 \\ 2 \\ -1 \\ 1 \end{bmatrix}$.

22. **(a)** Find two vectors that span the null space of $A = \begin{bmatrix} 1 & 1 & -2 & 0 \\ 3 & -1 & 2 & -4 \\ -1 & 2 & -4 & 3 \end{bmatrix}$.

(b) Use the result of part (a) to find the matrix that projects vectors onto the null space of A.

(c) Find two orthogonal vectors that span the null space of A.

(d) Use the result of (c) to find the matrix that projects vectors onto the null space of A. Compare this matrix with the one found in part (a).

(e) Find the vector u in the null space of A for which $\|\mathsf{u} - \mathsf{b}\|$ is a minimum, with $\mathsf{b} = \begin{bmatrix} 1 \\ 0 \\ -1 \\ 0 \end{bmatrix}$.

23. **(a)** [BB] Suppose one wants to project vectors in R^3 onto the column space of the matrix $A = \begin{bmatrix} 1 & -2 \\ 2 & -4 \\ 3 & -6 \end{bmatrix}$. Do the methods of this section apply? Explain how to find the matrix that projects vectors onto the column space of A.

(b) Find the matrix which projects vectors in R^2 onto the column space of $A = \begin{bmatrix} -2 & -1 & 0 & 1 \\ 1 & 1 & 1 & -1 \end{bmatrix}$.

(c) Find the matrix that projects vectors in R^3 onto the column space of $A = \begin{bmatrix} 1 & 0 & -1 \\ 0 & -1 & -1 \\ 1 & 3 & 2 \end{bmatrix}$.

24. Suppose A is a matrix with linearly independent columns.
Let $P = A(A^TA)^{-1}A^T$.

(a) [BB] Prove that $[(A^TA)^T]^{-1} = [(A^TA)^{-1}]^T$.

(b) Use Theorem 6.1.12 to prove that P is a projection matrix.

25. Let U be a subspace of a vector space V and let b be a vector in V. Suppose $\mathsf{p} \in U$ has the the property that $\mathsf{b}-\mathsf{p}$ is orthogonal to every vector in U. Show that $\|\mathsf{b}-\mathsf{p}\| \le \|\mathsf{b}-\mathsf{u}\|$ for every vector u in U. [*Hint:* $\mathsf{b}-\mathsf{u} = \mathsf{x}+\mathsf{y}$ with $\mathsf{x} = \mathsf{b}-\mathsf{p}$ and $\mathsf{y} = \mathsf{p}-\mathsf{u}$. Explain why the Theorem of Pythagoras applies and use it.]

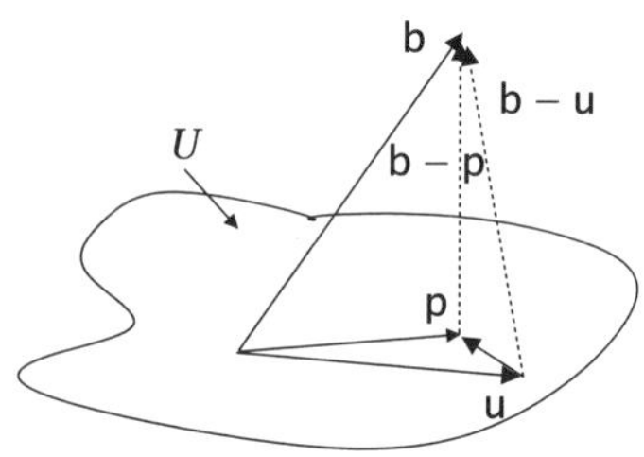

[With U = col sp A and $\mathsf{p} = A\mathsf{x}^+$, this exercise explains why the smallest $\|\mathsf{b}-A\mathsf{x}\|$ occurs when $\mathsf{b}-A\mathsf{x}$ is orthogonal to col sp A.]

Critical Reading

26. [BB] Let $P\colon \mathsf{R}^3 \to \mathsf{R}^3$ be the projection matrix that sends $\begin{bmatrix} x \\ y \\ z \end{bmatrix}$ to $\begin{bmatrix} x \\ y \\ 0 \end{bmatrix}$. Find the eigenvectors and eigenvalues of P without any calculation.

27. What are the possible values for the eigenvalues of a projection matrix P and why?

28. If P is a projection matrix which is not the zero matrix and if c is a scalar, can cP also be a projection matrix? State and prove a theorem.

29. [BB] Let P be that $n \times n$ matrix which projects vectors in R^n onto the subspace $U = \mathsf{R}^n$. What is P? Explain.

30. Prove that a projection matrix is unique; that is, if P_1 and P_2 are projection matrices with the same column space U, then $P_1 = P_2$.

31. Your linear algebra professor wants to construct some matrices A that satisfy $A^2 = A$. She knows all about projection matrices and could find suitable A using a formula like $A = B(B^TB)^{-1}B^T$, but such matrices are also symmetric. Find a method for constructing matrices A satisfying $A^2 = A$ which are **not** symmetric. Explain your answer with at least one example you consider interesting. [Hint: Consider that $A(A\mathsf{x}) = A\mathsf{x}$ for all vectors x.]

32. [BB] Let $A = \begin{bmatrix} 1 & 2 & 0 & 0 \\ 0 & 0 & 0 & 0 \\ 0 & 1 & 1 & 0 \end{bmatrix}$. **Without any calculation**, identify the column space of A and write down the matrix P that projects vectors in R^k onto the column space. A brief answer, in a sentence or two, should also reveal the value of k.

33. Suppose A has linearly independent columns and $A\mathsf{x} = \mathsf{b}$ has a real solution x_0. Show that the best solution to $A\mathsf{x} = \mathsf{b}$ in the sense of least squares is this vector x_0.

34. If a matrix A has linearly independent columns and $P = A(A^TA)^{-1}A^T$, show that the solution to $A\mathsf{x} = P\mathsf{b}$ is unique.

35. Given that $P = \frac{1}{26}\begin{bmatrix} 24 & 4 & -4 & -4 \\ 4 & 5 & -5 & 8 \\ -4 & -5 & 5 & -8 \\ -4 & 8 & -8 & 18 \end{bmatrix}$ is a projection matrix, find the column space of P.

36. Suppose A is an $m \times n$ matrix with linearly independent columns and $P = A(A^TA)^{-1}A^T$. Let y be any vector in R^m.

    (a) Without any calculation, explain why the system $A\mathsf{x} = P\mathsf{y}$ must have a solution.

    (b) Show that the solution is unique.

37. If P is a projection matrix, its column space and row space are the same (ignoring the cosmetic difference between a row and a column). Why?

38. If A has linearly independent columns, then $P = A(A^TA)^{-1}A^T$ is a projection matrix. This is Theorem 6.1.8. Conversely, is it the case that any projection matrix is $A(A^TA)^{-1}A^T$ for some A? Jerome answers

    "Yes. Let $A = P$. Then $P^TP = P^2 = P$, so $P(P^TP)^{-1}P^T = PP^{-1}P^T = IP^T = P$."

    (a) What is the flaw in Jerome's reasoning?

    (b) Is every P of the form $A(A^TA)^{-1}A^T$?

39. **(a)** Let P be a projection matrix and let x be a vector.

    i. If x is in the null space of P, prove that x is orthogonal to the column space of P.

    ii. If x is orthogonal to the column space of P, prove that x is in the null space of P.

    [Remark: These statements should be geometrically obvious. If you project a vector onto a plane and the vector is perpendicular to the plane, that vector goes to the zero vector.]

    (b) Find a 3×3 matrix whose null space is spanned by the vector $\mathsf{u} = \begin{bmatrix} 1 \\ 1 \\ 1 \end{bmatrix}$.

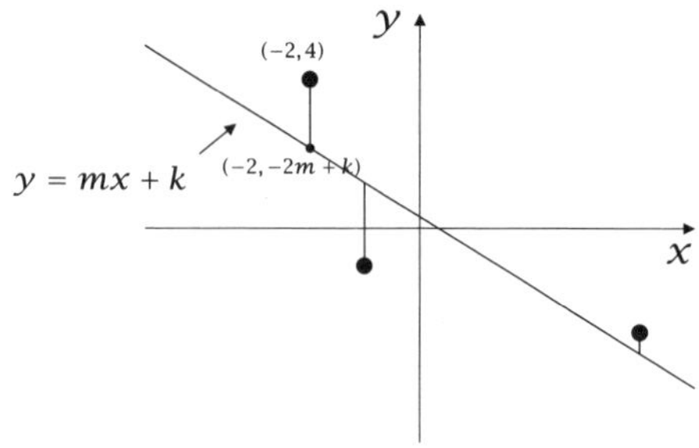

Figure 6.4

6.2 Application: Data Fitting

In this section, we return to the central idea of Section 6.1—how to find the "best solution" to a system that doesn't have a solution—and discuss a common application. Given a linear system $A\mathsf{x} = \mathsf{b}$ that may or may not have a solution, and assuming A has linearly independent columns, we have found that

$$\mathsf{x}^+ = (A^T A)^{-1} A^T \mathsf{b}$$

is the "best solution in the sense of least squares."

Suppose we want to put the "best" possible line through the three points

$$(-2, 4),\ (-1, -1),\ (4, -3). \tag{1}$$

See Figure 6.4. We'll make clear what we mean by "best" a little later.

The line has an equation of the form $y = mx + k$. If the points in question do in fact lie on a line, the coordinates (x, y) of each point would have to satisfy $y = mx + k$. So, we would have

$$\begin{aligned} 4 &= -2m + k \\ -1 &= -m + k \\ -3 &= 4m + k, \end{aligned}$$

which is $A\mathsf{x} = \mathsf{b}$, with

$$A = \begin{bmatrix} 1 & -2 \\ 1 & -1 \\ 1 & 4 \end{bmatrix}, \quad \mathsf{x} = \begin{bmatrix} k \\ m \end{bmatrix}, \quad \mathsf{b} = \begin{bmatrix} 4 \\ -1 \\ -3 \end{bmatrix}.$$

(The reason for writing the variables in the order k, m is explained in Exercise 10.) The system has no solution, of course; the points don't lie on a line. On the other hand, we can find the "best solution in the sense of least squares." We have

$$A^T A = \begin{bmatrix} 3 & 1 \\ 1 & 21 \end{bmatrix} \quad \text{and} \quad (A^T A)^{-1} = \frac{1}{62} \begin{bmatrix} 21 & -1 \\ -1 & 3 \end{bmatrix},$$

so the best solution to $A\mathsf{x} = \mathsf{b}$ is

$$\mathsf{x}^+ = \begin{bmatrix} k^+ \\ m^+ \end{bmatrix} = (A^TA)^{-1}A^T\mathsf{b}$$
$$= \frac{1}{62}\begin{bmatrix} 21 & -1 \\ -1 & 3 \end{bmatrix}\begin{bmatrix} 1 & 1 & 1 \\ -2 & -1 & 4 \end{bmatrix}\begin{bmatrix} 45 \\ -1 \\ -3 \end{bmatrix} = \begin{bmatrix} \frac{19}{62} \\ -\frac{57}{62} \end{bmatrix}.$$

Thus $k^+ = \frac{19}{62} \approx 0.305$, $m^+ = -\frac{57}{62} \approx -0.919$ and our "best" line is the one with equation $y = -.919x + .305$.

In what sense is this line "best?" Remember that $\mathsf{x} = \mathsf{x}^+$ minimizes $\|\mathsf{b} - A\mathsf{x}\|$. Now

$$\mathsf{b} - A\mathsf{x} = \begin{bmatrix} 4 \\ -1 \\ -3 \end{bmatrix} - \begin{bmatrix} -2m + k \\ -m + k \\ 4m + k \end{bmatrix}$$

so $\|\mathsf{b} - A\mathsf{x}\|^2$

$$= [4 - (-2m + k)]^2 + [-1 - (-m + k)]^2 + [-3 - (4m + k)]^2.$$

The terms being squared here,

$$4 - (-2m + k), \quad -1 - (-m + k), \quad -3 - (4m + k)$$

are (up to sign) the *deviations* or vertical distances of the given points from the line (see Figure 6.4). The "best" line, with equation $y = -.919x + .305$, is the one for which the sum of the squares of the deviations is least. We let the reader confirm that for this line, the deviations are (approximately) 1.855, −2.226 and −0.371 and so the minimum sum of squares of deviations is $1.885^2 + (-2.226)^2 + .371^2 = 8.534$. For no other line is the corresponding sum smaller.

We consider a more complicated situation. There may be reason to believe that only experimental error explains why the points

$$(-3, -.5),\ (-1, -.5),\ (1, .9),\ (2, 1.1),\ (3, 3.3) \qquad (2)$$

do not lie on a line. What line? We proceed as before.

If the points were to lie on the line with equation $y = mx + k$, we would have

$$\begin{aligned} -.5 &= -3m + k \\ -.5 &= -m + k \\ .9 &= m + k \\ 1.1 &= 2m + k \\ 3.3 &= 3m + k, \end{aligned}$$

x	y	$y^+ = m^+x + k^+$	$\text{dev} = y - y^+$	dev^2
-3	-0.5	-1.085	0.585	0.342
-1	-0.5	0.059	-0.559	0.312
1	0.9	1.203	-0.303	$.092$
2	1.1	1.775	-0.675	0.456
3	3.3	2.347	0.953	0.908
	Sum of squares of deviations =			2.110

Table 6.1

which is $A\mathsf{x} = \mathsf{b}$, with

$$A = \begin{bmatrix} 1 & -3 \\ 1 & -1 \\ 1 & 1 \\ 1 & 2 \\ 1 & 3 \end{bmatrix}, \quad \mathsf{x} = \begin{bmatrix} k \\ m \end{bmatrix} \quad \text{and} \quad \mathsf{b} = \begin{bmatrix} -.5 \\ -.5 \\ .9 \\ 1.1 \\ 3.3 \end{bmatrix}.$$

We have $A^TA = \begin{bmatrix} 5 & 2 \\ 2 & 24 \end{bmatrix}$ and $(A^TA)^{-1} = \frac{1}{116}\begin{bmatrix} 24 & -2 \\ -2 & 5 \end{bmatrix}$, so the best solution to $A\mathsf{x} = \mathsf{b}$ is

$$\mathsf{x}^+ = \begin{bmatrix} k^+ \\ m^+ \end{bmatrix} = (A^TA)^{-1}A^T\mathsf{b}$$

$$= \frac{1}{116}\begin{bmatrix} 24 & -2 \\ -2 & 5 \end{bmatrix}\begin{bmatrix} 1 & 1 & 1 & 1 & 1 \\ -3 & -1 & 1 & 2 & 3 \end{bmatrix}\begin{bmatrix} -.5 \\ -.5 \\ .9 \\ 1.1 \\ 3.3 \end{bmatrix} = \begin{bmatrix} .631 \\ .572 \end{bmatrix}.$$

Thus $k^+ = .631$, $m^+ = .572$, and our "best" line is the one with equation $y = .572x + .631$. We show the deviations of the given points from this line in Table 6.1. The sum of squares is 2.110. This number is pretty small, suggesting that the points are, on average, pretty close to a line. The line with equation $y = .572x + .631$ is "best" in the sense that 2.110 is least: no other line has the sum of squares of the deviations of the points from the line less than 2.110.

Other Curves

Perhaps we believe that the points listed in (2) should lie on a parabola, not a line. See Figure 6.5. If the parabola has equation $y = rx^2 + sx + t$, the

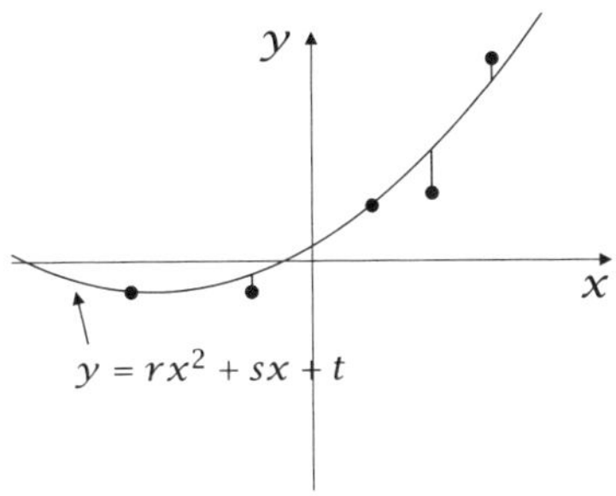

Figure 6.5

coordinates of each point would satisfy this equation; that is,

$$\begin{aligned}
-.5 &= r(-3)^2 - 3s + t \\
-.5 &= r(-1)^2 - s + t \\
.9 &= r + s + t \\
1.1 &= r(2^2) + 2s + t \\
3.3 &= r(3^2) + 3s + t,
\end{aligned}$$

which is $A\mathsf{x} = \mathsf{b}$, with

$$A = \begin{bmatrix} 1 & -3 & 9 \\ 1 & -1 & 1 \\ 1 & 1 & 1 \\ 1 & 2 & 4 \\ 1 & 3 & 9 \end{bmatrix}, \quad \mathsf{x} = \begin{bmatrix} t \\ s \\ r \end{bmatrix}, \quad \mathsf{b} = \begin{bmatrix} -.5 \\ -.5 \\ .9 \\ 1.1 \\ 3.3 \end{bmatrix}.$$

This time

$$A^T A = \begin{bmatrix} 1 & 1 & 1 & 1 & 1 \\ -3 & -1 & 1 & 2 & 3 \\ 9 & 1 & 1 & 4 & 9 \end{bmatrix} \begin{bmatrix} 1 & -3 & 9 \\ 1 & -1 & 1 \\ 1 & 1 & 1 \\ 1 & 2 & 4 \\ 1 & 3 & 9 \end{bmatrix} = \begin{bmatrix} 5 & 2 & 24 \\ 2 & 24 & 8 \\ 24 & 8 & 180 \end{bmatrix}$$

$(A^T A)^{-1} = \dfrac{1}{1876} \begin{bmatrix} 1064 & -42 & -140 \\ -42 & 81 & 2 \\ -140 & 2 & 29 \end{bmatrix}$, and the best solution in the sense of least squares is

$$\mathsf{x}^+ = \begin{bmatrix} t^+ \\ s^+ \\ r^+ \end{bmatrix} = (A^T A)^{-1} A^T \mathsf{b}$$

$$= (A^T A)^{-1} \begin{bmatrix} 1 & 1 & 1 & 1 & 1 \\ -3 & -1 & 1 & 2 & 3 \\ 9 & 1 & 1 & 4 & 9 \end{bmatrix} \begin{bmatrix} -.5 \\ -.5 \\ .9 \\ 1.1 \\ 3.3 \end{bmatrix} = \begin{bmatrix} .246 \\ .569 \\ .109 \end{bmatrix}.$$

So $t^+ = .246$, $s^+ = .569$, $r^+ = .109$. The "best" parabola has the equation $y = .109x^2 + .569x + .246$. This is the parabola for which the sum of the squares of the deviations of the points from the parabola is a minimum.

EXERCISES

Answers to exercises marked [BB] can be found at the **Back of the Book**.

1. Find the equation of the line that best fits each of the following sets of points in the sense of least squares.

   (a) [BB] $(0,0), (1,1), (3,12)$

   (b) $(3,2), (-2,-3), (2,4)$

   (c) [BB] $(-1,1), (0,0), (1,1), (2,0), (3,-8)$

   (d) $(-2,-2), (-1,-2), (0,-1), (2,0)$

   (e) $(2,3), (1,4), (1,-2), (-3,1), (-1,1)$

   (f) $(4,-3), (2,1), (1,-4), (-2,-2), (6,3), (-1,-2)$.

2. Find the equation of the parabola that best fits each of the following sets of points in the sense of least squares.

   (a) [BB] $(0,0), (1,1), (-1,2), (2,2)$

   (b) $(-1,-1), (0,4), (1,2), (2,3)$

   (c) $(-2,2), (-1,1), (0,0), (1,1), (2,2)$

   (d) $(1,-2), (2,1), (-1,1), (2,-2)$.

3. Find the cubic polynomial $f(x) = px^3 + qx^2 + rx + s$ that best fits the given points in the sense of least squares.

   (a) [BB] $(-2,0)$, $(-1,-\frac{1}{2})$, $(0,1)$, $(1,\frac{1}{2})$, $(2,-1)$

   (b) $(-2,0)$, $(-1,-2)$, $(0,0)$, $(1,2)$, $(2,-5)$

   (c) $(-1,-3)$, $(0,-\frac{1}{2})$, $(1,0)$, $(2,1)$, $(3,2)$.

4. The "best" quadratic polynomial through three given points, no two of which lie on a vertical line in the (x,y) plane, is in fact a perfect fit. Why?

5. Find a function of the form $f(x) = ax + b2^x$ that best fits the given points.

   (a) [BB] $(-1,1)$, $(0,0)$, $(2,-1)$ **(b)** $(-2,-3)$, $(-1,0)$, $(1,4)$

   (c) $(0,0)$, $(1,-1)$, $(2,2)$, $(3,-3)$.

6. **(a)** Show that the circle in the plane with center (x_0, y_0) and radius r has an equation of the form $ax + by + c = x^2 + y^2$.

   (b) Find the center and radius of the best circle, in the sense of least squares, which fits each of the following sets of points:

      i. [BB] $(0,0), (1,1), (3,-2)$;

      ii. $(0,0), (1,1), (3,-2), (-1,0)$;

iii. $(0,0), (1,0), (0,1), (1,1), (-1,-1)$;
iv. $(-1,2), (1,2), (1,-3), (-1,-1), (0,1)$.

7. According to the coach of a diving team, a person's score z on her third dive is a function $z = rx + sy + t$ of her scores x, y on her first two dives. Assuming this is the case,

	x	y	z
Gail	6	7	7
Michelle	8	6	8
Wendy	6	8	8
Melanie	7	4	9
Amanda	4	6	7

   (a) find a formula for z based on the scores shown;

   (b) estimate Brenda's score on her third dive, given that she scored 5 and 7, respectively, on her first two dives.

8. Find a curve with equation of the form $y = a \sin x + b \sin 2x$ that best fits the given sets of points.

   (a) [BB] $(-\frac{\pi}{3}, -4), (\frac{\pi}{6}, 2), (\frac{5\pi}{6}, -1)$

   (b) $(-\frac{7\pi}{6}, 0), (-\frac{\pi}{6}, 1), (\frac{\pi}{2}, 5)$

   (c) $(-\frac{5\pi}{3}, 1), (-\frac{2\pi}{3}, -7), (\frac{\pi}{6}, 0), (\frac{7\pi}{6}, -7)$.

9. **(a)** Explain how you might find the "best" plane with equation of the form $ax + by + cz = 1$ through a given set of points, (x_1, y_1, z_1), (x_2, y_2, z_2), ..., (x_n, y_n, z_n).

   (b) Describe a "sum of squares of deviations" test that could be used to estimate how good your plane is.

   (c) Find the "best" plane with equation of the form $ax + by + cz = 1$ through each of the following sets of points. Give the coefficients to two decimal places of accuracy and, in each case, find the sum of squares of the deviations. (This question assumes that some computing power is available.)

   i. [BB] $(2,4,6), (-1,0,12), (3,7,12), (4,1,-6)$
   ii. $(4,0,-3), (1,1,1), (3,2,1), (1,2,3)$
   iii. $(-2,1,3), (0,2,2), (1,5,2), (-1,-3,0), (4,0,-5)$
   iv. $(1,0,-1), (0,1,1), (1,0,1), (1,1,1), (2,2,2)$.

10. Suppose we want to find the line that best fits n points (x_1, y_1), (x_2, y_2), ..., (x_n, y_n) in the sense of least squares.

   (a) If the best line has equation $y = m^+x + k^+$, show that the constants m^+ and k^+ are the solutions to a system of the form $C\begin{bmatrix} k \\ m \end{bmatrix} = \mathsf{b}$ where

$$C = \begin{bmatrix} n & \sum x_i \\ \sum x_i & \sum x_i^2 \end{bmatrix} \text{ and } \mathsf{b} = \begin{bmatrix} \sum y_i \\ \sum x_i y_i \end{bmatrix}.$$

   (b) Use part (a) to find the line that best fits the points $(1,2)$, $(2,3)$, $(3,3)$, $(4,4)$, $(5,4)$.

   (c) Show that the system in (a) has a unique solution for any b if and only if the points (x_1, y_1), (x_2, y_2), ..., (x_n, y_n) do not lie on a vertical line.

Critical Reading

11. [With thanks to Jake Wells] In this section, we showed how to find the "best" line through a set of points in the plane, best in the sense that the sum of squares of vertical deviations of the points from the line is minimized. Suppose we wish instead to minimize the sum of squares of the horizontal deviations of a set of points $(x_1, y_1), \ldots, (x_n, y_n)$ from a line.

    (a) Find a procedure for accomplishing this.

    (b) In each case, find the equations of two lines, one that minimizes the sum of squares of vertical deviations and the other minimizing the sum of squares of horizontal deviations.

    i. $(-1,-1), (0,1), (1,2)$
    ii. $(-2,4), (-1,-1), (4,-3)$
    iii. $(-2,-1), (-1,1), (1,1), (3,2)$.

6.3 The Gram-Schmidt Algorithm and QR Factorization

When we say that a set $\{\mathsf{f}_1, \mathsf{f}_2, \ldots, \mathsf{f}_k\}$ of vectors is *orthogonal,* we mean orthogonal *pairwise*; that is, $\mathsf{f}_i \cdot \mathsf{f}_j = 0$ whenever $i \neq j$. If each of a set of orthogonal vectors has length 1, the vectors are called *orthonormal.*

6.3.1 Example. Any subset of the set $\{\mathsf{e}_1, \mathsf{e}_2, \ldots, \mathsf{e}_n\}$ of standard basis vectors in R^n is orthonormal, for example, $\left\{ \begin{bmatrix} 0 \\ 1 \\ 0 \end{bmatrix}, \begin{bmatrix} 0 \\ 0 \\ 1 \end{bmatrix} \right\}$ in R^3. ☺

It is useful to be able to find orthogonal sets of vectors, in order to project vectors onto subspaces, for example—see Section 6.1. We explain how to do so in this section, after recording a basic fact about orthogonal vectors. In R^2 or R^3, where vectors are easily pictured, two orthogonal vectors are certainly linearly independent since they have different directions. More generally, we have the following theorem.

6.3.2 Proposition. *An orthogonal set of nonzero vectors in* R^n *is linearly independent.*

Proof. Let $\{f_1, f_2, \ldots, f_k\}$ be an orthogonal set of vectors and suppose that

$$c_1 f_1 + c_2 f_2 + \cdots + c_k f_k = 0.$$

Taking the dot product of each side of this equation with f_1 gives

$$c_1(f_1 \cdot f_1) + c_2(f_1 \cdot f_2) + \cdots + c_k(f_1 \cdot f_k) = 0.$$

For $i \neq 1$, the vectors f_1 and f_i are orthogonal, so $f_1 \cdot f_i = 0$ and our equation is just $c_1(f_1 \cdot f_1) = 0$. Since $f_1 \neq 0$, we must have $c_1 = 0$. In a similar way, it follows that all the other scalars c_i are 0 too. ■

Linearly Independent to Orthogonal

In the first chapter of this book, in Section 1.4, we showed how to turn two linearly independent vectors u and v into orthogonal vectors e and f: let $e = u$ and $f = v - \text{proj}_\ell v$, where ℓ is the line spanned by u. See Figure 1.20 and examine Problem 1.4.4 once again.

READING CHECK 1. Let $U = \text{sp}\{u, v\}$ be the span of linearly independent vectors u and v and let e and f be the vectors just defined.

(a) Show that $f = v + au$ for some constant a.

(b) Use the result of (a) to show that e and f also span U.

In Section 1.4, we also discussed the projection of a vector w on the plane π spanned by two vectors u and v. If u, v and w are linearly independent, we can use this idea to produce three orthogonal vectors. We turn u and v into orthogonal e and f as before, then replace w by $g = w - \text{proj}_\pi w$—see Figure 1.21:

$$\begin{aligned} e &= u, \\ f &= v - \text{proj}_\ell v, \\ \text{and} \quad g &= w - \text{proj}_\pi w. \end{aligned}$$

The vectors e and f have the same span as u and v—see READING CHECK 1. In a similar way, one can show that the three vectors e, f, g have the same span as u, v, w. These ideas generalize to give a procedure for converting any linearly independent set of k vectors, $k \geq 2$, to an orthogonal set of k vectors that span the same subspace. The procedure, called the "Gram-Schmidt algorithm," is described in Table 6.2.

Gram-Schmidt Algorithm.[2] The Gram-Schmidt Algorithm is a procedure for turning a set of k linearly independent vectors $\mathsf{u}_1, \ldots, \mathsf{u}_k$ into a set of k orthogonal "subspace preserving" vectors $\mathsf{f}_1, \ldots, \mathsf{f}_k$.

Suppose $\{\mathsf{u}_1, \mathsf{u}_2, \ldots, \mathsf{u}_k\}$ is a set of linearly independent vectors in R^n. Let

$U_1 = \operatorname{sp}\{\mathsf{u}_1\}$ be the subspace spanned by vector u_1,
$U_2 = \operatorname{sp}\{\mathsf{u}_1, \mathsf{u}_2\}$ be the subspace spanned by vectors u_1 and u_2,
$U_3 = \operatorname{sp}\{\mathsf{u}_1, \mathsf{u}_2, \mathsf{u}_3\}$ be the subspace spanned by vectors u_1, u_2 and u_3,
$\vdots$
$U_k = \operatorname{sp}\{\mathsf{u}_1, \mathsf{u}_2, \ldots, \mathsf{u}_k\}$ be the subspace spanned by $\mathsf{u}_1, \mathsf{u}_2, \ldots, \mathsf{u}_k$.

Define vectors $\mathsf{f}_1, \mathsf{f}_2, \ldots, \mathsf{f}_k$ as follows:

$$\begin{aligned}
\mathsf{f}_1 &= \mathsf{u}_1 \\
\mathsf{f}_2 &= \mathsf{u}_2 - \frac{\mathsf{u}_2 \cdot \mathsf{f}_1}{\mathsf{f}_1 \cdot \mathsf{f}_1}\,\mathsf{f}_1 \\
\mathsf{f}_3 &= \mathsf{u}_3 - \frac{\mathsf{u}_3 \cdot \mathsf{f}_1}{\mathsf{f}_1 \cdot \mathsf{f}_1}\,\mathsf{f}_1 - \frac{\mathsf{u}_3 \cdot \mathsf{f}_2}{\mathsf{f}_2 \cdot \mathsf{f}_2}\,\mathsf{f}_2 \\
&\vdots \\
\mathsf{f}_k &= \mathsf{u}_k - \frac{\mathsf{u}_k \cdot \mathsf{f}_1}{\mathsf{f}_1 \cdot \mathsf{f}_1}\,\mathsf{f}_1 - \frac{\mathsf{u}_k \cdot \mathsf{f}_2}{\mathsf{f}_2 \cdot \mathsf{f}_2}\,\mathsf{f}_2 - \cdots - \frac{\mathsf{u}_k \cdot \mathsf{f}_{k-1}}{\mathsf{f}_{k-1} \cdot \mathsf{f}_{k-1}}\,\mathsf{f}_{k-1}.
\end{aligned}$$

Then $\{\mathsf{f}_1, \mathsf{f}_2, \ldots, \mathsf{f}_k\}$ is an orthogonal set of nonzero vectors with the property that

$$\begin{aligned}
\operatorname{sp}\{\mathsf{f}_1\} &= U_1 \\
\operatorname{sp}\{\mathsf{f}_1, \mathsf{f}_2\} &= U_2 \\
\operatorname{sp}\{\mathsf{f}_1, \mathsf{f}_2, \mathsf{f}_3\} &= U_3 \\
&\vdots \\
\operatorname{sp}\{\mathsf{f}_1, \mathsf{f}_2, \ldots, \mathsf{f}_k\} &= U_k.
\end{aligned}$$

Table 6.2

The key idea in the proof that the algorithm works is that if $\mathsf{f}_1, \mathsf{f}_2, \ldots, \mathsf{f}_t$ are orthogonal vectors spanning a subspace U, then, for any b, the projection of b on U,

$$\operatorname{proj}_U \mathsf{b} = \frac{\mathsf{b} \cdot \mathsf{f}_1}{\mathsf{f}_1 \cdot \mathsf{f}_1}\,\mathsf{f}_1 + \frac{\mathsf{b} \cdot \mathsf{f}_2}{\mathsf{f}_2 \cdot \mathsf{f}_2}\,\mathsf{f}_2 + \cdots + \frac{\mathsf{b} \cdot \mathsf{f}_t}{\mathsf{f}_t \cdot \mathsf{f}_t}\,\mathsf{f}_t,$$

has the property that $\mathsf{b} - \operatorname{proj}_U \mathsf{b}$ is orthogonal to every vector in U—see Figure 6.6 and Theorem 6.1.16. Why does this process lead to an orthogonal set of vectors as described?

[2]Published in 1883, the algorithm is named after Jørgen Pedersen Gram (1850–1916) and Erhard Schmidt (1876–1959).

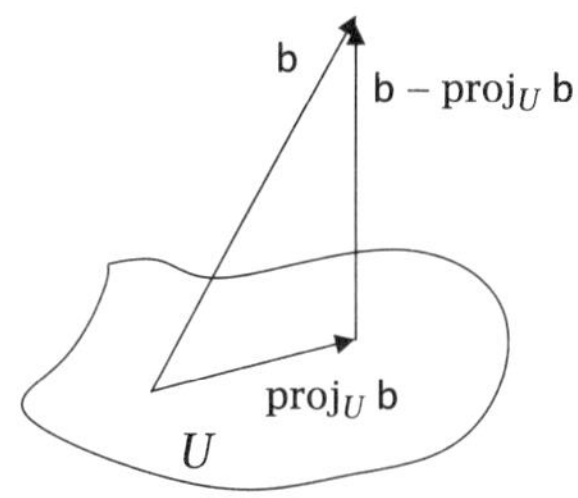

Figure 6.6: b is the projection on U if $\mathsf{b} - \operatorname{proj}_U \mathsf{b}$ is orthogonal to U.

First, since f_1 is a nonzero scalar multiple of u_1, the first subspace U_1 is certainly spanned by f_1. That $\operatorname{sp}\{\mathsf{f}_1, \mathsf{f}_2\} = U_2$ was exactly READING CHECK 1, but here is another argument that has the advantage that it shows us how to continue and prove that every U_i is spanned by $\mathsf{f}_1, \ldots, \mathsf{f}_i$.

We know that $\mathsf{f}_2 = \mathsf{u}_2 - \operatorname{proj}_{U_1} \mathsf{u}_2$ is orthogonal to f_1, by the definition of "projection," and it is also nonzero.

READING CHECK 2. Why is $\mathsf{f}_2 \neq 0$?

By Proposition 6.3.2, the orthogonal vectors f_1 and f_2 are linearly independent. Since they are contained in the 2-dimensional space U_2, they must also span this subspace; that is, $U_2 = \operatorname{sp}\{\mathsf{f}_1, \mathsf{f}_2\}$.

Notice that $\mathsf{f}_3 = \mathsf{u}_3 - \operatorname{proj}_{U_2} \mathsf{u}_3$, so f_3 is orthogonal to all vectors in U_2 (definition of "projection"), in particular, to both f_1 and f_2. It's also nonzero because u_3 is not in U_2. The orthogonal vectors $\mathsf{f}_1, \mathsf{f}_2, \mathsf{f}_3$ are linearly independent in the three-dimensional space U_3, so they span U_3: $U_3 = \operatorname{sp}\{\mathsf{f}_1, \mathsf{f}_2, \mathsf{f}_3\}$. Continuing in this way, we see that each U_i is spanned by the vectors $\mathsf{f}_1, \ldots, \mathsf{f}_i$: the Gram-Schmidt algorithm works as advertised.

6.3.3 Example. Suppose $\mathsf{u}_1 = \begin{bmatrix} 1 \\ 0 \\ -1 \\ 1 \end{bmatrix}$, $\mathsf{u}_2 = \begin{bmatrix} 1 \\ -2 \\ 1 \\ -1 \end{bmatrix}$ and $\mathsf{u}_3 = \begin{bmatrix} 2 \\ 0 \\ -1 \\ 0 \end{bmatrix}$.

The Gram-Schmidt algorithm sets

$$\mathsf{f}_1 = \mathsf{u}_1 = \begin{bmatrix} 1 \\ 0 \\ -1 \\ 1 \end{bmatrix}$$

and $$\mathsf{f}_2 = \mathsf{u}_2 - \frac{\mathsf{u}_2 \cdot \mathsf{f}_1}{\mathsf{f}_1 \cdot \mathsf{f}_1} \mathsf{f}_1 = \mathsf{u}_2 - \frac{-1}{3}\mathsf{f}_1 = \mathsf{u}_2 + \frac{1}{3}\mathsf{f}_1$$

$$= \begin{bmatrix} 1 \\ -2 \\ 1 \\ -1 \end{bmatrix} + \frac{1}{3}\begin{bmatrix} 1 \\ 0 \\ -1 \\ 1 \end{bmatrix} = \begin{bmatrix} \frac{4}{3} \\ -2 \\ \frac{2}{3} \\ -\frac{2}{3} \end{bmatrix}.$$

When working by hand, it will be easier to find f_3 if we multiply f_2 by $\frac{3}{2}$, obtaining a new $\mathsf{f}_2 = \begin{bmatrix} 2 \\ -3 \\ 1 \\ -1 \end{bmatrix}$ that is still orthogonal to f_1. Continuing,

$$\mathsf{f}_3 = \mathsf{u}_3 - \frac{\mathsf{u}_3 \cdot \mathsf{f}_1}{\mathsf{f}_1 \cdot \mathsf{f}_1}\mathsf{f}_1 - \frac{\mathsf{u}_3 \cdot \mathsf{f}_2}{\mathsf{f}_2 \cdot \mathsf{f}_2}\mathsf{f}_2 = \mathsf{u}_3 - \frac{3}{3}\mathsf{f}_1 - \frac{3}{15}\mathsf{f}_2 = \mathsf{u}_3 - \mathsf{f}_1 - \frac{1}{5}\mathsf{f}_2$$

$$= \begin{bmatrix} 2 \\ 0 \\ -1 \\ 0 \end{bmatrix} - \begin{bmatrix} 1 \\ 0 \\ -1 \\ 1 \end{bmatrix} - \frac{1}{5}\begin{bmatrix} 2 \\ -3 \\ 1 \\ -1 \end{bmatrix} = \begin{bmatrix} \frac{3}{5} \\ \frac{3}{5} \\ -\frac{1}{5} \\ -\frac{4}{5} \end{bmatrix},$$

which we would probably replace with $5\mathsf{f}_3$, giving as our final set of orthogonal vectors

$$\mathsf{f}_1 = \begin{bmatrix} 1 \\ 0 \\ -1 \\ 1 \end{bmatrix}, \quad \mathsf{f}_2 = \begin{bmatrix} 2 \\ -3 \\ 1 \\ -1 \end{bmatrix}, \quad \mathsf{f}_3 = \begin{bmatrix} 3 \\ 3 \\ -1 \\ -4 \end{bmatrix}.$$

6.3.4 Example. Suppose U is the subspace of R^5 with basis

$$\mathsf{u}_1 = \begin{bmatrix} 0 \\ 1 \\ 1 \\ 0 \\ 0 \end{bmatrix}, \quad \mathsf{u}_2 = \begin{bmatrix} 1 \\ 0 \\ 1 \\ -1 \\ 1 \end{bmatrix}, \quad \mathsf{u}_3 = \begin{bmatrix} 1 \\ 4 \\ -1 \\ 1 \\ -1 \end{bmatrix}, \quad \mathsf{u}_4 = \begin{bmatrix} 2 \\ 0 \\ 2 \\ 3 \\ 1 \end{bmatrix},$$

and we wish to find an orthogonal basis for U. The Gram–Schmidt algorithm sets

$$\mathsf{f}_1 = \mathsf{u}_1 = \begin{bmatrix} 0 \\ 1 \\ 1 \\ 0 \\ 0 \end{bmatrix}$$

$$\mathsf{f}_2 = \mathsf{u}_2 - \frac{\mathsf{u}_2 \cdot \mathsf{f}_1}{\mathsf{f}_1 \cdot \mathsf{f}_1}\mathsf{f}_1 = \mathsf{u}_2 - \frac{1}{2}\mathsf{f}_1 = \begin{bmatrix} 1 \\ 0 \\ 1 \\ -1 \\ 1 \end{bmatrix} - \frac{1}{2}\begin{bmatrix} 0 \\ 1 \\ 1 \\ 0 \\ 0 \end{bmatrix} = \begin{bmatrix} 1 \\ -\frac{1}{2} \\ \frac{1}{2} \\ -1 \\ 1 \end{bmatrix},$$

and it would be natural at this point to replace f_2 with $2\mathsf{f}_2$ giving a new $\mathsf{f}_2 = \begin{bmatrix} 2 \\ -1 \\ 1 \\ -2 \\ 2 \end{bmatrix}$. Continuing the algorithm,

$$\mathsf{f}_3 = \mathsf{u}_3 - \frac{\mathsf{u}_3 \cdot \mathsf{f}_1}{\mathsf{f}_1 \cdot \mathsf{f}_1}\mathsf{f}_1 - \frac{\mathsf{u}_3 \cdot \mathsf{f}_2}{\mathsf{f}_2 \cdot \mathsf{f}_2}\mathsf{f}_2 = \mathsf{u}_3 - \frac{3}{2}\mathsf{f}_1 - \frac{-7}{14}\mathsf{f}_2 = \mathsf{u}_3 - \frac{3}{2}\mathsf{f}_1 + \frac{1}{2}\mathsf{f}_2$$

$$= \begin{bmatrix} 1 \\ 4 \\ -1 \\ 1 \\ -1 \end{bmatrix} - \frac{3}{2}\begin{bmatrix} 0 \\ 1 \\ 1 \\ 0 \\ 0 \end{bmatrix} + \frac{1}{2}\begin{bmatrix} 2 \\ -1 \\ 1 \\ -2 \\ 2 \end{bmatrix} = \begin{bmatrix} 2 \\ 2 \\ -2 \\ 0 \\ 0 \end{bmatrix},$$

which might well be replaced by a new $\mathsf{f}_3 = \begin{bmatrix} 1 \\ 1 \\ -1 \\ 0 \\ 0 \end{bmatrix}$. Finally,

$$\begin{aligned} \mathsf{f}_4 &= \mathsf{u}_4 - \frac{\mathsf{u}_4 \cdot \mathsf{f}_1}{\mathsf{f}_1 \cdot \mathsf{f}_1}\mathsf{f}_1 - \frac{\mathsf{u}_4 \cdot \mathsf{f}_2}{\mathsf{f}_2 \cdot \mathsf{f}_2}\mathsf{f}_2 - \frac{\mathsf{u}_4 \cdot \mathsf{f}_3}{\mathsf{f}_3 \cdot \mathsf{f}_3}\mathsf{f}_3 \\ &= \mathsf{u}_4 - \frac{2}{2}\mathsf{f}_1 - \frac{1}{7}\mathsf{f}_2 - \frac{0}{3}\mathsf{f}_3 = \mathsf{u}_4 - \mathsf{f}_1 - \frac{1}{7}\mathsf{f}_2 \\ &= \begin{bmatrix} 2 \\ 0 \\ 2 \\ 3 \\ 1 \end{bmatrix} - \begin{bmatrix} 0 \\ 1 \\ 1 \\ 0 \\ 0 \end{bmatrix} - \frac{1}{7}\begin{bmatrix} 2 \\ -1 \\ 1 \\ -2 \\ 2 \end{bmatrix} = \begin{bmatrix} \frac{12}{7} \\ -\frac{6}{7} \\ \frac{6}{7} \\ \frac{23}{7} \\ \frac{5}{7} \end{bmatrix}, \end{aligned}$$

which might well be replaced by $\begin{bmatrix} 12 \\ -6 \\ 6 \\ 23 \\ 5 \end{bmatrix}$. ☺

QR Factorization

Let $A = \begin{bmatrix} 2 & 1 \\ 1 & 1 \\ -2 & 0 \end{bmatrix}$, label the columns of this matrix a_1 and a_2, respectively, and apply the Gram–Schmidt algorithm to these vectors. Thus $\mathsf{f}_1 = \mathsf{a}_1$ and

$$\mathsf{f}_2 = \mathsf{a}_2 - \frac{\mathsf{a}_2 \cdot \mathsf{f}_1}{\mathsf{f}_1 \cdot \mathsf{f}_1}\mathsf{f}_1 = \mathsf{a}_2 - \frac{3}{9}\mathsf{a}_1 = \mathsf{a}_2 - \frac{1}{3}\mathsf{a}_1 = \begin{bmatrix} 1 \\ 1 \\ 0 \end{bmatrix} - \frac{1}{3}\begin{bmatrix} 2 \\ 1 \\ -2 \end{bmatrix} = \begin{bmatrix} \frac{1}{3} \\ \frac{2}{3} \\ \frac{2}{3} \end{bmatrix}.$$

Multiplying by 3, we replace f_2 by $\begin{bmatrix}1\\2\\2\end{bmatrix}$, obtaining orthogonal vectors

$$f_1 = \begin{bmatrix}2\\1\\-2\end{bmatrix} \quad \text{and} \quad f_2 = \begin{bmatrix}1\\2\\2\end{bmatrix}$$

with $\operatorname{sp}\{f_1\} = \operatorname{sp}\{a_1\}$ and $\operatorname{sp}\{f_1, f_2\} = \operatorname{sp}\{a_1, a_2\}$. "Normalizing" f_1 and f_2, that is, dividing each vector by its length, gives orthonormal vectors

$$q_1 = \frac{1}{\|f_1\|}f_1 = \tfrac{1}{3}f_1 = \tfrac{1}{3}\begin{bmatrix}2\\1\\-2\end{bmatrix} \quad \text{and} \quad q_2 = \frac{1}{\|f_2\|}f_2 = \tfrac{1}{3}f_2 = \tfrac{1}{3}\begin{bmatrix}1\\2\\2\end{bmatrix}.$$

Now **express $\mathbf{a}_1$ and $\mathbf{a}_2$ first in terms of $\mathbf{f}_1$ and $\mathbf{f}_2$ and then in terms of $\mathbf{q}_1$ and $\mathbf{q}_2$.** We have

$$\begin{aligned} f_1 &= a_1 && \text{so} \quad a_1 = f_1 = 3q_1 \\ f_2 &= 3(a_2 - \tfrac{1}{3}f_1) = 3a_2 - f_1 && \text{so} \quad a_2 = \tfrac{1}{3}(f_1 + f_2) = q_1 + q_2. \end{aligned}$$

We obtain

$$\begin{bmatrix} a_1 & a_2 \\ \downarrow & \downarrow \end{bmatrix} = \begin{bmatrix} q_1 & q_2 \\ \downarrow & \downarrow \end{bmatrix}\begin{bmatrix}3 & 1\\0 & 1\end{bmatrix}$$

$$A = \begin{bmatrix}2 & 1\\1 & 1\\-2 & 0\end{bmatrix} = \begin{bmatrix}\frac{2}{3} & \frac{1}{3}\\ \frac{1}{3} & \frac{2}{3}\\ -\frac{2}{3} & \frac{2}{3}\end{bmatrix}\begin{bmatrix}3 & 1\\0 & 1\end{bmatrix}.$$

The matrix A has been factored $A = QR$ with $Q = \begin{bmatrix}\frac{2}{3} & \frac{1}{3}\\ \frac{1}{3} & \frac{2}{3}\\ -\frac{2}{3} & \frac{2}{3}\end{bmatrix}$ a matrix the same size as A, but with orthonormal columns, and $R = \begin{bmatrix}3 & 1\\0 & 1\end{bmatrix}$ a square upper triangular matrix.

6.3.5 Definition. A *QR factorization* of an $m \times n$ matrix A is a factorization $A = QR$ where Q is an $m \times n$ matrix with orthonormal columns and R is an $n \times n$ upper triangular matrix.

The following theorem, just illustrated, is a consequence of the Gram–Schmidt algorithm.

6.3.6 Theorem. *Any matrix with linearly independent columns has a QR factorization.*

READING CHECK 3. Suppose $A = QR$ is a QR factorization of an $m \times n$ matrix A has linearly independent columns.

(a) R also has linearly independent columns. Why?
[Hint: Use the definition of linear independence.]

(b) R is invertible. Why?

6.3.7 Problem. Find a QR factorization of $A = \begin{bmatrix} 0 & 1 & 1 \\ 1 & 0 & 4 \\ 1 & 1 & -1 \\ 0 & -1 & 1 \\ 0 & 1 & -1 \end{bmatrix}$. (See Example 6.3.4.)

Solution. The columns of A are the vectors $\mathsf{a}_1 = \begin{bmatrix} 0 \\ 1 \\ 1 \\ 0 \\ 0 \end{bmatrix}$, $\mathsf{a}_2 = \begin{bmatrix} 1 \\ 0 \\ 1 \\ -1 \\ 1 \end{bmatrix}$ and $\mathsf{a}_3 = \begin{bmatrix} 1 \\ 4 \\ -1 \\ 1 \\ -1 \end{bmatrix}$. In Example 6.3.4, we applied the Gram-Schmidt algorithm to obtain orthogonal vectors

$$\mathsf{f}_1 = \begin{bmatrix} 0 \\ 1 \\ 1 \\ 0 \\ 0 \end{bmatrix}, \quad \mathsf{f}_2 = \begin{bmatrix} 2 \\ -1 \\ 1 \\ -2 \\ 2 \end{bmatrix}, \quad \mathsf{f}_3 = \begin{bmatrix} 1 \\ 1 \\ -1 \\ 0 \\ 0 \end{bmatrix}.$$

Dividing each of these by its length, we obtain the orthonormal vectors

$$\mathsf{q}_1 = \frac{1}{\|\mathsf{f}_1\|}\mathsf{f}_1 = \frac{1}{\sqrt{2}}\mathsf{f}_1 = \frac{1}{\sqrt{2}}\begin{bmatrix} 0 \\ 1 \\ 1 \\ 0 \\ 0 \end{bmatrix}, \qquad \mathsf{q}_2 = \frac{1}{\|\mathsf{f}_2\|}\mathsf{f}_2 = \frac{1}{\sqrt{14}}\mathsf{f}_2 = \frac{1}{\sqrt{14}}\begin{bmatrix} 2 \\ -1 \\ 1 \\ -2 \\ 2 \end{bmatrix},$$

$$\mathsf{q}_3 = \frac{1}{\|\mathsf{f}_3\|}\mathsf{f}_3 = \frac{1}{\sqrt{3}}\mathsf{f}_3 = \frac{1}{\sqrt{3}}\begin{bmatrix} 1 \\ 1 \\ -1 \\ 0 \\ 0 \end{bmatrix}.$$

As before, we **express $\mathbf{a}_i$ in terms of $\mathbf{f}_i$ and then in terms of $\mathbf{q}_i$**. We previously had

$$\mathsf{f}_1 = \mathsf{a}_1 \qquad \text{so} \qquad \mathsf{a}_1 = \mathsf{f}_1 = \sqrt{2}\mathsf{q}_1,$$

$$\mathsf{f}_2 = 2(\mathsf{a}_2 - \tfrac{1}{2}\mathsf{f}_1) \qquad \text{so} \qquad \mathsf{a}_2 = \tfrac{1}{2}(\mathsf{f}_1 + \mathsf{f}_2) = \tfrac{\sqrt{2}}{2}\mathsf{q}_1 + \tfrac{\sqrt{14}}{2}\mathsf{q}_2,$$

$$\mathsf{f}_3 = \tfrac{1}{2}(\mathsf{a}_3 - \tfrac{3}{2}\mathsf{f}_1 + \tfrac{1}{2}\mathsf{f}_2) \qquad \text{so} \qquad \begin{aligned} \mathsf{a}_3 &= \tfrac{3}{2}\mathsf{f}_1 - \tfrac{1}{2}\mathsf{f}_2 + 2\mathsf{f}_3 \\ &= \tfrac{3\sqrt{2}}{2}\mathsf{q}_1 - \tfrac{\sqrt{14}}{2}\mathsf{q}_2 + 2\sqrt{3}\,\mathsf{q}_3. \end{aligned}$$

Thus

$$\begin{bmatrix} \mathsf{a}_1 & \mathsf{a}_2 & \mathsf{a}_3 \\ \downarrow & \downarrow & \downarrow \\ & & \end{bmatrix} = \begin{bmatrix} \mathsf{q}_1 & \mathsf{q}_2 & \mathsf{q}_3 \\ \downarrow & \downarrow & \downarrow \\ & & \end{bmatrix} \begin{bmatrix} \sqrt{2} & \frac{\sqrt{2}}{2} & \frac{3\sqrt{2}}{2} \\ 0 & \frac{\sqrt{14}}{2} & -\frac{\sqrt{14}}{2} \\ 0 & 0 & 2\sqrt{3} \end{bmatrix}$$

$$A = \begin{bmatrix} 0 & 1 & 1 \\ 1 & 0 & 4 \\ 1 & 1 & -1 \\ 0 & -1 & 1 \\ 0 & 1 & -1 \end{bmatrix} = \begin{bmatrix} 0 & \frac{2}{\sqrt{14}} & \frac{1}{\sqrt{3}} \\ \frac{1}{\sqrt{2}} & -\frac{1}{\sqrt{14}} & \frac{1}{\sqrt{3}} \\ \frac{1}{\sqrt{2}} & \frac{1}{\sqrt{14}} & -\frac{1}{\sqrt{3}} \\ 0 & -\frac{2}{\sqrt{14}} & 0 \\ 0 & \frac{2}{\sqrt{14}} & 0 \end{bmatrix} \begin{bmatrix} \sqrt{2} & \frac{\sqrt{2}}{2} & \frac{3\sqrt{2}}{2} \\ 0 & \frac{\sqrt{14}}{2} & -\frac{\sqrt{14}}{2} \\ 0 & 0 & 2\sqrt{3} \end{bmatrix}.$$

Orthogonal Matrices

This section has focused attention on matrices with orthonormal columns. Such matrices have some nice properties. Since the (i, j) entry of Q^TQ is the dot product of row i of Q^T and column j of Q, that is, the dot product of columns i and j of Q, and since these columns are orthonormal, the dot product is 0 if $i \neq j$ and 1 otherwise. In other words, $Q^TQ = I$. The converse is also true.

6.3.8 A matrix Q has orthonormal columns if and only if $Q^TQ = I$.

Recall that the dot product of vectors x and y is just the matrix product $\mathsf{x}^T\mathsf{y}$. If Q has orthonormal columns and x and y are vectors, the dot product of $Q\mathsf{x}$ and $Q\mathsf{y}$ is

$$Q\mathsf{x} \cdot Q\mathsf{y} = (Q\mathsf{x})^T(Q\mathsf{y}) = \mathsf{x}^TQ^TQ\mathsf{y} = \mathsf{x}^T\mathsf{y} = \mathsf{x} \cdot \mathsf{y}$$

because $Q^TQ = I$. This condition is usually summarized by saying that multiplication by an orthogonal matrix "preserves" the dot product. In the special case $\mathsf{y} = \mathsf{x}$, we get $Q\mathsf{x} \cdot Q\mathsf{x} = \mathsf{x} \cdot \mathsf{x}$. The right side is $\|\mathsf{x}\|^2$ and the left is $\|Q\mathsf{x}\|^2$, so $\|Q\mathsf{x}\| = \|\mathsf{x}\|$. Multiplication by Q also preserves the length of a vector.

6.3.9 Multiplication by a matrix with orthonormal columns preserves the dot product and the lengths of vectors, and hence also the angle between vectors.

Reading Check 4. Why does multiplication by an orthogonal matrix preserve the angle between (nonzero) vectors?

Now we turn our attention to **square** matrices with orthonormal columns. Such matrices have a special name.

6.3.10 Definition. An *orthogonal matrix* is a square matrix whose columns are orthonormal.[3]

Since an orthogonal matrix is square, the equation $Q^TQ = I$ says that Q is invertible (Corollary 4.5.11) with $Q^{-1} = Q^T$. In particular, $QQ^T = I$ too, and this says that the rows of Q are also orthonormal. (Is this obvious?)

6.3.11 Examples. • For any angle θ, the *rotation matrix*

$Q = \begin{bmatrix} \cos\theta & -\sin\theta \\ \sin\theta & \cos\theta \end{bmatrix}$ is orthogonal. Its columns are orthogonal and have length 1 because $\cos^2\theta + \sin^2\theta = 1$. For example, with $\theta = \frac{\pi}{4}$, we see that $\begin{bmatrix} \frac{1}{\sqrt{2}} & -\frac{1}{\sqrt{2}} \\ \frac{1}{\sqrt{2}} & \frac{1}{\sqrt{2}} \end{bmatrix}$ is orthogonal.

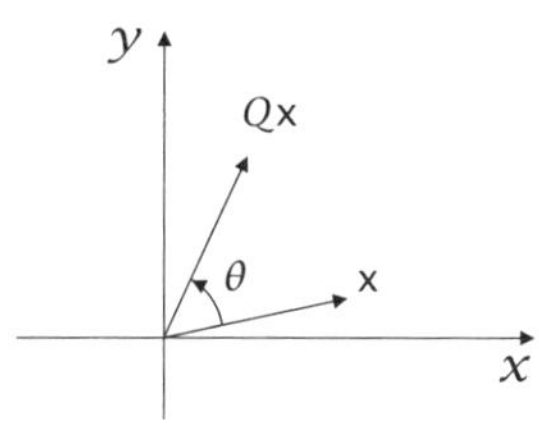

The name "rotation matrix" derives from the fact that if x is pictured in R^2 as an arrow based at the origin, then multiplication by Q rotates this arrow through the angle θ.

- Any permutation matrix is orthogonal because its columns are just the standard basis vectors $\mathsf{e}_1, \mathsf{e}_2, \ldots, \mathsf{e}_n$ in any order. For example, $P = \begin{bmatrix} 0 & 0 & 1 \\ 1 & 0 & 0 \\ 0 & 1 & 0 \end{bmatrix}$ is an orthogonal matrix.

6.3.12 Problem. Let w be a vector in R^n of length 1. Show that the *Householder matrix*[4] $H = I - 2\mathsf{w}\mathsf{w}^T$ is symmetric and orthogonal.

[3] It would be nice, and it would seem very natural, to use the term "orthonormal matrix" instead of "orthogonal matrix" to describe a square matrix with orthonormal columns, but there is no such luck. The term "orthogonal matrix" is firmly entrenched in the language of linear algebra. Neither you nor I are going to change it.

[4] After Alston Scott Householder (1904–1993), an American and mathematical biologist early in his career, but now best known for his contributions to linear algebra. His work was directly responsible for the development of many of the most effective algorithms in use today. A conference in honour of Householder has been held every three years since 1969. For more information about Householder, visit www-history.mcs.st-andrews.ac.uk/history/Mathematicians/Householder.html.

Solution. Symmetry follows from $H^T = I^T - 2(\mathsf{w}\mathsf{w}^T)^T = I - 2(\mathsf{w}^T)^T\mathsf{w}^T = I - 2\mathsf{w}\mathsf{w}^T = H$. Since H is symmetric, it is square, so to establish orthogonality, it is sufficient to show that $H^TH = I$. We compute

$$H^TH = (I - 2\mathsf{w}\mathsf{w}^T)(I - 2\mathsf{w}\mathsf{w}^T) = I - 2\mathsf{w}\mathsf{w}^T - 2\mathsf{w}\mathsf{w}^T + 4\mathsf{w}\underline{\mathsf{w}^T\mathsf{w}}\mathsf{w}^T.$$

Now $\mathsf{w}^T\mathsf{w} = \mathsf{w} \cdot \mathsf{w} = \|\mathsf{w}\|^2 = 1$ and so $H^TH = I - 4\mathsf{w}\mathsf{w}^T + 4\mathsf{w}\mathsf{w}^T = I$.

6.3.13 **Problem.** Let $\mathsf{w} = \frac{1}{\sqrt{3}}\begin{bmatrix} 1 \\ -1 \\ 1 \end{bmatrix}$. Compute the Householder matrix $H = I - 2\mathsf{w}\mathsf{w}^T$ and verify that its rows are orthonormal.

Solution. The matrix $H = I - \frac{2}{3}\begin{bmatrix} 1 \\ -1 \\ 1 \end{bmatrix}\begin{bmatrix} 1 & -1 & 1 \end{bmatrix}$

$$= \begin{bmatrix} 1 & 0 & 0 \\ 0 & 1 & 0 \\ 0 & 0 & 1 \end{bmatrix} - \frac{2}{3}\begin{bmatrix} 1 & -1 & 1 \\ -1 & 1 & -1 \\ 1 & -1 & 1 \end{bmatrix} = \frac{1}{3}\begin{bmatrix} 1 & 2 & -2 \\ 2 & 1 & 2 \\ -2 & 2 & 1 \end{bmatrix}.$$

The rows of H are orthonormal since they have length one and any two are orthogonal. A good way to check for orthonormal rows (especially with some computing power available) is to verify that $HH^T = I$, and this is the case here:

$$HH^T = \frac{1}{9}\begin{bmatrix} 1 & 2 & -2 \\ 2 & 1 & 2 \\ -2 & 2 & 1 \end{bmatrix}\begin{bmatrix} 1 & 2 & -2 \\ 2 & 1 & 2 \\ -2 & 2 & 1 \end{bmatrix} = \begin{bmatrix} 1 & 0 & 0 \\ 0 & 1 & 0 \\ 0 & 0 & 1 \end{bmatrix} = I.$$

Least Squares Revisited

In Section 6.1, we introduced the idea of *least squares* approximation. The idea was to find the "best solution" to a linear system $A\mathsf{x} = \mathsf{b}$ when, in fact, there weren't any solutions. "Best" meant that the vector x^+ minimizes $\|\mathsf{b} - A\mathsf{x}\|$. Since this number is (the square root of) a sum of squares, x^+ is best "in the sense of least squares."

We had an explicit formula for x^+ when the matrix A has linearly independent columns, namely,

$$\mathsf{x}^+ = (A^TA)^{-1}A^T\mathsf{b}.$$

(See Theorem 6.1.3.) Of course, if A has linearly independent columns, then $A = QR$ can be factored as the product of a matrix Q with orthonormal columns and an upper triangular matrix R. In this case, $A^TA = R^TQ^TQR = R^TR$ because $Q^TQ = I$, and so $\mathsf{x}^+ = (R^TR)^{-1}R^TQ^T\mathsf{b}$.

There is a world of advantage in replacing $(A^TA)^{-1}$ by $(R^TR)^{-1}$. The matrix R is invertible—see READING CHECK 3(b)—and so $(R^TR)^{-1} = R^{-1}(R^T)^{-1}$. This means

$$\mathsf{x}^+ = (R^TR)^{-1}R^TQ^T\mathsf{b} = R^{-1}(R^T)^{-1}R^TQ^T\mathsf{b} = R^{-1}Q^T\mathsf{b}.$$

Rewriting this as

$$R\mathsf{x}^+ = Q^T\mathsf{b}$$

we see that the desired vector x^+ is nothing but the solution to a triangular system of equations (which is easily solved by back substitution).

6.3.14 Example. In 6.1.4, we showed that the least squares solution to

$$\begin{aligned} x_1 + x_2 &= 2 \\ x_1 - x_2 &= 0 \\ 2x_1 + x_2 &= -4 \end{aligned}$$

is $\mathsf{x}^+ = \begin{bmatrix} -1 \\ 0 \end{bmatrix}$. This is $A\begin{bmatrix} x_1 \\ x_2 \end{bmatrix} = \mathsf{b}$, with $A = \begin{bmatrix} 1 & 1 \\ 1 & -1 \\ 2 & 1 \end{bmatrix}$ and $\mathsf{b} = \begin{bmatrix} 2 \\ 0 \\ -4 \end{bmatrix}$.

The QR factorization of A is

$$A = \begin{bmatrix} \frac{1}{\sqrt{6}} & \frac{2}{\sqrt{21}} \\ \frac{1}{\sqrt{6}} & -\frac{4}{\sqrt{21}} \\ \frac{2}{\sqrt{6}} & \frac{1}{\sqrt{21}} \end{bmatrix} \begin{bmatrix} \sqrt{6} & \frac{\sqrt{6}}{3} \\ 0 & \frac{\sqrt{21}}{3} \end{bmatrix},$$

so the desired vector x^+ is the solution to $R\mathsf{x} = Q^T\mathsf{b}$. We have

$$R\mathsf{x} = \begin{bmatrix} \sqrt{6} & \frac{\sqrt{6}}{3} \\ 0 & \frac{\sqrt{21}}{3} \end{bmatrix} \begin{bmatrix} x_1 \\ x_2 \end{bmatrix}$$

and

$$Q^T\mathsf{b} = \begin{bmatrix} \frac{1}{\sqrt{6}} & \frac{1}{\sqrt{6}} & \frac{2}{\sqrt{6}} \\ \frac{2}{\sqrt{21}} & -\frac{4}{\sqrt{21}} & \frac{1}{\sqrt{21}} \end{bmatrix} \begin{bmatrix} 2 \\ 0 \\ -4 \end{bmatrix} = \begin{bmatrix} -\frac{6}{\sqrt{6}} \\ 0 \end{bmatrix}.$$

The augmented matrix for the triangular system $R\mathsf{x} = Q^T\mathsf{b}$ is $\left[\begin{array}{cc|c} \sqrt{6} & \frac{\sqrt{6}}{3} & -\frac{6}{\sqrt{6}} \\ 0 & \frac{\sqrt{21}}{3} & 0 \end{array}\right]$. Back substitution gives

$$\tfrac{\sqrt{21}}{3}x_2 = 0, \text{ so } x_2 = 0; \quad \sqrt{6}x_1 + \tfrac{\sqrt{6}}{3}x_2 = -\tfrac{6}{\sqrt{6}}, \text{ so } x_1 = -\tfrac{6}{\sqrt{6}\sqrt{6}} = -1.$$

We obtain $\mathsf{x}^+ = \begin{bmatrix} -1 \\ 0 \end{bmatrix}$ as before. ☺

READING CHECK 5. Verify the QR factorization of A in Example 6.3.14.

Answers to Reading Checks

1. **(a)** $\mathsf{f} = \mathsf{v} - \operatorname{proj}_{\mathsf{u}} \mathsf{v} = \mathsf{f} - \dfrac{\mathsf{v} \cdot \mathsf{u}}{\mathsf{u} \cdot \mathsf{u}} \mathsf{u}$, so $a = -\dfrac{\mathsf{v} \cdot \mathsf{u}}{\mathsf{u} \cdot \mathsf{u}}$.

   (b) Any vector in $\operatorname{sp}\{\mathsf{e}, \mathsf{f}\}$ has the form $\mathsf{x} = c\mathsf{e} + d\mathsf{f}$. This is $c\mathsf{u} + d(\mathsf{v} + a\mathsf{u}) = (c + da)\mathsf{u} + e\mathsf{v}$, a linear combination of u and v, hence a vector of U. Conversely, if x is in U, then $\mathsf{x} = c\mathsf{u} + d\mathsf{v} = c\mathsf{e} + d(\mathsf{f} - a\mathsf{u}) = c\mathsf{e} + d(\mathsf{f} - a\mathsf{e}) = (c - da)\mathsf{e} + d\mathsf{f}$ is a linear combination of e and f. So $U = \operatorname{sp}\{\mathsf{e}, \mathsf{f}\}$.

2. $\operatorname{proj}_{U_1} \mathsf{u}_2 = c\mathsf{u}_1$ is a scalar multiple of u_1. If $\mathsf{f}_2 = 0$, then $\mathsf{u}_2 - c\mathsf{u}_1 = 0$, which says that u_1 and u_2 are linearly dependent.

3. **(a)** If a linear combination of the columns of R is 0, then $R\mathsf{x} = 0$ for some vector x. Thus $QR\mathsf{x} = 0$, so $A\mathsf{x} = 0$. Since the columns of A are linearly independent, $\mathsf{x} = 0$. Thus the columns of R are linearly independent.

   (b) The columns of R are linearly independent, so col rank $R = n = \operatorname{rank} R$.

4. $\cos\theta = \dfrac{\mathsf{u} \cdot \mathsf{v}}{\|\mathsf{u}\| \, \|\mathsf{v}\|}$, and multiplication by Q preserves $\mathsf{u} \cdot \mathsf{v}$, $\|\mathsf{u}\|$ and $\|\mathsf{v}\|$.

5. The columns of A are the vectors $\mathsf{a}_1 = \begin{bmatrix} 1 \\ 1 \\ 2 \end{bmatrix}$ and $\mathsf{a}_2 = \begin{bmatrix} 1 \\ -1 \\ 1 \end{bmatrix}$. We apply the Gram–Schmidt algorithm to these.

$$\mathsf{f}_1 = \mathsf{a}_1 = \begin{bmatrix} 1 \\ 1 \\ 2 \end{bmatrix}$$

$$\mathsf{f}_2 = \mathsf{a}_2 - \frac{\mathsf{a}_2 \cdot \mathsf{f}_1}{\mathsf{f}_1 \cdot \mathsf{f}_1} \mathsf{f}_1 = \mathsf{a}_2 - \tfrac{2}{6} \mathsf{f}_1 = \begin{bmatrix} 1 \\ -1 \\ 1 \end{bmatrix} - \tfrac{1}{3} \begin{bmatrix} 1 \\ 1 \\ 2 \end{bmatrix} = \begin{bmatrix} \frac{2}{3} \\ -\frac{4}{3} \\ \frac{1}{3} \end{bmatrix},$$

which we replace with $3\mathsf{f}_2$ to ease further calculations with this vector. Thus we take $\mathsf{f}_2 = \begin{bmatrix} 2 \\ -4 \\ 1 \end{bmatrix}$. Dividing f_1 and f_2 by their lengths, we obtain the orthonormal vectors

$$\mathsf{q}_1 = \frac{1}{\|\mathsf{f}_1\|} \mathsf{f}_1 = \frac{1}{\sqrt{6}} \mathsf{f}_1 = \begin{bmatrix} \frac{1}{\sqrt{6}} \\ \frac{1}{\sqrt{6}} \\ \frac{2}{\sqrt{6}} \end{bmatrix} \text{ and } \mathsf{q}_2 = \frac{1}{\|\mathsf{f}_2\|} \mathsf{f}_2 = \frac{1}{\sqrt{21}} \mathsf{f}_2 = \begin{bmatrix} \frac{2}{\sqrt{21}} \\ -\frac{4}{\sqrt{21}} \\ \frac{1}{\sqrt{21}} \end{bmatrix}.$$

Now

$$\begin{aligned} \mathsf{f}_1 &= \mathsf{a}_1 && \text{so that} && \mathsf{a}_1 = \mathsf{f}_1 = \sqrt{6}\mathsf{q}_1, \\ \mathsf{f}_2 &= 3(\mathsf{a}_2 - \tfrac{1}{3}\mathsf{f}_1) && \text{so that} && \mathsf{a}_2 = \tfrac{1}{3}\mathsf{f}_1 + \tfrac{1}{3}\mathsf{f}_2 = \tfrac{\sqrt{6}}{3}\mathsf{q}_1 + \tfrac{\sqrt{21}}{3}\mathsf{q}_2. \end{aligned}$$

Thus $Q = \begin{bmatrix} \mathsf{q}_1 & \mathsf{q}_2 \\ \downarrow & \downarrow \end{bmatrix} = \begin{bmatrix} \frac{1}{\sqrt{6}} & \frac{2}{\sqrt{21}} \\ \frac{1}{\sqrt{6}} & -\frac{4}{\sqrt{21}} \\ \frac{2}{\sqrt{6}} & \frac{1}{\sqrt{21}} \end{bmatrix}$ and $R = \begin{bmatrix} \sqrt{6} & \frac{\sqrt{6}}{3} \\ 0 & \frac{\sqrt{21}}{3} \end{bmatrix}$.

True/False Questions

Decide, with as little calculation as possible, whether each of the following statements is true or false and, if you say "false," explain your answer. (Answers are at the back of the book.)

1. An orthogonal set of nonzero vectors is a linearly independent set.

2. The Gram-Schmidt algorithm would convert the vectors $\mathsf{u}_1 = \begin{bmatrix} 1 \\ 2 \\ 3 \\ 4 \end{bmatrix}$, $\mathsf{u}_2 = \begin{bmatrix} 5 \\ 6 \\ 7 \\ 8 \end{bmatrix}$, $\mathsf{u}_3 = \begin{bmatrix} 2 \\ 4 \\ 6 \\ 8 \end{bmatrix}$ into three orthogonal vectors.

3. A vector space has an orthogonal basis if and only if it has an orthonormal basis.

4. A permutation matrix is an orthogonal matrix.

5. $\begin{bmatrix} \frac{1}{\sqrt{2}} & \frac{1}{\sqrt{3}} \\ 0 & \frac{1}{\sqrt{3}} \\ -\frac{1}{\sqrt{2}} & \frac{1}{\sqrt{3}} \end{bmatrix}$ is an orthogonal matrix.

6. The inverse of $\begin{bmatrix} \cos\theta & -\sin\theta \\ \sin\theta & \cos\theta \end{bmatrix}$ is $\begin{bmatrix} \cos(-\theta) & -\sin(-\theta) \\ \sin(-\theta) & \cos(-\theta) \end{bmatrix}$.

7. If $\mathsf{q}_1, \ldots, \mathsf{q}_k$ are orthonormal vectors in R^n, then these can be extended to a basis of R^n.

8. If $AA^T = I$, an identity matrix, then A has orthonormal rows.

9. If the columns of a square matrix are orthonormal, then its rows are orthonormal too.

10. The matrix $\begin{bmatrix} 1 & 0 \\ 0 & -1 \end{bmatrix}$ is a rotation matrix.

EXERCISES

*Answers to exercises marked [BB] can be found at the **Back of the Book**.*

1. Apply the Gram-Schmidt algorithm to each of the following sets of vectors.

   (a) [BB] $\mathsf{u}_1 = \begin{bmatrix} 0 \\ 1 \\ 3 \end{bmatrix}$, $\mathsf{u}_2 = \begin{bmatrix} -1 \\ 0 \\ 1 \end{bmatrix}$

   (b) $\mathsf{u}_1 = \begin{bmatrix} 1 \\ -1 \\ 1 \end{bmatrix}$, $\mathsf{u}_2 = \begin{bmatrix} 1 \\ 0 \\ 1 \end{bmatrix}$, $\mathsf{u}_3 = \begin{bmatrix} 1 \\ 1 \\ 2 \end{bmatrix}$

(c) $u_1 = \begin{bmatrix} 1 \\ 0 \\ 1 \\ 0 \end{bmatrix}, u_2 = \begin{bmatrix} 0 \\ 0 \\ 1 \\ 1 \end{bmatrix}, u_3 = \begin{bmatrix} 1 \\ 0 \\ 1 \\ 1 \end{bmatrix}$

(d) $u_1 = \begin{bmatrix} 1 \\ 0 \\ 2 \\ 1 \end{bmatrix}, u_2 = \begin{bmatrix} 0 \\ 1 \\ 0 \\ -3 \end{bmatrix}, u_3 = \begin{bmatrix} 2 \\ 1 \\ -3 \\ 1 \end{bmatrix}$.

2. Let $U = \operatorname{sp}\left\{\begin{bmatrix} 1 \\ 0 \\ 1 \\ 0 \end{bmatrix}, \begin{bmatrix} 1 \\ 1 \\ 1 \\ 0 \end{bmatrix}, \begin{bmatrix} 1 \\ 1 \\ 0 \\ 0 \end{bmatrix}\right\}$. Let $b = \begin{bmatrix} 2 \\ 0 \\ -1 \\ 3 \end{bmatrix}$.

   (a) Apply the Gram–Schmidt algorithm to exhibit an orthogonal basis for U.

   (b) Find the vector u in U for which $\|b - u\|$ is minimal.

3. [BB] Find a QR factorization of $A = \begin{bmatrix} 1 & 1 & 2 \\ 0 & -2 & 0 \\ -1 & 1 & -1 \\ 1 & -1 & 0 \end{bmatrix}$. Note that the Gram-Schmidt algorithm was applied to the columns of this matrix in Example 6.3.3.

4. Find a QR factorization of each matrix.

   (a) [BB] $A = \begin{bmatrix} 1 & 4 \\ 1 & 0 \end{bmatrix}$ **(b)** $A = \begin{bmatrix} 1 & 2 \\ 3 & 4 \end{bmatrix}$

   (c) [BB] $A = \begin{bmatrix} 1 & -1 & 2 \\ 0 & 1 & 1 \\ 1 & 1 & 1 \end{bmatrix}$ **(d)** $A = \begin{bmatrix} 1 & 3 & 1 \\ -1 & 1 & 0 \\ 0 & 2 & 1 \end{bmatrix}$

   (e) $A = \begin{bmatrix} -1 & 1 \\ 0 & 1 \\ 1 & -2 \end{bmatrix}$ **(f)** $A = \begin{bmatrix} 1 & 3 & 2 \\ -1 & 1 & 0 \\ 0 & 2 & 1 \\ 1 & 1 & 4 \end{bmatrix}$

   (g) $A = \begin{bmatrix} -1 & 0 \\ 2 & 1 \\ 1 & 1 \end{bmatrix}$ **(h)** $A = \begin{bmatrix} -2 & 2 \\ 1 & 0 \\ 0 & 1 \\ 2 & -1 \end{bmatrix}$

   (i) $A = \begin{bmatrix} 3 & -1 & 6 & -2 \\ 2 & -6 & 8 & 16 \\ 1 & 5 & -2 & -2 \\ 1 & -3 & 14 & 10 \\ -1 & 3 & 6 & 10 \end{bmatrix}$ **(j)** $A = \begin{bmatrix} -7 & 3 & 22 & 25 \\ -2 & 6 & -4 & 2 \\ -2 & -12 & 5 & 2 \\ -4 & 12 & -8 & 13 \\ -2 & 6 & 23 & 2 \\ -2 & 6 & -4 & 2 \end{bmatrix}$.

5. In each case, find a QR factorization of A and use this to find A^{-1}.

   (a) [BB] $A = \begin{bmatrix} 1 & 4 \\ 1 & 0 \end{bmatrix}$ [See Exercise 4(a).]

   (b) $A = \begin{bmatrix} 1 & -1 & 2 \\ 0 & 1 & 1 \\ 1 & 1 & 1 \end{bmatrix}$ [See Exercise 4(c).]

(c) $A = \begin{bmatrix} 0 & 3 & 2 \\ 3 & 5 & 5 \\ 4 & 0 & 5 \end{bmatrix}$ (d) $A = \begin{bmatrix} 0 & 1 & 1 & 0 \\ 1 & 0 & 4 & 1 \\ 1 & 1 & -1 & 1 \\ 0 & -1 & 1 & 1 \end{bmatrix}$.

6. Use a QR factorization of an appropriate matrix to find the best least squares solution to each of the given systems.

(a) [BB] $\begin{aligned} x + y &= 1 \\ 2x + 3y &= -1 \\ 2x + y &= 1 \end{aligned}$ (b) $\begin{aligned} x - y &= 1 \\ 2x + y &= -2 \\ x + 3y &= -3 \end{aligned}$

(c) $\begin{aligned} x \quad + z &= -1 \\ x + y + z &= 0 \\ 2x \quad &= 0 \\ x + y - z &= 3 \end{aligned}$ (d) $\begin{aligned} x + y + z &= 0 \\ x \quad + 2z &= -3 \\ -x + y \quad &= 1 \\ -x + y - z &= 0. \end{aligned}$

7. Suppose P is a projection matrix (see Section 6.1). Show that $Q = I - 2P$ is an orthogonal, symmetric matrix.

8. [BB] A matrix K is *skew-symmetric* if $K^T = -K$. Suppose K is a skew-symmetric matrix such that $I - K$ is invertible. Show that $(I + K)(I - K)^{-1}$ is orthogonal.

9. If Q is an orthogonal matrix and x is a nonzero vector such that $Q\mathsf{x} = c\mathsf{x}$, show that $c = \pm 1$.

10. (a) [BB] If Q is orthogonal and symmetric, show that $Q^2 = I$.
    (b) If Q is orthogonal and symmetric, show that $P = \frac{1}{2}(I - Q)$ is a projection matrix.

11. Show that the determinant of an orthogonal matrix is ± 1.

12. (a) [BB] If Q_1 is $m \times n$ with orthonormal columns and Q_2 is $n \times t$ with orthonormal columns, show that $Q_1 Q_2$ has orthonormal columns.
    (b) Is the product of orthogonal matrices orthogonal? Explain.

13. If Q is a real $n \times n$ matrix that preserves the dot product in R^n—that is, $Q\mathsf{x} \cdot Q\mathsf{y} = \mathsf{x} \cdot \mathsf{y}$ for any vectors x, y—is Q orthogonal? Explain.

14. Let $Q = \begin{bmatrix} \cos\theta & -\sin\theta \\ \sin\theta & \cos\theta \end{bmatrix}$, $\mathsf{x} = \begin{bmatrix} x \\ y \end{bmatrix}$ and $\mathsf{x}' = \begin{bmatrix} x' \\ y' \end{bmatrix} = Q\begin{bmatrix} x \\ y \end{bmatrix}$. Verify that $\|Q\mathsf{x}\| = \|\mathsf{x}\|$, consistent with 6.3.9.

15. If $Q^{-1}AQ$ is diagonal and Q is orthogonal, show that A is symmetric.

Critical Reading

16. In this section, we seem to have been finding QR factorizations for $m \times n$ matrices only when $m \geq n$. Why?

17. [BB] Suppose the Gram-Schmidt algorithm is applied to a set $\mathsf{u}_1, \mathsf{u}_2, \ldots, \mathsf{u}_k$ of vectors which are **not** linearly independent.

(a) The algorithm must break down. Why?

(b) Explain **how** the algorithm will break.

18. Apply the Gram–Schmidt algorithm to the vectors $u_1 = \begin{bmatrix} -1 \\ -2 \\ 1 \end{bmatrix}$, $u_2 = \begin{bmatrix} 3 \\ 1 \\ 2 \end{bmatrix}$, $u_3 = \begin{bmatrix} 0 \\ 1 \\ -1 \end{bmatrix}$. What goes wrong? Explain.

19. [BB] Any finite dimensional vector space has an orthogonal basis. Why?

20. If $q_1, q_2, \ldots, q_k$ are orthonormal vectors in R^n, then these vectors can be extended to an orthonormal basis of R^n. Why?

21. Let w be a vector in R^n of length 1 and $H = I - 2ww^T$ the corresponding Householder matrix. Let x be any vector in R^n and $y = Hx - x$.

    (a) [BB] Show that y is a scalar multiple of w.

    (b) [BB] Show that $x + \frac{1}{2}y$ is orthogonal to w.

    (c) Add the vectors y and $x + \frac{1}{2}y$ to the picture at the right. Your picture should make it appear that Hx is the reflection of x in a plane perpendicular to w.

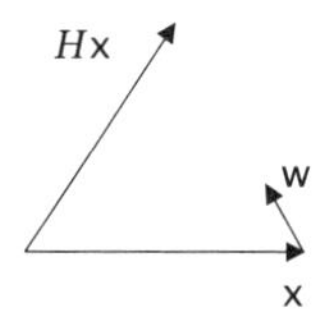

22. True or false: A Householder matrix is a projection matrix. (Justify your answer.)

23. Let V be the subspace of R^4 spanned by the vectors $u_1 = \begin{bmatrix} 1 \\ -1 \\ 0 \\ 0 \end{bmatrix}$, $u_2 = \begin{bmatrix} 1 \\ 0 \\ -1 \\ 0 \end{bmatrix}$, and $u_3 = \begin{bmatrix} 1 \\ 0 \\ 0 \\ -1 \end{bmatrix}$. Apply the Gram–Schmidt algorithm to find an **orthonormal** basis $\mathcal{B}$ of V. Is $\mathcal{B}$ the only orthonormal basis for V? Explain.

24. Roberta is having trouble finding the QR factorization of a 3×4 matrix. What would you say to her?

25. Does a matrix with orthonormal columns have a left inverse? a right inverse? Explain.

26. Suppose we want to solve the system $Ax = b$ and we factor $A = QR$. Then $QRx = b$. Multiplying each side by Q^T gives $Q^TQRx = Q^Tb$ and, since $Q^TQ = I$, we have $Rx = Q^Tb$. Since R is invertible, $x = R^{-1}Q^Tb$, so $Ax = b$ has a solution which is even unique. Is there a flaw in our reasoning?

27. (a) Suppose an $n \times n$ matrix A with nonzero determinant has two LU factorizations $A = LU = L_1U_1$, with L, L_1 lower triangular $n \times n$ matrices and U, U_1 upper triangular $n \times n$ matrices. Prove that $L = L_1$ and $U = U_1$.

(b) Show that an $m \times n$ matrix with linearly independent columns can be factored in exactly one way $A = QR$, Q with orthonormal columns and R upper triangular: The QR factorization is unique.

28. Suppose $A = QR$ is a QR factorization of a matrix A and we proceed to factor $R = Q_1R_1$, $R_1 = Q_2R_2$, and so on, continuing with QR factorizations of the triangular matrices R_i. Does the process ever stop? How? When?

[Hint: Exercise 27.]

6.4 Orthogonal Subspaces and Complements

We begin with some definitions and notation.

6.4.1 Definition. The *sum* of subspaces U and W of a vector space V is the set

$$U + W = \{\mathsf{u} + \mathsf{w} \mid \mathsf{u} \in U, \mathsf{w} \in W\}$$

of vectors that can be written as the sum of a vector in U and a vector in W. The *intersection* of U and W is the set

$$U \cap W = \{\mathsf{x} \in V \mid \mathsf{x} \in U \text{ and } \mathsf{x} \in W\}$$

of vectors that are in both U and W.

It is straightforward to show that sums and intersections of subspaces of a vector space V are also subspaces of V—see Exercise 8.

6.4.2 Example. Let $U = \{\begin{bmatrix} c \\ 2c \end{bmatrix} \mid c \in \mathsf{R}\}$ and $W = \{\begin{bmatrix} 0 \\ c \end{bmatrix} \mid c \in \mathsf{R}\}$ be the indicated subspaces of R^2. Then $U + W = \mathsf{R}^2$ because every $\mathsf{v} = \begin{bmatrix} x \\ y \end{bmatrix} \in \mathsf{R}^2$ can be written

$$\begin{bmatrix} x \\ y \end{bmatrix} = \begin{bmatrix} x \\ 2x \end{bmatrix} + \begin{bmatrix} 0 \\ y - 2x \end{bmatrix}$$

as the sum of a vector in U and a vector in W. If a vector $\mathsf{v} = \begin{bmatrix} x \\ y \end{bmatrix}$ is in the intersection of U and W, then $\mathsf{v} = \begin{bmatrix} x \\ y \end{bmatrix} = \begin{bmatrix} c \\ 2c \end{bmatrix} = \begin{bmatrix} 0 \\ d \end{bmatrix}$ for scalars c and d. So $x = c = 0$ and $y = 2c = d$, so $\mathsf{v} = \mathbf{0}$. In this case, $U \cap W = \{\mathbf{0}\}$. ☺

6.4.3 Example. Let U and W be the subspaces of R^3 defined by

$$U = \left\{\begin{bmatrix} x \\ y \\ z \end{bmatrix} \mid x + y - z = 0\right\} \quad \text{and} \quad W = \left\{\begin{bmatrix} x \\ y \\ z \end{bmatrix} \mid 2x - 3y + z = 0\right\}.$$

Then $U + W = \mathsf{R}^3$ since any $\begin{bmatrix} x \\ y \\ z \end{bmatrix}$ in R^3 can be written

$$\begin{bmatrix} x \\ y \\ z \end{bmatrix} = \begin{bmatrix} -y + \frac{1}{3}z \\ -x \\ -x - y + \frac{1}{3}z \end{bmatrix} + \begin{bmatrix} x + y - \frac{1}{3}z \\ x + y \\ x + y + \frac{2}{3}z \end{bmatrix}$$

as the sum of a vector in U and a vector in W.

What about $U \cap W$, the intersection of U and W? Each of these subspaces is a plane, the two planes are not parallel, so they intersect in a line. Solving the system

$$\begin{aligned} x + \ \ y - z &= 0 \\ 2x - 3y + z &= 0 \end{aligned}$$

gives $\begin{bmatrix} x \\ y \\ z \end{bmatrix} = t\begin{bmatrix} 2 \\ 3 \\ 5 \end{bmatrix}$, which is the equation of the line of intersection:

$$U \cap W = \left\{t\begin{bmatrix} 2 \\ 3 \\ 5 \end{bmatrix} \mid t \in \mathsf{R}\right\}.$$

:-)

READING CHECK 1. Explain why the planes in Example 6.4.3 are not parallel.

6.4.4 Example. If U is the subspace of R^4 spanned by $\mathsf{u}_1 = \begin{bmatrix} -1 \\ 1 \\ 2 \\ 3 \end{bmatrix}$ and $\mathsf{u}_2 = \begin{bmatrix} 2 \\ 0 \\ -1 \\ 3 \end{bmatrix}$ and W is the subspace of R^4 spanned by $\mathsf{w}_1 = \begin{bmatrix} 0 \\ 1 \\ -3 \\ 1 \end{bmatrix}$ and $\mathsf{w}_2 = \begin{bmatrix} 1 \\ 0 \\ 4 \\ 5 \end{bmatrix}$, then $U + W$ is spanned by $\{\mathsf{u}_1, \mathsf{u}_2, \mathsf{w}_1, \mathsf{w}_2\}$. To find a basis for $U + W$, we might put these vectors into the rows of a matrix and move to a row echelon matrix whose nonzero rows provide the basis we are after. Alternatively, and this will pay dividends later, we can form the matrix

$$A = \begin{bmatrix} \mathsf{u}_1 & \mathsf{u}_2 & \mathsf{w}_1 & \mathsf{w}_2 \\ \downarrow & \downarrow & \downarrow & \downarrow \\ & & & \end{bmatrix} = \begin{bmatrix} -1 & 2 & 0 & 1 \\ 1 & 0 & 1 & 0 \\ 2 & -1 & -3 & 4 \\ 3 & 3 & 1 & 5 \end{bmatrix} \tag{1}$$

whose columns are $\mathsf{u}_1, \mathsf{u}_2, \mathsf{w}_1, \mathsf{w}_2$, reduce this to row echelon form and then use the fact that the pivot columns are a basis for the column space of A (which is $U + W$)—see 4.4.21 of Section 4.3.

Row echelon form is $\begin{bmatrix} -1 & 2 & 0 & 1 \\ 0 & 1 & -1 & 2 \\ 0 & 0 & 1 & -1 \\ 0 & 0 & 0 & 0 \end{bmatrix}$. The pivot columns are columns one, two and three, so the first three columns of A—the vectors $\mathsf{u}_1, \mathsf{u}_2, \mathsf{w}_1$—are a basis for the column space, which is $U + W$.

Now we turn our attention to the intersection of U and W and begin by supposing y is in $U \cap W$. Then $\mathsf{y} = a_1\mathsf{u}_1 + a_2\mathsf{u}_2 = b_1\mathsf{w}_1 + b_2\mathsf{w}_2$ for certain scalars a_1, a_2, b_1, b_2. Thus $a_1\mathsf{u}_1 + a_2\mathsf{u}_2 - b_1\mathsf{w}_1 - b_2\mathsf{w}_2 = \mathsf{0}$, so $\mathsf{x} = \begin{bmatrix} a_1 \\ a_2 \\ -b_1 \\ -b_2 \end{bmatrix}$ is in the null space of A (using, once again, the famous formula in 2.1.33). Conversely, if $\mathsf{x} = \begin{bmatrix} x_1 \\ x_2 \\ y_1 \\ y_2 \end{bmatrix}$ is in the null space of A (it is useful to label the components of x this way), then $x_1\mathsf{u}_1 + x_2\mathsf{u}_2 + y_1\mathsf{w}_1 + y_2\mathsf{w}_2 = \mathsf{0}$, so $x_1\mathsf{u}_1 + x_2\mathsf{u}_2 = -y_1\mathsf{w}_1 - y_2\mathsf{w}_2$ is in $U \cap W$. So we can find $U \cap W$ by finding the null space of A. Looking at the row echelon form of A, we see that $y_2 = t$ is free, $y_1 = y_2 = t$, $x_2 = y_1 - 2y_2 = -t$ and $-x_1 = -2x_2 - y_2 = t$, so $x_1 = -t$ and $\mathsf{x} = \begin{bmatrix} -t \\ -t \\ t \\ t \end{bmatrix}$. Thus $-t\mathsf{u}_1 - t\mathsf{u}_2 + t\mathsf{w}_1 + t\mathsf{w}_2 = \mathsf{0}$. Vectors in $U \cap W$ have the form

$t(\mathsf{w}_1 + \mathsf{w}_2) = t(\mathsf{u}_1 + \mathsf{u}_2) = t\begin{bmatrix} 1 \\ 1 \\ 1 \\ 6 \end{bmatrix}$. The vector $\begin{bmatrix} 1 \\ 1 \\ 1 \\ 6 \end{bmatrix}$ is a basis for $U \cap W$. ☺

In this example, we had $\dim(U + W) = 3$ and $\dim(U \cap W) = 1$ and note that $\dim(U + W) + \dim(U \cap W) = 4 = 2 + 2 = \dim U + \dim W$. The reader should verify that this formula also holds in Examples 6.4.2 and 6.4.3. In fact, it is true in general.

6.4.5 Theorem. $\dim(U + W) + \dim(U \cap W) = \dim U + \dim W$ *for any subspaces* U *and* W *of* R^n.

Proof. Let $\{\mathsf{u}_1, \mathsf{u}_2, \ldots, \mathsf{u}_k\}$ be a basis for U and let $\{\mathsf{w}_1, \mathsf{w}_2, \ldots, \mathsf{w}_\ell\}$ be a basis for W. Let

$$A = \begin{bmatrix} \mathsf{u}_1 & \mathsf{u}_2 & \cdots & \mathsf{u}_k & \mathsf{w}_1 & \mathsf{w}_2 & \cdots & \mathsf{w}_\ell \\ \downarrow & \downarrow & & \downarrow & \downarrow & \downarrow & & \downarrow \\ & & & & & & & \end{bmatrix}$$

be the $n \times (k + \ell)$ matrix whose columns are as indicated. The column space of A is $U+W$, so $\dim(U+W) = \operatorname{rank} A$. If we could show that $\dim(U\cap W) = \operatorname{nullity} A$, then we would have $\dim(U+W)+\dim(U\cap W) = \operatorname{rank} A+\operatorname{nullity} A = k+\ell$, the number of columns of A. This would give the desired result because $k = \dim U$ and $\ell = \dim W$. To summarize, we can complete the proof by showing that $\dim(U\cap W) = \operatorname{nullity} A$. We do this by showing that the nice relationship between $U\cap W$ and the null space of A that we discovered in Example 6.4.4 holds in general.

So let y be a vector in $U\cap W$. Then

$$\mathsf{y} = x_1\mathsf{u}_1 + \cdots + x_k\mathsf{u}_k = y_1\mathsf{w}_1 + \cdots + y_\ell\mathsf{w}_\ell$$

is a linear combination of $\mathsf{u}_1,\ldots,\mathsf{u}_k$ and also a linear combination of $\mathsf{w}_1,\ldots,\mathsf{w}_\ell$, so $\mathsf{x} = \begin{bmatrix} x_1 \\ \vdots \\ x_k \\ -y_1 \\ \vdots \\ -y_\ell \end{bmatrix}$ is in the null space of A. On the other hand, if $\mathsf{x} = \begin{bmatrix} x_1 \\ \vdots \\ x_k \\ y_1 \\ \vdots \\ y_\ell \end{bmatrix}$ is in the null space of A, then $x_1\mathsf{u}_1 + \cdots + x_k\mathsf{u}_k + y_1\mathsf{w}_1 + \cdots + y_\ell\mathsf{w}_\ell = \mathsf{0}$, so

$$x_1\mathsf{u}_1 + \cdots + x_k\mathsf{u}_k = -y_1\mathsf{w}_1 - \cdots - y_\ell\mathsf{w}_\ell$$

is a vector in $U\cap W$. We have shown that the null space of A is precisely the set of vectors

$$\mathsf{x} = \begin{bmatrix} x_1 \\ \vdots \\ x_k \\ y_1 \\ \vdots \\ y_\ell \end{bmatrix}$$

with the property that $x_1\mathsf{u}_1 + \cdots + x_k\mathsf{u}_k = -y_1\mathsf{w}_1 - \cdots - y_\ell\mathsf{w}_\ell$ is in $U\cap W$. The function $T\colon \operatorname{null\,sp} A \to U\cap W$ defined by

$$\begin{bmatrix} x_1 \\ \vdots \\ x_k \\ y_1 \\ \vdots \\ y_\ell \end{bmatrix} \to x_1\mathsf{u}_1 + \cdots + x_k\mathsf{u}_k$$

establishes a one-to-one correspondence between null sp A and $U\cap W$. Since it is a linear transformation, T maps a basis for A to a basis for $U\cap W$, so these vector spaces have the same dimension—see Exercise 19 for details. ■

6.4.6 Remark. In Theorem 6.4.5, it was important that the given spanning sets $\mathcal{U} = \{\mathsf{u}_1,\mathsf{u}_2,\ldots,\mathsf{u}_k\}$ and $\mathcal{W} = \{\mathsf{w}_1,\mathsf{w}_2,\ldots,\mathsf{w}_\ell\}$ were bases for U and W, respectively, in order that the number of columns of A was $k + \ell = \dim U + \dim W$. If we are interested only in finding a basis for $U \cap W$, then it is sufficient that $\mathcal{U}$ and $\mathcal{W}$ be merely spanning sets for U and W, as illustrated in our next example where the vectors $\mathsf{w}_1, \mathsf{w}_2, \mathsf{w}_3$ are not a basis for W.

6.4.7 Example. Let U be the subspace of R^3 spanned by $\mathsf{u}_1 = \begin{bmatrix} 2 \\ 4 \\ -2 \end{bmatrix}$ and $\mathsf{u}_2 = \begin{bmatrix} 1 \\ 2 \\ 1 \end{bmatrix}$. Let W be the subspace of R^3 spanned by $\mathsf{w}_1 = \begin{bmatrix} 0 \\ 3 \\ -1 \end{bmatrix}$, $\mathsf{w}_2 = \begin{bmatrix} 1 \\ 0 \\ 1 \end{bmatrix}$ and $\mathsf{w}_3 = \begin{bmatrix} 3 \\ 6 \\ 1 \end{bmatrix}$. Let $A = \begin{bmatrix} 2 & 1 & 0 & 1 & 3 \\ 4 & 2 & 3 & 0 & 6 \\ -2 & 1 & -1 & 1 & 1 \end{bmatrix}$ be the matrix whose columns are $\mathsf{u}_1, \mathsf{u}_2, \mathsf{w}_1, \mathsf{w}_2, \mathsf{w}_3$. Row echelon form is $\begin{bmatrix} 2 & 1 & 0 & 1 & 3 \\ 0 & 2 & -1 & 2 & 4 \\ 0 & 0 & 3 & -2 & 0 \end{bmatrix}$. If $\mathsf{x} = \begin{bmatrix} x_1 \\ x_2 \\ y_1 \\ y_2 \\ y_3 \end{bmatrix}$ is in the null space of A, then $y_2 = t$ and $y_3 = s$ are free, $3y_1 = 2y_2$ so $y_1 = \frac{2}{3}y_2 = \frac{2}{3}t$,

$$2x_2 = y_1 - 2y_2 - 4y_3 = \tfrac{2}{3}t - 2t - 4s = -\tfrac{4}{3}t - 4s, \text{ so } x_2 = -\tfrac{2}{3}t - 2s,$$
$$2x_1 = -x_2 - y_2 - 3y_3 = \tfrac{2}{3}t + 2s - t - 3s = -\tfrac{1}{3}t - s, \text{ so } x_1 = -\tfrac{1}{6}t - \tfrac{1}{2}s$$

and $\mathsf{x} = \begin{bmatrix} -\frac{1}{6}t - \frac{1}{2}s \\ -\frac{2}{3}t - 2s \\ \frac{2}{3}t \\ t \\ s \end{bmatrix}$, so $(-\frac{1}{6}t - \frac{1}{2}s)\mathsf{u}_1 + (-\frac{2}{3}t - 2s)\mathsf{u}_2 + \frac{2}{3}t\mathsf{w}_1 + t\mathsf{w}_2 + s\mathsf{w}_3 = \mathsf{0}$.

Thus vectors in $U \cap W$ have the form

$$(\tfrac{1}{6}t + \tfrac{1}{2}s)\mathsf{u}_1 + (\tfrac{2}{3}t + 2s)\mathsf{u}_2$$
$$= \tfrac{2}{3}t\mathsf{w}_1 + t\mathsf{w}_2 + s\mathsf{w}_3 = \begin{bmatrix} t + 3s \\ 2t + 6s \\ \frac{1}{3}t + s \end{bmatrix} = (t + 3s)\begin{bmatrix} 1 \\ 2 \\ \frac{1}{3} \end{bmatrix},$$

showing that $U \cap W$ has basis the vector $\begin{bmatrix} 1 \\ 2 \\ \frac{1}{3} \end{bmatrix}$ and dimension 1. (Note that this is **not** the nullity of A.) Again $U + W$ is the column space of A, so $\dim(U + W) = \operatorname{rank} A = 3$ and $\dim(U + W) + \dim(U \cap W) = 3 + 1 = 4 = 2 + 2 = \dim U + \dim W$. (This number is no longer the number of columns in A.) ☺

READING CHECK 2. In Example 6.4.7, $\dim W = 2$. Why?

READING CHECK 3. With U and W as in Example 6.4.7, find a basis for $U + W$.

For the rest of this section, where the dot product plays a role, all vector spaces will be Euclidean or subspaces of a Euclidean vector space.

6.4.8 Definition. Subspaces U and W of R^n are *orthogonal* if $\mathsf{u} \cdot \mathsf{w} = 0$ for any vector u in U and any vector w in W.

6.4.9 Example. In R^3, the x-axis and the z-axis are orthogonal subspaces since if $\mathsf{u} = \begin{bmatrix} x \\ 0 \\ 0 \end{bmatrix}$ is on the x-axis and $\mathsf{w} = \begin{bmatrix} 0 \\ 0 \\ z \end{bmatrix}$ is on the z-axis, $\mathsf{u} \cdot \mathsf{w} = 0$. ☺

6.4.10 Example. The subspaces $U = \left\{ \begin{bmatrix} x_1 \\ x_2 \\ x_3 \\ x_4 \end{bmatrix} \in \mathsf{R}^4 \mid 2x_1 - x_2 + x_3 - x_4 = 0 \right\}$ and $W = \left\{ c \begin{bmatrix} 2 \\ -1 \\ 1 \\ -1 \end{bmatrix} \mid c \in \mathsf{R} \right\}$ are orthogonal since if $\mathsf{u} = \begin{bmatrix} x_1 \\ x_2 \\ x_3 \\ x_4 \end{bmatrix} \in U$ and $\mathsf{w} = c \begin{bmatrix} 2 \\ -1 \\ 1 \\ -1 \end{bmatrix} \in W$, then $\mathsf{u} \cdot \mathsf{w} = c(2x_1 - x_2 + x_3 - x_4) = 0$. ☺

6.4.11 Example. Let $U = \operatorname{sp}\{\mathsf{u}_1, \mathsf{u}_2\}$ be the subspace of R^4 spanned by $\mathsf{u}_1 = \begin{bmatrix} -1 \\ 1 \\ 2 \\ 1 \end{bmatrix}$ and $\mathsf{u}_2 = \begin{bmatrix} 2 \\ -1 \\ 0 \\ 1 \end{bmatrix}$ and $W = \operatorname{sp}\{\mathsf{w}\}$ the subspace spanned by $\mathsf{w} = \begin{bmatrix} 0 \\ -1 \\ 1 \\ -1 \end{bmatrix}$. Then U and W are orthogonal because $\mathsf{w} \cdot \mathsf{u}_1 = 0$ and $\mathsf{w} \cdot \mathsf{u}_2 = 0$, and these equations imply that any multiple of w is orthogonal to any linear combination of u_1 and u_2, that is, to any vector in U. ☺

6.4.12 The row space and the null space of a matrix are orthogonal.[5] To see why, let $A = \begin{bmatrix} \mathsf{r}_1 & \rightarrow \\ \mathsf{r}_2 & \rightarrow \\ \vdots & \\ \mathsf{r}_m & \rightarrow \end{bmatrix}$ be an $m \times n$ matrix with rows as indicated and let x be a

[5] Here and elsewhere in this section, we say "row" when we really mean the (column) vector that is its transpose. For example, we think of the rows of a 2×3 matrix as vectors in R^3.

vector in the null space of A. Since $A\mathsf{x} = \begin{bmatrix} \mathsf{r}_1 \cdot \mathsf{x} \\ \mathsf{r}_2 \cdot \mathsf{x} \\ \vdots \\ \mathsf{r}_m \cdot \mathsf{x} \end{bmatrix} = 0$, each $\mathsf{r}_i \cdot \mathsf{x} = 0$. Any r in the row space of A is a linear combination of the r_i, so $\mathsf{r} \cdot \mathsf{x} = 0$, which is what we wanted to show. Here's a specific illustration.

6.4.13 Example. Let $A = \begin{bmatrix} 1 & 4 & 0 \\ 2 & 3 & 10 \end{bmatrix}$. First we find the null space by reducing A to row echelon form: $A \to \begin{bmatrix} 1 & 4 & 0 \\ 0 & -5 & 10 \end{bmatrix}$. Thus, if $\begin{bmatrix} x \\ y \\ z \end{bmatrix}$ is in the null space of A, $z = t$ is free and $-5y = -10z = -10t$, so $y = 2t$ and $x = -4y = -8t$. The null space of A is the one-dimensional subspace

$$\text{null sp}\, A = \left\{ \mathsf{u} \in \mathsf{R}^3 \mid \mathsf{u} = t \begin{bmatrix} -8 \\ 2 \\ 1 \end{bmatrix} \right\}$$

and this is orthogonal to the row space of A since $\begin{bmatrix} -8 \\ 2 \\ 1 \end{bmatrix}$ is orthogonal to each row of A:

$$\begin{bmatrix} -8 \\ 2 \\ 1 \end{bmatrix} \cdot \begin{bmatrix} 1 \\ 4 \\ 0 \end{bmatrix} = -8 + 8 = 0, \quad \begin{bmatrix} -8 \\ 2 \\ 1 \end{bmatrix} \cdot \begin{bmatrix} 2 \\ 3 \\ 10 \end{bmatrix} = -16 + 6 + 10 = 0$$

and hence to any linear combination of these rows, that is, to any vector in the row space. ☺

6.4.14 Definition. The *orthogonal complement* of a subspace U of a Euclidean vector space V is the set of vectors in V that are orthogonal to every vector of U. The orthogonal complement of U is denoted $U^\perp$, read "U perp."

$$\boxed{U^\perp = \{\mathsf{v} \in V \mid \mathsf{v} \cdot \mathsf{u} = 0 \text{ for all } \mathsf{u} \in U\}}$$

6.4.15 Remarks. 1. The orthogonal complement of a subspace of V is also a subspace of V. See Exercise 10.

2. The assertion that subspaces U and W are orthogonal is precisely the assertion that each is contained in the orthogonal complement of the other: $U \subseteq W^\perp$; $W \subseteq U^\perp$. See Exercise 14.

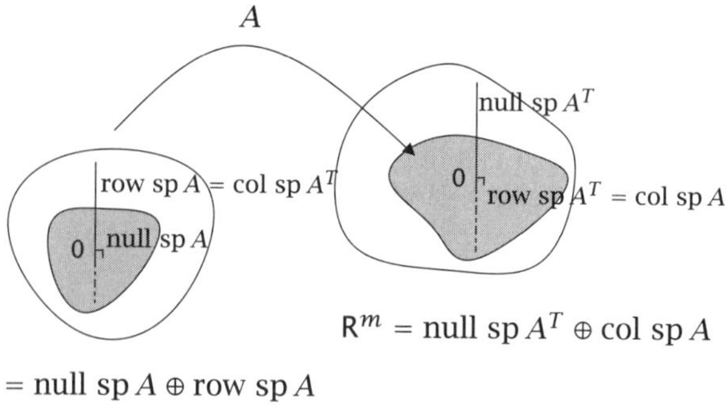

Figure 6.7: An $m \times n$ matrix A induces decompositions of $\mathbb{R}^n$ and $\mathbb{R}^m$ as shown.

6.4.16 Example. Let $V = \mathbb{R}^3$ and let U be the z-axis. Then $U^\perp$, the orthogonal complement of U, is the xy-plane: Those vectors of $\mathbb{R}^3$ which are orthogonal to the z-axis are precisely those lying in the xy-plane. ☺

Figure 6.7 illustrates the theorem that comes next.

6.4.17 Theorem. *The orthogonal complement of the row space of a matrix A is its null space. The orthogonal complement of the column space of A is the null space of A^T:*

$$(\text{row sp}\, A)^\perp = \text{null sp}\, A; \qquad (\text{col sp}\, A)^\perp = \text{null sp}\, A^T.$$

Proof. Let A be an $m \times n$ matrix. The second statement follows by applying the first statement to the matrix A^T, remembering that row sp A^T = col sp A. For the first statement, think of A in terms of its rows, $A = \begin{bmatrix} \mathsf{r}_1 & \rightarrow \\ \mathsf{r}_2 & \rightarrow \\ \vdots & \\ \mathsf{r}_m & \rightarrow \end{bmatrix}$. In 6.4.12, we showed null sp $A \subseteq (\text{row sp}\, A)^\perp$. On the other hand, $(\text{row sp}\, A)^\perp \subseteq$ null sp A because if x is in $(\text{row sp}\, A)^\perp$, then $\mathsf{r}_i \cdot \mathsf{x} = 0$ for each row r_i, so $A\mathsf{x} = \begin{bmatrix} \mathsf{r}_1 \cdot \mathsf{x} \\ \mathsf{r}_2 \cdot \mathsf{x} \\ \vdots \\ \mathsf{r}_m \cdot \mathsf{x} \end{bmatrix} = \mathbf{0}$, that is, x is in the null space of A. ■

6.4.18 Corollary. *If U is a subspace of $\mathbb{R}^n$ and* $\dim U = k$, *then* $\dim U^\perp = n - k$ *and* $(U^\perp)^\perp = U$.

Proof. Let $\mathsf{u}_1, \mathsf{u}_2, \ldots, \mathsf{u}_k$ be a basis for U and let $A = \begin{bmatrix} \mathsf{u}_1 & \mathsf{u}_2 & \cdots & \mathsf{u}_k \\ \downarrow & \downarrow & & \downarrow \end{bmatrix}$ be the $n \times k$ matrix with these vectors as columns. Thus $\operatorname{rank} A = k$ and $U = \operatorname{col\,sp} A$. Thus $U^\perp = (\operatorname{col\,sp} A)^\perp = \operatorname{null\,sp} A^T$. Since A^T is $k \times n$, $\operatorname{rank} A^T + \operatorname{nullity} A^T = n$. Since $\operatorname{rank} A^T = \operatorname{rank} A = k$, $\dim U^\perp = \operatorname{nullity} A^T = n - k$. Applying this fact to the subspace $U^\perp$, which has dimension $n - k$, we learn that $\dim(U^\perp)^\perp = n - (n - k) = k$. Since U is contained in $(U^\perp)^\perp$ and these subspaces have the same dimension, it follows that $U = (U^\perp)^\perp$, completing the proof. ■

READING CHECK 4. Why is U contained in $(U^\perp)^\perp$?

We can use Theorem 6.4.17 to find orthogonal complements.

6.4.19 Problem. Let U be the subspace of R^3 spanned by $\mathsf{u}_1 = \begin{bmatrix} 1 \\ 4 \\ 0 \end{bmatrix}$ and $\mathsf{u}_2 = \begin{bmatrix} 2 \\ 3 \\ 10 \end{bmatrix}$. Find $U^\perp$.

Solution. Let $A = \begin{bmatrix} 1 & 4 & 0 \\ 2 & 3 & 10 \end{bmatrix}$ be the matrix whose row space is U. This was the matrix that appeared in Example 6.4.13. There we found that the null space of A is the one-dimensional subspace of R^3 with basis $\begin{bmatrix} -8 \\ 2 \\ 1 \end{bmatrix}$. This is the orthogonal complement of the row space of A, that is, $U^\perp$. ♨

READING CHECK 5. Can you think of another way to find this vector $\begin{bmatrix} -8 \\ 2 \\ 1 \end{bmatrix}$?

6.4.20 Problem. Let U and W be as in 6.4.11. In that example, we showed that U and W are orthogonal subspaces. Does this mean that $W = U^\perp$?

Solution. The subspace U is the row space of $A = \begin{bmatrix} -1 & 1 & 2 & 1 \\ 2 & -1 & 0 & 1 \end{bmatrix}$, so the orthogonal complement of U is the null space of A. Row echelon form is $\begin{bmatrix} -1 & 1 & 2 & 1 \\ 0 & 1 & 4 & 3 \end{bmatrix}$, so, if $\mathsf{x} = \begin{bmatrix} x_1 \\ x_2 \\ x_3 \\ x_4 \end{bmatrix}$ is in $\operatorname{null\,sp} A$, then $x_3 = s$ and $x_4 = t$ are free, $x_2 = -4x_3 - 3x_4 = -4s - 3t$, $x_1 = x_2 + 2x_3 + x_4 = -2s - 2t$ and $\mathsf{x} = \begin{bmatrix} -2s - 2t \\ -4s - 3t \\ s \\ t \end{bmatrix}$. The orthogonal complement of U is two-dimensional, spanned by $\begin{bmatrix} -2 \\ -4 \\ 1 \\ 0 \end{bmatrix}$ and $\begin{bmatrix} -2 \\ -3 \\ 0 \\ 1 \end{bmatrix}$. We have $W \subseteq U^{\perp}$, but $W \neq U^{\perp}$.

Application: The Dual of a Code

In Section 2.2, we introduced the concept of a linear code, which we briefly review here. Let $F = \{0, 1\}$ and define addition and multiplication within F "modulo 2." Thus addition and multiplication within F work as usual with the one exception that $1 + 1 = 0$. The set F^n of column vectors over F is a vector space just like R^n except that components come from F instead of R. Previously, we defined a linear code as a subset C of F^n that is closed under addition. For any vector v and any scalar α, however, $\alpha\mathsf{v} = 0$ or $\alpha\mathsf{v} = \mathsf{v}$ (because α is 0 or 1), so C is also closed under scalar multiplication. Thus a code is actually a subspace of the vector space F^n. It is common to represent elements of C as sequences of 0s and 1s and to call any such sequence a "word," for example, the vector $\begin{bmatrix} 0 \\ 1 \\ 1 \\ 1 \end{bmatrix}$ becomes the word 0111.

Since a code is a vector space, the language of linear algebra can be introduced. For example, the code spanned by $\mathsf{u} = 10011$, $\mathsf{v} = 01001$, $\mathsf{w} = 00111$ is the set of linear combinations of these words with coefficients each 0 or 1. So, here are the words of C.

$0 = 0\mathsf{u} + 0\mathsf{v} + 0\mathsf{w}$	00000
$\mathsf{u} = 1\mathsf{u} + 0\mathsf{v} + 0\mathsf{w}$	10011
$\mathsf{v} = 0\mathsf{u} + 1\mathsf{v} + 0\mathsf{w}$	01001
$\mathsf{w} = 0\mathsf{u} + 0\mathsf{v} + 1\mathsf{w}$	00111
$\mathsf{u} + \mathsf{v} = 1\mathsf{u} + 1\mathsf{v} + 0\mathsf{w}$	11010
$\mathsf{u} + \mathsf{w} = 1\mathsf{u} + 0\mathsf{v} + 1\mathsf{w}$	10100
$\mathsf{v} + \mathsf{w} = 0\mathsf{u} + 1\mathsf{v} + 1\mathsf{w}$	01110
$\mathsf{u} + \mathsf{v} + \mathsf{w} = 1\mathsf{u} + 1\mathsf{v} + 1\mathsf{w}$	11101

Thinking of a code as a vector space, we can also talk about its dimension. The code just specified, for example, has basis $\{\mathsf{u}, \mathsf{v}, \mathsf{w}\}$ and hence dimension 3.

Suppose a code C has dimension k and generator matrix G. Thus G is an $n \times k$ matrix of rank k and $C = \operatorname{col\,sp} G$. The orthogonal complement $C^{\perp}$ of the subspace C is called the *dual code*. Since C is the row space of G^T, $C^{\perp}$ is null sp G^T (by Theorem 6.4.17). To repeat, the dual code $C^{\perp}$ is precisely the null space of G^T. This matrix G^T is called a *parity check* matrix for $C^{\perp}$. Note that $\dim C^{\perp} = n - k$, by Corollary 6.4.18, and G^T is $n - (n - k) \times n$.

Let H be a parity check matrix for C. Thus H is an $(n - k) \times n$ matrix and $C = \text{null sp}\, H$, which is $(\operatorname{col\,sp} H^T)^{\perp}$ by Theorem 6.4.17. So $C^{\perp} = \operatorname{col\,sp} H^T$ showing that H^T is a generator matrix for $C^{\perp}$.

We summarize.

6.4.21 Theorem. *Let C be a code with generator matrix G and parity check matrix H. Then the dual code $C^\perp$ has generator matrix H^T and parity check matrix G^T.*

We return to our example. Our code C, a subspace of F^5, has dimension 3 with basis $\mathsf{u} = 10011$, $\mathsf{v} = 01001$, $\mathsf{w} = 00111$. Thus a generator matrix for C is $G = \begin{bmatrix} 1 & 0 & 0 \\ 0 & 1 & 0 \\ 0 & 0 & 1 \\ 1 & 0 & 1 \\ 1 & 1 & 1 \end{bmatrix}$. This means that a word is in C if and only if it has the form $G\begin{bmatrix} x_1 \\ x_2 \\ x_3 \end{bmatrix} = \begin{bmatrix} x_1 \\ x_2 \\ x_3 \\ x_1 + x_3 \\ x_1 + x_2 + x_3 \end{bmatrix}$. So $\mathsf{x} = x_1x_2x_3x_4x_5$ is a code word if and only if $x_4 = x_1 + x_3$ and $x_5 = x_1 + x_2 + x_3$. Remember that arithmetic is performed modulo 2, so that $a + a = (1+1)a = 0$, that is, $-a = a$. Thus x is a code word if and only if

$$\begin{aligned} x_1 + x_3 + x_4 &= 0 \\ x_1 + x_2 + x_3 + x_5 &= 0, \end{aligned} \tag{2}$$

which is $H\mathsf{x} = 0$ with $H = \begin{bmatrix} 1 & 0 & 1 & 1 & 0 \\ 1 & 1 & 1 & 0 & 1 \end{bmatrix}$. The equations in (2) are called *parity check equations* and H is the *parity check matrix*. There are two ways to determine whether or not a word $\mathsf{x} = x_1x_2x_3x_4x_5$ is a code word: Is it of the form $G\mathsf{y}$? Is $H\mathsf{x} = 0$?

As we have shown, G^T is a parity check matrix for the dual code $C^\perp$; that is, $C^\perp$ is the null space of $G^T = \begin{bmatrix} 1 & 0 & 0 & 1 & 1 \\ 0 & 1 & 0 & 0 & 1 \\ 0 & 0 & 1 & 1 & 1 \end{bmatrix}$. Solving $G^T\mathsf{x} = 0$ with $\mathsf{x} = \begin{bmatrix} x_1 \\ x_2 \\ x_3 \\ x_4 \\ x_5 \end{bmatrix}$, we have x_4 and x_5 free,

$$\begin{aligned} x_3 &= x_4 + x_5 \\ x_2 &= x_5 \\ x_1 &= x_4 + x_5. \end{aligned}$$

So x is in $C^\perp$ if and only if

$$\mathsf{x} = \begin{bmatrix} x_4 + x_5 \\ x_5 \\ x_4 + x_5 \\ x_4 \\ x_5 \end{bmatrix} = x_4 \begin{bmatrix} 1 \\ 0 \\ 1 \\ 1 \\ 0 \end{bmatrix} + x_5 \begin{bmatrix} 1 \\ 1 \\ 1 \\ 0 \\ 1 \end{bmatrix} = H^T\mathsf{x},$$

confirming that H^T is a generator matrix for $C^\perp$.

Direct Sums

We conclude this section with a theorem that summarizes most of the important properties of orthogonal complements. One part requires familiarity with the concept of *direct sum*.

6.4.22 Definition. The sum $V = U + W$ of subspaces is *direct*, and we write $V = U \oplus W$, if every vector v in V can be written uniquely in the form $\mathsf{v} = \mathsf{u} + \mathsf{w}$, with $\mathsf{u} \in U$ and $\mathsf{w} \in W$; that is, if $\mathsf{v} = \mathsf{u}_1 + \mathsf{w}_1 = \mathsf{u}_2 + \mathsf{w}_2$ with u_1, u_2 in U and $\mathsf{w}_1, \mathsf{w}_2$ in W, then $\mathsf{u}_1 = \mathsf{u}_2$ and $\mathsf{w}_1 = \mathsf{w}_2$.

6.4.23 Lemma. *The sum $U + W$ of subspaces is direct if and only if the intersection $U \cap W = \{0\}$, the zero vector.*

Proof. ($\Longleftarrow$) Suppose that $U \cap W = \{0\}$. We have to show that if $\mathsf{u}_1 + \mathsf{w}_1 = \mathsf{u}_2 + \mathsf{w}_2$ with $\mathsf{u}_1, \mathsf{u}_2$ in U and $\mathsf{w}_1, \mathsf{w}_2$ in W, then $\mathsf{u}_1 = \mathsf{u}_2$ and $\mathsf{w}_1 = \mathsf{w}_2$. Now the equation $\mathsf{u}_1 + \mathsf{w}_1 = \mathsf{u}_2 + \mathsf{w}_2$ implies

$$\mathsf{u}_1 - \mathsf{u}_2 = \mathsf{w}_2 - \mathsf{w}_1.$$

The vector $\mathsf{u}_1 - \mathsf{u}_2$ is in U (because U is a subspace), and the vector $\mathsf{w}_2 - \mathsf{w}_1$ is in W (because W is a subspace). These vectors are equal, so we have a vector in U and in W. Since $U \cap W = \{0\}$, it must be that $\mathsf{u}_1 - \mathsf{u}_2 = \mathsf{w}_1 - \mathsf{w}_2 = 0$, so $\mathsf{u}_1 = \mathsf{u}_2$ and $\mathsf{w}_1 = \mathsf{w}_2$.

($\Longrightarrow$) Conversely, suppose $U + W$ is direct. We must show that $U \cap W = 0$. So we take a vector x that is in both U and W and show that $\mathsf{x} = 0$. The vector x is in both U and W. The zero vector is also in both U and W. This gives two ways to write x as the sum of a vector in U and a vector in W, namely, $\mathsf{x} = \mathsf{x} + 0 = 0 + \mathsf{x}$. By uniqueness, $\mathsf{x} = 0$. ■

6.4.24 Example. If $U = \left\{\begin{bmatrix} c \\ 2c \end{bmatrix} \mid c \in \mathsf{R}\right\}$ and $W = \left\{\begin{bmatrix} 0 \\ c \end{bmatrix} \mid c \in \mathsf{R}\right\}$ are as in Example 6.4.2, then $\mathsf{R}^2 = U \oplus W$ is a direct sum since we showed previously that $\mathsf{R}^2 = U + W$ and $U \cap W = \{0\}$. ☺

6.4.25 Example. If U and W are the subspaces of R^3 defined in Example 6.4.3, then $U + W$ is not direct, since the intersection of U and W is a line, not 0. ☺

6.4.26 Theorem. *Let U be a subspace of a (finite dimensional Euclidean) vector space V.*

1. $U \cap U^\perp = \{0\}$; *that is, if a vector x belongs to both U and $U^\perp$, then* $\mathsf{x} = 0$.

2. $V = U \oplus U^\perp$; *that is, every* $\mathsf{v} \in V$ *can be written uniquely in the form* $\mathsf{v} = \mathsf{u} + \mathsf{w}$ *with* $\mathsf{u} \in U$ *and* $\mathsf{w} \in U^\perp$.

3. $\dim V = \dim U + \dim U^\perp$.

Proof. 1. Let u be a vector which is in U and also in $U^\perp$. Since u is in $U^\perp$, it is orthogonal to every vector in U, in particular, to itself. Thus, $0 = \mathsf{u}\cdot\mathsf{u} = \|\mathsf{u}\|^2$, so $\mathsf{u} = 0$.

2. We have to show that any vector in V can be written uniquely as the sum $\mathsf{u}+\mathsf{w}$ of a vector u in U and a vector w in $U^\perp$. Uniqueness follows immediately from part 1 and Lemma 6.4.23, so we only have to show that for any v in V, there exist vectors u in U and w in U^T with $\mathsf{v} = \mathsf{u} + \mathsf{w}$.

So let v be in V and let $\mathsf{u} = \operatorname{proj}_U \mathsf{v}$ be the projection of v on U—see Theorem 6.1.16. By definition of projection, $\mathsf{v} - \mathsf{u} = \mathsf{v} - \operatorname{proj}_U \mathsf{v}$ is orthogonal to every vector in U; that is, $\mathsf{w} = \mathsf{v} - \mathsf{u}$ is in $U^\perp$. The obvious equation $\mathsf{v} = \mathsf{u} + (\mathsf{v} - \mathsf{u})$ expresses v as the sum of a vector in U and a vector in $U^\perp$.

3. Since $V = U + U^\perp$, $\dim V = \dim(U + U^\perp) = \dim U + \dim U^\perp - \dim(U \cap U^\perp)$ by Theorem 6.4.5. Since $U \cap U^\perp = \{0\}$, we obtain $\dim V = \dim U + \dim U^\perp$. (See Exercise 33 for a different proof.) ■

The most important part of Theorem 6.4.26 is part 2. If U is any subspace of a Euclidean space V, then V is the direct sum of U and its orthogonal complement: every v in V can be written uniquely as the sum of a vector in U and a vector in $\mathsf{U}^\perp$.

6.4.27 Example. If $V = \mathsf{R}^3$ and $U = \{\begin{bmatrix} 0 \\ 0 \\ z \end{bmatrix} \mid z \in \mathsf{R}\}$ is the z-axis, then $U^\perp = \{\begin{bmatrix} x \\ y \\ 0 \end{bmatrix} \mid x, y \in \mathsf{R}\}$ is the xy-plane and the assertion $V = U \oplus U^\perp$ simply expresses the fact that every vector $\begin{bmatrix} x \\ y \\ z \end{bmatrix}$ in R^3 can be written uniquely $\begin{bmatrix} 0 \\ 0 \\ z \end{bmatrix} + \begin{bmatrix} x \\ y \\ 0 \end{bmatrix}$ as the sum of a vector on the z-axis and a vector in the xy-plane. ☺

6.4.28 Problem. Let $V = \mathsf{R}^3$ and let $U = \left\{\mathsf{x} = \begin{bmatrix} x \\ y \\ z \end{bmatrix} \in \mathsf{R}^3 \mid 2x - y + z = 0\right\}$. Write $\mathsf{v} = \begin{bmatrix} x \\ y \\ z \end{bmatrix} \in V$ in the form $\mathsf{v} = \mathsf{u} + \mathsf{w}$ with $\mathsf{u} \in U$ and $\mathsf{w} \in U^\perp$.

Solution. Exactly as in the proof of the theorem, we take $\mathsf{u} = \operatorname{proj}_U \mathsf{v}$ and $\mathsf{w} = \mathsf{v} - \mathsf{u}$. An orthogonal basis for U is $\{\mathsf{f}_1, \mathsf{f}_2\}$ with $\mathsf{f}_1 = \begin{bmatrix} 0 \\ 1 \\ 1 \end{bmatrix}$ and $\mathsf{f}_2 = \begin{bmatrix} 1 \\ 1 \\ -1 \end{bmatrix}$.

So $\mathsf{u} = \operatorname{proj}_U \mathsf{v} = \dfrac{\mathsf{v}\cdot\mathsf{f}_1}{\mathsf{f}_1\cdot\mathsf{f}_1}\mathsf{f}_1 + \dfrac{\mathsf{v}\cdot\mathsf{f}_2}{\mathsf{f}_2\cdot\mathsf{f}_2}\mathsf{f}_2$

$$= \frac{y+z}{2}\begin{bmatrix}0\\1\\1\end{bmatrix} + \frac{x+y-z}{3}\begin{bmatrix}1\\1\\-1\end{bmatrix} = \begin{bmatrix}\frac{x+y-z}{3}\\ \frac{2x+5y+z}{6}\\ \frac{-2x+y+5z}{6}\end{bmatrix}.$$

Note that u is in U since $2(\frac{x+y-z}{3}) - \frac{2x+5y+z}{6} + \frac{-2x+y+5z}{6} = 0$. Finally,

$$\mathsf{w} = \mathsf{v} - \mathsf{u} = \begin{bmatrix}\frac{2x-y+z}{3}\\ \frac{-2x+y-z}{6}\\ \frac{2x-y+z}{6}\end{bmatrix} = \frac{2x-y+z}{6}\begin{bmatrix}2\\-1\\1\end{bmatrix}.$$

READING CHECK 6. The astute reader will not be surprised that the vector w in Problem 6.4.28 is a multiple of $\begin{bmatrix}2\\-1\\1\end{bmatrix}$. Why?

Answers to Reading Checks

1. The two normals, $\begin{bmatrix}1\\1\\-1\end{bmatrix}$ and $\begin{bmatrix}2\\-3\\1\end{bmatrix}$, are not parallel.

2. Let $A = \begin{bmatrix}\mathsf{w}_1 & \mathsf{w}_2 & \mathsf{w}_3\\ \downarrow & \downarrow & \downarrow\end{bmatrix} = \begin{bmatrix}0&1&3\\3&0&6\\-1&1&1\end{bmatrix}$. Row echelon form is $\begin{bmatrix}1&0&2\\0&1&3\\0&0&0\end{bmatrix}$. The pivot columns of A are columns one and two, so the first two columns of A, namely, w_1 and w_2, are a basis for col sp $A = W$.

3. We have seen that $\dim(U+W) = 3$. Since $U + W$ is a subspace of R^3, we have $U+W = \mathsf{R}^3$. One basis is $\{\mathsf{i},\mathsf{j},\mathsf{k}\}$. Another is $\{\mathsf{u}_1,\mathsf{u}_2,\mathsf{w}_1\}$, the set of pivot columns of A.

4. Let $\mathsf{u} \in U$ and let $\mathsf{x} \in U^\perp$. Then $\mathsf{u}\cdot\mathsf{x} = 0$, so u is orthogonal to $U^\perp$.

5. The subspace U is a plane in R^3, so its orthogonal complement is the one-dimensional subspace with basis the cross product $\mathsf{u}_1 \times \mathsf{u}_2$, which is $-5\begin{bmatrix}-8\\2\\1\end{bmatrix}$.

6. w is supposed to be orthogonal to U, which is a plane with normal $\begin{bmatrix}2\\-1\\1\end{bmatrix}$.

True/False Questions

Decide, with as little calculation as possible, whether each of the following statements is true or false and, if you say "false," explain your answer. (Answers are at the back of the book.)

1. In R^2, the sum of $U = \{\begin{bmatrix} x \\ x \end{bmatrix} \mid x \in \mathsf{R}\}$ and $W = \{\begin{bmatrix} x \\ 2x \end{bmatrix} \mid x \in \mathsf{R}\}$ is the xy-plane.
2. In R^3, the intersection of two (different) two-dimensional subspaces is a one-dimensional subspace.
3. If U and W are 3-dimensional subspaces of R^5, then $U \cap W \neq \{0\}$.
4. If U and W are subspaces of a vector space V and U is contained in $W^\perp$, then W is contained in $U^\perp$.
5. The sum of the xy- and yz-planes in R^3 is direct.
6. If U and W are orthogonal subspaces of a vector space V, then $\dim U + \dim W = \dim V$.
7. The orthogonal complement of the null space of a matrix is its row space.
8. If U is a subspace of a finite dimensional vector space V, there exists a subspace W such that $V = U + W$ and $U \cap W = \{0\}$.
9. If A is a symmetric matrix with column space U and $A\mathsf{x} = 0$, then x is in $U^\perp$.
10. If a code C has an $r \times s$ generator matrix, the dimension of C is r.

EXERCISES

Answers to exercises marked [BB] can be found at the Back of the Book.

1. In each of the following cases,
   - find a basis and the dimension of $U + W$,
   - find a basis and the dimension of $U \cap W$,
   - verify that $\dim(U + W) + \dim(U \cap W) = \dim U + \dim W$.

   (a) [BB] $V = \mathsf{R}^4$, $U = \left\{\begin{bmatrix} x-y \\ x+y \\ 2x-y \\ y \end{bmatrix} \mid x, y \in \mathsf{R}\right\}$; W is the subspace of R^4 with basis $\left\{\begin{bmatrix} -1 \\ 5 \\ 0 \\ 3 \end{bmatrix}, \begin{bmatrix} 0 \\ 0 \\ 1 \\ 0 \end{bmatrix}\right\}$.

(b) $V = \mathsf{R}^4$, $U = \left\{\begin{bmatrix} -2x+4y-3z \\ -x+y-z \\ x-7y+4z \\ x-5y+3z \end{bmatrix} \mid x,y \in \mathsf{R}\right\}$; W is the subspace of R^4 spanned by $\begin{bmatrix} 0 \\ 1 \\ 1 \\ 1 \end{bmatrix}$ and $\begin{bmatrix} 5 \\ 2 \\ 3 \\ 0 \end{bmatrix}$.

(c) [BB] $V = \mathsf{R}^5$, U has basis $\mathsf{u}_1 = \begin{bmatrix} -2 \\ 0 \\ 2 \\ 4 \\ 1 \end{bmatrix}$, $\mathsf{u}_2 = \begin{bmatrix} 3 \\ 5 \\ 2 \\ 4 \\ 1 \end{bmatrix}$, $\mathsf{u}_3 = \begin{bmatrix} 6 \\ 3 \\ 3 \\ -6 \\ -1 \end{bmatrix}$, and W has basis $\mathsf{w}_1 = \begin{bmatrix} 1 \\ 2 \\ -5 \\ 2 \\ 0 \end{bmatrix}$ and $\mathsf{w}_2 = \begin{bmatrix} 3 \\ 1 \\ 10 \\ -4 \\ 0 \end{bmatrix}$.

(d) $V = \mathsf{R}^4$, U has basis $\mathsf{u}_1 = \begin{bmatrix} 1 \\ 3 \\ 1 \\ 1 \end{bmatrix}$, $\mathsf{u}_2 = \begin{bmatrix} 1 \\ 5 \\ -1 \\ 2 \end{bmatrix}$ and W is spanned by $\mathsf{w}_1 = \begin{bmatrix} 2 \\ 2 \\ 3 \\ 1 \end{bmatrix}$, $\mathsf{w}_2 = \begin{bmatrix} 1 \\ 1 \\ 0 \\ 1 \end{bmatrix}$, $\mathsf{w}_3 = \begin{bmatrix} 0 \\ 6 \\ -3 \\ 2 \end{bmatrix}$, $\mathsf{w}_4 = \begin{bmatrix} 1 \\ -1 \\ 2 \\ 0 \end{bmatrix}$.

(e) $V = \mathsf{R}^5$, U is spanned by $\mathsf{u}_1 = \begin{bmatrix} 1 \\ 1 \\ 1 \\ 1 \\ 3 \end{bmatrix}$ and $\mathsf{u}_2 = \begin{bmatrix} 0 \\ -1 \\ 2 \\ 1 \\ 2 \end{bmatrix}$ and W is spanned by $\mathsf{w}_1 = \begin{bmatrix} 2 \\ 5 \\ -4 \\ -1 \\ 0 \end{bmatrix}$, $\mathsf{w}_2 = \begin{bmatrix} 0 \\ -2 \\ 6 \\ 3 \\ 6 \end{bmatrix}$, $\mathsf{w}_3 = \begin{bmatrix} -1 \\ 1 \\ 1 \\ 0 \\ -1 \end{bmatrix}$, $\mathsf{w}_4 = \begin{bmatrix} 1 \\ 0 \\ 3 \\ 2 \\ 5 \end{bmatrix}$.

2. In each case, you are given a vector space V and a subspace U. Find $U^\perp$.

(a) [BB] $V = \mathsf{R}^3$ and $U = \left\{c\begin{bmatrix} 2 \\ -3 \\ -7 \end{bmatrix} \mid c \in \mathsf{R}\right\}$

(b) $V = \mathsf{R}^3$ and $U = \left\{c\begin{bmatrix} 2 \\ -6 \\ 7 \end{bmatrix} \mid c \in \mathsf{R}\right\}$

(c) [BB] $V = \mathsf{R}^3$, U is the plane with equation $x - 2y + z = 0$

(d) $V = \mathsf{R}^3$, U is the plane with equation $2x + y - 3z = 0$

(e) [BB] $V = \mathsf{R}^3$, $\mathsf{u}_1 = \begin{bmatrix} 1 \\ 1 \\ -3 \end{bmatrix}$, $\mathsf{u}_2 = \begin{bmatrix} 1 \\ 2 \\ -1 \end{bmatrix}$ and $U = \operatorname{sp}\{\mathsf{u}_1, \mathsf{u}_2\}$

(f) $V = \mathsf{R}^3$, $\mathsf{u}_1 = \begin{bmatrix} -1 \\ 2 \\ 2 \end{bmatrix}$, $\mathsf{u}_2 = \begin{bmatrix} 4 \\ 0 \\ 1 \end{bmatrix}$, $U = \operatorname{sp}\{\mathsf{u}_1, \mathsf{u}_2\}$

(g) $V = \mathsf{R}^4$, $\mathsf{u}_1 = \begin{bmatrix} 2 \\ 0 \\ -5 \\ 6 \end{bmatrix}$, $\mathsf{u}_2 = \begin{bmatrix} 1 \\ 0 \\ -3 \\ 2 \end{bmatrix}$, $U = \operatorname{sp}\{\mathsf{u}_1, \mathsf{u}_2\}$

(h) $V = \mathsf{R}^5$, $\mathsf{u}_1 = \begin{bmatrix} -2 \\ 0 \\ 1 \\ 0 \\ 2 \end{bmatrix}$, $\mathsf{u}_2 = \begin{bmatrix} 5 \\ 4 \\ -4 \\ 0 \\ -1 \end{bmatrix}$, $\mathsf{u}_3 = \begin{bmatrix} 1 \\ 4 \\ -2 \\ 0 \\ 3 \end{bmatrix}$, $U = \operatorname{sp}\{\mathsf{u}_1, \mathsf{u}_2, \mathsf{u}_3\}$

(i) $V = \mathsf{R}^4$, $U = \left\{ \begin{bmatrix} x-y \\ x+y \\ 2x-y \\ y \end{bmatrix} \mid x, y \in \mathsf{R} \right\}$.

3. In each case, an orthogonal basis is given for a subspace U. Write the given vector $\mathsf{v} = \mathsf{u} + \mathsf{w}$ as the sum of a vector u in U and a vector w in $U^\perp$.

(a) [BB] $U = \operatorname{sp}\left\{ \begin{bmatrix} 1 \\ 1 \\ 1 \\ 1 \end{bmatrix}, \begin{bmatrix} 1 \\ 1 \\ -1 \\ -1 \end{bmatrix}, \begin{bmatrix} 1 \\ -1 \\ 1 \\ -1 \end{bmatrix} \right\}$; $\mathsf{v} = \begin{bmatrix} 2 \\ 0 \\ 1 \\ 6 \end{bmatrix}$

(b) $U = \operatorname{sp}\left\{ \begin{bmatrix} 1 \\ 0 \\ 0 \\ 1 \end{bmatrix}, \begin{bmatrix} -1 \\ 0 \\ 1 \\ 1 \end{bmatrix}, \begin{bmatrix} 1 \\ 0 \\ 2 \\ -1 \end{bmatrix} \right\}$; $\mathsf{v} = \begin{bmatrix} a \\ b \\ c \\ d \end{bmatrix}$.

4. Write $\mathsf{v} = \mathsf{u} + \mathsf{w}$ with u in U and w in $U^\perp$ in each of the following situations.

(a) [BB] $V = \mathsf{R}^2$, $\mathsf{v} = \begin{bmatrix} x \\ y \end{bmatrix}$, $U = \left\{ \begin{bmatrix} t \\ t \end{bmatrix} \mid t \in \mathsf{R} \right\}$

(b) $V = \mathsf{R}^3$, $\mathsf{v} = \begin{bmatrix} x \\ y \\ z \end{bmatrix}$, U is the span of $\begin{bmatrix} -2 \\ 1 \\ 0 \end{bmatrix}$

(c) [BB] $V = \mathsf{R}^3$, $\mathsf{v} = \begin{bmatrix} 3 \\ 1 \\ -1 \end{bmatrix}$, $U = \left\{ \begin{bmatrix} x \\ y \\ z \end{bmatrix} \mid 2x - y + 3z = 0 \right\}$

(d) $V = \mathsf{R}^3$, $\mathsf{v} = \begin{bmatrix} x \\ y \\ z \end{bmatrix}$, $U = \left\{ \begin{bmatrix} x \\ y \\ z \end{bmatrix} \mid 2x - y + 3z = 0 \right\}$

(e) $V = \mathsf{R}^4$, $\mathsf{v} = \begin{bmatrix} x_1 \\ x_2 \\ x_3 \\ x_4 \end{bmatrix}$, $U = \left\{ \begin{bmatrix} 2t-s \\ s+t \\ s-t \\ t \end{bmatrix} \mid t, s \in \mathsf{R} \right\}$.

5. [BB] Let U be the line in R^3 with equation $\begin{bmatrix} x \\ y \\ z \end{bmatrix} = t \begin{bmatrix} -2 \\ 5 \\ 1 \end{bmatrix}$.

(a) Find the matrix that projects vectors onto U.

(b) What is $U^\perp$?

(c) Find the matrix that projects vectors onto $U^\perp$.

(d) Find the vector w in $U^\perp$ that minimizes $\|\mathsf{w}-\mathsf{b}\|$ with $\mathsf{b}=\begin{bmatrix}-1\\2\\3\end{bmatrix}$.

6. (a) Find the matrix P that projects vectors in R^4 onto the orthogonal complement of the subspace U with equation $x_1+x_2-x_3+x_4=0$.

(b) Find the vector w in $U^\perp$ that minimizes $\|\mathsf{w}-\mathsf{b}\|$, with $\mathsf{b}=\begin{bmatrix}1\\1\\1\\1\end{bmatrix}$.

7. [BB] Let U be a subspace of R^n and v a vector in R^n. Explain why $\operatorname{proj}_U \mathsf{v}$ is unique; that is, if $\mathsf{v}=\mathsf{u}+\mathsf{w}$ with $\mathsf{u}\in U$ and $\mathsf{w}\in U^\perp$, then $\mathsf{u}=\operatorname{proj}_U\mathsf{v}$.

8. Suppose U and W are subspaces of R^n. Prove that $U+W$ [BB] and $U\cap W$ are also subspaces.

9. [BB] Let $\mathsf{u}_1,\mathsf{u}_2,\ldots,\mathsf{u}_k$ be vectors in R^n and let $U=\operatorname{sp}\{\mathsf{u}_1,\ldots,\mathsf{u}_k\}$. Show that $\mathsf{x}\in U^\perp$ if and only if $\mathsf{x}\cdot\mathsf{u}_i=0$ for all i.

10. If V is a vector space and U is a subspace of V, show that $U^\perp$ is a subspace of V.

11. [BB] Let U be the subspace of R^4 consisting of all vectors of the form $\begin{bmatrix}2s-t\\s\\t\\-5s+2t\end{bmatrix}$.

(a) Find a basis for U and a basis for $U^\perp$.

(b) Find the vector u in U that minimizes $\|\mathsf{u}-\mathsf{b}\|$, $\mathsf{b}=\begin{bmatrix}-1\\0\\0\\1\end{bmatrix}$, and a vector w in $U^\perp$ that minimizes $\|\mathsf{w}-\mathsf{b}\|$.

12. Let U be the subspace of R^4 that consists of all vectors of the form $\begin{bmatrix}2s\\-t\\4s-2t\\2t-2s\end{bmatrix}$.

(a) Find a basis for U and a basis for $U^\perp$.

(b) Find the vector u in U that minimizes $\|\mathsf{u}-\mathsf{b}\|$, $\mathsf{b}=\begin{bmatrix}1\\-1\\1\\-1\end{bmatrix}$ and the vector w in $U^\perp$ that minimizes $\|\mathsf{w}-\mathsf{b}\|$.

13. Let U be the subspace of R^4 that is the intersection of the planes with equations $x_1-x_3+2x_4=0$ and $2x_1+3x_2+x_3+x_4=0$.

(a) Find an orthogonal basis for U.

(b) Find a spanning set for $U^\perp$.

(c) Find the matrix P that projects vectors of R^4 onto U.

(d) Find the vector u in U that minimizes $\|\mathsf{u}-\mathsf{b}\|$, $\mathsf{b}=\begin{bmatrix}1\\2\\-1\\1\end{bmatrix}$.

14. Suppose U and W are subspaces of a finite dimensional vector space V.

(a) If $U \subseteq W$, show that $W^\perp \subseteq U^\perp$.

(b) [BB] Prove that the following three statements are equivalent:

i. U and W are orthogonal;

ii. $U \subseteq W^\perp$;

iii. $W \subseteq U^\perp$.

15. Let $F=\{0,1\}$ with addition and multiplication modulo 2 and let $V=F^3$. Let $\mathsf{u}=\begin{bmatrix}1\\1\\0\end{bmatrix}$ and $\mathsf{v}=\begin{bmatrix}1\\1\\1\end{bmatrix}$ and $U=\operatorname{sp}\{\mathsf{u},\mathsf{v}\}$. List the words in the dual code $C^\perp$. [This example shows that when the scalars are just 0 and 1, unlike the case of the reals, $U \cap U^\perp$ need not be 0.]

16. Let C be a code with generator matrix $G=\begin{bmatrix}1&0\\0&1\\1&0\\1&1\end{bmatrix}$. List the words of C and the words of the dual code $C^\perp$.

17. [BB] Let C be the code consisting of all linear combinations of 100011, 010101 and 001110.

(a) Find a generator matrix and a parity check matrix for C.

(b) Find a generator matrix and a parity check matrix for $C^\perp$.

(c) Is 110110 in C? Is this word in $C^\perp$?

18. Let $F=\{0,1\}$ with addition and multiplication modulo 2 and let C be the subspace of F^7 with basis 1000011, 0100101, 0010010 and 0001101.

(a) Find a generator matrix and a parity check matrix for C.

(b) Find a generator matrix and a parity check matrix for $C^\perp$.

(c) Is 1010010 in C? Is this word in $C^\perp$?

19. Let U and W be subspaces of R^n. Let $\{\mathsf{u}_1,\mathsf{u}_2,\ldots,\mathsf{u}_k\}$ be a basis for U and let $\{\mathsf{w}_1,\mathsf{w}_2,\ldots,\mathsf{w}_\ell\}$ be a basis for W. Let V be the set of vectors of the form $\begin{bmatrix}x_1\\\vdots\\x_k\\y_1\\\vdots\\y_\ell\end{bmatrix}$ that satisfy $x_1\mathsf{u}_1+\cdots+x_k\mathsf{u}_k=-y_1\mathsf{w}_1-\cdots-y_\ell\mathsf{w}_\ell$.

(a) [BB] Show that the function $T\colon V \to U \cap W$ defined by

$$T\left(\begin{bmatrix} x_1 \\ \vdots \\ x_k \\ y_1 \\ \vdots \\ y_\ell \end{bmatrix}\right) = x_1\mathsf{u}_1 + \cdots + x_k\mathsf{u}_k$$

is a linear transformation; that is, $T(\mathsf{x}+\mathsf{x}') = T(\mathsf{x}) + T(\mathsf{x}')$ for any vectors x, x' in V and $T(c\mathsf{x}) = cT(\mathsf{x})$ for any x in V and any scalar c.

(b) [BB] Show that T is *one-to-one*; that is, if $T(\mathsf{x}) = T(\mathsf{x}')$, then $\mathsf{x} = \mathsf{x}'$.

(c) [BB] Show that T is *onto*; that is, given a vector y in $U \cap W$, there exists an x in V such that $T(\mathsf{x}) = \mathsf{y}$.

(d) Let $\{\mathsf{v}_1, \mathsf{v}_2, \ldots, \mathsf{v}_t\}$ be a basis for V. Show that $\{T(\mathsf{v}_1), T(\mathsf{v}_2), \ldots, T(\mathsf{v}_t)\}$ is a basis for $U \cap W$. [Hence $\dim V = \dim(U \cap W)$.]

20. Explain why the orthogonal complement of the column space of a matrix A is the null space of A^T; symbolically, $(\text{col sp}\, A)^\perp = \text{null sp}\, A^T$.

Critical Reading

21. [BB] Suppose U and W are subspaces of some Euclidean space V with $U \cap W = 0$. Under what condition(s) will $U + W = V$?

22. True or false: If U and W are subspaces of V and $U \subseteq W$, then $U^\perp \subseteq W^\perp$.

23. Suppose U, W and S are subspaces of a Euclidean space V such that W is contained in S and $U \cap S = \{0\}$.

(a) Prove that $(U + W) \cap S = W$.

(b) Find $\dim(U + W + S)$.

24. [BB] If A is an $m \times n$ matrix with linearly independent columns and U is the column space of A^TA, find $U^\perp$.

25. Suppose an $n \times n$ matrix A satisfies $A^2 = A$. Show that $\mathsf{R}^n = \text{null sp}\, A \oplus \text{col sp}\, A$. [Hint: $\mathsf{x} = (\mathsf{x} - A\mathsf{x}) + A\mathsf{x}$.]

26. Let P be a projection matrix.

(a) Show that the eigenvalues of P are $\lambda = 0$ and $\lambda = 1$.

(b) Show that the eigenspaces of P are orthogonal complements.

27. Let A and B be matrices of the same size.

(a) Show that $\text{col sp}(A + B)$ is contained in $\text{col sp}\, A + \text{col sp}\, B$.

(b) Show that $\text{rank}(A + B) \le \text{rank}\, A + \text{rank}\, B$.

28. Suppose $\{f_1, f_2, f_3, f_4\}$ is an orthogonal basis for $V = R^4$. Let U be the subspace spanned by f_1 and f_2. Identify $U^\perp$ and explain your answer.

29. Is it possible for a 3×3 matrix to have row space spanned by i and $\begin{bmatrix} 1 \\ -2 \\ 1 \end{bmatrix}$ and $v = \begin{bmatrix} 1 \\ 1 \\ 1 \end{bmatrix}$ in the null space? Explain.

30. [BB] Let U and W be subspaces of a vector space V.
Prove that $(U + W)^\perp = U^\perp \cap W^\perp$.

31. Have another look at Problem 6.4.28. The vector w in $U^\perp$ is $\frac{2x-y+z}{6}\begin{bmatrix} 2 \\ -1 \\ 1 \end{bmatrix}$. It's not surprising that this is a multiple of $\begin{bmatrix} 2 \\ -1 \\ 1 \end{bmatrix}$ since this vector is normal to the plane U, but the coefficient contains the expression $2x - y + z$, which comes from the equation of the plane. Could this have been predicted? Do Problem 6.4.28 again, but this time computing w first, and then obtaining u as $v - w$. [Hint: Write $w = \alpha\begin{bmatrix} 2 \\ -1 \\ 1 \end{bmatrix}$ and find α.]

32. **(a)** Let V be a vector space. Suppose the vectors $u_1, u_2, \dots, u_k$ are a basis for a subspace U and the vectors $w_1, w_2, \dots, w_\ell$ are a basis for $U^\perp$. Show that $u_1, \dots, u_k, w_1, \dots, w_\ell$ are a basis for V.

(b) Use part (a) to show how to write a vector v in V in the form $v = u + w$ with u in U and w in $U^\perp$.

(c) Let $V = R^4$, $U = \{\begin{bmatrix} y \\ x + 2y \\ x \\ x + y \end{bmatrix} \mid x, y \in R\}$ and $v = \begin{bmatrix} 1 \\ 1 \\ 1 \\ 1 \end{bmatrix}$. Apply the result of part (b) to write v as the sum of a vector in U and a vector in $U^\perp$.

33. According to Theorem 6.4.26, if V is a finite dimensional vector space, say $\dim V = n$, and U is a subspace of V, then $\dim U + \dim U^\perp = n$. Give a proof of this result different from the one presented, which follows these lines. Let $\mathcal{U} = \{u_1, \dots, u_k\}$ be a basis for U and extend $\mathcal{U}$ to a basis $\mathcal{V}$ of V. Now apply the Gram-Schmidt process to $\mathcal{V}$, thus obtaining an orthogonal basis $\mathcal{F} = \{f_1, \dots, f_n\}$ of V.

(a) The vectors $f_1, \dots, f_k$ form a basis of U. Why?

(b) It suffices to show that $f_{k+1}, \dots, f_n$ form a basis for $U^\perp$. Why?

(c) The vectors $f_{k+1}, \dots, f_n$ are linearly independent. Why?

(d) The vectors $f_{k+1}, \dots, f_n$ span $U^\perp$. Why?

(e) The result follows. Why?

34. Let U be a subspace of $V = R^n$. Use the fact that the sum of the rank and nullity of an $m \times n$ matrix is n to give another proof that $\dim U + \dim U^\perp = \dim V$.

35. Let $V = \mathsf{R}^n$. Use the result $\dim U + \dim U^\perp = n$ (Theorem 6.4.26) to give another proof that the sum of the rank and nullity of an $m \times n$ matrix is n.

36. Theorem 6.4.26 presents some of the very few results in this book that depend on the fact that scalars are real numbers.

By contrast, let $F = \{0, 1\}$ denote the field of two elements and let $V = F^3$ with the usual dot product. Let $\mathsf{u} = \begin{bmatrix} 1 \\ 1 \\ 0 \end{bmatrix}$ and $U = \{\mathsf{0}, \mathsf{u}\}$.

(a) Prove that U is a subspace of V.

(b) List the vectors of $U^\perp$. Is $U \cap U^\perp = \{\mathsf{0}\}$?

(c) Give two reasons why $V \neq U \oplus U^\perp$.

6.5 The Pseudoinverse of a Matrix

In this section, we continue the discussion of "How to solve a linear system that doesn't have a solution!" that we started in Section 6.1.

Suppose we are given the linear system $A\mathsf{x} = \mathsf{b}$ with A an $m \times n$ matrix. If A has linearly independent columns, the $n \times n$ matrix A^TA is invertible and the vector $\mathsf{x}^+ = (A^TA)^{-1}A^T\mathsf{b}$ is the "best solution" to $A\mathsf{x} = \mathsf{b}$ in the sense that $\|\mathsf{b} - A\mathsf{x}^+\| \leq \|\mathsf{b} - A\mathsf{x}\|$ for any $\mathsf{x} \in \mathsf{R}^n$. In this case, x^+ is the (real) solution to $A\mathsf{x} = P\mathsf{b}$, where P is the projection matrix onto the column space of A, and this solution is unique. (See Figure 6.1.)

When the columns of A are not linearly independent, however, $A\mathsf{x} = P\mathsf{b}$ will still have a solution (since $P\mathsf{b}$ is in the column space of A), but no longer a unique one. In fact, when the columns of A are linearly dependent, there is a nonzero vector x_0 satisfying $A\mathsf{x}_0 = \mathsf{0}$, so if x_1 satisfies $A\mathsf{x} = P\mathsf{b}$, so does $\mathsf{x}_1 + \mathsf{x}_0$ because $A(\mathsf{x}_1 + \mathsf{x}_0) = A\mathsf{x}_1 + A\mathsf{x}_0 = P\mathsf{b} + \mathsf{0} = P\mathsf{b}$. Thus, in general, there is more than one vector x with $A\mathsf{x} = P\mathsf{b}$, more than one x with $\|\mathsf{b} - A\mathsf{x}\|$ minimal. How do we define x^+ in this situation?

We need to refine our convention: Amongst all solutions to $A\mathsf{x} = P\mathsf{b}$, let us agree that x^+ will be the shortest one.[6]

[6]The reader might well wonder if there could be several such vectors of minimal length. We shall soon see that this is not the case. There really is a unique shortest vector satisfying $A\mathsf{x} = P\mathsf{b}$.

6.5.1 Example. The column space of $A = \begin{bmatrix} \frac{1}{2} & 0 & 0 & 0 \\ 0 & 5 & 0 & 0 \\ 0 & 0 & 0 & 0 \end{bmatrix}$ is the xy-plane, $P = \begin{bmatrix} 1 & 0 & 0 \\ 0 & 1 & 0 \\ 0 & 0 & 0 \end{bmatrix}$ and $P\mathsf{b} = \begin{bmatrix} b_1 \\ b_2 \\ 0 \end{bmatrix}$ for $\mathsf{b} = \begin{bmatrix} b_1 \\ b_2 \\ b_3 \end{bmatrix}$. With $\mathsf{x} = \begin{bmatrix} x_1 \\ x_2 \\ x_3 \\ x_4 \end{bmatrix}$, the equation $A\mathsf{x} = P\mathsf{b}$ says $\begin{aligned} \tfrac{1}{2}x_1 &= b_1 \\ 5x_2 &= b_2 \\ 0 &= 0 \end{aligned}$, so $x_1 = 2b_1$, $x_2 = \frac{1}{5}b_2$, and x_3 and x_4 are arbitrary.

The solution is $\mathsf{x} = \begin{bmatrix} 2b_1 \\ \frac{1}{5}b_2 \\ x_3 \\ x_4 \end{bmatrix}$, a vector of length $\sqrt{4b_1^2 + \frac{1}{25}b_2^2 + x_3^2 + x_4^2}$. For given b, the shortest x is $\mathsf{x}^+ = \begin{bmatrix} 2b_1 \\ \frac{1}{5}b_2 \\ 0 \\ 0 \end{bmatrix}$. ☺

6.5.2 Example. The column space of $A = \begin{bmatrix} 1 & 1 & 1 \\ 1 & 1 & 1 \end{bmatrix}$ is the set of all scalar multiples of $\mathsf{v} = \begin{bmatrix} 1 \\ 1 \end{bmatrix}$. The matrix that projects $\mathsf{b} = \begin{bmatrix} b_1 \\ b_2 \end{bmatrix}$ on this space is $P = \dfrac{\mathsf{v}\mathsf{v}^T}{\mathsf{v} \cdot \mathsf{v}} = \frac{1}{2}\begin{bmatrix} 1 & 1 \\ 1 & 1 \end{bmatrix}$. (See the subsection "Projections onto Subspaces" in Section 6.1.) With $\mathsf{x} = \begin{bmatrix} x_1 \\ x_2 \\ x_3 \end{bmatrix}$, the equation $A\mathsf{x} = P\mathsf{b}$ gives $x_1 + x_2 + x_3 = \frac{1}{2}(b_1 + b_2)$, so $\mathsf{x} = \begin{bmatrix} \frac{1}{2}(b_1 + b_2) - x_2 - x_3 \\ x_2 \\ x_3 \end{bmatrix}$ and

$$\begin{aligned} \|\mathsf{x}\|^2 &= [\tfrac{1}{2}(b_1 + b_2) - x_2 - x_3]^2 + x_2^2 + x_3^2 \\ &= \tfrac{1}{4}(b_1 + b_2)^2 + 2x_2^2 + 2x_3^2 - (b_1 + b_2)(x_2 + x_3) + 2x_2x_3 = f(x_2, x_3) \end{aligned}$$

is a function of the two variables x_2, x_3. We use calculus to find the values of x_2 and x_3 that minimize this function.[7] The partial derivatives with respect to x_2 and x_3 are

$$\begin{aligned} \frac{\partial f}{\partial x_2} &= 4x_2 - (b_1 + b_2) + 2x_3 \\ \frac{\partial f}{\partial x_3} &= 4x_3 - (b_1 + b_2) + 2x_2. \end{aligned}$$

[7]If your calculus background makes this next part difficult, don't worry. A better method lies just ahead!

Setting these to 0 gives $x_2 = x_3 = \frac{1}{6}(b_1 + b_2)$. The Jacobian

$$\begin{vmatrix} \frac{\partial^2 f}{\partial x_2^2} & \frac{\partial^2 f}{\partial x_2 \partial x_3} \\ \frac{\partial^2 f}{\partial x_3 \partial x_2} & \frac{\partial^2 f}{\partial x_3^2} \end{vmatrix} = \begin{vmatrix} 4 & 2 \\ 2 & 4 \end{vmatrix} = 12$$

is positive, so we have identified a minimum of f, namely,

$$\begin{aligned} x_1 &= \tfrac{1}{2}(b_1 + b_2) - x_2 - x_3 = \tfrac{1}{2}(b_1 + b_2) - \tfrac{2}{6}(b_1 + b_2) = \tfrac{1}{6}(b_1 + b_2) \\ x_2 &= \tfrac{1}{6}(b_1 + b_2) \\ x_3 &= \tfrac{1}{6}(b_1 + b_2). \end{aligned}$$

The vector of shortest length is $\mathsf{x}^+ = \dfrac{1}{6}\begin{bmatrix} b_1 + b_2 \\ b_1 + b_2 \\ b_1 + b_2 \end{bmatrix}$. ☺

6.5.3 Example. The column space of $A = \begin{bmatrix} 1 & 1 & 2 \\ -4 & 1 & -3 \\ 3 & -2 & 1 \end{bmatrix}$ is the plane with equation $x + y + z = 0$ and the matrix that projects vectors onto this plane is $P = \dfrac{1}{3}\begin{bmatrix} 2 & -1 & -1 \\ -1 & 2 & -1 \\ -1 & -1 & 2 \end{bmatrix}$. With $\mathsf{x} = \begin{bmatrix} x_1 \\ x_2 \\ x_3 \end{bmatrix}$ and $\mathsf{b} = \begin{bmatrix} b_1 \\ b_2 \\ b_3 \end{bmatrix}$, the system $A\mathsf{x} = P\mathsf{b}$ is

$$\begin{bmatrix} 1 & 1 & 2 \\ -4 & 1 & -3 \\ 3 & -2 & 1 \end{bmatrix}\begin{bmatrix} x_1 \\ x_2 \\ x_3 \end{bmatrix} = \begin{bmatrix} \frac{2}{3}b_1 - \frac{1}{3}b_2 - \frac{1}{3}b_3 \\ -\frac{1}{3}b_1 + \frac{2}{3}b_2 - \frac{1}{3}b_3 \\ -\frac{1}{3}b_1 - \frac{1}{3}b_2 + \frac{2}{3}b_3 \end{bmatrix}.$$

Applying Gaussian elimination to the augmented matrix,

$$\left[\begin{array}{ccc|c} 1 & 1 & 2 & \frac{2}{3}b_1 - \frac{1}{3}b_2 - \frac{1}{3}b_3 \\ -4 & 1 & -3 & -\frac{1}{3}b_1 + \frac{2}{3}b_2 - \frac{1}{3}b_3 \\ 3 & -2 & 1 & -\frac{1}{3}b_1 - \frac{1}{3}b_2 + \frac{2}{3}b_3 \end{array}\right]$$

$$\to \left[\begin{array}{ccc|c} 1 & 1 & 2 & \frac{2}{3}b_1 - \frac{1}{3}b_2 - \frac{1}{3}b_3 \\ 0 & 5 & 5 & \frac{7}{3}b_1 - \frac{2}{3}b_2 - \frac{5}{3}b_3 \\ 0 & -5 & -5 & -\frac{7}{3}b_1 + \frac{2}{3}b_2 + \frac{5}{3}b_3 \end{array}\right] \to \left[\begin{array}{ccc|c} 1 & 1 & 2 & \frac{2}{3}b_1 - \frac{1}{3}b_2 - \frac{1}{3}b_3 \\ 0 & 1 & 1 & \frac{7}{15}b_1 - \frac{2}{15}b_2 - \frac{1}{3}b_3 \\ 0 & 0 & 0 & 0 \end{array}\right].$$

Thus x_3 is free,

$$\begin{aligned} x_2 &= \tfrac{7}{15}b_1 - \tfrac{2}{15}b_2 - \tfrac{1}{3}b_3 - x_3 \\ x_1 &= \tfrac{2}{3}b_1 - \tfrac{1}{3}b_2 - \tfrac{1}{3}b_3 - x_2 - 2x_3 \\ &= \tfrac{2}{3}b_1 - \tfrac{1}{3}b_2 - \tfrac{1}{3}b_3 - \tfrac{7}{15}b_1 + \tfrac{2}{15}b_2 + \tfrac{1}{3}b_3 + x_3 - 2x_3 \\ &= \tfrac{1}{5}b_1 - \tfrac{1}{5}b_2 - x_3, \end{aligned}$$

and

$$\mathsf{x} = \begin{bmatrix} \frac{1}{5}b_1 - \frac{1}{5}b_2 - x_3 \\ \frac{7}{15}b_1 - \frac{2}{15}b_2 - \frac{1}{3}b_3 - x_3 \\ x_3 \end{bmatrix}.$$

The length of this vector is a function of x_3, and one can use calculus to show that the minimum length occurs with $x_3 = \frac{2}{9}b_1 - \frac{1}{9}b_2 - \frac{1}{9}b_3$. We leave it to the reader to check that for this value of x_3, our desired vector is

$$\mathsf{x}^+ = \frac{1}{45}\begin{bmatrix} -b_1 - 4b_2 + 5b_3 \\ 11b_1 - b_2 - 10b_3 \\ 10b_1 - 5b_2 - 5b_3 \end{bmatrix}.$$

READING CHECK 1. Verify that the column space of the matrix A in Example 6.5.3 is the plane with equation $x + y + z = 0$. (You must do more than simply observe that the components of each column satisfy this equation!)

READING CHECK 2. Verify that the matrix that projects vectors onto the plane with equation $x + y + z = 0$ is the matrix P given in Example 6.5.3.

As we have seen, computing the length of a vector whose components involve variables and then finding the values of the variables that minimize this length can be rather formidable tasks. Fortunately, there is more than one way to find a vector x^+ of shortest length.

In Section 6.4, we showed that the null space and row space of a matrix are orthogonal complements. If A is $m \times n$, the row space of A and the null space of A are subspaces of R^n, so every vector in R^n can be written uniquely as the sum of a vector in the null space and a vector in the row space—see Theorems 6.4.17 and 6.4.26.

Let x_0 be any solution to

$$A\mathsf{x} = P\mathsf{b}, \tag{1}$$

where P is the matrix that projects vectors in R^m onto the column space of A. Thus $A\mathsf{x}_0 = P\mathsf{b}$. Write $\mathsf{x}_0 = \mathsf{x}_r + \mathsf{x}_n$ with x_r in the row space and x_n in the null space of A. Since $A\mathsf{x}_n = 0$, we have $A\mathsf{x}_0 = A(\mathsf{x}_r + \mathsf{x}_n) = A\mathsf{x}_r$, so x_r is also a solution to (1). By the Theorem of Pythagoras—Theorem 6.1.1—$\|\mathsf{x}_0\|^2 = \|\mathsf{x}_r\|^2 + \|\mathsf{x}_n\|^2 \geq \|\mathsf{x}_r\|^2$, so $\|\mathsf{x}_r\| \leq \|\mathsf{x}_0\|$. Since we are seeking a solution to (1) of shortest length, it follows that such a solution will lie in the row space of A. Wonderfully, there is just one such vector!

To see why, we suppose that x_r and x_r' are each solutions to $A\mathsf{x} = P\mathsf{b}$, both vectors lying in the row space. Then $A\mathsf{x}_r = A\mathsf{x}_r'$, so $A(\mathsf{x}_r - \mathsf{x}_r') = 0$. This says that $\mathsf{x}_r - \mathsf{x}_r'$ is in the null space of A as well as the row space. The intersection of a subspace and its orthogonal complement is 0. So $\mathsf{x}_r - \mathsf{x}_r' = 0$ and $\mathsf{x}_r = \mathsf{x}_r'$.

We summarize.

6.5.4 **Theorem.** *Suppose A is an $m \times n$ matrix and b is a vector in R^m. Then the best solution to $A\mathsf{x} = \mathsf{b}$—that is, the shortest vector among those minimizing $\|\mathsf{b} - A\mathsf{x}\|$—is the unique vector x^+ in the row space of A with the property that $A\mathsf{x}^+ = P\mathsf{b}$ is the projection of b on the column space of A.*

Since the map $\mathsf{b} \mapsto \mathsf{x}^+$ is a linear transformation and hence multiplication by a certain matrix—see Exercise 13—the "best solution" to $A\mathsf{x} = \mathsf{b}$ can be written in the form $\mathsf{x}^+ = A^+\mathsf{b}$ for a matrix A^+ called the *pseudoinverse* of A.[8]

6.5.5

The "best solution" to $A\mathsf{x} = \mathsf{b}$ is $\mathsf{x}^+ = A^+\mathsf{b}$.

We have seen that $A^+\mathsf{b} = \mathsf{x}^+$ is in the row space of A and that $AA^+\mathsf{b} = A\mathsf{x}^+ = P\mathsf{b}$ for all b, so $AA^+ = P$ is the projection matrix onto the column space of A. These two properties characterize A^+; that is, there is only one matrix with these two properties—see Exercise 14.

6.5.6 **Definition.** The *pseudoinverse* of an $m \times n$ matrix A is the matrix A^+ satisfying

1. $A^+\mathsf{b}$ is in the row space of A for all b, and
2. $AA^+ = P$ is the projection matrix from R^m onto the column space of A.

Equivalently, A^+ is defined by the condition that $A^+\mathsf{b}$ is the unique shortest vector x that minimizes $\|\mathsf{b} - A\mathsf{x}\|$.

The equation $\mathsf{x}^+ = A^+\mathsf{b}$ shows that if A is an $m \times n$ matrix, then A^+ is $n \times m$.

6.5.7 **Example.** In Example 6.5.1, we started with $A = \begin{bmatrix} \frac{1}{2} & 0 & 0 & 0 \\ 0 & 5 & 0 & 0 \\ 0 & 0 & 0 & 0 \end{bmatrix}$ and found

$$\mathsf{x}^+ = \begin{bmatrix} 2b_1 \\ \frac{1}{5}b_2 \\ 0 \\ 0 \end{bmatrix} = \begin{bmatrix} 2 & 0 & 0 \\ 0 & \frac{1}{5} & 0 \\ 0 & 0 & 0 \\ 0 & 0 & 0 \end{bmatrix} \begin{bmatrix} b_1 \\ b_2 \\ b_3 \end{bmatrix}, \text{ so } A^+ = \begin{bmatrix} 2 & 0 & 0 \\ 0 & \frac{1}{5} & 0 \\ 0 & 0 & 0 \\ 0 & 0 & 0 \end{bmatrix}.$$

:)

6.5.8 This example illustrates a general fact. If $A = \begin{bmatrix} D & 0 \\ 0 & 0 \end{bmatrix}$ is an $m \times n$ matrix, where $D = \begin{bmatrix} \mu_1 & & \\ & \ddots & \\ & & \mu_r \end{bmatrix}$ is an $r \times r$ matrix with nonzero diagonal entries

[8] A^+ is also known as the *Moore-Penrose* inverse.

and the three 0s represent zero matrices of the sizes needed for an $m \times n$ matrix, then A^+ is the $n \times m$ matrix

$$A^+ = \begin{bmatrix} D^{-1} & \mathbf{0} \\ \mathbf{0} & \mathbf{0} \end{bmatrix}, \quad \text{with} \quad D^{-1} = \begin{bmatrix} \frac{1}{\mu_1} & & \\ & \ddots & \\ & & \frac{1}{\mu_r} \end{bmatrix},$$

the bold 0s denoting blocks of 0s of the sizes needed for an $n \times m$ matrix.

READING CHECK 3. Find A^+, given $A = \begin{bmatrix} -2 & 0 \\ 0 & 3 \\ 0 & 0 \end{bmatrix}$.

6.5.9 Example. With $A = \begin{bmatrix} 1 & 1 & 2 \\ -4 & 1 & -3 \\ 3 & -2 & 1 \end{bmatrix}$ as in Example 6.5.3,

$$\mathsf{x}^+ = \frac{1}{45}\begin{bmatrix} -b_1 - 4b_2 + 5b_3 \\ 11b_1 - b_2 - 10b_3 \\ 10b_1 - 5b_2 - 5b_3 \end{bmatrix} = \frac{1}{45}\begin{bmatrix} -1 & -4 & 5 \\ 11 & -1 & -10 \\ 10 & -5 & -5 \end{bmatrix}\begin{bmatrix} b_1 \\ b_2 \\ b_3 \end{bmatrix},$$

so $A^+ = \frac{1}{45}\begin{bmatrix} -1 & -4 & 5 \\ 11 & -1 & -10 \\ 10 & -5 & -5 \end{bmatrix}$. ☺

6.5.10 Example. Here, we use the definition of "pseudoinverse" to find A^+ with $A = \begin{bmatrix} 1 & 1 & 1 \\ 1 & 1 & 1 \end{bmatrix}$, the matrix of Example 6.5.2. We require $A^+\mathsf{b}$ to be in the row space of A, so $A^+\mathsf{b} = t\begin{bmatrix} 1 \\ 1 \\ 1 \end{bmatrix}$ for some scalar t. We also want $A(A^+\mathsf{b}) = t\begin{bmatrix} 3 \\ 3 \end{bmatrix}$ to be the projection of $P\mathsf{b}$ of b on the column space of A. This is the matrix $P = \frac{1}{2}\begin{bmatrix} 1 & 1 \\ 1 & 1 \end{bmatrix}$ as before. The condition $A(A^+\mathsf{b}) = P\mathsf{b}$ says $t\begin{bmatrix} 3 \\ 3 \end{bmatrix} = \frac{1}{2}\begin{bmatrix} b_1 + b_2 \\ b_1 + b_2 \end{bmatrix}$, so $3t = \frac{1}{2}(b_1 + b_2)$ and $t = \frac{1}{6}(b_1 + b_2)$. We obtain

$$A^+\mathsf{b} = \frac{1}{6}(b_1 + b_2)\begin{bmatrix} 1 \\ 1 \\ 1 \end{bmatrix} = \begin{bmatrix} \frac{1}{6}(b_1 + b_2) \\ \frac{1}{6}(b_1 + b_2) \\ \frac{1}{6}(b_1 + b_2) \end{bmatrix} = \frac{1}{6}\begin{bmatrix} 1 & 1 \\ 1 & 1 \\ 1 & 1 \end{bmatrix}\begin{bmatrix} b_1 \\ b_2 \end{bmatrix}$$

so that $A^+ = \frac{1}{6}\begin{bmatrix} 1 & 1 \\ 1 & 1 \\ 1 & 1 \end{bmatrix}$, as before. To the author, this approach seems easier than the one used earlier with multi-variable calculus. ☺

Special Classes of Matrices

Certain classes of matrices have predictable pseudoinverses that do not have to be determined from the definition as in Example 6.5.10.

1. A is invertible.

Suppose first that A is an invertible $n \times n$ matrix. In this case, the column space of A is R^n and the matrix that projects vectors in R^n onto R^n is I, the identity matrix. (See Exercise 29 in Section 6.1.) The pseudoinverse A^+ of A has the property that AA^+ is projection onto the column space of A. Here, we have $AA^+ = I$ so $A^+ = A^{-1}$. The pseudoinverse is supposed to provide the "best solution" to $A\mathsf{x} = \mathsf{b}$. Here, this is certainly the case.

6.5.11 If A is invertible, $A^+ = A^{-1}$.

2. A is a projection matrix.

Suppose A is an $n \times n$ projection matrix. Then $A^2 = A$ and $A^T = A$, by Theorem 6.1.12. In this case, the matrix that projects vectors onto the column space of A is A itself, so the second property of A^+ given in 6.5.6 says $AA^+ = A$. Therefore, $A(A^+ - A) = 0$, since $A^2 = A$. This implies $A(A^+\mathsf{b} - A\mathsf{b}) = 0$ for any vector b. So $A^+\mathsf{b} - A\mathsf{b}$ is in the null space of A for any b. But this vector is also in the row space of A because both $A^+\mathsf{b}$ and $A\mathsf{b} = A^T\mathsf{b}$ are in the row space. Since the row space and the null space of A are orthogonal complements (so their intersection is 0), we obtain $A^+\mathsf{b} - A\mathsf{b} = 0$ for all b, so $A^+ = A$.

6.5.12 If A is a projection matrix, $A^+ = A$.

3. A has linearly independent columns.

Suppose the $m \times n$ matrix A has linearly independent columns. The results of Section 6.1 show that the matrix that projects vectors onto the column space of A is $P = A(A^TA)^{-1}A^T$. Thus $AA^+ = A(A^TA)^{-1}A^T$. Multiplying on the left by A^T gives $A^TAA^+ = A^TA(A^TA)^{-1}A^T = A^T$, so $A^+ = (A^TA)^{-1}A^T$.

6.5.13 If A has linearly independent columns, then $A^+ = (A^TA)^{-1}A^T$.

4. A has linearly independent rows.

Suppose the $m \times n$ matrix A has linearly independent rows. By Theorem 4.5.3, the columns of A span R^m. Thus the column space of A is R^m and the projection P onto the column space is I. Also $\operatorname{rank} A = m$. By Theorem 4.4.27, $\operatorname{rank} AA^T = \operatorname{rank} A$, so the $m \times m$ matrix AA^T is invertible. Now the argument is very similar to that used in the projection matrix case. We have

$I = AA^+ = AA^T(AA^T)^{-1}$, hence $A[(A^+ - A^T(AA^T)^{-1})\mathsf{b}] = \mathsf{0}$ for any vector b, so $(A^+ - A^T(AA^T)^{-1})\mathsf{b}$ is in the null space of A, for any b. It is also in the row space of A, so it's $\mathsf{0}$. We obtain $A^+ - A^T(AA^T)^{-1} = \mathsf{0}$.

6.5.14 If A has linearly independent rows, then $A^+ = A^T(AA^T)^{-1}$.

6.5.15 Remark. There are, of course, matrices not of the types we have considered here. To find the pseudoinverse in these cases, one can always resort to the definition. There are examples in the exercises. In particular, see Exercise 3. (Another way to find the pseudoinverse is described in Section 7.5.)

Answers to Reading Checks

1. The components of each column satisfy $x + y + z = 0$, so the column space is contained in the plane with this equation. The first two columns of A are linearly independent, so the column space has dimension at least 2. The plane has dimension 2, so the column space must be the plane.

2. We start with two linearly independent vectors in the plane. We could use two columns of the matrix, but $\begin{bmatrix} 1 \\ -1 \\ 0 \end{bmatrix}$ and $\begin{bmatrix} 1 \\ 0 \\ -1 \end{bmatrix}$ are simpler to handle. Let $B = \begin{bmatrix} 1 & 1 \\ -1 & 0 \\ 0 & -1 \end{bmatrix}$ be the matrix with these vectors as columns. Then $P = B(B^TB)^{-1}B^T$ is the desired projection matrix—see 6.1.6. We have

$$B^TB = \begin{bmatrix} 1 & -1 & 0 \\ 1 & 0 & -1 \end{bmatrix}\begin{bmatrix} 1 & 1 \\ -1 & 0 \\ 0 & -1 \end{bmatrix} = \begin{bmatrix} 2 & 1 \\ 1 & 2 \end{bmatrix}, \ (B^TB)^{-1} = \tfrac{1}{3}\begin{bmatrix} 2 & -1 \\ -1 & 2 \end{bmatrix} \text{ and}$$

$$\begin{aligned} P &= B(B^TB)^{-1}B^T \\ &= \frac{1}{3}\begin{bmatrix} 1 & 1 \\ -1 & 0 \\ 0 & -1 \end{bmatrix}\begin{bmatrix} 2 & -1 \\ -1 & 2 \end{bmatrix}\begin{bmatrix} 1 & -1 & 0 \\ 1 & 0 & -1 \end{bmatrix} = \frac{1}{3}\begin{bmatrix} 2 & -1 & -1 \\ -1 & 2 & -1 \\ -1 & -1 & 2 \end{bmatrix}. \end{aligned}$$

3. $A^+ = \begin{bmatrix} -\frac{1}{2} & 0 & 0 \\ 0 & \frac{1}{3} & 0 \end{bmatrix}$

True/False Questions

Decide, with as little calculation as possible, whether each of the following statements is true or false and, if you say "false," explain your answer. (Answers are at the back of the book.)

1. If A is an $m \times n$ matrix, the pseudoinverse A^+ of A is $n \times m$.

2. If A is an $m \times n$ matrix with pseudoinverse A^+, then AA^+ is the projection matrix onto the column space of A.

3. The pseudoinverse of $A = \begin{bmatrix} 5 & 0 & 0 \\ 0 & 7 & 0 \end{bmatrix}$ is $\begin{bmatrix} -5 & 0 & 0 \\ 0 & -7 & 0 \end{bmatrix}$.

4. If $A = \begin{bmatrix} 2 & 0 & 0 \\ 0 & -3 & 0 \\ 0 & 0 & 0 \\ 0 & 0 & 0 \end{bmatrix}$, then $(A^+)^+ = A$.

5. If $A = \begin{bmatrix} 1 & 0 \\ -2 & 1 \end{bmatrix}$, then $A^+ = \begin{bmatrix} 1 & 0 \\ 2 & 1 \end{bmatrix}$.

6. The pseudoinverse of $\begin{bmatrix} \cos\theta & -\sin\theta \\ \sin\theta & \cos\theta \end{bmatrix}$ is $\begin{bmatrix} \cos(-\theta) & -\sin(-\theta) \\ \sin(-\theta) & \cos(-\theta) \end{bmatrix}$.

7. The pseudoinverse of a matrix A with linearly independent rows is $A^+ = (A^TA)^{-1}A^T$.

8. The pseudoinverse of a matrix A with linearly independent columns is $A^+ = (A^TA)^{-1}$.

9. The pseudoinverse of the matrix $A = \begin{bmatrix} 1 & 2 & 3 & 4 & 5 \\ -2 & 0 & 1 & 1 & 7 \end{bmatrix}$ can be deduced from the formula $A^+ = A^T(AA^T)^{-1}$.

10. The pseudoinverse of a projection matrix is its inverse.

EXERCISES

Answers to exercises marked [BB] can be found at the Back of the Book.

1. [BB] **(a)** The matrix $A = \begin{bmatrix} 1 & 1 & 1 & 1 \\ 0 & 1 & -1 & 0 \end{bmatrix}$ has a right inverse. Why?

   (b) Why must AA^T be invertible?

   (c) Find the pseudoinverse of A.

2. **(a)** The matrix $A = \begin{bmatrix} 1 & 0 & 0 \\ 1 & 1 & 1 \\ 1 & -1 & 1 \\ 1 & 2 & 4 \end{bmatrix}$ has a left inverse. Why?

(b) Why must A^TA be invertible?

(c) Find the pseudoinverse of A.

3. [BB] Let $A = \begin{bmatrix} 1 & 1 & 2 \\ -4 & 1 & -3 \\ 3 & -2 & 1 \end{bmatrix}$ be the matrix of Example 6.5.3. Use Definition 6.5.6 to find A^+. (The calculations will be simpler if you first note that the row space of A is spanned by $\mathsf{u} = \begin{bmatrix} 1 \\ 0 \\ 1 \end{bmatrix}$ and $\mathsf{v} = \begin{bmatrix} 0 \\ 1 \\ 1 \end{bmatrix}$. Why is this the case?)

4. (a) Use your knowledge of single-variable calculus to find the pseudo-inverse of $A = \begin{bmatrix} 1 & 1 \\ 1 & 1 \\ 1 & 1 \end{bmatrix}$. (See Example 6.5.2.)

   (b) Find A^+ using its characterization in Definition 6.5.6.

5. Find the pseudoinverse of A and the "best solution" to $A\mathsf{x} = \mathsf{b}$ in each of the following cases.

   (a) [BB] $A = \begin{bmatrix} 1 & -1 & 0 \end{bmatrix}$; $\mathsf{b} = [10]$

   (b) $A = \begin{bmatrix} 1 \\ -2 \\ 0 \\ 2 \end{bmatrix}$; $\mathsf{b} = \begin{bmatrix} 1 \\ 1 \\ 1 \\ 1 \end{bmatrix}$

   (c) $A = \begin{bmatrix} -1 & 2 & 3 & 1 \\ 1 & 0 & 1 & 1 \end{bmatrix}$; $\mathsf{b} = \begin{bmatrix} -1 \\ 7 \end{bmatrix}$

   (d) [BB] $A = \begin{bmatrix} 1 & -1 \\ 1 & -1 \\ 1 & -1 \end{bmatrix}$; $\mathsf{b} = \begin{bmatrix} 2 \\ 0 \\ 1 \end{bmatrix}$

   (e) $A = \begin{bmatrix} 1 & 2 \\ 0 & 1 \\ -1 & 0 \end{bmatrix}$; $\mathsf{b} = \begin{bmatrix} 4 \\ -1 \\ 2 \end{bmatrix}$

   (f) $A = \begin{bmatrix} 1 & 2 \\ 2 & 4 \\ 3 & 6 \end{bmatrix}$; $\mathsf{b} = \begin{bmatrix} 1 \\ 1 \\ 3 \end{bmatrix}$

   (g) [BB] $A = \begin{bmatrix} 1 & -1 & 0 \\ 0 & 0 & 1 \\ 1 & -1 & 0 \end{bmatrix}$; $\mathsf{b} = \begin{bmatrix} -1 \\ 0 \\ 1 \end{bmatrix}$

   (h) $A = \begin{bmatrix} 1 & 0 & 1 & 0 \\ 0 & 0 & 1 & 1 \\ 1 & 0 & 1 & 1 \end{bmatrix}$; $\mathsf{b} = \begin{bmatrix} -5 \\ 1 \\ 2 \end{bmatrix}$.

6. (a) [BB] If a matrix A has linearly independent columns, its pseudoinverse is a left inverse of A. Why?

   (b) If a matrix A has linearly independent rows, its pseudoinverse is a right inverse of A. Why?

7. Let μ_1, μ_2, μ_3 be nonzero scalars and let $A = \begin{bmatrix} \mu_1 & 0 & 0 & 0 & 0 \\ 0 & \mu_2 & 0 & 0 & 0 \\ 0 & 0 & \mu_3 & 0 & 0 \\ 0 & 0 & 0 & 0 & 0 \end{bmatrix}$.

   Show that $A^+ = \begin{bmatrix} \frac{1}{\mu_1} & 0 & 0 & 0 \\ 0 & \frac{1}{\mu_2} & 0 & 0 \\ 0 & 0 & \frac{1}{\mu_3} & 0 \\ 0 & 0 & 0 & 0 \\ 0 & 0 & 0 & 0 \end{bmatrix}$, thus justifying the content of paragraph 6.5.8.

8. If $A = \begin{bmatrix} D & 0 \\ 0 & 0 \end{bmatrix}$ as in paragraph 6.5.8, show that $(A^+)^+ = A$.

Critical Reading

9. What is the pseudoinverse of a permutation matrix and why?

10. Find a simple formula for the pseudoinverse of a matrix A whose columns are orthonormal.

11. Suppose A is an $m \times n$ matrix with the property that $A\mathsf{x} = \mathsf{b}$ has at least one solution for any vector b in R^m. Find a formula for A^+.

12. Find a formula for the pseudoinverse of an $m \times n$ matrix of rank n.

13. [BB] Let A be an $m \times n$ matrix and let b be a vector in R^m. Show that the vector x^+ defined in Theorem 6.5.4 can be written in the form $\mathsf{x}^+ = B\mathsf{b}$ for some matrix B. (Hint: Show that the map $\mathsf{b} \mapsto \mathsf{x}^+$ is a linear transformation $\mathsf{R}^m \to \mathsf{R}^n$ and apply Theorem 5.1.13.)

14. Show that the two properties of the pseudoinverse given in Definition 6.5.6 characterize this matrix; that is, there is at most one matrix that has these two properties.

15. Suppose $A = \begin{bmatrix} \alpha_1 & \beta_2\alpha_1 & \beta_3\alpha_1 & \cdots & \beta_n\alpha_1 \\ \alpha_2 & \beta_2\alpha_2 & \beta_3\alpha_2 & \cdots & \beta_n\alpha_2 \\ \alpha_3 & \beta_2\alpha_3 & \beta_3\alpha_3 & \cdots & \beta_n\alpha_3 \\ \vdots & & & & \\ \alpha_m & \beta_2\alpha_m & \beta_3\alpha_m & \cdots & \beta_n\alpha_m \end{bmatrix}$ is an $m \times n$ matrix of rank1.
Find a formula for the pseudoinverse of A.

CHAPTER KEY WORDS AND IDEAS: Here are some technical words and phrases that were used in this chapter. Do you know the meaning of each? If you're not sure, check the glossary or index at the back of the book.

best solution in the sense of least squares
direct sum of subspaces
Gram–Schmidt algorithm
Householder matrix
intersection of subspaces
length of a vector
orthogonal complement
orthogonal matrix
orthogonal subspaces
orthogonal vectors
orthonormal vectors
permutation matrix
projection matrix
projection of a vector on a subspace
Pythagoras' Theorem
pseudoinverse
rotation matrix
sum of subspaces
symmetric matrix

Review Exercises for Chapter 6

1. Find the matrix P that projects vectors onto each of the following spaces.

   (a) the subspace of R^3 spanned by $\mathsf{v} = \begin{bmatrix} -2 \\ 7 \\ 1 \end{bmatrix}$

   (b) the plane spanned by the orthogonal vectors $\mathsf{e} = \begin{bmatrix} 1 \\ -2 \\ 1 \end{bmatrix}$ and $\mathsf{f} = \begin{bmatrix} -3 \\ 1 \\ 5 \end{bmatrix}$

   (c) the plane in R^3 with equation $x + 2y = 0$

   (d) the subspace U of R^4 consisting of vectors of the form $\begin{bmatrix} -t-s \\ 2t+s \\ s-t \\ s+2t \end{bmatrix}$

   (e) the column space of $A = \begin{bmatrix} 1 & 0 \\ 0 & 1 \\ -1 & 1 \end{bmatrix}$.

2. **(a)** The column space of $A = \begin{bmatrix} 3 & 1 \\ 2 & 0 \\ -1 & 1 \end{bmatrix}$ is a plane. Find the equation of this plane.

   (b) The system $A\mathsf{x} = \begin{bmatrix} 2 \\ 1 \\ 0 \end{bmatrix}$ has a solution. Without any calculation, explain why.

   (c) Find the matrix that projects vectors onto the column space of A.

(d) Find the best solution in the sense of least squares to the system

$$\begin{aligned} 3x_1 + x_2 &= -1 \\ 2x_1 \quad\quad &= 7 \\ -x_1 + x_2 &= 2. \end{aligned}$$

(e) What does "best solution in the sense of least squares" mean?

3. (a) Find the matrix P which projects vectors in R^3 onto the plane π with equation $2x - y + z = 0$.

(b) Find the vector u in π that minimizes $\|\mathsf{u} - \mathsf{b}\|$, with $\mathsf{b} = \begin{bmatrix} -3 \\ 1 \\ 1 \end{bmatrix}$.

4. Let A be an $m \times n$ matrix and let x and b be vectors with the property that $\mathsf{y} \cdot (A^T\mathsf{b} - A^TA\mathsf{x}) = 0$ for all vectors y. Prove that $\mathsf{b} - A\mathsf{x}$ is orthogonal to the column space of A.

5. Can a projection matrix have nonzero determinant? Explain.

6. If A is a 15×27 matrix and the dimension of its row space is 7, what is the dimension of the null space of A and why.

7. Let $\mathsf{u}_1 = \begin{bmatrix} 1 \\ 2 \\ 3 \end{bmatrix}$, $\mathsf{u}_2 = \begin{bmatrix} 1 \\ 0 \\ -1 \end{bmatrix}$, $\mathsf{u}_3 = \begin{bmatrix} -1 \\ -2 \\ 5 \end{bmatrix}$ and $\mathsf{u}_4 = \begin{bmatrix} 2 \\ 1 \\ -1 \end{bmatrix}$.
If the Gram–Schmidt algorithm were applied to these vectors, it would fail. Why?

8. Apply the Gram–Schmidt algorithm to $\mathsf{u}_1 = \begin{bmatrix} 1 \\ 0 \\ 1 \end{bmatrix}$, $\mathsf{u}_2 = \begin{bmatrix} 1 \\ 1 \\ 1 \end{bmatrix}$, $\mathsf{u}_3 = \begin{bmatrix} -1 \\ 0 \\ 1 \end{bmatrix}$.

9. (a) Apply the Gram–Schmidt algorithm to $\mathsf{a}_1 = \begin{bmatrix} 1 \\ 1 \\ 1 \\ 1 \end{bmatrix}$, $\mathsf{a}_2 = \begin{bmatrix} 2 \\ 1 \\ 0 \\ 1 \end{bmatrix}$, $\mathsf{a}_3 = \begin{bmatrix} -1 \\ 4 \\ 1 \\ 0 \end{bmatrix}$.

(b) Use the result of (a) to find a QR factorization of $A = \begin{bmatrix} 1 & 2 & -1 \\ 1 & 1 & 4 \\ 1 & 0 & 1 \\ 1 & 1 & 0 \end{bmatrix}$.

10. (a) Apply the Gram–Schmidt algorithm to $\mathsf{u}_1 = \begin{bmatrix} 1 \\ 0 \\ -1 \\ 0 \end{bmatrix}$, $\mathsf{u}_2 = \begin{bmatrix} 1 \\ 0 \\ 1 \\ 1 \end{bmatrix}$, $\mathsf{u}_3 = \begin{bmatrix} 1 \\ -1 \\ 0 \\ 0 \end{bmatrix}$.

(b) Find a QR factorization of $A = \begin{bmatrix} 1 & 1 & 1 \\ 0 & 0 & -1 \\ -1 & 1 & 0 \\ 0 & 1 & 0 \end{bmatrix}$.

(c) Use the result of (b) to find A^{-1}.

11. Let $A = \begin{bmatrix} -1 & 1 \\ 0 & 1 \\ 1 & -2 \end{bmatrix}$ and $B = \begin{bmatrix} 1 & -1 & 2 \\ 3 & 0 & 1 \end{bmatrix}$. One of these matrices has a QR factorization and the other does not. Which does not? Explain. Find a QR factorization of the other.

12. **(a)** Apply the Gram-Schmidt Algorithm to the vectors $\mathsf{a}_1 = \begin{bmatrix} -1 \\ 0 \\ 1 \\ 1 \end{bmatrix}$, $\mathsf{a}_2 = \begin{bmatrix} 0 \\ 1 \\ 1 \\ 1 \end{bmatrix}$.

    (b) Find the matrix which projects vectors in R^4 onto the span of a_1 and a_2.

    (c) Find a QR factorization of the matrix $A = \begin{bmatrix} -1 & 0 \\ 0 & 1 \\ 1 & 1 \\ 1 & 1 \end{bmatrix}$ whose columns are the vectors of part (a).

    (d) Explain how a QR factorization helps find the best solution (in the sense of least squares) to $A\mathsf{x} = \mathsf{b}$ and apply this method to find the best solution to
$$\begin{aligned} -x \qquad &= 1 \\ y &= 2 \\ x + y &= 2 \\ x + y &= 1. \end{aligned}$$

13. Find a QR factorization of $A = \begin{bmatrix} -1 & 0 & 1 \\ 0 & 1 & 1 \\ 1 & -1 & 2 \end{bmatrix}$.

14. Let $A = QR$ be a QR factorization of an $m \times n$ matrix A with linearly independent columns.

    (a) Show that the columns of R are linearly independent.

    (b) Assuming Q is $m \times n$, why is R invertible?

15. Does a matrix with orthonormal columns have a left inverse? a right inverse? Explain.

16. Let U be the subspace of R^4 spanned by $\mathsf{f}_1 = \begin{bmatrix} 1 \\ 1 \\ 1 \\ 1 \end{bmatrix}$ and $\mathsf{f}_2 = \begin{bmatrix} 1 \\ 1 \\ -1 \\ -1 \end{bmatrix}$.

    (a) Find the matrix that projects vectors in R^4 onto U.

    (b) Write $\mathsf{v} = \begin{bmatrix} 2 \\ 0 \\ 1 \\ 6 \end{bmatrix}$ as the sum of a vector in U and a vector in $U^\perp$.

    (c) Find the vector u in U that minimizes $\|\mathsf{u} - \mathsf{b}\|$, $\mathsf{b} = \begin{bmatrix} 2 \\ 0 \\ 1 \\ 6 \end{bmatrix}$.

    (d) Find the best solution in the sense of least squares to $\begin{aligned} x + y &= 2 \\ x + y &= 0 \\ x - y &= 1 \\ x - y &= 6. \end{aligned}$

17. Let A be an $m \times n$ matrix with linearly independent rows. Formulate a factorization theorem for A and prove that your statement is correct.

18. Let U be the subspace of R^4 spanned by $\mathsf{u}_1 = \begin{bmatrix} 1 \\ -1 \\ 2 \\ -3 \end{bmatrix}$ and $\mathsf{u}_2 = \begin{bmatrix} 0 \\ 1 \\ -1 \\ 0 \end{bmatrix}$. Find an orthogonal basis for U and an orthogonal basis for $U^\perp$.

19. Find an orthogonal basis for the subspace of R^4 defined by $x_1 + x_2 + x_3 + x_4 = 0$.

20. Let $U = \operatorname{sp}\{\mathsf{u}_1, \mathsf{u}_2, \mathsf{u}_3\}$, where $\mathsf{u}_1 = \begin{bmatrix} -1 \\ 0 \\ 1 \\ 0 \end{bmatrix}, \mathsf{u}_2 = \begin{bmatrix} 1 \\ 0 \\ 1 \\ 1 \end{bmatrix}, \mathsf{u}_3 = \begin{bmatrix} 1 \\ 0 \\ 0 \\ -1 \end{bmatrix}$.

(a) Use the Gram-Schmidt algorithm to find an orthogonal basis for U.

(b) Find the vector u in U that minimizes $\|\mathsf{u} - \mathsf{b}\|$, $\mathsf{b} = \begin{bmatrix} 1 \\ 1 \\ 1 \\ 0 \end{bmatrix}$.

21. Find $U^\perp$ in each of the following cases.

(a) $V = \mathsf{R}^3$, U is the span of $\mathsf{u}_1 = \begin{bmatrix} 2 \\ 2 \\ -4 \end{bmatrix}$ and $\mathsf{u}_2 = \begin{bmatrix} -2 \\ -1 \\ -3 \end{bmatrix}$

(b) $V = \mathsf{R}^4$, $U = \left\{ \begin{bmatrix} x - 2y \\ -x \\ -2y \\ x + 4y \end{bmatrix} \mid x, y \in \mathsf{R} \right\}$.

22. Write $\mathsf{v} = \mathsf{u} + \mathsf{w}$ with u in U and w in $U^\perp$ in each of the following situations.

(a) $V = \mathsf{R}^2$, $\mathsf{v} = \begin{bmatrix} x \\ y \end{bmatrix}$, $U = \left\{ \begin{bmatrix} -2t \\ 3t \end{bmatrix} \mid t \in \mathsf{R} \right\}$

(b) $V = \mathsf{R}^3$, $\mathsf{v} = \begin{bmatrix} x \\ y \\ z \end{bmatrix}$, U is the span of $\begin{bmatrix} 1 \\ -3 \\ 2 \end{bmatrix}$

(c) $V = \mathsf{R}^3$, $\mathsf{v} = \begin{bmatrix} x \\ y \\ z \end{bmatrix}$, $U = \left\{ \begin{bmatrix} x \\ y \\ z \end{bmatrix} \mid x + y - 3z = 0 \right\}$

(d) $V = \mathsf{R}^4$, $\mathsf{v} = \begin{bmatrix} 1 \\ 2 \\ -3 \\ -4 \end{bmatrix}$, $U = \left\{ \begin{bmatrix} -t + r \\ 2t + s + r \\ t - s + r \\ t - s + r \end{bmatrix} \mid t, s, r \in \mathsf{R} \right\}$

(e) $V = \mathsf{R}^4$, $\mathsf{v} = \begin{bmatrix} 2 \\ 1 \\ -1 \\ 1 \end{bmatrix}$, $U = \operatorname{sp}\left\{ \begin{bmatrix} 1 \\ -1 \\ 0 \\ 1 \end{bmatrix}, \begin{bmatrix} 1 \\ 0 \\ 1 \\ 0 \end{bmatrix} \right\}$.

23. Let A be an $m \times n$ matrix and let B an $n \times p$ matrix.

(a) Prove that nullity A + nullity $B \geq$ nullity AB. [Hint: Let $\mathsf{x}_1, \ldots, \mathsf{x}_k$ be a basis for null sp B and let $B\mathsf{y}_1, \ldots, B\mathsf{y}_\ell$ be a basis for null sp $A \cap$ col sp B. The vectors $\mathsf{x}_1, \ldots, \mathsf{x}_k, \mathsf{y}_1, \ldots, \mathsf{y}_\ell$ all lie in null sp AB, so it is sufficient to show that these are linearly independent. Explain and complete the proof.]

(b) Restate the inequality in part (a) as an assertion about rank A, rank B and rank AB.

24. Let A be a matrix with pseudoinverse A^+.

(a) In what sense is $\mathsf{x} = A^+\mathsf{b}$ the "best solution" to $A\mathsf{x} = \mathsf{b}$?

(b) Find the pseudoinverse of $A = \begin{bmatrix} -3 \\ 4 \end{bmatrix}$.

25. **(a)** Are the columns of $A = \begin{bmatrix} 1 & 1 & 1 \\ 1 & 2 & 3 \\ 1 & 3 & 6 \end{bmatrix}$ linearly independent?

 (b) Are the rows of A linearly independent?

 (c) Find the pseudoinverse of A.

26. **(a)** Matrix $A = \begin{bmatrix} -1 & 1 & 1 \\ 0 & 2 & 1 \\ 1 & 5 & 2 \end{bmatrix}$ fits into none of the four special classes of matrices identified in Section 6.5. Why not?

 (b) Find the pseudoinverse of A.

27. If A is an $m \times n$ matrix of rank m, what is AA^+ and why?

28. Suppose C is a code. What is the dual code $C^\perp$, and what is the connection between the generator and parity check matrices of C and $C^\perp$?

Chapter 7

The Spectral Theorem

7.1 Complex Vectors and Matrices

This section, and the next as well, can be omitted by those wishing to proceed as quickly as possible to a treatment of the diagonalization of real symmetric matrices, a topic that is treated without complex numbers in Section 7.3.

Readers of this section are referred to Appendix A for some common notation used with complex numbers and a discussion of the basic properties of this remarkable number system.

Complex Vector Spaces

Until this point of the text, the word "scalar" has been synonymous with "real number." In fact, it could equally well mean "complex number." The word "scalar" in the definition of vector space could mean "complex number" just as easily as "real number." Almost all the theorems and proofs about vector spaces that have appeared in this book work for any field of scalars. The primary differences between real and complex vector spaces involve the dot product.

Just as R^n is our basic and most important example of a *real* vector space (that is, the scalars are real numbers), so

$$\mathsf{C}^n = \left\{ \begin{bmatrix} z_1 \\ z_2 \\ \vdots \\ z_n \end{bmatrix} \mid z_1, z_2, \ldots, z_n \in \mathsf{C} \right\}$$

is the key example of a *complex* vector space (the scalars are complex numbers), but the definition of dot product in C^n has to be tweaked a little.

Let $\mathsf{v} = \begin{bmatrix} 1 \\ i \end{bmatrix}$ in $\mathbf{C}^2$. Defining the dot product as in $\mathbf{R}^n$, we get $\|\mathsf{v}\|^2 = \mathsf{v} \cdot \mathsf{v} = 1^2 + i^2 = 0$, which is surely not very desirable—a nonzero vector should surely have nonzero length! So we modify the usual definition.

7.1.1 Definitions. The *complex dot product* of vectors $\mathsf{z} = \begin{bmatrix} z_1 \\ \vdots \\ z_n \end{bmatrix}$ and $\mathsf{w} = \begin{bmatrix} w_1 \\ \vdots \\ w_n \end{bmatrix}$ in $\mathbf{C}^n$ is defined by

$$\mathsf{z} \cdot \mathsf{w} = \overline{z}_1 w_1 + \overline{z}_2 w_2 + \cdots + \overline{z}_n w_n.$$

Vectors z and w are *orthogonal* if $\mathsf{z} \cdot \mathsf{w} = 0$. The *length* of z is $\|\mathsf{z}\| = \sqrt{\mathsf{z} \cdot \mathsf{z}}$.

7.1.2 Examples. If $\mathsf{z} = \begin{bmatrix} 1 \\ i \end{bmatrix}$ and $\mathsf{w} = \begin{bmatrix} 2+i \\ 3-i \end{bmatrix}$, then

- $\mathsf{z} \cdot \mathsf{w} = 1(2+i) - i(3-i) = 2 + i - 3i - 1 = 1 - 2i$;
- $\mathsf{w} \cdot \mathsf{z} = (2-i)(1) + (3+i)(i) = 2 - i + 3i - 1 = 1 + 2i$;
- $\|\mathsf{z}\|^2 = \mathsf{z} \cdot \mathsf{z} = 1(1) - i(i) = 1 - i^2 = 2$;
- $\|\mathsf{w}\|^2 = \mathsf{w} \cdot \mathsf{w} = (2-i)(2+i) + (3+i)(3-i) = 4 + 1 + 9 + 1 = 15.$ ☺

7.1.3 Example. The vectors $\mathsf{z} = \begin{bmatrix} 2-i \\ 1+i \end{bmatrix}$ and $\mathsf{w} = \begin{bmatrix} 6+8i \\ 9-13i \end{bmatrix}$ are orthogonal because $\mathsf{z} \cdot \mathsf{w} = (2+i)(6+8i) + (1-i)(9-13i) = (4+22i) + (-4-22i) = 0.$ ☺

With our modified definition of dot product, notice that the length of $\mathsf{z} = \begin{bmatrix} 1 \\ i \end{bmatrix}$ is now $\sqrt{2}$, and no longer 0. In fact the new definition works as we'd like: nonzero vectors never have length 0.

7.1.4 Problem. With the definitions of dot product and length given in 7.1.1, show that nonzero vectors in $\mathbf{C}^n$ have nonzero length.

Solution. The square of the length of a vector $\mathsf{z} = \begin{bmatrix} z_1 \\ z_2 \\ \vdots \\ z_n \end{bmatrix}$ in $\mathbf{C}^n$ is

$$\|\mathsf{z}\|^2 = \mathsf{z} \cdot \mathsf{z} = \overline{z}_1 z_1 + \overline{z_2} z_2 + \cdots + \overline{z}_n z_n = |z_1|^2 + |z_2|^2 + \cdots + |z_n|^2. \quad (1)$$

For any complex number $z = a + bi$, $|z|^2 = a^2 + b^2$ is the sum of squares of two real numbers. Thus, equation (1) shows that $||\mathsf{z}^2||$ is the sum of squares of real numbers. Such a sum is nonnegative, and 0 if and only if each $z_i = 0$; that is, if and only if $\mathsf{z} = 0$. In particular, if $\mathsf{z} \neq 0$, then $\|\mathsf{z}\| \neq 0$. This is what we were asked to show. ☝

7.1.5 Remarks. The complex dot product has all the properties of the dot product in R^n listed in Theorem 1.2.20, with two exceptions.

i. As illustrated with the first two items in Example 7.1.2,

$$\mathsf{w} \cdot \mathsf{z} = \overline{\mathsf{z} \cdot \mathsf{w}}. \tag{2}$$

ii. While $\mathsf{z} \cdot (c\mathsf{w}) = c(\mathsf{z} \cdot \mathsf{w})$ as one might expect—see Exercise 9—if c is a factor of the first argument, it comes outside as $\overline{c}$, the complex conjugate of c:

$$(c\mathsf{z}) \cdot \mathsf{w} = \overline{c}(\mathsf{z} \cdot \mathsf{w}). \tag{3}$$

To see this, let $\mathsf{z} = \begin{bmatrix} z_1 \\ \vdots \\ z_n \end{bmatrix}$ and $\mathsf{w} = \begin{bmatrix} w_1 \\ \vdots \\ w_n \end{bmatrix}$. Then $c\mathsf{z} = \begin{bmatrix} cz_1 \\ \vdots \\ cz_n \end{bmatrix}$, so

$$\begin{aligned} (c\mathsf{z}) \cdot \mathsf{w} &= \overline{cz_1}w_1 + \overline{cz_2}w_2 + \cdots + \overline{cz_n}w_n \\ &= (\overline{c}\,\overline{z}_1)w_1 + (\overline{c}\,\overline{z}_2)w_2 + \cdots + (\overline{c}\,\overline{z}_n)w_n \\ &= \overline{c}(\overline{z}_1 w_1 + \overline{z}_2 w_2 + \cdots + \overline{z}_n w_n) = \overline{c}(\mathsf{z} \cdot \mathsf{w}). \end{aligned}$$

Matrices over C

If A is an $m \times n$ matrix, the *transpose* of A is the $n \times m$ matrix obtained from A by interchanging respective rows and columns and A is *symmetric* if $A^T = A$. For complex matrices, the analogue of "transpose" is *conjugate transpose* and the analogue of "symmetric" is *Hermitian*.[1]

7.1.6 Definitions. The *conjugate transpose* of a complex matrix A, denoted A^*, is the transpose of the matrix whose entries are the conjugates of those of A or, equivalently, the matrix obtained from A^T by taking the conjugates of its entries.

$$\boxed{A^* = \overline{A}^T = \overline{A^T}.}$$

Matrix A is *Hermitian* if $A^* = A$.

7.1.7 Example. If $A = \begin{bmatrix} 1+i & 3-2i \\ 2-4i & i \\ -5 & 5+7i \end{bmatrix}$, then $A^* = \begin{bmatrix} 1-i & 2+4i & -5 \\ 3+2i & -i & 5-7i \end{bmatrix}$. The 2×2 matrix $A = \begin{bmatrix} -3 & 4-i \\ 4+i & 7 \end{bmatrix}$ is Hermitian. So is $\begin{bmatrix} 1 & 2 & 3 \\ 2 & 4 & 5 \\ 3 & 5 & 6 \end{bmatrix}$. ⌣̈

[1]After the French mathematician Charles Hermite (1822–1901).

7.1.8 Remark. The last matrix here illustrates an important point. If A is real, then $A^* = A^T$. Thus a real matrix is Hermitian if and only if it is symmetric, so all statements about Hermitian matrices apply also to real symmetric matrices.

Since $(A^T)^T = A$ and $(AB)^T = B^T A^T$ for any matrices A and B (see the last part of Section 2.1) it follows that if A and B have complex entries

7.1.9
$$\boxed{(AB)^* = B^* A^* \text{ and } (A^*)^* = A.}$$

We have made great use of the fact that a dot product in R^n can be viewed as the product of a row matrix and a column matrix: if x and y are vectors in R^n, then $\mathsf{x} \cdot \mathsf{y} = \mathsf{x}^T \mathsf{y}$. A similar sort of thing happens in C^n.

Suppose $\mathsf{z} = \begin{bmatrix} z_1 \\ \vdots \\ z_n \end{bmatrix}$ and $\mathsf{w} = \begin{bmatrix} w_1 \\ \vdots \\ w_n \end{bmatrix}$ are vectors in C^n. Then

$$\mathsf{z} \cdot \mathsf{w} = \overline{z}_1 w_1 + \overline{z}_2 w_2 + \cdots + \overline{z}_n w_n = [\overline{z}_1 \; \overline{z}_2 \; \ldots \; \overline{z}_n] \begin{bmatrix} w_1 \\ w_2 \\ \vdots \\ w_n \end{bmatrix} = \mathsf{z}^* \mathsf{w}.$$

7.1.10
$$\boxed{\mathsf{z} \cdot \mathsf{w} = \mathsf{z}^* \mathsf{w}.}$$

7.1.11 Remark. The decision to define $\mathsf{z} \cdot \mathsf{w}$ as we did for complex vectors z and w—see 7.1.1—was a matter of taste. We wanted to obtain the formula $\mathsf{z} \cdot \mathsf{w} = \mathsf{z}^* \mathsf{w}$ because of its analogy with $\mathsf{x} \cdot \mathsf{y} = \mathsf{x}^T \mathsf{y}$ for real vectors. In other textbooks, you may well see

$$\mathsf{z} \cdot \mathsf{w} = z_1 \overline{w}_1 + z_2 \overline{w}_2 + \cdots + z_n \overline{w}_n,$$

which involves the complex conjugates of the components of w rather than z.

READING CHECK 1. Let $\mathsf{z} = \begin{bmatrix} z_1 \\ z_2 \\ \vdots \\ z_n \end{bmatrix}$ and $\mathsf{w} = \begin{bmatrix} w_1 \\ w_2 \\ \vdots \\ w_n \end{bmatrix}$ be complex vectors. The author defined

$$\mathsf{z} \cdot \mathsf{w} = \overline{z}_1 w_1 + \overline{z}_2 w_2 + \cdots + \overline{z}_n w_n,$$

but the student remembered

$$\mathsf{z} \odot \mathsf{w} = z_1 \overline{w}_1 + z_2 \overline{w}_2 + \cdots + z_n \overline{w}_n.$$

Do the student and author agree, or is there a difference?

Suppose A is an $n \times n$ matrix and λ is a scalar. Recall that a vector z is an *eigenvector* of A corresponding to the *eigenvalue* λ if $\mathsf{z} \neq 0$ and $A\mathsf{z} = \lambda \mathsf{z}$.

In Theorem 3.5.14, we saw that eigenvectors that correspond to different eigenvalues of a matrix are linearly independent. Now we shall see that in the case of Hermitian (and this includes real symmetric) matrices, eigenvectors corresponding to different eigenvalues are even orthogonal.

7.1.12 Theorem. (Properties of Hermitian matrices) *Let A be a Hermitian matrix. Then*

1. $\mathsf{z} \cdot A\mathsf{z}$ *is real for any* $\mathsf{z} \in \mathbb{C}^n$;
2. *all eigenvalues of A are real;*
3. *eigenvectors corresponding to different eigenvalues are orthogonal.*

As a special case, if A is a real symmetric matrix, then the eigenvalues of A are real and eigenvectors corresponding to different eigenvalues are orthogonal.

Proof. The final sentence follows from statements 2 and 3 and the fact that a real symmetric matrix is Hermitian, so we are left having to establish statements 1, 2 and 3.

1. A complex number is real if and only if it equals its conjugate, so we prove that $\mathsf{z} \cdot A\mathsf{z} = \overline{\mathsf{z} \cdot A\mathsf{z}}$. Now $\mathsf{z} \cdot A\mathsf{z} = \mathsf{z}^* A\mathsf{z}$ is a scalar, so it equals its transpose. Using this fact and $A^* = A$ gives what we want:

$$\overline{\mathsf{z} \cdot A\mathsf{z}} = \overline{\mathsf{z}^* A\mathsf{z}} = \overline{(\mathsf{z}^* A\mathsf{z})^T} = (\mathsf{z}^* A\mathsf{z})^* = \mathsf{z}^* A^* (\mathsf{z}^*)^* = \mathsf{z}^* A\mathsf{z} = \mathsf{z} \cdot A\mathsf{z}.$$

2. Suppose λ is an eigenvalue of A; thus $A\mathsf{z} = \lambda\mathsf{z}$ for some vector $\mathsf{z} \neq 0$. Now

$$\mathsf{z} \cdot A\mathsf{z} = \mathsf{z} \cdot \lambda\mathsf{z} = \lambda(\mathsf{z} \cdot \mathsf{z}) = \lambda \|\mathsf{z}\|^2 .$$

By part 1, $\mathsf{z} \cdot A\mathsf{z}$ is real. Also $\|\mathsf{z}\|^2$ is real, so $\lambda = \dfrac{\mathsf{z} \cdot A\mathsf{z}}{\|\mathsf{z}\|^2}$ is real.

3. Let λ and μ be eigenvalues of A corresponding, respectively, to eigenvectors z and w, and suppose that $\lambda \neq \mu$. We compute $\mathsf{z} \cdot A\mathsf{w}$ in two ways. On the one hand,

$$\mathsf{z} \cdot A\mathsf{w} = \mathsf{z} \cdot (\mu\mathsf{w}) = \mu(\mathsf{z} \cdot \mathsf{w}),$$

whereas, on the other hand, remembering that $A^* = A$,

$$\mathsf{z} \cdot A\mathsf{w} = \mathsf{z}^* A\mathsf{w} = \mathsf{z}^* A^* \mathsf{w} = (A\mathsf{z})^* \mathsf{w} = (A\mathsf{z}) \cdot \mathsf{w} = (\lambda\mathsf{z}) \cdot \mathsf{w} = \overline{\lambda}(\mathsf{z} \cdot \mathsf{w}).$$

By part 2, $\overline{\lambda} = \lambda$, so $\mu(\mathsf{z} \cdot \mathsf{w}) = \lambda(\mathsf{z} \cdot \mathsf{w})$ giving $(\mu - \lambda)(\mathsf{z} \cdot \mathsf{w}) = 0$. Since $\mu - \lambda \neq 0$, we obtain $\mathsf{z} \cdot \mathsf{w} = 0$ as desired. ∎

A real $n \times n$ matrix may have no real eigenvalues or it may have some that are real and others that are not. The matrix $A = \begin{bmatrix} 0 & -1 \\ 1 & 0 \end{bmatrix}$ has no real eigenvalues since its characteristic polynomial is

$$\begin{vmatrix} -\lambda & -1 \\ 1 & -\lambda \end{vmatrix} = \lambda^2 + 1.$$

The matrix $A = \begin{bmatrix} 1 & 0 & 0 \\ 0 & 0 & -1 \\ 0 & 1 & 0 \end{bmatrix}$ has one real eigenvalue and two eigenvalues that are not real since its characteristic polynomial is

$$\begin{vmatrix} 1-\lambda & 0 & 0 \\ 0 & -\lambda & -1 \\ 0 & 1 & -\lambda \end{vmatrix} = (1-\lambda)(\lambda^2+1).$$

It is remarkable that if a matrix is real and **symmetric**, its eigenvalues must all be real.

7.1.13 Example. The matrix $A = \begin{bmatrix} 1 & -2 \\ -2 & -2 \end{bmatrix}$ is real symmetric, so its eigenvalues must be real. We investigate.

The characteristic polynomial of A is

$$\det(A - \lambda I) = \begin{vmatrix} 1-\lambda & -2 \\ -2 & -2-\lambda \end{vmatrix} = \lambda^2 + \lambda - 6 = (\lambda+3)(\lambda-2).$$

The eigenvalues $\lambda = -3, 2$ are real. Since they are different, the corresponding eigenvectors should be orthogonal. Let $\mathsf{x} = \begin{bmatrix} x_1 \\ x_2 \end{bmatrix}$ be an eigenvector for $\lambda = -3$. Gaussian elimination applied to $A - \lambda I$ with $\lambda = -3$ gives

$$\begin{bmatrix} 4 & -2 \\ -2 & 1 \end{bmatrix} \to \begin{bmatrix} 1 & -\frac{1}{2} \\ 0 & 0 \end{bmatrix},$$

so $x_2 = t$ is free, $x_1 = \frac{1}{2}t$ and $\mathsf{x} = \begin{bmatrix} \frac{1}{2}t \\ t \end{bmatrix} = t\begin{bmatrix} \frac{1}{2} \\ 1 \end{bmatrix}$. If $\mathsf{x} = \begin{bmatrix} x_1 \\ x_2 \end{bmatrix}$ is an eigenvector for $\lambda = 2$, Gaussian elimination applied to $A - \lambda I$, with $\lambda = 2$, gives

$$\begin{bmatrix} -1 & -2 \\ -2 & -4 \end{bmatrix} \to \begin{bmatrix} 1 & 2 \\ 0 & 0 \end{bmatrix},$$

so $x_2 = t$ is free, $x_1 = -2t$, and $\mathsf{x} = \begin{bmatrix} -2t \\ t \end{bmatrix} = t\begin{bmatrix} -2 \\ 1 \end{bmatrix}$. The eigenspaces are each one-dimensional with bases $\begin{bmatrix} 1 \\ 2 \end{bmatrix}$ and $\begin{bmatrix} -2 \\ 1 \end{bmatrix}$ and indeed orthogonal. ☺

7.1.14 Example. We find the eigenvalues and corresponding eigenvectors of the Hermitian matrix $A = \begin{bmatrix} 1 & 1+i \\ 1-i & 2 \end{bmatrix}$. The determinant of $A - \lambda I$ is

$$\begin{aligned} \begin{vmatrix} 1-\lambda & 1+i \\ 1-i & 2-\lambda \end{vmatrix} &= (1-\lambda)(2-\lambda) - (1+i)(1-i) \\ &= (2 - 3\lambda + \lambda^2) - (1+1) = \lambda^2 - 3\lambda = \lambda(\lambda - 3), \end{aligned}$$

so the eigenvalues of A are $\lambda = 0, 3$ (both real, of course). To find the eigenvectors for $\lambda = 0$, we solve the homogeneous system $(A - \lambda I)\mathsf{z} = \mathsf{0}$ with $\lambda = 0$:

$$\begin{bmatrix} 1 & 1+i \\ 1-i & 2 \end{bmatrix} \to \begin{bmatrix} 1 & 1+i \\ 0 & 0 \end{bmatrix},$$

so if $\mathsf{z} = \begin{bmatrix} z_1 \\ z_2 \end{bmatrix}$, we have $z_2 = t$ free and $z_1 = -(1+i)z_2 = -(1+i)t$. Thus $\mathsf{z} = \begin{bmatrix} -(1+i)t \\ t \end{bmatrix} = t\begin{bmatrix} -(1+i) \\ 1 \end{bmatrix}$. To find the eigenvectors for $\lambda = 3$, we solve the homogeneous system $(A - \lambda I)\mathsf{z} = \mathsf{0}$ with $\lambda = 3$:

$$\begin{bmatrix} -2 & 1+i \\ 1-i & -1 \end{bmatrix} \to \begin{bmatrix} -2 & 1+i \\ 0 & 0 \end{bmatrix},$$

so with $\mathsf{z} = \begin{bmatrix} z_1 \\ z_2 \end{bmatrix}$, we again have $z_2 = t$ free and $2z_1 = (1+i)z_2$, so $z_1 = \frac{1+i}{2}z_2 = \frac{1+i}{2}t$. Thus $\mathsf{z} = \begin{bmatrix} \frac{1+i}{2}t \\ t \end{bmatrix} = t\begin{bmatrix} \frac{1+i}{2} \\ 1 \end{bmatrix}$. Note that any eigenvector for $\lambda = 0$ is orthogonal to any eigenvector for $\lambda = 3$ (as asserted in Theorem 7.1.12) because

$$\begin{bmatrix} -(1+i) \\ 1 \end{bmatrix} \cdot \begin{bmatrix} \frac{1+i}{2} \\ 1 \end{bmatrix} = -(1-i)\frac{1+i}{2} + 1(1) = -\frac{(1-i)(1+i)}{2} + 1 = 0.$$

(Don't forget the definition (7.1.1) of dot product in the complex case!)

Unitary Matrices

Remember that *orthonormal vectors* are orthogonal vectors of length 1, and an *orthogonal* matrix is a real square matrix with orthonormal columns. The analogue of this concept for complex matrices is "unitary matrix."

7.1.15 Definition. A matrix over C is *unitary* if it is square with orthonormal columns.

Recall that a real matrix Q has orthonormal columns if and only if $Q^TQ = I$—see 6.3.8. Similarly, a complex matrix U has orthonormal columns if and only if $U^*U = I$. If Q is square, then $Q^TQ = I$ implies $Q^T = Q^{-1}$; similarly, if U is square, then $U^*U = I$ implies $U^* = U^{-1}$.

7.1.16 Example. The matrix $U = \begin{bmatrix} \frac{1}{\sqrt{2}} & \frac{1+i}{2} \\ \frac{i}{\sqrt{2}} & \frac{1-i}{2} \end{bmatrix}$ has columns the vectors $\mathsf{z}_1 = \begin{bmatrix} \frac{1}{\sqrt{2}} \\ \frac{i}{\sqrt{2}} \end{bmatrix}$ and $\mathsf{z}_2 = \begin{bmatrix} \frac{1+i}{2} \\ \frac{1-i}{2} \end{bmatrix}$. These are unit vectors because

$$\|\mathsf{z}_1\|^2 = \frac{1}{\sqrt{2}}\frac{1}{\sqrt{2}} + \frac{-i}{\sqrt{2}}\frac{i}{\sqrt{2}} = \frac{1}{2} + \frac{1}{2} = 1$$

and

$$\|\mathsf{z}_2\|^2 = \frac{1-i}{2}\frac{1+i}{2} + \frac{1+i}{2}\frac{1-i}{2} = \frac{2}{4} + \frac{2}{4} = 1.$$

They are orthogonal because

$$\mathsf{z}_1 \cdot \mathsf{z}_2 = \frac{1}{\sqrt{2}}\frac{1+i}{2} + \frac{-i}{\sqrt{2}}\frac{1-i}{2} = \frac{1+i}{2\sqrt{2}} + \frac{-i+i^2}{2\sqrt{2}} = 0,$$

so the columns are orthonormal and, since U is square, it is unitary. Equivalently, we note that $U^*U = \begin{bmatrix} \frac{1}{\sqrt{2}} & \frac{1}{\sqrt{2}} \\ \frac{1-i}{\sqrt{2}} & \frac{1+i}{2} \end{bmatrix} \begin{bmatrix} \frac{1}{\sqrt{2}} & \frac{1+i}{2} \\ \frac{i}{\sqrt{2}} & \frac{1-i}{2} \end{bmatrix} = \begin{bmatrix} 1 & 0 \\ 0 & 1 \end{bmatrix} = I$. :-)

7.1.17 Example. The matrix $Q = \begin{bmatrix} \frac{1}{\sqrt{2}} & -\frac{1}{\sqrt{2}} \\ \frac{1}{\sqrt{2}} & \frac{1}{\sqrt{2}} \end{bmatrix}$ has orthonormal columns because

$$Q^TQ = \begin{bmatrix} \frac{1}{\sqrt{2}} & \frac{1}{\sqrt{2}} \\ -\frac{1}{\sqrt{2}} & \frac{1}{\sqrt{2}} \end{bmatrix} \begin{bmatrix} \frac{1}{\sqrt{2}} & -\frac{1}{\sqrt{2}} \\ \frac{1}{\sqrt{2}} & \frac{1}{\sqrt{2}} \end{bmatrix} = \begin{bmatrix} 1 & 0 \\ 0 & 1 \end{bmatrix} = I$$

and, since it is square and real, Q is orthogonal. :-)

Have another look at the definition of the complex dot product given in 7.1.1. If all the components of z are real, then each $\overline{z}_i = z_i$ and $\mathsf{z} \cdot \mathsf{w}$ is just the usual dot product in R^n. Thus two vectors in R^n are orthogonal with respect to the complex dot product if and only if they are orthogonal in R^n with respect to the usual dot product there. In particular, a unitary matrix whose entries are all real numbers is an orthogonal matrix, so anything we can prove about unitary matrices holds equally for real orthogonal matrices.

7.1.18 Theorem (Properties of Unitary Matrices). *Let U be a unitary matrix.*

1. *U preserves dot products and lengths: $U\mathsf{z} \cdot U\mathsf{w} = \mathsf{z} \cdot \mathsf{w}$ and $\|U\mathsf{z}\| = \|\mathsf{z}\|$ for any vectors z and w in C^n.*

2. *The eigenvalues of U have modulus* 1*:* $|\lambda| = 1$ *for any eigenvalue* λ.
3. *Eigenvectors corresponding to different eigenvalues are orthogonal.*

Proof. 1. Since $U^*U = I$, we have $U\mathsf{z} \cdot U\mathsf{w} = (U\mathsf{z})^*(U\mathsf{w}) = \mathsf{z}^*U^*U\mathsf{w} = \mathsf{z}^*\mathsf{w} = \mathsf{z} \cdot \mathsf{w}$, so U preserves the dot product. It follows that U also preserves length since $\|U\mathsf{z}\|^2 = U\mathsf{z} \cdot U\mathsf{z} = \mathsf{z} \cdot \mathsf{z} = \|\mathsf{z}\|^2$.

2. Let λ be an eigenvalue of U and z a corresponding eigenvector; thus $\mathsf{z} \neq 0$ and $U\mathsf{z} = \lambda\mathsf{z}$. By part 1, $U\mathsf{z} \cdot U\mathsf{z} = \mathsf{z} \cdot \mathsf{z}$, so $(\lambda\mathsf{z}) \cdot (\lambda\mathsf{z}) = \mathsf{z} \cdot \mathsf{z}$. Thus $\overline{\lambda}\lambda(\mathsf{z} \cdot \mathsf{z}) = \mathsf{z} \cdot \mathsf{z}$. Since $\mathsf{z} \cdot \mathsf{z} = \|\mathsf{z}\|^2 \neq 0$, we get $\overline{\lambda}\lambda = 1$, so $|\lambda|^2 = 1$ and $|\lambda| = 1$.

3. This is very similar to part 2. Let λ and μ be different eigenvalues of U corresponding to the eigenvectors z and w, respectively. Then

$$\mathsf{z} \cdot \mathsf{w} = U\mathsf{z} \cdot U\mathsf{w} = (\lambda\mathsf{z}) \cdot (\mu\mathsf{w}) = (\overline{\lambda}\mu)(\mathsf{z} \cdot \mathsf{w}),$$

so $(\overline{\lambda}\mu - 1)(\mathsf{z} \cdot \mathsf{w}) = 0$ and we can conclude $\mathsf{z} \cdot \mathsf{w} = 0$, as desired, if we can show that $\overline{\lambda}\mu \neq 1$. But if $\overline{\lambda}\mu = 1$, multiplying by λ gives $\lambda\overline{\lambda}\mu = \lambda$, so $|\lambda|^2\mu = \lambda$ (because $|\lambda| = 1$) implying $\mu = \lambda$, which is not true, so we are done. ■

Answers to Reading Checks

1. The two versions of dot product are not the same. Since $\overline{\overline{z}w} = z\overline{w}$, the author's and student's versions are complex conjugates: $\mathsf{z} \odot \mathsf{w} = \overline{\mathsf{z} \cdot \mathsf{w}}$.

True/False Questions

Decide, with as little calculation as possible, whether each of the following statements is true or false and, if you say "false," explain your answer. (Answers are at the back of the book.)

1. If $z = \begin{bmatrix} 1 \\ i \end{bmatrix}$ and $w = \begin{bmatrix} -1 \\ 2i \end{bmatrix}$, then $z \cdot w = -3$.
2. If $A = \begin{bmatrix} 1 & i \\ 1-i & 2i \end{bmatrix}$, then $A^* = \begin{bmatrix} 1 & 1+i \\ -i & 2i \end{bmatrix}$.
3. The diagonal entries of a Hermitian matrix are real.
4. All the eigenvalues $\begin{bmatrix} 1 & 1-i & 3-2i \\ 1+i & -5 & -i \\ 3+2i & i & 0 \end{bmatrix}$ are real.
5. A real orthogonal matrix is unitary.
6. If U is a unitary matrix and $\mathsf{z} \cdot \mathsf{w} = 0$, then $U\mathsf{z} \cdot U\mathsf{w} = 0$.
7. If the eigenvalues of a matrix U are $-1, \frac{1 \pm i}{\sqrt{2}}$, the matrix is unitary.

8. A rotation matrix is orthogonal.

9. Matrix $U = \begin{bmatrix} \frac{1}{\sqrt{3}} & -\frac{i}{\sqrt{2}} \\ \frac{i}{\sqrt{3}} & 0 \\ \frac{1}{\sqrt{3}} & \frac{i}{\sqrt{2}} \end{bmatrix}$ is unitary.

10. Matrix $H = \begin{bmatrix} 2 & 1-3i & 2+i \\ 1+3i & 6 & 7 \\ 2-i & 7 & 9 \end{bmatrix}$ is Hermitian.

EXERCISES

*Answers to exercises marked [BB] can be found at the **Back of the Book**.*

1. Find $A+B$, $A-B$, $2A-B$, AB, and BA (where possible) in each of the following cases.

   (a) [BB] $A = \begin{bmatrix} 1+i & 1 \\ 2-2i & -3i \end{bmatrix}$, $B = \begin{bmatrix} 1-i & 3i \\ -3 & -i \end{bmatrix}$

   (b) $A = \begin{bmatrix} 3+i & i \\ 4+i & 0 \end{bmatrix}$, $B = \begin{bmatrix} -1 & -3-3i \\ -i & -1 \end{bmatrix}$

   (c) [BB] $A = \begin{bmatrix} i & 2+i \\ 2-2i & -5 \end{bmatrix}$, $B = \begin{bmatrix} i & 2-i \\ 1+3i & 0 \\ -3 & -i \end{bmatrix}$

   (d) $A = \begin{bmatrix} -i & 3+3i \\ -2+3i & -7 \end{bmatrix}$, $B = \begin{bmatrix} 1 \\ i \end{bmatrix}$

   (e) $A = \begin{bmatrix} 1+i & 2i & -1 \\ 3-i & i & 1-i \\ 1 & 1-2i & -i \end{bmatrix}$, $B = \begin{bmatrix} 1+i & 2-i \\ -i & 4 \\ 1+2i & 3-2i \end{bmatrix}$

   (f) $A = \begin{bmatrix} -1+2i & 2 & 3 \\ 3-i & 0 & -6+2i \\ i & -i & i \end{bmatrix}$, $B = \begin{bmatrix} 1 & -6-6i & -3i \\ i & 2 & 6 \\ 4+4i & 2i & 3+i \end{bmatrix}$.

2. Let $A = \begin{bmatrix} 1-i & 3i \\ -3 & -i \end{bmatrix}$. Find $\det A$ and $\det(4iA)$.

3. Find the conjugate transpose of each matrix.

   (a) $A = \begin{bmatrix} 2 & 1-i \\ 3+4i & 3-i \\ i & 4+7i \end{bmatrix}$ **(b)** $A = \begin{bmatrix} 2i & 1-i & 5-i \\ 3 & 4+i & 7-2i \end{bmatrix}$

4. [BB] One of the matrices below is unitary, another is Hermitian, and two others are conjugate transposes. Which is which?

$$A = \begin{bmatrix} 1+i & 1-i \\ -1+i & -1-i \end{bmatrix}, \quad B = \tfrac{1}{2}\begin{bmatrix} 1+i & 1-i \\ 1-i & 1+i \end{bmatrix},$$

$$C = \tfrac{1}{\sqrt{2}}\begin{bmatrix} 1-i & -1-i \\ 1+i & -1+i \end{bmatrix}, \quad D = \tfrac{1}{\sqrt{2}}\begin{bmatrix} 1 & i \\ i & -1 \end{bmatrix}, \quad E = \tfrac{1}{\sqrt{2}}\begin{bmatrix} 1 & i \\ -i & 1 \end{bmatrix}.$$

5. **(a)** Find a unit vector in the direction of $\mathsf{z} = \begin{bmatrix} -i \\ 2 \\ 1+i \\ 2-3i \end{bmatrix}$.

(b) Find a vector of length 2 in direction opposite $\mathsf{z} = \begin{bmatrix} 2+i \\ 1-2i \\ 4 \\ -i \end{bmatrix}$.

(c) Find two unit vectors parallel to $\mathsf{z} = \begin{bmatrix} 3+4i \\ -3i \\ 1-i \\ 2 \end{bmatrix}$.

6. In each case, find $\|\mathsf{u}\|$, $\|\mathsf{v}\|$, a unit vector in the direction of u and $\mathsf{u}\cdot\mathsf{v}$. If u and v are orthogonal, say so.

(a) [BB] $\mathsf{u} = \begin{bmatrix} 1+2i \\ i \end{bmatrix}, \mathsf{v} = \begin{bmatrix} 3-i \\ 1-i \end{bmatrix}$ **(b)** $\mathsf{u} = \begin{bmatrix} 3-4i \\ 2+i \end{bmatrix}, \mathsf{v} = \begin{bmatrix} 1-2i \\ 1+i \end{bmatrix}$

(c) $\mathsf{u} = \begin{bmatrix} 2+3i \\ 3+2i \end{bmatrix}, \mathsf{v} = \begin{bmatrix} 5+i \\ -5+i \end{bmatrix}$ **(d)** $\mathsf{u} = \begin{bmatrix} 2-i \\ 1+3i \\ -i \end{bmatrix}, \mathsf{v} = \begin{bmatrix} 1+i \\ 2+3i \\ 12i \end{bmatrix}$

(e) $\mathsf{u} = \begin{bmatrix} 1-i \\ 1+2i \\ -i \end{bmatrix}, \mathsf{v} = \begin{bmatrix} 2+i \\ -1-4i \\ i \end{bmatrix}$.

7. [BB] In this section, we defined the length $\|\mathsf{z}\|$ of $\mathsf{z} = \begin{bmatrix} z_1 \\ z_2 \\ \vdots \\ z_n \end{bmatrix}$ by $\|\mathsf{z}\|^2 = \mathsf{z}\cdot\mathsf{z}$, "$\cdot$" denoting the complex dot product. Show that $\|\mathsf{z}\|$ is a real number.

8. Let c be a complex number and let z be a vector in C^n. Show that $\|c\mathsf{z}\| = |c|\,\|\mathsf{z}\|$.

9. **(a)** [BB] Use equations (2) and (3) to show that $\mathsf{z}\cdot(c\mathsf{w}) = c(\mathsf{z}\cdot\mathsf{w})$ for any complex number c and any vectors z and w in C^n.

(b) Let c denote a complex number and z,w vectors in C^n. Find a complex number q such that $z\cdot(c\mathsf{w}) = (q\mathsf{z})\cdot\mathsf{w}$. (Assume $\mathsf{z}\cdot\mathsf{w} \neq 0$.)

10. [BB] Suppose $\mathsf{z} = \begin{bmatrix} z_1 \\ \vdots \\ z_n \end{bmatrix}$ is a vector in C_n and $\overline{\mathsf{z}}$ is the vector $\begin{bmatrix} \overline{z_1} \\ \vdots \\ \overline{z_n} \end{bmatrix}$ whose components are the complex conjugates of the components of z. Show that z and $\overline{\mathsf{z}}$ have the same length.

11. Given nonzero vectors u and v in C^n, the *projection* of u on v is the vector $\mathsf{p} = c\mathsf{v}$ with the property that $\mathsf{u}-\mathsf{p}$ is orthogonal to v.

(a) [BB] Find a formula for the projection of u on v.

(b) Find the projection of $\mathsf{u} = \begin{bmatrix} 1 \\ -i \\ 0 \end{bmatrix}$ on $\mathsf{v} = \begin{bmatrix} 3+2i \\ -i \\ 2 \end{bmatrix}$ [BB] and v on u.

(c) Find two orthogonal vectors that span the same subspace as u and v.

12. Find the eigenvalues and the corresponding eigenspaces of each of the following matrices over the complex numbers.

(a) [BB] $A = \begin{bmatrix} 2 & -3 \\ 1 & -1 \end{bmatrix}$ (b) $A = \begin{bmatrix} -1 & 4+2i \\ 4-2i & 7 \end{bmatrix}$

(c) $A = \begin{bmatrix} 2 & -1 & -2 \\ -2 & 3 & 3 \\ 2 & -1 & -1 \end{bmatrix}$ (d) $A = \begin{bmatrix} 0 & 1 & 0 \\ 0 & 0 & 1 \\ -52 & 3 & 0 \end{bmatrix}$.

13. A real matrix is *skew-symmetric* if $A^T = -A$.

(a) [BB] Show that $\mathsf{x}^T A\mathsf{x} = 0$ for any real skew-symmetric $n \times n$ matrix A and any vector x in R^n. [Hint: $\mathsf{x}^T A\mathsf{x}$ is 1×1.]

(b) Where is the fact that x is real used in part (a)?

(c) Prove that the only real eigenvalue of a real skew-symmetric matrix is 0.

14. A complex matrix A *skew-Hermitian* if $A^* = -A$. Let A be a complex $n \times n$ matrix.

(a) Show that $A + A^*$ is Hermitian.

(b) Show that $A - A^*$ is skew-Hermitian.

(c) Show that $A = S + K$ is the sum of a Hermitian matrix S and a skew-Hermitian matrix K.

15. Let K be a skew-Hermitian matrix. (See Exercise 14.)

(a) Show that K^2 and iK are Hermitian.

(b) Show that any eigenvalue λ of K is pure imaginary, that is, $\lambda = ib$ for some real number b.

16. [BB] Let A be an $n \times n$ matrix that commutes with its conjugate transpose and let z be a vector in C^n. Prove that $\|A\mathsf{z}\| = \|A^*\mathsf{z}\|$.

17. Let A and B be $n \times n$ complex matrices. True or false, and justify your answer.

(a) [BB] If A and B are unitary, so is AB.

(b) [BB] If A and B are symmetric, so is AB.

(c) If A and B are Hermitian, so is AB.

(d) If A and B each commute with their conjugate transposes, so does AB.

18. (a) If A is a square complex matrix, show that $(A^*)^{-1} = (A^{-1})^*$. [Hint: Remember that square matrices X and Y are inverses if and only if $XY = I$.]

(b) If U is a unitary matrix, show that U^{-1} is also unitary.

19. In each case, give an example or explain why no example exists.

(a) [BB] a Hermitian matrix that is not symmetric

(b) a symmetric matrix that is not Hermitian

(c) a 3×3 matrix, all entries real, that has no real eigenvalues.

20. (a) Show that the orthogonal complement of the column space of a complex matrix A is the null space of A^*: symbolically, $(\operatorname{col} \operatorname{sp} A)^\perp = \operatorname{null} \operatorname{sp} A^*$.

(b) Use (a) to show that the null space of A and the column space of A^* are orthogonal complements.

21. Prove by elementary methods that the eigenvalues of a real 2×2 symmetric matrix $A = \begin{bmatrix} a & b \\ b & c \end{bmatrix}$ are real.

Critical Reading

22. The eigenvalues of a permutation matrix have modulus 1. Why?

7.1.19 Definition. A subspace U of R^n or C^n is said to *invariant* under an $n \times n$ matrix A if $AU \subseteq U$ (AU is contained in U).

23. [BB] Let U be an eigenspace of a matrix A.

    (a) Show that U is invariant under A.

    (b) If A is Hermitian (and vector spaces are complex), show that $U^\perp$ is also invariant under A.

24. If A and B are matrices that commute and U is an eigenspace of A, show that U is invariant under B.

25. Let A be a complex $n \times n$ matrix and let U be a subspace of C^n that is invariant under A. Show that $U^\perp$ is invariant under A^*.

26. (a) Show that any 2×2 Hermitian matrix over the complex numbers is similar to a diagonal matrix. [Hint: The matrix A has an eigenvector $\mathbf{z}_1$. Why? Let U be the eigenspace spanned by $\mathbf{z}_1$ and write $\mathsf{C}^2 = U \oplus U^\perp$. Since $\dim U^\perp = 1$ (why?) $U^\perp$ has a basis $\mathbf{z}_2$. Using the result of Exercise 25, we see that the vector $\mathbf{z}_2$ is also an eigenvector of A. Why? Now let P be the matrix whose columns are $\mathbf{z}_1$ and $\mathbf{z}_2$.]

    (b) The matrix P satisfying $P^{-1}AP = D$, which exists by part (c), can be chosen unitary. Why?

    [This exercise outlines a proof of the "Spectral Theorem" for 2×2 matrices and can be generalized to arbitrary matrices by a technique known as "mathematical induction." See Theorems 7.2.8 and 7.2.16.]

7.2 Unitary Diagonalization

In this section, we prove that every Hermitian matrix is similar to a real diagonal matrix and, moreover, that the diagonalizing matrix can be chosen to be unitary.

Since this section depends heavily on the notions of eigenvalue, eigenvector and diagonalizability, the reader should be very comfortable with the material in Sections 3.4 and 3.5 before continuing.

Unless we state specifically to the contrary, we assume in what follows that all matrices have entries that are complex numbers. (This includes the possibility that all entries are real, of course.)

A Theorem of Schur[2]

We first introduced the concept of similarity in Section 3.5. Square matrices A and B are *similar* if there exists an invertible matrix P such that $P^{-1}AP = B$. An $n \times n$ matrix A is *diagonalizable* if it is similar to a diagonal matrix, equivalently, A has n linearly independent eigenvectors—see Theorem 3.5.9. Not all matrices are diagonalizable, by any means, but every matrix can be triangularized.

7.2.1 **Theorem (Schur).** *Any $n \times n$ matrix over C is similar to a triangular matrix via a unitary matrix; that is, there exists a unitary matrix U such that $U^*AU = T$ is upper triangular.*

Proof. We provide a proof in the case $n = 3$ with an argument that can be generalized in a straightforward manner.

Let A be a 3×3 matrix over C. By the Fundamental Theorem of Algebra, the characteristic polynomial of A has a root, so A has an eigenvalue λ_1. Let z_1 be a corresponding eigenvector of length 1. Now z_1 is part of basis for C^3 (any linearly independent set of vectors can be extended to a basis). Applying the Gram-Schmidt algorithm to this basis, we obtain an orthonormal basis z_1, u, v for C^3. Form the (unitary) matrix $U_1 = \begin{bmatrix} \mathsf{z}_1 & \mathsf{u} & \mathsf{v} \\ \downarrow & \downarrow & \downarrow \\ & & \end{bmatrix}$ whose columns are these vectors. Then

$$AU_1 = A\begin{bmatrix} \mathsf{z}_1 & \mathsf{u} & \mathsf{v} \\ \downarrow & \downarrow & \downarrow \\ & & \end{bmatrix} = \begin{bmatrix} A\mathsf{z}_1 & A\mathsf{u} & A\mathsf{v} \\ \downarrow & \downarrow & \downarrow \\ & & \end{bmatrix} = \begin{bmatrix} \lambda_1\mathsf{z}_1 & A\mathsf{u} & A\mathsf{v} \\ \downarrow & \downarrow & \downarrow \\ & & \end{bmatrix}$$

and

$$U_1^*AU_1 = \begin{bmatrix} \bar{\mathsf{z}}_1^T & \rightarrow \\ \bar{\mathsf{u}}^T & \rightarrow \\ \bar{\mathsf{v}}^T & \rightarrow \end{bmatrix}\begin{bmatrix} \lambda_1\mathsf{z}_1 & A\mathsf{u} & A\mathsf{v} \\ \downarrow & \downarrow & \downarrow \\ & & \end{bmatrix} = \begin{bmatrix} \lambda_1 & \star & \star \\ 0 & & \\ 0 & \multicolumn{2}{c}{B} \end{bmatrix} \tag{1}$$

[2]Issai Schur (1875–1941), who lived most of his life in Germany and is especially known for his work in group theory and number theory, is considered to be one of the world's greatest mathematicians. The author is a fifth generation student of Schur's, through Brauer, Bruck, Kleinfeld and Anderson.

because z_1 of length 1 gives $\overline{\mathsf{z}}_1^T(\lambda_1\mathsf{z}_1) = \lambda_1(\mathsf{z}_1^*\mathsf{z}_1) = \lambda_1\|\mathsf{z}_1\|^2 = \lambda_1$ and z_1 orthogonal to u and v gives $\overline{\mathsf{u}}^T\mathsf{z}_1 = 0 = \overline{\mathsf{v}}^T\mathsf{z}_1$. We don't know anything about the second or third column of $U_1^*AU_1$, but we shall be interested in the 2×2 matrix B in the lower right corner.

Let λ_2 be an eigenvalue of B and then, just as in equation (1), find a unitary 2×2 matrix P so that

$$P^*BP = \begin{bmatrix} \lambda_2 & \star \\ 0 & \lambda_3 \end{bmatrix}.$$

Let $U_2 = \begin{bmatrix} 1 & 0 \quad 0 \\ 0 & \\ 0 & P \end{bmatrix}$. Then U_2 is unitary, $U_2^* = \begin{bmatrix} 1 & 0 \quad 0 \\ 0 & \\ 0 & P^* \end{bmatrix}$ and

$$\begin{aligned} U_2^*(U_1^*AU_1)U_2 &= \begin{bmatrix} 1 & 0 \quad 0 \\ 0 & \\ 0 & P^* \end{bmatrix}\begin{bmatrix} \lambda_1 & \star \quad \star \\ 0 & \\ 0 & B \end{bmatrix}\begin{bmatrix} 1 & 0 \quad 0 \\ 0 & \\ 0 & P \end{bmatrix} \\ &= \begin{bmatrix} \lambda_1 & \star \quad \star \\ 0 & \\ 0 & P^*BP \end{bmatrix} = \begin{bmatrix} \lambda_1 & \star & \star \\ 0 & \lambda_2 & \star \\ 0 & 0 & \lambda_3 \end{bmatrix} = T \end{aligned}$$

is upper triangular. Moreover this is U^*AU with $U = U_1U_2$. This finishes the proof because U is unitary: $U^* = (U_1U_2)^* = U_2^*U_1^* = U_2^{-1}U_1^{-1} = (U_1U_2)^{-1} = U^{-1}$. ■

7.2.2 Example. We demonstrate the proof of Schur's Theorem using the matrix $A = \begin{bmatrix} 5 & 8 & 16 \\ 5 & 0 & 9 \\ -3 & -5 & -10 \end{bmatrix}$. The characteristic polynomial of A is $-(\lambda-1)(\lambda+3)^2$, so the eigenvalues are 1 and -3. An eigenvalue for $\lambda = 1$ is $\mathsf{z} = \begin{bmatrix} -2 \\ -1 \\ 1 \end{bmatrix}$. We extend to the basis $\mathsf{z}, \mathsf{e}_1, \mathsf{e}_2$ and then apply the Gram-Schmidt process to produce an orthonormal basis

$$\mathsf{z}_1 = \begin{bmatrix} \frac{-2}{\sqrt{6}} \\ \frac{-1}{\sqrt{6}} \\ \frac{1}{\sqrt{6}} \end{bmatrix}, \quad \mathsf{u} = \begin{bmatrix} \frac{1}{\sqrt{3}} \\ \frac{-1}{\sqrt{3}} \\ \frac{1}{\sqrt{3}} \end{bmatrix}, \quad \mathsf{v} = \begin{bmatrix} 0 \\ \frac{1}{\sqrt{2}} \\ \frac{1}{\sqrt{2}} \end{bmatrix}.$$

(We expect the reader to verify!) The matrix $U_1 = \begin{bmatrix} \frac{-2}{\sqrt{6}} & \frac{1}{\sqrt{3}} & 0 \\ \frac{-1}{\sqrt{6}} & \frac{-1}{\sqrt{3}} & \frac{1}{\sqrt{2}} \\ \frac{1}{\sqrt{6}} & \frac{1}{\sqrt{3}} & \frac{1}{\sqrt{2}} \end{bmatrix}$ is unitary and $U_1^*AU_1 = \begin{bmatrix} 1 & -8\sqrt{2} & -12\sqrt{3} \\ 0 & -3 & 0 \\ 0 & \sqrt{6} & -3 \end{bmatrix}$, which is $\begin{bmatrix} 1 & -8\sqrt{2} \quad -12\sqrt{3} \\ 0 & \\ 0 & B \end{bmatrix}$ with $B = \begin{bmatrix} -3 & 0 \\ \sqrt{6} & -3 \end{bmatrix}$. An eigenvalue of B is -3 (in fact, this is the only eigenvalue) and $\mathsf{e}_2 = \begin{bmatrix} 0 \\ 1 \end{bmatrix}$ is a corresponding eigenvector. An obvious extension of e_2 to an orthonormal basis of R^2 is $\mathsf{e}_2, \mathsf{e}_1$. Thus $P = \begin{bmatrix} 0 & 1 \\ 1 & 0 \end{bmatrix}$ is unitary and $P^*BP = \begin{bmatrix} -3 & \sqrt{6} \\ 0 & -3 \end{bmatrix}$. With $U_2 = \begin{bmatrix} 1 & 0 \;\; 0 \\ 0 & \\ 0 & P \end{bmatrix} = \begin{bmatrix} 1 & 0 & 0 \\ 0 & 0 & 1 \\ 0 & 1 & 0 \end{bmatrix}$, we have

$$\begin{aligned} U_2^* U_1^* A U_1 U_2 &= U_2^* \begin{bmatrix} 1 & -8\sqrt{2} \;\; -12\sqrt{3} \\ 0 & \\ 0 & B \end{bmatrix} U_2 \\ &= \begin{bmatrix} 1 & -8\sqrt{2} \;\; -12\sqrt{3} \\ 0 & \\ 0 & P^*BP \end{bmatrix} = \begin{bmatrix} 1 & -8\sqrt{2} & -12\sqrt{3} \\ 0 & -3 & \sqrt{6} \\ 0 & 0 & -3 \end{bmatrix}. \end{aligned}$$

This is U^*AU with $U = U_1U_2$ unitary. :-)

Schur's Theorem leads to what has always struck the author as one of the most beautiful (and surprising) theorems of linear algebra: Every matrix satisfies its characteristic polynomial!

We demonstrate first with an example.

7.2.3 Example. Let $A = \begin{bmatrix} -1 & 2 \\ 3 & 5 \end{bmatrix}$. The characteristic polynomial of A is

$$\begin{vmatrix} -1-\lambda & 2 \\ 3 & 5-\lambda \end{vmatrix} = -5 - 4\lambda + \lambda^2 - 6 = \lambda^2 - 4\lambda - 11.$$

Evaluating this polynomial at the matrix A gives

$$A^2 - 4A - 11I = \begin{bmatrix} 7 & 8 \\ 12 & 31 \end{bmatrix} - 4\begin{bmatrix} -1 & 2 \\ 3 & 5 \end{bmatrix} - 11\begin{bmatrix} 1 & 0 \\ 0 & 1 \end{bmatrix} = \begin{bmatrix} 0 & 0 \\ 0 & 0 \end{bmatrix}.$$:-)

As another prelude to the theorem we have mentioned—Theorem 7.2.5—let us try to understand why a 3×3 triangular matrix might satisfy its characteristic polynomial.

7.2.4 Let $T = \begin{bmatrix} \lambda_1 & a & b \\ 0 & \lambda_2 & c \\ 0 & 0 & \lambda_3 \end{bmatrix}$ be an upper triangular 3×3 matrix. The determinant of a triangular matrix is the product of its diagonal entries, so the characteristic polynomial of T is

$$\det(T - \lambda I) = \begin{vmatrix} \lambda_1 - \lambda & a & b \\ 0 & \lambda_2 - \lambda & c \\ 0 & 0 & \lambda_3 - \lambda \end{vmatrix} = (\lambda_1 - \lambda)(\lambda_2 - \lambda)(\lambda_3 - \lambda).$$

Our goal is to discover why this polynomial, evaluated at T, should be 0, that is, why T should satisfy the equation

$$(\lambda_1 I - T)(\lambda_2 I - T)(\lambda_3 I - T) = 0.$$

The factors of the left side are the matrices

$$\lambda_1 I - T = \begin{bmatrix} 0 & -a & -b \\ 0 & \lambda_1 - \lambda_2 & -c \\ 0 & 0 & \lambda_1 - \lambda_3 \end{bmatrix} = A_1$$

$$\lambda_2 I - T = \begin{bmatrix} \lambda_2 - \lambda_1 & -a & -b \\ 0 & 0 & -c \\ 0 & 0 & \lambda_2 - \lambda_3 \end{bmatrix} = A_2$$

$$\lambda_3 I - T = \begin{bmatrix} \lambda_3 - \lambda_1 & -a & -b \\ 0 & \lambda_3 - \lambda_2 & -c \\ 0 & 0 & 0 \end{bmatrix} = A_3$$

and we wish to show that $A_1 A_2 A_3 = 0$. Remember how matrix multiplication works:

$$A \begin{bmatrix} \mathsf{b}_1 & \mathsf{b}_2 & \mathsf{b}_3 \\ \downarrow & \downarrow & \downarrow \end{bmatrix} = \begin{bmatrix} A\mathsf{b}_1 & A\mathsf{b}_2 & A\mathsf{b}_3 \\ \downarrow & \downarrow & \downarrow \end{bmatrix}.$$

Remember too that each vector $A\mathsf{b}_i$ is a linear combination of the columns of A, with coefficients the components of b_i. Consider then the product $A_1 A_2$. The first column is the product of A_1 and the first column of A_2:

$$\begin{bmatrix} 0 & -a & -b \\ 0 & \lambda_1 - \lambda_2 & -c \\ 0 & 0 & \lambda_1 - \lambda_3 \end{bmatrix} \begin{bmatrix} \lambda_2 - \lambda_1 \\ 0 \\ 0 \end{bmatrix} = (\lambda_2 - \lambda_1) \begin{bmatrix} 0 \\ 0 \\ 0 \end{bmatrix} = \begin{bmatrix} 0 \\ 0 \\ 0 \end{bmatrix}.$$

The second column is the product of A_1 and the second column of A_2:

$$\begin{bmatrix} 0 & -a & -b \\ 0 & \lambda_1 - \lambda_2 & -c \\ 0 & 0 & \lambda_1 - \lambda_3 \end{bmatrix} \begin{bmatrix} -a \\ 0 \\ 0 \end{bmatrix} = -a \begin{bmatrix} 0 \\ 0 \\ 0 \end{bmatrix} = \begin{bmatrix} 0 \\ 0 \\ 0 \end{bmatrix}.$$

The first two columns of $A_1 A_2$ are 0, so $A_1 A_2 = \begin{bmatrix} 0 & 0 & \mathsf{x} \\ 0 & 0 & \downarrow \\ 0 & 0 & \end{bmatrix}$ for some column x.

Now consider the product of $A_1 A_2$ and A_3. The first column is

$$\begin{bmatrix} 0 & 0 & \mathsf{x} \\ 0 & 0 & \downarrow \\ 0 & 0 & \end{bmatrix} \begin{bmatrix} \lambda_3 - \lambda_1 \\ 0 \\ 0 \end{bmatrix} = (\lambda_3 - \lambda_1) \begin{bmatrix} 0 \\ 0 \\ 0 \end{bmatrix} = \begin{bmatrix} 0 \\ 0 \\ 0 \end{bmatrix}.$$

The second column is

$$\begin{bmatrix} 0 & 0 & \mathsf{x} \\ 0 & 0 & \downarrow \\ 0 & 0 & \end{bmatrix} \begin{bmatrix} -a \\ \lambda_3 - \lambda_2 \\ 0 \end{bmatrix} = -a \begin{bmatrix} 0 \\ 0 \\ 0 \end{bmatrix} + (\lambda_3 - \lambda_2) \begin{bmatrix} 0 \\ 0 \\ 0 \end{bmatrix} + 0\mathsf{x} = \begin{bmatrix} 0 \\ 0 \\ 0 \end{bmatrix}.$$

The third column is

$$\begin{bmatrix} 0 & 0 & \mathsf{x} \\ 0 & 0 & \downarrow \\ 0 & 0 & \end{bmatrix}\begin{bmatrix} -b \\ -c \\ 0 \end{bmatrix} = -b\begin{bmatrix} 0 \\ 0 \\ 0 \end{bmatrix} + -c\begin{bmatrix} 0 \\ 0 \\ 0 \end{bmatrix} + 0\mathsf{x} = \begin{bmatrix} 0 \\ 0 \\ 0 \end{bmatrix}.$$

Each column of $(A_1A_2)A_3$ is 0. So $A_1A_2A_3 = 0$: The triangular matrix T satisfies its characteristic polynomial.

Having examined the special case of a triangular matrix, we state the result for arbitrary (square) matrices.

7.2.5 Theorem (Cayley–Hamilton).[3] *Any (square) matrix satisfies its characteristic polynomial.*

Proof. First, we remark that the reasoning applied to a 3×3 triangular matrix in Example 7.2.4 generalizes to triangular matrices of any size. Specifically, if T is upper triangular with $\lambda_1, \lambda_2, \ldots, \lambda_n$ on the diagonal, then

- $\lambda_1 I - T$ has first column 0,
- $(\lambda_1 I - T)(\lambda_2 I - T)$ has first two columns 0 and, in general,
- $(\lambda_1 I - T) \cdots (\lambda_k I - T)$ has first k columns 0.

Thus all n columns of the matrix $(\lambda_1 I - T)(\lambda_2 I - T) \cdots (\lambda_n I - T)$ are 0, so the matrix is 0. Remembering that the determinant of a triangular matrix like $T - \lambda I$ is the product of the diagonal entries, we see that T satisfies its characteristic polynomial

$$\det(T - \lambda I) = (\lambda_1 - \lambda)(\lambda_2 - \lambda) \cdots (\lambda_n - \lambda). \tag{2}$$

Thus the theorem we are trying to prove is true for upper triangular matrices.

Now let A be any $n \times n$ matrix. Schur's Theorem says that there is a unitary matrix U with

$$U^{-1}AU = \begin{bmatrix} \lambda_1 & & \star \\ & \lambda_2 & \\ & & \ddots \\ 0 & & & \lambda_n \end{bmatrix} = T$$

upper triangular. The characteristic polynomial of A is the characteristic polynomial of T (see READING CHECK 1), which is the polynomial in (2), so we have only to show that

$$(\lambda_1 I - A)(\lambda_2 I - A) \cdots (\lambda_n I - A) = 0.$$

[3]After Arthur Cayley (1821–1896) and Sir William Rowan Hamilton (1805–1865).

Since $U^{-1}AU = T$, we have $A = UTU^{-1}$, so $\lambda_1 I - A = \lambda_1 I - UTU^{-1} = U(\lambda_1 I - T)U^{-1}$. Similarly, each factor $\lambda_i I - A = U(\lambda_i I - T)U^{-1}$, so that

$$\begin{aligned}(\lambda_1 I - A)(\lambda_2 I - A)\cdots(\lambda_n I - A) \\ &= U(\lambda_1 I - T)U^{-1}U(\lambda_2 I - T)U^{-1}\cdots U(\lambda_n I - T)U^{-1} \\ &= U(\lambda_1 I - T)(\lambda_2 I - T)\cdots(\lambda_n I - T)U^{-1},\end{aligned}$$

which is 0 because the factor between U and U^{-1} is the characteristic polynomial of the triangular matrix T evaluated at T, which we have already seen is 0. ■

READING CHECK 1. Show that similar matrices have the same characteristic polynomial.

Hermitian Matrices are Unitarily Diagonizable

The subject of diagonalizability was introduced in Section 3.5. To say that a matrix A is *diagonalizable* is to say that A is similar to a diagonal matrix; that is, there is an equation of the form $P^{-1}AP = D$ with D a diagonal matrix. When this occurs, the equation $AP = PD$ shows that the diagonal entries of D are eigenvalues of A and the columns of P are eigenvectors corresponding in order to the diagonal entries of D—see 3.5.3. Theorem 3.5.14 stated that an $n \times n$ real matrix A is diagonalizable if and only if A has n linearly independent eigenvectors, hence if and only if R^n has a basis of eigenvectors of A. Arguments similar to those presented there would show that an $n \times n$ complex matrix is diagonalizable if and only if C^n has a basis of eigenvectors.

7.2.6 **Theorem.** *An $n \times n$ matrix A with real (or complex) entries is diagonalizable if and only if R^n (respectively, C^n) has a basis of eigenvectors of A.*

In this section, we show that there is always a basis of eigenvectors if A is Hermitian. Thus Hermitian matrices are diagonalizable, in fact, unitarily diagonalizable; that is, if A is a Hermitian matrix, then there is a **unitary** matrix U such that $U^{-1}AU = D$ is diagonal. Thus the eigenvectors of A can be chosen orthonormal, so C^n has an orthonormal basis of eigenvectors of A.

7.2.7 **Definition.** A matrix A is *unitarily diagonalizable* if there exists a unitary matrix U such that $U^{-1}AU$ (which is U^*AU) is a diagonal matrix.

The name "Spectral Theorem," by which the next theorem is generally known, is derived from the word "spectrum," a term some people use for the set of eigenvalues of a matrix.

7.2.8 Theorem (Spectral Theorem). *Any Hermitian matrix is unitarily diagonalizable, in fact, if A is Hermitian, there exists a unitary matrix U such that $U^{-1}AU = D$ is a real diagonal matrix.*

Proof. Suppose A is an $n \times n$ Hermitian matrix. Schur's Theorem gives a unitary matrix U and upper triangular matrix T such that $U^*AU = T$. Since A is Hermitian, $A = A^*$, so $T^* = U^*A^*(U^*)^* = U^*AU = T$. Now T is upper triangular, so T^* is lower triangular. The only way an upper triangular matrix and a lower triangular matrix can be equal is if the matrix in question is a diagonal matrix. Thus $T^* = T$ implies that T is diagonal. Since the diagonal entries of T^* are the complex conjugates of the diagonal entries of T, each diagonal entry of T equals its complex conjugate and so is real. This completes the proof. ∎

7.2.9 Example. Let $A = \begin{bmatrix} 2 & 1-i \\ 1+i & 3 \end{bmatrix}$. Since A is Hermitian, the Spectral Theorem says that A is unitarily diagonalizable. The characteristic polynomial of A is

$$\begin{vmatrix} 2-\lambda & 1-i \\ 1+i & 3-\lambda \end{vmatrix} = 6 - 5\lambda + \lambda^2 - 2 = \lambda^2 - 5\lambda + 4 = (\lambda-1)(\lambda-4),$$

so the eigenvalues of A are $\lambda = 1$ and $\lambda = 4$. If $\mathsf{z} = \begin{bmatrix} z_1 \\ z_2 \end{bmatrix}$ is an eigenvector for $\lambda = 1$, then $(A - \lambda I)\mathsf{z} = \mathbf{0}$. Using Gaussian elimination,

$$\begin{bmatrix} 1 & 1-i \\ 1+i & 2 \end{bmatrix} \to \begin{bmatrix} 1 & 1-i \\ 0 & 0 \end{bmatrix},$$

so $z_2 = t$ is free, $z_1 = -(1-i)t$ and $\mathsf{z} = \begin{bmatrix} -(1-i)t \\ t \end{bmatrix}$. The eigenspace has basis $\mathsf{f}_1 = \begin{bmatrix} -1+i \\ 1 \end{bmatrix}$. If $\mathsf{z} = \begin{bmatrix} z_1 \\ z_2 \end{bmatrix}$ is an eigenvector for $\lambda = 4$, then $(A - \lambda I)\mathsf{z} = \mathbf{0}$. Using Gaussian elimination,

$$\begin{bmatrix} -2 & 1-i \\ 1+i & -1 \end{bmatrix} \to \begin{bmatrix} 1 & -\frac{1-i}{2} \\ 0 & 0 \end{bmatrix},$$

so $z_2 = t$ is free, $z_1 = \frac{1-i}{2}t$ and $\mathsf{z} = \begin{bmatrix} \frac{1-i}{2}t \\ t \end{bmatrix}$. The eigenspace has basis $\mathsf{f}_2 = \begin{bmatrix} 1-i \\ 2 \end{bmatrix}$. Since f_1 and f_2 are eigenvectors of a Hermitian matrix corresponding to different eigenvalues, they must be orthogonal (see Theorem 7.1.12) and they are:

$$\mathsf{f}_1 \cdot \mathsf{f}_2 = (-1-i)(1-i) + 1(2) = -(1+i)(1-i) + 2 = -2 + 2 = 0.$$

The square of the length of f_1 is $(-1-i)(-1+i)+1(1) = -(1+i)(-1+i)+1 = 2+1 = 3$, so $\mathsf{u}_1 = \begin{bmatrix} \frac{-1+i}{\sqrt{3}} \\ \frac{1}{\sqrt{3}} \end{bmatrix}$ is a unit eigenvector for $\lambda = 1$. Similarly, $\|\mathsf{f}_2\|^2 = (1+i)(1-i)+2(2) = 2+4 = 6$, so $\mathsf{u}_2 = \begin{bmatrix} \frac{1-i}{\sqrt{6}} \\ \frac{2}{\sqrt{6}} \end{bmatrix}$ is a unit eigenvector for $\lambda = 4$. The matrix

$$U = \begin{bmatrix} \mathsf{u}_1 & \mathsf{u}_2 \\ \downarrow & \downarrow \end{bmatrix} = \begin{bmatrix} -\frac{1+i}{\sqrt{3}} & \frac{1-i}{\sqrt{6}} \\ \frac{1}{\sqrt{3}} & \frac{2}{\sqrt{6}} \end{bmatrix}$$

is unitary and $U^*AU = D = \begin{bmatrix} 1 & 0 \\ 0 & 4 \end{bmatrix}$, the real diagonal matrix whose diagonal entries are the eigenvalues corresponding to u_1 and u_2, respectively. ☺

Real Symmetric Matrices are Orthogonally Diagonalizable

7.2.10 **Definition.** A real $n \times n$ matrix A is *orthogonally diagonalizable* if there exists a (real) orthogonal matrix Q such that $Q^{-1}AQ$ (which is Q^TAQ) is a (real) diagonal matrix.

When $Q^{-1}AQ = D$ is diagonal, the columns of Q are eigenvectors for A. Thus, if an $n \times n$ matrix A is orthogonally diagonalizable, we have not just n linearly independent eigenvectors, but n orthonormal eigenvectors. An $n \times n$ matrix A is orthogonally diagonalizable if and only if R^n has a basis of **orthonormal** eigenvectors for A.

7.2.11 **Example.** In Example 7.1.13, we saw that the matrix $A = \begin{bmatrix} 1 & -2 \\ -2 & -2 \end{bmatrix}$ has two orthogonal eigenvectors, $\mathsf{f}_1 = \begin{bmatrix} 1 \\ 2 \end{bmatrix}$ and $\mathsf{f}_2 = \begin{bmatrix} -2 \\ 1 \end{bmatrix}$, which comprise a basis for R^2, so A is orthogonally diagonalizable. Replacing f_1 with $\mathsf{q}_1 = \frac{1}{\|\mathsf{f}_1\|}\mathsf{f}_1$ and f_2 with $\mathsf{q}_2 = \frac{1}{\|\mathsf{f}_2\|}\mathsf{f}_2$, we obtain an orthogonal matrix

$$Q = \begin{bmatrix} \mathsf{q}_1 & \mathsf{q}_2 \\ \downarrow & \downarrow \end{bmatrix} = \begin{bmatrix} \frac{1}{\sqrt{5}} & -\frac{2}{\sqrt{5}} \\ \frac{2}{\sqrt{5}} & \frac{1}{\sqrt{5}} \end{bmatrix}$$

whose columns are eigenvectors corresponding to -3 and 2, respectively. Thus $Q^{-1}AQ = Q^TAQ = \begin{bmatrix} -3 & 0 \\ 0 & 2 \end{bmatrix}$. ☺

Remembering that a real symmetric matrix is Hermitian, the Spectral Theorem has an almost immediate, and important, corollary.

7.2.12 Theorem (Principal Axes Theorem). *A real symmetric matrix is orthogonally diagonalizable.*

Proof. Suppose A is a real symmetric matrix. By the Spectral Theorem, there is a unitary matrix U such that $U^{-1}AU = D$ is real diagonal. The columns of U are eigenvectors of A. Since A is real and the eigenvalues are real, the eigenspace corresponding to the real λ, which is obtained by solving $(A - \lambda I)\mathsf{x} = \mathsf{0}$, consists of vectors with real components. Thus all eigenvectors are real. So U is real, and then the fact that $U^T = \overline{U^T} = U^* = U^{-1}$ says U is orthogonal. ∎

7.2.13 Example. Let A be the real symmetric matrix $A = \begin{bmatrix} 1 & -2 & 0 \\ -2 & 0 & 2 \\ 0 & 2 & -1 \end{bmatrix}$. The characteristic polynomial of A is $-\lambda(\lambda^2 - 9)$, so the eigenvalues for A are $\lambda = 0$ and $\lambda = \pm 3$. Since these are different, eigenvectors corresponding to $0, 3, -3$ respectively must be orthogonal, so A is orthogonally similar to one of the six diagonal matrices with diagonal entries $0, 3, -3$, for instance, $D = \begin{bmatrix} 0 & 0 & 0 \\ 0 & 3 & 0 \\ 0 & 0 & -3 \end{bmatrix}$. To find the orthogonal matrix Q with $Q^{-1}AQ = D$, we find eigenvectors corresponding to $\lambda = 0$, $\lambda = 3$ and $\lambda = -3$, respectively, and *normalize*, that is, divide each by its length to obtain eigenvectors of length 1. We find that the eigenspace corresponding to $\lambda = 0$ consists of multiples of $\mathsf{f}_1 = \begin{bmatrix} 2 \\ 1 \\ 2 \end{bmatrix}$, so let $\mathsf{q}_1 = \frac{\mathsf{f}_1}{\|\mathsf{f}_1\|} = \begin{bmatrix} \frac{2}{3} \\ \frac{1}{3} \\ \frac{2}{3} \end{bmatrix}$. The eigenspace corresponding to $\lambda = 3$ consists of multiples of $\mathsf{f}_2 = \begin{bmatrix} -2 \\ 2 \\ 1 \end{bmatrix}$, so let $\mathsf{q}_2 = \frac{\mathsf{f}_2}{\|\mathsf{f}_2\|} = \begin{bmatrix} -\frac{2}{3} \\ \frac{2}{3} \\ \frac{1}{3} \end{bmatrix}$. The eigenspace corresponding to $\lambda = -3$ consists of multiples of $\mathsf{f}_3 = \begin{bmatrix} -1 \\ -2 \\ 2 \end{bmatrix}$, so let $\mathsf{q}_3 = \frac{\mathsf{f}_3}{\|\mathsf{f}_3\|} = \begin{bmatrix} -\frac{1}{3} \\ -\frac{2}{3} \\ \frac{2}{3} \end{bmatrix}$. With $Q = \begin{bmatrix} \mathsf{q}_1 & \mathsf{q}_2 & \mathsf{q}_3 \\ \downarrow & \downarrow & \downarrow \\ & & \end{bmatrix} = \begin{bmatrix} \frac{2}{3} & -\frac{2}{3} & -\frac{1}{3} \\ \frac{1}{3} & \frac{2}{3} & -\frac{2}{3} \\ \frac{2}{3} & \frac{1}{3} & \frac{2}{3} \end{bmatrix}$, we have

$$Q^{-1}AQ = Q^TAQ = D = \begin{bmatrix} 0 & 0 & 0 \\ 0 & 3 & 0 \\ 0 & 0 & -3 \end{bmatrix}.$$

☺

7.2.14 Example. The characteristic polynomial of the real symmetric matrix $A = \begin{bmatrix} 0 & 2 & -1 \\ 2 & 3 & -2 \\ -1 & -2 & 0 \end{bmatrix}$ is $-(\lambda+1)^2(\lambda-5)$, so the eigenvalues of A are $\lambda = 5$ and $\lambda = -1$. The eigenspace for $\lambda = 5$ consists of multiples of $\mathsf{f}_1 = \begin{bmatrix} -1 \\ -2 \\ 1 \end{bmatrix}$. The eigenspace for $\lambda = -1$ consists of linear combinations of $\mathsf{u} = \begin{bmatrix} 1 \\ 0 \\ 1 \end{bmatrix}$ and $\mathsf{v} = \begin{bmatrix} -2 \\ 1 \\ 0 \end{bmatrix}$. Each of these vectors is orthogonal to f_1, as must be the case—eigenvectors of a real symmetric matrix that correspond to different eigenvalues are orthogonal. On the other hand, u and v are not orthogonal (why should they be?), but this is just an inconvenience, not a genuine obstacle. Applying the Gram-Schmidt algorithm in its simplest form, we can convert u and v to orthogonal eigenvectors by taking $\mathsf{f}_2 = \mathsf{u}$ and

$$\mathsf{f}_3 = \mathsf{v} - \frac{\mathsf{v}\cdot\mathsf{f}_2}{\mathsf{f}_2\cdot\mathsf{f}_2}\mathsf{f}_2 = \begin{bmatrix} -2 \\ 1 \\ 0 \end{bmatrix} - \frac{-2}{2}\begin{bmatrix} 1 \\ 0 \\ 1 \end{bmatrix} = \begin{bmatrix} -2 \\ 1 \\ 0 \end{bmatrix} + \begin{bmatrix} 1 \\ 0 \\ 1 \end{bmatrix} = \begin{bmatrix} -1 \\ 1 \\ 1 \end{bmatrix}.$$

Now the set $\mathsf{f}_1, \mathsf{f}_2, \mathsf{f}_3$ comprise an orthogonal basis of eigenvectors. Dividing each of these by its length, we obtain

$$\mathsf{q}_1 = \begin{bmatrix} \frac{1}{\sqrt{6}} \\ \frac{2}{\sqrt{6}} \\ -\frac{1}{\sqrt{6}} \end{bmatrix}, \quad \mathsf{q}_2 = \begin{bmatrix} \frac{1}{\sqrt{2}} \\ 0 \\ \frac{1}{\sqrt{2}} \end{bmatrix}, \quad \mathsf{q}_3 = \begin{bmatrix} -\frac{1}{\sqrt{3}} \\ \frac{1}{\sqrt{3}} \\ \frac{1}{\sqrt{3}} \end{bmatrix}.$$

The matrix

$$Q = \begin{bmatrix} \mathsf{q}_1 & \mathsf{q}_2 & \mathsf{q}_3 \\ \downarrow & \downarrow & \downarrow \\ & & \end{bmatrix} = \begin{bmatrix} -\frac{1}{\sqrt{6}} & \frac{1}{\sqrt{2}} & -\frac{1}{\sqrt{3}} \\ -\frac{2}{\sqrt{6}} & 0 & \frac{1}{\sqrt{3}} \\ \frac{1}{\sqrt{6}} & \frac{1}{\sqrt{2}} & \frac{1}{\sqrt{3}} \end{bmatrix}$$

is orthogonal and $Q^{-1}AQ = D = \begin{bmatrix} 5 & 0 & 0 \\ 0 & -1 & 0 \\ 0 & 0 & -1 \end{bmatrix}$ is the diagonal matrix whose diagonal entries are eigenvalues for q_1, q_2 and q_3, respectively. ☺

Unitarily Diagonalizable If and Only If Normal

7.2.15 The converse of the Spectral Theorem is the statement that a unitarily diagonalizable matrix is Hermitian. This is not true! The matrix $V = \begin{bmatrix} 0 & -1 \\ 1 & 0 \end{bmatrix}$ is not Hermitian. The eigenvalues of V are $\pm i$. Corresponding eigenvectors are $\begin{bmatrix} 1 \\ -i \end{bmatrix}$ and $\begin{bmatrix} 1 \\ i \end{bmatrix}$. These vectors are orthogonal and of length $\sqrt{2}$. Dividing by their lengths and setting

$$U = \begin{bmatrix} \frac{1}{\sqrt{2}} & \frac{1}{\sqrt{2}} \\ -\frac{i}{\sqrt{2}} & \frac{i}{\sqrt{2}} \end{bmatrix},$$

we have U unitary and $U^*VU = D = \begin{bmatrix} i & 0 \\ 0 & -i \end{bmatrix}$ diagonal. Thus V is unitarily diagonalizable, but not Hermitian.

So, while every Hermitian matrix is unitarily diagonalizable, not every unitarily diagonalizable matrix is Hermitian. It would be satisfying to have a theorem characterizing unitarily diagonalizable matrices, that is, a theorem that says

> "A matrix is unitarily diagonalizable if and only if … "

We conclude this section with just such a theorem.

7.2.16 Theorem. *A matrix A is unitarily diagonalizable if and only if it commutes with its conjugate transpose, that is, if and only if $AA^* = A^*A$.*

Proof. ($\Longrightarrow$) Assume that A is unitarily diagonalizable. Then there exists a unitary matrix U and a diagonal matrix D such that $U^{-1}AU = D$. Thus $A = UDU^{-1}$. Since U is unitary, $U^{-1} = U^*$, so $A = UDU^*$ and $A^* = (U^*)^*D^*U^* = U\overline{D}U^*$ because D is diagonal (and hence equal to its transpose). Since $U^*U = I$, $AA^* = UDU^*U\overline{D}U^* = UD\overline{D}U^*$, whereas $A^*A = U\overline{D}U^*UDU^* = U\overline{D}DU^*$. The desired result $AA^* = A^*A$ follows because $D\overline{D} = \overline{D}D$.

($\Longleftarrow$) Conversely, suppose $AA^* = A^*A$. By Schur's Theorem, there exists a unitary matrix U such that $U^{-1}AU = T$ is upper triangular. We shall show that T is diagonal, completing the proof. First we show that $TT^* = T^*T$. Why is this?

Since $U^{-1} = U^*$, $T = U^*AU$ and $T^* = U^*A^*(U^*)^* = U^*A^*U$. Since $U^*U = I$, $TT^* = U^*AUU^*A^*U = U^*AA^*U$ and, similarly, $T^*T = U^*A^*AU$. Since $AA^* = A^*A$, we obtain $TT^* = T^*T$, as claimed.

Next, for any vector z, $\|T\mathsf{z}\|^2 = T\mathsf{z} \cdot T\mathsf{z} = (T\mathsf{z})^*(T\mathsf{z}) = \mathsf{z}^*T^*T\mathsf{z}$ and, similarly, $\|T^*\mathsf{z}\|^2 = (T^*\mathsf{z})^*(T^*\mathsf{z}) = \mathsf{z}^*TT^*\mathsf{z}$. Since $TT^* = T^*T$, we have $\|T\mathsf{z}\|^2 = \|T^*\mathsf{z}\|^2$. In particular,

$$\|T\mathsf{e}_i\|^2 = \|T^*\mathsf{e}_i\|^2 \tag{3}$$

for each of the standard basis vectors $\mathsf{e}_1, \mathsf{e}_2, \ldots, \mathsf{e}_n$. Now T has the form

$$T = \begin{bmatrix} \lambda_1 & a_2 & a_3 & \cdots & a_n \\ 0 & \lambda_2 & b_3 & \cdots & b_n \\ 0 & 0 & \lambda_3 & \cdots & \star \\ \vdots & \vdots & & \ddots & \vdots \\ 0 & 0 & & \cdots & \lambda_n \end{bmatrix}, \quad \text{so} \quad T^* = \begin{bmatrix} \overline{\lambda}_1 & 0 & 0 & \cdots & 0 \\ \overline{a}_2 & \overline{\lambda}_2 & 0 & & 0 \\ \overline{a}_3 & \overline{b}_3 & \overline{\lambda}_3 & & 0 \\ \vdots & \vdots & \vdots & \ddots & \\ \overline{a}_n & \overline{b}_n & \star & \cdots & \overline{\lambda}_n \end{bmatrix}.$$

We examine the consequences of (3) in the cases $i = 1, 2, \ldots, n$. The vector $Te_1 = \begin{bmatrix} \lambda_1 \\ 0 \\ \vdots \\ 0 \end{bmatrix}$ is the first column of T, and $T^*e_1 = \begin{bmatrix} \overline{\lambda}_1 \\ \overline{a}_2 \\ \overline{a}_3 \\ \vdots \\ \overline{a}_n \end{bmatrix}$ is the first column of T^*. The lengths of these vectors is the same, so

$$|\lambda_1|^2 = |\lambda_1|^2 + |a_2|^2 + |a_3|^2 + \cdots + |a_n|^2,$$

implying $|a_2|^2 + |a_3|^2 + \cdots + |a_n|^2 = 0$, so $a_2 = a_3 = \cdots = a_n = 0$. At this point,

$$T = \begin{bmatrix} \lambda_1 & 0 & 0 & \cdots & 0 \\ 0 & \lambda_2 & b_3 & \cdots & b_n \\ 0 & 0 & \lambda_3 & \cdots & \star \\ \vdots & \vdots & & \ddots & \vdots \\ 0 & 0 & & \cdots & \lambda_n \end{bmatrix}, \qquad T^* = \begin{bmatrix} \overline{\lambda}_1 & 0 & 0 & \cdots & 0 \\ 0 & \overline{\lambda}_2 & 0 & & 0 \\ 0 & \overline{b}_3 & \overline{\lambda}_3 & & 0 \\ \vdots & \vdots & \vdots & \ddots & \\ 0 & \overline{b}_n & \star & \cdots & \overline{\lambda}_n \end{bmatrix}$$

and T is starting to look like the diagonal matrix we want it to be. The vector $Te_2 = \begin{bmatrix} 0 \\ \lambda_2 \\ 0 \\ \vdots \\ 0 \end{bmatrix}$ and $T^*e_2 = \begin{bmatrix} 0 \\ \overline{\lambda}_2 \\ \overline{b}_3 \\ \vdots \\ \overline{b}_n \end{bmatrix}$. Equating the lengths of these vectors, we get

$$|\lambda_2|^2 = |\lambda_2|^2 + |b_3|^2 + \cdots + |b_n|^2,$$

so $b_3 = \cdots = b_n = 0$. Continuing to examine the consequences of (3) in the cases $i = 3, 4, \ldots, n$ shows that T is diagonal. ■

7.2.17 Corollary. *Any unitary matrix is unitarily diagonalizable.*

Proof. If U is a unitary matrix, $UU^* = U^*U\ (= I)$. ■

7.2.18 Definition. A matrix A is *normal* if and only if $AA^* = A^*A$.

Thus Theorem 7.2.16 says that a matrix is unitarily diagonalizable if and only if it is normal, Corollary 7.2.17 says that a unitary matrix is normal, and Theorem 7.2.8 is clear because a Hermitian matrix is also normal.

READING CHECK 2. Why is a Hermitian matrix normal?

[The converse is false. See Exercise 12.]

Answers to Reading Checks

1. Suppose $B = P^{-1}AP$. Using basic properties of the determinant, the characteristic polynomial of B is $\det(P^{-1}AP - \lambda I) = \det P^{-1}(A - \lambda I)P = \frac{1}{\det P}\det(A - \lambda I)\det P = \det(A - \lambda I)$, which is the characteristic polynomial of A.

2. $A^* = A$, so $AA^* = A^*A = A^2$.

True/False Questions

Decide, with as little calculation as possible, whether each of the following statements is true or false and, if you say "false," explain your answer. (Answers are at the back of the book.)

1. Matrices $A = \begin{bmatrix} 1 & 0 \\ 0 & 1 \end{bmatrix}$ and $B = \begin{bmatrix} 3 & 0 \\ 0 & 3 \end{bmatrix}$ are similar.

2. The matrix $A = \begin{bmatrix} 1 & 1 \\ 0 & 1 \end{bmatrix}$ satisfies $A^2 = 2A - I$.

3. The matrix $A = \begin{bmatrix} 1 & 2i \\ -2i & 3 \end{bmatrix}$ is unitarily diagonalizable.

4. If A is a Hermitian matrix with no eigenvalue 0, then A is invertible.

5. If D is a diagonal matrix and Q is a real orthogonal matrix, then QDQ^T is also diagonal.

6. If A is a real symmetric matrix, there is an orthogonal matrix Q such that Q^TAQ is a diagonal matrix.

7. A Hermitian matrix is normal.

8. A normal matrix is Hermitian.

9. A normal matrix is unitary.

10. A unitary matrix is normal.

EXERCISES

Answers to exercises marked [BB] can be found at the **Back of the Book.**

1. For each real symmetric matrix A, find an orthogonal matrix Q and a real diagonal matrix D such that $Q^TAQ = D$.

   (a) [BB] $A = \begin{bmatrix} 2 & 3 \\ 3 & -6 \end{bmatrix}$ **(b)** $A = \begin{bmatrix} 17 & 24 \\ 24 & 3 \end{bmatrix}$

   (c) [BB] $A = \begin{bmatrix} -2 & 1 & 1 \\ 1 & -2 & 1 \\ 1 & 1 & -2 \end{bmatrix}$ **(d)** $A = \begin{bmatrix} 1 & -3 & 2 \\ -3 & 1 & -2 \\ 2 & -2 & 2 \end{bmatrix}$

(e) $A = \begin{bmatrix} 5 & 2 & 2 \\ 2 & 2 & -4 \\ 2 & -4 & 2 \end{bmatrix}$ **(f)** $A = \begin{bmatrix} -1 & 1 & 2 & 1 \\ 1 & -1 & -2 & -1 \\ 2 & -2 & -4 & -2 \\ 1 & -1 & -2 & -1 \end{bmatrix}$

(g) $A = \begin{bmatrix} 0 & 2 & 1 & 1 \\ 2 & 0 & -1 & -1 \\ 1 & -1 & 0 & -2 \\ 1 & -1 & -2 & 0 \end{bmatrix}$ **(h)** $A = \begin{bmatrix} 3 & -1 & -1 & 1 \\ -1 & 3 & 1 & -1 \\ -1 & 1 & 3 & -1 \\ 1 & -1 & -1 & 3 \end{bmatrix}$.

The characteristic polynomials of the matrices in (f), (g) and (h), respectively, are $-\lambda^4 - 7\lambda^3 = -\lambda^3(\lambda+7)$, $\lambda^4 - 12\lambda^2 + 16\lambda = \lambda(\lambda+4)(\lambda-2)^2$ and $\lambda^4 - 12\lambda^3 + 48\lambda^2 - 80\lambda + 48$.

2. For each real symmetric matrix A,

   i. find an orthogonal matrix Q such that Q^TAQ is diagonal;

   ii. use the result of part i to find A^5.

   (a) $A = \begin{bmatrix} 3 & 0 & -1 \\ 0 & -1 & 0 \\ -1 & 0 & 3 \end{bmatrix}$ **(b)** $A = \begin{bmatrix} 3 & 0 & 2 \\ 0 & 1 & 0 \\ 2 & 0 & 3 \end{bmatrix}$.

3. Find the eigenvalues and corresponding eigenspaces of each Hermitian matrix.

   (a) $A = \begin{bmatrix} 1 & i \\ -i & 1 \end{bmatrix}$ **(b)** $A = \begin{bmatrix} 1 & 0 & 1+i \\ 0 & 5 & 0 \\ 1-i & 0 & 0 \end{bmatrix}$.

4. For each Hermitian matrix A, find a unitary matrix U and a diagonal matrix D such that $U^*AU = D$.

   (a) [BB] $A = \begin{bmatrix} 1 & -i \\ i & 1 \end{bmatrix}$ **(b)** $A = \begin{bmatrix} 3 & 2+i \\ 2-i & 7 \end{bmatrix}$

   (c) $A = \begin{bmatrix} 3 & 1+i \\ 1-i & 2 \end{bmatrix}$ **(d)** $A = \begin{bmatrix} 1 & 1+i \\ 1-i & 3 \end{bmatrix}$

   (e) $A = \begin{bmatrix} 1 & i & 0 \\ -i & 0 & 1 \\ 0 & 1 & 1 \end{bmatrix}$ **(f)** $A = \begin{bmatrix} 1 & 0 & 1-i \\ 0 & 2 & 0 \\ 1+i & 0 & 0 \end{bmatrix}$.

5. In each case, determine whether or not A is diagonalizable over R and over C. Justify your answers.

   (a) [BB] $A = \begin{bmatrix} 2 & -3 \\ 1 & 1 \end{bmatrix}$ **(b)** $A = \begin{bmatrix} 3 & 1 & -1 \\ 2 & 2 & -1 \\ 2 & 2 & 0 \end{bmatrix}$

   (c) $A = \begin{bmatrix} 0 & -1 & 1 \\ 1 & 1 & 0 \\ -1 & 0 & 1 \end{bmatrix}$ **(d)** $A = \begin{bmatrix} 1 & 1 & 1 \\ -1 & 3 & -1 \\ -1 & 2 & 0 \end{bmatrix}$.

6. [BB] Let A be a real skew-symmetric matrix, that is, a matrix satisfying $A^T = -A$. Prove that there exists a real orthogonal matrix Q and a matrix B with $A = QBQ^T$ and B^2 is diagonal. [Hint: A^TA is real symmetric.]

7. A matrix K is *skew-Hermitian* if $K^* = -K$. Prove that a skew-Hermitian matrix is unitarily diagonalizable.

8. For each of the following matrices, find a unitary matrix U and an upper triangular matrix T such that $U^*AU = T$. [Mimic the proof of Schur's Theorem.]

   (a) [BB] $A = \begin{bmatrix} 2 & 2 \\ -1 & 0 \end{bmatrix}$ **(b)** $A = \begin{bmatrix} 6 & 9 \\ -1 & 12 \end{bmatrix}$

   (c) $A = \begin{bmatrix} 1 & -1 & 1 \\ 1 & -1 & 1 \\ 1 & -1 & 1 \end{bmatrix}$ **(d)** $A = \begin{bmatrix} 0 & 1 & 0 \\ 1 & 0 & -3 \\ 0 & 1 & 0 \end{bmatrix}$.

9. Prove that the determinant of a Hermitian matrix is a real number.

10. **(a)** Is $A = \begin{bmatrix} 0 & -1 \\ 1 & 0 \end{bmatrix}$ Hermitian?

    (b) Is A normal?

    (c) Is A unitarily diagonalizable?

    Justify your answers.

11. **(a)** Is $A = \begin{bmatrix} 1 & 0 & -i \\ 0 & 1 & 1 \\ i & -1 & 2 \end{bmatrix}$ normal?

    (b) Is A unitarily diagonalizable?

    (c) Is A similar to a diagonal matrix?

12. **(a)** This section contains an example of a normal matrix that is not Hermitian. Find it.

    (b) Show that for any real numbers a and b, $A = \begin{bmatrix} a & b \\ -b & a \end{bmatrix}$ is normal but not symmetric.

    (c) Show that if A is a real 2×2 normal matrix that is not symmetric, then A looks like the matrix in (a).

13. Determine with as few calculations as possible whether or not the given matrix is diagonalizable over $\mathbf{C}$.

    (a) [BB] $A = \begin{bmatrix} 1 & 0 & 1 \\ 0 & 1 & -1 \\ 1 & 0 & 0 \end{bmatrix}$ **(b)** $A = \begin{bmatrix} 1 & 1 & 0 \\ 0 & 1 & 1 \\ 0 & 0 & -1 \end{bmatrix}$

    (c) $A = \begin{bmatrix} \frac{1}{\sqrt{2}} & \frac{1}{\sqrt{2}} & 0 \\ 0 & 0 & 1 \\ -\frac{1}{\sqrt{2}} & \frac{1}{\sqrt{2}} & 0 \end{bmatrix}$ **(d)** $A = \begin{bmatrix} i & i & 0 \\ -i & i & i \\ 0 & -i & i \end{bmatrix}$.

14. Prove that a matrix A is normal if and only if $A+A^*$ commutes with $A-A^*$.

15. [BB] Suppose A is a normal $n \times n$ (complex) matrix. Prove that null sp A = null sp A^*. [Hint: Let $\mathbf{z} \in \mathbf{C}^n$ and compute $(A^*\mathbf{z})\cdot(A^*\mathbf{z})$.]

16. Suppose A is a normal $n \times n$ (complex) matrix. Prove that col sp A = col sp A^* [Hint: Exercise 20(b) of Section 7.1 and Exercise 15.]

17. Suppose A is a symmetric $n \times n$ matrix and $\mathbf{q}_1, \mathbf{q}_2, \ldots, \mathbf{q}_n$ are orthonormal eigenvectors corresponding to eigenvalues $\lambda_1, \lambda_2, \ldots, \lambda_n$ respectively.

(a) Prove that $A = \lambda_1 q_1 q_1^T + \lambda_2 q_2 q_2^T + \cdots + \lambda_n q_n q_n^T$. [Hint: Call the matrix on the right B. It is sufficient to show that $Ax = Bx$ for any x. Why. Also use (2) of Section 6.1 to rewrite an expression like $(qq^T)x$.]

(b) Any real symmetric matrix is a linear combination of projection matrices. Why?

[The decomposition of A given in part (a) is its *spectral decomposition.*]

18. We noted that the converse of the Spectral Theorem is false. On the other hand, the converse of the Principal Axes Theorem is true. Prove this fact; namely, prove that if a real matrix is orthogonally diagonalizable, then A is symmetric.

Critical Reading

19. Think of a way to show that any square complex matrix is unitarily similar to a **lower** triangular matrix. (Use Schur's Theorem without changing the proof.)

20. Suppose the only eigenvalue of a real symmetric matrix A is 3. What can you say about A?

21. [BB] Suppose A and B are real symmetric matrices. Why are all the roots of $\det(A + B - \lambda I)$ real?

22. If U is a unitary matrix and every eigenvalue of U is either $+i$ or $-i$, find U^2 and justify your answer.

23. [BB] Let A be a (real) $n \times n$ matrix and suppose that $x^T Ax > 0$ for all vectors x.

(a) Let λ be a real eigenvalue of A. Show that $\lambda > 0$.

(b) If A is symmetric, show that A has a square root; that is, there exists a matrix B such that $A = B^2$.

(c) If A is symmetric, show that A can be factored in the form $A = RR^T$ for some matrix R.

24. (a) Let A be a real symmetric matrix with nonnegative eigenvalues. Show that A can be written in the form $A = RR^T$ for some matrix R. (This is a partial converse of the elementary fact that RR^T is always symmetric.) [Hint: A diagonal matrix with real nonnegative diagonal entries has a square root.]

(b) Let A and B be real symmetric matrices and suppose the eigenvalues of B are all positive. Show that the roots of the polynomial $\det(A - \lambda B)$ are real. [Hint: First explain why $\det(A - \lambda B) = 0$ gives a nonzero x with $Ax = \lambda Bx$. Then use the result of 24(a) to factor $B = RR^T$ with R invertible.]

25. (a) What are the possible eigenvalues of a projection matrix?

(b) Under what conditions does a projection matrix have all possible eigenvalues?

Explain your answers.

26. [BB] **(a)** Find the characteristic polynomial of $A = \begin{bmatrix} 3 & 3 & -4 \\ -4 & -3 & 5 \\ 2 & 4 & 0 \end{bmatrix}$.

(b) Evaluate $A^6 + 4A^5 - 5A^4 - 50A^3 - 76A^2 - 10A + 50I$.

27. (a) Let $A = \begin{bmatrix} 0 & -1 \\ 1 & -2 \end{bmatrix}$. Why is A invertible? Show that A^{-1} is a polynomial in A. (In fact, in this case, there are scalars a and b such that $A^{-1} = aA + bI$.)

(b) Repeat part (a) for $A = \begin{bmatrix} 0 & 0 & -2 \\ 1 & 0 & 3 \\ 0 & 1 & -5 \end{bmatrix}$. (This time, A^{-1} is a quadratic polynomial in A.)

(c) Suppose A is an $n \times n$ invertible matrix. Explain why A^{-1} is a polynomial in A.

7.3 Real Symmetric Matrices

This section offers an alternative approach to the diagonalization of symmetric matrices whose entries are all real numbers for readers who wish to avoid the arguments involving complex numbers that were used in previous sections. If you are familiar with the material in Sections 7.1 and 7.2, what we present here will be completely redundant.

Since this section depends heavily upon the notions of eigenvalue, eigenvector and diagonalizability, the reader should be comfortable with the material in Sections 3.4 and 3.5 before continuing.

Recall that a matrix A is *diagonalizable* if there exists an invertible matrix P and a diagonal matrix D such that $P^{-1}AP = D$. When this occurs, the columns of P are eigenvectors of A and the diagonal entries of D are eigenvalues in the order that corresponds to the order of the columns of A; that is, the $(1,1)$ entry of D is an eigenvalue and the first column of P is a corresponding eigenvector, and so on. Invertibility of P is equivalent to linearly independent columns, so an $n \times n$ matrix is diagonalizable if and only if it has n linearly independent eigenvectors. (See Theorem 3.5.9.) Since n linearly independent vectors in R^n form a basis, this theorem is clear.

7.3.1 Theorem. *An $n \times n$ real matrix A is diagonalizable if and only if R^n has a basis consisting of eigenvectors of A.*

As the name suggests, a matrix is *orthogonally diagonalizable* if it is diagonalizable and the *diagonalizing matrix* P can be chosen orthogonal.

7.3.2 Definition. A matrix A is *orthogonally diagonalizable* if there exist an orthogonal matrix Q and a diagonal matrix D such that $Q^{-1}AQ = D$, equivalently, such that $Q^TAQ = D$.

READING CHECK 1. Why "equivalently" in this definition?

Remembering that the basis to which reference is made in Theorem 7.3.1 consists of the columns of the diagonalizing matrix, the following corollary is immediate.

7.3.3 Corollary. *A real $n \times n$ matrix A is orthogonally diagonalizable if and only if R^n has an orthogonal basis of eigenvectors of A.*

7.3.4 Example. The characteristic polynomial of $A = \begin{bmatrix} 1 & -2 \\ -2 & -2 \end{bmatrix}$ is

$$\det(A - \lambda I) = \begin{vmatrix} 1-\lambda & -2 \\ -2 & -2-\lambda \end{vmatrix} = \lambda^2 + \lambda - 6 = (\lambda+3)(\lambda-2).$$

The eigenvalues are $\lambda = -3$ and $\lambda = 2$. The eigenspace for $\lambda = -3$ consists of multiples of $\mathsf{f}_1 = \begin{bmatrix} 1 \\ 2 \end{bmatrix}$. The eigenspace for $\lambda = 2$ consists of multiples of $\mathsf{f}_2 = \begin{bmatrix} -2 \\ 1 \end{bmatrix}$. The eigenvectors f_1 and f_2 are orthogonal and constitute a basis for R^2. Thus A is orthogonally diagonalizable. Normalizing f_1 and f_2, that is, dividing each by its length so as to produce unit eigenvectors $\mathsf{q}_1 = \begin{bmatrix} \frac{1}{\sqrt{5}} \\ \frac{2}{\sqrt{5}} \end{bmatrix}$ and $\mathsf{q}_2 = \begin{bmatrix} -\frac{2}{\sqrt{5}} \\ \frac{1}{\sqrt{5}} \end{bmatrix}$, the matrix

$$Q = \begin{bmatrix} \mathsf{q}_1 & \mathsf{q}_2 \\ \downarrow & \downarrow \end{bmatrix} = \begin{bmatrix} \frac{1}{\sqrt{5}} & -\frac{2}{\sqrt{5}} \\ \frac{2}{\sqrt{5}} & \frac{1}{\sqrt{5}} \end{bmatrix}$$

is orthogonal and $Q^TAQ = D = \begin{bmatrix} -3 & 0 \\ 0 & 2 \end{bmatrix}$, the diagonal matrix whose diagonal entries are eigenvalues in order corresponding to the columns of Q: the first diagonal entry is an eigenvalue for q_1 and the second is an eigenvalue for q_2. ⌣̈

Properties of Symmetric Matrices

Symmetric matrices are quite remarkable objects. The characteristic polynomial of $A = \begin{bmatrix} 0 & 1 \\ -1 & 0 \end{bmatrix}$ (which is not symmetric) is $\begin{vmatrix} -\lambda & -1 \\ 1 & -\lambda \end{vmatrix} = \lambda^2 + 1$, which has no real roots, so A has no real eigenvalues. The characteristic polynomial of $A = \begin{bmatrix} 0 & 1 & 0 \\ 0 & 0 & 1 \\ 6 & -10 & 5 \end{bmatrix}$ (also not symmetric) is $(1-\lambda)(\lambda^2 - 4\lambda + 6)$, which has just one real root, $\lambda = 1$, so A has just one real eigenvalue.

On the other hand, **all the eigenvalues of a symmetric matrix are always real.** This is easy to show for 2×2 matrices—see Exercise 4—and not hard in general, though the proof involves complex matrices and is omitted here (see Theorem 7.1.12).

Remember that eigenvectors corresponding to different eigenvalues are linearly independent (see Theorem 3.5.14). For eigenvectors of symmetric matrices, even more is true.

7.3.5 **Theorem.** *If A is a real symmetric matrix, then eigenvectors that correspond to different eigenvalues are orthogonal.*

Proof. Let x and y be eigenvectors for A with corresponding eigenvalues λ and μ, and suppose $\lambda \neq \mu$. We compute $\mathsf{x}^T A\mathsf{y}$ in two different ways. On the one hand, since $A\mathsf{y} = \mu\mathsf{y}$, we have $\mathsf{x}^T A\mathsf{y} = \mu\mathsf{x}^T\mathsf{y} = \mu(\mathsf{x} \cdot \mathsf{y})$. On the other hand, since $A = A^T$,

$$\mathsf{x}^T A\mathsf{y} = \mathsf{x}^T A^T\mathsf{y} = (A\mathsf{x})^T\mathsf{y} = (\lambda\mathsf{x})^T\mathsf{y} = \lambda(\mathsf{x}^T\mathsf{y}) = \lambda(\mathsf{x} \cdot \mathsf{y}).$$

Thus $\lambda(\mathsf{x} \cdot \mathsf{y}) = \mu(\mathsf{x} \cdot \mathsf{y})$, so $(\lambda - \mu)(\mathsf{x} \cdot \mathsf{y}) = 0$. This gives $\mathsf{x} \cdot \mathsf{y} = 0$ because $\lambda - \mu \neq 0$. Thus x and y are orthogonal, as claimed. ■

The symmetric 2×2 matrix A in Example 7.3.4 had two different eigenvalues, -3 and 2, so in the light of Theorem 7.3.5, it is not accidental that corresponding eigenvectors are orthogonal. In general, if a symmetric $n \times n$ matrix has n different eigenvalues, corresponding eigenvectors are orthogonal and, since there are n of them, they form an orthogonal basis for R^n. Normalizing produces an orthonormal basis of eigenvectors, so the following corollary is immediate.

7.3.6 **Corollary.** *A real symmetric $n \times n$ matrix with n different eigenvalues is orthogonally diagonalizable.*

We have presented a quite straightforward proof of this corollary, not difficult reasoning that explains why a symmetric $n \times n$ matrix with n different eigenvalues is orthogonally diagonalizable. It is much harder to see why this is true when there are fewer than n different eigenvalues. The 3×3 symmetric matrix $A = \begin{bmatrix} 0 & 2 & -1 \\ 2 & 3 & -2 \\ -1 & -2 & 0 \end{bmatrix}$ has three orthogonal eigenvectors, so it's orthogonally diagonalizable, yet it has just two eigenvalues. The proof of the "Principal Axes Theorem," first stated as Corollary 7.2.12, is beyond the scope of this section.

7.3.7 **Theorem (Principal Axes Theorem).** *A real symmetric matrix is orthogonally diagonalizable.*

7.3.8 Example. Consider the real symmetric matrix $A = \begin{bmatrix} 0 & 2 & -1 \\ 2 & 3 & -2 \\ -1 & -2 & 0 \end{bmatrix}$ introduced above, whose eigenvalues are $\lambda = 5$ and $\lambda = -1$. The eigenspace for $\lambda = 5$ consists of multiples of $\mathsf{f}_1 = \begin{bmatrix} -1 \\ -2 \\ 1 \end{bmatrix}$. The eigenspace for $\lambda = -1$ consists of linear combinations of $\mathsf{u} = \begin{bmatrix} 1 \\ 0 \\ 1 \end{bmatrix}$ and $\mathsf{v} = \begin{bmatrix} -2 \\ 1 \\ 0 \end{bmatrix}$. Each of these vectors is orthogonal to f_1, as must be the case—eigenvectors corresponding to different eigenvalues of a symmetric matrix are orthogonal. On the other hand, u and v are not orthogonal (no reason they should be), but this is just an inconvenience, not a genuine obstacle. Applying the Gram–Schmidt Algorithm in its simplest form, we can convert u and v to orthogonal eigenvectors by setting $\mathsf{f}_2 = \mathsf{u}$ and

$$\mathsf{f}_3 = \mathsf{v} - \frac{\mathsf{v} \cdot \mathsf{f}_2}{\mathsf{f}_2 \cdot \mathsf{f}_2}\,\mathsf{f}_2 = \begin{bmatrix} -2 \\ 1 \\ 0 \end{bmatrix} - \frac{-2}{2}\begin{bmatrix} 1 \\ 0 \\ 1 \end{bmatrix} = \begin{bmatrix} -2 \\ 1 \\ 0 \end{bmatrix} + \begin{bmatrix} 1 \\ 0 \\ 1 \end{bmatrix} = \begin{bmatrix} -1 \\ 1 \\ 1 \end{bmatrix}.$$

The eigenvectors $\mathsf{f}_1, \mathsf{f}_2, \mathsf{f}_3$ constitute an orthogonal basis which we can turn into an orthonormal basis by *normalizing*, dividing each vector by its length. We obtain

$$\mathsf{q}_1 = \begin{bmatrix} -\frac{1}{\sqrt{6}} \\ -\frac{2}{\sqrt{6}} \\ \frac{1}{\sqrt{6}} \end{bmatrix}, \quad \mathsf{q}_2 = \begin{bmatrix} \frac{1}{\sqrt{2}} \\ 0 \\ \frac{1}{\sqrt{2}} \end{bmatrix}, \quad \mathsf{q}_3 = \begin{bmatrix} -\frac{1}{\sqrt{3}} \\ \frac{1}{\sqrt{3}} \\ \frac{1}{\sqrt{3}} \end{bmatrix}.$$

The matrix

$$Q = \begin{bmatrix} \mathsf{q}_1 & \mathsf{q}_2 & \mathsf{q}_3 \\ \downarrow & \downarrow & \downarrow \end{bmatrix} = \begin{bmatrix} -\frac{1}{\sqrt{6}} & \frac{1}{\sqrt{2}} & -\frac{1}{\sqrt{3}} \\ -\frac{2}{\sqrt{6}} & 0 & \frac{1}{\sqrt{3}} \\ \frac{1}{\sqrt{6}} & \frac{1}{\sqrt{2}} & \frac{1}{\sqrt{3}} \end{bmatrix}$$

is orthogonal, and $Q^TAQ = D = \begin{bmatrix} 5 & 0 & 0 \\ 0 & -1 & 0 \\ 0 & 0 & -1 \end{bmatrix}$, the diagonal matrix whose entries are eigenvalues for q_1, q_2 and q_3, respectively. ☺

7.3.9 Remark. Why the name "Principal Axes Theorem?" A set of n mutually orthogonal coordinate axes in R^n—like the x-, y- and z-axes in R^3—are traditionally called "principal axes," these corresponding to orthogonal vectors like i, j and k. The Principal Axes Theorem says that a symmetric $n \times n$ matrix A has n orthogonal eigenvectors, so R^n has a set of "principal axes," each the span of an eigenvector.

Answers to Reading Checks

1. If Q is orthogonal, $Q^{-1} = Q^T$.

True/False Questions

Decide, with as little calculation as possible, whether each of the following statements is true or false and, if you say "false," explain your answer. (Answers are at the back of the book.)

1. If A is a matrix, P is an invertible matrix and D is a diagonal matrix with $P^{-1}AP = D$, then the columns of P are eigenvectors of A.
2. If A is a matrix, Q is an orthogonal matrix and D is a diagonal matrix with $Q^{-1}AQ = D$, then $Q^TAQ = D$.
3. There exists a real symmetric 4×4 matrix with characteristic polynomial $\lambda^4+\lambda$.
4. If D is a diagonal matrix and Q is a real orthogonal matrix, then QDQ^T is also diagonal.
5. If $P^{-1}AP = D$ for some matrix P and some diagonal matrix D, then $A^2 = D^2$.
6. If the eigenvalues of a matrix A are all real, then A is symmetric.
7. If a 3×3 matrix A has three orthogonal eigenvectors, then A is symmetric.
8. A real symmetric matrix can be orthogonally diagonalized.
9. The standard basis vectors of R^n determine a set of principal axes for R^n.
10. If $\lambda = 1$ is the only eigenvalue of a real symmetric $n \times n$ matrix A, then $A = I$, the identity matrix.

EXERCISES

Answers to exercises marked [BB] can be found at the Back of the Book.

1. For each real symmetric matrix A, find an orthogonal matrix Q and a real diagonal matrix D such that $Q^TAQ = D$.

   (a) [BB] $A = \begin{bmatrix} 2 & 3 \\ 3 & -6 \end{bmatrix}$ **(b)** $A = \begin{bmatrix} 17 & 24 \\ 24 & 3 \end{bmatrix}$

   (c) [BB] $A = \begin{bmatrix} -2 & 1 & 1 \\ 1 & -2 & 1 \\ 1 & 1 & -2 \end{bmatrix}$ **(d)** $A = \begin{bmatrix} 1 & -3 & 2 \\ -3 & 1 & -2 \\ 2 & -2 & 2 \end{bmatrix}$

   (e) $A = \begin{bmatrix} 5 & 2 & 2 \\ 2 & 2 & -4 \\ 2 & -4 & 2 \end{bmatrix}$ **(f)** $A = \begin{bmatrix} -1 & 1 & 2 & 1 \\ 1 & -1 & -2 & -1 \\ 2 & -2 & -4 & -2 \\ 1 & -1 & -2 & -1 \end{bmatrix}$

   (g) $A = \begin{bmatrix} 0 & 2 & 1 & 1 \\ 2 & 0 & -1 & -1 \\ 1 & -1 & 0 & -2 \\ 1 & -1 & -2 & 0 \end{bmatrix}$ **(h)** $A = \begin{bmatrix} 3 & -1 & -1 & 1 \\ -1 & 3 & 1 & -1 \\ -1 & 1 & 3 & -1 \\ 1 & -1 & -1 & 3 \end{bmatrix}$.

   The characteristic polynomials of the matrices in (f), (g) and (h), respectively, are $-\lambda^4 - 7\lambda^3 = -\lambda^3(\lambda+7)$, $\lambda^4 - 12\lambda^2 + 16\lambda = \lambda(\lambda+4)(\lambda-2)^2$ and $\lambda^4 - 12\lambda^3 + 48\lambda^2 - 80\lambda + 48$.

2. For each real symmetric matrix A,

   i. find an orthogonal matrix Q such that Q^TAQ is diagonal;

   ii. use the result of part i to find A^5.

(a) [BB] $A = \begin{bmatrix} 1 & 2 \\ 2 & 1 \end{bmatrix}$ **(b)** $A = \begin{bmatrix} 3 & 0 & -1 \\ 0 & -1 & 0 \\ -1 & 0 & 3 \end{bmatrix}$

(c) $A = \begin{bmatrix} 3 & 0 & 2 \\ 0 & 1 & 0 \\ 2 & 0 & 3 \end{bmatrix}$ **(d)** $A = \begin{bmatrix} -1 & -4 & 4 & 0 \\ -4 & -5 & 0 & -4 \\ 4 & 0 & -5 & -4 \\ 0 & -4 & -4 & -1 \end{bmatrix}$.

3. [BB] State the converse of the Principal Axes Theorem. Is this true? Explain.

4. Show that the eigenvalues of a 2×2 (real) symmetric matrix are always real numbers.

5. [BB] If A is a real symmetric $n \times n$ matrix with nonnegative eigenvalues, show that $\mathsf{x}^T A \mathsf{x} \geq 0$ for all vectors $\mathsf{x} \in \mathsf{R}^n$.

6. Let A be a (real) $n \times n$ matrix with the property that $\mathsf{x}^T A \mathsf{x} \geq 0$ for all vectors x.

   (a) Let λ be a real eigenvalue of A. Show that $\lambda \geq 0$.

   (b) If A is symmetric, show that A has a square root; that is, there exists a matrix B such that $A = B^2$.

   (c) If A is symmetric, show that A can be factored in the form $A = RR^T$ for some matrix R.

7. Let A be a real symmetric matrix with nonnegative eigenvalues. It is a fact that $\mathsf{x}^T A \mathsf{x} \geq 0$ for all vectors x of compatible size. (See Exercise 5.) Suppose A and B are real symmetric matrices of the same size with nonnegative eigenvalues. Show that all eigenvalues of $A + B$ are nonnegative. [Hint: Exercise 6.]

Critical Reading

8. What can be said about a real symmetric matrix with precisely one eigenvalue?

9. Suppose A and B are real symmetric matrices.

   (a) [BB] Is $A + B$ orthogonally diagonalizable? Explain.

   (b) Why are all the roots of $\det(A + B - \lambda I)$ real?

10. Let A be a (real) $n \times n$ matrix.

   (a) The matrix $A^T A$ is diagonalizable over R. Why?

   (b) Show that the eigenvalues of $A^T A$ are nonnegative.

(c) Let $\mathsf{v}_1, \mathsf{v}_2, \ldots$ be eigenvectors of A^TA satisfying $\mathsf{v}_{k+1} = A\mathsf{v}_k$. If all the eigenvalues of A^TA are less than 1, show that $\|\mathsf{v}_1\|^2 \geq \|\mathsf{v}_2\|^2 \geq \|\mathsf{v}_3\|^2 \geq \cdots$.

11. Suppose A is a *skew-symmetric* matrix, that is $A^T = -A$. Let x and y be eigenvectors of A corresponding to eigenvalues λ and μ respectively. Show that either x and y are orthogonal or $\lambda = -\mu$.

12. [BB] Let A be a real skew-symmetric matrix. Prove that there exist a real orthogonal matrix Q and a matrix B with B^2 diagonal and $A = QBQ^T$.

    [Hint: A^TA is real symmetric.]

13. **(a)** Find all the possible eigenvalues of a projection matrix.

    (b) Does every projection matrix have all possible eigenvalues?

    Explain your answers.

14. Let v be an eigenvector of a real symmetric $n \times n$ matrix A and let U be the one-dimensional subspace Rv. Show that $U^\perp$ is invariant under multiplication by A. (See the glossary for the definition of "invariant subspace.")

15. **(a)** [BB] Let A be a real symmetric matrix with nonnegative eigenvalues. Show that A can be written in the form $A = RR^T$ for some matrix R. (This is a partial converse to the elementary fact that RR^T is always symmetric.) [Hint: A diagonal matrix with real nonnegative diagonal entries has a square root.]

    (b) Let A and B be real symmetric matrices and suppose the eigenvalues of B are all positive. Show that the roots of the polynomial $\det(A - \lambda B)$ are real. [Hint: First explain why $\det(A - \lambda B) = 0$ gives a nonzero x with $A\mathsf{x} = \lambda B\mathsf{x}$. Then use the result of part (a) to factor $B = RR^T$ with R invertible.]

7.4 Application: Quadratic Forms, Conic Sections

7.4.1 Definition. A *quadratic form* in n variables $x_1, x_2, \ldots, x_n$ is a linear combination of the products x_ix_j, that is, a linear combination of squares x_1^2, x_2^2, ..., x_n^2 and *cross terms* x_1x_2, x_1x_3, ..., x_1x_n, x_2x_3, ..., x_2x_n, ..., $x_{n-1}x_n$.

7.4.2 Examples. • $q = x^2 - y^2 + 4xy$ and $q = x^2 + 3y^2 - 2xy$ are quadratic forms in x and y;

- $q = -4x_1^2 + x_2^2 + 4x_3^2 + 6x_1x_3$ is a quadratic form in x_1, x_2 and x_3;
- the most general quadratic form in x_1, x_2, x_3 is

$$a_1x_1^2 + a_2x_2^2 + a_3x_3^2 + a_{12}x_1x_2 + a_{13}x_1x_3 + a_{23}x_2x_3.$$

:-)

In the study of quadratic forms, our starting point is the observation that any quadratic form can be written in the form $\mathsf{x}^TA\mathsf{x}$ where A is a symmetric matrix and $\mathsf{x} = \begin{bmatrix} x_1 \\ \vdots \\ x_n \end{bmatrix}$. The matrix A is obtained by putting the coefficient of x_i^2 in the ith diagonal position and splitting equally the coefficient of x_ix_j between the (i, j) and (j, i) positions; that is, putting one half the coefficient in each position.

7.4.3 Example. Suppose $q = x_1^2 - x_2^2 + 4x_1x_2$. The coefficients of x_1^2 and x_2^2 are 1 and -1, respectively, so we put these, in order, into the two diagonal positions of a matrix A. The coefficient of x_1x_2 is 4, which we split equally between the $(1, 2)$ and $(2, 1)$ positions, putting a 2 in each place.

So $A = \begin{bmatrix} 1 & 2 \\ 2 & -1 \end{bmatrix}$ and, with $\mathsf{x} = \begin{bmatrix} x_1 \\ x_2 \end{bmatrix}$, the reader might check that $\mathsf{x}^TA\mathsf{x} = x_1^2 - x_2^2 + 4x_1x_2 = q$:

$$\begin{bmatrix} x_1 & x_2 \end{bmatrix} \begin{bmatrix} 1 & 2 \\ 2 & -1 \end{bmatrix} \begin{bmatrix} x_1 \\ x_2 \end{bmatrix} = x_1^2 - x_2^2 + 4x_1x_2.$$

Here are some other examples.

7.4.4 Examples.

- $$4x^2 - 7xy + y^2 = \begin{bmatrix} x & y \end{bmatrix} \begin{bmatrix} 4 & -\frac{7}{2} \\ -\frac{7}{2} & 1 \end{bmatrix} \begin{bmatrix} x \\ y \end{bmatrix}$$

- $$4x_1^2 + x_2^2 + 4x_3^2 + 6x_1x_3 - 10x_2x_3 = \begin{bmatrix} x_1 & x_2 & x_3 \end{bmatrix} \begin{bmatrix} -4 & 0 & 3 \\ 0 & 1 & -5 \\ 3 & -5 & 4 \end{bmatrix} \begin{bmatrix} x_1 \\ x_2 \\ x_3 \end{bmatrix}$$

- $$2x_1^2 - 3x_2^2 + 7x_3^2 = \begin{bmatrix} x_1 & x_2 & x_3 \end{bmatrix} \begin{bmatrix} 2 & 0 & 0 \\ 0 & -3 & 0 \\ 0 & 0 & 7 \end{bmatrix} \begin{bmatrix} x_1 \\ x_2 \\ x_3 \end{bmatrix}.$$

Look carefully at the last example. When a quadratic form $q = \lambda_1x_1^2 + \lambda_2x_2^2 + \cdots + \lambda_nx_n^2$ has no cross terms, the matrix A giving $q = \mathsf{x}^TA\mathsf{x}$ is a diagonal matrix, with diagonal entries the coefficients of $x_1^2, x_2^2, \ldots, x_n^2$:

$$\lambda_1x_1^2 + \lambda_2x_2^2 + \cdots + \lambda_nx_n^2 = \begin{bmatrix} x_1 & x_2 & \cdots & x_n \end{bmatrix} \begin{bmatrix} \lambda_1 & 0 & \cdots & 0 \\ 0 & \lambda_2 & & 0 \\ \vdots & & \ddots & \vdots \\ 0 & 0 & \cdots & \lambda_n \end{bmatrix} \begin{bmatrix} x_1 \\ x_2 \\ \vdots \\ x_n \end{bmatrix}. \quad (1)$$

In general, the matrix A in $\mathsf{x}^TA\mathsf{x}$ is not diagonal, but it is symmetric, so it can be orthogonally diagonalized (this is the Principal Axes Theorem). So, there exists an orthogonal matrix Q such that $Q^TAQ = D$ is a real diagonal matrix. Set $\mathsf{y} = Q^T\mathsf{x}$ $(= \mathsf{Q}^{-1}\mathsf{x})$. Then $\mathsf{x} = Q\mathsf{y}$ and $\mathsf{x}^TA\mathsf{x} = \mathsf{y}^TQ^TAQ\mathsf{y} = \mathsf{y}^TD\mathsf{y}$. If the diagonal entries of D are $\lambda_1, \lambda_2, \ldots, \lambda_n$, the form $q = \mathsf{x}^TA\mathsf{x}$ becomes $q = \mathsf{y}^TD\mathsf{y} = \lambda_1 y_1^2 + \lambda_2 y_2^2 + \cdots + \lambda_n y_n^2$.

7.4.5 Examples. • The quadratic form $q = x^2 - y^2 + 4xy = \mathsf{x}^TA\mathsf{x}$ with $A = \begin{bmatrix} 1 & 2 \\ 2 & -1 \end{bmatrix}$. The eigenvalues of A are $\lambda_1 = \sqrt{5}$, $\lambda_2 = -\sqrt{5}$, so $q = \mathsf{x}^TA\mathsf{x} = \mathsf{y}^TD\mathsf{y} = \sqrt{5}y_1^2 - \sqrt{5}y_2^2$ with $\mathsf{y} = Q^T\mathsf{x}$ for a certain matrix Q.

- The quadratic form $q = -4x_1^2 + x_2^2 + 4x_3^2 + 6x_1x_3 = \mathsf{x}^TA\mathsf{x}$ with $A = \begin{bmatrix} -4 & 0 & 3 \\ 0 & 1 & 0 \\ 3 & 0 & 4 \end{bmatrix}$. The eigenvalues of A are $1, 5, -5$, so $q = y_1^2 + 5y_2^2 - 5y_3^2$, with $\mathsf{y} = Q^T\mathsf{x}$ for a certain matrix Q. ⌣̈

Conic Sections

Remember that the *coordinates* of a vector x relative to a basis $\{\mathsf{v}_1, \mathsf{v}_2, \ldots, \mathsf{v}_n\}$ are the scalars $c_1, c_2, \ldots, c_n$ required as coefficients when x is written as a linear combination of $\mathsf{v}_1, \ldots, \mathsf{v}_n$. The coordinates of $\begin{bmatrix} -3 \\ 4 \end{bmatrix}$ relative to the standard basis are -3 and 4 because $\begin{bmatrix} -3 \\ 4 \end{bmatrix} = -3\begin{bmatrix} 1 \\ 0 \end{bmatrix} + 4\begin{bmatrix} 0 \\ 1 \end{bmatrix} = -3\mathsf{e}_1 + 4\mathsf{e}_2$.

Let's examine the equation $\mathsf{y} = Q^T\mathsf{x}$ in more detail with $\mathsf{y} = \begin{bmatrix} y_1 \\ y_2 \end{bmatrix}$ and $Q = \begin{bmatrix} \mathsf{q}_1 & \mathsf{q}_2 \\ \downarrow & \downarrow \end{bmatrix}$. Since $Q^T = Q^{-1}$, $\mathsf{x} = Q\mathsf{y} = y_1\mathsf{q}_1 + y_2\mathsf{q}_2$, so x has coordinates y_1 and y_2 relative to the basis $\{\mathsf{q}_1, \mathsf{q}_2\}$. See Figure 7.1.

The set $\mathcal{Q} = \{\mathsf{q}_1, \mathsf{q}_2\}$ is an orthonormal basis for R^2 just like the standard basis $\{\mathsf{e}_1, \mathsf{e}_2\}$. A graph in R^2 has an equation relative to the standard basis and another relative to $\mathcal{Q}$—same graph, different equations.

For example, suppose $\mathsf{q}_1 = \begin{bmatrix} \frac{1}{\sqrt{2}} \\ \frac{1}{\sqrt{2}} \end{bmatrix}$ and $\mathsf{q}_2 = \begin{bmatrix} -\frac{1}{\sqrt{2}} \\ \frac{1}{\sqrt{2}} \end{bmatrix}$. The relationship between the coordinates x_1, x_2 of a vector relative to the standard basis and its coordinates y_1, y_2 relative to $\mathcal{Q}$ is expressed by $\mathsf{x} = Q\mathsf{y}$:

$$\begin{bmatrix} x_1 \\ x_2 \end{bmatrix} = \begin{bmatrix} \frac{1}{\sqrt{2}} & -\frac{1}{\sqrt{2}} \\ \frac{1}{\sqrt{2}} & \frac{1}{\sqrt{2}} \end{bmatrix} \begin{bmatrix} y_1 \\ y_2 \end{bmatrix},$$

so

$$x_1 = \tfrac{1}{\sqrt{2}}y_1 - \tfrac{1}{\sqrt{2}}y_2 \quad \text{and} \quad x_2 = \tfrac{1}{\sqrt{2}}y_1 + \tfrac{1}{\sqrt{2}}y_2.$$

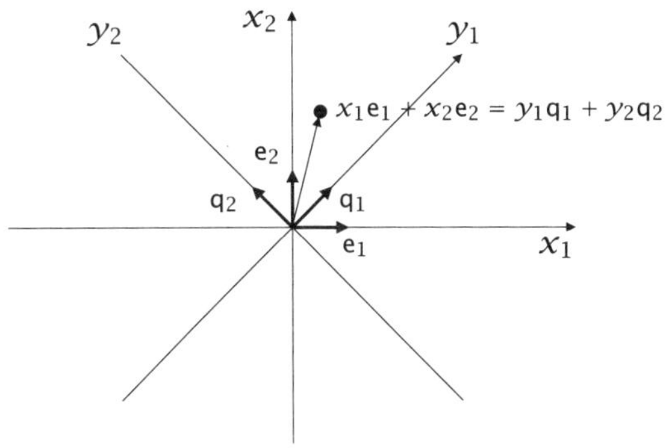

Figure 7.1: The coordinates of (x_1, x_2) relative to the orthonormal basis $\{\mathsf{q}_1, \mathsf{q}_2\}$ are (y_1, y_2).

The ellipse with equation $\frac{x_1^2}{16} + \frac{x_2^2}{9} = 1$ relative to the standard basis is shown in Figure 7.2. Relative to the basis $\mathcal{Q}$, it has equation

$$\tfrac{1}{16}\left(\tfrac{1}{\sqrt{2}}y_1 - \tfrac{1}{\sqrt{2}}y_2\right)^2 + \tfrac{1}{9}\left(\tfrac{1}{\sqrt{2}}y_1 + \tfrac{1}{\sqrt{2}}y_2\right)^2 = 1,$$

that is, $25y_1^2 + 25y_2^2 + 14y_1y_2 = 288$. This example is only for the purposes of illustration; if we had to choose between the equations $\frac{x_1^2}{16} + \frac{x_2^2}{9} = 1$ and $25y_1^2 + 25y_2^2 + 14y_1y_2 = 288$, most of us would choose the former, without the cross terms.

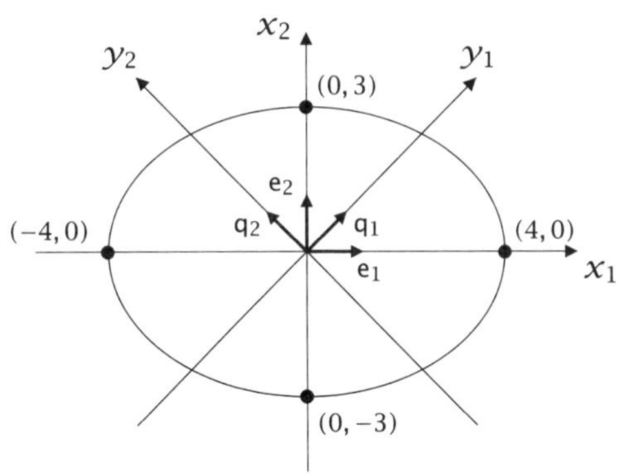

Figure 7.2: The ellipse with equation $\frac{x_1^2}{16} + \frac{x_2^2}{9} = 1$ relative to the standard basis.

On the other hand, if a quadratic form happens to come to us with some nonzero cross terms, it would be nice to find a new set of "principal axes" relative to which the cross terms disappear.

7.4.6 Problem. Identify the graph of $5x_1^2 - 4x_1x_2 + 8x_2^2 = 36$.

Solution. The given equation is $q = 36$, where $q = 5x_1^2 - 4x_1x_2 + 8x_2^2 = \mathsf{x}^T A\mathsf{x}$ with

$$\mathsf{x} = \begin{bmatrix} x_1 \\ x_2 \end{bmatrix} \quad \text{and} \quad A = \begin{bmatrix} 5 & -2 \\ -2 & 8 \end{bmatrix}.$$

The eigenvalues of A are $\lambda_1 = 4$ and $\lambda_2 = 9$, so, relative to a certain new orthonormal basis $\mathcal{Q}$, our equation is $4y_1^2 + 9y_2^2 = 36$. The graph is an ellipse, in standard position relative to the basis $\mathcal{Q} = \{\mathsf{q}_1, \mathsf{q}_2\}$. See Figure 7.3. The vectors

$$\mathsf{q}_1 = \frac{1}{\sqrt{5}} \begin{bmatrix} 2 \\ 1 \end{bmatrix} \quad \text{and} \quad \mathsf{q}_2 = \frac{1}{\sqrt{5}} \begin{bmatrix} -1 \\ 2 \end{bmatrix}$$

are eigenvectors of A corresponding to $\lambda_1 = 4$ and $\lambda_2 = 9$, respectively.

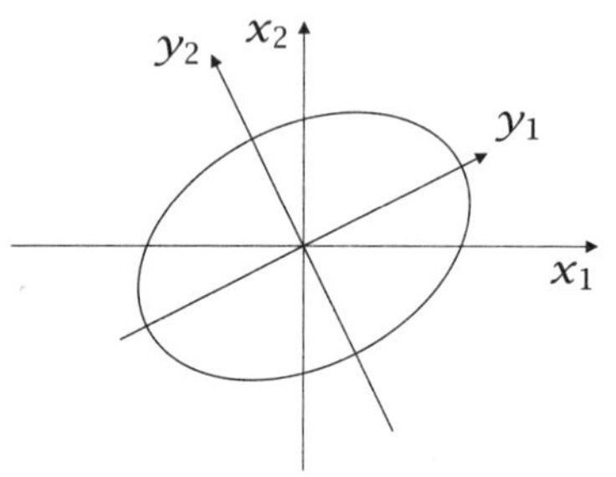

Figure 7.3: The ellipse with equation $4y_1^2 + 9y_2^2 = 1$.

EXERCISES

*Answers to exercises marked [BB] can be found at the **Back** of the **Book**.*

1. Write each of the given quadratic forms as $\mathsf{x}^T A\mathsf{x}$ for some symmetric matrix A.

   (a) [BB] $q = 2x^2 - 3xy + y^2$

   (b) $q = x_1^2 - 2x_2^2 + 3x_3^2 - x_1x_3$

   (c) $q = x_2^2 + 4x_3^2 + x_1x_2 + 4x_1x_3 - 7x_2x_3$

   (d) $q = 7y_1^2 - 3y_2^2 + 5y_3^2 - 8y_1y_3 + 14y_2y_3 - y_1y_2$.

2. Let $\mathsf{q}_1 = \frac{1}{5}\begin{bmatrix} 3 \\ -4 \end{bmatrix}$ and $\mathsf{q}_2 = \frac{1}{5}\begin{bmatrix} 4 \\ 3 \end{bmatrix}$. Let (x_1, x_2) be the coordinates of the point P in the Euclidean plane relative to the standard basis and (y_1, y_2) the coordinates of P relative to the orthogonal basis $\mathcal{Q} = \{\mathsf{q}_1, \mathsf{q}_2\}$.

(a) Write down the equations that relate x_1, x_2 to y_1, y_2.

(b) Find the equations of each of the following quadratic forms in terms of y_1 and y_2.

i. [BB] $q = 25x_1^2 - 50x_2^2$

ii. $q = x_1^2 + x_2^2$

iii. $q = x_1x_2$

iv. $q = 2x_1^2 + 7x_2^2$.

3. Rewrite the given equations so that relative to some orthonormal basis the cross terms disappear. (You are not being asked to find the new basis.) Try to identify the graph.

(a) [BB] $6x^2 - 4xy + 3y^2 = 1$

(b) $3x^2 + y^2 + 3z^2 + 2xz = 1$

(c) $5y^2 + 4z^2 + 4xy + 8xz + 12yz = 3$

(d) $4x^2 + 4y^2 - 8z^2 - 10xy - 4yz = 1$.

4. [BB] Find an orthonormal basis of R^3 relative to which the quadratic form $q = 3x_1^2 + 4x_2^2 + x_3^2 - 4x_2x_3$ has no cross terms. Express q relative to this new basis.

5. Answer Exercise 4 again using each of the following quadratic forms.

(a) $q = x_1^2 + x_2^2 + 5x_3^2 - 2x_1x_3 + 4x_2x_3$

(b) $q = 2x_1^2 + 2x_3^2 - 2x_1x_2 - 6x_1x_3 - 2x_2x_3$

(c) $q = 2x_1^2 - 13x_2^2 - 10x_3^2 + 20x_1x_2 - 12x_1x_3 + 24x_2x_3$

(d) $q = 2x_1 + x_2^2 + 2x_3^2 + 2x_1x_2 + 2x_2x_3$.

6. Show that the graph (in the plane) of the equation $ax^2 + bxy + cy^2 = 1$ is an ellipse if $b^2 - 4ac < 0$ and a hyperbola if $b^2 - 4ac > 0$. What is the graph if $b^2 - 4ac = 0$?

[Remark: An equation of the given type may not have a graph at all. Consider, for example, $-x^2 - y^2 = 1$. In this exercise, assume that the values of a, b and c are such that no "vacuous" case occurs.]

7.5 The Singular Value Decomposition

Not every real matrix A has a factorization $A = P^{-1}DP$ with D diagonal, as we well know. In this section, however, we show that every matrix over R (not necessarily square) does have a factorization $Q^{-1}DP$ with D diagonal and both P and Q orthogonal. This "Singular Value Decomposition" has become extremely important. To learn about applications in politics, to crystals and in quantum information, for example, we refer the reader to the article "The Extraordinary SVD," by Carla Martin and Mason Porter, which appeared in the *American Mathematical Monthly* **10** (2012), pp. 838–851.

Suppose A is a real $m \times n$ matrix. The $n \times n$ matrix A^TA is symmetric, so by the Principal Axes Theorem—Theorem 7.2.12—R^n has a basis $\mathsf{x}_1, \dots, \mathsf{x}_n$ of orthonormal eigenvectors of A^TA. Let $\lambda_1, \dots, \lambda_n$ be corresponding eigenvalues, so that $A^TA\mathsf{x}_i = \lambda_i\mathsf{x}_i$ for $i = 1, 2, \dots, n$.

Now $\|\mathsf{x}_i\|^2 = \mathsf{x}_i \cdot \mathsf{x}_i = 1$ for each i, so

$$(A^TA\mathsf{x}_i) \cdot \mathsf{x}_i = (\lambda_i\mathsf{x}_i) \cdot \mathsf{x}_i = \lambda_i(\mathsf{x}_i \cdot \mathsf{x}_i) = \lambda_i,$$

while, on the other hand,

$$\begin{aligned}(A^TA\mathsf{x}_i) \cdot \mathsf{x}_i &\overset{\downarrow}{=} (A^TA\mathsf{x}_i)^T\mathsf{x}_i \\ &= \mathsf{x}_i^TA^TA\mathsf{x}_i = (A\mathsf{x}_i)^T(A\mathsf{x}_i) = (A\mathsf{x}_i) \cdot (A\mathsf{x}_i) = \|A\mathsf{x}_i\|^2.\end{aligned}$$

READING CHECK 1. Explain the first equality here, the one marked with the arrow.

So

$$\lambda_i = \|A\mathsf{x}_i\|^2 \quad \text{for } i = 1, 2, \dots, n, \tag{1}$$

which implies that $\lambda_i \geq 0$ for all i. Now rearrange the λ_i, if necessary, so that $\lambda_1 > 0$, $\lambda_2 > 0$, ..., $\lambda_r > 0$, and $\lambda_{r+1} = \lambda_{r+2} = \cdots = \lambda_n = 0$. Set

$$\sigma_i = \sqrt{\lambda_i} \quad \text{and} \quad \mathsf{y}_i = \frac{1}{\sigma_i}A\mathsf{x}_i \qquad \text{for } i = 1, 2, \dots, r.$$

The y_i are vectors in R^m and each has length 1 because

$$\|\mathsf{y}_i\| = \frac{1}{\sigma_i}\|A\mathsf{x}_i\| = \frac{1}{\sigma_i}\sqrt{\lambda_i} = 1.$$

In fact, the y_i are orthonormal because

$$\begin{aligned}\mathsf{y}_i \cdot \mathsf{y}_j &= \frac{1}{\sigma_i\sigma_j}A\mathsf{x}_i \cdot A\mathsf{x}_j \\ &= \frac{1}{\sigma_i\sigma_j}\mathsf{x}_i^TA^TA\mathsf{x}_j = \frac{1}{\sigma_i\sigma_j}\mathsf{x}_i^T\lambda_j\mathsf{x}_j = \frac{\lambda_j}{\sigma_i\sigma_j}\mathsf{x}_i \cdot \mathsf{x}_j\end{aligned} \tag{2}$$

which is 0 if $i \neq j$ (because the x_i are orthonormal).

Now extend the orthonormal vectors $\mathsf{y}_1, \dots, \mathsf{y}_r$ to an orthonormal basis y_1, $\mathsf{y}_2, \dots, \mathsf{y}_m$ of R^m (extend to any basis and then apply the Gram–Schmidt Algorithm) and set

$$Q_1 = \begin{bmatrix} \mathsf{y}_1 & \cdots & \mathsf{y}_m \\ \downarrow & & \downarrow \end{bmatrix} \quad \text{and} \quad Q_2 = \begin{bmatrix} \mathsf{x}_1 & \cdots & \mathsf{x}_n \\ \downarrow & & \downarrow \end{bmatrix}.$$

Then Q_1 is an orthogonal $m \times m$ matrix, Q_2 is an orthogonal $n \times n$ matrix and

$$\Sigma = Q_1^TAQ_2 = \begin{bmatrix} \mathsf{y}_1^T & \rightarrow \\ \vdots & \\ \mathsf{y}_m^T & \rightarrow \end{bmatrix}\begin{bmatrix} A\mathsf{x}_1 & \cdots & A\mathsf{x}_n \\ \downarrow & & \downarrow \end{bmatrix}$$

is an $m \times n$ matrix, the (i,j) entry of which is $\mathsf{y}_i^T(A\mathsf{x}_j)$. If $j \le r$, then $\mathsf{y}_j = \dfrac{1}{\sigma_j}A\mathsf{x}_j$ and

$$\mathsf{y}_i^T(A\mathsf{x}_j) = \mathsf{y}_i^T(\sigma_j \mathsf{y}_j) = \sigma_j \mathsf{y}_i \cdot \mathsf{y}_j = \begin{cases} \sigma_j & \text{if } i = j \\ 0 & \text{if } i \ne j. \end{cases}$$

If $j > r$, $\left\|A\mathsf{x}_j\right\|^2 = \lambda_j = 0$, so $A\mathsf{x}_j = \mathsf{0}$ and $\mathsf{y}_i^T(A\mathsf{x}_j) = 0$. Thus

$$\Sigma = \begin{bmatrix} \sigma_1 & 0 & \cdots & 0 & 0 & & 0 \\ 0 & \sigma_2 & & & \vdots & & \vdots \\ \vdots & & \ddots & & & & \\ 0 & & & \sigma_r & & & \\ 0 & & & & 0 & & \\ \vdots & & & & & \ddots & \\ 0 & & \cdots & & & & 0 \end{bmatrix}.$$

Since $Q_1^T = Q_1^{-1}$ and $Q_2^T = Q_2^{-1}$, the equation $\Sigma = Q_1^T A Q_2$ implies $A = Q_1 \Sigma Q_2^T$.

All this establishes the following theorem.

7.5.1 Theorem. *Any $m \times n$ (real) matrix A can be factored $A = Q_1 \Sigma Q_2^T$, where Q_1 is an orthogonal $m \times m$ matrix, Q_2 is an orthogonal $n \times n$ matrix and $\Sigma = \begin{bmatrix} D & \mathbf{0} \\ \mathbf{0} & \mathbf{0} \end{bmatrix}$ is a diagonal $m \times n$ matrix with $D = \begin{bmatrix} \sigma_1 & & \\ & \ddots & \\ & & \sigma_r \end{bmatrix}$, a diagonal $r \times r$ matrix with positive diagonal entries σ_i.*

[The boldface **0**s in the representation of Σ denote zero matrices of the sizes needed to make Σ $m \times n$.]

7.5.2 Remark. A factorization $A = Q_1 \Sigma Q_2^T$ as in Theorem 7.5.1 is called a *singular value decomposition* of A. The nonzero diagonal entries of Σ, namely, $\sigma_1, \ldots, \sigma_r$, are called the *singular values* of A. The orthogonal matrices Q_1 and Q_2 are not unique, but the singular values of A are uniquely determined as the eigenvalues of the matrix $A^T A$. (See Exercise 5.)

7.5.3 Example. Let $A = \begin{bmatrix} -2 \\ 1 \\ 2 \end{bmatrix}$. Then $A^T A = \begin{bmatrix} -2 & 1 & 2 \end{bmatrix} \begin{bmatrix} -2 \\ 1 \\ 2 \end{bmatrix} = [9]$. The vector $\mathsf{x}_1 = [1]$ is a basis for R^1 and it's a unit eigenvector for $A^T A$ corresponding to eigenvalue $\lambda_1 = 9$. Set $\sigma_1 = \sqrt{\lambda_1} = 3$ and $\mathsf{y}_1 = \frac{1}{\sigma_1}A\mathsf{x}_1 = \frac{1}{3}\begin{bmatrix} -2 \\ 1 \\ 2 \end{bmatrix}$. Extend

y_1 to an orthonormal basis of R^3, for example,

$$\mathsf{y}_1 = \frac{1}{3}\begin{bmatrix} -2 \\ 1 \\ 2 \end{bmatrix}, \quad \mathsf{y}_2 = \frac{1}{\sqrt{5}}\begin{bmatrix} 0 \\ 2 \\ -1 \end{bmatrix}, \quad \mathsf{y}_3 = \frac{1}{\sqrt{45}}\begin{bmatrix} 5 \\ 2 \\ 4 \end{bmatrix}.$$

Set

$$Q_1 = \begin{bmatrix} -\frac{2}{3} & 0 & \frac{5}{\sqrt{45}} \\ \frac{1}{3} & \frac{2}{\sqrt{5}} & -\frac{1}{\sqrt{5}} \\ \frac{2}{3} & -\frac{1}{\sqrt{5}} & \frac{4}{\sqrt{45}} \end{bmatrix}, \quad Q_2 = [1] \quad \text{and} \quad \Sigma = \begin{bmatrix} 3 \\ 0 \\ 0 \end{bmatrix}.$$

Then $A = Q_1 \Sigma Q_2^T$ is a singular value decomposition. The lone singular value of A is 3.

7.5.4 Example. Suppose $A = \begin{bmatrix} 1 & 1 \\ 1 & 0 \\ 0 & 1 \end{bmatrix}$. Then $A^T A = \begin{bmatrix} 2 & 1 \\ 1 & 2 \end{bmatrix}$ and

$$\mathsf{x}_1 = \tfrac{1}{\sqrt{2}}\begin{bmatrix} 1 \\ -1 \end{bmatrix}, \quad \mathsf{x}_2 = \tfrac{1}{\sqrt{2}}\begin{bmatrix} 1 \\ 1 \end{bmatrix}$$

provide an orthonormal basis of eigenvectors of $A^T A$ with corresponding eigenvalues $\lambda_1 = 1$, $\lambda_2 = 3$.

Let $\sigma_1 = \sqrt{\lambda_1} = 1$, $\sigma_2 = \sqrt{\lambda_2} = \sqrt{3}$ and $\Sigma = \begin{bmatrix} 1 & 0 \\ 0 & \sqrt{3} \\ 0 & 0 \end{bmatrix}$, then set

$$\mathsf{y}_1 = \frac{1}{\sigma_1} A\mathsf{x}_1 = \frac{1}{\sqrt{2}}\begin{bmatrix} 0 \\ 1 \\ -1 \end{bmatrix} \quad \text{and} \quad \mathsf{y}_2 = \frac{1}{\sigma_2} A\mathsf{x}_2 = \frac{1}{\sqrt{6}}\begin{bmatrix} 2 \\ 1 \\ 1 \end{bmatrix}.$$

Together with $\mathsf{y}_3 = \frac{1}{\sqrt{3}}\begin{bmatrix} -1 \\ 1 \\ 1 \end{bmatrix}$, we obtain an orthonormal basis $\mathsf{y}_1, \mathsf{y}_2, \mathsf{y}_3$ for R^3. Letting

$$Q_1 = \begin{bmatrix} 0 & \frac{2}{\sqrt{6}} & -\frac{1}{\sqrt{3}} \\ \frac{1}{\sqrt{2}} & \frac{1}{\sqrt{6}} & \frac{1}{\sqrt{3}} \\ -\frac{1}{\sqrt{2}} & \frac{1}{\sqrt{6}} & \frac{1}{\sqrt{3}} \end{bmatrix} \quad \text{and} \quad Q_2 = \begin{bmatrix} \frac{1}{\sqrt{2}} & \frac{1}{\sqrt{2}} \\ -\frac{1}{\sqrt{2}} & \frac{1}{\sqrt{2}} \end{bmatrix}$$

we have

$$\underset{A}{\begin{bmatrix} 1 & 1 \\ 1 & 0 \\ 0 & 1 \end{bmatrix}} = \underset{Q_1}{\begin{bmatrix} 0 & \frac{2}{\sqrt{6}} & -\frac{1}{\sqrt{3}} \\ \frac{1}{\sqrt{2}} & \frac{1}{\sqrt{6}} & \frac{1}{\sqrt{3}} \\ -\frac{1}{\sqrt{2}} & \frac{1}{\sqrt{6}} & \frac{1}{\sqrt{3}} \end{bmatrix}} \underset{\Sigma}{\begin{bmatrix} 1 & 0 \\ 0 & \sqrt{3} \\ 0 & 0 \end{bmatrix}} \underset{Q_2^T,}{\begin{bmatrix} \frac{1}{\sqrt{2}} & -\frac{1}{\sqrt{2}} \\ \frac{1}{\sqrt{2}} & \frac{1}{\sqrt{2}} \end{bmatrix}}$$

a singular value decomposition of A. The singular values of A are 1 and $\sqrt{3}$.

The Pseudoinverse Again

A singular value decomposition for a matrix A leads to a nice formula for the pseudoinverse A^+, a matrix we first met in Section 6.5:

7.5.5

$$\boxed{\text{If } A = Q_1\Sigma Q_2^T, \text{ then } A^+ = Q_2\Sigma^+Q_1^T.}$$

The formula is easily remembered. Since Q_1 and Q_2 are orthogonal, $A = Q_1\Sigma Q_2^{-1}$, so if A and Σ were invertible, we would have $A^{-1} = Q_2\Sigma^{-1}Q_1^{-1} = Q_2\Sigma^{-1}Q_1^T$. In general, without any assumptions about invertibility, one simply replaces the inverses of A and Σ by their pseudoinverses. But why is 7.5.5 true?

You might remember that A^+ provides the best solution to $A\mathsf{x} = \mathsf{b}$ in the sense that $\mathsf{x}^+ = A^+\mathsf{b}$ is the vector of shortest length amongst those vectors x minimizing $\|\mathsf{b} - A\mathsf{x}\|$. Given $A = Q_1\Sigma Q_2^T$ and in view of 6.3.9—multiplication by the orthogonal matrix Q_1^T preserves length—we have

$$\|\mathsf{b} - A\mathsf{x}\| = \left\|\mathsf{b} - Q_1\Sigma Q_2^T\mathsf{x}\right\| = \left\|Q_1^T\mathsf{b} - Q_1^T(Q_1\Sigma Q_2^T\mathsf{x})\right\| = \left\|Q_1^T\mathsf{b} - \Sigma Q_2^T\mathsf{x}\right\|$$

so, in particular,

$$\|\mathsf{b} - A\mathsf{x}^+\| = \left\|Q_1^T\mathsf{b} - \Sigma\mathsf{y}\right\| \tag{3}$$

with $\mathsf{y} = Q_2^T\mathsf{x}^+$. The left side of (3) is a minimum, hence so is the right. Moreover x^+ is the shortest vector giving this minimum, hence so is y because $\|\mathsf{y}\| = \|\mathsf{x}\|$ (orthogonal matrices preserve length). It follows that $\mathsf{y} = \mathsf{y}^+ = Q_2^T\mathsf{x}^+$, while a basic property of Σ^+ is that it gives the best solution to $\Sigma\mathsf{y} = Q_1^T\mathsf{b}$; that is, $\mathsf{y}^+ = \Sigma^+Q_1^T\mathsf{b}$. We conclude that $\mathsf{x}^+ = Q_2\mathsf{y}^+ = Q_2\Sigma^+Q_1^T\mathsf{b}$, so $A^+ = Q_2\Sigma^+Q_1^T$, as asserted.

Formula 7.5.5 makes it quite easy to discover a number of properties of the pseudoinverse. The proof of the following theorem makes use of the fact that if $\Sigma = \begin{bmatrix} \sigma_1 & & & \\ & \ddots & & \mathbf{0} \\ & & \sigma_r & \\ & \mathbf{0} & & \mathbf{0} \end{bmatrix}$ is $m \times n$ with $\sigma_1, \ldots, \sigma_r$ all nonzero, then Σ^+ is the $n \times m$ matrix $\Sigma^+ = \begin{bmatrix} \frac{1}{\sigma_1} & & & \\ & \ddots & & \mathbf{0} \\ & & \frac{1}{\sigma_r} & \\ & \mathbf{0} & & \mathbf{0} \end{bmatrix}$ (see paragraph 6.5.8), so $(\Sigma^+)^+ = \Sigma$.

(As always, the boldface $\mathbf{0}$s indicate zero matrices of the size needed to complete a matrix of the right size.)

7.5.6 Theorem. (Properties of the Pseudoinverse) *The pseudoinverse A^+ of an $m \times n$ matrix A has the following properties:*

1. A^+ *is* $n \times m$;
2. $A^{++} = A$;
3. $\operatorname{rank} A^+ = \operatorname{rank} A$;
4. $\operatorname{col\,sp} A^+ = \operatorname{row\,sp} A$;
5. $\operatorname{row\,sp} A^+ = \operatorname{col\,sp} A$;
6. A^+A *is the projection matrix onto the row space of* A.

Proof. 1. We have $A = Q_1\Sigma Q_2^T$ with Q_1 an orthogonal $m \times m$ matrix, Σ $m \times n$ and Q_2 an orthogonal $n \times n$ matrix. Thus $A^+ = Q_2\Sigma^+Q_1^T$ is $n \times m$.

2. With $A = Q_1\Sigma Q_2^T$, we have $A^+ = Q_2\Sigma^+Q_1^T$, so $(A^+)^+ = Q_1(\Sigma^+)^+Q_2^T = Q_1\Sigma Q_2^T = A$ since $(\Sigma^+)^+ = \Sigma$.

3. For this part, we remember that multiplication by an invertible matrix preserves rank—see Exercise 4 in Section 4.5. Thus $\operatorname{rank} A = \operatorname{rank} Q_1\Sigma Q_2^T = \operatorname{rank}\Sigma$ and, similarly, $\operatorname{rank} A^+ = \operatorname{rank}\Sigma^+$. Since the rank of a diagonal matrix with r nonzero entries is r, we have $\operatorname{rank}\Sigma = r = \operatorname{rank}\Sigma^+$.

4. A vector in the column space of A^+ has the form $A^+\mathsf{b}$, so it's in the row space of A. (This is part of the definition of the pseudoinverse—see Definition 6.5.6.) Thus $\operatorname{col\,sp} A^+ \subseteq \operatorname{row\,sp} A$. Since the dimensions of these two spaces are equal by part 3, these spaces are the same.

5. Applying part 4 to the matrix A^+ gives $\operatorname{row\,sp} A^+ = \operatorname{col\,sp}(A^+)^+ = \operatorname{col\,sp} A$.

6. The definition of the pseudoinverse specifies that $A(A^+\mathsf{b})$ is the projection of b on the column space of A. Applying this fact to the matrix A^+ tells us that $A^+(A^+)^+\mathsf{b} = A^+A\mathsf{b}$ is the projection of b on the column space of A^+, which is the row space of A. ■

Answers to Reading Checks

1. For any vectors u and v (in the same Euclidean space), we know that $\mathsf{u} \cdot \mathsf{v} = \mathsf{u}^T\mathsf{v}$.

True/False Questions

Decide, with as little calculation as possible, whether each of the following statements is true or false and, if you say "false," explain your answer. (Answers are at the back of the book.)

1. For any (real) symmetric $n \times n$ matrix A, there exists an orthonormal basis of R^n consisting of eigenvectors of A.

2. If u, v, w are orthonormal vectors in R^4, there exists a vector of length 1 orthogonal to each of u, v, w.

3. If A is a matrix and Q_1 and Q_2 are orthogonal matrices with $\Sigma = Q_1^T A Q_2$, then $A = Q_1 \Sigma Q_2^{-1}$.

4. The singular values of a matrix A are the eigenvalues of $A^T A$.

5. If $A = Q_1 \Sigma Q_2^T$ is a singular value decomposition, then $A^+ = Q_1^+ \Sigma^+ Q_2$.

EXERCISES

Answers to exercises marked [BB] can be found at the **Back** *of the* **Book.**

1. Find a singular value decomposition of each of the following matrices and use this to find the pseudoinverse of each matrix as well.

   (a) [BB] $A = \begin{bmatrix} 1 & -1 \\ 0 & 0 \\ 1 & -1 \end{bmatrix}$ **(b)** $A = \begin{bmatrix} 1 & 1 & -1 \\ 1 & 3 & 1 \end{bmatrix}$

   (c) [BB] $A = \begin{bmatrix} 1 & 2 & 3 \end{bmatrix}$ **(d)** $A = \begin{bmatrix} 1 & -1 \\ -1 & 1 \\ 2 & -2 \end{bmatrix}$

   (e) $A = \begin{bmatrix} 1 & -1 & 0 \\ 0 & 0 & 1 \\ 1 & -1 & 0 \end{bmatrix}$ **(f)** $A = \begin{bmatrix} 0 & -1 & 3 \\ 0 & 2 & 0 \\ 0 & 0 & 2 \end{bmatrix}$.

2. Prove that the determinant of a square matrix A is plus or minus the product of the singular values of A.

3. [BB] Let A be an $m \times n$ matrix. Prove that $(A^+)^T = (A^T)^+$.

4. **(a)** [BB] Prove that $AA^+A = A$ for any matrix A.

   (b) Use part (a) to prove that $A^+ = A^+AA^+$.

   (c) Prove that AA^+ is a projection matrix.

5. Let $A = Q_1 \Sigma Q_2^T$ be a factorization of the $m \times n$ matrix A with Q_1 and Q_2 orthogonal and $\Sigma = \begin{bmatrix} \sigma_1 & & 0 \\ & \ddots & \\ & & \sigma_r \\ 0 & & 0 \end{bmatrix}$.

   (a) [BB] Show that the columns of Q_2 must be eigenvectors of $A^T A$.

   (b) Show that the columns of Q_1 must be eigenvectors of AA^T.

   (c) Show that the nonzero diagonal entries of Σ are the square roots of the nonzero (hence positive) eigenvalues of $A^T A$.

   [This exercise shows that the singular values of a matrix are unique.]

6. Let $A = Q_1 \Sigma Q_2^T$ be a singular value decomposition of an $m \times n$ matrix A.

(a) Show that the matrices A^TA and $\Sigma^T\Sigma$ are similar.

(b) Explain why A^TA and $\Sigma^T\Sigma$ have the same eigenvalues.

(c) Explain why A^TA and AA^T have the same nonzero eigenvalues. [See Exercise 7 for another proof of this surprising fact.]

7. (a) [BB] Show that the singular values of a matrix and its transpose are the same.

   (b) For any $m \times n$ matrix A, the matrices AA^T and A^TA have the same set of nonzero eigenvalues. Why?

8. (a) Let $A = Q_1\Sigma Q_2^T$ be a singular value decomposition of an $m \times n$ matrix A. The columns of Q_2 are orthonormal eigenvectors of A^TA. Show that the first r columns of Q_1 are orthonormal eigenvectors of AA^T.

   (b) Let A be a real symmetric matrix with positive eigenvalues. Explain why the singular values of A are precisely its eigenvalues.

9. Let $A = Q_1\Sigma Q_2^T$ be a singular value decomposition of an $m \times n$ matrix A of rank r with $Q_1 = \begin{bmatrix} \mathsf{y}_1 & \cdots & \mathsf{y}_m \\ \downarrow & & \downarrow \end{bmatrix}$ and $Q_2 = \begin{bmatrix} \mathsf{x}_1 & \cdots & \mathsf{x}_n \\ \downarrow & & \downarrow \end{bmatrix}$ as in the text.

   (a) [BB] The first r columns of Q_1 are orthonormal eigenvectors of AA^T and precisely those with nonzero eigenvalues—see Exercise 8. Show that these are also a basis for the column space of A. Show that the last $m - r$ columns of Q_1 form a basis for the null space of A^T.

   (b) The first r columns of Q_2 are orthonormal eigenvectors of A^TA and precisely those with nonzero eigenvalues. Show that these are also a basis for the row space of A. Show that the last $n-r$ columns of Q_2 are a basis for the null space of A. [Hint: Establish the second statement first!]

10. Recall the $A^+ = A^{-1}$ if A is invertible—see Section 6.5—and in at least one respect, the formula $(A^+)^+ = A$, the pseudoinverse behaves like the inverse in general. Determine whether or not $(AB)^+ = B^+A^+$ by considering the matrices $A = \begin{bmatrix} 1 & 2 & 3 \end{bmatrix}$ and $B = \begin{bmatrix} -2 \\ 1 \\ 2 \end{bmatrix}$.

11. Explain how to obtain a singular value decomposition of A^{-1} from a singular value decomposition of an invertible matrix A.

Critical Reading

12. Your job is to produce a 2×3 matrix A with two singular values, both $\sqrt{5}$. How would you do this? Develop a procedure and illustrate by finding the singular value decomposition of a suitable matrix.

CHAPTER KEY WORDS AND IDEAS: Here are some technical words and phrases that were used in this chapter. Do you know the meaning of each? If you're not sure, check the glossary or index at the back of the book.

Cayley–Hamilton Theorem
commute
conjugate transpose of a matrix
diagonalizable
dot product (complex)
eigenvalue
eigenvector
Fundamental Theorem of Algebra
Hermitian
normalize
normal matrix
orthogonally diagonalizable
orthogonal matrix
orthonormal vectors
principal axes
pseudoinverse
Schur's Theorem
similar matrices
singular value decomposition
singular values
Spectral Theorem
spectrum
transpose of a matrix
unitarily diagonalizable
unitary

Review Exercises for Chapter 7

1. Determine if each of the following matrices is Hermitian, unitary, or normal.

(a) $A = \begin{bmatrix} 1 & -i \\ i & i \end{bmatrix}$ **(b)** $A = \begin{bmatrix} 2i & -1+i \\ 1+i & i \end{bmatrix}$

(c) $A = \begin{bmatrix} 1 & i \\ -i & 2 \end{bmatrix}$ **(d)** $A = \frac{1}{2}\begin{bmatrix} 1+i & \sqrt{2} \\ 1-i & \sqrt{2}i \end{bmatrix}$

(e) $A = \begin{bmatrix} 2i & 1+i \\ -1+i & i \end{bmatrix}$ **(f)** $A = \begin{bmatrix} 7 & 2+i & -i \\ 2-i & -4 & 3-4i \\ i & 3+4i & -2 \end{bmatrix}$.

2. Fill in the missing entries in the table, which is intended to show the analogues in C^n of terminology in R^n.

R^n	C^n
$\mathsf{x} \cdot \mathsf{y} = x_1y_1 + x_2y_2 + \cdots$	$\mathsf{z} \cdot \mathsf{w} = ?$
transpose A^T	?
?	$\mathsf{z} \cdot \mathsf{w} = \mathsf{z}^*\mathsf{w}$
symmetric: $A^T = A$	?
?	unitary $U^* = U^{-1}$
orthogonally diagonalizable $Q^TAQ = D$	?
?	$\mathsf{z}^*\mathsf{w}$

3. Suppose A and B are $n \times n$ matrices with the same set of n linearly independent eigenvectors. Prove that $AB = BA$.

4. Suppose the only eigenvalue of a real symmetric matrix A is -7. What can you say about A?

5. Is a permutation matrix diagonalizable? orthogonally diagonalizable? Explain.

6. Is the matrix $\begin{bmatrix} \frac{1+2i}{\sqrt{7}} & \frac{3-i}{\sqrt{35}} \\ \frac{1+i}{\sqrt{7}} & -\frac{5}{\sqrt{35}} \end{bmatrix}$ unitary? Explain.

7. **(a)** Show that the product of commuting Hermitian matrices is Hermitian.

   (b) Establish the converse of part (a): If AB is Hermitian, then A and B commute.

8. "A symmetric matrix (whose entries are not necessarily real) is diagonalizable." Is this statement true or false? Explain. [Hint: $A = \begin{bmatrix} 3 & i \\ i & 1 \end{bmatrix}$.]

9. The characteristic polynomial of $A = \begin{bmatrix} 9 & -2 & 6 \\ -2 & 9 & 6 \\ 6 & 6 & 7 \end{bmatrix}$ is $-(\lambda - 11)^2(\lambda + 11)$. Explain why there exists an orthogonal matrix Q such that Q^TAQ is diagonal. What are the diagonal entries of D? Find Q such that $Q^TAQ = \begin{bmatrix} 11 & 0 & 0 \\ 0 & -11 & 0 \\ 0 & 0 & 11 \end{bmatrix}$.

10. **(a)** The matrix $A = \begin{bmatrix} 1 & 0 & -1 \\ 0 & 0 & 0 \\ -1 & 0 & 1 \end{bmatrix}$ is orthogonally diagonalizable. Why?

    (b) Find an orthogonal matrix Q and a diagonal matrix D such that $Q^TAQ = D$.

11. Repeat Exercise 10 with $A = \begin{bmatrix} 3 & 0 & 2 \\ 0 & 1 & 0 \\ 2 & 0 & 3 \end{bmatrix}$.

12. In each case, find an orthogonal matrix Q and a real diagonal matrix D so that $Q^TAQ = D$.

    (a) $A = \begin{bmatrix} 5 & 4 & -4 \\ 4 & 5 & 4 \\ -4 & 4 & 5 \end{bmatrix}$

    (b) $A = \begin{bmatrix} -5 & -2 & -4 & -9 \\ -2 & 2 & 4 & 2 \\ -4 & 4 & 8 & 4 \\ -9 & 2 & 4 & -5 \end{bmatrix}$ The characteristic polynomial is $\lambda^4 - 196\lambda^2$.

    (c) $A = \begin{bmatrix} 7 & -2 & 5 \\ -2 & 4 & 2 \\ 5 & 2 & 7 \end{bmatrix}$

    (d) $A = \begin{bmatrix} 3 & 3 & -1 & -1 \\ 3 & -5 & 3 & 3 \\ -1 & 3 & -3 & 5 \\ -1 & 3 & 5 & -3 \end{bmatrix}$ The characteristic polynomial is $(\lambda - 4)^2(\lambda + 8)^2$.

13. What can be said about a real symmetric matrix with precisely one eigenvalue? State a proposition and prove it.

14. Find the eigenvalues and corresponding eigenvectors of $A = \begin{bmatrix} 2 & 3-3i \\ 3+3i & 5 \end{bmatrix}$. Verify that eigenvectors corresponding to different eigenvalues are orthogonal. Why should this be the case?

15. Prove that a matrix that is unitarily similar to a real diagonal matrix is Hermitian.

16. **(a)** Show that if A and Q are orthogonal matrices, so is AQ.

    (b) If Q is an orthogonal matrix and x is a vector, show that $\|Q\mathsf{x}\| = \|\mathsf{x}\|$.

    (c) For any $\mathsf{x} \in \mathsf{R}^n$ and any scalar t, show that $\|t\mathsf{x}\| = |t|\,\|\mathsf{x}\|$.

    (d) If λ is an eigenvalue of an orthogonal matrix, then $\lambda = \pm 1$.

17. Let A be a real symmetric matrix. Show that the following three conditions are equivalent:

    i. A is orthogonal;

    ii. If λ is an eigenvalue of A, then $\lambda = \pm 1$;

    iii. $A^2 = I$.

18. Answer true or false and justify your choice.

    (a) $A = \begin{bmatrix} 1 & i \\ -i & 1+i \end{bmatrix}$ is Hermitian.

    (b) If U^*AU is a diagonal matrix and U is unitary, then A is normal.

    (c) If x and y are different eigenvectors of a normal matrix, then x is orthogonal to y.

    (d) All eigenvalues of a Hermitian matrix are real.

    (e) If U is a unitary matrix, then $U - 2I$ is invertible.

    (f) For any $n \times n$ matrix A, the eigenvalues of $A + A^*$ are all real.

    (g) $A = \begin{bmatrix} 1 & -1 & 1 \\ 1 & -1 & 1 \\ 1 & -1 & 1 \end{bmatrix}$ is similar to a diagonal matrix with at least one 0 on the diagonal.

19. **(a)** Find a symmetric matrix A such that $q = 2x^2 + y^2 + 2z^2 + 2xy + 2yz = \mathsf{x}^T A\mathsf{x}$.

    (b) Find an orthonormal basis for R^3 relative to which q has no cross terms. What is q relative to this new basis?

20. In each case, find a singular value decomposition of A and use this to find the pseudoinverse of A.

    (a) $A = \begin{bmatrix} 1 & 0 & -1 \\ 0 & -1 & 2 \end{bmatrix}$ **(b)** $A = \dfrac{1}{3}\begin{bmatrix} -2 & -2 & -1 \\ -1 & -4 & 1 \\ 3 & 0 & 3 \end{bmatrix}$.

21. Let A^+ denote the pseudoinverse of a matrix A. Prove that A^+A is symmetric. [Hint: Singular value decomposition.]

Appendix A: Complex Numbers

Zero and the negative integers were invented so that we could solve equations such as $x + 5 = 5$ and $x + 7 = 0$, at a time when the natural numbers $1, 2, 3, \ldots$ were the only "numbers" known. The rational numbers (common fractions) were introduced so that equations such as $5x - 7 = 0$ would have solutions. The real numbers include the rationals, but also solutions to equations like $x^2 - 2 = 0$ as well as numbers like π and e, which are not solutions to any polynomial equations (with integer coefficients). The real numbers are much more numerous than the rationals, but there are still not enough to provide solutions to equations like $x^2 + 1 = 0$. It is a remarkable fact that by adding to the real numbers the solution to this one equation, our inability to solve polynomial equations disappears. It is common to label i a number satisfying $i^2 + 1 = 0$. It then turns out that every polynomial with real coefficients has a root of the form $a + bi$ with a and b real. Such a number is called *complex* and the amazing fact we have just stated—every polynomial with real coefficients has a complex root—is known as the "Fundamental Theorem of Algebra."

A *complex number* is a number of the form $a + bi$, where a and b are real numbers. The *real part* of $a + bi$ is a; the *imaginary part* is b. The set of all complex numbers is denoted C.

For example, $2 - 3i$, $\frac{1}{2} + \frac{17}{5}i$, $-16 = -16 + 0i$ and $\pi = \pi + 0i$ are complex numbers, the last two examples serving to illustrate that any real number is also a complex number (with imaginary part 0): We write $\mathsf{R} \subseteq \mathsf{C}$.

Keeping in mind that $i^2 = -1$, complex numbers can be added and multiplied in the more or less obvious way:

$$(a + bi) + (c + di) = (a + c) + (b + d)i,$$
$$(a + bi)(c + di) = (ac - bd) + (ad + bc)i.$$

A-1 Examples.

- $(2 - 3i) + (4 + i) = 6 - 2i$
- $i(4 + 5i) = -5 + 4i$
- $(4 + 3i)(2 - i) = 8 - 4i + 6i - 3i^2 = 11 + 2i$ ☺

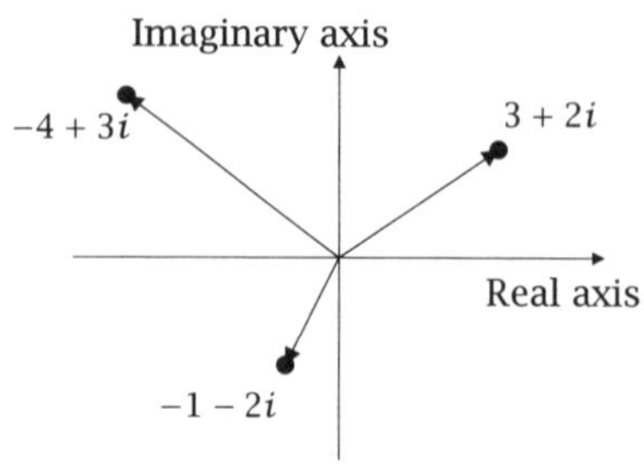

Figure A.1: The complex plane

The Complex Plane

The familiar Cartesian plane can be used to picture complex numbers by associating the point (a, b) with the complex number $a + bi$. See Figure A.1. It is common to picture $a + bi$ as the vector $\begin{bmatrix} a \\ b \end{bmatrix}$ too. Results from Section 1.1 show that addition of complex numbers can be viewed geometrically as vector addition according to the parallelogram rule: the sum of $a + bi$ and $c + di$ is the diagonal of the parallelogram with sides the vectors $\begin{bmatrix} a \\ b \end{bmatrix}$ and $\begin{bmatrix} c \\ d \end{bmatrix}$—see Figure 1.5. When the Cartesian plane is used to picture complex numbers, it is called the *complex plane*. The x-axis is called the *real axis* and the y-axis is the *imaginary axis*.

Complex Conjugation

A-2 **Definition.** The *conjugate* of the complex number $a + bi$ is the complex number $a - bi$. We denote the conjugate of $a + bi$ by $\overline{a + bi}$.

If $z = a + bi$, then $\overline{z} = a - bi$.

A-3 **Example.** $\overline{1 + 2i} = 1 - 2i$; $\overline{2 - 3i} = 2 + 3i$; $\overline{-5} = -5$. ☺

If $z = a = a + 0i$ is real, then $\overline{z} = a - 0i = a = z$. Conversely, if $z = a + bi$ and $\overline{z} = z$, then $a - bi = a + bi$, so $2bi = 0$ implying $b = 0$.

A-4

A complex number z is real if and only if $\overline{z} = z$.

Suppose we multiply $z = a + bi$ by its conjugate. We get

$$z\overline{z} = (a + bi)(a - bi) = a^2 - b^2 i^2 = a^2 + b^2.$$

Representing z as the vector $\mathsf{v} = \begin{bmatrix} a \\ b \end{bmatrix}$, the product $z\overline{z}$ is just $\|\mathsf{v}\|^2$, the square of the length of v.

A-5 Definition. The *modulus* of the complex number $z = a + bi$ is $\sqrt{a^2 + b^2}$ and denoted $|z|$.

If $z = a + bi$, then $\lvert z\rvert = \sqrt{a^2 + b^2}$. Thus $\lvert z\rvert^2 = a^2 + b^2 = z\overline{z}$.

A-6 Example. If $z = 2 - 3i$, $|z|^2 = z\overline{z} = (2 - 3i)(2 + 3i) = 2^2 + 3^2 = 13$. Thus $|z| = \sqrt{13}$. ☺

When z is real, the modulus of z is just its *absolute value* because $z = a + 0i$ real implies $|z| = \sqrt{a^2 + 0^2} = |a|$, the absolute value of $z = a$ in the usual sense.

Here are some basic properties of complex conjugation and modulus.

A-7 Proposition. *If z and w are complex numbers, then*

1. $\overline{\overline{z}} = z$,
2. $\overline{zw} = \overline{z}\,\overline{w}$ *and*
3. $|zw| = |z||w|$.

Proof. 1. Let $z = a + bi$. Then $\overline{z} = a - bi$, so $\overline{\overline{z}} = \overline{a - bi} = a + bi = z$.

2. If $z = a + bi$ and $w = c + di$, then $zw = (a + bi)(c + di) = (ac - bd) + (ad + bc)i$, so $\overline{zw} = (ac - bd) - (ad + bc)i$. On the other hand, $\overline{z} = a - bi$, $\overline{w} = c - di$ and $(bi)(di) = bdi^2 = -bd$, so $\overline{z}\,\overline{w} = (a - bi)(c - di) = (ac - bd) - (ad + bc)i = (ac - bd) - (ad + bc)i$.

3. We have $|zw|^2 = zw\overline{zw} = zw\overline{z}\,\overline{w} = z\overline{z}w\overline{w} = |z|^2|w|^2$. Taking the square root of both sides gives $|zw| = |z||w|$. ■

A-8 Example. Let $z = 2+3i$ and $w = 4-7i$. Then $zw = 29-2i$, so $\overline{zw} = 29+2i$. On the other hand, $\overline{z}\,\overline{w} = (2 - 3i)(4 + 7i) = 29 + 2i$. Thus $\overline{zw} = \overline{z}\,\overline{w}$. ☺

Division of Complex Numbers

We learn early in life how to add and multiply real numbers and we have now seen how to add and multiply complex numbers as well. What about division? Any nonzero real number b has a *multiplicative inverse* $\frac{1}{b}$, which is a real number satisfying $b\left(\frac{1}{b}\right) = 1$. If $a + bi$ is a nonzero complex number, we can write down $\frac{1}{a+bi}$ with the expectation that $(a+bi)\frac{1}{a+bi} = 1$, but is $\frac{1}{a+bi}$ a complex number? Remember that complex numbers are numbers of the form $x + yi$.

Here the notion of complex conjugation comes to the rescue. If $z = a + bi$, then $z\overline{z} = a^2 + b^2$ is a **real number** which is zero if and only if $a = b = 0$, that is, if and only if $z = 0$. So if $z \neq 0$, $\frac{1}{z\overline{z}}$ is a real number and

$$\frac{1}{z} = \frac{1}{z\overline{z}}\,\overline{z} = \frac{1}{a^2+b^2}\overline{z} = \frac{a}{a^2+b^2} - \frac{b}{a^2+b^2}i.$$

Thus $\frac{1}{z}$ is indeed a complex number, with real part $\frac{a}{a^2+b^2}$ and imaginary part $\frac{-b}{a^2+b^2}$.

A-9 Any nonzero complex number z has a multiplicative inverse $\frac{1}{z\overline{z}}\overline{z}$.

A-10 Examples.

- $\frac{1}{2+3i} = \frac{1}{(2+3i)(2-3i)}(2-3i) = \frac{1}{13}(2-3i) = \frac{2}{13} - \frac{3}{13}i$;
- $\frac{1}{2-i} = \frac{1}{(2-i)(2+i)}(2+i) = \frac{1}{5}(2+i) = \frac{2}{5} + \frac{1}{5}i.$ 🙂

With multiplicative inverses in hand, we now have a way to **divide** one complex number by another nonzero complex number. Just as $\frac{a}{b} = a\left(\frac{1}{b}\right)$, so $\frac{w}{z} = w\left(\frac{1}{z}\right) = w\frac{\overline{z}}{z\overline{z}} = \frac{w}{z}\frac{\overline{z}}{\overline{z}}$.

A-11

$$\frac{w}{z} = \frac{w}{z}\frac{\overline{z}}{\overline{z}}$$

A-12 Example. $\frac{1+2i}{3-i} = \frac{1+2i}{3-i}\frac{3+i}{3+i} = \frac{1}{10}(1+7i).$ 🙂

The Polar Form of a Complex Number

Just as addition of complex numbers has a nice geometrical interpretation (using the parallelogram rule for vector addition), so does multiplication. To describe this, we begin with the observation that if the point (a, b) in the plane is at distance r from the origin, then

$$\begin{aligned} a &= r\cos\theta \\ \text{and} \quad b &= r\sin\theta, \end{aligned} \tag{1}$$

where θ is the angle between the positive x-axis and the ray from $(0,0)$ to (a, b). As usual, we agree that the counterclockwise direction is the positive direction for measuring angles. Thus $a + bi = r\cos\theta + (r\sin\theta)i = r(\cos\theta + i\sin\theta)$—see Figure A.2. With the agreement that $0 \le \theta < 2\pi$, the representation $z = r(\cos\theta + i\sin\theta)$ is called the *polar form* of the complex number z. By comparison, we call the representation $z = a + bi$ its *standard form.*

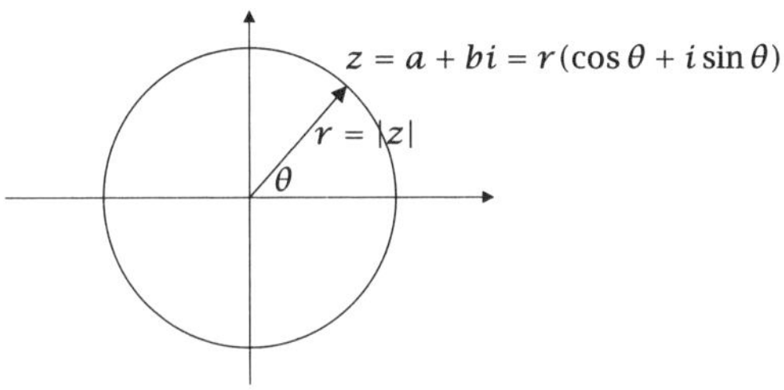

Figure A.2

A-13

$z = a + bi$	**standard form**
$z = r(\cos\theta + i\sin\theta),\ 0 \le \theta < 2\pi$	**polar form**

Notice that when $z = r(\cos\theta + i\sin\theta)$ is written in polar form, the number r is just $|z|$, the modulus of z. The angle θ is called its *argument.*

It is useful to be able to convert between the standard and polar forms of a complex number. Equations (1) show how to convert from polar form to standard form.

If $z = r(\cos\theta + i\sin\theta)$, then $z = a + bi$ with $a = r\cos\theta$ and $b = r\sin\theta$.

Conversely, given $z = a + bi$, the standard form of z, the equations

$$r = |z| = \sqrt{a^2 + b^2}, \quad \cos\theta = \frac{a}{r} = \frac{a}{\sqrt{a^2 + b^2}}, \quad \sin\theta = \frac{b}{r} = \frac{b}{\sqrt{a^2 + b^2}}$$

show how to convert from standard to polar form.

A-14 Examples. • If $z = 2+2i$, we have $r = |z| = \sqrt{2^2+2^2} = \sqrt{8} = 2\sqrt{2}$, $\cos\theta = \frac{2}{2\sqrt{2}} = \frac{1}{\sqrt{2}}$ and $\sin\theta = \frac{2}{2\sqrt{2}} = \frac{1}{\sqrt{2}}$, so $\theta = \frac{\pi}{4}$ and $z = 2\sqrt{2}\left(\cos\frac{\pi}{4} + i\sin\frac{\pi}{4}\right)$.

- If $z = -1+i\sqrt{3}$, then $r = |z| = \sqrt{1+3} = 2$, $\cos\theta = -\frac{1}{2}$ and $\sin\theta = \frac{\sqrt{3}}{2}$. So $\theta = \frac{2\pi}{3}$ and the polar form of z is $2\sqrt{2}\left(\cos\frac{2\pi}{3} + i\sin\frac{2\pi}{3}\right)$.
- If $z = 4-5i$, then $r = \sqrt{16+25} = \sqrt{41}$, $\cos\theta = \frac{4}{\sqrt{41}}$ and $\sin\theta = -\frac{5}{\sqrt{41}}$. A calculator gives $\arccos\frac{4}{\sqrt{41}} \approx .896$ rads $\approx 51°$. Since $\sin\theta < 0$, θ is in quadrant IV, so $\theta = 2\pi - .896 \approx 5.387$ rads and the polar form of z is $\sqrt{41}(\cos 5.387 + i\sin 5.387)$. ☺

Now let $z = r(\cos\alpha + i\sin\alpha)$ and $w = s(\cos\beta + i\sin\beta)$ and observe that

$$\begin{aligned} zw &= rs[(\cos\alpha + i\sin\alpha)(\cos\beta + i\sin\beta)] \\ &= rs[(\cos\alpha\cos\beta - \sin\alpha\sin\beta) + i(\cos\alpha\sin\beta + \sin\alpha\cos\beta)] \\ &= rs[\cos(\alpha+\beta) + i\sin(\alpha+\beta)] \end{aligned}$$

using the so-called *addition rules* for sines and cosines:

$$\cos(\alpha+\beta) = \cos\alpha\cos\beta - \sin\alpha\sin\beta$$

and

$$\sin(\alpha+\beta) = \sin\alpha\cos\beta + \cos\alpha\sin\beta.$$

A-15 To multiply complex numbers, multiply moduli and add arguments: $zw = rs[\cos(\alpha+\beta) + i\sin(\alpha+\beta)]$.

See Figure A.3.

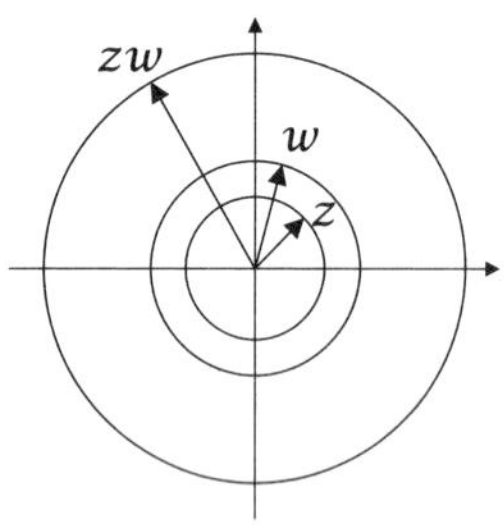

Figure A.3: To multiply complex numbers, multiply moduli and add arguments.

Similarly,

A-16 To divide complex numbers, divide moduli and subtract arguments: $\dfrac{z}{w} = \dfrac{r}{s}\left[\cos(\alpha-\beta) + i\sin(\alpha-\beta)\right]$.

A-17 Example. Suppose $z = 6\left(\cos\frac{\pi}{3} + i\sin\frac{\pi}{3}\right)$ and $w = 3\left(\cos\frac{\pi}{4} + i\sin\frac{\pi}{4}\right)$. Then

$$zw = 18\left[\cos\left(\tfrac{\pi}{3} + \tfrac{\pi}{4}\right) + i\sin\left(\tfrac{\pi}{3} + \tfrac{\pi}{4}\right)\right] = 18\left(\cos\tfrac{7\pi}{12} + i\sin\tfrac{7\pi}{12}\right);$$

and $$\frac{z}{w} = 2\left[\cos\left(\tfrac{\pi}{3} - \tfrac{\pi}{4}\right) + i\sin\left(\tfrac{\pi}{3} - \tfrac{\pi}{4}\right)\right] = 2\left(\cos\tfrac{\pi}{12} + i\sin\tfrac{\pi}{12}\right).$$

The powers z^n of a complex number z are products, of course, so if $z = r(\cos\theta + i\sin\theta)$, multiplying moduli and adding arguments gives

$$z^2 = r^2(\cos 2\theta + i\sin 2\theta)$$
$$z^3 = zz^2 = r^3(\cos 3\theta + i\sin 3\theta)$$

and, in general

$$z^n = r^n(\cos n\theta + i\sin n\theta).$$

This last statement is known as *De Moivre's Formula.*[1]

A-18 If $z = r(\cos\theta + i\sin\theta)$, then $z^n = r^n(\cos n\theta + i\sin n\theta)$.

A-19 Problem. Find z^{17} if $z = -1 + i\sqrt{3}$.

Solution. The modulus of z is $r = 2$ and the argument is $\theta = \frac{2\pi}{3}$. The polar form of z is $z = 2\left(\cos\frac{2\pi}{3} + i\sin\frac{2\pi}{3}\right)$. By De Moivre's formula,

$$\begin{aligned} z^{17} &= 2^{17}\left(\cos\tfrac{34\pi}{3} + i\sin\tfrac{34\pi}{3}\right) \\ &= 2^{17}\left(\cos\tfrac{4\pi}{3} + i\sin\tfrac{4\pi}{3}\right) \\ &= 2^{17}\left(-\frac{1}{2} - i\frac{\sqrt{3}}{2}\right) = -2^{16} - (2^{16}\sqrt{3})i. \end{aligned}$$

Obviously, polar form is the natural way to multiply complex numbers. Who would contemplate computing $(-1 + i\sqrt{3})^{17}$ as the notation suggests,

$$(-1 + \sqrt{3}i)^{17} = \underbrace{(-1 + \sqrt{3}i)(-1 + \sqrt{3}i)\cdots(-1 + \sqrt{3}i)}_{\text{17 factors}}\ ?$$

On the other hand, standard form is the obvious way to add complex numbers. Who would convert to polar form if asked to add $-1 + i\sqrt{3}$ and $2 - 2i$?

[1]Abraham de Moivre (1667–1754), geometer and probabilist, is probably best known for this formula.

A-20

To add or subtract complex numbers, use standard form. To multiply or divide complex numbers, use polar form.

A-21 **Remark.** In a calculus course, the student has probably been introduced to a number denoted e, which is the base e of the natural logarithms. Perhaps you have seen the famous *Taylor expansion* for e^x.

$$e^x = 1 + x + \frac{x^2}{2!} + \frac{x^3}{3!} + \frac{x^4}{4!} + \cdots$$

which implies such approximations as

$$e^1 = 1 + 1 + \frac{1}{2} + \frac{1}{6} + \frac{1}{24} + \cdots \approx 2.708$$

(in fact, $e \approx 2.718281828459\ldots$) and

$$\sqrt{e} = e^{1/2} = 1 + \frac{1}{2} + \frac{\frac{1}{4}}{2} + \frac{\frac{1}{8}}{6} + + \frac{\frac{1}{16}}{24} + \cdots \approx 1.648.$$

In the theory of complex variables, the complex valued function e^{ix} is defined by $e^{ix} = \cos x + i \sin x$. Using this formula, many of the definitions and results of this section become "obvious." Specifically, if $z = r(\cos\alpha + i\sin\alpha)$ and $w = s(\cos\beta + i\sin\beta)$, then $z = re^{i\alpha}$ and $w = se^{i\beta}$, so

$$zw = re^{i\alpha}se^{i\beta} = rse^{i(\alpha+\beta)}$$

(see A-15) while

$$\frac{z}{w} = \frac{re^{i\alpha}}{se^{i\beta}} = \frac{r}{s}e^{i\alpha}e^{-i\beta} = \frac{r}{s}e^{i(\alpha-\beta)}$$

(see A-16). De Moivre's formula for the nth power of z becomes simply

$$z^n = (re^{i\theta})^n = r^n e^{in\theta} = r^n(\cos n\theta + i \sin n\theta).$$

Roots of Polynomials

We are now ready to discuss the Fundamental Theorem of Algebra, which was mentioned briefly at the start of this section. This theorem was proven by Karl Friedrich Gauss (1777–1855) in 1797 when he was just 20 years of age!

A-22 **Theorem (Fundamental Theorem of Algebra).** *Every polynomial with real coefficients has a root in the complex numbers.*

Actually, this theorem says much more. Let $f(x)$ be a polynomial with real coefficients. The Fundamental Theorem says there is a (complex) root r_1, so we can factor $f(x)$

$$f(x) = (x - r_1)q_1(x),$$

where $q_1(x)$ is a polynomial of degree one less than the degree of $f(x)$. If $q_1(x)$ is not a constant, the Fundamental Theorem says that $q_1(x)$ also has a complex root, so $q_1(x) = (x - r_2)q_2(x)$ and

$$f(x) = (x - r_1)(x - r_2)q_2(x),$$

with $q_2(x)$ a polynomial of degree two less than the degree of $f(x)$. Continuing in this way, we eventually have

$$f(x) = (x - r_1)(x - r_2)\cdots(x - r_n)q(x)$$

where $q(x)$ has degree 0; that is, $q(x) = c$ is a constant. Thus the Fundamental Theorem of Algebra implies that a polynomial $f(x)$ of degree n factors completely over $\mathbb{C}$ as the product of linear factors:

$$f(x) = c(x - r_1)(x - r_2)\cdots(x - r_n).$$

Each r_i is a root of $f(x)$, so $f(x)$ has n roots, $r_1, r_2, \ldots, r_n$, providing we count *multiplicities* (double roots twice, triple roots three times, and so on).

A-23 Examples. • $x^2 - 3x - 4 = (x + 1)(x - 4)$ is a polynomial of degree two; it has two roots;

- the polynomial $x^4 - 1 = (x^2 - 1)(x^2 + 1) = (x - 1)(x + 1)(x - i)(x + i)$ has degree four and four roots;
- the polynomial $x^2 - 6x + 9 = (x - 3)^2$ has degree two and two roots because we count the double root 3 twice: 3 is a root of "multiplicity" two. ☺

The polynomial $f(z) = z^n - a$ has n roots called the *nth roots of a*. To discover what these are, write a and z in polar form:

$$a = r(\cos\theta + i\sin\theta)$$
$$z = s(\cos\alpha + i\sin\alpha).$$

If $f(z) = 0$, then $z^n = a$, so

$$s^n(\cos n\alpha + i\sin n\alpha) = r(\cos\theta + i\sin\theta).$$

Equating moduli and arguments, we have $s^n = r$ and $n\alpha = \theta + 2k\pi$ for some integer k; that is,

$$s = \sqrt[n]{r} \quad \text{and} \quad \alpha = \frac{\theta + 2k\pi}{n}.$$

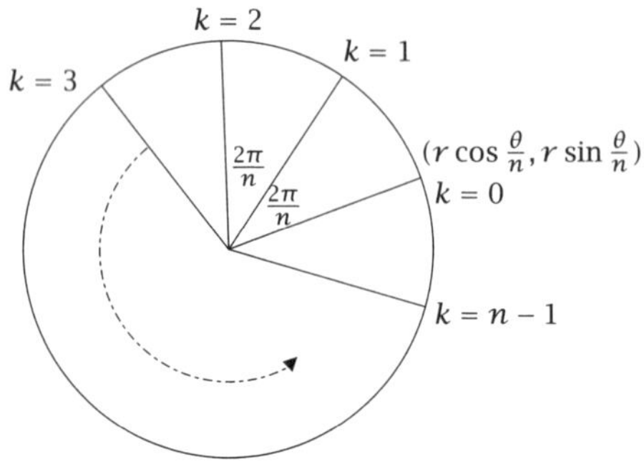

Figure A.4: The n solutions to $z^n = r$.

A-24

> The n solutions to $z^n = r(\cos\alpha + i\sin\alpha)$ are
>
> $z = \sqrt[n]{r}\left(\cos\frac{\theta+2k\pi}{n} + i\sin\frac{\theta+2k\pi}{n}\right), \qquad k = 0, 1, 2, \ldots, n-1.$

These solutions correspond to points distributed evenly around the circumference of the circle with centre $(0, 0)$ and radius r, as shown in Figure A.4.

A-25 **Problem.** Find all complex numbers z such that $z^4 = -64$.

Solution. In polar form, $-64 = 64(\cos\pi + i\sin\pi)$. The solutions to $z^4 = 64(\cos\pi + i\sin\pi)$ are

$$\sqrt[4]{64}\left(\cos\frac{\pi+2k\pi}{4} + i\sin\frac{\pi+2k\pi}{4}\right), \quad k = 0, 1, 2, 3.$$

Since $\sqrt[4]{64} = (2^6)^{1/4} = 2^{3/2} = 2\sqrt{2}$, these solutions are

$$z_1 = 2\sqrt{2}(\cos\tfrac{\pi}{4} + i\sin\tfrac{\pi}{4}) = 2\sqrt{2}\left(\tfrac{\sqrt{2}}{2} + i\tfrac{\sqrt{2}}{2}\right) = 2(1+i) \qquad (k=0)$$

$$z_2 = 2\sqrt{2}(\cos\tfrac{3\pi}{4} + i\sin\tfrac{3\pi}{4}) = 2\sqrt{2}\left(-\tfrac{\sqrt{2}}{2} + i\tfrac{\sqrt{2}}{2}\right) = 2(-1+i) \qquad (k=1)$$

$$z_3 = 2\sqrt{2}(\cos\tfrac{5\pi}{4} + i\sin\tfrac{5\pi}{4}) = 2(-1-i) \qquad (k=2)$$

$$z_4 = 2\sqrt{2}(\cos\tfrac{7\pi}{4} + i\sin\tfrac{7\pi}{4}) = 2(1-i). \qquad (k=3)$$

A-26 **Problem.** Find all complex numbers z such that $z^3 = -1 + i\sqrt{3}$.

Solution. In polar form, $-1+i\sqrt{3}=2\left(\cos\frac{2\pi}{3}+i\sin\frac{2\pi}{3}\right)$, so the solutions to $z^3=-1+i\sqrt{3}$ are $\sqrt[3]{2}\left(\cos\frac{\frac{2\pi}{3}+2k\pi}{3}+i\sin\frac{\frac{2\pi}{3}+2k\pi}{3}\right)$, $k=0,1,2$; that is,

$$z_1=\sqrt[3]{2}\left(\cos\tfrac{2\pi}{9}+i\sin\tfrac{2\pi}{9}\right)\approx .965+.810i \qquad (k=0)$$

$$z_2=\sqrt[3]{2}\left(\cos\tfrac{8\pi}{9}+i\sin\tfrac{8\pi}{9}\right)\approx -1.184+.431i \qquad (k=1)$$

$$z_3=\sqrt[3]{2}\left(\cos\tfrac{14\pi}{9}+i\sin\tfrac{14\pi}{9}\right)\approx .289-1.241i. \qquad (k=2)$$

☝

The roots of $z^n=1$ are called *nth roots of unity.*

A-27 Problem. Find the two square roots of unity.

Solution. We wish to solve $z^2=1\cos 0+i\sin 0$. The solutions are $\cos\frac{0+2k\pi}{2}+i\sin\frac{0+2k\pi}{2}$, $k=0,1$; that is,

$$z_1=\cos 0+i\sin 0=1 \qquad (k=0)$$

$$z_2=\cos\pi+i\sin\pi=-1. \qquad (k=1)$$

There wasn't anything new here! ☝

A-28 Problem. Find the three cube roots of unity.

Solution. We wish to solve $z^3=1=\cos 0+i\sin 0$. The solutions are $\cos\frac{0+2k\pi}{3}+i\sin\frac{0+2k\pi}{3}$, $k=0,1,2$; that is,

$$z_1=\cos 0+i\sin 0=1 \qquad (k=0)$$

$$z_2=\cos\tfrac{2\pi}{3}+i\sin\tfrac{2\pi}{3}=-\tfrac{1}{2}+i\sin\tfrac{\sqrt{3}}{2} \qquad (k=1)$$

$$z_3=\cos\tfrac{4\pi}{3}+i\sin\tfrac{4\pi}{3}=-\tfrac{1}{2}-i\sin\tfrac{\sqrt{3}}{2}. \qquad (k=2)$$

☝

True/False Questions

Decide, with as little calculation as possible, whether each of the following statements is true or false and, if you say "false," explain your answer. (Answers are at the back of the book.)

1. The integer 7 is a complex number.
2. The Fundamental Theorem of Algebra says that every polynomial with real coefficients has a root in the complex numbers.
3. If $\overline{z} = z$, then z is a real number.
4. If z is a complex number, $|z| = z\overline{z}$.
5. $|3 + 4i| = 5$.
6. If $z = a + bi \neq 0$ with a and b real, then there exist real numbers c and d such that $\frac{1}{z} = c + di$.
7. $1 + i = 2(\cos\frac{\pi}{4} + i\sin\frac{\pi}{4})$.
8. If $z = 2(\cos\frac{\pi}{3} + i\sin\frac{\pi}{3})$ and $w = 3(\cos\frac{\pi}{4} + i\sin\frac{\pi}{4})$, then $zw = 12(\cos\frac{7\pi}{12} + i\sin\frac{7\pi}{12})$.
9. De Moivre's Formula says that if $z = r(\cos\theta + i\sin\theta)$ and n is a positive integer, then $z^n = r^n(\cos n\theta + i\sin n\theta)$.
10. The polynomial $z^5 + 1$ has five distinct roots.

EXERCISES

Answers to exercises marked [BB] can be found at the Back of the Book.

1. Write each of the following complex numbers in polar form. Do not approximate any angles with decimals.

   (a) [BB] $1 + \sqrt{3}i$ **(b)** $-7i$ **(c)** $5 - 5i$ **(d)** $-\sqrt{3} - i$.

2. Write each complex number in polar form (with any approximations accurate to three decimal places).

   (a) [BB] $2+3i$ **(b)** $-2+3i$ **(c)** $-2-3i$ **(d)** $2-3i$.

3. Find the polar form of each of the following complex numbers,

   (a) [BB] $1 + 6i$ **(b)** $3 - 4i$ **(c)** $-7 + i$ **(d)** $-2 - 9i$,

   giving each angle to three decimal place accuracy.

4. Express each of the following complex numbers in standard form.

   (a) [BB] $\dfrac{3-4i}{1+i}$ **(b)** $\dfrac{1+2i}{1-4i}$ **(c)** $\dfrac{-2-2i}{-3-3i}$

   (d) $\dfrac{3+3i}{2-i}$ **(e)** $\dfrac{-i}{4+2i}$ **(f)** $\dfrac{-2+3i}{1-i}$.

5. [BB] Let $z = 2 - 2i$. Express z, $\overline{z}$, z^2 and $1/z$ in standard and polar forms.

6. Answer Exercise 5 with $z = \sqrt{3} - i$.

7. [BB] Let $z = 1 - 2i$ and $w = 3 + i$. Express zw, $\frac{z}{w}$, $z^2 - w^2$, and $\frac{z+w}{z-w}$ in standard form.

8. Answer Exercise 7 with $z = 3 + 3i$ and $w = 2 - 4i$.

9. [BB] Let $z = i$ and $w = 2+i$, then express $\frac{z\overline{w} + z^2}{w\overline{z} + w^2}$ as a complex number in standard form.

10. Let z and w be complex numbers. Show that the complex conjugate of $z + w$ is $\overline{z} + \overline{w}$.

11. **(a)** [BB] Show that the real and imaginary parts of a complex number z are, respectively, $\frac{1}{2}(z + \overline{z})$ and $\frac{1}{2i}(z - \overline{z})$.

    (b) Prove that a complex number z is real if and only if $\overline{z} = z$.

12. **(a)** Compute $(2 + i)(3 + i)$.

    (b) Convert $2 + i$ and $3 + i$ to polar form and compute the product in (a) again (giving your answer in polar form).

    (c) Recall that $\arctan\theta$ means the angle whose tangent is θ. Use the result of (b) to determine the value of $\arctan\frac{1}{2} + \arctan\frac{1}{3}$.

Critical Reading

13. Find a condition which is both necessary and sufficient for the product of two complex number numbers to be real.

Appendix B: Show and Prove

Gillian came to my office one day seeking help with a homework assignment. She circled the word "show" in one problem and said, "You know, I guess I don't know what that little word means." Words like **show** and **prove** in a question imply that some understanding and writing will be required to answer it, not just a number that can be easily checked with a text. In this appendix, we attempt to make "show and prove" problems less difficult than some people think they are.

The author has heard "proof" defined as "a convincing communication that answers why." Proofs are **communications**. When you prove something, you aren't done until you can explain it to others. To give a proof, you have to convince the other person that something is so. This is usually done with a sequence of sentences, **each one following clearly and logically from the previous**, which explain **why** the desired result is true.

Most mathematical theorems and most show and prove problems are statements of the form "if $\mathcal{A}$, then $\mathcal{B}$," where $\mathcal{A}$ and $\mathcal{B}$ are other statements. "If a given vector is a scalar multiple of another vector, then the vectors are linearly dependent" is one example.

In any show and prove problem, it's a good first step to determine what $\mathcal{A}$ and $\mathcal{B}$ are. In our example,

$\mathcal{A}$ is "One vector is a scalar multiple of another"
and $\mathcal{B}$ is "The vectors are linearly dependent."

The statement $\mathcal{A}$ (which could also be a collection of several statements) is called the *hypothesis*; $\mathcal{B}$ is called the *conclusion.*

READING CHECK 1. A student is asked to prove the following:

"A homogeneous system of equations always has a solution."

State this in the form "If $\mathcal{A}$, then $\mathcal{B}$." What is the hypothesis? What is the conclusion?

To prove "If $\mathcal{A}$, then $\mathcal{B}$" requires writing down a sequence of statements, **each of which follows logically from the preceding**, starting with $\mathcal{A}$ (usually) and finishing with $\mathcal{B}$.

B-1 **Problem.** Suppose a vector u is a scalar multiple of another vector v. Prove that u and v are linearly dependent.

Solution.

1. Let u and v be vectors and suppose u is a scalar multiple of v.
2. Then $\mathsf{u} = c\mathsf{v}$ for some scalar c. (This is the definition of "scalar multiple.")
3. Then $1\mathsf{u} - c\mathsf{v} = \mathsf{0}$. (This is basic algebra.)
4. This is a linear combination of u and v that equals the zero vector and in which not all the coefficients are 0 (because the coefficient of u is $1 \neq 0$).
5. Thus u and v are linearly dependent. (This is the definition of "linear independence.")

In this proof, and elsewhere in this appendix, we often number the statements in proofs so that it is easy to refer to them. While we are not recommending that you number the statements in your proofs, we do suggest that you try writing proofs in single statements, one under the other (rather than in a paragraph), making sure that every statement you write down is a logical consequence of previous statements. If you find it difficult to know when to justify something and when justification is not necessary, remember the rule: "If in doubt, justify."

When Jennifer enrolls for a German course, she understands that she'll have to memorize a lot of vocabulary. It is very difficult to translate a sentence if you do not know how to translate the words! So it is with mathematics. If you don't know what the words mean, you aren't going to be able to translate a mathematical sentence into language you understand and you certainly will not be able to solve show or prove problems.

B-2 Before attempting to show or prove something, make sure you know the meaning of every word in the statement.

We would have had no chance of solving Problem B-1 had we not known what "scalar multiple" and "linearly dependent" mean. And we certainly couldn't start the next problem without knowing what is meant by "linear combination."

B-3 **Problem.** Suppose u, v and w are vectors and u is a scalar multiple of v. Show that u is a linear combination of v and w.

Solution.

1. Since u is a scalar multiple of v, there is some scalar c so that u $= c$v.

2. Thus u $= c$v $+$ 0w. This is a linear combination of v and w. ♁

This solution required only two lines, but each line required knowledge of linear algebra vocabulary. First, we used the definition of "scalar multiple" to interpret the hypothesis "u is a scalar multiple of v." Then, after a moment's thought and remembering the definition of "linear combination," adding the term "0w" gave the desired conclusion.

B-4 **Problem.** If X is a matrix, show that the product XX^T of X and its transpose is symmetric.

The hypothesis is "X is a matrix." Not much! The desired conclusion is "XX^T is symmetric." We should know what "symmetric" means. We check the index. "Symmetric matrix" is defined on p. 210 (and also in the glossary). A matrix A is *symmetric* if and only if $A^T = A$. Here, the matrix A is XX^T. This is the matrix for which we must show that $A^T = A$.

Solution.

1. The transpose of XX^T is $(XX^T)^T = (X^T)^T X^T$ because the transpose of AB is $B^T A^T$.

2. $(X^T)^T = X$.

3. Therefore $(XX^T)^T = XX^T$ as desired. ♁

B-5 **Problem.** Suppose A is an $m \times n$ matrix such that the matrix $A^T A$ is invertible. Let b be any vector in R^n. Show that the linear system Ax = b has at most one solution.

It's pretty clear how we should begin.

Solution.

1. Suppose Ax = b has two solutions.

2. Let these solutions be x_1 and x_2.

3. Thus $A\mathsf{x}_1 = \mathsf{b}$ and $A\mathsf{x}_2 = \mathsf{b}$.

4. Therefore $A\mathsf{x}_1 = A\mathsf{x}_2$.

We interrupt our proof to make a couple of comments. The first few lines were "obvious," we suggest, what the author calls "follow your nose" steps.

But what should come next? So far, we haven't used the most important part of the hypothesis, that $A^T A$ is invertible. Looking at line 4 and realizing we have to get $A^T A$ into the game, we get an idea.

5. Multiply by A^T! We get $A^T A\mathbf{x}_1 = A^T A\mathbf{x}_2$.
6. So $(A^T A)^{-1} A^T A\mathbf{x}_1 = (A^T A)^{-1} A^T A\mathbf{x}_2$, because $A^T A$ is invertible.
7. So $I\mathbf{x}_1 = I\mathbf{x}_2$, since $(A^T A)^{-1}(A^T A) = I$ is the identity matrix.
8. So $\mathbf{x}_1 = \mathbf{x}_2$, as desired. 👍

B-6 Problem. Suppose A is an invertible matrix. Show that the determinant of A is not 0.

Solution.

1. Let A be an invertible matrix.
2. Then $AB = I$, the identity matrix, for some matrix B. (This comes from the definition of "invertible.")
3. Then $\det AB = \det I = 1$.
4. Thus $(\det A)(\det B) = 1$ (because $\det AB = (\det A)(\det B)$).
5. Thus $\det A$ cannot be 0; otherwise, its product with $\det B$ would be 0. 👍

In an attempt to prove "if $\mathcal{A}$, then $\mathcal{B}$," an all too common error is for a student to assume at the very beginning that the desired statement $\mathcal{B}$ is true and, after a few steps, having reached a statement that is true, to cry "Eureka" and claim to have confirmed that $\mathcal{B}$ is true. See Problem 6.1.5 in Section 6.1 for one example showing the fallacy of this line of reasoning. Here is another. It involves the complex number i which, as the reader may know, satisfies $i^2 = -1$.

B-7 Example. Prove that $1 = -1$. (You're going to love what follows.)

1. Let's begin with what we are to prove. We write down $1 = -1$.
2. Since $i^2 = -1$, we have $1 = i^2$.
3. Dividing by i, $\frac{1}{i} = \frac{i^2}{i} = \frac{i}{1}$.
4. Thus $\frac{\sqrt{1}}{\sqrt{-1}} = \frac{\sqrt{-1}}{\sqrt{1}}$, since $1 = \sqrt{1}$ and $i = \sqrt{-1}$.

5. $\sqrt{\frac{1}{-1}} = \sqrt{\frac{-1}{1}}$.
6. $\sqrt{-1} = \sqrt{-1}$.
7. $-1 = -1$, squaring both sides.

We have reached a correct statement, so the first statement must have been correct. Maybe not! This is horrible logic. We started with what we were asked to prove! ☺

Let u and v be vectors. The Cauchy–Schwarz inequality is the statement $|\mathsf{u} \cdot \mathsf{v}| \leq \|\mathsf{u}\| \, \|\mathsf{v}\|$, while the triangle inequality asserts $\|\mathsf{u} + \mathsf{v}\| \leq \|\mathsf{u}\| + \|\mathsf{v}\|$.

To answer a homework exercise that read

> Use the Cauchy–Schwarz inequality to prove the triangle inequality,

the author once saw this "proof:"

Solution. (Not correct.) $\|\mathsf{u} + \mathsf{v}\| \leq \|\mathsf{u}\| + \|\mathsf{v}\|$. (The author was not impressed to see the student begin with what she was asked to prove!)

$\|\mathsf{u} + \mathsf{v}\|^2 \leq (\|\mathsf{u}\| + \|\mathsf{v}\|)^2$.

$\|\mathsf{u} + \mathsf{v}\|^2 \leq \|\mathsf{u}\|^2 + 2\,\|\mathsf{u}\|\,\|\mathsf{v}\| + \|\mathsf{v}\|^2$.

$(\mathsf{u} + \mathsf{v}) \cdot (\mathsf{u} + \mathsf{v}) \leq \mathsf{u} \cdot \mathsf{u} + 2\,\|\mathsf{u}\|\,\|\mathsf{v}\| + \mathsf{v} \cdot \mathsf{v}$.

$\mathsf{u} \cdot \mathsf{u} + 2\mathsf{u} \cdot \mathsf{v} + \mathsf{v} \cdot \mathsf{v} \leq \mathsf{u} \cdot \mathsf{u} + 2\,\|\mathsf{u}\|\,\|\mathsf{v}\| + \mathsf{v} \cdot \mathsf{v}$.

This statement is true because $\mathsf{u} \cdot \mathsf{v} \leq \|\mathsf{u}\|\,\|\mathsf{v}\|$ (Cauchy–Schwarz). Eureka! 👍

Again, the student started with what was to be proved, reached another statement that was true and concluded the first was correct. But the student assumed the first statement was correct when she began the proof!

To the author, the disappointing part of this alleged "proof" is that it contains the ingredients of a completely correct proof, that consists essentially in writing down the argument in reverse order.

Solution.

$2\mathsf{u} \cdot \mathsf{v} \leq 2\,\|\mathsf{u}\|\,\|\mathsf{v}\|$. This is Cauchy-Schwarz.

Add $\mathsf{u} \cdot \mathsf{u} + \mathsf{v} \cdot \mathsf{v}$ to each side to get $\mathsf{u} \cdot \mathsf{u} + 2\mathsf{u} \cdot \mathsf{v} + \mathsf{v} \cdot \mathsf{v} \leq \mathsf{u} \cdot \mathsf{u} + 2\,\|\mathsf{u}\|\,\|\mathsf{v}\| + \mathsf{v} \cdot \mathsf{v}$.

$(\mathsf{u} + \mathsf{v}) \cdot (\mathsf{u} + \mathsf{v}) \leq \mathsf{u} \cdot \mathsf{u} + 2\,\|\mathsf{u}\|\,\|\mathsf{v}\| + \mathsf{v} \cdot \mathsf{v}$.

$\|\mathsf{u} + \mathsf{v}\|^2 \leq \|\mathsf{u}\|^2 + 2\,\|\mathsf{u}\|\,\|\mathsf{v}\| + \|\mathsf{v}\|^2$.

$\|\mathsf{u} + \mathsf{v}\|^2 \leq (\|\mathsf{u}\| + \|\mathsf{v}\|)^2$

The result now follows by taking the square root of each side. 👍

Some questions ask you to prove that one set X is contained in another set Y. To do this, you have to show that anything in X is also in Y.

B-8 Problem. Suppose A and B are matrices for which the product AB is defined. Show that the null space of B is contained in the null space of AB.

If you are not sure what "null space" means, refer to the index (or glossary). The null space of a matrix A is the set of all vectors x for which $A\mathsf{x} = 0$.

There is only one way to begin the solution.

1. Let x be in the null space of B.
2. Therefore $B\mathsf{x} = 0$.

The second line is more or less obvious—show the marker we know what it means for something to belong to the null space of B! But now what to do? If you are unsure, you can always try moving to the conclusion of the proof and working upward. The last line must be "Therefore the null space of B is contained in the null space of AB," and this would follow if the previous line read "So x is in the null space of AB." In turn, this statement would follow from a previous statement saying "$(AB)\mathsf{x} = 0$." (We again assure the marker that we know the meaning of "null space.") So we have the last three lines of our proof.

$n-2$. $(AB)\mathsf{x} = 0$.

$n-1$. So x is in the null space of AB.

n. Therefore the null space of B is contained in the null space of AB.

Altogether, what do we have?

1. Let x be in the null space of B.

2. Therefore $B\mathsf{x} = 0$.

⋮

$n-2$. $(AB)\mathsf{x} = 0$.

$n-1$. So x is in the null space of AB.

n. Therefore the null space of B is contained in the null space of AB.

It remains only to fill in the missing lines. In this case, however, there are no missing lines! Line $n-2$ follows from line 2 upon multiplication by A, using the fact that matrix multiplication is associative; $A(B\mathsf{x}) = (AB)\mathsf{x}$. Our proof is complete. ☝

Sometimes you are asked to prove that two sets X and Y are the same. There are two things you have to do this time. You have to show that anything in X is in Y and then that anything in Y is in X.

B-9 Problem. Suppose A and B are matrices for which the product AB is defined. If A is invertible, show that the null space of B is the null space of AB.

Solution. First we show that the null space of B is contained in the null space of AB.

1. Take a vector x in the null space of B.
2. Then $B\mathsf{x} = 0$, by definition of "null space of B."
3. So $A(B\mathsf{x}) = A(0) = 0$.
4. So $(AB)\mathsf{x} = 0$, because matrix multiplication is associative.
5. So x is in the null space of AB, by definition of "null space of AB."

Now we show that the null space of AB is contained in the null space of B.

1. Let x be a vector in the null space of AB.
2. Then $(AB)\mathsf{x} = 0$, by definition of "null space of AB."
3. Therefore $A(B\mathsf{x}) = 0$, since matrix multiplication is associative.
4. Therefore $A^{-1}[A(B\mathsf{x})] = A^{-1}0 = 0$, using the given information that A^{-1} exists.
5. Therefore $B\mathsf{x} = 0$, because $A^{-1}A = I$ is the identity and $IX = X$ for any matrix X.
6. Therefore x is in the null space of B, by definition of "null space of B." ☝

Some time ago, the author put this question on a linear algebra exam.

> Suppose A and B are matrices and $BA = 0$. Show that the column space of A is contained in the null space of B.

Here is how one student answered:

1. Let y be in the column space of A.
2. Therefore $A\mathsf{x} = \mathsf{y}$ for some x.
3. Let x be in the null space of B.

   ⋮

n. Therefore A is in the null space of B.

My student got off to a great start. We want to show that the column space of A is contained in the null space of B, so we should start with something in the column space of A and show that we know what this means. The first two lines are great, but from where did statement 3 come? It sure does not follow from anything preceding. And the last line is **not** what we wanted to show! The last three lines had to be these:

$n-2$. $B\mathsf{y} = 0$.

$n-1$. So y is in the null space of B.

n. Therefore the column space of A is contained in the null space of B.

Here is a correct solution.

Solution.

1. Let y be in the column space of A.
2. Therefore $A\mathsf{x} = \mathsf{y}$ for some x.
3. Multiplying by B gives $B\mathsf{y} = B(A\mathsf{x}) = (BA)\mathsf{x}$.
4. $B\mathsf{y} = 0$ because $BA = 0$.
5. So y is in the null space of B.
6. Therefore the column space of A is contained in the null space of B. 👍

The first two lines were automatic, as were the last three. We had only to insert line 3 to close the gap between beginning and end.

Answers to Reading Checks

1. "If a system of equations is homogeneous, then it has a solution." The hypothesis is that a system of equations is homogeneous; the conclusion is that it has a solution.

Appendix C: Things I Must Remember

Over the years, students who have finished their linear algebra course have told me there were some important things they wished they had known when they started to use this book. There are indeed some recurring themes in the approach to linear algebra adopted here, and we suggest that many users of this text would find it helpful to consult this short note regularly. Points are raised in order of appearance in this book; page references point to the spot where the concept was first introduced.

1. Memorize definitions (the glossary at the back of this book was put there to assist you with this). If you don't know what "similar" matrices are or what it means for a set of vectors to "span" a subspace, or if you don't know that virtually any argument intended to establish the "linear independence" of vectors $\mathsf{u}, \mathsf{v}, \mathsf{w}, \ldots$ **must** begin "Suppose $a\mathsf{u} + b\mathsf{v} + c\mathsf{w} + \cdots = 0$," then you are in trouble!

FROM CHAPTER ONE

2. Vectors in this book are always written as columns. Euclidean n-space, denoted R^n, is the set of all n-dimensional vectors, that is, the set of all vectors with n components. For example, $\begin{bmatrix} 1 \\ 7 \end{bmatrix}$ is a two-dimensional vector, that is, a vector in R^2 (also called *the plane*) and $\begin{bmatrix} -3 \\ 2 \\ 6 \end{bmatrix}$ is a three-dimensional vector, that is, a vector in R^3 (also called 3*-space*). *p. 8*

FROM CHAPTER TWO

3. Since vectors are columns, vectors in R^n can also be regarded as matrices. For example, is $\begin{bmatrix} 1 \\ 2 \\ 3 \end{bmatrix}$ a vector or a 3×1 matrix? Often it doesn't matter; sometimes it's convenient to change points of view. The dot product of vectors of u and v in R^n is also the matrix

product $\mathsf{u}^T\mathsf{v}$, thinking of u and v as $n \times 1$ matrices. For example, if $\mathsf{u} = \begin{bmatrix} 1 \\ 2 \\ 3 \end{bmatrix}$ and $\mathsf{v} = \begin{bmatrix} -5 \\ 4 \\ 8 \end{bmatrix}$, then

$$\mathsf{u} \cdot \mathsf{v} = 1(-5) + 2(4) + 3(8) = 27 = \begin{bmatrix} 1 & 2 & 3 \end{bmatrix} \begin{bmatrix} -5 \\ 4 \\ 8 \end{bmatrix} = \mathsf{u}^T\mathsf{v}.$$

In particular, the square of the length of a vector u is $\|\mathsf{u}\|^2 = \mathsf{u} \cdot \mathsf{u} = \mathsf{u}^T\mathsf{u}$. If u is a *unit vector* (that is, $\|\mathsf{u}\| = 1$), then the $n \times n$ matrix $Y = \mathsf{u}\mathsf{u}^T$ satisfies $Y^2 = Y$ because

$$Y^2 = (\mathsf{u}\mathsf{u}^T)(\mathsf{u}\mathsf{u}^T) = \mathsf{u}(\mathsf{u}^T\mathsf{u})\mathsf{u}^T = \mathsf{u}\mathsf{u}^T = Y$$

since $\mathsf{u}^T\mathsf{u} = \mathsf{u} \cdot \mathsf{u} = \|\mathsf{u}\|^2 = 1$. *p. 99*

4. When presented with the product AB of matrices, it is often useful to view B as a series of columns. If $B = \begin{bmatrix} \mathsf{b}_1 & \mathsf{b}_2 & \dots & \mathsf{b}_p \\ \downarrow & \downarrow & & \downarrow \end{bmatrix}$ has columns $\mathsf{b}_1, \mathsf{b}_2, \dots, \mathsf{b}_p$, the product AB is the matrix

$$AB = A \begin{bmatrix} \mathsf{b}_1 & \mathsf{b}_2 & \dots & \mathsf{b}_p \\ \downarrow & \downarrow & & \downarrow \end{bmatrix} = \begin{bmatrix} A\mathsf{b}_1 & A\mathsf{b}_2 & \dots & A\mathsf{b}_p \\ \downarrow & \downarrow & & \downarrow \end{bmatrix}$$

whose columns are $A\mathsf{b}_1, A\mathsf{b}_2, \dots, A\mathsf{b}_p$. Suppose $A = \begin{bmatrix} 1 & -2 & 0 \\ -2 & 0 & 2 \\ 0 & 2 & -1 \end{bmatrix}$, $\mathsf{x} = \begin{bmatrix} 2 \\ 1 \\ 2 \end{bmatrix}$, $\mathsf{y} = \begin{bmatrix} 1 \\ 1 \\ 1 \end{bmatrix}$ and $\mathsf{z} = \begin{bmatrix} -2 \\ 2 \\ 1 \end{bmatrix}$. You should check that $A\mathsf{x} = \mathsf{0}$, $A\mathsf{y} = \begin{bmatrix} -1 \\ 0 \\ 1 \end{bmatrix}$ and $A\mathsf{z} = 3\mathsf{z}$. This means that if $B = \begin{bmatrix} \mathsf{x} & \mathsf{y} & \mathsf{z} \\ \downarrow & \downarrow & \downarrow \end{bmatrix} = \begin{bmatrix} 2 & 1 & -2 \\ 1 & 1 & 2 \\ 2 & 1 & 1 \end{bmatrix}$, then $AB = \begin{bmatrix} A\mathsf{x} & A\mathsf{y} & A\mathsf{z} \\ \downarrow & \downarrow & \downarrow \end{bmatrix} = \begin{bmatrix} 0 & -1 & -6 \\ 0 & 0 & 6 \\ 0 & 1 & 3 \end{bmatrix}$.

p. 102

5. The product $A\mathsf{x}$ of an $m \times n$ matrix A and a vector x in R^n is a linear combination of the columns of A with coefficients the components of the vector. For example,

$$\begin{bmatrix} 1 & 4 & 7 \\ 2 & 5 & 8 \\ 3 & 6 & 9 \end{bmatrix} \begin{bmatrix} x_1 \\ x_2 \\ x_3 \end{bmatrix} = x_1 \begin{bmatrix} 1 \\ 2 \\ 3 \end{bmatrix} + x_2 \begin{bmatrix} 4 \\ 5 \\ 6 \end{bmatrix} + x_3 \begin{bmatrix} 7 \\ 8 \\ 9 \end{bmatrix}.$$

So, if b is a vector in R^m, the linear system $A\mathsf{x} = \mathsf{b}$ has a solution if and only if b is a linear combination of the columns of A—that is, if and only if b is in the column space of A. The column space of a matrix is the set of all vectors $A\mathsf{x}$. *p. 107*

6. The inverse of the product of invertible matrices A and B is the product of the inverses **order reversed**: $(AB)^{-1} = B^{-1}A^{-1}$. Transpose works the same way: the transpose of the product of A and B is the product of the transposes **order reversed**: $(AB)^T = B^T A^T$. *pp. 130 and 133*

7. Most students know a symmetric matrix when they see one—how about $\begin{bmatrix} 1 & 2 & -7 \\ 2 & -5 & 6 \\ -7 & 6 & 0 \end{bmatrix}$, for instance? On the other hand, knowing what a symmetric matrix "looks like" doesn't help much with exercises that involve the concept. Could you prove that AA^T is symmetric for any matrix A? Once again, we ask you to take seriously our request to **learn definitions**. A matrix X is *symmetric* if $X^T = X$. Thus, $X = AA^T$ is symmetric because $X^T = (AA^T)^T = (A^T)^T A^T = AA^T = X$. *p. 210*

FROM CHAPTER THREE

8. A (square) matrix A is invertible if and only if $\det A \neq 0$. Thus, if $\det A \neq 0$, then $A\mathsf{x} = \mathsf{b}$ has a solution for any vector b, and a unique one, namely, $\mathsf{x} = A^{-1}\mathsf{b}$. In particular, the homogeneous system $A\mathsf{x} = \mathsf{0}$ has only the trivial solution, $\mathsf{x} = \mathsf{0}$. For example, the system $\begin{bmatrix} -1 & 2 \\ 3 & 4 \end{bmatrix}\begin{bmatrix} x \\ y \end{bmatrix} = \begin{bmatrix} 0 \\ 0 \end{bmatrix}$ has only the trivial solution $\begin{bmatrix} x \\ y \end{bmatrix} = \begin{bmatrix} 0 \\ 0 \end{bmatrix}$ because $\det \begin{bmatrix} -1 & 2 \\ 3 & 4 \end{bmatrix} = -10 \neq 0$. *p. 255*

9. To find eigenvalues and eigenvectors of a (square) matrix A,
   - Compute the characteristic polynomial $\det(A - \lambda I)$.
   - Set $\det(A - \lambda I) = 0$. The solutions of this equation are the eigenvalues.
   - For each eigenvalue λ, solve the homogeneous system $(A - \lambda I)\mathsf{x} = \mathsf{0}$. The solution is the eigenspace corresponding to λ.

   p. 285

FROM CHAPTER FOUR

10. The column space of a matrix is an important concept throughout this book, and most students remember the definition: The column space of a matrix is the set of linear combinations of the columns. Deep understanding and comfort with the idea, however, seems to take a while to develop. Do you see immediately why the column space of $A = \begin{bmatrix} 2 & 0 & 1 & 4 & 0 \\ 3 & 1 & 0 & 5 & 0 \\ 1 & 0 & 0 & -7 & 1 \end{bmatrix}$ is R^3, for example? Look at columns two, three and five, which are the standard

basis vectors for R^3. Every vector $\begin{bmatrix} x \\ y \\ z \end{bmatrix}$ in R^3 is a linear combination of the standard basis vectors, hence a linear combination of the columns of A:

$$\begin{bmatrix} x \\ y \\ z \end{bmatrix} = 0\begin{bmatrix} 2 \\ 3 \\ 1 \end{bmatrix} + y\begin{bmatrix} 0 \\ 1 \\ 0 \end{bmatrix} + x\begin{bmatrix} 1 \\ 0 \\ 0 \end{bmatrix} + 0\begin{bmatrix} 4 \\ 5 \\ -7 \end{bmatrix} + z\begin{bmatrix} 0 \\ 0 \\ 1 \end{bmatrix}.$$

Can you see that the column space of $\begin{bmatrix} -1 & 3 & 5 \\ 2 & 4 & 9 \\ 0 & 0 & 0 \end{bmatrix}$ is the xy-plane? How about the column space of $A = \begin{bmatrix} -1 & 3 & 5 \\ 2 & 4 & 9 \\ 3 & 1 & 4 \end{bmatrix}$? Can you see that this is the plane with equation $x - y + z = 0$? Remember that a vector b is in the column space if and only if it is a linear combination of the columns of A, that is, if and only if $\mathsf{b} = A\mathsf{x}$ for some x. So, solve the system $A\mathsf{x} = \mathsf{b}$:

$$\left[\begin{array}{ccc|c} -1 & 3 & 5 & b_1 \\ 2 & 4 & 9 & b_2 \\ 3 & 1 & 4 & b_3 \end{array}\right] \to \left[\begin{array}{ccc|c} 1 & -3 & -5 & -b_1 \\ 0 & 10 & 19 & b_2 + 2b_1 \\ 0 & 10 & 19 & b_3 + 3b_1 \end{array}\right]$$
$$\to \left[\begin{array}{ccc|c} 1 & -3 & -5 & -b_1 \\ 0 & 10 & 19 & b_2 + 2b_1 \\ 0 & 0 & 0 & (b_3 + 3b_1) - (b_2 + 2b_1) \end{array}\right].$$

The system has a solution if and only if $(b_3 + 3b_1) - (b_2 + 2b_1)$; that is, if and only if $b_1 - b_2 + b_3 = 0$. This is the case if and only if b is in the plane with equation $x - y + z = 0$. *p. 334*

11. The row space of a matrix A is the set of linear combinations of the rows of A. The three elementary row operations do not affect the row space, so to investigate the span of some vectors, put them into the rows of a matrix and reduce to row echelon form. For example, if $\mathsf{u}_1 = \begin{bmatrix} 1 \\ 3 \\ 1 \\ -1 \\ 1 \end{bmatrix}$, $\mathsf{u}_2 = \begin{bmatrix} -1 \\ -1 \\ 3 \\ 1 \\ 3 \end{bmatrix}$ and $\mathsf{u}_3 = \begin{bmatrix} 1 \\ 4 \\ 3 \\ -1 \\ 3 \end{bmatrix}$, and we want to find the dimension of the subspace of R^5 spanned by these vectors, apply Gaussian elimination to $A = \begin{bmatrix} 1 & 3 & 1 & -1 & 1 \\ -1 & -1 & 3 & 1 & 3 \\ 1 & 4 & 3 & -1 & 3 \end{bmatrix}$:

$$A \to \begin{bmatrix} 1 & 3 & 1 & -1 & 1 \\ 0 & 2 & 4 & 0 & 4 \\ 0 & 1 & 2 & 0 & 2 \end{bmatrix} \to \begin{bmatrix} 1 & 3 & 1 & -1 & 1 \\ 0 & 1 & 2 & 0 & 2 \\ 0 & 0 & 0 & 0 & 0 \end{bmatrix} = U.$$

The row space of A is the row space of U, that is, the span of

$\begin{bmatrix} 1 & 3 & 1 & -1 & 1 \end{bmatrix}$ and $\begin{bmatrix} 0 & 1 & 2 & 0 & 2 \end{bmatrix}$. So the subspace of R^5 spanned by $\mathsf{u}_1, \mathsf{u}_2$ and u_3 is also spanned by $\mathsf{v}_1 = \begin{bmatrix} 1 \\ 3 \\ 1 \\ -1 \\ 1 \end{bmatrix}$ and $\mathsf{v}_2 = \begin{bmatrix} 0 \\ 1 \\ 2 \\ 0 \\ 2 \end{bmatrix}$. Since these vectors are linearly independent, the subspace has dimension 2. *p. 357*

FROM CHAPTER FIVE

12. Suppose $\mathcal{V} = \{\mathsf{v}_1, \mathsf{v}_2, \ldots, \mathsf{v}_n\}$ and $\mathcal{W}$ are bases for vector spaces V and W and $T: V \to W$ is a linear transformation. The matrix of T relative to $\mathcal{V}$ and $\mathcal{W}$ is the matrix

$$M_{\mathcal{W} \leftarrow \mathcal{V}}(T) = \begin{bmatrix} [T(\mathsf{v}_1)]_{\mathcal{W}} & [T(\mathsf{v}_2)]_{\mathcal{W}} & \cdots & [T(\mathsf{v}_n)]_{\mathcal{W}} \\ \downarrow & \downarrow & & \downarrow \\ & & & \end{bmatrix}$$

whose columns are the coordinate vectors of $T(\mathsf{v}_1)$, $T(\mathsf{v}_2), \ldots$, $T(\mathsf{v}_n)$ relative to $\mathcal{W}$. This matrix converts the coordinate vector of x relative to $\mathcal{V}$ to the coordinate vector of $T(\mathsf{x})$ relative to $\mathcal{W}$: $[T(\mathsf{x})]_{\mathcal{W}} = M_{\mathcal{W} \leftarrow \mathcal{V}}(T)\mathsf{x}_{\mathcal{V}}$. *p. 433*

13. Given bases $\mathcal{V} = \{\mathsf{v}_1, \mathsf{v}_2, \ldots, \mathsf{v}_n\}$ and $\mathcal{V}'$ for a vector space, to change coordinates of a vector x from $\mathcal{V}$ to $\mathcal{V}'$, multiply the coordinate vector of x relative to $\mathcal{V}$ by the change of coordinates matrix

$$P_{\mathcal{V}' \leftarrow \mathcal{V}} = \begin{bmatrix} [\mathsf{v}_1]_{\mathcal{V}'} & [\mathsf{v}_2]_{\mathcal{V}'} & \cdots & [\mathsf{v}_n]_{\mathcal{V}'} \\ \downarrow & \downarrow & & \downarrow \\ & & & \end{bmatrix} : \qquad \mathsf{x}_{\mathcal{V}'} = P_{\mathcal{V}' \leftarrow \mathcal{V}} \mathsf{x}_{\mathcal{V}}.$$

Sometimes $P_{\mathcal{V}' \leftarrow \mathcal{V}}$ seems hard to find, but $P_{\mathcal{V} \leftarrow \mathcal{V}'}$ is easy. In this case, remember that $P_{\mathcal{V}' \leftarrow \mathcal{V}} = P^{-1}_{\mathcal{V} \leftarrow \mathcal{V}'}$. *p. 444*

FROM CHAPTER SIX

14. To find the matrix P that projects vectors onto a subspace U, find an orthogonal basis $\mathsf{f}_1, \mathsf{f}_2, \ldots, \mathsf{f}_n$ for U. Then

$$P = \frac{\mathsf{f}_1 \mathsf{f}_1^T}{\mathsf{f}_1 \cdot \mathsf{f}_1} + \frac{\mathsf{f}_2 \mathsf{f}_2^T}{\mathsf{f}_2 \cdot \mathsf{f}_2} + \cdots + \frac{\mathsf{f}_n \mathsf{f}_n^T}{\mathsf{f}_n \cdot \mathsf{f}_n}.$$

In particular, to project onto a one-dimensional subspace U spanned by a (nonzero) vector u, use $P = \dfrac{\mathsf{u}\mathsf{u}^T}{\mathsf{u} \cdot \mathsf{u}}$. *p. 470*

FROM CHAPTER SEVEN

15. A singular value decomposition of an $m \times n$ matrix A is a factorization $A = Q_1 \Sigma Q_2^T$, where Q_1 is an orthogonal $m \times m$ matrix, Q_2 is an orthogonal $n \times n$ matrix and $\Sigma = \begin{bmatrix} D & 0 \\ 0 & 0 \end{bmatrix}$ is a diagonal $m \times n$ matrix with $D = \begin{bmatrix} \mu_1 & & \\ & \ddots & \\ & & \mu_r \end{bmatrix}$, a diagonal $r \times r$ matrix with positive diagonal entries μ_i. The matrix $Q_2 = \begin{bmatrix} \mathsf{x}_1 & \cdots & \mathsf{x}_n \\ \downarrow & & \downarrow \end{bmatrix}$ has columns an orthonormal set of eigenvalues for $A^T A$. The first r columns of $Q_1 = \begin{bmatrix} \mathsf{y}_1 & \cdots & \mathsf{y}_m \\ \downarrow & & \downarrow \end{bmatrix}$ are orthonormal eigenvectors of AA^T corresponding to eigenvalues $\mu_1, \ldots, \mu_r$. The remaining columns are obtained by extending the first r to an orthonormal basis of R^m.
p. 582

Answers to True/False and BB Exercises

Section 1.1—True/False

1. False. $\overrightarrow{AB} = \begin{bmatrix} -4 \\ 3 \end{bmatrix}$ (the coordinates of B less those of A).
2. True. $\begin{bmatrix} 0 \\ 0 \end{bmatrix} = 0\begin{bmatrix} 1 \\ 2 \end{bmatrix}$.
3. True. This is 1.1.27.
4. False. A linear combination of $\begin{bmatrix} 1 \\ 0 \\ 1 \end{bmatrix}$ and $\begin{bmatrix} 2 \\ 0 \\ -1 \end{bmatrix}$ is a vector of the form $a\begin{bmatrix} 1 \\ 0 \\ 1 \end{bmatrix} + b\begin{bmatrix} 2 \\ 0 \\ -1 \end{bmatrix} = \begin{bmatrix} a+2b \\ 0 \\ a-b \end{bmatrix}$. Since $\begin{bmatrix} 1 \\ 2 \\ 3 \end{bmatrix}$ does not have second component 0, it is not such a linear combination.
5. True. $4\mathsf{u} - 9\mathsf{v} = 2(2\mathsf{u}) + (-\frac{9}{7})(7\mathsf{v})$.
6. True.
7. True. $0 = 0\mathsf{u} + 0\mathsf{v} + 0\mathsf{w}$.
8. True. The vectors are not parallel.
9. True. These vectors, also labeled $\mathsf{e}_1, \mathsf{e}_2, \mathsf{e}_3$, are the standard basis for R^3. See 1.1.32.
10. True. Since every vector in R^3 is a linear combination of $\mathsf{i}, \mathsf{j}, \mathsf{k}$, the same is true of any set containing i, j and k. If $\mathsf{x} = \begin{bmatrix} x_1 \\ x_2 \\ x_3 \end{bmatrix}$, then $\mathsf{x} = x_1\mathsf{e}_1 + x_2\mathsf{e}_2 + x_3\mathsf{e}_3 + 0\mathsf{u}$.

Exercises 1.1

1. (a) $\overrightarrow{AB} = \begin{bmatrix} 3 \\ 3 \end{bmatrix}$.
2. (a) $B = (0, 6)$.

3. $x = -\frac{1}{4}u = -7w$; x is not a scalar multiple of v.

7. The zero vector, u_4, is parallel to each of the others. Since we've agreed that 0 has no direction, it doesn't make sense to say it has the same or opposite direction as another vector. The remaining pairs of parallel vectors are u_2, u_5 (same direction) and u_1, u_3 (opposite direction).

8. (a) $4\begin{bmatrix}2\\-3\end{bmatrix}+2\begin{bmatrix}3\\1\end{bmatrix}=\begin{bmatrix}14\\-10\end{bmatrix}$, (c) $3\begin{bmatrix}2\\1\\3\end{bmatrix}-2\begin{bmatrix}1\\0\\-5\end{bmatrix}-4\begin{bmatrix}0\\-1\\2\end{bmatrix}=\begin{bmatrix}4\\7\\11\end{bmatrix}$.

9. (a) $a = 2, b = 17$.

10. (a) $x = \frac{3}{5}u + \frac{1}{5}v$; $y = -\frac{2}{5}u + \frac{1}{5}v$.

11. (a) $a = 5, b = -3$.

12. (a) $v = 0v_1 + 0v_2 + (-1)v_3 + 0v_4 + 0v_5$.

13. (a) $p = -4u + 7v + 5w$.

14. (a)

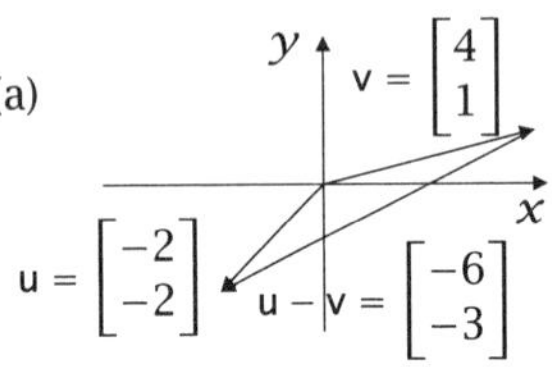

15. (a)

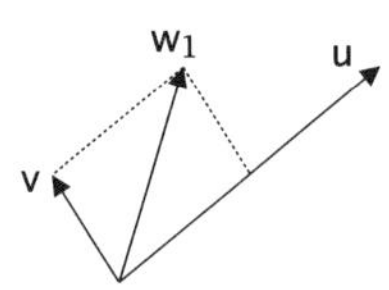

It appears that $w_1 = \frac{1}{2}u + v$.

16. (a) $\begin{bmatrix}7\\7\end{bmatrix}=3\begin{bmatrix}2\\3\end{bmatrix}-\begin{bmatrix}-1\\2\end{bmatrix}$.

18. (a) No.

19. $\begin{bmatrix}1\\0\\0\end{bmatrix} = -v_1 + \frac{1}{2}v_2 - \frac{1}{2}v_3$.

20. (a) There are many ways to express $\begin{bmatrix}0\\0\end{bmatrix}$ as a linear combination of $\begin{bmatrix}1\\4\end{bmatrix}$ and $\begin{bmatrix}-2\\-8\end{bmatrix}$. For example, $\begin{bmatrix}0\\0\end{bmatrix}=0\begin{bmatrix}1\\4\end{bmatrix}+0\begin{bmatrix}-2\\-8\end{bmatrix}=2\begin{bmatrix}1\\4\end{bmatrix}+1\begin{bmatrix}-2\\-8\end{bmatrix}=4\begin{bmatrix}1\\4\end{bmatrix}+2\begin{bmatrix}-2\\-8\end{bmatrix}=20\begin{bmatrix}1\\4\end{bmatrix}+10\begin{bmatrix}-2\\-8\end{bmatrix}$.

21. (a) $0 = 0u + 0v$ is in the plane spanned by any set of vectors.

22. (a) No: Vector u is not in the span of the other vectors.

23. (a) No.

26. (a) $u=\begin{bmatrix}2\\2\end{bmatrix}$; $v=\begin{bmatrix}2\\-1\end{bmatrix}$, (b) i. $\overrightarrow{OD}=\begin{bmatrix}4\\1\end{bmatrix}=u+v$.

27. We wish to show that $\overrightarrow{DE} = \frac{1}{2}\overrightarrow{BC}$. Now $\overrightarrow{AD} = \frac{1}{2}\overrightarrow{AB}$ and $\overrightarrow{AE} = \frac{1}{2}\overrightarrow{AC}$, so $\overrightarrow{DE} = \overrightarrow{DA} + \overrightarrow{AE} = -\frac{1}{2}\overrightarrow{AB} + \frac{1}{2}\overrightarrow{AC} = \frac{1}{2}(\overrightarrow{AC} - \overrightarrow{AB}) = \frac{1}{2}\overrightarrow{BC}$.

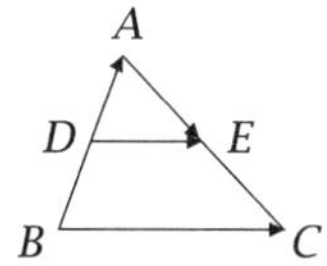

32. There are three possibilities for $D(x, y)$.

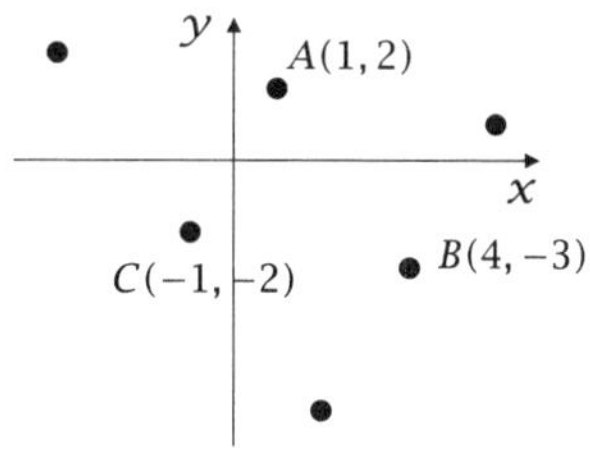

If $\overrightarrow{CA} = \overrightarrow{DB}$, then $\begin{bmatrix} 2 \\ 4 \end{bmatrix} = \begin{bmatrix} 4-x \\ -3-y \end{bmatrix}$ and D is $(2, -7)$.

If $\overrightarrow{DA} = \overrightarrow{CB}$, then $\begin{bmatrix} 1-x \\ 2-y \end{bmatrix} = \begin{bmatrix} 5 \\ -1 \end{bmatrix}$ and D is $(-4, 3)$.

If $\overrightarrow{AD} = \overrightarrow{CB}$, then $\begin{bmatrix} x-1 \\ y-2 \end{bmatrix} = \begin{bmatrix} 5 \\ -1 \end{bmatrix}$ and D is $(6, 1)$.

36. 1. Closure under addition: Let $\mathsf{u} = \begin{bmatrix} x \\ y \end{bmatrix}$ and $\mathsf{v} = \begin{bmatrix} z \\ w \end{bmatrix}$ be vectors.

Then $\mathsf{u} + \mathsf{v} = \begin{bmatrix} x+z \\ y+w \end{bmatrix}$ is a vector.

2. Commutativity of addition: Let $\mathsf{u} = \begin{bmatrix} x \\ y \end{bmatrix}$ and $\mathsf{v} = \begin{bmatrix} z \\ w \end{bmatrix}$ be vectors.

Then $\mathsf{u} + \mathsf{v} = \begin{bmatrix} x+z \\ y+w \end{bmatrix}$ and $\mathsf{v} + \mathsf{u} = \begin{bmatrix} z+x \\ w+y \end{bmatrix}$. Since $z + x = x + z$ and $w + y = y + w$, $\mathsf{v} + \mathsf{u} = \mathsf{u} + \mathsf{v}$.

5. Negatives: If $\mathsf{u} = \begin{bmatrix} x \\ y \end{bmatrix}$, then $-\mathsf{u} = \begin{bmatrix} -x \\ -y \end{bmatrix}$, so $\mathsf{u} + (-\mathsf{u}) = \begin{bmatrix} x \\ y \end{bmatrix} + \begin{bmatrix} -x \\ -y \end{bmatrix} = \begin{bmatrix} x-x \\ y-y \end{bmatrix} = \begin{bmatrix} 0 \\ 0 \end{bmatrix} = \mathsf{0}$. Similarly, $(-\mathsf{u}) + \mathsf{u} = \mathsf{0}$.

41. A linear combination of $\begin{bmatrix} 1 \\ \frac{3}{2} \\ 0 \end{bmatrix}$ and $\begin{bmatrix} 0 \\ 3 \\ 6 \end{bmatrix}$ is a vector of the form $a\begin{bmatrix} 1 \\ \frac{3}{2} \\ 0 \end{bmatrix} + b\begin{bmatrix} 0 \\ 3 \\ 6 \end{bmatrix}$. This is $\frac{1}{2}\begin{bmatrix} 2 \\ 3 \\ 0 \end{bmatrix} + 3b\begin{bmatrix} 0 \\ 1 \\ 2 \end{bmatrix}$, which is a linear combination of $\begin{bmatrix} 2 \\ 3 \\ 0 \end{bmatrix}$ and $\begin{bmatrix} 0 \\ 1 \\ 2 \end{bmatrix}$.

42. (a) $a(2\mathsf{u}) + b(-3\mathsf{v}) = (2a)\mathsf{u} + (-3b)\mathsf{v}$.

48. (a) Three points X, Y and Z are collinear if and only if the vector $\overrightarrow{XY}$ is a scalar multiple of the vector $\overrightarrow{XZ}$ (or $\overrightarrow{YZ}$). (There are other similar responses that are equally correct.)

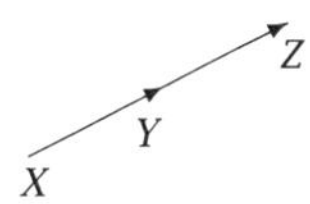

51. (a) We are given that $\mathsf{v} = c\mathsf{u}$ or $\mathsf{u} = c\mathsf{v}$ for some scalar c. Suppose $\mathsf{v} = c\mathsf{u}$. This can be rewritten as $1\mathsf{v} - c\mathsf{u} = \mathsf{0}$. This expresses $\mathsf{0}$ as a linear combination of u and v in a nontrivial way since at least one coefficient is not zero (the coefficient of v). The case $\mathsf{u} = c\mathsf{v}$ is similar.

Section 1.2—True/False

1. False: $|c|$ times the length of v.
2. False. A unit vector has length 1; this vector has length $\sqrt{3}$.
3. False. Try $\mathsf{u} = \mathsf{i}$ and $\mathsf{v} = \mathsf{j}$.
4. True. The dot product of these vectors is 0.
5. False. True if and only if the vectors are unit vectors.
6. True. The Cauchy-Schwarz inequality says $|\mathsf{u}\cdot\mathsf{v}| \le \|\mathsf{u}\|\,\|\mathsf{v}\|$. Since $\mathsf{u}\cdot\mathsf{v} \le |\mathsf{u}\cdot\mathsf{v}|$, the given statement is also true.
7. False. It makes no sense to talk about the length of a scalar, like $\mathsf{u}\cdot\mathsf{v}$. The Cauchy-Schwarz inequality says $|\mathsf{u}\cdot\mathsf{v}| \le \|\mathsf{u}\|\,\|\mathsf{v}\|$, absolute value on the left, not length.
8. False. If u and v are unit vectors, $\mathsf{u}\cdot\mathsf{v}$ is the cosine of the angle between them and this cannot be more than 1.
9. True. First note that neither vector is $\mathsf{0}$ because the vectors are not orthogonal. Since $\mathsf{u}\cdot\mathsf{v} = \|\mathsf{u}\|\,\|\mathsf{v}\|\cos\theta$ and $\|\mathsf{u}\|\,\|\mathsf{v}\| > 0$, the sign of the dot product is the sign of $\cos\theta$. If this is positive, the angle between u and v is acute; if $\cos\theta$ is negative, θ is obtuse.
10. False. The "Triangle Inequality" is so-named because the sum of the lengths of any two sides of a triangle can never be less than the length of the third: $\|a\| + \|b\| \ge \|a+b\|$.

Exercises 1.2

1. $\mathsf{u} = \pm\dfrac{3}{\sqrt{6}}\begin{bmatrix}1\\-2\\1\end{bmatrix}$.

3. (a) $-\frac{2}{3}\begin{bmatrix}1\\2\\2\end{bmatrix} = \begin{bmatrix}-\frac{2}{3}\\-\frac{4}{3}\\-\frac{4}{3}\end{bmatrix}$.

5. (a) $\|\mathsf{u}\| = \sqrt{3^2+4^2+0^2} = 5$; $\|\mathsf{u}+\mathsf{v}\| = \sqrt{54}$; $\left\|\frac{\mathsf{w}}{\|\mathsf{w}\|}\right\| = 1$,

   (b) $\frac{1}{\|\mathsf{u}\|}\mathsf{u} = \frac{1}{5}\mathsf{u} = \begin{bmatrix}\frac{3}{5}\\\frac{4}{5}\\0\end{bmatrix}$, (c) $\begin{bmatrix}-\frac{8}{3}\\-\frac{4}{3}\\-\frac{8}{3}\end{bmatrix}$.

6. (a) These are the points on the circle with radius 1 centred at P.

7. (a) $\theta = \frac{\pi}{2}$.

   (b) $\cos\theta = \dfrac{1}{5\sqrt{2}} \approx .1414$, so $\theta \approx \arccos(.1414) \approx 1.43$ rads $\approx 82°$.

9. (a) not defined, (b) $\|\mathsf{y}\| = \sqrt{6}$.

10. (a) $\mathsf{u}\cdot\mathsf{v} = \frac{3\sqrt{2}}{2}$.

11. (a) No: $-1 \le \cos\theta \le +1$ for any θ.

13. (a) $(\mathsf{u}-\mathsf{v})\cdot(2\mathsf{u}-3\mathsf{v}) = 53$, (c) $\|\mathsf{u}+\mathsf{v}\|^2 = 50$.

15. $\|2\mathsf{u}-\mathsf{v}\| = 7$.

16. (a) $k = -1$ or $k = 2$.

19. (a) $c = \pm\frac{4}{3}\sqrt{6}$.

21. $(\mathsf{u}+k\mathsf{v})\cdot\mathsf{u} = \mathsf{u}\cdot\mathsf{u} + k\mathsf{v}\cdot\mathsf{u} = \mathsf{u}\cdot\mathsf{u}$ since $\mathsf{v}\cdot\mathsf{u} = 0$. Since $\mathsf{u} \ne 0$, $\mathsf{u}\cdot\mathsf{u} \ne 0$, so $(\mathsf{u}+k\mathsf{v})\cdot\mathsf{u} \ne 0$.

23. $C = (-1-\sqrt{2}, \sqrt{2})$ and $C = (-1+\sqrt{2}, -\sqrt{2})$.

26. 1. Commutativity: $\mathsf{u}\cdot\mathsf{v} = xz + yw$ and $\mathsf{v}\cdot\mathsf{u} = zx + wy = xz + yw$, so $\mathsf{u}\cdot\mathsf{v} = \mathsf{v}\cdot\mathsf{u}$.

    5. For any vector $\mathsf{v} = \begin{bmatrix} x \\ y \end{bmatrix}$, $\mathsf{v}\cdot\mathsf{v} = x^2 + y^2$ is the sum of squares of real numbers. Since $a^2 \ge 0$ for any real number a, $x^2 + y^2 \ge 0$ and positive if either $x \ne 0$ or $y \ne 0$; that is, $x^2 + y^2 = 0$ if and only if $x = y = 0$. This says precisely that $\|\mathsf{v}\| = 0$ if and only if $\mathsf{v} = 0$.

29. (a) $\mathsf{u}\cdot\mathsf{v} = -6+5 = -1$, $\|\mathsf{u}\| = \sqrt{5}$ and $\|\mathsf{v}\| = \sqrt{34}$. The Cauchy-Schwarz inequality says that $1 \le \sqrt{170}$.

    (d) $\mathsf{u}\cdot\mathsf{v} = -3$, $\|\mathsf{u}\| = \sqrt{5}$, $\|\mathsf{v}\| = \sqrt{10}$. Cauchy-Schwarz says $3 \le \sqrt{5}\sqrt{10} \approx 7.1$. Since $\mathsf{u}+\mathsf{v} = \begin{bmatrix} -2 \\ 1 \\ -2 \end{bmatrix}$, $\|\mathsf{u}+\mathsf{v}\| = \sqrt{9} = 3$. The triangle inequality says $3 \le \sqrt{5} + \sqrt{10} \approx 5.4$.

32. Let $\mathsf{u} = \begin{bmatrix} a \\ b \end{bmatrix}$ and $\mathsf{v} = \begin{bmatrix} \cos\theta \\ \sin\theta \end{bmatrix}$. Then $\mathsf{u}\cdot\mathsf{v} = a\cos\theta + b\sin\theta$, $\|\mathsf{u}\|^2 = a^2 + b^2$ and $\|\mathsf{v}\|^2 = \cos^2\theta + \sin^2\theta = 1$. Squaring the Cauchy-Schwarz inequality gives $(\mathsf{u}\cdot\mathsf{v})^2 \le \|\mathsf{u}\|^2\,\|\mathsf{v}\|^2$, the given inequality.

34. (a) Let $\mathsf{u} = \begin{bmatrix} \sqrt{3}a \\ b \end{bmatrix}$ and $\mathsf{v} = \begin{bmatrix} \sqrt{3}c \\ d \end{bmatrix}$. We have $\mathsf{u}\cdot\mathsf{v} = 3ac + bd$, $\|\mathsf{u}\| = \sqrt{3a^2 + b^2}$ and $\|\mathsf{v}\| = \sqrt{3c^2 + d^2}$, so the result follows by squaring the Cauchy-Schwarz inequality: $(\mathsf{u}\cdot\mathsf{v})^2 \le \|\mathsf{u}\|^2\,\|\mathsf{v}\|^2$.

37. A better approach would be to say that $\mathsf{u} = \frac{4}{7}\mathsf{v}$, with $\mathsf{v} = \begin{bmatrix} -2 \\ 1 \\ 5 \end{bmatrix}$, so $\|\mathsf{u}\| = \frac{4}{7}\|\mathsf{v}\| = \frac{4}{7}\sqrt{30}$.

41. (a) The diagonals are $\overrightarrow{AC} = \mathsf{u}+\mathsf{v}$ and $\overrightarrow{DB} = \mathsf{u}-\mathsf{v}$.

43. $\frac{1}{\|\mathsf{u}\|}\mathsf{u} + \frac{1}{\|\mathsf{v}\|}\mathsf{v}$.

Section 1.3—True/False

1. False. A line (in the xy-plane) is the **graph** of such an equation.
2. False. $\mathsf{u} \times \mathsf{v}$ is a vector and the absolute value of a vector is meaningless.
3. False. $\mathsf{u} \cdot \mathsf{v}$ is a number and the length of a number is meaningless.
4. True. Since $\|\mathsf{u} \times \mathsf{v}\| = \|\mathsf{u}\| \, \|\mathsf{v}\| \sin\theta$, the given statement says $\sin\theta = 1$, so $\theta = \frac{\pi}{2}$. The vectors are orthogonal.
5. True. The coordinates of the point satisfy the equation.
6. False. The line has direction normal to the plane, so the line is orthogonal to the plane and certainly hits it!
7. True. The normal vectors $\begin{bmatrix} -2 \\ 3 \\ -1 \end{bmatrix}$ and $\begin{bmatrix} 4 \\ -6 \\ 2 \end{bmatrix}$ are parallel.
8. False. If $\mathsf{u} \times \mathsf{v} = 0$, then u and v are parallel.
9. True. $\mathsf{i} \times (\mathsf{i} \times \mathsf{j}) = \mathsf{i} \times \mathsf{k} = -\mathsf{j}$.
10. True. If the dot product is 0, the direction and normal are orthogonal.

Exercises 1.3

1. (a) For example, linear combinations of $\begin{bmatrix} 1 \\ 0 \\ -3 \end{bmatrix}$ and $\begin{bmatrix} 0 \\ 1 \\ 2 \end{bmatrix}$.

2. (a) $4x + 2y - z = 5$ is one equation.

3. (a) $\mathsf{u}\times\mathsf{v} = \begin{bmatrix} 3 \\ 5 \\ 7 \end{bmatrix}$, $(\mathsf{u}\times\mathsf{v})\cdot\mathsf{u} = \begin{bmatrix} 3 \\ 5 \\ 7 \end{bmatrix} \cdot \begin{bmatrix} 1 \\ -2 \\ 1 \end{bmatrix} = 3-10+7 = 0$ and $(\mathsf{u}\times\mathsf{v})\cdot\mathsf{v} = \begin{bmatrix} 3 \\ 5 \\ 7 \end{bmatrix} \cdot \begin{bmatrix} 3 \\ 1 \\ -2 \end{bmatrix} = 9 + 5 - 14 = 0$. $\mathsf{v} \times \mathsf{u} = \begin{bmatrix} -3 \\ -5 \\ -7 \end{bmatrix} = -(\mathsf{u} \times \mathsf{v})$.

4. (a) $\mathsf{u} \times \mathsf{v} = 2\begin{bmatrix} 1 \\ 3 \\ -1 \end{bmatrix}$; $(\mathsf{u} \times \mathsf{v}) \times \mathsf{w} = \begin{bmatrix} -18 \\ -2 \\ -24 \end{bmatrix}$.

   $\mathsf{v} \times \mathsf{w} = \begin{bmatrix} -3 \\ 12 \\ -4 \end{bmatrix}$; $\mathsf{u} \times (\mathsf{v} \times \mathsf{w}) = \begin{bmatrix} -16 \\ -11 \\ -21 \end{bmatrix}$.

   Since the cross product is not an associative operation, there is no reason why these vectors should be the same.

6. One vector orthogonal to both u and v is $\mathsf{u}\times\mathsf{v} = \begin{bmatrix} 6 \\ -3 \\ 0 \end{bmatrix}$. Any multiple of this vector, for instance $\begin{bmatrix} 2 \\ -1 \\ 0 \end{bmatrix}$, is another.

9. (a) $\mathsf{u}\times\mathsf{v} = \begin{vmatrix} u_2 & u_3 \\ v_2 & v_3 \end{vmatrix}\mathsf{i} - \begin{vmatrix} u_1 & u_3 \\ v_1 & v_3 \end{vmatrix}\mathsf{j} + \begin{vmatrix} u_1 & u_2 \\ v_1 & v_2 \end{vmatrix}\mathsf{k}$
$= (u_2v_3 - u_3v_2)\mathsf{i} - (u_1v_3 - u_3v_1)\mathsf{j} + (u_1v_2 - u_2v_1)\mathsf{k}.$
$\mathsf{v}\times\mathsf{u} = \begin{vmatrix} v_2 & v_3 \\ u_2 & u_3 \end{vmatrix}\mathsf{i} - \begin{vmatrix} v_1 & v_3 \\ u_1 & u_3 \end{vmatrix}\mathsf{j} + \begin{vmatrix} v_1 & v_2 \\ u_1 & u_2 \end{vmatrix}\mathsf{k}$
$= (v_2u_3 - v_3u_2)\mathsf{i} - (v_1u_3 - v_3u_1)\mathsf{j} + (v_1u_2 - v_2u_1)\mathsf{k} = -(\mathsf{u}\times\mathsf{v}).$

10. (a) i. $\mathsf{u}\times\mathsf{v} = \begin{bmatrix} 5 \\ -1 \\ -1 \end{bmatrix}$, so $\|\mathsf{u}\times\mathsf{v}\| = \sqrt{27}$. Now $\|\mathsf{u}\| = \sqrt{14}$, $\|\mathsf{v}\| = \sqrt{2}$ and $\mathsf{u}\cdot\mathsf{v} = 1$. Since $27 + 1 = 14(2)$, we have verified the formula.

ii. $\cos\theta = \frac{1}{\sqrt{28}}$ and $\sin\theta = \frac{\sqrt{27}}{\sqrt{28}}$.

12. (a) $a\mathsf{u} + b\mathsf{v} = \begin{bmatrix} a-b \\ 2b \\ 4a \end{bmatrix}$. A normal is $\mathsf{u}\times\mathsf{v} = \begin{bmatrix} -8 \\ -4 \\ 2 \end{bmatrix}$. The vector $\begin{bmatrix} 4 \\ 2 \\ -1 \end{bmatrix}$ is also a normal. An equation is $4x + 2y - z = 0$.

13. (a) $\overrightarrow{AB} = \begin{bmatrix} 1 \\ -2 \\ 2 \end{bmatrix}$ and $\overrightarrow{AC} = \begin{bmatrix} -2 \\ 1 \\ -7 \end{bmatrix}$. A normal vector is $\overrightarrow{AB}\times\overrightarrow{AC} = \begin{bmatrix} 12 \\ 3 \\ -3 \end{bmatrix}$.
Take $\mathsf{n} = \begin{bmatrix} 4 \\ 1 \\ -1 \end{bmatrix}$. An equation of the plane is $4x + y - z = 6$.

14. (a) Since $\overrightarrow{AB} = \begin{bmatrix} 1 \\ 1 \\ 0 \end{bmatrix} = \overrightarrow{CD}$, $ABCD$ is a parallelogram. Since $\overrightarrow{AB}\times\overrightarrow{AC} = \begin{bmatrix} 4 \\ -4 \\ -5 \end{bmatrix}$, the area is $\left\|\overrightarrow{AB}\times\overrightarrow{AC}\right\| = \sqrt{57}$.

(b) The area of the triangle is $\frac{1}{2}\sqrt{57}$.

16. (a) Two sides of the triangle are $\overrightarrow{AB} = \begin{bmatrix} 6 \\ 1 \end{bmatrix}$ and $\overrightarrow{AC} = \begin{bmatrix} 7 \\ -6 \end{bmatrix}$. Think of these as lying in 3-space. The triangle has area one half the area of the parallelogram with sides $\mathsf{u} = \begin{bmatrix} 6 \\ 1 \\ 0 \end{bmatrix}$ and $\mathsf{v} = \begin{bmatrix} 7 \\ -6 \\ 0 \end{bmatrix}$. This is one half the length of $\mathsf{u}\times\mathsf{v} = \begin{bmatrix} 0 \\ 0 \\ -43 \end{bmatrix} = \dfrac{43}{2}$.

18. (a) The direction of the line is $\begin{bmatrix} -1 \\ 1 \end{bmatrix}$, so the slope is -1. An equation is $x + y = 3$.

21. (a) Setting $t = 0$ and then $t = 1$, we see that $A(1,0,2)$ and $B(-3,3,3)$ are also on the plane. The plane contains the arrows from P to A and from P to B, so it is parallel to the vectors $\begin{bmatrix} -2 \\ -1 \\ 3 \end{bmatrix}$ and $\begin{bmatrix} -6 \\ 2 \\ 4 \end{bmatrix}$. A

normal vector is $\mathsf{n} = \begin{bmatrix} -10 \\ -10 \\ -10 \end{bmatrix}$. The vector $\begin{bmatrix} 1 \\ 1 \\ 1 \end{bmatrix}$ is equally good. An equation of the plane is $x + y + z = 3$.

22. (a) We need triples (x, y, z) that satisfy both equations. Setting $x = 0$ gives $y + z = 5$ and $-y + z = 1$, so $z = 3$ and $y = 2$. This gives the point $A(0, 2, 3)$. Setting $y = 0$ gives $2x + z = 5$ and $x + z = 1$, so $x = 4$ and $z = -3$. This gives the point $B(4, 0, -3)$. Setting $z = 0$ gives $2x + y = 5$ and $x - y = 1$, so $x = 2$ and $y = 1$. This gives the point $C(2, 1, 0)$. Many other points are possible, of course.

(b) **Solution 1.** The line has direction $\overrightarrow{AB} = \begin{bmatrix} 4 \\ -2 \\ -6 \end{bmatrix}$ and hence also $\begin{bmatrix} 2 \\ -1 \\ -3 \end{bmatrix}$. An equation is $\begin{bmatrix} x \\ y \\ z \end{bmatrix} = \begin{bmatrix} 0 \\ 2 \\ 3 \end{bmatrix} + t \begin{bmatrix} 2 \\ -1 \\ -3 \end{bmatrix}$.

Solution 2. The line has direction orthogonal to the normal vector of each plane, so the cross product of the normal vectors gives a direction vector. This cross product is $\begin{bmatrix} 2 \\ -1 \\ -3 \end{bmatrix}$. As in Solution 1, we still need to find a point on the line; $A(0, 2, 3)$ will do. An equation is $\begin{bmatrix} x \\ y \\ z \end{bmatrix} = \begin{bmatrix} 0 \\ 2 \\ 3 \end{bmatrix} + t \begin{bmatrix} 2 \\ -1 \\ -3 \end{bmatrix}$, as before.

[Solutions to this question may look different from the one we have obtained, of course. A correct answer must be an equation of a line whose direction vector is a multiple of $\begin{bmatrix} 2 \\ -1 \\ -3 \end{bmatrix}$ and passes through a point whose coordinates satisfy the equation of each plane.]

23. (a) The line has direction orthogonal to the normal vector of each plane, so $\begin{bmatrix} 0 \\ 3 \\ -1 \end{bmatrix}$ is a direction vector. Putting $z = 0$ in the two given equations gives $x + 2y = 5$, $x - y = -1$, whose solution is $x = 1$, $y = 2$. Thus $(1, 2, 0)$ is on the line and an equation is $\begin{bmatrix} x \\ y \\ z \end{bmatrix} = \begin{bmatrix} 1 \\ 2 \\ 0 \end{bmatrix} + t \begin{bmatrix} 0 \\ 3 \\ -1 \end{bmatrix}$.

[Solutions to this question may look different from the one we have obtained, of course. A correct answer must be an equation of a line whose direction vector is a multiple of $\begin{bmatrix} 0 \\ 3 \\ -1 \end{bmatrix}$ and that passes through a point whose coordinates satisfy the equation of each plane.]

25. (a) If (x, y, z) is on the line, then, for some t, $x = 1 + 2t$, $y = -2 + 5t$, $z = 3 - t$. If such a point is on the plane, then $4 = x - 3y + 2z = (1 + 2t) - 3(-2 + 5t) + 2(3 - t) = 13 - 15t$, so $15t = 9$, $t = \frac{3}{5}$ and $(x, y, z) = (\frac{11}{5}, 1, \frac{12}{5})$.

26. (a) $\begin{bmatrix} 2 \\ -3 \\ 5 \end{bmatrix}$ is a normal and hence orthogonal to π.

27. (a) Since $ad - bc = 0$, we have $ad = bc$. If $a \neq 0$ and $d \neq 0$, then $b \neq 0$, so $\frac{c}{a} = \frac{d}{b} = k$. This says $c = ka$ and $d = kb$, so $\begin{bmatrix} c \\ d \end{bmatrix} = k\begin{bmatrix} a \\ b \end{bmatrix}$. Suppose $a = 0$ and $d \neq 0$. Then $b = 0$ or $c = 0$. In the case $b = 0$, then $\begin{bmatrix} a \\ b \end{bmatrix} = \begin{bmatrix} 0 \\ 0 \end{bmatrix} = 0\begin{bmatrix} c \\ d \end{bmatrix}$, while if $c = 0$ (and $b \neq 0$), then $\begin{bmatrix} c \\ d \end{bmatrix} = \begin{bmatrix} 0 \\ d \end{bmatrix} = \frac{d}{b}\begin{bmatrix} 0 \\ b \end{bmatrix} = \frac{d}{b}\begin{bmatrix} a \\ b \end{bmatrix}$. Suppose $a \neq 0$ and $d = 0$. Again, $b = 0$ or $c = 0$. In the case $b = 0$, then $\begin{bmatrix} c \\ d \end{bmatrix} = \begin{bmatrix} c \\ 0 \end{bmatrix} = \frac{c}{a}\begin{bmatrix} a \\ 0 \end{bmatrix} = \frac{c}{a}\begin{bmatrix} a \\ b \end{bmatrix}$, while if $c = 0$ (and $b \neq 0$), then $\begin{bmatrix} c \\ d \end{bmatrix} = \begin{bmatrix} 0 \\ 0 \end{bmatrix} = 0\begin{bmatrix} a \\ b \end{bmatrix}$. Finally, if $a = d = 0$, then $b = 0$ or $c = 0$. If $b = 0$, then $\begin{bmatrix} a \\ b \end{bmatrix} = \begin{bmatrix} 0 \\ 0 \end{bmatrix} = 0\begin{bmatrix} c \\ d \end{bmatrix}$, while if $c = 0$ (and $b \neq 0$), then $\begin{bmatrix} c \\ d \end{bmatrix} = \begin{bmatrix} 0 \\ d \end{bmatrix} = \frac{d}{b}\begin{bmatrix} 0 \\ b \end{bmatrix} = \frac{d}{b}\begin{bmatrix} a \\ b \end{bmatrix}$. In every case, one of the two given vectors is a multiple of the other.

28. (a) If the lines intersect at (x, y, z), then (x, y, z) is on both lines. So there is a t such that $\begin{bmatrix} x \\ y \\ z \end{bmatrix} = \begin{bmatrix} 1 \\ 0 \\ -2 \end{bmatrix} + t\begin{bmatrix} -3 \\ 1 \\ 1 \end{bmatrix}$ and an s such that $\begin{bmatrix} x \\ y \\ z \end{bmatrix} = \begin{bmatrix} -4 \\ 1 \\ 1 \end{bmatrix} + s\begin{bmatrix} 11 \\ -3 \\ -5 \end{bmatrix}$. We find that $t = -2, s = 1$ is a solution. The lines intersect where $\begin{bmatrix} x \\ y \\ z \end{bmatrix} = \begin{bmatrix} 1 \\ 0 \\ -2 \end{bmatrix} - 2\begin{bmatrix} -3 \\ 1 \\ 1 \end{bmatrix} = \begin{bmatrix} -4 \\ 1 \\ 1 \end{bmatrix} + 1\begin{bmatrix} 11 \\ -3 \\ -5 \end{bmatrix} = \begin{bmatrix} 7 \\ -2 \\ -4 \end{bmatrix}$; that is, at the point $(7, -2, -4)$.

32. Since the lines are orthogonal, the direction vectors must have dot product 0, so $b = 3$. Since the lines intersect, there are values of t and s such that
$$\begin{aligned} 2 - t &= 3 + s \\ -1 + 2t &= 1 - s \\ 3 + t &= a + bs = a + 3s. \end{aligned}$$
We get $a = 18$.

Section 1.4—True/False

1. False. The projection of a vector u onto a nonzero vector v is a multiple of v.

2. False. If u and v are orthogonal, the projection of u onto v is the zero vector.

3. False.
4. True.
5. True.
6. False. The vector $\begin{bmatrix} 1 \\ -1 \\ 1 \end{bmatrix}$ is not in the plane.
7. False in general. The vectors e and f must be **orthogonal**. See Theorem 1.4.10.
8. True.
9. True. The point $(-2, 3, 1)$ is on the plane.
10. False. The line is perpendicular to the plane since its direction is a normal to the plane.

Exercises 1.4

1. When the angle θ between u and v is **obtuse:** $\frac{\pi}{2} < \theta < \pi$.
2. (a) $\mathsf{u} = c\mathsf{v}$, so $\frac{\mathsf{u}\cdot\mathsf{v}}{\mathsf{v}\cdot\mathsf{v}} = \frac{c\mathsf{v}\cdot\mathsf{v}}{\mathsf{v}\cdot\mathsf{v}}$ and $\mathrm{proj}_{\mathsf{v}}\,\mathsf{u} = c\mathsf{v} = \mathsf{u}$.
3. (a) $\mathrm{proj}_{\mathsf{v}}\,\mathsf{u} = \frac{4}{11}\begin{bmatrix} 3 \\ 1 \\ 1 \end{bmatrix}$, $\mathrm{proj}_{\mathsf{u}}\,\mathsf{v} = \frac{2}{3}\begin{bmatrix} 1 \\ 2 \\ -1 \end{bmatrix}$.
4. (a) Use the procedure described by the diagram in the text for Problem 1.4.5. The distance is $\frac{2}{3}\left\|\begin{bmatrix} -1 \\ 2 \\ -1 \end{bmatrix}\right\| = \frac{2}{3}\sqrt{6}$ and the closest point is $A(-\frac{1}{3}, \frac{2}{3}, \frac{5}{3})$.
5. Let Q be any point on the plane. Then use the fact that $\overrightarrow{PA}$ is the projection of $\overrightarrow{PQ}$ on n.
6. (a) i. The distance is $\frac{2}{3}\sqrt{3}$.
   ii. The closest point is $A = (\frac{8}{9}, \frac{29}{9}, -\frac{2}{9})$.
7. (a) Any two nonzero orthogonal vectors whose components satisfy $2x - 3y + 4z = 0$ is a correct answer, for instance, $\begin{bmatrix} -9 \\ 5 \\ 18 \end{bmatrix}$ and $\begin{bmatrix} 2 \\ 0 \\ 1 \end{bmatrix}$.
   (d) Any two nonzero orthogonal vectors whose components satisfy $2x - 3y + 4z = 0$ is a correct answer, for instance, $\begin{bmatrix} 3 \\ 10 \\ 6 \end{bmatrix}$ and $\begin{bmatrix} 2 \\ 0 \\ -1 \end{bmatrix}$.
9. (a) A correct answer consists of two (nonzero) orthogonal vectors whose components satisfy $2x - y + z = 0$, for example, $\begin{bmatrix} 0 \\ 1 \\ 1 \end{bmatrix}$ and $\begin{bmatrix} 1 \\ 1 \\ -1 \end{bmatrix}$.
   (b) $\mathrm{proj}_{\pi}\,\mathsf{w} = \frac{1}{3}\begin{bmatrix} 1 \\ 4 \\ 2 \end{bmatrix}$.

12. (a) $\text{proj}_\pi \mathbf{w} = \begin{bmatrix} 1 \\ 0 \\ 2 \end{bmatrix}$, (b) With $\mathbf{w} = \begin{bmatrix} x \\ y \\ z \end{bmatrix}$, $\text{proj}_\pi \mathbf{w} = \frac{1}{6}\begin{bmatrix} 2x-2y+2z \\ -2x+5y+z \\ 2x+y+5z \end{bmatrix}$.

14. (a) $\text{proj}_\pi \mathbf{w} = \frac{1}{6}\begin{bmatrix} -13 \\ 17 \\ 2 \end{bmatrix}$.

15. (a) This is exactly as in the proof of Theorem 1.4.10. Write $\mathbf{w} = a\mathbf{e} + b\mathbf{f}$. Then $\mathbf{w} \cdot \mathbf{e} = a\mathbf{e} \cdot \mathbf{e}$, so $a = \frac{\mathbf{w} \cdot \mathbf{e}}{\mathbf{e} \cdot \mathbf{e}}$. Similarly $b = \frac{\mathbf{w} \cdot \mathbf{f}}{\mathbf{f} \cdot \mathbf{f}}$, so $\mathbf{w} = \frac{\mathbf{w} \cdot \mathbf{e}}{\mathbf{e} \cdot \mathbf{e}}\mathbf{e} + \frac{\mathbf{w} \cdot \mathbf{f}}{\mathbf{f} \cdot \mathbf{f}}\mathbf{f}$.

18. (a) The picture certainly makes Don's theory appear plausible and the following argument establishes that it is indeed correct. Don takes for the projection the vector $\mathbf{p} = \mathbf{w} - \text{proj}_\mathbf{n} \mathbf{w}$.

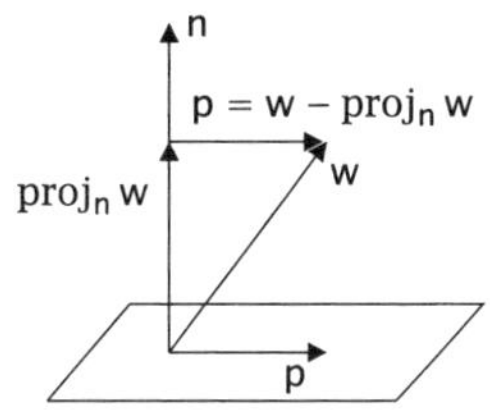

By definition of "projection on a plane," we must show that $\mathbf{w} - \mathbf{p}$ is orthogonal to every vector in π. This is immediately clear since $\mathbf{w} - \mathbf{p} = \mathbf{w} - (\mathbf{w} - \text{proj}_\mathbf{n} \mathbf{w}) = \text{proj}_\mathbf{n} \mathbf{w}$ is a normal to the plane.

(b) A normal to π is $\mathbf{n} = \begin{bmatrix} 3 \\ -2 \\ 1 \end{bmatrix}$. Thus $\text{proj}_\mathbf{n} \mathbf{w} = \frac{\mathbf{w} \cdot \mathbf{n}}{\mathbf{n} \cdot \mathbf{n}}\mathbf{n} = \frac{2}{14}\begin{bmatrix} 3 \\ -2 \\ 1 \end{bmatrix} = \frac{1}{7}\begin{bmatrix} 3 \\ -2 \\ 1 \end{bmatrix}$ and $\mathbf{p} = \mathbf{w} - \text{proj}_\mathbf{n} \mathbf{w} = \begin{bmatrix} 1 \\ 1 \\ 1 \end{bmatrix} - \frac{1}{7}\begin{bmatrix} 3 \\ -2 \\ 1 \end{bmatrix} = \frac{1}{7}\begin{bmatrix} 4 \\ 9 \\ 6 \end{bmatrix}$.

21. (a) The lines have directions $\mathbf{d}_1 = \begin{bmatrix} 4 \\ 3 \\ 1 \end{bmatrix}$ and $\mathbf{d}_2 = \begin{bmatrix} 2 \\ 6 \\ 7 \end{bmatrix}$. Neither vector is a scalar multiple of the other, so the lines are not parallel. Suppose they intersect at (x, y, z). Then there would exist t and s such that

$$\begin{aligned} x &= 3 + 4t &&= 2 + 2s \\ y &= 3t &&= -1 + 6s \\ z &= -1 + t &&= -3 + 7s. \end{aligned}$$

The third equation says $-3 + 3t = -9 + 21s$. Since $3t = -1 + 6s$, we have $-3 - 1 + 6s = -9 + 21s$, so $15s = 5$, $s = \frac{1}{3}$ and (third equation) $t = -2 + 7s = \frac{1}{3}$ too. But $s = t = \frac{1}{3}$ do not satisfy the first equation. No s, t exist, so the lines do not intersect.

(b) An equation is $15x - 26y + 18z = 27$.

(c) The shortest distance between the lines is $\frac{1}{49}\sqrt{1225}$.

24. (b) The shortest distance is realized along a line with direction perpendicular to each line. Thus $\mathsf{d} = \begin{bmatrix} 2 \\ 3 \\ 6 \end{bmatrix}$, the cross product of direction vectors of ℓ_1 and ℓ_2, is a direction vector.

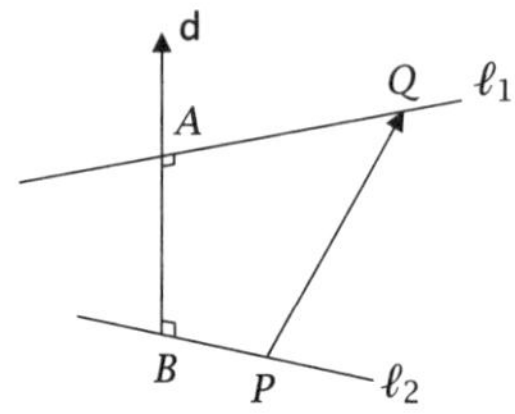

Take any point Q on ℓ_1 and any point P on ℓ_2, say $P(-2,-1,-1)$ and $Q(3,0,5)$. The projection of PQ on d is $\overrightarrow{BA}$, a vector whose length is the desired shortest distance between ℓ_1 and ℓ_2. We have $\operatorname{proj}_{\mathsf{d}} \overrightarrow{PQ} == \frac{49}{49}\mathsf{d} = \mathsf{d}$, so the shortest distance is $\|\mathsf{d}\| = \sqrt{49} = 7$.

25. (a) The line has direction $\mathsf{d} = \begin{bmatrix} 2 \\ 1 \\ 3 \end{bmatrix}$, the plane has normal $\mathsf{n} = \begin{bmatrix} -2 \\ 1 \\ 1 \end{bmatrix}$. Since $\mathsf{d} \cdot \mathsf{n} = 0$, the line is perpendicular to a normal to the plane and hence parallel to the plane.

(b) Two possibilities are $P = (3,5,-4)$ and $Q = (0,1,2)$.

(c) The distance we want is the length of the projection of $\overrightarrow{PQ}$ on n. We have $\overrightarrow{PQ} = \begin{bmatrix} -3 \\ -4 \\ 6 \end{bmatrix}$ and $\mathsf{n} = \begin{bmatrix} -2 \\ 1 \\ 1 \end{bmatrix}$, so $\operatorname{proj}_{\mathsf{n}} \overrightarrow{PQ} = \dfrac{8}{6}\begin{bmatrix} -2 \\ 1 \\ 1 \end{bmatrix}$, of length $\frac{4}{3}\sqrt{6}$, the required distance.

Section 1.5—True/False

1. False. The statement $0\mathsf{u} + 0\mathsf{v} = \mathsf{0}$ is true for any vectors u and v and has nothing to do with linear independence.
2. False. Saying "$a\mathsf{u} + b\mathsf{v} = \mathsf{0}$ where a and b are 0" is the same as saying $0\mathsf{u} + 0\mathsf{v} = \mathsf{0}$, which is true for any vectors u and v and has nothing to do with linear independence.
3. True. $0\mathsf{u} + 0\mathsf{v} + (58)\mathsf{0}$ is a nontrivial linear combination of u, v and 0 that equals the zero vector.
4. False. If the vectors are linearly dependent, there is such an equation with **not all** the scalars $c_i = 0$. "Not all" does not mean "none."
5. False. One of them is a **linear combination** of the others.
6. True.
7. True. For instance, $\mathsf{v}_1 = \frac{3}{2}\mathsf{v}_2 - \frac{1}{2}\mathsf{v}_3$.
8. False. Let $\mathsf{u} = \mathsf{0}$, let v be any nonzero vector and let w any nonzero vector that is not a scalar multiple of v.
9. False: For any vectors u, v, w—linearly dependent or not—$0\mathsf{u} + 0\mathsf{v} + 0\mathsf{w} = \mathsf{0}$.

10. False. If you state at the outset that $c_1 = 0$, $c_2 = 0$, ..., $c_m = 0$, then it is abundantly clear that $c_1\mathsf{u}_1 + c_2\mathsf{u}_2 + \cdots + c_m\mathsf{u}_m = \mathsf{0}$, but this is not what linear independence is all about. Linear independence means that if you happen to stumble across a linear combination of the vectors which equals the zero vector, then you can **conclude** that the coefficients are all 0. This is very different from **assuming** at the outset that the coefficients are 0.

Exercises 1.5

1. (a) linearly independent, (f) linearly dependent: $-7\mathsf{v}_1 - 2\mathsf{v}_2 + \mathsf{v}_3 = \mathsf{0}$.
2. **First line:** Suppose $c_1\mathsf{v}_1 + c_2\mathsf{v}_2 + \cdots + c_k\mathsf{v}_k = \mathsf{0}$ for scalars $c_1, c_2, \ldots, c_k$. **Last line:** Therefore $c_1 = 0$, $c_2 = 0$, ..., $c_k = 0$.
4. Suppose $a\mathsf{e} + b\mathsf{f} = \mathsf{0}$ for scalars a and b. Taking the dot product with e and using $\mathsf{f} \cdot \mathsf{e} = 0$ gives $a\mathsf{e} \cdot \mathsf{e} = 0$. Since $\mathsf{e} \neq \mathsf{0}$, we know that $\mathsf{e} \cdot \mathsf{e} \neq 0$, so $a = 0$. Similarly, taking the dot product with f gives $b = 0$.
8. If you know, please tell the author.
9. Suppose $a\mathsf{u} + b\mathsf{v} + c\mathsf{x} = \mathsf{0}$. We must show $a = b = c = 0$. Now $a\mathsf{u} + b\mathsf{v} + c\mathsf{x} + 0\mathsf{y} = \mathsf{0}$. Since $\mathsf{u}, \mathsf{v}, \mathsf{x}, \mathsf{y}$ are linearly independent, all coefficients are 0. In particular, $a = b = c = 0$.

Section 2.1—True/False

1. False. Take A to be 2×3 and B to be 3×2. Then AB is 2×2 and BA is 3×3, both of which are square.
2. True.
3. False. For example, let $A = \begin{bmatrix} 1 & 0 \end{bmatrix}$ and $B = \begin{bmatrix} 0 \\ 1 \end{bmatrix}$.
4. False. For example, let $A = \begin{bmatrix} 1 & 0 \end{bmatrix}$, $B = \begin{bmatrix} 0 \\ 1 \end{bmatrix}$ and $X = \begin{bmatrix} 1 & 1 \\ 1 & 1 \end{bmatrix}$.
5. False. The $(2, 1)$ entry is 2, the entry in row two, column 1.
6. True. Since $\mathsf{b} = \mathsf{e}_3$, the third standard basis vector, $A\mathsf{b}$ is the third column of A.
7. True. This is the very important fact 2.1.33.
8. False. This says matrix multiplication is **associative**.
9. True. The hypothesis implies that $A\mathsf{e}_i = \mathsf{0}$, where $\mathsf{e}_1, \mathsf{e}_2, \ldots, \mathsf{e}_n$ denote the standard basis vectors. But $A\mathsf{e}_i$ is column i of A, so each column of A is $\mathsf{0}$. Thus $A = 0$.
10. False. $A(2\mathsf{x}_0) = 2A\mathsf{x}_0 = 2\mathsf{b} \neq \mathsf{b}$.

Exercises 2.1

1. (a) $x = 1, y = -2$.
2. (a) No x, y, a, b exist.
3. (a) $\begin{bmatrix} 1 & 2 & 3 \\ 1 & 2 & 3 \end{bmatrix}$.
6. The matrix P must be square.
10. (a) $A + B = \begin{bmatrix} 4 & -2 & 5 \\ -4 & 0 & 6 \end{bmatrix}$; $2A^T - B$ is not defined.
11. (a) $x = 4$, $y = 4$ and $z = 5$, (c) No such x, y, z exist.
13. $AB = [30]$, $BA = \begin{bmatrix} 8 & 8 & 8 \\ 10 & 10 & 10 \\ 12 & 12 & 12 \end{bmatrix}$; neither A^TB nor AB^T is determined; $A^TB^T = \begin{bmatrix} 8 & 10 & 12 \\ 8 & 10 & 12 \\ 8 & 10 & 12 \end{bmatrix}$, $B^TA^T = [30]$.
15. (a) $AB = \begin{bmatrix} 1 & -7 \\ 8 & -4 \end{bmatrix}$, $BA = \begin{bmatrix} -5 & -5 & 12 \\ 6 & 2 & -8 \\ -2 & 1 & 0 \end{bmatrix}$, $A^TB^T = (BA)^T = \begin{bmatrix} -5 & 6 & -2 \\ -5 & 2 & 1 \\ 12 & -8 & 0 \end{bmatrix}$, $B^TA^T = (AB)^T = \begin{bmatrix} 1 & 8 \\ -7 & -4 \end{bmatrix}$.
16. (a) The $(1, 1)$ and $(2, 2)$ entries of AB are, respectively, $ax + by + cz$ and $du + ev + fw$.
22. (a) $\begin{bmatrix} -1 & 2 & 1 \\ 0 & 1 & 8 \end{bmatrix} \begin{bmatrix} 3 \\ 4 \\ -1 \end{bmatrix}$.
23. (a) $A\mathsf{x} = \begin{bmatrix} 7 \\ 12 \end{bmatrix} = -4\begin{bmatrix} 1 \\ -2 \end{bmatrix} + 4\begin{bmatrix} 2 \\ 0 \end{bmatrix} + \begin{bmatrix} 3 \\ 4 \end{bmatrix}$.
24. (a) $\begin{bmatrix} 2 & -1 \\ 3 & 5 \\ -1 & 1 \\ 2 & -5 \end{bmatrix} \begin{bmatrix} x \\ y \end{bmatrix} = \begin{bmatrix} 10 \\ 7 \\ -3 \\ 1 \end{bmatrix}$.
25. $\begin{bmatrix} 1 & 1 & 1 \\ 4 & 2 & 1 \end{bmatrix} \begin{bmatrix} a \\ b \\ c \end{bmatrix} = \begin{bmatrix} 4 \\ 8 \end{bmatrix}$.
32. $A(B + C) = AB + AC = BA + CA = (B + C)A$.
35. "No." With $A = \begin{bmatrix} 3 & -1 \\ 0 & -2 \end{bmatrix}$, we find $A - 3I = \begin{bmatrix} 0 & -1 \\ 0 & -5 \end{bmatrix}$ and $A + 2I = \begin{bmatrix} 5 & -1 \\ 0 & 0 \end{bmatrix}$, so that $(A - 3I)(A + 2I) = \begin{bmatrix} 0 & 0 \\ 0 & 0 \end{bmatrix}$. Yet $A \neq 3I$ and $A \neq -2I$.
37. (b) $A = \begin{bmatrix} A_1 & A_2 \end{bmatrix}$ is partitioned as a 1×2 matrix with blocks $A_1 = \begin{bmatrix} 1 & 2 \\ 4 & 5 \\ 7 & 8 \\ 0 & 1 \end{bmatrix}$ and $A_2 = \begin{bmatrix} 3 & 4 \\ 6 & 7 \\ 9 & 0 \\ 2 & 3 \end{bmatrix}$.

(d) $A = \begin{bmatrix} A_1 & A_2 & A_3 \\ A_4 & A_5 & A_6 \end{bmatrix}$ is partitioned as a 2×3 matrix with blocks $A_1 = \begin{bmatrix} 1 \\ 4 \end{bmatrix}$, $A_2 = \begin{bmatrix} 2 & 3 \\ 5 & 6 \end{bmatrix}$, $A_3 = \begin{bmatrix} 4 \\ 7 \end{bmatrix}$, $A_4 = \begin{bmatrix} 7 \\ 0 \end{bmatrix}$, $A_5 = \begin{bmatrix} 8 & 9 \\ 1 & 2 \end{bmatrix}$, $A_6 = \begin{bmatrix} 0 \\ 3 \end{bmatrix}$.

39. In (a), $A = \begin{bmatrix} A_1 & A_2 \end{bmatrix}$ is partitioned as a 1×2 matrix. The only compatible partitionings of B when A is 1×2 occur as i, iii and iv. In i, $AB = \begin{bmatrix} A_1B_1 + A_2B_2 \end{bmatrix}$, but $A_1 = \begin{bmatrix} 1 \\ 0 \\ 4 \end{bmatrix}$ is 3×1 and $B_1 = \begin{bmatrix} 2 & 3 & 2 & -2 \end{bmatrix}$ is 1×4; the block sizes are not compatible so AB is not defined. In iii, $AB = \begin{bmatrix} A_1B_1 + A_2B_2 & A_1B_2 + A_2B_4 \end{bmatrix}$,

$$A_1B_1 + A_2B_3 = \begin{bmatrix} 1 \\ 0 \\ 4 \end{bmatrix} [2] + \begin{bmatrix} 2 \\ -2 \\ 4 \end{bmatrix} [1] = \begin{bmatrix} 4 \\ -2 \\ 12 \end{bmatrix},$$

$$A_1B_2 + B_2B_4 = \begin{bmatrix} 1 \\ 0 \\ 4 \end{bmatrix} \begin{bmatrix} 3 & 2 & -2 \end{bmatrix} + \begin{bmatrix} 2 \\ -2 \end{bmatrix} \begin{bmatrix} 0 & 2 & 4 \end{bmatrix}$$

$$= \begin{bmatrix} 3 & 2 & -2 \\ 0 & 0 & 0 \\ 12 & 8 & -8 \end{bmatrix} + \begin{bmatrix} 0 & 4 & 8 \\ 0 & -4 & -8 \\ 0 & 8 & 16 \end{bmatrix} = \begin{bmatrix} 3 & 6 & 6 \\ 0 & -4 & -8 \\ 12 & 16 & 8 \end{bmatrix},$$

so $AB = \left[\begin{array}{c|ccc} 4 & 3 & 6 & 6 \\ -2 & 0 & -4 & -8 \\ 12 & 12 & 16 & 8 \end{array}\right]$ is defined.

In case iv, $AB = \begin{bmatrix} A_1B_1 + A_2B_4 & A_1B_2 + A_2B_5 & A_1B_3 + A_2B_6 \end{bmatrix}$,

$$A_1B_1 + B_2B_4 = \begin{bmatrix} 1 \\ 0 \\ 4 \end{bmatrix} [2] + \begin{bmatrix} 2 \\ -2 \\ 4 \end{bmatrix} [1] = \begin{bmatrix} 4 \\ -2 \\ 12 \end{bmatrix},$$

$$A_1B_2 + A_2B_5 = \begin{bmatrix} 1 \\ 0 \\ 4 \end{bmatrix} \begin{bmatrix} 3 & 2 \end{bmatrix} + \begin{bmatrix} 2 \\ -2 \\ 4 \end{bmatrix} \begin{bmatrix} 0 & 2 \end{bmatrix}$$

$$= \begin{bmatrix} 3 & 2 \\ 0 & 0 \\ 12 & 8 \end{bmatrix} + \begin{bmatrix} 0 & 4 \\ 0 & -4 \\ 0 & 8 \end{bmatrix} = \begin{bmatrix} 3 & 6 \\ 0 & -4 \\ 12 & 16 \end{bmatrix},$$

$$A_1B_3 + A_2B_6 = \begin{bmatrix} 1 \\ 0 \\ 4 \end{bmatrix} [-2] + \begin{bmatrix} 2 \\ -2 \\ 4 \end{bmatrix} [4] = \begin{bmatrix} 6 \\ -8 \\ 8 \end{bmatrix}$$

and $AB = \left[\begin{array}{c|cc|c} 4 & 3 & 6 & 6 \\ -2 & 0 & -4 & -8 \\ 12 & 12 & 16 & 8 \end{array}\right]$ as before.

43. (a) The dot product of a vector y and a vector $A\mathsf{x}$ is the matrix product of y^T and $A\mathsf{x}$.

45. (a) The first column of AB is A times the first column of B; this is $b_{11}\mathsf{a}_1$. The second column of AB is A times the second column of B; this is $b_{12}\mathsf{a}_1 + b_{22}\mathsf{a}_2$. In general, column k of AB is A times column k of B; this is $b_{1k}\mathsf{a}_1 + b_{2k}\mathsf{a}_2 + \cdots + b_{kk}\mathsf{a}_k$.

Exercises 2.2

1. (a) 10011 is in C.
2. (a) This word is not in C.
3. $H = \begin{bmatrix} 1 & 0 & 1 & 1 & 0 \\ 1 & 1 & 1 & 0 & 1 \end{bmatrix}$ is one possibility.
5. We use the parity check matrix given in (4).
   (a) The message is 1010. (c) The message is 1111.

Section 2.3—True/False

1. False. If C is invertible, multiplying $AC = BC$ on the right by C^{-1} gives $A = B$, but not necessarily otherwise.
2. True.
3. False.
4. False, unless $AB = BA$ (which is seldom the case).
5. True. See 2.3.13.
6. False. $X = BA^{-1}$. (In general, this is different from $A^{-1}B$ unless A^{-1} and B commute, which is unlikely.)
7. True.
8. True. AB is 2×5.
9. False. If A is 2×3 and B is 3×2, then A^TB^T is defined and 3×3, and B^TA^T is defined and 2×2.
10. False.

Exercises 2.3

1. (a) These matrices are inverses.
3. (a) $A^{-1} = \begin{bmatrix} -3 & -5 \\ -1 & -2 \end{bmatrix}$. The system is $A\mathsf{x} = \mathsf{b}$ with $\mathsf{x} = \begin{bmatrix} x \\ y \end{bmatrix}$ and $\mathsf{b} = \begin{bmatrix} 3 \\ 7 \end{bmatrix}$, so the solution is $\mathsf{x} = A^{-1}\mathsf{b} = -\begin{bmatrix} 3 & 5 \\ 1 & 2 \end{bmatrix}\begin{bmatrix} 3 \\ 7 \end{bmatrix} = \begin{bmatrix} -44 \\ -17 \end{bmatrix}$.

6. The inverse of cA is $B = \frac{1}{c}A^{-1}$. To prove this, it suffices to prove that $(cA)B = I$. Using part 4 of Theorem 2.1.29 to move $\frac{1}{c}$ to the front, we have $(cA)B = (cA)(\frac{1}{c}A^{-1}) = c\frac{1}{c}AA^{-1} = 1I = I$.
9. (a) $(AB)^T = (BA)^T = A^TB^T$.
11. (a) $AB = \begin{bmatrix} 1 & 0 \\ 0 & 1 \end{bmatrix}$, $BA = \begin{bmatrix} 5 & -20 & 2 \\ 2 & -9 & 1 \\ 10 & -50 & 6 \end{bmatrix}$. (b) A is not invertible.
13. (a) Multiplying $AC = BC$ on the right by C^{-1} gives $ACC^{-1} = BCC^{-1}$, so $AI = BI$ and $A = B$.
16. We are asked to show that the inverse of A^T is $X = (A^{-1})^T$. To show that two $n \times n$ matrices are inverses, we compute their product (in either order) and expect to get I. Here, we compute $A^TX = A^T(A^{-1})^T = (A^{-1}A)^T$ (because $(U^TV^T = (VU)^T) = I^T = I$.
20. It suffices to show that $(I - BA)[I + B(I - AB)^{-1}A] = I$. For this, we compute

$$\begin{aligned}(I - BA)[I + B(I - AB)^{-1}A] &= I + B(I - AB)^{-1}A - BA - BAB(I - AB)^{-1}A \\ &= I - BA + B[(I - AB)^{-1} - AB(I - AB)^{-1}]A \\ &= I - BA + B[(I - AB)(I - AB)^{-1}]A \\ &= I - BA + BIA = I - BA + BA = I.\end{aligned}$$

23. (a) Taking the inverse of each side of the given equation, we get $2A = \begin{bmatrix} 1 & 2 \\ 3 & 4 \end{bmatrix}$, so $A = \frac{1}{2}\begin{bmatrix} 1 & 2 \\ 3 & 4 \end{bmatrix} = \begin{bmatrix} \frac{1}{2} & 1 \\ \frac{3}{2} & 2 \end{bmatrix}$.
24. (a) $X = B^{-1} + A^{-1}$.
27. (a) The inverse is $\begin{bmatrix} A^{-1} & 0 \\ -C^{-1}BA^{-1} & C^{-1} \end{bmatrix} = \begin{bmatrix} 5 & 3 & 0 & 0 \\ 2 & 1 & 0 & 0 \\ -38 & -\frac{118}{5} & \frac{7}{5} & -\frac{9}{5} \\ 6 & \frac{19}{5} & -\frac{1}{5} & \frac{2}{5} \end{bmatrix}$.

    (e) The inverse is $\begin{bmatrix} -2 & -1 & 0 & 0 \\ \frac{3}{2} & \frac{1}{2} & 0 & 0 \\ 3 & 2 & \frac{1}{3} & -\frac{4}{3} \\ -\frac{5}{2} & -\frac{3}{2} & -\frac{1}{6} & \frac{7}{6} \end{bmatrix}$.
30. There exists a matrix X such that $(AB)X = I$. Thus $A(BX) = I$, so A is invertible. Now B is invertible because $B = A^{-1}(AB)$ is the product of invertible matrices.

Section 2.4—True/False

1. False. The variable x_3 does not appear to the first power.
2. True. The second system is the first after application of the elementary row operation $E3 \to E3 - 2(E1)$.

3. False. A system of linear equations has no solution, one solution, or infinitely many solutions.
4. True.
5. True, because the two systems are equivalent.
6. False. The last row corresponds to the equation $0 = 1$, so there are no solutions.
7. False. Free variables correspond to the columns **without** pivots.
8. True. See Problem 2.4.23.
9. False. A pivot cannot be 0.
10. False. Think about $\left[\begin{array}{cc|c} 1 & 0 & 3 \\ 0 & 1 & 2 \\ 0 & 0 & 1 \end{array}\right]$.

Exercises 2.4

1. (a) not in row echelon form.
2. (a) $a = 2$.
3. (a) $\begin{bmatrix} 1 & -1 & -2 \\ 2 & -3 & -5 \\ -1 & 4 & 5 \end{bmatrix} \to \begin{bmatrix} 1 & -1 & -2 \\ 0 & -1 & -1 \\ 0 & 0 & 0 \end{bmatrix}$.
   The pivots are 1 and -1. The pivot columns are columns one and two.
4. (a) The pivot columns are columns one, two, five and six.
7. (a) $A = \begin{bmatrix} -2 & 1 & 5 \\ -8 & 7 & 19 \end{bmatrix}$, $\mathsf{x} = \begin{bmatrix} x_1 \\ x_2 \\ x_3 \end{bmatrix}$ and $\mathsf{b} = \begin{bmatrix} -10 \\ -42 \end{bmatrix}$.
   $$\mathsf{x} = \begin{bmatrix} \frac{8}{3}t + \frac{14}{3} \\ \frac{1}{3}t - \frac{2}{3} \\ t \end{bmatrix} = \begin{bmatrix} \frac{14}{3} \\ -\frac{2}{3} \\ 0 \end{bmatrix} + t\begin{bmatrix} \frac{8}{3} \\ \frac{1}{3} \\ 1 \end{bmatrix}.$$
9. (a) $\left[\begin{array}{cccc|c} ① & -2 & 3 & -1 & 5 \end{array}\right]$. The free variables are x_2, x_3 and x_4.
   (b) $\begin{bmatrix} x_1 \\ x_2 \\ x_3 \\ x_4 \end{bmatrix} = \begin{bmatrix} 5 + 2t - 3s + r \\ t \\ s \\ r \end{bmatrix} = \begin{bmatrix} 5 \\ 0 \\ 0 \\ 0 \end{bmatrix} + t\begin{bmatrix} 2 \\ 1 \\ 0 \\ 0 \end{bmatrix} + s\begin{bmatrix} -3 \\ 0 \\ 1 \\ 0 \end{bmatrix} + r\begin{bmatrix} 1 \\ 0 \\ 0 \\ 1 \end{bmatrix}$.
12. (a) $\mathsf{x} = \begin{bmatrix} 1 \\ \frac{7}{3} - t \\ t \\ \frac{1}{3} \end{bmatrix} = \begin{bmatrix} 1 \\ \frac{7}{3} \\ 0 \\ \frac{1}{3} \end{bmatrix} + t\begin{bmatrix} 0 \\ -1 \\ 1 \\ 0 \end{bmatrix}$. There are infinitely many solutions.
13. (a) The solution is $\begin{bmatrix} -1 \\ 2 \\ 0 \end{bmatrix}$.

(c) The solution is $\begin{bmatrix} x \\ y \\ z \end{bmatrix} = \begin{bmatrix} 4+t-2s \\ t \\ s \end{bmatrix} = \begin{bmatrix} 4 \\ 0 \\ 0 \end{bmatrix} + t\begin{bmatrix} 1 \\ 1 \\ 0 \end{bmatrix} + s\begin{bmatrix} -2 \\ 0 \\ 1 \end{bmatrix}$.

(e) There is no solution.

(i) The solution is $\begin{bmatrix} 4-\frac{15}{2}t \\ 1-5t \\ t \\ -1 \end{bmatrix} = \begin{bmatrix} 4 \\ 1 \\ 0 \\ -1 \end{bmatrix} + t\begin{bmatrix} -\frac{15}{2} \\ -5 \\ 1 \\ 0 \end{bmatrix}$.

14. $\begin{bmatrix} 1 \\ 6 \\ -4 \end{bmatrix} = 1\begin{bmatrix} 2 \\ 4 \\ 6 \end{bmatrix} + 1\begin{bmatrix} 3 \\ 7 \\ -1 \end{bmatrix} - \begin{bmatrix} 4 \\ 5 \\ 9 \end{bmatrix}$.

15. (a) $\begin{bmatrix} 8 \\ -11 \\ -3 \end{bmatrix} = 3\begin{bmatrix} 2 \\ -1 \\ 5 \end{bmatrix} - 2\begin{bmatrix} -1 \\ 4 \\ 9 \end{bmatrix}$.

16. (a) $\mathsf{v} = 2\mathsf{v}_1 + \mathsf{v}_2 + 0\mathsf{v}_3$ is a linear combination of $\mathsf{v}_1, \mathsf{v}_2, \mathsf{v}_3$.

17. The answer is "no:" the given vectors span the plane with equation $x - z = 0$.

19. (a) $\mathsf{x} = \begin{bmatrix} 3b_1 - 5b_2 \\ -b_1 + 2b_2 \end{bmatrix}$.

(b) The results of part (a) show that $A\begin{bmatrix} 3b_1 - 5b_2 \\ 2b_2 - b_1 \end{bmatrix} = \mathsf{b}$. So $\begin{bmatrix} b_1 \\ b_2 \end{bmatrix} = (3b_1 - 5b_2)\begin{bmatrix} 2 \\ 1 \end{bmatrix} + (2b_2 - b_1)\begin{bmatrix} 5 \\ 3 \end{bmatrix}$.

21. (a) The normals to the planes are not parallel.

(b) The line of intersection has an equation $\begin{bmatrix} x \\ y \\ z \end{bmatrix} = \begin{bmatrix} -2t+5 \\ 4t-8 \\ t \end{bmatrix} = \begin{bmatrix} 5 \\ -8 \\ 0 \end{bmatrix} + t\begin{bmatrix} -2 \\ 4 \\ 1 \end{bmatrix}$.

23.
$$\begin{aligned} 10a + 7b + c &= -149 \\ -6a - b + c &= -37 \\ -4a - 7b + c &= -65. \end{aligned}$$

29. (i) No solution if $k \neq -1$. (ii) Under no conditions is there a unique solution.

(iii) Infinitely many solutions if $k = -1$.

31. (a) $b + 3a \neq 0$.

(b) Under no circumstances does the system have a unique solution.

(c) $b + 3a = 0$.

35. (a) $2a + b + c = 0$, (b) No.

39. (b) The points are not coplanar.

Exercises 2.5

1. (a) $I = \frac{E}{R} = \frac{4}{8} = \frac{1}{2}$ amps. (e) $I_1 = \frac{11}{9}, I_2 = \frac{5}{9}, I_3 = \frac{2}{3}$.

4. (a) The required single resistance is $\dfrac{V}{I} = \dfrac{V}{\frac{V}{R_1} + \frac{V}{R_2}} = \dfrac{1}{\frac{1}{R_1} + \frac{1}{R_2}} = \dfrac{R_1 R_2}{R_1 + R_2}$.

Section 2.6—True/False

1. True.
2. False. Column four is a nonpivot column, so there is a free variable, so the system has infinitely many solutions.
3. True. We are given that $A\mathsf{u} = \mathsf{b}$ and $A\mathsf{v} = \mathsf{b}$, so $A(\mathsf{u} - \mathsf{v}) = A\mathsf{u} - A\mathsf{v} = \mathsf{b} - \mathsf{b} = \mathsf{0}$.
4. True. Since $A\mathsf{x}$ is a linear combination of the columns of A, this is the definition of linear dependence.
5. True. See 2.6.4.

Exercises 2.6

1. (a) The solution is $\begin{bmatrix} 0 \\ 0 \\ 0 \end{bmatrix}$, the zero vector.

3. (a) $2\begin{bmatrix} 1 \\ 0 \\ 1 \end{bmatrix} + \begin{bmatrix} -1 \\ 1 \\ 0 \end{bmatrix} - \begin{bmatrix} 1 \\ 1 \\ 2 \end{bmatrix} = \begin{bmatrix} 0 \\ 0 \\ 0 \end{bmatrix}$.

   (b) The result of (a) shows that $A\mathsf{x} = \mathsf{0}$, where $\mathsf{x} = \begin{bmatrix} -2 \\ -1 \\ 1 \end{bmatrix}$. If A were invertible, $A^{-1}A\mathsf{x} = A^{-1}\mathsf{0}$; that is, $\mathsf{x} = \mathsf{0}$, which is not true. Thus A is not invertible.

4. (i) $c - b + a \neq 0$.

   (ii) Not possible. $x = y = z = 0$ is a solution.

   (iii) $c - b + a = 0$.

6. (a) The solution is $\mathsf{x} = \begin{bmatrix} -\frac{13}{3} - 5t \\ t \\ \frac{1}{3} \end{bmatrix} = \begin{bmatrix} -\frac{13}{3} \\ 0 \\ \frac{1}{3} \end{bmatrix} + t\begin{bmatrix} -5 \\ 1 \\ 0 \end{bmatrix}$, which is $\mathsf{x}_p + \mathsf{x}_h$ with $\mathsf{x}_p = \begin{bmatrix} -\frac{13}{3} \\ 0 \\ \frac{1}{3} \end{bmatrix}$ and $\mathsf{x}_h = t\begin{bmatrix} -5 \\ 1 \\ 0 \end{bmatrix}$.

7. (a) $A = \begin{bmatrix} 1 & -2 & 3 & 1 & 3 & 4 \\ -3 & 6 & -8 & 2 & -11 & -15 \\ 1 & -2 & 2 & -4 & 6 & 9 \\ -2 & 4 & -6 & -2 & -6 & -7 \end{bmatrix}$, $\mathsf{x} = \begin{bmatrix} x_1 \\ x_2 \\ x_3 \\ x_4 \\ x_5 \\ x_6 \end{bmatrix}$ and $\mathsf{b} = \begin{bmatrix} -1 \\ 2 \\ 3 \\ 1 \end{bmatrix}$.

   (b) The solution is $\mathsf{x} = \begin{bmatrix} x_1 \\ x_2 \\ x_3 \\ x_4 \\ x_5 \\ x_6 \end{bmatrix} = \begin{bmatrix} -30 + 2t + 14s \\ t \\ 6 - 5s \\ s \\ 5 \\ -1 \end{bmatrix} = \begin{bmatrix} -30 \\ 0 \\ 6 \\ 0 \\ 5 \\ -1 \end{bmatrix} + t \begin{bmatrix} 2 \\ 1 \\ 0 \\ 0 \\ 0 \\ 0 \end{bmatrix} +$ $s \begin{bmatrix} 14 \\ 0 \\ -5 \\ 1 \\ 0 \\ 0 \end{bmatrix}$, which is $\mathsf{x}_p + \mathsf{x}_h$ with $\mathsf{x}_p = \begin{bmatrix} -30 \\ 0 \\ 6 \\ 0 \\ 5 \\ -1 \end{bmatrix}$ and $\mathsf{x}_h = t \begin{bmatrix} 2 \\ 1 \\ 0 \\ 0 \\ 0 \\ 0 \end{bmatrix} + s \begin{bmatrix} 14 \\ 0 \\ -5 \\ 1 \\ 0 \\ 0 \end{bmatrix}$.

   (c) $\begin{bmatrix} -1 \\ 2 \\ 3 \\ 1 \end{bmatrix} = -30 \begin{bmatrix} 1 \\ -3 \\ 1 \\ -2 \end{bmatrix} + 0 \begin{bmatrix} -2 \\ 6 \\ -2 \\ 4 \end{bmatrix} + 6 \begin{bmatrix} 3 \\ -8 \\ 2 \\ -6 \end{bmatrix} + 0 \begin{bmatrix} 1 \\ 2 \\ -4 \\ -2 \end{bmatrix} + 5 \begin{bmatrix} 3 \\ -11 \\ 6 \\ -6 \end{bmatrix} -$ $\begin{bmatrix} 4 \\ -15 \\ 9 \\ -7 \end{bmatrix}$

9. (a) Linearly dependent: $\mathsf{v}_1 - \mathsf{v}_2 + \mathsf{v}_3 + 0\mathsf{v}_4 = \mathsf{0}$.

Section 2.7—True/False

1. False. If an elementary row operation applied to the identity matrix puts a 2 in the $(2,2)$ position, the operation is "multiply row two by 2". This operation leaves a 0, not a 1, in the $(2,3)$ position.
2. True. E is elementary, corresponding to the elementary row operation which subtracts twice row three from row two. The inverse of E adds twice row three to row two.
3. True. The elementary matrix E is obtained from the 3×3 identity matrix I by an elementary row operation that does not change rows two or three of I.
4. False. For example, $\begin{bmatrix} 0 & 1 \\ 1 & 0 \end{bmatrix}$ is not lower triangular.
5. True. This is Theorem 2.7.12.
6. False. First solve $L\mathsf{y} = \mathsf{b}$ for y, then $U\mathsf{x} = \mathsf{y}$ for x.
7. True: $1 + 3 + 5 + \cdots + (2k - 1) = k^2$.

8. True. The given matrix is a permutation matrix, so its inverse is its transpose.

9. True.

10. True. By definition, the pivots of A are the pivots of U, these being the first nonzero entries in the nonzero rows of A. Since A is invertible, U is invertible, so these entries must be on the diagonal.

Exercises 2.7

1. (a) The second row of EA is the sum of the second row and four times the third row of A; all other rows of EA are the same as those of A.

   (b) elementary matrix.

   (c) $E^{-1} = \begin{bmatrix} 1 & 0 & 0 \\ 0 & 1 & -4 \\ 0 & 0 & 1 \end{bmatrix}$.

2. (a) $\begin{bmatrix} 1 & 0 & 0 \\ 0 & 1 & 0 \\ 0 & -1 & 1 \end{bmatrix}$.

3. (a) $E = \begin{bmatrix} 1 & 0 & 0 \\ 0 & 1 & 0 \\ -6 & 0 & 1 \end{bmatrix}$.

4. EA is A but with its first row multiplied by -2; FA is A, but with row two replaced by the sum of row two and three times row one.

6. (a) $E = \begin{bmatrix} 1 & 0 & 0 \\ 4 & 1 & 0 \\ 0 & 0 & 1 \end{bmatrix}$, $F = \begin{bmatrix} 1 & 0 & 0 \\ 0 & 1 & 0 \\ 0 & -3 & 1 \end{bmatrix}$, $EF = \begin{bmatrix} 1 & 0 & 0 \\ 4 & 1 & 0 \\ 0 & -3 & 1 \end{bmatrix}$, $FE = \begin{bmatrix} 1 & 0 & 0 \\ 4 & 1 & 0 \\ -12 & -3 & 1 \end{bmatrix}$.

   (b) $E^{-1} = \begin{bmatrix} 1 & 0 & 0 \\ -4 & 1 & 0 \\ 0 & 0 & 1 \end{bmatrix}$, $F^{-1} = \begin{bmatrix} 1 & 0 & 0 \\ 0 & 1 & 0 \\ 0 & 3 & 1 \end{bmatrix}$, $(EF)^{-1} = \begin{bmatrix} 1 & 0 & 0 \\ -4 & 1 & 0 \\ -12 & 3 & 1 \end{bmatrix}$,

   $(FE)^{-1} = \begin{bmatrix} 1 & 0 & 0 \\ -4 & 1 & 0 \\ 0 & 3 & 1 \end{bmatrix}$.

9. $L = \begin{bmatrix} 1 & 0 & 0 \\ -\frac{2}{3} & 1 & 0 \\ 0 & \frac{1}{7} & 1 \end{bmatrix}$ and $U = \begin{bmatrix} -3 & 3 & 6 \\ 0 & 7 & 14 \\ 0 & 0 & 2 \end{bmatrix}$.

11. (a) $A = \begin{bmatrix} -3 & 6 \\ 8 & -5 \end{bmatrix} \to \begin{bmatrix} -3 & 6 \\ 0 & 11 \end{bmatrix} = MA$ with $M = \begin{bmatrix} 1 & 0 \\ \frac{8}{3} & 1 \end{bmatrix}$.

12. (a) $L = E_1E_2E_3$ with $E_1 = \begin{bmatrix} 1 & 0 & 0 \\ -2 & 1 & 0 \\ 0 & 0 & 1 \end{bmatrix}$, $E_2 = \begin{bmatrix} 1 & 0 & 0 \\ 0 & 1 & 0 \\ 3 & 0 & 1 \end{bmatrix}$ and $E_3 = \begin{bmatrix} 1 & 0 & 0 \\ 0 & 1 & 0 \\ 0 & 5 & 1 \end{bmatrix}$. $L^{-1} = E_3^{-1}E_2^{-1}E_1^{-1} = \begin{bmatrix} 1 & 0 & 0 \\ 2 & 1 & 0 \\ -13 & -5 & 1 \end{bmatrix}$.

13. (a) $U = E_1E_2$ with $E_1 = \begin{bmatrix} 1 & 0 & 0 \\ 0 & 1 & -5 \\ 0 & 0 & 1 \end{bmatrix}$ and $E_2 = \begin{bmatrix} 1 & 2 & 0 \\ 0 & 1 & 0 \\ 0 & 0 & 1 \end{bmatrix}$.

$$U^{-1} = E_2^{-1}E_1^{-1} = \begin{bmatrix} 1 & -2 & 0 \\ 0 & 1 & 0 \\ 0 & 0 & 1 \end{bmatrix} \begin{bmatrix} 1 & 0 & 0 \\ 0 & 1 & 5 \\ 0 & 0 & 1 \end{bmatrix} = \begin{bmatrix} 1 & -2 & -10 \\ 0 & 1 & 5 \\ 0 & 0 & 1 \end{bmatrix}.$$

14. $A = \begin{bmatrix} -1 & 1 & -2 \\ 2 & 1 & 7 \end{bmatrix} \to \begin{bmatrix} -1 & 1 & -2 \\ 0 & 3 & 3 \end{bmatrix} = EA = U$ with $E = \begin{bmatrix} 1 & 0 \\ 2 & 1 \end{bmatrix}$.

Thus $A = E^{-1}U = LU$ with $L = E^{-1} = \begin{bmatrix} 1 & 0 \\ -2 & 1 \end{bmatrix}$.

16. (a) $A \to \begin{bmatrix} 2 & 1 & 0 \\ 0 & 4 & 2 \\ 0 & 0 & 5 \end{bmatrix} = U$ with $E = \begin{bmatrix} 1 & 0 & 0 \\ 0 & 1 & 0 \\ -3 & 0 & 1 \end{bmatrix}$. (b) $L = \begin{bmatrix} 1 & 0 & 0 \\ 0 & 1 & 0 \\ 3 & 0 & 1 \end{bmatrix}$.

(c) $A\mathsf{v} = -2\begin{bmatrix} 2 \\ 0 \\ 6 \end{bmatrix} + 3\begin{bmatrix} 1 \\ 4 \\ 3 \end{bmatrix} + 5\begin{bmatrix} 0 \\ 2 \\ 5 \end{bmatrix}$.

17. (a) A is elementary, so its inverse is the elementary matrix that "undoes" A: $A^{-1} = \begin{bmatrix} 1 & -3 \\ 0 & 1 \end{bmatrix}$.

(e) A is elementary, so its inverse is the elementary matrix that "undoes" A: $A^{-1} = \begin{bmatrix} 1 & 0 & 0 \\ 0 & 1 & 2 \\ 0 & 0 & 1 \end{bmatrix}$.

18. $x = 10$.

20. (a) $M = \begin{bmatrix} 1 & 0 \\ -a & 1 \end{bmatrix}$.

21. (a) The first row of EA is the same as the first row of A. The second row of EA is (Row 2 of A) − (Row 1 of A); the third row of EA is (Row 3 of A) + 2 (Row 1 of A).

The first row of DA is 4 (Row 1 of A); the second row of DA is the same as the second row of A, and the third row of DA is −(Row 3 of A).

The rows of PA, in order, are the third, first and second rows of A.

(b) $P^{-1} = P^T = \begin{bmatrix} 0 & 1 & 0 \\ 0 & 0 & 1 \\ 1 & 0 & 0 \end{bmatrix}$.

24. Ian is right. This matrix has no LU factorization. Lynn's first step corresponds to multiplication by $E = \begin{bmatrix} 1 & 1 \\ 0 & 1 \end{bmatrix}$, which is not lower triangular. A lower triangular matrix can be obtained only if the row operations used in the transformation from A to U change entries on or below the main diagonal.

28. (a) $A = LU$ with $L = \begin{bmatrix} 1 & 0 \\ -\frac{1}{2} & 1 \end{bmatrix}$ and $U = \begin{bmatrix} 2 & 4 \\ 0 & 8 \end{bmatrix} = U$.

(f) There is no LU factorization.

29. (a) First we solve $L\begin{bmatrix} y_1 \\ y_2 \end{bmatrix} = \begin{bmatrix} -2 \\ 9 \end{bmatrix}$. We get $y_1 = -1$ and $y_2 = 3$. Now we solve $U\begin{bmatrix} x \\ y \end{bmatrix} = \begin{bmatrix} -1 \\ 3 \end{bmatrix}$. We get $x = -\frac{5}{2}$, $y = 3$. Thus $\begin{bmatrix} x \\ y \end{bmatrix} = \begin{bmatrix} -\frac{5}{2} \\ 3 \end{bmatrix}$.

(b) $\begin{bmatrix} -2 \\ 9 \end{bmatrix} = -\frac{5}{2}\begin{bmatrix} 1 \\ 6 \end{bmatrix} + 3\begin{bmatrix} 1 \\ 8 \end{bmatrix}$.

35. (a) $P = \begin{bmatrix} 0 & 1 \\ 1 & 0 \end{bmatrix}$, $L = I$, $U = \begin{bmatrix} 1 & 2 & 1 \\ 0 & 3 & 1 \end{bmatrix}$.

Section 2.8—True/False

1. False.
2. True.
3. True. $\begin{bmatrix} 0 & 0 \\ 0 & 0 \end{bmatrix} = \begin{bmatrix} 1 & 0 \\ 0 & 1 \end{bmatrix}\begin{bmatrix} 0 & 0 \\ 0 & 0 \end{bmatrix}\begin{bmatrix} 1 & 0 \\ 0 & 1 \end{bmatrix}$.
4. False. $\begin{bmatrix} 0 & 0 \\ 0 & 1 \end{bmatrix} \neq \begin{bmatrix} 1 & 0 \\ a & 1 \end{bmatrix}\begin{bmatrix} 0 & 0 \\ 0 & 1 \end{bmatrix}\begin{bmatrix} 1 & b \\ 0 & 1 \end{bmatrix}$ for any a and b, as shown in the text.
5. False. $\begin{bmatrix} 0 & 1 \\ 1 & 0 \end{bmatrix}\begin{bmatrix} 2 & 0 \\ 0 & 1 \end{bmatrix} = \begin{bmatrix} 0 & 1 \\ 2 & 0 \end{bmatrix}$.

Exercises 2.8

1. (a) $L = \begin{bmatrix} 1 & 0 & 0 \\ -2 & 1 & 0 \\ -3 & 3 & 1 \end{bmatrix}$, $D = \begin{bmatrix} -2 & 0 & 0 \\ 0 & 2 & 0 \\ 0 & 0 & -52 \end{bmatrix}$, $U = \begin{bmatrix} 1 & -\frac{1}{2} & 3 \\ 0 & 1 & 10 \\ 0 & 0 & 1 \end{bmatrix}$.

2. (a) $L = \begin{bmatrix} 1 & 0 & 0 \\ -\frac{2}{3} & 1 & 0 \\ 0 & \frac{1}{7} & 1 \end{bmatrix}$, $D = \begin{bmatrix} -3 & 0 & 0 \\ 0 & 7 & 0 \\ 0 & 0 & 2 \end{bmatrix}$, $U = \begin{bmatrix} 1 & -1 & -2 \\ 0 & 1 & 2 \\ 0 & 0 & 1 \end{bmatrix}$.

3. (a) $L = \begin{bmatrix} 1 & 0 \\ -\frac{1}{2} & 1 \end{bmatrix}$, $D = \begin{bmatrix} 2 & 0 \\ 0 & 8 \end{bmatrix}$ and $U = \begin{bmatrix} 1 & 2 \\ 0 & 1 \end{bmatrix}$.

(f) No LU, hence no LDU factorization.

5. (a) A matrix X is symmetric if $X^T = X$. We are given that $A^T = A$ and $B^T = B$. Since $(AB)^T = B^TA^T = BA$ which is not, in general, the same as AB, AB is not symmetric, in general.

7. $(A+B)^T = A^T + B^T = -A - B = -(A+B)$. Since the transpose of $A+B$ is its additive inverse, $A+B$ is skew-symmetric.

9. (a) $D = \begin{bmatrix} 2 & 0 \\ 0 & -\frac{7}{2} \end{bmatrix}$, $U = \begin{bmatrix} 1 & \frac{5}{2} \\ 0 & 1 \end{bmatrix}$, $L = U^T = \begin{bmatrix} 1 & 0 \\ \frac{5}{2} & 1 \end{bmatrix}$.

14. (a) Since L is lower triangular, L^T is upper triangular, so we can express $L^T = DU$ as the product of a diagonal matrix D and an upper triangular matrix U with 1s on its diagonal. Thus $L = (DU)^T = U^T D^T$ with U^T lower triangular with 1s on its diagonal and $D^T = D$ diagonal. With $L = \begin{bmatrix} 2 & 0 & 0 \\ 4 & -3 & 0 \\ 6 & 7 & 4 \end{bmatrix}$, we have $L^T = \begin{bmatrix} 2 & 4 & 6 \\ 0 & -3 & 7 \\ 0 & 0 & 4 \end{bmatrix} = DU_1$ with $D = \begin{bmatrix} 2 & 0 & 0 \\ 0 & -3 & 0 \\ 0 & 0 & 4 \end{bmatrix}$ and $U_1 = \begin{bmatrix} 1 & 2 & 3 \\ 0 & 1 & -\frac{7}{3} \\ 0 & 0 & 1 \end{bmatrix}$. Thus $L = U_1^T D$ with $U_1^T = \begin{bmatrix} 1 & 0 & 0 \\ 2 & 1 & 0 \\ 3 & -\frac{7}{3} & 1 \end{bmatrix}$ and $D = \begin{bmatrix} 2 & 0 & 0 \\ 0 & -3 & 0 \\ 0 & 0 & 4 \end{bmatrix}$.

This problem shows that to factor a lower triangular matrix L as $L = D_1 L_1$, where L_1 has 1s on the diagonal, you get the diagonal matrix D_1 by factoring the diagonal entries of L from its **columns**.

Section 2.9—True/False

1. True.
2. False: the pivots are not all 1.
3. False. Both $\begin{bmatrix} -1 & 0 & 2 \\ 0 & 2 & 14 \end{bmatrix}$ and $\begin{bmatrix} 1 & 0 & -2 \\ 0 & 1 & 7 \end{bmatrix}$ are row echelon forms of $A = \begin{bmatrix} -1 & 0 & 2 \\ 4 & 2 & 6 \end{bmatrix}$.
4. True.
5. False.
6. True. $(AB)^{-1} = B^{-1}A^{-1}$.
7. True: A has to be square and $A[B(AB)^{-1}] = I$, so $A^{-1} = B(AB)^{-1}$.
8. True. Since E and F are elementary, they are invertible, and the product of invertible matrices is invertible.
9. False. $(ABC)^{-1} = C^{-1}B^{-1}A^{-1}$ and there is no reason why $(ABC)(A^{-1}B^{-1}C^{-1})$ should be I.
10. False. This holds only if A and B are square. (See Problem 2.9.2.)

Exercises 2.9

1. (a) Yes. A is square and $AB = I$.
2. (a) $\begin{bmatrix} 1 & 0 & 0 \\ 0 & 1 & 0 \\ 0 & 0 & 1 \\ 0 & 0 & 0 \end{bmatrix}$.

3. (a) The inverse is $\begin{bmatrix} 1 & \frac{1}{3} \\ -2 & -1 \end{bmatrix}$. (d) The inverse is $\begin{bmatrix} -\frac{1}{4} & 0 & \frac{1}{2} \\ \frac{1}{2} & -\frac{1}{4} & -\frac{1}{4} \\ -\frac{1}{4} & \frac{1}{2} & 0 \end{bmatrix}$.

4. (a) $A = \begin{bmatrix} 1 & -2 & 2 \\ 2 & 1 & 1 \\ 1 & 0 & 1 \end{bmatrix}$, $\mathsf{x} = \begin{bmatrix} x_1 \\ x_2 \\ x_3 \end{bmatrix}$, $\mathsf{b} = \begin{bmatrix} 3 \\ 0 \\ -2 \end{bmatrix}$.

   (b) $A^{-1} = \begin{bmatrix} 1 & 2 & -4 \\ -1 & -1 & 3 \\ -1 & -2 & 5 \end{bmatrix}$, so $\mathsf{x} = A^{-1}\mathsf{b} = \begin{bmatrix} 11 \\ -9 \\ -13 \end{bmatrix}$.

   (c) $A\begin{bmatrix} 11 \\ -9 \\ -13 \end{bmatrix} = \begin{bmatrix} 3 \\ 0 \\ -2 \end{bmatrix}$, so $\begin{bmatrix} 3 \\ 0 \\ -2 \end{bmatrix} = 11\begin{bmatrix} 1 \\ 2 \\ 1 \end{bmatrix} - 9\begin{bmatrix} -2 \\ 1 \\ 0 \end{bmatrix} - 13\begin{bmatrix} 2 \\ 1 \\ 1 \end{bmatrix}$.

7. $A = \begin{bmatrix} 1 & 2 \\ 3 & 0 \end{bmatrix}\begin{bmatrix} 5 & 1 \\ -1 & 1 \end{bmatrix}^{-1} = \begin{bmatrix} 1 & 2 \\ 3 & 0 \end{bmatrix}\frac{1}{6}\begin{bmatrix} 1 & -1 \\ 1 & 5 \end{bmatrix} = \frac{1}{6}\begin{bmatrix} 3 & 9 \\ 3 & -3 \end{bmatrix} = \frac{1}{2}\begin{bmatrix} 1 & 3 \\ 1 & -1 \end{bmatrix}$.

10. (a) $A = \begin{bmatrix} 0 & 1 \\ 1 & 0 \end{bmatrix}\begin{bmatrix} 2 & 0 \\ 0 & 1 \end{bmatrix}\begin{bmatrix} 1 & 0 \\ 0 & -1 \end{bmatrix}\begin{bmatrix} 1 & \frac{1}{2} \\ 0 & 1 \end{bmatrix}$.

11. (a) The columns are linearly independent, so the matrix is invertible.

12. (a) You compute XY (or YX). If $XY = I$ (or $YX = I$), then X and Y are inverses.

14. No, it is not. For instance, $\begin{bmatrix} 1 & 0 \\ 0 & 1 \end{bmatrix} = \begin{bmatrix} 1 & a \\ 0 & 1 \end{bmatrix}\begin{bmatrix} 1 & -a \\ 0 & 1 \end{bmatrix}$ for any a.

Section 3.1—True/False

1. False. Strike out the first row and second column of the matrix. The determinant of what is left is 2, not −2.
2. True. The $(1,2)$ cofactor is $(-1)^{1+2}m_{12}$.
3. True. The scalar is $\det A$.
4. False. If $I = \begin{bmatrix} 1 & 0 \\ 0 & 1 \end{bmatrix}$, $-2I = \begin{bmatrix} -2 & 0 \\ 0 & -2 \end{bmatrix}$, so $\det(-2I) = 4$.
5. False. A Laplace expansion along the second row gives $+3$.
6. False. Only a square matrix has a determinant.
7. False. Let $A = \begin{bmatrix} 2 & 1 \\ 4 & 3 \end{bmatrix}$ and $B = -A$. Then $\det A = \det B = 2$, but $\det(A+B) = 0$.
8. False. For example, consider $A = \begin{bmatrix} 1 & 0 & 0 \\ 0 & 0 & 0 \\ 0 & 0 & 0 \end{bmatrix}$.
9. True. Since $\det A \neq 0$, A is invertible, the solution to $A\mathsf{x} = \mathsf{b}$ is $\mathsf{x} = A^{-1}\mathsf{b}$.
10. True. A square matrix A is invertible if and only if $\det A \neq 0$.

Exercises 3.1

1. (a) i. $M = \begin{bmatrix} 9 & -7 \\ -4 & 2 \end{bmatrix}$, $C = \begin{bmatrix} 9 & 7 \\ 4 & 2 \end{bmatrix}$, $AC^T = -10I = C^T A$.

ii. $AC^T = (\det A)I$, so $\det A = -10$. iii. $A^{-1} = -\frac{1}{10}\begin{bmatrix} 9 & 4 \\ 7 & 2 \end{bmatrix}$.

(c) i. $M = \begin{bmatrix} -26 & -12 & 4 \\ -13 & -6 & 2 \\ 13 & 6 & -2 \end{bmatrix}$, $C = \begin{bmatrix} -26 & 12 & 4 \\ 13 & -6 & -2 \\ 13 & -6 & -2 \end{bmatrix}$,

$AC^T = \begin{bmatrix} 0 & 0 & 0 \\ 0 & 0 & 0 \\ 0 & 0 & 0 \end{bmatrix} = C^T A$.

ii. $AC^T = (\det A)I$, so $\det A = 0$.

iii. A is not invertible since $\det A = 0$. See Example 3.1.16.

3. (a) $\det A = 2\begin{vmatrix} -1 & 2 \\ 1 & 1 \end{vmatrix} + \begin{vmatrix} 1 & 2 \\ 3 & 1 \end{vmatrix} + 3\begin{vmatrix} 1 & -1 \\ 3 & 1 \end{vmatrix} = 2(-3) + (-5) + 3(4) = 1$.

5. (a) $M = \begin{bmatrix} -2 & 41 & 7 \\ 3 & 1 & 2 \\ 3 & 26 & 2 \end{bmatrix}$, (b) $C = \begin{bmatrix} -2 & -41 & 7 \\ -3 & 1 & -2 \\ 3 & -26 & 2 \end{bmatrix}$.

(c) $\begin{bmatrix} 4 \\ 1 \\ 7 \end{bmatrix} \cdot \begin{bmatrix} 3 \\ -26 \\ 2 \end{bmatrix} = 12 - 26 + 14 = 0$. This is the dot product of the second row of A and the third column of C^T. Since $AC^T = \det(A)I$, entries not on the diagonal are 0.

(d) $\begin{bmatrix} 2 \\ 4 \\ -3 \end{bmatrix} \cdot \begin{bmatrix} -41 \\ 1 \\ -26 \end{bmatrix} = -82 + 4 + 78 = 0$. This is the dot product of the second row of C^T and the first column of A. Since $C^T A = \det(A)I$, entries not on the diagonal are 0.

(e) $\begin{bmatrix} 0 \\ 1 \\ 1 \end{bmatrix} \cdot \begin{bmatrix} -41 \\ 1 \\ -26 \end{bmatrix} = 1 - 26 = -25$. Since $C^T A = (\det A)I$, each diagonal entry is $\det A$, in particular, the $(2,2)$ entry. So $\det A = -25$.

7. (a) $c_{13} = 3$, $c_{21} = 2$, $c_{32} = 1$, (b) $\det A = 2$.

(c) A is invertible since $\det A \neq 0$. (d) $A^{-1} = \dfrac{1}{2}\begin{bmatrix} -1 & 2 & -1 \\ -1 & 0 & 1 \\ 3 & -2 & 1 \end{bmatrix}$.

9. (a) $x = -1$ or $x = -2$.

10. Let $A = \begin{bmatrix} -1 & 2 & x \\ 0 & 3 & y \\ 2 & -2 & z \end{bmatrix}$.

(a) $c_{13} = -6$, $c_{23} == 2$, $c_{33} = -3$.

(b) Since $C = \begin{bmatrix} 19 & 10 & -6 \\ -14 & -11 & 2 \\ -2 & 5 & -3 \end{bmatrix}$, $AC^T = \begin{bmatrix} \det A & 0 & 0 \\ 0 & \det A & 0 \\ 0 & 0 & \det A \end{bmatrix}$

$= \begin{bmatrix} 1 - 6x & -8 + 2x & 12 - 3x \\ 30 - 6y & -33 + 2y & 15 - 3y \\ 18 - 6z & -6 + 2z & -14 - 3z \end{bmatrix}$. So $x = 4$, $y = 5$, $\det A = -23$.

(c) $A^{-1} = -\frac{1}{23}\begin{bmatrix} 19 & -14 & -2 \\ 10 & -11 & 5 \\ -6 & 2 & -3 \end{bmatrix}$.

14. $A = \begin{bmatrix} 4 & -2 \\ -7 & -1 \end{bmatrix}$.

16. (a) $A = \begin{bmatrix} 1 & 2 \\ -1 & 3 \end{bmatrix}$.

Section 3.2—True/False

1. True. The determinant of a triangular matrix is the product of its diagonal entries. If no diagonal entry is 0, the determinant is not 0, so the matrix is invertible.
2. False. The matrix does not have a determinant because it is not square.
3. False in general. If A is $n \times n$, $\det(-A) = (-1)^n \det A$. This is $-\det A$ if and only if n is odd.
4. False: $\det \frac{1}{5}\begin{bmatrix} 4 & 3 \\ 1 & 2 \end{bmatrix} == (\frac{1}{5})^2 \det \begin{bmatrix} 4 & 3 \\ 1 & 2 \end{bmatrix} = \frac{1}{25}5 = \frac{1}{5}$.
5. True. "Singular" means "not invertible."
6. True. $AC^T = (\det A)I = 5I$, so $\det(AC^T) = (\det A)(\det C^T) = 125$ and $\det C^T = \dfrac{125}{\det A} = 25$.
7. False. Interchanging two rows, for example, changes the sign of the determinant.
8. False. The matrix $A = \begin{bmatrix} 1 & 2 \\ 1 & 2 \end{bmatrix}$ has determinant 0.
9. False. The determinant is **not** a linear function, although it is linear **in each row**.
10. True. $\det AB = (\det A)(\det B) = (\det B)(\det A) = \det BA$.

Exercises 3.2

1. For a fixed vector u in R^3,

   i. $\mathsf{u} \times (\mathsf{x}_1 + \mathsf{x}_2) = (\mathsf{u} \times \mathsf{x}_1) + (\mathsf{u} \times \mathsf{x}_2)$, and

   ii. $\mathsf{u} \times (c\mathsf{x}) = c(\mathsf{u} \times \mathsf{x})$

   for all vectors $\mathsf{x}, \mathsf{x}_1, \mathsf{x}_2$ in R^3 and all scalars c.

3. (a) 0 (two equal rows).
5. $\det A = -\frac{4}{243}$, $\det B = \frac{1}{2}$, $\det AB = -\frac{2}{243}$.
7. $\det A = 60$, $\det A^{-1} = \frac{1}{60}$, $\det A^2 = 3600$.

11. $A^{-1} = \frac{1}{\det A} \operatorname{adj} A = -\frac{1}{2} \operatorname{adj} A$, so $\operatorname{adj} A = 2A^{-1}$. It follows that $\det(15A^{-1} - 6 \operatorname{adj} A) = \frac{81}{2}$.

16. (a) $-1\begin{vmatrix} 1 & 1 & 3 \\ 1 & 1 & 2 \\ 3 & -1 & 2 \end{vmatrix} - 2\begin{vmatrix} -1 & 1 & 0 \\ 1 & 1 & 2 \\ 3 & -1 & 2 \end{vmatrix} - \begin{vmatrix} -1 & 1 & 0 \\ 1 & 1 & 3 \\ 1 & 1 & 2 \end{vmatrix} = -1(-4) - 2(0) - (2) = 2.$

(b) $\det A = \begin{vmatrix} -1 & -1 & 1 & 0 \\ 2 & 1 & 1 & 3 \\ 0 & 1 & 1 & 2 \\ 1 & 3 & -1 & 2 \end{vmatrix} = \begin{vmatrix} -1 & -1 & 1 & 0 \\ 0 & -1 & 3 & 3 \\ 0 & 1 & 1 & 2 \\ 0 & 2 & 0 & 2 \end{vmatrix}$

$= \begin{vmatrix} -1 & -1 & 1 & 0 \\ 0 & -1 & 3 & 3 \\ 0 & 0 & 4 & 5 \\ 0 & 0 & 6 & 8 \end{vmatrix} = 4\begin{vmatrix} -1 & -1 & 1 & 0 \\ 0 & -1 & 3 & 3 \\ 0 & 0 & 1 & \frac{5}{4} \\ 0 & 0 & 6 & 8 \end{vmatrix}$

$= 4\begin{vmatrix} -1 & -1 & 1 & 0 \\ 0 & -1 & 3 & 3 \\ 0 & 0 & 1 & \frac{5}{4} \\ 0 & 0 & 0 & \frac{1}{2} \end{vmatrix} = 4(\tfrac{1}{2}) = 2.$

18. (a) Let $A = \begin{bmatrix} 1 & 3 \\ -2 & 4 \end{bmatrix}$ be the matrix whose columns are the given vectors. Since $\det A = 10 \neq 0$, the vectors are linearly independent.

20. (a) The determinant is 7. (d) The determinant is 60.

23. $\det A = \pm 35$.

24. (a) $a = 0$ or $a = -1$.

27. (a) -18. (c) -62. (e) 0.

28. The determinant is 6.

30. The determinant is -12.

36. $\det \begin{bmatrix} 1 & x & x^2 & x^3 \\ x & x^2 & x^3 & 1 \\ x^2 & x^3 & 1 & x \\ x^3 & 1 & x & x^2 \end{bmatrix} = \det \begin{bmatrix} 1 & x & x^2 & x^3 \\ 0 & 0 & 0 & 1-x^4 \\ 0 & 0 & 1-x^4 & x-x^5 \\ 0 & 1-x^4 & x-x^5 & x^2-x^6 \end{bmatrix}$

$= -\det \begin{bmatrix} 1 & x & x^2 & x^3 \\ 0 & 1-x^4 & x-x^5 & x^2-x^6 \\ 0 & 0 & 1-x^4 & x-x^5 \\ 0 & 0 & 0 & 1-x^4 \end{bmatrix} = -(1-x^4)^3 = (x^4-1)^3.$

42. Suppose $A = LDU$. Each of L and U is triangular with 1s on the diagonal, so $\det L = \det U = 1$, the product of the diagonal entries in each case. Thus $\det A = \det L \det D \det U = \det D$. Using just the third elementary row operation to reduce

$A = \begin{bmatrix} 2 & -1 & 4 & 1 \\ 1 & 1 & -10 & -2 \\ 4 & 0 & -7 & 6 \\ 6 & -3 & 0 & 1 \end{bmatrix}$ to an upper triangular matrix, we have

$$\begin{bmatrix} 2 & -1 & 4 & 1 \\ 1 & 1 & -10 & -2 \\ 4 & 0 & -7 & 6 \\ 6 & -3 & 0 & 1 \end{bmatrix} \to \begin{bmatrix} 2 & -1 & 4 & 1 \\ 0 & \frac{3}{2} & -12 & -\frac{5}{2} \\ 0 & 2 & -15 & 4 \\ 0 & 0 & -12 & -2 \end{bmatrix} \to \begin{bmatrix} 2 & -1 & 4 & 1 \\ 0 & \frac{3}{2} & -12 & -\frac{5}{2} \\ 0 & 0 & 1 & \frac{22}{3} \\ 0 & 0 & -12 & -2 \end{bmatrix}$$

$$\to \begin{bmatrix} 2 & -1 & 4 & 1 \\ 0 & \frac{3}{2} & -12 & -\frac{5}{2} \\ 0 & 0 & 1 & \frac{22}{3} \\ 0 & 0 & 0 & 86 \end{bmatrix} = U' = DU \text{ with } D = \begin{bmatrix} 2 & 0 & 0 & 0 \\ 0 & \frac{3}{2} & 0 & 0 \\ 0 & 0 & 1 & 0 \\ 0 & 0 & 0 & 86 \end{bmatrix}, \text{ so}$$

$\det A = \det D = 2(\frac{3}{2})(1)(86) = 258$.

44. We must show that $\det A = 0$. Let the columns of A be $a_1\mathsf{u} + b_1\mathsf{v}$, $a_2\mathsf{u} + b_2\mathsf{v}$, $a_3\mathsf{u} + b_3\mathsf{v}$ for scalars $a_1, b_1, a_2, b_2, a_3, b_3$. Write $\det(\mathsf{w}_1, \mathsf{w}_2, \mathsf{w}_3)$ for the determinant of a matrix whose columns are the vectors $\mathsf{w}_1, \mathsf{w}_2, \mathsf{w}_3$. Then

$$\begin{aligned} \det A &= \det(a_1\mathsf{u} + b_1\mathsf{v}, a_2\mathsf{u} + b_2\mathsf{v}, a_3\mathsf{u} + b_3\mathsf{v}) \\ &= \det(a_1\mathsf{u}, a_2\mathsf{u} + b_2\mathsf{v}, a_3\mathsf{u} + b_3\mathsf{v}) \\ &\quad + \det(b_1\mathsf{v}, a_2\mathsf{u} + b_2\mathsf{v}, a_3\mathsf{u} + b_3\mathsf{v}) \end{aligned}$$

by linearity of the determinant in the first column. Using linearity of the determinant in columns two and three, eventually we obtain

$$\begin{aligned} \det A = {} & \det(a_1\mathsf{u}, a_2\mathsf{u}, a_3\mathsf{u}) + \det(a_1\mathsf{u}, a_2\mathsf{u}, b_3\mathsf{v}) + \det(a_1\mathsf{u}, b_2\mathsf{v}, a_3\mathsf{u}) \\ & + \det(a_1\mathsf{u}, b_2\mathsf{v}, b_3\mathsf{v}) + \det(b_1\mathsf{v}, a_2\mathsf{u}, a_3\mathsf{u}) + \det(b_1\mathsf{v}, a_2\mathsf{u}, b_3\mathsf{v}) \\ & + \det(b_1\mathsf{v}, b_2\mathsf{v}, a_3\mathsf{u}) + \det(b_1\mathsf{v}, b_2\mathsf{v}, b_3\mathsf{v}) \end{aligned}$$

and each of these eight determinants is 0 because in each case, either one column is 0 or one column is a multiple of another. In the second determinant—$\det(a_1\mathsf{u}, a_2\mathsf{u}, b_3\mathsf{v})$—for example, if $a_1 \neq 0$, the second column is $\dfrac{a_2}{a_1}$ times the first.

46. We have $\frac{1}{2}(I - A)A = I$, so $(I - A)A = 2I$. Taking the determinant of each side gives $\det(I - A)\det A = 2^n$. If the product of two numbers is a power of 2, each number itself is a power of 2, so the result follows.

Exercises 3.3

1. The (i,i) entry of A^2 is the number of walks of length two from v_i to v_i. Such a walk is of the form $v_i v_j v_i$ and this exists if and only if $v_i v_j$ is an edge.
2. The $(1,2)$ entry of A^4 is the number of walks of length four from v_1 to v_2. This number is two since the only such walks are $v_1 v_5 v_3 v_1 v_2$ and $v_1 v_3 v_5 v_1 v_2$.
3. (a) The (i,j) entry of A^2 is the dot product of row i of A with column j of A. This is the number of positions k for which a_{ik} and a_{kj} are both 1. This is the number of vertices v_k for which $v_i v_k$ and $v_k v_j$ are edges and this is the number of walks of length two from v_i to v_j since any such walk is of the form $v_i v_k v_j$.

5. $\mathcal{T}$ is connected, so you can get from u to v within the tree. Follow this route in the larger graph and return to u along the new edge. This gives a cycle.

6. (a) 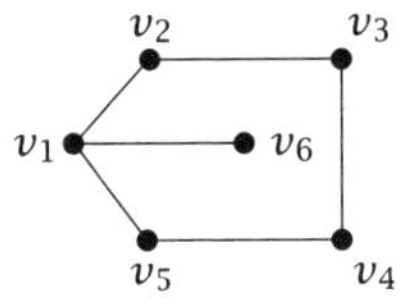

7. (a) The adjacency matrix is $A = \begin{bmatrix} 0 & 1 & 0 & 0 \\ 1 & 0 & 1 & 1 \\ 0 & 1 & 0 & 1 \\ 0 & 1 & 1 & 0 \end{bmatrix}$. The matrix defined in Kirchhoff's Theorem is $M = \begin{bmatrix} 1 & -1 & 0 & 0 \\ -1 & 3 & -1 & -1 \\ 0 & -1 & 2 & -1 \\ 0 & -1 & -1 & 2 \end{bmatrix}$. There are three spanning trees, 3 being the value of any cofactor of M.

8. (a) $\mathcal{K}_4$: 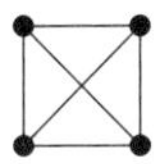

10. (a) $A_1 = \begin{bmatrix} 0 & 1 & 0 & 1 & 1 \\ 1 & 0 & 1 & 0 & 1 \\ 0 & 1 & 0 & 1 & 1 \\ 1 & 0 & 1 & 0 & 1 \\ 1 & 1 & 1 & 0 & 0 \end{bmatrix}$ and $A_2 = \begin{bmatrix} 0 & 1 & 1 & 1 & 0 \\ 1 & 0 & 1 & 0 & 1 \\ 1 & 1 & 0 & 1 & 0 \\ 1 & 0 & 1 & 0 & 1 \\ 1 & 1 & 0 & 1 & 0 \end{bmatrix}$. The graphs are isomorphic as we see by relabelling the vertices of $\mathcal{G}_1$.

Let $P = \begin{bmatrix} 0 & 1 & 0 & 0 & 0 \\ 0 & 0 & 1 & 0 & 0 \\ 0 & 0 & 0 & 1 & 0 \\ 0 & 0 & 0 & 0 & 1 \\ 1 & 0 & 0 & 0 & 0 \end{bmatrix}$ be the permutation matrix that has the rows of the identity in the order 23451. Then $A_2 = P^T A_1 P$.

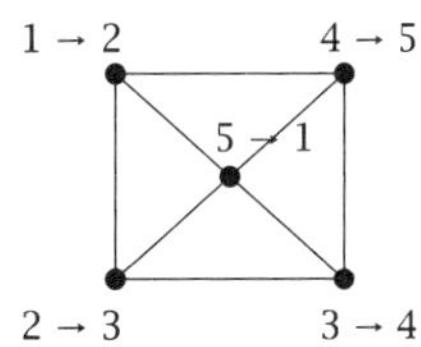

(e) $A_1 = \begin{bmatrix} 0 & 1 & 0 & 1 & 0 & 1 \\ 1 & 0 & 1 & 0 & 1 & 0 \\ 0 & 1 & 0 & 1 & 0 & 1 \\ 1 & 0 & 1 & 0 & 1 & 0 \\ 0 & 1 & 0 & 1 & 0 & 1 \\ 1 & 0 & 1 & 0 & 1 & 0 \end{bmatrix}$ and $A_2 = \begin{bmatrix} 0 & 1 & 0 & 0 & 1 & 1 \\ 1 & 0 & 1 & 1 & 0 & 0 \\ 0 & 1 & 0 & 1 & 0 & 1 \\ 0 & 1 & 1 & 0 & 1 & 0 \\ 1 & 0 & 0 & 1 & 0 & 1 \\ 1 & 0 & 1 & 0 & 0 & 1 \end{bmatrix}$, respectively. No such permutation matrix P exists because the graphs are not isomorphic: graph $\mathcal{G}_2$ has several *triangles*—156, 234, for instance—but graph $\mathcal{G}_1$ contains no such triangles.

11. (a) The given matrices are the adjacency matrices of the graphs $\mathcal{G}_1, \mathcal{G}_2$ shown. The graphs are not isomorphic ($\mathcal{G}_2$ has more edges than $\mathcal{G}_1$), so no such permutation matrix exists.

$\mathcal{G}_1$: 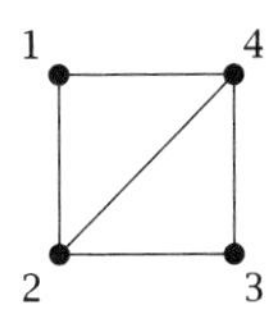

$\mathcal{G}_2$; 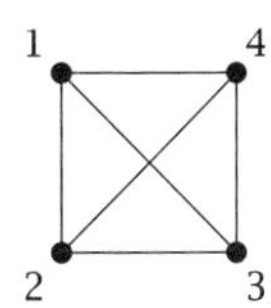

Section 3.4—True/False

1. True. See discussion of the PageRank algorithm.
2. Not quite true. An eigenvector is, by definition, not zero, so the eigenvectors of A are the **nonzero** solutions to $(A - \lambda I)\mathsf{x} = 0$.
3. True.
4. True.
5. False. See our "Final Thought," on p. 288.
6. True. If 0 is an eigenvalue of A, then there exists a nonzero vector x with $A\mathsf{x} = 0\mathsf{x} = 0$. This cannot happen if A is invertible.
7. False. The eigenvalues of the triangular matrix $A = \begin{bmatrix} 0 & 1 \\ 0 & 0 \end{bmatrix}$ are all 0, but $A \neq 0$.
8. False. The only eigenvalue of $\begin{bmatrix} 1 & 1 \\ 0 & 1 \end{bmatrix}$ is 1.
9. True. Given $A\mathsf{x} = 3\mathsf{x}$, we have $A^2\mathsf{x} = A(A\mathsf{x}) = A(3\mathsf{x}) = 3A\mathsf{x} = 9\mathsf{x}$.
10. False.

Exercises 3.4

1. Yes, because $A\mathsf{x} = 4\mathsf{x}$.

3. (a) $A - \lambda I = \begin{bmatrix} 1-\lambda & 2 \\ 2 & 4-\lambda \end{bmatrix}$ has determinant $(1-\lambda)(4-\lambda) - 4 = \lambda^2 - 5\lambda = \lambda(\lambda - 5)$. Thus $\lambda = 0$ and $\lambda = 5$ are eigenvalues, while -1, 1 and 3 are not.

   (d) When A is larger than 2×2, it is not so easy to find $\det(A - \lambda I)$ and its roots, so here we answer the question by computing $\det(A - \lambda I)$ for each given value of λ and determining whether or not this matrix has 0 determinant. We have matrix $A - \lambda I = \begin{bmatrix} 5-\lambda & -7 & 7 \\ 4 & -3-\lambda & 4 \\ 4 & -1 & 2-\lambda \end{bmatrix}$ and discover that $\lambda = 1$ is an eigenvalue, $\lambda = 2$ is not an eigenvalue, $\lambda = 4$ is not an eigenvalue, $\lambda = 5$ is an eigenvalue, $\lambda = 6$ is not an eigenvalue.

5. (a) $A\begin{bmatrix} 2 \\ 3 \end{bmatrix} = \begin{bmatrix} 8 \\ 12 \end{bmatrix} = 4\begin{bmatrix} 2 \\ 3 \end{bmatrix}$, so $\begin{bmatrix} 2 \\ 3 \end{bmatrix}$ is an eigenvector corresponding to $\lambda = 4$.

   (b) The characteristic polynomial of A is $\lambda^2 - 3\lambda - 4$.

   (c) The eigenvalues of A are 4 and -1.

7. (a) The characteristic polynomial of A is $\lambda^2 - 6\lambda - 27 = (\lambda - 9)(\lambda + 3)$, so the eigenvalues are $\lambda = 9$ and $\lambda = -3$. The eigenspace corresponding to $\lambda = 9$ is the set of vectors of the form of $\mathsf{x} = \begin{bmatrix} x_1 \\ x_2 \end{bmatrix} = \begin{bmatrix} 2t \\ t \end{bmatrix} =$

$t\begin{bmatrix}2\\1\end{bmatrix}$. The eigenspace corresponding to $\lambda = -3$ is the set of vectors of the form of $\mathsf{x} = \begin{bmatrix}x_1\\x_2\end{bmatrix} = \begin{bmatrix}-t\\t\end{bmatrix} = t\begin{bmatrix}-1\\1\end{bmatrix}$.

(d) The characteristic polynomial of $A = \begin{bmatrix}1 & -2 & 3\\2 & 6 & -6\\1 & 2 & -1\end{bmatrix}$ is $-\lambda^3 + 6\lambda^2 - 12\lambda + 8 = -(\lambda - 2)^3$. The only eigenvalue of A is $\lambda = 2$. The corresponding eigenspace consists of vectors of the form $\begin{bmatrix}-2s+3t\\s\\t\end{bmatrix} = s\begin{bmatrix}-2\\1\\0\end{bmatrix} + t\begin{bmatrix}3\\0\\1\end{bmatrix}$.

9. (a) Since multiplication by A is reflection in a line, any vector on this line will be fixed by A, so we seek solutions to $A\mathsf{x} = \mathsf{x}$ and discover that our matrix reflects vectors in the y-axis. Thus any nonzero vector on the y-axis is an eigenvector corresponding to $\lambda = 1$ and any nonzero vector on the x-axis is an eigenvector corresponding to $\lambda = -1$.

10. (a) Given $A\mathsf{x} = \lambda\mathsf{x}$, we have $A(a\mathsf{x}) = a(A\mathsf{x}) = a(\lambda\mathsf{x}) = \lambda(a\mathsf{x})$, so $a\mathsf{x}$ is also in U.

(b) Given $A\mathsf{x} = \lambda\mathsf{x}$ and $A\mathsf{y} = \lambda\mathsf{y}$, we have $A(\mathsf{x}+\mathsf{y}) = A\mathsf{x}+A\mathsf{y} = \lambda\mathsf{x}+\lambda\mathsf{y} = \lambda(\mathsf{x}+\mathsf{y})$, so $\mathsf{x}+\mathsf{y}$ is also in U.

13. Let $A = \begin{bmatrix}\mathsf{a}_1 & \mathsf{a}_2\\ \downarrow & \downarrow\end{bmatrix}$. We are given that $A\begin{bmatrix}2\\2\end{bmatrix} = 6\begin{bmatrix}2\\2\end{bmatrix} = \begin{bmatrix}12\\12\end{bmatrix}$. By 2.1.33, $\begin{bmatrix}12\\12\end{bmatrix} = 2\mathsf{a}_1 + 2\mathsf{a}_2$, so $\mathsf{a}_1 + \mathsf{a}_2 = \begin{bmatrix}6\\6\end{bmatrix}$.

15. Let λ be an eigenvalue of A and let $\mathsf{x} \neq \mathsf{0}$ be a corresponding eigenvector. Then $A\mathsf{x} = \lambda\mathsf{x}$. Multiplying on the left by P gives $PA\mathsf{x} = \lambda P\mathsf{x}$. Since $PA = BP$, we get $BP\mathsf{x} = \lambda P\mathsf{x}$, that is, $B(P\mathsf{x}) = \lambda(P\mathsf{x})$. Since P is invertible, $P\mathsf{x} \neq \mathsf{0}$. So λ is an eigenvalue of B with corresponding eigenvector $P\mathsf{x}$.

20. (a) Yes. Since $A\mathsf{v} = \lambda\mathsf{v}$, $(5A)\mathsf{v} = 5\lambda\mathsf{v} = (5\lambda)\mathsf{v}$, so v is an eigenvector of $5A$ with eigenvalue 5λ.

Section 3.5—True/False

1. True. $A = 4I$, so $B = P^{-1}AP$ means $B = 4P^{-1}I = 4I = A$.
2. True. The characteristic polynomial is the determinant of $\begin{bmatrix}1-\lambda & 2 & 3\\0 & 4-\lambda & 5\\0 & 0 & 6-\lambda\end{bmatrix}$. This matrix is triangular, so its determinant is the product of its diagonal entries.

3. True. The eigenvalues of the diagonal matrix $A^{10} = \begin{bmatrix} 1 & 0 \\ 0 & 2^{10} \end{bmatrix}$ are its diagonal entries.
4. False. The matrix is triangular: its eigenvalues are its diagonal entries, 2 and -14.
5. False. Similar matrices have the same trace (sum of diagonal entries), but the traces here are 1 and 3, respectively.
6. True. Any diagonal matrix is diagonalizable: Take : $P = I$.
7. The identity matrix $\begin{bmatrix} 1 & 0 \\ 0 & 1 \end{bmatrix}$ is diagonalizable with 1 as its only eigenvalue.
8. True. This is part of Theorem 3.5.14.
9. False. The 3×3 matrix in Example 3.5.12 is diagonalizable, but it has just two different eigenvalues.
10. False. The symmetric matrix $A = \begin{bmatrix} 0 & 1 \\ 1 & 0 \end{bmatrix}$ has eigenvalues ± 1.

Exercises 3.5

1. (a) $A = I$. (b) No, by part (a). The given matrix is not I.
   (c) A is not diagonalizable.
3. (a) $\det B = -2$; the characteristic polynomial is $\lambda^2 - 5\lambda - 2$, the characteristic polynomial of A, the easier matrix.
5. The 2×2 matrix A has two distinct eigenvalues. It is similar to both $\begin{bmatrix} 2 & 0 \\ 0 & 7 \end{bmatrix}$ and $\begin{bmatrix} 7 & 0 \\ 0 & 2 \end{bmatrix}$.
7. A has three different eigenvalues because the characteristic polynomial of A, $(5-\lambda)(\lambda+2)(\lambda-1)$, has distinct roots. One of six diagonal matrices to which A is similar is $\begin{bmatrix} 1 & 0 & 0 \\ 0 & -2 & 0 \\ 0 & 0 & 5 \end{bmatrix}$.
9. (a) $\lambda = 2$ is an eigenvalue because $A - 2I = \begin{bmatrix} -1 & 2 \\ 2 & -4 \end{bmatrix}$ is not invertible. The corresponding eigenspace is the set of scalar multiples of $\begin{bmatrix} 2 \\ 1 \end{bmatrix}$.
   (b) The characteristic polynomial of A is $(\lambda - 2)(\lambda + 3) = \lambda^2 + \lambda - 6$.
   (c) A is diagonalizable because A is 2×2 with two different eigenvalues.
   (d) The columns of P should be eigenvectors corresponding to -3 and 2, respectively. So $P = \begin{bmatrix} -1 & 2 \\ 1 & 2 \end{bmatrix}$.
   (e) The columns of Q are eigenvectors of A corresponding to eigenvalues -3 and 2, respectively. So $Q^{-1}AQ = \begin{bmatrix} -3 & 0 \\ 0 & 2 \end{bmatrix}$.

11. (a) Given $P^{-1}AP = D$, we have $A = PDP^{-1}$, so $A^2 = (PDP^{-1})(PDP^{-1}) = PD^2P^{-1}$, $A^3 = A^2A = (PD^2P^{-1})(PDP^{-1}) = PD^3P^{-1}$ and, in general, $A^k = PD^kP^{-1}$. This makes the calculation of A^k easy since the powers D^k of the diagonal matrix D are easy: Just raise the diagonal entries of D to the power k.

(b) The characteristic polynomial of A is $(1-\lambda)(3-\lambda)$, so A has two distinct eigenvalues, $\lambda = 1, 3$, and is diagonalizable. For $\lambda = 1$, an eigenvector is $\begin{bmatrix} 1 \\ 0 \end{bmatrix}$ and for $\lambda = 3$, $\begin{bmatrix} 1 \\ 1 \end{bmatrix}$. With $P = \begin{bmatrix} 1 & 1 \\ 0 & 1 \end{bmatrix}$, we have $P^{-1}AP = D = \begin{bmatrix} 1 & 0 \\ 0 & 3 \end{bmatrix}$, so $A = PDP^{-1}$ and $A^{10} = PD^{10}P^{-1} = \begin{bmatrix} 1 & 3^{10}-1 \\ 0 & 3^{10} \end{bmatrix}$.

14. (a) The characteristic polynomial of A is $-\lambda^3 + 5\lambda^2 - 2\lambda - 8$ $= -(\lambda-4)(\lambda+1)(\lambda-2)$.

(b) The 3×3 matrix A has three different eigenvalues, 4, -1 and 2.

(c) The desired matrix P is a matrix whose columns are eigenvectors corresponding to 4, -1 and 2, **in that order**. One possibility is $P = \begin{bmatrix} 1 & 0 & -1 \\ -1 & -1 & 1 \\ 1 & 1 & 0 \end{bmatrix}$.

16. (a) The characteristic polynomial of A is $(1-\lambda)(3-\lambda)$, so A has two distinct eigenvalues and hence is diagonalizable. For $\lambda = 1$, $\mathsf{x} = \begin{bmatrix} 1 \\ -1 \end{bmatrix}$ is an eigenvector and, for $\lambda = 3$, $\mathsf{x} = \begin{bmatrix} 0 \\ 1 \end{bmatrix}$. For $P = \begin{bmatrix} 1 & 0 \\ -1 & 1 \end{bmatrix}$ we have $P^{-1}AP = D = \begin{bmatrix} 1 & 0 \\ 0 & 3 \end{bmatrix}$.

(d) The characteristic polynomial of A is $(2-\lambda)(\lambda+1)^2$. There are two eigenvalues, $\lambda = -1$ and $\lambda = 2$. For $\lambda = 2$, the eigenspace is spanned by $\begin{bmatrix} 1 \\ 1 \\ 1 \end{bmatrix}$. For $\lambda = -1$, the eigenspace is spanned by $\begin{bmatrix} 0 \\ 0 \\ 1 \end{bmatrix}$. There are just two linearly independent eigenvectors. The matrix is not diagonalizable.

(h) The characteristic polynomial of A is $(1-\lambda)(\lambda^2-4\lambda+3) = (1-\lambda)(\lambda-1)(\lambda-3)$, so A has eigenvalues $\lambda = 1$ and $\lambda = 3$. The eigenvectors for $\lambda = 1$ are multiples of $\begin{bmatrix} -1 \\ 0 \\ 1 \end{bmatrix}$. Eigenvectors for $\lambda = 3$ are multiples of $\begin{bmatrix} 1 \\ 0 \\ 1 \end{bmatrix}$. Since A does not have three linearly independent eigenvectors, it is not diagonalizable.

17. (a) The eigenvalues of A are 1 and 2 with corresponding eigenvectors, respectively, $\begin{bmatrix} 1 \\ 1 \end{bmatrix}$ and $\begin{bmatrix} 2 \\ 1 \end{bmatrix}$. Let $P = \begin{bmatrix} 1 & 2 \\ 1 & 1 \end{bmatrix}$. Then $P^{-1}AP = D = \begin{bmatrix} 1 & 0 \\ 0 & 2 \end{bmatrix}$.

(b) $D = D_1^2$ with $D_1 = \begin{bmatrix} 1 & 0 \\ 0 & \sqrt{2} \end{bmatrix}$. Now $P^{-1}AP = D_1^2$, so $A = PD_1^2P^{-1} = (PD_1P^{-1})(PD_1P^{-1}) = B^2$, with $B = PD_1P^{-1} = \begin{bmatrix} -1+2\sqrt{2} & 2-2\sqrt{2} \\ -1+\sqrt{2} & 2-\sqrt{2} \end{bmatrix}$.

20. The characteristic polynomial of A is $\lambda^2 - 150\lambda - 4375 = (\lambda-175)(\lambda+25)$, A has two eigenvalues, $\lambda = 175$ and $\lambda = -25$. The eigenspace for $\lambda = 175$ is the set of multiples of $\begin{bmatrix} -4 \\ 3 \end{bmatrix}$. The eigenspace for $\lambda = -25$ is the set of multiples of $\begin{bmatrix} 3 \\ 4 \end{bmatrix}$. The matrix $P = \begin{bmatrix} -4 & 3 \\ 3 & 4 \end{bmatrix}$ has orthogonal columns and $P^{-1}AP = \begin{bmatrix} 175 & 0 \\ 0 & -25 \end{bmatrix}$.

24. (a) For example, $P_1 = \begin{bmatrix} -1 & -13 & 26 \\ 1 & 11 & -22 \\ 0 & 2 & -4 \end{bmatrix}$, $P_2 = \begin{bmatrix} -13 & -1 & 26 \\ 11 & 1 & -22 \\ 2 & 0 & -4 \end{bmatrix}$,

$P_3 = \begin{bmatrix} -1 & -13 & 2 \\ 1 & 11 & -2 \\ 0 & 2 & 0 \end{bmatrix}$.

(b) None of the matrices P_i is invertible. In each case, one column is a multiple of another, so the determinant is 0.

(c) No P and D exist because there do not exist three linearly independent eigenvectors. The matrix A is not diagonalizable.

32. (a) Can't be done: Such a matrix A must be diagonalizable (because the eigenvalues are distinct), so the matrix P whose columns are $\mathsf{x}_1, \mathsf{x}_2, \mathsf{x}_3$ must be invertible.

Exercises 3.6

1. (a) The first eight terms are $0, 1, 1, 3, 5, 11, 21, 45$.

   (b) a_n is the second component of $\mathsf{v}_n = \frac{1}{3}\left[-\begin{bmatrix} (-1)^{n+1} \\ (-1)^n \end{bmatrix} + \begin{bmatrix} 2^{n+1} \\ 2^n \end{bmatrix}\right]$, which is $\frac{1}{3}[(-1)^{n+1} + 2^n]$.

4. (a) $a_n = -8 + 4(2^n)$.

Section 3.7—True/False

1. False. The entries in the first column do not have sum 1.
2. True. Each entry of $A^T\mathsf{u}$ is the dot product of u and a column of A. This is twice the sum of the entries of a column of A, which is 2.

3. False; for example, $\begin{bmatrix} .5 & 0 \\ .5 & 1 \end{bmatrix}$.
4. False. The term "regular" applies only to Markov matrices and this matrix is not Markov.
5. True. The matrix is Markov.
6. False. If x is an eigenvector of A, then $c\mathsf{x}$ is also an eigenvector for any scalar c.
7. True. This is the steady state vector.
8. False. The eigenspace corresponding to $\lambda = 1$ is a line.
9. True. Each column is the steady state vector.
10. True.

Exercises 3.7

1. (a) With $P = \begin{bmatrix} 3 & 1 \\ 8 & 1 \end{bmatrix}$, we have $P^{-1}AP = \begin{bmatrix} 1 & 0 \\ 0 & \frac{1}{12} \end{bmatrix}$.

(b) $A^k = PD^kP^{-1} = \frac{1}{11}\begin{bmatrix} 3 & 1 \\ 8 & -1 \end{bmatrix}\begin{bmatrix} 1 & 0 \\ 0 & \frac{1}{12^k} \end{bmatrix}\begin{bmatrix} 1 & 1 \\ 8 & -3 \end{bmatrix}$

$$= \begin{bmatrix} \frac{3}{11} + \frac{8}{11}\frac{1}{12^k} & \frac{3}{11} - \frac{3}{11}\frac{1}{12^k} \\ \frac{8}{11} - \frac{8}{11}\frac{1}{12^k} & \frac{8}{11} + \frac{3}{11}\frac{1}{12^k} \end{bmatrix}.$$

(c) $A \to B = \begin{bmatrix} \frac{3}{11} & \frac{3}{11} \\ \frac{8}{11} & \frac{8}{11} \end{bmatrix}$.

(d) Since A is a regular Markov matrix, A will converge to a matrix each of whose columns is x_∞, the unique eigenvector corresponding to 1 whose components sum to 1. The components of $t\begin{bmatrix} 3 \\ 8 \end{bmatrix}$ have sum $3t + 8t = 11t$. We want $11t = 1$, so $t = \frac{1}{11}$ and $B = \begin{bmatrix} \frac{3}{11} & \frac{3}{11} \\ \frac{8}{11} & \frac{8}{11} \end{bmatrix}$, as before.

2. (a) $A^4 = \begin{bmatrix} \frac{1}{8} & \frac{1}{16} & \frac{1}{16} & \frac{1}{16} \\ \frac{1}{8} & \frac{3}{16} & \frac{1}{8} & \frac{1}{8} \\ \frac{1}{2} & \frac{1}{2} & \frac{9}{16} & \frac{1}{2} \\ \frac{1}{4} & \frac{1}{4} & \frac{1}{4} & \frac{5}{16} \end{bmatrix}$. All the entries are positive, so A is regular.

The steady state vector is $\frac{4}{15}\mathsf{x} = \dfrac{1}{15}\begin{bmatrix} 1 \\ 2 \\ 8 \\ 4 \end{bmatrix}$.

4. The eventual proportion of the original $x_0 + y_0 + z_0$ students in Dr. G's class is $\frac{55}{102} \approx .539$. The corresponding proportions for Dr. L and Dr. P are, respectively, $\frac{12}{102} \approx .118$ and $\frac{35}{102} \approx .343$. Notice that these fractions are just the components of v_∞—see READING CHECK 2.

6. The eventual distribution is $\frac{2}{9} \approx 22\%$ of the initial total population in British Columbia, $\frac{1}{3} \approx 33\%$ in Ontario and $\frac{4}{9} \approx 44\%$ in Newfoundland, proportions independent of the initial distribution. They are the components of the steady state vector.

Section 4.1—True/False

1. False. It defines a function from $\mathsf{R}^\ell \to \mathsf{R}^k$.
2. True. Every vector in the xz-plane is a linear combination of the columns: $\begin{bmatrix} x \\ 0 \\ z \end{bmatrix} = -x \begin{bmatrix} -1 \\ 0 \\ 0 \end{bmatrix} + \frac{z}{2} \begin{bmatrix} 0 \\ 0 \\ 2 \end{bmatrix} + 0 \begin{bmatrix} 4 \\ 0 \\ 5 \end{bmatrix} + 0 \begin{bmatrix} 1 \\ 0 \\ 1 \end{bmatrix}$.
3. True. The matrix A is $m \times n$ with $m = 3 < 4 = n$, so we apply Theorem 4.1.4.
4. True. The hypothesis says each of A and B is invertible (with inverses $\frac{1}{c}B$ and $\frac{1}{c}A$, respectively), so $\operatorname{rank} A = \operatorname{rank} B = n$.
5. True. $\mathsf{x} = \frac{1}{2}A\mathsf{x}$ and $A\mathsf{x}$ is a linear combination of the columns of A.
6. False. The system has **at most one** solution, by Theorem 4.1.17.
7. True. Let the columns of A be $\mathsf{a}_1, \ldots, \mathsf{a}_n$ and suppose $c_1\mathsf{a}_1 + \cdots + c_n\mathsf{a}_n = \mathsf{0}$. By 2.1.33, this is $A\mathsf{c} = \mathsf{0}$ with $\mathsf{c} = \begin{bmatrix} c_1 \\ \vdots \\ c_n \end{bmatrix}$.
8. True. The null space of A consists of the vectors x that satisfy $A\mathsf{x} = \mathsf{0}$. These are exactly the vectors x satisfying $U\mathsf{x} = \mathsf{0}$. Passing to U is the way we solve systems of equations!
9. False. The matrices $A = \begin{bmatrix} 1 & 1 \\ 1 & 1 \end{bmatrix}$ and $U = \begin{bmatrix} 1 & 1 \\ 0 & 0 \end{bmatrix}$ have different column spaces.
10. True. "Singular" means "not invertible."

Exercises 4.1

1. $m = 7, n = 9$.

3. (a) The null space consists of all scalar multiples of $\begin{bmatrix} -4 \\ 2 \\ 7 \\ 1 \end{bmatrix}$. Rank A = 3 since there are three pivot columns.

4. The column space is the set of linear combinations of the three columns. The first two columns (for example) are not parallel, so any vector is a linear combination of them and hence of all three columns.

5. (a) Yes: $\mathsf{b} = \frac{1}{2}\begin{bmatrix} 2 \\ 2 \end{bmatrix} + \frac{2}{3}\begin{bmatrix} 0 \\ 3 \end{bmatrix}$.

6. (a) No.

7. (a) No: $A\mathsf{u} \neq \mathsf{0}$.

   (b) Yes: b = 0(column one)+0(column two)+0(column three)+0(column four)+0(column five) + 1(column six).

   (c) The null space is the span of $\begin{bmatrix} 2 \\ 1 \\ 0 \\ 0 \\ 0 \\ 0 \end{bmatrix}$, $\begin{bmatrix} -1 \\ 0 \\ 2 \\ 1 \\ 0 \\ -0 \end{bmatrix}$ and $\begin{bmatrix} 4 \\ 0 \\ -5 \\ 0 \\ -2 \\ 1 \end{bmatrix}$.

   (d) The column space is the set of all $\mathsf{b} = \begin{bmatrix} b_1 \\ b_2 \\ b_3 \end{bmatrix}$ such that $A\mathsf{x} = \mathsf{b}$, so we solve the system starting with the augmented matrix $[A \mid \mathsf{b}]$ $= \left[\begin{array}{cccccc|c} 6 & -12 & -5 & 16 & -2 & 53 & b_1 \\ -3 & 6 & 3 & -9 & 1 & 29 & b_2 \\ -4 & 8 & 3 & -10 & 1 & 33 & b_3 \end{array}\right]$. We use the same sequence of elementary row operations as in part (c) and reach the stage $\left[\begin{array}{cccccc|c} 6 & -12 & -5 & 16 & -2 & -53 & b_1 \\ 0 & 0 & 1 & -2 & 0 & 5 & 2b_2 + b_1 \\ 0 & 0 & 0 & 0 & -\frac{1}{3} & -\frac{2}{3} & (b_3 + \frac{2}{3}b_1) + \frac{(}{2}b_2 + b_1) \end{array}\right]$ where we can see (without going further) that there are infinitely many solutions for any b. The column space is R^3.

   (e) Three pivots, so the rank is 3.

9. (a) For $A\mathsf{x} = \mathsf{v}$, the general solution is

$$\mathsf{x} = \begin{bmatrix} x_1 \\ x_2 \\ x_3 \\ x_4 \\ x_5 \end{bmatrix} = \begin{bmatrix} \frac{15}{4} - 3t - \frac{7}{26}s \\ \frac{21}{4} + 2t - \frac{3}{26}s \\ t \\ -\frac{7}{4} - \frac{1}{26}s \\ s \end{bmatrix} = \begin{bmatrix} \frac{15}{4} \\ \frac{21}{4} \\ 0 \\ -\frac{7}{4} \\ 0 \end{bmatrix} + t\begin{bmatrix} -3 \\ 2 \\ 1 \\ 0 \\ 0 \end{bmatrix} + s\begin{bmatrix} -\frac{7}{26} \\ -\frac{3}{26} \\ 0 \\ -\frac{1}{26} \\ 1 \end{bmatrix}.$$

10. (a) The column space is the plane with equation $2x - y + 3z = 0$. The null space consists of multiples of $\begin{bmatrix} \frac{3}{5} \\ -\frac{14}{5} \\ 1 \end{bmatrix}$, equivalently, multiples of $\begin{bmatrix} 3 \\ -14 \\ 5 \end{bmatrix}$.

11. (a) The rank is 1.

12. The given matrix is elementary, so invertible. col sp $A = \mathsf{R}^3$, null sp $A = \{\mathsf{0}\}$.

14. $A = \begin{bmatrix} 1 & 2 & 3 & 4 \\ 2 & 4 & 6 & 8 \\ 3 & 6 & 9 & 12 \end{bmatrix}$.

16. (a) The null space is all vectors of the form $\begin{bmatrix} -\frac{13}{2}t \\ 3t \\ t \end{bmatrix} = t\begin{bmatrix} -\frac{13}{2} \\ 3 \\ 1 \end{bmatrix}$.

(b) rank $A = 2$.

(c) The solution is $\mathsf{x} = \begin{bmatrix} -\frac{11}{2} - \frac{13}{2}t \\ 4 + 3t \\ t \end{bmatrix} = \begin{bmatrix} -\frac{11}{2} \\ 4 \\ 0 \end{bmatrix} + t\begin{bmatrix} -\frac{13}{2} \\ 3 \\ 1 \end{bmatrix}$.

(d) $\begin{bmatrix} 1 \\ 6 \\ -4 \end{bmatrix} = -\frac{11}{2}\begin{bmatrix} 2 \\ 4 \\ 0 \end{bmatrix} + 4\begin{bmatrix} 3 \\ 7 \\ -1 \end{bmatrix}$.

(e) The column space is the plane with equation $2x - y - z = 0$.

19. (a) Since $A\mathsf{x} = 0$, $(BA)\mathsf{x} = B(A\mathsf{x}) = B0 = 0$ too.

20. Let the first two columns of A be i and j and let the third be $\begin{bmatrix} a \\ b \\ 0 \end{bmatrix}$ with any a and b.

21. (a) Yes, since $A(B\mathsf{x}) = 0$.

24. (a) rank $A = 2$.

(b) The system has a solution if and only if $c = 1$ or $c = -2$. If $c = 1$,
$\mathsf{x} = \begin{bmatrix} 4 - 2t - s \\ -5 - t - 2s \\ t \\ s \end{bmatrix} = \begin{bmatrix} 4 \\ -5 \\ 0 \\ 0 \end{bmatrix} + t\begin{bmatrix} -2 \\ -1 \\ 1 \\ 0 \end{bmatrix} + s\begin{bmatrix} -1 \\ -2 \\ 0 \\ 1 \end{bmatrix}$.
If $c = -2$, $\mathsf{x} = \begin{bmatrix} 1 - 2t - s \\ 4 - t - 2s \\ t \\ s \end{bmatrix} = \begin{bmatrix} 1 \\ 4 \\ 0 \\ 0 \end{bmatrix} + t\begin{bmatrix} -2 \\ -1 \\ 1 \\ 0 \end{bmatrix} + s\begin{bmatrix} -1 \\ -2 \\ 0 \\ 1 \end{bmatrix}$.

28. If b is in col sp AB, then $\mathsf{b} = (AB)\mathsf{x}$ for some vector x. Since $\mathsf{b} = A(B\mathsf{x})$, b is in col sp A. Thus col sp $AB \subseteq$ col sp A. On the other hand, if b is in col sp A, then $\mathsf{b} = A\mathsf{x}$ for some vector x, so $\mathsf{b} = (ABA)\mathsf{x} = AB(A\mathsf{x})$ is in col sp AB. Thus col sp $A \subseteq$ col sp AB, so these column spaces are the same.

Section 4.2—True/False

1. False. A subspace must contain the zero vector and this line does not.
2. True. Any scalar multiple of a vector with equal components has equal components.
3. True. The column space of the given matrix is R^2.
4. True. The null space of a matrix is a subspace.

5. True. The given set is the null space of the matrix $A - 2I$.
6. True. The second matrix is a row echelon form of the first.
7. False. The column space of the second matrix is the xy-plane, but the column space of the first matrix contains vectors with nonzero third component.
8. False. The zero vector is in any subspace.
9. False. Let $\mathsf{u} = \begin{bmatrix} 1 \\ 1 \end{bmatrix}$ and $\mathsf{v} = \begin{bmatrix} 0 \\ 0 \end{bmatrix}$. Then $\{\mathsf{u}, \mathsf{v}\}$ is a linearly dependent set, but u is not a multiple of v.
10. True.

Exercises 4.2

1. (a) Closed under addition and scalar multiplication.
3. (a) Not a subspace because it is not closed under addition. Equally, because it is not closed under scalar multiplication and because it does not contain the zero vector.
5. $\mathsf{u}_1 - \mathsf{u}_2 = \mathsf{u}_1 + (-1)\mathsf{u}_2$ and U is closed under scalar multiplication and addition.
6. (a) Vector v is **not** in the span of the others: the equation $\mathsf{v} = a\mathsf{v}_1 + b\mathsf{v}_2$ has no solution.
10. If U is contained in W, then $U \cup W = W$ is a subspace. Similarly, if W is contained in U, then $U \cup W = U$ is a subspace. Conversely, suppose that $U \cup W$ is a subspace and, following the hint, suppose that neither U nor W is contained in the other. Then there is a vector u that is in U, but not in W, and a vector w that is in W but not in U. Consider the vector $\mathsf{u} + \mathsf{w}$ which, by assumption, is in the subspace $U \cup W$. If $\mathsf{u} + \mathsf{w} = \mathsf{u}_1$ is in U, then $\mathsf{w} = \mathsf{u}_1 - \mathsf{u}$ is in U since U is a subspace, contradicting the fact that w was not in U. So $\mathsf{u} + \mathsf{w}$ is not in U and, similarly, not in W. This contradicts the fact that $U \cup W$ is a subspace, so our assumption that neither U nor W was contained in the other must be wrong.

Section 4.3—True/False

1. False. A spanning set of R^4 needs at least four vectors.
2. True. Every vector in R^4 is a linear combination of the first four vectors.
3. False. A linearly independent set in R^3 contains at most three vectors.
4. True.
5. False. A finite dimensional vector space has a finite **basis**, but an infinite number of vectors: if v is in the vector space, so are all the vectors $c\mathsf{v}$, for any real number c.

6. True. The matrix has rank 3, so its column space is R^3 and the three given vectors are the standard basis for R^3.
7. True. The nonzero rows of row echelon form constitute a basis.
8. True. There are n columns, each being a vector in R^m. In R^m, any $n > m$ vectors are linearly dependent.
9. False. A vector space of positive dimension has infinitely many subspaces; for example, any line through 0 is a subspace.
10. True, by Proposition 4.4.14. The possible dimensions are $0, 1, 2, 3, 4, 5, 6$.

Exercises 4.3

1. (a) Any three vectors in R^2 are linearly dependent. Here, $\mathsf{v}_3 = 2\mathsf{v}_1 - \mathsf{v}_2$.

3. (a) i. Here is row echelon form: $\begin{bmatrix} 1 & -1 & 0 & -2 \\ 0 & 1 & \frac{1}{4} & 4 \\ 0 & 0 & -\frac{5}{2} & -20 \end{bmatrix}$. The set $\left\{ \begin{bmatrix} 0 \\ -2 \\ -8 \\ 1 \end{bmatrix} \right\}$ spans the null space and is linearly independent.

ii. The nonzero vectors in a row echelon form of A form are linearly independent and span the row space of A: $\begin{bmatrix} 1 & -1 & 0 & -2 \end{bmatrix}$, $\begin{bmatrix} 0 & 1 & \frac{1}{4} & 4 \end{bmatrix}$ and $\begin{bmatrix} 0 & 0 & -\frac{5}{2} & -20 \end{bmatrix}$.

iii. Here is row echelon for A^T: $\begin{bmatrix} 1 & -3 & -2 \\ 0 & 1 & 1 \\ 0 & 0 & 1 \\ 0 & 0 & 0 \end{bmatrix}$. The nonzero rows form a linearly independent set of vectors whose transposes span the column space of A: $\begin{bmatrix} 1 \\ -3 \\ -2 \end{bmatrix}$, $\begin{bmatrix} 0 \\ 1 \\ 1 \end{bmatrix}$ and $\begin{bmatrix} 0 \\ 0 \\ 1 \end{bmatrix}$.

4. (a) Not a basis.
5. $b \neq 2a$.

8. W is the null space of $A = \begin{bmatrix} 1 & 2 & 3 & 5 \\ 1 & 0 & 1 & 3 \\ 2 & 3 & 5 & 9 \end{bmatrix}$. The vectors $\begin{bmatrix} -1 \\ -1 \\ 1 \\ 0 \end{bmatrix}$ and $\begin{bmatrix} -3 \\ -1 \\ 0 \\ 1 \end{bmatrix}$ are a basis: $\dim W = 2$.

12. Two vectors span a vector space of dimension at most 2, so $n \leq 2$.
14. (a) Take any two vectors, neither of which is a multiple of the other (so these are linearly independent). These vectors will span a plane. For a third vector, take any linear combination of the first two.

(b) I would cry. This cannot be done.

Section 4.4—True/False

1. True. The matrix has row rank = column rank = 3.
2. True. n vectors span R^n if and only if they are linearly independent.
3. True. Once we have row echelon form for a matrix A, we can identify the pivot columns, and these constitute a basis for the column space of A.
4. False. The first two columns of each matrix give bases for the respective column spaces, but these spaces are different. The vectors $\begin{bmatrix}1\\5\\9\end{bmatrix}$ and $\begin{bmatrix}2\\6\\10\end{bmatrix}$ give a basis for the column space of A.
5. False. See the comments after 4.4.20.
6. False: $\begin{bmatrix}0 & 0\\1 & 2\end{bmatrix}$.
7. True. The column rank is 1, so the row rank is 1 too.
8. False. An $m \times n$ matrix has rank at most the smaller of m and n.
9. True. $\operatorname{rank} AA^T = \operatorname{rank} A \leq 3$.
10. True. $\operatorname{rank} A = 2$, $\operatorname{rank} B = 3$ and $\operatorname{rank} AB \leq \operatorname{rank} A$.

Exercises 4.4

1. The span of the given vectors is the column space of

$$A = \begin{bmatrix}-1 & 0 & 2 & -4 & 4\\0 & 1 & 1 & -4 & 4\\1 & 1 & 1 & -6 & -2\\1 & 0 & -1 & 1 & -5\end{bmatrix}$$ which, by 4.4.21, has basis the pivot columns: columns one, two and three.

4. (a) The most obvious answer is $\mathsf{e}_1 = \begin{bmatrix}1\\0\\0\end{bmatrix}$, $\mathsf{e}_2 = \begin{bmatrix}0\\1\\0\end{bmatrix}$, $\mathsf{e}_3 = \begin{bmatrix}0\\0\\1\end{bmatrix}$.

   (c) V has dimension 3. The first three vectors are linearly independent and hence comprise a basis for V.

5. (a) A basis is $\{\begin{bmatrix}1\\-1\\3\\2\end{bmatrix}, \begin{bmatrix}0\\3\\-5\\-1\end{bmatrix}\}$: $\dim V = 2$.

   (b) Any two are linearly independent and hence comprise a basis for V.

7. (a) Either condition—linear independence or spanning—is equivalent to basis, by Corollaries 4.4.3 and 4.4.10.

8. (a) $U = \begin{bmatrix}1 & -1 & 0 & 2 & 3 & 1\\0 & 1 & 4 & -2 & 2 & 3\\0 & 0 & 0 & 1 & -5 & 1\\0 & 0 & 0 & 0 & 0 & 0\end{bmatrix}$.

(b) A basis is the set of nonzero rows of U,
$\begin{bmatrix}1 & -1 & 0 & 2 & 3 & 1\end{bmatrix}$, $\begin{bmatrix}0 & 1 & 4 & -2 & 2 & 3\end{bmatrix}$ and $\begin{bmatrix}0 & 0 & 0 & 1 & -5 & 1\end{bmatrix}$.

(c) The row rank of U is $\dim \operatorname{row\,sp} U = 3$. Since U and A have the same row spaces, this is also the row rank of A.

(d) The pivot columns of U are columns one, two and four, that is, $\begin{bmatrix}1\\0\\0\\0\end{bmatrix}$, $\begin{bmatrix}-1\\1\\0\\0\end{bmatrix}$ and $\begin{bmatrix}2\\-2\\1\\0\end{bmatrix}$. Now $a\begin{bmatrix}1\\0\\0\\0\end{bmatrix}+b\begin{bmatrix}-1\\1\\0\\0\end{bmatrix}+c\begin{bmatrix}2\\-2\\1\\0\end{bmatrix}=\begin{bmatrix}0\\0\\0\\0\end{bmatrix}$ implies $\begin{bmatrix}a-b+2c\\b-2c\\c\\0\end{bmatrix}=\begin{bmatrix}0\\0\\0\\0\end{bmatrix}$ and back substitution gives $c=b=a=0$. So these columns are linearly independent.

(e) The pivot columns of A are columns one, two and four. So we must show that $\mathsf{a}_1=\begin{bmatrix}1\\3\\5\\2\end{bmatrix}$, $\mathsf{a}_2=\begin{bmatrix}-1\\-2\\-4\\-1\end{bmatrix}$ and $\mathsf{a}_4=\begin{bmatrix}2\\4\\9\\2\end{bmatrix}$ are linearly independent.

$$a\begin{bmatrix}1\\3\\5\\2\end{bmatrix}+b\begin{bmatrix}-1\\-2\\-4\\-1\end{bmatrix}+c\begin{bmatrix}2\\4\\9\\2\end{bmatrix}=\begin{bmatrix}0\\0\\0\\0\end{bmatrix} \text{ implies } \begin{aligned} a-b+2c&=0\\ 3a-2b+4c&=0\\ 5a-4b+9c&=0\\ 2a-b+2c&=0\end{aligned}$$

and

$$\begin{bmatrix}1&-1&2\\3&-2&4\\5&-4&9\\2&-1&2\end{bmatrix}\to\begin{bmatrix}1&-1&2\\0&1&-2\\0&1&-1\\0&1&-2\end{bmatrix}\to\begin{bmatrix}1&-1&2\\0&1&-2\\0&0&1\\0&0&0\end{bmatrix}$$

gives $c=b=a=0$, so $\mathsf{a}_1,\mathsf{a}_2,\mathsf{a}_4$ are linearly independent.

(f) The "other" columns are $\mathsf{a}_2=\begin{bmatrix}0\\4\\4\\4\end{bmatrix}$, $\mathsf{a}_5=\begin{bmatrix}3\\11\\12\\8\end{bmatrix}$ and $\mathsf{a}_6=\begin{bmatrix}1\\6\\9\\5\end{bmatrix}$.
We have $\mathsf{a}_3=4\mathsf{a}_1+4\mathsf{a}_2$, $\mathsf{a}_5=5\mathsf{a}_1-8\mathsf{a}_2-5\mathsf{a}_4$ and $\mathsf{a}_6=4\mathsf{a}_1+5\mathsf{a}_2+\mathsf{a}_4$.

(g) Since columns a_1, a_2 and a_4 form a basis for the column space, the column space has dimension 3. The column rank of A is 3, the same as the row rank.

(h) The rank of A is its row rank; the rank of A^T is its column rank. These numbers are the same for any matrix.

(i) The nullity of A is $n - \operatorname{rank} A = 6 - 3 = 3$.

10. (a) A basis for the row space is $\begin{bmatrix}1 & 0 & 1 & -1\end{bmatrix}$, $\begin{bmatrix}0 & 1 & -1 & 2\end{bmatrix}$ and $\begin{bmatrix}0 & 0 & 0 & 1\end{bmatrix}$. The dimension is 3.

(b) Columns one, two and four comprise a basis for the column space of A: namely, $\begin{bmatrix} -1 \\ 0 \\ 1 \\ 2 \\ 0 \end{bmatrix}$, $\begin{bmatrix} 0 \\ 1 \\ 1 \\ -1 \\ -1 \end{bmatrix}$ and $\begin{bmatrix} 1 \\ 2 \\ 1 \\ 1 \\ 0 \end{bmatrix}$. The dimension is 3.

(c) A basis for the null space is $\begin{bmatrix} -1 \\ 1 \\ 1 \\ 0 \end{bmatrix}$. The nullity is 1.

12. (a) This part uses heavily 2.6.4. Suppose the columns of U are linearly independent. If some linear combination of the columns of A is 0, we have $A\mathsf{x} = 0$ for some x. Thus $EA\mathsf{x} = 0$, so $U\mathsf{x} = 0$. Since the columns of U are linearly independent, $\mathsf{x} = 0$, so the columns of A are linearly independent.

(b) Suppose the columns of A span R^m and let y be a vector in R^m. Then $E^{-1}\mathsf{y} = A\mathsf{x}$ for some x, so $\mathsf{y} = EA\mathsf{x} = U\mathsf{x}$. This says y is a linear combination of the columns of U, so these span R^m too.

13. (a) B has (column) rank n, so it is invertible by Theorem 4.1.17.

14. (a) Via Gaussian elimination, $A \to \begin{bmatrix} 1 & 1 \\ 0 & 1 \\ 0 & 0 \\ 0 & 0 \end{bmatrix} = U$.

i. The null space is $\{0\}$ and nullity $A = 0$.

ii. The row space has basis $\begin{bmatrix} 1 & 1 \end{bmatrix}$ and $\begin{bmatrix} 0 & 1 \end{bmatrix}$. The row rank of A is 2.

iii. The pivot columns of A comprise a basis for the column space, here both columns: $\begin{bmatrix} 1 \\ 1 \\ 2 \\ 1 \end{bmatrix}$ and $\begin{bmatrix} 1 \\ -1 \\ 0 \\ 2 \end{bmatrix}$. The column rank is 2.

iv. $n = 2$, the number of columns, and $2 = 2+0 = \operatorname{rank} A + \operatorname{nullity} A$.

(d) $A \to \begin{bmatrix} 1 & 0 & 1 & 1 \\ 0 & 1 & 1 & -1 \\ 0 & 0 & 0 & 0 \end{bmatrix} = U$.

i. The vectors $\begin{bmatrix} -1 \\ -1 \\ 1 \\ 0 \end{bmatrix}$ and $\begin{bmatrix} -1 \\ 1 \\ 0 \\ 1 \end{bmatrix}$ are a basis for the null space: nullity $A = 2$.

ii. The rows $\begin{bmatrix} 1 & 0 & 1 & 1 \end{bmatrix}$ and $\begin{bmatrix} 0 & 1 & 1 & -1 \end{bmatrix}$ are a basis for the two space. The row rank is 2.

iii. The pivot columns, columns one and two, comprise a basis for the column space: $\begin{bmatrix} 1 \\ 1 \\ 0 \end{bmatrix}$ and $\begin{bmatrix} 0 \\ 1 \\ 1 \end{bmatrix}$. The column rank is 2.

iv. $n = 4$, the number of columns of A, and $4 = 2 + 2 = \operatorname{rank} A + \operatorname{nullity} A$.

23. The row rank of the matrix is 1, so the column rank is 1. Let v be any nonzero column of A. Then v is a basis for the column space, so every column is a multiple of v.

25. (a) xy^T is $n \times n$. Let $\mathsf{y} = \begin{bmatrix} y_1 \\ y_2 \\ \vdots \\ y_n \end{bmatrix}$. Then $\mathsf{xy}^T = \begin{bmatrix} \mathsf{x} \\ \downarrow \\ \ \end{bmatrix} \begin{bmatrix} y_1 & y_2 & \cdots & y_n \end{bmatrix} =$

$\begin{bmatrix} y_1\mathsf{x} & y_2\mathsf{x} & \cdots & y_n\mathsf{x} \\ \downarrow & \downarrow & & \downarrow \\ & & & \end{bmatrix}$. Every column is a multiple of x, so the column space is spanned by x. Its dimension is 1; the rank is 1.

Section 4.5—True/False

1. False. Take $A = 0$ and B any singular matrix.
2. True. In the proof of part 3 of Theorem 4.5.3, we even showed how to construct a right inverse.
3. True.
4. True.
5. False. The system has a unique solution **if a solution exists.**
6. True. The rows of A are linearly independent.
7. False. The columns of A are linearly dependent (three vectors in R^2).
8. False in general. The hypothesis says that A has a right inverse, $\frac{1}{c}B$, so rank $A = m$, and that B has a left inverse, $\frac{1}{c}A$, so rank $B = n$.
9. True.
10. True.

Exercises 4.5

1. The rank 2, which is neither the number of columns nor the number of rows.

3. (a) i. The rank of A is 2, the number of rows, so A has a right inverse.

ii. No. If A has a right inverse and also a left inverse, then A is square, by Corollary 4.5.7.

iii. The general right inverse for A is $B = \begin{bmatrix} -\frac{1}{7} + \frac{1}{7}t & \frac{2}{7} + \frac{1}{7}s \\ \frac{3}{7} - \frac{3}{7}t & \frac{1}{7} - \frac{3}{7}s \\ t & s \end{bmatrix}$.

iv. Setting $t = s = 0$ gives $\begin{bmatrix} -\frac{1}{7} & \frac{2}{7} \\ \frac{3}{7} & \frac{1}{7} \\ 0 & 0 \end{bmatrix}$.

(b) i. The rank of A is 2, the number of columns, so A has a left inverse.

ii. No. If A has a left inverse and also a right inverse, then A is square, by Corollary 4.5.7.

iii. The most general right inverse for A^T is $C^T = \begin{bmatrix} 2 & -1 \\ 1+2t & -1+2s \\ t & s \end{bmatrix}$,

so the most general left inverse for A is $C = \begin{bmatrix} 2 & 1+2t & t \\ -1 & -1+2s & s \end{bmatrix}$.

iv. Setting $t = s = 0$ gives $\begin{bmatrix} 2 & 1 & 0 \\ -1 & -1 & 0 \end{bmatrix}$.

4. (a) By Theorem 4.4.28, rank $AB \le$ rank A. Using the very same fact, rank A = rank$(AB)B^{-1} \le$ rank AB, so rank AB = rank A.

5. (a) There exist matrices C and D such that $CA = I$ and $DB = I$. Then $(DC)(AB) = D(CA)B = DIB = DB = I$, so AB has left inverse DC.

7. (a) Yes. Since the $n \times n$ matrix A^TA is invertible, its rank is n by Corollary 4.5.11, so the rank of A is n by Theorem 4.4.27 and the columns of A are linearly independent by Theorem 4.5.5.

8. (a) Suppose $AB = I$ with B unique. Suppose $A\mathsf{x} = 0$ and let $X = \begin{bmatrix} \mathsf{x} & \mathsf{x} & \cdots & \mathsf{x} \\ \downarrow & \downarrow & & \downarrow \end{bmatrix}$. Then $A(B + X) = AB + AX = I$ (since $AX = 0$), so $B + X$ is also a right inverse for A. Thus $B + X = B$, so $X = 0$ and $\mathsf{x} = 0$. It follows that the columns of A are linearly independent, so A has a left inverse. So A is square and invertible by Corollary 4.5.7.

9. Since the rows of U are linearly independent, U has a right inverse.

Section 5.1—True/False

1. False. $\sqrt{x+y} \neq \sqrt{x} + \sqrt{y}$.
2. False. $T(0) \neq 0$.
3. True.
4. True.
5. True. The matrix is $\begin{bmatrix} \cos\theta & -\sin\theta \\ \sin\theta & \cos\theta \end{bmatrix}$ with $\theta = \frac{\pi}{4}$.
6. True.
7. False. $P(\mathsf{j}) = \begin{bmatrix} 1 \\ -1 \end{bmatrix}$ is not on the line.
8. False. We have $P(\mathsf{i}) = \begin{bmatrix} 1 \\ 1 \end{bmatrix}$. Let $A = (1,0)$ and $B = (1,1)$. Then $\overrightarrow{AB} = \begin{bmatrix} 0 \\ 1 \end{bmatrix}$ should be orthogonal to $\begin{bmatrix} 1 \\ 1 \end{bmatrix}$, but it is not.
9. True. $T(\mathsf{u} + \mathsf{v}) = T(\mathsf{u}) + T(\mathsf{v})$.

10. False. The person who says this has forgotten that "closed" is a word that applies to subspaces, not functions. A linear transformation is a function that **preserves** addition and scalar multiplies.

Exercises 5.1

1. (a) Since $T(0) = \begin{bmatrix} 0 \\ 1 \\ 0 \end{bmatrix} \neq 0$, T is not a linear transformation.

(b) Let $\mathsf{x} = \begin{bmatrix} x_1 \\ x_2 \\ x_3 \end{bmatrix}$ and $\mathsf{y} = \begin{bmatrix} y_1 \\ y_2 \\ y_3 \end{bmatrix}$ be vectors in R^3. Then $\mathsf{x} + \mathsf{y} = \begin{bmatrix} x_1 + y_1 \\ x_2 + y_2 \\ x_3 + y_3 \end{bmatrix}$, so

$$T(\mathsf{x}+\mathsf{y}) = \begin{bmatrix} (x_1 + y_1) + (x_2 + y_2) \\ (x_2 + y_2) + (x_3 + y_3) \end{bmatrix} = \begin{bmatrix} x_1 + x_2 + y_1 + y_2 \\ x_2 + x_3 + y_2 + y_3 \end{bmatrix}$$

$$= \begin{bmatrix} x_1 + x_2 \\ x_2 + x_3 \end{bmatrix} + \begin{bmatrix} y_1 + y_2 \\ y_2 + y_3 \end{bmatrix} = T(\mathsf{x}) + T(\mathsf{y}).$$

Also, for any scalar c, $c\mathsf{x} = c\begin{bmatrix} x_1 \\ x_2 \\ x_3 \end{bmatrix} = \begin{bmatrix} cx_1 \\ cx_2 \\ cx_3 \end{bmatrix}$, so

$$T(c\mathsf{x}) = \begin{bmatrix} cx_1 + cx_2 \\ cx_2 + cx_3 \end{bmatrix} = c\begin{bmatrix} x_1 + x_2 \\ x_2 + x_3 \end{bmatrix} = cT(\mathsf{x}).$$

Thus T is a linear transformation.

3. $T\left(\begin{bmatrix} x \\ y \end{bmatrix}\right) = T\left(x\begin{bmatrix} 1 \\ 0 \end{bmatrix} + y\begin{bmatrix} 0 \\ 1 \end{bmatrix}\right) = xT\left(\begin{bmatrix} 1 \\ 0 \end{bmatrix}\right) + yT\left(\begin{bmatrix} 0 \\ 1 \end{bmatrix}\right) = x\begin{bmatrix} 3 \\ -1 \\ 1 \end{bmatrix} + y\begin{bmatrix} 4 \\ 5 \\ -2 \end{bmatrix} = \begin{bmatrix} 3x + 4y \\ -x + 5y \\ x - 2y \end{bmatrix}$.

5. $T\left(\begin{bmatrix} x \\ y \end{bmatrix}\right) = T\left((3x+2y)\begin{bmatrix} 1 \\ -1 \end{bmatrix} + (x+y)\begin{bmatrix} -2 \\ 3 \end{bmatrix}\right) = (3x+2y)T\left(\begin{bmatrix} 1 \\ -1 \end{bmatrix}\right) + (x+y)T\left(\begin{bmatrix} -2 \\ 3 \end{bmatrix}\right) = (3x + 2y)\begin{bmatrix} 1 \\ 2 \\ 3 \end{bmatrix} + (x + y)\begin{bmatrix} -2 \\ -4 \\ 7 \end{bmatrix} = \begin{bmatrix} x \\ 2x \\ 16x + 13y \end{bmatrix}$.

8. (a) No. Let $\mathsf{v} = \begin{bmatrix} 1 \\ 0 \\ -2 \end{bmatrix}$. Since $T(\mathsf{v}) = \begin{bmatrix} 1 \\ 2 \\ 3 \end{bmatrix}$, if T were linear, $T(2\mathsf{v}) = 2T(\mathsf{v}) = \begin{bmatrix} 2 \\ 4 \\ 6 \end{bmatrix}$; that is, $T\left(\begin{bmatrix} 2 \\ 0 \\ -4 \end{bmatrix}\right) = \begin{bmatrix} 2 \\ 4 \\ 6 \end{bmatrix}$.

9. (a) $A = \begin{bmatrix} 3 & 0 \\ 1 & -1 \\ -1 & 5 \end{bmatrix}$.

10. (a) $A = \begin{bmatrix} \cos\theta & -\sin\theta \\ \sin\theta & \cos\theta \end{bmatrix} = \begin{bmatrix} \frac{\sqrt{3}}{2} & -\frac{1}{2} \\ \frac{1}{2} & \frac{\sqrt{3}}{2} \end{bmatrix}$.

11. (a) $A = \begin{bmatrix} 1 & 0 \\ 0 & -1 \end{bmatrix}$; (c) $A = \begin{bmatrix} 0 & 1 \\ 1 & 0 \end{bmatrix}$.

13. (a) $A = \frac{1}{6}\begin{bmatrix} 1 & 2 & -1 \\ 2 & 4 & -2 \\ -1 & -2 & 1 \end{bmatrix}$.

14. (a) $A = \frac{1}{14}\begin{bmatrix} 5 & 3 & -6 \\ 3 & 13 & 2 \\ -6 & 2 & 10 \end{bmatrix}$.

15. (a) $A = \begin{bmatrix} 7 & -12 \\ 14 & -23 \end{bmatrix}$.

18. $BA = I$ is the identity matrix and $B = A^{-1}$.

Section 5.2—True/False

1. True: $\mathsf{x} = \mathsf{j}$.
2. True.
3. False.
4. False. For any x, $(f \circ g)(x) = f(g(x)) = f(x) = 1$.
5. False. For any x, $(g \circ f)(x) = g(f(x)) = g(1) = 1$.
6. False. If $f, g\colon \mathsf{R} \to \mathsf{R}$ are defined by $f(x) = 2$ and $g(x) = \frac{1}{x}$ for all x, then $(f \circ g)(x) = 2$, while $(g \circ f)(x) = \frac{1}{2}$.
7. True. See Theorem 5.2.1.
8. True. Rotation through 2θ is achieved with the composition TT, for which the matrix is A^2. See Example 5.2.4.
9. True. $T(T\mathsf{x}) = (TT)\mathsf{x}$ and the matrix for TT is $A^2 = I$.
10. True. The matrix of TS is $AB = \begin{bmatrix} 4 & 5 \\ 6 & 3 \end{bmatrix}$.

Exercises 5.2

1. (a) i. $ST\left(\begin{bmatrix} x \\ y \end{bmatrix}\right) = \begin{bmatrix} -x - 2y \\ -3x - 3y \\ 6x + y \\ x - 3y \end{bmatrix}$.

ii. The matrix for ST is $\begin{bmatrix} -1 & -2 \\ -3 & -3 \\ 6 & 1 \\ 1 & -3 \end{bmatrix}$.

iii. The matrix for S is $A = \begin{bmatrix} 1 & 1 & -1 \\ 0 & 1 & -1 \\ 1 & 0 & 1 \\ 0 & 1 & 0 \end{bmatrix}$, the matrix for T is $B = \begin{bmatrix} 2 & 1 \\ 1 & -3 \\ 4 & 0 \end{bmatrix}$, the matrix for ST is AB.

3. (a) $ST\left(\begin{bmatrix} x \\ y \end{bmatrix}\right) = \begin{bmatrix} y - x \\ -y + 2x \end{bmatrix}$; $TS\left(\begin{bmatrix} x \\ y \end{bmatrix}\right) = \begin{bmatrix} -x + 2y \\ x - y) \end{bmatrix}$.

(b) $A = \begin{bmatrix} 1 & -1 \\ -1 & 2 \end{bmatrix}, B = \begin{bmatrix} 0 & 1 \\ 1 & 0 \end{bmatrix}, C = \begin{bmatrix} -1 & 1 \\ 2 & -1 \end{bmatrix}, D = \begin{bmatrix} -1 & 2 \\ 1 & -1 \end{bmatrix}$.

(c) $AB = C$, $BA = D$.

5. The matrix of S is $A = \begin{bmatrix} 3 & 0 & -1 \\ 4 & 2 & 0 \end{bmatrix}$. Thus $S\left(\begin{bmatrix} x \\ y \\ z \end{bmatrix}\right) = \begin{bmatrix} 3 & 0 & -1 \\ 4 & 2 & 0 \end{bmatrix}\begin{bmatrix} x \\ y \\ z \end{bmatrix} = \begin{bmatrix} 3x - z \\ 4x + 2y \end{bmatrix}$.

7. (a) $(ST)\left(\begin{bmatrix} x \\ y \end{bmatrix}\right) = \begin{bmatrix} x - y \\ x + y \end{bmatrix}$; $(TU)\left(\begin{bmatrix} x \\ y \end{bmatrix}\right) = \begin{bmatrix} 5x \\ -x \end{bmatrix}$.

(b) $[(ST)U]\left(\begin{bmatrix} x \\ y \end{bmatrix}\right) = \begin{bmatrix} -x \\ 5x \end{bmatrix}$; $[S(TU)]\left(\begin{bmatrix} x \\ y \end{bmatrix}\right) = \begin{bmatrix} -x \\ 5x \end{bmatrix}$.

Exercises 5.3

1. (a) $\begin{bmatrix} 0 & 1 & 0 \\ 1 & 0 & 0 \\ 0 & 0 & 1 \end{bmatrix}$.

3. Scaling in both the x- and y-directions by the given factors is accomplished by multiplying first by $\begin{bmatrix} k_1 & 0 \\ 0 & 1 \end{bmatrix}$ and then by $\begin{bmatrix} 1 & 0 \\ 0 & k_2 \end{bmatrix}$ (or the other way around). Since these matrices commute and the product is $\begin{bmatrix} k_1 & 0 \\ 0 & k_2 \end{bmatrix}$, multiplication by this one matrix accomplishes both scalings.

Section 5.4—True/False

1. True.

2. False. Let $v_1 = \begin{bmatrix} 0 \\ 1 \end{bmatrix}$ and $v_2 = \begin{bmatrix} 1 \\ 0 \end{bmatrix}$. Since $\begin{bmatrix} -2 \\ 3 \end{bmatrix} = 3v_1 - 2v_2$, the coordinate vector is $\begin{bmatrix} 3 \\ -2 \end{bmatrix}$.
3. True: $\begin{bmatrix} -2 \\ 3 \end{bmatrix} = 1\begin{bmatrix} -2 \\ 3 \end{bmatrix} + 0\begin{bmatrix} 5 \\ 7 \end{bmatrix}$.
4. True. $v_1 = 1v_1 + 0v_2$.
5. True. $\begin{bmatrix} 3 \\ -4 \end{bmatrix} = -1v_1 + 1v_2$.
6. True: $\mathrm{id}(v_1) = v_1$ has coordinate vector $\begin{bmatrix} 1 \\ 0 \end{bmatrix}$ relative to $\mathcal{V}$ and $\mathrm{id}(v_2) = v_2$ has coordinate vector $\begin{bmatrix} 0 \\ 1 \end{bmatrix}$ relative to $\mathcal{V}$.
7. True, as in Question 6.
8. False. The second column of $M_{\mathcal{E}\leftarrow\mathcal{V}}(\mathrm{id})$ should be the coordinate vector of $\mathrm{id}(v_2)$ relative to $\mathcal{E}$, and this is the coordinate vector of v_2 relative to $\mathcal{E}$ which is $\begin{bmatrix} 1 \\ 1 \end{bmatrix}$.
9. True. The first column should be the coordinate vector of $\mathrm{id}(e_1)$ relative to $\mathcal{V}$, and it is: $e_1 = 1v_1 + 0v_2$. The second column should be the coordinate vector of $\mathrm{id}(e_2)$ relative to $\mathcal{V}$ and it is: $e_2 = -1v_1 + 1v_2$.
10. True. The first column should be the coordinate vector of $\mathrm{id}(v_1)$ relative to $\mathcal{V}$, and it is: $v_1 = 1v_1 + 0v_2$. The second column should be the coordinate vector of $\mathrm{id}(v_2)$ relative to $\mathcal{V}$ and it is: $v_2 = 0v_1 + 1v_2$.

Exercises 5.4

1. $\frac{1}{17}x_1 + \frac{3}{17}x_2$ and $-\frac{5}{17}x_1 + \frac{2}{17}x_2$.

4. (a) $T\left(\begin{bmatrix} x_1 \\ x_2 \end{bmatrix}\right) = \begin{bmatrix} x_1 + 2x_2 \\ 3x_1 + 4x_2 \end{bmatrix}$, (b) $T\left(\begin{bmatrix} x_1 \\ x_2 \end{bmatrix}\right) = \begin{bmatrix} x_1 + x_2 \\ 3x_1 + x_2 \end{bmatrix}$,

(c)] $T\left(\begin{bmatrix} x_1 \\ x_2 \end{bmatrix}\right) = \begin{bmatrix} 4x_1 + 6x_2 \\ 3x_1 + 4x_2 \end{bmatrix}$, (d) $T\left(\begin{bmatrix} x_1 \\ x_2 \end{bmatrix}\right) = \begin{bmatrix} 4x_1 + 2x_2 \\ 3x_1 + x_2 \end{bmatrix}$.

6. (a) $M_{\mathcal{E}\leftarrow\mathcal{E}}(T) = \begin{bmatrix} 1 & 4 \\ 2 & 7 \end{bmatrix}$, (b) $M_{\mathcal{E}\leftarrow\mathcal{V}}(T) = \begin{bmatrix} -3 & 2 \\ -5 & 3 \end{bmatrix}$,

(c) $M_{\mathcal{V}\leftarrow\mathcal{E}}(T) = \begin{bmatrix} -5 & -18 \\ -3 & -11 \end{bmatrix}$, (d) $M_{\mathcal{V}\leftarrow\mathcal{V}}(T) = \begin{bmatrix} 13 & -8 \\ 8 & -5 \end{bmatrix}$.

Section 5.5—True/False

1. False. Any invertible matrix is the matrix of the identity transformation relative to suitable bases. See Proposition 5.5.3.

2. True. Any invertible matrix is a matrix for the identity.
3. True. $P_{\mathcal{V}\to\mathcal{V}'} = P^{-1}_{\mathcal{V}'\to\mathcal{V}}$. Here $P^2 = I$, so $P^{-1} = P$.
4. True. Let $\mathsf{x} = a\mathsf{v}_1 + b\mathsf{v}_2$. Then $\mathsf{x}_\mathcal{V} = \begin{bmatrix} a \\ b \end{bmatrix}$ and the equation $P_{\mathcal{W}\leftarrow\mathcal{V}}\begin{bmatrix} a \\ b \end{bmatrix} = \begin{bmatrix} c \\ d \end{bmatrix}$ is $P_{\mathcal{W}\leftarrow\mathcal{V}}\mathsf{x}_\mathcal{V} = \begin{bmatrix} c \\ d \end{bmatrix}$, which says that the coordinate vector of x relative to $\mathcal{W}$ is $\begin{bmatrix} c \\ d \end{bmatrix}$, that is, $\mathsf{x} = c\mathsf{w}_1 + d\mathsf{w}_2$.
5. True. This is Definition 5.5.5.
6. True.
7. False. The correct matrix is the inverse of the one given.
8. True: $[T(\mathsf{x})]_\mathcal{V} = M_{\mathcal{V}\leftarrow\mathcal{E}}\mathsf{x}_\mathcal{E}$.
9. True: $M_{\mathcal{V}\leftarrow\mathcal{E}}(T) = P_{\mathcal{V}\leftarrow\mathcal{E}}M_{\mathcal{E}\leftarrow\mathcal{E}}(T)$.
10. True, since $\det P^{-1}AP = \det A$.

Exercises 5.5

1. (a) $M_{\mathcal{E}\leftarrow\mathcal{E}}(\mathrm{id}) = \begin{bmatrix} 1 & 0 \\ 0 & 1 \end{bmatrix}$, (c) $M_{\mathcal{E}\leftarrow\mathcal{V}}(\mathrm{id}) = \begin{bmatrix} -1 & 2 \\ 1 & 0 \end{bmatrix}$.

2. (a) $P_{\mathcal{V}\leftarrow\mathcal{E}} = P^{-1} = \begin{bmatrix} 0 & 1 & 0 \\ 0 & 0 & 1 \\ \frac{1}{3} & -\frac{1}{3} & 0 \end{bmatrix}$.

   (b) The coordinates are the components of $P_{\mathcal{V}\leftarrow\mathcal{E}}\begin{bmatrix} 2 \\ 1 \\ -1 \end{bmatrix} = \begin{bmatrix} 1 \\ 2 \\ -\frac{2}{3} \end{bmatrix}$.

6. If $\mathsf{x}_\mathcal{V} = \begin{bmatrix} 3 \\ 1 \\ -2 \end{bmatrix}$, then $\mathsf{x} = \mathsf{x}_\mathcal{E} = P_{\mathcal{E}\leftarrow\mathcal{V}}\mathsf{x}_\mathcal{V} = \begin{bmatrix} 1 \\ -1 \\ 2 \end{bmatrix}$. If $\mathsf{y}_\mathcal{V} = \begin{bmatrix} 4 \\ 3 \\ -4 \end{bmatrix}$, then $\mathsf{y} = \mathsf{y}_\mathcal{E} = P_{\mathcal{E}\leftarrow\mathcal{V}}\mathsf{y}_\mathcal{V} = \begin{bmatrix} 0 \\ 2 \\ 3 \end{bmatrix}$. So $T(\mathsf{x}) = \mathsf{y}$ because $T(\mathsf{x}) = \begin{bmatrix} -x_1 + x_2 + x_3 \\ x_3 \\ 2x_1 - 3x_2 - x_3 \end{bmatrix}$ implies $T\left(\begin{bmatrix} 1 \\ -1 \\ 2 \end{bmatrix}\right) = \begin{bmatrix} 0 \\ 2 \\ 3 \end{bmatrix}$.

7. (a) $T(\mathsf{u}) = \begin{bmatrix} -109 \\ -24 \end{bmatrix}$, $T(\mathsf{v}) = \begin{bmatrix} 712 \\ 157 \end{bmatrix}$ and $T(\mathsf{w}) = \begin{bmatrix} 1155 \\ 255 \end{bmatrix}$.

   (b) $T(\mathsf{u}) = -109\begin{bmatrix} -1 \\ 1 \end{bmatrix} - 24\begin{bmatrix} 4 \\ -5 \end{bmatrix} = \begin{bmatrix} 13 \\ 11 \end{bmatrix}$,

   $T(\mathsf{v}) = 712\begin{bmatrix} -1 \\ 1 \end{bmatrix} + 157\begin{bmatrix} 4 \\ -5 \end{bmatrix} = \begin{bmatrix} -84 \\ -73 \end{bmatrix}$,

   $T(\mathsf{w}) = 1155\begin{bmatrix} -1 \\ 1 \end{bmatrix} + 255\begin{bmatrix} 4 \\ -5 \end{bmatrix} = \begin{bmatrix} -135 \\ -120 \end{bmatrix}$.

10. (a) $M_{\mathcal{V}\leftarrow\mathcal{V}}(T) = \frac{1}{2}\begin{bmatrix} 3 & -1 & 6 \\ -3 & -3 & -6 \\ 1 & 2 & 4 \end{bmatrix}$.

11. (a) $\mathsf{u} = \begin{bmatrix} -8 \\ 4 \\ 16 \end{bmatrix}, \mathsf{v} = \begin{bmatrix} 0 \\ 1 \\ 1 \end{bmatrix}, \mathsf{w} = \begin{bmatrix} -4 \\ -2 \\ 7 \end{bmatrix}$.

13. (a) $P_{\mathcal{E}\leftarrow\mathcal{V}} = \begin{bmatrix} 1 & 1 & -1 \\ 1 & -1 & 0 \\ 1 & 0 & 1 \end{bmatrix}$.

16. (a) $A = M_{\mathcal{E}\leftarrow\mathcal{E}}(T) = \begin{bmatrix} 2 & -1 \\ 0 & 3 \end{bmatrix}$; (b) $B = M_{\mathcal{V}\leftarrow\mathcal{V}}(T) = \frac{1}{7}\begin{bmatrix} 11 & -15 \\ 2 & 24 \end{bmatrix}$.

    (c) We have $B = P_{\mathcal{V}\leftarrow\mathcal{E}} A P_{\mathcal{E}\leftarrow\mathcal{V}}$. Let $P = P_{\mathcal{E}\leftarrow\mathcal{V}} = \begin{bmatrix} 1 & -2 \\ 1 & 5 \end{bmatrix}$.

    Then $P_{\mathcal{V}\leftarrow\mathcal{E}} = P^{-1} = \frac{1}{7}\begin{bmatrix} 5 & 2 \\ -1 & 1 \end{bmatrix}$

    so $B = P^{-1}AP = \frac{1}{7}\begin{bmatrix} 5 & 2 \\ -1 & 1 \end{bmatrix}\begin{bmatrix} 2 & -1 \\ 0 & 3 \end{bmatrix}\begin{bmatrix} 1 & -2 \\ 1 & 5 \end{bmatrix}$.

19. $B = \begin{bmatrix} -68 & 59 \\ -55 & 72 \end{bmatrix}$.

21. $B = \frac{1}{2}\begin{bmatrix} -4 & -8 & -4 \\ 3 & 6 & 2 \\ -3 & -4 & 0 \end{bmatrix}$.

Section 6.1—True/False

1. False. The length of $\begin{bmatrix} 1 \\ 1 \\ 1 \\ 1 \end{bmatrix}$ is $\sqrt{4} = 2$.
2. True. $\mathsf{x} \cdot \mathsf{x}$ is a sum of squares of real numbers.
3. False: $\mathsf{u} \cdot \mathsf{v} = \mathsf{u}^T\mathsf{v}$.
4. False. The columns of A must be linearly independent.
5. False. It's the vector x that makes $\|\mathsf{b} - A\mathsf{x}\|$ as small as possible.
6. True.
7. True. Multiply $P^2 = P$ on both sides by P^{-1}.
8. True. This is just READING CHECK 6 in the case that $\mathsf{v} \cdot \mathsf{v} = 1$.
9. False. The vectors u and v must be orthogonal.
10. True. The vector b clearly has the property that it is in U and $\mathsf{b} - \mathsf{b}$ is orthogonal to U.

Exercises 6.1

1. A is not square matrices, so A^T is not square. The step $(AA^T)^{-1} = (A^T)^{-1}A^{-1}$ is invalid.

3. (a) $\mathsf{x}^+ = \frac{1}{14}\begin{bmatrix} 31 \\ -2 \end{bmatrix}$.

   (b) This is "best" in the sense that $\|\mathsf{b} - A\mathsf{x}^+\|$ is minimal as x varies over all $\mathsf{x} \in \mathsf{R}^2$.

   Equally good would be to say that the vector $A\mathsf{x}^+ = \dfrac{1}{14}\begin{bmatrix} 29 \\ 33 \\ 60 \end{bmatrix}$ is the vector in the column space of A which minimizes $\|\mathsf{b} - A\mathsf{x}\|$.

4. (a) The best solution is $\mathsf{x}^+ = \frac{1}{58}\begin{bmatrix} 11 \\ 32 \end{bmatrix}$.

5. (a) No, (d) Yes.

6. (a) $P = \dfrac{1}{35}\begin{bmatrix} 11 & 14 & 2 & 8 \\ 14 & 21 & -7 & 7 \\ 2 & -7 & 29 & 11 \\ 8 & 7 & 11 & 9 \end{bmatrix}$.

7. (a) There is a solution if and only if $b_1 - b_2 + 3b_3 - 2b_4 = 0$.

   (b) Let $\mathsf{a}_1, \mathsf{a}_2, \mathsf{a}_3, \mathsf{a}_4$ be the columns of A. Suppose $c_1\mathsf{a}_1 + c_2\mathsf{a}_2 + c_3\mathsf{a}_3 = \mathsf{0}$.

   This is $A\mathsf{c} = \mathsf{0}$ with $\mathsf{c} = \begin{bmatrix} c_1 \\ c_2 \\ c_3 \end{bmatrix}$. The reader should check that $A\mathsf{c} = \mathsf{0}$ implies $\mathsf{c} = 0$.

   (c) The columns of A are linearly independent.

   (d) The components of b do not satisfy $b_1 - b_2 + 3b_3 - 2b_4 = 0$, the condition found in part (a).

   (e) $\mathsf{x}^+ = -\begin{bmatrix} \frac{4}{15} \\ \frac{3}{5} \\ \frac{1}{5} \end{bmatrix}$.

   (f) The vector $\mathsf{b} - A\mathsf{x}^+$ has minimal length amongst all vectors of the form $\mathsf{b} - A\mathsf{x}$. Alternatively, $A\mathsf{x}^+$ is the projection of b on col sp A.

   (g) $P = \dfrac{1}{15}\begin{bmatrix} 14 & 1 & -3 & 2 \\ 1 & 14 & 3 & -2 \\ -3 & 3 & 6 & 6 \\ 2 & -2 & 6 & 11 \end{bmatrix}$.

   (h) $P^2 = P$.

   (i) The solution is $\mathsf{x} = -\begin{bmatrix} \frac{4}{15} \\ \frac{3}{5} \\ \frac{1}{5} \end{bmatrix}$, which is x^+, as it should be.

10. Let $A = c_1P_1 + c_2P_2 + \cdots + c_nP_n$ be a linear combination of projection matrices $P_1, P_2, \ldots, P_n$. Since $P_1^T = P_1$, $P_2^T = P_2$, and so on, $A^T = c_1P_1^T + c_2P_2^T + \cdots + c_nP_n^T = c_1P_1 + c_2P_2 + \cdots P_n = A$. Thus A is symmetric.

13. For $P = AA^T$, $P^2 = AA^TAA^T = AIA^T = P$ and $P^T = (A^T)^TA^T = AA^T = P$.

16. (a) i. $\mathsf{x}^+ = \frac{1}{3}\begin{bmatrix} 2 \\ -1 \end{bmatrix}$.

ii. $A\mathsf{x}^+ = \frac{1}{3}\begin{bmatrix} 3 \\ -2 \\ 1 \\ 5 \end{bmatrix}$ is the vector in the column space of A which minimizes $\|\mathsf{b} - A\mathsf{x}\|$.

iii. $P = \frac{1}{3}\begin{bmatrix} 1 & 0 & 1 & 1 \\ 0 & 1 & 1 & -1 \\ 1 & 1 & 2 & 0 \\ 1 & -1 & 0 & 2 \end{bmatrix}$.

iv. Gaussian elimination on the augmented matrix $[A|P\mathsf{b}]$ proceeds

$$\left[\begin{array}{cc|c} 1 & -1 & 1 \\ 0 & 2 & -\frac{2}{3} \\ 1 & 1 & \frac{1}{3} \\ 1 & -3 & \frac{5}{3} \end{array}\right] \to \left[\begin{array}{cc|c} 1 & -1 & 1 \\ 0 & 2 & -\frac{2}{3} \\ 0 & 2 & -\frac{2}{3} \\ 0 & -2 & \frac{2}{3} \end{array}\right] \to \left[\begin{array}{cc|c} 1 & -1 & 1 \\ 0 & 1 & -\frac{1}{3} \\ 0 & 0 & 0 \\ 0 & 0 & 0 \end{array}\right],$$

so $x_2{}^+ = -\frac{1}{3}$, $x_1{}^+ = 1 + x_2{}^+ = \frac{2}{3}$ and $\mathsf{x}^+ = \begin{bmatrix} \frac{2}{3} \\ -\frac{1}{3} \end{bmatrix}$ in agreement with our answer to part i.

17. **Method 1:** We form the matrix $A = \begin{bmatrix} 1 & 1 \\ 0 & 1 \\ 1 & 0 \end{bmatrix}$ whose columns are the given vectors. We wish to project vectors onto the column space of A. We have $A^TA = \begin{bmatrix} 2 & 1 \\ 1 & 2 \end{bmatrix}$ and $(A^TA)^{-1} = \frac{1}{3}\begin{bmatrix} 2 & -1 \\ -1 & 2 \end{bmatrix}$. The desired projection matrix is $P = A(A^TA)^{-1}A^T = \frac{1}{3}\begin{bmatrix} 1 & 1 \\ 0 & 1 \\ 1 & 0 \end{bmatrix} 3 \begin{bmatrix} 2 & -1 \\ -1 & 2 \end{bmatrix} \begin{bmatrix} 1 & 0 & 1 \\ 1 & 1 & 0 \end{bmatrix} = \frac{1}{3}\begin{bmatrix} 2 & 1 & 1 \\ 1 & 2 & -1 \\ 1 & -1 & 2 \end{bmatrix}$.

Method 2: We replace the given vectors by two **orthogonal** vectors which span the same plane, taking, for example, $\mathsf{e} = \begin{bmatrix} 1 \\ 0 \\ 1 \end{bmatrix}$ and $\mathsf{f} = \begin{bmatrix} 1 \\ 2 \\ -1 \end{bmatrix}$. Then

$$P = \frac{\mathsf{e}\mathsf{e}^T}{\mathsf{e}\cdot\mathsf{e}} + \frac{\mathsf{f}\mathsf{f}^T}{\mathsf{f}\cdot\mathsf{f}} = \frac{1}{2}\begin{bmatrix} 1 & 0 & 1 \\ 0 & 0 & 0 \\ 1 & 0 & 1 \end{bmatrix} + \frac{1}{6}\begin{bmatrix} 1 & 2 & -1 \\ 2 & 4 & -2 \\ -1 & -2 & 1 \end{bmatrix} = \frac{1}{3}\begin{bmatrix} 2 & 1 & 1 \\ 1 & 2 & -1 \\ 1 & -1 & 2 \end{bmatrix}$$

as before.

19. $P = \frac{1}{6}\begin{bmatrix} 1 & 0 & -1 & -2 \\ 0 & 1 & 2 & -1 \\ -1 & 2 & 5 & 0 \\ -2 & -1 & 0 & 5 \end{bmatrix}$.

20. (a) For example, U is spanned by the vectors $\mathsf{u}_1 = \begin{bmatrix} 1 \\ 1 \\ 0 \\ 0 \end{bmatrix}$, $\mathsf{u}_2 = \begin{bmatrix} 0 \\ 0 \\ 1 \\ 0 \end{bmatrix}$ and

$$\mathbf{u}_3 = \begin{bmatrix} -1 \\ 0 \\ 0 \\ 1 \end{bmatrix}.$$

21. (a) $P = \frac{1}{35}\begin{bmatrix} 1 & -3 & 5 \\ -3 & 9 & -15 \\ 5 & -15 & 25 \end{bmatrix}$. (c) $P = \frac{1}{66}\begin{bmatrix} 65 & -4 & 7 \\ -4 & 50 & 28 \\ 7 & 28 & 17 \end{bmatrix}$.

(e) $P = \frac{1}{30}\begin{bmatrix} 26 & 2 & -10 \\ 2 & 29 & 5 \\ -10 & 5 & 5 \end{bmatrix}$.

23. (a) Since the columns of A are linearly dependent, the methods of this section don't apply directly. The object, however, is to project vectors onto the column space of A, which is the span of $\mathbf{v} = \begin{bmatrix} 1 \\ 2 \\ 3 \end{bmatrix}$. The answer is $P = \frac{\mathbf{v}\mathbf{v}^T}{\mathbf{v}\cdot\mathbf{v}} = \frac{1}{14}\begin{bmatrix} 1 \\ 2 \\ 3 \end{bmatrix}\begin{bmatrix} 1 & 2 & 3 \end{bmatrix} = \frac{1}{14}\begin{bmatrix} 1 & 2 & 3 \\ 2 & 4 & 6 \\ 3 & 6 & 9 \end{bmatrix}$.

24. (a) Let $B = A^TA$. We are asked to prove that the inverse of B^T is $(B^{-1})^T$. We show that the product of these matrices is the identity matrix. Remembering that $X^TY^T = (XY)^T$, we simply compute $B^T(B^{-1})^T = (B^{-1}B)^T = I^T = I$.

26. Vectors of the form $\begin{bmatrix} 0 \\ 0 \\ z \end{bmatrix}$ are eigenvectors corresponding to the eigenvalue $\lambda = 0$. Vectors of the form $\begin{bmatrix} x \\ y \\ 0 \end{bmatrix}$ are eigenvectors corresponding to $\lambda = 1$.

29. $P = I$ is the $n \times n$ identity matrix. By definition, $\mathbf{b} - P\mathbf{b}$ must be orthogonal to the column space of P for all $\mathbf{b}$, and the column space of P is the space onto which P projects; here this is all of R^n. The only vector in R^n which is orthogonal to R^n is the zero vector. So $\mathbf{b} - P\mathbf{b} = \mathbf{0}$ for all $\mathbf{b}$, and this says $P = I$.

32. The column space is the xz-plane. $P = \begin{bmatrix} 1 & 0 & 0 \\ 0 & 0 & 0 \\ 0 & 0 & 1 \end{bmatrix}$.

Exercises 6.2

1. (a) The "best" line has equation $y = \frac{59}{14}x - \frac{9}{7}$.

   (c) The "best" line has equation $y = -\frac{9}{5}x + \frac{3}{5}$.

2. (a) The best parabola has equation $y = \frac{3}{4}x^2 - \frac{13}{20}x + \frac{9}{20}$.

3. (a) The best cubic is $f(x) = -\frac{1}{4}x^3 - \frac{2}{7}x^2 + \frac{3}{4}x + \frac{4}{7}$.

5. (a) The best function is $f(x) = -\frac{17}{20}x + \frac{1}{6}2^x \approx 0.85x + 0.17(2^x)$.

6. (b) i. The center of the best circle is $(\frac{17}{10}, -\frac{7}{10})$ and the radius is $\frac{13}{5\sqrt{2}}$.

8. (a) The best curve has equation $y = \frac{7\sqrt{3}+9}{6}\sin x + \frac{13\sqrt{3}-9}{18}\sin 2x$ $\approx 3.52\sin x + 0.75\sin 2x$.

9. (c) i. The best plane has equation $0.51x - 0.27y + 0.13z = 1$. The sum of squares of deviations is 7.35.

Section 6.3—True/False

1. True.
2. False. The given vectors are not linearly independent.
3. True. An orthonormal basis is an orthogonal basis. On the other hand, an orthogonal basis can be converted to an orthonormal basis by dividing each vector by its length.
4. True. A permutation matrix is square with orthonormal columns since its columns are the standard basis vectors $\mathsf{e}_1, \mathsf{e}_2, \ldots, \mathsf{e}_n$ in some order.
5. False. This matrix has orthonormal columns but it is not square.
6. True. The given matrix is rotation through θ; its inverse is rotation through $-\theta$.
7. True. Orthonormal vectors are linearly independent and any linearly independent set of vectors can always be extended to a basis.
8. True.
9. True. See comments after Definition 6.3.10.
10. False. The $(1,1)$ and $(2,2)$ entries should be $\cos\theta$ but here these entries are not equal.

Exercises 6.3

1. (a) $\mathsf{f}_1 = \mathsf{u}_1 = \begin{bmatrix} 0 \\ 1 \\ 3 \end{bmatrix}, \mathsf{f}_2 = \begin{bmatrix} -1 \\ -\frac{3}{10} \\ \frac{1}{10} \end{bmatrix}$.

3. $Q = \begin{bmatrix} \frac{1}{\sqrt{3}} & \frac{2}{\sqrt{15}} & \frac{3}{\sqrt{35}} \\ 0 & -\frac{3}{\sqrt{15}} & \frac{3}{\sqrt{35}} \\ -\frac{1}{\sqrt{3}} & \frac{1}{\sqrt{15}} & -\frac{1}{\sqrt{35}} \\ \frac{1}{\sqrt{3}} & -\frac{1}{\sqrt{15}} & -\frac{4}{\sqrt{35}} \end{bmatrix}$ and $R = \begin{bmatrix} \sqrt{3} & -\frac{\sqrt{3}}{3} & \sqrt{3} \\ 0 & 2\frac{\sqrt{15}}{3} & \frac{\sqrt{15}}{5} \\ 0 & 0 & \frac{\sqrt{35}}{5} \end{bmatrix}$.

4. (a) $Q = \begin{bmatrix} \frac{1}{\sqrt{2}} & \frac{1}{\sqrt{2}} \\ \frac{1}{\sqrt{2}} & -\frac{1}{\sqrt{2}} \end{bmatrix}$ and $R = \begin{bmatrix} \sqrt{2} & 2\sqrt{2} \\ 0 & 2\sqrt{2} \end{bmatrix}$.

(c) $Q = \begin{bmatrix} \frac{1}{\sqrt{2}} & -\frac{1}{\sqrt{3}} & \frac{1}{\sqrt{6}} \\ 0 & \frac{1}{\sqrt{3}} & \frac{2}{\sqrt{6}} \\ \frac{1}{\sqrt{2}} & \frac{1}{\sqrt{3}} & -\frac{1}{\sqrt{6}} \end{bmatrix}$ and $R = \begin{bmatrix} \sqrt{2} & 0 & \frac{3\sqrt{2}}{2} \\ 0 & \sqrt{3} & 0 \\ 0 & 0 & \frac{\sqrt{6}}{2} \end{bmatrix}$.

5. (a) $Q = \begin{bmatrix} \frac{1}{\sqrt{2}} & \frac{1}{\sqrt{2}} \\ \frac{1}{\sqrt{2}} & -\frac{1}{\sqrt{2}} \end{bmatrix}$ and $R = \begin{bmatrix} \sqrt{2} & 2\sqrt{2} \\ 0 & 2\sqrt{2} \end{bmatrix}$, so $A^{-1} = R^{-1}Q^{-1} = R^{-1}Q^T =$ $\begin{bmatrix} \frac{\sqrt{2}}{2} & -\frac{\sqrt{2}}{2} \\ 0 & \frac{\sqrt{2}}{4} \end{bmatrix} \begin{bmatrix} \frac{1}{\sqrt{2}} & \frac{1}{\sqrt{2}} \\ \frac{1}{\sqrt{2}} & -\frac{1}{\sqrt{2}} \end{bmatrix} = \begin{bmatrix} 0 & 1 \\ \frac{1}{4} & -\frac{1}{4} \end{bmatrix}$.

6. (a) $Q = \begin{bmatrix} \frac{1}{3} & 0 \\ \frac{2}{3} & \frac{1}{\sqrt{2}} \\ \frac{2}{3} & -\frac{1}{\sqrt{2}} \end{bmatrix}$ and $R = \begin{bmatrix} 3 & 3 \\ 0 & \sqrt{2} \end{bmatrix}$. The desired vector x^+ is the solution to $R\mathsf{x} = Q^T\mathsf{b} = \begin{bmatrix} \frac{1}{3} & \frac{2}{3} & \frac{2}{3} \\ 0 & \frac{1}{\sqrt{2}} & -\frac{1}{\sqrt{2}} \end{bmatrix} \begin{bmatrix} 1 \\ -1 \\ 1 \end{bmatrix} = \begin{bmatrix} \frac{1}{3} \\ -\frac{2}{\sqrt{2}} \end{bmatrix}$. The augmented matrix is $\left[\begin{array}{cc|c} 3 & 3 & \frac{1}{3} \\ 0 & \sqrt{2} & -\frac{2}{\sqrt{2}} \end{array}\right]$. Back substitution gives $x_2 = -1$ and $x_1 = \frac{1}{9} - x_2 = \frac{10}{9}$. The best solution is $\mathsf{x}^+ = \begin{bmatrix} \frac{10}{9} \\ -1 \end{bmatrix}$.

8. Let $Q = (I+K)(I-K)^{-1}$. It suffices to show that $Q^TQ = I$. Recalling that $(XY)^T = Y^TX^T$ for matrices X and Y and noting that $(X^{-1})^T = (X^T)^{-1}$ if X is invertible, we have $Q^TQ = [(I-K)^T]^{-1}(I+K)^T(I+K)(I-K)^{-1}$. Now $(I-K)^T = I^T - K^T = I+K$ and $(I+K)^T = I^T + K^T = I-K$. Also, the matrices $I+K$ and $I-K$ commute. Thus $Q^TQ = (I+K)^{-1}(I-K)(I+K)(I-K)^{-1} = (I+K)^{-1}(I+K)(I-K)(I-K)^{-1} = I$. So Q is orthogonal.

10. (a) Since Q is orthogonal, $Q^T = Q^{-1}$ and since Q is symmetric, $Q^T = Q$. Thus $Q = Q^{-1}$, so $Q^2 = I$.

12. (a) It suffices to show that $(Q_1Q_2)^T(Q_1Q_2) = I$. Since $Q_1^TQ_1 = I$ and $Q_2^TQ_2 = I$, we have $(Q_1Q_2)^T(Q_1Q_2) = Q_2^TQ_1^TQ_1Q_2 = Q_2^TQ_2 = I$.

17. (a) The subspace spanned by $\mathsf{u}_1, \mathsf{u}_2, \ldots, \mathsf{u}_k$ has dimension less than k, so it cannot contain k orthogonal (and hence linearly independent) vectors.

(b) Since $\mathsf{u}_1, \mathsf{u}_2, \ldots, \mathsf{u}_k$ are linearly dependent, some u_t is a linear combination of the previous vectors in the list. Let U be the span of these vectors; that is, $U = \operatorname{sp}\{\mathsf{u}_1, \ldots, \mathsf{u}_{t-1}\}$. Since u_t is in U, the projection of u_t on U is u_t and the next vector $\mathsf{f}_t = \mathsf{u}_t - \operatorname{proj}_U \mathsf{u}_t$ will be $\mathbf{0}$.

19. Let $\mathsf{x}_1, \ldots, \mathsf{x}_n$ be any basis and apply the Gram-Schmidt algorithm.

21. (a) $\mathsf{y} = H\mathsf{x} - \mathsf{x} = (I - 2\mathsf{w}\mathsf{w}^T)\mathsf{x} - \mathsf{x} = I\mathsf{x} - 2\mathsf{w}\mathsf{w}^T\mathsf{x} - \mathsf{x} = -2\mathsf{w}(\mathsf{w} \cdot \mathsf{x}) = -(2\mathsf{w} \cdot \mathsf{x})\mathsf{w}$.

(b) Using the result of (a), we have $\mathsf{x} + \frac{1}{2}\mathsf{y} = \mathsf{x} - (\mathsf{w} \cdot \mathsf{x})\mathsf{w}$. Thus $[\mathsf{x} + \frac{1}{2}\mathsf{y}] \cdot \mathsf{w} = \mathsf{x} \cdot \mathsf{w} - (\mathsf{w} \cdot \mathsf{x})(\mathsf{w} \cdot \mathsf{w}) = 0$ because $\mathsf{w} \cdot \mathsf{w} = \|\mathsf{w}\|^2 = 1$.

Section 6.4—True/False

1. True. The vectors span a subspace of dimension 2.
2. True. A two-dimensional subspace of R^3 is a plane and the intersection of two planes is a line.
3. True. Since $\dim(U + W) \le 5$, the formula $\dim(U + W) + \dim(U \cap W) = \dim U + \dim W = 6$ implies $\dim(U \cap W) \neq 0$.
4. True. Either statement is equivalent to the assertion that $\mathsf{u} \cdot \mathsf{w} = 0$ for any u in U and any w in W.
5. False. These planes have nonzero intersection (the y-axis).
6. False. Let $V = \mathsf{R}^3$, let U be the x-axis and W the y-axis.
7. True. In Theorem 6.4.17, we showed that $R^\perp = \text{null sp}\, A$, so $(\text{null sp}\, A)^\perp = (R^\perp)^\perp = R$.
8. True: $W = U^\perp$.
9. True. $A\mathsf{x} = 0$ says x is in the null space of A, which is the orthogonal complement of the row space of A. Since $A^T = A$, the row space of A is U, its column space.
10. False. The columns of the generator matrix are a basis for C: $\dim C = s$.

Exercises 6.4

1. (a) A basis for $U + W$ is $\mathsf{u}_1 = \begin{bmatrix} 1 \\ 1 \\ 2 \\ 0 \end{bmatrix}$, $\mathsf{u}_2 = \begin{bmatrix} -1 \\ 1 \\ -1 \\ 1 \end{bmatrix}$, $\mathsf{w}_1 = \begin{bmatrix} -1 \\ 5 \\ 0 \\ 3 \end{bmatrix}$. A basis for $U \cap W$ is $\begin{bmatrix} 1 \\ -5 \\ -1 \\ -3 \end{bmatrix}$. Since $\dim U = \dim W = 2$, $\dim(U + W) + \dim(U \cap W) = 3 + 1 = 2 + 2 = 4$.

   (c) We have $U + W = U$, so W is contained in U.
   For $U + W$, $\begin{bmatrix} -2 \\ 0 \\ 2 \\ 4 \\ 1 \end{bmatrix}$, $\begin{bmatrix} 3 \\ 5 \\ 2 \\ 4 \\ 1 \end{bmatrix}$, $\begin{bmatrix} 6 \\ 3 \\ 3 \\ -6 \\ -1 \end{bmatrix}$. For $U \cap W = W$, $\begin{bmatrix} 1 \\ 2 \\ -5 \\ 2 \\ 0 \end{bmatrix}$ and $\begin{bmatrix} 3 \\ 1 \\ 10 \\ -4 \\ 0 \end{bmatrix}$.
   The result is easy because $\dim(U + W) = \dim U$ and $\dim(U \cap W) = \dim W$.

2. (a) $U^\perp$ is the plane with equation $2x - 3y - 7z = 0$.

   (c) $U^\perp$ is the one-dimensional subspace consisting of multiples of the normal $\begin{bmatrix} 1 \\ -2 \\ 1 \end{bmatrix}$.

   (e) $U^\perp$ is the set of multiples of $\begin{bmatrix} 5 \\ -2 \\ 1 \end{bmatrix}$.

3. (a) $\mathsf{u} = \frac{1}{4}\begin{bmatrix} 1 \\ 7 \\ 11 \\ 17 \end{bmatrix}$, $\mathsf{w} = \mathsf{v} - \mathsf{u} = \frac{7}{4}\begin{bmatrix} 1 \\ -1 \\ -1 \\ 1 \end{bmatrix}$.

4. (a) $\mathsf{u} = \operatorname{proj}_U \mathsf{v} = \frac{x+y}{2}\begin{bmatrix} 1 \\ 1 \end{bmatrix} = \begin{bmatrix} \frac{x+y}{2} \\ \frac{x+y}{2} \end{bmatrix}$ and $\mathsf{w} = \mathsf{v} - \mathsf{u} = \begin{bmatrix} \frac{x-y}{2} \\ \frac{-x+y}{2} \end{bmatrix}$.

   (c) Since U is two-dimensional and $U^\perp$ is one-dimensional with basis $\mathsf{b} = \begin{bmatrix} 2 \\ -1 \\ 3 \end{bmatrix}$, it is easier to project onto $U^\perp$. We have $\mathsf{w} = \operatorname{proj}_{\mathsf{b}} \mathsf{v} = \frac{1}{7}\begin{bmatrix} 2 \\ -1 \\ 3 \end{bmatrix}$, the desired vector in $U^\perp$, and $\mathsf{u} = \mathsf{v} - \mathsf{w} = \frac{1}{7}\begin{bmatrix} 19 \\ 8 \\ -10 \end{bmatrix}$.

5. (a) The matrix is $\frac{1}{30}\begin{bmatrix} 4 & -10 & -2 \\ -10 & 25 & 5 \\ -2 & 5 & 1 \end{bmatrix}$.

   (b) $U^\perp$ is the plane with equation $-2x + 5y + z = 0$.

   (c) $P = \frac{1}{30}\begin{bmatrix} 26 & 10 & 2 \\ 10 & 5 & -5 \\ 2 & -5 & 29 \end{bmatrix}$.

   (d) The vector we seek is the projection of $\begin{bmatrix} -1 \\ 2 \\ 3 \end{bmatrix}$ on the plane: the answer is $P\begin{bmatrix} -1 \\ 2 \\ 3 \end{bmatrix} = \frac{1}{2}\begin{bmatrix} 0 \\ -1 \\ 5 \end{bmatrix}$.

7. $\mathsf{x} = \operatorname{proj}_U \mathsf{v}$ is in U and $\mathsf{y} = \mathsf{v} - \operatorname{proj}_U \mathsf{v}$ is in $U^\perp$. Thus $\mathsf{v} = \mathsf{x} + \mathsf{y}$ is one way to write v as the sum of a vector in U and another in $U^\perp$. Part 2 of Theorem 6.4.26 says this is the only way.

8. Since U and W are subspaces, each contains 0. Thus $0 = 0 + 0 \in U + W$. So $U + W$ is not empty. Suppose $\mathsf{v}_1 = \mathsf{u}_1 + \mathsf{w}_1$ and $\mathsf{v}_2 = \mathsf{u}_2 + \mathsf{w}_2$ are each in $U + W$, $\mathsf{u}_1, \mathsf{u}_2 \in U$, $\mathsf{w}_1, \mathsf{w}_2 \in W$. Since U is a subspace $\mathsf{u}_1 + \mathsf{u}_2 \in U$ and, since W is a subspace, $\mathsf{w}_1 + \mathsf{w}_2 \in W$. Thus $\mathsf{v}_1 + \mathsf{v}_2 = (\mathsf{u}_1 + \mathsf{u}_2) + (\mathsf{w}_1 + \mathsf{w}_2) \in U + W$, showing that $U + W$ is closed under addition. Let $\mathsf{v} = \mathsf{u} + \mathsf{w}$, $\mathsf{u} \in U$, $\mathsf{w} \in W$ belong to $U + W$ and let c be a scalar. Since U is a subspace, $c\mathsf{u} \in U$ and, since W is a subspace, $c\mathsf{w} \in W$. Thus $c\mathsf{v} = (c\mathsf{u}) + (c\mathsf{w}) \in U + W$, showing that $U + W$ is also closed under scalar multiplication. Thus $U + W$ is a subspace.

9. Certainly if $\mathsf{x} \in U^\perp$, then each $\mathsf{x} \cdot \mathsf{u}_i = 0$ since $\mathsf{u}_i \in U$. Conversely, if $\mathsf{x} \cdot \mathsf{u}_i = 0$ for all i and u is any vector in U, then $\mathsf{u} = a_1\mathsf{u}_1 + \cdots + a_k\mathsf{u}_k$ for some scalars $a_1, \ldots, a_k$. So $\mathsf{x} \cdot \mathsf{u} = a_1\mathsf{x} \cdot \mathsf{u}_1 + a_2\mathsf{x} \cdot \mathsf{u}_k + \cdots + a_k\mathsf{x} \cdot \mathsf{u}_k = 0$, which shows that $\mathsf{x} \in U^\perp$.

11. (a) U has basis $\mathsf{u}_1 = \begin{bmatrix} 2 \\ 1 \\ 0 \\ -5 \end{bmatrix}$, $\mathsf{u}_2 = \begin{bmatrix} -1 \\ 0 \\ 1 \\ 2 \end{bmatrix}$.

   $U^\perp$ has basis $\mathsf{v}_1 = \begin{bmatrix} 1 \\ -2 \\ 1 \\ 0 \end{bmatrix}$ and $\mathsf{v}_2 = \begin{bmatrix} 2 \\ 1 \\ 0 \\ 1 \end{bmatrix}$.

(b) $\mathsf{u} = \operatorname{proj}_U \mathsf{b} = \begin{bmatrix} -\frac{1}{2} \\ -\frac{1}{6} \\ \frac{1}{6} \\ \frac{7}{6} \end{bmatrix}$; $\mathsf{w} = \operatorname{proj}_{U^\perp} \mathsf{b} = \begin{bmatrix} -\frac{1}{2} \\ \frac{1}{6} \\ -\frac{1}{6} \\ -\frac{1}{6} \end{bmatrix}$.

14. (b) (i $\Longrightarrow$ ii) Assume U and W are orthogonal. Let u be a vector in U. We wish to show that u is in $W^\perp$, that is, that $\mathsf{u} \cdot \mathsf{w} = 0$ for any w in W. This is true because U and W are orthogonal.

(ii $\Longrightarrow$ iii) Assume $U \subseteq W^\perp$. By part (a), $(W^\perp)^\perp \subseteq U$. Now use $(W^\perp)^\perp = W$.

(iii $\Longrightarrow$ i) Assume $W \subseteq U^\perp$. We wish to show that U and W are orthogonal, that is, $\mathsf{u} \cdot \mathsf{w} = 0$ for any u in U and any w in W. This is true because w is in $U^\perp$.

17. (a) $G = \begin{bmatrix} 1 & 0 & 0 \\ 0 & 1 & 0 \\ 0 & 0 & 1 \\ 0 & 1 & 1 \\ 1 & 0 & 1 \\ 1 & 1 & 0 \end{bmatrix}$ is a generator matrix for C. A parity check matrix is

$H = \begin{bmatrix} 0 & 1 & 1 & 1 & 0 & 0 \\ 1 & 0 & 1 & 0 & 1 & 0 \\ 1 & 1 & 0 & 0 & 0 & 1 \end{bmatrix}$.

(b) A generator matrix for $C^\perp$ is $H^T = \begin{bmatrix} 0 & 1 & 1 \\ 1 & 0 & 1 \\ 1 & 1 & 0 \\ 1 & 0 & 0 \\ 0 & 1 & 0 \\ 0 & 0 & 1 \end{bmatrix}$. A parity check matrix for $C^\perp$ is $G^T = \begin{bmatrix} 1 & 0 & 0 & 0 & 1 & 1 \\ 0 & 1 & 0 & 1 & 0 & 1 \\ 0 & 0 & 1 & 1 & 1 & 0 \end{bmatrix}$.

(c) The word 110110 is in C and in $C^\perp$.

19. (a) Let $\mathsf{x} = \begin{bmatrix} x_1 \\ \vdots \\ x_k \\ y_1 \\ \vdots \\ y_\ell \end{bmatrix}$ and $\mathsf{x}' = \begin{bmatrix} x_1' \\ \vdots \\ x_k' \\ y_1' \\ \vdots \\ y_\ell' \end{bmatrix}$. Then $\mathsf{x} + \mathsf{x}' = \begin{bmatrix} x_1 + x_1' \\ \vdots \\ x_k + x_k' \\ y_1 + y_1' \\ \vdots \\ y_\ell + y_\ell' \end{bmatrix}$, so

$$\begin{aligned} T(\mathsf{x} + \mathsf{x}') &= (x_1 + x_1')\mathsf{u}_1 + \cdots + (x_k + x_k')\mathsf{u}_k \\ &= (x_1\mathsf{u}_1 + \cdots + x_k\mathsf{u}_k) + (x_1'\mathsf{u}_1 + \cdots + x_k'\mathsf{u}_k) = T(\mathsf{x}) + T(\mathsf{x}'). \end{aligned}$$

Let c be a scalar. Then $c\mathsf{x} = \begin{bmatrix} cx_1 \\ \vdots \\ cx_k \\ cy_1 \\ \vdots \\ cy_\ell \end{bmatrix}$, so $T(c\mathsf{x}) = cx_1\mathsf{u}_1 + \cdots +$

$cx_k\mathsf{u}_k = c(x_1\mathsf{u}_1 + \cdots + x_k\mathsf{u}_k) = cT(\mathsf{x})$.

(b) Let $\mathsf{x} = \begin{bmatrix} x_1 \\ \vdots \\ x_k \\ y_1 \\ \vdots \\ y_\ell \end{bmatrix}$ and $\mathsf{x}' = \begin{bmatrix} x_1' \\ \vdots \\ x_k' \\ y_1' \\ \vdots \\ y_\ell' \end{bmatrix}$ with

$$\begin{aligned} x_1\mathsf{u}_1 + \cdots + x_k\mathsf{u}_k &= -y_1\mathsf{w}_1 - \cdots - y_\ell\mathsf{w}_\ell \quad \text{and} \\ x_1'\mathsf{u}_1 + \cdots + x_k'\mathsf{u}_k &= -y_1'\mathsf{w}_1 - \cdots - y_\ell'\mathsf{w}_\ell. \end{aligned} \qquad (*)$$

Suppose $T(\mathsf{x}) = T(\mathsf{x}')$. Then $x_1\mathsf{u}_1 + \cdots + x_k\mathsf{u}_k = x_1'\mathsf{u}_1 + \cdots + x_k'\mathsf{u}_k$ so $(x_1 - x_1')\mathsf{u}_1 + \cdots + (x_k - x_k')\mathsf{u}_k = \mathsf{0}$. Since $\mathsf{u}_1, \ldots, \mathsf{u}_k$ are linearly independent, this implies $x_1 = x_1'$, ..., $x_k = x_k'$. From (*), we obtain

$$-y_1\mathsf{w}_1 - \cdots - y_\ell\mathsf{w}_\ell = -y_1'\mathsf{w}_1 - \cdots - y_\ell'\mathsf{w}_\ell$$

so that

$$(y_1 - y_1')\mathsf{w}_1 + \cdots + (y_\ell - y_\ell')\mathsf{w}_\ell = \mathsf{0}.$$

Since $\mathsf{w}_1, \ldots, \mathsf{w}_\ell$ are linearly independent, this implies $y_1 = y_1'$, ..., $y_\ell = y_\ell'$. Thus $\mathsf{x} = \mathsf{x}'$.

(c) Let $\mathsf{y} = x_1\mathsf{u}_1 + \cdots + x_k\mathsf{u}_k = y_1\mathsf{w}_1 + \cdots + y_\ell\mathsf{w}_\ell$ be a vector in $U \cap W$.

Then $\mathsf{x} = \begin{bmatrix} x_1 \\ \vdots \\ x_k \\ -y_1 \\ \vdots \\ -y_\ell \end{bmatrix}$ is in V and $\mathsf{y} = T(\mathsf{x})$.

21. $U + W = V$ if and only if $\dim U + \dim W = \dim V$.

24. $U^\perp = \{\mathsf{0}\}$.

30. Let x be in $(U+W)^\perp$. Then $\mathsf{x} \cdot (\mathsf{u} + \mathsf{w}) = 0$ for every $\mathsf{u} \in U$ and every $\mathsf{w} \in W$. In particular, with $\mathsf{w} = \mathsf{0}$, we see that $\mathsf{x} \cdot \mathsf{u} = 0$ for every $\mathsf{u} \in U$. Thus $\mathsf{x} \in U^\perp$ and, similarly, $\mathsf{x} \in W^\perp$, so $\mathsf{x} \in U^\perp \cap W^\perp$.

    On the other hand, suppose $\mathsf{x} \in U^\perp \cap W^\perp$. Then $\mathsf{x} \cdot \mathsf{u} = 0$ for every $\mathsf{u} \in U$ and $\mathsf{x} \cdot \mathsf{w} = 0$ for every $\mathsf{w} \in W$, so $\mathsf{x} \cdot (\mathsf{u} + \mathsf{w}) = \mathsf{x} \cdot \mathsf{u} + \mathsf{x} \cdot \mathsf{w} = 0$ for every $\mathsf{u} + \mathsf{w} \in U + W$. Thus $\mathsf{x} \in (U+W)^\perp$.

Section 6.5—True/False

1. True.
2. True: $P\mathsf{b} = A\mathsf{x}^+ = AA^+\mathsf{b}$.
3. False: $A^+ = \begin{bmatrix} \frac{1}{5} & 0 & 0 \\ 0 & \frac{1}{7} & 0 \end{bmatrix}$.

4. True: $A^+ = \begin{bmatrix} \frac{1}{2} & 0 & 0 \\ 0 & -\frac{1}{3} & 0 \\ 0 & 0 & 0 \\ 0 & 0 & 0 \end{bmatrix}$—see paragraph 6.5.8—and the same observation gives $(A^+)^+ = A$.
5. True. A is invertible.
6. True. The given matrix is rotation through θ, so its pseudoinverse is its inverse, which is rotation through $-\theta$.
7. False. The given formula is that for a matrix with independent **columns.**
8. False. It's $(A^TA)^{-1}A^T$.
9. True. Matrix A has linearly independent rows.
10. False. Most projection matrices are not invertible. If P is a projection matrix, $P^+ = P$.

Exercises 6.5

1. (a) The rank is 2, which is the number of rows.

   (b) The 2×2 matrix AA^T has linearly independent rows.

   (c) $A^+ = \begin{bmatrix} \frac{1}{4} & 0 \\ \frac{1}{4} & \frac{1}{2} \\ \frac{1}{4} & -\frac{1}{2} \\ \frac{1}{4} & 0 \end{bmatrix}$.

3. $A^+ = \dfrac{1}{45}\begin{bmatrix} -1 & -4 & 5 \\ 11 & -1 & -10 \\ 10 & -5 & -5 \end{bmatrix}$.

5. (a) $A^+ = \dfrac{1}{2}\begin{bmatrix} 1 \\ -1 \\ 0 \end{bmatrix}$, $\mathsf{x}^+ = \begin{bmatrix} 5 \\ -5 \\ 0 \end{bmatrix}$.

   (d) $A^+ = \frac{1}{6}\begin{bmatrix} 1 & 1 & 1 \\ -1 & -1 & -1 \end{bmatrix}$, $\mathsf{x}^+ = \frac{1}{2}\begin{bmatrix} 1 \\ -1 \end{bmatrix}$.

   (g) $A^+ = \dfrac{1}{4}\begin{bmatrix} 1 & 0 & 1 \\ -1 & 0 & 1 \\ 0 & 4 & 0 \end{bmatrix}$, $\mathsf{x}^+ = \dfrac{1}{2}\begin{bmatrix} 0 \\ 1 \\ 0 \end{bmatrix}$.

6. (a) If A has linearly independent columns, $A^+ = (A^TA)^{-1}A^T$, so $A^+A = (A^TA)^{-1}(A^TA) = I$.

13. The vector x^+ is the unique vector in the row space of A with the property that $A\mathsf{x}^+ = P\mathsf{b}$ is the projection of b on the column space of A. Let $T\colon \mathsf{R}^m \to \mathsf{R}^n$ be defined by $T(\mathsf{b}) = \mathsf{x}^+$. First we show that T preserves vector addition. Given vectors b_1 and b_2, let x_1^+ and x_2^+ be the unique vectors in the row space of A with the property that $A\mathsf{x}_1^+ = P\mathsf{b}_1$ and $A\mathsf{x}_2^+ = P\mathsf{b}_2$. To show that $T(\mathsf{b}_1 + \mathsf{b}_2) = T(\mathsf{b}_1) + T(\mathsf{b}_2)$, we must show that the unique vector x^+ in the row space of A with the property that

$A\mathbf{x}^+ = P(\mathbf{b}_1 + \mathbf{b}_2)$ is $\mathbf{x}_1^+ + \mathbf{x}_2^+$. Now $A(\mathbf{x}_1^+ + \mathbf{x}_2^+) = A\mathbf{x}_1^+ + A\mathbf{x}_2^+ = P\mathbf{b}_1 + P\mathbf{b}_2 = P(\mathbf{b}_1 + \mathbf{b}_2)$. So $\mathbf{x} = \mathbf{x}_1^+ + \mathbf{x}_2^+$ satisfies $A\mathbf{x} = P(\mathbf{b}_1 + \mathbf{b}_2)$. Since this vector is also in the row space of A (the row space is closed under vector addition), we have what we want. To see that T preserves scalar multiplication, let $\mathbf{b}$ be a vector in R^m and let c be a scalar. Let $\mathbf{x}^+ = T(\mathbf{b})$ be the unique vector in the row space of A satisfying $A\mathbf{x}^+ = P\mathbf{b}$. To show $T(c\mathbf{b}) = cT(\mathbf{b})$, we must show that the unique vector $\mathbf{x}$ in the row space of A that satisfies $A\mathbf{x} = P(c\mathbf{b})$ is $\mathbf{x} = c\mathbf{x}^+$. This follows because $c\mathbf{x}^+$ is in the row space of A (the row space is closed under scalar multiplication) and $A(c\mathbf{x}^+) = cA\mathbf{x}^+ = cP\mathbf{b} = P(c\mathbf{b})$.

Section 7.1—True/False

1. False: $z \cdot w = 1(-1) + \bar{i}2i = -1 - 2i^2 = +1$.
2. False: $A^* = \begin{bmatrix} 1 & 1+i \\ -i & -2i \end{bmatrix}$.
3. True.
4. True. The matrix is Hermitian.
5. True.
6. True. Unitary matrices preserve the dot product.
7. False. The eigenvalues of a unitary matrix are real.
8. True. The columns of $\begin{bmatrix} \cos\theta & -\sin\theta \\ \sin\theta & \cos\theta \end{bmatrix}$ are orthonormal.
9. False. The matrix is not square.
10. True.

Exercises 7.1

1. (a) $A + B = \begin{bmatrix} 2 & 1+3i \\ -1-2i & -4i \end{bmatrix}$, $A - B = \begin{bmatrix} 2i & 1-3i \\ 5-2i & -2i \end{bmatrix}$,

$2A - B = \begin{bmatrix} 1+3i & 2-3i \\ 7-4i & -5i \end{bmatrix}$, $AB = \begin{bmatrix} -1 & -3+2i \\ 5i & 3+6i \end{bmatrix}$,

$BA = \begin{bmatrix} 8+6i & 10-i \\ -5-5i & -6 \end{bmatrix}$.

(c) $BA = \begin{bmatrix} 1-6i & -11+7i \\ -3+i & -1+7i \\ -2-5i & -6+2i \end{bmatrix}$; $A+B, A-B, 2A-B, AB$ are not defined.

4. A and C are conjugate transposes, B is unitary, E is Hermitian.
6. (a) $\|\mathbf{u}\| = \sqrt{6}$, $\|\mathbf{v}\| = \sqrt{12}$, and a unit vector in the direction of $\mathbf{u}$ is $\frac{1}{\sqrt{6}}\begin{bmatrix} 1+2i \\ i \end{bmatrix}$. The vectors are not orthogonal.

7. $\mathsf{z}\cdot\mathsf{w} = \overline{z}_1 z_1 + \overline{z}_2 z_2 + \cdots + \overline{z}_n z_n = |z_1|^2 + |z_2|^2 + \cdots + |z_n|^2$ is the sum of the squares of the moduli of complex numbers and the modulus of the complex number $z = a + bi$ is the real number $|z|^2 = a^2 + b^2$.

9. (a)
$$\begin{aligned} \mathsf{z}\cdot(c\mathsf{w}) &= \overline{(c\mathsf{w})\cdot\mathsf{z}} && \text{by (2)}\\ &= \overline{\overline{c}(\mathsf{w}\cdot\mathsf{z})} && \text{by (3)}\\ &= \overline{\overline{c}}\,\overline{\mathsf{w}\cdot\mathsf{z}} && \text{by Proposition A-7}\\ &= c\,\overline{\mathsf{w}\cdot\mathsf{z}} && \text{by Proposition A-7}\\ &= c(\mathsf{z}\cdot\mathsf{w}) && \text{using (2) again.} \end{aligned}$$

10. Since $\overline{\mathsf{z}} = \begin{bmatrix} \overline{z}_1 \\ \vdots \\ \overline{z}_n \end{bmatrix}$, $\|\overline{\mathsf{z}}\|^2 = \overline{\overline{z}}_1\overline{z}_1 + \cdots + \overline{\overline{z}}_n\overline{z}_n$. Since $\overline{\overline{z}}_i = z_i$, we have $\|\overline{\mathsf{z}}\|^2 = z_1\overline{z}_1 + \cdots + z_n\overline{z}_n$, and since $z_i\overline{z}_i = \overline{z}_i z_i$, we have $\|\overline{\mathsf{z}}\|^2 = \overline{z}_1 z_1 + \cdots + \overline{z}_n z_n$, which is precisely $\|\mathsf{z}\|^2$.

11. (a) We want $(\mathsf{u} - c\mathsf{v})\cdot\mathsf{v} = 0$, so $\mathsf{u}\cdot\mathsf{v} - \overline{c}(\mathsf{v}\cdot\mathsf{v}) = 0$ and $\overline{c} = \dfrac{\mathsf{u}\cdot\mathsf{v}}{\mathsf{v}\cdot\mathsf{v}}$. Thus $c = \dfrac{\overline{\mathsf{u}\cdot\mathsf{v}}}{\mathsf{v}\cdot\mathsf{v}} = \dfrac{\mathsf{v}\cdot\mathsf{u}}{\mathsf{v}\cdot\mathsf{v}}$. So the projection of u on v is the vector $\mathsf{p} = \dfrac{\mathsf{v}\cdot\mathsf{u}}{\mathsf{v}\cdot\mathsf{v}}\mathsf{v}$.

(b) The projection of u on v is $\dfrac{4-2i}{18}\mathsf{v} = \dfrac{2-i}{9}\begin{bmatrix} 3+2i \\ -i \\ 2 \end{bmatrix} = \dfrac{1}{9}\begin{bmatrix} 8+i \\ -1-2i \\ 4-2i \end{bmatrix}$.

12. (a) The eigenvalues are $\lambda = \frac{1\pm\sqrt{-3}}{2} = \frac{1}{2}(1 \pm \sqrt{3}i)$.
The eigenspace corresponding to $\lambda = \frac{1}{2}(1+\sqrt{3}i)$ consists of vectors of the form $t\begin{bmatrix} 3+\sqrt{3}i \\ 2 \end{bmatrix}$. The eigenspace corresponding to $\lambda = \frac{1}{2}(1-\sqrt{3}i)$ consists of vectors of the form $t\begin{bmatrix} 3-\sqrt{3}i \\ 2 \end{bmatrix}$.

13. (a) The hint gives this away! Since $\mathsf{x}^TA\mathsf{x}$ is 1×1, this matrix equals its transpose. Thus $\mathsf{x}^TA\mathsf{x} = (\mathsf{x}^TA\mathsf{x})^T = \mathsf{x}^TA^T\mathsf{x} = -\mathsf{x}^TA\mathsf{x}$, so $2\mathsf{x}^TA\mathsf{x} = 0$, so $\mathsf{x}^TA\mathsf{x} = 0$.

16. $\|A\mathsf{z}\|^2 = (A\mathsf{z})\cdot(A\mathsf{z}) = (A\mathsf{z})^*(A\mathsf{z}) = \mathsf{z}^*A^*A\mathsf{z} = \mathsf{z}^*AA^*\mathsf{z} = (A^*\mathsf{z})^*(A^*\mathsf{z}) = \|A^*\mathsf{z}\|^2$.

17. (a) True, (b) False.

19. (a) $\begin{bmatrix} 0 & i \\ -i & 0 \end{bmatrix}$.

23. (a) Let u be a vector in U. We must show that $A\mathsf{u}$ is also in U. This follows from $A\mathsf{u} = \lambda\mathsf{u}$ for some scalar λ and the fact that $\lambda\mathsf{u}$ is in U because U is a subspace.

(b) Let $\mathsf{w} \in U^\perp$. We have to show that $A\mathsf{w}$ is also in $U^\perp$, that is, we have to show that $A\mathsf{w}$ is orthogonal to any vector in U. So let z be a vector U. Then z is an eigenvector of A, corresponding to λ, say. We have

$$\begin{aligned} A\mathsf{w}\cdot\mathsf{z} &= (A\mathsf{w})^*\mathsf{z}\\ &= \mathsf{w}^*A^*\mathsf{z}\\ &= \mathsf{w}^*A\mathsf{z} \quad \text{because } A^* = A\\ &= \mathsf{w}^*\lambda\mathsf{z} = \lambda(\mathsf{w}^*\mathsf{z}) = \lambda(\mathsf{w}\cdot\mathsf{z}) = 0, \end{aligned}$$

because z is in U and w is in $U^\perp$.

Section 7.2—True/False

1. False. $A = I$, so any matrix similar to A is also I: $P^{-1}IP = I$.
2. True. The characteristic polynomial of A is $(1-\lambda)^2 = 1 - 2\lambda + \lambda^2$, and this equation is satisfied by A (by the Cayley-Hamilton theorem).
3. True. A is Hermitian.
4. True. We have $U^*AU = D$ for some unitary matrix U and diagonal matrix D. The diagonal entries of D are the eigenvalues of A, so U is invertible. Thus $A = UDU^*$ is also invertible.
5. False. Let A be a real symmetric matrix that is not diagonal. There exists a diagonal matrix and an orthogonal matrix Q such that $Q^{-1}AQ = D$. Then $A = QDQ^{-1} = QDQ^T$ is not diagonal.
6. True. This is the Principal Axes Theorem.
7. True. If H is Hermitian, $HH^* = HH^* = H^2$.
8. False. It was shown in this section that $V = \begin{bmatrix} 0 & -1 \\ 1 & 0 \end{bmatrix}$ is unitarily diagonalizable, hence normal, but V is not Hermitian.
9. False. $A = \begin{bmatrix} 1 & i \\ -i & 1 \end{bmatrix}$ is Hermitian, hence normal, but it is not unitary because its columns do not have modulus 1 (neither are its columns orthogonal).
10. True. $U^* = U^{-1}$ implies $UU^* = U^*U = I$.

Exercises 7.2

1. (a) $Q = \begin{bmatrix} \frac{3}{\sqrt{10}} & -\frac{1}{\sqrt{10}} \\ \frac{1}{\sqrt{10}} & \frac{3}{\sqrt{10}} \end{bmatrix}$, $D = \begin{bmatrix} 3 & 0 \\ 0 & -7 \end{bmatrix}$.

   (c) $Q = \begin{bmatrix} \frac{1}{\sqrt{3}} & -\frac{1}{\sqrt{2}} & -\frac{1}{\sqrt{6}} \\ \frac{1}{\sqrt{3}} & 0 & \frac{2}{\sqrt{6}} \\ \frac{1}{\sqrt{3}} & \frac{1}{\sqrt{2}} & -\frac{1}{\sqrt{6}} \end{bmatrix}$, $D = \begin{bmatrix} 0 & 0 & 0 \\ 0 & -3 & 0 \\ 0 & 0 & -3 \end{bmatrix}$.

4. (a) $U = \begin{bmatrix} \frac{i}{\sqrt{2}} & -\frac{i}{\sqrt{2}} \\ \frac{1}{\sqrt{2}} & \frac{1}{\sqrt{2}} \end{bmatrix}$, $D = \begin{bmatrix} 0 & 0 \\ 0 & 2 \end{bmatrix}$.

5. (a) $\det(A - \lambda I) = (2-\lambda)(1-\lambda) + 3 = \lambda^2 - 3\lambda + 5$ has no real roots, so A is not diagonalizable over $\mathbb{R}$. Over $\mathbb{C}$, however, the characteristic polynomial has the distinct roots $\lambda = \dfrac{3 \pm i\sqrt{11}}{2}$, so A is diagonalizable.

6. Since A^TA is a real symmetric matrix, there exists an orthogonal matrix Q such that $Q^TA^TAQ = D$ is diagonal. Thus $-Q^TA^2Q = D$. Since $Q^T = Q^{-1}$, $Q^TA^2Q = Q^{-1}A^2Q = Q^{-1}AQQ^{-1}AQ = (Q^{-1}AQ)^2 = (Q^TAQ)^2$. Let $B = Q^TAQ$. Then $B^2 = -D$ is diagonal and $A = QBQ^T$.

8. (a) $U = \begin{bmatrix} -\frac{1+i}{\sqrt{3}} & \frac{1+i}{\sqrt{6}} \\ \frac{1}{\sqrt{3}} & \frac{2}{\sqrt{6}} \end{bmatrix}$, $T = \begin{bmatrix} 1+i & -\frac{3}{\sqrt{2}} + \frac{i}{\sqrt{2}} \\ 0 & 1-i \end{bmatrix}$.

13. (a) The characteristic polynomial of this 3×3 matrix is $-(\lambda - 1)(\lambda^2 - \lambda - 1)$ has three different roots, so the eigenvalues are all different. So the matrix is diagonalizable over C.

15. We have $(A^*z) \cdot (A^*z) = (A^*z)^*A^*z = z^*(A^*)^*A^*z = z^*AA^*z = z^*A^*Az$ using $AA^* = A^*A$. Thus $(A^*z) \cdot (A^*z) = (Az) \cdot (Az)$, so $(A^*z) \cdot (A^*z) = 0$ if and only if $(Az) \cdot (Az) = 0$. Now $z \in \text{null sp} A$ if and only if $Az = 0$, and this occurs if and only if $\|Az\|^2 = (Az) \cdot (Az) = 0$. Similarly, $z \in \text{null sp} A^*$ if and only if $(A^*z) \cdot (A^*z) = 0$. So we have the desired result.

21. The given polynomial is the characteristic polynomial of $A + B$. Since A and B are symmetric, so is $A + B$. So the characteristic polynomial of $A + B$ has real roots.

23. (a) Let x be an eigenvector corresponding to λ. Then $x^T Ax = x^T(\lambda x) = \lambda \|x\|^2$. Since $x \neq 0$, $\|x\|^2 > 0$ and $x^T Ax > 0$, so $\lambda > 0$.

    (b) There exists an orthogonal matrix Q such that $Q^{-1}AQ = D$ is a diagonal matrix. The diagonal entries of D are the eigenvalues of A, so these are positive, hence each is the square of a real number. It follows that $D = E^2$ for some matrix E and then $A = QDQ^{-1} = QE^2Q^{-1} = (QEQ^{-1})(QEQ^{-1}) = B^2$, with $B = QEQ^{-1}$.

    (c) There exists an orthogonal matrix Q such that $Q^{-1}AQ = D$ is a diagonal matrix with positive entries on the diagonal. As in (b), $D = E^2$ for some matrix E, so $A = QDQ^{-1} = QE^2Q^T = (QE)(EQ^T)$. This is RR^T with $R = QE$ since $R^T = E^TQ^T = EQ^T$ (because E is diagonal).

26. (a) The characteristic polynomial is $-\lambda^3 + 9\lambda + 10$.

    (b) By the Cayley-Hamilton Theorem, A satisfies its characteristic polynomial; thus $A^3 = 9A + 10$. Using this fact, the given polynomial is

$$-6A + 10I = \begin{bmatrix} -8 & -18 & 24 \\ 24 & 28 & -30 \\ -12 & -24 & 10 \end{bmatrix}.$$

Section 7.3—True/False

1. True.
2. True. Since Q is orthogonal, $Q^{-1} = Q^T$.
3. False. The eigenvalues of a real symmetric matrix are all real, but the roots of the given polynomial are not.
4. False. Let A be any real symmetric matrix that is not diagonal. There exist a diagonal matrix D and an orthogonal matrix Q such that $Q^TAQ = D$. Then $A = QDQ^T$ is not diagonal.
5. False. Since $P^{-1}AP = D$, $D^2 = P^{-1}APP^{-1}AP = P^{-1}A^2P$, and there is no reason why this should be A^2.
6. False. The eigenvalues of the upper triangular matrix $A = \begin{bmatrix} 1 & 2 \\ 0 & 3 \end{bmatrix}$ are real (its diagonal entries), but A is not symmetric.

7. True. The hypothesis implies that $Q^TAQ = D$ is diagonal, where Q is the orthogonal matrix whose columns are the given vectors divided by their lengths. So $A = QDQ^T$ is symmetric. (Do you see why?)
8. True. This is the Principal Axes Theorem.
9. True.
10. True. We have $Q^{-1}AQ = I$ for some matrix Q.

Exercises 7.3

1. (a) $Q = \begin{bmatrix} \frac{3}{\sqrt{10}} & -\frac{1}{\sqrt{10}} \\ \frac{1}{\sqrt{10}} & \frac{3}{\sqrt{10}} \end{bmatrix}$, $D = \begin{bmatrix} 3 & 0 \\ 0 & -7 \end{bmatrix}$.

   (c) $Q = \begin{bmatrix} \frac{1}{\sqrt{3}} & -\frac{1}{\sqrt{2}} & -\frac{1}{\sqrt{6}} \\ \frac{1}{\sqrt{3}} & 0 & \frac{2}{\sqrt{6}} \\ \frac{1}{\sqrt{3}} & \frac{1}{\sqrt{2}} & -\frac{1}{\sqrt{6}} \end{bmatrix}$, $D = \begin{bmatrix} 0 & 0 & 0 \\ 0 & -3 & 0 \\ 0 & 0 & -3 \end{bmatrix}$.

2. (a) i. $Q = \begin{bmatrix} -\frac{1}{\sqrt{2}} & \frac{1}{\sqrt{2}} \\ \frac{1}{\sqrt{2}} & \frac{1}{\sqrt{2}} \end{bmatrix}$, $D = \begin{bmatrix} -1 & 0 \\ 0 & 3 \end{bmatrix}$.

   ii. $A^5 = \begin{bmatrix} 121 & 122 \\ 122 & 121 \end{bmatrix}$.

3. The converse of the Principal Axes Theorem says that if a real matrix is orthogonally diagonalizable, then A is symmetric. This is true, and here is the argument. Suppose $Q^{-1}AQ = D$ is diagonal for some orthogonal matrix Q. Thus $A = QDQ^{-1} = QDQ^T$, so $A^T = (Q^T)^T D^T Q^T = QDQ^T = A$, since the diagonal matrix D is its own transpose.

5. There exist an orthogonal matrix Q and a diagonal matrix D with nonnegative diagonal entries $\lambda_1, \lambda_2, \ldots, \lambda_n$ such that $Q^TAQ = D$. Thus $A = QDQ^T$ and for any x, $\mathsf{x}^TA\mathsf{x} = \mathsf{x}^TQDQ^T\mathsf{x}$. Set $\mathsf{y} = Q^T\mathsf{x} = \begin{bmatrix} y_1 \\ \vdots \\ y_n \end{bmatrix}$.

   Then $\mathsf{x}^TQ = \mathsf{y}^T$ and $\mathsf{x}^TA\mathsf{x} = \mathsf{y}^TD\mathsf{y} = \lambda_1 y_1^2 + \lambda_2 y_2^2 + \cdots + \lambda_n y_n^2$ and this is nonnegative because each $\lambda_i \geq 0$ and each $\mathsf{y}_i^2 \geq 0$.

9. (a) Since A and B are symmetric, so is $A + B$, so $A + B$ is orthogonally diagonalizable.

12. Since A^TA is a real symmetric matrix, there exists an orthogonal matrix Q such that $Q^TA^TAQ = D$ is diagonal. Thus $-Q^TA^2Q = D$. Since $Q^T = Q^{-1}$, $Q^TA^2Q = Q^{-1}A^2Q = Q^{-1}AQQ^{-1}AQ = (Q^{-1}AQ)^2 = (Q^TAQ)^2$. Let $B = Q^TAQ$. Then $B^2 = Q^TA^2Q = -D$ is diagonal and $A = QBQ^{-1} = QBQ^T$.

15. (a) There exist an orthogonal matrix Q and a diagonal matrix D with nonnegative diagonal entries such that $Q^TAQ = D$. Now D has a square root, so $D = D_1^2$ for some diagonal matrix D_1. We have $A = QDQ^T = QD_1^2Q^T = (QD_1)(D_1Q^T)$. Let $R = QD_1$. Then $R^T = D_1^TQ^T = D_1Q^T$ (because D_1 is diagonal). So $A = RR^T$ as required.

Exercises 7.4

1. (a) $q = \mathsf{x}^T A \mathsf{x}$ with $\mathsf{x} = \begin{bmatrix} x \\ y \end{bmatrix}$ and $A = \begin{bmatrix} 2 & -\frac{3}{2} \\ -\frac{3}{2} & 1 \end{bmatrix}$.

2. (b) i. $q = -23y_1^2 + 72y_1y_2 - 2y_2^2$.

3. (a) $2x^2 + 7y^2 = 1$. The graph is an ellipse.

4. $\mathsf{q}_1 = \frac{1}{\sqrt{5}}\begin{bmatrix} 0 \\ 1 \\ 2 \end{bmatrix}$, $\mathsf{q}_2 = \begin{bmatrix} 1 \\ 0 \\ 0 \end{bmatrix}$ $\mathsf{q}_3 = \frac{1}{\sqrt{5}}\begin{bmatrix} 0 \\ -2 \\ 1 \end{bmatrix}$ and $q = 3y_2^2 + 5y_3^2$.

Section 7.5—True/False

1. True. This is the Principal Axes Theorem.
2. True. Extend $\{\mathsf{u}, \mathsf{v}, \mathsf{w}\}$ to a basis $\{\mathsf{u}, \mathsf{v}, \mathsf{w}, \mathsf{t}\}$ of R^4 and apply the Gram-Schmidt Algorithm to these four vectors.
3. True, since $Q_1^T = Q_1^{-1}$.
4. False. The singular values of A are the **square roots** of the **nonzero** eigenvalues of A^TA.
5. False: $A^+ = Q_2\Sigma^+Q_1^T$.

Exercises 7.5

1. (a) $A = Q_1\Sigma Q_2^T$ with $Q_1 = \begin{bmatrix} \frac{1}{\sqrt{2}} & \frac{1}{\sqrt{2}} & 0 \\ 0 & 0 & 1 \\ \frac{1}{\sqrt{2}} & -\frac{1}{\sqrt{2}} & 0 \end{bmatrix}$, $Q_2 = \begin{bmatrix} \frac{1}{\sqrt{2}} & \frac{1}{\sqrt{2}} \\ -\frac{1}{\sqrt{2}} & \frac{1}{\sqrt{2}} \end{bmatrix}$ and

   $\Sigma = \begin{bmatrix} 2 & 0 \\ 0 & 0 \\ 0 & 0 \end{bmatrix}$. $A^+ = Q_2\Sigma^+Q_1^T = \begin{bmatrix} \frac{1}{4} & 0 & \frac{1}{4} \\ -\frac{1}{4} & 0 & -\frac{1}{4} \end{bmatrix}$.

   (c) $A = Q_1\Sigma Q_2^T$ with $Q_1 = [1]$, $Q_2 = \begin{bmatrix} \frac{1}{\sqrt{14}} & -\frac{2}{\sqrt{5}} & -\frac{3}{\sqrt{70}} \\ \frac{2}{\sqrt{14}} & \frac{1}{\sqrt{5}} & -\frac{6}{\sqrt{70}} \\ \frac{3}{\sqrt{14}} & 0 & \frac{5}{\sqrt{70}} \end{bmatrix}$

   and $\Sigma = \begin{bmatrix} \sqrt{14} & 0 & 0 \end{bmatrix}$. $A^+ = \frac{1}{14}\begin{bmatrix} 1 \\ 2 \\ 3 \end{bmatrix}$.

3. Let $A = Q_1\Sigma Q_2^T$ be a singular value decomposition of A. Then $A^+ = Q_2\Sigma^+Q_1^T$, by 7.5.5, so $(A^+)^T = Q_1(\Sigma^+)^TQ_2^T$. Now $A^T = Q_2\Sigma^TQ_1^T$ is a singular value decomposition of A^T, so 7.5.5 says that $(A^T)^+ = Q_1(\Sigma^T)^+Q_2^T$ and the result follows because $(\Sigma^+)^T = (\Sigma^T)^+$.

4. (a) Write $A = Q_1\Sigma Q_2^T$. Then $A^+ = Q_2\Sigma^+Q_1^T$, so
   $AA^+A = Q_1\Sigma Q_2^TQ_2\Sigma^+Q_1^TQ_1\Sigma Q_2^T = Q_1\Sigma Q_2^T = A$, using $Q_1^TQ_1 = I$, $Q_2^TQ_2 = I$ and $\Sigma\Sigma^+\Sigma = \Sigma$.

5. (a) From $A = Q_1\Sigma Q_2^T$, we have $A^TA = Q_2\Sigma^T Q_1^T Q_1 \Sigma Q_2^T = Q_2\Sigma^T\Sigma Q_2^{-1}$ because Q_1 and Q_2 are orthogonal. Hence $A^TAQ_2 = Q_2\Sigma^T\Sigma$. Let $B = A^TA$ and note that $\Sigma^T\Sigma$ is a diagonal matrix D. We have an equation of the form $BP = PD$ with $P = Q_2$. The result now follows exactly as shown at the start of Section 3.5 (see paragraph 3.5.3).

7. (a) Let $A = Q_1\Sigma Q_2^T$ be a singular value decomposition of A. Then $A^T = Q_2\Sigma^T Q_1^T$ is a singular value decomposition of A^T, so the singular values of A^T are the positive elements on the diagonal of Σ^T, but these are the positive elements on the diagonal of Σ, which are the singular values of A.

9. (a) We have $\mathsf{y}_i = \frac{1}{\sigma_i} A\mathsf{x}_i$ for $i = 1, 2, \ldots, r$. Thus $\mathsf{y}_1, \mathsf{y}_2, \ldots, \mathsf{y}_r$ are r orthonormal (hence linearly independent) vectors in col sp A, which has dimension r, so they form a basis. Now it follows that $\mathsf{y}_{r+1}, \ldots, \mathsf{y}_m$ are a basis for the orthogonal complement of col sp A, which is $(\text{null sp}\, A)^\perp$ by Theorem 6.4.17.

Appendix A—True/False

1. True: $7 = 7 + 0i$.
2. True.
3. True.
4. False: $z\bar{z} = |z|^2$.
5. True.
6. True. The inverse of nonzero complex number is also a complex number.
7. False: $1 + i = \sqrt{2}(\cos\frac{\pi}{4} + i\sin\frac{\pi}{4})$.
8. True. Multiply moduli and add arguments.
9. True.
10. True. The equation $z^5 + 1 = 0$ has five roots, spaced equally around the unit circle. They are all different.

Exercises Appendix A

1. (a) $1 + \sqrt{3}i = 2(\cos\frac{\pi}{3} + i\sin\frac{\pi}{3}$.
2. (a) $2 + 3i = \sqrt{13}(\cos .983 + i\sin .983)$.
3. (a) $1 + 6i = \sqrt{37}(\cos 1.406 + i\sin 1.406)$.
4. (a) $\dfrac{3-4i}{1+i} = -\dfrac{7}{2} - \dfrac{7}{2}i$.
5. $z = 2 - 2i = 2[\cos(-\frac{\pi}{4}) + i\sin(-\frac{\pi}{4})]$; $\bar{z} = 2 + 2i = 2\sqrt{2}\cos\frac{\pi}{4} + i\sin\frac{\pi}{4})$; $z^2 = -8i = 8[\cos(-\frac{\pi}{2}) + i\sin(-\frac{\pi}{2})]$; $\dfrac{1}{z} = \dfrac{\bar{z}}{z\bar{z}} = \frac{1}{8}(2 + 2i) = \frac{1}{4} + \frac{1}{4}i = \frac{1}{2\sqrt{2}}(\cos\frac{\pi}{4} + i\sin\frac{\pi}{4})$.

7. $zw = 1(3+i) - 2i(3+i) = 3 + i - 6i + 2 = 5 - 5i$; $\dfrac{z}{w} = \frac{1}{10}(1 - 7i)$, $z^2 - w^2 = -11 - 10i$, $\dfrac{z+w}{z-w} = \frac{1}{13}(-5 + 14i)$.

9. $\dfrac{z\overline{w} + z^2}{w\overline{z} + w^2} = \dfrac{1}{5} + \dfrac{2}{5}i$.

11. (a) Write $z = a + bi$. Then $\overline{z} = a - bi$, so $z + \overline{z} = 2a$. The real part of z is $a = \frac{1}{2}(z + \overline{z})$. Also, $z - \overline{z} = 2bi$, so the imaginary part of z is $b = \frac{1}{2i}(z - \overline{z})$.

Glossary

If you do not know what the words mean, it is impossible to read anything with understanding. This fact, which is completely obvious to students of German or Russian, is often lost on mathematics students, but it is just as applicable. What follows is a vocabulary list of all the technical terms discussed in this book together with the page where each was first introduced. In most cases, each definition is followed by an example.

Additive inverse: The additive inverse of the number 5 is -5 because $5 + (-5) = 0$. The additive inverse of a vector v, denoted $-\mathsf{v}$ ("minus v") is the vector with the property that $\mathsf{v} + (-\mathsf{v}) = \mathbf{0}$. The additive inverse of a matrix A, denoted $-A$ ("minus A"), is the matrix with the property that $A + (-A) = 0$. For example, the additive inverse of $\begin{bmatrix} 2 \\ -3 \end{bmatrix}$ is $\begin{bmatrix} -2 \\ 3 \end{bmatrix}$. The additive inverse of $A = \begin{bmatrix} -1 & 3 \\ -2 & 4 \end{bmatrix}$ is $-A = \begin{bmatrix} 1 & -3 \\ 2 & -4 \end{bmatrix}$. *pp. 7, 97*

Angle between vectors: If u and v are nonzero vectors in R^n, the angle between u and v is that angle θ, $0 \le \theta \le \pi$, whose cosine satisfies $\cos\theta = \dfrac{\mathsf{u}\cdot\mathsf{v}}{\|\mathsf{u}\|\,\|\mathsf{v}\|}$. For example, if $\mathsf{u} = \begin{bmatrix} 1 \\ 2 \\ 1 \\ 0 \\ 3 \end{bmatrix}$ and $\mathsf{v} = \begin{bmatrix} 1 \\ -1 \\ 1 \\ 1 \\ 1 \end{bmatrix}$, then $\cos\theta = \dfrac{\mathsf{u}\cdot\mathsf{v}}{\|\mathsf{u}\|\,\|\mathsf{v}\|} = \dfrac{3}{\sqrt{15}\sqrt{5}} = \dfrac{\sqrt{3}}{5} \approx .346$, so $\theta \approx 1.217$ rads $\approx 70°$. *p. 30*

Basis: (Plural "bases") A basis of a vector space V is a linearly independent set of vectors that spans V. For example, the standard basis vectors $\mathsf{e}_1, \mathsf{e}_2, \ldots, \mathsf{e}_n$ in R^n form a basis for R^n, and the polynomials $1, x, x^2, \ldots$ form a basis for the vector space $\mathsf{R}[x]$ of polynomials in x with real coefficients. *p. 365*

Change of Coordinates Matrix: The change of coordinates matrix from $\mathcal{V}$ to $\mathcal{V}'$, where $\mathcal{V}$ and $\mathcal{V}'$ are bases for a vector space V, is the matrix of the identity linear transformation relative to these bases. It is denoted $P_{\mathcal{V}' \leftarrow \mathcal{V}}$. Thus $P_{\mathcal{V}' \leftarrow \mathcal{V}} = M_{\mathcal{V}' \leftarrow \mathcal{V}}(\mathrm{id})$. For example, let $\mathcal{E} = \{\mathsf{e}_1, \mathsf{e}_2\}$ be the standard basis of R^2 and let $\mathcal{V} = \{\mathsf{v}_1, \mathsf{v}_2\}$ be the basis with $\mathsf{v}_1 = \begin{bmatrix} -1 \\ 2 \end{bmatrix}$ and $\mathsf{v}_2 = \begin{bmatrix} 2 \\ -3 \end{bmatrix}$. The change of coordinates matrix $\mathcal{V}$ to $\mathcal{E}$ is $P = P_{\mathcal{E} \leftarrow \mathcal{V}} =$

$M_{\mathcal{E}\leftarrow\mathcal{V}}(\text{id}) = \begin{bmatrix} -1 & 2 \\ 2 & -3 \end{bmatrix}$. The change of coordinates matrix $\mathcal{E}$ to $\mathcal{V}$ is
$P_{\mathcal{V}\leftarrow\mathcal{E}} = P^{-1}_{\mathcal{E}\leftarrow\mathcal{V}} = \begin{bmatrix} 3 & 2 \\ 2 & 1 \end{bmatrix}$. *p. 443*

Closed under: See *Subspace.*

Cofactor: See *Minor.*

Column space: The column space of a matrix A is the set of all linear combinations of the columns of A. It is denoted col sp A. For example, the column space of $A = \begin{bmatrix} 1 & 0 \\ 0 & 1 \end{bmatrix}$ is R^2, the entire Euclidean plane, while the column space of $A = \begin{bmatrix} 1 & 3 & 7 \\ 2 & 6 & 14 \end{bmatrix}$ is the set of scalar multiples of $\begin{bmatrix} 1 \\ 2 \end{bmatrix}$. Since a linear combination of the columns of A is a vector of the form $A\mathsf{x}$, the column space of A is the set of all vectors of the form $A\mathsf{x}$. So, to determine whether or not a given vector b is in the column space of A, decide whether or not there is an x with $A\mathsf{x} = \mathsf{b}$. *p. 334*

Complex conjugate: The conjugate of the complex number $a + bi$ is the complex number $a - bi$, denoted $\overline{a + bi}$. For example, $\overline{1 + 2i} = 1 - 2i$, $\overline{-3i} = +3i$ and $\overline{-5} = -5$. *p. 594*

Component (of a vector): See *Vector.*

Conjugate transpose: See *Transpose.*

Coordinate vector: The coordinate vector of a vector x relative to a basis $\mathcal{V} = \{\mathsf{v}_1, \mathsf{v}_2, \ldots, \mathsf{v}_n\}$ is the vector $\mathsf{x}_{\mathcal{V}} = \begin{bmatrix} x_1 \\ x_2 \\ \vdots \\ x_n \end{bmatrix}$ whose components are the unique numbers satisfying $\mathsf{x} = x_1\mathsf{v}_1 + x_2\mathsf{v}_2 + \cdots + x_n\mathsf{v}_n$. For example, if $\mathsf{x} = \begin{bmatrix} x_1 \\ x_2 \end{bmatrix}$ is a vector in R^2, and $\mathcal{E} = \{\mathsf{e}_1, \mathsf{e}_2\}$ is the standard basis, then $\mathsf{x}_{\mathcal{E}} = \begin{bmatrix} x_1 \\ x_2 \end{bmatrix}$ because $\mathsf{x} = x_1\mathsf{e}_1 + x_2\mathsf{e}_2$. If $\mathcal{V} = \{\mathsf{v}_1, \mathsf{v}_2\}$ with $\mathsf{v}_1 = \begin{bmatrix} -1 \\ 2 \end{bmatrix}$ and $\mathsf{v}_2 = \begin{bmatrix} 2 \\ -3 \end{bmatrix}$, then $\mathsf{x}_{\mathcal{V}} = \begin{bmatrix} 3x_1 + 2x_2 \\ 2x_1 + x_2 \end{bmatrix}$ because $\mathsf{x} = (3x_1 + 2x_2)\mathsf{v}_1 + (2x_1 + x_2)\mathsf{v}_2$. *p. 429*

Cross product: The cross product of vectors $\mathsf{u} = \begin{bmatrix} u_1 \\ u_2 \\ u_3 \end{bmatrix}$ and $\mathsf{v} = \begin{bmatrix} v_1 \\ v_2 \\ v_3 \end{bmatrix}$ is a vector whose calculation we remember by the scheme

$$\mathsf{u} \times \mathsf{v} = \begin{vmatrix} \mathsf{i} & \mathsf{j} & \mathsf{k} \\ u_1 & u_2 & u_3 \\ v_1 & v_2 & v_3 \end{vmatrix}$$

$$\text{so that } \mathsf{u} \times \mathsf{v} = \begin{vmatrix} u_2 & u_3 \\ v_2 & v_3 \end{vmatrix} \mathsf{i} - \begin{vmatrix} u_1 & u_3 \\ v_1 & v_3 \end{vmatrix} \mathsf{j} + \begin{vmatrix} u_1 & u_2 \\ v_1 & v_2 \end{vmatrix} \mathsf{k}.$$

For example, with $\mathsf{u} = \begin{bmatrix} 1 \\ 3 \\ 2 \end{bmatrix}$ and $\mathsf{v} = \begin{bmatrix} 0 \\ 2 \\ -1 \end{bmatrix}$, we have

$$\mathsf{u} \times \mathsf{v} = \begin{vmatrix} \mathsf{i} & \mathsf{j} & \mathsf{k} \\ 1 & 3 & 2 \\ 0 & -2 & 1 \end{vmatrix} = \begin{vmatrix} 3 & 2 \\ -2 & 1 \end{vmatrix} \mathsf{i} - \begin{vmatrix} 1 & 2 \\ 0 & 1 \end{vmatrix} \mathsf{j} + \begin{vmatrix} 1 & 3 \\ 0 & -2 \end{vmatrix} \mathsf{k}$$

$$= 7\mathsf{i} - 1\mathsf{j} + (-2)\mathsf{k} = \begin{bmatrix} 7 \\ -1 \\ -2 \end{bmatrix}.$$

p. 48

Determinant: The determinant of the 2×2 matrix $A = \begin{bmatrix} a & b \\ c & d \end{bmatrix}$ is the number $ad - bc$. It is denoted $\det A$ or $\begin{vmatrix} a & b \\ c & d \end{vmatrix}$. For $n > 2$, the determinant of an $n \times n$ matrix A is defined by the equation $AC^T = (\det A)I$, where C is the matrix of cofactors of A. *pp. 48, 240*

Diagonal: See *Main diagonal.*

Diagonalizable: A matrix A is diagonalizable if it is similar to a diagonal matrix; that is, if there exist an invertible matrix P and a diagonal matrix D such that $P^{-1}AP = D$. For example, $A = \begin{bmatrix} -11 & 18 \\ -6 & 10 \end{bmatrix}$ is diagonalizable because $P^{-1}AP = D$, with $P = \begin{bmatrix} 2 & 3 \\ 1 & 2 \end{bmatrix}$ and $D = \begin{bmatrix} -2 & 0 \\ 0 & 1 \end{bmatrix}$. A real $n \times n$ matrix A is orthogonally diagonalizable if there exists a (real) orthogonal matrix Q such that $Q^{-1}AQ = Q^TAQ = D$ is a (real) diagonal matrix. For example, let $A = \begin{bmatrix} 1 & -2 \\ -2 & -2 \end{bmatrix}$. The matrix $Q = \begin{bmatrix} \frac{1}{\sqrt{5}} & -\frac{2}{\sqrt{5}} \\ \frac{2}{\sqrt{5}} & \frac{1}{\sqrt{5}} \end{bmatrix}$ is orthogonal and $Q^TAQ = \begin{bmatrix} -3 & 0 \\ 0 & 2 \end{bmatrix} = D$. A complex matrix A is unitarily diagonalizable if there exists a unitary matrix U such that $U^{-1}AU = U^*AU$ is diagonal. For example, let $A = \begin{bmatrix} 2 & 1-i \\ 1+i & 3 \end{bmatrix}$. The matrix $U = \begin{bmatrix} \frac{-1+i}{\sqrt{3}} & \frac{1-i}{\sqrt{6}} \\ \frac{1}{\sqrt{3}} & \frac{2}{\sqrt{6}} \end{bmatrix}$ is unitary and $U^{-1}AU = U^*AU = \begin{bmatrix} 1 & 0 \\ 0 & 4 \end{bmatrix}$.

pp. 294, 559, 561

Dot product: The dot product of vectors $\mathsf{x} = \begin{bmatrix} x_1 \\ x_2 \\ \vdots \\ x_n \end{bmatrix}$ and $\mathsf{y} = \begin{bmatrix} y_1 \\ y_2 \\ \vdots \\ y_n \end{bmatrix}$ is the number $\mathsf{x} \cdot \mathsf{y} = x_1y_1 + x_2y_2 + \cdots + x_ny_n$. For example, with $\mathsf{x} = \begin{bmatrix} -1 \\ 2 \end{bmatrix}$ and $\mathsf{y} = \begin{bmatrix} 4 \\ -3 \end{bmatrix}$, we have $\mathsf{x} \cdot \mathsf{y} = -1(4) + 2(-3) = -10$. If $\mathsf{x} = \begin{bmatrix} -1 \\ 0 \\ 2 \end{bmatrix}$ and $\mathsf{y} = \begin{bmatrix} 1 \\ 2 \\ 3 \end{bmatrix}$, then $\mathsf{x} \cdot \mathsf{y} = -1(1) + 0(2) + 2(3) = 5$. The complex dot product of vectors

$\mathsf{z} = \begin{bmatrix} z_1 \\ \vdots \\ z_n \end{bmatrix}$ and $\mathsf{w} = \begin{bmatrix} w_1 \\ \vdots \\ w_n \end{bmatrix}$ in C^n is $\mathsf{z} \cdot \mathsf{w} = \overline{z}_1 w_1 + \overline{z}_2 w_2 + \cdots + \overline{z}_n w_n$. For example, if $\mathsf{z} = \begin{bmatrix} 1 \\ i \end{bmatrix}$ and $\mathsf{w} = \begin{bmatrix} 2+i \\ 3-i \end{bmatrix}$, then $\mathsf{z} \cdot \mathsf{w} = 1(2+i) - i(3-i) = 2+i-3i-1 = 1-2i$ and $\mathsf{w} \cdot \mathsf{z} = (2-i)(1) + (3+i)(i) = 2-i+3i-1 = 1+2i$.
pp. 29, 542

Eigenspace; Eigenvalue; Eigenvector: An eigenvalue of a (square) matrix is a number λ with the property that $A\mathsf{x} = \lambda \mathsf{x}$ for some nonzero vector x. The vector x is called an eigenvector of A corresponding to λ. The set of all solutions to $A\mathsf{x} = \lambda\mathsf{x}$ is called the eigenspace of A corresponding to λ. For example, for $A = \begin{bmatrix} 1 & 3 \\ 2 & -4 \end{bmatrix}$, we have

$$A\begin{bmatrix} 3 \\ 1 \end{bmatrix} = \begin{bmatrix} 1 & 3 \\ 2 & -4 \end{bmatrix}\begin{bmatrix} 3 \\ 1 \end{bmatrix} = \begin{bmatrix} 6 \\ 2 \end{bmatrix} = 2\mathsf{x},$$

so $\begin{bmatrix} 3 \\ 1 \end{bmatrix}$ is an eigenvector of A corresponding to the eigenvalue $\lambda = 2$. The eigenspace corresponding to $\lambda = 2$ is the set of multiples of $\begin{bmatrix} 3 \\ 1 \end{bmatrix}$. You find this by solving the homogeneous system $(A - 2I)\mathsf{x} = \mathsf{0}$, which is a useful way to rewrite $A\mathsf{x} = 2\mathsf{x}$.
p. 281

Elementary matrix: An elementary matrix is a (square) matrix obtained from the identity matrix by a single elementary row operation. For example, $E = \begin{bmatrix} 0 & 1 & 0 \\ 1 & 0 & 0 \\ 0 & 0 & 1 \end{bmatrix}$ is an elementary matrix, obtained from the 3×3 identity matrix I by interchanging rows one and two. The matrix $E = \begin{bmatrix} 1 & 0 & 0 \\ 0 & 3 & 0 \\ 0 & 0 & 1 \end{bmatrix}$ is elementary; it is obtained from I by multiplying row two by 3.
p. 185

Elementary row operations: The three elementary row operations on a matrix are

1. $R \leftrightarrow R'$: Interchange two rows;
2. $R \to cR$: Multiply a row by a scalar $c \neq 0$;
3. $R \to R - cR'$: Replace a row by that row less a multiple of another row.
p. 144

Euclidean n-space: Euclidean n-space is the set of all n-dimensional vectors. It is denoted R^n. Euclidean 2-space, $\mathsf{R}^2 = \{\begin{bmatrix} x \\ y \end{bmatrix} \mid x, y \in \mathsf{R}\}$ is more commonly called the Euclidean plane and Euclidean 3-space $\mathsf{R}^3 = \{\begin{bmatrix} x \\ y \\ z \end{bmatrix} \mid x, y, z \in \mathsf{R}\}$ is often called simply 3-space.
p. 8

Finite dimension: A vector space V is finite dimensional (has finite dimension) if $V = \{\mathsf{0}\}$ or if V has a basis of n vectors for some $n \geq 1$. A vector space that is not finite dimensional is infinite dimensional. The dimension of a finite dimensional vector space V is the number of elements in any basis of V. For example, the set $\mathbb{R}[x]$ of polynomials over $\mathbb{R}$ in a variable x, with usual addition and scalar multiplication, is an infinite dimensional vector space, while $\mathbb{R}^n$ has finite dimension n. Planes through the origin in $\mathbb{R}^3$ have dimension 2, while lines through the origin have dimension 1. *p. 366*

Hermitian: A complex matrix A is Hermitian if $A = A^*$, its conjugate transpose. For example, the matrix $A = \begin{bmatrix} -3 & 4-i \\ 4+i & 7 \end{bmatrix}$ is Hermitian, as is $\begin{bmatrix} 1 & 2 & 3 \\ 2 & 4 & 5 \\ 3 & 5 & 6 \end{bmatrix}$. *p. 543*

Homogeneous system: A system of linear equations is homogeneous if it has the form $A\mathsf{x} = 0$, that is, the vector of constants to the right of = is the zero vector. For example, the system

$$\begin{aligned} 3x_1 + 2x_2 - x_3 &= 0 \\ 2x_1 - 5x_2 + 7x_3 &= 0, \end{aligned}$$

which is $A\mathsf{x} = 0$ with $A = \begin{bmatrix} 3 & 2 & -1 \\ 2 & -5 & 7 \end{bmatrix}$ and $\mathsf{x} = \begin{bmatrix} x_1 \\ x_2 \end{bmatrix}$, is homogeneous. *p. 177*

Identity matrix: An identity matrix is a square matrix with 1s on the diagonal and 0s everywhere else. For example, $\begin{bmatrix} 1 & 0 \\ 0 & 1 \end{bmatrix}$ and $\begin{bmatrix} 1 & 0 & 0 \\ 0 & 1 & 0 \\ 0 & 0 & 1 \end{bmatrix}$ are identity matrices, the first being the 2×2 identity matrix and the second the 3×3 identity matrix. *p. 105*

Inconsistent: See *Linear equation/Linear system.*

Intersection of subspaces: The intersection of subspaces U and W of a vector space V is the set $U \cap W = \{\mathsf{x} \in V \mid \mathsf{x} \in U \text{ and } \mathsf{x} \in W\}$ of vectors that are in both U and W. For example, if $U = \left\{\begin{bmatrix} c \\ 2c \end{bmatrix} \in \mathbb{R}^2\right\}$ and $W = \left\{\begin{bmatrix} 0 \\ c \end{bmatrix} \in \mathbb{R}^2\right\}$, then $U \cap W = \{0\}$ since $\mathsf{v} = \begin{bmatrix} x \\ y \end{bmatrix}$ in both U and W implies $\mathsf{v} = \begin{bmatrix} x \\ y \end{bmatrix} = \begin{bmatrix} c \\ 2c \end{bmatrix} = \begin{bmatrix} 0 \\ d \end{bmatrix}$ for scalars c and d. So $x = c = 0$ and $y = 2c = d = 0$ meaning $\mathsf{v} = 0$. *p. 503*

Invariant subspace: A subspace U of $\mathbb{R}^n$ or $\mathbb{C}^n$ is invariant under multiplication by an $n \times n$ matrix A if AU is contained in U, symbolically, $AU \subseteq U$. For example, the xy-plane is invariant under multiplication by $A = \begin{bmatrix} 2 & -3 & 1 \\ -1 & 5 & 1 \\ 0 & 0 & 8 \end{bmatrix}$ because $A\begin{bmatrix} x \\ y \\ 0 \end{bmatrix} = \begin{bmatrix} 2x-3y \\ -x+5y \\ 0 \end{bmatrix}$ is in the xy-plane for any $\begin{bmatrix} x \\ y \\ 0 \end{bmatrix}$ in that plane. *p. 553*

Inverse: See *Invertible matrix, Right inverse* and *Left inverse.*

Invertible matrix: A square matrix A is invertible (or has an inverse) if there is another matrix B such that $AB = I$ and $BA = I$. The matrix B is called the inverse of A and we write $B = A^{-1}$. For example, the matrix $A = \begin{bmatrix} 1 & -2 \\ 2 & -3 \end{bmatrix}$ has an inverse, namely, $B = \begin{bmatrix} -3 & 2 \\ -2 & 1 \end{bmatrix}$, since

$$AB = \begin{bmatrix} 1 & -2 \\ 2 & -3 \end{bmatrix}\begin{bmatrix} -3 & 2 \\ -2 & 1 \end{bmatrix} = \begin{bmatrix} 1 & 0 \\ 0 & 1 \end{bmatrix} = I$$

and

$$BA = \begin{bmatrix} -3 & 2 \\ -2 & 1 \end{bmatrix}\begin{bmatrix} 1 & -2 \\ 2 & -3 \end{bmatrix} = \begin{bmatrix} 1 & 0 \\ 0 & 1 \end{bmatrix} = I.$$

From theory, we know that if $AB = I$ with A and B square, then BA must also be I so, in point of fact, we had only to compute one of the above two products to be sure that A and B are invertible. The matrix $A = \begin{bmatrix} 1 & 2 \\ 2 & 4 \end{bmatrix}$ is not invertible because, if we let $B = \begin{bmatrix} x & y \\ z & w \end{bmatrix}$, then $AB = \begin{bmatrix} x+2z & y+2w \\ 2x+4z & 2y+4w \end{bmatrix}$. Since the second row of AB is twice the first, AB can never be I. *p. 127*

LDU factorization: An LDU factorization of a matrix A is a representation of $A = LDU$ as the product of a lower triangular matrix L with 1s on the diagonal, a diagonal matrix D and a row echelon matrix U with 1s on the diagonal. For example,

$$\begin{bmatrix} 1 & 2 & 3 \\ 5 & 14 & 7 \\ 9 & 10 & 0 \\ 0 & 2 & 3 \end{bmatrix} = \begin{bmatrix} 1 & 0 & 0 & 0 \\ 5 & 1 & 0 & 0 \\ 9 & -2 & 1 & 0 \\ 0 & \frac{1}{2} & -\frac{7}{43} & 1 \end{bmatrix}\begin{bmatrix} 1 & 0 & 0 & 0 \\ 0 & 4 & 0 & 0 \\ 0 & 0 & -43 & 0 \\ 0 & 0 & 0 & 1 \end{bmatrix}\begin{bmatrix} 1 & 2 & 3 \\ 0 & 1 & -2 \\ 0 & 0 & 1 \\ 0 & 0 & 0 \end{bmatrix}$$

is an LDU factorization. *p. 208*

Left inverse: A left inverse of an $m \times n$ matrix A is an $n \times m$ matrix C such that $CA = I_n$, the $n \times n$ identity matrix. For example, any matrix C of the form $C = \begin{bmatrix} 1-5t & t & t \\ -2-5s & 1+s & s \end{bmatrix}$ is a left inverse of $A = \begin{bmatrix} 1 & 0 \\ 2 & 1 \\ 3 & -1 \end{bmatrix}$ since $CA = I_2$, the 2×2 identity matrix. *p. 388*

Length: The length of $\mathsf{x} = \begin{bmatrix} x_1 \\ x_2 \\ \vdots \\ x_n \end{bmatrix}$ is $\|\mathsf{x}\| = \sqrt{\mathsf{x} \cdot \mathsf{x}} = \sqrt{x_1^2 + x_2^2 + \cdots + x_n^2}$. For example, if $\mathsf{x} = \begin{bmatrix} 2 \\ 1 \\ 2 \end{bmatrix}$, then $\|\mathsf{x}\| = \sqrt{4+1+4} = 3$. The rule is the same for complex vectors (but you must use the complex dot product). For instance, if $\mathsf{z} = \begin{bmatrix} 1 \\ i \end{bmatrix}$, then $\|\mathsf{z}\|^2 = \mathsf{z} \cdot \mathsf{z} = 1(1) - i(i) = 1 - i^2 = 2$, so $\|\mathsf{z}\| = \sqrt{2}$. *pp. 27, 542*

Linear combination: A linear combination of k vectors $\mathsf{u}_1, \mathsf{u}_2, \ldots, \mathsf{u}_k$ is a vector of the form $c_1\mathsf{u}_1 + c_2\mathsf{u}_2 + \cdots + c_k\mathsf{u}_k$, where $c_1, \ldots, c_k$ are scalars. For example, $\begin{bmatrix} -5 \\ 9 \end{bmatrix}$ is a linear combination of $\mathsf{u} = \begin{bmatrix} -2 \\ 3 \end{bmatrix}$ and $\mathsf{v} = \begin{bmatrix} -1 \\ 1 \end{bmatrix}$

since $\begin{bmatrix}-5\\9\end{bmatrix} = 4\begin{bmatrix}-2\\3\end{bmatrix} - 3\begin{bmatrix}-1\\1\end{bmatrix} = 4\mathsf{u} - 3\mathsf{v}$ and $\begin{bmatrix}2\\-6\end{bmatrix}$ is a linear combination of $\mathsf{u}_1 = \begin{bmatrix}-2\\3\end{bmatrix}$, $\mathsf{u}_2 = \begin{bmatrix}6\\-5\end{bmatrix}$ and $\mathsf{u}_3 = \begin{bmatrix}4\\5\end{bmatrix}$ since $\begin{bmatrix}2\\-6\end{bmatrix} = 3\begin{bmatrix}-2\\3\end{bmatrix} + 2\begin{bmatrix}6\\-5\end{bmatrix} - \begin{bmatrix}4\\5\end{bmatrix} = 3\mathsf{u}_1 + 2\mathsf{u}_2 + (-1)\mathsf{u}_3$. A linear combination of matrices $A_1, A_2, \ldots, A_k$ (all of the same size) is a matrix of the form $c_1A_1 + c_2A_2 + \cdots + c_kA_k$, where $c_1, c_2, \ldots, c_k$ are scalars. *pp. 10, 106*

Linear equation/Linear system: A linear equation is an equation of the form

$$a_1x_1 + a_2x_2 + \cdots + a_nx_n = b,$$

where $a_1, a_2, \ldots, a_n$ and b are real numbers and $x_1, x_2, \ldots, x_n$ are variables. A set of one or more linear equations is called a linear system. To solve a linear system means to find values of the variables that make each equation true. A system that has no solution is called inconsistent. *p. 141*

Linear transformation: A linear transformation is a function $T\colon V \to W$ from a vector space V to another vector space W that satisfies

1. $T(\mathsf{u} + \mathsf{v}) = T(\mathsf{u}) + T(\mathsf{v})$
2. $T(c\mathsf{u}) = cT(\mathsf{u})$

for all vectors u, v in V and all scalars c. We describe these two properties by saying that T "preserves" addition and scalar multiplication. The most important example of a linear transformation in this book is left multiplication by a matrix, which is the function $T\colon \mathsf{R}^n \to \mathsf{R}^m$ defined by $T(\mathsf{x}) = A\mathsf{x}$ with A an $m \times n$ matrix. The function $T\colon \mathsf{R}^1 \to \mathsf{R}^1$ defined by $T(x) = 2x$ is a linear transformation because $T(x + y) = 2(x + y) = 2x + 2y = T(x) + T(y)$ and $T(cx) = 2(cx) = c(2x) = cT(x)$ for all vectors x, y in $\mathsf{R}^1 = \mathsf{R}$ and all scalars c. The function sin: $\mathsf{R} \to \mathsf{R}$ does not define a linear transformation because sine satisfies neither of the required properties. *p. 401*

Linearly dependent: See *Linearly independent.*

Linearly independent: A set of vectors $\mathsf{x}_1, \mathsf{x}_2, \ldots, \mathsf{x}_n$ is linearly independent if the only linear combination of them that equals the zero vector is the trivial one, where all the coefficients are 0:

$$c_1\mathsf{x}_1 + c_2\mathsf{x}_2 + \cdots + c_n\mathsf{x}_n = \mathsf{0} \quad \text{implies} \quad c_1 = c_2 = \cdots = c_n = 0.$$

Vectors that are not linearly independent are linearly dependent; that is, there exist scalars $c_1, c_2, \ldots, c_n$, not all 0, with $c_1\mathsf{x}_1 + c_2\mathsf{x}_2 + \cdots + c_n\mathsf{x}_n = \mathsf{0}$. The vectors $\mathsf{x}_1 = \begin{bmatrix}1\\2\end{bmatrix}$ and $\mathsf{x}_2 = \begin{bmatrix}-1\\0\end{bmatrix}$ are linearly independent (solve the equation $c_1\mathsf{x}_1 + c_2\mathsf{x}_2 = \mathsf{0}$ and obtain $c_1 = c_2 = 0$) whereas the vectors $\mathsf{u}_1 = \begin{bmatrix}1\\2\end{bmatrix}$, $\mathsf{u}_2 = \begin{bmatrix}1\\0\end{bmatrix}$ and $\mathsf{u}_3 = \begin{bmatrix}2\\4\end{bmatrix}$ are linearly dependent because $2\mathsf{u}_1 + 0\mathsf{u}_2 - 1\mathsf{u}_3 = \mathsf{0}$. *pp. 77, 361*

LU Factorization: An LU factorization of a matrix A is the representation of $A = LU$ as the product of a square lower triangular matrix L with 1s on the diagonal and a row echelon matrix U. For example, $\begin{bmatrix}1&2\\4&6\end{bmatrix} = \begin{bmatrix}1&0\\4&1\end{bmatrix}\begin{bmatrix}1&2\\0&-2\end{bmatrix}$ is an LU factorization of $A = \begin{bmatrix}1&2\\4&6\end{bmatrix}$. *p. 187*

Main diagonal: The main diagonal of an $m \times n$ matrix $A = [a_{ij}]$ is the list of elements $a_{11}, a_{22}, a_{33}, \ldots$. For instance, the main diagonal of $\begin{bmatrix} 1 & 2 & 3 \\ 4 & 5 & 6 \\ 7 & 8 & 9 \end{bmatrix}$ is 1, 5, 9 and the main diagonal of $\begin{bmatrix} 7 & 9 & 3 \\ 6 & 5 & 4 \end{bmatrix}$ is 7, 5. A matrix is diagonal if and only if its only nonzero entries lie on the main diagonal. For example, the matrix $\begin{bmatrix} 1 & 0 & 0 \\ 0 & -7 & 0 \\ 0 & 0 & 2 \end{bmatrix}$ is diagonal. So is $\begin{bmatrix} -2 & 0 & 0 \\ 0 & 7 & 0 \end{bmatrix}$.

p. 145

Markov matrix: A Markov matrix is a matrix $A = [a_{ij}]$ with $0 \le a_{ij} \le 1$ for all i, j and the entries in each column adding to 1. For example, $\begin{bmatrix} \frac{1}{2} & \frac{1}{4} \\ \frac{1}{2} & \frac{3}{4} \end{bmatrix}$ is Markov and so is $\begin{bmatrix} 0 & \frac{2}{3} \\ 1 & \frac{1}{3} \end{bmatrix}$. *p. 313*

Matrix: A matrix is a rectangular array of numbers enclosed in square brackets. If there are m rows and n columns in the array, the matrix is called $m \times n$ (read "m by n") and said to have size $m \times n$. For example, $A = \begin{bmatrix} 2 & 1 & -1 \\ 3 & -2 & 6 \\ 1 & 0 & -5 \end{bmatrix}$ is a 3×3 ("three by three") matrix, while $B = \begin{bmatrix} 1 & 2 & 3 \\ 4 & 5 & 6 \end{bmatrix}$ is 2×3 and $\mathsf{x} = \begin{bmatrix} x_1 \\ x_2 \\ x_3 \end{bmatrix}$ is 3×1. The numbers in the matrix are called its entries. We always use a capital letter to denote a matrix and the corresponding lower case letter, with subscripts, for its entries. Thus if $A = [a_{ij}]$, the $(2, 3)$ entry is a_{23}. If $A = \begin{bmatrix} 2 & 1 & 3 \\ 3 & 2 & 8 \\ 9 & 0 & 1 \end{bmatrix}$, then $a_{32} = 0$. If $A = [2i - j]$ is a 3×4 matrix, the $(1, 3)$ entry of A is $2(1) - 3 = -1$.

p. 94

Matrix of a linear transformation: The matrix of a linear transformation $T\colon V \to W$ with respect to bases $\mathcal{V} = \{\mathsf{v}_1, \mathsf{v}_2, \ldots, \mathsf{v}_n\}$ of V and $\mathcal{W}$ for W is the matrix

$$M_{\mathcal{W} \leftarrow \mathcal{V}}(T) = \begin{bmatrix} [T(\mathsf{v}_1)]_{\mathcal{W}} & [T(\mathsf{v}_2)]_{\mathcal{W}} & \cdots & [T(\mathsf{v}_n)]_{\mathcal{W}} \\ \downarrow & \downarrow & & \downarrow \\ & & & \end{bmatrix}$$

whose columns are the coordinate vectors of $T(\mathsf{v}_1)$, $T(\mathsf{v}_2), \ldots, T(\mathsf{v}_n)$ relative to $\mathcal{W}$. For example, let $V = W = \mathsf{R}^2$ and let $T\colon V \to V$ be the linear transformation defined by

$$T\left(\begin{bmatrix} x_1 \\ x_2 \end{bmatrix}\right) = \begin{bmatrix} x_2 \\ x_1 \end{bmatrix}.$$

Let $\mathcal{E} = \{\mathsf{e}_1, \mathsf{e}_2\}$ be the standard basis for V and let

$$\mathcal{V} = \{\mathsf{v}_1, \mathsf{v}_2\}, \quad \text{where} \quad \mathsf{v}_1 = \begin{bmatrix} -1 \\ 2 \end{bmatrix} \text{ and } \mathsf{v}_2 = \begin{bmatrix} 2 \\ -3 \end{bmatrix}.$$

Then $\mathcal{V}$ is also a basis for V. Since

$$T(\mathbf{e}_1) = T\left(\begin{bmatrix}1\\0\end{bmatrix}\right) = \begin{bmatrix}0\\1\end{bmatrix} \text{ and } T(\mathbf{e}_2) = T\left(\begin{bmatrix}0\\1\end{bmatrix}\right) = \begin{bmatrix}1\\0\end{bmatrix},$$

the matrix of T relative to $\mathcal{E}$ and $\mathcal{E}$ is $M_{\mathcal{E}\leftarrow\mathcal{E}}(T) = \begin{bmatrix}0&1\\1&0\end{bmatrix}$. Since

$$T\left(\begin{bmatrix}-1\\2\end{bmatrix}\right) = \begin{bmatrix}2\\-1\end{bmatrix} = 2\mathbf{e}_1 + (-1)\mathbf{e}_2$$

so that $[T(\mathbf{v}_1)]_{\mathcal{E}} = \begin{bmatrix}2\\-1\end{bmatrix}$ and

$$T\left(\begin{bmatrix}2\\-3\end{bmatrix}\right) = \begin{bmatrix}-3\\2\end{bmatrix} = -3\mathbf{e}_1 + 2\mathbf{e}_2$$

so that $[T(\mathbf{v}_2)]_{\mathcal{E}} = \begin{bmatrix}-3\\2\end{bmatrix}$, the matrix of T with respect to $\mathcal{V}$ and $\mathcal{E}$ is $M_{\mathcal{E}\leftarrow\mathcal{V}}(T) = \begin{bmatrix}2&-3\\-1&2\end{bmatrix}$. *p. 433*

Minor: The (i, j) minor of a square matrix A, denoted m_{ij}, is the determinant of the $(n-1)\times(n-1)$ matrix obtained from A by deleting row i and column j. The (i, j) cofactor of A is $(-1)^{i+j}m_{ij}$ and is denoted c_{ij}. For example, let $A = \begin{bmatrix}1&2&3\\0&-2&2\\3&7&-4\end{bmatrix}$. The $(1,1)$ minor is the determinant of the matrix $\begin{bmatrix}-2&2\\7&-4\end{bmatrix}$, which is what remains when we remove row one and column one of A. So $m_{11} = -2(-4) - 2(7) = -6$ and $c_{11} = (-1)^{1+1}m_{11} = (-1)^2 m_{11} = -6$. The $(-1)^{i+j}$ in the definition of cofactor has the affect of either leaving the minor alone or changing its sign according to the pattern $\begin{bmatrix}+&-&+&-&\cdots\\-&+&-&+&\cdots\\\vdots&\vdots&&&\end{bmatrix}$. Thus the $(2,3)$ minor of A is the determinant of $\begin{bmatrix}1&2\\3&7\end{bmatrix}$, which is what remains after removing row two and column three of A, $m_{23} = 1$, while $c_{23} = -1$ since there is a $-$ in the $(2,3)$ position of the pattern. *p. 237*

Modulus of a complex number: The modulus of the complex number $z = a + bi$ is $\sqrt{a^2+b^2}$. It is denoted $|z|$. For example, if $z = 2-3i$, $|z|^2 = z\bar{z} = (2-3i)(2+3i) = 2^2 + 3^2 = 13$, so $|z| = \sqrt{13}$. *p. 595*

Null space: The null space of a matrix A is the set of all vectors $\mathbf{x}$ that satisfy $A\mathbf{x} = \mathbf{0}$. It is denoted null sp A. For example, the null space of $A = \begin{bmatrix}1&0\\0&1\end{bmatrix}$ is $\left\{\begin{bmatrix}0\\0\end{bmatrix}\right\}$, while the null space of $A = \begin{bmatrix}1&0&2\\0&1&3\end{bmatrix}$ is the set of vectors of the form $\begin{bmatrix}-2t\\-3t\\t\end{bmatrix} = t\begin{bmatrix}-2\\-3\\1\end{bmatrix}$. [Solve the homogeneous system $A\mathbf{x} = \mathbf{0}$!] A vector $\mathbf{x}$ is in the null space of A if and only if $A\mathbf{x} = \mathbf{0}$. *p. 332*

Nullity: The nullity of a matrix is the dimension of its null space, that is, the number of vectors in a basis. For example, the nullity of $A = \begin{bmatrix} 1 & -3 & -2 & 4 \\ 2 & 0 & 2 & 2 \\ 0 & 4 & 4 & -4 \end{bmatrix}$ is 2. (Solve $A\mathsf{x} = 0$ and discover that the two vectors $\begin{bmatrix} -1 \\ -1 \\ 1 \\ 0 \end{bmatrix}$ and $\begin{bmatrix} -1 \\ 1 \\ 0 \\ 1 \end{bmatrix}$ form a basis for the null space.) *p. 375*

Opposite direction: Vectors u and v have opposite direction if $\mathsf{u} = c\mathsf{v}$ for some scalar $c < 0$. For example, $\mathsf{u} = \begin{bmatrix} 2 \\ -3 \end{bmatrix}$ and $\mathsf{v} = \begin{bmatrix} -3 \\ \frac{9}{2} \end{bmatrix}$ have opposite direction because $\mathsf{u} = -\frac{2}{3}\mathsf{v}$ with the scalar $-\frac{2}{3} < 0$. *p. 3*

Orthogonal: A set $\{\mathsf{f}_1, \mathsf{f}_2, \dots, \mathsf{f}_k\}$ of vectors is orthogonal if the vectors are pairwise orthogonal: $\mathsf{f}_i \cdot \mathsf{f}_j = 0$ when $i \neq j$. If each f_i also has length 1, then the set is called orthonormal. For example, any set like $\{\begin{bmatrix} 0 \\ 1 \\ 0 \end{bmatrix}, \begin{bmatrix} 0 \\ 0 \\ 1 \end{bmatrix}\}$, a subset of the set of standard basis vectors, is orthonormal. The four vectors $\mathsf{f}_1 = \begin{bmatrix} 1 \\ 1 \\ 1 \\ -1 \end{bmatrix}$, $\mathsf{f}_2 = \begin{bmatrix} 1 \\ 0 \\ 1 \\ 2 \end{bmatrix}$, $\mathsf{f}_3 = \begin{bmatrix} -1 \\ 0 \\ 1 \\ 0 \end{bmatrix}$, $\mathsf{f}_4 = \begin{bmatrix} -1 \\ 3 \\ -1 \\ 1 \end{bmatrix}$ form an orthogonal set in R^4. They can be converted to an orthonormal set $\mathsf{q}_1, \mathsf{q}_2, \mathsf{q}_3, \mathsf{q}_4$ by normalizing, that is, dividing each by its length:

$$\mathsf{q}_1 = \frac{1}{2}\begin{bmatrix} 1 \\ 1 \\ 1 \\ -1 \end{bmatrix}, \quad \mathsf{q}_2 = \frac{1}{\sqrt{6}}\begin{bmatrix} 1 \\ 0 \\ 1 \\ 2 \end{bmatrix}, \quad \mathsf{q}_3 = \frac{1}{\sqrt{2}}\begin{bmatrix} -1 \\ 0 \\ 1 \\ 0 \end{bmatrix}, \quad \mathsf{q}_4 = \frac{1}{\sqrt{12}}\begin{bmatrix} -1 \\ 3 \\ -1 \\ 1 \end{bmatrix}.$$

pp. 32, 486

Orthogonal complement: Let U be a subspace of R^n. The orthogonal complement of U, denoted $U^\perp$ ("U perp"), is the set of all vectors in R^n that are orthogonal to all vectors of U:

$$U^\perp = \{\mathsf{v} \in \mathsf{R}^n \mid \mathsf{v} \cdot \mathsf{u} = 0 \text{ for all } \mathsf{u} \text{ in } U\}.$$

For example, with $U = \{\begin{bmatrix} 0 \\ 0 \\ z \end{bmatrix} \mid z \in \mathsf{R}\}$, the z-axis, $U^\perp$ is the set of vectors of the form $\begin{bmatrix} x \\ y \\ 0 \end{bmatrix}$, the xy-plane. Subspaces U and W are orthogonal complements if each is the orthogonal complement of the other; for example, the null space and the row space of a matrix are orthogonal complements. *p. 509*

Orthogonal matrix: An orthogonal matrix is a square matrix with orthonormal columns. For example, the rotation matrix $Q = \begin{bmatrix} \cos\theta & -\sin\theta \\ \sin\theta & \cos\theta \end{bmatrix}$ is orthogonal for any angle θ. Any permutation matrix is orthogonal because its columns are just the standard basis vectors $\mathsf{e}_1, \mathsf{e}_2, \ldots, \mathsf{e}_n$ in some order. One such matrix is $P = \begin{bmatrix} 0 & 0 & 1 \\ 1 & 0 & 0 \\ 0 & 1 & 0 \end{bmatrix}$. *p. 495*

Orthogonal subspaces: Subspaces U and W of R^n are orthogonal if $\mathsf{u} \cdot \mathsf{w} = 0$ for any vector u in U and any vector w in W. For example, in R^3, the x-axis and the z-axis are orthogonal subspaces since if $\mathsf{u} = \begin{bmatrix} x \\ 0 \\ 0 \end{bmatrix}$ is on the x-axis and $\mathsf{w} = \begin{bmatrix} 0 \\ 0 \\ z \end{bmatrix}$ is on the z-axis, then $\mathsf{u} \cdot \mathsf{w} = 0$. The subspaces

$$U = \left\{ \begin{bmatrix} u_1 \\ u_2 \\ u_3 \\ u_4 \end{bmatrix} \in \mathsf{R}^4 \mid 2u_1 - u_2 + u_3 - u_4 = 0 \right\} \text{ and } W = \left\{ c \begin{bmatrix} 2 \\ -1 \\ 1 \\ -1 \end{bmatrix} \mid c \in \mathsf{R} \right\}.$$

are orthogonal because if $\mathsf{u} = \begin{bmatrix} u_1 \\ u_2 \\ u_3 \\ u_4 \end{bmatrix} \in U$ and $\mathsf{w} = c \begin{bmatrix} 2 \\ -1 \\ 1 \\ -1 \end{bmatrix} \in W$, then $\mathsf{u} \cdot \mathsf{w} = c(2u_1 - u_2 + u_3 - u_4) = 0$. *p. 508*

Orthogonally diagonalizable: See *Diagonalizable.*

Orthonormal: See *Orthogonal.*

Pairwise orthogonal: See *Orthogonal.*

Parallel: Vectors u and v are parallel if one is a scalar multiple of the other, that is, if $\mathsf{u} = c\mathsf{v}$ or $\mathsf{v} = c\mathsf{u}$ for some scalar c. For example, $\mathsf{u} = \begin{bmatrix} -2 \\ 1 \end{bmatrix}$ and $\mathsf{v} = \begin{bmatrix} 6 \\ -3 \end{bmatrix}$ are parallel because $\mathsf{v} = -3\mathsf{u}$. The zero vector is parallel to any vector u because $0 = 0\mathsf{u}$. *p. 3*

Permutation matrix: A permutation matrix is a matrix whose rows are the standard basis vectors, in some order; equivalently, a matrix whose columns are the standard basis vectors, in some order. For example, $\begin{bmatrix} 0 & 1 \\ 1 & 0 \end{bmatrix}$, $\begin{bmatrix} 0 & 1 & 0 \\ 0 & 0 & 1 \\ 1 & 0 & 0 \end{bmatrix}$ and $\begin{bmatrix} 0 & 0 & 1 & 0 \\ 0 & 1 & 0 & 0 \\ 0 & 0 & 0 & 1 \\ 1 & 0 & 0 & 0 \end{bmatrix}$ are all permutation matrices. *p. 197*

Pivot: A pivot of a row echelon matrix is the first nonzero entry in a nonzero row and a pivot column is a column containing a pivot. For example, the pivots of $\begin{bmatrix} 0 & -3 & 1 & 2 & 1 \\ 0 & 0 & 2 & 0 & -3 \\ 0 & 0 & 0 & 0 & 8 \end{bmatrix}$ are -3, 2 and 8 and the pivot columns are columns two, three and five. If U is a row echelon form of a matrix A, the pivot columns of A are the pivot columns of U. If U was obtained without multiplying any row by a scalar, then the pivots of A are the pivots of U. For example,

$$A = \begin{bmatrix} -2 & 4 & 3 \\ 3 & -6 & 0 \\ 4 & -8 & 1 \end{bmatrix} \to \begin{bmatrix} -2 & 4 & 3 \\ 0 & 0 & \frac{9}{2} \\ 0 & 0 & 7 \end{bmatrix} \to \begin{bmatrix} -2 & 4 & 3 \\ 0 & 0 & \frac{9}{2} \\ 0 & 0 & 0 \end{bmatrix} = U$$

so the pivots of A are -2 and $\frac{9}{2}$, and the pivot columns of A are columns one and three. *p. 146*

Projection: The projection of a vector u on a (nonzero) vector v is the vector $\mathsf{p} = \operatorname{proj}_{\mathsf{v}} \mathsf{u}$ parallel to v with the property that $\mathsf{u} - \mathsf{p}$ is orthogonal to v. One can show that $\operatorname{proj}_{\mathsf{v}} \mathsf{u} = \frac{\mathsf{u} \cdot \mathsf{v}}{\mathsf{v} \cdot \mathsf{v}} \mathsf{v}$. For example, if $\mathsf{u} = \begin{bmatrix} 1 \\ -3 \\ 4 \end{bmatrix}$ and $\mathsf{v} = \begin{bmatrix} -3 \\ 1 \\ 2 \end{bmatrix}$, $\operatorname{proj}_{\mathsf{v}} \mathsf{u} = \frac{1}{7}\mathsf{v} = \begin{bmatrix} -\frac{3}{7} \\ \frac{1}{7} \\ \frac{2}{7} \end{bmatrix}$. The projection of a vector b onto a plane π is the vector $\mathsf{p} = \operatorname{proj}_\pi \mathsf{b}$ in π with the property that $\mathsf{b} - \mathsf{p}$ is orthogonal to every vector in π. If the plane π is spanned by orthogonal vectors e and f, one can show that $\operatorname{proj}_\pi \mathsf{b} = \frac{\mathsf{b}\cdot\mathsf{e}}{\mathsf{e}\cdot\mathsf{e}} \mathsf{e} + \frac{\mathsf{b}\cdot\mathsf{f}}{\mathsf{f}\cdot\mathsf{f}} \mathsf{f}$. For example, the plane π with equation $3x - 2y + z = 0$ is spanned by the orthogonal vectors $\mathsf{e} = \begin{bmatrix} 5 \\ 6 \\ -3 \end{bmatrix}$ and $\mathsf{f} = \begin{bmatrix} 0 \\ 1 \\ 2 \end{bmatrix}$. If $\mathsf{b} = \begin{bmatrix} 5 \\ -3 \\ -7 \end{bmatrix}$, $\operatorname{proj}_\pi \mathsf{b} = \frac{\mathsf{b} \cdot \mathsf{e}}{\mathsf{e} \cdot \mathsf{e}} \mathsf{e} + \frac{\mathsf{b} \cdot \mathsf{f}}{\mathsf{f} \cdot \mathsf{f}} \mathsf{f}$

$$= \frac{28}{70} \begin{bmatrix} 5 \\ 6 \\ -3 \end{bmatrix} - \frac{17}{5} \begin{bmatrix} 0 \\ 1 \\ 2 \end{bmatrix} = \frac{2}{5} \begin{bmatrix} 5 \\ 6 \\ -3 \end{bmatrix} - \frac{17}{5} \begin{bmatrix} 0 \\ 1 \\ 2 \end{bmatrix} = \begin{bmatrix} 2 \\ -1 \\ -8 \end{bmatrix}.$$

In general, the projection of a vector b on a subspace U is the vector $\mathsf{p} = \operatorname{proj}_U \mathsf{b}$ with the property that $\mathsf{b} - \mathsf{p}$ is orthogonal to U. If $\mathsf{f}_1, \mathsf{f}_2, \ldots, \mathsf{f}_n$ is an orthogonal basis for U, then

$$\operatorname{proj}_U \mathsf{b} = \frac{\mathsf{b} \cdot \mathsf{f}_1}{\mathsf{f}_1 \cdot \mathsf{f}_1} \mathsf{f}_1 + \frac{\mathsf{b} \cdot \mathsf{f}_2}{\mathsf{f}_2 \cdot \mathsf{f}_2} \mathsf{f}_2 + \cdots + \frac{\mathsf{b} \cdot \mathsf{f}_n}{\mathsf{f}_n \cdot \mathsf{f}_n} \mathsf{f}_n.$$

pp. 66, 69, 470

Projection matrix: A (necessarily square) matrix P is a projection matrix if and only if $\mathsf{b} - P\mathsf{b}$ is orthogonal to the column space of P for all vectors b. For example, $P = \frac{1}{3} \begin{bmatrix} 2 & -1 & 1 \\ -1 & 2 & 1 \\ 1 & 1 & 2 \end{bmatrix}$ is a projection matrix. (This is most easily verified by noticing that P is symmetric and checking that $P^2 = P$. These two properties characterize projection matrices.) *p. 465*

Pseudoinverse: The pseudoinverse of a matrix A is the matrix A^+ satisfying

- $A^+\mathsf{b}$ is in the row space of A for all b, and
- $A(A^+\mathsf{b})$ is the projection of b on the column space of A.

Equivalently, A^+ is defined by the condition that $A^+\mathsf{b} = \mathsf{x}^+$ is the shortest vector amongst those minimizing $\|\mathsf{b} - A\mathsf{x}\|$. Thanks to x^+, the system $A\mathsf{x} = \mathsf{b}$ always has a solution, namely, $\mathsf{x} = A^+\mathsf{b}$, called the best solution to $A\mathsf{x} = \mathsf{b}$ in the sense of least squares.

For example, if $A = \begin{bmatrix} -1 & 0 & 0 \\ 0 & \frac{1}{2} & 0 \\ 0 & 0 & 0 \end{bmatrix}$, $\mathsf{x}^+ = \begin{bmatrix} -b_1 \\ 2b_2 \\ 0 \end{bmatrix} = \begin{bmatrix} -1 & 0 & 0 \\ 0 & 2 & 0 \\ 0 & 0 & 0 \end{bmatrix} \begin{bmatrix} b_1 \\ b_2 \\ b_3 \end{bmatrix}$,
so $A^+ = \begin{bmatrix} -1 & 0 & 0 \\ 0 & 2 & 0 \\ 0 & 0 & 0 \end{bmatrix}$. *p. 528*

Quadratic form: A quadratic form in n variables $x_1, x_2, \ldots, x_n$ is a linear combination of terms of the form $x_i x_j$; that is, a linear combination of $x_1^2, x_2^2, \ldots, x_n^2$ and cross terms like x_1x_2, x_2x_5 and x_4x_9. For example, $q = x^2 - y^2 + 4xy$ and $q = 3y^2 - 2xy$ are quadratic forms in x and y, while $q = -4x_1^2 + x_2^2 + 4x_3^2 + 6x_1x_3$ is a quadratic form in x_1, x_2 and x_3. *p. 576*

Rank: The rank of a matrix is the number of pivot columns. For example, the rank of $A = \begin{bmatrix} 4 & 0 & 2 \\ 0 & 1 & 3 \end{bmatrix}$ is 2, while the rank of $A = \begin{bmatrix} 1 & 2 & 3 \\ 2 & 4 & 6 \end{bmatrix}$ is 1. *p. 338*

Reduced row echelon form: See *Row echelon form.*

Right inverse: A right inverse for an $m \times n$ matrix A is an $n \times m$ matrix B such that $AB = I_m$, the $m \times m$ identity matrix. For example, any matrix B of the form $B = \begin{bmatrix} t - \frac{5}{3} & s + \frac{2}{3} \\ \frac{4}{3} - 2t & -\frac{1}{3} - 2s \\ t & s \end{bmatrix}$ is a right inverse of $A = \begin{bmatrix} 1 & 2 & 3 \\ 4 & 5 & 6 \end{bmatrix}$ since $AB = I_2$. *p. 388*

Row echelon form: A matrix is in row echelon form if it is upper triangular with the following properties:

1. all rows consisting entirely of 0s are at the bottom,
2. the pivots, which are the leading nonzero entries in nonzero rows, step from left to right as you read down the matrix, and
3. every entry below a pivot is 0.

For example, $\begin{bmatrix} 7 & 1 & 3 & 4 & 5 \\ 0 & 0 & 5 & 3 & 1 \\ 0 & 0 & 0 & 0 & 1 \end{bmatrix}$ and $\begin{bmatrix} 0 & 0 & 1 & 1 & 5 \\ 0 & 0 & 0 & 0 & 1 \\ 0 & 0 & 0 & 0 & 0 \end{bmatrix}$ are in row echelon form, while $\begin{bmatrix} 1 & 0 & 3 & 2 \\ 0 & -1 & 2 & 4 \\ 0 & 1 & 0 & 1 \end{bmatrix}$ and $\begin{bmatrix} 1 & 0 & 3 & 2 \\ 0 & 0 & 1 & 4 \\ 0 & 1 & 0 & 2 \end{bmatrix}$ are not. A matrix is in reduced row echelon form if it is in row echelon form, each pivot is a 1, and all entries above and below each pivot are 0. For example, $\begin{bmatrix} 1 & 0 \\ 0 & 1 \end{bmatrix}$, $\begin{bmatrix} 1 & 0 & -2 \\ 0 & 1 & 1 \end{bmatrix}$ and $\begin{bmatrix} 1 & 0 & 0 & 0 \\ 0 & 0 & 1 & 0 \\ 0 & 0 & 0 & 1 \end{bmatrix}$ are all in reduced row echelon form.

pp. 145, 223

Same direction: Vectors u and v have the same direction if $\mathsf{u} = c\mathsf{v}$ with $c > 0$. For example, $\mathsf{u} = \begin{bmatrix} -2 \\ 4 \\ 6 \end{bmatrix}$ and $\mathsf{v} = \begin{bmatrix} -1 \\ 2 \\ 3 \end{bmatrix}$ have the same direction because $\mathsf{v} = \frac{1}{2}\mathsf{u}$ with the scalar $\frac{1}{2} > 0$. *p. 3*

Similar: Matrices A and B are similar if there is an invertible matrix P such that $B = P^{-1}AP$. For example, $A = \begin{bmatrix} 1 & 2 \\ 3 & 4 \end{bmatrix}$ and $B = \begin{bmatrix} -55 & -97 \\ 34 & 60 \end{bmatrix}$ are similar because

$$B = \begin{bmatrix} -55 & -97 \\ 34 & 60 \end{bmatrix} = \begin{bmatrix} 2 & -5 \\ -1 & 3 \end{bmatrix}\begin{bmatrix} 1 & 2 \\ 3 & 4 \end{bmatrix}\begin{bmatrix} 3 & 5 \\ 1 & 2 \end{bmatrix} = P^{-1}AP$$

with $P = \begin{bmatrix} 3 & 5 \\ 1 & 2 \end{bmatrix}$. *p. 292*

Singular matrix: A square matrix is singular if it is not invertible, a condition equivalent to having determinant 0. For example, $A = \begin{bmatrix} 2 & 3 \\ 4 & 6 \end{bmatrix}$ is singular. *p. 255*

Singular values: The singular values of a matrix A are the positive diagonal entries $\sigma_1, \ldots, \sigma_r$ of $D = \begin{bmatrix} \sigma_1 & & \\ & \ddots & \\ & & \sigma_r \end{bmatrix}$ when A has been factored $A = Q_1 \Sigma Q_2^T$ with Q_1 and Q_2 orthogonal matrices and $\Sigma = \begin{bmatrix} D & 0 \\ 0 & 0 \end{bmatrix}$. For example,

$$Q_1 = \begin{bmatrix} 0 & \frac{2}{\sqrt{6}} & -\frac{1}{\sqrt{3}} \\ \frac{1}{\sqrt{2}} & \frac{1}{\sqrt{6}} & \frac{1}{\sqrt{3}} \\ -\frac{1}{\sqrt{2}} & \frac{1}{\sqrt{6}} & \frac{1}{\sqrt{3}} \end{bmatrix} \text{ and } Q_2 = \begin{bmatrix} \frac{1}{\sqrt{2}} & \frac{1}{\sqrt{2}} \\ -\frac{1}{\sqrt{2}} & \frac{1}{\sqrt{2}} \end{bmatrix}$$

are orthogonal matrices and $A = \begin{bmatrix} 1 & 1 \\ 1 & 0 \\ 0 & 1 \end{bmatrix} = Q_1 \Sigma Q_2^T$ with $\Sigma = \begin{bmatrix} 1 & 0 \\ 0 & \sqrt{3} \\ 0 & 0 \end{bmatrix}$, so the singular values of A are 1 and $\sqrt{3}$. *p. 583*

Size: See *Matrix*.

Span, spanned by: The span of vectors $\mathsf{u}_1, \mathsf{u}_2, \ldots, \mathsf{u}_n$ is the set of all linear combinations of $\mathsf{u}_1, \mathsf{u}_2, \ldots, \mathsf{u}_n$. It is denoted $\operatorname{sp}\{\mathsf{u}_1, \ldots, \mathsf{u}_n\}$. We say that the space $\operatorname{sp}\{\mathsf{u}_1, \ldots, \mathsf{u}_n\}$ is spanned by the vectors $\mathsf{u}_1, \ldots, \mathsf{u}_n$. For example, the span of i and j in R^2 is all R^2 since any vector $\mathsf{u} = \begin{bmatrix} a \\ b \end{bmatrix}$ in R^2 is $a\mathsf{i} + b\mathsf{j}$. We say that R^2 is spanned by i and j. On the other hand, the span of $\mathsf{u} = \begin{bmatrix} -1 \\ 2 \end{bmatrix}$ and $\mathsf{v} = \begin{bmatrix} -3 \\ 6 \end{bmatrix}$ is the set of scalar multiples of u since $a\mathsf{u} + b\mathsf{v} = \begin{bmatrix} -a - 3b \\ 2a + 6b \end{bmatrix} = (a + 3b)\begin{bmatrix} -1 \\ 2 \end{bmatrix}$. The span of two nonparallel vectors is a plane. For example, the span of $\begin{bmatrix} 1 \\ 2 \end{bmatrix}$ and $\begin{bmatrix} -3 \\ 4 \end{bmatrix}$ is the xy-plane. Any plane through the origin in R^3 is spanned by any two nonparallel vectors it contains. For example, the plane with equation $2x - y + 3z = 0$ is spanned by $\mathsf{u} = \begin{bmatrix} 1 \\ 2 \\ 0 \end{bmatrix}$ and $\mathsf{v} = \begin{bmatrix} 0 \\ 3 \\ 1 \end{bmatrix}$. *pp. 12, 13*

Standard basis vectors: The standard basis vectors in R^n are the n-dimensional vectors $\mathsf{e}_1, \mathsf{e}_2, \ldots, \mathsf{e}_n$ where e_i has ith component 1 and all other components 0; that is,

$$e_1 = \begin{bmatrix}1\\0\\0\\0\\ \vdots\\0\end{bmatrix}, \quad e_2 = \begin{bmatrix}0\\1\\0\\0\\ \vdots\\0\end{bmatrix}, \quad e_3 = \begin{bmatrix}0\\0\\1\\0\\ \vdots\\0\end{bmatrix}, \quad \ldots, \quad e_n = \begin{bmatrix}0\\0\\0\\ \vdots\\0\\1\end{bmatrix}.$$

In R^2, the standard basis vectors $e_1 = \begin{bmatrix}1\\0\end{bmatrix}$ and $e_2 = \begin{bmatrix}0\\1\end{bmatrix}$ are also denoted i and j, respectively, and, in R^3, the standard basis vectors $e_1 = \begin{bmatrix}1\\0\\0\end{bmatrix}$, $e_2 = \begin{bmatrix}0\\1\\0\end{bmatrix}$ and $e_3 = \begin{bmatrix}0\\0\\1\end{bmatrix}$ are also denoted i, j and k, respectively. *p. 15*

Subspace: A subspace of a vector space V is a subset U of V that is itself a vector space under the operations of addition and scalar multiplication in V. It is useful to note that a nonempty subset U is a subspace if and only if it is

- closed under addition: If u and v are in U, so is u + v, and
- closed under scalar multiplication: If u is in U and c is a scalar, then cu is also in U.

For example, the set of vectors of the form $\begin{bmatrix}x\\y\\0\end{bmatrix}$ is a subspace of R^3 as is the plane with equation $x + y + 2z = 0$. The set of polynomials in x with real coefficients and constant term 0 is a subspace of $R[x]$, the vector space of all polynomials in x (with real coefficients). *p. 350*

Sum of subspaces: The sum of subspaces U and W of R^n is the set $U+W = \{u+w \mid u \in U, w \in W\}$ of vectors in R^n that can be written as the sum of a vector in U and a vector in W. For example, if $U = \left\{\begin{bmatrix}c\\2c\end{bmatrix} \in R^2\right\}$ and $W = \left\{\begin{bmatrix}0\\c\end{bmatrix} \in R^2\right\}$, then $U + W = R^2$ because every $v = \begin{bmatrix}x\\y\end{bmatrix} \in R^2$ can be written $\begin{bmatrix}x\\y\end{bmatrix} = \begin{bmatrix}x\\2x\end{bmatrix} + \begin{bmatrix}0\\y-2x\end{bmatrix}$ as the sum of a vector in U and a vector in W. A sum $U+W$ is direct, and we write $U \oplus W$, if every vector in $U + W$ can be written in just one way in the form u + w. For example, the representation of $R^2 = U + W$ just given is direct because if $u_1 = \begin{bmatrix}x_1\\2x_1\end{bmatrix}$ and $u_2 = \begin{bmatrix}x_1\\2x_1\end{bmatrix}$ are in U, and $w_1 = \begin{bmatrix}0\\y_1-2x_1\end{bmatrix}$ and $w_2 = \begin{bmatrix}0\\y_1-2x_1\end{bmatrix}$ are in W, and $u_1 + w_1 = u_2 + w_2$, then (it is easily seen that) $x_1 = x_2$ and $y_1 = y_2$, so $u_1 = u_2$ and $w_1 = w_2$.

In general, it is not hard to see that $U + W$ is direct if and only if the intersection $U \cap W = 0$. For example, if $U = \left\{\begin{bmatrix}x\\y\\z\end{bmatrix} \mid x+y-z = 0\right\}$ and

$W = \left\{ \begin{bmatrix} x \\ y \\ z \end{bmatrix} \mid 2x - 3y + z = 0 \right\}$, then $\mathsf{R}^3 = U + W$ since any $\begin{bmatrix} x \\ y \\ z \end{bmatrix}$ in R^3 can be written

$$\begin{bmatrix} x \\ y \\ z \end{bmatrix} = \begin{bmatrix} -y + \frac{1}{3}z \\ -x \\ -x - y + \frac{1}{3}z \end{bmatrix} + \begin{bmatrix} x + y - \frac{1}{3}z \\ x + y \\ x + y + \frac{2}{3}z \end{bmatrix}$$

as the sum of a vector in U and a vector in W, but the sum is not direct because (for example), the nonzero vector $\begin{bmatrix} 3 \\ 4 \\ 5 \end{bmatrix}$ is in both U and W.

p. 503

Symmetric: A matrix A is symmetric if $A = A^T$, the transpose of A. For example, $\begin{bmatrix} 1 & 2 \\ 2 & 4 \end{bmatrix}$ is a symmetric matrix, as is $\begin{bmatrix} 1 & 2 & 3 \\ 2 & 7 & -8 \\ 3 & -8 & 5 \end{bmatrix}$. For any matrix A, both A^TA and AA^T are symmetric. *p. 210*

Transpose: The transpose of an $m \times n$ **real** matrix A is the $n \times m$ matrix whose rows are the columns of A in the same order. The transpose of A is denoted A^T. For example, if $A = \begin{bmatrix} 1 & 2 & 3 \\ 4 & 5 & 6 \\ 7 & 8 & 9 \end{bmatrix}$, then $A^T = \begin{bmatrix} 1 & 4 & 7 \\ 2 & 5 & 8 \\ 3 & 6 & 9 \end{bmatrix}$ and if $A = \begin{bmatrix} -1 & 2 \\ 3 & 4 \\ 0 & -5 \end{bmatrix}$, then $A^T = \begin{bmatrix} -1 & 3 & 0 \\ 2 & 4 & -5 \end{bmatrix}$. The conjugate transpose of a complex matrix A, denoted A^*, is the transpose of the matrix whose entries are the conjugates of those of A: $A^* = \overline{A}^T = \overline{A^T}$. For example, if $A = \begin{bmatrix} 1+i & 3-2i \\ 2-4i & i \\ -5 & 5+7i \end{bmatrix}$, then $A^* = \begin{bmatrix} 1-i & 2+4i & -5 \\ 3+2i & -i & 5-7i \end{bmatrix}$.

pp. 95, 543

Triangular: A matrix is upper triangular if all entries below the main diagonal are 0 and lower triangular if all entries above the main diagonal are 0. For example, $U = \begin{bmatrix} 1 & -1 & 2 \\ 0 & 1 & 1 \end{bmatrix}$ is upper triangular, while $L = \begin{bmatrix} -1 & 0 \\ 2 & 3 \end{bmatrix}$ is lower triangular. Triangular means either upper triangular or lower triangular. *p. 145*

Unit vector: See *Vector.*

Unitarily diagonalizable: See *Diagonalizable.*

Unitary matrix: A (complex) matrix is unitary if it is square and its columns are orthonormal. For example, $U = \begin{bmatrix} \frac{1}{\sqrt{2}} & \frac{1+i}{2} \\ \frac{i}{\sqrt{2}} & \frac{1-i}{2} \end{bmatrix}$ is unitary. *p. 547*

Vector: An n-dimensional vector ($n \geq 1$) is a column of n numbers enclosed in brackets. The numbers are called the components of the vector. For example, $\begin{bmatrix} 1 \\ 2 \end{bmatrix}$ is a two-dimensional vector with components 1 and 2, and $\begin{bmatrix} -1 \\ 2 \\ 2 \\ 5 \end{bmatrix}$ is a four-dimensional vector with components $-1, 2, 2$ and 5.

A unit vector is a vector of length 1. For example, $\begin{bmatrix} \frac{1}{\sqrt{2}} \\ \frac{1}{\sqrt{2}} \end{bmatrix}$ and $\begin{bmatrix} \frac{1}{\sqrt{6}} \\ -\frac{2}{\sqrt{6}} \\ \frac{1}{\sqrt{6}} \end{bmatrix}$ are unit vectors. *pp. 1, 8, 28*

Vector space: A vector space is a nonempty set V of objects called vectors together with two operations called addition and scalar multiplication. For any two vectors u and v in V, there is another vector $\mathsf{u} + \mathsf{v}$ in V called the sum of u and v. For any vector u and any scalar c, there is a vector $c\mathsf{u}$ called a scalar multiple of v. Addition and scalar multiplication have the following properties:

1. (Closure under addition) If u and v are in V, so is $\mathsf{u} + \mathsf{v}$.
2. (Commutativity of addition) $\mathsf{u} + \mathsf{v} = \mathsf{v} + \mathsf{u}$ for all $\mathsf{u}, \mathsf{v} \in V$.
3. (Associativity of addition) $\mathsf{u} + (\mathsf{v} + \mathsf{w}) = (\mathsf{u} + \mathsf{v}) + \mathsf{w}$ for all $\mathsf{u}, \mathsf{v}, \mathsf{w} \in V$.
4. (Zero) There is an element 0 in V with the property that $0 + \mathsf{v} = \mathsf{v}$ for any $\mathsf{v} \in V$.
5. (Additive inverse) For each $\mathsf{v} \in V$, there is another vector $-\mathsf{v}$ ("minus v"), called the additive inverse of v, with the property that $\mathsf{v} + (-\mathsf{v}) = 0$.
6. (Closure under scalar multiplication) If $\mathsf{v} \in V$ and c is a scalar, then $c\mathsf{v}$ is in V.
7. (Scalar associativity) $c(d\mathsf{v}) = (cd)\mathsf{v}$ for all $\mathsf{v} \in V$ and all scalars c, d.
8. (One) $1\mathsf{v} = \mathsf{v}$ for any $\mathsf{v} \in V$.
9. (Distributivity)
   - $c(\mathsf{v} + \mathsf{w}) = c\mathsf{v} + c\mathsf{w}$ for all $\mathsf{v}, \mathsf{w} \in V$ and any scalar c.
   - $(c + d)\mathsf{v} = c\mathsf{v} + d\mathsf{v}$ for all $\mathsf{v} \in V$ and all scalars c, d.

The most important vector space in this book is R^n, Euclidean n-space, with addition and scalar multiplication defined componentwise. The set $\mathsf{R}[x]$ of polynomials in x with real coefficients and the set $M_{mn}(\mathsf{R})$ of $m \times n$ matrices with real entries are other examples. *p. 348*

Index